Vehicle Aerodynamics

PT-49

Edited by
Dr. V. Sumantran and Dr. Gino Sovran

Published by
Society of Automotive Engineers, Inc.
400 Commonwealth Drive
Warrendale, PA 15096-0001
U.S.A.
Phone (412) 776-4841
Fax: (412) 776-5760

ISBN 1-56091-594-3

SAE Order No. PT-49

Preface

This volume is primarily an assemblage of published papers selected to illustrate current activity in the field of vehicle aerodynamics. In its broadest sense, vehicle aerodynamics encompasses many different and interesting aspects of the airflow around and through a vehicle. Many of these aspects are addressed, including wind tunnel testing, on-road testing, computational simulations, and selected examples of aerodynamic development in a vehicle's design and development process. Today, the aerodynamic development of road vehicles also requires a system-level integration with other disciplines such as thermal analysis (where one deals with airflow for engine cooling) and acoustics (in the face of growing importance of flow-induced noise).

While there is clear evidence that interest in aerodynamics among automakers continues to grow, this is by no means a new field. To imply otherwise would be unfair to the significant body of work in this subject that has accompanied the evolution of automobiles. Over the last decade, there has been significant progress made in both our understanding of the technology and in our ability to create products that continually advance the level of aerodynamic efficiency of the cars and trucks made around the world. This volume offers a window to such progress.

This collection of papers does not purport to represent either a comprehensive coverage or a critical assessment of road-vehicle aerodynamic technology. It is limited by what is available and there are consequent gaps in the technical coverage. Furthermore, in the highly competitive and commercial auto industry, where proprietary considerations are important, current publications may not represent state-of-the-art technology.

There has been no re-editing of the papers included in this volume, so there is some lack of uniformity in presentation. The papers were chosen to balance representation of the various activities that constitute the subject. It is hoped that the information contained will help engineers and aerodynamicists in their efforts to confront not only the challenges of today, but those that will be faced in the future as well.

V. Sumantran Ph.D.
GM Research & Development Center

Gino Sovran Ph.D.
GM Research & Development Center (Retired)

Table of Contents

Section I
Historical Background/Overview

Section II
Wind Tunnel Testing
Testing Methodology

Ground-Simulation

Measurement Techniques

Section III
On-Road Testing

Section IV
Computational Fluid Dynamics

Section V
Vehicle Aerodynamic Development and Drag Reduction

Section VI
Aerodynamics of Engine Cooling

Section VII
Vehicle Aerodynamics and Aeroacoustics

Section I
Historical Background/ Overview

Aerodynamics of Road Vehicles

Wolf-Heinrich Hucho
Ostring 48, D-6231, Schwalbach (Ts), Germany

Gino Sovran
General Motors Research and Environmental Staff

1. INTRODUCTION

In fluid mechanical terms, road vehicles are bluff bodies in very close proximity to the ground. Their detailed geometry is extremely complex. Internal and recessed cavities which communicate freely with the external flow (i.e. engine compartment and wheel wells, respectively) and rotating wheels add to their geometrical and fluid mechanical complexity. The flow over a vehicle is fully three-dimensional. Boundary layers are turbulent. Flow separation is common and may be followed by reattachment. Large turbulent wakes are formed at the rear and in many cases contain longitudinal trailing vortices.

As is typical for bluff bodies, drag (which is a key issue for most road vehicles—but far from the only one) is mainly pressure drag. This is in contrast to aircraft and ships, which suffer primarily from friction drag. The avoidance of separation or, if this is not possible, its control are among the main objectives of vehicle aerodynamics.

With regard to their geometry, road vehicles comprise a large variety of configurations (Figure 1). Passenger cars, vans, and buses are closed, single bodies. Trucks and race cars can be of more than one body. Motorcycles and some race cars have open driver compartments. With the race car being the only exception, the shape of a road vehicle is not primarily

Figure 1 With respect to geometry, road vehicles comprise a wide variety of shapes. Race cars and, even more so, motorcycles have to be studied with the driver in place.

determined by the need to generate specific aerodynamic effects—as, for instance, an airplane is designed to produce lift.

To the contrary, a road vehicle's shape is primarily determined by functional, economic and, last but not least, aesthetic arguments. The aerodynamic characteristics are not usually generated intentionally; they are the consequences of, but not the reason for, the shape. These "other than aerodynamic" considerations place severe constraints on vehicle aerodynamicists. For example, there are good reasons for the length of a vehicle being a given. Length for a passenger car is a measure of its size, and thus its class. To place a car in a specific market niche means recognizing length as an invariant in design. Furthermore, mass and cost are proportional to length. In the same sense all the other main dimensions of a vehicle, such as width and height (which define frontal area), are frozen very early in the design process. Even the details of a car's proportions are prescribed to close limits for reasons of packaging and aesthetics (Figure 2). Of course, some maneuvering room must be left to the aerodynamicists (the hatched regions). Otherwise, they would do no more than just measure the aerodynamic characteristics of configurations designed by others.

Depending on the specific purpose of each type of vehicle, the objectives of aerodynamics differ widely. While low drag is desirable for all road vehicles, other aerodynamic properties are also significant. Negative lift is decisive for the cornering capability of race cars, but is of no importance for trucks. Cars and, even more so, vans are sensitive to cross wind, but heavy trucks are not. Wind noise should be low for cars and buses, but is of no significance for race cars.

While the process of weighing the relative importance of a set of needs from various disciplines is generally comparable to that in other branches of applied fluid mechanics, the situation in vehicle aerodynamics is unique in that an additional category of arguments has to be taken into account: art, fashion, and taste. In contrast to technical and economic factors, these additional arguments are subjective in nature and cannot be quantified.

Exterior design (the term "styling" that was formerly used is today usually avoided) has to be recognized as extremely important. "Design is what sells" rules the car market worldwide. While design gives technical requirements a form that is in accord with fashion, the fundamental nature

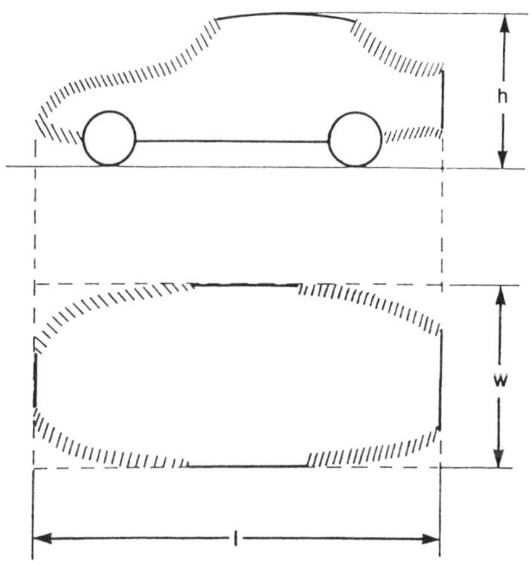

Figure 2 Right from the beginning of the development of a new vehicle, its main dimensions and detailed proportions are frozen. The limited maneuvering room for aerodynamics is identified by the hatched lines.

of fashion is change. Consequently, although vehicle aerodynamics is getting better and better, it is not progressing toward a single ultimate shape as in the case, for instance, of subsonic transport aircraft. To the contrary, it must come to terms with new shapes again and again.

There is no question, however, that aerodynamics does influence design. The high trunk typical of notchback cars with low drag is the most striking example. Despite the fact that it tends to look "bulky," it had to be accepted by designers because of its favorable effect on drag—and the extra luggage space it provides. Today's cars are streamlined more than ever, and an "aero-look" has become a styling feature of its own.

2. HISTORY

When the carriage horse was replaced by a thermal engine more than 100 years ago, nobody thought about aerodynamics. The objective of the body shell of the now horseless carriage was, as before, to shelter the driver and passengers from wind, rain, and mud. The idea of applying aerodynamics to road vehicles came up much later, after flight technology had made considerable progress. For both airships and aircraft, streamlined shapes were developed which lowered drag significantly, thus permitting higher cruising speeds with any given (limited) engine power.

The early attempts (Figure 3) to streamline cars were made according to aeronautical practice and by adapting shapes from naval architecture. These failed for two reasons. First, the benefits of aerodynamics were simply not needed. Bad roads and low engine power only permitted moderate driving speeds. Second, the approach of directly transplanting (with almost no change) shapes which had been developed for aeronautical and marine purposes was not appropriate. These streamlined shapes could be accommodated only if some important details of car design were subordinated, e.g. engine location, or the layout of the passenger compartment.

The long road from those days to today's acceptance of aerodynamics in the automobile industry has been described in great detail (Kieselbach 1982a,b, 1983; Hucho 1987b). From this history, only those events which were decisive will be highlighted here. Acknowledging the danger of being superficial, only five will be identified.

1. The recognition that the pattern of flow around half a body of revolution is changed significantly when that half body is brought close to the ground (Klemperer 1922, Figure 4).
2. The truncation of a body's rear end (Koenig-Fachsenfeld et al 1936, Kamm et al 1934, Figure 4).

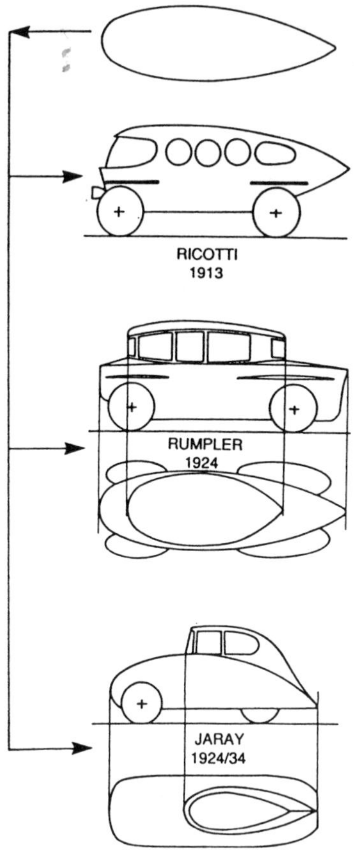

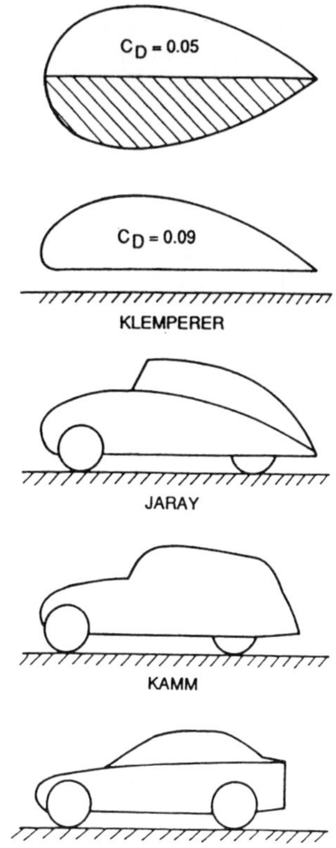

Figure 3 (*left*) The early attempts to apply aerodynamics to road vehicles consisted of the direct transfer of shapes originating from aeronautical and marine practice. The resulting shapes differed widely from those of contemporary cars and were rejected by the buying public. This unsuitable transfer procedure was very embarrassing for later attempts to introduce aerodynamics into vehicles.

Figure 4 (*right*) Klemperer (1922) recognized that the flow over a body of revolution, which is axisymmetric in free flight, changed drastically and lost symmetry when the body came close to the ground. By modifying its shape, however, he was able to reduce the related drag increase. Despite their extreme length, flow separates from the rear of streamlined cars. By truncating the rear shortly upstream of the location where separation would take place, shapes of acceptable length were generated with no drag penalty. This idea was first proposed by Koenig-Fachsenfeld for buses, and was transferred to cars by Kamm.

3 The introduction of "detail-optimization" into vehicle development (Figure 5, Hucho et al 1976).
4. The deciphering of the detailed flow patterns at car rear ends (Section 4.1).
5. The application of "add-ons" like underbody air dams, fairings, and wings to passenger cars, trucks, and race cars.

With these five steps, aerodynamics has been adapted to road vehicles, rather than road-vehicle configurations being determined by the demands of aerodynamics. The shape of cars changed in an evolutionary rather than a revolutionary manner over the years (Figure 6), and at first for reasons other than aerodynamic ones. Taste, perhaps influenced by the fascinating shapes of aircraft, called for smooth bodies with integrated headlamps and fenders (the "pontoon body"), and production technology made them possible. Better flow over the car and thus lower drag was only a spinoff. But, finally, the two oil crises of the 1970s generated great pressure for improving fuel economy drastically, and provided a break-

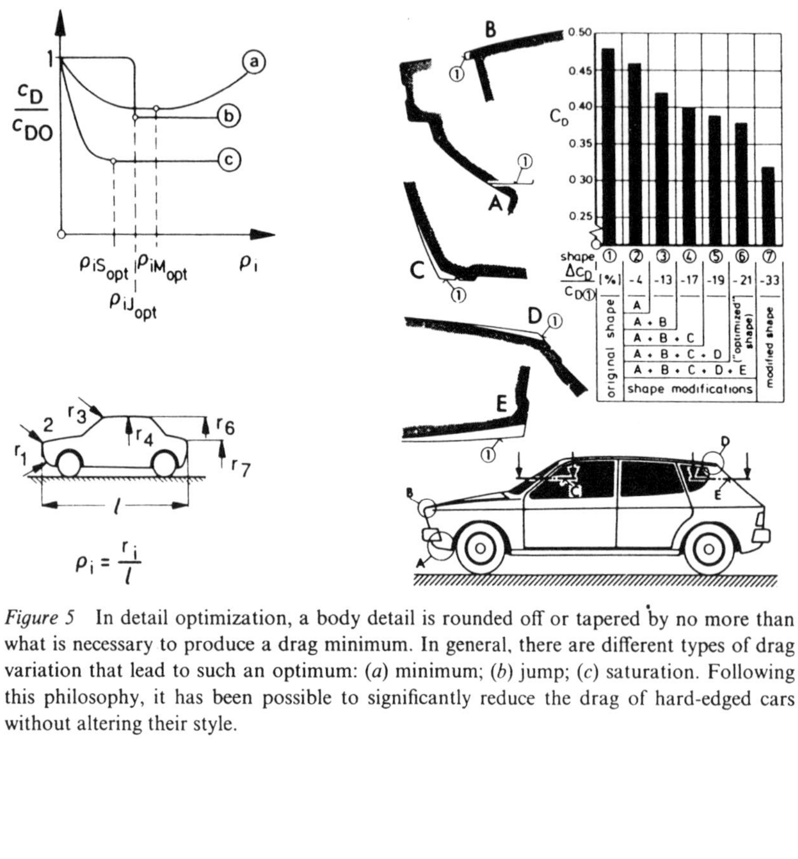

Figure 5 In detail optimization, a body detail is rounded off or tapered by no more than what is necessary to produce a drag minimum. In general, there are different types of drag variation that lead to such an optimum: (*a*) minimum; (*b*) jump; (*c*) saturation. Following this philosophy, it has been possible to significantly reduce the drag of hard-edged cars without altering their style.

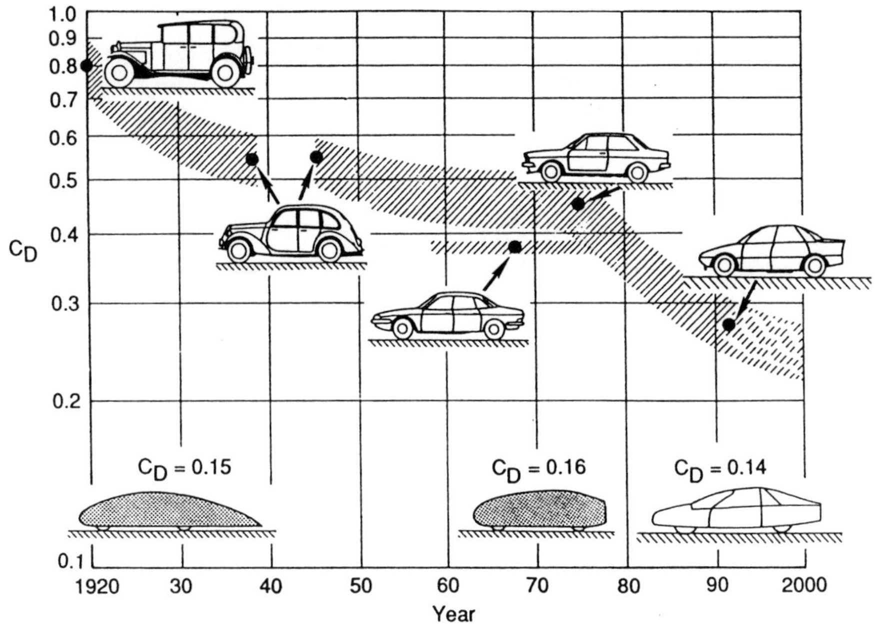

Figure 6 The drag history of cars. Using a logarithmic scale for drag emphasizes how difficult it is to achieve very low drag values. Research has been far ahead of what has been realized in production.

through for vehicle aerodynamics. Since then, drag coefficients have come down dramatically. This has been a major contributor to the large improvements in fuel economy that have been realized.

Research in road-vehicle aerodynamics has always been far ahead of practical application. Drag values demonstrate this. A drag coefficient as low as $C_D = 0.15$ was demonstrated for a body with wheels as early as 1922 (Klemperer, Figure 6), but it took more than 40 years to reproduce this value with an actual car—and then only with a research vehicle. Nevertheless, blaming the automobile industry for not taking advantage of the full potential of this technology is not justified, since the "concept" of any car is influenced by a variety of factors (Figure 7) which, collectively, are summarized by the term "market." However, the long-known potential for reducing drag (which relates to one of these factors) is now being exploited more and more. How far this trend will go depends on the future course of fuel prices and, perhaps, of emissions regulation (e.g. the possible regulation of CO_2 to control global warming).

In the following, the subject of road-vehicle aerodynamics will be treated

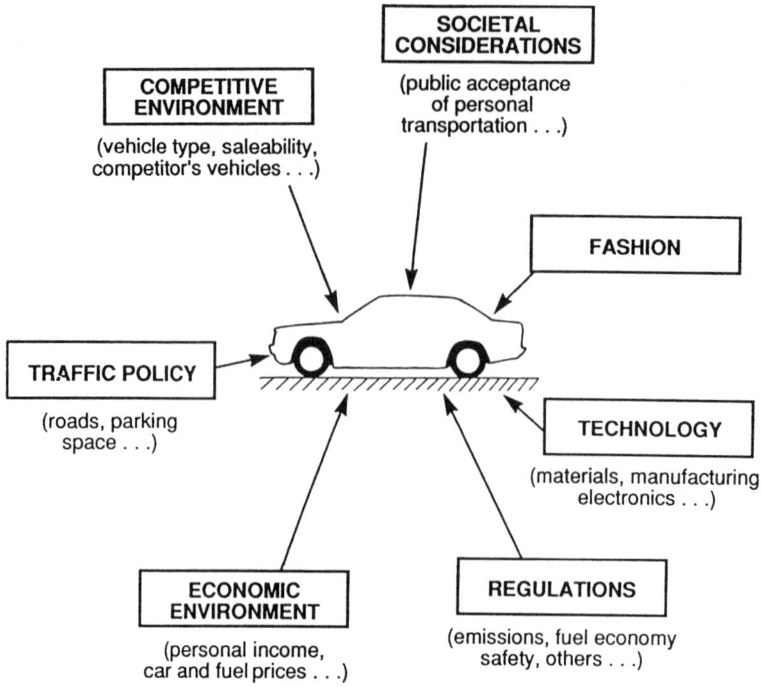

Figure 7 The concept of a car is influenced by many requirements of very different nature. A careful balance between them is required by the market.

in four sections. In the first (Section 3), the way that aerodynamics influences the operation of vehicles will be described, and without delving into the related fluid-mechanic mechanisms. These mechanisms will be discussed in a second section (Section 4). Then (Section 5), the aerodynamic-development process and the tools that are used in it will be described. In the final section (Section 6), open issues that remain and future trends will be discussed.

3. VEHICLE ATTRIBUTES AFFECTED BY AERODYNAMICS

3.1 *Performance and Fuel Economy*

The motivation for allowing aerodynamics to influence ·the shape of vehicles, if not their style, is the market situation, and this changes with time. Fuel economy and, increasingly, global warming are the current key arguments for low drag worldwide. In Europe, particularly Germany, top speed is still considered an important sales feature despite the rapidly increasing traffic density which largely prohibits fast driving even in the absence of speed limits.

Vehicle fuel consumption is a matter of demand and supply. On the demand side are the mechanical energies required for propulsion and by accessories. On the supply side is the efficiency with which this energy can be generated by the powerplant and delivered to the points of application. The influence of aerodynamics on this demand-supply relationship is through the drag force, which affects the propulsive part of the demand side.

Commercial aircraft, trains, ships, and highway trucks typically operate at a relatively constant cruising speed. In typical automobile driving, however, vehicle speed varies with time or distance. An analysis of the factors affecting automobile fuel economy can best be made if the driving pattern is prescribed. In the U.S., two speed variations of particular relevance are the Environmental Protection Agency (EPA) Urban and Highway schedules that are the basis for the fuel-economy and exhaust-emissions regulations. They represent the two major types of driving, and a combination of their fuel consumptions is used for regulatory purposes. In Europe, regulation is based on the Euromix cycle, a combination of a simple urban schedule and two constant-speed cruising conditions.

At any instant, the tractive force required at the tire/road interface of a car's driving wheels is (Sovran & Bohn 1981)

$$F_{\mathrm{TR}} = \underbrace{R+D}_{\text{Road Load}} + \underbrace{M\frac{dV}{dt}}_{\text{Inertia}} + \underbrace{Mg\sin\theta}_{\text{Grade}},$$

where F_{TR} is the tractive force, R the tire rolling resistance, D the aerodynamic drag, M the vehicle mass, g the acceleration of gravity, and θ the inclination angle of the road.

The corresponding tractive power is

$$P_{\mathrm{TR}} = F_{\mathrm{TR}} V,$$

and the tractive energy required for propulsion during any given driving period is

$$E_{\mathrm{TR}} = \int_0^T P_{\mathrm{TR}}\, dt$$

for positive values of the integrand. The main reason that fuel is consumed in an automobile is to provide this tractive energy.

Writing an equation for instantaneous fuel consumption, integrating it over a total driving duration, and using the mean-value theorem to introduce appropriate averages for some of the integrands, the following fundamental equation for the average fuel-consumed-per-unit-distance-traveled, $\tilde{g}$, can be obtained (Sovran 1983):

$$\tilde{g} = \underbrace{\frac{k}{\eta_b \eta_d S}[E_{TR} + E_{ACC}]}_{\substack{\text{propulsion} \\ (P_{TR} > 0)}} + \underbrace{g_u}_{\substack{\text{braking and} \\ \text{idle}\,(P_{TR} \leq 0)}},$$

where k is a fuel-dependent constant, η_b is the average engine efficiency during propulsion, η_d is the average drivetrain efficiency, S is the total distance traveled, E_{ACC} is the energy required by vehicle accessories, and g_u is the fuel consumption during idling and braking. Typical units are gallons per mile in the U.S. and liters per 100 kilometers in Europe.

Aerodynamic drag is responsible for part of E_{TR}. However, E_{TR} is only part of the propulsive fuel consumption which is only part of the total fuel consumption. The impact of drag on total vehicle fuel consumption therefore depends on the relative magnitudes of these contributions.

For the U.S. driving schedules, E_{TR} can be described accurately (Sovran & Bohn 1981) by linear equations of the form

$$\frac{E_{TR}}{S} = [\underbrace{\alpha r_0}_{\text{Tire}} + \underbrace{\gamma}_{\text{Inertia}}]M + \underbrace{\beta(C_D A)}_{\text{Aero}},$$

where the vehicle descriptors M, C_D, A, and r_0 are the mass, drag coefficient, frontal area, and tire-rolling-resistance coefficient, respectively, and α, β, and γ are known constants which are different for each schedule. For a typical mid-size American car, drag is responsible for 18% of E_{TR} on the Urban schedule and 51% on the Highway.

The energy part (square brackets) of the propulsion term in the fuel-consumption equation is dominated by E_{TR}, which contributes $\simeq$94% for both Urban and Highway for the midsize car. The propulsion term itself accounts for $\simeq$81% of the total fuel consumption on the Urban schedule and $\simeq$96% on the Highway.

These quantifications permit the influence coefficient relating a percentage change in $C_D A$ to a percentage change in $\tilde{g}$ to be established. In general, this coefficient is vehicle as well as driving-schedule dependent (Sovran 1983), but for the midsize car being considered they are $\simeq$0.14 and $\simeq$0.46 for Urban and Highway, respectively. For the Euromix cycle, a typical influence coefficient for cars powered by spark-ignition engines is $\simeq$0.3, while for diesel engines it is $\simeq$0.4 (Emmelmann 1987b). In all cases, these values presume that the drivetrain gearing is rematched so that the road load power-requirement curve runs through the engine's brake-specific-fuel-consumption map in the same manner at the lower drag as at the higher drag.

If no other changes are made in a vehicle, the benefits of reduced drag are actually threefold: reduced fuel consumption, increased acceleration capability, and increased top speed. When maximum fuel-economy benefit is the objective the increased acceleration and top-speed capabilities can be converted to additional reductions in fuel consumption. Conversion of the increased acceleration capability is accomplished by regearing the drivetrain, as discussed above. Conversion of the increased top speed requires a reduction in installed engine power, and a corresponding percentage reduction in vehicle mass so that the acceleration capability of the vehicle is not diminished.

The preceding discussions have presumed the absence of ambient wind while driving. In the presence of wind a vehicle's wind speed is generally different than its ground speed, and its yaw angle is generally not zero. This affects the operating drag force, and therefore vehicle fuel economy (Sovran 1984). On the average, the result is a reduction in fuel economy.

3.2 *Handling*

While traveling along a road, a vehicle experiences more than just drag. The resultant aerodynamic force has components in all six degrees of freedom (Figure 8). In principle, they all influence a vehicle's dynamics. Traditionally, test engineers distinguish between a vehicle's behavior in still

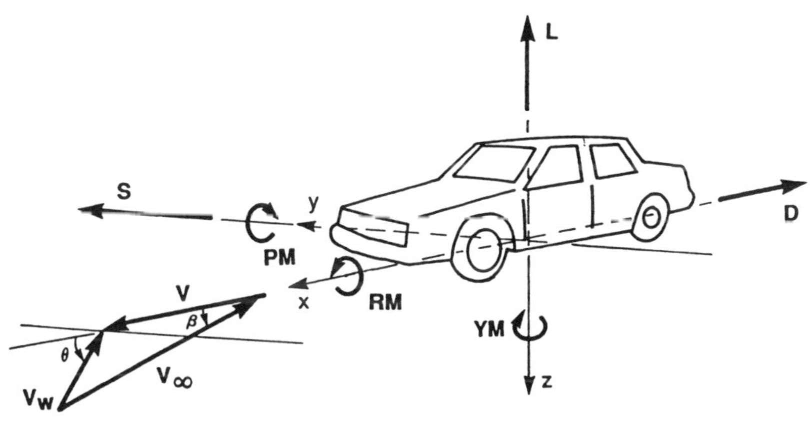

Figure 8 Aerodynamic forces and moments acting on a vehicle. Definitions of yaw angle β and the coordinate system; the latter is different from that used for flight dynamics.

air (handling), and its behavior in the presence of a crosswind (crosswind sensitivity).

The flow over a vehicle moving through still air is nominally symmetric about the vehicle's plane of symmetry. Lift, pitching moment and, of course, drag are therefore the only aerodynamic components. Unless special measures are taken, the vertical force on a bluff body close to the ground is positive, i.e. it tends to lift the vehicle. The accompanying reduction in load on the tires is, in principle, disadvantageous to handling. This is because the maximum side force that a tire can generate decreases when wheel load is reduced. However, the effect is negligible for most vehicles except race cars—at least at reasonable driving speeds. For a typical European car with a typical lift coefficient of 0.3, lift amounts to less than 3% of the vehicle's weight at a speed of 60 mph, and only 10% at 120 mph.

It is the pitching moment rather than total lift that counts in vehicle dynamics because it changes the load distribution between the front and rear axles, which alters the steering properties of a vehicle. With increasing speed, a negative (nose down) pitching moment promotes a tendency to oversteer, which is undesirable. But, this effect is hardly noticeable to the average driver. Nevertheless some European manufacturers design their cars for as low a rear-axle lift as possible, even if it involves a drag penalty.

In contrast, race cars live on negative lift. Their cornering capability has been improved dramatically by aerodynamic downforce. This downforce has to be balanced against the accompanying increase in drag, with the nature of the balance depending on the racetrack. High-speed courses with only a few bends, like LeMans (France), call for cars with low drag because top speed and fuel economy are decisive (the latter determines the number of fuel stops); speed while cornering is less important. On the other hand, courses with numerous bends, like Brands Hatch (Great Britain), require high downforce (Flegl & Rauser 1992) for short lap times.

3.3 *Crosswind Sensitivity*

In a crosswind, and while passing another vehicle in still air, the flow around a vehicle becomes asymmetric and so a side force, a yawing moment, and a rolling moment are produced. Also, the components of drag, lift, and pitching moment are altered, and normally they are increased. From both experience and numerical simulations of vehicle dynamics, it is known that only two of these components are significant to a vehicle's behavior in crosswind; these are yawing moment and side force.

The yawing moment referred to a vehicle's center of gravity gives a first indication of sensitivity to crosswind. For almost any vehicle, the yawing moment is unstable, i.e. it tends to twist the vehicle further away from the wind. As a result, the angle of yaw, the yawing moment, and the side force are increased even further.

In recent years, the center of gravity of passenger cars has, on average, moved steadily forward. This is mainly for two reasons. First, rear-engine cars have become rare; second, the move to front-wheel drive has resulted in vertical load being shifted from the rear to the front axle. Consequently, while the yawing moment referred to the center of the wheelbase has remained relatively constant, that referred to the center of gravity has decreased. As a result, the matter of crosswind sensitivity seems to have ceased to be of concern to the driving public.

Nevertheless, crosswind sensitivity remains a subject for research. Concern is focused on the influence of the various aerodynamic and vehicle parameters on a driver's reaction in sudden crosswind gusts (Sorgatz & Buchheim 1982). Driving simulators allow for changing all these parameters independently of each other (Willumeit et al 1991). On-road driving, computations, and tests on the simulator have all led to a result which is important for aerodynamic development: Yawing moment is more disturbing than side force.

There are two reasons why crosswind sensitivity might require future attention in production vehicles: decreasing vehicle weight and increasing yaw moment. Currently, the weight of passenger cars is increasing rather than decreasing. But new government regulations on fuel economy and emissions will soon force vehicle engineers to build lighter cars, because this makes significant improvements in fuel economy possible (Piech 1992). With regard to yawing moment, it is increased by the rounded rear-end shape which is in fashion at the present time, and it will go up even further unless countermeasures are taken.

In principle, with active kinematics for the rear axle (Donges et al 1990) the twist induced by crosswind can be compensated by an appropriate steering angle of the rear wheels; this would produce dynamic stability.

3.4 *Functionals*

The flow over a vehicle not only produces aerodynamic forces and moments, but also many other effects that can be summarized under the term *functionals*. During the development of a new vehicle they require at least the same attention as the forces and moments. These functional effects are:

FORCES ON BODY PARTS Vehicle bodies are made up of large and comparatively flat panels, and these have to withstand considerable aerodynamic loading. Hoods, doors, and frameless windows have to be tight under all conditions. Modern lightweight structures are prone to flutter. Special attention has to be paid to add-on parts like air shields and the various types of spoilers.

WIND NOISE The more that the formerly dominant noise sources (i.e. engine and tires) have been attenuated, the more that wind noise has become objectionable inside vehicles. Passenger cars and buses are of particular concern. In 1983, the vehicle speed at which wind noise was of the same annoyance as the noise from those other sources was 100 mph (Buchheim et al 1983); in 1992 it is only 60 mph. Open sun roofs and side windows can also cause low-frequency noise (booming) which is extremely annoying.

BODY-SURFACE WATER FLOW AND SOILING Water flow on a body's surfaces can impede visibility. Water streaks and droplets accumulating on the forward side windows prevent a clear view into the outside rearview mirrors. Soil diminishes the function of headlights and taillights. The sides of vans and buses are often used for advertising purposes and therefore need to be kept clean.

INTERIOR FLOW SYSTEMS Several interior flow systems pass through a vehicle. For passenger and sports cars the design of engine-cooling ducting has become extremely difficult because of increased engine power and less underhood space. Race cars are the most demanding. According to Flegl & Rauser (1992) up to 12 separate flow ducts have to be provided; these are for the "air box" which conducts the combustion air to the engine, several coolers for water, oil, and the turbocharged air, brake cooling, and cockpit ventilation. Passenger vehicles require ducting for proper ventilation and heating of the passenger compartment. Buses require high rates of air exchange free of drafts.

4. AERODYNAMIC CHARACTERISTICS

4.1 *Drag*

The physics of drag can be addressed from two different perspectives: 1. from that of the vehicle; 2. from that of the fluid through which it moves. From Newton's third law, the forces on the body and on the stream are equal and opposite to each other.

VEHICLE PERSPECTIVE The drag that a fluid stream exerts on a vehicle is the integral, over all surfaces exposed to the stream, of the local streamwise component of the normal (pressure) and tangential (skin friction) surface forces, i.e. $D = D_p + D_f$. The physics of drag is in the pressure (D_p) and friction (D_f) components. Measurement of only total drag (D) per se does not provide information on their relative magnitudes, nor on their distributions over the body surfaces. However, a degree of understanding of the origin and nature of drag contributions can be and has been gained by making systematic, parametric changes in the body surfaces. This has even been accomplished in the course of normal vehicle development (e.g. detail optimization, Section 5.1).

Direct evaluation of D_p and D_f requires knowledge of the detailed stress distributions over all vehicle surfaces. To obtain this experimentally is very difficult, except for very simple shapes, even in a reasearch sense. However, direct evaluations of components of D_p have been made in selected regions of a body. This is feasible in regions where the surface pressure is reasonably uniform, e.g. in base regions where the flow is fully separated. In general, however, the direct evaluation of D_p and D_f for bodies with the complex geometries of typical road vehicles is not practical. On the other hand, detailed surface-stress distributions are the specific output of computational fluid dynamics (CFD). Adequately validated computational codes can therefore contribute greatly to better understanding of road-vehicle aerodynamics.

STREAM PERSPECTIVE The wind-axis drag acting on a vehicle can be determined by applying the streamwise momentum equation to a large control volume containing the vehicle. One of the most significant results of the past 20 years of R&D in vehicle aerodynamics has been the identification of streamwise trailing vortices as a dominant feature of wakes. These originate in a number of different regions on a car body, and from the fluid-stream perspective are the producers (or the consequences?) of drag. In recognition of this, the momentum equation can be written in a form that expresses drag in terms of three integrals across the fluid stream downstream of the body: the downstream defects in stagnation pressure and streamwise dynamic head, and the dynamic head of the crossflow velocities in the wake which can be interpreted as a vortex drag. With uniform swirl-free inlet flow (characterized by V_∞ and $p_{t,\infty}$),

$$D = \iint_A (p_{t,\infty} - p_t)\,da + \frac{\rho}{2}\iint_A (U^2 - u^2)\,da + \frac{\rho}{2}\iint_A (v^2 + w^2)\,da,$$

where u, v, and w are the local x, y, and z velocity components, respectively, in the downstream cross-section, and p_t the corresponding stagnation (total) pressure.

This method of drag evaluation and breakdown requireş data from extensive and detailed traverses behind vehicles. By reducing the crossflow velocities to a vorticity field, vortex structures are better discriminated, facilitating the process of tracking them back to their origin. The method is used primarily in research (e.g. Ahmed 1981, Ahmed et al 1984, Bearman 1984, Onorato et al 1984, Hackett & Sugavanam 1984). However, it has also been reduced to a fairly routine test procedure in one particular automotive tunnel (Cogotti 1987, 1989), where it is used as an aid in car-shape development. With it, the vortical structures in the wake are identified and their drag contributions evaluated. By tracing them back to their origin, local body-shape modifications can be explored for reducing drag. An inherent premise of this approach is that drag is minimized by minimizing vortex drag.

VORTEX DRAG In the early years of vehicle aerodynamics the direct application of concepts and knowledge from the aircraft field seemed reasonable, and this tendency has persisted even to this day (although not usually with practicing vehicle aerodynamicists). The wing that makes flight possible is a lifting body, streamlined (i.e. without extensive flow separation), and of large aspect ratio (i.e. with nominally 2-D flow, but with 3-D end effects). Conceptual models such as circulation and induced drag have proved to be very useful for it. On the other hand, the typical road vehicle is not designed to produce lift, is bluff (i.e. has regions of locally separated flow and a large wake), and is of very small aspect ratio in planview (width/length $\simeq 1/3$ for cars). Consequently, its flow field is highly 3-D (nearly all end effect), not a perturbation of a nominally 2-D one. The concepts of circulation and induced drag (i.e. a drag component related to lift) are therefore in question for such flows (e.g. Jones 1978, Hucho 1978). Nevertheless, airfoil thinking is still used (although not usually by workers in the automobile industry) to suggest that the inadvertent lift of passenger cars produces an induced-drag component.

Trailing streamwise vortices are generated at the tips of wings by the pressure difference between pressure and suction surfaces that produces lift (Figure 9a, where a wing is represented by a flat plate with an automobile-like aspect ratio of 1/3 at an angle of attack). The strength of these vortices is lift dependent. If two identical such wings at equal and opposite angles of attack are connected together (Figure 9b) each generates tip vortices, but the net lift of the system is zero. The drag represented by the vortices is clearly not related to net lift. However, identifiable lifting-body components exist, so the vortex-related drag can be attributed to the lift of the individual wings.

Trailing vortices can also be generated by nonlifting solid bodies at zero angle of attack (Figure 9c and 9d). They are generated by local pressure gradients at edges that are oblique to the local flow. The net lift is zero and there are no identifiable lifting-body components, so the drag associated with these vortices cannot be related to a lift. This demonstrates that

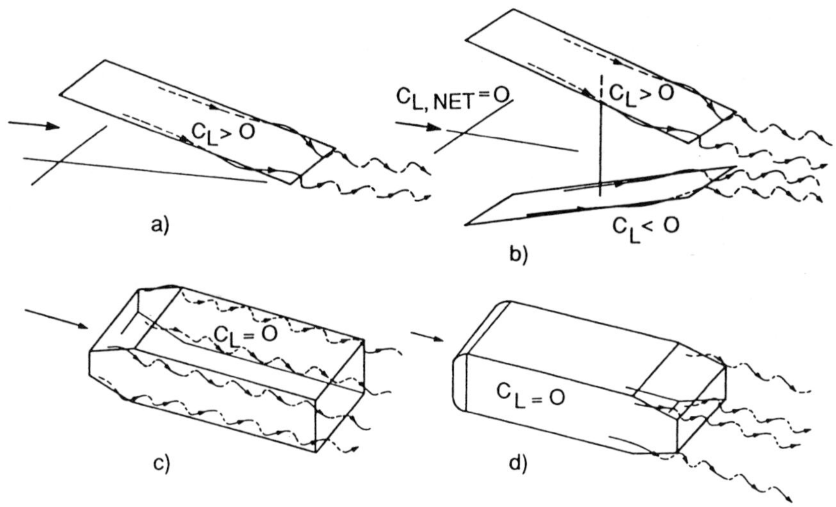

Figure 9 Longitudinal streamwise vortices (schematic) generated by lifting and nonlifting bodies.

the induced-drag concept for the contribution of trailing vortices to drag is not universal. The geometries and vortices of Figures 9c and 9d are typical of road vehicles.

CRITICAL AFTERBODY GEOMETRY The flow over the front part of any body moving subsonically through the air is easier to manage than that over the rear. This is especially true for automobiles now that free-standing headlights, radiators, and fenders are no longer in fashion. For any current car that has received aerodynamic attention, the contribution of the forebody to drag is usually small. This does not even require streamline shaping, just careful attention to detail. The major aerodynamic problem is at the rear.

The flow over and from the upper-rear surface of vehicles is particularly interesting. The influence of the slant angle of this surface on drag has been extensively investigated over the past 20 years (e.g. Janssen & Hucho 1975; Morel 1978a,b; Ahmed 1984). The nature of the flow behavior is illustrated in Figure 10, where the slant angle is that measured from the horizontal. As the rear roof of the simple body is slanted from the squareback baseline (point A, Figure 10a), trailing vortices are formed at its lateral edges which are drag producing. However, the downwash generated between them promotes attached flow on the central portion of the slanted surface, generating a pressure recovery that is drag reducing. The net effect is a drag reduction.

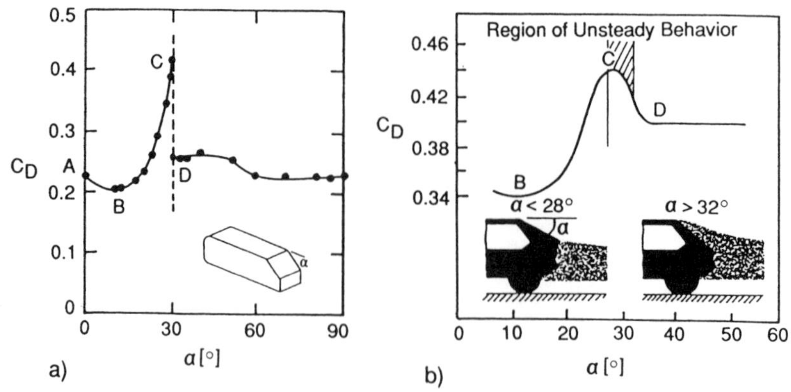

Figure 10 Dependence of drag, C_D, on the slant angle α of the upper-rear surface of fastbacks, and the existence of a critical phenomenon; (*a*) simple body, (*b*) automobile.

As the angle continues to increase the competing mechanisms grow individually, and the net drag reaches a minimum at $\simeq 15°$ (point B). It is to be noted that vortex drag is not a minimum at this point. At a sufficiently larger angle the drag reduction has decreased to zero, and further increases in angle result in a net increase in drag even though nominally attached flow is maintained on the centerline of the slanted surface.

At point C the drag is a maximum, and the trailing vortices are of maximum strength. At a slightly greater angle (the critical angle) the vortices lift off and/or burst, and the induced downwash is no longer able to produce attached flow on the centerline. The flow detaches abruptly, and a fully-separated flow develops on the slanted surface (point D). Its drag level is comparable to that of the initial square-back configuration. For this simple body with sharp rear edges, the transition is discontinuous and unidirectional. The flow patterns corresponding to C and D have been measured, and that of C is shown in Figure 11 (Ahmed 1984).

The same general behavior is observed for an actual automobile, where it was originally discovered (Figure 10b, Janssen & Hucho 1975). This has rounded edges in the slanted-roof area, and the flow pattern at transition can be bistable, switching slowly (with a period of many seconds) and randomly between that of C and D. As the slant angle increases the bistability eventually disappears, and only the D-pattern persists. For angles greater than this, the drag of the D-pattern is relatively insensitive to angle.

A drag maximum as a function of rear-upper-body geometry has also been observed for notchback cars (e.g. Figure 12, Nouzawa et al 1990). The relevant characteristic angle (β) is that between the horizontal and the

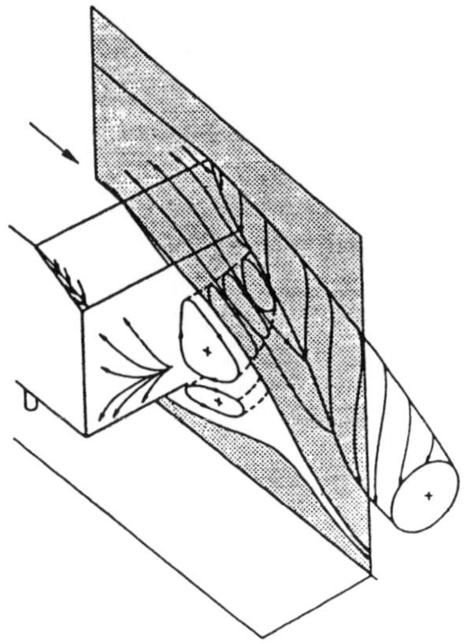

Figure 11 Wake flow pattern for a large subcritical upper-rear slant angle of fastbacks.

line connecting the end of the roof to the end of the trunk deck. For rear-window angles (α) that are $\geq 25°$, a drag maximum occurs at $\beta \simeq 25°$. At this maximum the wake pattern consists of a separation bubble behind the rear window, and trailing vortices. Measurements of instantaneous

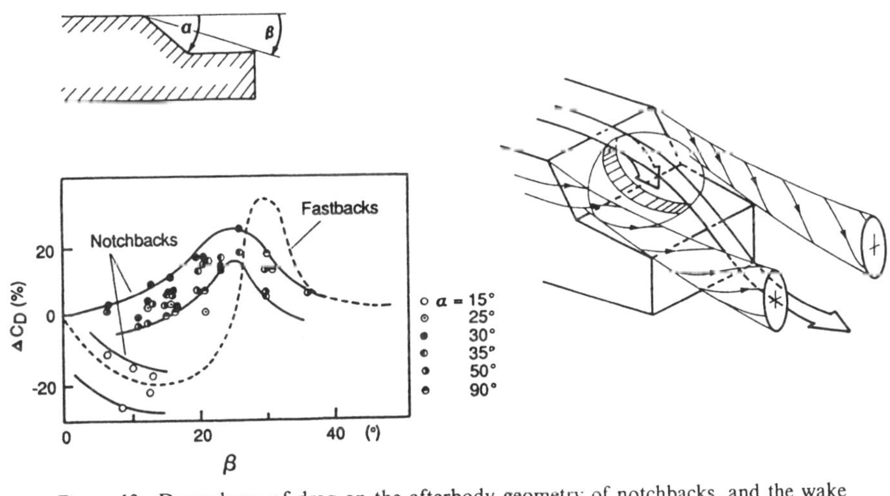

Figure 12 Dependence of drag on the afterbody geometry of notchbacks, and the wake flow pattern for high subcritical drag.

velocity in the wake and pressure on the trunk deck indicate significant bubble and wake unsteadiness (Nouzawa et al 1992). Corresponding measurements of the instantaneous drag force show that its fluctuations correlate with the unsteadiness of the separation bubble.

In vehicle design, the generic type of rear geometry (i.e. squareback, fastback, or notchback) is selected by the stylist, not the aerodynamicist. The choice is based on function and design theme, as well as aesthetics. The role of the aerodynamicist is to achieve low drag for the configuration that has been selected.

EFFECT OF AMBIENT WIND On the average, there is always an ambient wind and the wind is not aligned with the road along which a vehicle is traveling. Therefore, road vehicles operate at a nonzero angle of yaw, on the average (Figure 13). It is the aerodynamic-force component along the direction of travel that resists vehicle motion, so this is the relevant drag. This is in the body-axis frame of reference (when a vehicle is at zero pitch angle). Some typical variations of C_D with yaw angle are shown in Figure

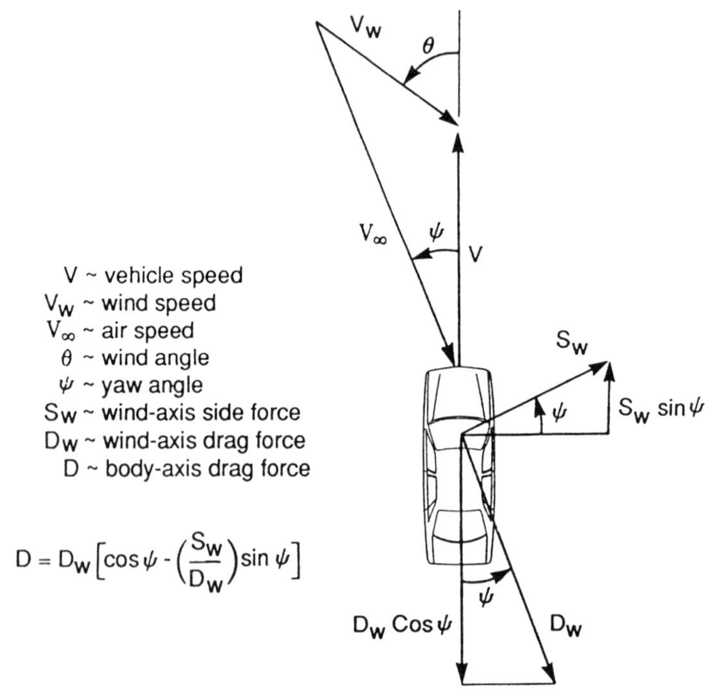

Figure 13 Body-axis drag at nonzero yaw angle.

14. Driving conditions with ψ large enough to reach maximum C_D are rare, except in gusts, and average values of ψ are said to be less than 10° for typical driving.

As is shown in Figure 13, the aerodynamics is in the wind-axis reference frame. The variation of body-axis drag (D) with yaw angle depends on the rate of increase of wind-axis drag (D_w) and the rate of decrease of the bracketed term containing the wind-axis ratio of side force to drag (S_w/D_w). Interpreting the plan view of the vehicle as the cross-section of a bluff airfoil, this ratio is analogous to the lift/drag ratio of an airfoil. The larger S_w/D_w, the greater the rate of decrease of the bracketed term and hence the smaller the rate of increase of C_D. In an extreme case, C_D can actually decrease with ψ, i.e. the wind can contribute an aerodynamic push. This has been accomplished with an experimental, small, commuter-type, high-fuel-economy vehicle having a truncated-airfoil planform (Retzlaff & Hertz 1990).

OTHER DRAG CONTRIBUTIONS The bodies of road vehicles are characterized by two types of open cavity that communicate freely with the external flow field. One is the engine compartment, with throughflow against internal resistances. Its drag characteristics are well understood by vehicle aerodynamicists (Wiedemann 1986). The other type consists of the wheel wells; these also contain large rotating tires that pump air. The wells contribute to drag, and the exposed wheels themselves generate bluff-body drag.

A measure of the state of the art in vehicle aerodynamics is the typical C_D of production cars. In the U.S. in 1975, this was about 0.55. Today, it is about 0.35. This large improvement reflects not only the skill of the aerodynamicists, but also the greatly increased concern about vehicle fuel

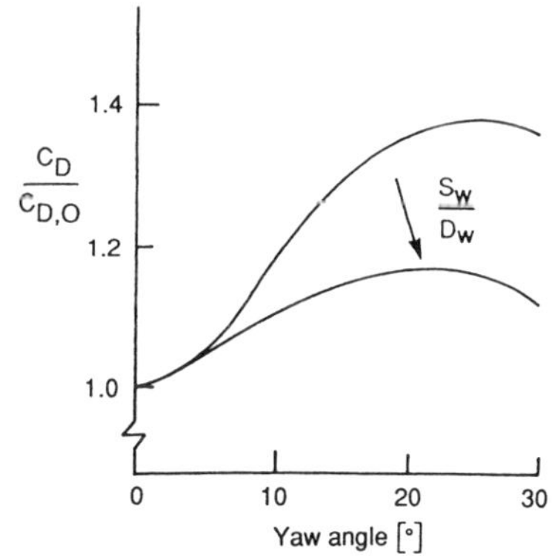

Figure 14 Typical variations of body-axis drag coefficient (normalized) with yaw angle.

economy. In Europe where fuel economy has always been important, a significant reduction in C_D has also taken place. In 1975 a typical value for production cars was 0.46. Today, several cars are below 0.30, and one car (Emmelmann et al 1990) is listed at 0.26.

4.2 *Lift and Pitching Moment*

The lift on a body close to the ground is determined by two counteracting effects (e.g. Stollery & Burns 1969). Although a symmetrical body has no lift when in free air, when close to the ground the body and the ground-plane—or in the terms of potential flow, the body and its mirror image—form a narrowing channel (Venturi nozzle) between them in which the local flow is accelerated and then decelerated. This results in an average negative gauge pressure on the body's underside, tending to generate a downforce. At the same time, flow resistance in the channel displaces more flow to the upper side of the body creating an effective positive camber of the body. This makes a negative contribution to the upper-surface pressure, tending to produce an upward force.

Lift effects are pronounced on race cars, because they are generated intentionally. Two means are used to produce a large downforce.

1. A Venturi channel in the body's underside. The effectiveness of this measure is enhanced by preventing lateral flow with side skirts. A large downforce can be generated with only a small penalty in drag.
2. Negatively cambered wings at the car's front and rear, even equipped with double flaps. In this case, there is a large increase in drag due to the induced drag of the wings, and mainly from the wing at the rear (Figure 15, Rauser & Eberius 1987).

Both means for producing large downforce have been developed to such an extent that lift coefficients on the order of $C_L = -3$ have been achieved with Formula One cars (Wright 1983). This has permitted transverse accelerations beyond a driver's endurability. Therefore, in 1982 regulations prescribed only flat underbodies, and side skirts were ruled out. Now, lift coefficients on the order of $C_L = -2$ are typical, producing a downforce at 300 km/h that is approximately twice a car's weight.

To be most effective, downforce-generating elements should produce only a minimum increase in drag. One means of accomplishing this would be to make the flaps of the wings movable. In a curve, they could be deflected to a high angle to generate maximum downforce, while in the straights they could be swiveled to zero deflection for minimum drag. However, for safety reasons, movable body parts have been prohibited on race cars. A means of circumventing this regulation would be to use a jet-flap for downforce/drag control, but this has not yet been tried.

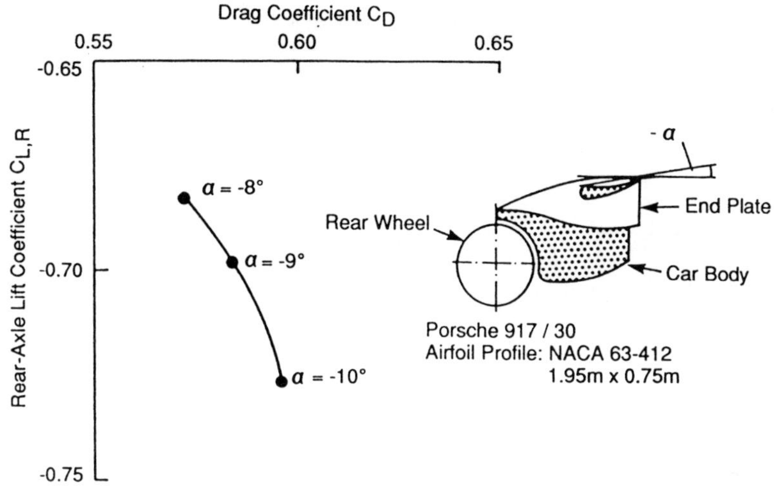

Figure 15 Rear-axle lift versus drag for three different angles of attack of the rear wing of a race car.

Passenger and sports cars use knowledge gained with race cars. Front lift is reduced by an underbody air dam; rear lift is reduced by either a rear spoiler (sometimes shaped as a wing), an underbody diffuser at the rear, or by a combination of the two. A few sports cars are equipped with a movable rear wing which is withdrawn at low speed to improve visibility to the rear.

4.3 *Yawing Moment and Side Force*

The generation of side force and yawing moment on cars is very much the same as the generation of lift and pitching moment on a wing at angle of attack (Figure 16, Squire & Pankhurst 1952, Barth 1960). When viewed from above, a car body can be regarded as a blunt airfoil that can generate lift (side force for the car) when its angle of attack (yaw angle for the car) is greater than zero. However, to generate a pressure diagram comparable to that for a wing the car needs a yaw angle much greater than the angle of attack required by the wing. In addition to its poor cross-sectional shape for a lifting body, this is because the aspect ratio of the car when viewed from the side (height over body length) is < 1, while that of the wing in Figure 16 is $\gg 1$. The center of pressure of the area enclosed by the pressure distributions on the lee- and windward sides is located forward of the midpoint of the car's length, and therefore a yawing moment is generated.

Similar to airfoils, cars are aerodynamically unstable, i.e. the yawing

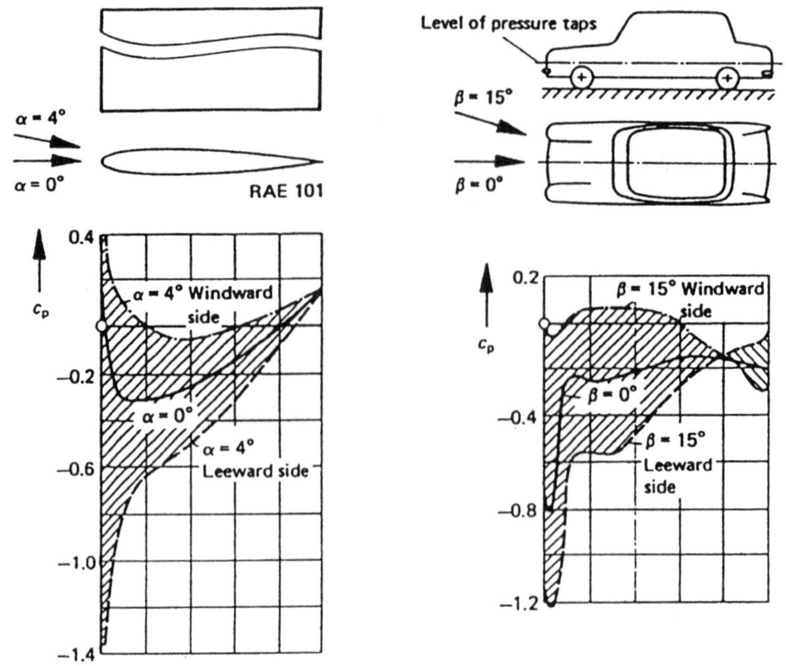

Figure 16 Comparison of the pressure distribution around an airfoil and a horizontal section of a car.

moment tends to turn the car away from the wind. Both their yawing moment and their side force increase almost linearly with yaw angle up to $\beta = 20°$, and side force is linear to even larger angles (Figure 17, Emmelmann 1987a). The rate of increase is mainly determined by the side-view cross-sectional shape of a vehicle. Generally, a higher yawing moment goes with a lower side force, and vice versa.

Under yawed conditions, a rolling moment is also generated, lift is increased, and pitching moment is altered. In the speed range of passenger cars, the influence of these components on vehicle dynamics is minor. However, the subjective perception of a driver has to be considered, and this is currently being investigated in a vehicle simulator (H. Goetz, private communication).

Measures for reducing aerodynamic instability, or even reversing it to stability, have been known for a long time (Koenig-Fachsenfeld 1951). The only really effective method is the tail fin. Race cars make use of it,

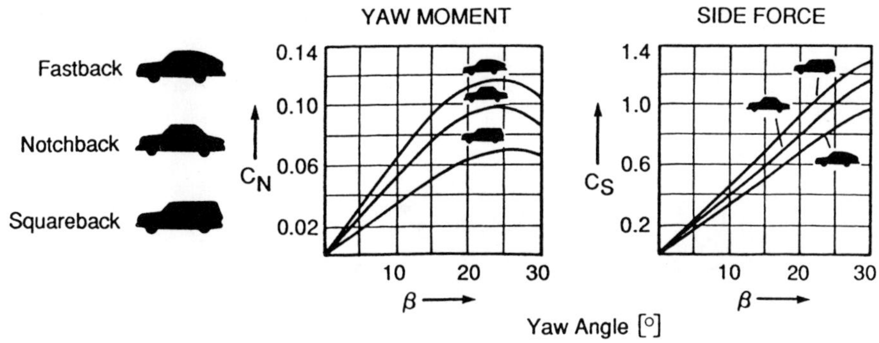

Figure 17 Yaw-moment and side-force coefficients for a family of cars with different rear-end shapes.

since the endplates of their rear wing act as such. But for passenger cars, airplane-like tail fins cannot be applied for practical reasons.

With the comparatively small shape modifications usually feasible in car development, only a limited reduction of yawing moment is possible (Sorgatz & Buchheim 1982, Gilhaus & Renn 1986). Rounded rear-end contours are unfavorable for crosswind sensitivity. Under yawed conditions the flow around the windward side's rear corner remains attached longer for a rounded corner than for a sharp edge, producing high negative pressure at this location and increased yawing moment. Some European cars have a sharp trailing edge on the rear pillars of the passenger-compartment "greenhouse" (the C pillars) to reduce yawing moment (Gilhaus & Hoffmann 1992). The accompanying increase in drag at yaw is tolerable because the large angles at which it occurs are infrequent.

Implicitly, all the preceding considerations are based on the assumption of an idealized wind, i.e. steady, with a spatially uniform velocity. In reality, this is never the case. Natural wind has a boundary layer character. When its velocity profile is combined vectorially with a vehicle's forward speed, a skewed oncoming relative-wind profile is generated (Figure 18, Hucho 1974) which cannot be reproduced either in a wind tunnel or by a side-wind facility at a test track.

Because of varying landscape along a road, the strength and local direction of wind can vary with distance along it. Furthermore, natural wind is gusty. Therefore, steady-state data obtained in test facilities cannot be exactly applicable to actual windy-day driving. The transient side force and yawing moment computed with an analytical model based on slender-body theory (Hucho & Emmelmann 1973) are significantly

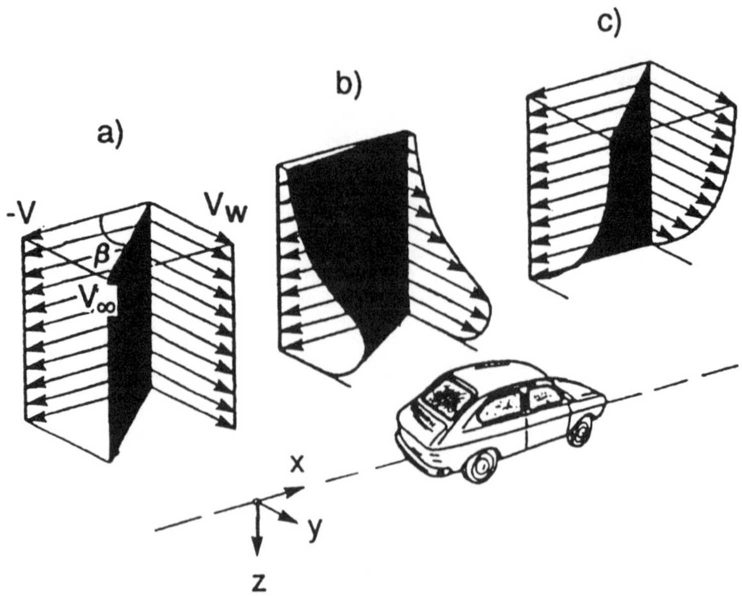

Figure 18 Wind profiles: (*a*) in the wind tunnel, (*b*) on a crosswind track, and (*c*) in natural wind.

larger than their steady state values, the magnitude of the overshoots depending on the gradient of the crosswind velocity along the direction of travel, i.e. on the gradient at the edge of the simulated wind gust.

4.4 *Detailed Surface Flows*

Detailed surface flows are investigated for a number of reasons: to properly locate openings for air inlets and outlets; to ascertain the forces on particular body parts; to determine the sources of wind noise; and to find means for controlling water flow and soil deposition on surfaces.

The openings for ducted air flows are placed in regions of high pressure. Therefore, the inlet for engine-cooling air has moved downward over the years along with the lower hoodlines, and is now close to the stagnation point. Consequently, large inlet grilles have become unnecessary and, to the complaint of many buyers, cars have lost their distinctive faces.

The inlet for passenger-compartment ventilation air is normally placed in the cowl area at the base of the windshield, and most commonly in the central part. The disadvantage of this location is that the ram-air volume flow rate depends on driving speed and so, therefore, does the flow velocity inside the passenger compartment. Placing the inlet away from the centerline in an area of zero pressure coefficient avoids this speed dependency, but then a continuously running fan is needed to achieve desired interior airflow levels.

Outlet openings were formerly placed at locations of high negative surface pressure. But, due to the high flow-field velocities at such locations the local wind noise was high and was communicated to the inside of a car. Ventilation outlets are now placed in regions of low to moderate negative pressure to avoid this problem. However, the larger outlet areas required must be hidden from view; for example, behind the rear bumper.

Knowledge of the aerodynamic forces acting on body parts such as doors and the hood is needed to properly select the position of hinges and locks. Flexible structures tend to flutter, and so locations for reinforcement also have to be determined.

Wind noise, as perceived inside the passenger compartment, is receiving increasing attention. The basic phenomena are now well understood (Stapleford & Carr 1971, George 1989). However, based on buyer complaints, the application of this knowledge to specific cars still seems to be a major difficulty.

Investigations of wind noise have to deal with three factors: sources, paths, and receivers. Sources may be such things as a leaking window seal, a large-scale separation bubble on the hood, high local flow velocities, or a small-scale separation in a gap between two doors (Figure 19, Hucho 1987b). By comparing the actual flow with one where separation has been avoided by artificial means, such as by sealing door gaps, the contribution of individual noise sources can be identified (Buchheim et al 1983, Ogata et al 1987). Aerodynamic noise can also be generated by the separated flow over protuberances on a car body, such as side rearview mirrors and radio antennae (Figure 20, Ogata et al 1987). Paths have to be treated according to their function. Panels can be dampened with absorbing materials, but windows can't. Openings for ventilation must remain open. Highly turbulent separated flow from the front part of a vehicle reattaches further downstream and may excite vibrations of the structure in the reattachment area; the wake of the side-mounted rearview mirror is a typical example. The receivers are human beings. Their individual sensitivity to noise (level, spectrum, discrete tones) must be accommodated.

It might be expected that low-drag cars would produce low wind noise, but no such correlation exists for today's cars. Even aerodynamically-well-designed cars have many flow separations (of large and small.scale), areas of high local velocity, and leaking seals. Further advances are needed in the ability to identify noise sources. Recent research on high-speed trains

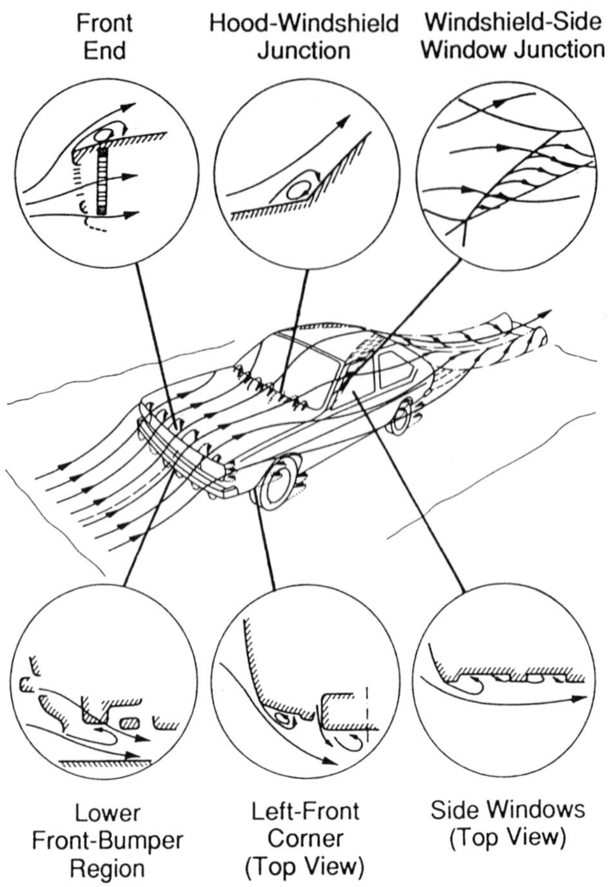

Figure 19 Flow around a car, and major locations of flow separation.

(Mackrodt & Pfitzenmaier 1987) should be applicable to cars. Progress in analytical modeling of sources and paths is also a necessity (Haruna et al 1992).

The air through which a vehicle moves is not always dry and clean. Water drops (either directly from rain or from splash and spray whirled up from the road) and dirt particles (either wet or dry) are heavier than air and so do not follow streamlines. The resulting two-phase flow has to be managed with consideration for both safety and aesthetics. Two points of view deserve consideration. Whirled-up water and dirt are annoying and dangerous not only for the vehicle generating them, but also for other vehicles in its vicinity. The latter problem is still very much neglected.

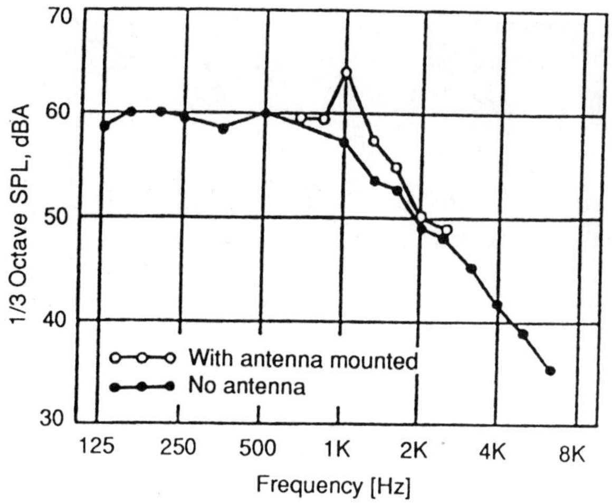

Figure 20 Aerodynamic noise generated by a radio antenna

Water streaks that spill over from the windshield to the A-pillar (the post between windshield and front door) are broken up into droplets by an A-pillar vortex. These droplets tend to collect on the forward part of the side windows, obscuring a driver's view in the side-mounted rearview mirrors. By integrating a rain gutter into the A-pillar, the water from the windshield can be trapped and led down and away. Similarly, water from the roof can be trapped and kept away from the rear window, either by a trapping strip in the rubber seal (Piatek 1987) or, in the case of a hatchback door, by controlling the gap between roof and door (Figure 21). If these traps are carefully faired into a body's contour, they will not cause an increase in drag.

The deposition of dirt is more difficult to prevent. Those parts of the body close to the stagnation point are the most susceptible to dirt deposition, e.g. the headlamps. Up to now, no aerodynamic alternative to the mechanical cleaning of headlamps, either by high-pressure water spraying or by brushing, has been found. Dirt deposition on the rear window of a squareback car can be reduced by a guide vane at the end of the roof. However, the air-curtain effect has to be balanced against the increase in drag that accompanies it. The sides of delivery vans can be kept free of soiling, at least partly, by impeding the dirty water that spills out of the front wheel houses. This can be achieved by generating a low pressure underneath the vehicle that reverses the direction of the wheel-house outflow from outboard to inboard.

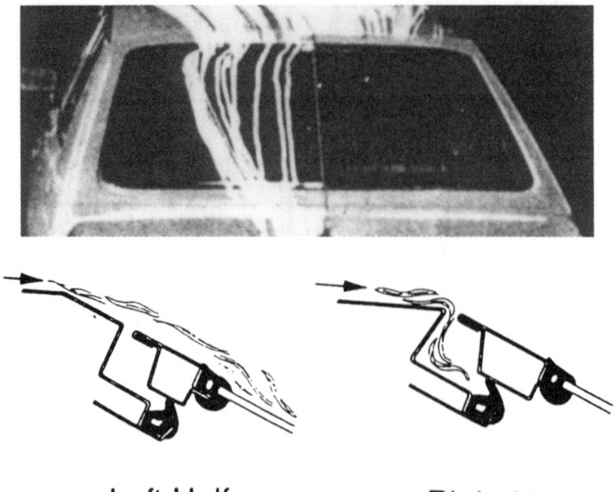

Figure 21 The gap between roof and hatch door can be used as a water trap. On the left side the large trapping gap has been sealed for the photograph, on the right side it is open and functional.

Splash and spray whirl-up is also a safety hazard for other vehicles. This is especially true for trucks and should no longer be ignored, particularly since solutions are already on hand. On wet roads, trucks with their almost uncovered wheels can raise opaque curtains of water higher than the eye-point of a neighboring car's driver (Figure 22, Goehring & Kraemer 1987). Up to now, the only way to impede this has been to completely cover all nonsteering wheels. This measure is even rewarded with a drag reduction, mainly under crosswind conditions, and is also useful in preventing cars from going under the truck in a crash.

Figure 22 Splash and spray can be kept under a vehicle by shielding the non-steered wheels. (*Left*) Wheels uncovered; the passing car can hardly be seen. (*Right*) Wheels and sides shielded, as well as the rear of the truck. (Photos courtesy of Mercedes-Benz.)

4.5 *Internal Flow Systems*

The flows in internal and recessed cavities that communicate freely with the external flow contribute to drag and have been discussed in Section 4.1. The flow in the passenger compartment does not make a significant drag contribution, but it has to be carefully designed for passenger comfort. High air-exchange rates have to be realized, but with a flow free from drafts. Consequently, flow velocities are kept low. The influence of buoyancy has to be recognized because of significant differences in temperature between heating and cooling. During heating, a temperature gradient is preferred (warm feet, cool head). During cooling, the conditioned air is directed to the face and chest. By making use of the physiological effect that a temperature lower than ambient can be simulated by blowing ambient air at moderate velocity over passengers, the use of air conditioning per se can be avoided in moderate climates (Gengenbach 1987).

5. DEVELOPMENT PROCESS AND TOOLS

5.1 *Aerodynamic Development of a Vehicle*

The aerodynamic development of a typical road vehicle follows a procedure different from that used for other types of fluid-dynamic machines, e.g. aircraft, compressors, or turbines. While the latter are designed to meet specific aerodynamic performance criteria—otherwise they wouldn't work—the aerodynamic properties of a road vehicle are largely by-products of the design process.

The typical design of a fluid-dynamic machine is accomplished in three major steps:

1. The main overall dimensions of the machine, be it the planform and span of an aircraft wing or the diameter and number of stages of a turbine or compressor, are determined by using nondimensional coefficients (derived from similarity laws) for which optimum values are known from accumulated experience. Typical coefficients are ones such as the lift coefficient (C_L) of a wing during aircraft cruise, or the specific speed (N_S) of a fan.
2. The detailed design of the machine's components follows an iterative process between analytical design and experimental verification. Computational fluid dynamics (CFD) has accelerated the convergence of this process significantly; furthermore, it permits interference effects among the components to be taken into account.
3. The separately optimized components are put together in a prototype system which is then tested. If the final device is very large, this is

preceded by testing with small-scale models. When design performance targets are not achieved, modifications are made.

The development of a road vehicle is accomplished differently. Aerodynamic development is performed in a closed loop containing aesthetic, packaging, and aerodynamic considerations. Both the number of iterations necessary and the quality of the final result depend on the ability of the aerodynamicist to recognize the intentions of the exterior designer (stylist), and to find solutions within the designer's limits of acceptability—which are extremely difficult to define because of their subjective nature.

In general, the aerodynamic development of a road vehicle can proceed from two different and opposite starting points. Typically, the designer makes the first proposal for a vehicle configuration. Then the aerodynamicist tries to improve the shape, according to his goals as described in Sections 3 and 4. One strategy which has proved to be very effective is the so-called detail-optimization (Hucho et al 1976). In this approach, details like corner and edge radii, and taper or boat-tailing are modified in small increments in order to find optimum values (Figure 5). As long as the drag coefficient of vehicles was still fairly high, say $C_D = 0.40$, the modifications necessary were frequently so small that aesthetics was hardly affected (Janssen & Hucho 1975).

The opposite approach starts with a simple body of very low drag which has the desired overall dimensions of the subsequent car. This so-called basic body is modified step by step, in cooperation with the exterior designer, until a final shape materializes which has both low drag and the desired appearance. Drag-coefficients below $C_D = 0.30$ have been achieved this way (Buchheim et al 1981). This is called *shape-optimization*.

Today the actual aerodynamic development process is usually a mixture of these two extremes. The original design of the stylist is influenced by his general knowledge of vehicle aerodynamics, more or less, and the shape modifications which are eventually necessary are constrained within rather narrow limits.

5.2 *Test Facilities*

Aerodynamic development per se is done almost entirely experimentally, guided by empiricism based on the knowledge described in Section 4. The primary test facility is the wind tunnel. Water tunnels and towing tanks are sometimes also used, but only as supplements. They lend themselves to flow visualization (Williams et al 1991), and also have the capability for reproducing the relative motion between vehicle and road, including wheel rotation (Larsson et al 1989, Aoki et al 1992). However, only those towing tanks can be used which have an extremely level and smooth bottom on which the submerged test vehicle can roll.

All vehicle characteristics influenced by aerodynamics are finally evaluated on the road. This also permits tests that cannot be made in a wind tunnel, e.g. measuring a vehicle's behavior in a cross wind.

Historically, the wind tunnel testing of automobiles started with small-scale models. Scales like 1:4 or 1:5 were preferred in Europe, the somewhat larger 3/8-scale in the U.S. The advantages of small-scale testing are that the models are cheaper than full-scale ones, are easy to handle, and can be quickly modified. Furthermore, only small wind tunnels are needed, and these are more generally available and can be rented at moderate cost.

For two reasons, small-scale testing was eventually only rarely used. At first, this was because test results from partial-scale models very often did not reproduce full-scale values with the accuracy needed. This deficiency was partly due to a lack of geometric similarity in the models, and partly to the unpredictable effects of Reynolds number. However, geometric similarity is not a fundamental problem but rather a matter of the skill and care of the model maker. Also, a Reynolds-number gap on the order of two can be bridged by artificially increasing the turbulence level of the wind tunnel (Wiedemann & Ewald 1989). Consequently, small-scale testing has again come into favor and is used by some car manufacturers with great success.

The second and even stronger objection to small-scale testing is non-aerodynamic in nature. Vehicle exterior design is done in full scale, because shapes in small scale cannot be adequately assessed aesthetically. Therefore, a full-scale model always exists, and if it is built on a realistic chassis, such as the one from the preceding model year, it can also be used as the wind-tunnel model.

5.3 *Limitations of Wind-Tunnel Testing*

Two major limitations exist when cars are tested in tunnels. First, due to their pronounced bluntness, cars disturb the flow in the test section of a wind tunnel significantly, the situation being similar to that with aircraft models under extreme stall (e.g. fighter aircraft during combat maneuvers). Second, the relative motion between vehicle and road and the rotation of the wheels are very difficult to reproduce, and are therefore usually neglected. However, great progress has recently been made in overcoming both of these limitations.

BLOCKAGE As a carryover from aircraft terminology, the ratio of a car's frontal area to the cross-sectional area at the tunnel-nozzle exit is called the blockage ratio. For road vehicles, this ratio can be extremely large compared to aircraft. Blockage ratios up to 20% are sometimes used. Thermal tests (e.g. engine cooling) are done with blockage ratios even higher than this. For a long time, a blockage ratio on the order of 5% was said to be appropriate for the aerodynamic testing of cars, this figure being borrowed from aeronautical practice. For a typical car, this would lead to a test-section cross-sectional area of 40 m^2. Very few automotive wind tunnels are this large; 25 m^2 is a typical value for the wind tunnels built recently in Europe.

With regard to the boundaries of the test section airstream, two different types of configuration are used. The closed-throat test section with solid walls is preferred in the U.S. The open-throat test section with freestream boundaries is common in Europe. The latter is characterized more precisely as only three-quarters open, the remaining quarter of the boundary being the ground floor which represents the road. A third type of boundary configuration which lies between the open- and closed-throat versions is one with slotted walls. Some automotive wind tunnels in Europe use this type (Eckert et al 1990).

Testing in an open test section is affected by a body as bluff as a car in three different ways:

1. The determination of the wind speed may be in error. If the model is too close to the nozzle's exit its flow field extends upstream far enough to modify the pressure at the location of the downstream pressure taps inside the nozzle (Figure 23, Kuenstner et al 1992).
2. The flow into the collector may be disturbed if the length (L) of the test section is insufficient and the cross-sectional area (A_C) of the collector is too small compared to the nozzle's cross-sectional area (A_N) (Figure 24, v. Schultz-Hausmann & Vagt 1988).
3. The streamlines in the vicinity of a vehicle are diverted more than in an air stream of infinite cross-section. Consequently, the drag is lower. Only this effect is subject to a blockage correction, not the preceding two.

While the first two effects are now well understood, blockage corrections for the third which are generally valid for open test sections still do not exist (SAE 1990). They are often said to be small and are therefore, for simplicity, neglected. It is doubtful whether this holds for blockage ratios of 10% or more.

The effect of body bluffness on closed-throat tunnel testing is primarily one of model blockage, although a possible influence on the wind-speed measurement cannot be ignored. In contrast to open throats, the streamline divergence around a model is less than that in an infinite stream. This causes the measured values of C_D to be larger than those in an infinite stream, rather than smaller, and the magnitude of the discrepancy is significantly greater than that for open test sections.

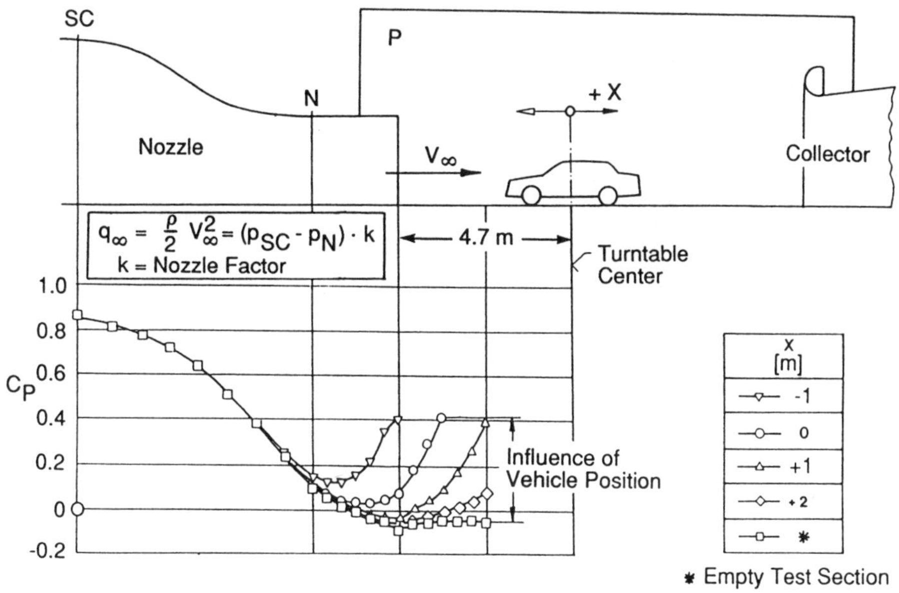

Figure 23 Pressure distribution on the centerline of the floor inside the nozzle of the BMW 10 m² acoustic wind tunnel. If the model is positioned too close to the nozzle's exit, the static pressure at location N is increased.

All classical blockage corrections are all-inclusive in nature, i.e. the velocity head of the oncoming wind is first corrected, and then the coefficients corresponding to all the measured forces, moments, and pressures are adjusted using this corrected head. The assumption underlying any concept of data correction (it is, more properly, data adjustment) is that the detailed nature of the flow field (in a nondimensional sense) is unchanged by blockage. This does not necessarily hold if the blockage ratio is large, or if flow separations and reattachments are affected.

Blockage correction for road vehicles in closed test sections has been the subject of an ongoing cooperative study by aerodynamicists active in vehicle testing (SAE 1991). From the many procedures considered, three have been established as acceptable for zero-yaw-angle testing in this application. These are: a wall-pressure signature method (Hackett & Wilsden 1975), a less comprehensive wall-pressure method (Hensel 1951), and a semi-empirical method (Mercker 1986). All three give approximately the same blockage correction, even at nonzero yaw angle. The pressure-signature method is the most comprehensive, requires the least subjective user input, and is readily applicable to vehicles with unknown flow patterns. It has the disadvantage of being demanding in terms of instru-

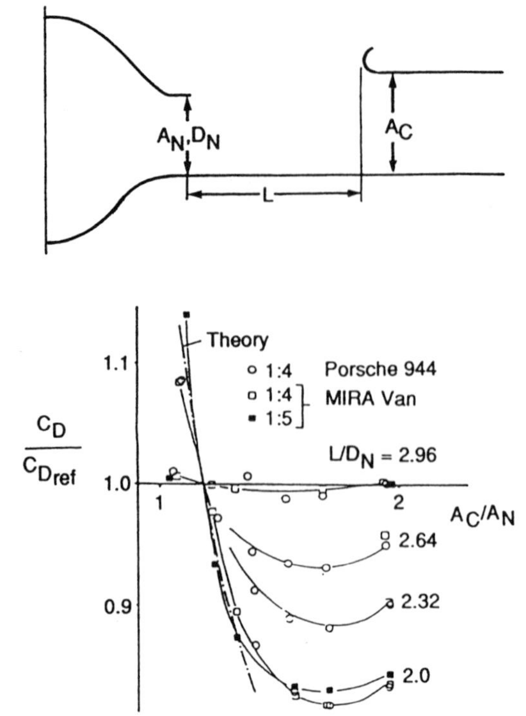

Figure 24 Influence of test-section length L (relative to the hydraulic diameter D_N of the nozzle) and the ratio of collector area (A_C) to nozzle area (A_N) on measured drag; Porsche 22 m^2 wind tunnel, open test section.

mentation and computation. The method has been modified and extended for tractor-trailer trucks by Schaub et al (1990), and the improvement is applicable to automobiles.

ROAD REPRESENTATION AND WHEEL ROTATION From the numerous suggestions for simulating the road in a wind tunnel (Figure 25) the following four have been reduced to practice (Mercker & Wiedemann 1990): a solid floor without any boundary layer control (*a*), with tangential blowing (*h*), or with boundary layer suction, (*d* and *g*); a moving belt in combination with an upstream scoop or a suction slot to remove the oncoming boundary layer (*c*).

The simplest and most common way to represent the road is with a solid and fixed ground floor. This permits a vehicle to be easily connected to a balance under the floor through four pads, one under each wheel, which can be adjusted for wheelbase and track. The wheels of a vehicle are not rotated during conventional testing. As long as the displacement thickness

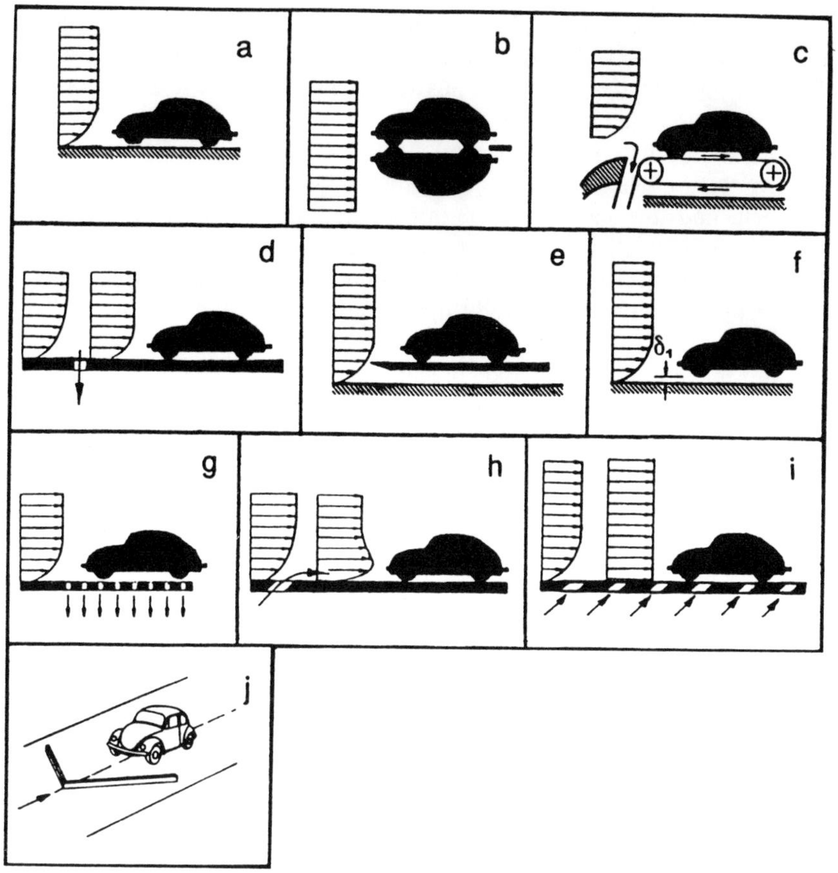

Figure 25 Various possibilities for representing the road in a wind tunnel.

of the floor boundary layer, as measured in an empty test section, is less than 10% of the vehicle's ground clearance, this kind of ground simulation is adequate for passenger-car development at drag-coefficient levels of $\simeq 0.40$ and above (Hucho et al 1975).

The moving belt represents (in principle at least) an almost perfect way to simulate the road. There are a surprising number of moving-belt wind tunnels in the world used for automotive testing. Nearly all of them (20, by one count) are for partial-scale models, and many are used for testing Formula One race cars. Quite a number are at educational institutions. There are only two full-scale facilities.

In moving-belt tunnels the belt's ability to support the weight of the test vehicle is limited. Therefore, in nearly all of them the vehicle is suspended from a force balance located above the test section, usually by a vertical strut. This strut is shielded from aerodynamic loads by a streamlined fairing. In a very few facilities the vehicle has an internal balance and is supported from the rear by a sting. The resultant aerodynamic interference of the strut or sting can be significant, and must be quantified by calibration.

Nonrotating wheels over a moving belt require a gap between wheels and belt, and this leads to errors in drag and lift (Beauvais et al 1968). Consequently, in all moving-belt facilities the wheels of the test vehicle are in contact with the belt, and driven by it. While the two-fold effect increases the fidelity with which the wind tunnel simulates on-road driving, it also confounds the aerodynamic measurements. The wheels are either attached to the vehicle in a manner that prevents their vertical loads from being transmitted to the body, or detached and independently supported in a manner that permits the streamwise force on them to be measured. In either case, the lift of the wheels is not measured. Furthermore vehicle drag can be inferred only if appropriate correction is made for the rolling resistance of the tires. The latter is determined by a tare test with the belt moving but the wind off.

A means for measuring the lift of a complete vehicle (i.e. including the contributions of the wheels) as well as that of only its body has been developed for full-scale cars. The vehicle is fitted with an internal balance and supported from the rear on a sting (Figure 26, Mercker et al 1991).

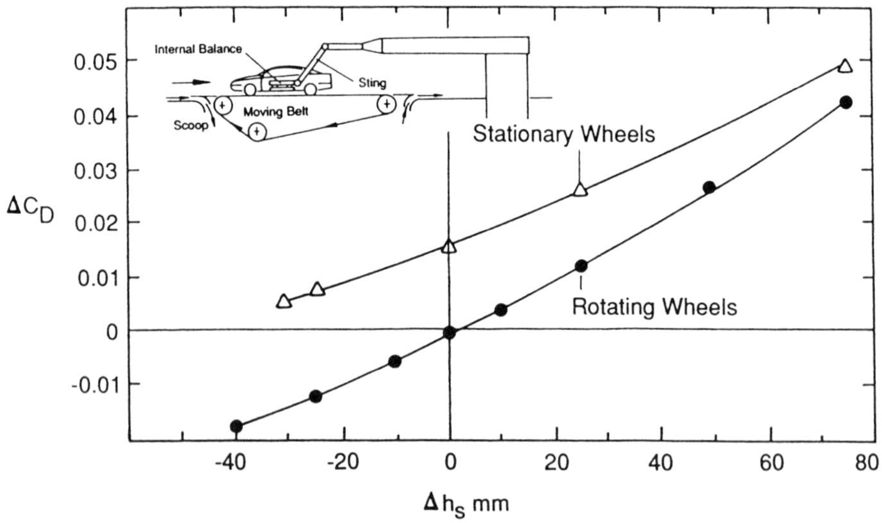

Figure 26 Influence of wheel rotation on drag, as measured for an Opel Calibra over a moving belt in the German-Dutch wind tunnel (DNW): Δh_s is the difference in underbody clearance height from the design level.

The suspension of all four wheels is replaced by a pneumatic system which permits accurate control of the vertical load at each wheel. This is necessary to permit measurement of the total vehicle lift, and to properly simulate the tire rolling resistance which depends on vertical load. The technique requires considerable model preparation, and setting up the test and performing the necessary calibrations requires much valuable time in the test section. Therefore, it is not well suited for routine testing, and is primarily restricted to research.

The influence of wheel rotation per se on the drag of passenger cars has been evaluated using this technique (Figure 26, Mercker et al 1991). Rotating wheels make a smaller contribution to vehicle drag than stationary ones. In addition, wheel rotation increases vehicle lift significantly.

It is argued that tangential blowing over a fixed floor comes close to a moving-floor simulation (Figure 27, Mercker & Knape 1989). In order to cope with wheel rotation, special wheel pads are under consideration. Either a miniature moving belt or a pair of little rollers integrated into the pads under each wheel may provide a practical solution.

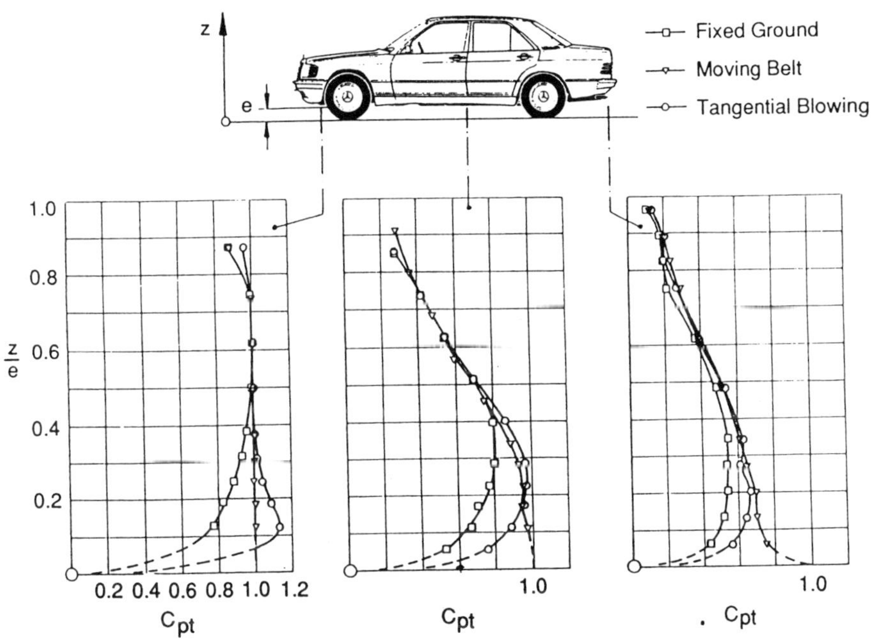

Figure 27 Dynamic-head distribution [$C_{pt} = (P_t - p_\infty)/q_\infty$] in the centerplane under a vehicle in a wind tunnel with three different types of ground plane simulation: fixed ground, moving belt, and tangential blowing.

In some fixed-floor automotive wind tunnels, boundary layer suction is applied, most frequently as concentrated suction. Air is sucked off through a strip of perforated sheet metal at the nozzle exit. The alternative, distributed suction, is used in only one tunnel (Vagt & Wolff 1987, Eckert et al 1992). In this case air is sucked through a large porous area of the ground floor in front of and under the car. The use of several suction chambers permits spatial variation of the suction rate over the surface of the floor.

In comparison to tangential blowing, suction has several disadvantages. If the boundary layer is to be reduced by the same amount, a much larger volume of air must be removed than is added when blowing is employed. Consequently, the negative angle of attack of the oncoming flow induced by the suction may exceed a tolerable value for car tests, say 0.2°. Forces and moments on a car are very sensitive to angle of attack.

With distributed suction, it is also not clear how much air should be sucked off, nor where. In comparison to the natural floor boundary layer in an empty test section, the thickness of the boundary layer on the ground underneath a moving vehicle is small (Hucho et al 1975), but not known a priori. Therefore, it is difficult to define the proper suction rate and distribution. In the case of tangential blowing, it has been empirically determined that the blowing rate required for zero displacement thickness in an empty tunnel at the location of the front wheels of a car will result in aerodynamic measurements in good agreement with those obtained with a moving belt (Mercker & Knape 1989).

5.4 *New Test Facilities*

Aerodynamic noise can be investigated in a wind tunnel only if there is a sufficient difference in sound-pressure level between the noise produced by the flow around (and through) a vehicle and the background noise of the tunnel. The formerly required 10 dB(A) difference (Buchheim et al 1983) may be sufficient for objective measurement of the external sound pressure field, but it is inadequate for subjective assessment of wind noise inside a car. Aero-acoustic wind tunnels have now been built which provide a 30 dB(A) sound-pressure-level difference (Ogata et al 1987, Hucho 1989a). Existing tunnels are now being equipped with additional sound dampers, even at the expense of a lower maximum wind speed, to make them better suited for aero-acoustic investigations.

Thermal tests make up the other half of vehicle aerodynamics. Adequate comfort for passengers and proper functioning of the engine and accessories have to be guaranteed under all operating conditions. One approach to thermal testing is to perform all aerodynamic test work in the same facility (Moerchen 1968). The other is to use separate specialized facilities for each kind of test. The latter has turned out to be the more efficient. Facilities for thermal tests can tolerate a less accurate simulation of the flow field around a car, so they can be built much smaller. Furthermore, the number of aerodynamic and thermal tests has grown so large that more than one facility is needed in any case. For air conditioning work in passenger cars, climatic tunnels with a test-section area between 6 and 10 m^2 are suitable (Buchheim et al 1986). An airstream of even smaller cross-section can be used for engine-cooling tests, and 2 to 4 m^2 are sufficient (Basshuysen et al 1989).

5.5 *Measurement Techniques*

The motivations for improving measurement techniques have been, and still are, threefold: time saving, increased accuracy, and more-detailed information for reaching the ever-higher aerodynamic targets.

Routinely, the frontal area of every model has to be determined. Formerly, this was done by projecting the shadow of a vehicle on a screen using a spotlight a long distance in front of it, and integrating the area of the shadow by hand. This procedure has now been replaced by several fully automated techniques using laser light (Buchheim et al 1987) which are far faster and more accurate.

For a long time, only three kinds of measurement were performed during routine vehicle-shape development: forces and moments; pressure distribution on specific sectors of the surface; and flow visualization by various techniques (Hucho & Janssen 1977). Information on the flow field was rather limited, leaving shape modification to the imagination and experience of the test engineer.

Deeper insight into flow mechanisms should provide a more rational basis for development work, but this requires additional types of data. Boundary layer profiles on the external surfaces and underhood velocity distributions are now being measured by laser-Doppler anemometry (LDA) (Buchheim et al 1987, Cogotti & Berneburg 1991). Wake surveys are being made with multi-hole probes (Cogotti 1987, 1989) and evaluated on the basis of momentum considerations (see Section 4.1).

5.6 *Computational Fluid Dynamics*

The primary reason for the automobile industry's interest in numerical methods is to save time during product development. The ability to quickly react to the ever-changing needs of the market has even higher priority than cost saving, which is very important in its own right. Consequently, all numerical methods need to satisfy two conditions—the first one is necessary, but only the second is sufficient:

- They have to reproduce the related physics with adequate accuracy;
- They have to be faster than experiment.

Only in some narrow niches does computational fluid dynamics as applied to vehicles fulfill both these conditions today. CFD is still more a subject of research than a development tool.

What accuracy means in this context becomes evident from the following requirement: CFD must be able to discriminate a change in drag as small as $\Delta C_D = 0.002$ (Ahmed 1992). Also, computing time has to be competitive with the wind tunnel (Hucho 1989b). The time needed to carry out one specific modification to a model and make the related measurement in a wind tunnel is only from five to ten minutes in partial scale, and from ten to fifteen minutes in full scale.

However, when it finally becomes practical, CFD will have much to offer. First of all, it can generate information before a testable model even exists. Secondly, numerical methods are not necessarily burdened with the limitations of the wind tunnel. For example, computational space can be made large enough to eliminate blockage effects, although not without cost. Also, simulation of the relative motion between vehicle and road and rotation of the wheels are comparatively easy to accommodate. Finally, once the equations have been solved, there is much more information available than from a routine experiment.

Exterior designers would be happy if reliable predictions of aerodynamic characteristics could be made right now. More and more, their creative design process is taking place on the computer. The selection of candidate designs is being made from the computer screen, and aerodynamic data (e.g. C_D) could contribute to these pre-hardware decisions, but only if CFD is able to generate results on time. Ultimately, it should only be necessary to build one or two clay models to finalize a configuration.

Consequently, a large effort is under way to improve CFD, and this is documented in numerous recent publications (e.g. Ahmed 1992, Kobayashi & Kitoh 1992). Four different routes are being followed in the application to road vehicles. They are based on:

- Laplace's equation,
- Reynolds-averaged Navier-Stokes equations (RANS),
- instantaneous Navier-Stokes equations, called direct numerical simulation (DNS),
- zonal models (hybrids).

The Laplace equation is being solved by the well-known panel method; only the surface of the model (and the road) have to be discretized. The vortex-lattice method has only been applied infrequently to cars (Stafford

1973), but was recently used to simulate the trailing vortices at their slanted base (Hummel & Ramm 1992). By definition, only nonviscous flow can be treated with the panel method, and hence no information on drag can be expected. Results are improved if the wake is modeled (Ahmed & Hucho 1977). Its geometry has to be taken from experiment, or from intuition. Pressure distribution can be predicted fairly well on all surfaces where the flow is attached. Forces and moments on body panels like doors, hood, and windows can be calculated. The panel method has been successfully applied to the development of the German high-speed train ICE (Inter City Express, Mackrodt et al 1980). The shape of the train's front end had to be designed to create the smallest possible pressure wave, in order to avoid problems when entering tunnels, passing through stations, or when two trains pass going in opposite directions.

The Reynolds-averaged Navier-Stokes equations (RANS) need a turbulence model for closure, and there is much controversy in the fluid mechanics community as to which one is the most appropriate. In engineering applications the k-ε model is the one most widely used. Close to a wall the empirical logarithmic law of the wall is usually applied, but has only been validated for plane two-dimensional boundary layers. As previously described, the flow over a road vehicle is anything but 2-D.

A very good result has been reported with the RANS concept (Hutchings & Pien 1988). Drag was predicted correctly for a generic car model; however, there were significant differences between computed and measured pressure distributions on the upper and lower rear surfaces. This highlights the need to validate more than the prediction of overall aerodynamic forces, since these might fortuitously camouflage the presence of large, but compensating, errors in local flows.

To date, none of the RANS codes has reproduced the large changes in flow pattern and drag that take place at a critical slant angle of the upper-rear surface of vehicles (see Section 4.1). A computation of such critical behavior has, however, been made using the full instantaneous Navier-Stokes equations (Figure 28, Tsuboi et al 1988). The longitudinal slender cylinder with slanted base of Morel (1978a) was used, and it was located out of ground effect. This configuration was used by Morel to explore the drastic change in drag at a slant angle of $\simeq 30^\circ$ discovered during the development of the Volkswagen Rabbit (Figure 10b, Janssen & Hucho 1975). Numerical results (drag coefficient and flow pattern) agree well with the measurements. Whether or not all the turbulence scales at the test Reynolds number were adequately resolved with the computational grid employed is an open question.

This result has stimulated numerous applications of DNS to actual vehicle development problems in Japan. With almost 10^6 grid points for

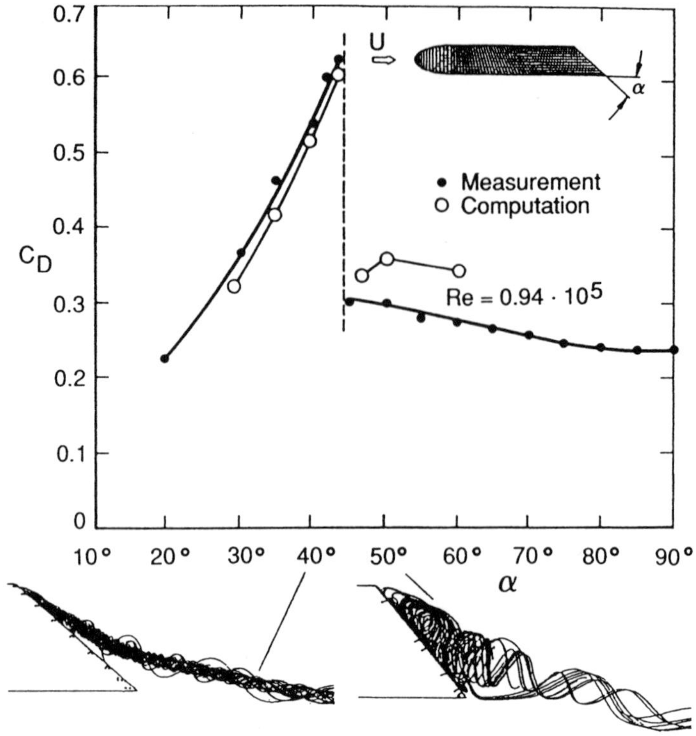

Figure 28 Drag versus rear slant angle α computed with a DNS code and compared to measurements at the same Reynolds number (measurement: Morel 1978a; computation: Tsuboi et al 1988). Bottom: computed flowpaths of the vortices emanating from the sharp edge of the slant.

a half model, DNS has been able to discriminate the effect of several aerodynamic devices (spoilers, flaps) on the drag and lift of a sports car (Figure 29, Kataoka et al 1991). Grid generation is said to require only three days, and CPU-time for a single configuration between 10 and 20 hours on a supercomputer. Calculated drag compares to measurement within 5%.

Hybrid computational schemes need some a priori knowledge of the flow field. The field around a model is partitioned (Larsson et al 1991) into zones of attached flow and zones where separation is expected to occur. The flow in the first type of zone is computed iteratively with a panel method and a boundary layer code. The second zone is treated with either RANS or DNS. An alternative approach is to compute the attached flow with an Euler code and a boundary layer code, and to model the wake.

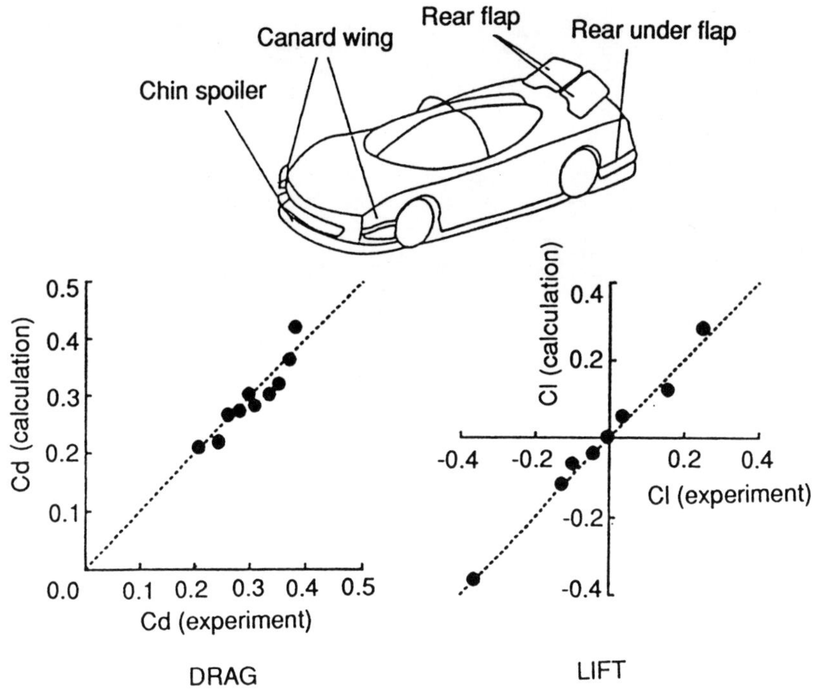

Figure 29 Comparison of computed (DNS) and measured drag and lift coefficients for a sports car with various aerodynamic devices.

Vorticity is generated within the boundary layer, and its transport can be computed with the Euler equations. A promising result has been achieved for two-dimensional flow around a circular cylinder (Krukow & Stricker 1990). No 3-D results and no details of the wake model have been published.

CFD has also been applied to internal flow systems, whose common feature is that they are nonisothermal. Underhood flow has been computed with RANS using the k-ε turbulence model (see e.g. Kuriyama 1988). Interaction between the interior and exterior flows has been treated by simultaneous computation of both flow fields (Ono et al 1992). The flow pattern inside the passenger compartment of cars and buses attracts increasing attention. For moderate air fluxes the velocities are sufficiently small that a laminar treatment is appropriate (Figure 30, Glober & Rumez 1990); for greater fluxes, turbulent flows have been computed (e.g. Ishihara et al 1991). Temperature and velocity distributions inside the compartment have been used to compute the thermal comfort of passengers, with

Figure 30 Streamlines inside a passenger compartment without passengers; laminar flow.

humans being described by a discretized thermo-physiological model (Dick & Stricker 1988).

CFD lends itself to the development of corrections for wind-tunnel data, but has been used for that purpose in only a few cases. However, the contours of interference-free streamline-shaped tunnel walls, either fixed (L. G. Stafford 1979, personal communication) or adjustable (Goenka 1990), have been computed. The geometries of the vehicles in the test section were comparatively simple because only their far-field flow was needed. A blockage ratio of up to 20% can be accommodated with properly shaped walls.

5.7 *Empirical Methods for Drag Estimation*

Early in the recent history of automative aerodynamics it was perceived as a shortcoming that the aerodynamic characteristics of a vehicle could only be established by measurement. The Motor Industry Research Association (MIRA) tried to overcome this with a rating method for drag (White 1967). It made use of nine descriptors of a vehicle's shape. Each geometric feature, such as the front end or the A-pillar, had to be subjectively rated with regard to its drag contribution. The drag of cars was high in those days, the coefficient ranging from 0.50 to 0.60. Freestanding headlamps and fenders were still in vogue. The method had the virtue of identifying those areas that were most important for drag, but it was unsuited for shapes with lower drag. Later, when CFD was still not recognized as being applicable to vehicle design, MIRA made a second attempt at a rating method (Carr 1987). Similar to the first one, drag is broken down according to the location of generation: forebody, afterbody, underbody, wheel and wheel wells, protuberances (e.g. mirrors), and engine cooling system. The contribution of each of these is determined using empirical functions which correlate geometric parameters (similar to those shown in Figure 5) that can be read from a vehicle drawing with local drag. Some 50 generic details are taken into account, 27 empirical (confidential) constants are applied, and no subjective rating is required. Validation of the method with 26 cars (Rose 1984) resulted in a fairly high accuracy of $\pm 5\%$. The method could be integrated into a CAD system, but its limits of applicability are difficult to define. Also, a major drawback is its linearity: No interaction between the various regions of a car body is allowed for. Work is currently under way at MIRA to extend this rating method to lift and lateral coefficients.

An attempt to develop an expert system for drag prediction failed despite promising first results (Buchheim et al 1989), and has been abandoned.

6. OPEN ISSUES AND FUTURE TRENDS

Using drag coefficient as the measure of the state of the art in vehicle aerodynamics, further progress appears to be possible. Today, well-tuned cars have drag coefficients of about 0.30. However lower values are possible (Figure 6), and have already been realized with production cars (Emmelman et al 1990). Whether or not they are feasible for any particular car is more a question of consistency with the vehicle's design concept than it is of aerodynamic capability.

Drag will be of even greater importance for some of the specialty vehicles likely to be on the market in the future (e.g. electric, hybrid). The greatest problem of electric vehicles—limited driving range—is directly impacted by drag. Drag coefficients less than 0.20 are feasible, and electrics can provide the incentive for achieving them. However, the development process will be of increased difficulty because there appears to be significant interaction between the local flows in different areas of a car body at these low drag levels.

If future fuel economy regulation forces cars to become much lighter, the other components of the resultant aerodynamic force will become more important. Directional stability can be improved by reducing lift, mainly at the rear axle. Sensitivity to crosswinds must be kept under control by reducing yawing moment and side force.

The management of underhood airflow will become more and more difficult. Fuel economy regulation is forcing more compact body shapes, and a current styling feature is lower hood lines. Together, they are dictating smaller, and hence more tightly packed, engine compartments. At the same time, the airflow requirement is increasing. Firstly, ozone-layer depletion has mandated a worldwide elimination of the chlorofluorocarbon refrigerant (freon) currently used in automotive air conditioning systems. (in 1991, $\simeq$95% of the new cars sold by U.S. manufacturers had air conditioning). Its environmentally friendly replacement is a less effective refrigerant and requires more airflow for condenser cooling. Secondly, higher underhood temperatures are having a deleterious effect on the durability of engine and vehicle accessories in the engine compartment, and so these temperatures must be controlled. The combination of increased airflow requirement, higher-resistance air path, and under-bumper air inlets will make thermal management of the underhood compartment even more difficult than it is today. Consequently, it will require an even larger fraction of the aerodynamic effort expended in new-vehicle development.

Other parts of the flow field will also call for closer attention. Splash and spray have to be kept under control—not only for trucks where technically feasible solutions are waiting for application. Wind noise has to be reduced as the engine and tires become quieter.

The designers of vehicle shape need more room in which to maneuver. Sometimes they create new aesthetic shapes that are unfavorable for low drag. The required aerodynamic compensation has to come from places that do not affect appearance. Such regions are the vehicle's underside, the wheels, and the wheel wells.

From experiments in both research (Cogotti 1983) and development (Mercker et al 1991) it can be concluded that wheels and wheel wells contribute approximately half the drag of a low-drag car. Major contributors are the front wheels. Due to an outwardly spreading flow under the front part of a vehicle, the local flow approaches the front wheels at a large yaw angle, resulting in high wheel drag. Wake measurements (Cogotti 1987) also confirm this. Because cars must have wheels, and unconventional wheel arrangements are unacceptable, the only way to improve the situation is to suitably control the lateral spreading of the flow under a vehicle. However, only very modest attempts at this have been made so far (Emmelmann et al 1990).

With respect to wind-tunnel test techniques, ground simulation has to be improved for routine testing. Miniature moving belts or mini-dynos, one under each wheel, together with tangential blowing seems to be a feasible approach. Better boundary corrections (adjustments) for open-jet test sections are to be expected by using CFD.

Computer technology per se continues to improve at an almost astonishing rate. More and more aerodynamic computations (CFD) will become possible. They will definitely have a role in research. The question is the degree to which they will be feasible in vehicle development (accuracy, cost, time). How will CFD stack up against intelligent experimentation, e.g. as exemplified by the detail-optimization technique? Will it cope better with the increasingly interactive nature of the flow fields associated with very-low-drag shapes which is making experimental implementation of that technique more difficult? In principle, DNS can do it all. But are computational grids fine enough to discriminate all the scales of turbulent flow at realistic operating Reynolds numbers possible and/or feasible? Even if they are, is the huge amount of detailed data generated actually needed for good aerodynamic design?

Despite some fascinating published results from CFD, the technology is still applicable in only limited areas during the daily course of aerodynamic engineering. From the authors' point of view, the major reservations on existing CFD codes are that, with the exception of the panel method, their accuracy and limits are not well known. Many published comparisons with experiment have been made for body geometries having flows that are too complex. Consequently, it is not possible to determine whether the differences between experiment and CFD are due to the physics of the equations (e.g. turbulence model), numerical characteristics (e.g. mesh geometry, numerical algorithm, computation scheme), or a combination of the two.

As has been common practice in other areas of fluid dynamics, computational codes should be tested—and developed—by comparing their results with either exact (analytical) solutions (which don't exist for complex turbulent flows) or with experiments, but one should start with simple flows. Only when these are fully understood is the step to more complex flows justified. Configurations like cars and trucks have extremely complex flow fields. They should be tackled at the end of a code development process, not at the beginning.

Literature Cited

Ahmed, S. R. 1981. Wake structure of typical automobile shapes. *Trans. ASME, J. Fluids Engrg.* 103: 162–69

Ahmed, S. R. 1984. Influence of base slant on the wake structure and drag of road vehicles. *Trans. ASME, J. Fluids Engrg.* 105: 429–34

Ahmed, S. R. 1992. Numerische Verfahren. See Hucho 1992, Chap. 14

Ahmed, S. R., Hucho, W.-H. 1977. The calculation of the flow field past a van with the aid of a panel method. *SAE Pap. 770 390*

Ahmed, S. R., Ramm, R., Faltin, G. 1984. Some salient features of the time averaged ground vehicle wake. *SAE Pap. 840 300*

Aoki, K., Miyata, H., Kanai, M., Hanaoka, Y., Zhu, M. 1992. A water-basin test technique for the aerodynamic design of road vehicles. *SAE Pap. 920 348*

Barth, R. 1960. Der Einfluss unsymmetrischer Stroemung auf die Luftkraefte an Fahrzeugmodellen und aehnlichen Koerpern. *Automobiltech. Z.* 62. 80–95

Basshuysen, R. v., Chemnitzer, E., Stock, D. 1989. Der neue 3-m2- Klimawindkanal mit Allrad-Rollenpruefstand bei Audi. *Automobiltech. Z.* 91: 646–52

Bearman, P. W. 1984. Some observations on road vehicle wakes. *SAE Pap. 840 301*

Beauvais, F. N., Tignor, S. C., Turner, T. R. 1968. Problems of ground simulation in automotive aerodynamics. *SAE Pap. 680 121*

Buchheim, R., Deutenbach, K.-R., Lueckhoff, H.-J. 1981. Necessity and premises

for reducing the aerodynamic drag of future passenger cars. *SAE Pap. 810 185*
Buchheim, R., Dobrzynski, W., Mankau, H., Schwabe, D. 1983. Vehicle interior noise related to external aerodynamics. *Int. J. Veh. Des.* SP 3
Buchheim, R. Schwabe, D., Roehe, H. 1986. Der neue 6 m2- Klimawindkanal von Volkswagen. *Automobiltech. Z.* 88: 211–18, 389–92
Buchheim, R., Durst, F., Beeck, M. A., Hentschel, W., Piatek, R., Schwabe, D. 1987. Advanced experimental techniques and their application to automotive aerodynamics. *SAE Pap. 870 244*
Buchheim, R., Knorr, G., Mankau, H., Tang, T. 1989. Expertensystem zur Unterstuetzung der Aerodynamikoptimierung von Fahrzeugen. *GMD-Stud.* 160: 60–83
Carr, G. W. 1987. New MIRA drag reduction prediction method for cars. *Auto. Eng.* June/July 1987: 34–38
Cogotti, A. 1983. Aerodynamic characteristics of car wheels. *Int. J. Veh. Des.* SP 3: 173–96
Cogotti, A. 1987. Flow-field surveys behind three squareback car models using a new fourteen-hole probe. *SAE Pap. 870 243*
Cogotti, A. 1989. A strategy for optimum surveys of passenger-car flow fields. *SAE Pap. 890 374*
Cogotti, A., Berneburg, H. 1991. Engine compartment airflow investigations using a laser-doppler velocimeter. *SAE Pap. 910 308*
Dick, A., Stricker, R. 1988. Zur Bewertung inhomogenen Klimas im Pkw durch ein thermophysiologisches Insassenmodell. *VDI-Bere.* 699: 247–64, Duesseldorf
Donges, E., Mueller, B., Seifdenfuss, T. 1990. Design of the BMW-safety concept for active rear-axle kinematics. *23rd FISITA Congr., Torino,*Pap. 905 059
Eckert, W., Vagt, J.-D., Wolff, B. 1990. Die Messtrecke mit geschlitzten Waenden im Porsche Windkanal. *Automobiltech. Z.* 92: 286–97
Eckert, W., Singer, N., Vagt, J.-D. 1992. The Porsche wind tunnel floor-boundary-layer control—a comparison with road data and results from moving belt. *SAE Pap. 920 346*
Emmelmann, H.-J. 1987a. Driving stabilty in side winds. See Hucho 1987, Chap. 5
Emmelmann, H.-J. 1987b. Performance of cars and light vans. See Hucho 1987, Chap. 3
Emmelmann, H.-J., Berneburg, H., Schulze, J. 1990. Aerodynamic development of the Opel Calibra. *SAE Pap. 900 327*
Flegl. H., Rauser, M. 1992. Hochleistungsfahrzeuge. See Hucho 1992, Chap. 7
Gengenbach, W. 1987. Heating, ventilation and air conditioning of motor vehicles. See Hucho 1987, Chap. 10
George, A. R. 1989. Automobile aeroacoustics. *AIAA Pap.* 89–1067, San Antonio, TX
Gilhaus, A., Renn, V. 1986. Drag and driving-stability related aerodynamic forces and their interdependence—results of measurements on 3/8 scale basic car shapes. *SAE Pap. 860 211*
Gilhaus, A., Hoffmann, R. 1992. Richtungsstabilitaet. See Hucho 1992, Chap. 5
Glober, S., Rumez, W. 1990. Berechnung der Stroemung und Temperaturverteilung in Fahrzeugkabinen mit einem Navier-Stokes-Verfahren. *VDI-Ber.* 816: 335–44
Goehring, E., Kraemer, W. 1987. Seitliche Fahrgestellverkleidungen fuer Nutzfahrzeuge. *Automobiltech. Z.* 89: 481–88; 659–66
Goenka, L. N. 1990. A numerical model to determine vehicle pressure-simulation accuracy in a slotted-wall automotive wind tunnel. *SAE Pap. 900 188*
Hackett, J. E., Wilsden, D. J. 1975. Determination of low speed wake blockage correction via tunnel wall static pressure measurements. *AGARD Fluid Dynamics Panel, Symp. on Wind Tunnel Des. Testing Techniques, London*
Hackett, J. E., Sugavanam, A. 1984. Evaluation of a complete wake integral for the drag of a car-like shape. *SAE Pap. 840 577*
Haruna, S., Kamimoto, I., Okamoto, S. 1992. Estimation method for automobile aerodynamic noise. *SAE Pap. 920 205*
Hensel, R. W. 1951. Rectangular wind tunnel corrections using the velocity ratio method. *NACA TN* 2372
Hucho, W.-H. 1974. *Versuchstechnik in der Fahrzeugaerodynamik. Koll. Industrieaerodynamik, Teil* 3: *Aerodynamik von Strassenfahrzeugen.* Aachen
Hucho, W.-H. 1978. The aerodynamic drag of cars—current understanding, unresolved problems and future prospects. See Sovran et al 1978, pp. 7–44
Hucho, W.-H., ed. 1987a. *The Aerodynamics of Road Vehicles.* London: Butterworth
Hucho, W.-H. 1987b. Aerodynamic drag of passenger cars. See Hucho 1987a, Chap. 4
Hucho, W.-H. 1989a. Neuartiger Akustik-Windkanal bei BMW. *Auto. Rev.* 41/5.10.1989: 43–45
Hucho, W.-H. 1989b. Numerischer Windkanal—Stromlinienautos aus dem Supercomputer? *Comput. Technol.* 1989/10: 44–62
Hucho, W.-H., ed. 1992. *Aerodynamik des Automobils.* Duesseldorf: VDI. In press
Hucho, W.-H., Emmelmann, H.-J. 1973.

Theoretical prediction of the aerodynamic derivatives of a vehicle in cross wind gusts. *SAE Pap. 730 232*

Hucho, W.-H., Janssen, L. J., Schwarz, G. 1975. The wind tunnel's ground floor boundary layer—its interference with the flow underneath cars. *SAE Pap. 750 066*

Hucho, W.-H., Janssen, L. J., Emmelmann, H.-J. 1976. The optimization of body details—a method for reducing the aerodynamic drag of vehicles. *SAE Pap. 760 185*

Hucho, W.-H., Janssen, L. J. 1977. Flow Visualization techniques in vehicle aerodynamics. *Int. Symp. on Flow Visualization, Tokyo*

Hummel, D., Ramm, G. 1992. A panel method for the computation of the flow around vehicles including side-edge vortices and wakes. *Proc. ATA-Conf. Innovation and Reliability in Auto. Des. and Testing, Florence*

Hutchings, B. J., Pien, W. 1988. Computation of three-dimensional vehicle aerodynamics using FLUENT/BFC. See Marino 1988, pp. 233–55

Ishihara, Y., Hara, J., Sakamoto, H., Kamemoto, K., Okamoto, H. 1991. Determination of flow velocity distribution in a vehicle interior using visualization and computation techniques. *SAE Pap. 910 310*

Janssen, L. J., Hucho, W.-H. 1975. Aerodynamische Entwicklung von VW-Golf und VW-Scirocco. *Automobiltech. Z.* 77: 1–5

Jones, R. T. 1978. Discussion. See Sovran et al 1978, pp. 40–44

Kamm, W., Schmid, C., Riekert, P., Huber, L. 1934. Einfluss der Reichsautobahn auf die Gestaltung der Kraftfahrzeuge. *Automobiltech. Z.* 37: 341–54

Kataoka, T., China, H., Nakagawa, K., Yanagimoto, K., Yoshida, M. 1991. Numerical simulation of road vehicle aerodynamics and effect of aerodynamic devices. *SAE Pap. 910 597*

Kieselbach, R. J. F. 1982a. *Streamline Cars in Germany. Aerodynamics in the Construction of Passenger Vehicles* 1900–1945. Stuttgart: Kohlhammer Ed. Auto und Verk.

Kieselbach, R. J. F. 1982b. *Streamline Cars in Europe and USA. Aerodynamics in the Construction of Passenger Vehicles* 1900–1945. Stuttgart: Kohlhammer Ed. Auto und Verk.

Kieselbach, R. J. F. 1983. *Aerodynamically Designed Commercial Vehicles* 1931–1961 *Built on the Chassis of: Daimler Benz, Krupp, Opel, Ford.* Stuttgart: Kohlhammer Ed. Auto und Verk.

Klemperer, W. 1922. Luftwiderstandsuntersuchungen an Automodellen. *Z. Flugtech. Motorluftschiffahrt.* 13: 201–6

Kobayashi, T., Kitoh, K. 1992. A review of CFD methods and their application to automobile aerodynamics. *SAE Pap. 920 338*

Koenig-Fachsenfeld, R. v. 1951. *Aerodynamik des Kraftfahrzeugs.* Frankfurt: Umschau

Koenig-Fachsenfeld, R. v., Ruehle, D., Eckert, A., Zeuner, M. 1936. Windkanalmessungen an Omnibusmodellen. *Automobiltech. Z.* 39: 143–49

Krukow, G., Stricker, R. 1990. Einsatzpotential gekoppelter Verfahren zur Simulation von 3D-Fahrzeugstroemungen. *VDI Ber.* 816: 367–77, Duesseldorf

Kuenstner, R., Deutenbach, K.-R., Vagt, J.-D. 1992. Measurement of reference dynamic pressure in open-jet automotive wind tunnels. *SAE Pap. 920 344*

Kuriyama, T. 1988. Numerical simulation on three dimensional flow and heat transfer in the engine compartment using "STREAM." See Marino 1988, pp. 283–92

Larsson, L., Nilsson, L. U., Berndtsson, A., Hammar, L., Knutson, K., Danielson, H. 1989. A study of ground simulation-correlation between wind-tunnel and water-basin tests of a full-scale car. *SAE Pap. 890 368*

Larsson, L., Broberg, L., Janson, C.-E. 1991. A zonal method for predicting external automoblie aerodynamics. *SAE Pap. 910 595*

Mackrodt, P.-A., Steinheuer, J., Stoffers, G. 1980. Entwicklung aerodynamisch optimaler Formen fuer das Rad/Schiene-Versuchsfahrzeug II. *Arch. Eisenbahntech.* 35: 67–77

Mackrodt, P.-A., Pfitzenmaier, E. 1987. Aerodynamik und Aeroakustik fuer Hochgeschwindigkeitszuege. *Phys. Unserer Zeit.* 18: 65–76

Marino, C., ed. 1988. Supercomputer applications in automotive research and development. *Proc. 2nd Int. Conf. on Supercomputing Appl. in the Auto. Ind., Seville.* Minneapolis: Cray Res.

Mercker, E. 1986. A blockage correction for automotive testing in a wind tunnel with closed test section. *J. Wind Eng. Ind. Aerodyn.* 22: 149–67

Mercker, E., Knape, H. W. 1989. Ground simulation with moving belt and tangential blowing for full scale automotive testing in a wind tunnel. *SAE Pap. 890 248*

Mercker, E., Wiedemann, J. 1990. Comparison of different ground simulation techniques for use in automotive wind tunnels. *SAE Pap. 900 321*

Mercker, E., Breuer, N., Berneburg, H., Emmelmann, H.-J. 1991. On the aerodynamic interference due to the rolling wheels of passenger cars. *SAE Pap. 910 311*

Moerchen, W. 1968. The climatic wind tunnel of Volkswagenwerk AG. *SAE Pap. 680 120*

Morel, T. 1978a. The effect of base slant on the flow pattern and drag of three-dimensional bodies with blunt ends. See Sovran et al. 1978, pp. 191–226

Morel, T. 1978b. Aerodynamic drag of bluff body shapes characteristic of hatch-back cars. *SAE Pap. 780 267*

Nouzawa, T., Haruna, S., Hiasa, K., Nakamura, T., Sato, H. 1990. Analysis of wake pattern for reducing aerodynamic drag of notchback model. *SAE Pap. 900 318*

Nouzawa, T., Hiasa, K., Nakamura, T., Kawamoto, A. 1992. Unsteady-wake analysis of the aerodynamic drag of a notchback model with critical afterbody geometry. *SAE Pap. 920 202*

Ogata, N., Iida, N., Fuji, Y. 1987. Nissan's low-noise full scale wind tunnel. *SAE Pap. 870 250*

Ono, K., Himeno, R., Fujitani, K., Uematsu, Y. 1992. Simultaneous computation of the external flow around a car body and the internal flow through its engine compartment. *SAE Pap. 920 342*

Onorato, M., Costelli, A. F., Garrone, A. 1984. Drag measurement through wake analysis. *SAE Pap. 840 302*

Piatek, R. 1987. Operation, safety and comfort. See Hucho 1987a, Chap. 6

Piech, F. 1992. 3 Liter/100 km im Jahr 2000. *Automobiltech. Z.* 94: 20–23

Rauser, M., Eberius, J. 1987. Verbesserung der Fahrzeugaerodynamik durch Unterbodengestaltung. *Automobiltech. Z.* 89: 535–42

Retzlaff, R. N., Hertz, P. B. 1990. Airfoil plan-view body shapes to reduce drag at yaw. *SAE Pap. 900 314*

Rose, M. J. 1984. Appraisal and modification of an empirical method for predicting the aerodynamic drag of cars. *MIRA Released Rep.* 1984/1

SAE 1990. Aerodynamic testing of road vehicles: open-jet wind tunnel boundary interference. *SAE Inform. Rep. J* 2071

SAE 1991. Aerodynamic testing of road vehicles: closed-test-section wind tunnel boundary interference. *SAE Inform. Rep. J* 2085

Schaub, U. W., Olson, M. E., Raimondo, S. 1990. Correction of wind tunnel force data for yawed full and half-scale truck models using a modified pressure-signature method. *SAE Pap. 900 187*

Sorgatz, U., Buchheim, R. 1982. Untersuchungen zum Seitenwindverhalten zukuenftiger Fahrzeuge. *Automobiltech. Z.* 84: 11–18

Sovran, G. 1983. Tractive-energy-based formulae for the impact of aerodynamics on fuel economy over the EPA driving schedules. *SAE Pap. 830 304*

Sovran, G. 1984. The effect of ambient wind on a road vehicle's aerodynamic work requirement and fuel consumption. *SAE Pap. 840 298*

Sovran, G., Morel, T., Mason, W. T., eds. 1978. *Aerodynamic Drag Mechanisms of Bluff Bodies and Road Vehicles.* New York: Plenum

Sovran, G., Bohn, M. S. 1981. Formulae for the tractive-energy requirements of vehicles driving the EPA schedules. *SAE Pap. 810 184*

Squire, H. B., Pankhurst, R. C. 1952. *ARC CP* 80

Stafford, L. G. 1973. A numerical method for the calculation of the flow around a motor vehicle. *Proc. Adv. Road Veh. Aerodyn. BHRA Fluid Eng., Cranfield,* pp. 167–83

Stapleford, W. R., Carr, G. W. 1971. Aerodynamic noise in road vehicles. *MIRA Rep.* 1971/2

Stollery, J. L., Burns, W. K. 1969. Forces on bodies in the presence of the ground. *Proc. 1st Symp. on Road Veh. Aerodynam.,* London, Nov. 6–7

Tsuboi, K., Shirayama, S., Oana, M., Kuwahara, K. 1988. Computational study of the effect of base slant. See Marino 1988, pp. 257–72

Vagt, J.-D., Wolff, B. 1987. Das neue Messzentrum fuer Aerodynamik—Zwei neue Windkanaele bei Porsche. Teil 1: *Automobiltech. Z.* 89: 121–29; Teil 2: *Automobiltech. Z.* 89: 183–89

von Schulz-Hausmann, K., Vagt, J.-D. 1988. Influence of test section length and collector area on measurements in 3/4-open-jet automotive wind tunnels. *SAE Pap. 880 251*

White, R. G. S. 1967. A rating method for assessing vehicle aerodynamic drag coefficients. *MIRA Rep.* 167/9

Wiedemann, J. 1986. Optimierung der Kraftfahrzeugdurchstroemung zur Steigerung des aerodynamischen Abtriebes. *Automobiltech. Z.* 88: 429–31

Wiedemann, J., Ewald, B. 1989. Turbulence

manipulation to increase effective Reynolds numbers in vehicle aerodynamics. *AIAA J.* 27: 763–69
Williams. J. E., Hackett, J. E. Oler, J. W. Hamar, L. 1991. Water flow simulation of automotive underhood airflow phenomena. *SAE Pap. 910 307*
Willumeit, H. P., Matheis, A., Mueller, K. 1991. Korrelation von Untersuchungsergebnissen zur Seitenwindempfindlichkeit eines Pkw im Fahrsimulator und Prueffeld. *Automobiltech. Z.* 93: 28–35
Wright, P. G. 1983. The influence of aerodynamics on the design of Formula-One racing cars. *Int. J. Veh. Des.*, SP 3: 158–72

Section II
Wind Tunnel Testing

Testing Methodology

Section II
Wind Tunnel Testing

Wind tunnels have served as the primary tool of aerodynamicists since the beginning of active interest in the behavior of the flow around road vehicles. Today, amid the growing use of other tools, they still remain invaluable, and the scope of wind tunnel testing has grown far beyond early expectations.

In the aerodynamic development of road vehicles, a practicing engineer is called upon to integrate aerodynamics into a framework involving other related disciplines. For example, the flow around a vehicle impacts not only its aerodynamic characteristics, but also the ability to cool the engine and the flow-induced noise generated by the air-rush around the passenger compartment. The need to assess the impact of aerodynamics on these related systems broadens the nature of wind tunnel testing. This imposes further requirements on the wind tunnel facilities. As a consequence, the design, construction and commissioning of a new wind tunnel is a very complex task and requires considerable resources and understanding. In view of this challenge, the growing number of different types of wind tunnels being constructed or upgraded for road vehicle aerodynamics reflects the importance still accorded this discipline.

Wind tunnels, in the broadest sense, try to simulate the aerodynamic conditions that a vehicle would experience in its operating environment. However, the aerodynamics generated in a wind tunnel are not identical, but merely equivalent (to some degree) to the aerodynamics experienced "on-road." For example, in a wind tunnel, both the far-field boundary and the ground boundary are represented in some convenient and yet useful manner. Likewise, the test-section flow velocity (around a stationary vehicle) attempts to re-create the relative motion between a moving vehicle and the surrounding (stagnant) environment. Where differences exist between wind tunnel and road, it is important to know how they influence the results that are generated by wind tunnel testing. This has been extensively studied for many years and continues to be an area of enduring interest. It has resulted in many useful ways of dealing with the lack of true replication of on-road conditions.

The aerodynamic forces and moments experienced by a vehicle in its operating condition are often of crucial interest to aerodynamicists. Force and moment balances have served as the most direct tool for determining these quantities. Much progress has been made in minimizing their sources of error. Moreover, the motivation to understand the causes of dynamic loads and to create more opportunities for the diagnosis of flow conditions has led to significant advances in the kinds of tools that are available. These tools, many of them derived from computer-based devices and methods, have helped shed light on many previously unknown flow phenomena, both local and global. The plethora of these tools ranges from simple subjective flow visualization to very complex quantitative measurement systems. They have been used in applications for both external and internal flow situations.

In the final analysis, however, the utility of any special wind tunnel test methodology or any measurement technique is linked to the cost (in its broadest sense) incurred through its use. Many wind tunnel test procedures involve very expensive wind tunnel modifications (such as a moving ground-plane) or complex test-vehicle preparation. Some measurement techniques involve expensive instrumentation and prolonged data acquisition duration. As a consequence, their use is usually limited to those kinds of wind tunnel tests where the information being sought justifies such effort and cost.

940414

The Effect on Aerodynamic Drag of the Longitudinal Position of a Road Vehicle Model in a Wind Tunnel Test Section

K. P. Garry
Cranfield Institute of Technology

S. B. Wallis
Ford Motor Co.

K. R. Cooper and A. Fediw
National Research Council

D. J. Wilsden
Lockheed Aeronautical Systems Co.

ABSTRACT

It is not generally appreciated that the longitudinal position of a road vehicle model in a wind tunnel can have a significant influence on its measured aerodynamic drag. This paper explores the influence of the proximity of the end of the test section on measured aerodynamic drag, where the 'end' of the test section is defined by the start of the first diffuser or the end of a separate groundboard. Both flat plates and three-dimensional, automotive shapes were tested in three different model-scale and full-scale wind tunnels. It was found that the drag began to change from its upstream, undisturbed value when a vehicle model was closer than a distance of four times the square root of its base area from the end of the test section and that large changes occur when a vehicle model was closer than twice the square-root of its base area to the end of the test section. The effect is attributed to base pressure changes in the proximity of the diffuser or of the end of a groundboard. It is suggested that this effect may have contributed to some of the lack of correlation observed between wind tunnels and that it may be partially responsible for the lower drag levels measured in short-test-section, open-jet wind tunnels, when compared with closed-wall wind tunnels.

INTRODUCTION

The influences of wind tunnel boundaries on the aerodynamic forces measured on models have been and are a subject of active research [1,2]*. In virtually all instances, these investigations have focussed on the constraint posed on the flow over a model by the lateral boundaries of the wind tunnel, whether they were solid walls, an open jet, or porous surfaces. Little attention has been paid to the upstream and downstream boundaries of the test section. It is the intention of this paper to demonstrate that the proximity of the base of an automobile to the downstream boundary of the test section may cause significant errors in its drag measurement and to provide a general guide to the longitudinal placement of models to avoid this problem.

To effect this purpose, wind tunnel tests were performed in the 9mx9m wind tunnel of the National Research Council of Canada (NRC 9m), the 0.9mx0.9m Pilot Wind Tunnel of the NRC (NRC Pilot Tunnel), and the 5mx7m Low Speed Wind Tunnel of the Lockheed Aeronautical Systems Company (Lockheed LSWT). The tests were jointly sponsored by the Ford Motor Company, the National Research Council of Canada, and the Lockheed Aeronautical Systems Company.

STATEMENT OF THE PROBLEM

Comparisons of aerodynamic forces and moments measured on the same models in different wind tunnels [3,4] have shown that their absolute magnitudes usually do not agree, with aerodynamic drag differences as large as ±7.5% from the mean of the tunnels having been observed for long, heavy-duty truck models [3] after correction for wind tunnel blockage. The variations between wind tunnels were not random and the data from any pair of wind tunnels correlated well [4], in the sense that their relationship was linear but with a slope other than unity. This behaviour suggests that there were consistent, physical explanations for the observed behaviour. It was postulated in [3] that these measurement differences might result from: calibration errors, flow non-uniformity, longitudinal pressure gradients, inadequate blockage corrections, and model longitudinal-position/test-section-length effects.

Of these possibilities, the last seemed the most likely, supported by preliminary measurements of the effect of model position by Garry [5] and by some of the measurements reported in the SAE Information Report on open-jet boundary interference [6].

* numbers in square parenthesis refer to references at the end of the paper

The measurements in [5] showed an effect of proximity to the end of a groundboard on drag at the 5% level for a truck-like shape. As the end of the groundboard was moved closer to the model, the drag was first observed to rise above the undisturbed value measured away from the end of the groundboard, followed by a drop to below the undisturbed value when the end of the groundboard was close to the rear face of the model. These measurements also showed that a body with a more streamlined rear end showed a larger drag rise before the final drop. This preliminary work did not cover a sufficiently large range of position from the end of the groundboard to be certain that the region over which significant changes could occur had been fully surveyed. Further, the model position in this study was changed by removing pieces from the trailing edge of the groundboard, rather than moving the model, offering the possibility that some of the observed variation could be due to changing groundboard length. Finally, there was the possibility that the results were tunnel-specific and could not be generalized.

The measurements presented in Figures 2.15 and 4.4 of [6] showed that the drag measured on simplified automobile shapes decreased as a model was moved downstream in the test section of an open jet tunnel. The drag reductions were presented as a function of test section length (Fig. 2.15) and as a function of distance from the nozzle (Fig. 4.4) but these reductions may, in fact, result from proximity to the diffuser at the end of the typically short, open-jet test sections.

A new series of experiments were planned in several wind tunnels with different test section geometries and ground simulations to better understand this behaviour.

While the flow mechanisms that produced the drag differences reported in [3,4] are not known with certainty, it is possible that changes in the base pressure due to proximity to the end of the test section could have produced them, based on the results from [5,6]. Most bluff shapes typical of road vehicles have separated flow regions at their rear ends that encompass some or all of the rear faces of the vehicles at zero yaw angle. These complex, three-dimensional separations vary with yaw angle, usually increasing in size with increasing angle. The separating shear layers produce a low-pressure, closed, separation bubble containing recirculating flow. Since the pressures in this bubble are a major contributor to base drag, it was postulated that the too-close proximity of downstream tunnel boundaries could lead to changes in the developing wake flow, changing the base pressure and producing the observed drag behaviour. The pressure changes could come from effects of the diffuser pressure field or from leakage flows entering the wake bubble from floor discontinuities like that found at the end of a separate groundboard or a floor-mounted breather. The position at which the bubble is disturbed by end-of-test-section flow effects should depend on the size of the separation bubble which will relate to the size of the separation region on the base of the vehicle and, thereby, to the shape of the rear of the vehicle.

Because the size of the separation bubble is determined by the size of the vertical or near-vertical rear surfaces, the tunnel positions over which significant effects can occur should relate to a characteristic dimension of this area. Several possibilities can be suggested, including: separation-region height, separation-region width, or the square-root of separation-region area. In practice, these areas will be the base areas of trucks, vans, or other square-backed vehicles and the trunk areas of fast-back or notch-back automobiles.

EXPERIMENTAL APPROACH

A series of models were selected that produced flow fields that were typical of automotive shapes. The models consisted of flat plates of various areas and aspect ratios, simplified block automotive shapes with square-back and fast-back configurations, and two full-sized light trucks - a van and a pick-up. The models were positioned at a series of locations relative to the end of the test section, where the 'end' was generally defined by some physical limit that included: the start of the first diffuser, the start of a floor-mounted, downstream breather slot, or the end of a separate groundboard. Aerodynamic drag and/or base pressure were measured as a function of longitudinal position in each test section.

WIND TUNNEL DESCRIPTIONS - Three wind tunnels were used to provide a range of test section configurations and diffuser angles, improving the generality of the results. These wind tunnels had test locations that were on the tunnel floor or on separate, full-width groundboards - the two most common test installations. The Lockheed LSWT had a floor test location only, while the two NRC tunnels could test on their floors or on groundboards. These full-width groundboards occupied part of the length of the test sections and ended 1/3 of the tunnel width ahead of the first diffusers in both cases. The upper surfaces of the groundboards were 14.7 % of the tunnel width above the floor, resulting in reduced test section areas. The NRC tunnels also had breather slots in the floor that could be closed downstream of the models to investigate this effect. The pertinent test section geometries are summarized in Table 1.

MODEL DESCRIPTIONS - The sharp-edged, flat plate models can be divided into two groups - those built for the full scale tunnels (Lockheed LSWT, NRC 9m tunnel) and those built for the Pilot Tunnel. The three-dimensional models were provided by full-scale light trucks in the large tunnels and by block models in the small tunnel. The dimensions of these models are summarized in Table 2. Photographs of the models are presented in Figure 1.

The flat plate model tested in the full scale tunnels was a 860 mm-square plate attached to a mounting frame through three load cells to provide drag measurement. Base pressure was measured at four points on the back face and averaged pneumatically. The plate could be mounted at either of two heights above the ground, providing 90 mm or 483 mm clearance from the bottom of

Table 1: Test Section Geometries						
Tunnel	**Ground Simulation**	**Test Section. (h,w,ℓ) m**	**Test-Section Area, C, m²**	**Diffuser Cone Angle**	**Contraction Ratio**	**Floor Breather**
Lockheed LSWT	floor	5.0, 7.1, 12.5	35.1	6.0°	7.0	no
NRC 9m	floor	9.1, 9.1, 22.9	81.9	3.0°	6.0	yes
NRC 9m	groundboard	7.9, 9.1, 16.3	71.2	3.0°	6.0	no
NRC pilot	floor	0.91, 0.91, 2.29	0.823	3.0°	6.0	yes
NRC pilot	groundboard	0.79, 0.91, 1.62	0.706	3.0°	6.0	no

the plate to the tunnel floor. The upper location removed the plate from ground effect while the lower position placed the plate in ground effect. Most of the measurements were made with the plate low, although one set of measurements was made on the groundboard in the NRC 9m tunnel with the plate high. The plate model was bolted to the floor or to the groundboard at a series of predetermined locations in each wind tunnel.

Both square and rectangular flat plate models were tested in the Pilot Tunnel measurements. The five square plates had different areas while the three rectangular plates had fixed area and varying height-to-width (aspect) ratios. The rectangular plates were used to allow the significant plate dimension - height, width, or root area - to be determined. The plates were attached to a strut connected to the front face of each plate and were mounted at a clearance height of 15 % of the height of each plate. The leading strut was used to allow the plate to be positioned close to the end of the groundboard or the test section. The strut was attached to a balance outside the wind tunnel that was connected to a sliding mount that provided rapid position change. Base pressure was measured as in the full-scale tunnels.

Full-scale vehicles were tested in both full-scale wind tunnels. Only base pressure was measured in the NRC 9m tunnel while both base and front-face pressures were measured in the Lockheed LSWT. The vehicles were driven up the lengths of the test sections with the tunnels running to survey the position effect. The NRC test was done with the breather open and closed.

A simple block model was tested in the NRC Pilot Tunnel to provide a configuration similar to that used in the truck tests in the large tunnels. This rectangular block had rounded leading edges to provide attached flow to the base. Its rear shape could be changed from a square-back configuration to a 27° fast-back configuration by removing the upper-rear corner block. The length of the slanted portion was such that the vertical rear surface was one-half that of the square-back configuration.

TEST PROCEDURE - The models were positioned as far back in each test section as possible and then moved forward to the standard testing location - the reference location - and then toward the inlet of the test section. Drag and/or base pressure were recorded at a wind speed of approximately 30 m/s at each position. This speed was chosen because it provided good signal levels and because it was a typical test speed for the trucks. Most of the tests were performed at zero yaw angle, except for two cases in the Pilot Tunnel where the square-back and the fast-back models were also tested at 14° yaw angle.

PRESENTATION OF RESULTS

All the measurements showed the same, consistent trends - that the base pressure increased and the drag decreased as the base of a model approached the end of the test section.

FLAT PLATES - The general behaviour of the flat plates was best typified by the measurements made on the flat plate models. Figure 2 shows the base pressure coefficient measured for the square plate models mounted on the groundboard in the NRC Pilot Tunnel as a function of absolute position in the test section. The origin of coordinates was at the trailing edge of the groundboard and x was measured positively upstream. The reference model location was at the turntable centre, at x=1050 mm. It was evident that each plate had a similar variation in base pressure behaviour as it was moved downstream from the reference location toward the end of the groundboard at x=0. Initially, the base pressure was negative and constant. As the plate approached the end of the groundboard, the base pressure increased sharply before decreasing once again when close to the end of the groundboard.

The aerodynamic drag was measured at the same time as the base pressures and showed parallel trends, as can be seen in Figure 3. The drag was initially constant far upstream of the end of the groundboard, followed by a sudden decrease as the plate was moved downstream and then a slight rise as the end of the groundboard was reached. The drag drop and the base pressure rise occurred farther upstream for the successively larger plates, suggesting that the observed behaviour might collapse on longitudinal position when non-dimensionalized by a relevant plate dimension - height, width, or square-root area.

The rectangular plate data demonstrated that a collapse was possible, using the *square root of plate area* to non-dimensionalize the longitudinal position, as shown in Figures 4 and 5. Here, the base pressure

Table 2: Model Geometries

Model	Test Section	w,h,ℓ mm	w:h①	Clearance z/h②	Frontal Area, S, m2	Base Area A, m²	$A^{1/2}$ mm	Blockage S/C③
Full-scale Calibration Plate	LSWT	864.0, 864.0, -	1:1	0.10	0.746	0.746	864.0	0.021
	9 m, floor	864.0, 864.0, -	1:1	0.10	0.746	0.746	864.0	0.009
	9 m, gb ④	864.0, 864.0, -	1:1	0.10	0.746	0.746	864.0	0.010
E150 Van	LSWT	1651.0, 1854.2, 5385.0	1.12:1	0.26	3.67	3.67	1915.7	0.105
F150 Pick-up	9 m, floor	1879.6, 1371.6, 5283.2	1.37:1	0.35	2.79	2.79	1670.3	0.034
Square Plate Models	0.9 m, gb	78.0, 78.0, -	1:1	0.15	0.0061	0.0061	77.8	0.009
		129.0, 129.0, -	1:1	0.15	0.0166	0.0166	128.8	0.024
		162.0, 162.0, -	1:1	0.15	0.0262	0.0262	168.9	0.032
		189.0, 189.0, -	1:1	0.15	0.0357	0.0357	188.9	0.051
		230.0, 230.0, -	1:1	0.15	0.0530	0.0530	230.2	0.064
Square Plate Models	0.9 m, floor	129.0, 129.0, -	1:1	0.15	0.0166	0.0166	128.8	0.020
		162.0, 162.0, -	1:1	0.15	0.0262	0.0262	168.9	0.032
		230.0, 230.0, -	1:1	0.15	0.0530	0.0530	230.2	0.064
Rectangular Plate Models	0.9 m, gb	114.5, 228.8, -	1:2	0.15	0.0262	0.0262	168.9	0.032
		228.8, 114.5, -	2:1	0.15	0.0262	0.0262	168.9	0.032
		280.2, 93.5, -	3:1	0.15	0.0262	0.0262	168.9	0.032
Square-back Block	0.9 m, gb	179.0, 138.0, 345.0	1.3:1	0.15	0.0247	0.0247	157.2	0.035
	0.9 m, floor	179.0, 138.0, 345.0	1.3:1	0.15	0.0247	0.0247	157.2	0.030
Fast-back Block	0.9 m, gb	179.0, 138.0, 345.0	1.3:1	0.15	0.0247	0.0124	111.1	0.035
	0.9 m, floor	179.0, 138.0, 345.0	1.3:1	0.15	0.0247	0.0124	111.1	0.030

① ratio of body width to body height ② ratio of underbody clearance to body height
③ ratio of model frontal area, S, to test section area, C ④ groundboard

coefficients and the drag coefficients of all eight plates have been normalized by the upstream values at the reference location and longitudinal position has been normalized by the square-root of plate area. The plotted data show some dependence of the magnitudes of the normalized drag and base pressure coefficient ratios with plate size, probably as a result of the interference caused by the common mounting strut. In Figure 5, the drag is seen to begin dropping when the plate was approximately $x=4A^{1/2}$ from the end of the groundboard and reached a minimum for all plates at $x\approx1.4\ A^{1/2}$. The main drag change occurred when the plates were closer to the end of the groundboard than $x=2A^{1/2}$.

An identical trend was seen for the full-scale calibration plate on the groundboard of the NRC 9m wind tunnel, as shown in Figure 6. The plate was tested at two heights and it can be seen that the influence of the approach of the end of the groundboard was greater when the plate was in the lower position, where the drag reduction was double that in the higher position.

When the full-scale calibration plate was mounted on the floor of the Lockheed LSWT a similar trend was observed - the base pressure increased as the end of the test section (the start of the first diffuser) was approached, Figure 7. In this case, the pressure rise was monotonic with position, not showing the final decrease measured with the groundboard.

AUTOMOTIVE SHAPES - Figure 7 also shows the changes in base pressure measured in the Lockheed LSWT on the base of a full-scale Ford Econoline van and in the NRC 9m tunnel on a Ford pick-up truck with a cap. Both trucks have a significantly different flow field than the plate although all three have similar wake structures. The normalized pressure coefficient variation with position of the base of the Econoline van from the diffuser was nearly identical to that for the flat plate. The measurements in the NRC 9m tunnel showed a similar pressure reductions but of about one-half the magnitude of those in the Lockheed LSWT.

The front face pressure on the Econoline van in the Lockheed LSWT remained constant at all positions.

Additional measurements were made on the floor and on the groundboard in the NRC Pilot Tunnel using small-scale, block models that represented common automotive shapes. The measurements on these bodies are presented in Figures 8 and 9, respectively. The block models had either a square or a slanted rear end and a rounded front.

Only the square-back configuration of the block model was tested on the floor with the floor-mounted breather vents open and closed, Figure 8. The normalized drag coefficient distribution showed the characteristic drag drop and compared closely with the pressure data from the Lockheed LSWT in Figure 7. It can be seen that the presence of an open breather slot on the floor downstream of the model caused little change in the overall behaviour.

Figure 9 shows that the drag drop close to the end of the groundboard also occurred for fast-back configurations and that the behaviour at 14° yaw angle was basically the same as at zero for both rear geometries. Unlike the flat plate measurements, the drag rises prior to the final drop, as was observed in [5]. The data do not collapse as well as for the previous cases but still show drag changes starting at $x \approx 4A^{1/2}$. The major drag changes occur for $x \leq 2A^{1/2}$, where the reference area used for the fast-back configuration was one-half that used for the square-back. The square-back drag showed a larger drag drop than the fast-back drag. This may be attributable to the strong vortical flows in the wake of the fast-back geometry being less sensitive to base pressure changes.

DISCUSSION

Irrespective of the test section or the model geometry employed, all the measurements showed one consistent trend - that the positioning of an automotive model too near to the end of the test section caused significant base pressure and drag changes. At times, the base pressure decreased and the drag increased but the dominant tendency was for the base pressure to rise and for the drag to decrease. The mechanism of these changes is not fully understood but could relate to changes in the near-wake structure induced by leakage flows from beneath a groundboard and from the pressure rise that commences just upstream of the start of a diffuser.

Most groundboards will carry some lift, a result of the circulation that occurs because of the influence of mounting struts and other hardware that are below the groundboard, in the lower flow path. Thus, the pressures below a groundboard are often positive. When the lower flow can bleed into the near-wake of a model because it is too near the end of the groundboard they may increase the base pressure and cause the drag reductions observed. As the plate moves to the end of the groundboard and away from it, the flow will return to the undisturbed case and the drag will rise toward the upstream value, a trend that was seen in the flat plate data on the groundboard.

That the near-wake structure is changed when approaching the end of the groundboard can be seen in the NRC water tunnel flow visualization photographs of the block models in Figure 10. The visualization was done using aluminium particles illuminated by a strobed plane of light on the model longitudinal centre line to trace the flow. The upper photograph shows the model away from the influence of the end of the groundboard at $x/A^{1/2}=5.0$. In the middle photograph at $x/A^{1/2}=1.5$, the point at which the drag drop was maximum, evidence of distortion of the curvature of the flow over the rear of the model and an interaction of this flow with the end of the groundboard can be seen. Flow can also be seen to enter the wake of the model from below the groundboard. The reduced flow curvature in the wake region shown in the middle photograph, in comparison to that of the upper photograph, may be an indicator of the higher base pressures produced as the end of the groundboard is approached. The lower photograph shows an even stronger flow from below the groundboard into the base region of the model.

The drag reductions on the plates and the other shapes that were observed as the first diffuser was approached must be caused by the rising static pressure in the diffuser increasing the base pressures of these models. Thus, one would expect some effect of diffuser angle, as can be seen in Figure 7 where the base pressure change in the Lockheed LSWT with a diffuser cone angle of 6° was greater than that in the NRC 9m tunnel which has a cone angle of 3°. Because most wind tunnels are designed for efficient diffusion and the minimum circuit length, they will have similar diffuser angles to these and should produce similar base pressure and drag changes to those observed.

An extreme example of the diffuser effect was observed when the simple block models were tested in the 0.6mx0.6m insert test section of the NRC Pilot Tunnel which has a larger diffuser cone angle of 10°. The presence of the large model near the end of the test section caused the first diffuser to stall, and the drag to drop to one-half the value measured just before stall. The higher pressures downstream in the diffuser propagated up the separated region in the diffuser and caused a large change in the base pressure of the model. This set of measurements was not presented because of the flow separation induced in the large-angle diffuser but they serve to underline the effect.

CONCLUDING REMARKS

Wind tunnel measurements on automotive shapes were made in five test sections of three wind tunnels to determine the effect of model longitudinal position on aerodynamic drag. It was found that a position that placed the base of a model closer to the end of the test section than four times the square-root of its separated base area could lead to drag measurement errors that were of the order of 10% to 20%, depending on vehicle configuration and position. This effect was ascribed to changes in base pressure due to the influence of the diffuser pressure field or due to interference flows from below the groundboard disturbing the near wake.

Although these measurements were made in closed-wall wind tunnels, the same diffusion occurs in open-jet tunnels and so they should be affected similarly. Most full-scale, open-jet wind tunnels have shorter test sections than closed tunnels and so may be more prone to model-position-induced measurement errors.

This proximity effect may explain part of the lack of correlation found between the same models in different wind tunnels and may also partially explain the tendency of open-jet wind tunnels to measure lower drag values than closed-wall wind tunnels, even after all other boundary effects are corrected for.

It was recommended that the base of automotive models should never be closer than twice the square-root of the base area to the end of a groundboard or the start of the first diffuser, $x=2A^{1/2}$, and it was further recommended that a more conservative separation of $x=4A^{1/2}$ be utilized.

RECOMMENDATIONS

The measurements presented in this paper show a consistent trend that supports a guideline on the positioning of models in a wind tunnel. The guideline was developed in closed-wall wind tunnels but should also be applicable to open jet tunnels. The recommendation is: ***that the base of a road vehicle model in a wind tunnel should be at least four times the square-root of its base area, $x \geq 4A^{1/2}$, from the end of the test section, defined by the end of the groundboard or the start of the first diffuser, and should never be closer than $x=2A^{1/2}$.***

REFERENCES

1. Garner, H.C.; Rogers, E.W.E.; Acum, W.E.A. and Maskell, E.C. 'Subsonic Wind Tunnel Wall Corrections'. AGARDograph 109, Oct. 1966.

2. 'Aerodynamic Testing of Road Vehicles - Closed-Test-Section Wind Tunnel Boundary Interference'. SAE Information Report J2085, 1994.

3. Cooper, K.R.; Gerhardt, H.J.; Whitbread, R.; Garry, K.P. and Carr, G.W. 'A Comparison of Aerodynamic Drag Measurements on Model Trucks in Closed-Jet and Open-Jet Wind Tunnels'. 6th Colloquium on Industrial Aerodynamics, Aachen, Germany, 1985.

4. Cooper, K.R.; 'A Note on the Correlation of Automotive Force and Moment Measurements Between the Low Speed Wind Tunnels at Lockheed and at the National Research Council'. Report issued to the Ford Motor Co by the NRC, July 18, 1990.

5. Garry, K.P. 'Preliminary Wind Tunnel Tests on the Influence of Ground Board Length on the Aerodynamic Drag of Simple Commercial Vehicle Models'. College of Aeronautics, CoA Report No. 8619, Cranfield, UK, 1986.

6. 'Aerodynamic Testing of Road Vehicles - Open Jet Wind Tunnel Boundary Interference'. SAE Information Report J2071, 1993.

flat plates and the simple block model with alternate slanted rear in the NRC 0.9 m pilot tunnel

block model on the groundboard in the NRC 0.9 m pilot tunnel at 14° yaw angle looking downstream

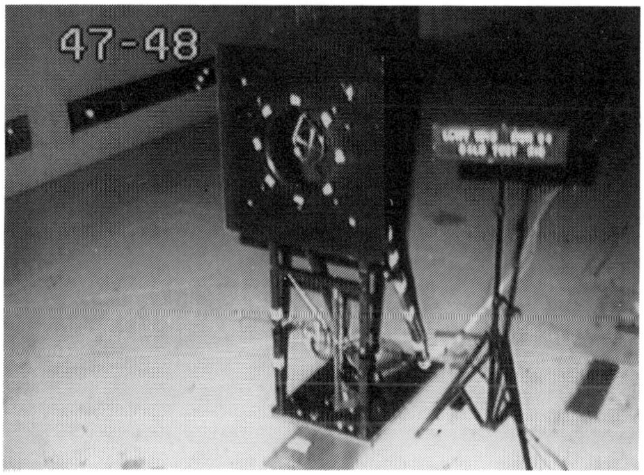

flat plate model attached to its stand in the Lockheed LSWT

Econoline van in the Lockheed LSWT at the downstream end of the test section

pick-up truck with cap on the floor of the NRC 9 m wind tunnel

FIG. 1: Model Photographs

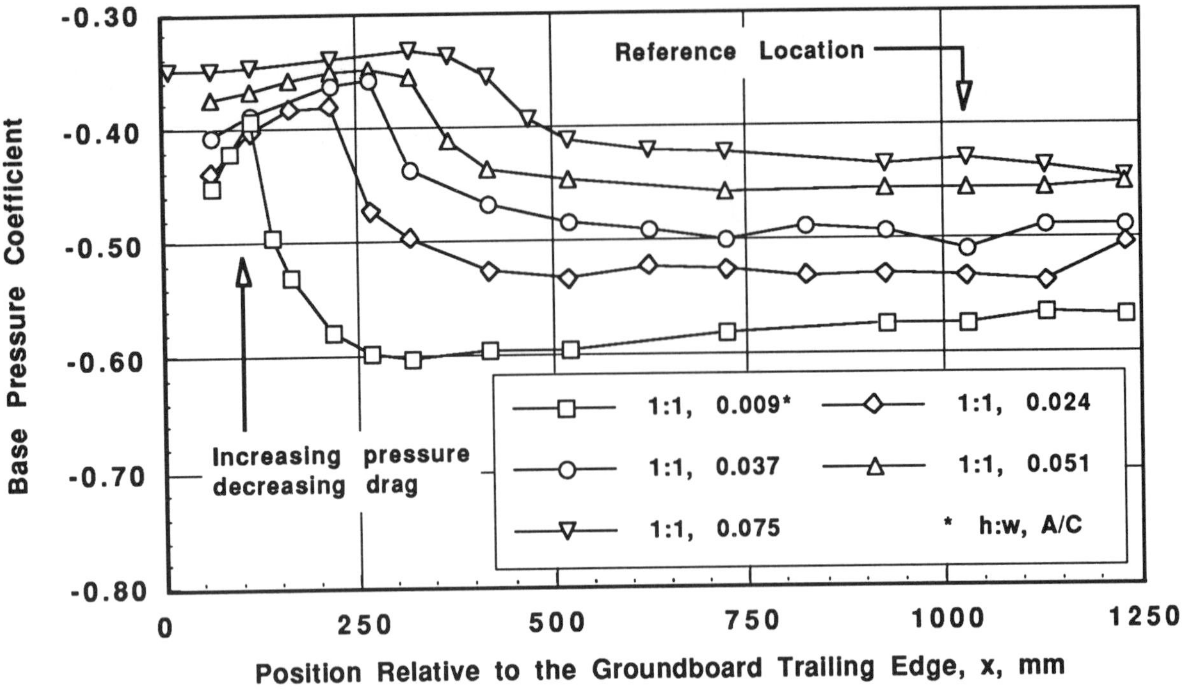

FIG. 2: Effect of Longitudinal Position on the Base Pressure of Square Flat Plates on the Groundboard of the NRC Pilot Tunnel

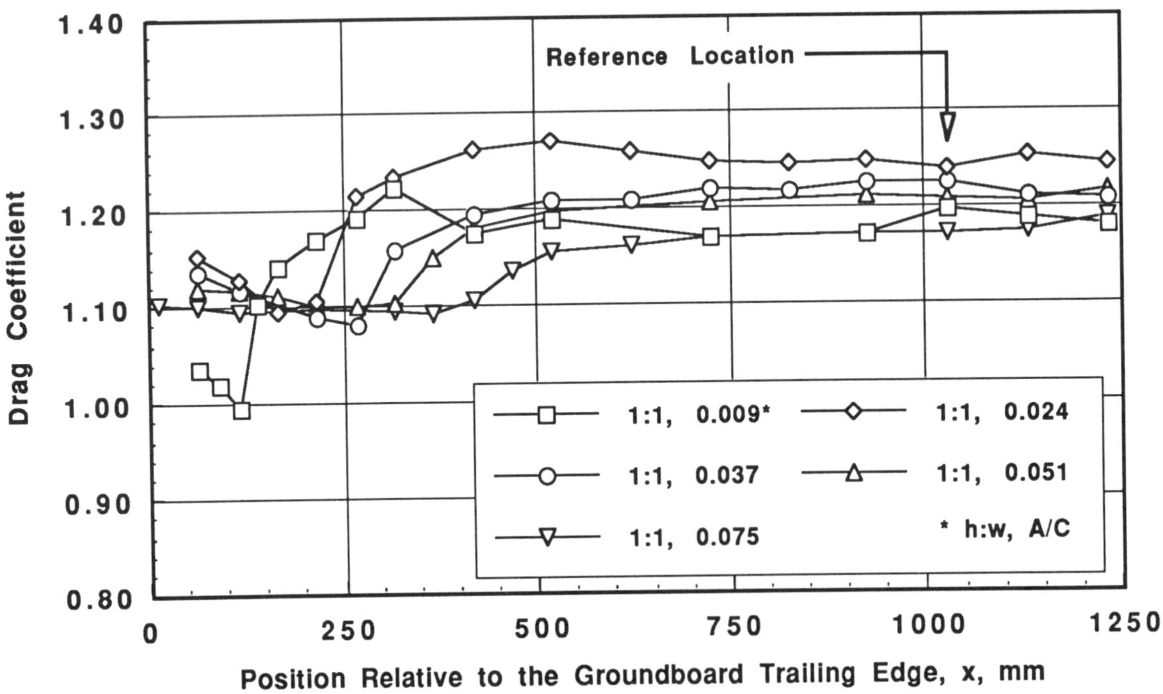

FIG. 3: Effect of Longitudinal Position on the Drag of Square Flat Plates on the Groundboard of the NRC Pilot Tunnel

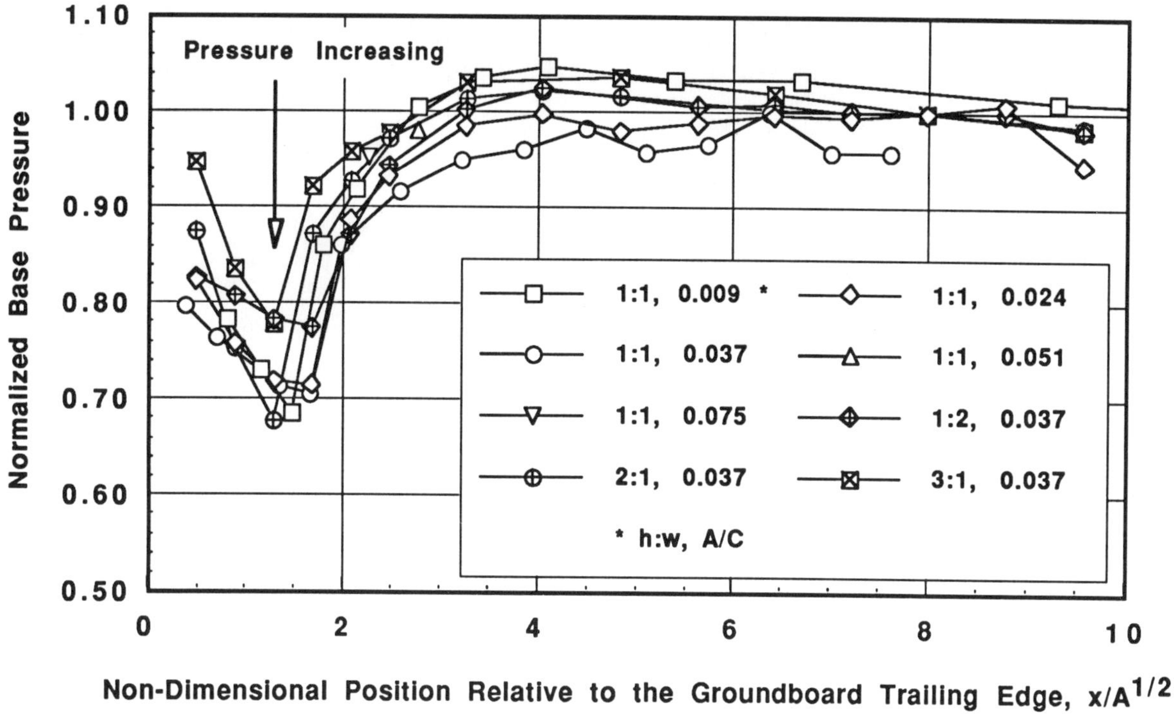

FIG. 4: Collapse of the Normalized Base Pressure Coefficient Distributions for Square and Rectangular Plates on the Groundboard of the NRC Pilot Tunnel

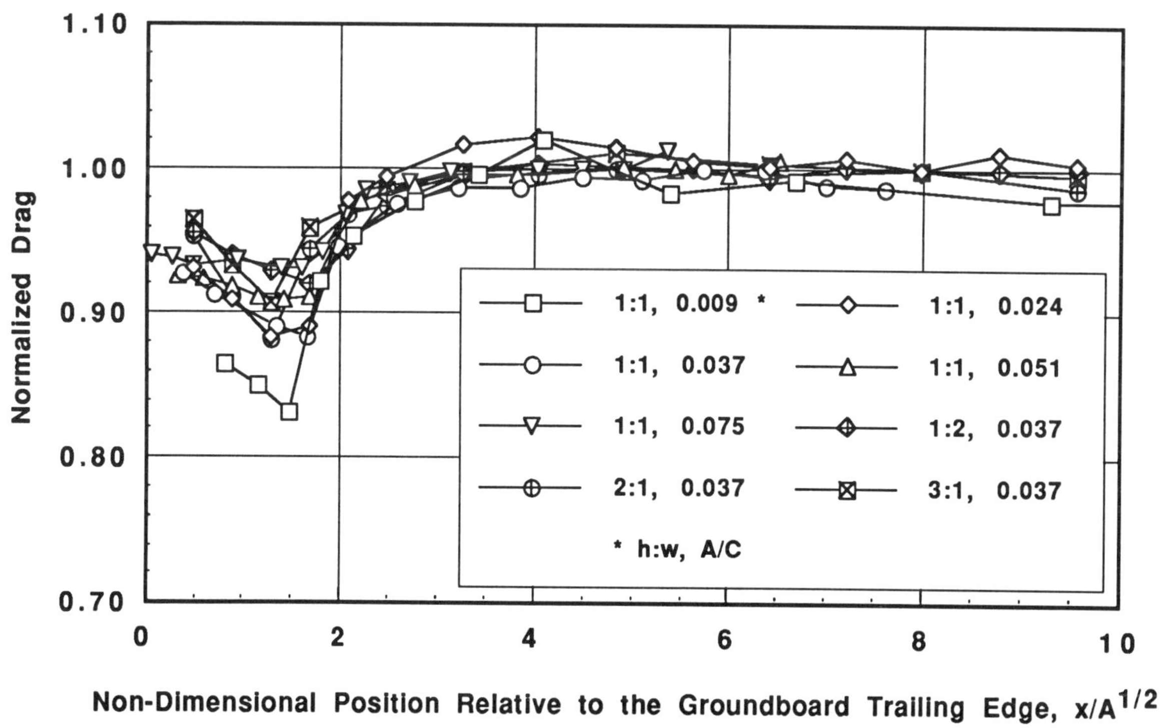

FIG. 5: Collapse Of The Normalized Drag Coefficient Distributions for Square and Rectangular Plates on the Groundboard of the NRC Pilot Tunnel

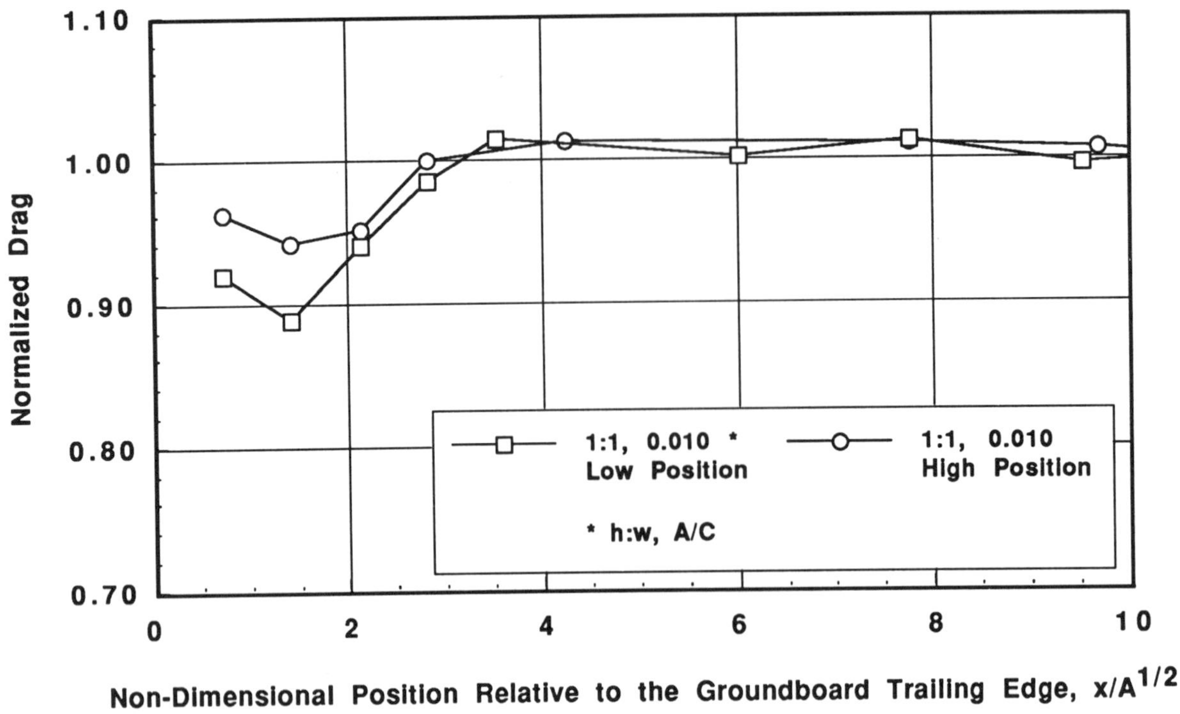

FIG. 6: Normalized Drag Coefficient Distributions for a Square Plate on the Groundboard of the NRC 9m Tunnel

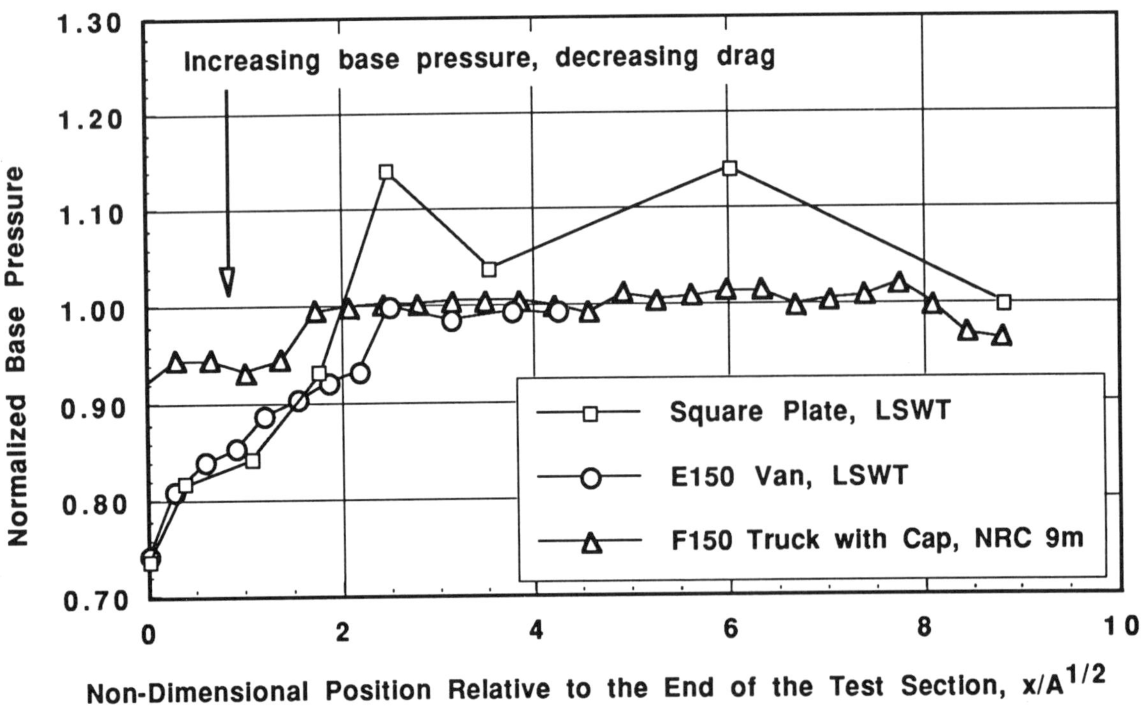

FIG. 7: Normalized Base Pressure Distributions for an Econoline Van and a Square Plate in the Lockheed LSWT and for a Pick-up Truck in the NRC 9m Tunnel

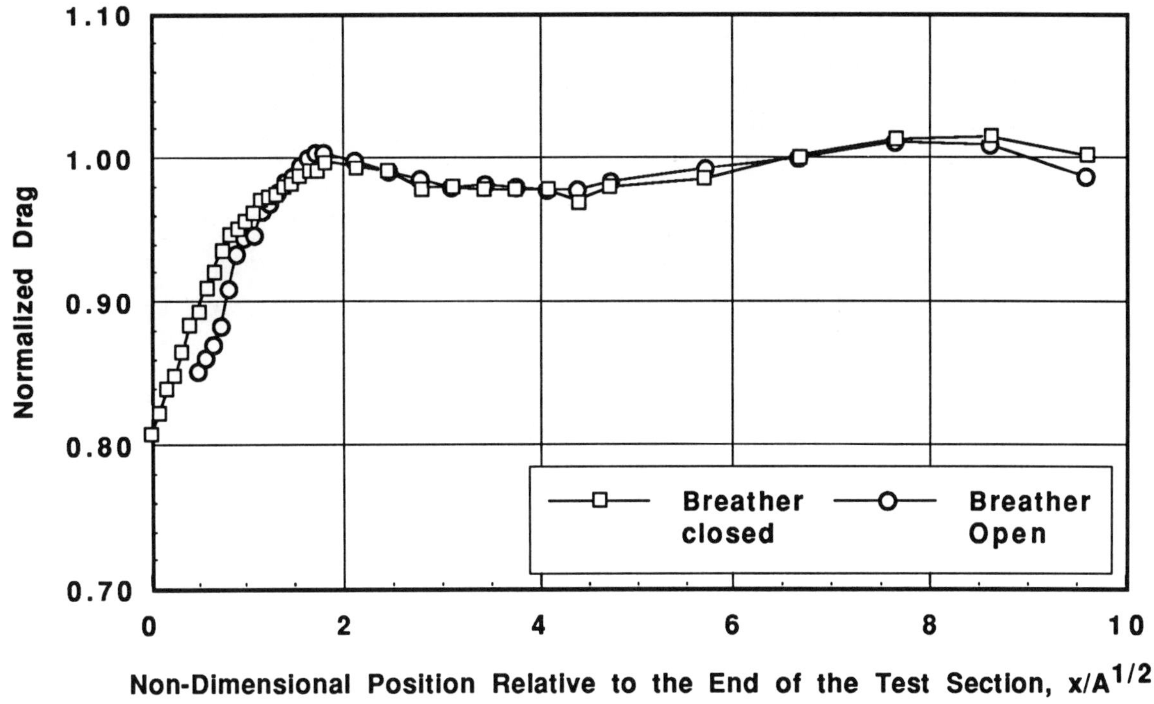

FIG. 8: Normalized Drag Coefficient Distributions for a Rectangular Body on the Floor of the NRC Pilot Tunnel

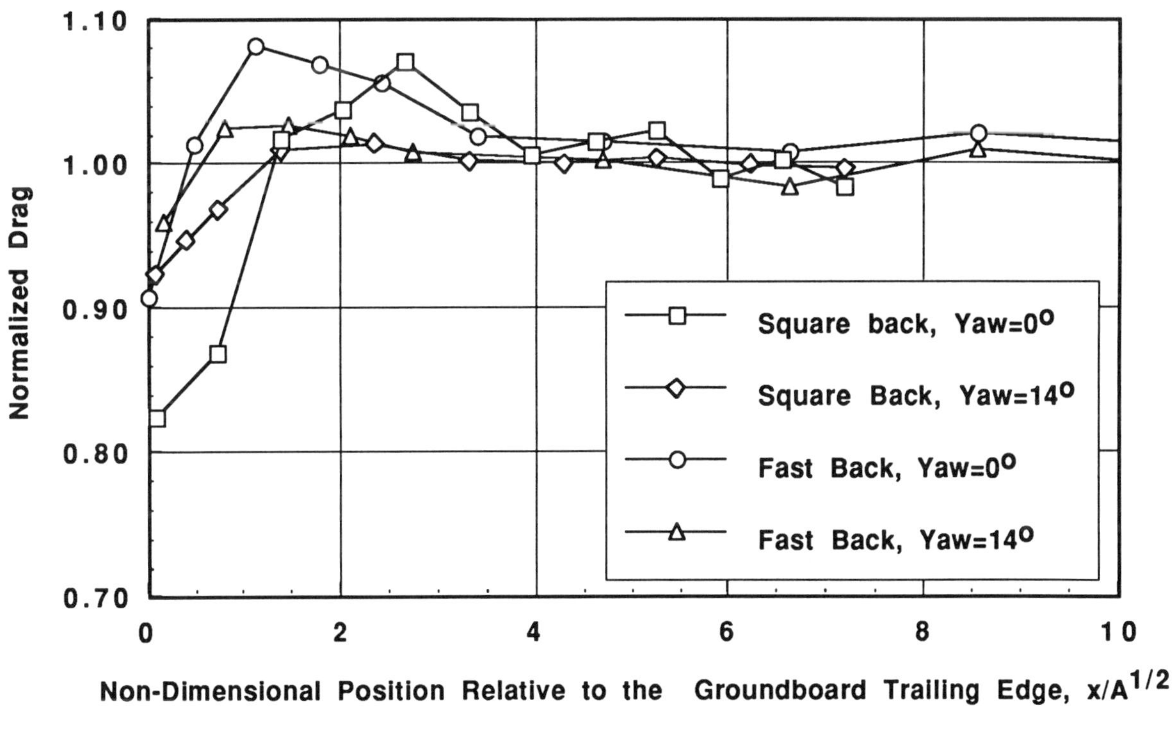

FIG. 9: Normalized Drag Coefficient Distributions for a Rectangular Body on the Groundboard of the NRC Pilot Tunnel

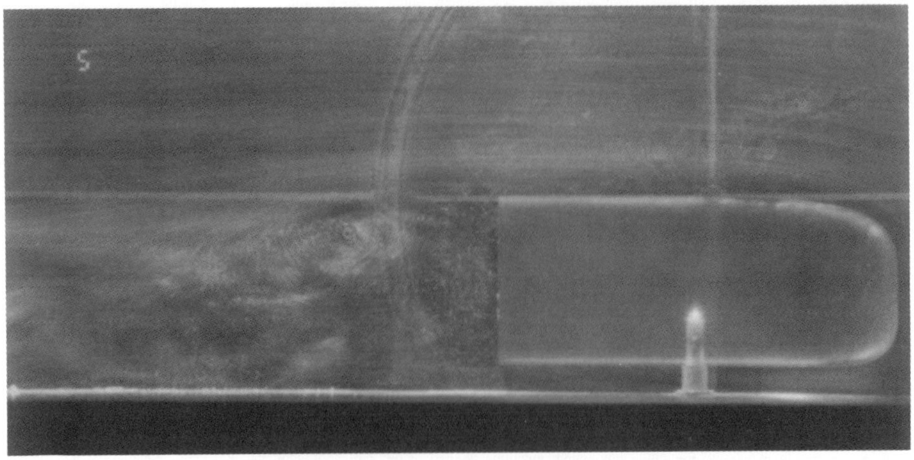

$x/A^{1/2}=5.0$, flow over the model centre plane visualized using aluminium particles.

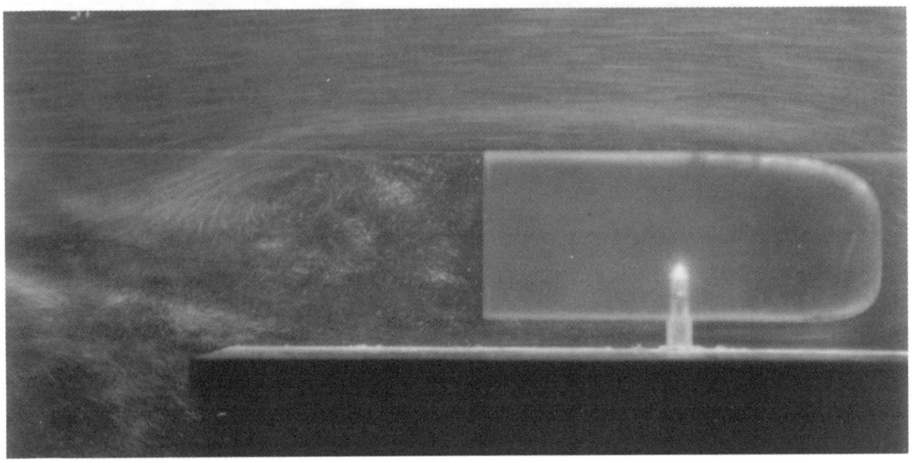

$x/A^{1/2}=1.5$, model at position of lowest drag. Note distortion of flow aft of model base and flow entering wake from below groundboard

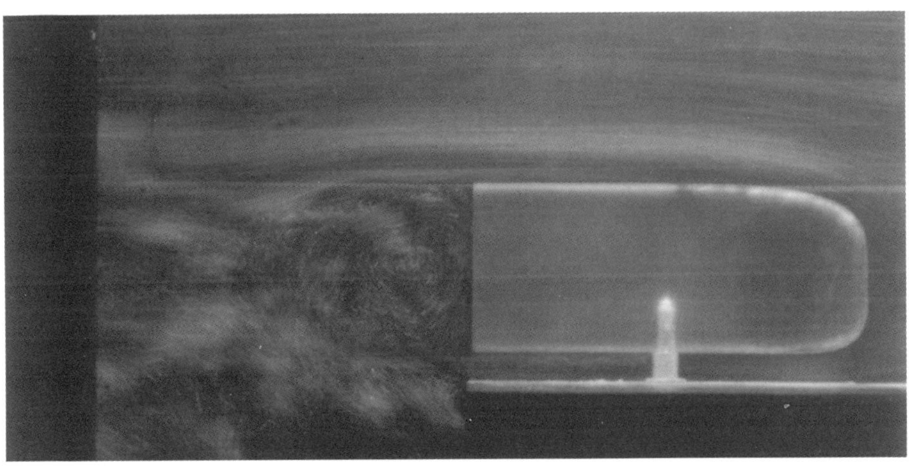

$x/A^{1/2}=0.0$, note stronger flow from below groundboard into wake.

FIG. 10: Vizualization of the Flow over a Simple Automotive Shape on a Groundboard in the NRC Water Tunnel

920344

Measurement of Reference Dynamic Pressure in Open-Jet Automotive Wind Tunnels

Rudi Kuenstner
University of Stuttgart

Klaus-Rainer Deutenbach
BMW AG

Jorg-Dieter Vagt
Porsche AG

ABSTRACT

In automotive open-jet wind tunnels reference velocity is usually measured in terms of a static pressure difference between two different cross-sectional areas of the tunnel. Most commonly used are two sections within the nozzle (Method 1: ΔP-Nozzle). Sometimes, the reference velocity is deduced from the static pressure difference between settling chamber and plenum (Method 2: ΔP-Plenum).

Investigations in three full-scale open-jet automotive wind tunnels have clearly shown that determination of reference dynamic pressure according to ΔP-Plenum is physically incorrect. Basically, all aerodynamic coefficients, including drag coefficient, obtained by this method are too low. For test objects like cars and vans it was found that the error Δc_D depends on the test object's drag blockage in an open-jet wind tunnel. On the other hand, correct determination of dynamic pressure according to ΔP-Nozzle for car and van testing requires at least a certain minimum distance from balance center to reference-tap location upstream of the nozzle exit.

INTRODUCTION

Testing in open-jet wind tunnels requires a clear understanding of at least two main influencing parameters, which can be hidden within the test results and affect their comparability. The first one is given by the way of determing the reference dynamic pressure, and the second one by the set up of the collector.

The determination of reference dynamic pressure or reference velocity is of fundamental importance, because it is a reference quantity for all dimensionless force and moment coefficients of a test object. Moreover, it is related to all static pressure coefficients measured on the test objects surface, and it is related to all pressure coefficients or velocity ratios in the flow around the test object.

In European full-scale automotive wind tunnels a test section with a 3/4-open jet has been used predominantly. It has been believed that this kind of test section simulates the condition "car on road" best. In order to avoid ambient influences, the open test section is usually surrounded by a more-or-less large plenum chamber. In most cases the static plenum pressure corresponds with the atmospheric one by means of a connecting duct. It has been thought, so far, that the static plenum pressure is the same as that of the open jet in an empty test section, and that there is no change when a test object is in place. Due to this mainly historical reason, reference dynamic pressure in open-jet wind tunnels has been determined partly to date by the static pressure difference between settling chamber and plenum (Method 2: ΔP-Plenum). First investigations of P. Müllenbach and K.-R. Deutenbach in 1989 [1][1] showed that the pressure level in the complete tunnel airpath is significantly changed when a test object is in place and, therefore, the control of mass flow rate through the nozzle is affected by the test object itself.

In aeronautical testing, traditionally executed in wind tunnels with closed jet, the reference speed had to be determined within the tunnel airpath from the very beginning. From this, some open-jet automotive wind tunnels adopted method 1: ΔP-Nozzle. This method measures the average velocity in terms of the static pressure difference between two sections within the nozzle. The first one is in the settling chamber and the second one is before the nozzle exit. Basically, due to stability, continuity and homogeneity reasons the test-section length of open jets must be significantly shorter than those of comparable closed jets. Therefore, when using this method in open-jet facilities the static pressure close to the nozzle

[1]Numbers in brackets designate references at end of paper.

exit can be influenced by the test object itself, which can affect the comparability of test results in open-jet automotive wind tunnels. Other main influencing parameters are the collector set up, longitudinal static pressure gradient, flow angularity and floor boundary-layer thickness.

DETERMINATION OF REFERENCE DYNAMIC PRESSURE

Tunnel speed is usually defined as the mean speed or average flow velocity in an empty tunnel in that region of the test section that will be occupied by the test object when (for constant intake conditions) the rate of mass flow is the same as with the test object. Generally, this mass flow can be measured in terms of static pressure difference between two sections of the tunnel with different cross sectional areas, Fig. 1. Method 1, ΔP-Nozzle, uses two sections within the nozzle, whereas method 2, ΔP-Plenum, employs the static pressure difference between settling chamber and plenum (at a specific location).

Unfortunately, method 2 with plenum pressure can lead to serious errors, because the pressure level in the whole tunnel airpath changes significantly when a test object is in place. This is equivalent to an additional pressure loss in the tunnel circuit. Fig. 2 shows schematically the effect of an additional pressure loss, generated by a test vehicle, on calibration method 2 (plenum pressure). Since the plenum pressure is determined by the surrounding condition (the atmosphere, by a connecting duct), it is independent of the test object. Thus, it is the same for the empty and occupied test section. The additional pressure loss should initially raise the pressure difference ΔP-Plenum. But, due to the calibration method used, ΔP-Plenum and ΔP-Fan are kept constant by reducing fan speed, and thus the volume flow rate.

The amount of reduction of volume flow rate clearly depends on the amount of the additional pressure loss, which is determined by the size and shape of the test object and by the tunnel size. The drag coefficient is calculated with the true force acting on the car and the dynamic pressure QA (due to the calibration method) instead of the really existing dynamic pressure QB. It is due to the fact that the static plenum pressure does not experience the static pressure rise in the tunnel airpath caused by the test object, as it is shown in the chapter "Test Results". The resulting drag coefficient then is too low. Consequently, the determination of reference dynamic pressure according to the plenum-pressure calibration procedure (method 2) is physically incorrect.

The situation is different when using method 1 (ΔP-Nozzle), see Fig. 3. This method uses information which is directly related to the flow or is directly proportional to the volume flow rate. The volume flow rate itself is kept the same for the empty test section and the test section occupied by a car. This is simply done by increasing ΔP-Fan according to ΔP-Car in order to maintain constant ΔP-Nozzle.

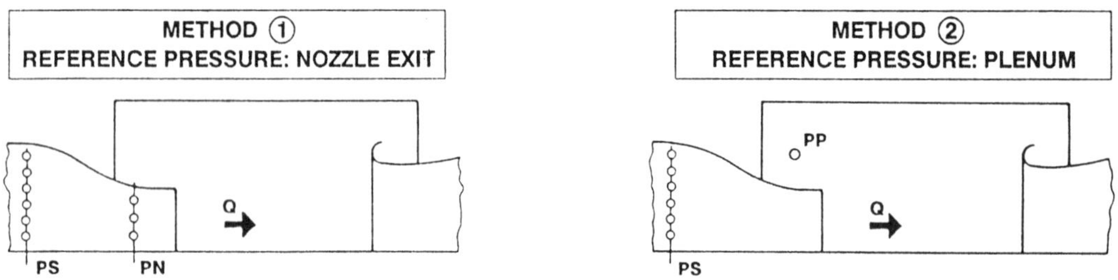

$$Q1 = \Delta P_{Nozzle} \cdot KF1$$

$$Q1 = (PS - PN) \cdot KF1$$

Q = Dynamic pressure in test section

PS = Static wall pressure in settling chamber

PN = Static wall pressure before nozzle exit

KF1 = Tunnel factor, based on the static wall pressure before the nozzle exit and determined in the empty test section by a pitot-static tube

$$Q2 = \Delta P_{Plenum} \cdot KF2$$

$$Q2 = (PS - PP) \cdot KF2$$

Q = Dynamic pressure in test section

PS = Static wall pressure in settling chamber

PP = Static pressure at specific location in the plenum

KF2 = Tunnel factor, based on the plenum static pressure and determined in the empty test section by a pitot-static tube

Fig. 1: Comparison of different methods of determination of reference dynamic pressure

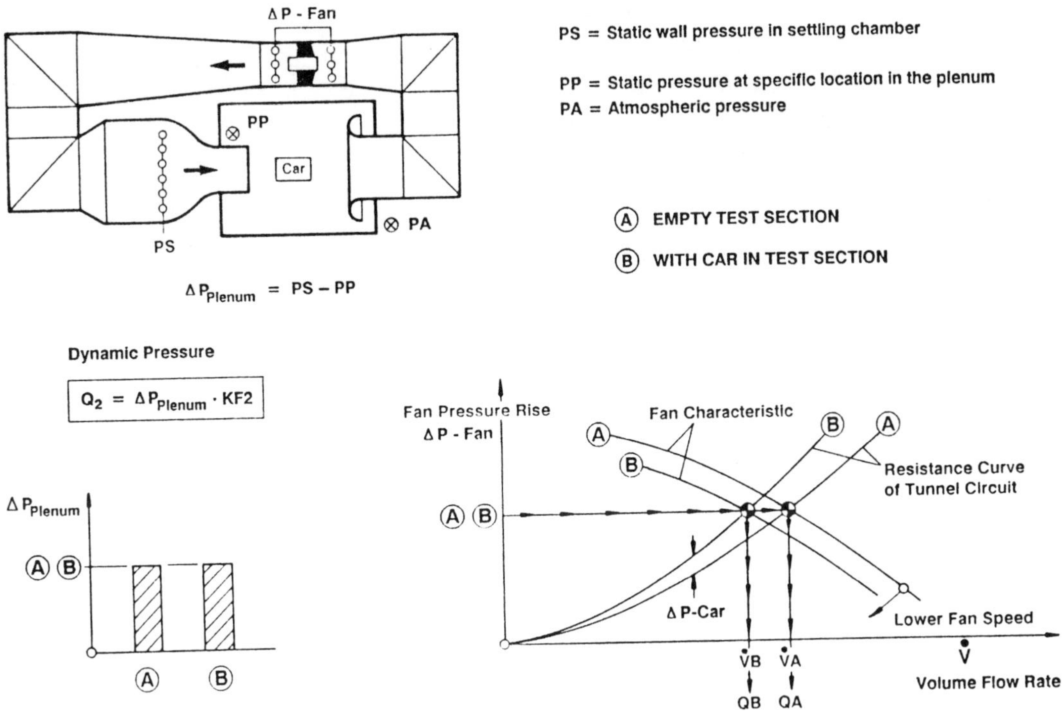

Fig. 2: Interaction between fan characteristic and measurement of dynamic pressure with method 2 ΔP-Plenum

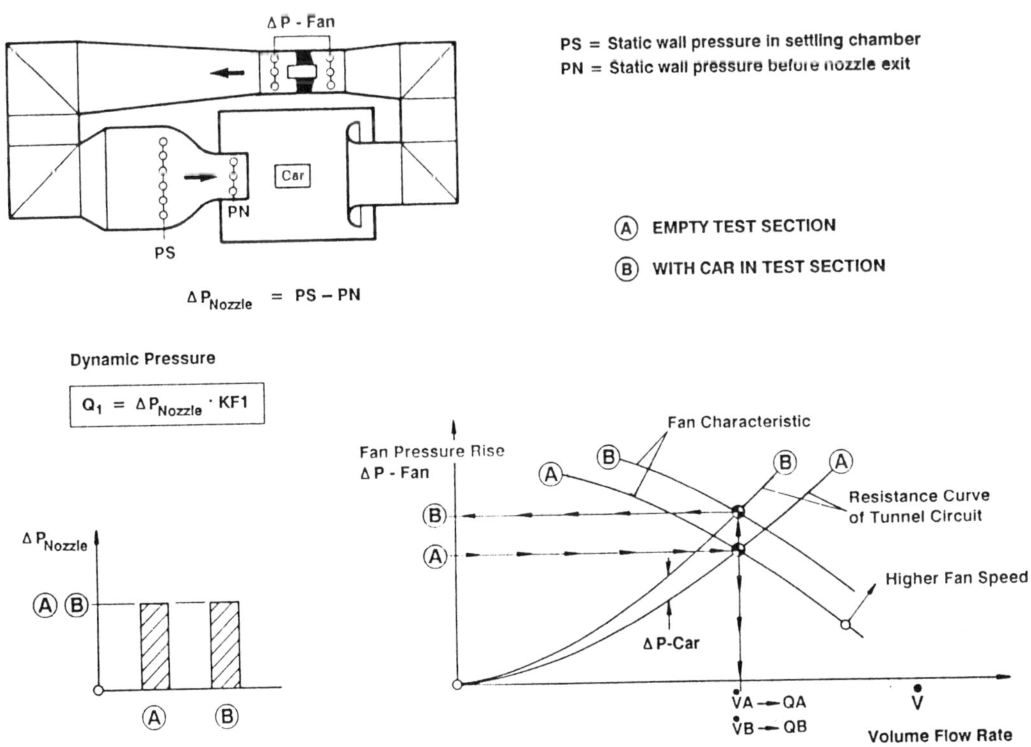

Fig. 3: Interaction between fan characteristic and measurement of dynamic pressure with method 1 ΔP-Nozzle

EQUIVALENT SYSTEM

For the determination of static pressure coefficients c_p, the local static pressures cannot be determined anymore as a difference from the static plenum pressure. The difference pressure should use the static pressure of the undisturbed flow. But during test operation with a car in the test section, the undisturbed static pressure of the flow cannot be determined directly any longer. It must be related to a flow equivalent pressure – in this case the settling chamber's static pressure. This leads to a "Static Pressure Offset KP" that has to be obtained in the empty test section, see Fig. 4. This indirectly relates the test object's local static pressure to that of the undisturbed flow in the empty test section. By plotting the negative KP-factor versus dynamic pressure according to ΔP-Nozzle, a nearly constant value results.

THE WIND TUNNELS

The test results reported in this paper are from three German full-scale automotive wind tunnels:

1. The Open-Jet Wind Tunnel of the Institute of Internal Combustion Engines and Motor Vehicle Engineering (IVK) of Stuttgart University. It is of Göttingen type (single return) with a nozzle exit area of 22.5 m^2 (5.8 x 3.87 m^2) and a test section length of 9.5 m, details in [2]. No correction procedure was applied to the test results.

2. The Open-Jet Wind Tunnel of BMW AG in München. It is of Göttingen type (double return) with a nozzle exit area of 20 m^2 (5.77 x 3.47 m^2) and a test section length of 10.0 m, details in [3]. The Open-Jet Wind Tunnel of BMW was originally designed with a slotted-wall test section, but it was changed to open jet in 1985. No correction procedure was applied to the test results.

3. The Wind Tunnel of Porsche AG in Weissach. It is of Göttingen type (single return) with a nozzle exit area of 22.3 m^2 (6.2 x 3.6 m^2) and a test section length of 13.5 m, details in [4]. The Porsche test section can be changed from open-jet to slotted-wall, but all results in this paper refer to open jet. No correction procedure was applied to the test results.

THE TEST VEHICLES

The automotive wind tunnels of IVK, BMW and Porsche are only suitable for full-scale testing of cars and vans. Consequently, such types of vehicles served as test objects, see Table 1. In the IVK Wind Tunnel production cars from compact class up to luxury class were used, and a relatively large delivery van with high roof (frontal area 4.08 m^2, blockage 18.2%). BMW used a luxury class car and a van, and Porsche a sports car and a van.

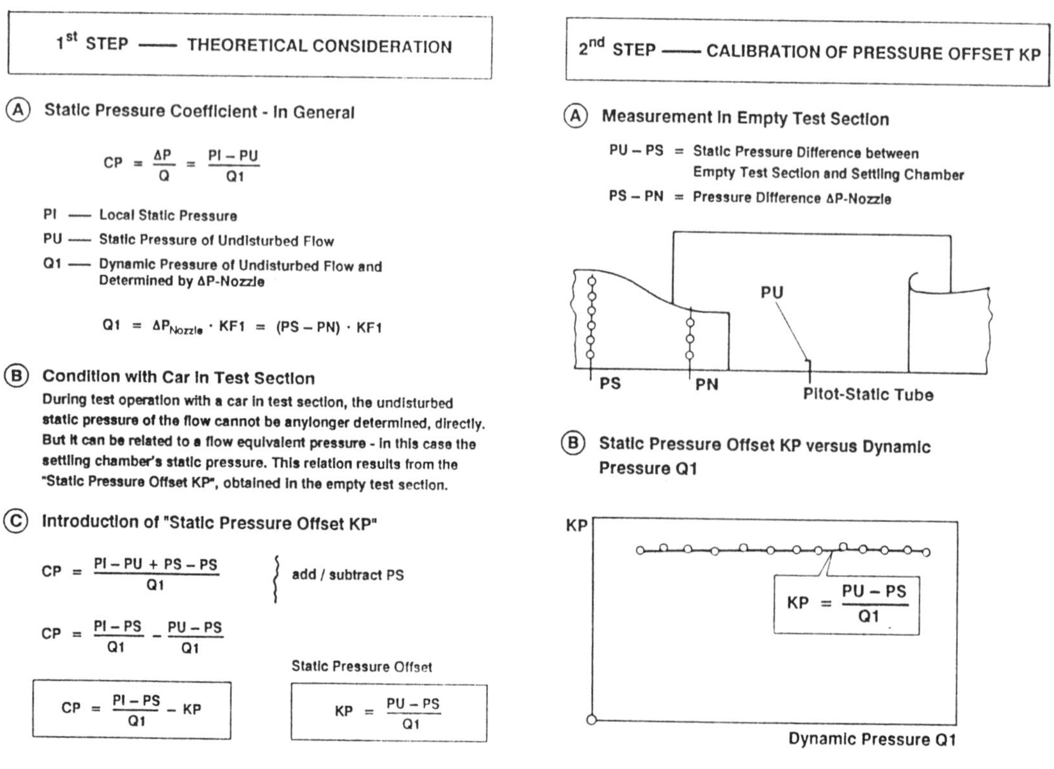

Fig. 4: Procedure for correct pressure coefficients and definition of static pressure offset KP

A/AN – Blockage
L – Car Length A – Frontal Area of Car AN – Nozzle Exit Area

Open Jet Wind Tunnel	Test Vehicle				
	Compact Car	Middle Class Car	Luxury Class Car	Sports Car	Delivery Van
	L in m A in m2 A/AN in %	L in m A in m2 A/AN in %	L in m A in m2 A/AN in %	L in m A in m2 A/AN in %	L in m A in m2 A/AN in %
IVK AN=22.5 m2	4.00 1.86 8.3	4.69 2.08 9.3	5.18 2.19 9.8		5.19 4.08 18.2
Porsche AG AN=22.3 m2				4.53 1.99 8.9	4.75 3.90 17.5
BMW AG AN=20.0 m2			5.02 2.13 10.6		4.57 3.07 15.3

Table 1: Data of test vehicles used in IVK, Porsche and BMW Wind Tunnels

TEST RESULTS

The tests in all three wind tunnels lead to similar results and the same conclusions. In order to simplify matters the test results of IVK are shown as an example, and all results refer to a test speed of 140 kph.

Fig. 5 shows the test set up in the IVK Open-Jet Automotive Wind Tunnel. In this wind tunnel and for this investigation, reference dynamic pressure was determined according to both methods. For method 2 (ΔP-Plenum) there is a ring tube on the nozzle's outer surface. Additionally, three pressure taps A, B, C were fixed to the plenum's front and side wall at the nozzle's mid height. Because of a duct to the atmosphere, the plenum pressure corresponds with the atmospheric one. Furthermore, as a control, the pressure difference between the plenum ring tube and the atmosphere was measured using two taps D and E outside the wind tunnel building.

For method 1 (ΔP-Nozzle) there are different cross sections of reference pressure taps upstream of the nozzle exit in the nozzle ceiling at locations 0.39 m, 0.85 m, 1.30 m, 1.75 m, 2.50 m and 3.25 m from the nozzle exit.

In order to determine the upstream stagnation influences of the test vehicles, an array of pressure taps on the centerline of the wind tunnel floor was used. It started in the settling chamber's static pressure ring on the wall, went through the contraction, and ended at the turntable's front end (pressure taps 1 – 21).

Fig. 6 shows coefficients of static difference pressure of the three taps A, B, C on the plenum walls and of the two atmospheric pressure taps D and E to the plenum ring tube, both for the empty and occupied (3 cars, 1 delivery van) test section. The graph shows the static plenum pressures not being affected by the different test objects. There is a small data scatter due to ambient wind outside the wind tunnel building, but there is no significant pressure rise relative to the empty test section caused by the test objects. The lower static pressures for the plenum-taps B and C as against tap A and the ones at the ring tube are due to the recirculating flow (secondary flow) in the plenum.

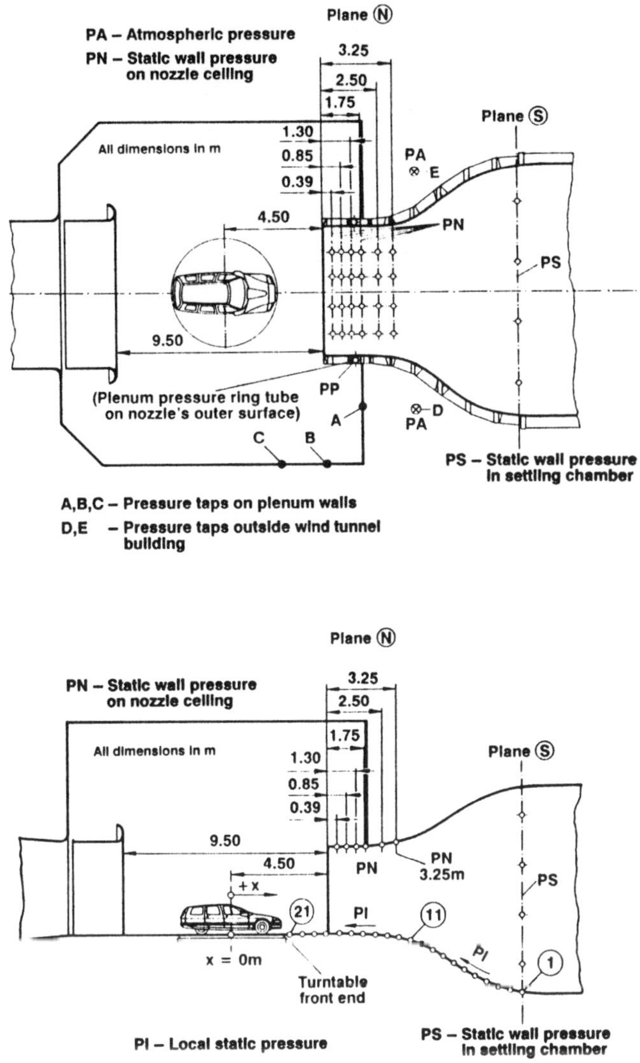

Fig. 5: Location of reference-pressure taps and local static pressure taps in IVK Wind Tunnel (top, plan view; bottom, side view)

Fig. 7 shows the static pressure coefficients on the tunnel floor's centerline between settling chamber and turntable front end. The test section is occupied by the middle-class car. The reference-tap location for dynamic pressure determination according to ΔP-Nozzle is 3.25 m upstream of the nozzle exit in the nozzle ceiling. The local static pressure differences are taken both to plenum (broken line) and to settling chamber (full line, static pressure coefficients determined with static pressure offset KP). The difference between the two curves shows a pressure rise Δc_{pCar} in the whole tunnel airpath, including the open jet. It is simply due to the fact that the static plenum pressure, here used in the difference pressure, does not experience the pressure rise in the tunnel

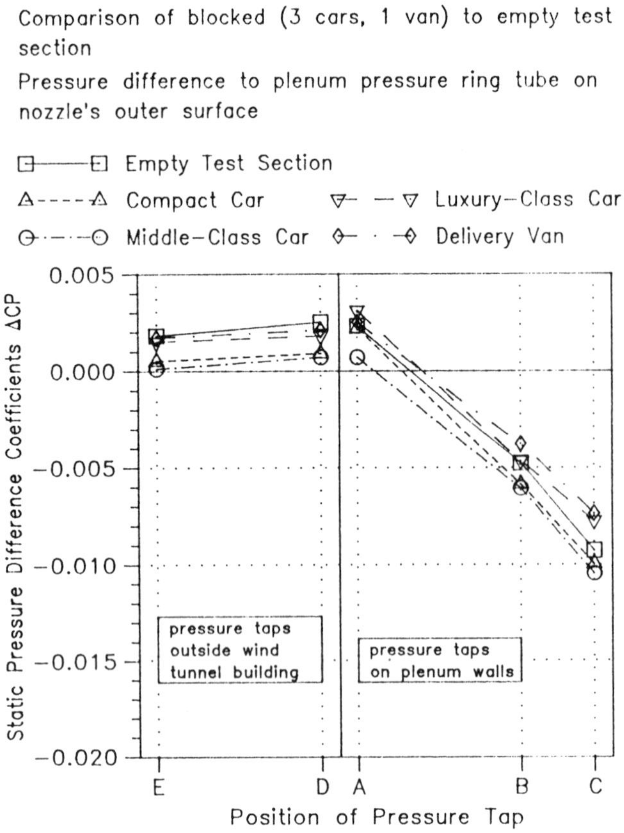

Fig. 6: Comparison of occupied to empty test section for pressure taps on plenum walls and in atmosphere

airpath and caused by the car. The local static pressures on the tunnel floor, however, experience the pressure rise and consequently, those pressure coefficients are too high.

Fig. 8 quantifies, for the open-jet wind tunnels of IVK, Porsche and BMW at constant volume flow rate the static pressure at specific locations in the tunnel airpath with and without test objects. The dynamic pressure was determined by using method ΔP-Nozzle. Constant volume flow rate means constant pressure difference ΔP-Nozzle. It is only achievable if ΔP-Fan is increased according to the test object's pressure loss. This pressure rise is between 2.2% (compact car) and 3.3% (luxury class car) in the IVK Wind Tunnel. Porsche has for a sports car 3% and BMW for a luxury class car (different from the one used in the IVK) 2.7%. For the larger delivery vans the pressure rise is 7.2% at IVK, 7% at Porsche and 5.3% at BMW. A dynamic pressure determination according to ΔP-Plenum would lead to higher dynamic pressures with exactly those increments as stated before. All resulting coefficients including drag coefficient would then be too low by the same rate. Consequently, the determination of reference dynamic pressure according to ΔP-Plenum is physically incorrect.

Fig. 9 again shows for the IVK Wind Tunnel pressure coefficients on the tunnel floor's centerline both for the empty test section and with a middle-class car in place. All pressure coefficients were determined with the equiva-

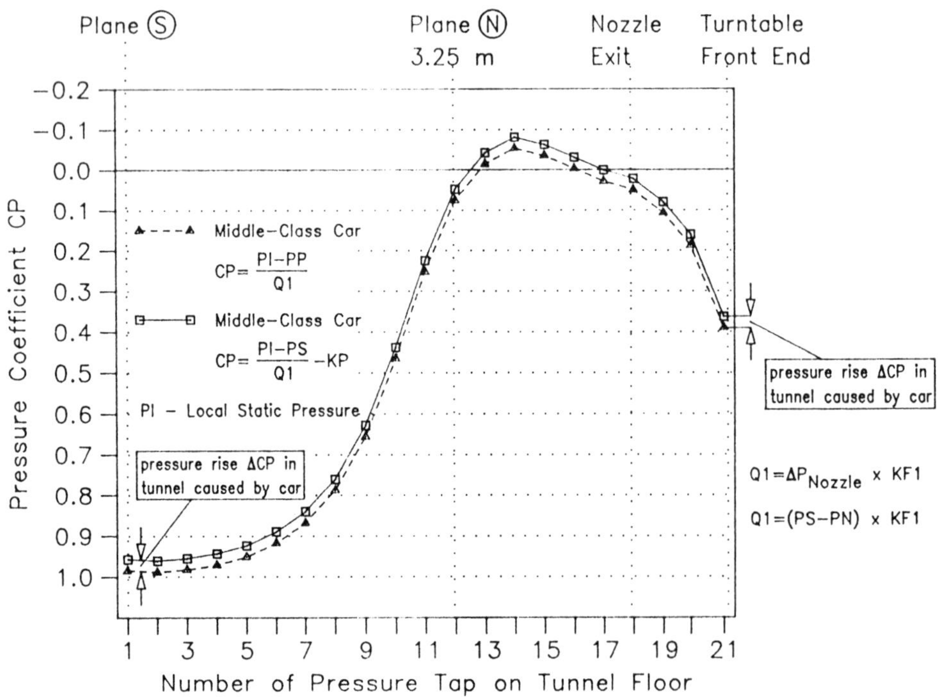

Fig. 7: Pressure rise in tunnel caused by car, IVK Wind Tunnel

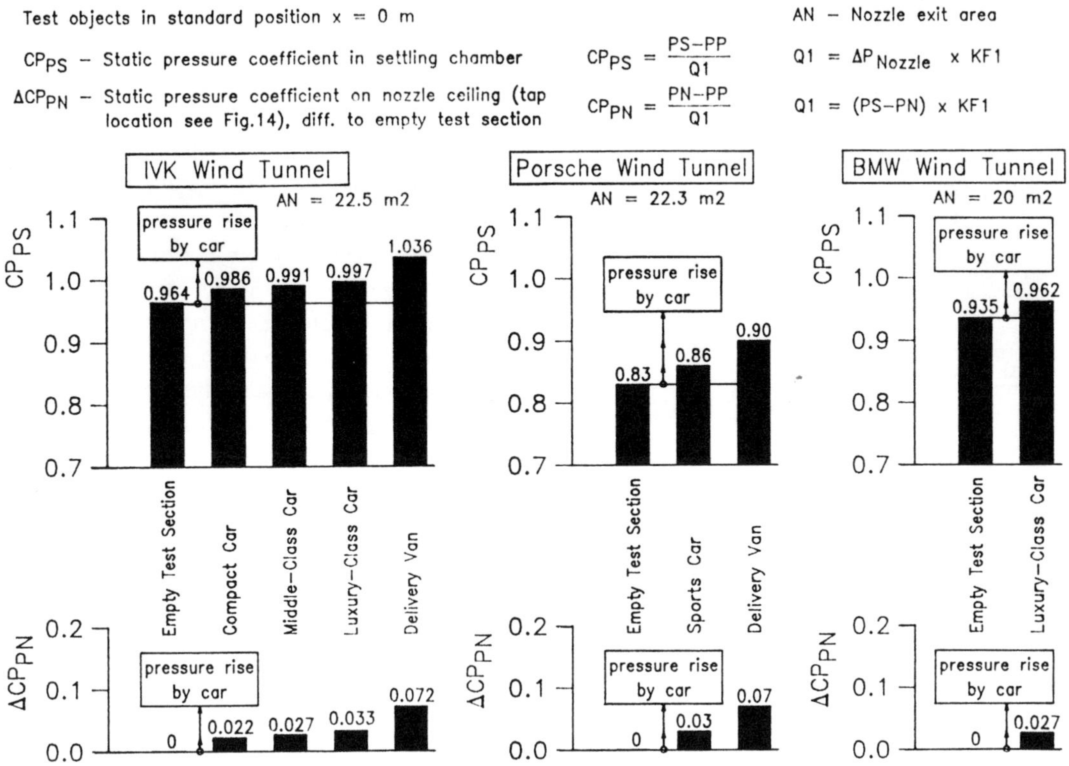

Fig. 8: Pressure rise in settling chamber and on nozzle ceiling for constant dynamic pressure

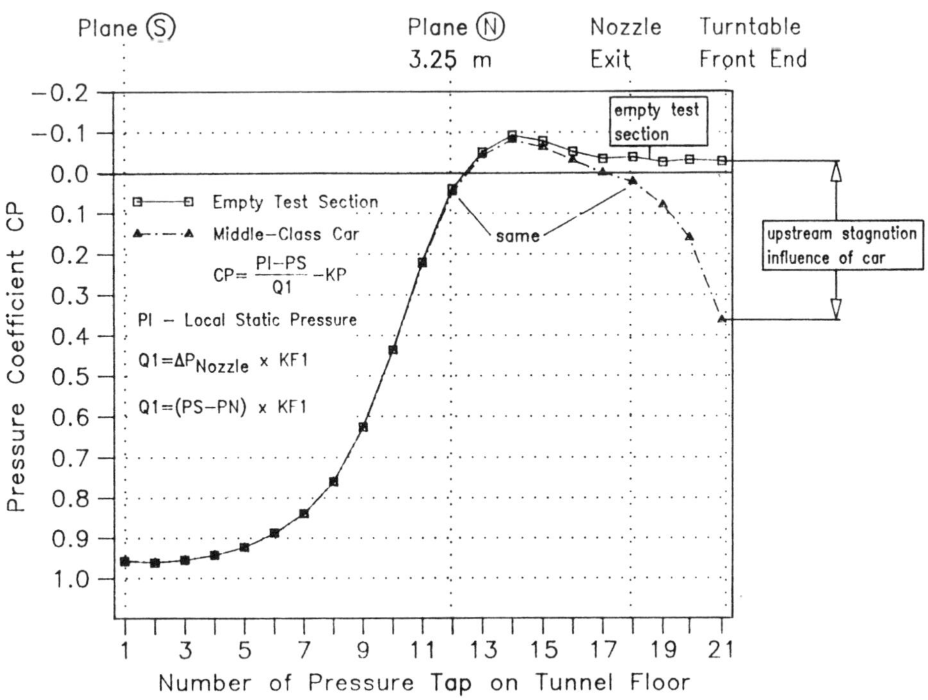

Fig. 9: Upstream stagnation influence of a car in the IVK Wind Tunnel

lent system and with static pressure offset KP. Reference pressure PN on the nozzle ceiling was taken 3.25 m before the exit. In front of the car the influence of the car's pressure field can be seen. This influence does not disappear, even in the sensitive floor area, until 3.25 m upstream of the nozzle exit. Basically, the physical limit of method ΔP-Nozzle is given by the requirement that the pressure taps be located far enough away from the front of the test object that they are not, or are only marginally, influenced by it. In cross section 3.25 m, the pressure difference to the settling chamber is the same as in the strongly influenced area at the nozzle exit. This is due to an overshoot of the velocity along the nozzle contour before the exit, as Fig. 10 schematically shows. A wind tunnel nozzle provides a location within it where the pressure difference to the settling chamber is the same as at the nozzle exit.

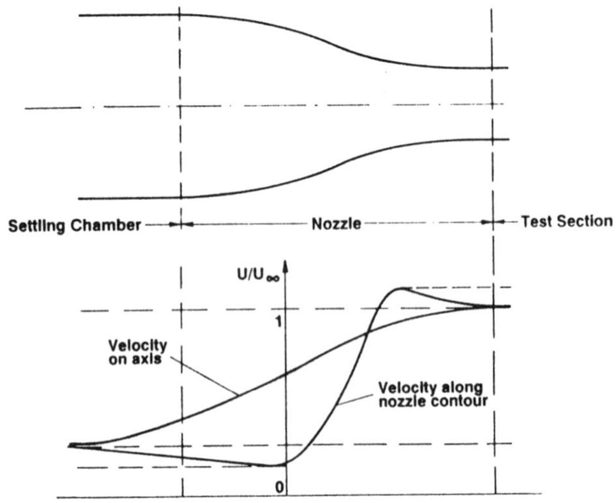

Fig. 10: Schematic velocity development in a nozzle

Figs. 11 and 12 show the upstream influences of the middle-class car and the delivery van on the IVK nozzle-floor pressures. Reference-pressure taps for Q determination are 3.25 m upsteam in the nozzle ceiling, and pressure coefficients are calculated with static pressure offset KP. The vehicles on the turntable are in the standard position (x = 0 m), 1 m upstream (x = +1 m) and 1 m downstream (x = −1 m). Only in the case of the van (frontal area 4.08 m^2, blockage 18.2%) in the upstream position is there an appreciable influence at floor point 3.25 m.

Fig. 13 shows, for three cars and a van in the standard position, the influence on reference pressure taps in the nozzle ceiling (reference-tap location 3.25 m for Q determination). From this Figure a minimum distance of 2.5 m upstream of the nozzle exit is required, especially for the van. The demand for a reference-tap location on the nozzle ceiling that is either not or only slightly in-

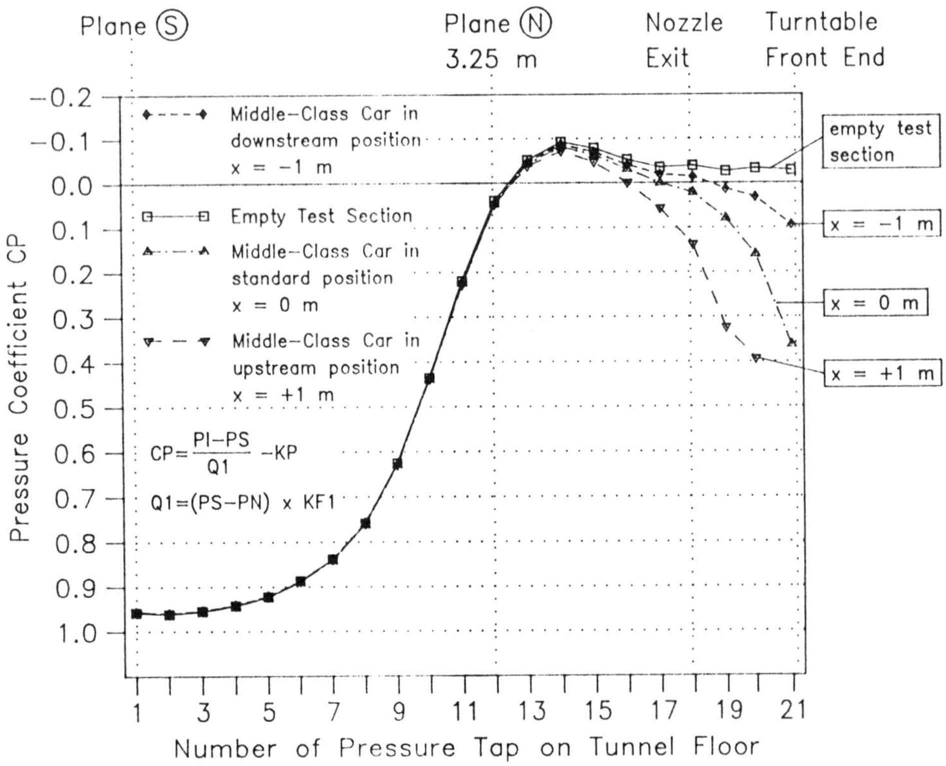

Fig. 11: Influence of position of a middle-class car in the test section on the pressure distribution on the tunnel floor

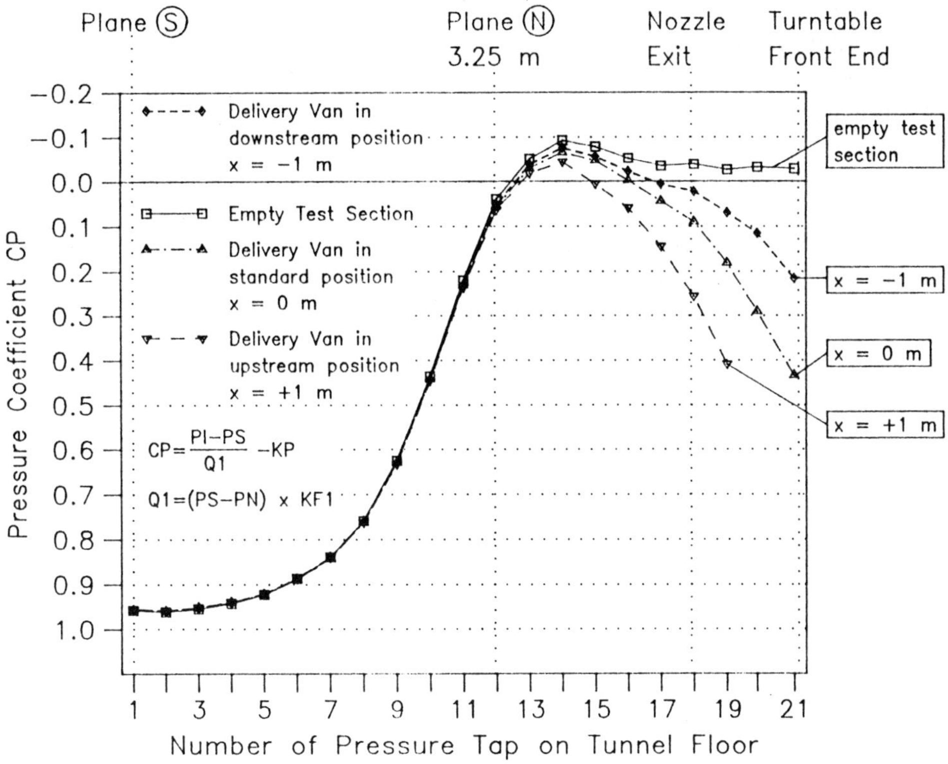

Fig. 12: Influence of position of a van in the test section on the pressure distribution on the tunnel floor

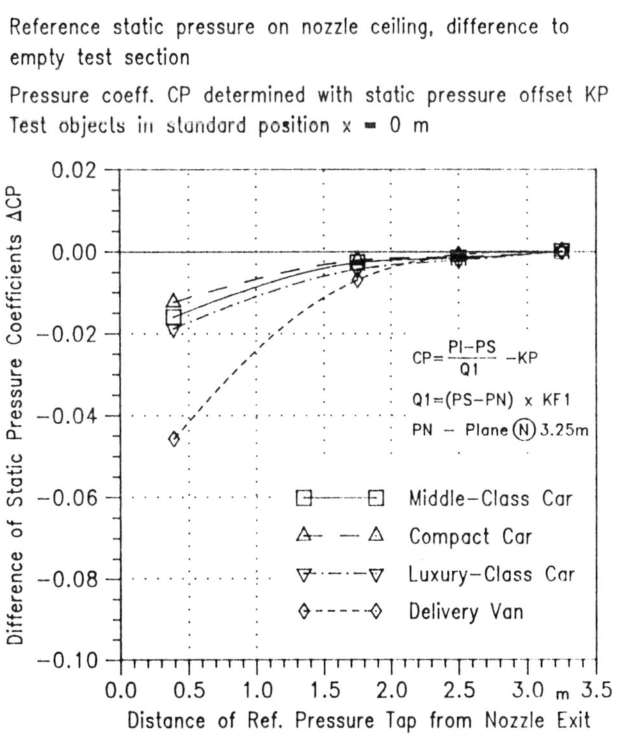

Fig. 13: Influence of various test vehicles on reference static pressure on nozzle ceiling, IVK Wind Tunnel

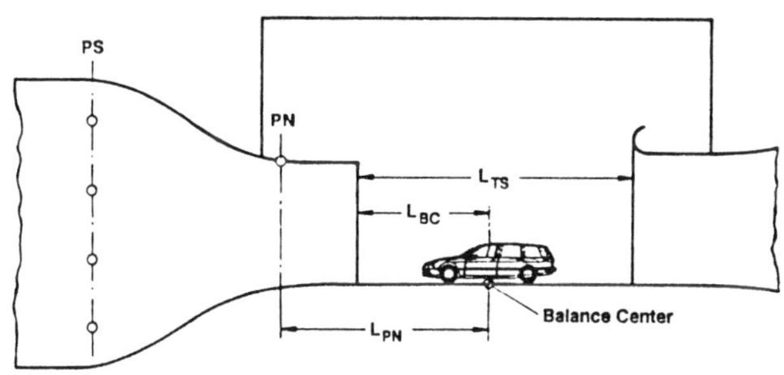

AN – Nozzle Exit Area

Open-Jet Wind Tunnel	AN /m2/	L_{TS} /m/	L_{BC} /m/	L_{PN} /m/
IVK	22.5	9.5	4.5	7.75
Porsche AG	22.3	13.5	5.5	7.26
BMW AG	20.0	10.0	4.7	7.47

Fig. 14: Realized distances of reference-pressure taps in nozzle ceiling from balance center for correct car and van testing in IVK, Porsche and BMW Wind Tunnels

fluenced, leads in the IVK Wind Tunnel for car and van testing to a minimum longitudinal distance of 7 m from reference-pressure tap in the nozzle ceiling to the turntable center (2.5 m in the nozzle and 4.5 m from nozzle exit to turntable center). Fig. 14 shows for the IVK, BMW and Porsche Wind Tunnels the realized distances of the reference pressure taps in the nozzle ceiling from the balance center for correct car and van testing. Because of the restricted test section length of open-jet wind tunnels, such minimum distances mean that any direct measurement of reference dynamic pressure in an open jet by means of a pitot-static tube in front of a vehicle is in error when the vehicle is in place. Investigations in this respect in the IVK Wind Tunnel showed only the total head being correct, whereas the measured static pressure in the test section in any location was influenced by the test vehicles. Figs. 11 – 13 confirm this fact.

Reference static pressure on nozzle ceiling, difference to empty test section

Pressure coeff. CP determined with static pressure offset KP

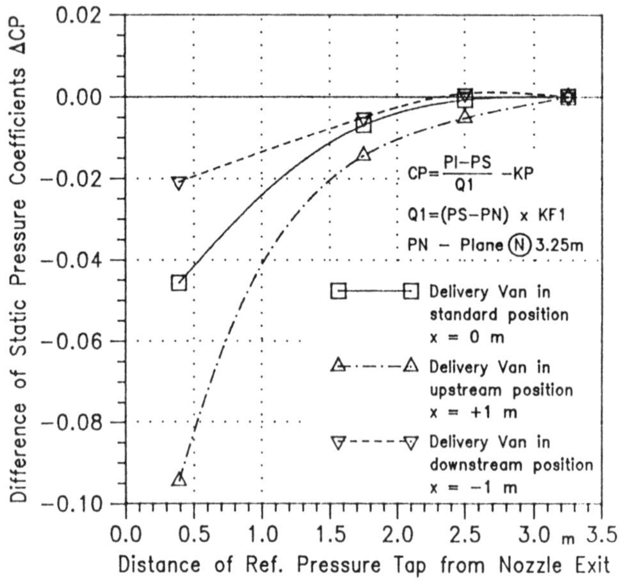

Fig. 15: Influence of van position in the test section on the reference static pressure on the nozzle ceiling, IVK Wind Tunnel

Fig. 15 shows for the delivery van the influence of its upstream and downstream positions x = +1 m and x = −1 m on the various pressure taps upstream of the nozzle exit. The downstream van position shows at the reference-tap locations more than 1.75 m ahead of the nozzle exit very little change compared to the standard position, but in the van's upstream position those reference taps react sensitively.

Fig. 16 shows for three cars and the van the drag coefficients versus reference-tap location on the nozzle ceiling from 0.39 m upstream of nozzle exit up to 3.25 m. Q determination according to ΔP-Nozzle was used for

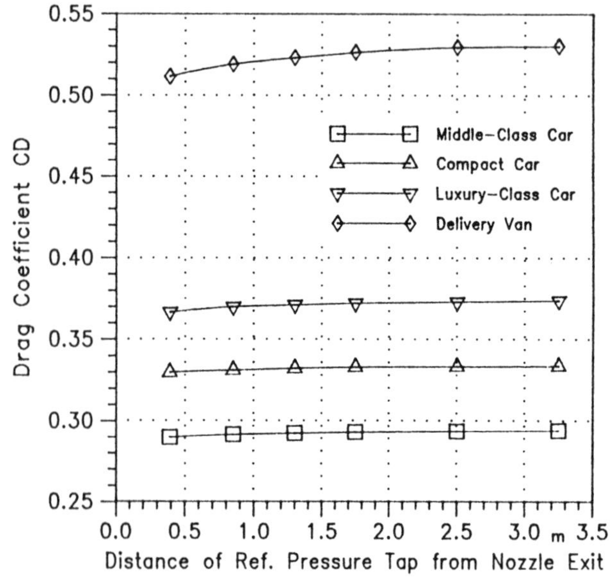

Fig. 16: Influence of the reference-pressure tap location in the nozzle ceiling on the drag coefficients of various cars and a van, IVK Wind Tunnel

each reference-tap section in plane N (0.39 m, ... , 3.25 m). The car curves level out far earlier than the van, of course. The decrease of the drag coefficients with reference-pressure tap's decreasing distance from the nozzle exit is rather surprising, because it has been thought that the static reference-pressure at the nozzle ceiling is raised by the upstream stagnation zone of a car, and thus the pressure difference to the settling chamber is smaller and that results in a higher drag coefficient. The correct physical explanation for the lower drag coefficients that results by shifting the static reference pressure towards the nozzle exit is that the settling chamber's still uninfluenced mass flow has also to pass the nozzle exit for continuity reasons. If there exists now in the lower middle region of the nozzle exit flow a test object's upstream stagnation zone (retarded flow), there must be higher local velocities in the remaining sections both at the nozzle ceiling and at the side walls in order to maintain the mass flow. Consequently, lower static wall pressures are impressed on the ceiling near the nozzle exit as compared to the empty test section. Therefrom results a higher pressure difference to the settling chamber and therewith a lower drag coefficient will be obtained than when using uninfluenced wall areas more upstream on the nozzle ceiling for the reference taps.

Fig. 17 shows for the cars and the van the deviation from the "uninfluenced" drag coefficient obtained with the reference-pressure tap location 3.25 m before the nozzle exit. If a deviation of no more than 0.001 in drag coefficient can be tolerated, it requires again a minimum upstream distance of 2.5 m from the nozzle exit for car and van testing. From the turntable center it amounts again to a 7 m longitudinal distance, or in other

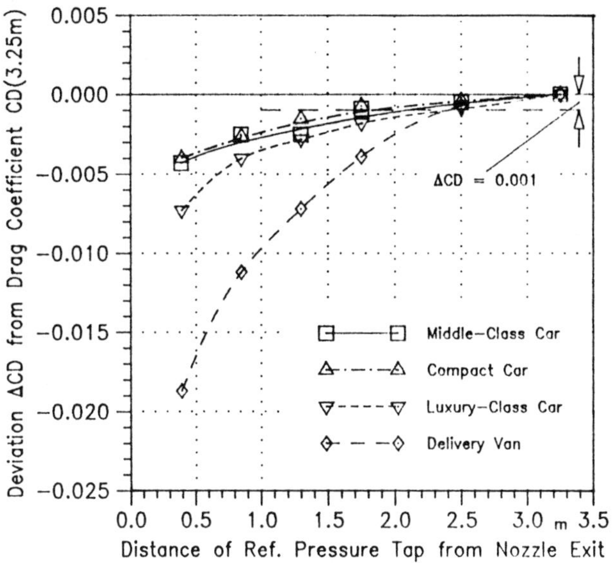

Fig. 17: Δc_D of three cars and a van from the "uninfluenced" drag coefficient $c_D(3.25\ m)$ and versus reference-pressure tap location in the nozzle ceiling

Open-Jet Wind Tunnel	Category of Test Vehicle: Compact Car	Middle-Class Car	Upper Middle-Class Car	Luxury-Class Car	Delivery Van
IVK AN = 22.5 m2	△	⊡		▽	◇
BMW Technik GmbH AN = 10 m2				⊙	
BMW AG AN = 20 m2				◆	▲
Mercedes-Benz AN = 32.6 m2		⊕	×		+

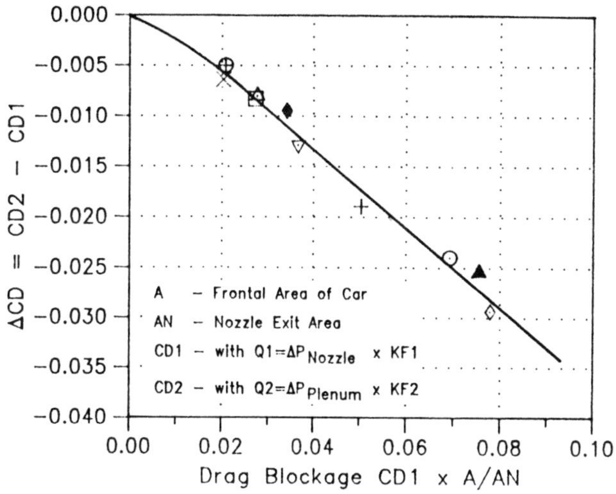

Fig. 18: Error Δc_D by applying the dynamic-pressure determination method ΔP-Plenum

words: a vehicle length upstream from the test object's front end.

For non or low lifting vehicle configurations at zero yaw the additional pressure loss in the tunnel airpath that the fan has to overcome is due to the drag blockage c_D x A/A_N of the vehicle. The quantity drag blockage relates the test object's drag force to the available thrust of the open jet. The difference in Q determination between the methods ΔP-Nozzle and ΔP-Plenum again traces back to the necessary ΔP_{Fan} to overcome the pressure loss ΔP_{Car} in the tunnel . Since the Q difference determines all coefficient differences, the Δc_D between the two methods of Q determination depends on the test object's drag blockage in the wind tunnel. Therefore, by means of drag blockage, results from different open-jet wind tunnels can be compared. The curve in Fig. 18 shows how the error in drag coefficient, caused by the different methods of Q determination depends on the test object's drag blockage. The results from different-sized open-jet automotive wind tunnels are well correlated. Generally, the higher the car drag and the smaller the wind tunnel, the greater the error Δc_D.

In Fig. 19 a simplified car range with a minimum drag area $c_D \times A = 0.44\ m^2$ and a maximum one of 0.88 m^2 is considered. For a medium-size full-scale open-jet wind tunnel with a nozzle exit area of 20 m^2 there results the considerable error range of $\Delta c_D = -0.007 \ldots -0.015$ for car testing. For a 10 m^2 full-scale wind tunnel there results a big error with a wide range of $\Delta c_D = -0.015 \ldots -0.033$. A large open-jet wind tunnel with 40 m^2 nozzle exit area shows on the contrary rather negligible deviations $\Delta c_D = -0.003 \ldots -0.007$ for this simplified consideration of a production car range.

	Car Drag	CD1 x A	=	CD1 x A
Car Range	Min. Drag	0.26 x 1.7	=	0.44 m2
	Average	0.33 x 2.0	=	0.66 m2
	Max. Drag	0.40 x 2.2	=	0.88 m2

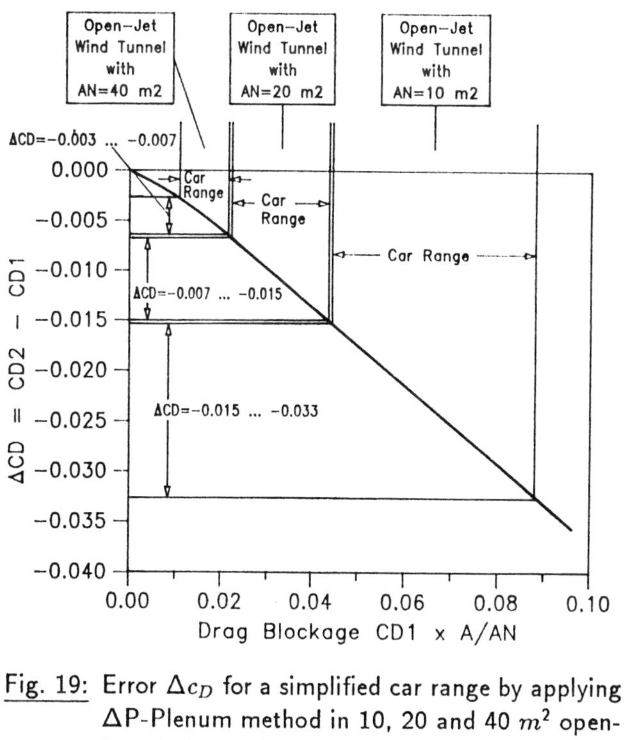

Fig. 19: Error Δc_D for a simplified car range by applying ΔP-Plenum method in 10, 20 and 40 m^2 open-jet wind tunnels

Figs. 20 and 21 show static pressure coefficients on the centerline of a BMW reference car (luxury class car). The local pressure coefficients were determined on road, in BMW's 20 m^2 wind tunnel (Fig. 20) and in BMW's Acoustic Open-Jet Wind Tunnel (Fig. 21) with a nozzle exit area of 10 m^2. In both wind tunnels the local pressure differences were taken both to the plenum (ref. plenum) and to the settling chamber in context with the static pressure offset KP. Reference dynamic pressure was determined with method ΔP-Nozzle, in each case. The reference quantity on road was a pitot-static tube in front of the car, and this test set up was calibrated in both wind tunnels. The measurements from the road correspond best to those of the wind tunnel obtained with the static pressure offset KP, and confirm the correctness of this new method. The pressure rise with the reference quantity being the plenum is in the 10 m^2 wind tunnel significantly higher than in the 20 m^2 tunnel. Both graphs confirm Fig. 19, as well. Since the error Δc_D is due to the wind tunnel fan's additional pressure rise required to overcome the car's pressure loss in the tunnel circuit, the pressure rise in a 10 m^2 wind tunnel must be considerably higher than in a 20 m^2 wind tunnel, according to Fig. 19.

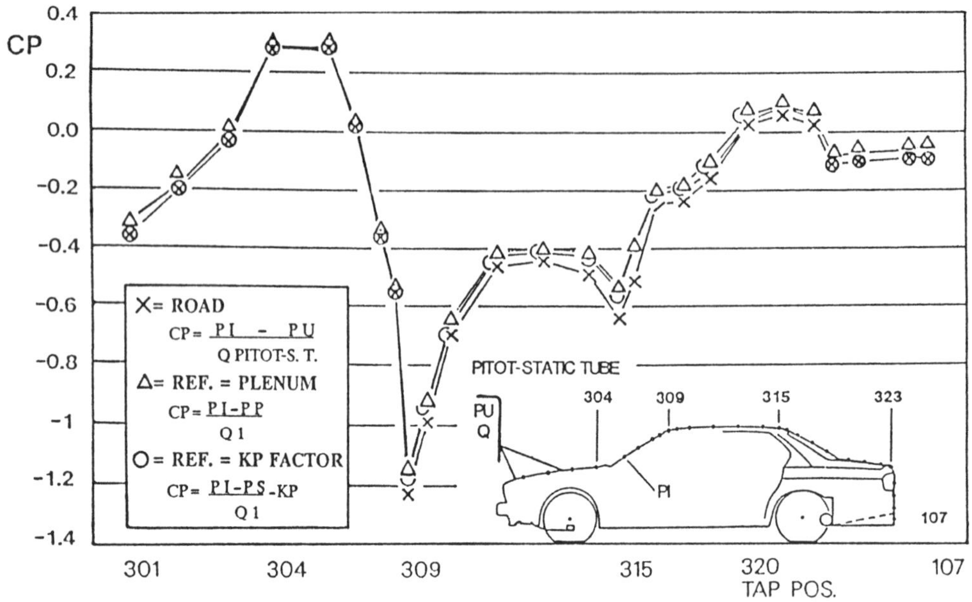

Fig. 20: Pressure coefficients on a car centerline obtained on road and in the BMW 20 m^2 Open-Jet Wind Tunnel

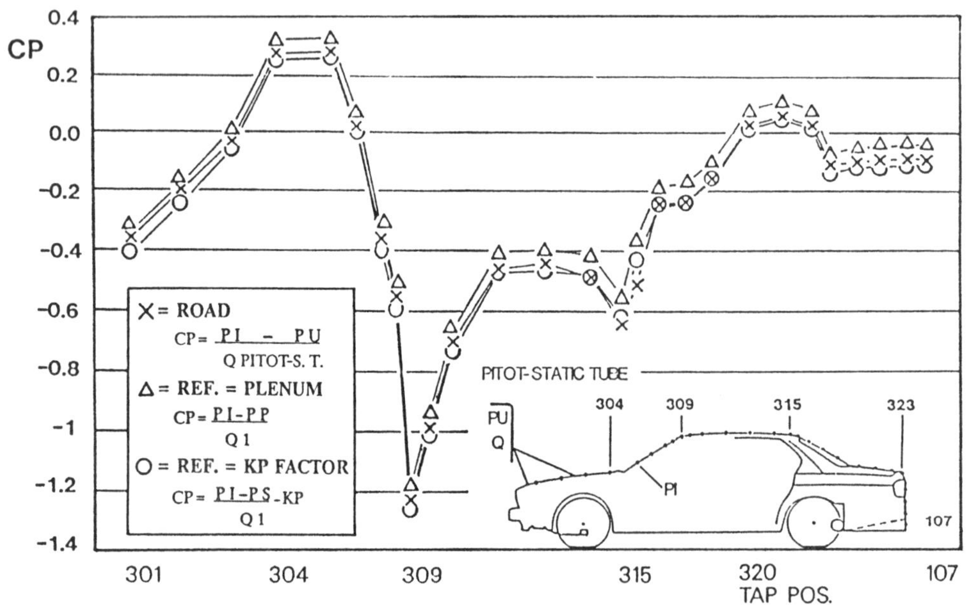

Fig. 21: Pressure coefficients on a car centerline obtained on road and in the BMW 10 m^2 Open-Jet Wind Tunnel

SUMMARY AND CONCLUSIONS

Two methods for the determination of reference dynamic pressure in open-jet automotive wind tunnels were considered theoretically, and force and pressure measurements were carried out in three full-scale open-jet wind tunnels. By using pressure taps on the tunnel floor from settling chamber to turntable, in the nozzle ceiling, in the plenum and in the atmosphere, and by testing different cars and vans, it was investigated what changes in the tunnel-airpath pressure occur relative to the condition of the empty test section, and how test vehicles can influence the reference-pressure taps in the nozzle. Finally, pressure measurements on a car surface on the road were compared to those in wind tunnels obtained with different reference pressures. From these investigations and considerations the following conclusions may be drawn:

1. Q measurement in open-jet wind tunnels using the pressure difference between settling chamber and plenum is physically incorrect, because the plenum pressure does not experience the fan's additional pressure rise required to overcome the test object's pressure loss in the tunnel airpath in order to maintain constant mass flow rate. Thus, all resulting (drag) coefficients are too low.

2. For the determination of pressure coefficients c_p, the static plenum pressure cannot be used as reference pressure. The local static pressure must be related to a flow-equivalent pressure – the settling chamber's static pressure – by introducing the static pressure offset KP obtained in the empty test section. This relates the test object's local static pressure to that of the undisturbed flow in the empty test section, but indirectly.

3. For correct Q determination according to the method ΔP-Nozzle, the reference pressure taps should be in the nozzle ceiling, preferably, to minimize the influence of the test object. When using several of them in any given cross-section, they should have approximately the same pressure level. Therefore, reference pressure taps should not be close to the corner to the side wall.

4. In the case of car and van testing the reference static pressure-taps in the nozzle ceiling should be at least 7 m longitudinal distance from the turntable center or, in other words, a vehicle length upstream from the vehicle's front end. This also means that any reference velocity measurement in the open jet by means of a pitot-static tube in front of a vehicle is in error because the presence of the test object influences the static pressure in the test section at all locations.

5. The test object's upstream stagnation zone can be determined by an array of pressure taps on the tunnel floor's centerline. Therefrom, a suitable location for the reference pressure taps in the nozzle ceiling may be derived. A wind tunnel nozzle has an overshoot of velocity along the contour before the nozzle exit. It provides a location where the pressure difference to the settling chamber is the same as at the nozzle exit and thus, the accuracy of dynamic pressure measurement is not affected.

6. The difference in drag coefficient between the two methods of Q determination depends on the test object's drag blockage in the wind tunnel. By means of this reference quantity different wind tunnels can be compared. Generally, the higher the car drag and the smaller the open-jet automotive wind tunnel, the greater the error in drag coefficient when obtained by Q measurement according to ΔP-Plenum. The indicated drag coefficient is too small.

ACKNOWLEDGEMENTS

The authors wish to express their thanks to all those people who were involved in the test campaigns that have led to this paper. Special thanks also go to Peter Müllenbach from BMW AG, Lutz J. Janssen and Norbert Lindener from BMW Technik GmbH for supporting this paper with some basic ideas and helpful suggestions. Our thanks are also extended to Alfred Kuhn from Mercedes-Benz AG for making available some wind tunnel results.

REFERENCES

[1] Müllenbach, P.; Deutenbach, K.-R.; "Determination of Dynamic Pressure and Reference Pressure in Automobile Wind Tunnels with Open Test Section", BMW Internal Report, 1989

[2] Potthoff,J.; "Die neue Kraftfahrzeug-Windkanalanlage der Universität Stuttgart am Institut für Verbrennungsmotoren und Kraftfahrwesen", Tag. Aerodynamik des Kraftfahrzeugs, Haus der Technik e.V., Essen, 3.–4. November 1987

[3] Egle, S.; Herzum, N.; Hofele, G.; Konitzer, H.; "Wind Tunnel for Aerodynamic Research", SAE Paper 820372, 1982

[4] Vagt, J.-D.; Wolff, B.; "Das neue Meßzentrum für Aerodynamik – Zwei neue Windkanäle bei Porsche", Teil 1 ATZ Automobiltechnische Zeitschrift 89(1987)3, Teil 2 ATZ Automobiltechnische Zeitschrift 89(1987)4

910311

On the Aerodynamic Interference Due to the Rolling Wheels of Passenger Cars

E. Mercker and N. Breuer
German-Dutch Wind Tunnel (DNW)

H. Berneburg and H. J. Emmelmann
Technical Development Center of General Motors Europe

ABSTRACT

Wind tunnel tests were carried out with a full-scale passenger car over a moving belt. The suspension system of the vehicle was redesigned in such a way that drag and lift forces could be measured whilst the wheels were rolling on the moving ground. The measurements were carried out with an internal balance installed inside the vehicle. Additionally, total-pressure-deficit contour plots were reduced from wake-rake measurements behind the front and rear wheels in order to identify the origin of different bound vortices generated at the wheels. It was found from these tests that rolling wheels have a large aerodynamic influence on passenger cars. They decrease the drag and increase the lift forces in comparison to fixed wheels. This has been established in an absolute and a relative sense by investigating different vehicle configurations.

THE AERODYNAMIC INTERFERENCE of rolling wheels has been a more or less neglected subject in automotive testing in wind tunnels. Although, numerous tests have been carried out with racing cars, only a few have been carried out with passenger cars, and these used models in small-scale. An excellent paper on this subject was given by J.Hackett et al. in 1987 [1]*. The results revealed that the influence of rotating wheels on drag and lift is rather limited, at least for passenger cars. This conclusion was drawn from a comparison of integrated wake measurements. Direct measurements of the forces acting on a vehicle with rolling wheels were not carried out. Probably due to this conclusion, no continuation of experiments with rolling wheels has apparently taken place.

Some basic studies were carried out by A.Cogotti [2] on full-scale isolated wheels and wheels fitted to a low-drag passenger car of 1:2-scale. This test campaign also yielded only a moderate influence on the horizontal and vertical forces of the vehicle. The difference, for example, in drag coefficient between a fixed wheel and a rolling wheel amounted to only 5 drag counts (one drag count is equivalent to a drag coefficient of 1/1000). Unfortunately, during this test a tunnel-floor boundary was present, and the relative movement of a car over a road was not simulated. Whether these had a decisive effect on the results can only be answered by extended experiments.

Both results, mentioned above, are contradictory to the findings of E.Mercker and H.W.Knape [3]. They carried out experiments on a full-scale vehicle over a moving belt including some limited tests with rolling wheels. The forces acting on the vehicle were measured with an internal balance. It turned out that, compared to tests with fixed wheels, the drag coefficient could decrease by 27 drag counts if the wheels were rolling. Due to this large difference, it is suspected by the present authors that a systematic off-set takes place between measurements with fully equipped production cars, as in our case, and small-scale models. Whether this is due to Reynolds number effects or other influences can not be answered without additional tests in both scales.

In order to prove that the previous results measured with a full-scale car over a moving belt with wheels rolling were not a unique result of the specific test vehicle, a new test campaign was in-

*) Numbers in parentheses designate references at end of paper.

itiated in cooperation with the TDC of GME. In order to gain some more understanding of the interference effects of rolling wheels on a car body, plans were made to provide a capability for measuring drag and lift forces with an internal balance whilst the wheels are in contact with and driven by a moving belt. Furthermore, two wake-rakes equipped with a number of total-pressure probes were installed behind the front and the rear wheels in order to measure the total-pressure deficit of the flow in the wakes of the wheels. It was anticipated that they could distinguish between the interferences originating from the front and rear wheels.

Clearly, the emphasis of this test campaign was put on the effect of rolling wheels, and not on moving-ground simulation. Thus, a comparison is made between fixed wheels and rolling wheels. But in order to attain the closest resemblance to road driving, all tests were carried out with a moving belt. The inevitable gap between moving ground and fixed wheels therefore had to be calibrated.

TEST EQUIPMENT

THE WIND TUNNEL- The experiments reported in this paper were executed in the German-Dutch Wind Tunnel (DNW) in Emmeloord, The Netherlands.

At DNW, three closed test sections of different sizes are available. The medium-size 8x6 m^2 section was chosen for this investigation. The test section is 20 m long and is operated under atmospheric conditions by air breathers at the downstream end. The maximum attainable wind speed is 420 km/h. Details about the flow quality in the test section can be taken from [4]. Except for the pressure data, all measured model data is corrected for blockage effects. The correction procedure applied is the volume-ratio method described by E.Mercker [5].

THE MOVING BELT- The moving belt of DNW was initially designed for testing aircraft models during take-off and landing. In the meantime, this test device has also been used for automotive testing almost on a routine basis. The moving belt used during these experiments had a width of 5.2 m and a length of 7.6 m. With a different belt, the useable width can be extended to 6.3 m. The belt speed range of 5 m/s to 50 m/s can be continuously varied.

During operation the flatness of the belt is monitored by a video camera taking pictures from a vertical laser light sheet crossing from the front right wheel to the left rear wheel underneath the vehicle and throwing a shadow on the leeward side of the wind tunnel wall. Depending on the height of the belt above the ground the shadow is increased, and can be measured by a ruler. Only during the start-up phase of the wind tunnel during which the static pressure, in the whole test section is considerably different from ambient pressure the belt tends to lift from the ground. But during the measurement of model data, even under the most severe condition when the vehicle was producing a down-force coefficient of CL=-0.5, the belt remained flat on the ground and no lifting of any kind was discernible.

In order to cope with the approach tunnel-floor boundary layer in front of the moving belt, the device is equipped with a boundary layer removal scoop which is installed in front of the main idling roller. The scoop and the moving-belt surface are installed in the test section 200 mm above the tunnel floor, which is well above the maximum boundary layer thickness occurring during this test campaign. Details about the calibration procedure and the aerodynamic characteristics of the moving belt as well as the way a possible flow angularity at the leading edge of the removal scoop is monitored can be found in [3].

THE TEST VEHICLE- The newly developed Opel Calibra Coupe was selected for the full-scale measurements with rolling wheels over a moving belt. During the development phase of this car the vehicle was tested in various wind tunnels throughout Europe, using the specific high technologies, available at these test facilities to the advantage of producing a low-drag passenger car of CD=0.26. As a consequence, a pretty clear picture of the general flow pattern around the car was available, see [6], although none of the previous tests had been carried out with rolling wheels. Due to the large proportions of the internal balance, which along with the aerodynamic forces and moments must also support the vehicles weight, it was necessary to modify the inside of the vehicle. Furthermore, in order to allow a support sting to be connected to the internal balance the rear window was replaced by a perspex window with an opening just large enough to allow passage of the sting. A sketch of the arrangement of the vehicle, the sting and the sting-support mechanism over the moving belt is displayed in Fig.1.

In addition, the suspension of all four wheels was reconstructed in order to cope with the challenge of propelling the wheels by the belt while still measuring the horizontal and vertical forces acting on the vehicle, i.e., to construct a de-

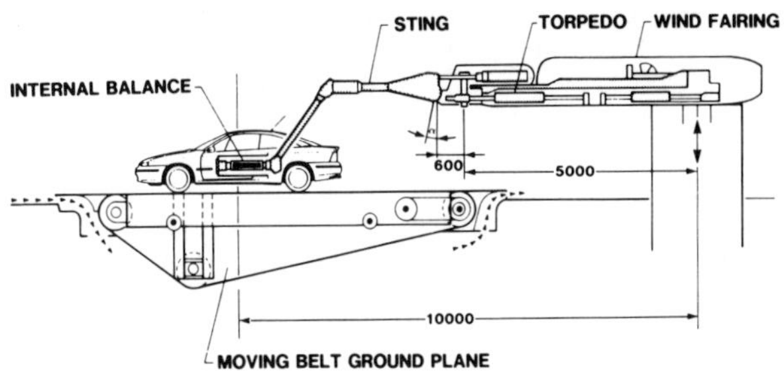

Fig.1: The Opel Calibra with internal balance and sting support mechanism.

vice which provides defined forces into the belt through the wheels under all flow conditions of the wind tunnel stream. Details of this construction will be described in the following chapter.

In order to evaluate the aerodynamic interference effects of the sting and sting mechanism on the aerodynamic forces acting on the vehicle, a calibration procedure was carried out during a previous test campaign with the same type of car [6]. For those tests the external balance with a fixed floor was used. The sting and sting mechanism were present in the test section but disconnected from the internal balance. In a second step, the complete support mechanism was removed from the test section. A comparison of the two measurements revealed a decrease in drag coefficient of 3 drag counts when the support was in place. Apparently the vehicle is propelled by the presence of the sting. This is a characteristic of the Opel Calibra and has something to do with where and under what angle the sting penetrates the rear window. Other cars might show the opposite effect of an increased drag force. In [3], for instance, an increase of 15 drag counts was measured for a square-back type car when the sting and support mechanism was present. The lift coefficients at the front and rear wheels of the Opel Calibra were changed by $CL_R=CL_F= -0.01$ due to the presence of the sting mechanism. Thus, all drag and lift coefficients have been corrected for this effect.

The vehicle of this test campaign was tested with 29 different configurations. These can be classified with respect to 6 main investigation objectives:

The influence of rolling wheels on:

- different underfloor covers
- fully and partly covered wheel houses
- open and closed cooling air flow
- different wheel sizes
- different tire treads
- different spoilers

Furthermore, two configurations were tested with different standing heights of the vehicle. Otherwise, all configurations were tested at the design value of standing height.

THE WHEEL SUSPENSION- The wheel suspension of the Opel Calibra had to be modified for the test purposes. An attempt was made to integrate all major original parts of the suspension system into this new design in order not to change the flow conditions of this part of the vehicle in comparison to the road-going production car. Nevertheless, the drive axles, the brake callipers, the shock absorbers and the springs were removed. All original bushings of the wheel suspension levers were exchanged by low-friction bearings, and the steering column was secured in order to prevent an unwanted angle of steer during wind tunnel testing, see Fig.2.

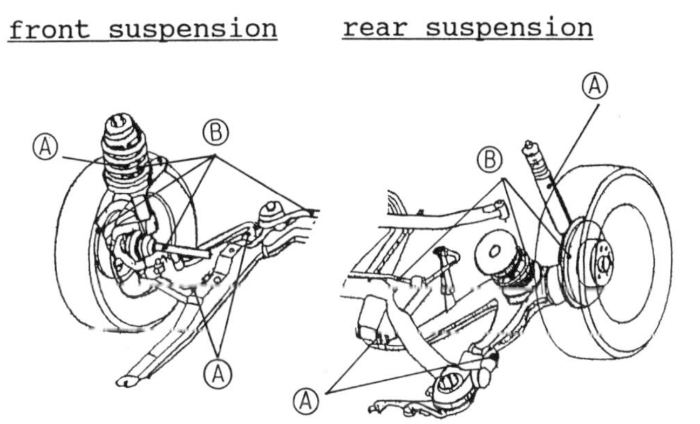

Fig.2: The front and rear suspension systems. Parts denoted with A were removed. Parts denoted with B were exchanged.

By removing shock absorbers and springs from the suspension system and locking the lower ball-head of the wishbone-arm, a wheel could move freely in the vertical direction on a certain camber. A sketch of the principle of the suspension system is displayed in Fig.3.

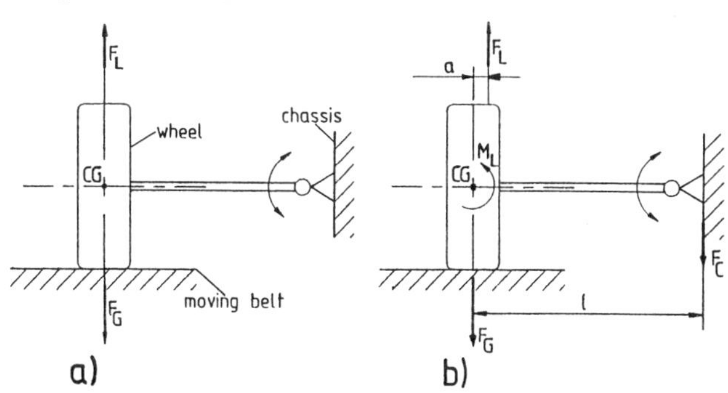

Fig.3: Schematic sketch of the wheel suspension system and the acting forces and moments.

This system would have been sufficient for measuring the total drag force of the vehicle including the wheels. Furthermore, the lift force of the vehicle body excluding the vertical forces acting on the wheels can be measured, because this suspension system is not able to transfer any vertical forces to the vehicle body. The latter is true under the assumption that the lift force F_L(positive or negative) acting on the wheel is in line with the weight force F_G of the wheel and its suspension system, because no additional forces will be introduced into the vehicle through the bearings (except friction forces which are assumed to be small in comparison to the aerodynamic forces involved). All forces created by aerodynamic lift at the wheels are supported by the belt, see Fig.3a.

But even, under the assumption that the weight force F_G and lift force F_L are not in line (a moment M_L will then be created around the centre of gravity CG which results in a fractional-force F_C acting on the vehicle) the error is presumably small because both forces will, most likely, be acting close together somewhere at the centre of the tire, see Fig.3b (the gross weight is created by the tire and the wheel rim, and the lifting surface is provided by the carcass; for $a \ll l$ and $F_C = a/l * F_L$ the fractional force will be small).

The procedure to measure lift and drag is the following: With the rolling wheels resting on the running belt a tare-measurement of the weight and friction forces is taken with the internal balance. There after, the wind tunnel is started and using the internal balance the aerodynamic forces acting on the vehicle can be measured. As mentioned above, no vertical forces from the vehicle will be transferred into the suspension system and vice versa. Thus, only the lift on the vehicle body will be measured.

Due to possible lift forces at the wheels the loading of the wheel bearings will be changed and thus the horizontal forces acting on the vehicle will be altered as well. In a calibration test prior to the wind tunnel test it was established that, for instance, for a vertical load change of 130 N applied to one rolling wheel, the drag force changed by less than 1 N. This variation can be regarded to be rather small and therefore, the influence of varying wheel loadings due to lift on drag can be neglected.

This procedure is common practice in scale-model testing but firstly, the lift forces acting directly on the wheels can not be measured and, secondly, the procedure buries some uncertainties concerning the absolute value of the lift of the car body, due to the reasons mentioned above.

Our full-scale test brought up another problem. For scale model testing the weight of the wheels is not a problem, but for full-scale tests the weight of the wheels (about 32 kg each) was by far too much. Rolling on the belt, the wheels would have destroyed the belt after a few minutes of operation due to frictional heating. Therefore, pneumatic pressure cylinders were built into the vehicle in place of the shock absorbers. The cylinders, which are commercially available, had about the same dimensions as the shock absorbers and thus no major alterations to the wheelhouse had to be made. The secondary part of the air cylinders was pressurized with 8 bar from DNW's air-distribution-system. Due to the large buffer capacity of this system, leakage through the packing at the shaft was continuously compensated. The maximum pressure of the cylinders was 10 bar, and they could hold a weight of approximately 50 kg each for a whole day without any severe leakage between the primary and secondary side of the piston.

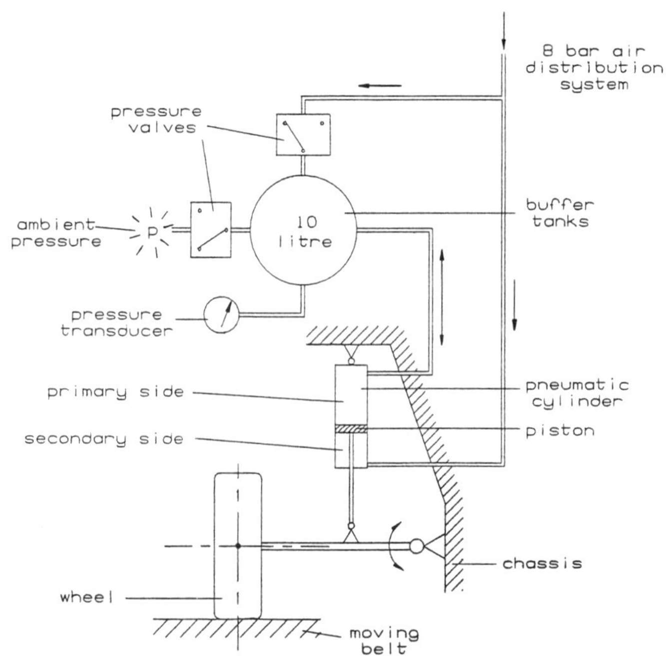

Fig.4: Schematic sketch of the suspension system with pneumatic cylinder and the pressure distribution system inside the car.

From the sketch in Fig.4, showing the principles of this test suspension system, it can be seen that the wheels are lifted, due to the way the cylinders are installed in the vehicle. Increasing the pressure on the primary side of the cylinder, on the other hand, will create a down-force. At 8 bar the force at the piston is zero and the full weight of the wheels is present at the floor. Thus, in an easy manner this system provides the opportunity to adjust the weight of the wheels as desired. Furthermore, as long as the pressures on both sides of the piston are kept constant the resulting down-forces at the wheels are independent of the position of the piston inside the cylinder making it easy to adjust the standing height of the car above the ground without creating any extra forces. Also, vibrations of small amplitude of the whole model due to varying windloads will not falsify the measurement.

In order to control the pressure on the primary side of the cylinders, 4 buffer tanks of 10 liter each were installed inside the drivers compartment and connected to the cylinders. Moving the piston a complete stroke of 200 mm from bottom to top of the cylinder created a pressure increase inside the closed system of approximately 0.1 bar, which results in a force of about 8 N. Considering that it is most unlikely that the vehicle height over the ground will be varied over such a large distance during one wind-tunnel run, this force can be neglected.

Monitoring of the pressure in the cylinder took place through a 20-bar electronic pressure transducer for each wheel. The pressure itself was controlled by two simple electric pressure valves connected to the buffer tanks and to the air distribution system on one side, and to the buffer tanks and ambient pressure on the other side. The latter was used to release the pressure in the system, see Fig.4. With this arrangement it was possible to adjust the residual weight of the wheels to any desired level. During testing this level was set to about 6 kg.

The procedure for measuring the drag and lift forces was as follows:

<u>First step:</u> The belt was brought up to speed with the wheels rolling on the belt; the wind was turned off and a tare-measurement of the weight and friction forces acting on the internal balance was taken.

<u>Second step:</u> The pressure in the cylinders was decreased continuously until the wheels were lifted from the ground one after the other. After any particular wheel was lifted from the ground a force measurement with the internal balance was taken. In this way the residual weight of each individual wheel was determined.

<u>Third step:</u> The pressure in the cylinders was increased until it reached the same level as during the tare-measurement. This could easily be monitored by the installed pressure transducers and verified by a measurement with the internal balance (the vertical and horizontal forces should then again be zero).

<u>Fourth step:</u> The wind was turned on and another measurement was taken with the internal balance. This measurement yields the total drag force of the car body including the wheels, as well as the lift force on the car body alone. As explained above, the lift force might be falsified by a fractional portion of the lift force acting on the wheels.

<u>Fifth and final step:</u> Similar to step 2, the residual weights forces of the wheels are measured again. However, in this case they will differ by the existence of additional lift forces acting on the wheel. In other words, the difference in weight of the wheels under wind-on and wind-off conditions is an aerodynamic lift force.

Therefore, starting with one wheel, the pressure in the relating cylinder was decreased step by step until the wheel was lifting from ground. When the point of floating was reached the pressure was increased again by another step causing the wheel to lower and touching the running belt again. Upon completion, a measurement with the internal balance was made and the result was compared with the one attained with the same wheel but without wind (step 2). The difference of both measurements yielded the lift of the wheel. The step width for increasing or decreasing the pressure inside the cylinder determines the accuracy of this procedure. The smaller the chosen step width, the smaller the residual forces introduced into the belt will be. These forces cannot be measured by the internal balance and thus bias the result. For our test a step width of 1000 pa was chosen, yielding a resolution of 0.8 N for the vertical forces.

For monitoring purpose four video cameras were installed; one for each wheel. With the aid of these devices the rotational speed of the wheels could be checked.

The procedure described above may seems confusing. Once the exercise has been carried out a few times, however, it is quite straightforward. Furthermore, assuming symmetrical flow over the vehicle in a crosswise direction the procedure can be reduced to a measurement at one front and one rear wheel only. The lift forces will in that case be the same on each pair of wheels.

In an extensive calibration program prior to the wind tunnel test it was established that the maximum error was about 31 N in the vertical direction, which is established by the resolution capabilities of the internal balance and the step width of the pressure variation inside the cylinders. For the internal balance the digital rounding error is equivalent to 7 N. As this error can occur for all four wheels due to the way the total lift is reduced, a variation of the lift coefficient of 0.017 is possible at a test speed of 140 km/h.

WIND TUNNEL TESTS

DRAG AND LIFT MEASUREMENTS

The basic body- One of the major goals of this investigation was to evaluate and to differentiate, in general, between the aerodynamic interference effects created by the fixed and the rolling wheels of a passenger car. Therefore it was interesting to establish, first the drag and lift forces of the body-shell alone without any wheels. For this purpose the wheels were removed and the underbody of the Opel Calibra was completely covered with flat sheet panels. Additionally the wheelhouses were closed with specially made inserts and side dams, see Fig.5.

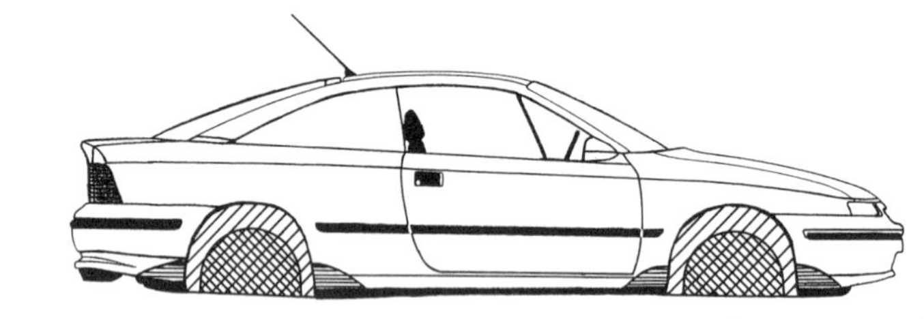

Fig.5: The Opel Calibra with wheel inserts and side dams.

Fig.6 displays the drag force and the lift coefficient for this vehicle as a function of ground clearance between the front spoiler and the moving belt. It indicates that drag is decreased with increased distance from the ground, which is plausible because the interference originating from the induced boundary layer on the moving belt and the boundary layer on the underbody of the vehicle is decreased. Furthermore, the effective angle of attack of the flow will be altered. This will produce a different flow on the top and bottom surfaces of the vehicle changing drag and lift respectively. For a larger ground clearance, which was beyond our measuring range, the drag force will most likely approach asymptotically a constant value.

On the other hand, for small ground clearances the drag force decrease again, after a maximum is reached. This effect can be explained by borrowing the image concept from potential flow theory: A vehicle brought into the vicinity of a ground plane comes under the influence of its image beneath the ground. As the ground clearance tends to zero the stagnation point of the vehicle will move from somewhere at the centre of the front towards the plane of symmetry (ground board). The vehicle with an outer flow passing over the body and an underbody flow will then appear as a vehicle and its image with an outer flow and no through-flow. In this case, the frontal area of the complete body has doubled, whereas the drag coefficient will not increase as much (this can be concluded from bluff body flow of different aspect ratio). Thus, as in our experiment, only the forces acting on the upper half (vehicle) of the symmetrical body (vehicle plus image) was measured the drag coefficient has to decrease with decreasing ground clearance.

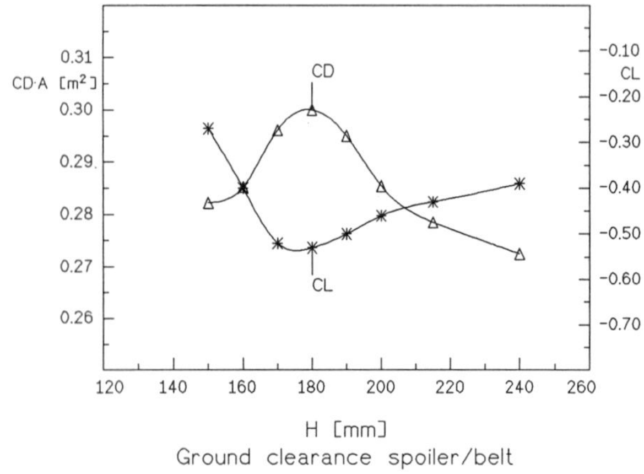

Fig.6: Drag force and lift coefficient versus ground clearance for the basic body without wheels and with a flat underfloor.

However, in viscous flow a reversed trend of increasing drag might occur again for very small ground clearance (not measured during our experiment). This is caused by flow separation from the ground in front of the vehicle generating a so-called 'horseshoe vortex'. This vortex creates drag and thus the drag coefficient of the vehicle will be increased with increasing vortex strength. For the drag force, the same tendency as displayed in Fig.6 can be observed by keeping the standing height of the vehicle constant and varying the length of the front spoiler, see [7].

The lift force reveals just the opposite behaviour. With decreasing ground clearance the suction pressure underneath the vehicle is increased as

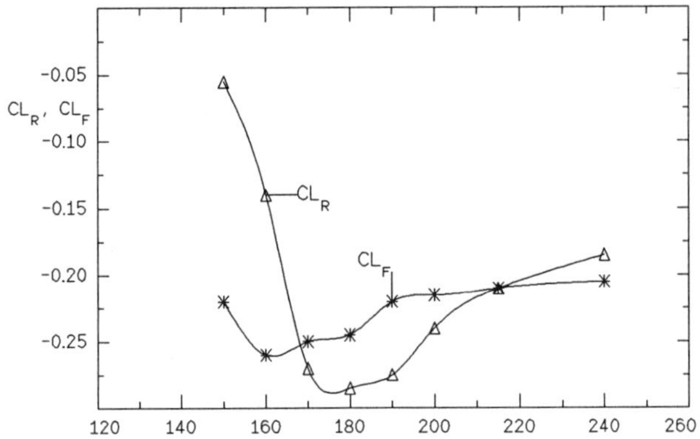

Fig.7: Lift coefficient at the front and rear axle versus different ground clearance for the basic body. CL_F= front lift; CL_R= rear lift

velocity increases due to continuity. However, due to the expanding interference of the horseshoe vortex the free passage will progressively be blocked and the mass flux reduced, which in turn decreases the down-force. This flow mechanism can also be deduced from Fig.7 which shows the lift distribution at the front and rear axle. The steep increase in lift at the rear axles for small ground clearances shows how much the down-force is diminished in this part of the underfloor due to the confined through-flow. Unfortunately, very small ground clearances could not be achieved because, due to viscous pumping, the vehicle and its suspension system were excited in its natural frequency and the car vibrated in a resonant frequency with rather high amplitudes. When moving the car higher above the ground, pumping was greatly reduced.

The influence of the wheel gap- A direct force measurement with fixed wheels over a moving ground is only possible with a gap between tire and belt. In scale testing and also during full-scale measurements (see [3]) it has been established that this gap can be of decisive influence on the level of the drag and lift coefficients. Therefore a calibration test for the gap influence was executed.

For this test the standing heights of the vehicle were kept constant. With a small pulley for each wheel installed inside the driver's compartment, the gap at the wheels was adjusted. Fig.8a and 8b show the drag area (drag force) and lift coefficient as a function of gap clearance for the basic configuration, with flat underbody and wheel inserts installed. Fig.9a and 9b show the same kind

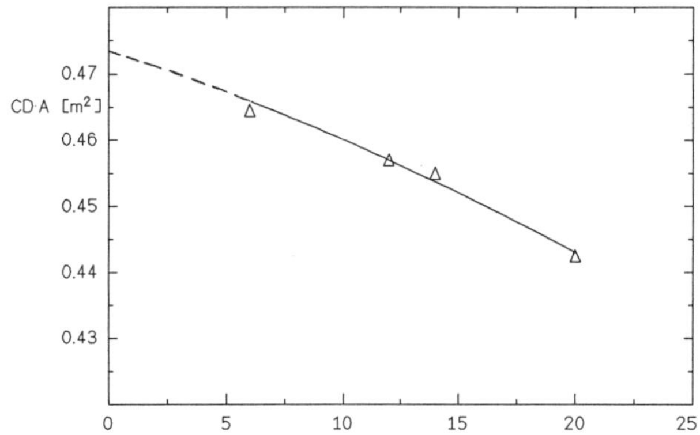

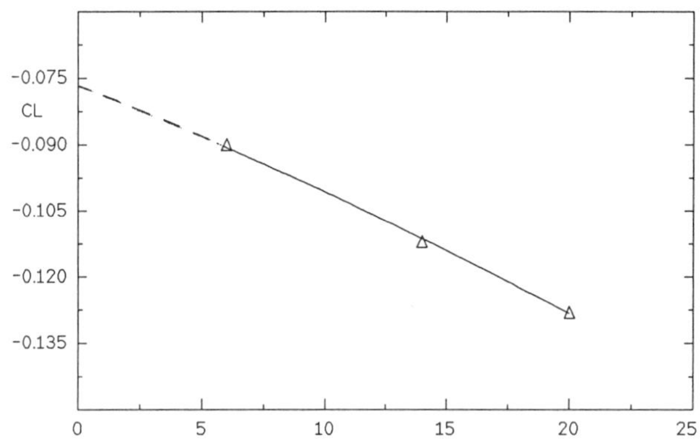

Fig.8a,b: Influence of wheel gap clearance h for fixed wheel on drag force CD*A and lift coefficient CL (flat underfloor, wheelhouse inserts).

of measurements but for a vehicle configuration equivalent to a production car with rough underbody, cooling air flow etc.. Extrapolating these curves to vanishing gap clearance reveals the desired drag and lift coefficient without gap interference. The gap clearance during testing was kept constant and was adjusted to 12 mm for all four wheels. From Fig.8 and Fig.9 it can be taken that the difference in drag area is then 0.019 and 0.021 respectively in comparison to the extrapolated value for the two different configurations.

With respect to a frontal area of 1.884 m^2 of the car this is equivalent to 10 and 11 drag counts. For the lift coefficient a difference of about 0.03 is yielded. Although, the slopes of the curves in the above mentioned figures reveal a different gradient, which means, that the influence of gap clearance depends on the vehicle configuration, all drag and lift data with fixed wheels was corrected with a constant value of 0.011

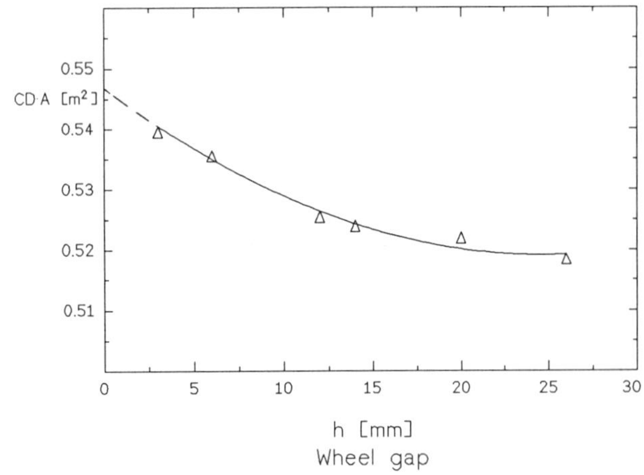

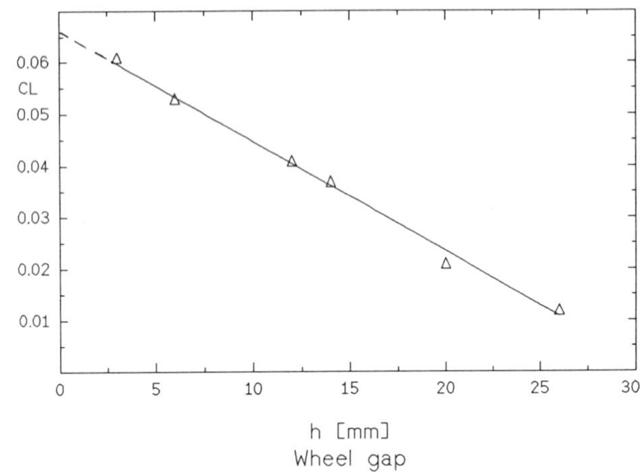

Fig.9a,b: Influence of wheel gap clearance h for fixed wheel on drag force CD*A and lift coefficient CL for the production vehicle.

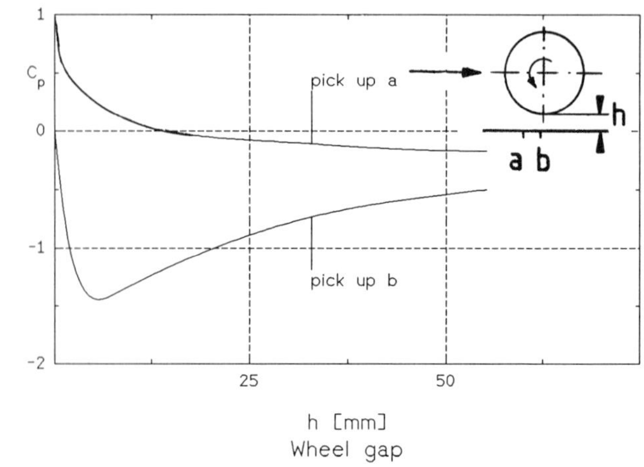

Fig.10: Static pressure at the ground under and in front of an isolated wheel versus gap clearance, [2].

for the drag coefficients and 0.030 for the lift coefficients. Furthermore, from Fig.8 and Fig.9 it can be taken that a polynomial extrapolation was chosen. From a physical point of view the curves should rather have a horizontal tangent for vanishing gap clearance. However, such an extrapolation would change the result by one or two drag counts only and thus has been neglected.

It is interesting to note that both drag and lift increase when gap clearance is decreased. The reason for the effect of increased drag can be seen again in a horseshoe-vortex in front of each wheel which progressively increases in strength when the gap is decreased. This flow mechanism creates drag. The phenomenon itself is very well known in the aerodynamics of high-rise buildings. It will also occur when the floor is moving relative to the model in an automotive wind tunnel.

The slope of the lift curve can also be explained straightforwardly. However, some pressure measurements carried out by A.Cogotti [2] might be of some additional help. In Fig.10 these results are reproduced, showing the static pressure in front of an isolated fixed wheel and at the centre of the gap as a function of the gap height. The latter curve of the pressure coefficient reveals the same trend as the lift curve of Fig.6. With decreasing clearance the suction pressure is initially decrease because the velocity is increased due to continuity. But at the same time, a stagnation area is built up progressively in front of the wheels, ending in a horseshoe vortex which yields reversed flow in front of the wheel. As a consequence, the gap flow will be disturbed and the mass flux through the gap will be decreased. Simultaneously the down-force will be decreased. When the gap clearance, vanishes completely, a positive pressure finally remains at the forward part of the wheel, and due to the wake flow the aft region will experience a smaller negative pressure (see [2]). Thus, for small gap clearances, the integral of the pressure distribution over the lower part of the wheel will be positive, creating a lift force.

In principle, the same mechanism takes place with rolling wheels. However, due to the rotation the stagnation point of the flow on the forward part of the wheel will be shifted upwards and the frontal down-wash will be diminished. As a consequence, the drag force will be decreased whilst the lift is increased.

This qualitative description of the flow pattern will be discussed in more detail in a following chapter. However, it is worthwhile to note that a down-force originating from the Magnus-effect is apparently not existing. The reason for this can be recognized in the presence of the ground board and thus the horseshoe vortex. Nevertheless, due to a crossflow from underneath the vehicle the overall vertical forces on the wheels can be negative, because the downgoing front and upgoing rear part might be exposed to a wake flow generated at the edge of the inner tread of the wheel.

<u>The influence of wheelhouse covers</u>- In a next step, small underbody panels were removed in order to install four rolling wheels to the vehicle. The inserts at the sides of the wheel arches remained in place for the first measurement. In a second and third step they were half opened and finally completely removed.

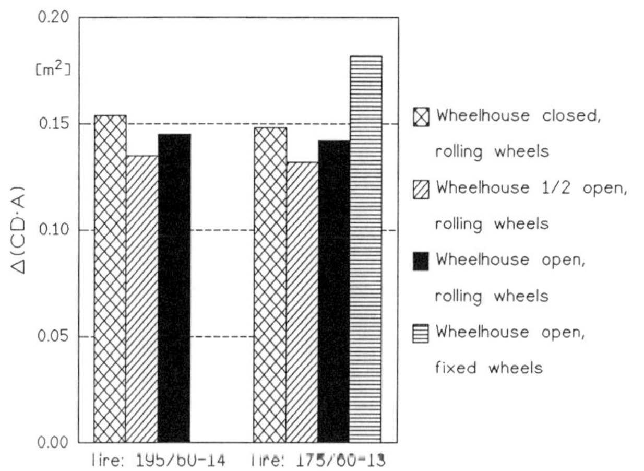

Fig.11: Increase of drag force due to rolling and fixed wheels for different wheelhouse openings in comparison to the basic body.
$\Delta(CD*A)=D_W-D_{bb}$;
D_W =CD*A with wheels
D_{bb} =CD*A for basic body (no wheels)
note: tire 175/60-13 was also tested with fixed wheels.

Fig.11 shows the difference in drag force compared with the basic model without wheels for these three configurations. The measurements were also carried out with two different wheel sizes. For a vehicle frontal area of 1.884 m^2 an average increase in drag coefficient of about 75 counts occurred. Apparently, the wheels and the openings in the underfloor panel contribute significantly to the overall drag coefficient of a vehicle. As this increase is over proportional to the increase in frontal area of the car due to the wheels, it becomes clear that it is important to study the wake flow of the wheels and their interference effect on the car body in more detail when optimizing the drag.

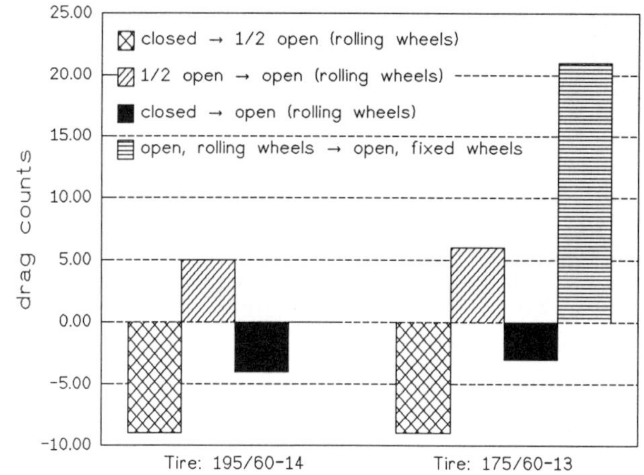

Fig.12: Change of drag due to changes of the wheelhouse opening for rolling wheels, and due to rolling wheels versus fixed wheels for the open wheelhouse.

Fig.12, which is a crossplot of Fig.11, shows the change in drag for the various wheelhouse inserts. The lowest drag was measured the wheelhouse half open. The reason for this may be identified in a suppressed bound vortex shed from the upper part of the rotating wheel when the wheel arch is open. Details about this mechanism can be concluded from the work of L.J.Oswald and A.L.-Browne [8] who investigated the airflow field around an operating tire. However, the largest increase in drag was measured when the wheels were fixed. A considerable increase of 21 drag counts was reduced from the measurements in comparison to rolling wheels with the open wheelhouse.

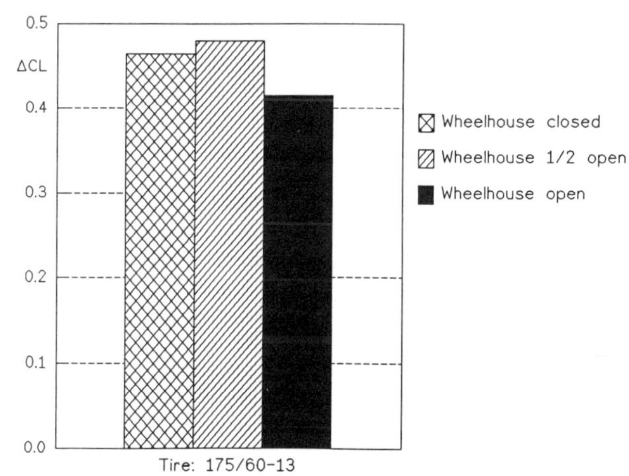

Fig.13: Increase of total lift coefficient due to rolling wheels in comparison to the basic configuration for different wheelhouse openings. $\Delta CL = CL_{rw} - CL_{bb}$
index: rw= rolling wheel,
bb= basic body (no wheels)

Fig.13 shows the change in lift when rolling wheels are fitted to the basic body. A dramatic increase of the overall lift force is discernable. However, from Fig.14 it can be concluded that the gross increase is created by the interference flow with the vehicle shell that is generated by the wheel. The contribution of the wheel itself is expressed by a moderate down-force depending on the wheelhouse opening. When analyzing the lift forces in a fashion described above, it is to be noted that it was the front wheels which created negative lift. The rear wheels reacted indifferently for most configurations (the weight force of the wheels with and without wind was identical).

Fig.14: Total lift coefficient CL_T and Lift coefficient of vehicle body CL_C for different wheelhouse openings with rolling wheels.
$CL_T=CL_C+CL_W$;
index: T=total lift,
C=body lift,
W=wheel lift.

<u>The influence of different floor sections</u>- The next few configurations were investigated with respect to different underbody panels and front spoilers. Fig.15 displays the difference in drag between fixed and rolling wheels when floor panels were removed step by step. The largest difference was attained by adding the production front spoiler. An increase of almost 23 drag counts was measured when the rolling wheels were fixed. Fig.15 also contains the absolute drag values for the configurations with fixed wheels. A comparison of both drag levels reveals that there is no correlation between the absolute value for one configuration and the change in this level due to rolling wheels. Thus, subtracting a certain amount of drag force for the effect of rolling wheels during standard measurements with fixed wheels does not seem to be feasible. Nevertheless, it is suspected, that the effect of rolling wheels will be diminished without the presence of ground movement or other ground simulation techniques such as boundary layer control.

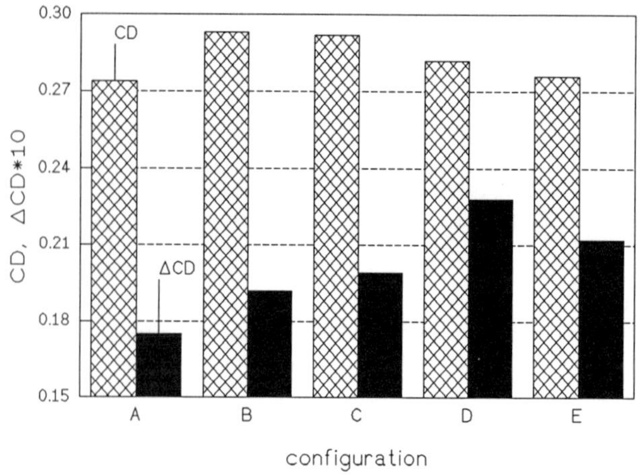

Fig.15: The effect of rolling wheels on drag for different configurations in comparison to fixed wheels. -Also drag coefficient for fixed wheels.
$\Delta CD=CD_{fw}-CD_{rw}$;
index: fw=fixed wheel
rw=rolling wheel
Configurations:
A - with complete underfloor
B - underfloor removed but with engine cover
C - same as B but side dams removed, Fig.5
D - same as C but production spoiler
E - same as D but engine compartment cover removed

The variation of lift for the different configurations is given in Fig.16. Whereas for a fully covered floor section a down-force was measured, a positive lift force was yielded when the floor panels were removed. From Fig.16 it can be further concluded that the overall lift force is generally increased with the wheels rolling.

Analysis of the lift measurements also revealed that the vertical forces on the individual wheels increased when the car had a production underbody, see Fig.17. For the production car, for instance (conf. I), when a front spoiler was added the entire level of the vertical forces at the axles was raised by a certain amount. In comparison to the vehicle with smooth underbody the data analysis revealed that the lift at the front axle was now unchanged, and positive at the rear axle, but the pitch moment was constant in both cases.

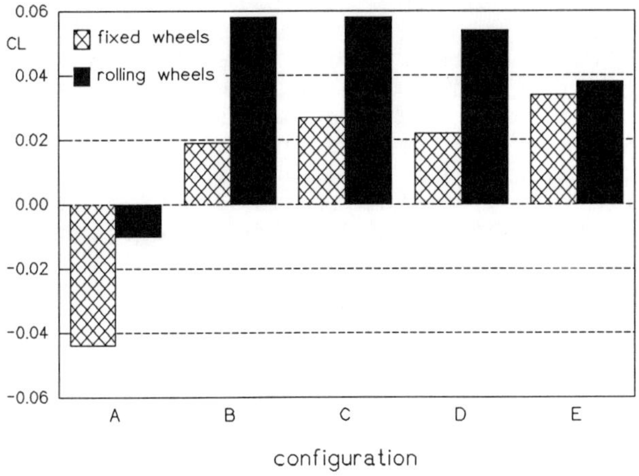

Fig.16: Total lift coefficient for fixed and rolling wheels. Configuration, see Fig.15.

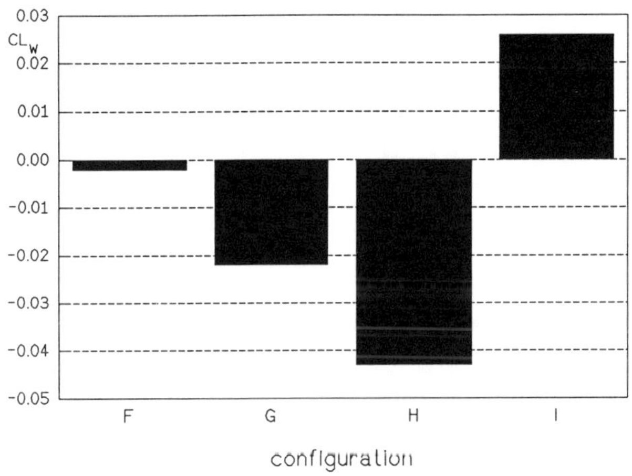

Fig.17: Lift coefficient of the rolling wheels for different configurations.
$CL_W = CL_T - CL_C$
index: W=wheel, T=total, C=body
Configuration:
F - smooth underfloor, wheelhouse closed
G - same as F but wheelhouse 1/2 open
H - same as G but wheelhouse open
I - rough underfloor, wheelhouse open

<u>The influence of cooling-air flow</u>- During the aerodynamic development of a car the influence of cooling-air flow on the drag and lift forces is an important aspect. Often, a comparison of aerodynamic forces is made between configurations with a closed and with an open air intake in front of the radiator in order to asses the drag penalty due to the through-flow. Furthermore, sometimes identical cars are available with different engine sizes which demand a different

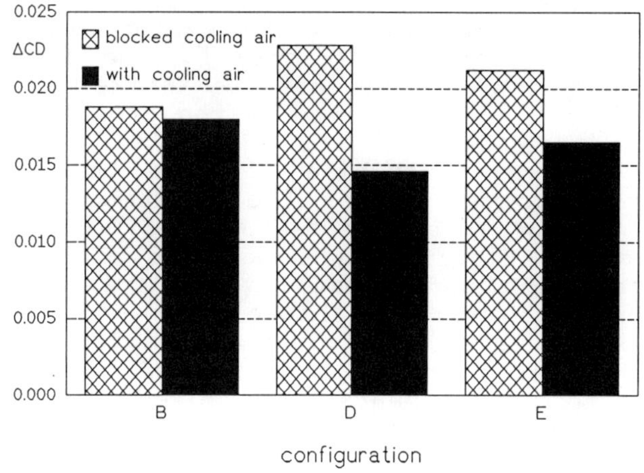

Fig.18: The effect of rolling wheels on drag for different cooling-air flow in comparison to fixed wheels.
$\Delta CD = CD_{fw} - CD_{rw}$;
index: fw=fixed wheel,
rw=rolling wheel
Configuration, see Fig.15.

cooling-air rates. In order to yield a lower drag the intake is often partially blocked by the car manufacturer.

Fig.18 shows, for different configurations, the change in drag due to rolling wheels in comparison to standard measurements with fixed wheels. The same vehicle configurations were investigated with open and closed cooling-air intake. Especially, for the configuration where the production front spoiler was added a rather large difference between closed and open intake is visible. In this case, with blocked cooling-air flow a decrease of almost 23 drag counts was measured due to rolling wheels. But this influence is reduced down to 15 drag counts for an open intake, which again demonstrates the dependence of the results on the car configuration.

Possibly, the reason for this behaviour can be seen in the yaw angle of the attack-flow at the wheels. During the development of the Opel Calibra it was demonstrated (see [6]) that, depending on the design of the complete front spoiler the angle of attack of the approach flow of the wheels could be influenced to a large extent, and thus drag and lift. With the cooling-air flow blocked the mode of action of the spoiler has probably been changed in comparison to the configuration with cooling-air flow. This resulted in an increased crossflow at the wheels and consequently the drag was increased. In a following chapter, this fact will be illustrated by some total-pressure measurements with two wake-rakes fitted to the car behind the front and rear wheels.

From this result it can be further concluded that only a partial drag increase is generated directly by the flow through the radiator and its intake. The other part is exhibited by the crossflow at the wheels.

It is worthwhile to note that a direct comparison of the influence of cooling-air flow leads to very different results. Fig.19 shows that for rolling wheels the difference between open and closed cooling-air intake remains at a constant level of 27 drag counts for all three configurations investigated, whereas the measurements with fixed wheels reveal a large variation. Apparently, the crossflow effect is less significant with the wheels rolling. This proves again how difficult it is for the development engineer to come to the right decisions without the wheels turning.

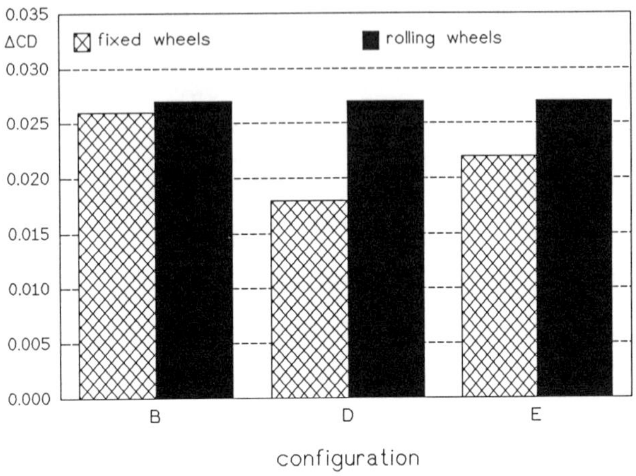

Fig.19: Change in drag coefficient due to different cooling-air flows for fixed and rolling wheels.
$\Delta CD = CD_{oc} - CD_{cc}$;
index: oc=open cooling air,
cc=closed cooling air
Configuration, see Fig.15.

With respect to lift forces, the situation seems to be less dramatic. Although, the lift is increased with open cooling-air flow, the differences between rolling and fixed wheels remained constant. Therefore these result is not displayed in a figure.

Influence of the tire-width-In order to investigate the aerodynamic interference of the surrounding wheelhouse upon the rolling wheels, tires of different width but with a constant diameter were mounted to the vehicle. All tires had the same tread and the tire width could be varied over the range 165 mm to 205 mm. It can be reported from the measured results that the displacement of volume inside the wheelhouse by wider tires is only of moderate influence. Taking into account that the frontal area of the vehicle was increased with wider tires, the difference between the smallest and the widest tire was only 9 drag counts. Whether this difference occurs entirely due to the displacement effect of the tires inside the wheelhouse, or due to a different wake-interference effect between the wheels on the body shell cannot be answered without extended tests. Most likely, several flow-parameters were changed simultaneously by varying the width of the wheels. The most effective test would be to vary the internal wheelhouse dimensions while keeping the tires constant (see [2]). Some explanation of the flow mechanism involved will be given in a following chapter where the wake flow of the tires will be analyzed.

Influence of tire tread- Four sets of tires were investigated with respect to different treads. All tires had the same dimensions and were fitted to the same type of rim. This investigation was executed with rolling wheels only. Fig.20 shows the different kind of treads for three tires arbitrarily selected from the dealers stock. The fourth set of tires was a so-called 'slick' without any tread. As expected, the slick produced the smallest drag force; tire No.3 yielded the largest. However, the maximum difference measured was merely 5 drag counts. Interestingly, the tire with the highest drag created the lowest lift force, which is advantageous for driving in wet weather conditions.

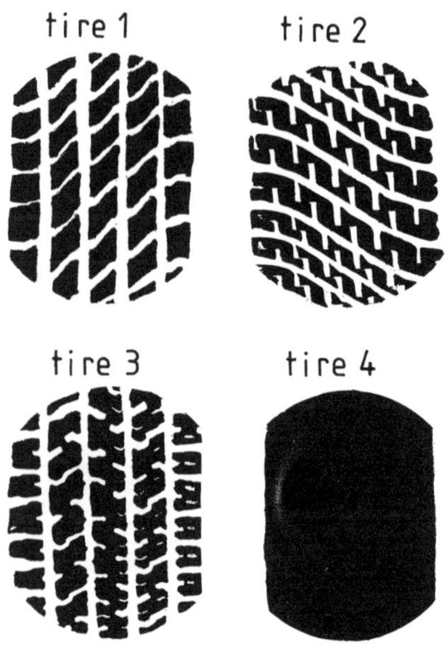

Fig.20: Sketch of different tire treads. Tire 4 is a so-called 'slick'

It is worthwhile to refer to the fact that the wheels were not fully loaded during our moving belt test. Thus, during road-driving the result might be different due to deformation of the tire carcass.

Influence of standing height- The vehicle configuration similar to the production car was tested with different standing heights, H_S, above the moving ground. This was achieved by adjusting the sting mechanism in height. Fig.21 reveals the difference between rolling and fixed wheels. Again, the gap clearance occurring for the test with fixed wheels was corrected with a value extracted from Fig.9. The difference between both configurations decreased with larger standing heights. It was merely 7 drag counts for the highest level above ground, but this increased to over 20 drag counts for the smallest level. The reason for this can be explained by the different vortices straddling the upper and lower part of the wheel. This will be discussed in the following chapter using some wake-rake measurements.

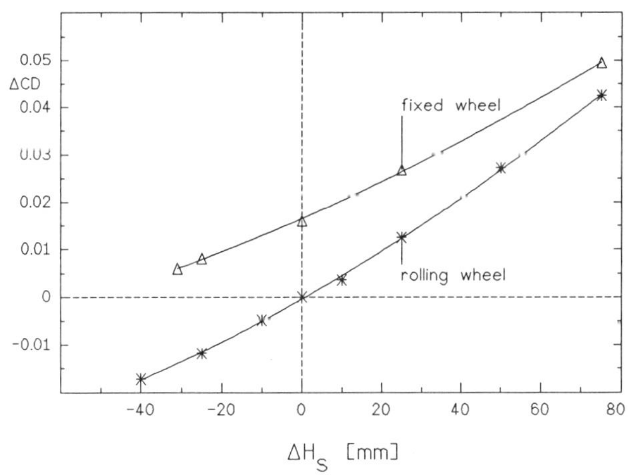

Fig.21: Effect of standing hight H_S on drag for fixed and rolling wheels (reference value is the production vehicle with rolling wheels).
*: $\Delta CD = CD_{rw} - CD_{rw}(0)$ and
Δ: $\Delta CD = CD_{fw} - CD_{rw}(0)$;
index: rw=rolling wheels,
fw=fixed wheels
rw(0)=rolling wheels measured at production height.

WAKE MEASUREMENTS

As already mentioned above, two rakes provided with 25 and 43 total-head probes, respectively, were mounted to the vehicle. These gave sufficient resolution to generate contour-plots of the total-pressure deficit behind the wheels.

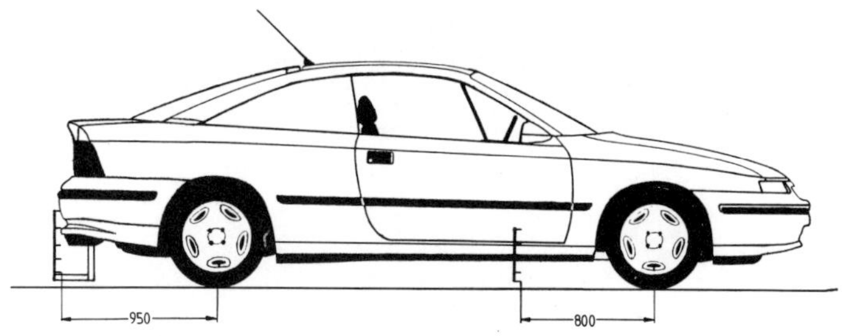

Fig.22: Position of the wake-rakes
note: only one rake on each side

A sketch of the locations and dimensions of the rakes with respect to the car is shown in Fig.22.

Although measuring vorticity and crossflow-vectors would have been beneficial for the interpretation of the flow phenomena, it was thought in the present test that knowledge of the deficit of total-head would be sufficient to understand the basic flow mechanism involved in the aerodynamics of rolling wheels. This was concluded from measurements carried out by J.E.Hackett et al.[1]. From their measurements it can be taken that the region of maximum vorticity is approximately equivalent to the region of maximum total-pressure deficit (vorticity can be evaluated as the ratio of vortex strength Γ to unit volume flux $\dot{Q}$). Our total-head probes were insensitive to yaw angles up to ±25°. For larger flow angles the contour plots can only be used on a comparative basis.

In the following, a simple flow model is introduced in order to elucidate some of the rather complicated flow field mechanisms around a wheel. Fig.23a shows schematically the front view of a non-rotating, isolated wheel exposed to a parallel flow. In the centre of the cross section the stagnation point of the flow is indicated. Due to the fact that our wheel is larger in height than in width, the flow rate $\dot{Q}$ towards the sides will be greater than towards the top and the

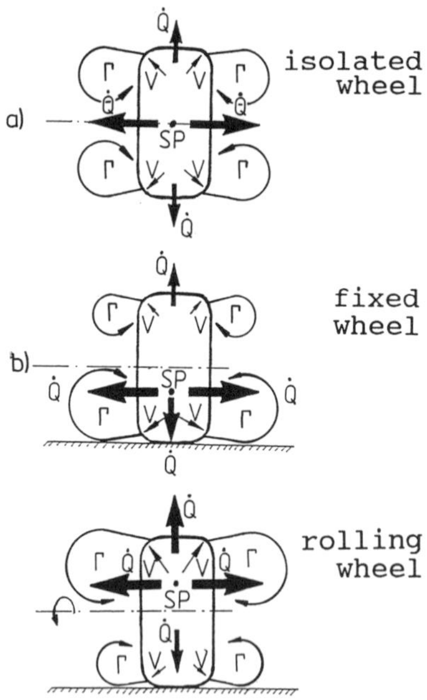

Fig.23: Flow model for the vortex flow around isolated, fixed and rolling wheel.
$\dot{Q}$ = flow rate
Γ = vortex strength
SP = stagnation point
V = flow vector

bottom. Except in the axis of symmetry, an oblique flow V towards the outer edge of the wheel will be experienced. As the flow component in the horizontal direction is larger than in the vertical direction, there is a flow vector horizontally, away from the tire in the corners of the plane. Thus, the flow will roll-up into two vortex pairs on each side of the wheel with a vortex strength Γ (the sense of rotation of these vortices is opposite to those encountered frequently in textbooks on vehicle aerodynamics). The vorticity of all four vortices will be the same.

Fig.23b shows the same non-rotating wheel fixed to a ground board. The same flow mechanism takes place, except that the lower part of the flow is blocked by the floor. The flow which, in our first example, could escape at the bottom of the wheel will now be deflected to the sides, increasing the vortex strength of the ground vortices. This will increase the drag and lower the lift, because the base pressure at the aft part of the wheel will be decreased. Apart from this mechanism, a typical horseshoe-vortex in front of the wheel will originate from viscous flow separation at the ground. The two ground vortices on each side of the wheel and the horseshoe vortex having the same sense of rotation will merge and straddle the wheel. Adopting potential flow theory again, the stagnation point also moves downwards due to the image of the wheel beneath the ground board.

Finally, Fig.23c displays the wheel rolling on the ground board. In an isolated mode, like the wheel in Fig.23a, rotation will cause the stagnation point to shift upwards. This mechanism is known as the Magnus-effect and generates a down-force. Although, the Magnus-effect will somehow be disturbed by the presence of the ground board, it still can be assumed that the stagnation point is moved upwards to some extent when the wheel is rotating. Therefore, less flow will escape at the top part of the wheel and most of the flow is deflected towards the sides, strengthening the upper bound vortices.Consequently the down-wash in front of the wheel will decrease and thus the strength of vortices at the ground will be debilitated. Most of the vorticity experienced in this region will originate from the horseshoe vortex. In opposition to the fixed wheel this fact will decrease the drag and higher the lift, because the base pressure at the down going aft part of the wheel will be increased.

We are aware of the fact that this flow model is arguable because the sense of the upper bound-vortices depends on the ratio of width to height of the wheel. However, the sense of the ground vortices and the horseshoe-vortex in front of the wheel is predetermined and cannot be altered. For this reason, wide tires, such as those used for Formula-1 racing cars, will bring a third vortex on each side of the wheel into operation in order to balance the upper and lower bound vortices. Furthermore, the wheels of our vehicle were more or less shielded by the wheelhouse. In accordance with reference [8], the symmetry of the upper vortex pair will then be highly disturbed depending on the geometry of the wheelhouse. Beyond that, all assumptions are made for a uniform approach flow. Thus, it is difficult to say how the mechanism will be changed for the rear wheels which are exposed to the wake flow of the front wheels. It is most likely that, in principle, the same flow mechanism will take place, but according to the work of A.J. Scibor-Rylski [9] the effects will be diminished.

<u>The wake flow for different standing heights</u>- For the front wheels, Fig.24 shows the total-pressure-deficit contours for different standing heights of the vehicle with rolling and with fixed wheels. Only some contour values are given in this and the following figures but each increment is constant.

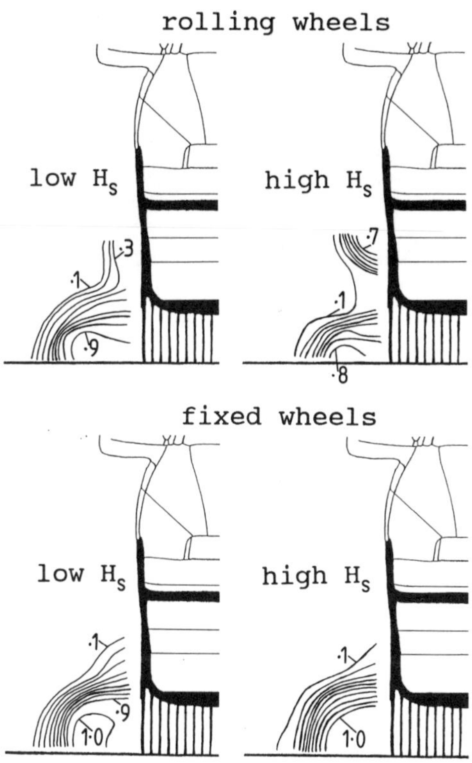

Fig.24: Total-pressure-deficit contour plots at the front wheel for different standing heights: rolling wheel - fixed wheel Sense of vortex rotation: clockwise for lower bound vortex, anti-clockwise for upper bound vortex.

As mentioned above, the deficit in total head can reasonably be expected to represent a developing vortex. Therefore, the contour-plot for the low standing height reveal a strong vortex at the ground when the wheels were fixed. Presumably, the strength of this vortex would be even more increased for no gap clearance. For the rolling wheel the vorticity of the ground vortex is greatly reduced because of a reduced downwash. With low standing heights no upper bound-vortex will become visible because of the shielding-effect of the wheelhouse. The comparison of drag forces in Fig.21 revealed the largest differences for these standing height.

For high standing heights the situation is somewhat different. The plot for rolling wheels now yields part of the upper bound vortex because of the reduced shielding-effect of the wheelhouse and a weakened ground vortex. For fixed wheels, though, a weak upper bound-vortex is probably generated; no total-pressure deficit has been measured because of a remaining shielding-effect of the wheelhouse. This is in accordance with the simple flow model explained above. The upper bound vortex for rolling wheels, on the contrary, generates an additional drag force, and thus the difference in drag between rolling and fixed wheel is reduced. This fact has been elucidated in Fig.21.

<u>The wake flow for different spoilers</u>- Fig.25 shows the wake flow for the front wheels and the influence of rolling wheels for different spoiler configurations. The so-called basic spoiler was a spoiler with a flange of 10 mm spanning the full frontal width of the vehicle. This spoiler was compared with the spoiler integrated into the production version of the car which had additional spoiler corners.

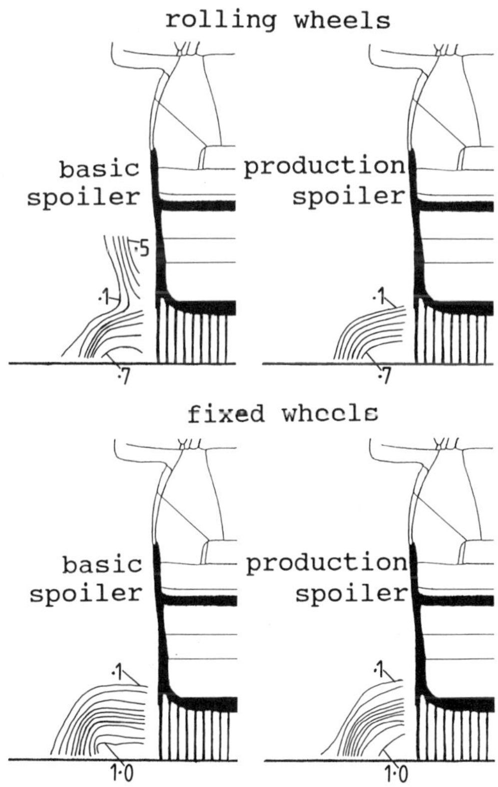

Fig.25: Total-pressure-deficit contour plots at front wheel for different spoilers.

With rolling wheels the upper bound vortex is visible when the basic spoiler is installed, but this vortex is completely suppressed when the production version was fitted. Combined with a slightly reduced ground vortex, this fact results in a drag reduction of 13 drag counts with the production spoiler.

In opposition to this result, the experiments with fixed wheels revealed no upper bound vortex, but an increased ground vortex. A reduction of 9 drag counts was yielded when the production spoiler was installed, which can be anticipated by the reduced size of the inner total-pressure-deficit contour. The total difference between rolling and fixed wheel for the production spoiler sums up to 22 drag counts in favour of the rolling wheels, demonstrating again that the results are varied not only in a relative sense, but also in an absolute sense when the wheels are rolling.

Fig.26 reveals the total-pressure-deficit contours behind the rear wheel. This plot contains not only the deficit of the rear wheels but also the wake history of the front wheels, making it more difficult to distinguish different effects. Evidently, the vortices originating from the front wheels have merged with each other on their way downstream. But with the knowledge of the flow pattern in front of the vehicle it is probable that the mechanism at the rear wheels is generally similar.

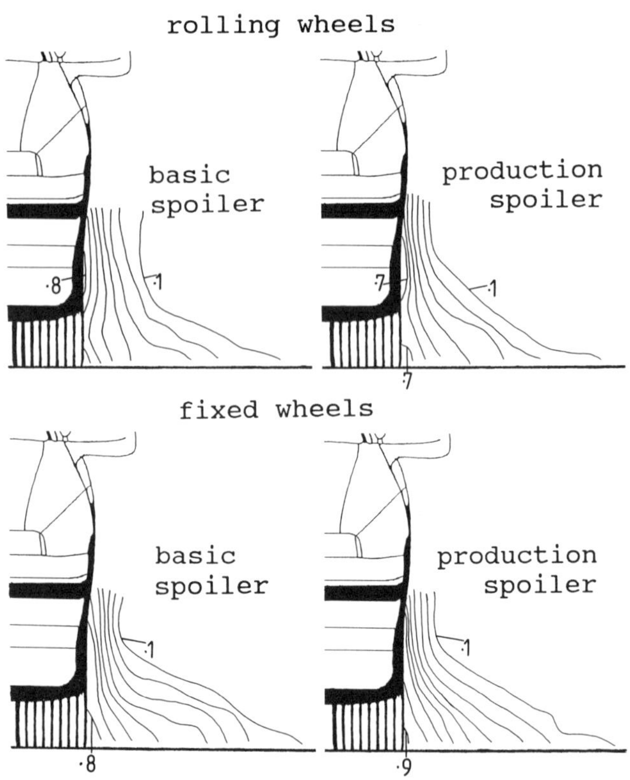

Fig.26: Total-pressure-deficit contour plots at rear wheel for different spoilers.

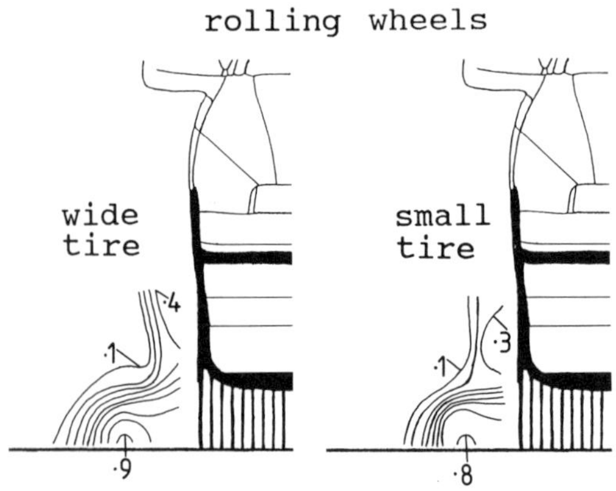

Fig.27: Total-pressure-deficit contour plot at front rolling wheel for different tire widths.

The wake flow for wheels of different widths- Finally, Fig.27 shows wake-rake measurements for rolling wheels of different widths. Apparently as a consequence of greater displacement of the flow by the wider tire, a vortex of increased strength is exhibited at the ground. Furthermore, due to the fact that for wider tires the edge of the carcass will be closer to the outer part of the wheelhouse, the upper bound vortex can be measured. As already mentioned above, the increase in drag coefficient was measured to be 9 counts between the smallest and the widest tire. It is strongly suspected that this was created by the vortices and not by some other mechanism taking place inside the wheelhouse.

SUMMARY and CONCLUSIONS

Wake and force measurements were carried out in a wind tunnel on a full-scale passenger car. The experiments were executed with a moving belt. Therefore the vehicle was suspended by a sting and the force measurements were made with an internal balance installed inside the car. With different vehicle configurations, the effect of rolling wheels was investigated in comparison to fixed wheels. A wake-rake analysis behind a front and rear wheels was included in the test program in order to gain a better understanding of the flow mechanism around wheels.

From these experiments it was found:

1) For all the configurations tested, a large reduction in drag was measured for rolling wheels. The amount depended on the particular vehicle configuration. These results are contrary to the findings of other researchers.

2) The lift on the vehicle increased with rolling wheels in comparison to fixed wheels. The reason for this can be seen in an increased base pressure at the aft part of the wheels when they are rolling. This is a consequence of the stagnation point shifting upwards in front of the wheel and debilitating the lower bound vortices.

3) With the aid of a simple flow model a picture could be drawn of the overall flow mechanism at the wheels. This was supported by some force and wake measurements behind the wheels when the standing height of the vehicle or the front spoiler was varied.

4) The investigation of rolling wheels varying in width yielded a notable influence on drag. It is strongly suggested that this result is related to the upper and lower bound vortices of the wheels. These are strengthened due to increased displacement of the flow and reduced shielding by the wheelhouse when the width of the wheels is increased.

5) A large increase in drag and lift forces was measured when wheels were mounted to the basic vehicle body. The lowest drag was measured with a half-open wheelhouse, for which the upper bound vortex was suppressed.

6) On the other hand, the half-open-wheelhouse configuration created more lift than an open or a completely closed wheelhouse. Most of the lift was generated at the car body. The wheels themselves create only moderate vertical forces. Which could be negative when the wheels were exposed to crossflow. For the production version of the car a moderate positive vertical force was measured at the wheels because with an optimized spoiler the crossflow angle was greatly reduced, and presumably the base pressure of the wheel was increased.

7) Finally, experiments with different tire-treads revealed a moderate influence on drag when the wheels were rolling. Only 5 drag counts lower drag was measured between a slick tire and the worst of three tires.

ACKNOWLEDGEMENT

The authors wish to express their thanks to all those people who helped to execute the test campaign and to accomplish this paper. Special appreciation is extended to Jelke Veldman and his crew, Chris Groenendijk and Jaap Menger from DNW. They prepared the car for testing, build the wake rakes and redesigned the suspension system. Our thanks are also extended to Hermann Holthusen from DNW who produced the plots and who rescued us, with great virtuosity from computer problems during the writing of this paper. In addition, special thanks are extended to Dr.Gino Sovran from GM who reviewed the paper. Due to his comments and constructive criticism the paper has been greatly improved.

REFERENCES

[1] J.E.Hackett, J.B.Baker, J.E.Williams, S.B.Wallis: On the influence of ground movement and wheel rotation in tests on modern car shapes.
SAE paper 870245, 1987

[2] A.Cogotti: Aerodynamic characteristics of car wheels. Impact of aerodynamics on vehicle design.
Int.J. of vehicle design.SP3, London 1983.

[3] E.Mercker, H.W.Knape: Ground simulation with moving belt and tangential blowing for full-scale automotive testing in a wind tunnel.
SAE Paper 890367, 1989

[4] M.Seidel: Construction 1976-1980 - Design, manufacturing, calibration of DNW, 1982

[5] E.Mercker: A blockage correction for automotive testing in a wind tunnel with closed test section.
Journal of Wind Eng.and Ind. Aerodyn. 22, 1986

[6] H.J.Emmelmann, H.Berneburg, J.Schulze: The Aerodynamic development of the Opel Calibra.
SAE Paper 900327, 1990

[7] H.J.Emmelmann, H.Berneburg: Aerodynamic Development of the Opel Calibra in the DNW.
Proc. Colloquium 'Ten years testing at DNW', Oct. 1990

[8] L.J.Oswald, A.L.Browne: The airflow field around an operating tire and its effect on power loss.
SAE Paper 810166, 1981

[9] A.J.Scibor-Rylski: Road vehicle aerodynamics.
2nd edition, Pentech Press, 1984

920345

Aerodynamics of Trucks in Wind Tunnels: The Importance of Replicating Model Form, Model Detail, Cooling System and Test Conditions

M. Eugene Olson
Navistar International Transportation Corp.

Uwe W. Schaub
National Research Council Canada

ABSTRACT

Drag coefficient comparisons are made for two half-size truck models with variable length, grain-haul height, closed trailers, and with identical frontal areas and length/volume distributions.

One, a simple block model, with sharp edges, but characterized by correct overall dimensions, was designed to simulate only the major physical features of the truck, such as wheels, fenders, hood, cab and trailer. The second model accurately replicated body shape and hardware exposed to external and cooling air streams.

Testing was conducted in the 9m x 9m NRCC, solid- wall wind tunnel of the National Research Council Canada (NRCC) over a velocity range of 48 to 193 kph (30 to 120 mph). Both models were yawed through ±14° with full-length and truncated trailer bodies.

The measured coefficient data were corrected by adjusting the tunnel dynamic pressure, yaw angle, and the horizontal buoyancy effect using a simple modified pressure-signature correction method. This method was previously shown to result in correlating the wind-averaged drag coefficient data of the half-scale detailed model and actual full-scale vehicle within a 1% to 2% margin for the 7 mph wind speed and 55 mph road speed (7 and 55 mph) and 7 and 30 mph conditions, respectively.

The half-scale model, representing a 2.2% frontal-area blockage in the wind tunnel was observed to be independent of Reynolds number effects inside the range of 2.2×10^6 to 4.4×10^6. Increasing trailer body length increased drag significantly. Engine cooling airflow was found to be the source of 3.8% of the total wind-averaged drag coefficient at a Reynolds number of 2.2×10^6, the ram and fan airflow contributions being 66% and 34% respectively. The difference in the wind-averaged drag due to model form and detail was inversely proportional to body length: a minimum of 2% for the full-length model and up to 11% for a 38% shortened model with the block model having the higher drag.

1.0 INTRODUCTION

The present paper is the 3rd in a series of papers emanating from a much larger investigative program into various aspects of tractor-trailer aerodynamic testing and design. The program included three wind tunnel entries with full-scale and half-scale tractor-trailer systems featuring functioning engine cooling airflow and air intake systems.

A primary concern during the entire program was the relative accuracy of the data. Could it be corrected with sufficient accuracy to allow a qualitative analysis of the various aspects of the program? A critical requirement was that the full-scale and half-scale results must provide a high degree of correlation.

In support of the total program, various areas of concern were investigated, such as:

1. identification of similarity parameters (primarily Reynolds number) and their limitations;

2. size limitations in cross-sectional area, volume and overall length, and

3. identification of form and detail requirements and limitations as they apply to long three-dimensional bluff bodies.

Analysis in these areas proved to be difficult without adequate correction to the data.

Aerodynamic testing of full-scale vehicles in wind tunnels is a well-established practice in the automotive industry because it assures testing with the correct scaling numbers and it is convenient; but, until recently the measured forces were only useful for A to B comparisons since the absolute values were always subject to unknown tunnel wall-interference and ground effects. Most of the experience to date relates to small and short vehicles such as cars and vans. The problem of compensating accurately for solid wall wind tunnel interference appears to have been resolved through the use of a technique known as the pressure-signature method, typically as developed by Hackett et al. (1) and, more recently by Schaub et al. (2). It is especially appropriate for high-blockage long three-dimensional bodies because it accounts for both the solid and wake blockage-induced interference velocities and the horizontal buoyancy force due to the generated axial pressure gradient. The second method (2) accounts also for any side-force effects generated by yawing the model and was developed in support of this program. Ground effects remain an unresolved unknown.

In an earlier paper (2), it was shown that the calculated tunnel-wall interference corrections were surprisingly large for the present half-scale models which presented only a 2.2% frontal area blockage in the 9m x 9m NRCC wind tunnel: the wind-averaged drag and side-force coefficients were reduced by about 6% and 4%, respectively. Appreciable increases in yaw angle of up to 3% were also indicated. The study demonstrated that the modified-pressure-signature correction method resulted in correlating the full and half-scale wind-averaged drag coefficient data within a 1% and 2% margin of repeatability for the 7 mph wind speed and 55 mph road speed (7 & 55 mph) and 7 & 30 mph condition, respectively (3). The study also demonstrated the limitations of the method: clearly indicating when the model was too large, too long or in too close a proximity to the walls.

Before testing models in solid-wall wind tunnels, there are important tunnel interference effects related to the installation of the model that must be considered, as reported in Reference 4, including model-generated disturbances that propagate upstream and that may influence the tunnel's 'q'-measuring system, the development of the model wake downstream of the test section, the interaction between the model and the tunnel floor boundary layer, and the possibility of break-down of the mainstream flow when a long model is yawed. These effects, and how they relate to the present model, have been discussed amply (4).

The above considerations deal with the appropriateness of testing a given size model in a solid-wall wind tunnel; there are, however, additional requirements to ensure that the actual data are representative of full-scale vehicle aerodynamic performance. As mentioned previously, these include Reynolds number effects, the importance of modeling surface and volumetric details, the influence of model fineness ratio (L/D), and the importance of modeling the cooling airflow on force measurements. They are all well-known requirements, but not so well-known or understood as they pertain to long three-dimensional bluff bodies.

The present paper addresses these subjects. The study was made at half scale in order to provide ease of configurational changes (vs. full scale), optimization of test time, testing over a Reynolds number range including full-scale vehicle Reynolds number, and because the chosen model size assured convenient replication of body details.

Two half-scale models were built to support these investigations. One was a highly detailed duplication of the full-scale truck including the engine compartment and engine-cooling airflow system. It is generally referred to as the reference or detailed model. The other, a simple block model, was used to address the ever-recurring question of how much surface detail is required and to ensure flow separation independent of Reynolds number. It was characterized by a generally simple shape, correct geometric volume distribution, zero surface detail and sharp edges.

2.0 EXPERIMENTAL METHOD

2.1 DESCRIPTION OF TEST FACILITY - The tests were conducted in the 9m x 9m NRCC wind tunnel, which features a 23m long test section, a single eight bladed fan with a large hub/tip radius ratio, seven flow-straightening vanes, two one-'q' screens and a dense heat exchanger matrix inside the settling chamber, as depicted in Fig. 1.

2.2 DESCRIPTION OF MODELS - An International XF 4370 tractor with a standard van-box-style semi trailer, modified (channelled) to grain-haul height was chosen as the vehicle to be modeled at half scale, see Fig. 2. Two half-scale models were built to simulate this vehicle: a reference model which replicated the actual vehicle in exacting detail (±1.5 mm at full scale), and a second one featuring approximately the correct volume distribution, but lacking all of the small parts detail.

The overall dimensions of the half-scale reference model truck are given in Fig. 3. While the reference model was a scaled copy of the selected full-scale vehicle, see Fig. 4, details difficult to replicate in model scale and of relative aerodynamic insignificance to the prime objectives of the program were neutralized on both the full- and half-scale models. As an example, tire tread on the full-scale model was taped over and then smooth tread tire simulation duplicated on the half-scale model. Tire tread in itself was not considered aerodynamically significant to the program objectives, just duplication of the full-scale into half-scale. This was done to simplify half-scale model build and reduce costs. The tractor and trailer were rigidly connected to prevent movement at the fifth wheel.

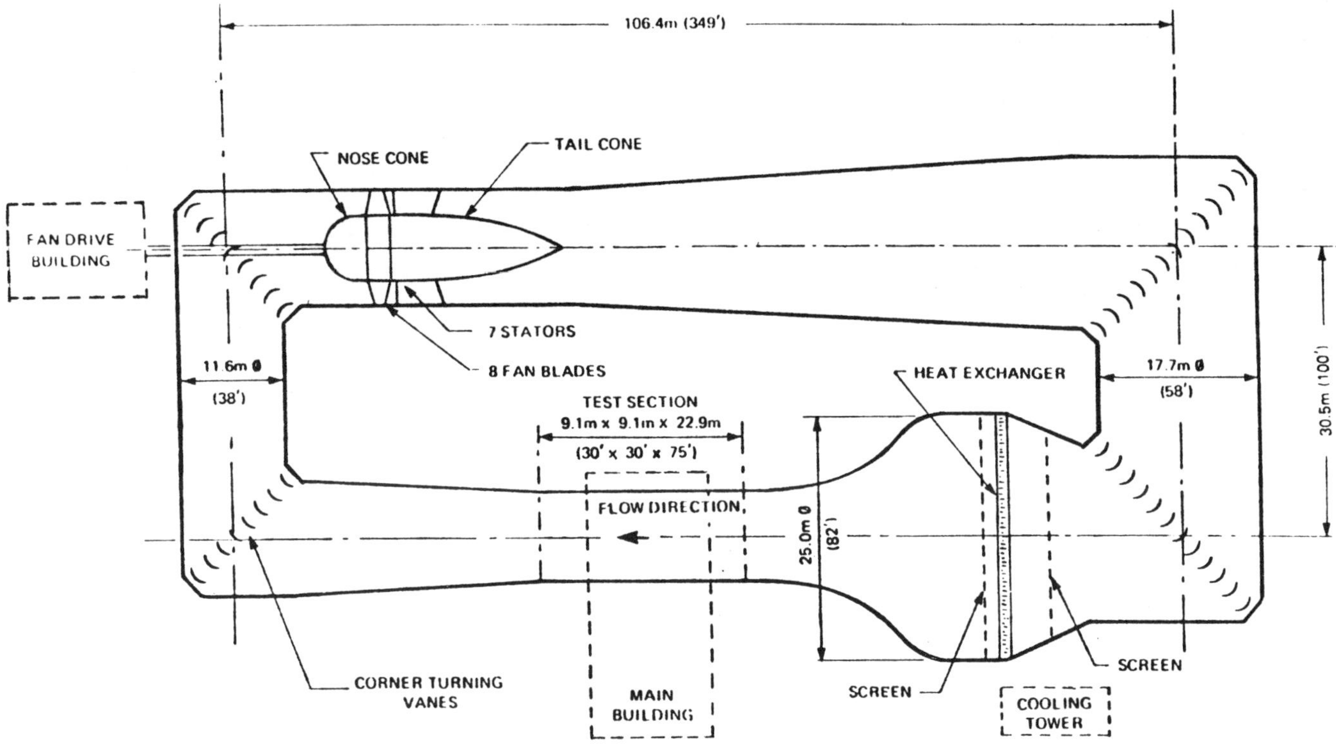

Figure 1 9m x 9m NRCC Wind Tunnel Circuit

Figure 2 Full-Scale Test Vehicle in the 9m x 9m NRCC Wind Tunnel

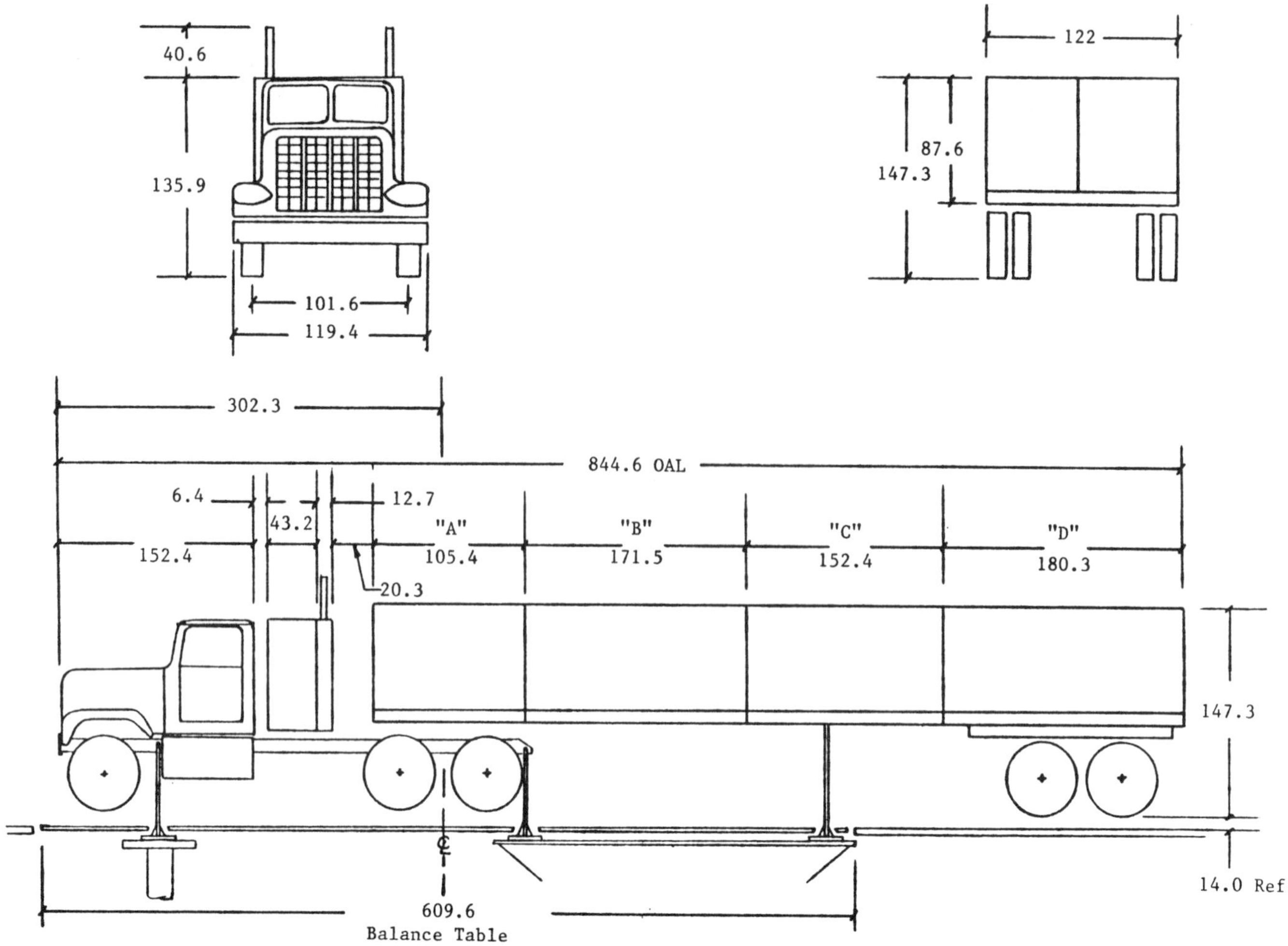

Figure 3 Half-Scale Reference Model Dimensions (cm)

Figure 4 Half-Scale Reference Model in the 9m x 9m NRCC Wind Tunnel

To facilitate testing over the same Reynolds number range with both full-and half-scale models, the tunnel speed had to be doubled for the half-scale tests. The study included investigating the vehicle drag changes due to both ram-only and combined ram-and-fan cooling effects. There was, therefore, a requirement for a scaled engine cooling system package, including the radiator, fan and shroud.

To match the full-scale velocity ratio of the average cooling-air velocity (V_a) at the radiator exit (or radiator face velocity) to the approaching free stream velocity (V_∞) at a full-scale Reynolds number, but in half-scale, would require doubling the radiator face velocity. This, however, would result in the cooling mass flow not being scaled correctly. To acquire a scaled match of both V_a/V_∞ and mass flow would require testing at the same free-stream velocity. Therefore, a half-scale radiator, designed and built by Blackstone Corporation, was designed to generate the same frictional characteristics, and thereby the same pressure-loss coefficient, over the same velocity range as the full-scale over its normal velocity range. The cooling fan, designed, built, and tested by Allied Automotive, was geometrically scaled: same hub/tip radius ratio, percent tip clearance, number of blades, blade aspect ratio, and circumferential spacing on the hub spider. The blade pitch angles, nominally set to full-scale specifications, were subsequently tweaked to achieve basically similar aerodynamic performance, as verified by means of in-shroud tests inside the supplier's standard airflow test stand. These verification tests showed that the half-scale fan speeds turned out to be nominally twice as large as for full-scale; however, the power draw was higher than expected from fan-scaling rules. Within the model, the fan was driven by a variable speed d.c. motor imbedded inside a simulated engine block.

The engine compartment, auxiliaries and other parts were also modeled in exacting detail (Fig. 5). Engine air was drawn from the grille through a scaled air-route system, at a scaled flow rate by a vacuum cleaner located inside the trailer. The half-scale axles and suspensions were reproduced accurately and were adjusted at the correct operating height. The half-scale trailer was built using structural detail, fasteners, and fabrication techniques as used for the full-scale trailer. To enable studying the vehicle length effect on drag measurement and on tunnel interference, the half-scale trailer was sectioned, as illustrated in Fig. 3, into four units A, B, C, and D.

The block model was a half-scale volume-approximation of the reference vehicle with the correct frontal-area distribution along its length. Only the major body shapes were represented: engine housing, fenders, cab, trailer box, and the wheels. The wheels were cylinders of correct diameter and width. All body parts were of the simplest possible shape representation, with sharp edges, see Fig. 6.

Figure 5 Half-Scale Detailed Model Engine Compartment

Figure 6 Half-Scale Block Model in the 9m x 9m NRCC Wind Tunnel

Both models were supported fully on the tunnel balance via six slender vertical struts, suitably protected from the airstream by full-height wind shields and from engaging with non-metric surfaces by means of electric anti-fouling circuits.

2.3 TEST METHOD - Basic force and moment testing was conducted per SAE Recommended Practice J1252 (3) and the drag results were wind-averaged. The models were tested at wind tunnel speeds of 48, 73, 97, 145 and 193 kph. The test yaw angle range was ±14° with individual settings in steps of 4°. Tunnel speed and cooling fan speed were also varied as needed at constant yaw settings. Zero-yaw settings were determined by nulling vehicle side force. The detailed model was tested with closed-grille cooling and engine-air intakes (reference model tests), with open-grille intakes (tests with a passive cooling system), and with a powered cooling fan operating over a range of fan speeds (tests with an active cooling system).

Extensive calibrations were made with both full- and half-scale models in still ambient air in order to associate radiator airflow, and hence radiator face velocity, with cooling fan speed. The half-scale model cooling-airflow was calibrated by means of an orifice flow meter and matched to the full-scale airflows through use of arrays of propeller anemometers placed inside the fan/radiator shrouds, as illustrated in Fig. 7. Correct cooling airflow conditions were then established for the chosen velocity ratio V_a/V_∞ conditions by setting appropriate fan speeds.

Figure 7 Anemometer System - Half-Scale Radiator and Fan Shroud

3.0 ANALYSIS AND DISCUSSION

3.1 EFFECT OF TUNNEL SPEED ON MEASURED FORCES - When testing models in wind tunnels at insufficient Reynolds numbers there is always a concern about the correctness of the measured force data. The full-scale Reynolds number (based on maximum width) at a road speed of 97 kph (60 mph) is 4.4×10^6, so the half-scale model was tested over a Reynolds number range of 1.1×10^6 to 4.4×10^6 (48 to 193 kph in half-scale). Fig. 8 illustrates corrected drag coefficient (C_D) data of the reference model and the block model at yaw angles of zero and 14°, spanning the tunnel speed range of 48 to 193 kph. Figs. 9 and 10 show typical drag coefficient polars for the two models in this speed range.

These data, together with the wind-averaged drag coefficient data presented in Table 1 for the full-length reference and block models at the tunnel speeds of 97 and 145 kph, exhibit a random scatter of 1% or less, well within the observed band of repeatability and the estimated margin of experimental uncertainty (2). It

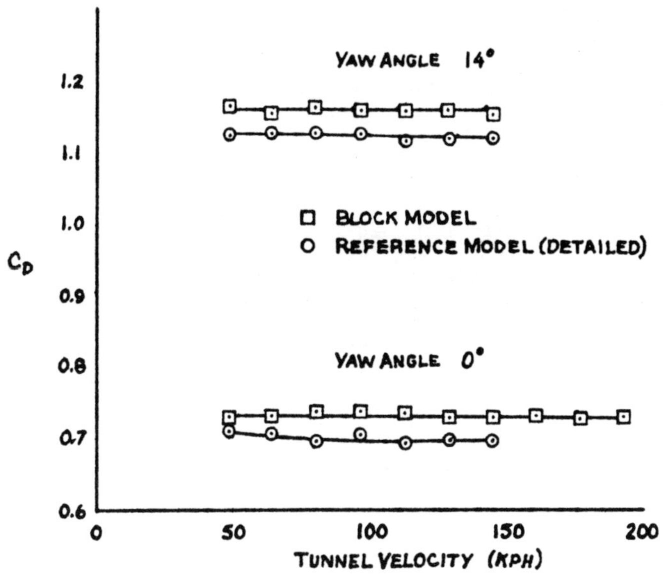

Figure 8 **Drag Coefficients of the Full-Length (ABCD) Reference and Block Models as a Function of Tunnel Velocity**

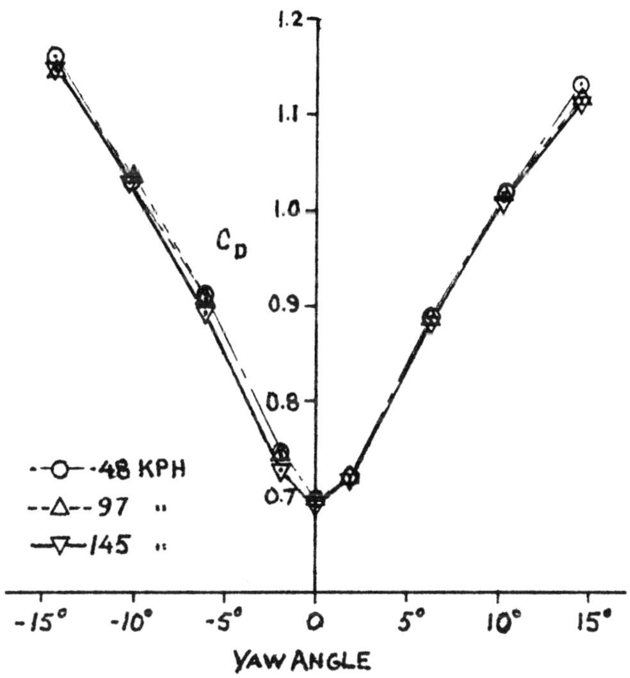

Figure 9 **Drag-Coefficient Polar of the Reference Model (ABCD)**

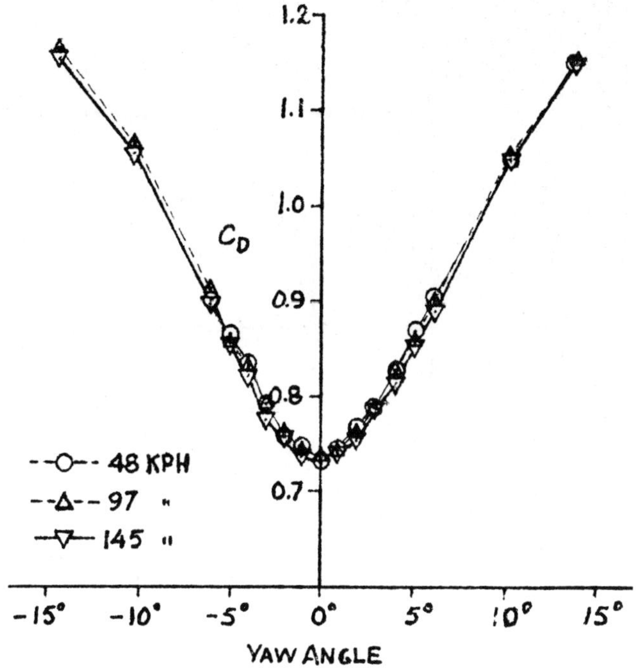

Figure 10 **Drag-Coefficient Polar of the Block Model (ABCD)**

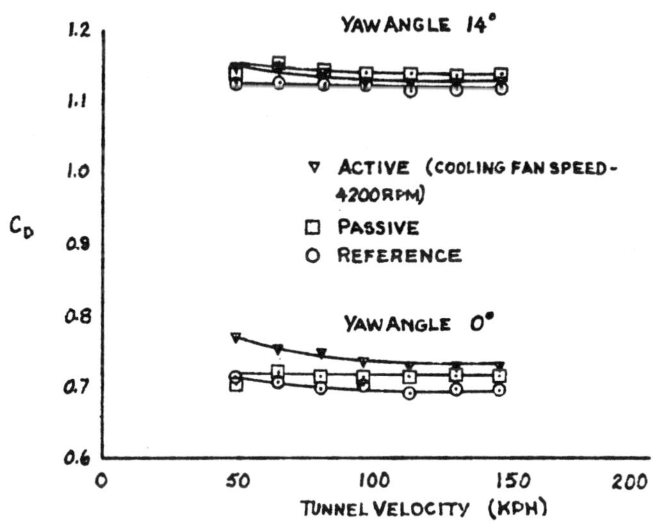

Figure 11 **Effect of Cooling Air on Drag Coefficient as a Function of Tunnel Velocity**

appears therefore that there is no discernible dependency on Reynolds number inside the range 2.2 x 10^6 to 4.4 x 10^6. Previous internal Navistar research had indicated that 1.7 x 10^6 is the lower limit for effective Reynolds number simulation.

Fig. 11 on the other hand, suggests a dependency on Reynolds number at the lower velocities (Reynolds number < 2.2 x 10^6) for both the active and passive cooling systems. The passive cooling drag coefficient variation (as well as the reference model data) fall within the estimated margin of experimental uncertainty and are not significant. Because tunnel speed varies and cooling fan rpm is constant (fan-induced cooling airflow rate essentially constant), what looks like a dependency in the active cooling system case, is actually a relative increase in cooling drag at the lower speeds. It can readily be seen that as the tunnel speed decreases, the ratio of the radiator face velocity to the free stream tunnel velocity (V_a/V_∞) increases and

Table 1 Effect of Tunnel Speed, Cooling and Hardware Detail on Wind-Averaged Drag Coefficient

TUNNEL VELOCITY (KPH)	WIND AVERAGE (MPH)	REF. MODEL	EFFECT OF LACK OF DETAIL		EFFECT OF COOLING			
					PASSIVE		ACTIVE	
		$\overline{C}_D$	$\overline{C}_D$	$\frac{\overline{C}_D}{\overline{C}_{D_{ref}}}-1$	$\overline{C}_D$	$\frac{\overline{C}_D}{\overline{C}_{D_{ref}}}-1$	$\overline{C}_D$	$\frac{\overline{C}_D}{\overline{C}_{D_{ref}}}-1$
48	7 & 30	1.011	1.026	1.5%	1.024	1.3%	1.049	3.8%
	7 & 55	0.854	0.868	1.7%	0.864	1.2%	0.900	5.4%
97	7 & 30	1.004	1.034	3.0%	1.022	1.8%	1.031	2.7%
	7 & 55	0.849	0.871	2.6%	0.870	2.5%	0.881	3.8%
145	7 & 30	0.997	1.024	2.7%	1.022	2.5%	1.024	2.7%
	7 & 55	0.841	0.861	2.4%	0.870	3.5%	0.873	3.9%

hence the ratio of cooling drag to total drag (a ratio of forces)*.

During the testing, considerable difficulty was experienced in obtaining reliable and repeatable data at tunnel speeds less than approximately 100 kph (2.2 x 10^6 Reynolds number). Variations were occurring that, while small, seemed to be something other than just a random unsteadiness in the flow. What had appeared as random unsteadiness, thought to be caused by both excessive blockage and instrumentation limitation during full-scale testing, now showed similar characteristics but with definitive, recurring level shifts at half-scale. This seemed to indicate that the half-scale reference model may have been experiencing a temporally and spatially varying local-flow-separation because of the low Reynolds number.

Considerable uncorrected drag level differences existed between the full- and half-scale models (2), and at the time, whether any of the difference might be due to a Reynolds number problem was unknown. Reinstalling the full-scale model, for additional testing to determine local separation location was time and cost prohibitive. It was therefore decided to construct a sharp-edged, half-scale block model which would have well-defined separation under all conditions. This model, however, experienced the same characteristic variations in the data as the half-scale reference model. These questions relating to low speed testing were not resolved and analysis was therefore restricted to data obtained at a Reynolds number of 2.2 x 10^6 and higher.

3.2 EFFECT OF TRAILER LENGTH - Trailer-length effect was investigated and is depicted for the reference and block models in Fig. 12. These results show that drag coefficient decreases with decreasing trailer length. The first reduction in trailer length ((ABCD-ACD)/ABCD is 20%) produced less of a decrease in C_D than the second reduction in length ((ACD-AD)/ACD is 23%). In general, the reductions in C_D were greater at

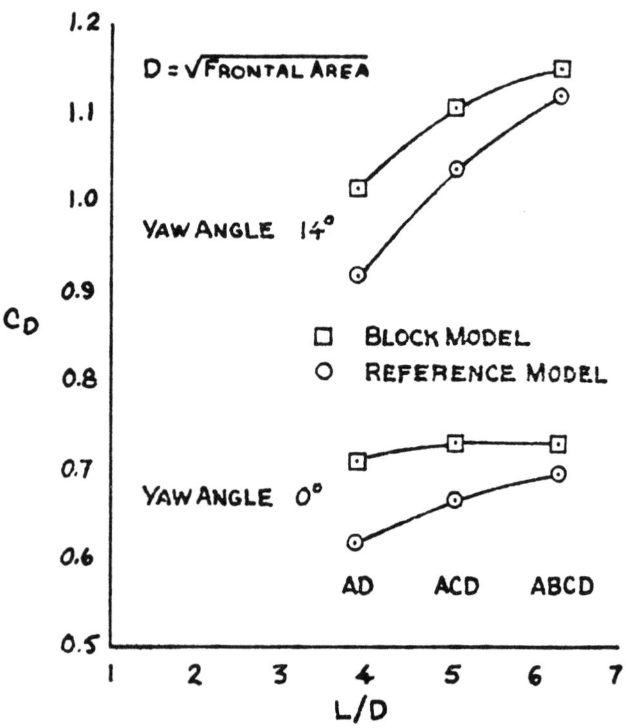

Figure 12 Influence of Vehicle Length on Drag Coefficient

---

*The relationship between the velocity ratio and the ratio of forces may be established on simplifying the latter with continuity applied to the captured cooling airflow stream tube. The cooling system drag expression (momentum) used was derived in Reference 6; but, it is important to note that this simple expression does not account for any adjustment in the external flow around the truck (which causes changes in the total drag) due to the flow removed through the cooling system air path.

Table 2 Wind-Averaged Drag Coefficient Data for Three Vehicle-Length Configurations of the Reference and Block Models and Percent Increase over the Reference Model

MODEL	WIND AVERAGE (MPH)	VEHICLE LENGTH CONFIGURATION					
		ABCD		ACD		AD	
		$\overline{C}_D$	$\frac{\overline{C}_D}{\overline{C}_{D_{ref}}}-1$	$\overline{C}_D$	$\frac{\overline{C}_D}{\overline{C}_{D_{ref}}}-1$	$\overline{C}_D$	$\frac{\overline{C}_D}{\overline{C}_{D_{ref}}}-1$
REF.	7 & 30	1.006		0.935		0.839	
	7 & 55	0.851		0.794		0.728	
BLOCK	7 & 30	1.034	2.8%	1.004	7.4%	0.939	11.9%
	7 & 55	0.871	2.4%	0.852	7.3%	0.808	11.0%

yaw than at zero yaw; for example, the reductions in C_D at a yaw angle of 14° were between two and three times as great as at a yaw angle of 0°. This is, however, not surprising because reduction in trailer length not only reduces surface friction, but also the size of the resulting wake. The corresponding wind-averaged drag coefficients for the different model lengths are given in Table 2.

The fact that C_D decreased with decreasing trailer length more for the detailed reference model than for the block model (whose flow separations were assumed to be well-defined and constant) underlines the role of friction and flow separation along the sides of the trailer box and front-end flow separations.

The results confirm (7) that minimum drag coefficient results when using short bodies (small L/D) with aerodynamic forebodies where the afterbody airflow is at least parallel to the free stream and experiences no flow separation before the cut-off. The difference in drag is primarily friction drag. This is not true for short bodies with non-aerodynamic forebodies where the flow separates and continues to expand for a distance resulting in increasing base pressure drag; for example, at L/D ratios less than two, the drag rapidly increases as the flat-plate limit is approached.

3.3 EFFECT OF ENGINE COOLING ON MEASURED FORCES

3.3.1 Ram-Only Cooling - Wind tunnel models are normally tested either with completely closed intakes or with ram pressure being the only prime mover of cooling air, i.e. under passive cooling conditions. In the present study, the front-end cooling-air path of the detailed model was very carefully matched to (and experimentally verified with) full scale. For reasons mentioned in the previous section, cooling drag comparisons are discussed primarily on the basis of the 97 kph and 145 kph results, which showed consistent trends and were repeatable; however, the 48 kph data are also presented visually. As Fig. 11 illustrates that, for both zero and 14° of yaw, passive cooling increases drag by a nearly constant amount of $\Delta C_D = 0.02$, or approximately 3% and 2%, respectively. Table 1 shows that passive cooling increases the wind-averaged drag from 1.8% and 2.5% to 2.5% and 3.5% for the 7 & 30 and 7 & 55 mph conditions, respectively. This seems to indicate that the passive-cooling drag increment is larger at small yaw angles, but for practical testing purposes one may assume it to be nearly constant and independent of yaw angle within the limits of testing.

3.3.2 Combined Ram and Fan Cooling - Both the approach flow and the cooling fan can be prime movers of cooling air. Active cooling in the half-scale detailed model was generated by operating the scaled cooling fan at a fan speed required to generate the radiator face velocity needed to simulate the ratio of full-scale operational radiator face velocity to approach flow velocity. Fig. 11 shows only the 4200 rpm data which was dynamically correct only at a total system Reynolds number of 2.2×10^6 (96 kph) at half scale.

In terms of the wind-averaged drag contribution shown in Table 1, the drag coefficient increase was 2.7% for the 7 & 30 condition and 3.8% for the 7 & 55 mph case. The latter being greater because the 7 & 55 average does not include large yaw angles, where the cooling drag was observed to be minimal. If it is assumed there is no or insignificant interaction between ram and fan air, then the ram air contribution was 68% and 66% for the wind-averaged cooling drag at 7 & 30 and 7 & 55 respectively, and 32% and 34% for the fan contribution.

Clearly, the vehicle drag is increased significantly by both passive and active cooling. By drawing cooling air from the approach flow and then discharging it without streamwise kinetic energy relative to the vehicle, streamwise momentum is destroyed and additional drag is generated. The increase in drag coefficient due to passive cooling is zero for a stationary vehicle, and is seen to increase directly with wind speed. When friction effects can be neglected, it is evident from force-balance considerations alone that a fan operating inside the engine bay, while drawing air from still surroundings and discharging downward, cannot generate vehicle forces in the horizontal plane. When there is approach flow the constant-speed cooling-fan drag contribution (additional to the ram-only contribution) is seen to be highest at low vehicle speeds and to shrink as vehicle speed builds up. The cause for this trend was discussed previously in Section 3.1.

3.4 BLOCK MODEL VERSUS DETAILED MODEL - Table 2 suggests maximum possible drag differences associated with the lack of replicating hardware detail and corner rounding. Not reproducing any of the hardware detail, but maintaining the correct volume distribution, introduced the following 7 & 30 wind-averaged drag penalties: 11.9% for AD, 7.4% for ACD, and 2.8% for ABCD. In terms of the 7 & 55 wind-averaged drag the penalties for AD, ACD, and ABCD were: 11.0%, 7.3%, and 2.4%, respectively.

The present study is ultimately concerned with drag reductions of only the full-length vehicle ABCD, and therefore it is relevant to observe that omissions of detail could at worst cause a drag penalty of only 2%. On the other hand, if it is assumed that the aerodynamic effect of rounding the corners on the detailed model is constant and independent of trailer length, then the maximum benefit of the rounded corners on this model could only be 2% or less. This means that the basic shape of the detailed model is not much more aerodynamic than the block model. This would mean that the hardware-detail effect on the ABCD long-body flow field is masked or buried in the overall flow field of a not very aerodynamic body. This is consistent with previous research on non-aerodynamic tractor-trailer systems. This previous research, although not part of the present program, has shown that the hardware detail has an aerodynamic effect on increasingly aerodynamic long-body tractor-trailer shapes (the more aerodynamically clean - the more the effect of detail).

These results illustrate some interesting facts:

1. The detailed model was not very aerodynamic.

2. If hardware detail accounted only for 11% of the drag at most, vehicle shape must account either directly or indirectly for at least 90%.

3. The accurate reproduction of hardware detail seems to be most important on short-bodied vehicles. Therefore it would seem necessary to model hardware detail with much greater accuracy for short vehicles, such as cars or vans, than for long-bodied vehicles such as long highway vehicles and trains.

4. The advantage of minor hardware "streamlining" features are difficult to demonstrate on a vehicle that is long and not aerodynamically clean. Alternately, it is hard to justify the accurate reproduction of hardware details on blunt or "rough" aerodynamic shapes.

4.0 CONCLUSIONS

1. There was no significant variation in drag coefficient with Reynolds number inside the range 2.2×10^6 to 4.4×10^6.

2. Drag coefficient increased with trailer length. Such increases were greatest at large yaw angles because of a larger wake and an increasing friction effect.

3. Cooling drag contributions were a maximum at zero yaw and significant out as far as 14° of yaw angle.

4. The 7 & 30 wind-averaged cooling drag added 2.7% to the basic $\bar{C}_D$, of which 68% was due to ram alone and 32% due to the fan. The 7 & 55 wind-averaged cooling drag added 3.8% to the basic $\bar{C}_D$, of which 66% was due to the ram alone and 34% due to the fan. At zero yaw, cooling drag added 4.9%.

5. The cooling fan's absolute contribution to vehicle drag decreased with yaw angle.

6. At a Reynolds number of 2.2×10^6 and a matched fan speed of 4200 rpm, ram air caused twice as large a contribution to cooling drag as the cooling fan.

7. These half-scale block-model tests indicated that form and/or hardware details reduced the wind-averaged drag of the basic shape as follows: ABCD - 3%, ACD - 7%, AD - 11%.

REFERENCES

1. Hackett, J.E., Wilsden, D.J., Lilley, D.E. "Estimation of Tunnel Blockage from Wall Pressure Signatures: A Review and Data Correlation." N.A.S.A. CR-152, 241. March 1979.

2. Schaub, U.W., Olson, M.E., Raimondo, S. "Correction of Wind Tunnel Force Data for Yawed Full and Half-Scale Truck Models Using a Modified Pressure-Signature Method." SAE 900187. February 1990.

3. "SAE Wind Tunnel Test Procedure for Trucks and Buses." SAE Recommended Practice. J1252. 1979.

4. Waudby-Smith, P.M. and Rainbird, W.J. "Some Principles of Automotive Aerodynamic Testing in Wind Tunnels with Examples from Slotted Wall Test Section Facilities." SAE 850284. February 1985.

5. Schaub, U.W., Nishimura, Y., Olson, M.E., Creswick, G.E. "A Full Scale Class 8 Conventional Tractor-Trailer in the 9x9m Wind Tunnel." SAE 881876. November 1988.

6. Hucho, Wolf-Heinrich. "Aerodynamik des Automobils." Vogel-Verlag. 1981. 4: 156-157

7. Götz, Hans. "Commercial Vehicles", Chapter 8 of Aerodynamics of Road Vehicles, Butterworths, 1987. Hucho, Wolf-Heinrich, Editor.

APPENDIX

NOMENCLATURE

C_D Drag coefficient of model based on frontal area

$\overline{C}_D$ Wind-averaged drag coefficient of model based on 7 mph wind speed and 30 mph vehicle speed (7 & 30 mph, or 11 & 48 kph) or 7 & 55 mph (or 11 & 88 kph)

D Characteristic linear dimension of the model, here the square root of the frontal area

L Length of model

q Tunnel average dynamic pressure at the test section measuring station

V_a Average cooling-air velocity at the radiator-exit, also called radiator face velocity

V_∞ Tunnel velocity or vehicle speed

Δ Denotes a change in a quantity

Section II
Wind Tunnel Testing

Ground Simulation

900321

Comparison of Different Ground Simulation Techniques for Use in Automotive Wind Tunnels

Edzard Mercker
German-Dutch Wind Tunnel (DNW)

Jochen Wiedemann
AUDI AG

ABSTRACT*

The range of applicability and the physical restrictions for the use of ground-simulation techniques in automotive wind tunnels are elucidated. The techniques considered are the moving-belt technique, as well as boundary layer control techniques like tangential blowing and distributed normal suction for use in wind tunnels with stationary ground boards. Attention has to be paid to the question of whether the flow to be simulated is of boundary layer or Couette type. In the case of boundary layer flow, interaction of the ground-floor boundary layer with the inviscid flow in the gap between a vehicle and the road can be fully simulated by introducing a negative transpiration velocity along the stationary ground plane. In practise however, angularity effects on the external flow result from mismatched control parameters. Very small relative ground clearances give rise to the formation of a Couette flow between the road and the vehicle. For this type of gap flow interaction with the rest of the flow field around the car vanishes asymptotically. When performing tests on a stationary ground board, mass and momentum have to be added to the gap flow to meet the characteristics of the moving-ground case. This can be achieved by tangential blowing.

IN AN AUTOMOTIVE WINDTUNNEL a steady turbulent boundary layer always develops along the ground. This boundary layer usually is considerably thicker than the unsteady boundary layer generated at the road under a moving car. Thus, underbody regions which encounter inviscid or at least high-momentum flow in real life might be within the ground-floor boundary layer of the test section. If parts of the car (tires, suspension, diffusors) penetrate into these regions the aerodynamic coefficients will be falsified. Furthermore, the displacement effect of the wind tunnel boundary layer differs from the displacement effect of the boundary layer along the road. The latter has been investigated by Beese [1]**, who showed that for a 2-D wing profile in ground effect the displacement thickness may even attain negative values. The boundary layer then entrains fluid from the inviscid flow at its edge instead of displacing mass. The displacement effect on a stationary ground introduces a flow angularity into the air stream which may lead to considerable errors when testing ground-effect racing cars with small ground clearances. But even conventional passenger cars might be affected as well. There is a strong trend in road vehicle aerodynamics to smooth the undersides of the vehicles, and thus correct prediction of the efficiency of costly underbody shields is essential.

The only way to avoid the interference effects mentioned above is to simulate the relative motion between road and vehicle, and this is almost perfectly achieved by means of a so-called moving belt. This technique, however, requires the removal of the approaching boundary layer in front of the belt in order to avoid boundary layer relaxation effects on the belt.

---

* Dedicated to Prof.Dr.-Ing. K.Gersten on his 60th birthday.

**Numbers in parentheses designate references at end of paper.

Several moving-belt facilities for testing small-scale models are available in Europe. Tests on scale models executed with moving-belt devices in comparison with alternative simulation techniques are presented in papers by Wiedemann [2] [3], Carr [4] and Bearman et al. [5]. A moving belt for full-scale passenger cars is in operation at the German-Dutch Windtunnel (DNW) [6]. An alternative approach to realistic ground simulation has been discussed by Larsson et al. [7], who performed tests on a full-scale car in a large towing tank, thus enabling relative motion between car and ground. However, full-scale tests of both methods require time-consuming preparations of the model. Therefore, alternative methods of ground simulation are desirable which allow testing on a stationary ground plane with an external underfloor balance.

In road vehicle aerodynamics it is attempted to keep the integral quantities (δ_1, δ_2) of the empty-test-section boundary layer as small as possible in order to minimize interference effects. This can be achieved by means of a boundary layer scoop in front of the vehicle, or a porous or slotted upstream section of the ground floor where hard suction is applied ("Basic Suction System"). In both cases a "new" boundary layer of reduced thickness develops downstream. An alternative approach involves application of tangential blowing. In this case a thin high-momentum wall jet is blown into the lower part of the boundary layer. Depending on the momentum ratio between jet and boundary layer, the displacement thickness of the combination can be adjusted to zero. Even negative values at a given downstream location can be attained. Tangential blowing is in operation on a routine basis at DNW. Both control techniques are discussed in a paper by Berndtsson et al. [8].

To prevent the boundary layer from further growth downstream of the control unit, distributed suction through a perforated ground plane is applied in some wind tunnels. This method also evokes some technical difficulties. Careful sealing between suction chamber and balance is required in order to avoid erroneous force measurements. Furthermore, the perforation of the ground plane has to be finely divided to approximate a uniformly "porous" surface without any discrete sinks. In order not to affect the distribution of the suction velocity by the imposed pressure field of the car it is desirable to maintain so-called choked conditions in the perforation.

Many experimental studies have been carried out so far on the comparison of the above-mentioned simulation techniques. Most of these investigations have been basically phenomenological, and little theoretical work is available on the subject. Wiedemann [3] discussed simulation criteria based on theoretical considerations which were deduced from wings in ground effect for typical 2-D flow cases. Furthermore, orders of magnitude of the simulation parameters were assessed and compared with experiments. In the present study, the definition of moving-ground simulation criteria with reference to automotive applications is further enhanced by means of a detailed analysis of the physics of the flow field.

THE THREE ZONES OF AUTOMOTIVE GROUND SIMULATION

The objective of ground simulation in stationary-ground-board wind tunnels is to generate the correct interactions between vehicle and ground even though the kinematic boundary condition at the ground plane is violated. This requires some sort of flow control, which is usually restricted to the close vicinity of the ground and which is commonly termed "Boundary Layer Control" (BLC). When defining the requirements for ground simulation by flow control, it has to be considered that a vehicle in ground proximity generates flow regimes in the underbody region which do not necessarily exhibit boundary layer properties.

According to the two-dimensional flow field classification by [1] which has been applied to road vehicle aerodynamics by [3], the flow field about a body in the proximity of a moving ground can be split up into

- regions of asymptotically inviscid flow, i.e. the oncoming flow.

- regions where (to a first-order approximation) Prandtl's boundary layer equations hold. These regions extend along the body contour and along those parts of the ground plane where the inviscid velocity differs from the velocity of the ground, i.e. from U_∞,

- and regions which exhibit fully developed or creeping-flow behaviour. These flows can be described, at least to a first-order approximation, by the equations of hydrodynamic lubrication theory.

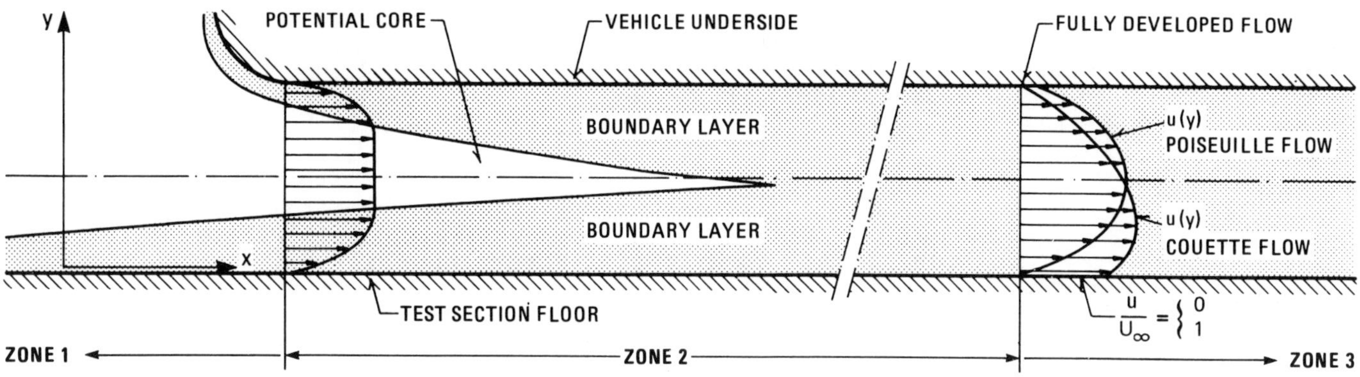

Fig. 1 Three Zones of Ground Simulation.

Thus, ground simulation is restricted to regions of viscous flow which can be subdivided into three zones, see Fig. 1:

ZONE 1: Boundary Layer Flow

ZONE 2: Intermediate Zone where a transition from boundary layer flow to fully developed flow takes place.

ZONE 3: Fully Developed Flow

The following considerations focus on the properties of the three zones where ground simulation may be applied.

SIMULATION REQUIREMENTS

ZONE 1 (BOUNDARY LAYER FLOW) - Boundary layer flow can be encountered in front of a vehicle and in sections underneath it. In a test section with stationary ground a turbulent ground-floor boundary layer always develops, the properties of which differ from those of the boundary layer on a moving ground. In this zone, to first-order approximation, the bias of the non-moving ground is restricted to the immediate wall proximity, i.e. to the boundary layer. However, interactions between the boundary layer and the free stream, which are second-order effects in the terminology of higher order boundary layer theory [9], also take place. The matching condition between the first-order boundary layer flow and the second-order external flow is given by the displacement effect of the boundary layer which results in an additional velocity component v_e perpendicular to the wall at $y=\delta$, see Schlichting [9], p.145 . The velocity v_e van be computed according to Eq. (1) which is derived in [3].

$$v_e(x) = U(x)\frac{d\delta_1(x)}{dx} + \delta_1(x)\frac{dU(x)}{dx} + v_W(x) \quad (1)$$

Here v_e serves as a boundary condition for the second-order external flow. Therefore, distortions of the boundary layer flow will lead to incorrect V-components in the inviscid flow field, thus introducing flow angularity. Eq. (2) gives the boundary-layer-induced flow angle at the edge of the boundary layer:

$$\alpha_e(x) = arc\tan\left[\frac{v_e(x)}{U(x)}\right] \quad (2)$$

With the help of Eq. (1), it is possible in principle, to calculate the distribution of the transpiration velocity $v_w(x)$ at the stationary ground, which is required to achieve the "correct" $v_e(x)$ of the flow along a moving ground. However, the v_e for a moving ground and $\delta_1(x)$, the displacement thickness of the transpired boundary layer, have to be known. Whereas the latter can be obtained as a function of $v_w(x)$ from iterative boundary layer calculations, $v_e(x)$ at the edge of the boundary layer with moving ground is unknown and can only be calculated from Eq. (1), if $\delta_1(x)$ for the moving ground is known. An exact determination of $v_e(x)$ therefore appears unrealistic, and only order-of-magnitude estimations can be made for certain categories of flows, like lift or downforce configurations [3]. Even if the calculations could yield the required distribution of the transpiration velocity $v_w(x)$ at the ground wall, a considerable technical difficulty would exist because a porous ground board, with a capability for adjusting $v_w(x)$ to required distributions, would be needed. These problems lead to some compromises with respect to simulation accuracy. A pragmatic approach for approximating $v_w(x)$ is attained by installing a number of floor sections of constant suction velocity but with the velocity varying from section to section. For some typical flow cases it has been discussed in [3] how to select the v_w for the different sections. However, in most industrial applications of distributed suction a constant suction velocity for all sections is selected. Typical magnitudes of $-v_w/U_\infty$ range from 0.002 to 0.006.

In the following, suction is discussed with respect to so-called laminar asymptotic suction layers. In such asymptotic flows boundary layers properties are constant in the streamwise direction. The justification for assuming asymptotic conditions is given by Iglisch's [11] 2-D considerations for laminar flow along a flat plate at zero incidence, which are presented in [9]. The constant shape factor $H_{12}=\delta_1/\delta_2=2$ of the exponential velocity profile of the asymptotic suction layer is approximated within 0.5 % if

$$\xi = \left(\frac{-v_w}{U_\infty}\right)^2 \cdot Re_l \geq 5.12 \qquad (3)$$

According to Eq. (3), a suction rate of $v_w/U_\infty=-0.005$, a free stream velocity $U_\infty=40$ m/s and $\nu=15\cdot10^{-6}$ m²/s yields asymptotic boundary layer properties after a flow length l of only 0.077 m from the leading edge. Considering that the length of typical suction boards is of the order of 5 m, it appears reasonable to assume for simplicity that asymptotic suction exists at every point on the suction board. Hereby it is presupposed that the approach boundary layer in front of the suction board has been removed with the aid of a scoop or a basic suction system. The latter is a short porous section in the ground plane upstream of the model and the distributed suction area, where hard suction is applied. Typical rates for basic suction with porous surfaces are $v_w/U_\infty=-0.030$.

The justification for considering laminar flow can be found in [9] where Ulrich's [12] results were reprocessed. It was found that above a critical suction rate of $-v_w/U_\infty=1.2\cdot10^{-4}$ the flow will always be stable, i.e. laminar, independently of the initial suction length. However, this rate is more than one order of magnitude smaller than commonly used suction rates and thus a laminar boundary layer in an empty test section can be expected for sucked boundary layers. To first-order approximation, this will not be altered by the non-zero pressure gradient at the ground imposed by a vehicle. This has been shown (e.g. by Gersten et al. [13] and Wiedemann [14]) with the help of asymptotic methods. Due to the asymptotic character of the boundary layer, the first two terms on the right side of Eq. (1) vanish identically. Therefore, v_e equals v_w. To satisfy continuity, however, such flows must induce a flow angle in the external inviscid flow. In the limiting case of an infinite suction plane this angle remains constant in the entire flow field, independently of vertical distance from the ground plane. Therefore, for a constant free stream velocity Eq. (2) reduces to

$$\alpha = arc\tan\left[\frac{v_w}{U_\infty}\right] \qquad (4)$$

A more sophisticated evaluation of the flow angularity due to suction can be performed on the basis of realistic wind tunnel dimensions and with consideration of the basic suction which usually precedes distributed suction. The flow in the test section can be composed by superposing a parallel flow of velocity U_∞ and a source distribution q along the ground plane y=0 simulating the effect of boundary layer suction along the ground plane. For constant suction in the flow direction q will be constant as well. Because of symmetry, q equals $2\cdot v_w$. Eq. (5) gives the velocity component v at any location (x,y) in the flow field as the integral of all contributions from source elements $q(x')\cdot dx'$ at x=x' on the ground.

$$v(x,y) = \frac{1}{2\pi} \int_{x'_1}^{x'_2} q(x') \cdot \frac{y}{(x-x')^2+y^2}\, dx' \qquad (5)$$

Integration of the source inductions along the x-axis yields

$$v(x,y) = \frac{q}{2\pi}\; arc\tan\left(\frac{x'-x}{y}\right)_{x'_1}^{x'_2} \qquad (6)$$

The primed quantities denote the location of the source elements. The effect of finite test-section dimensions in the y-direction has been neglected.

For this analysis it is assumed that basic suction (index "BS") is applied in the range $-0.7 \leq (x/L)_{BS} < -0.5$

and distributed suction (index "DS") for $-0.5 \leq (x/L)_{DS} \leq 0.5$. In other words, the length of the section where basic suction is applied is 20 % of the length for distributed suction. It is important to note that no gap is assumed between the basic suction surface and the distributed suction surface. L denotes the diameter of the turntable and x=0 is at the center of the turntable.

Fig. 2 shows results of the evaluation of Eq. (6) for basic suction. The results reveal a considerable vertical influence of basic suction, even at a non-dimensional height of y/L=1 above the ground. Furthermore, at the front edge of the turntable (x/L=-0.5) at a height of y/L=0.1, which is approximately the height of the cooling-air intake of passenger cars, the v-component amounts to 35 % of the value at the wall. The displacement effect of the boundary layer developing downstream of the basic suction board has been neglected in this computation.

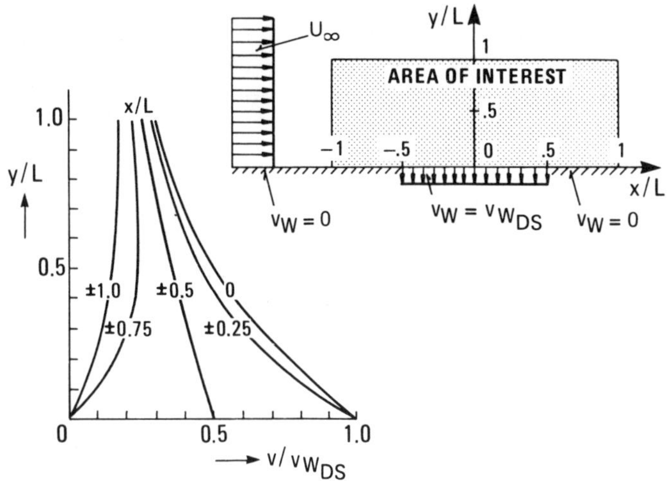

Fig. 3 Vertical Distribution of the v-Component in the Potential Flow Field Due to Distributed Suction for $-0.5 \leq (x/L)_{DS} \leq 0.5$.

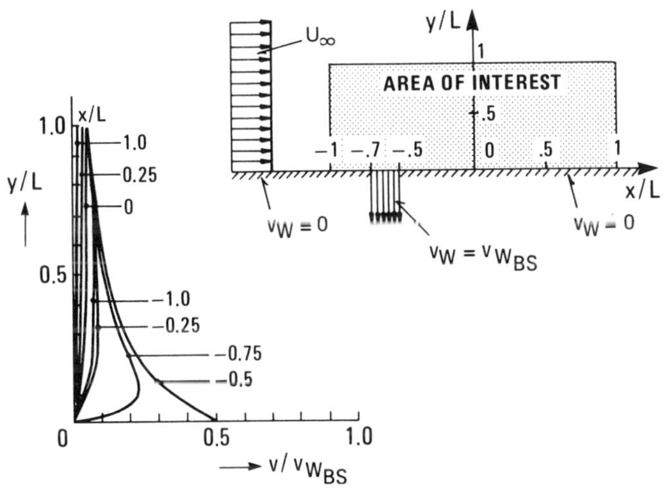

Fig. 2 Vertical Distribution of the v-Component in the Potential Flow Field Due to Basic Suction for $-0.7 \leq (x/L)_{BS} < -0.5$.

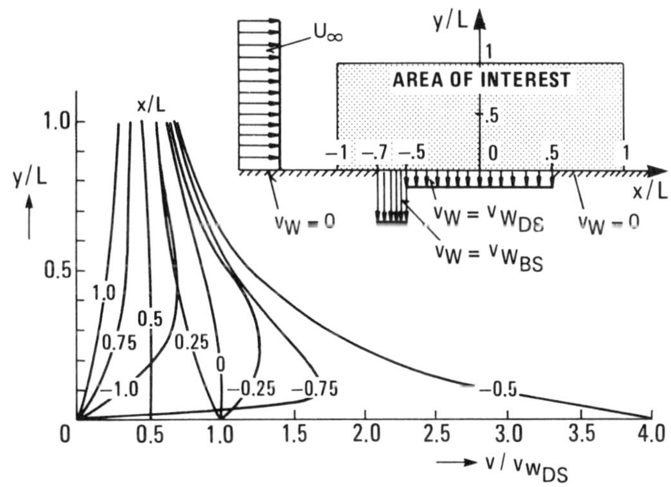

Fig. 4 Vertical Distribution of the v-Component in the Potential Flow Field Due to Basic Suction in Conjunction with Distributed Suction According to $-0.7 \leq (x/L)_{BS} < -0.5$ and $-0.5 \leq (x/L)_{DS} \leq 0.5$; $(v_{W\,BS}/v_{W\,DS} = 7)$.

Fig. 3 displays results for distributed suction. It has been assumed that the approach boundary layer has been removed without introducing a v-component into the flow field. As can be seen in Fig. 3, the decay of the v-component between x/L=-0.5 and x/L=+0.5 takes place slowly with lateral distance.

In combining basic and distributed suction, the ratio $v_{W\,BS}/v_{W\,DS}$ is assumed to be 7, which is an averaged value for the quoted ranges of each suction type. The range of integration of Eq. (5) is divided into two sections, $-0.7 \leq (x/L)_{BS} < -0.5$ and $-0.5 \leq (x/L)_{DS} \leq 0.5$. Results of such composite calculations are displayed in Fig. 4. The velocity v(x,y) is scaled with the value of distributed suction $v_{W\,DS}$. Again, the v-component does not decrease by more than 20 % within the first 10 % of y/L above the ground. Considering that L is of the order of 5 m in a full-scale facility, there will be little decrease of flow angularity due to suction within 0.5 m above the ground. At some locations upstream of the origin (center of turntable) but within the distributed-suction area $v/v_{W\,DS}$ even increases after starting from unity at the wall (e.g. at x/L=-0.25). This is due to the close proximity of the basic-suction area which affects points at non-zero wall distance in this region. The position x/L=-0.5 is of great interest for automotive testing, as it designates approximately the front of a car. A large overshoot of $v/v_{W\,DS}$ is discernible which is equivalent to a large flow angularity. This, again, occurs due to the proximity of the basic-suction area. Therefore, if basic suction is to be applied to remove the approach boundary layer, it is suggested that its downstream extension end before x/L=-0.75. This can be deduced from the curve x/L=-0.25, which shows a much smaller influence from basic suction than x/L=-0.5 does.

Another approach to flow control is tangential blowing. In [8] this method is primarily discussed with respect to momentum which is injected into the boundary layer along the ground by a jet. With this method a reduction of the displacement thickness and the momentum thickness is attained. Therefore, a scoop or a basic-suction system is not necessary to cope with the approach boundary layer. The wall jet entrains and accelerates outer fluid and therefore compensates, or even overcompensates, for the displacement effect of the residual ground-floor boundary layer. In contrast to distributed suction with a constant suction rate, no asymptotic boundary layer develops with tangential blowing.

From an experiment at DNW [8], where a jet with a uniform blowing rate of $(U_0/U_\infty)=5.85$ was introduced 3.75 m upstream of the center of the turntable, it is revealed that the displacement thickness of the composite layer (jet plus boundary layer) can be tuned to be zero at the front-wheel position of a car in the test section by adjusting the momentum of the jet appropriately. An increase of 5 mm over a length of 2 m downstream of the front wheel position was measured for the displacement thickness δ_1. According to Eq. (1) this leads to a positive mean displacement velocity of $v_e/U_\infty=0.0025$ in the empty test section between x=-2 m and x=0. Eq. (2) then yields an equivalent flow angle of $\alpha=0.143^0$ which can be regarded as rather small.

A general view of angularity effects due to tangential blowing can be obtained by considering the total displacement of the composite layer, consisting of contributions from the boundary layer and the wall jet. In the following analysis use is made of a suggestion from Gartshore et al. [15] to convert the velocity profile of the composite layer into the profile of an equivalent "simple" wall jet which is discharged into a field of constant velocity U_∞, see Fig. 5. A momentum balance serves as a criterion for the calculation of the equivalent wall jet.

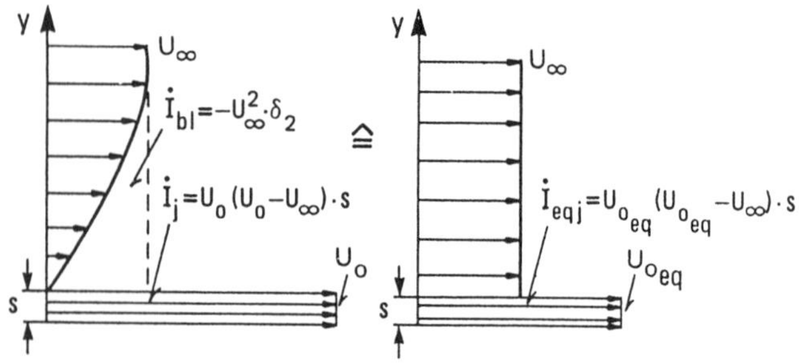

Fig. 5 Equivalent "Simple" Wall Jet.

$$-U_\infty^2\delta_2+U_0\left(U_0-U_\infty\right)s=U_{0_{eq}}\left(U_{0_{eq}}-U_\infty\right)s \qquad (7)$$

The left-hand side of Eq. (7) denotes the momentum of the composite layer, whereas the right-hand side denotes the momentum of the equivalent "simple" wall jet (Fig. 5). Reprocessing the data in [8] yields the momentum thickness at the jet discharge slot: $\delta_2(x_s=0)=14.8$ mm. The corresponding equivalent jet would have a velocity ratio of $(U_{0eq}/U_\infty)=4.223$ at the slot, from Eq. (7).

The displacement of the wall jet is obtained by integration of the continuity equation from the wall (y=0) to the edge of the jet (index "e"), see Fig. 6.

$$v_{ej}=-\frac{dQ}{dx}+U_\infty\frac{db}{dx}=-\frac{d}{dx}\left(Q_I+Q_{II}+Q_{III}\right)+U_\infty\frac{db}{dx} \qquad (8)$$

As indicated by Eq. (8) and Fig. 6 the total volume flux of the jet Q can be subdivided into the wall region (I), where boundary layer flow according to the 1/10 - power law can be expected, and the free-shear-flow regions (II and III), see Thomas [16].

In region II, the change of volume flux due to a velocity overshoot is calculated with the help of an entrainment approach, see Turner [17]:

$$\frac{dQ_{II}}{dx} = -0.084\,\frac{U_{max}-U_\infty}{U_\infty} \qquad (9)$$

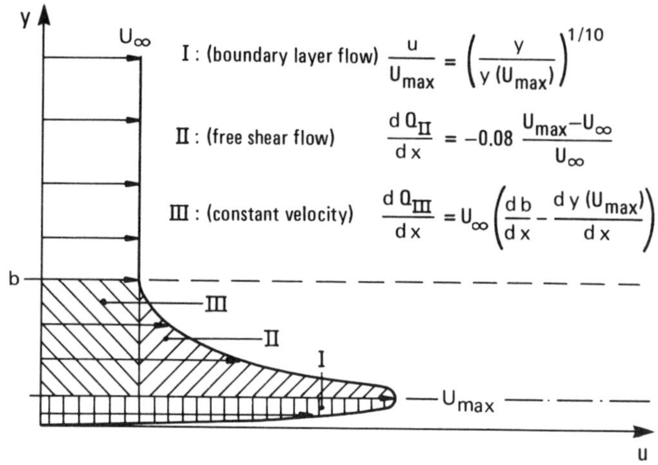

Fig. 6 Wall Jet with Constant External Velocity U_∞.

The relation between $(U_{max}-U_\infty)/U_\infty$ and (x_j/s), the non-dimensional distance from the virtual origin of the jet (i.e. the location where the width of the jet b equals zero), is given in [16] as

$$\frac{U_{max} - U_\infty}{U_\infty} = C\left(\frac{x_j}{s}\right)^\sigma \tag{10}$$

Here, σ is a function of the ratio of the uniform jet discharge velocity U_0 to U_∞, and is given by

$$\sigma = -0.6\left(1-\frac{U_\infty}{U_{0eq}}\right)^{-1} \tag{11}$$

The streamwise variation of volume flux dQ_{III}/dx in region III can be computed according to:

$$\frac{dQ_{III}}{dx} = U_\infty\left(\frac{db}{dx}-\frac{dy(U_{max})}{dx}\right) \tag{12}$$

Here, $y(U_{max})$ denotes the location of the velocity maximum, which can be calculated, according to [16] from

$$\frac{y(U_{max})}{s} = 0.011\left(\frac{x_j}{s}\right) \tag{13}$$

Substituting Eqs. (9) to (13) into Eq. (8) yields

$$\frac{v_{ej}}{U_\infty} = -(0.01\sigma+0.09)\cdot C\cdot\left(\frac{x_j}{s}\right)^\sigma + 0.001 \tag{14}$$

The relation between x_j and the downstream distance x_s from the jet discharge slot can be taken from [16]. The slot height is denoted by s.

$$\frac{x_j}{s} = \frac{x_s}{s} + 20 \tag{15}$$

Eq. (14) contains the unknown constant C. For the present analysis C can be evaluated by utilizing the experimentally observed change in displacement thickness δ_1 of the composite layer between the slot location and an arbitrarily chosen position in the test section. In [8] it was revealed that at DNW the displacement thickness δ_1 decreased to zero at x=-1.96 m, which is equivalent to a position of x_s=1.79 m downstream of the slot. The δ_1 of the approach boundary layer at the slot was processed to be 19.0 mm at a tunnel speed of U_∞=40 m/s. Due to the sudden momentum addition by the jet, a negative jet displacement thickness $\delta_{1j}=-(U_{0eq}-U_\infty)/U_\infty)\cdot s=-3.2$ mm is attained which has to be added to δ_1 of the approach boundary layer. Therefore, the total change in displacement thickness between the slot ($x=-x_s=-3.75$ m) and x=-1.96 m (the location where δ_1=0) amounts to $\Delta\delta_1$=-15.8 mm. Integrating Eq. (14) in x-direction and substituting the left-hand side by $\Delta\delta_1$=-15.8 mm yields C=9.208.

From Eq. (14) there is a negative term accounting for the entrainment of the outer part of the wall jet, and a positive contribution due to the displacement of the near-wall region. Evaluation of Eq. (14) at x=-1.96 m (x_j/s=1810) yields $v_e/U_\infty=-1.08\cdot10^{-3}$, whereas at x=0 v_e/U_∞ equals $-1.7\cdot10^{-4}$. However, these results overestimate the entrainment of the composite layer compared to experimental findings at DNW suggesting a mean value of $v_e/U_\infty=+2.5\cdot10^{-3}$ between these boundaries. This discrepancy is due to the different streamwise evolutions of the displacement thickness in theory and experiment, which are displayed in Fig. 7. The theoretical curves were computed from the integrated form of Eq. (14) and compared with experimental results in [8]. The difference between both curves becomes most obvious at large distances downstream of the slot, whereas there is a closer agreement between theory and experiment between the slot and x_j/s=1810 (x=-1.96 m), because of the common boundary values at these locations. Consequently, v-components evaluated from Eq. (14) will be most accurate in front of a vehicle, but still, their values will be slightly negatively biased. The observed discrepancies between theory and experiment may presumably be attributed to the influence of the composite

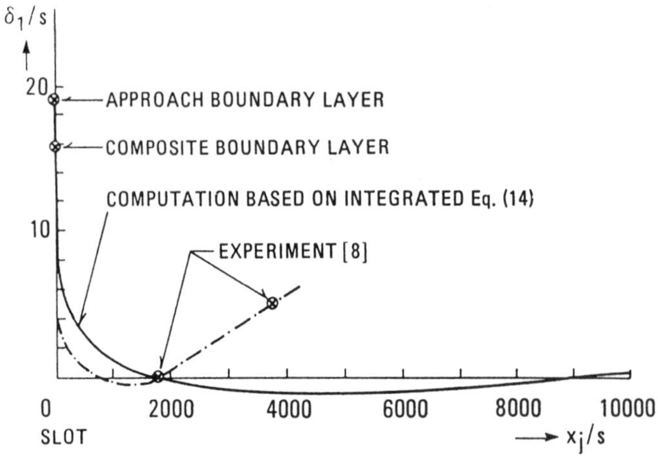

Fig. 7 Computed Displacement Thickness of Composite Layer at DNW with Tangential Blowing.

boundary layer being approximated by the equivalent "simple" wall jet.

To evaluate the interaction of the entrainment with the rest of the flow field, the wall jet is substituted by an appropriate sink distribution along the plane y=0, yielding the same v_e. If $q(x)=2\ v_c(x)$ is substituted into Eq. (5) in an analogous way as for constant suction, the v-component can be calculated for the entire 2-D flow field. The result of such a calculation is shown in Fig. 8. Unlike Figs. 2 to 4, the abscissa in Fig. 8 is scaled with U_∞ instead of v_w because for tangential blowing v is a function of x according to Eq. (14). The vertical axis is scaled with the diameter of the turntable; $L_{DNW}=5$ m at DNW. It can be seen that the induced v-components are small, even at

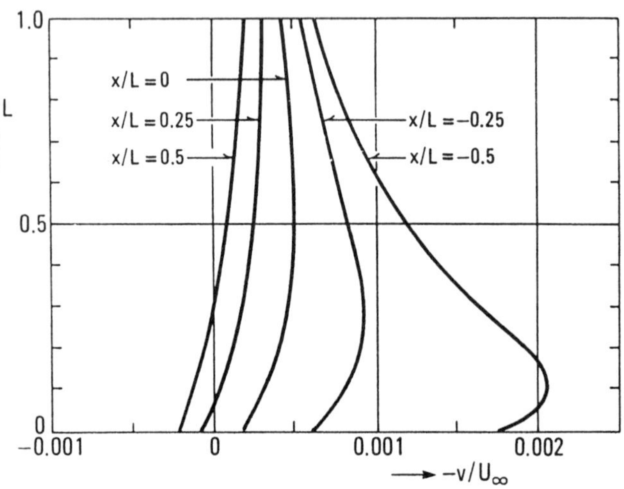

Fig. 8 Vertical Distribution of the v-Component in the Potential Flow Field Due to Tangential Blowing for $U_0/U_\infty=5.85$. Slot Location $x_s/L=-0.75$.

the upstream edge of the turntable (x/L=-0.5), which would be close to the front wheel position of a car in the test section.

Boundary Layer Control by constant suction and by tangential blowing can be compared with respect to the distortion they introduce into the inviscid flow field. However, it should be kept in mind that the results for tangential blowing are based on boundary conditions at DNW, and thus suffer from a loss of generality. Conversely, the results attained with distributed suction are independent of boundary layer effects, due to the asymptotic character of the suction layer.

Fig. 9 displays the results of such a comparison. Instead of showing the v-components, the abscissa is the flow angle α. However, the conversion of the suction results, which are independent of suction intensity (Eq. 5), into flow angles requires stretching the abscissa of Fig. 3 and 4 with $v_{w\ DS}/U_\infty$. For the comparison displayed in Fig. 9 a value of $v_{w\ DS}/U_\infty=-0.005$ $(-\alpha=0.286°)$ was chosen, which represents common practise in automotive tunnels with suction planes.

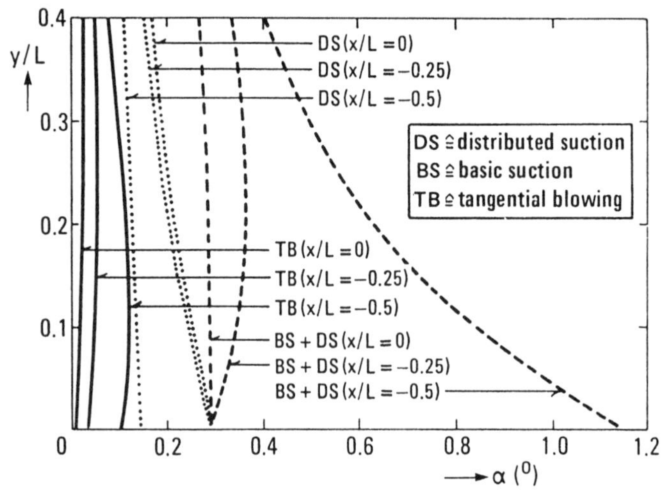

Fig. 9 Vertical Distribution of Flow Angularity Due to Distributed Suction, Combined Basic and Distributed Suction, and Tangential Blowing.

Apparently, at all streamwise locations where y/L<0.05, the flow-induced angles of combined basic and distributed suction differ by an order of magnitude from those evaluated for tangential blowing. At larger distances from the ground the difference decreases for some values of x/L. At x/L=-0.25 even a considerable increase of angularity in y-direction is discernible for the combined basic and distributed suction. This occurs due to the downstream influence of the basic suction, the intensity of which is $v_{BS}/v_{DS}=7$ in the present example. As the results for suction are proportional to $v_{W\,DS}/U_\infty$, a variation of this parameter allows proportional shifting parallel to the abscissa for those curves applicable to suction. The lateral (y-direction) distribution of α remains unaffected, however. Distributed suction without basic suction reveals considerably reduced angularity which, nevertheless, is still larger than for tangential blowing, for the magnitudes selected for this example. As stated earlier, the present analysis tends to overestimate the negative angularity of tangential blowing. The results clearly show that angularity due to BLC affects the entire flow field and is not restricted to the near-wall region. This finding is crucial because biased force measurements will result from such "angles of attack".

For tangential blowing, different α -distributions and magnitudes can be obtained by changing the slot height s and the discharge velocity U_0. For constant momentum flux and air density of the discharged jet, U_0 will change proportionally to $(s)^{-0.5}$ and, according to Eqs. (7), (10) and (11), the streamwise distribution of U_{max} will be changed, thus affecting the constant C and the exponent σ in Eq. (14).

ZONE 2 (INTERMEDIATE ZONE) - The gap between a road vehicle and the ground can be considered as a channel intake. In Zone 2 the flow undergoes transition from boundary layer flow with a potential core to fully developed flow. Fig. 1 qualitatively shows the evolution of the velocity profile. At the entrance, boundary layers develop along the ground and the underside of the vehicle, regardless of the presence of a moving belt. However, the ground-floor boundary layer along a moving ground will rapidly diminish in the upstream direction, since U approaches U_∞ at all points of the flow field upstream of the car. Depending on the boundary condition at the wall, the flow has a tendency to form the velocity profile of a Poiseuille flow (stationary ground) or a Couette flow (moving ground) respectively. In a 2-D flow the continuity equation causes the flow to accelerate and thus the pressure under a car to decrease. In a 3-D case, however, cross flow is created and, due to streamline curvature, a positive pressure gradient might be attained.

Unlike Zone 1, viscous effects become first-order effects at any location within the gap. They are no longer restricted to the wall proximity, thus classical boundary layer approximations, assuming small boundary layer dimensions and small velocity components in the lateral direction in comparison to the external flow properties, are violated in this region. The wall boundary condition of the streamwise velocity distribution will directly affect the pressure distribution in this zone. Furthermore, a transpiration velocity at the wall will lead to first-order effects in the remaining core flow. Therefore, the term "boundary layer control" is not adequate in this zone and should be replaced by "flow control". Whereas, for Zone 1 flow angularity is the most important property to look at, Zone 2 requires momentum effects to be considered. In contrast to external flows, the internal flow in the gap suffers from a lack of mass and momentum transfer from inviscid flow regimes. A moving ground plane, however, continuously feeds momentum into the flow compensating for viscous losses. Simultaneously, the underside volume flux is increased compared to that for a stationary ground plane. As the volume flow rate through the gap is a function of the external pressure distribution and vice versa, the base pressure may be altered by the moving belt. Consequently, the volume flux is changed until a state of equilibrium is attained.

The above discussion suggests that proper ground simulation by flow control demands momentum supply to the gap flow. In principle, this requirement can be met by tangential blowing. However, unlike the moving belt, the wall jet does not continuously provide momentum. Thus, the correct momentum flux can be met at only a single streamwise position. Further downstream with decreasing momentum the entrainment capability of the jet changes into a displacement one as the gap flow undergoes transition into a Poiseuille flow, analogous to uncontrolled underbody flows. The possibility, with tangential blowing, of adjusting the momentum at any desired streamwise location to desired values offers the opportunity to properly optimize the rear underfloor diffusors of cars, which tend to create flow separation on stationary ground boards. Such flow separation will not occur with a moving belt because due to the non-slip condition a flow sheet close to the surface always moves in the direction of the belt, thus giving rise to free recirculation as was shown in [18]. The same effect can be attained with tangential blowing by adjusting the wall jet according to a criterion which assures positive wall shear stress, see [19].

Ground simulation by distributed suction, on the contrary, continuously withdraws volume flux from the gap flow. The justification for mass removal has been adopted from aspects of boundary layer flow which are not applicable, because in the gap, both viscous layers merge and undergo transition into a fully developed flow. Therefore, displacement effects vanish asymptotically, and so the required suction velocity should tend to zero for increasing x. Thus, constant suction will not enable the convective terms of the Navier-Stokes equations to vanish. In case constant suction is applied, due to continuity, volume flux has to be supplied from crosswise direction, hence creating 3-D effects. In a 2-D flow, however, a suction rate of $v_w/U_\infty=-0.005$ and a relative gap height of h/L=0.01 (L: vehicle length) lead to the removal of 50 % of the volume flux across the entrance of the gap. This may cause interactions with the external flow field. Analogous to such continuity considerations, distributed suction creates momentum losses according to the so-called sink drag, rather than supplying momentum to the gap flow.

In principle, any flow control creates interactions with the external flow which may lead to falsified forces and moments on the upper side of a vehicle. This is schematically illustrated in Fig. 10, where ground simulation by stationary ground, by moving ground, by tangential blowing and by distributed suction are contrasted. As a consequence of the difference in volume flux through the gap with the different simulation techniques, variations of the location of the stagnation point on the body surface occur. Experimental verification will experience this effect as a change in downforce, for instance. Additionally, an increased streamwise gradient of volume flux and momentum flux is created into the gap for the case of distributed suction.

ZONE 3 (FULLY DEVELOPED FLOW) - Depending on the ground-plane boundary condition the flow in Zone 3 either results in a Poiseuille flow (stationary ground) or in a Couette flow (moving ground). In both cases the left-hand side of the Navier-Stokes equations will be zero because the convective terms vanish identically for fully developed flow.

$$u\frac{\partial u}{\partial x} = 0 \quad ; \quad v\frac{\partial u}{\partial y} = 0 \qquad (16)$$

By using the method of matched asymptotic expansions Beese [1] showed that a first-order approximation to fully developed flow is attained by expansion of the Navier-Stokes equations with the perturbation parameter

$$Re_L \cdot \left(\frac{h}{L}\right)^2 = Re^* \to 0 \qquad (17)$$

Thus, even for flows where the convective terms do not vanish identically, basically the same equations of motion can be obtained asymptotically with vanishing Re^*. According to [1] the maximum tolerable value of Re^* depends on the streamwise variation of the ground clearance h. For variations smaller than 10 %, Re^* may be as big as 10. A result of the first-order approximation is that the pressure p(x) is an analytical function of h(x) as well as a function of the pressures p_1 and p_2 at the entrance and at the exit of the gap respectively. The flow around the vehicle can merely influence the boundary values p_1 and p_2. These however, tend to be independent of the gap height as h approaches zero because the gap volume flux becomes too small to affect the external flow.

Fig. 10 Internal - External Flow Interactions.

When considering flow control for Zone 3 it will be important to know at which location in the downstream direction fully developed flow will occur. For laminar flow it has been shown in [9] that the inlet length l of a 2-D channel flow with stationary walls is given by

$$\frac{l}{L} = 0.04 \cdot Re^{*} \tag{18}$$

Assume that fully developed flow occurs at the end of the car, (l/L)=1; with a car length L=4.5 m, a free stream velocity of 33 m/s, and $\nu=15 \cdot 10^{-6}$ m^2/s, a gap height of h=7.1 mm can be deduced by combining Eq. (17) and (18). This ground clearance is considerably smaller than even for racing cars. Hence, for laminar flow Zone 3 is unlikely to occur.

For turbulent flow which generally can be experienced at the rough underside of a road vehicle, the inlet length is much smaller compared to laminar flow. A rough estimate based on the properties of a turbulent pipe flow can elucidate this fact. A gap height and width of 0.15 m and 1.8 m respectively, is assumed for a car. For a relative wall roughness of $k/D_h=0.1$ and a flow velocity of 33 m/s an inlet length of l/L=0.19 for 2-D flow can be calculated according to Truckenbrodt [10]. Apparently, after 20 % of the vehicle length the flow will be fully developed for the chosen example. Considering the different boundary conditions with a stationary ground or a moving ground, and both exhibiting much smaller wall roughness than what has been assumed, this result may be altered. Thus, it seems to be difficult to predict the occurrence and the extent of Zone 3. However, the location of Zone 3 is important to know because in a strict sense flow control has to vanish in this region in order to meet the demands of fully developed flow.

Momentum and mass-conservation aspects are of vital importance in this flow regime. Whereas in Zone 2 the volume flux and pressure gradient are in accordance with the external flow, in Zone 3 there is no effect of the volume flux on the external pressure distribution. Thus, the same pressure gradient will be attained for a Poiseuille flow and a Couette flow. Therefore, a fundamental momentum consideration can be executed. By integrating the respective velocity profiles, the momentum-flux ratio $\dot{I}_C/\dot{I}_P$ of the Couette flow to the Poiseuille flow is yielded. For vanishing values of the gap height h the following relation is obtained:

$$\lim_{h \to 0} \frac{\dot{I}_C}{\dot{I}_P} = \lim_{h \to 0} \frac{U_\infty \cdot \eta^2}{\frac{1}{40} \cdot h^4 \cdot \left(\frac{dp}{dx}\right)^2} = \infty \tag{19}$$

Apparently, flow control has to supply momentum to Zone 3. This can be achieved by tangential blowing. However, it has to be realized that the need for vanishing convective terms for fully developed flow cannot be fulfilled simultaneously. This means that flow control by tangential blowing sustains the transitional Zone 2 in order to cope with the momentum requirements.

Distributed suction, on the other hand, continuously removes mass and momentum, thus preventing the flow from ever becoming fully developed. Additionally, the momentum requirements are violated. To satisfy continuity, a vanishing relative gap height leads to infinite velocities within the gap, if finite suction rates are applied, as may be deduced from basic 2-D mass-flux considerations.

$$\lim_{h \to 0} \frac{U_{gap}}{v_W} = \lim_{h \to 0} \frac{L}{h} = \infty \tag{20}$$

Consequently, infinite interactions with the external flow field occur. This contradicts the above statement that Zone 3 does not affect the external flow field. Thus, distributed suction is not suited to simulating fully developed flow regions.

EVALUATION OF EXPERIMENTAL DATA WITH REFERENCE TO THE ZONAL FLOW MODEL

Considerable experimental work has been carried out in the past on the effect of ground simulation for automotive testing. Unfortunately, these studies have provided little insight into the physics of the flow because they were basically phenomenological. In the following, some examples of this experimental work will be discussed with reference to the zonal flow model of Fig. 1.

It was shown in [8] that ground simulation, either by tangential blowing or basic suction, can considerably increase the Reynolds number sensitivity of the flow past a vehicle. Fig. 11 displays some data reproduced from [8].

Provided that the flow on the upper side of a vehicle is not significantly affected by a change in free stream velocity from 100 km/h to 200 km/h, the change in c_D must be due to the vehicle's underside. This however, excludes the occurrence of a Poiseuille flow (Zone 3) for the following reasons.

- In fully developed flow, viscous drag is proportional to dp/dx in the gap. As was mentioned above, p_1 and p_2, the pressures at the entrance and the exit of the gap are imposed by the external inviscid flow and are proportional to U_∞^2. Therefore, $c_{D\ gap}$, the viscous drag of the gap flow, is independent of the velocity level.

- On the other hand, in a Zone 3 flow along a moving wall the shear stresses are proportional to $C_1 U_\infty + C_2 U_\infty^2$ which implies that $c_{D\ gap} \propto C_3 + C_4/U_\infty$. This relation, at least qualitatively, matches the curves of Fig. 11.

These results suggest that improved ground simulation along fixed walls by means of decreasing the momentum thickness delays the formation of fully developed flow and approximates the effect of a Couette flow profile. This conjecture is supported by the experiments of Mercker and Knape [6]. They measured total pressure profiles under a passenger car when applying different ground simulation techniques, tangential blowing and moving belt. The results attained at the rear bumper are reproduced in Fig. 12. Whereas for the fixed-ground profile a considerable momentum deficit is discernible relative to the moving-belt case, tangential blowing closely matches the moving belt profile, except in the proximity of the wall. This close matching appears plausible because in both cases momentum has been transferred from the wall region to the flow in the gap.

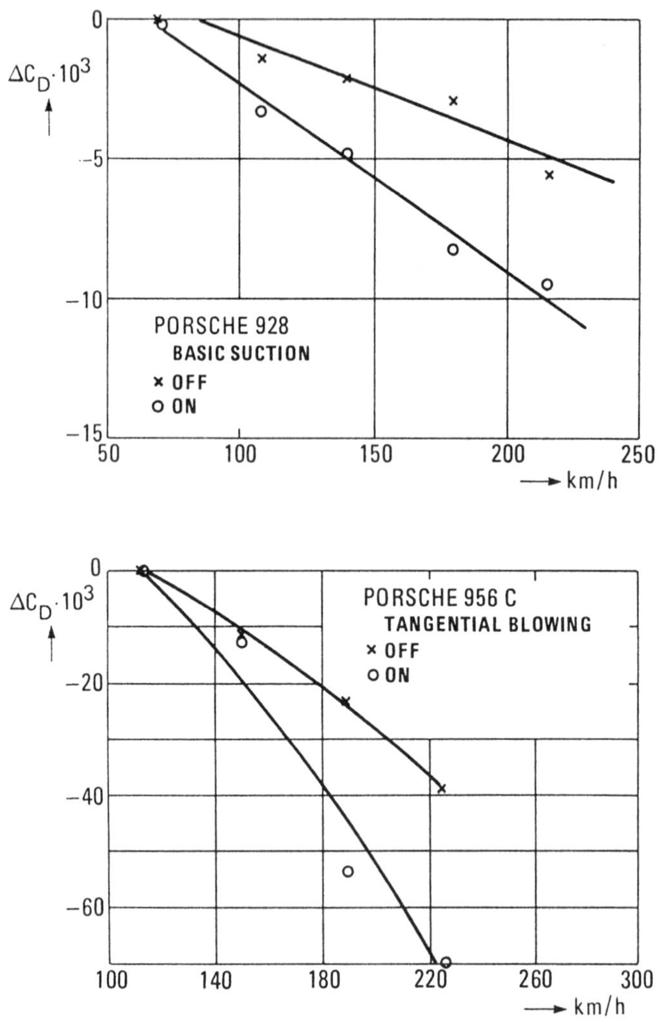

Fig. 11 Influence of Reynolds Number on the Flow Past a Vehicle with and without Ground Simulation; after [8].

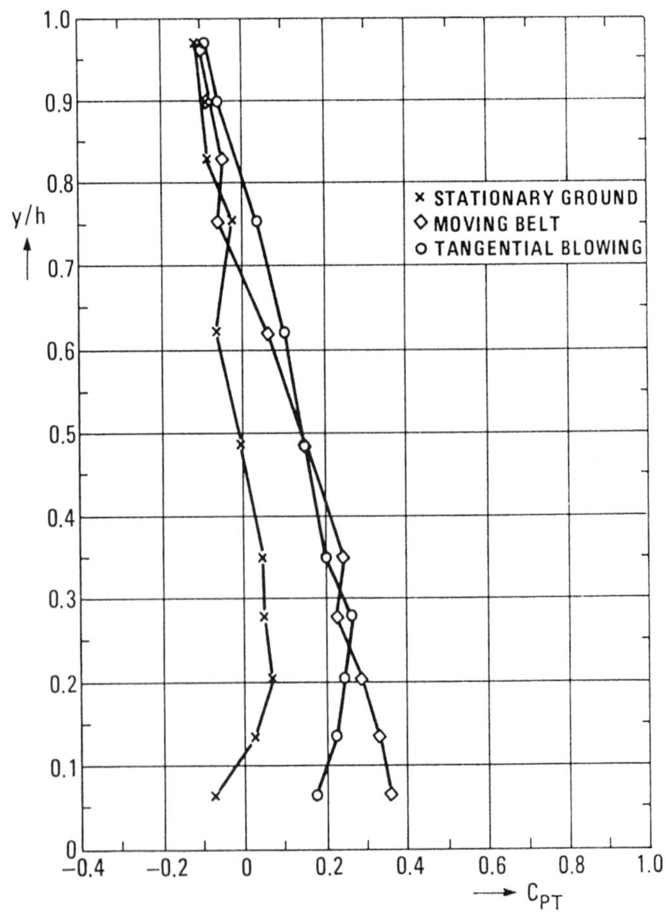

Fig. 12 Total-Pressure Profiles at the Rear-Bumper Position of a Passenger Car; after [6].

Carr [4] performed experiments with scale models (1:6-1:4) on the comparison of ground simulation by distributed suction and moving belt. The investigations included different types of passenger cars as well as sports cars and a formula one racing car. In each case, except for the formula one car, it was found that the change in lift coefficient from the fixed-ground case due to distributed suction (v_w/U_∞=-0.0015) was much greater than the corresponding effect of the moving belt. This may be due to flow-control-induced angularity effects or sink effects in the suction case which result from overcompensation of the diminishing displacement effect underneath the car. On racing cars with very steep rear diffusors, however, flow recirculation is likely to occur on a moving ground plane, whereas there may be flow separation along a stationary ground. Both flow types differ considerably, as has been shown by Telionis [18] and, in general, lead to different forces on the model. However, by adequately adjusting the suction rate, the same change in lift as experienced on a moving belt can be generated, though there is stationary-ground flow separation instead of the recirculation along the moving ground [3]. But since the correct lift change is not known *a priori* this shows the impossibility of a "once for all adjustment" of the suction rate to suit every simulation requirement.

Fig. 13 shows pressure distributions along the underside of a vehicle for different ground simulations.

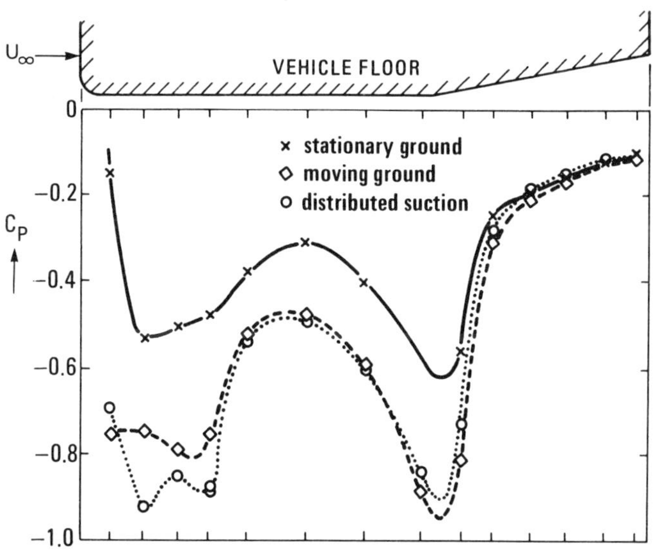

Fig. 13 Pressure Distributions with Different Ground-Simulation Techniques; after [4].

This diagram has been reproduced from [4]. It can be seen that the greatest pressure differences between moving belt and suction occur near the contour radii. Due to curvature there are strong streamwise pressure gradients which cause gradients of the displacement thickness leading to local changes of v_e in Eq. (1), thus requiring a streamwise change of the suction rate v_w. However, a constant suction rate was used during this experiment. Nevertheless, it is important to note, that in this example distributed suction reveals a considerable improvement of the flow simulation compared to the conventional stationary ground.

SUMMARY AND CONCLUSIONS

Three possible zones of flow have been identified in the gap between vehicle floor and road. These zones require different flow controls in order to simulate a moving ground in a wind tunnel with stationary floor.

In ZONE 1, the flow exhibits boundary layer characteristics. With the aid of some fundamental calculations it is shown, that applying distributed boundary layer suction may result in an unwanted flow angularity effect in front of a vehicle which decays slowly with vertical distance from the floor. This effect will be amplified if basic suction, which is used in order to reduce the approach boundary layer, precedes the distributed suction. Flow control by tangential blowing reveals flow angularity effects which are an order of magnitude smaller than for combined basic and distributed boundary layer suction. This is due to the fact that flow angularity created by the entrainment flow of the wall jet is partly compensated by the displacement effect of the wall boundary layer.

In the intermediate zone (Zone 2), where transition from boundary layer flow to fully developed flow takes place, ground simulation demands momentum supply to the gap flow. In principle, this requirement can be met by tangential blowing, but the correct momentum flux is attained at one single streamwise position only. In contrast, distributed suction withdraws volume flux from the gap flow, and thus momentum loss is created. The removal of mass flux creates interactions with the external flow which may lead to falsified forces and moments on the upper side of a vehicle.

In ZONE 3, fully developed flow occurs. Hence, flow control has to vanish, in a strict sense, because in this zone the convective terms of the

Navier-Stokes equations vanish identically. The volume flux through the gap has no influence on the external pressure distribution and thus, the same pressure gradient will be effective for a Poiseuille flow and a Couette flow in the gap. Nevertheless, the momentum ratio between Couette flow and Poiseuille flow tends to infinity for vanishing gap height. Consequently, flow control should supply momentum, which can be achieved by tangential blowing. This way, however, flow control terminates in a Poiseuille flow, although delayed, and is analogous to uncontrolled flow. In order to meet momentum and mass conservation aspects, the condition of vanishing convective terms has to be abandoned. Distributed suction, however, does not lead to any type of developed flow. Furthermore, it removes mass and momentum from the gap flow, as mentioned above, and thus is not recommended for ground simulation in Zone 3.

In conclusion, the results show that among the techniques considered, tangential blowing meets the simulation requirements of the different zones best. It should be mentioned, however, that this technique requires costly technical effort in full scale facilities.

NOMENCLATURE

b	width of the jet
C	constant
C_1	constant
C_2	constant
c_D	drag coefficient
D_h	hydraulic diameter
h	gap height under a car
H_{12}	shape factor; $H_{12}=\delta_1/\delta_2$
$\dot{I}$	momentum flux
k	wall roughness
l	flow length after which asymptotic or fully developed conditions are attained
L	diameter of turntable, length of car
p	pressure
p_1	pressure at the entrance of the gap
p_2	pressure at the exit of the gap
q	source strength per unit length
Q	volume flux
Re_l	Reynolds number based on l
Re_L	Reynolds number based on L
Re^*	perturbation parameter
s	slot height
u	streamwise velocity component
U	streamwise velocity component outside of the boundary layer
U_{gap}	average streamwise velocity in the gap
U_{max}	maximum streamwise velocity in the jet
U_0	jet discharge velocity
U_∞	free stream velocity
v	normal velocity component
v_e	normal velocity component at the outer edge of the boundary layer
v_w	normal blowing (positive) or suction (negative) velocity on the ground plane
x	streamwise coordinate (x=0 at center of turntable)
x_j	distance from virtual origin of the jet
x_s	distance from blowing slot
x'	location of source element
x'_1	upstream edge of suction board
x'_2	downstream edge of suction board
y	coordinate normal to the ground plane
α	vertical flow angle
δ_1	displacement thickness
δ_2	momentum thickness
η	viscosity
ν	kinematic viscosity

SUBSCRIPTS

BS basic suction

DS distributed suction

e edge of boundary layer, edge of wall jet

eq equivalent simple wall jet

j due to jet

TB tangential blowing

REFERENCES

[1] BEESE, E., Untersuchungen zum Einfluß der Reynolds-Zahl auf die aerodynamischen Beiwerte von Tragflügelprofilen in Bodennähe. Dissertation, Fakultät für Maschinenbau, Ruhr- Universität Bochum, (1982).

[2] WIEDEMANN, J., Grenzen und Möglichkeiten der Modelltechnik innerhalb der Kraftfahrzeug-Aerodynamik. Symposium No. T-30-905-056-7 Aerodynamik des Kraftfahrzeugs, Haus der Technik, Essen, FRG, (1987).

[3] WIEDEMANN, J., Some Basic Investigations into the Principles of Ground Simulation Techniques in Automotive Aerodynamics. SAE Paper 890369, Detroit, (1989).

[4] CARR, G.W., A Comparison of the Ground-Plane-Suction and Moving-Belt Ground-Representation Techniques. SAE Paper 880249, Detroit (1988).

[5] BEARMAN, P.W., DE BEER, D., HAMIDY, E., AND HARVEY, J.K., The Effect of a Moving Floor on Wind-Tunnel Simulation of Road Vehicles. SAE Paper 880245, Detroit, (1988).

[6] MERCKER, E. and KNAPE, H.W., Ground Simulation with Moving Belt and Tangential Blowing for Full-Scale Automotive Testing in a Wind Tunnel. SAE Paper 890367, Detroit (1989).

[7] LARSSON, L., HAMMAR, L., NILSSON, L.U., BERNDTSSON, A., KNUTSON, K. and DANIELSON, H., A Study of Ground Simulation - Correlation between Wind-Tunnel and Water-Basin Tests of a Full-Scale Car. SAE Paper 890368, Detroit, (1989).

[8] BERNDTSSON, A., ECKERT, W.T. and MERCKER,E., The Effect of Groundplane Boundary Layer Control on Automotive Testing in a Wind Tunnel. SAE Paper 880248, Detroit (1988).

[9] SCHLICHTING, H., Boundary-Layer Theory. Seventh Edition. McGraw-Hill, New York, (1979).

[10] TRUCKENBRODT, E., Fluidmechanik, Band 1, Springer-Verlag, Berlin, Heidelberg, New York, (1980).

[11] IGLISCH, R., Exakte Berechnungen der laminaren Reibungsschicht an der längsangeströmten ebenen Platte mit homogener Absaugung. Schriften dt. Akad. d. Luftfahrtforschung, 8 B, No. 1 (1944); NACA RM 1205 (1949).

[12] ULRICH, A., Theoretische Untersuchungen über die Widerstandsersparnis durch Laminarhaltung mit Absaugung. Schriften dt. Akad. d. Luftfahrtforschung, 8 B, No. 2 (1944).

[13] GERSTEN, K., PAPENFUSS, H.D., GROSS, J.F., Second Order Boundary Layer Flow With Hard Suction, AIAA Journal, Vol. 15, pp. 1750-1755, (1977).

[14] WIEDEMANN, J., Der Einfluß von Ausblasen und Absaugen an durchlässigen Wänden auf Strömungen bei großen Reynoldszahlen. Dissertation, Fakultät für Maschinenbau, Ruhr-Universität Bochum, (1983).

[15] GARTSHORE, I.S. and NEWMAN, B.G., The Turbulent Wall Jet in an Arbitrary Pressure Gradient. Aero. Quart. 20, pp 25-56, (1969).

[16] THOMAS, F., Untersuchungen über die Grenzschicht an einer Wand stromabwärts von einem Ausblasespalt. Abhandlungen der Braunschweigischen Wissenschaftlichen Gesellschaft, Vol 15, (1963). Editor: Bogen., H.J., Verlag Friedr. Vieweg&Sohn, Braunschweig.

[17] TURNER, J.S., Buoyancy Effects in Fluids. Cambridge University Press, (1973).

[18] TELIONIS, P.D., Unsteady Viscous Flows. Springer Series in Computational Physics. Springer Verlag, New York, (1981).

[19] HACKETT, J.E. and BOLES, R.A., Moving-Ground Simulation by Tangential Blowing. J. Aircraft, Vol. 16, No. 12, Article No. 78-814R, Dec. (1979).

950996

Ground Effect Simulation for Full-Scale Cars in the Pininfarina Wind Tunnel

Antonello Cogotti
Industrie Pininfarina S.p.A.

ABSTRACT

The paper refers to a research program which is presently in progress at Pininfarina. It deals with a new ground effect simulation system for full-scale car testing in an automotive wind tunnel.
This research program is carried out in the framework of the "Progetto Finalizzato Trasporti 2" of the Italian C.N.R.

The first part of the paper makes an analysis of:

- The various ground effect simulation systems already existing in the world.
- The results which can be achieved using these systems.
- Their main limitations and drawbacks and some considerations about the possibility of using these systems for routine aerodynamic development work on full-scale vehicles.

The second part of the paper describes the new Ground Effect Simulation System (GESS) which is presently in operation in the wind tunnel of the Pininfarina Aerodynamic and Aeroacoustic Research Center.

Its layout is quite different from the ground effect simulation systems existing in other full-scale automotive wind tunnels. It includes several subsystems, that is to say:

- A Moving Belt (MB) "between the wheels" integrated in the balance-turntable system.
- A Tangential Blowing System (TBS) in front of the MB.
- Two Basic Suction Systems (BSS), one immediately in front of the TBS and the second between the TBS and the MB.
- A System for Rotating the car Wheels (RWS) through four rollers integrated in the balance pads.
- A System for Lifting a car (LS) through the wheels and/or the rockers.
  Also the LS is integrated in the balance pads.

The new Pininfarina GESS is therefore more complex than the existing systems, however, it gives several advantages, in particular:

- The relative car-to-ground motion is well reproduced along the most important part of the car, that is to say under the full length of the car.
- In general the overall system is much easier to use than the most of the existing full-scale ground simulation systems.
  Only minor preparation of the car is necessary, in order to conduct tests with a full simulation of the ground motion and wheel rotation.

The third part of the paper reports some preliminary results measured on a full-scale car in various conditions with fixed ground/fixed wheels, rotating wheels only, moving ground only and moving ground/rotating wheels.
Although these values are likely to be "car-dependent", they give a first indication of the possible differences which can be expected when using the new Ground Effect Simulation System.

INTRODUCTION

Aerodynamic optimization of road vehicles is usually carried out by most of the car companies by working on full-scale models, in automotive wind tunnels.
The flow-field simulation, which is achieved in this type of facility is in general, good if not excellent. However, some areas of the aerodynamic field exist where the floor simulation, compared with the road, is less satisfactory.
The most important of these areas is the one under the car, between the floor of the test section and the car underbody. This is due to the fact that in automotive wind tunnels, neither the relative motion between the car and the ground, nor the wheel rotation, is usually reproduced.
This lack of simulation of the aerodynamic field, already recognized in the past, was regarded as not very important since the car underbody was not optimized from an aerodynamic standpoint.
In these conditions, underbody flow is typically separated and therefore a good simulation of this flow is not really very important.
In reality, aerodynamic development of passengers cars, in these last 20 years, has concentrated mainly on the most visible parts of the car body, disregarding modifications to underbody areas for various reasons, like: costs, mechanical and thermal problems and, last but not least, the lower importance, from a sale standpoint.
However, in recent years, the aerodynamic optimization of the car body upper surfaces, both as shape and details (i.e. flush-glass, etc.), has achieved levels which are increasingly difficult to improve upon.
On the other hand, a significant potential for the improvement of car aerodynamic characteristics does exist by working on the system "underbody-wheelhousings-wheels". The flow maps in fig. 1, measured behind a 20-year old notchback (Cd = 0.45) and a modern notchback (Cd = 0.29) show clearly that energy losses in the underbody area are not too different and that, in any case, this area might be greatly improved. However, to do this work correctly, that is to say to find out aerodynamic solutions which are really valid both in the wind tunnel and on the road, it is necessary that the aerodynamic flow, close to these parts, corresponds closely to that existing on the road.
A correct simulation of the aerodynamic flow, in the area of the system "underbody-wheels", requires in particular that, the relative motion between floor and underbody, as well as the wheel rotation, is reproduced in the wind tunnel, as it happens on the road.
The first idea is obviously to use a large moving belt under the car. However, although moving belts of various sizes have been already built in the past, they are practical enough to be used only in case of tests on small scale models, while their use for full-scale models of cars is relatively impractical.

The research program which is in progress at Pininfarina, in the framework of the "Progetto Finalizzato Trasporti 2" (PFT-2) of the Italian "Consiglio Nazionale delle Ricerche" (CNR), has the aim of finding out technical solutions and of building, in the wind tunnel of the Pininfarina Aerodynamic and Aeroacoustic Research Center., a system which is able to:

- provide a good simulation of the aerodynamic flow in the underbody areas of full-scale cars so as to achieve a good simulation of the ground effect for these road vehicles;
- be in the same time practical enough, so that this new system can be currently used in the wind tunnel for the routine development work of every new car models.

PART I

DESCRIPTION OF EXISTING GROUND EFFECT SIMULATION SYSTEMS.

An analysis of the existing systems for the simulation of the ground effect both in full-scale and small-scale wind tunnels was made at the beginning of this research program. It shows that such systems can be grouped in the following 5 categories [see also ref. (1)]:

1. Suction of the boundary layer at the beginning of the test section and/or immediately in front of the car.
   This system, often named "basic suction", removes most of the incoming boundary layer. Immediately after the suction inlet, the boundary layer starts to grow up again. A device of this type was installed in Pininfarina in 1980. Its main characteristics can be summarised as follows (2).
   The air inlet is located 0.60 m from the nozzle end. It has a width of 3.0 m and a length of 0.23 m.
   The air outlets are located at the end of the test section, far enough from the car's rear end, so as not to disturb the car flow field or its wake.

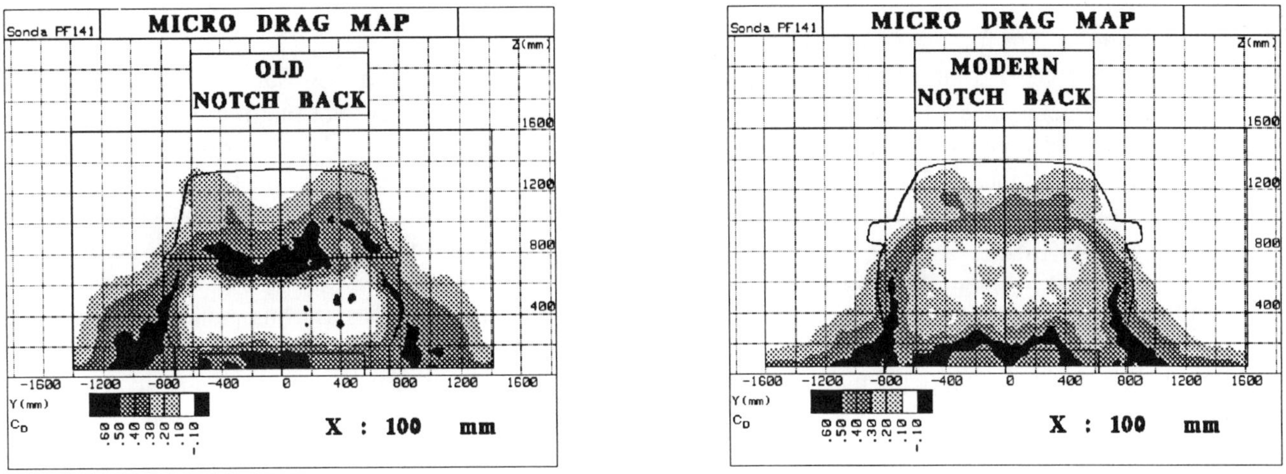

Fig.1 - "Micro-drag" maps showing the momentum losses measured behind an old notchback (C_D=0.45) and a modern notchback (C_D=0.29) (from ref. 3).

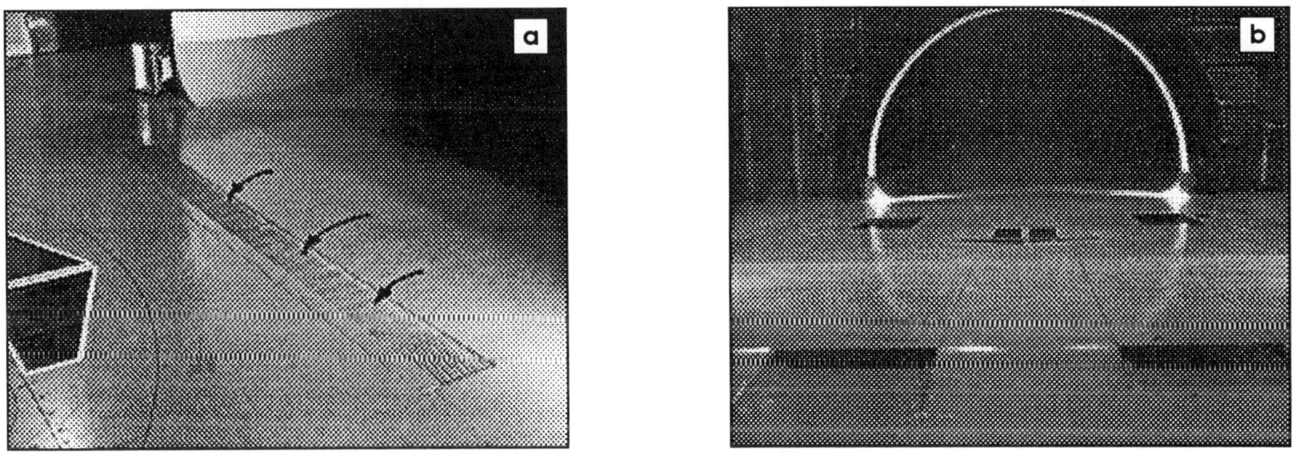

Fig.2A - Basic suction system, Pininfarina wind tunnel (from ref. 2)
a) suction slot at the beginning of the test section
b) airflow outlets at the end of the test section

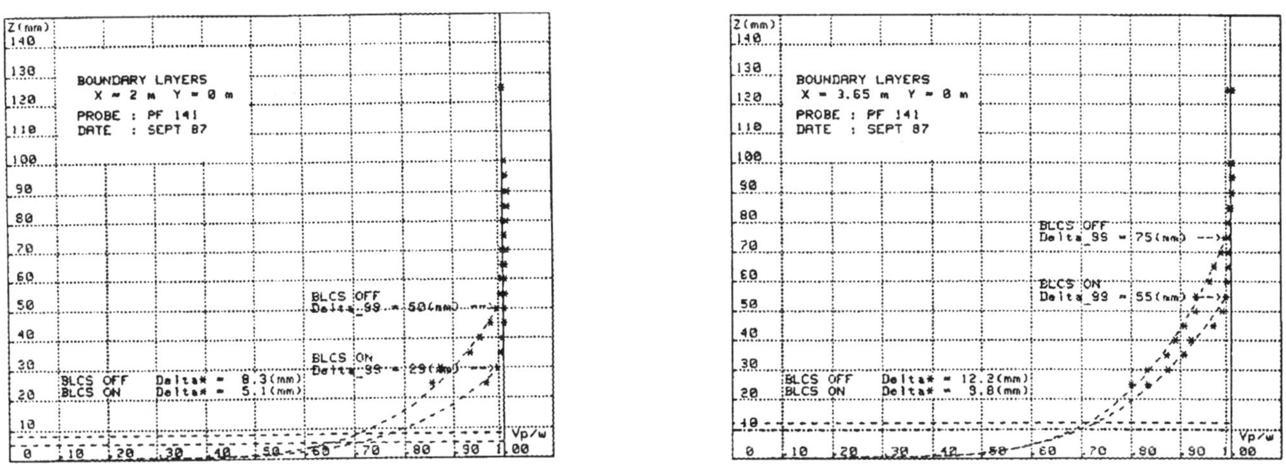

Fig.2B - Boundary layer at X=2 m and X=3.675 m from the beginning of the test section, with and without suction of boundary layer (from ref. 2).

By using this system, the boundary layer thickness δ_{99} at 2 m from the nozzle end, at V = 100 Km/h is reduced from 50 to 29 mm and the displacement thickness δ^* from 8.3 to 5.1 mm.
Fig. 2a shows the boundary layer suction system: respectively, the air inlet and the two air outlets.
Fig. 2b shows the boundary layer velocity profile at X = 2 m and X = 3.675 m (= turntable center) with and without the suction system in operation.
Fig. 2c shows the evolution of the boundary layer after the air suction inlet with and without the system in operation.
The main advantage of this technique, which is common to many wind tunnels, is that it is a relatively simple technique and it does not interfere with the wind tunnel work.
In practice, the operation of the boundary layer suction system is automatic and fully transparent to the wind tunnel user: the blowers, which suck the boundary layer, are linked to the main fan motor. The air volume, which is removed from the boundary layer, increases proportionally to the wind velocity in the test section.
However, this system has a number of limitations, namely:

- The boundary layer is usually not completely removed and, in any case, its suction cannot be increased beyond a certain limit, otherwise flow angularities are introduced in the main flow in the test section.
- Secondly, the boundary layer starts increasing up immediately after the suction inlet, along the car underbody.

Therefore, good correspondence with the road is not achieved.
Tests made over the past 14 years have shown that, for passenger cars with a conventional underbody, ground clearance in the range of 150-200 mm and stationary wheels, the suction of the boundary layer does not produce any significant change in the aerodynamic coefficients. Typically, for a car model, drag coefficient increases by $\Delta Cd = +0.005$ and lift coefficients also do not significantly change. The increase of Cd is probably due to the extra flow impinging on the (stationary) front wheel lower parts.
However, in the case of passenger cars with a ground clearance lower than 150 mm, boundary layer suction results in an improvement in the flow under the car. In particular the "horse-shoe" vortices, which can be seen in the wind tunnel and do not exist on the road, are removed or strongly reduced.
Fig. 3 shows a flow survey behind a car with a ground clearance of the front dam = 65 mm, with and without boundary layer suction (3).
It can be noticed that in the case of suction, the "horse-shoe" vortices disappear. Similar tests in other wind tunnels (4) also confirm these conclusions.
In any case, besides this reduction or removal of the "horse-shoe" vortices, the simple basic suction of the boundary layer is not able to reproduce, to a satisfactory level, the flow conditions existing under a car moving on the road.

2. <u>Distributed suction of the boundary layer over the complete floor of the test section.</u>

A system of this type was built by Porsche in its full-scale wind tunnel, in addition to a basic suction system, located at the beginning of the test section (5), (6), (7).
It is a relatively complex system, as the whole surface of the turntable and the other parts of the wind tunnel floor must be "porous", so that a suction can be applied to the entire test section floor.

The main advantage of this system is that it does not interfere with the wind tunnel work, nor does it require any additional preparation to the car or to the wind tunnel (at least as long as wheels are kept stationary). In other terms, this system is practically transparent to the wind tunnel users.
A second important advantage is that this system controls the boundary layer along the entire length of the car.
However, some possible inconveniences of this system, beside its cost and complexity, are:

- removing the boundary layer does not necessarily mean that the boundary condition (velocity at the floor level corresponding to the free stream velocity) is reproduced . This may represent a drawback mainly in the "diffuser-like" area at the rear of some cars.
- it may be not easy to define how much flow volume has to be removed ;
- instead of adding the missing flow volume and momentum to the boundary layer, a certain flow volume is removed. Someone maintains that this is not the best approach to restore the flow conditions existing on the road (31).

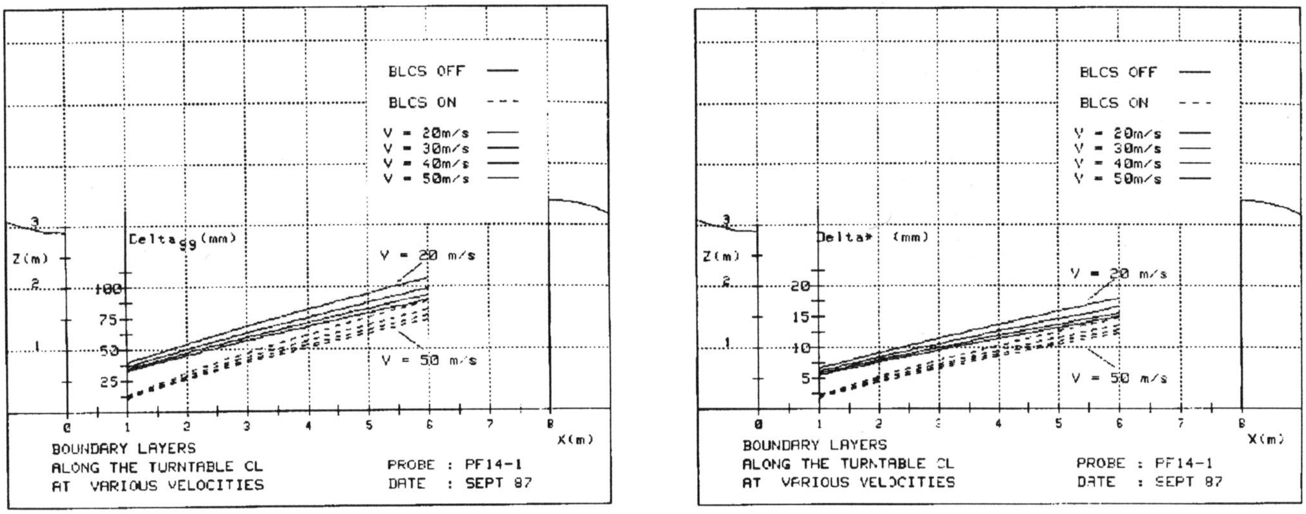

Fig. 2C - Boundary layer development downstream of the suction slot, with and without the basic suction system in operation.

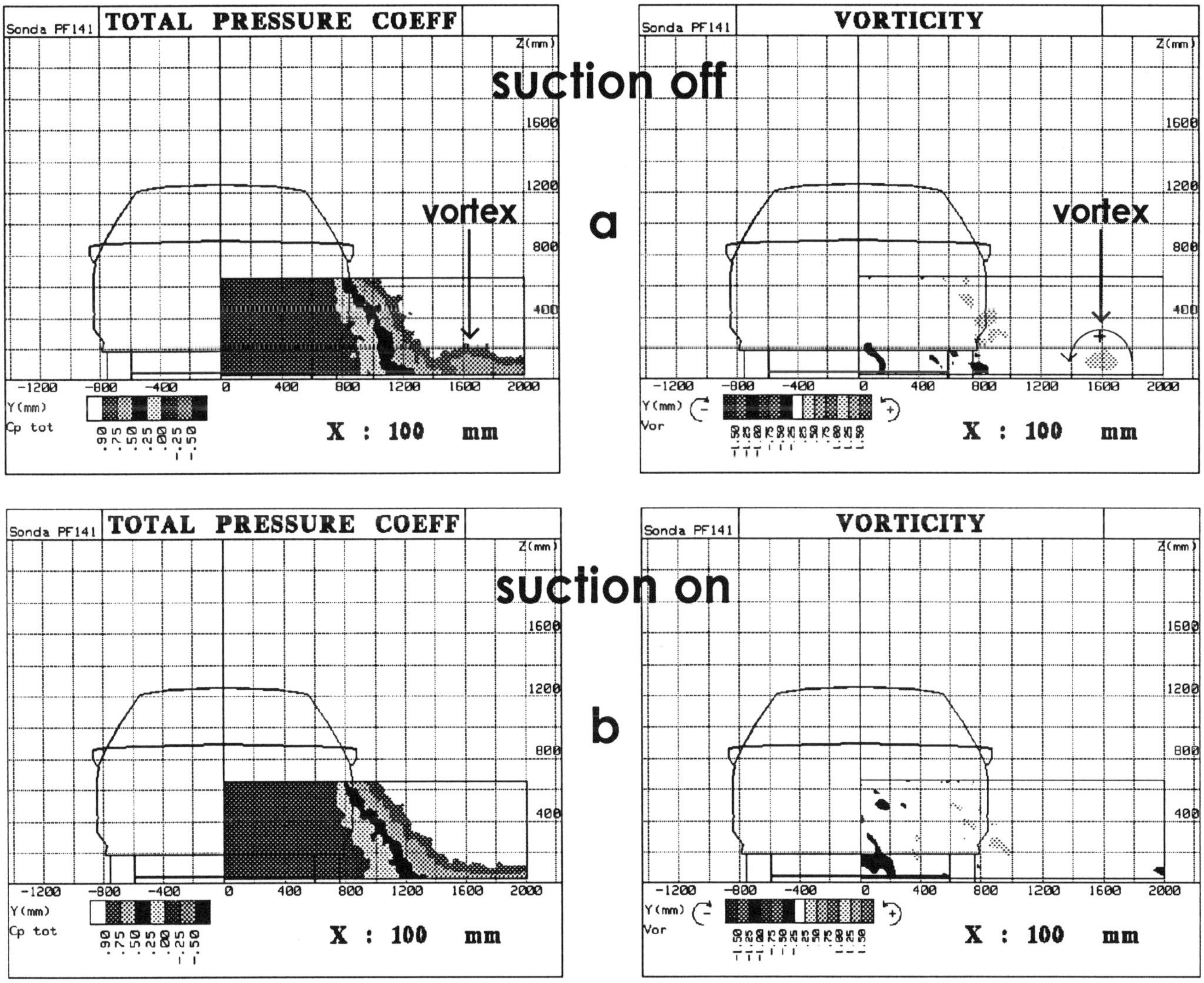

Fig.3 - Pressure and vorticity maps, 100 mm downstream of a full-scale car, having ground clearance of the front dam of 65 mm, with and without the basic suction system in operation. Vorticity ($\omega = \delta Vy/\delta z - \delta Vz/\delta y$) is calculated from Vy and Vz values, with grid spacing = 25 mm.

Additional means are necessary to rotate the car wheels, but that is not necessarily a problem.

3. Moving belts.

Moving belts have been used for many years in a number of wind tunnels for small scale-model testing (8), (9), (10), (11), (12), (13), (14).
In general, a moving belt, much wider and larger than the scale model is used (fig. 4).
In these wind tunnels, the balance is usually installed at the top of the test section. Then, the model is supported by the balance through some vertical struts. These struts are usually faired and also attached to the "non-metric" part of the wind tunnel balance. Therefore, their aerodynamic drag is not measured by the balance.
In some cases, the small scale model is supported by a sting (12), that is to say a longitudinal horizontal strut which supports the scale model from the rear.
In this case, the balance is often installed inside the model between the model and the sting. The model's wheels can stay in contact with the moving belt, which, in this case, provides their rotation. However, this causes an additional drag (rolling drag), which has to be taken into account.
As an alternative, wheels are separated from the model body and supported by external arms. In this way, the wheel rolling resistance is not measured by the balance. However, the aerodynamic forces, acting on the wheels, are no longer measured by the balance, and this leads to other errors which are not negligible.
In general, indications provided by these moving belts are very helpful on condition that the moving belt system is correctly designed and built.

In particular:

- The belt must not lift or produce waviness, when it moves at high speed.
- The upstream flow which reaches the beginning of the belt must have an almost perfectly straight velocity profile, that is to say a perfectly flat boundary layer.
  Otherwise, the belt carries downstream the incoming boundary layer.

A moving belt system for small scale models is not directly transferable to full-scale cars or models. In fact, regardless of same technological difficulties which can still exist when building moving belts of big size (i.e. weight, stiffness, etc.), the following additional problems have to be considered:

- In full-scale automotive wind tunnels, the balance is usually installed under the test section floor. A car is usually supported through the 4 wheels by 4 small-sized pads linked to the balance.
  Consequently the room required by a conventional large moving belt is already occupied by the balance and the simultaneous installation of these 2 systems is physically impossible.

- A car-body is built so that it can easily withstand concentrated loads (corresponding to its own weight and other additional loads like passenger weight, inertial loads, aerodynamic forces, etc.) only if these loads are applied through the car wheels, or to the points of attachment of the suspension system or , to a minor extent, to some specific points along the rockers. These points are the only ones suitable to support a car. Therefore, it is impossible, as in the case of small scale models, to support a car, for instance, through a point of attachment on the roof or on the rear end. These parts are relatively weak, and in some cases (the rear end), are designed so as to be collapsible in case of an accident. They are not suitable to carry concentrated loads unless these are very low.

Taking into consideration these problems, two aerodynamic facilities (DNW and FIAT) have put in operation, in these last years, two moving belts for full-scale car testing.

1. The first moving belt is in use at the DNW wind tunnel. It is a large moving belt (15) (16) much larger and wider than a full-scale car (fig. 5).
   The problem of the physical incompatibility between the moving floor and balance is, in this case, avoided because this wind tunnel is mainly an aeronautical wind tunnel in which a sting-balance system is also available. A car, supported by the sting through the backlight, is positioned over the belt. However to support a car in this way, it is necessary to reinforce the car body by inserting a frame inside the passenger compartment. Consequently, to make a test of this type, a car has to be heavily modified.

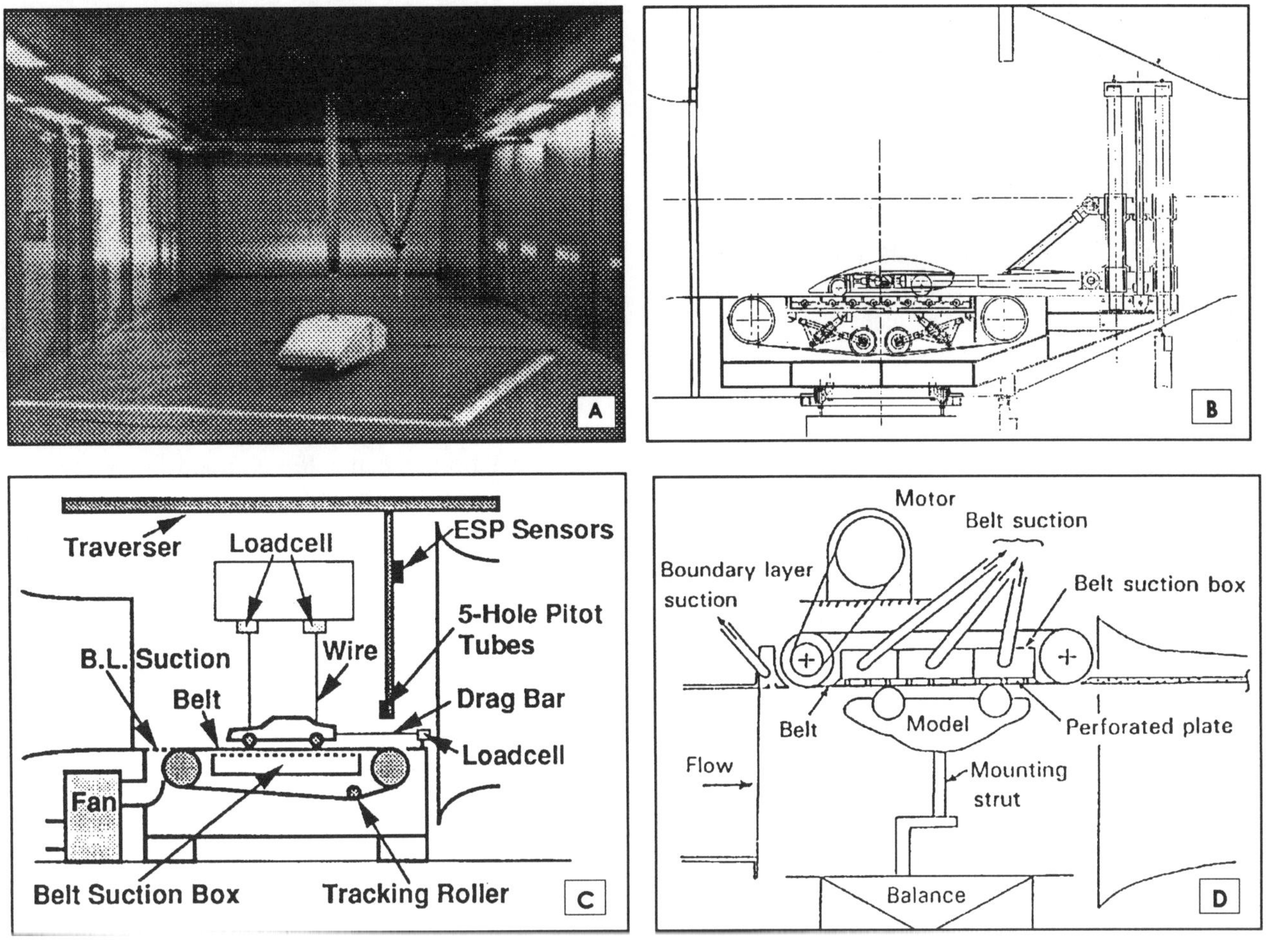

Fig.4 - Examples of moving belts for reduced scale models:
Fig.4A - Imperial College (ref.11)
Fig.4B - Sardou (ref.12)
Fig.4C - Nissan (ref.14)
Fig.4D - Jari (ref.13)

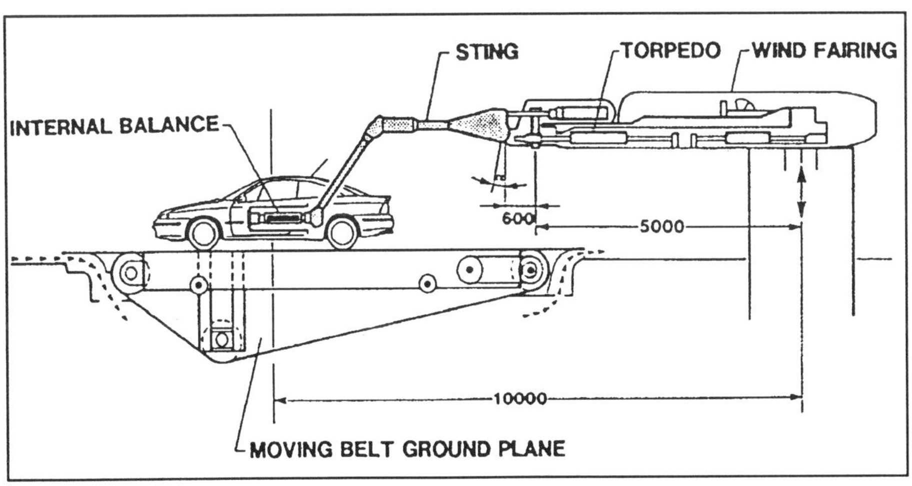

Fig.5 - DNW wind tunnel.
Full-scale moving belt and supporting sting for a full-scale car (from ref.15).

2. The second moving belt is one built by Fiat (17). In this case (when needed), the moving belt is installed over the test section floor. It is therefore necessary to build a new floor all around the belt, which is about 500 mm higher than the ordinary floor. In this case, with the balance under the test section floor (as in every automotive wind tunnel), the car can be supported by 4 struts which come from the balance, turn around the belt sides and support the car through the rockers.

It is therefore clear that in both cases, DNW and Fiat, these tests require an important preparation of the car and moreover, at least for Fiat, an additional time to install the moving belt, the new floor, etc.
This means that tests of this type can be carried out from time to time, but certainly cannot be done as a routine everyday work.
However, it is important to acknowledge the contribution these two facilities have given in understanding the various aspects of ground effect simulation. At the same time, an extensive knowledge of the flow conditions existing , for ground vehicles, in the presence of floor motion and rotating wheels has already been achieved.

4. Tangential blowing of the boundary layer.

A further system, which has been used as an alternative to the boundary layer suction or to a moving belt system, consists of "blowing" the boundary layer tangential to the floor, upstream to the car (fig. 6) ref. (16), (18), (19).
In practice, since a boundary layer is characterised by a lack of flow volume and momentum, one can consider the possibility of restoring the missing flow volume and/or momentum by injecting a thin high-speed jet tangent to the test section floor.
Some systems of this type have already been built and some of them are already operational in full-scale wind tunnels (DNW, Lockheed).
From a theoretical standpoint, this system should be better than the suction system as it tries to restore in the boundary layer, the volume and momentum which is missing.
However, also in this case some limitations and problems exist:

- Blowing is only effective for a limited distance, after that the boundary layer starts again to increase.
- It is difficult to decide in advance how much to blow. If blowing is too weak a velocity deficit in the boundary layer still remains, if blowing is too strong one risks to "blow" the car front wheels, therefore increasing their aerodynamic drag.
- No blowing system exists which can provide a good flow simulation along the whole underbody.
  It seems that the best trade-off is to use a blowing level which produces a flat velocity profile immediately in front of front wheels. Afterwards the velocity profile starts to deteriorate, that is to say a new boundary layer grows under the car.
- A multiple blowing (along transversal lines) might be considered, but the setting up of this system become more and more complex.
- Last but not least, as for the suction, additional means for rotating the 4 wheels have to be provided.

5. Tests in hydrodynamic basins.

A further technique for a ground effect simulation on full-scale cars has been proposed, in recent years. It consists in towing a car along the floor of a hydrodynamic basin through a dynamometric carriage (20), (21), (22) and fig. 7.
The advantage of this technique is obvious.
The correspondence with the road is almost perfect. Furthermore, due to the water high density (compared with the air), tests are carried out at very low speeds and this makes flow visualisations easy to perform (as an example 2.7 m/s in water corresponds to 38.9 m/s = 140 Km/h in air).
The inconveniences related to this technique are numerous: In particular, the long preparation on the car and the difficulties of doing modifications to the car under the water.
It is therefore doubtful that this system can be used in the future except, maybe, for comparison tests with aerodynamic wind tunnels or for visualisation tests which may become a bit easier in the water.

All that has been further debated during the round table at the end of the CNR-PININFARINA Workshop on "Wind Tunnel Simulation of Ground Effect", hold in Turin on May 12, 1994 (Chairman Dr. G. Sovran, Speakers Dr. J.Wiedemann, G. Carr, J. Vagt, M. Yoshida, K.Cooper, A. Cogotti),

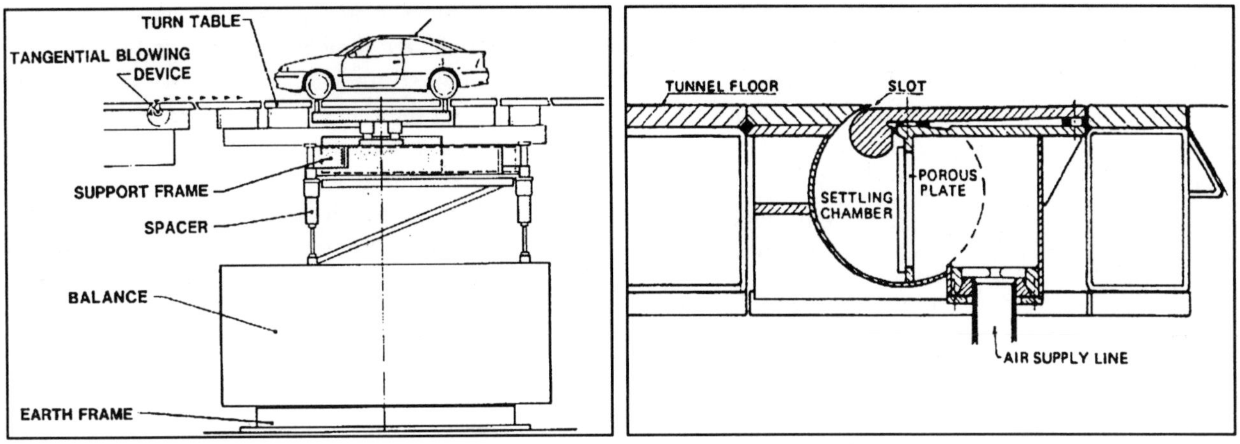

Fig.6 - DNW wind tunnel.
Tangential blowing system for full-scale cars (from ref. 16).

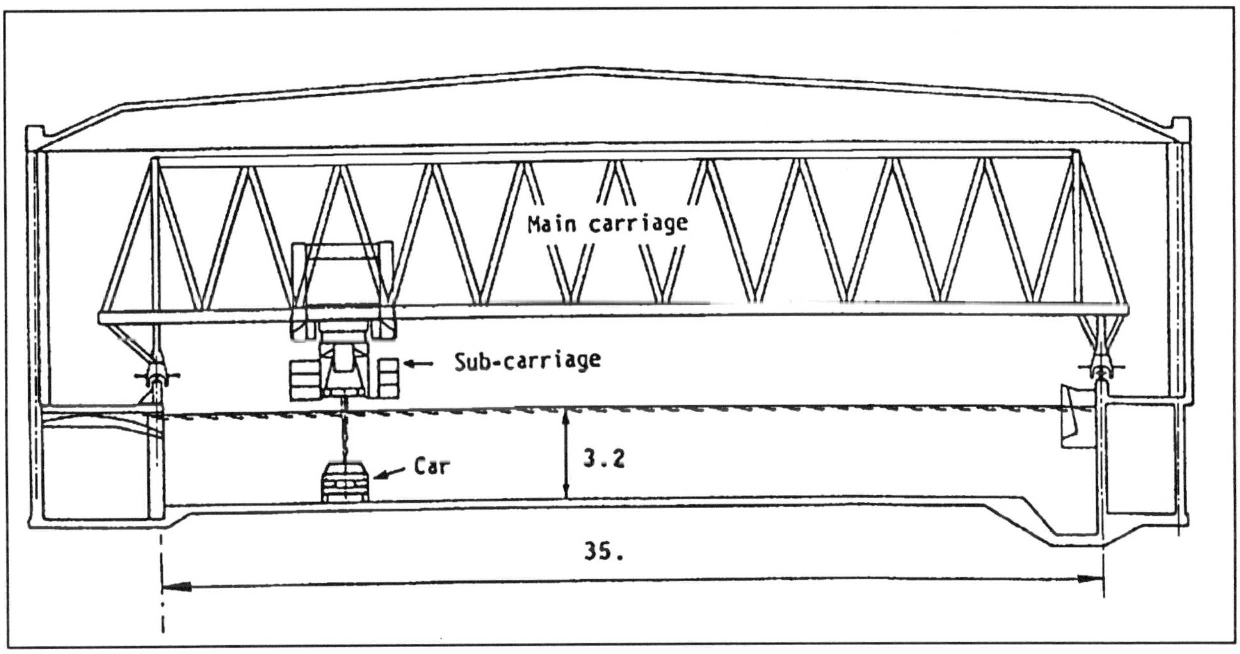

Fig.7 - Test basin of the SSPA Maritime Dynamics Laboratory for hydrodynamics tests on a submerged vehicle (from ref. 21).

[Participant list in Appendix 1; see also ref. (23), (24), (25), (1)].
The conclusions drawn at the end of the round table discussion have confirmed the results of the analysis of the existing ground effect simulation techniques , made during the first year of research, that is to say:

- To be really able to reproduce the kinematics and fluid dynamic conditions existing on the road in proximity of the ground, the boundary layer along the test section floor must be controlled along the full length of the car underbody.
  Consequently :
- "Simple" Systems like "Basic Suction" or "Tangential Blowing" are not adequate, at least in principle, because they can control the boundary layer at a given station only. And furthermore the criterion to adopt for their operation (that is to say, how much "to blow", how much "to suck"), is always questionable.
- The system which can better reproduce the kinematics and fluid dynamic conditions existing on the road in proximity of the ground, also in case of yaw tests, is that based on a moving belt.
- On the other hand, the choice of a moving belt of big size presents problems of practicality due to the difficulties of coexistence with the aerodynamic balance.

That is why, considering all that, and also considering that this work is made within the framework of a joint research program CNR-Pininfarina, a decision was taken by Pininfarina , during the first year of research, to reconsider the whole matter in a more comprehensive way.
In particular, considering that none of the existing systems seemed to be fully satisfactory either from an aerodynamic standpoint or for practical reasons, it was decided that it was worth making the effort to design something new, even more complex, that could reach a better trade-off between a good flow simulation and an acceptable practicality level.

<u>PART II</u>

DESCRIPTION OF THE NEW GROUND EFFECT SIMULATION SYSTEM.

The new Ground Effect Simulation System (GESS) which has been realised in the wind tunnel of the Pininfarina Aerodynamic and Aeroacoustic Research Center is based on the following considerations:

- The most practical way to support a car is through its 4 wheels. Four additional slender vertical struts can be used to support the car through the rockers. Any other supporting system creates severe practical problems not acceptable in the routine work. Therefore, the new system was designed so that cars can continue to be supported through the wheels, directly on the balance pads, already in use for conventional wind tunnel tests.

- At the same time, it was decided that the new system has to be based, as much as possible, on a moving belt. That's because this device can provide, in principle, the closest boundary condition at the floor level along the whole car underbody compared to on-road conditions.

- As a consequence of these two points, it was decided that the best choice was a moving belt of reduced width, which extends between the car wheels.
  After an analysis of the typical tracks of existing cars, a moving belt width of 1 m was chosen.
  This means that small cars with very narrow tracks cannot be tested. However, all the existing cars, from segment "1" on (starting, for instance, from Fiat Uno or Punto , Peugeot 106 or 205, Opel Corsa), can be tested.
  Fig. 8 shows the layout of a segment "1" type car over the belt, in case of a full-scale car as well as scale models from 1:2.5 to 1:5.

- A further consideration concerns the installation of the moving belt over the test section floor, as for the Fiat moving belt. This type of installation is critical from a practical standpoint, as the preparation time is relatively long. Therefore, it has been decided that the moving belt has to be installed inside the balance and its turntable. In this way, the moving belt is integrated in the balance and permanently installed. Plus the belt itself is at the same level of the wind tunnel floor.
  The preparation time will be almost nil and this should results in a test time not too different from conventional tests.

- The former decision limits the belt overall length. The belt length must be a bit shorter than the turntable diameter. However, since the moving belt can rotate with the balance and the turntable, it will be possible to perform measurements in side wind conditions. This is an additional

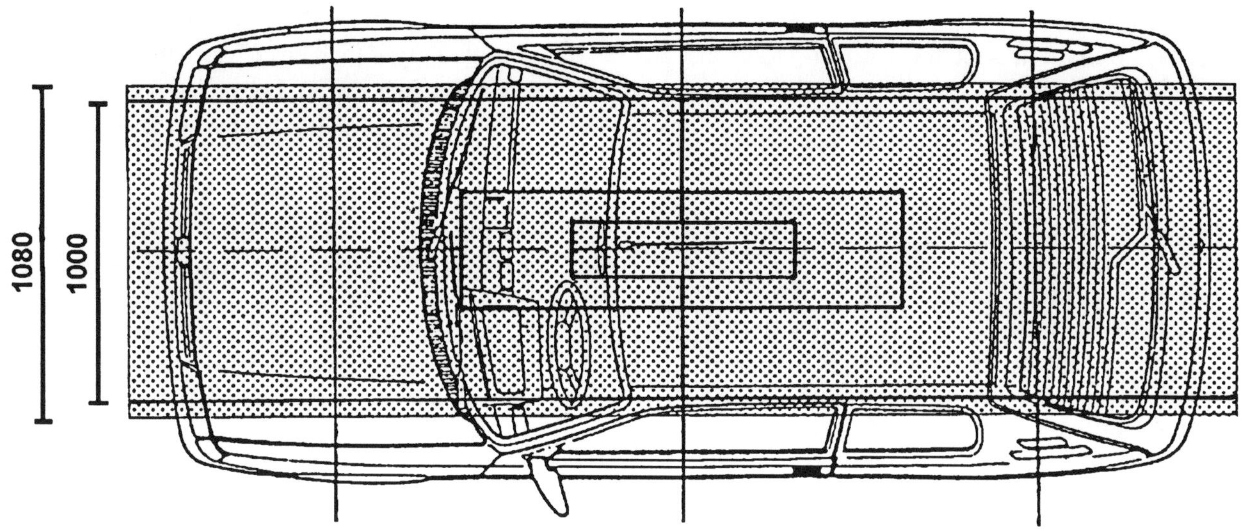

Full-scale car

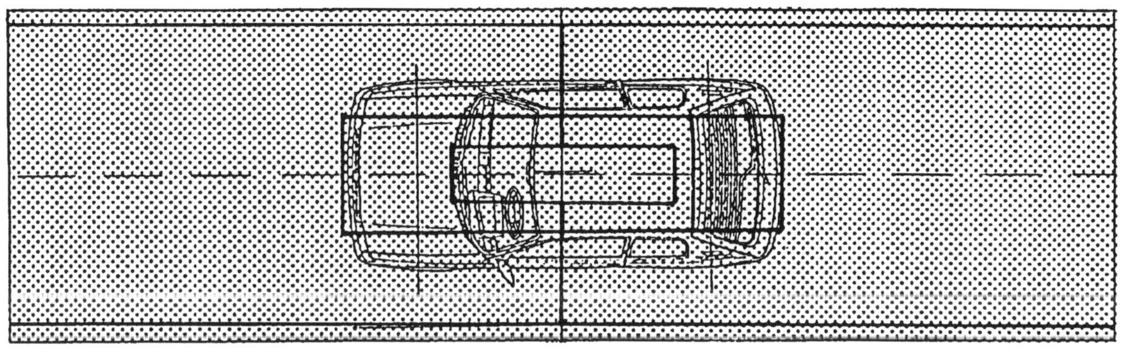

1:2.5 scale model

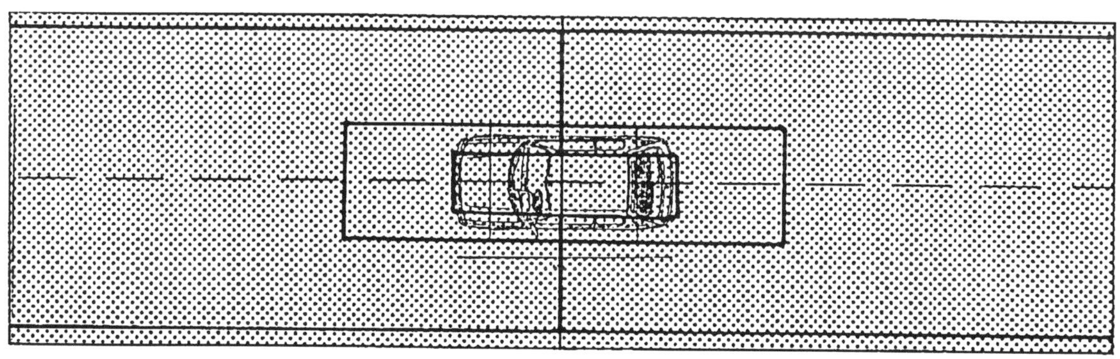

1:5 scale model

Fig. 8 - Car & moving belt layout in the case of:
a) full-scale car : moving belt "between the wheels"
b) 1:2.5 scale model over the belt
c) 1:5 scale model over the belt

feature, which can be rather important considering the contemporary effect of the wheel rotation in presence of side wind .

- Since the belt is relatively narrow and short, it is necessary to add some other system to "cover" the floor areas which are not affected by the moving belt.
  To achieve this aim, a Tangential Blowing System (TBS) has to be installed at the end of the nozzle. This blowing system provides a flat velocity profile of the boundary layer immediately in front of the moving belt. Then the moving belt carries this velocity profile along the underbody. In this way, it is not necessary to "push" the blowing system too much because its task is to cover only the area upstream of the moving belt and not the full test section floor. The risk of "blowing the front wheels", in this way, is totally avoided.

- It is well known from previous experience, both Pininfarina and other wind tunnels, that wheel rotation is a key-factor to produce good simulation of the flow under a car.
  Therefore, additional means which can provide the wheel rotation are integrated in the balance upper frame.

- Last but not least, keeping in mind the difficulty of carrying out modifications to the car underbody, it was decided to integrate in the balance pads a Lifting System (LS) . It can lift the car over the floor , without moving the car from the balance pads

The moving belt was designed and built during the first year of the research program (1992-93).
During the second year of this program, the moving belt was carefully tested in an "over the floor layout" (1). In the meantime, the necessary modification to the balance and to the turntable was designed.
At the end of the second year of research (summer 1994), the upper frame of the balance was replaced by a new one built on purpose and the moving belt was installed in the new balance/turntable system.
At the beginning of the third year of research (Sept.-Oct. 94), the blowing system, as well as the system for wheel rotation, has been installed.
After a careful set-up of these various subsystems , the new Ground Effect Simulation System (GESS) has been officially put in service in the second half of 1994.

An overall view of its final layout is shown in fig. 10A. In summary it includes (figs. 10B, C, D):

- A Moving Belt (MB) between the wheels
- A Tangential Blowing System (TBS) in front of the MB
- Two Basic Suction Systems .
  The first (BSS1) in front of the TBS , the second already existing (BSS2) , between the TBS and the MB.
- A System for the Wheel Rotation (WRS), integrated in the balance pads.
- A System for Lifting the Car (LS) , which is also integrated in the balance pads.

Details on the GESS subsystems and some results of the tests already made in the years 1993-94 are reported in the following paragraphs.

THE MOVING BELT

The Moving Belt (MB) is obviously the most important subsystem. Its main characteristics and performances are summarised hereunder.
The belt has a width of 1.00 m and a length of 3.8 m (between roller axes). The "working" part of the belt , on the turntable surface is about 3.9 m long.
The machine's overall width and length are only a bit bigger, see fig 9.
These dimensions have been chosen on the base of typical car tracks and of the existing turntable diameter.
The overall height of the moving belt is 320 mm only, and its total weight is also very low (less than 1700 kg) .
This means that the design of this new moving belt is very compact with only two rollers of relatively small diameter.
The front roller is the steering roller while the rear roller provides to drive the belt.
The compactness of the whole system is important for two reasons:

- First, the moving belt has to be integrated in the balance/turntable system , therefore its size, including its height, must be necessarily reduced.
- Second , this reduced height was important to allow the use of the moving belt in an "over-the-floor layout" during the second year of the research program. That possibility was essential to check the performances of the moving belt before starting any expensive modification to the balance and to the turntable.

Fig.9A - Overall view of the moving belt built in the first year of research, during the final assembly.

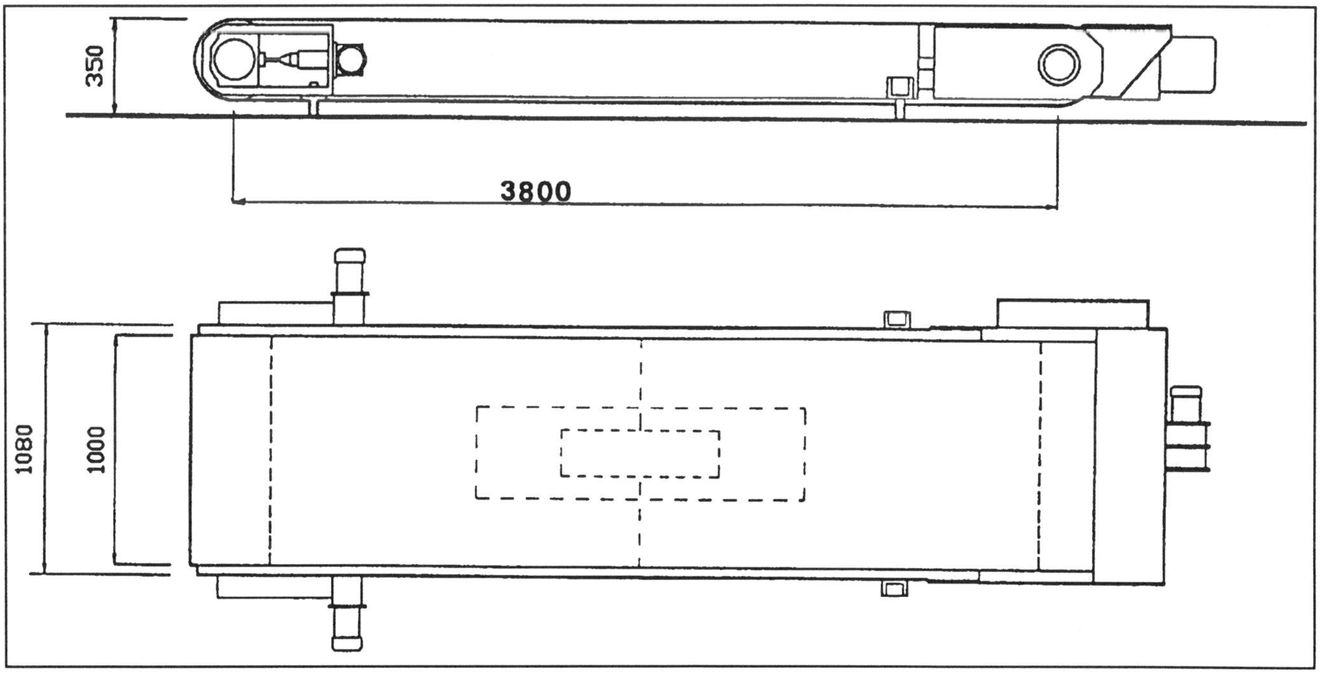

Fig.9B - Schematic layout of the moving belt.

Max. speed of the belt is 55 m/s (200 k/h), with a possibility of increasing this speed to 70 m/s , should that become necessary in the future.
The moving belt is furthermore equipped with 5 independent chambers , located immediately under the belt. Through them, a distributed suction can be applied to the belt to avoid a belt lift in case of strong negative pressures under the car. And anyway the belt lift is monitored, at present, by 3 displacement sensors, located in the front and in the center of the belt.
Also the belt temperature is kept under control by a water cooling system and monitored by some temperature sensors.

THE TANGENTIAL BLOWING SYSTEM

The tangential blowing system (TBS) is located 2.52 m ahead of the moving belt, on the floor nozzle (see fig. 10B).
Its main characteristics are the following:
- jet width = 4 m
- jet thickness = adjustable from 0 to10 mm.
- jet velocity = adjustable from 0 to 200 m/s

THE BASIC SUCTION SYSTEMS

At present, 2 Basic Suction Systems (BSS1 and BSS2) are available.
The first (BSS1) is located .68 m in front of the TBS. (see fig.10 B) . Its inlet is 4 m wide and about 25 mm. high.
The inlet of the BSS1 is on the floor of the nozzle where the static pressure coefficient is 0.4.
The BSS1 outlet is located in an area of Cp static $\cong 0$.
Therefore there is an automatic suction of the incoming boundary layer without any need to install additional fans.
The main aim of BSS1 is to remove as much as possible of the floor boundary layer so that the tangential blowing can be kept at a minimum level.
The second (BSS2) is in reality the old system built in 1980 and already mentioned at the beginning of this paper.
Tests made in the second half of 1994, have clearly shown the importance of using BSS2 together with BSS1 and the TB system , to achieve an almost perfect velocity profile in the boundary layer approaching the moving belt.

THE WHEEL ROTATION SYSTEM (WRS)

A system for the rotation of the car wheels has been designed and integrated in the balance pads. It is made, for each pad, of a roller having a diameter of 200 mm. and a width of 50 mm. (fig. 10C).
Max. tangential speed of the roller is 55 m/s.

THE CAR LIFTING SYSTEM (LS)

A system for lifting the car through the balance pads has also been designed and integrated in the balance system. It is made by 4 + 4 hydraulic lifters, which can lift the car through the balance pads and /or the rockers by 290 mm. This height , added to a typical car-to-ground clearance of 100 - 200 mm, is enough to work easily on the car underbody without removing the car from the balance (fig.10D).

THE BALANCE

The balance upper frame has been modified by Pfister so as to be able to accommodate the moving belt and the other subsystem (RWS and LS) previously mentioned. In particular, the balance has been transformed from a 6-component to a 7-component-type by installing a new moment frame with 4 load cells for the vertical loads. Furthermore, the old load cells for the drag and side forces (horizontal loads) have been replaced by new load cells . The "dead-weights" existing in the original balance to increase the accuracy of measurement, have been removed because they are no longer necessary after the new load cell installation.
The calibration made after these changes has shown that the balance now has an accuracy, better than ±20 g. (0.2 N) on the drag component and similar accuracy for the other components.

PART III

PRELIMINARY RESULTS

The moving belt was designed and built during the first year (92-93) of the CNR-PFT2 Research Program.
In the second year (93-94) , the moving belt has been carefully tested in an "over-the-floor layout" .
To do that a new floor , 350 mm height , was built over the original test section floor , around the four sides of the moving belt.
The existing Basic Suction System was modified and a Scoop was also created ahead of the suction area.

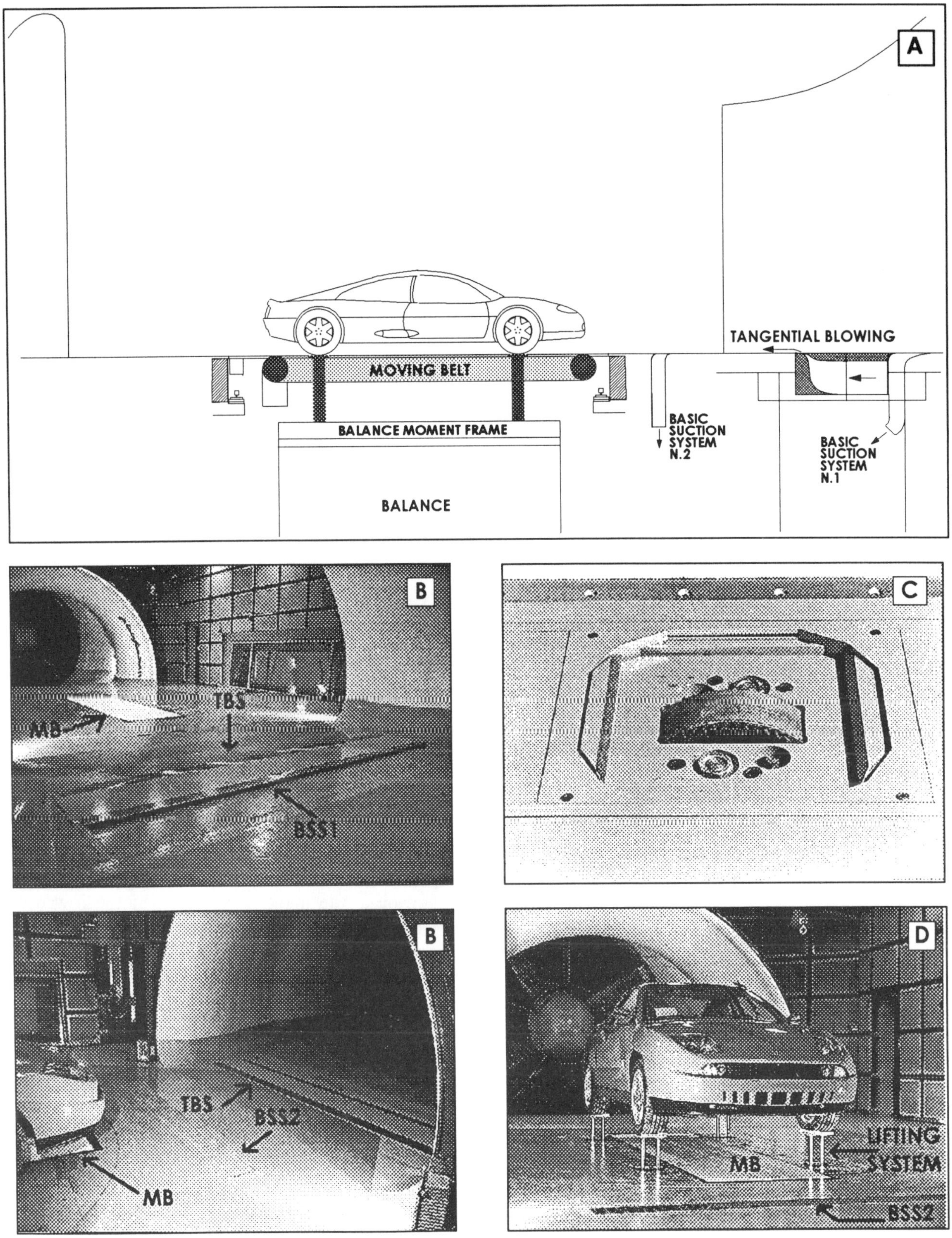

Fig.10 : A - Final layout of the Pininfarina Ground Effect Simulation System.
B - Overall view of the Moving Belt, the Tangential Blowing and the Suction Systems
C - The rollers for the Wheel Rotation (WRS)
D - The car Lifting System (LS)

The layout of this temporary "over-the-floor" solution is shown in fig. 11A.
The result of this new floor optimization work is summarised by the velocity profiles, (fig. 11B) where the two conditions , Moving Belt "On" and Moving Belt "Off" are compared.
The velocity profile with Moving Belt "On" is rather good , but not perfect, as a small velocity deficit is still present.
This was considered acceptable for the "over-the-floor" solution; however, it confirmed that , for the final solution , i.e. the moving belt integrated into the turntable, a tangential blowing system was necessary to be able to produce an uniform velocity profile in front of the belt.

Using this "over-the-floor" layout tests have been performed on car models of various scales (1:5, 1:3 and 1:1).

Tests on scale models

A 1:5 scale model is shown in fig.12, and a 1:3 scale model in fig. 13.
In these tests, wheels are rotated by the belt .
The car models were supported on the balance by four side arms through the four wheels. This way of supporting the model is new for scale models. Usually scale models tested on moving belts are supported from the top and/or from the back.
This new supporting system has several advantages, namely:

- It is rather compact , and easy to install.
- The four side arms interfere with the flow around the model in an area (the central part of the four wheels) where the local flow is already highly disturbed and turbulent.
- The flow over the roof , which is very important for the wake development or for the efficiency of rear wings and spoilers, is not disturbed at all.
- The area behind the model is totally free, and that is also important so as not to preclude flow investigations in this area with the "14-hole probe" technique (26) or with a Laser Doppler Velocimeter (27).

Its main drawback is the stiffness, which may be not enough in case of heavy models or strong vertical aerodynamic loads.

Some results measured on these scale models , are reported in fig.15 to show the difference between the two test conditions, "S" i.e. Stationary wheels and floor and " W+G" i.e. rotating Wheels and moving Ground.

In general , it can be seen that , in this specific case, due to the combined effect of floor motion and of wheels rotation, Drag coefficients are a bit lower , while Lift coefficients are always lower. In particular the 1:3 scale model car has a greater drag sensitivity which can be due to its relatively lower ground clearance as well as to its relatively wider tires.
These differences can also be explained by the analysis of the model wakes measured in the two different test conditions "S" and "W+G" (figs. 14A and B).
In particular , in the "W+G" condition, the flow in the underbody area is accelerated by the wheel rotation and at the same time is not more slowed down by the stationary floor. These effects are mainly important for the 1:3 scale model car.
Therefore, in the lower part of the wake maps velocities are higher, as well as the total and static pressures. Mainly due to this increase of total pressure CpTot, drag is reduced , according to the expression (26):

$$Cd*S = \int(1-CpTot)*dS - \int(1-Vx/Vw)^2*dS + \int(Vy/Vw)^2*dS + \int(Vz/Vw)^2*dS$$

The increase of down force (negative lift) caused by the floor motion and by the wheel rotation cannot be explained only by the analysis of the wake flow. In fact, lift depends also (or mostly) on the static pressure distribution along the underbody, and these pressures cannot be seen directly on the wake maps.
However, the higher velocities in the lower wake indirectly confirm the existence of accelerated flows under the car which, in turn, means lower static pressures ,that is to say, more down force.

Tests on full-scale cars

Results reported hereunder were measured during an extensive test session carried out on a same car . Therefore it is likely that these results are somewhat "specific" for this car. They should be taken only as an indication of the differences which can be expected when using the GESS.
About 80 different configurations were tested. The car body was modified mainly in the underbody area (fairings of various shape and size), in the front and rear bumper , as well in other parts like wheel disks, rockers, rear spoilers and wings.

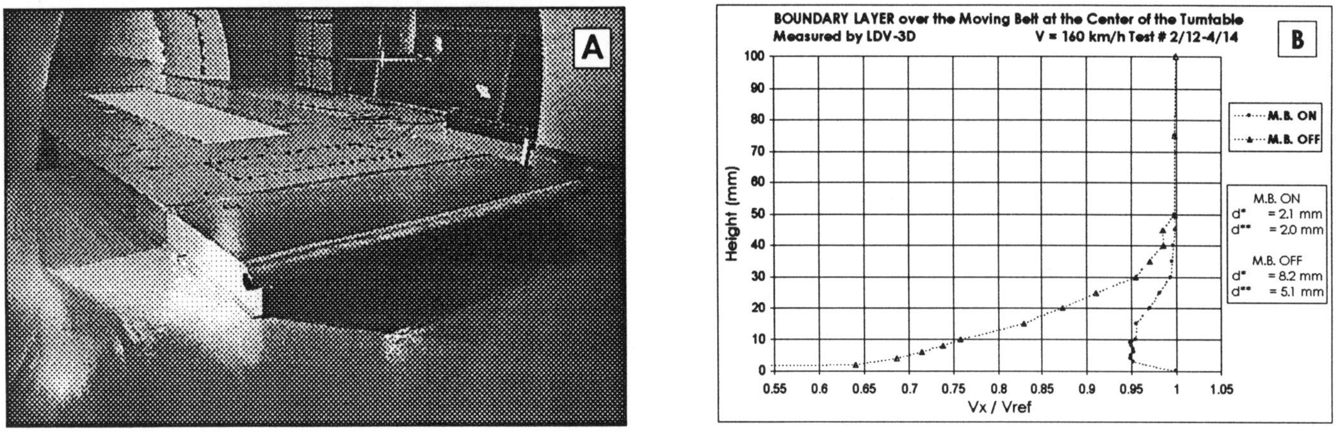

Fig.11: A - Layout of the moving belt in the wind tunnel over the test section floor, with the new floor all around thebelt and boundary layer suction in front of the belt. Temporary installation during the second year of research (1993-94)
B - Velocity profile over the belt at X=turntable center, with the moving belt ON and OFF, basic suction ON, no tangential blowing

Fig.12 - A 1:5 scale model teste with moving floor and rotating wheels

Fig.13 - A 1:3 scale model tested with moving floor and rotating wheels

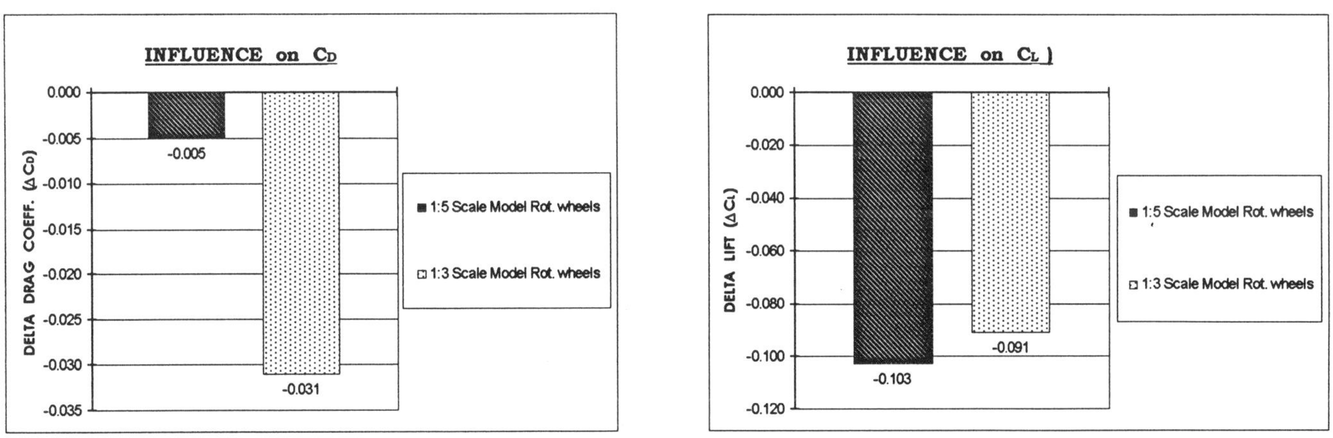

Fig.15 - Change of drag and lift coefficients caused by moving floor and wheel rotation (1:5 and 1:3 scale models)

FIXED FLOOR & STATIONARY WHEELS **MOVING FLOOR & ROTATING WHEELS**

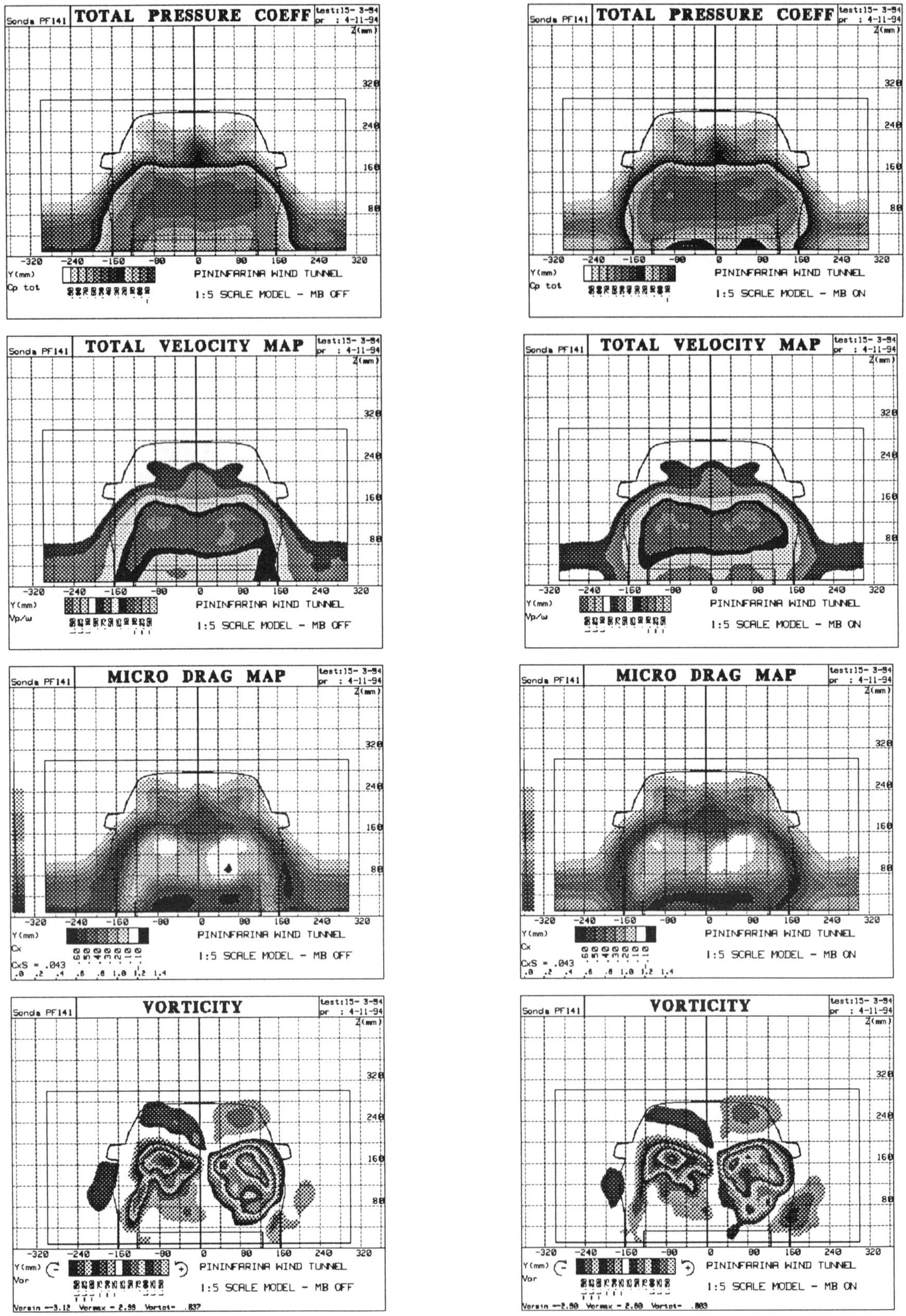

Fig.14A - A wake image behind a 1:5 scale model in the two conditions of:
- moving floor and rotating wheels
- fixed floor and stationary wheels

FIXED FLOOR & STATIONARY WHEELS **MOVING FLOOR & ROTATING WHEELS**

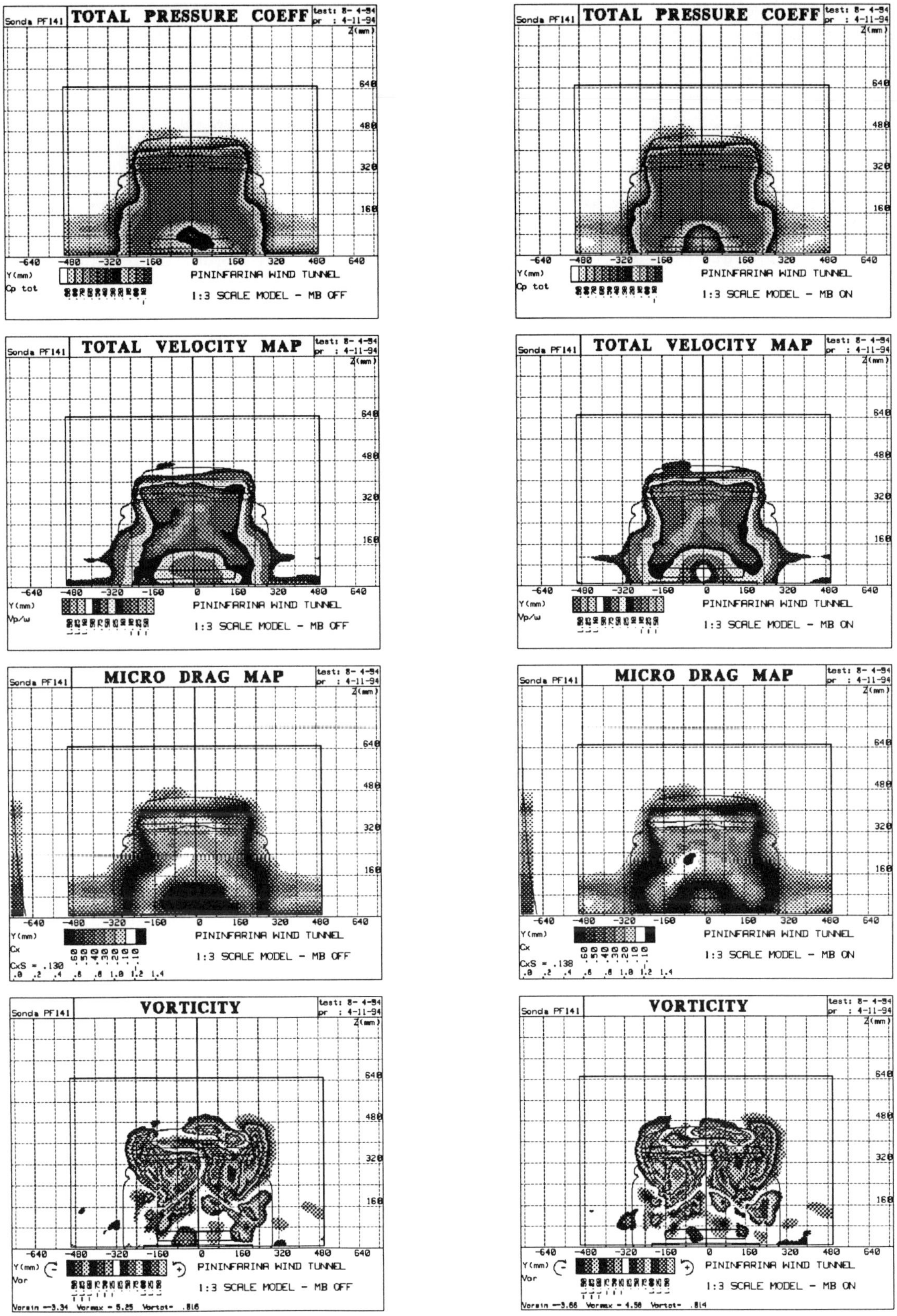

Fig.14B - A wake image behind a 1:3 scale model in the two conditions of:
- moving floor and rotating wheels
- fixed floor and stationary wheels

Each configuration was tested at least in the two conditions , "S" and "W+G".
In some case also the intermediate conditions , "W" i.e. "rotating Wheels only" and "G" i.e. "moving Ground only" , were tested. Some statistics is reported also for these tests , however due to the limited number of available data , these indications are less reliable.
Fig. 16 A shows the distribution of the Cd differences, Cd, between the two test conditions "W+G" and "S", measured for the various car configurations.
One can see that, on average , the "W+G" test condition gives a ΔCd value which is lower by .003 , that is to say a ΔCd = -.003.
However these Cd differences may vary between limits (-.011 to +.005) which are rather large.
And that confirms once more (see also 28, 29) that the combined effect of the wheel rotation and of the ground motion cannot be taken in any case , as a constant term to subtract to the Cd value measured in "Stationary" condition.
In reality the combined effect of Wheel rotation and ground motion is the result of two effects , which produce , for this specific car, a ΔCd = -.011 to 0 for the wheel rotation only , and a ΔCd = +.006 to +.012 for the ground motion only.
Similarly for the lift (total, front and rear), Fig.16A/B shows the distribution of the differences, ΔCL, ΔCLF and ΔCLR , between the two test conditions "W+G" and "S", measured for the various car configurations.
On average the differences are the followings:
ΔCL = -.015 (range -.032 to +.021)
Δ CLF = -.005 (range -.016 to +.007)
ΔCLR = -.010 (range -.021 to +.010)

The wheel rotation alone brings differences of:
ΔCL = +.023 (range +.000 to +.038)
Δ CLF = +.012 (range -.002 to +.023)
ΔCLR = +.011 (range +.002 to +.015)

while the ground motion alone has given
ΔCL = -.010 (range -.000 to -.020)
ΔCLF = -.006 (range -.000 to -.012)
ΔCLR = -.004 (range -.000 to -.007)

And the comments already made for the Drag Coefficient can be applied to the Lift coefficients and probably to the others coefficients not mentioned in this paper.

ADDITIONAL COMMENTS ON THE TEST RESULTS

A general consensus probably exists on the fact that the combined effect of Wheel rotation and Ground motion is not a "constant term" which can be added or subtracted to the values measured in "Stationary" condition.
However , the analysis of these first results may suggest the conclusion that , for a conventional passenger car, without any important "ground effect", the differences which can be expected with the use of the GESS, are small and apparently not very important. This conclusion may change completely, when the following points are taken into consideration:

1. The small modifications which are usually carried out on a car during the conventional aerodynamic development work (such as small spoilers, lips, rocker fairings, underbody covers, etc.) , often reduces the drag by only some counts, for instance ΔC_d = -0.003.
   These parts have a cost which is often significant, furthermore, they may not be well accepted by the stylists. Therefore, the ratio cost-benefit may be critical and the decision whether to put these parts into production is also rather difficult.
   The same parts tested using the GESS, that is to say with a very realistic simulation of the road condition, give C_d values slightly different compared with the static test. This small C_d difference between the "stationary" and the "dynamic" test, is often in the same order of magnitude of the C_d improvement given by a modification during the "stationary" test. This means that a C_d improvement measured during a "stationary" test, may "disappear" when the same test is repeated using the GESS. And that can completely change the decision whether to put these parts into production or not.
   Some typical examples are:
   - a small rear spoiler, which gave a ΔC_d = -.003 in “ stationary” conditions , gave only -0.001 with moving ground/rotating wheels. The same spoiler re-tested on the same car with a different underbody fairing gave a ΔC_d = - 0.004 in the stationary test, and 0 with moving ground/rotating wheels.
   - a rear wing, which gave a ΔC_d = - 0.007 with fixed ground/fixed wheels, gave only -0.003 with moving ground/rotating wheels;

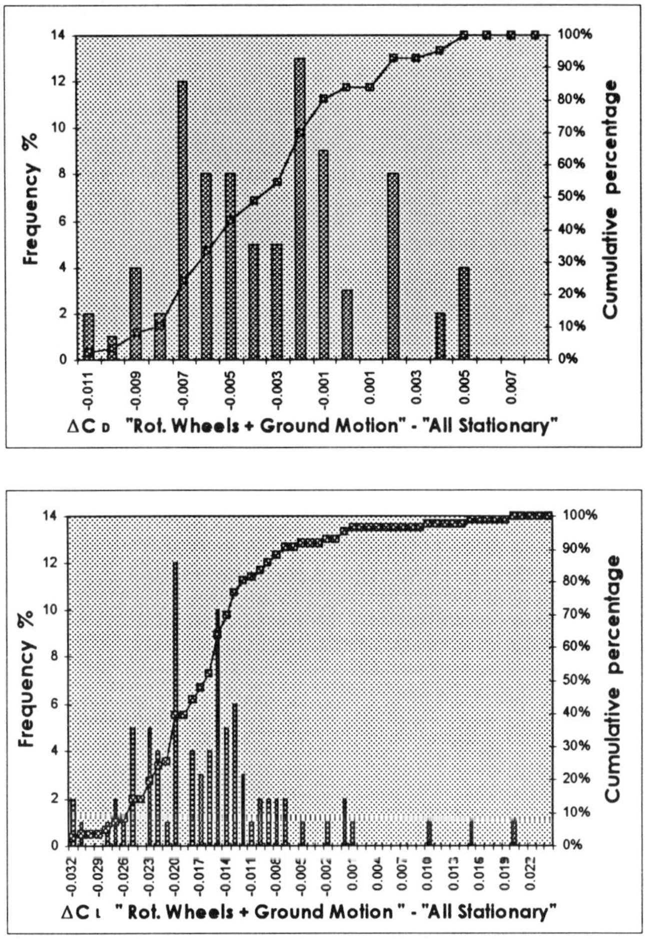

Fig.16A: ΔC_D and ΔC_L values between the two configurations:
"Rot. Wheels + Ground Motion" and
"All stationary"

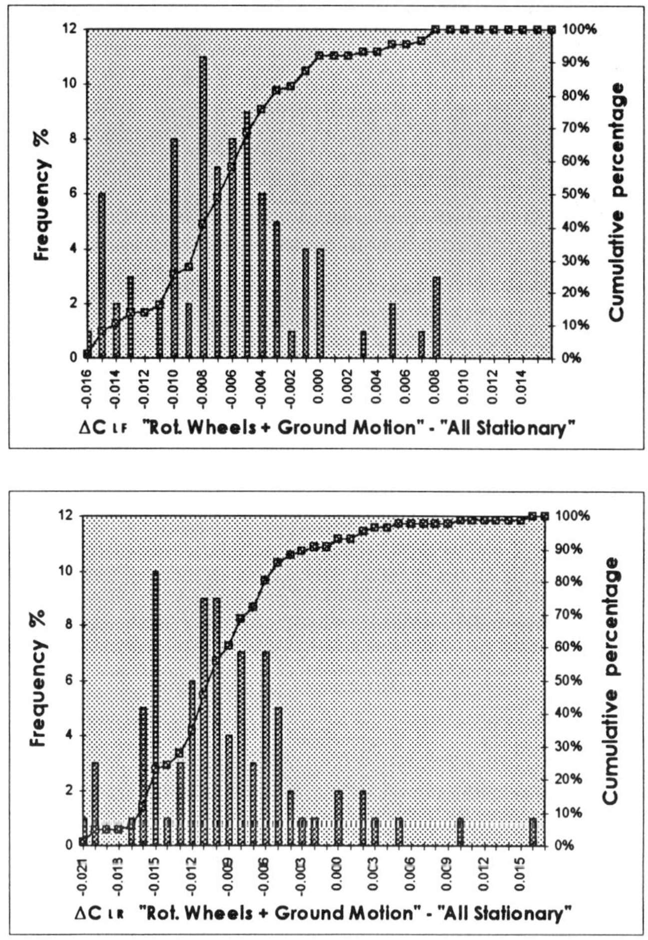

Fig.16B: ΔC_{LF} and ΔC_{LR} values between the two configurations:
"Rot. Wheels + Ground Motion" and
"All stationary"

- an underbody fairing, which gave a poor ΔC_d = -0.002 with fixed ground/fixed wheels, gave -0.005 with moving ground/rotating wheels;

It is clear that in these cases the final decision about these parts will be quite different . In particular the cost of parts which do not give any true aerodynamic benefit will be saved.

2. A second even more important point is that, if the car aerodynamic development work is made using the GESS, the overall final result can be rather different. In fact it will be the sum of a number of small modifications, each one made in a slightly different way. This aspect will be explored more in detail in the future.

3. A third important area which will benefit from, this improved simulation technique, is the cooling system. It is known (23, 28) that the efficiency of this system is dependent on the characteristics of the local flow (velocity module and flow angularities). And these characteristics are somewhat dependent on the Wheel rotation and on the relative ground motion.

4. A further subject which may benefit from the GESS is the correlation between wind tunnel and road.
   Up to now the differences between road tests and wind tunnel results have often been justified by the different test conditions (fixed ground/fixed wheels) existing in the wind tunnel..
   In the future it will be possible to separate these wheels/ground effects from others effects such as nozzle or collector interference, blockage correction, etc. which are wind tunnel specific.
   A correlation of the wind tunnel results (measured with the GESS) and road results will be carried out in the future, by starting a co-operation with automotive Companies which are able to perform coast-down tests with a reasonably good level of accuracy.

CONCLUSION

A "Ground Effect Simulation System" of new design was designed and built in the Pininfarina Aerodynamic and Aeroacoustic Research Center in the framework of a joint research program CNR-Pininfarina .

This new system has the aim of reaching an excellent qualitative level in the simulation of the flow conditions existing under a car running on the road, while being at the same time almost "transparent " to the wind tunnel user.

The major benefits which are expected are the following:

- a more accurate aerodynamic development work on every new car model by working in full scale, on its underbody and on its cooling system.
- a better cost-estimate of the aerodynamic benefit which can be given by add-on parts, spoilers, wings, fairings etc.
- a possibility of a better correlation with the road tests as well a further chance of understanding the interference effects existing in the wind tunnel.

REFERENCES

1) A. Cogotti, "Ground Effect Simulation for full-scale cars in the Pininfarina Wind Tunnel", CNR-Pininfarina Workshop on Wind Tunnel Simulation of Ground Effect, Turin, May 12, 1994

2) A. Cogotti, "La Galleria del Vento Pininfarina", Status Report - V Meeting EADE, Paris, October 1987

3) A.Cogotti, "Prospects for Aerodynamic Research in the Pininfarina Wind Tunnel", FISITA paper 905149, XIII FISITA Congress, Torino, May, 1990

4) J.E.Hackett, J.E. Williams, J.B. Baker, S.B. Wallis, "On the Influence of Ground Movement and Wheel Rotation in Tests on Modern Car Shapes", SAE paper 870245, Intl' SAE Congress & Exposition, Detroit, February 1987

5) A. Berndtsson, W.T. Eckert, E. Mercker, "The Effect of Groundplane Boundary Layer Control on Automotive Testing in a Wind Tunnel", SAE paper 880248, Intl' SAE Congress & Exposition, Detroit, February - March 1988

6) W. Eckert, N. Singer, J.D. Vagt, "The Porsche Wind Tunnel Floor-Boundary-Layer Control - A Comparison with Road Data and Results from Moving Belt", SAE paper 920346, Intl' SAE Congress & Exposition, Detroit, February 1992

7) G.W. Carr, W. Eckert, "A Further Evaluation of the Ground-Plane Suction Method for Ground Simulation in Automotive Wind Tunnels", SAE paper 940418, Intl' SAE Congress & Exposition, Detroit, February-March 1994

8) B.Bonis, F.B. Quagliotti, "Methodology for Testing Vehicles in Ground Effect", 6th Colloquium on Industrial Aerodynamics, Aachen, June 1985

9) B.Bonis, F.B. Quagliotti, "A Study of the Validity of Fixed-Ground Testing for Formula 1 Racing Cars", SAE paper 860092, Intl' SAE Congress & Exposition, Detroit, February 1986

10) B. Bonis, F.B. Quagliotti, G.P. Dallara, "Is the Moving Belt Ground Plane really necessary for the Automobile Testing in the Wind Tunnel?", SAE paper 870719, Intl' SAE Congress & Exposition, Detroit, February 1987

11) P.W. Bearmann, D. De Beer, E. Hamidy, J.K. Harvey, "The Effect of a Moving Floor on Wind-Tunnel Simulation of Road Vehicles", SAE paper 880245, Intl' SAE Congress & Exposition, Detroit, February-March 1988

12) M.Sardou, "The Sensitivity of Wind-Tunnel Data to a High-Speed Moving Ground for Different Types of Road Vehicles", SAE paper 880246, Intl' SAE Congress & Exposition, Detroit, February-March 1988

13) T.Imaizumi, "The Effect of Moving Ground on the Aerodynamic Characteristics", JSAE magazine, July 1991

14) I.Yamaguchi, K. Nakatsuru, M. Takagi, Y. Shimpo, "The Effect of a Moving Belt on Underfloor Airflow in a Scale Wind Tunnel Test", JSAE paper 902258

15) H.J.Emmelmann, H. Berneburg, "Aerodynamic Development of the Opel Calibra in the DNW", from "Ten Years of Testing at DNW - 1980-1990", Colloquium at the occasion of ten years operational phase of DNW, Noordoostpolder, October 1990

16) E.Mercker, J. Wiedemann, "Comparison of Different Ground Simulation Techniques for Use in Automotive Wind Tunnels", SAE paper 900321, Intl' SAE Congress & Exposition, Detroit February-March 1990

17) C.Berta, "Full-Scale Moving Belt in Fiat Aerodynamic Wind Tunnel", FISITA paper 905147, XIII FISITA Congress, Torino, May, 1990

18) J.E.Hackett, R.A. Bolest, "Moving-Ground Simulation by Tangential Blowing", AIAA paper 78-814, 10th Aerodynamic Testing Conference, San Diego, CA, April 1978

19) P.Müllenbach, K. R. Deutenbach, "Influence of Boundary Layer Control Systems to the Flow Fields around Passenger Cars", Aachen, 1989

20) M.L.Vorwaller, G.J. Germane, "Aerodynamic Drag Studies on Rolling Vehicles by Underwater Tow Testing", SAE paper 860091, Intl' SAE Congress & Exposition, Detroit, February 1986

21) L. Larsson, L. Hammar, L.U. Nilsson, A. Berndtsson, K. Ktutson, H. Danielson, "A Study of Ground Simulation-Correlation between Wind-Tunnel and Water-Basin Tests of a Full-Scale Car", SAE paper 890368, Intl' SAE Congress & Exposition, February-March 1989.

22) K.Aoki, H.Miyata, M. Kanai, Y. Hanaoka, M. Zhu, "A Water-Basin Test Techniques for the Aerodynamic Design of Road Vehicles", SAE paper 920348, Intl' SAE Congress & Exposition, Detroit, February 1992

23) J. Wiedemann, "On the Importance of Ground Simulation for the Optimization of the Underside of Passenger Cars", CNR-Pininfarina Workshop on Wind Tunnel Simulation of Ground Effect, Turin, May 12, 1994

24) G.W. Carr, "Some Experimental Comparisons Between Different Methods for Ground-Plane Boundary-Layer Control", CNR-Pininfarina Workshop on Wind Tunnel Simulation of Ground Effect, Turin, May 12, 1994

25) K.R. Cooper, "Approaches to Ground Simulation in North America", CNR-Pininfarina Workshop on Wind Tunnel Simulation of Ground Effect, Turin, May 12, 1994

26) A. Cogotti, "A Strategy for Optimum Surveys of Passenger-Car Flow Fields", SAE paper 890374, Intl' SAE Congress & Exposition, Detroit, February-March 1989

27) H. Berneburg, A. Cogotti, "Development and Use of LDV and Other Airflow Measurement Techniques as a Basis for the Improvement of Numerical Simulation of Engine Compartment Air Flows", SAE paper 930294, Intl' SAE Congress & Exposition, Detroit, March 1993

28) E. Mercker, H. Soja, J. Wiedemann, "Experimental Investigation on the Influence of Various Ground Simulation Techniques on a Passenger Car", Vehicle Aerodynamics - Two Day Conference, Loughborough University of Technology, UK, July 1994

29) K.R. Cooper, A.A. Fediw, K.P.Garry, S. Wiodarski, "Development of a Moving Ground System for the Study of Vehicle Aerodynamics", Vehicle Aerodynamics - Two Day Conference, Loughborough University of Technology, UK, July 1994

30) A. Cogotti, "The Pininfarina Ground Effect Simulation System", 10th EADE Meeting, Sindelfingen, October 1994

31) Private communications at the SAE RVAC Ground Simulation Standards Committee (1991-1994)

APPENDIX 1

CNR-PININFARINA WORKSHOP
Turin, May 12, 1994

LIST OF PARTICIPANTS

Company	Name
ABARTH	Ing. S. Limone
BECCIO	Mr. S. Beccio
BENZING	Ing. E. Benzing
BERTONE	Mr. E. Pagliano
BERTONE	Mr. L. D'Ambrosio
BIZZARRINI	Ing. G. Bizzarrini
BMW AG	Mr. K. R. Deutenbach
BMW AG	Mr. H. Kerschbaum
BWM Technik GmbH	Mr. N. Lindener
BREDA C.F.	Ing. M. Pecorini
BUGATTI	Ing. M. Forghieri
C.I.R.A.	Mr. B. Sciarretta
CHRYSLER Corp.	Mr. M. Gleason
DALLARA	Ing. D. Zinelli
FERRARI	Ing. R. Fedeli
FIAT Auto s.p.a.	Ing. C. Berta
FIAT Auto s.p.a.	Dr. P. Bergamini
FIAT Auto s.p.a.	Arch. R. Hamnett
FIAT Auto s.p.a.	Ing. B. Bonis
FIAT C.R.F.	Ing. M. Casella
FIAT C.R.F.	Ing. D. Vitali
FORD AG	Mr. A. Gilhaus
FORTI CORSE	Ing. D. Caronna
GENERAL MOTORS	Dr. G. Sovran
GRONDONA	Ing. G. Borello
IDEA	Ing. R. Grosso
ITALDESIGN s.p.a.	Mr. G. Borge
ITALDESIGN s.p.a.	Mr. R. Chiesa
IVK, UNIVERSITAET	Mr. R. Kuenstner
ADAM OPEL AG	Dr. H. Berneburg
PININFARINA S & R	Ing. M. Vernacchia
POLITECNICO Torino	Prof. A. Morelli
PSA Peugeot/Citroen	Mr. R. Marazzato
PSA Peugeot/Citroen	Mr. M. Zaccariotto
VOLKSWAGEN AG	Mr. H. Mankau

SPEAKERS

Company	Name
MIRA	Mr. G. Carr
PORSCHE AG	Dr. J. D. Vagt
MITSUBISHI	Mr. M. Yoshida
AUDI AG	Dr. J. Wiedemann
N.R.C.	Mr. K.R. Cooper
IND. PININFARINA	Ing. A. Cogotti

Consiglio Nazionale delle Ricerche (CNR)
Prof. C. Caputo
Prof. A. Scognamiglio
Ing. V. Delle Site
Ing. O. Landolfi

Industrie Pininfarina
Ing. A. Pininfarina
Ing. L. Borgarelli
Mr. E. Dogliani
Ing. R. Trivero
Mr. A. Bianco
Mr. A. Catania
Mr. M. Simioni
Mr. L. Gentile
Mr. C. La Fauci
Mr. P. Grillo

940418

A Further Evaluation of the Ground-Plane Suction Method for Ground Simulation in Automotive Wind Tunnels

Geoffrey W. Carr
MIRA

Winfried Eckert
Porsche AG

ABSTRACT

For wind-tunnel measurements of drag and lift on passenger and racing-car models, correlations have been established between results obtained with a stationary ground-plane, with and without boundary-layer-suction, and a moving ground-plane.

Compared with the stationary-ground-plane results without suction, the application of suction caused significant reductions in lift. These reductions were smaller, though, than those obtained with the moving-ground-plane, particularly in the case of rear-axle lift force. The influence of ground-plane suction on drag was also less than that of the moving ground-plane, which induced small drag increases.

1. INTRODUCTION

Whilst it is generally accepted that, in principle, a moving ground plane provides the perfect simulation of the road or track surface in an automotive wind tunnel, it is also widely recognised that practical difficulties can preclude the use of this technique in some facilities.

This is particularly true in wind tunnels used for full-scale vehicle tests, where a stationary or 'fixed' ground plane is normally used, often in conjunction with some form of boundary-layer thinning system.

Firstly, the complexity of the vehicle mounting arrangements required are a major deterrent to the use of a moving-ground system. Drastic modifications may be necessary to the body structure of the test vehicle, and a special wheel suspension system may be needed, as described in (1)*. Secondly, in order to examine the cross-wind characteristics of the vehicle, it is normally required to measure the aerodynamic forces over a range of yaw angles. This then necessitates yawing not only the vehicle and its mounting system but also the moving ground plane,which, in practice, is normally a moving belt. Yawing the moving belt introduces considerable mechanical complexity, together with a need to remove the upstream boundary layer along the windward side of the belt as well as at the leading edge. The resulting simulation, however, is no longer ideal for the case where a cross wind is added to the motion of the vehicle.

These problems can be reduced by compromise arrangements such as that described in (2), where a removable moving-belt rig is installed astride the turntable and the underfloor six-component balance. Zero-yaw tests can then be conducted with the moving belt and with the vehicle mounted on four cantilever struts attached to the underfloor balance. Measurements at other yaw angles can be made with the moving belt removed. Such an arrangement, however, involves time-consuming installation and removal of the moving-belt rig.

Alternatively, as proposed in (3), a narrower moving-belt rig can be installed permanently in the turntable between the normal wheel mounting pads. However, the advantage of the rotation of the wheels, which results when these are allowed to run on the moving belt, is then lost. But so also is the disadvantage of the complexity of the necessary wheel suspension modifications.

*Numbers in parentheses designate references at end of paper.

In view of the difficulties associated with full-scale moving-belt tests, the relative simplicity of some of the alternative fixed-ground simulation methods encourages consideration of possible compromise solutions. Both tangential blowing (4,5) and distributed ground-plane suction (6) have been advocated as methods for ground-plane representation. Both appear to involve less mechanical complexity than the moving-belt method, plus simpler vehicle mounting arrangements. Clearly, neither method provides ideal representation of the flow close to the ground-plane. Therefore, as argued on a theoretical basis in (7), for example, neither can be expected to give true values for the underbody pressures and velocities (and hence the lift and drag forces) associated with vehicles with smooth, profiled underbodies in fairly close proximity to the ground.

These shortcomings might be considered acceptable, however, if a consistent relationship could be established between the measured forces with one of the above-mentioned fixed-ground methods and the corresponding measurements with a moving-ground plane. The present study was conducted with the objective of finding such a relationship, especially for Formula 1 - type racing cars.

2. PREVIOUS WORK

Of the two alternative methods, distributed suction was chosen for the present study in preference to tangential blowing. Although it has been argued theoretically (7) that ground plane simulation by tangential blowing introduces smaller errors, particularly in the onset flow, the need to compromise between the conflicting optimum levels of blowing for the front and rear ends of the vehicle is seen as a major disadvantage. In order to produce a flow field representative of road/track conditions at the rear of the vehicle, where an underbody diffuser would be located (and flow separation on the tunnel floor might occur), it is necessary to 'over-blow' to an extent which creates unrealistically high drag forces on the front wheels.

On the other hand, with distributed suction, the excessive angularity of the onset flow calculated in (7) can be reduced to acceptable levels by allowing sufficient distance between the area of distributed suction and the short region of 'basic' suction which precedes it and which removes the upstream boundary layer. There remains the disadvantage that the boundary layer is not eliminated at any point in the vicinity of the vehicle. However, this disadvantage is mitigated by the fact that the residual boundary layer is asymptotic and relatively thin beneath the vehicle and its wake. This boundary layer has also been found to be resistant to the adverse pressure gradient generated by an underbody diffuser (8).

Another theoretical objection to distributed suction, that at small ground clearances it can directly influence and reduce the vehicle underbody pressures, was not substantiated by the experimental work reported in (8). Although such an effect must occur in the limiting case, when the ground clearance tends to zero, no effect was measured at a clearance equivalent to 25mm at full scale.

For these reasons, distributed suction remained the preferred choice, following the earlier work of (8), for installation in the MIRA Full-Scale Wind Tunnel. However, the need for further evaluation of this technique had been noted in (8), and the present work was designed to meet that need. The comparison of force and pressure measurements in (8) had shown generally good agreement between data obtained in the MIRA Model Wind Tunnel with interchangeable moving-belt and distributed-suction installations. A few significant differences were found, though, in the case of a Formula 1 racing car model, and these were assumed to be attributable to wheel rotation effects on the moving-belt data.

More recently, with the aim of verification of further improved suction adjustment criteria, measurements made with the distributed-suction systems in the Porsche Full-Scale and Model Wind Tunnels were compared with corresponding measurements on the road and in a moving-ground model wind tunnel, respectively (9). A Group C - type racing car and its 1:4 scale model formed the test objects in this investigation, which covered a wide range of configurations in both cases. Good agreement was found in the comparisons of incremental coefficients of drag and front and rear lift.

3. PRESENT APPROACH

3.1 WIND TUNNELS - For the present study, two model wind tunnels were used, both of which were designed primarily for 1:4 scale car models. The Porsche Model Wind Tunnel provided the distributed-suction data and the MIRA Model Wind Tunnel the corresponding moving-belt data.

The MIRA Model Wind Tunnel, which is fully described in (10) and shown in side elevation in Fig.1, has a closed-jet test section 2.06m wide by 1.00m high. The moving-ground rig is illustrated in more detail in Fig. 2. It has a belt length of 3.0m between roller centres with a width of 1.05m. The local air velocity remains within $\pm$ 1 per cent of free-stream down to the belt surface.

Models are suspended over the belt by a rigid vertical arm, which attaches via a pivot to a 2-component load cell housed centrally within the model. A vertical link attached

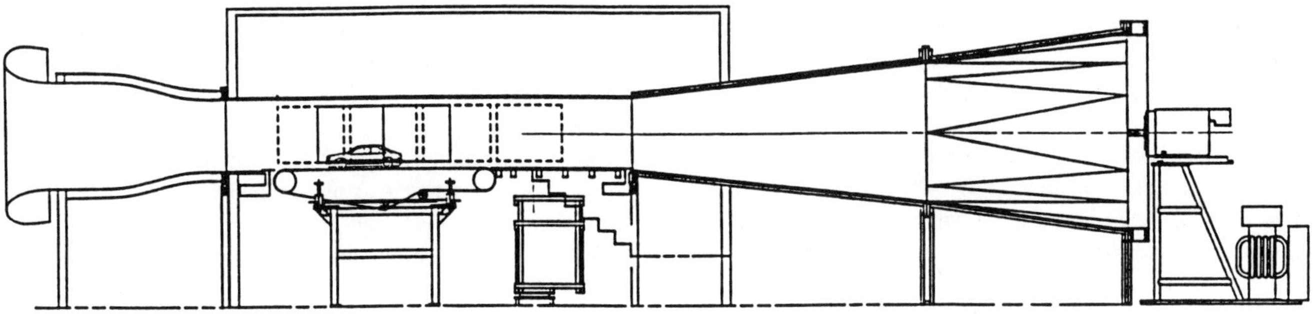

Fig.1 MIRA Model Wind Tunnel

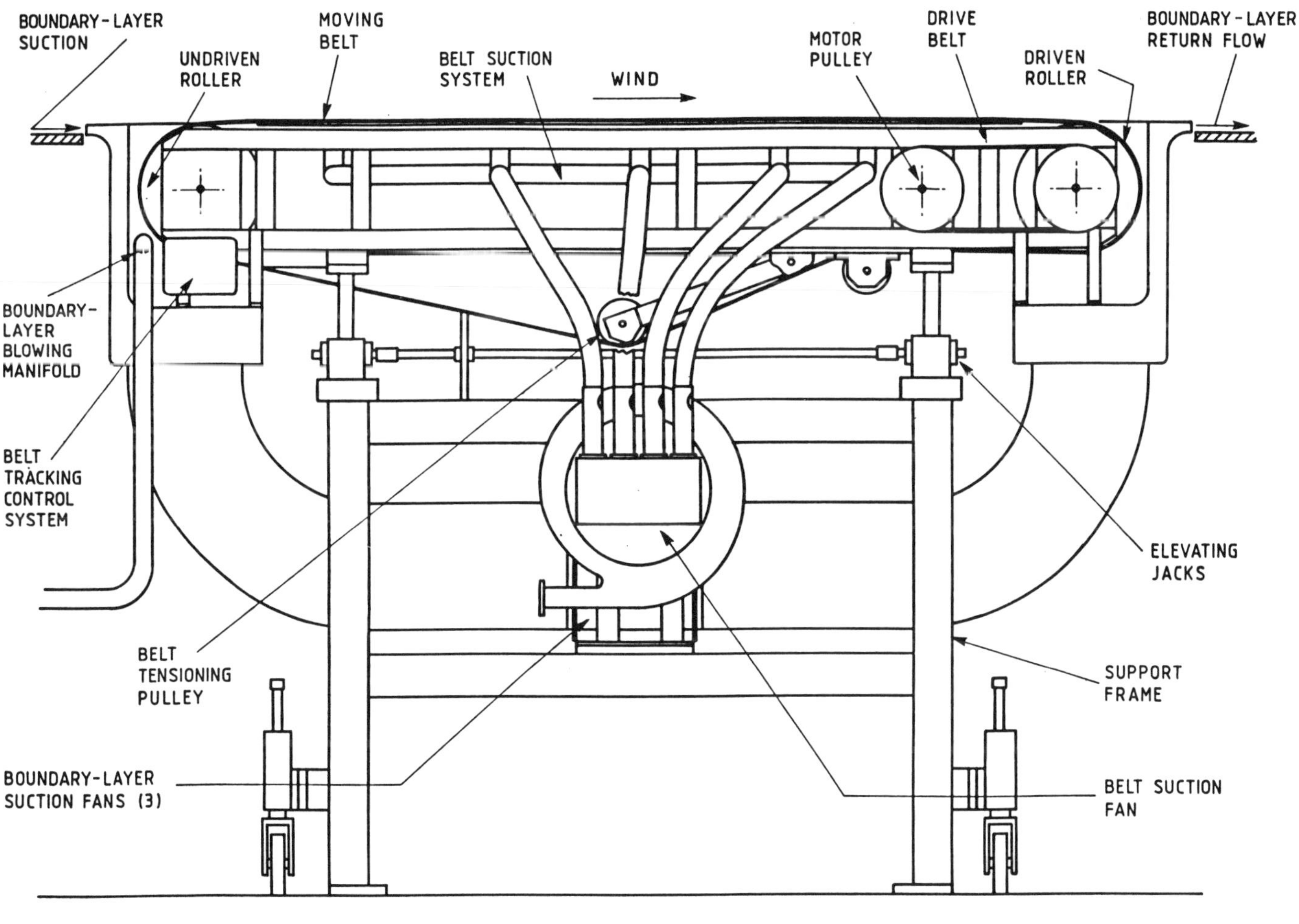

Fig.2 MIRA Moving-Ground Rig

to the rear of the model restrains it in pitch and incorporates a load cell for the third component of aerodynamic force. The front and rear supports can be individually adjusted, under computer control, to vary the model ride heights, i.e., the ground clearance and the pitch angle. The model wheels are normally allowed to run on the belt and so rotate.

The Porsche Model Wind Tunnel has been fully described in (6) and(11). It has a nozzle 0.9m high x 1.55m wide leading into a test section which can be of either slotted-wall or open-jet configuration. In (12), tests were also conducted with the slotted walls sealed to provide a closed-jet test section. This precedent was followed for the present work in order to improve the comparability with the MIRA MWT measurements.

Full details of the complete basic and distributed-suction system are given in (9), and a schematic illustration is reproduced in Fig.3. The effective velocity ratios through the basic and distributed-suction areas are, respectively, 0.036 and 0.002 to 0.0003. The resulting displacement thickness of the boundary layer is constant at approximately 0.8mm over the area covered by the distributed suction. The associated flow pitch angle is approximately -0.2 deg. at the start of the turntable. Beneath the turntable is a 6-component balance, and models are mounted on four pins passing through clearance holes in the turntable.

For both wind tunnels, the boundary layer velocity profiles are shown in Fig.4. The fixed-ground and moving-ground profiles in the MIRA tunnel are compared with the Porsche tunnel profiles with and without suction.

3.2 MODELS - Two models were used in this study. The first was the 1:4 scale version of the MIRA Reference Model, the design of which is shown in Fig.5. Interchangeable rear-end sections allow four different configurations to be represented, but in the present work only square-back and fast-back forms were tested.

Additional variations were introduced as follows:-

(i) by adding three transverse roughness strips to the underbody,

(ii) by removing the wheels,

(iii) by eliminating the rear underbody diffuser.

Removal of the wheels exposed the 16mm-diameter supports which were used in the Porsche MWT to locate the model on the mounting pins. Elimination of the diffuser was effected by adding a wedge-shaped piece to create a full-length flat undersurface.

Whereas the first model was expected to confirm previous findings, the second was intended to be the main source of new information. This was a 1:4 scale model of a 1991 Formula 1 racing car (Fig.6). The exposed wheels of the model were mounted separately from the body in both tunnels, using transverse arms attached to mounting blocks at the sides of the test section. In the Porsche MWT, the wheels rested on the tunnel floor and did not rotate. Rotation was effected in the MIRA MWT by running the wheels on the moving belt. For direct comparison with the Porsche results, however, it was also necessary to run with the wheels stationary and the belt moving. This was achieved by raising the wheels fractionally clear of the belt, locking them on their spindles and adding a thin, low-friction seal to the bottom of each wheel.

In both tunnels, the forces on the wheels were not measured. The results thus consist of body forces only (in the presence of the wheels). These forces were transmitted to the underfloor balance in the Porsche MWT through four mounting pins of streamlined section, which supported the model at the ends of rudimentary suspension arms, just inboard of the wheels.

To create four different test configurations of the Formula 1 model two changes were made to the underbody:-

(i) by eliminating the step under the raised nose,

(ii) by eliminating the rear diffuser.

In both cases, a piece was added to make a full-length flat underbody.
Additional variations in the aerodynamic characteristics of this model were achieved by covering a range of five different ride-height combinations.

4. TEST PROCEDURE

In both wind tunnels and with both models, all measurements were made at 25 m/s nominal wind speed and zero yaw angle. The measurements covered drag and front and rear lift forces.

The MIRA Reference Model was mounted in the MIRA MWT by means of the overhead suspension arrangement already described. A 2mm- clearance was maintained between the belt surface and the (fixed) wheels or, when the wheels were removed, the supports which were then exposed. In the Porsche MWT, this model was mounted with the same clearance under the wheels. It was supported, however, by means of the four pins connected

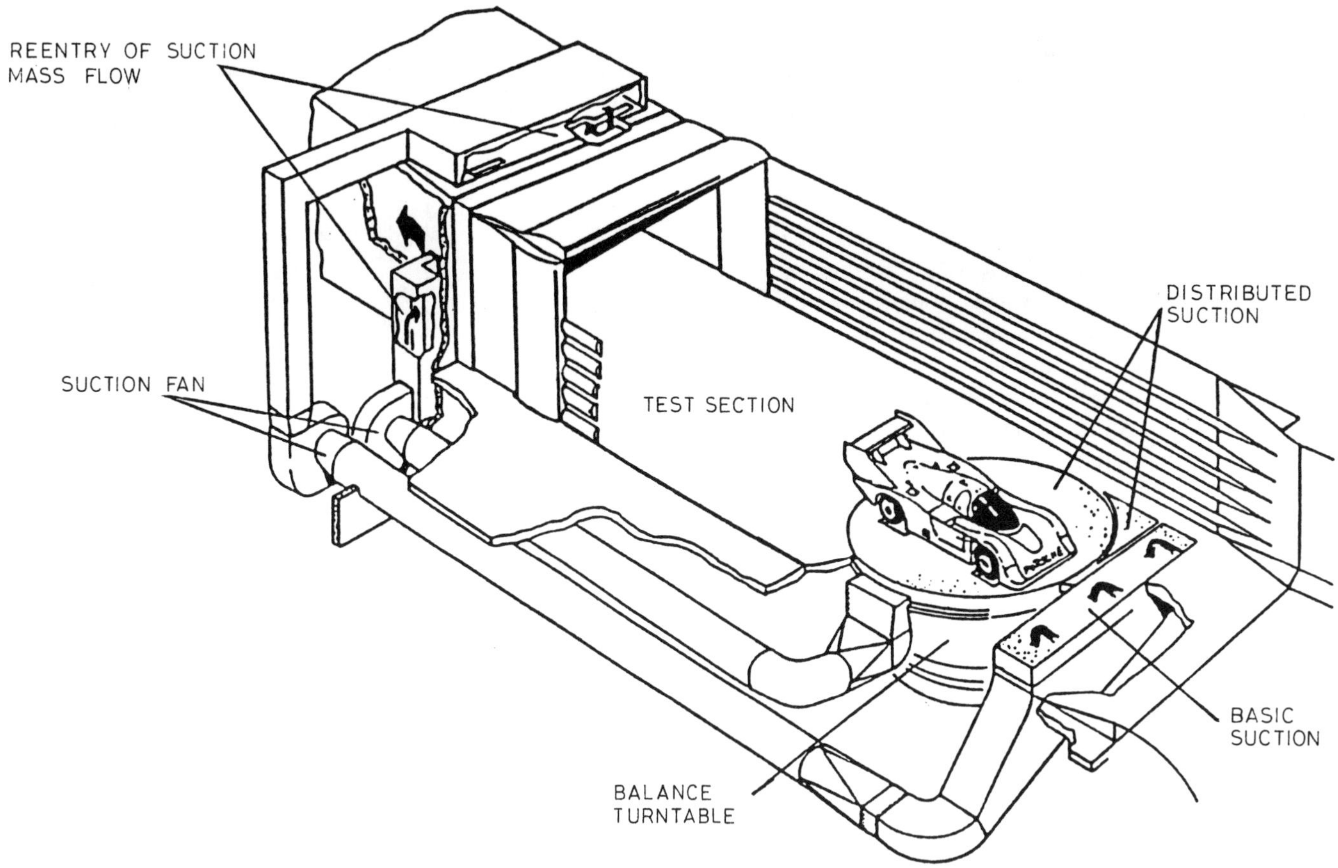

Fig.3 Porsche Model Wind Tunnel, showing test section and ground-plane suction system

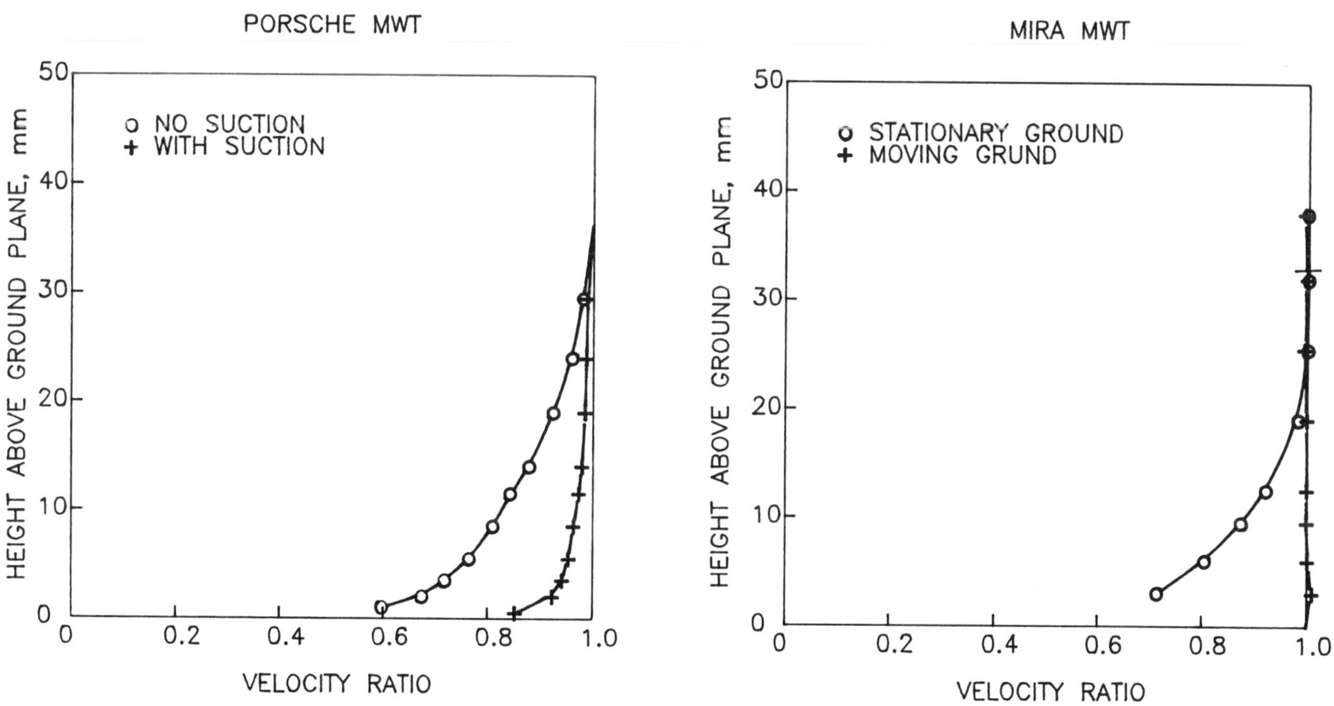

Fig.4 Boundary-layer velocity profiles

to the underfloor balance. To determine the effects of the front mounting strut in the MIRA MWT, additional measurements were made with a replica of this strut positioned above the model but not in contact with it. It was not feasible, however, to simulate the Porsche mounting pins in the MIRA MWT tests.

The body of the F.1 model was mounted from the overhead suspension struts in the MIRA MWT, and as previously mentioned, the wheels were mounted separately on transverse arms from the sides of the test section. The wheels were similarly mounted in the Porsche MWT, but the body was mounted on the four pins as described above in Section 3.2. Although the replica of the MIRA MWT front strut was originally intended to be used also with this model in the Porsche facility, this plan was subsequently abandoned because of the time taken in adjusting the 'strut' for each change of model ride heights.

The full programme of tests covered with the two models in the two tunnels is given in Tables 1 and 2. For the tests in the MIRA MWT with the belt stationary, the gaps at the front and the sides of the belt were sealed, and the upstream boundary-layer removal system and the belt suction system were both in operation.

TABLE 2 - Test Programme for Formula 1 Model

Configuration No.	Front Underbody	Rear Underbody
1	Flat	Flat
2	Flat	With diffuser
3	Raised	Flat
4	Raised	With diffuser

All configurations tested at front/rear ride height combinations of 3/3, 6/6, 6/10, 10/10, 10/6 mm.

All configurations tested with stationary wheels, in Porsche MWT with and without suction on and in MIRA MWT with and without belt moving. Also tested, with rotating wheels, in MIRA MWT with belt moving.

TABLE 1 - Test Programme for MIRA Reference Model

Unless otherwise stated all configurations were tested in Porsche MWT with and without suction on and in MIRA MWT with and without belt moving.

Configuration No.	Rear End	Wheels *	Underbody	Diffuser
1	Fast-back	On	Rough	With
2	Square-back	On	Rough	With
3	Square-back	Off	Rough	With
4	Fast-back	Off	Rough	With
5	Fast-back	Off	Smooth	With
6	Square-back	Off	Smooth	With
7	Square-back	Off	Smooth	Without
8	Fast-back	Off	Smooth	Without
9	Fast-back	On	Smooth	Without
10	Square-back	On	Smooth	Without
11	Square-back	On	Smooth	With
12	Fast-back	On	Smooth	With
13	Fast-back	On	Rough	Without
14	Square-back	On	Rough	Without
15+1	Square-back	On	Rough	Without
16+	Fast-back	On	Rough	Without

* All 'wheels off' tests were with 16mm diameter struts in place of wheels.

\+ Tested only in Porsche MWT, with dummy strut above model roof.

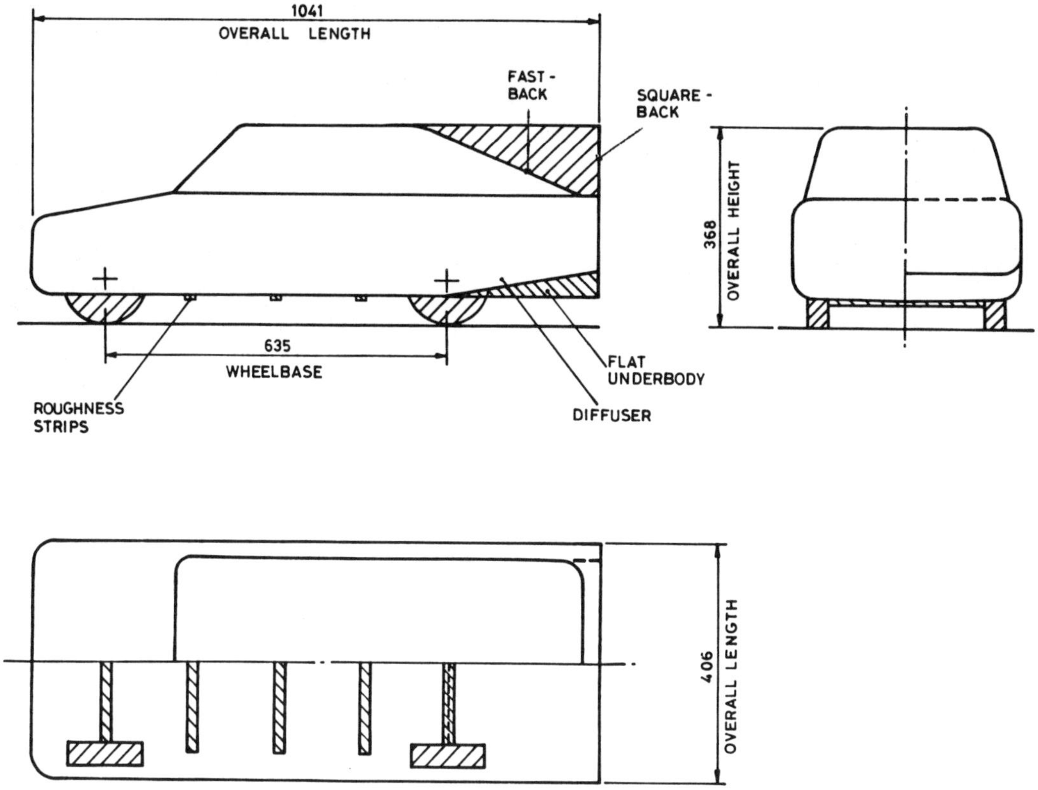

Fig.5 MIRA Reference Car Model
Shaded areas denote removable parts

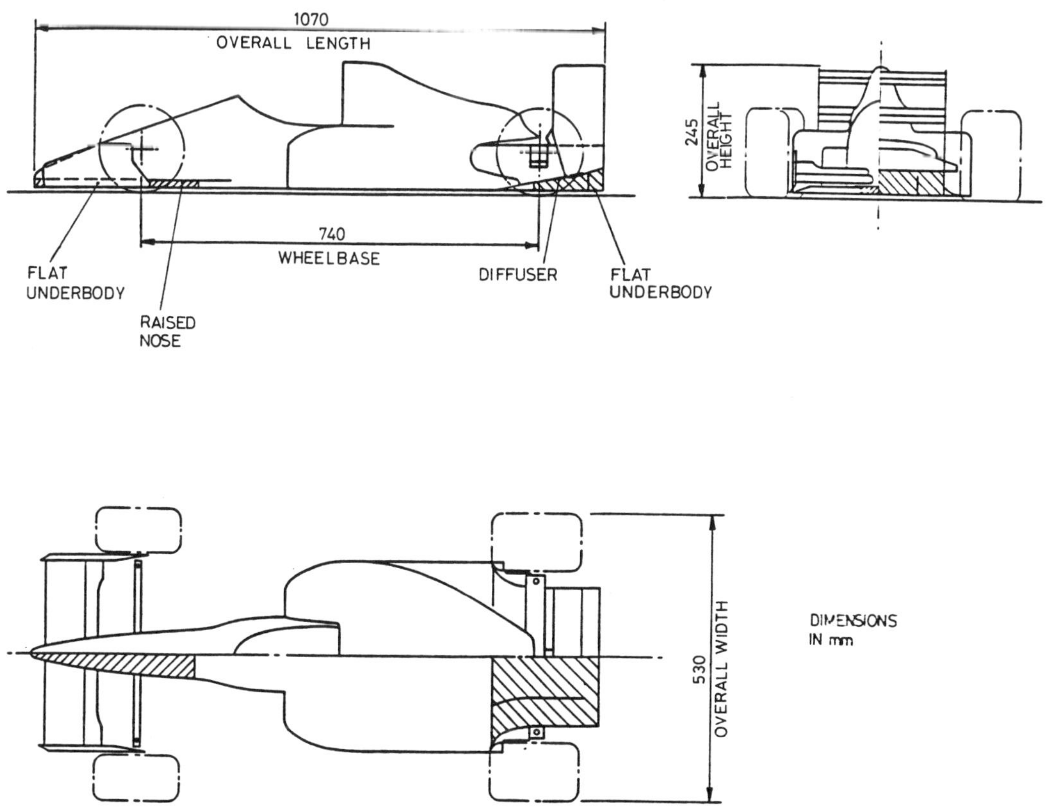

Fig.6 Formula 1 Model
Shaded areas denote removable parts

5. RESULTS AND DISCUSSION

In the presentation and comparison of results, the approach taken has been to examine the correlation between the various sets of data and to seek a simple relationship. This approach has been followed in determining the respective effects of:

(i) ground-plane suction (Porsche MWT)

(ii) ground-plane motion (MIRA MWT)

(iii) wheel rotation (MIRA MWT: F.1 model only)

The results have been corrected for wind-tunnel blockage in the case of the MIRA Reference Model, but not the F.1 Model. The continuity method was used for the corrections, as is the standard practice in the MIRA MWT. The resulting corrections were -16.0 per cent in the Porsche MWT and -11.0 per cent in the MIRA MWT. Blockage corrections were ignored in the case of the F.1 model because of its smaller frontal area and the uncertainty concerning the magnitude of blockage effects on open-wheeled racing car models (12).

Although the effects of the differences in model-mounting arrangements between the two tunnels were not quantified, it is believed that these would only affect significantly the correlation of data between the tunnels. The required effects of ground-plane suction, or ground-plane motion, or wheel rotation, in one tunnel are considered unlikely to be influenced significantly.

The repeatability of results obtained with the moving ground surface was generally less good than for the other measurements. This may have been due, in some cases, to very slight undetected local lifting of the moving belt. However, the deficiencies in repeatability were not sufficient to cast significant uncertainty on the general observations and conclusions made in the following sections.

5.1 MIRA REFERENCE MODEL - The drag and lift coefficients (C_D and C_L) for the reference passenger car model in its 14 configurations are presented in Figs. 7 and 8. In addition to the total lift coefficients, the front and rear-axle components (C_{LF} and C_{LR}) are also presented.

Fig. 7 shows the effects of the ground-plane suction in the Porsche MWT. For C_D, the effect was very small, with a mean 1:1 relationship between values with suction on and suction off and with a residual scatter of + 0.010/- 0.007. A tendency can be observed in Fig. 7 for C_D to be reduced more by the application of suction in the case of the lower-drag versions of the fast-back and square-back configurations (i.e., with smooth underbody and no wheels).

Fig. 7 also shows that the effect of suction on the total lift coefficient was a reduction in C_L which varied apparently randomly between zero and about 0.05. The division of C_L into front and rear-axle components reveals that, in general, a reduction of approximately 0.015 in C_{LF} was induced by the application of suction. This reduction was practically independent of the level of C_{LF} over the range from -0.16 to + 0.09. In three cases, C_{LR} was also reduced, by about 0.03, these being three out of the four configurations which included both the wheels and the diffuser. In the remaining cases, C_{LR} was practically unchanged (within $\pm$ 0.01) by operation of the suction system.

Fig. 8 provides comparable data on the effects of a moving ground surface. It shows greater changes in C_D than were recorded due to suction (Fig. 7). In all except two cases, C_D was increased by approximately 5 per cent. The exceptions were the two rear-end configurations with wheels on, a smooth underbody and no diffuser, and for these the increase was only about 1 per cent.

The effect of the moving ground surface on C_L (Fig.8) was generally a greater reduction than had been found with the application of suction (Fig.7). Moreover, in this case, the reduction increased with decreasing C_L. An average reduction equivalent to approximately ($0.06 - 0.04\ C_L$) can be deduced from Fig.8, though with appreciable scatter.

This reduction in lift was divided fairly evenly between the front and rear-axle components. For C_{LF} the reduction could mostly be represented fairly accurately by ($0.018 - 0.07\ C_{LF}$). As with C_D, the exceptions were the two configurations with wheels on, a smooth underbody and no diffuser. These showed considerably larger reductions in C_{LF}. Thus, the moving ground-plane caused a reduction in C_{LF} which was greater by approximately 7 per cent of ($-C_{LF}$) than that induced by ground-plane suction.

Fig.8 also shows that for C_{LR} the scatter was slightly greater than for C_{LF}, and the mean effect was a reduction of approximately ($0.042 - 0.042\ C_{LR}$). Since C_{LR} had mostly not been significantly affected by the ground-plane suction, this reduction then constitutes the difference between the effects of the two techniques on C_{LR}.

In summary, the reference model results showed straight-line relationships between the coefficients obtained with and without each of the two ground-plane simulation techniques. The scatter about these straight lines was mostly small and apparently random, indicating that the relationships, and

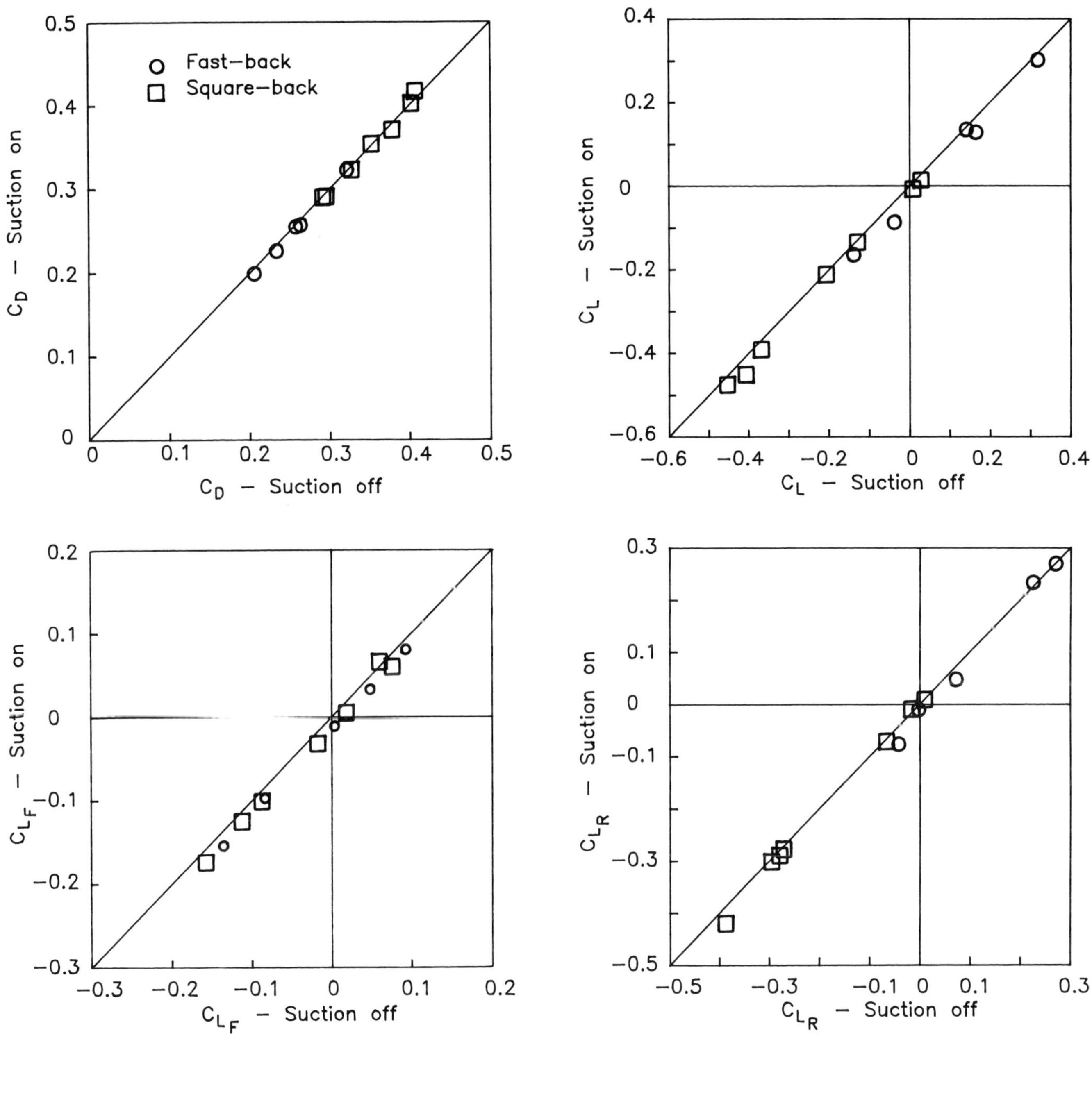

Fig.7 Effects of ground-plane suction
MIRA Reference Car Model in Porsche MWT

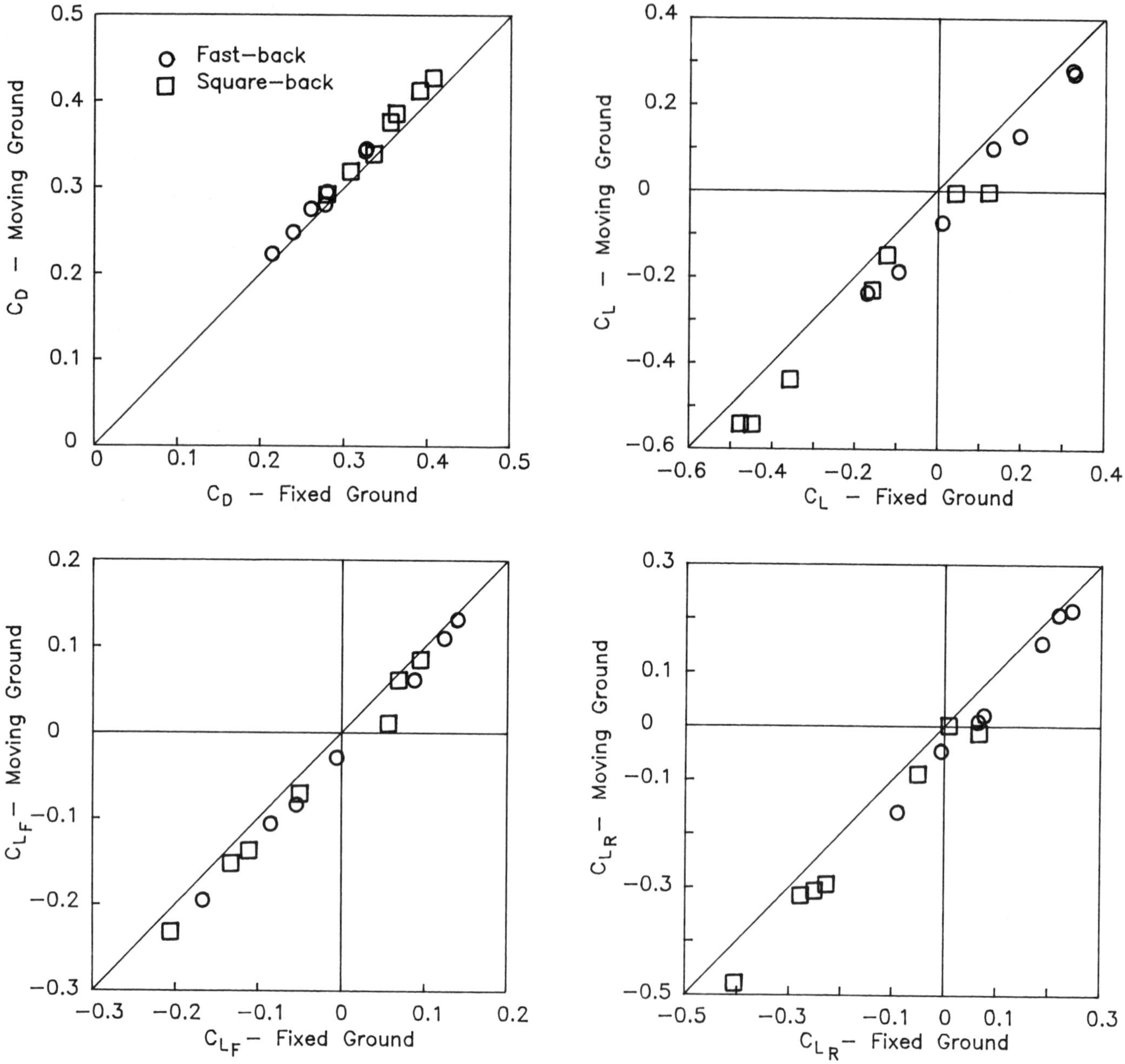

Fig.8 Effects of moving ground-plane
MIRA Reference Car Model in MIRA MWT

hence the effects, were largely independent of the model configuration.

The only significant effect observed consistently due to ground-plane suction was a small incremental reduction in the front lift coefficient C_{LF}. In contrast, the effects of the moving ground plane were seen in all of the coefficients recorded. A percentage increase in drag coefficient was accompanied by a combination of incremental and percentage reductions in lift coefficient. The lift reduction, which was shared fairly equally between front and rear axles, could thus be related to the mean underbody velocity, since this is also a direct function of C_L.

5.2 FORMULA 1 MODEL - The results of the drag and lift measurements on the F.1 model in its 20 ride-height/underbody-configuration combinations are presented in coefficient form in Figs.9 to 11. Again, front and rear-axle lift coefficients are included.

The effects of ground-plane suction in the Porsche MWT are shown in Fig.9. As in the case of the MIRA reference model, the influence of the suction system on the drag coefficient was small. However, the mean change in C_D was larger for the F.1 model at + 0.006, i.e., an increase of 1.7 per cent. Since the wheels were excluded from the measurements on this model, the increased drag cannot be attributed to the reduction in the proportion of the wheels immersed in the boundary layer when the boundary-layer thickness was reduced by the suction. It is therefore suggested instead that this small drag increase could be due to the extra vortex drag resulting from the increase in the magnitude of the (negative) lift coefficient which is shown in Fig.9.

The greater negative lift due to suction shown in Fig.9 appears to consist of an average percentage increase in $-C_L$ of approximately 16 per cent. As with the drag coefficients, there is considerable scatter about this mean increase. Fig.9, however, also reveals that the scatter in the C_L values is not entirely random. Firstly, although the increase in $-C_{LF}$, which ranges from 16 to 38 per cent, appears to be independent of the underbody configuration, it does vary with ride heights. In particular, the smallest increases are found to be associated with the greatest front ride height of 10mm (irrespective of which of the two rear ride heights of 6 and 10mm applied).

On the other hand the magnitude of the effects of suction on the rear lift coefficient was mainly related to the underbody configuration and less to the ride heights. The two configurations without a diffuser showed negligible changes, while the two with a diffuser exhibited increases in $-C_{LR}$ from 5 to 17 per cent. Within these smaller ranges of scatter, a trend could be discerned such that the lower ride heights gave the greater increases in $-C_{LR}$.

The effects of the moving ground plane in the MIRA MWT on the drag and lift coefficients (for the body only) are shown in Fig.10. Generally, the effects were large and exhibited considerable scatter. Firstly, the drag coefficients were increased by 0.013 to 0.050 (4 to 17 per cent), despite the exclusion of the wheels from the measurements. The magnitude of these increases may again be attributable to the increases in induced drag resulting from the large increases in (negative) lift coefficient shown in Fig.10.

The increases in C_L can mostly be described by the approximate expression ($-0.10 + 0.36\ C_L$). Considerably greater increases in downforce were measured, however, at the minimum ride heights with the two underbody configurations without a diffuser. The additional downforce was generated fairly equally at the front and rear of the model, as Fig.10 shows. The mean increase in $-C_{LF}$ was 30 per cent, with $\pm$ 9 per cent of apparently random scatter about the mean. The associated increases in $-C_{LR}$ could be expressed as ($-0.10 + 0.31\ C_L$), with outlying data points as shown for $-C_L$ in Fig.10.

The effects of wheel rotation on the body drag and lift forces are presented in Fig.11. The drag coefficients, which were increased first by the moving ground plane (as discussed above), were increased further by allowing the wheels to rotate. As Fig.11 shows, increases ranging from 4 to 13 per cent were recorded. The smallest increases occurred with the completely flat underbody and the largest with the raised nose and the rear diffuser.

Wheel rotation also augmented the effects of the moving ground plane on the lift coefficient. Fig.11 shows a substantial increase in $-C_L$, which can be represented by ($-0.11 + 0.05\ C_L$). From Fig.11, it can also be seen that nearly all of this additional downforce acted in the region of the rear axle. At the front axle position the effects of wheel rotation had a mostly negligible effect on the local component of lift.

As with the reference model, a linear regression appeared to be valid for each of the comparisons of coefficients for the F.1 model. In the case of the latter, however, the scatter about the regression lines was mostly not negligible. Some of this scatter was evidently related to variations in the underbody configuration, and some was associated with changes in ride heights. It should be noted here that the ride heights were only 10mm or less, whereas the reference model ride height was much greater, at 50mm.

Unlike the reference model, the F.1 model exhibited large

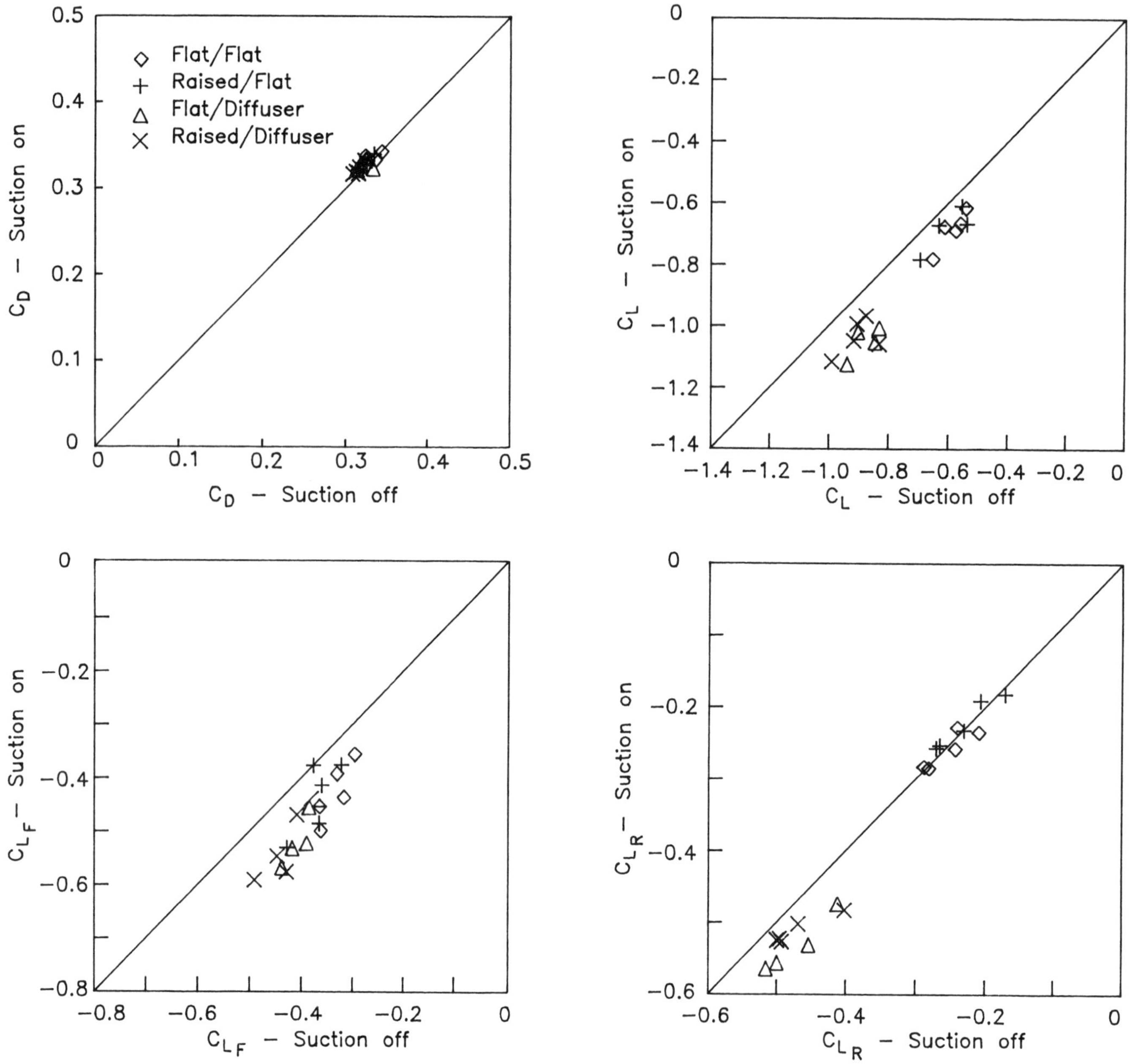

Fig.9 Effects of ground-place suction
Formula 1 model in Porsche MWT

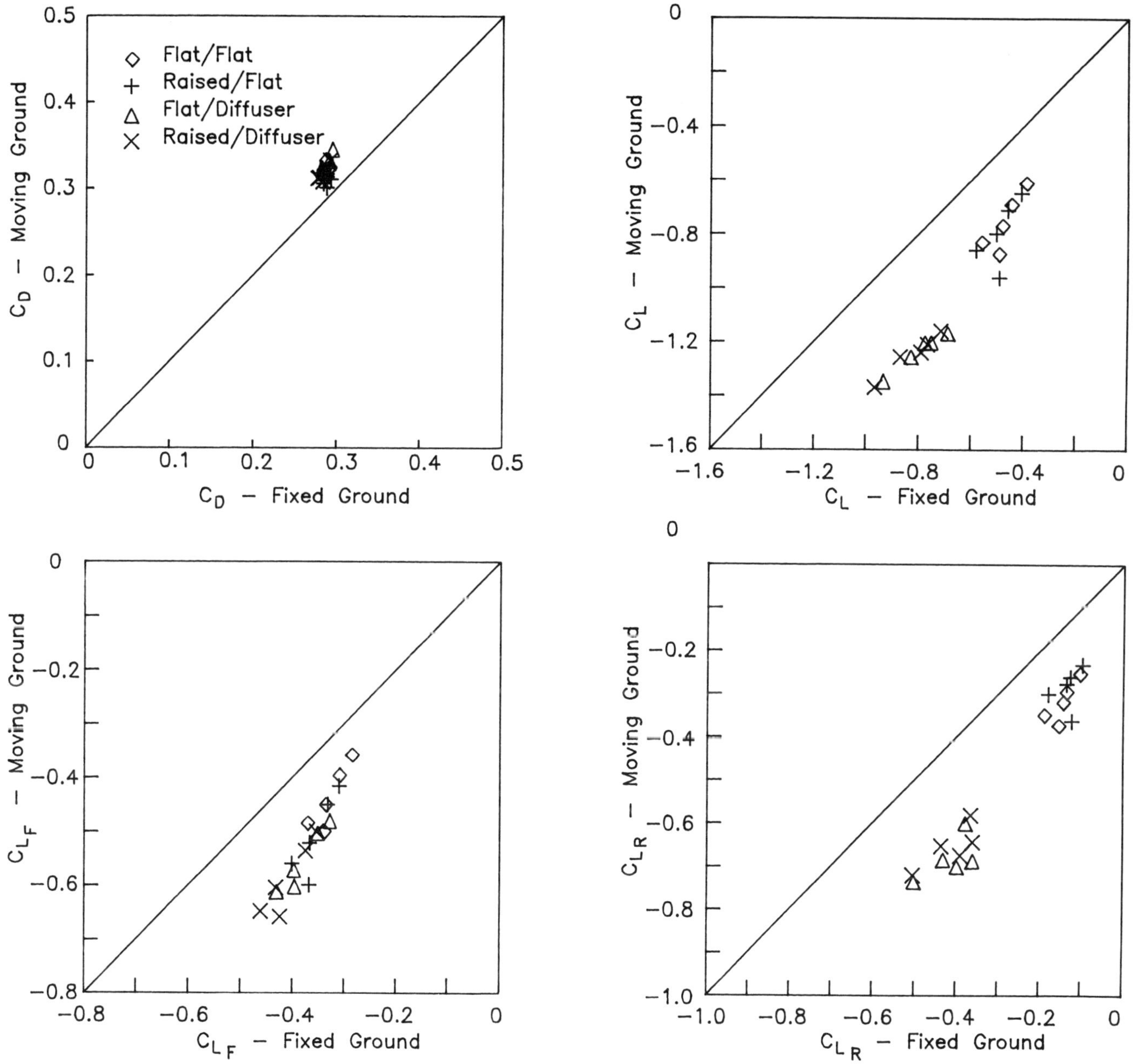

Fig.10 Effects of moving ground-plane
Formula 1 model in MIRA MWT (with wheels stationary)

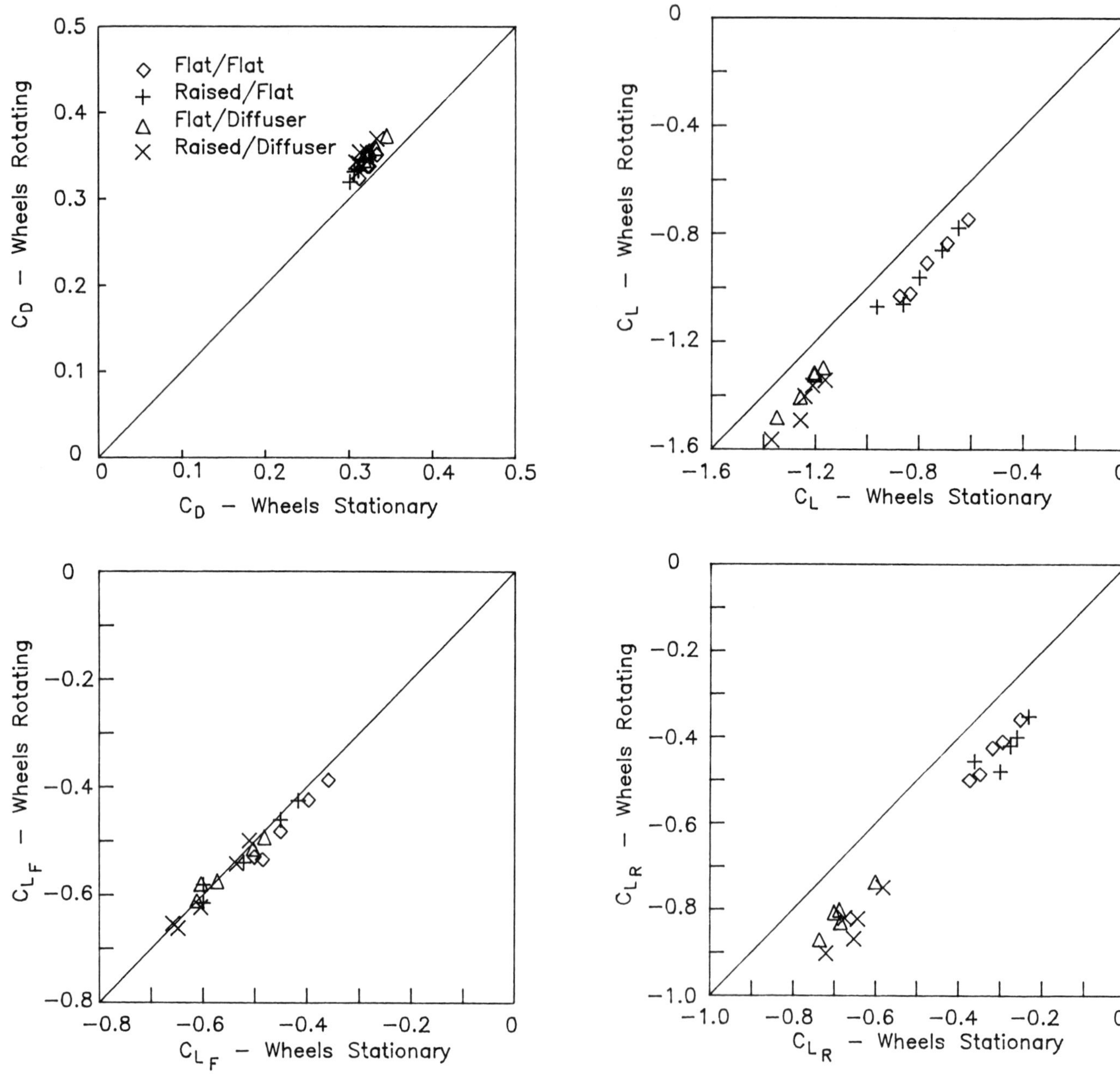

Fig.11 Effects of wheel rotation
Formula 1 model in MIRA MWT (with moving ground-plane)

negative lift coefficients in every test condition, and the reductions in lift due to both ground-plane suction and the moving ground plane were correspondingly larger. However, the relationships between the respective effects of the two ground simulation techniques showed some similarity for the two models. Firstly, their drag coefficients were increased more by the moving ground plane than by ground-plane suction. Likewise, their lift coefficients - front, rear and total - were reduced more by the moving ground plane. Furthermore, for both models, the reduction in front-axle lift relative to the reduction in rear-axle lift was greater with ground-plane suction than with the moving ground plane.

This last feature of the comparison between the two ground simulation techniques means that, in addition to the smaller reductions in lift with ground-plane suction, there was also a difference in the ratio of front lift to rear lift. This is particularly important in wind-tunnel testing of racing cars, since high-speed stability is strongly dependent on this ratio (or the centre-of-pressure location which it defines).

6. CONCLUSIONS

From these comparative measurements between ground-plane suction and the generally-accepted ideal simulation by a moving ground plane, the following conclusions are drawn:-

- The motion of the ground plane substantially increases the downforce (or reduces the lift) at both front and rear axle positions, even on a passenger car model with a rough, flat underbody. The drag is also slightly increased.

- Distributed suction through the ground plane also gives increased downforce (or reduced lift), particularly at the front-axle position, but generally to a lesser extent than the moving ground plane. The increases in drag were small to negligible.

- A linear relationship can be determined between the respective changes in the aerodynamic forces due to the two ground simulation techniques. In some cases, however, there is considerable scatter about the regression line, and the relationship becomes dependent on underbody configuration and/or ride heights.

- On an open-wheeled racing-car model, wheel rotation generates markedly higher downforce on the body, particularly at the rear-axle position, and significantly greater body drag.

7. ACKNOWLEDGMENT

The support of the Rover Group Ltd in this study is gratefully acknowledged.

8. REFERENCES

1. Mercker, E., et al. On the Aerodynamic Interference due to the Rolling Wheels of Passenger Cars. SAE Paper 910311, Detroit, 1991.

2. Berta, C. Full-Scale Moving Belt in Fiat Aerodynamic Wind Tunnel. SAE Paper 905147, 1990.

3. Wiedemann, J., et al. Audi Aero-Acoustic Wind Tunnel. Proc. Royal Aero. Soc. Conference: Wind Tunnels and Wind Tunnel Techniques. Southampton, Sept. 1992, pp.56.1 - 56.14.

4. Mercker, E., and Knape, H.W. Ground Simulation with Moving Belt and Tangential Blowing for Full-Scale Automotive Testing in a Wind Tunnel. SAE Paper 890367, Detroit, 1989.

5. Hackett, J.E., and Boles, R.A. Moving Ground Simulation by Tangential Blowing. J. Aircraft, Vol.16, No.12, Article No. 78-814 R, Dec.1979.

6. Vagt, J.-D., and Wolff, B. Das neue Messzentrum für Aerodynamik Zwei neue Windkanäle bei Porsche - Teil 2. ATZ 89 (1987) 4, pp.183-189.

7. Mercker, E., and Wiedemann, J. Comparison of Different Ground Simulation Techniques for Use in Automotive Wind Tunnels. SAE Paper 900321, Detroit, 1990.

8. Carr, G.W. A Comparison of the Ground-Plane-Suction and Moving-Belt Ground-Representation Techniques. SAE Paper 880249, Detroit, 1988.

9. Eckert, W., Singer, N., and Vagt, J.-D. The Porsche Wind-Tunnel Floor-Boundary-Layer Control - A Comparison with Road Data and Results from Moving Belt. SAE Paper 920346, Detroit, 1992.

10. MIRA Aerodynamic Test Facilities: Users' Handbook, Oct. 1993.

11. Vagt, J.-D., and Wolff, B. Das neue Messzentrum für Aerodynamik - Zwei neue Windkanäle bei Porsche - Teil 1. ATZ 89 (1987)3, pp.121-129.

12. Eckert, W., and Müller, R. Influence of Different Wind Tunnel Test Sections - Open Jet, Slotted Walls,

Solid Walls - on Aerodynamic Test Results of Passenger Cars and a Racing Car. Proc. 3rd. Int. Conf. : Innovation and Reliability in Automotive Design and Testing. Florence, April 1992, pp.1567-1577.

Section II
Wind Tunnel Testing

Measurement Techniques

920347

Experimental Techniques for the Aerodynamic Development of Convertible Cars

Antonello Cogotti
Industrie Pininfarina S.p.A.

ABSTRACT

The aerodynamic development of a convertible passenger car requires the optimization of additional characteristics, when compared to those which are considered for normal passenger cars.
To be more specific, this means that , in addition to the usual requirements of hard-top passenger cars, (i.e. low values of the aerodynamic coefficients Cd, Cl, Cmy, etc., optimization of the cooling flow and achievement of low aerodynamic noise), it is important to also check the following:

- that the car with the soft top in the closed position, shows:
  . reduced soft-top ballooning
  . an acceptable increase in interior noise relative to a hard top
  . aerodynamic coefficient values which are not too different when compared with the hard-top version

- that the car in the open configuration (soft top down) shows:
  . an acceptable increase of the aerodynamic coefficients when compared with the hard-top version, with side windows both raised and lowered
  . a good comfort level inside the passenger compartment, i.e., air flow velocities close to zero near the passengers' heads.

To achieve this,in some cases,it has been necessary to set up new instrumentation for these measurements, i.e., such as for measurement of soft-top ballooning.
In other cases, the standard techniques (flow mapping with the 14-hole probe, acoustic dummies, etc.) are used, although in a different way than for the development of normal passenger cars.
This paper describes these techniques and the way they are employed for the aerodynamic development of convertible cars.

CONVERTIBLE CARS have always played an important role in the history of Pininfarina. Many of them became famous and contributed, in a positive way, to the development and growth of the Company.
Among them,it is worth mentioning (fig.1):
Giulietta spyder - Alfa Romeo
504 Cabriolet - Peugeot
365 GTS Daytona spyder - Ferrari
Spyder Europa - Fiat
348 TS - Ferrari
A.R. spyder - Alfa Romeo
205 Cabrio - Peugeot
Allanté - Cadillac

A convertible-car project requires additional experience in design and engineering compared with that normally required for conventional cars.
From a structural standpoint, loads and stresses are distributed on the body in a different way, and that requires a different design of the chassis.
The design of the soft top and the relevant mechanical elements (opening and closing devices) also requires a specific knowledge which is often not available (or no longer available) in the engineering department of the major car companies.
From the aerodynamic standpoint,a convertible car has, in addition to the typical requirements of a conventional passenger car (Cd, Cl, Cmy.....low values, cooling flow optimization, etc.), a need for characteristics which mainly relate to the following points:

- for the car with closed soft-top:
  . optimization of the aerodynamic coefficients

Fig. 1: Some of the convertible cars designed and built by Pininfarina.
Giulietta spyder - Alfa Romeo - 1962 / 504 Cabriolet - Peugeot - 1969
365 GTS Daytona spyder - Ferrari - 1969 / Spyder Europa - Fiat - 1982
348 TS - Ferrari - 1989 / A.R. spyder - Alfa Romeo - 1990
205 Cabrio Roland Garros - Peugeot - 1990 / Allanté MY 91 - Cadillac - 1991

. reduction of soft-top ballooning
. management and optimization of internal noise

- for the car in the open configuration (soft top lowered):
  . optimization of the aerodynamic coefficients, with side windows both raised and lowered
  . improvement of comfort in the passenger compartment through the analysis of the flow in the region of the passengers' heads.

In the following paragraphs, examples of measurements carried out to verify the achievement of these additional requirements for the convertible cars designed by Pininfarina are reported.
Also, information about the experimental techniques which are used to achieve these measurements is given.

Fig. 2: Scanning of the soft-top profile at various wind velocities

CONVERTIBLE CARS WITH THE SOFT TOP CLOSED.

The first characteristic of a car with a soft top instead of a more conventional hard top, is that the upper part of the car has a certain amount of deformation, which is furthermore dependent on car velocity.
It is therefore a task of the Engineering Department to design a soft top and the relevant supporting frame so as to ensure minimum deformation of the soft-top shape during running conditions: this is necessary for both aesthetic and aerodynamic reasons.
This is achieved both by working on the supporting frame and by choosing carefully the soft-top fabric, the type of sewing between the various parts, as well as the type and extension of the links between the fabric and the supporting frame.
In case that, as often happens, various mechanical solutions or types of fabric can be used, the choice of the best solution or improvement of the chosen solution is carried out in the wind tunnel after evaluation of:
- the soft-top deformation
- the change of Cd produced by a given soft top and compared for instance with the hard-top version
- the aerodynamic noise inside the car

The soft-top deformation is measured by using the traversing system in the wind tunnel, together with an optical sensor for distance measurements. This scans the soft-top shape, for instance along the longitudinal center line (and in some cases along other lines), with the wind blowing at velocities from 100 to 160 Km/h (Fig.2).

As a result, profiles of the soft-top center-line at the various wind velocities are obtained and they are compared with the profiles measured in conditions of no-wind.
The displacement of the car body during the measurements with the wind are taken into account, and the results are presented in fig. 3.
From these figures, it is easy to observe the progressive lifting (ballooning) of the soft top with increasing wind speed.
The profile shown in Fig. 3A indicates a soft top where the fabric is linked to a certain number of cross arches,in order to reduce the ballooning.
The profile shown in Fig. 3B indicates a soft top without cross arches.
Therefore, the amount of ballooning is greater; however the longitudinal profile remains more uniform.
These two possible solutions present different aesthetic qualities, different difficulties in engineering and production, and different costs.

From the point of view of drag excessive ballooning of the fabric can increase the aerodynamic drag coefficient and worsen the aesthetics; therefore, it is preferable to look for designs which prevent too much deformation of the soft top, even at very high speeds.
Usually a soft top, even in the closed position, increases the Cd value compared with the hard-top version by at least ΔCd = 0.015 - 0.030.

With regard to evaluation of the aero-acoustic characteristics,in case of a two-seater convertible car, this is done by installing two dummies on the two (front) seats, and in case of a four-seater convertible car, by installing a dummy on the driver's seat (left side) and a second dummy on the rear seat (right side). Bruel & Kjaer head & torsos are used as dummies.
Measurements are made on the following configurations:

- basic car, that is to say a car with soft top closed
- basic car without some of the extension components which are more easily removable (rearview mirror and wind-shield wipers); these, being very close to the driver's head, are likely to be the source of undesirable aerodynamic noise
- basic car taped along the soft-top edges to the windshield upper edge,side glass, backlite contour, etc., to stop the infiltration noise.

A soft top has reduced acoustic insulation compared with the hard-top version. Furthermore, the possibility of noise infiltration through the sealing around the soft-top contour is increased.
The presence of a backlite made of plastic produces an additional path of noise entry, compared with one made of glass.
The following figures show examples of SPL (A weighted) measurements made inside the car at V = 100 and 140 Km/h,for different conditions of soft-top sealing, etc.
Figs. 4 and 5 show the noise inside a two- and a four-seater convertible car.
Fig. 6 compares the interior noise of two two-seater convertible cars: the first has a glass backlite and a conventional soft top, while the second has a plastic backlite and a better sound insulated soft top. As a result the first car has a lower interior noise at the lower sound frequencies, due to the glass backlite, while the second car has a lower interior noise at the higher sound frequencies, due to the better sound insulation of the soft top.
An additional acoustic measurement is usually carried out by measuring the RASTI (Rapid Speech Transmission Index).
In this case, an acoustic signal emitted by one of the two dummies (by mean of a loud speaker incorporated in its mouth) is received and measured by the second dummy (by one of its microphone/ear) in order to estimate the effect of background noise inside the passenger compartment on the quality of speech transmission between the two dummies.
The RASTI index varies from 0 to 1, where 1 means a perfect acoustic transmission, while 0 means that, due to the back-ground noise, the signal emitted by the first dummy cannot be properly understood by the second dummy.
Fig. 7 shows an example of the RASTI measured at various wind speeds in two different convertible cars.
The first diagram regards the 500 Hz band, the second diagram the 2 KHz band. In a certain way, they can be considered as representative of two different kind of voices. From these diagrams, it can be seen that the speech transmission index at the 500 Hz band is lower for the car B (plastic backlite) than the car A (glass backlite). RASTI values for the 2 KHz band are approximately the same for the two cars; car B is slightly better due to the better sound insulation of the soft top.

CONVERTIBLE CARS WITH THE SOFT TOP OPEN.

When driving with the soft top open,(i.e. folded down), the aerodynamic characteristics of convertible cars are obviously very different from those with the soft top closed.

First of all, the air flow separates at the windshield upper edge and at its sides.
This generates a highly turbulent separated flow which, in the best case, reattaches at the end of the deck lid.
This flow dissipates a large amount of energy, therefore often producing a large increasing of drag.
Furthermore, it may produce quite annoying vortices for the car passengers.
It is therefore important,during the aerodynamic development of a convertible car, to minimize these negative effects by taking into account various geometric factors; for instance:

- the windshield slant angle
- the inward sloping of the windshield pillars and side windows
- the central roll bar (which may be used not only for passenger protection in case of roll over, but also for aerodynamic reasons)
- a rear-end height sufficient to reattach and guide the airflow over it.

Usually a car with soft top open shows a drag increase between 10% and 25% compared with the same car with the soft top closed (see fig. 8).
This drag increase can be easily explained by an analysis of the car wake.
In Fig. 9, the pressure, velocity, micro drag and vorticity maps for the same car in the two different configurations (soft top open/closed) are compared.
The big loss of total pressure behind the windshield,and the corresponding increase of micro drag due to the energy dissipated by the separated flow, can be seen.

The second important characteristics of a car with soft top open regards, as already mentioned, the level of comfort in the passenger compartment; it may noteceably worsen due to the intensity of vortices near the passengers' heads.
In Pininfarina, the analysis of the comfort level is done by mapping the flow inside the passenger compartment usually along two vertical planes (car center line and center of the driver seat) and, furthermore, along the horizontal plane which cuts the passenger compartment at the height of the dummy's ear.
These flow maps are usually made with the dummies on board and with the side windows in at least two positions, fully raised and fully lowered.
Maps in Fig. 10 show the flow characteristics in the passenger compartment of two convertible cars with soft top open.
From these flow maps and mainly from the velocity map, it is quite easy to evaluate the characteristics of the separated flow, the reattachment area, the existence of important reverse flows close to the passenger head and the way the flow enters from the car sides to the passenger comppartment.

SUMMARY

The need for more accurate aerodynamic development of new convertible cars being designed and engineered for Pininfarina has led to the setting up of new wind-tunnel experimental techniques specifically conceived for this type of car.

These techniques allow the evaluation and subsequent optimization of the aerodynamic coefficients of these cars, soft-top ballooning, internal aerodynamic noise, and comfort level inside the car when it is driven with the soft top open.

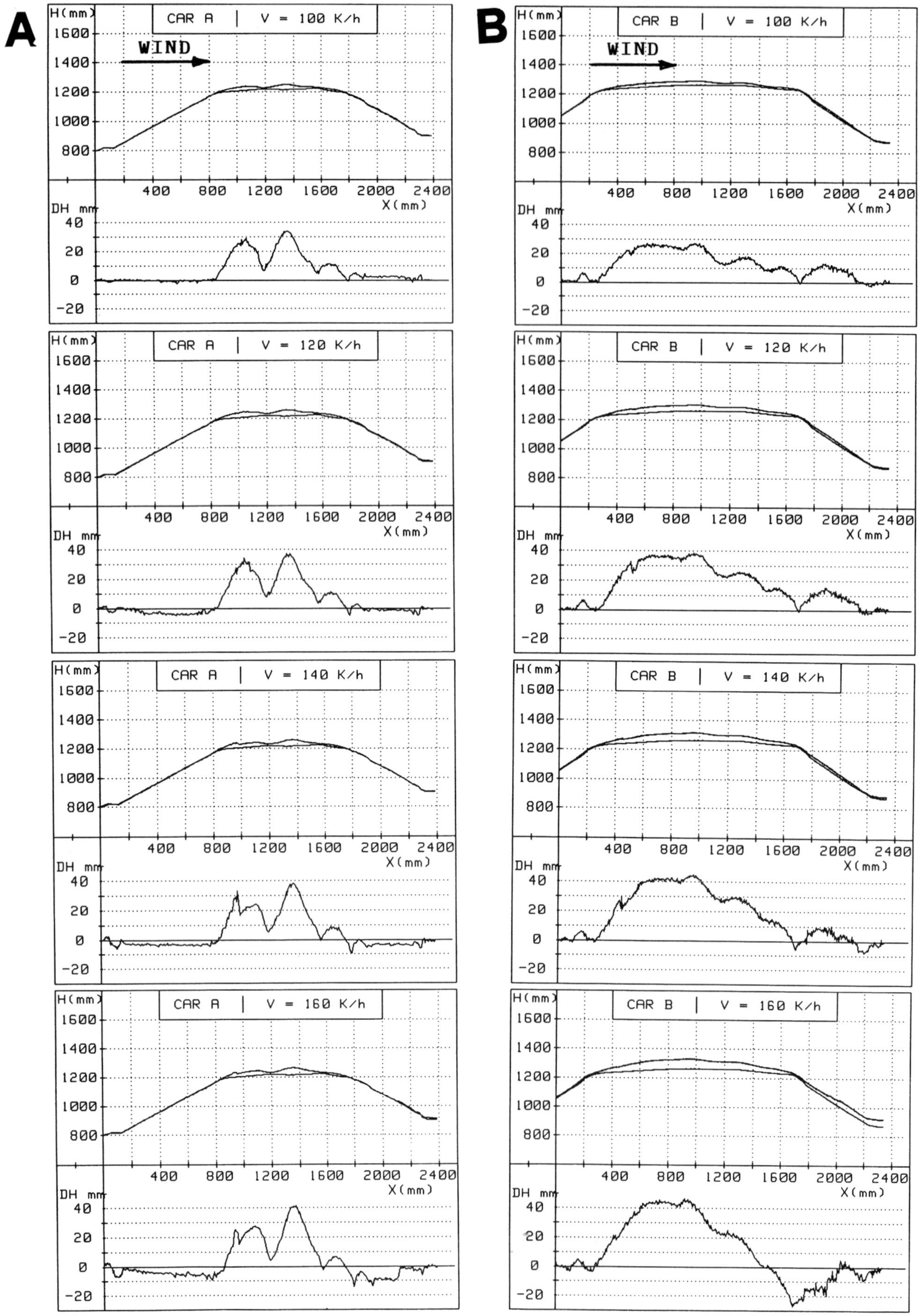

Figs. 3A,3B: ballooning of a soft top with (A) and without (B) cross arches, at velocities from 100 to 160 Km/h. The upper part of each diagram compares the centerline profile at V = 0 Km/h and at the test speed. The lower part shows the difference Dh between the two profiles (scale enlarged).

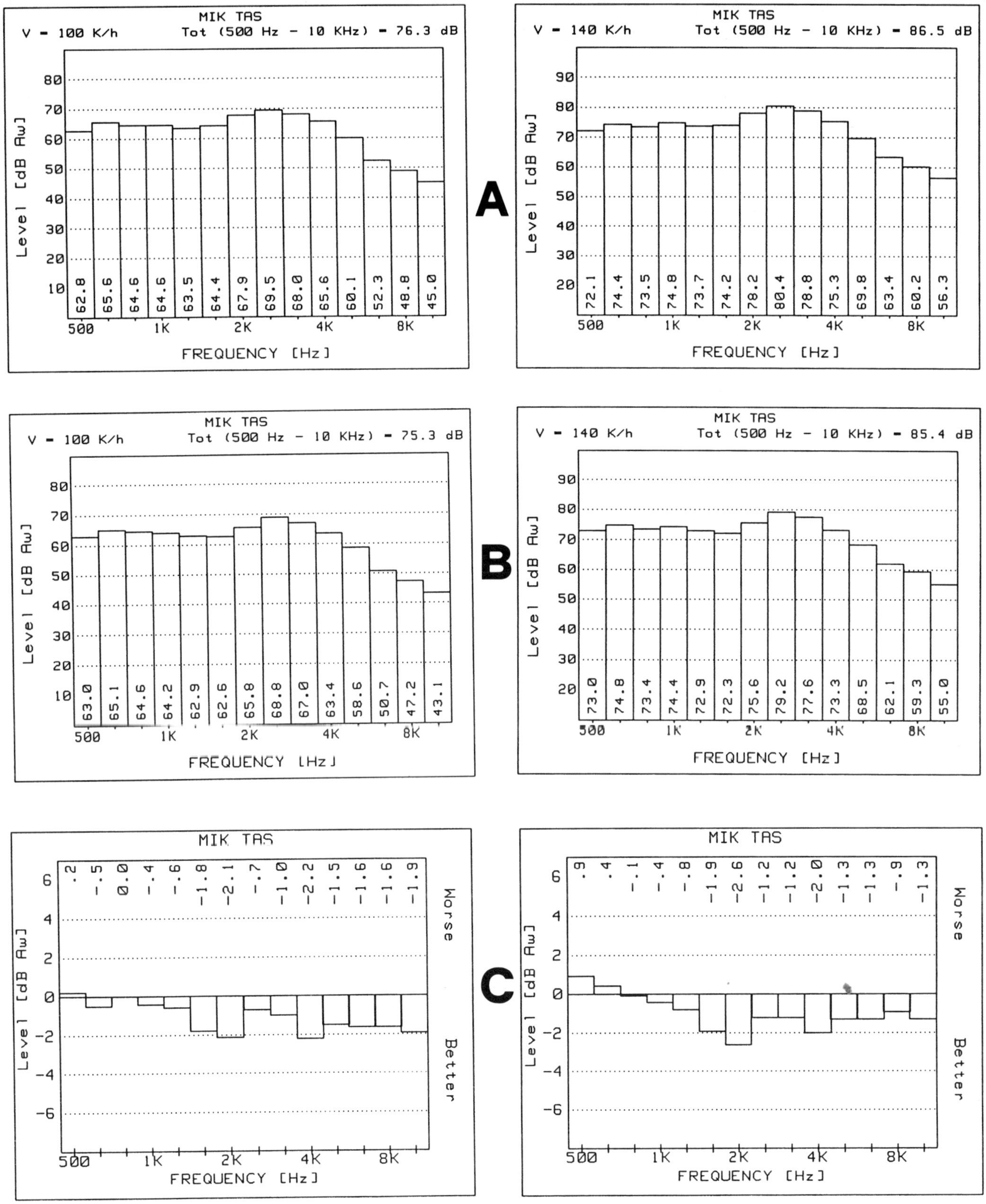

Fig. 4: SPL (Aw) values measured at the driver's seat position by a B&K dummy (left ear) at V = 100 (column at left) and 140 Km/h (column at right).

A – two-seater convertible car – glass backlite

B – as "A", plus tape around the soft-top contour

C – noise reduction due to the improved sealing

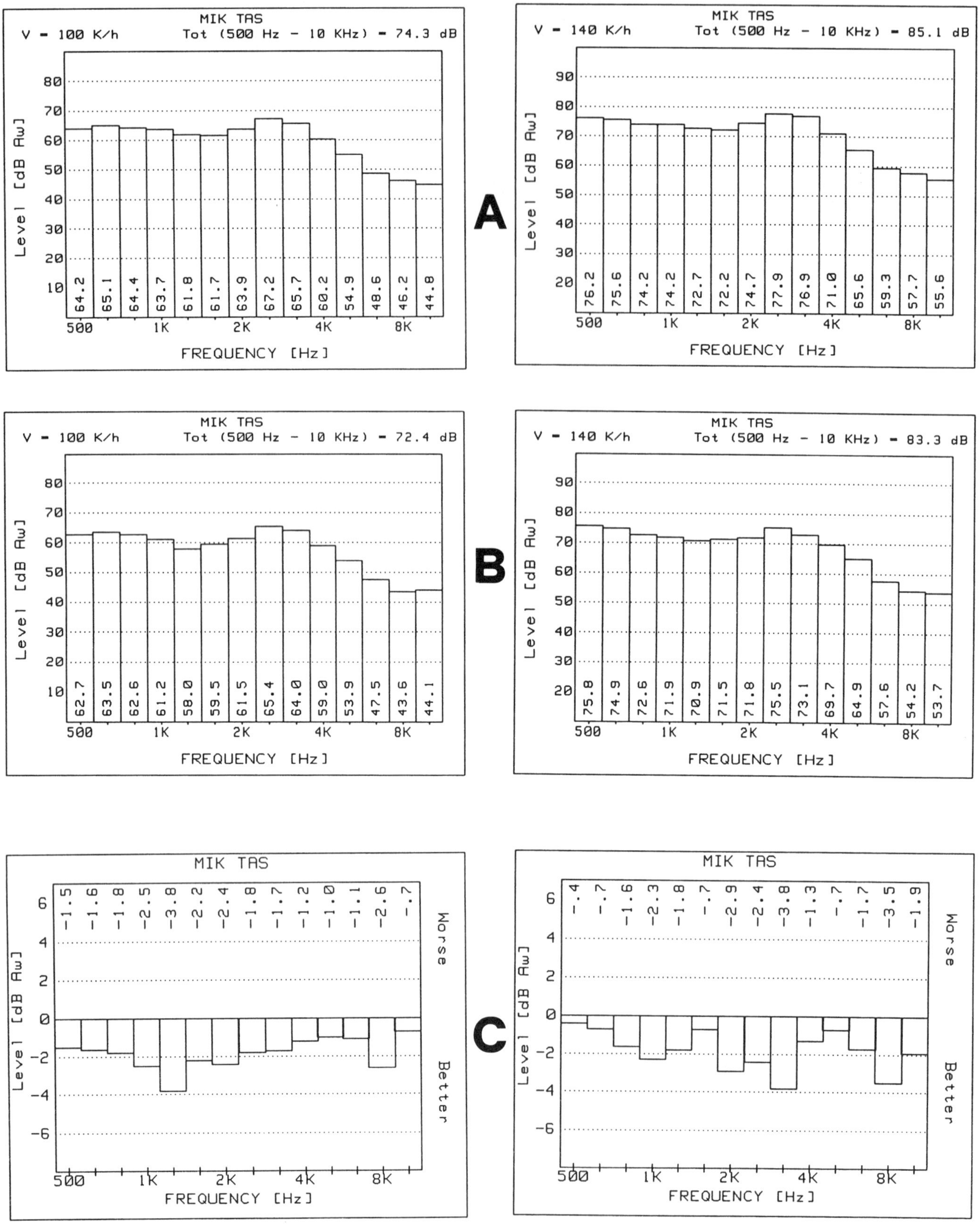

Fig. 5: SPL (Aw) values measured at the driver's seat position by a B&K dummy (left ear) at V = 100 (column at left) and 140 Km/h (column at right).

A - four-seater convertible car

B - as "A", plus tape around the soft-top contour

C - noise reduction due to the better sealing

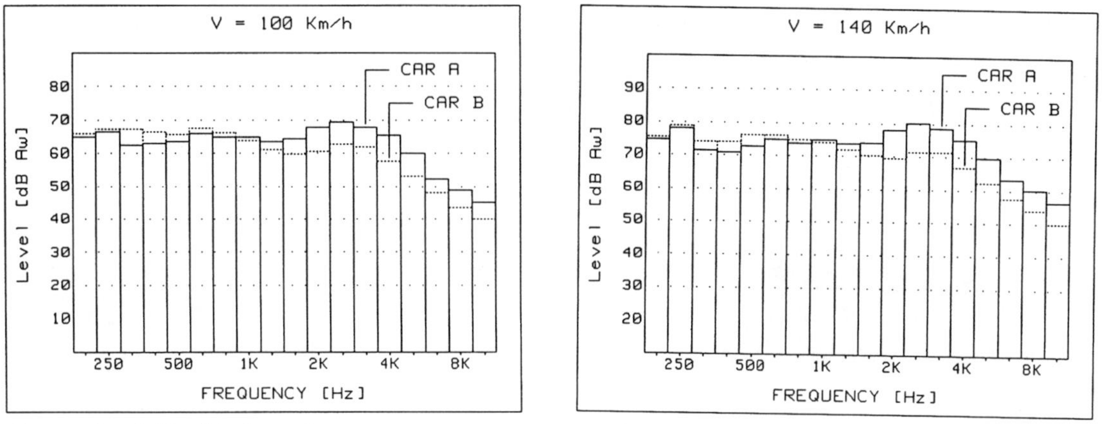

Fig. 6: SPL (Aw) values. Driver seat - dummy left ear
A - two seater - glass backlite
B - two seater - plastic backlite

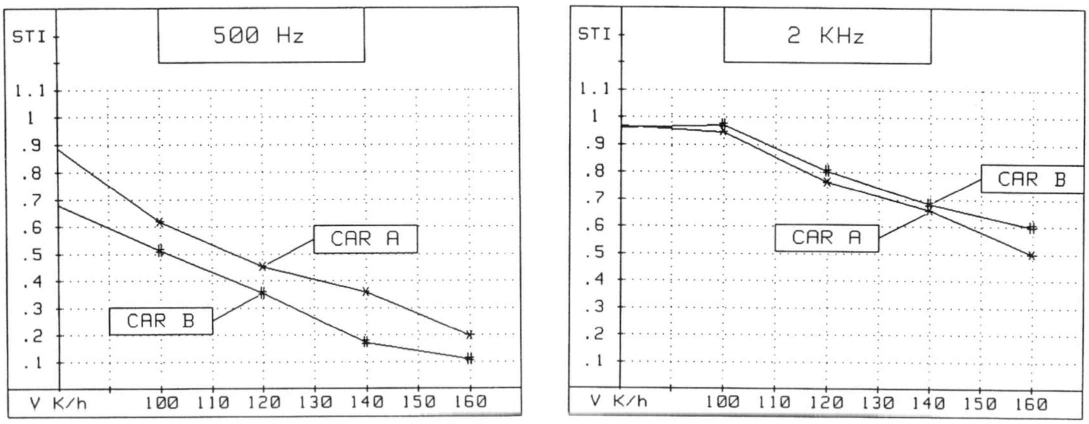

Fig. 7: RASTI INDEX
A - two seater - glass backlite
B - two seater - plastic backlite

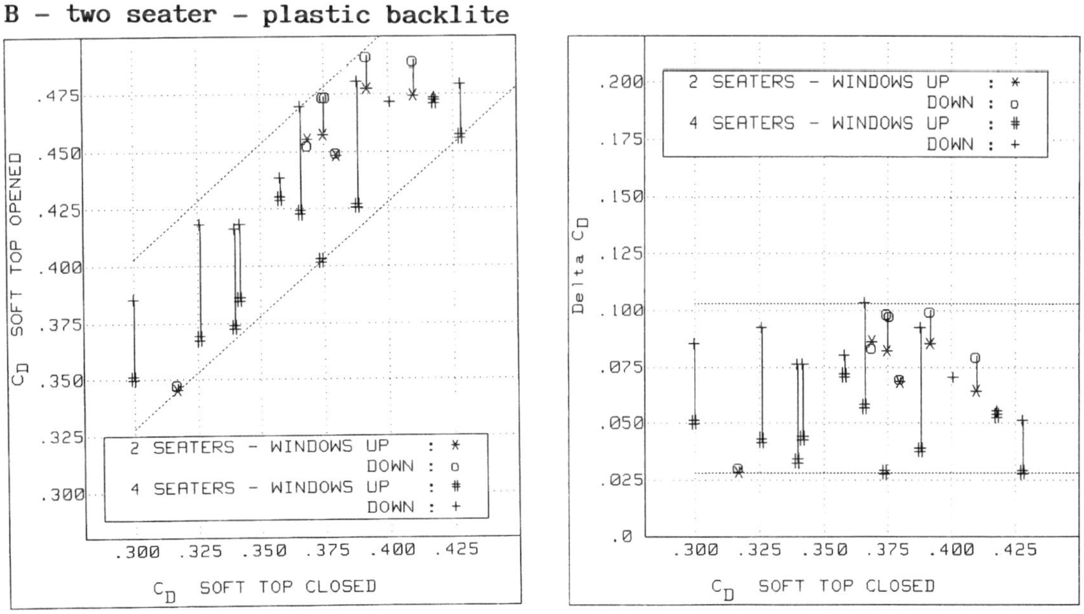

Fig. 8: increase of CD for 2 and 4-seaters convertible cars caused by opening of soft top and by lowering the side windows (courtesy of Adam Opel AG and Pininfarina Industries)
Delta CD = CD (soft top open) - CD (soft top closed)

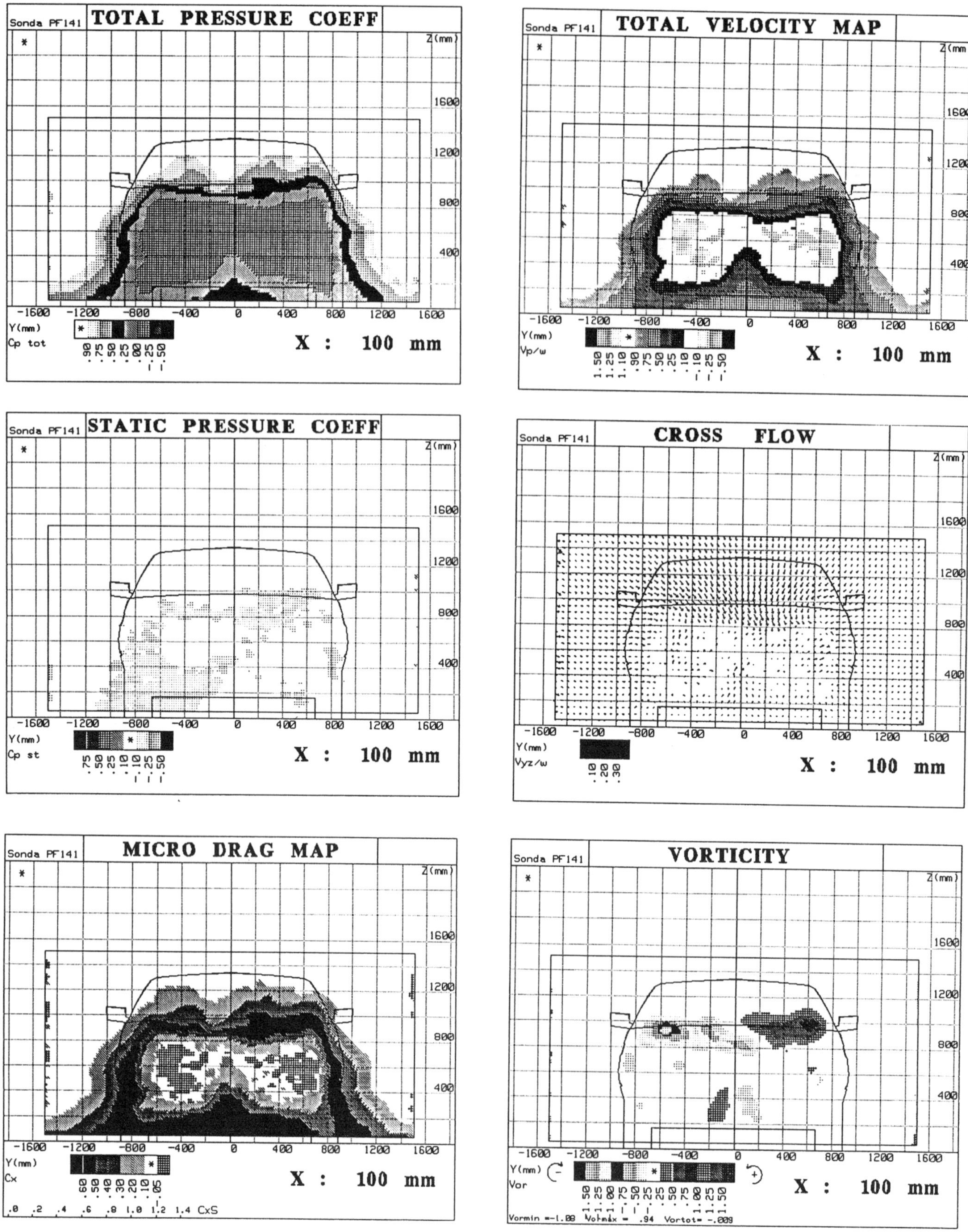

Fig. 9A: Flow field behind a convertible car ;
Soft top closed

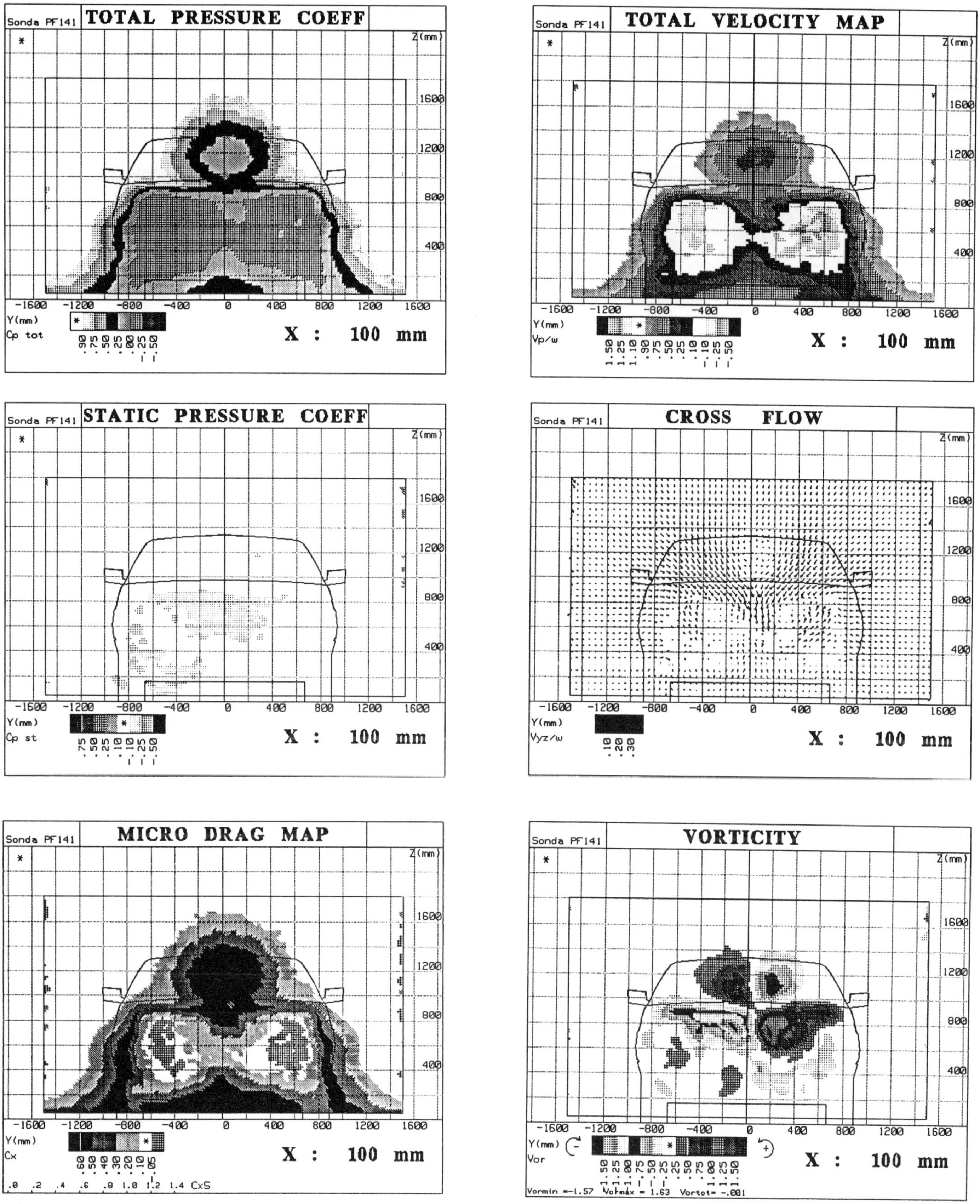

Fig. 9B: Flow field behind a convertible car;
Soft top open - Side windows up

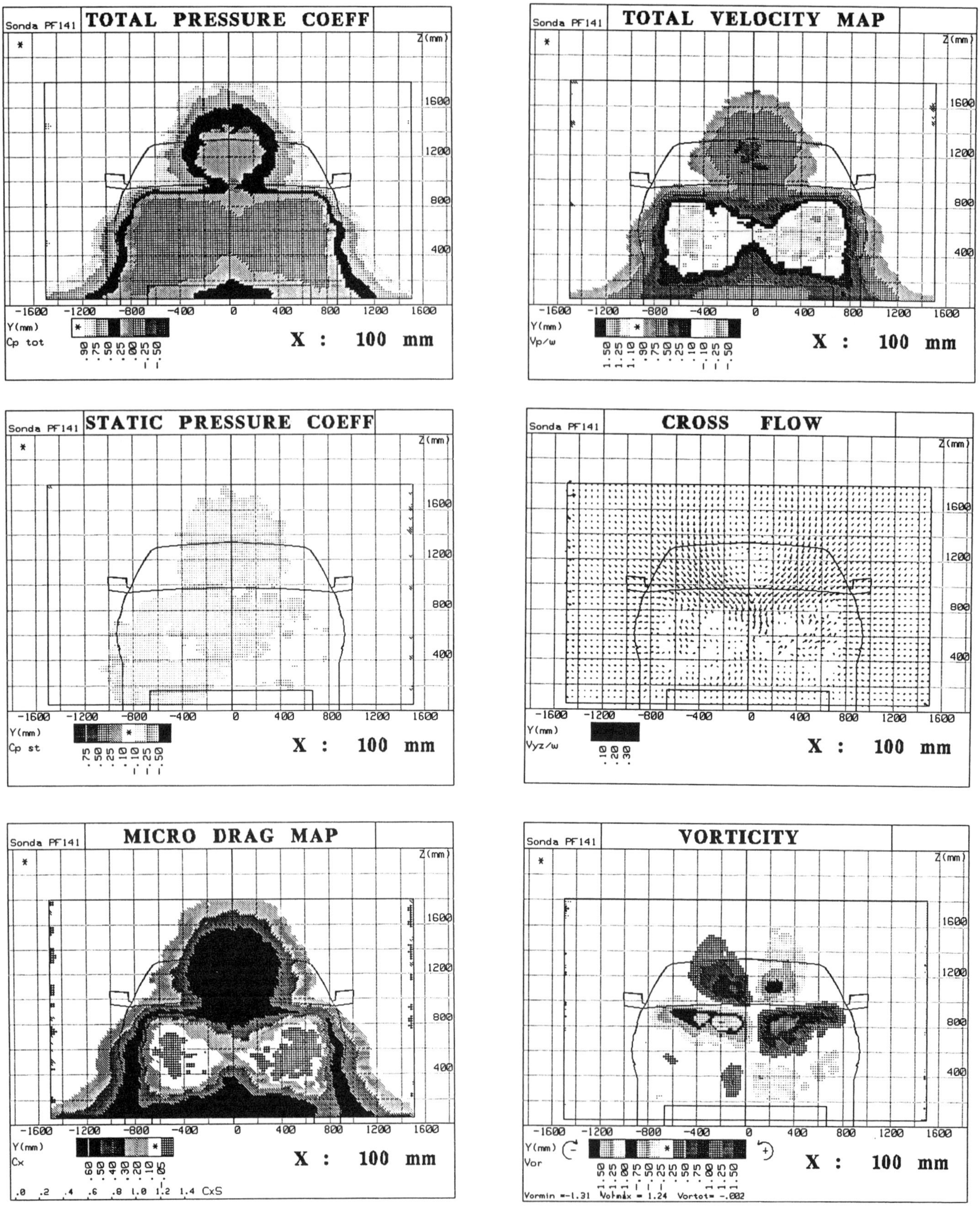

Fig. 9C: Flow field behind a convertible car;
Soft top open - Side windows down

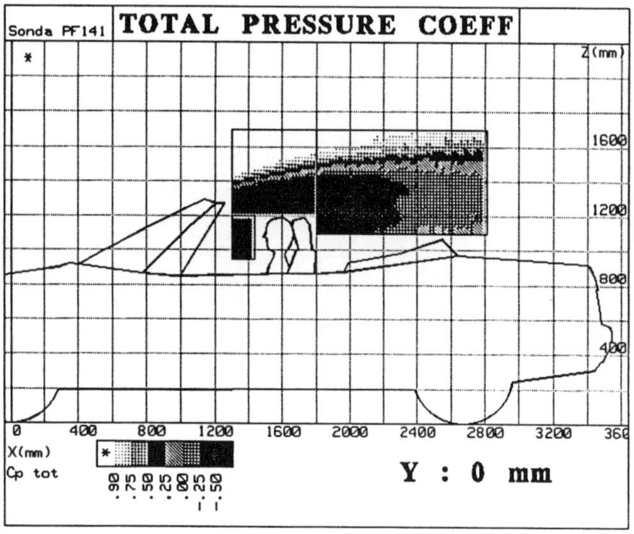

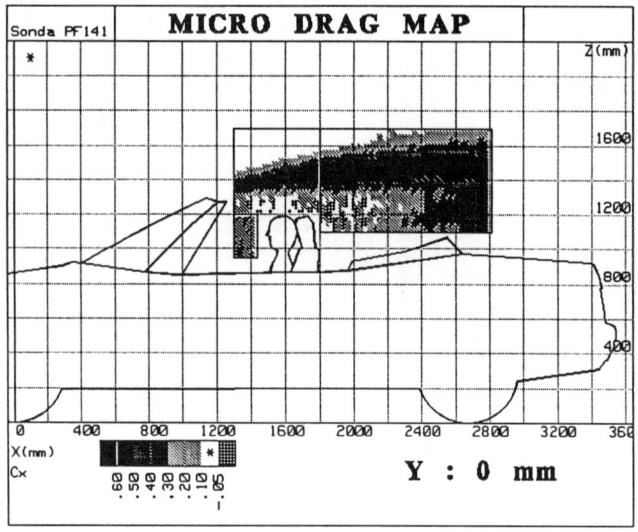

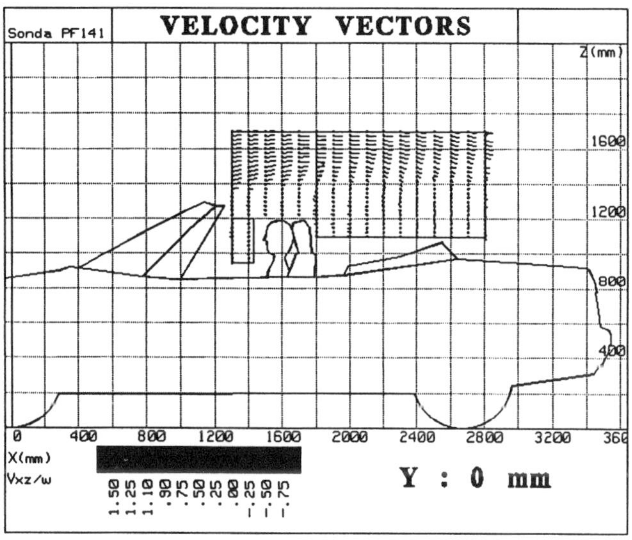

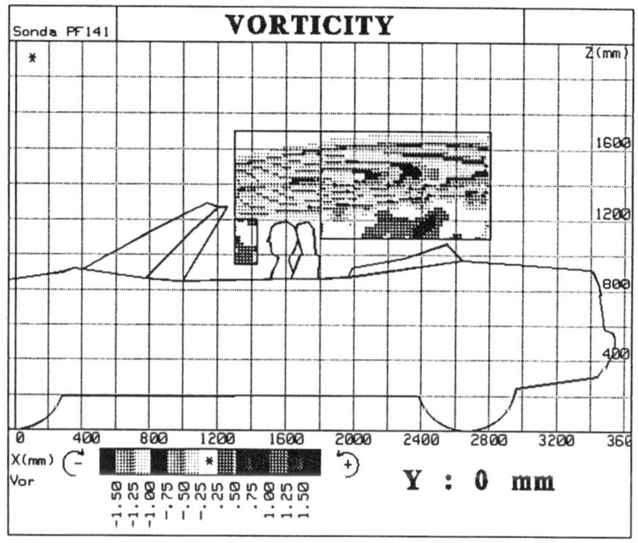

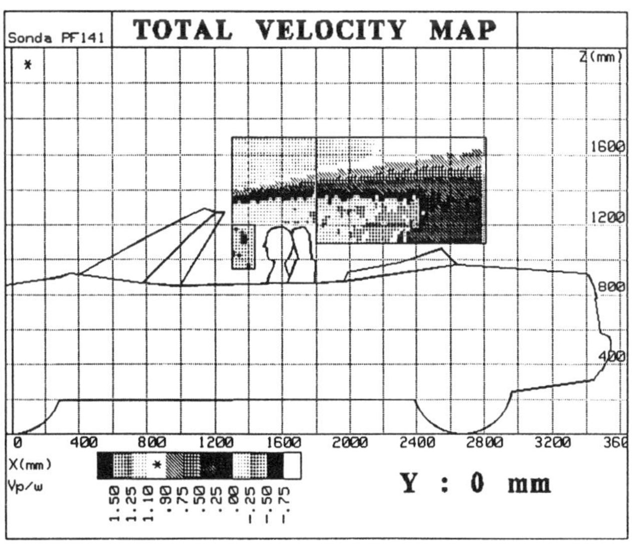

Fig. 10A: Flow maps at Y = 0 mm (center-line longitudinal vertical plane); two-seater car with soft top open and side windows lowered

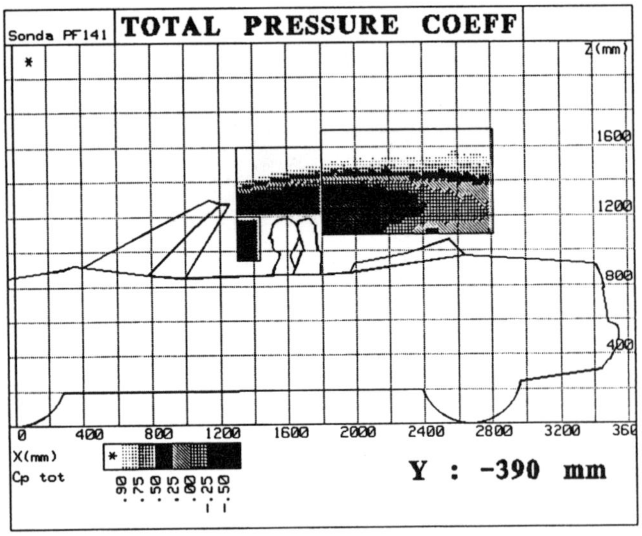

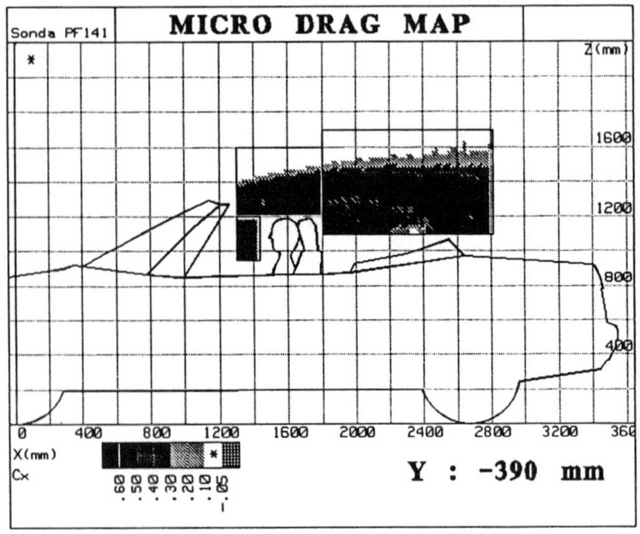

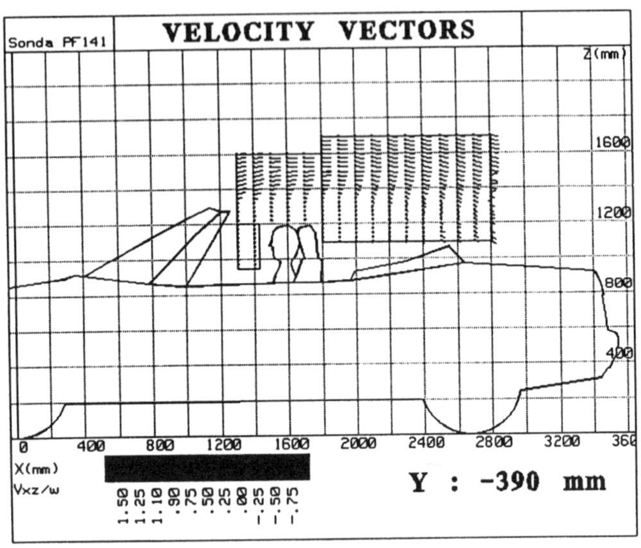

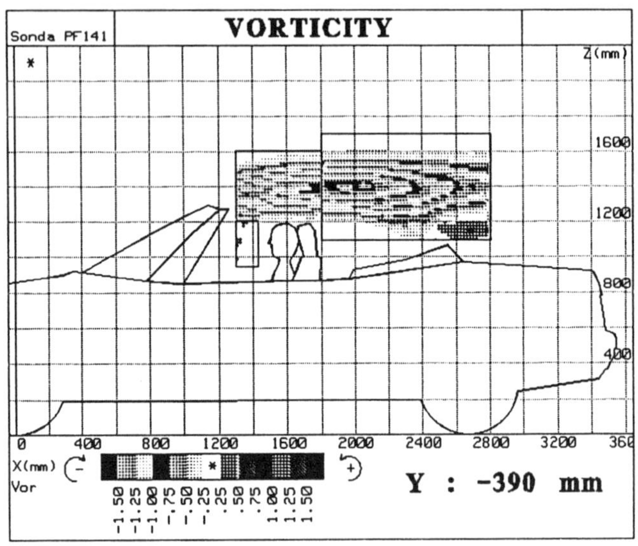

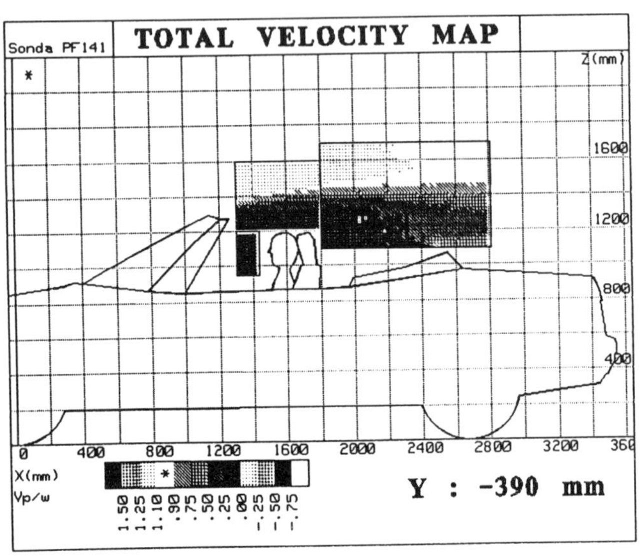

Fig. 10B: Flow maps at Y = -390 mm. Two-seater car with soft top open and side windows lowered.

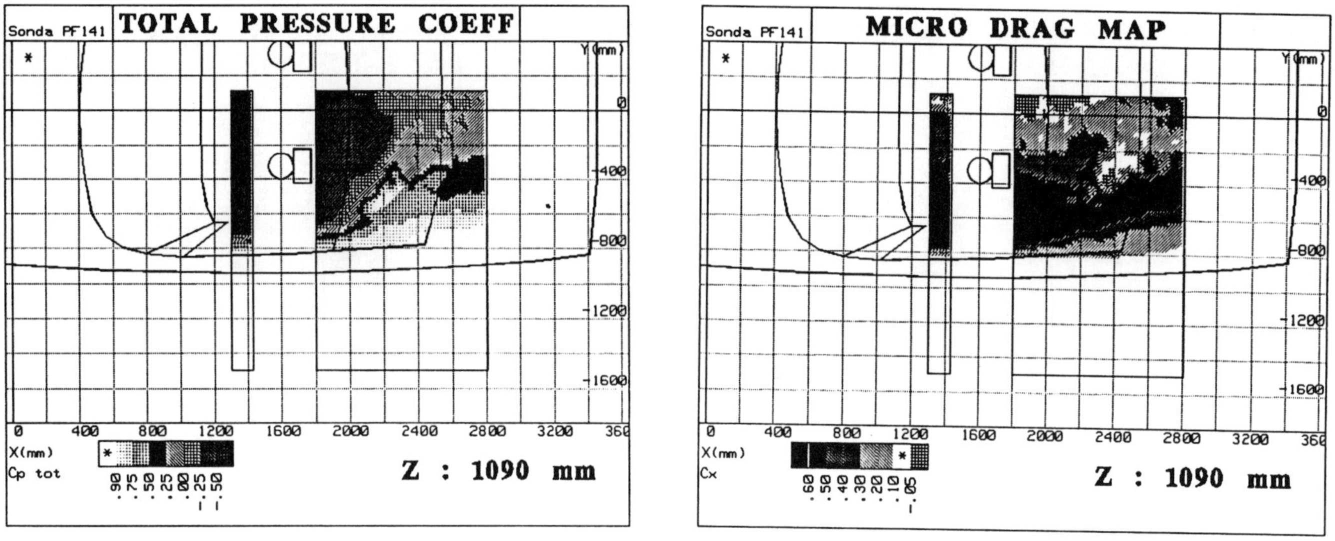

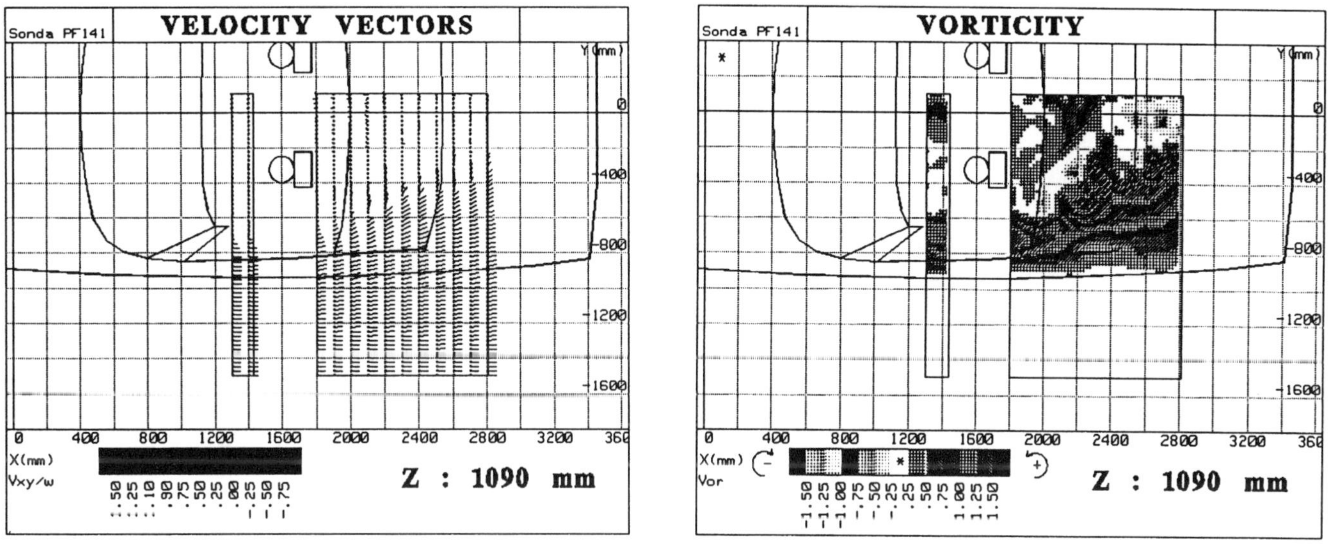

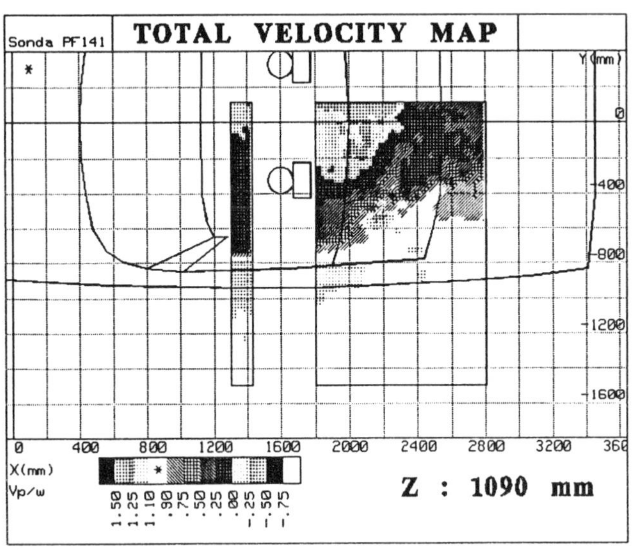

Fig. 10C: Flow maps at Z = 1.09 m (dummy ear height).
Two seater car with soft top open and side windows down

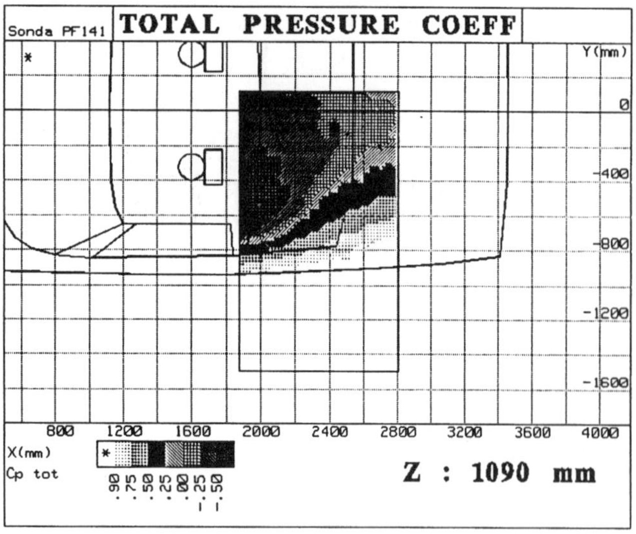

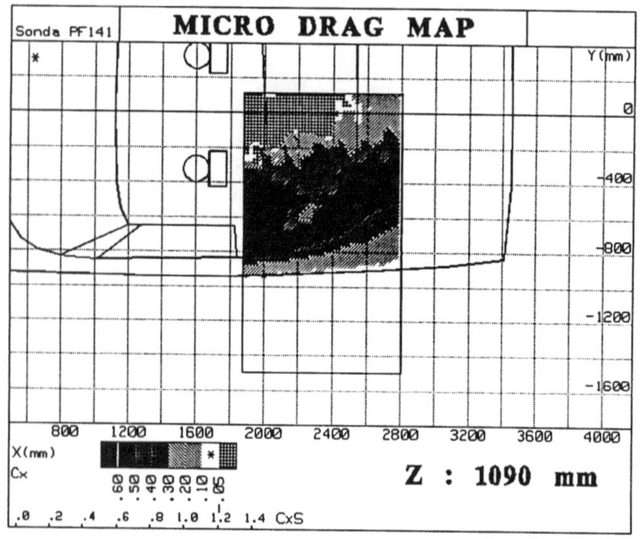

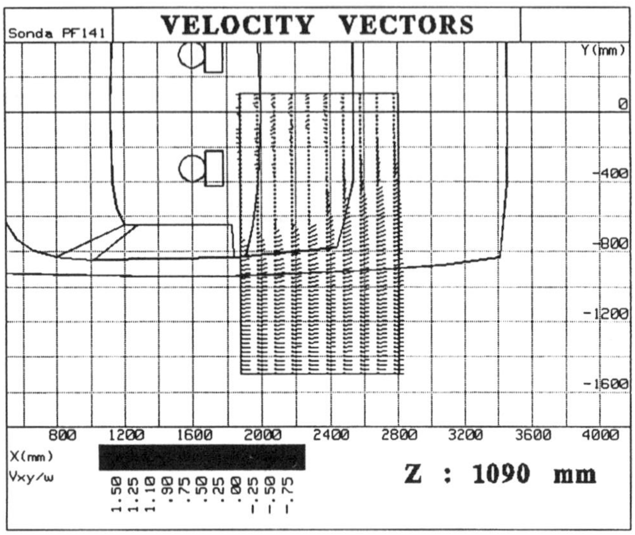

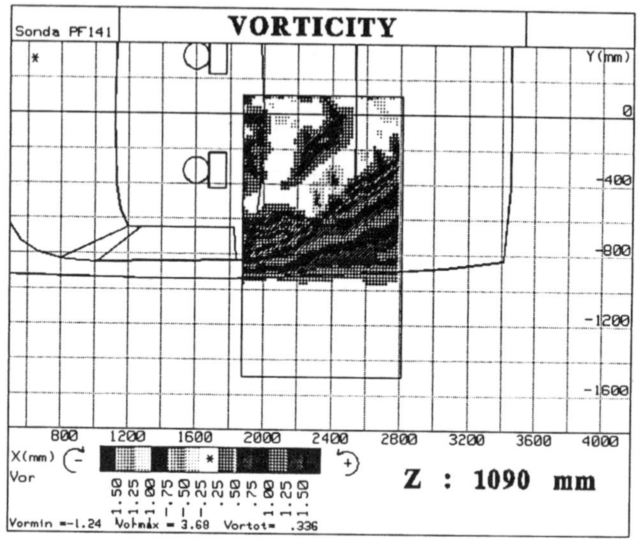

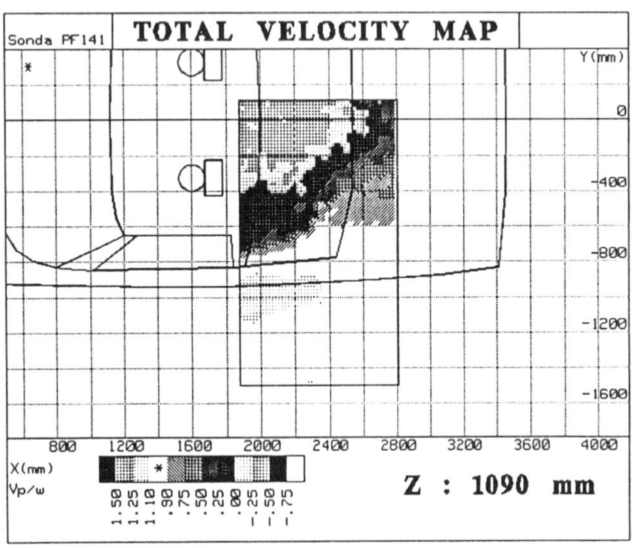

Fig. 10D: Flow maps at Z = 1.09 m (dummy ear height).
Two-seater car with soft top open and side windows up.

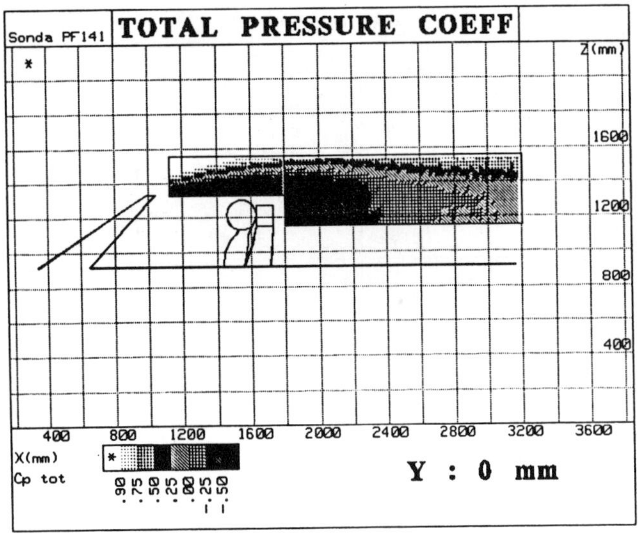

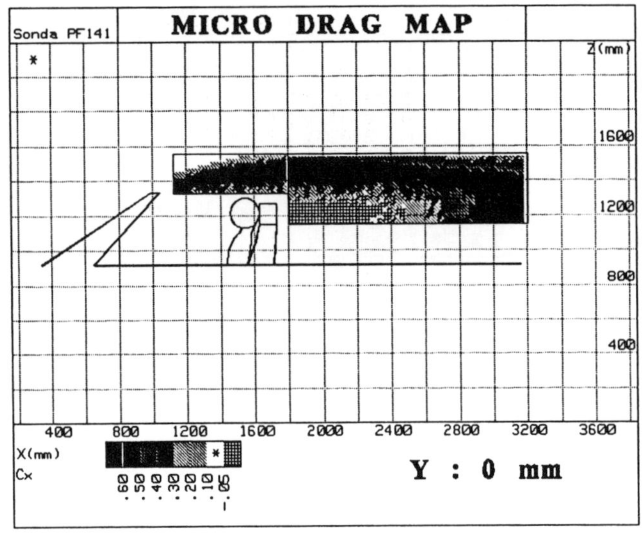

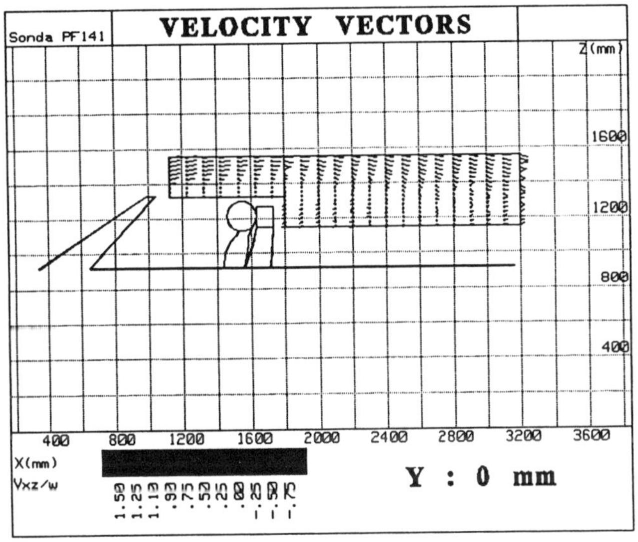

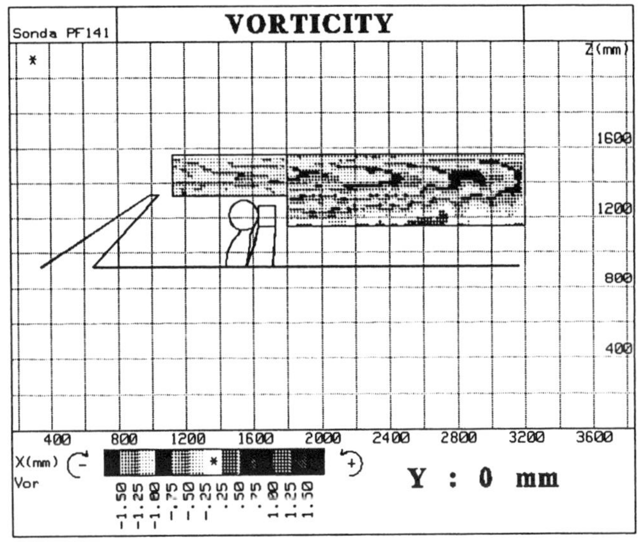

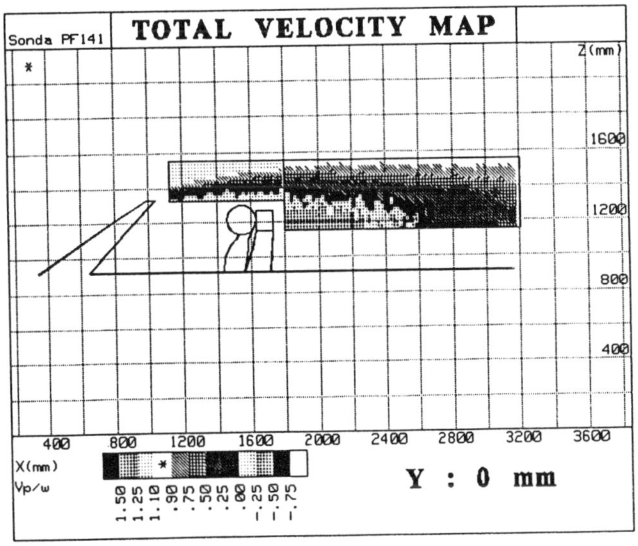

Fig. 10E: Flow maps at Y = 0 mm (center-line longitudinal vertical plane). Four-seater car with open soft top and side windows lowered

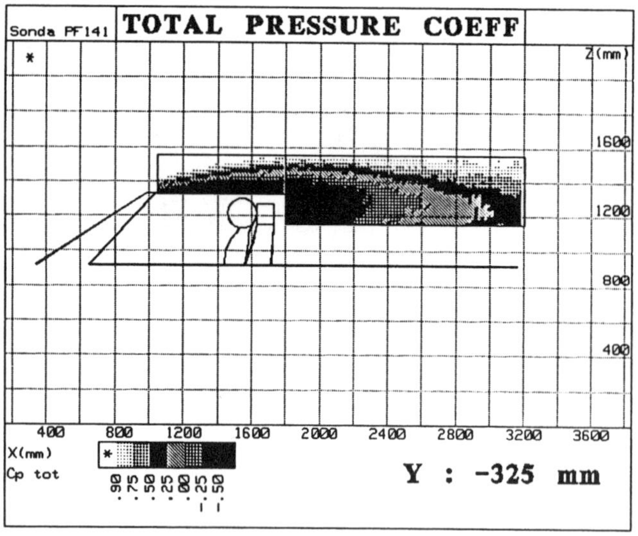

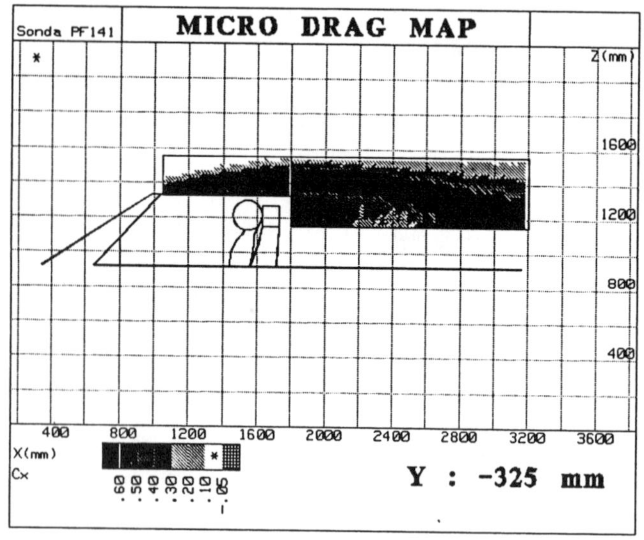

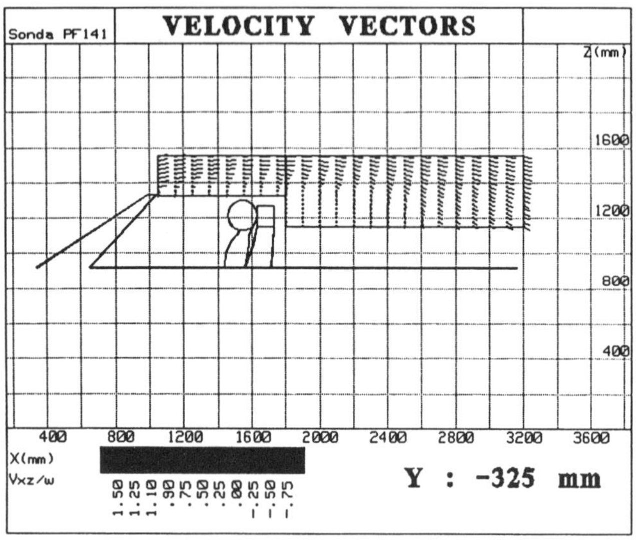

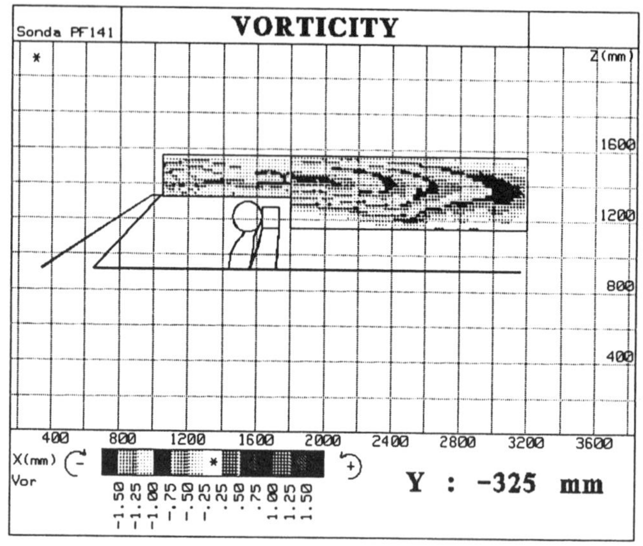

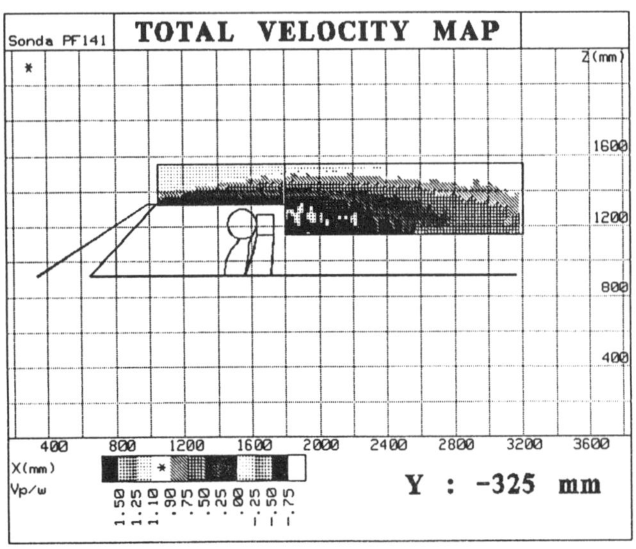

Fig. 10F: Flow maps at Y = -325 mm. Four-seater car with open soft top and side windows lowered.

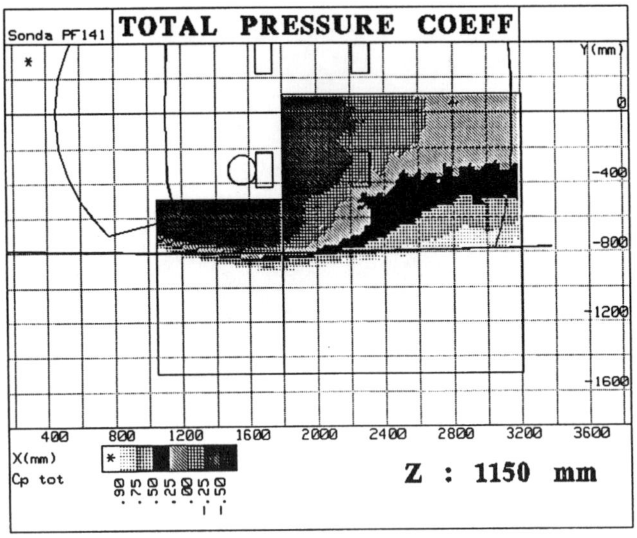

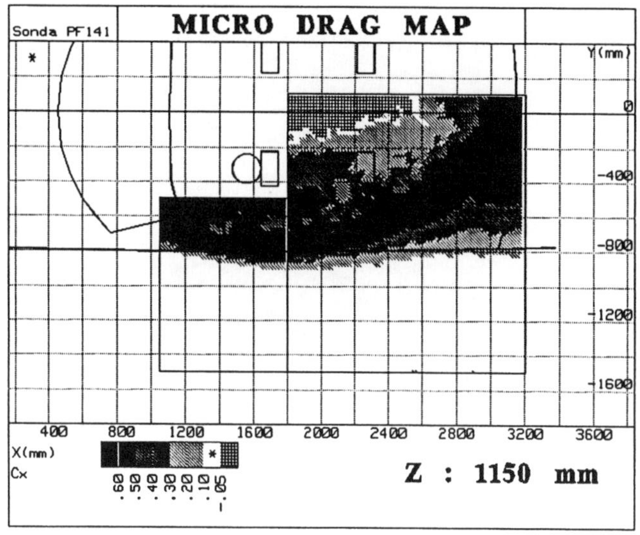

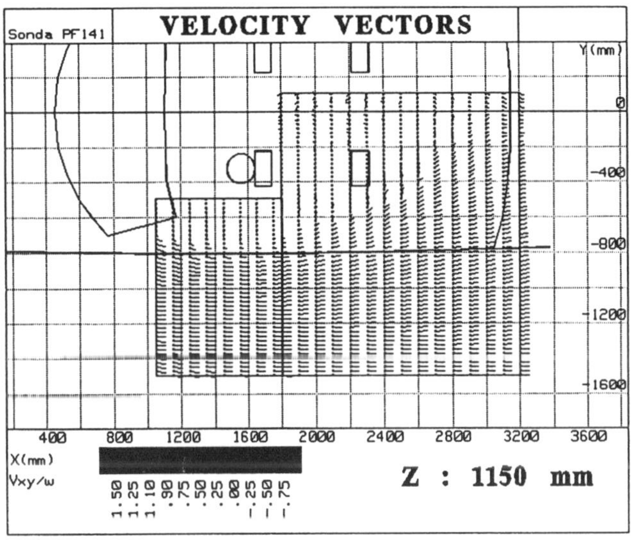

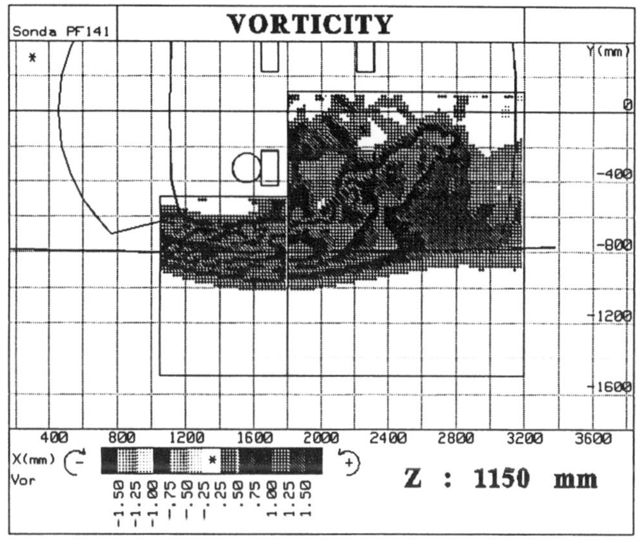

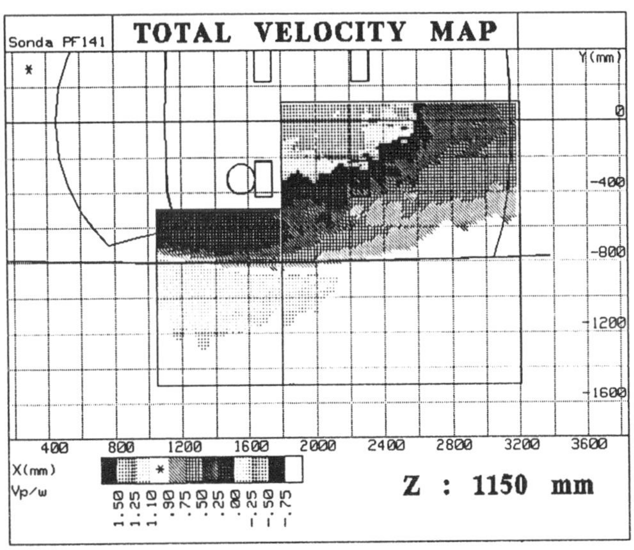

Fig. 10G: Flow maps at Z = 1.15 m (dummy ear height).
Four-seater car with open soft top and side windows down

910314

Aerodynamics of Road Vehicles in Tunnels — Flow Visualization Using the Laser Light Sheet Method and Its Digital Image Processing

Kiyoyuki Minato and Hano Ryu
Japan Automobile Research Institute, Inc. (JARI)

Toshio Kobayashi
Institute of Industrial Science, University of Tokyo

ABSTRACT

For the past several years, construction of Japanese road network has maintained a rapid pace, but those roads, having many automotive tunnels, are affected by mountainous topography. The air pollution in tunnels by exhaust gases from cars, especially those from large-size vehicles, has become problematical in Japan.

The purpose of this paper is to clarify the flow behavior, especially wake structure behind trucks, and their aerodynamics characteristics at overtaking in road vehicle tunnels with digital image processing technique and three dimensinoal numerical analysis.

Recently the high performance of the road vehicle, especially in high speed, is strongly desired. The improvement of vehicle aerodynamics and the clarification of its effect on vehicular motion are gradually being done. Although many works and data on the vehicle aerodynamics are published [1,2,3,4], a few argue relations between the flow around the vehicle, especially the wake structure, and the aerodynamic characteristics of the vehicle [5,6,7,8].

Japanese roads and their traffic are characterized by their topographical nature due to mountainous land. Exhaust gas in road vehicle tunnels is assumed to be transported partly with the air flow of vehicles and partly removed by exhaust fans of the tunnels. In an overtaking process in a tunnel, the drag coefficient reaches a maximum when the front end of the overtaking vehicle reaches the rear end of a vehicle being overtaken and the flow pattern behind the vehicle being overtaken shows a specific pattern. The lack of reports on overtaking process in road vehicle tunnels prompted us to investigate it [9,10,11,12].

The purpose of this paper is to clarify the effect of wake structure behind the road vehicle and its aerodynamic characteristics at overtaking in road vehicle tunnels. The drag and pressure around road vehicles at overtaking were measured using force sensors and pressure tranducer. Flows in the wake, especially those at overtaking, were observed by a laser light sheet method and digital image processing technique [13,14]. Results were then compared with numerical results obtained by the finite volume method with the $\kappa - \epsilon$ model. The numerical results agreed well with those of the experiment.

EXPERIMENTAL EQUIPMENT AND PROCEDURES

The facility used for this series of tests was the low-speed wind tunnel at JARI. The tunnel is open and single-return type with a test section of 600 mm in height, 600 mm in width and 1200 mm in length and maximum speed of 50 m/sec. The tunnel model used was a 1/40 scale model of ENASAN tunnel, one of the most famous long road tunnels in Japan. The model was made of transparent acrylic resin. The tunnel model was 212 mm in width, 110 mm in height and 2000 mm in length. The model was mounted on a beam in the wind tunnel. In this experiment rectangular

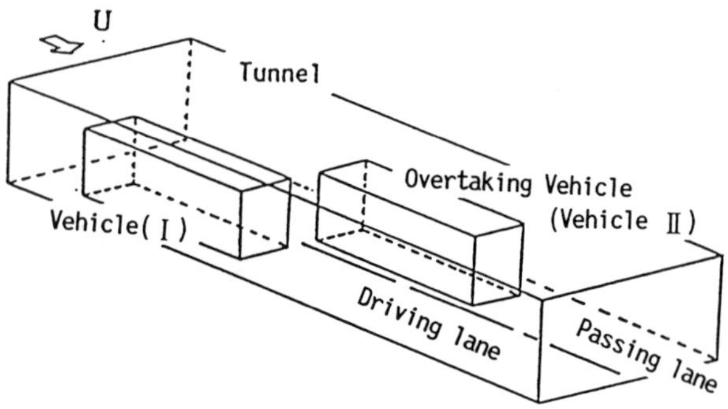

Fig 1. Overtaking Situation in a tunnel

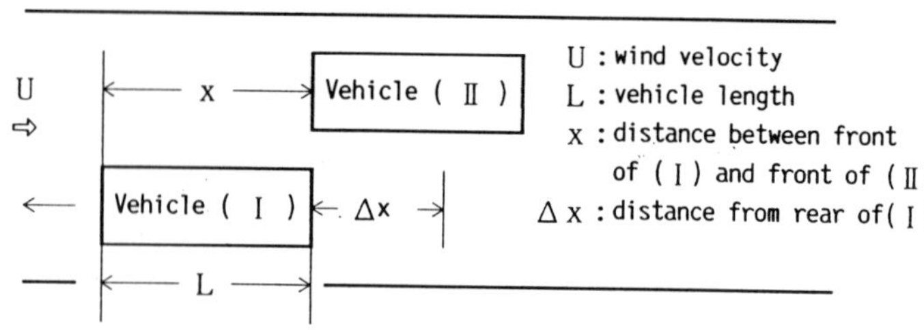

Fig 2. Geometry of Overtaking Situation

parallelepipeds of 1/40 scale (width : 62 mm, height : 72 mm, length : 300 mm) were used as large-size trucks. Figure 1 shows overtaking situation in a tunnel. The definitions of geometry are shown in Figure 2. The aerodynamic coefficients were determined from 30-second measurements of the balance output. The pressure distribution along vehicle centerlines section was measured by a pressure transducer. Experiments were carried out at a wind velocity of 20 m/sec and Reynolds number of 95000.

FLOW VISUALIZATION

Flow patterns of various overtaking situations were made visible by the smoke method. The smoke was injected from a small tube. For trace particles, kerosene was heated and vaporized by an electric heater at the front of the tunnel. To visualize the detailed characterristics of turbulent flow around the vehicles in tunnels, a laser light sheet was used. This system can clearly visualize details of the turbulent flow around vehicles and is helpful for clarifying the mechanism of wake structures. The source of the light sheet is an argon-ion laser (power : 4W , thickness of light sheet : 10 mm). The visualized flow is photographed by a CCD camera and recorded on video tape.

IMAGE PROCESSING TECHNIQUE

As to the image processing technique, the information of a flow field is directly image-input by using the image input system consisting of CCD cameras (570× 485 image elements) and a image processing board (512 × 512 image elements, 256 gradations/pixel). This study tries to evaluate quantitatively flow patterns by the image processing of the spatial data of superimposed input images. The image data obtained from visualized photographs are often accompanied by nonuniformity in the concentration due to uneven illumination at the time of taking photographs. To overcome these demerits, the nonuniformity in the concentration distribution is made smooth by superimposing the input image. In this study, a mean image is calculated from 10 visualized images around a vehicle body in the same conditions, and the mean brigthness is obtained as a flow pattern. The definitions of the mean brightness are given as follows :

$$\text{mean brightness}(\overline{I(x,y)}) = \frac{1}{M} \Sigma (I(x,y))$$

$I(x,y)$ = brightness on the pixcl

M = number of images

Numerical Method

Now, we treat the numerical flow field around the vehicles as three dimensional unsteady incompressible viscous flow. The equation for $\kappa - \epsilon$ turbulent model are written as follows :

Continuity equation :

$$\frac{\partial U_i}{\partial x_i} = 0$$

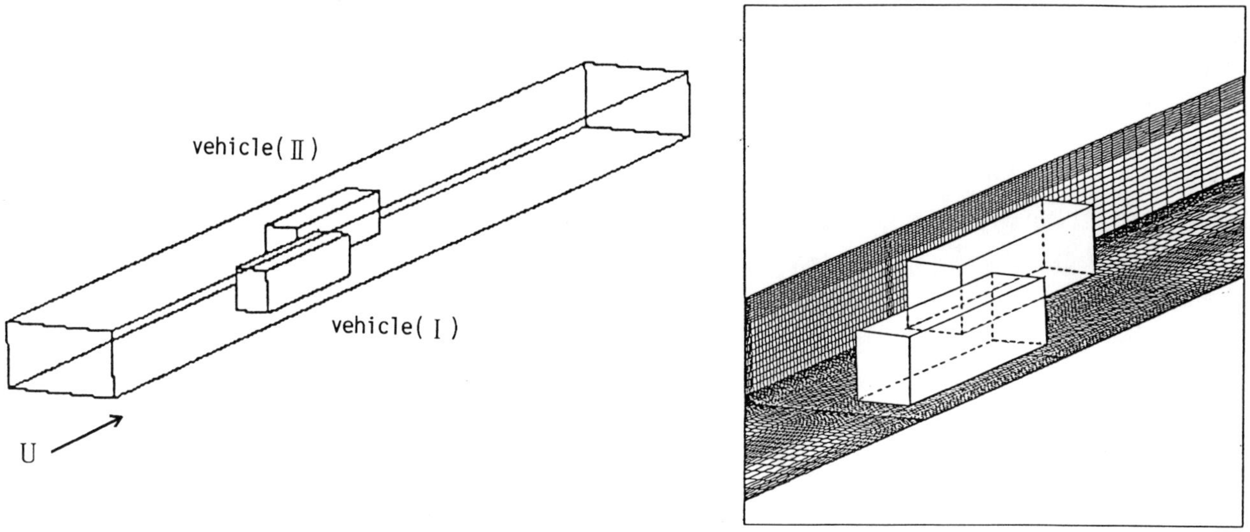

Fig 3. Grid distributions global view and near the vehicles

Navier-Stokes equation :

$$\frac{\partial U_i}{\partial t} + \frac{\partial}{\partial x_j}(U_i U_j) = -\frac{\partial P}{\partial x_i} + \frac{\partial}{\partial x_j}\left\{-\overline{u_i' u_j'} + \frac{1}{Re}\left(\frac{\partial U_i}{\partial x_j} + \frac{\partial U_j}{\partial x_i}\right)\right\}$$

where, velocity u_i, length x_i, pressure p, and time t are non-dimensionalized by the inlet velocity U_{in}, ρU_{in}^2 (ρ = density), vehicle hight H, and time H/U_{in}, respectively. The Reynolds number Re is defined by $U_{in}H/\nu$ (ν = kinematic viscosity). The Reynolds stress $-\overline{u_i' u_j'}$ is expressed as follows in the $\kappa-\varepsilon$ model:

$$-\overline{u_i' u_j'} = \nu_t\left(\frac{\partial U_i}{\partial x_j} + \frac{\partial U_j}{\partial x_i}\right) - \frac{2}{3}\kappa\delta_{ij}$$

The eddy kinematic viscosity ν_t is defined by

$$\nu_t = C\mu\,\kappa^2/\varepsilon$$

Assuming local homogeneous turbulence, the final $\kappa-\varepsilon$ equations are written as follows :

$$\frac{\partial \kappa}{\partial t} + \frac{\partial}{\partial x_i}(\kappa U_j) = \frac{\partial}{\partial x_j}\left\{\left(\frac{\nu_t}{\sigma_1} + \frac{1}{Re}\right)\frac{\partial \kappa}{\partial x_j}\right\} + \nu_t\left(\frac{\partial U_i}{\partial x_j} + \frac{\partial U_j}{\partial x_i}\right)\frac{\partial U_i}{\partial x_j} - \varepsilon$$

$$\frac{\partial \varepsilon}{\partial t} + \frac{\partial}{\partial x_j}(\varepsilon U_j) = \frac{\partial}{\partial x_j}\left\{\left(\frac{\nu_t}{\sigma_2} + \frac{1}{Re}\right)\frac{\partial \varepsilon}{\partial x_j}\right\} + C_1\frac{\varepsilon}{\kappa}\nu_t\left(\frac{\partial U_i}{\partial x_j} + \frac{\partial U_j}{\partial x_i}\right)\frac{\partial U_i}{\partial x_j} - C_2\frac{\kappa\varepsilon}{\nu_t}$$

where $C\mu$, C_1, C_2, σ_1 and σ_2 are numerical constants.

Calculations in this paper concern the turbulent flow with a Reynolds number of 95000 (U_{in} = 20.0 m/s , H = 72mm) around a three-dimensional vehicle corresponding to experimentl conditions in the wind tunnel. Meshes are arranged at equal and unequal intervals near and away from the vehicle, respectively, with a total number of meshs of 140 × 32 × 30. Figure 3 schematically illustrates the mesh arrangement used for the calculations. It should also be noted that the mesh is finer near the vehicle where the flow is studied in detail.

Based on the staggered mesh system central differencing is applied with regard to time and space.

It should be noted that upwind differencing is applied for the convective terms of κ and ε transport equations. With regard to the boundary conditions for the periphery of the calculational domain, constant values and zero gradients in the flow direction are imposed on the inlet and outlet planes, respectively. Solid boundary conditions are imposed on the surface of the vehicle, on the ground, upper and side planes of the tunnel, where the log-law is applied for the parallel and normal components of velocity to the wall. This paper employs Laufer's experimental values ($\kappa = 3.2\times10^{-3}$, $\varepsilon = 7.1\times10^{-4}$) and Launder and Spalding's recommended values ($C\mu = 0.09$, $C_1 = 1.44$, $C_2 = 0.1728$, $\sigma_1 = 1.0$, $\sigma_2 = 1.3$) as the κ and ε values on the inlet plane and the numerical constants in the turbulence model, respectively.

RESULTS AND DISCUSSIONS

Figure 4 shows the changes of drag coefficient during an overtaking process in a tunnel. In this experiment only the vehicle which is overtaken by the other was set on the balance. In an overtaking process drag coefficient increased gradually and reached a maximum when the front end of the overtaking vehicle reaches the rear end of a vehicle being overtaken. Figures 5 and 6 show the pressure distribution on both sides of the body in a horizontal plane and those along their longitudinal centerline at overtaking process, respectively. Most

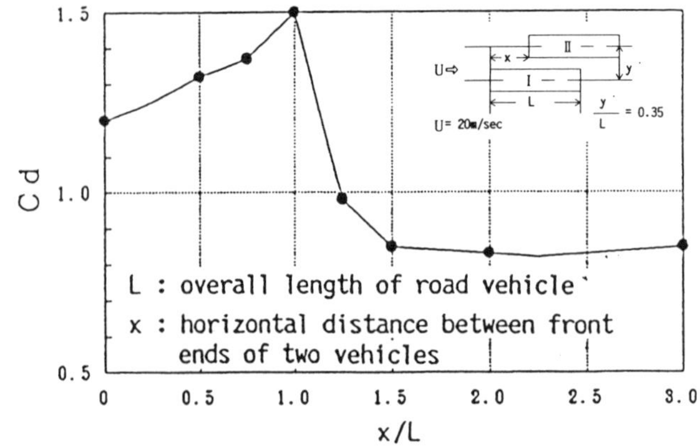

Fig 4. Development of C_D at overtaking (Vehicle I)

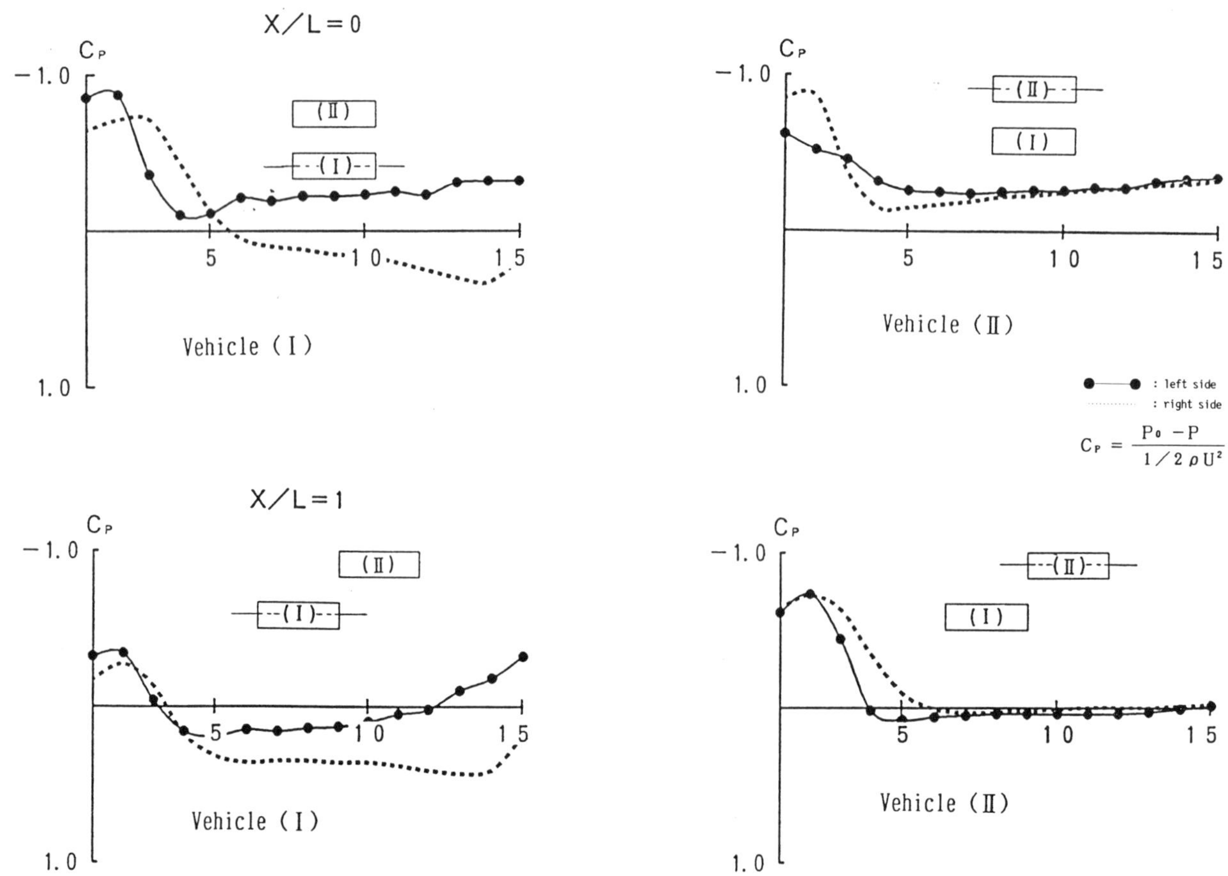

$$C_P = \frac{P_0 - P}{1/2\,\rho U^2}$$

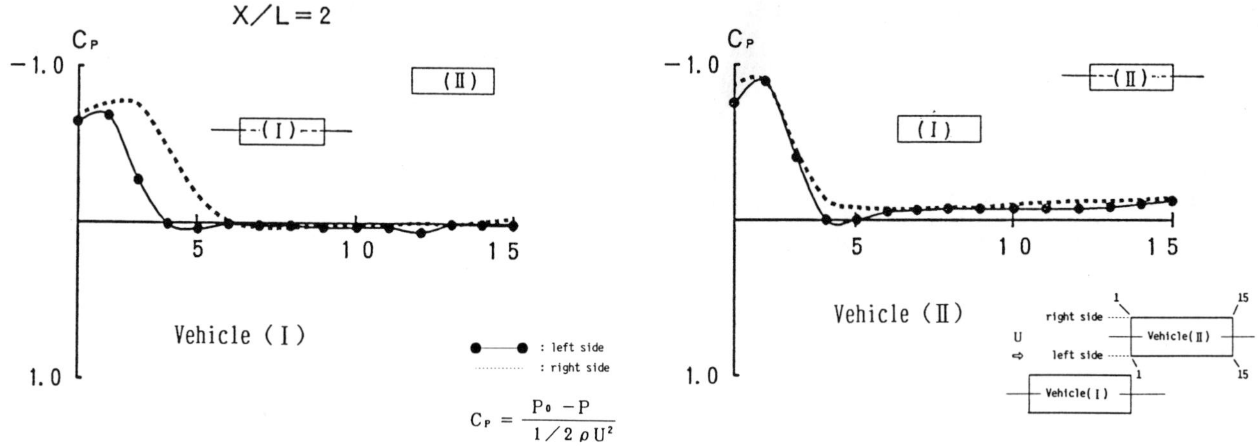

Fig 5. Pressure distribution in horizontal section

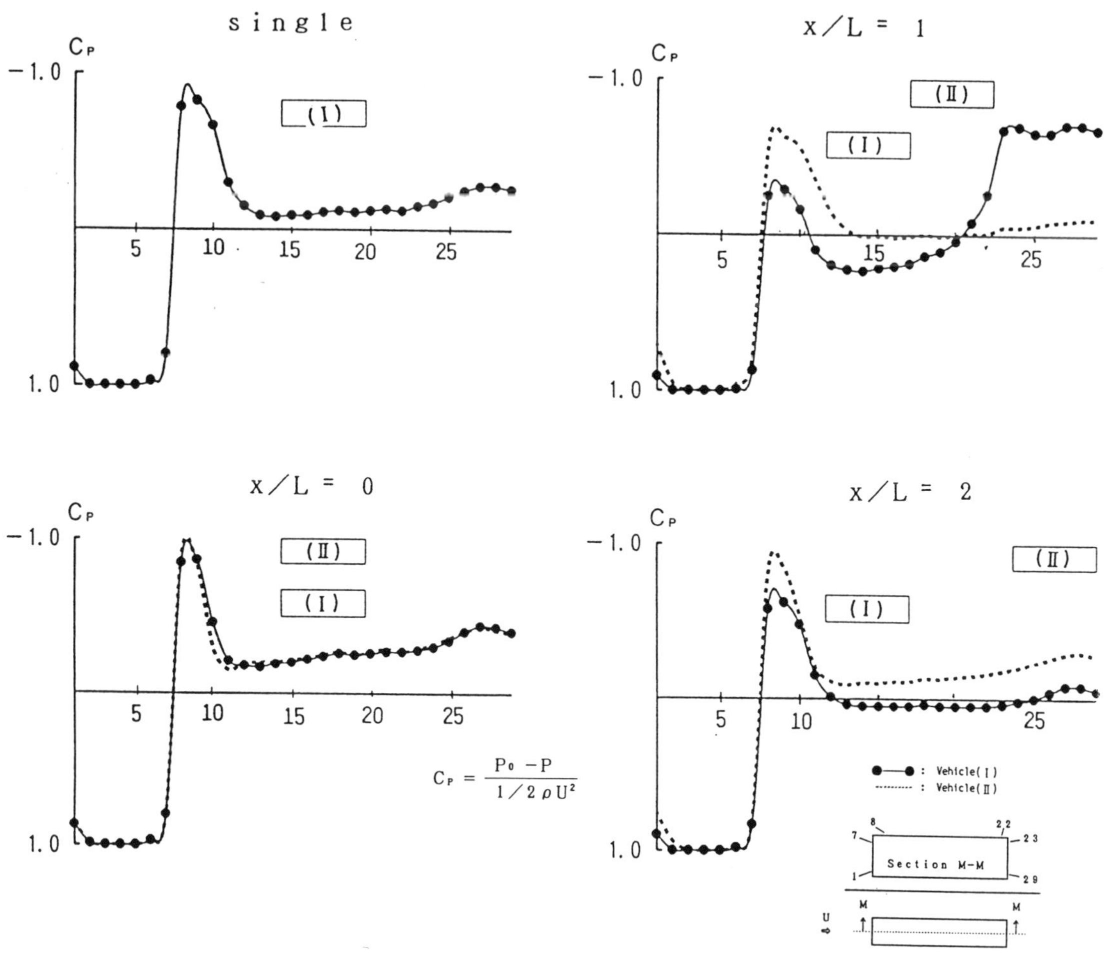

Fig 6. Pressure distribution along longitudinal centeline at overtaking process

dramatic changes of pressure occur on the outer side, roof and rear of overtaken vehicle at the begining of overtaking. An interesting result to be deduced from figs.4 to 6 is the change of drag coefficient and pressure distribution of overtaken vehicle when overtaken by the other. As the overtaking vehicle approaches, pressure distribution around overtaken vehicle body rises and then flow separation of overtaken vehicle is promoted very much. Figure 7 shows visualized flows around vehicles at overtaking in the tunnel. When vehicles are single, relatively large vortices are easily formed in the lower portion of wake of trucks near the rear surface of the vehicle, generating negative pressure regions. A pair of strong vortices rotating in opposite directions to one another is also formed near the upper surface of

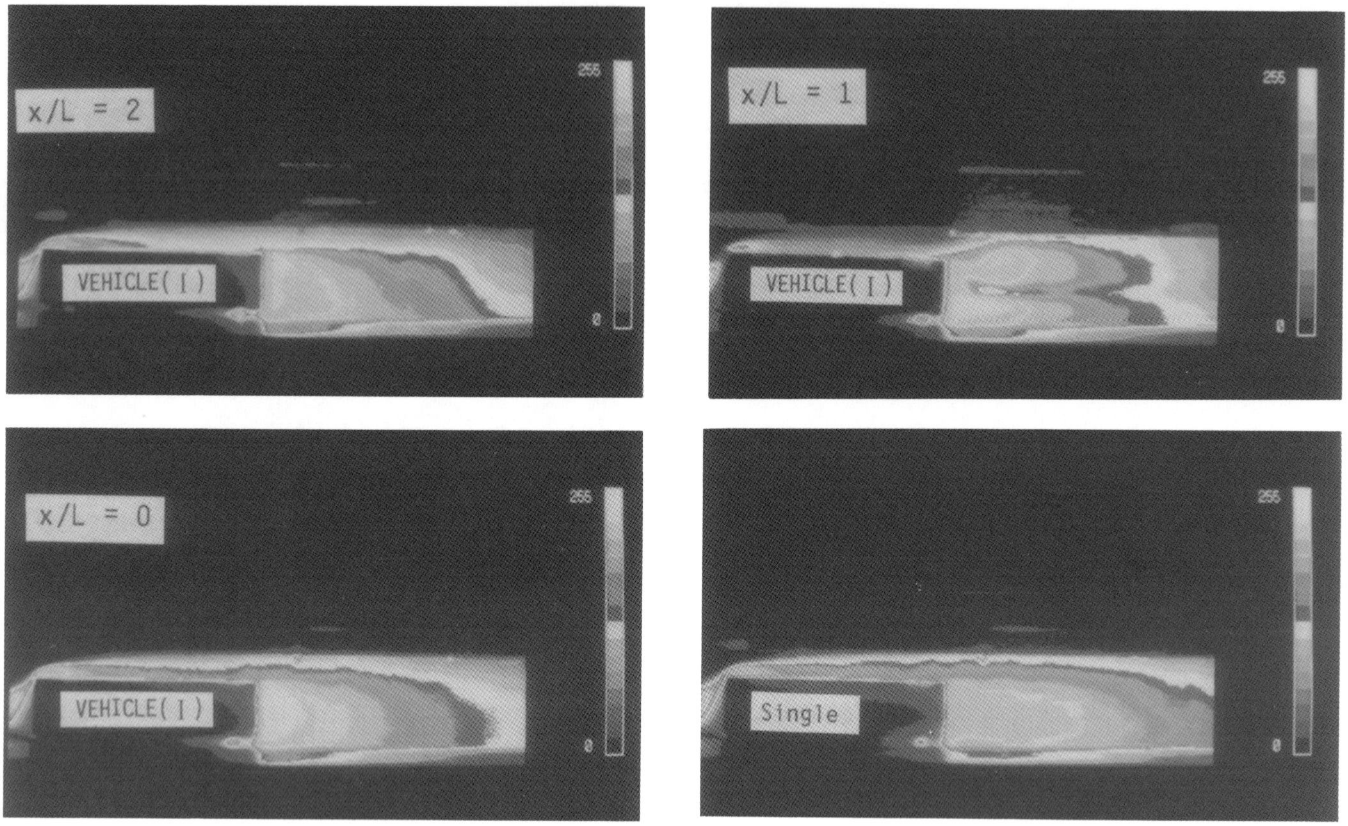

Fig 7. Wake patterns at overtaking by laser light sheet

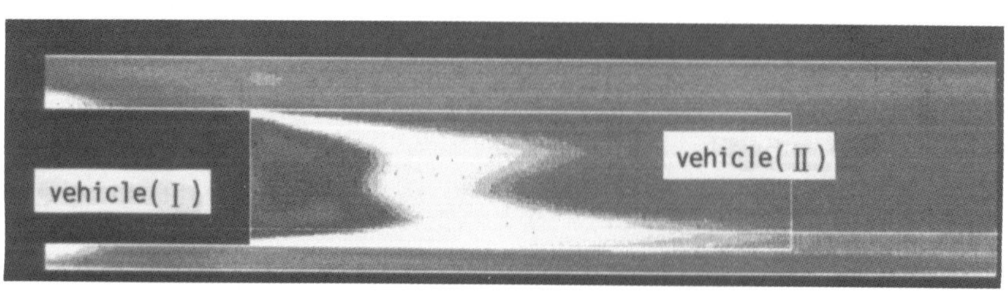

Fig 8. Wake patterns at overtaking by Simulation

the roof, flowing diagonally up from the sides of the vehicle. At overtaking, however the wake strucuture changes drastically. The flow pattern, which was separated, is shown in Figure 7. Figure 8 shows the velocity distribution at overtaking. The results of calcuration shows excellent agreement with experimental results. The sequence of frames in Figure 9 shows the flow patterns at vertical section behind the overtaken vehicle at overtaking. These figures are of interest showing formation process of wake structure as the overtaking vehicle approches the rear of the overtaken vehicle. These wake flow patterns were formed by the mixture of very high velocity flows streaming through the gap between the wall of tunnel and the vehicle and the slow velocity flows streaming through the gap between two vehicles.

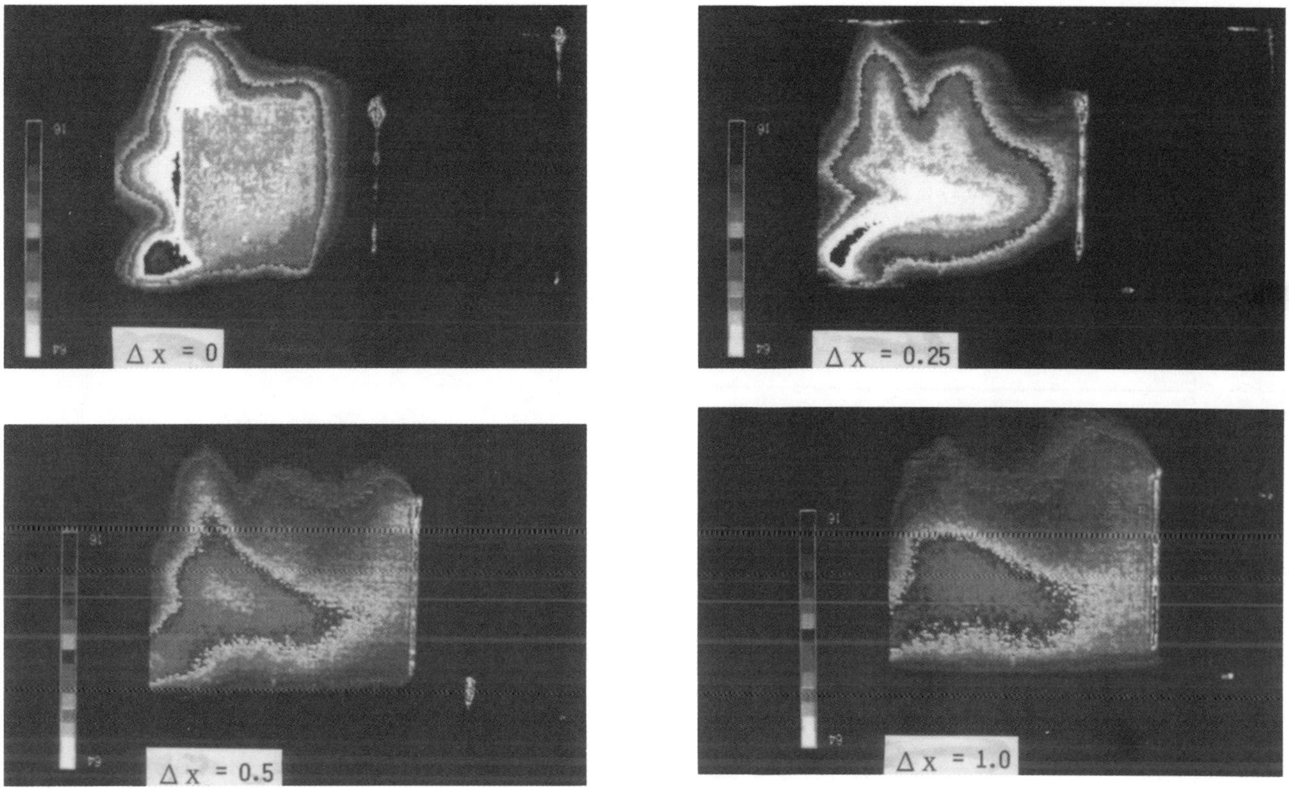

Fig 9. Process of wake patterns at cross-section behind vehicle(I)

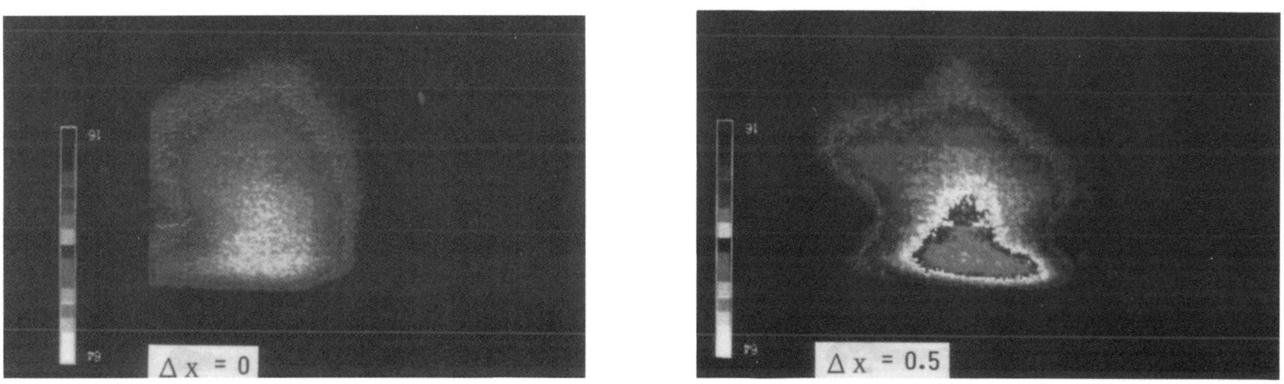

Fig 10. Process of wake patterns at cross-section behind single vehicle

Figure 10 shows flow patterns on vertical plane behind the single vehicle in a tunnel. Figure 11 shows computer generated streaklines at overtaking. These streaklines are made in terms of over-view, side-view and rear-view planes. Figure 12 showa time line at overtaking. These results show the wake structure change drastically. The results of the calcuration are in good agreement with experimental results.

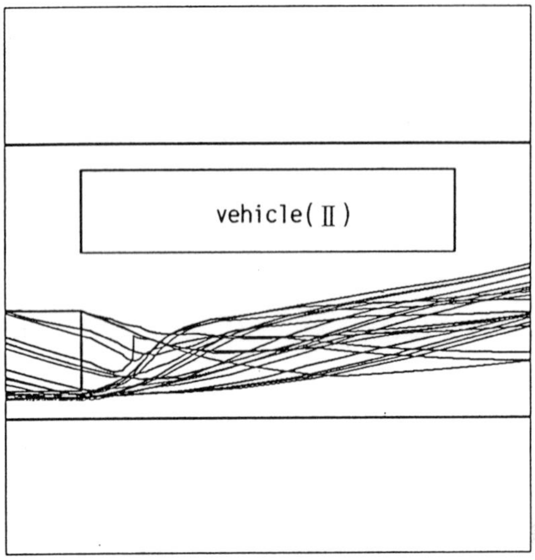

over-view sections

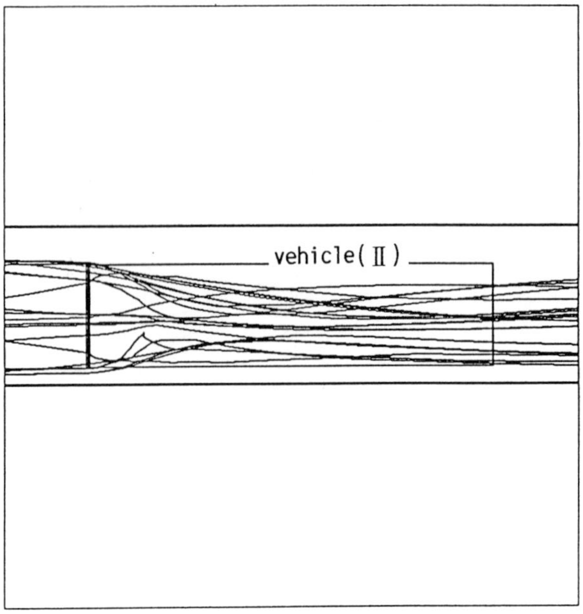

side-view sections

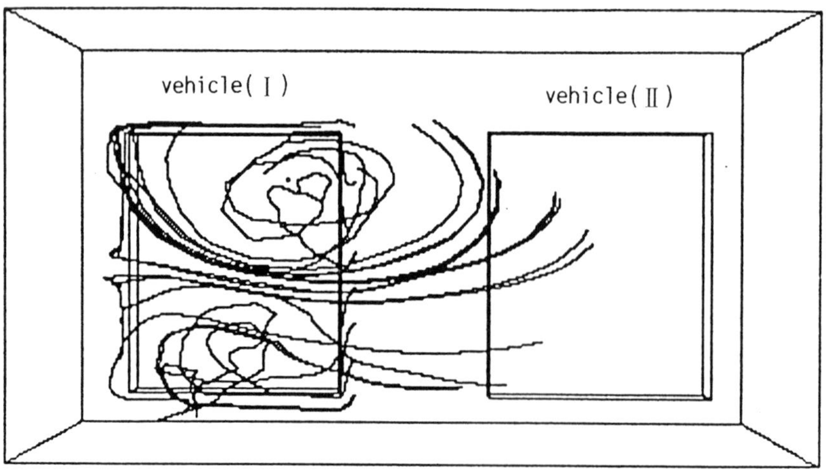

rear-view

Fig 11. The computer generated streak lines at overtaking

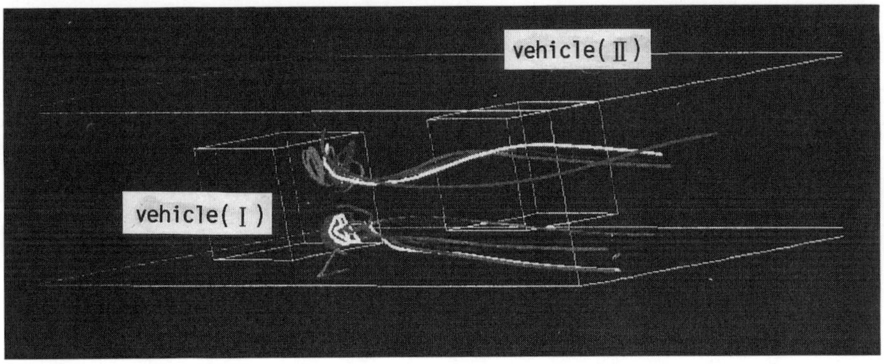

Fig 12. The computer generated time lines at overtaking

CONCLUSIONS

The effects of overtaking in tunnel were studied by a laser ligth sheet method and digital image processing technique. Experimented results were also compared with numerical results. The present investigation indicated that effects of interest are experienced when the overtaking vehicle approches the end of the overtaken vehicle. In the process of overtaking, the drag coefficient of aerodynamic force components increased sharply and the wake structure changed drastically. The numerical results agreed well with those of the experiment.

REFERENCES

1. R.Himeno et al., Numerical Analysis of the Airflow around Automobiles Using Multi-block Structured Grids, SAE Paper 900319,1990
2. k.kitoh et al., Effects of Boundary Conditions on Numerical Turbulent Flow around a Two-Dimensional Vehicle by k- ε Model, IIS Annual Report of Group 1986,pp.67-72.
3. G.W.Carr., Aerodynamic Testing and Data Adjustments Testing Methods and Procedures, SAE Technical Information Report J2084
4. T.Morel., The Effect of Base Slant on the Flow Pattern and Drag of Three-Dimensional Bodies, Proceeding of 1976 GM Symposium, 1976
5. Ahmed,S.R., Wake structure of typical automobile shapes, Transactions of the ASME, Journal of Fluids Engineering, Vol 103,1981
6. Cogotti,A., Wake Survey of Different Car Body Shapes with Coloured Isopressure Maps, SAE Paper 840299, 1984
7. Cogotti,A., Recent Advances in Flow Filed Mapping Techniques, SAE Paper 870718(1987)
8. Ahmed,S.R et al., Some Salient Features of the Time Averaged Ground Vehicle Wake, SAE Paper 840300, 1984
9. Ahmes,S.R, Hucho,W.H., The Calculation of the Flow Field Past a Van with the Aid of a Panel Method, SAE Paper 770390, 1977
10. Hucho,W.H et al., The Optimization of Body Details-A Method for Reducin the Aerodynamic Drag of Road Vehicles, SAE Paper 760185, 1976
11. F.N.Beauvais., Aerodynamic Inputs to a Parked Vehicle Caused by a Passing Bus,SAE Paper 700037, 1970
12. J.Steinheuer., Calculation of Unsteady Pressures During Passing and Tunnel Entrance of Trains, Aerodynamics of Transportation, ASME 1979
13. T.Kobayashi et al., An Image Processing Technique for Determing Two-Dimensional Flow Fields with Reverse Flow, Journal of The Visualization Society of Japan, Vol5-17(1985)(in Japanese)
14. T.Kobayashi,Y.Yoshitake., Development of Digital Image Processing System for Pathline Pictures, Transaction of the JSAME, Vol51-466(1985)(in Japanese)

930297

Airflow Measurement Around Passenger Car Models Using a Two-Channel Laser Coppler Velocimeter

Hiroshi Sato and Michitoshi Takagi
Nissan Motor Co., Ltd.

ABSTRACT

A two-channel LDV system is used to obtain accurate airflow measurements around scale models of passenger cars in wind tunnel tests at the Nissan Research Center. A 2-watt argon-ion laser is employed as the light source. The main optical unit and probe head are connected by optical fibers. The probe head consists of a compact LDV probe with a beam expander and focusing lens with a long focal length can be easily traversed. A new type of signal processor, performing a digital autocorrelation function, is employed to process the Doppler signals. Mean airflow velocities and turbulence intensities are calculated by a micro computer to evaluate the flow fields. The results of preliminary experiments conducted with this system indicate that the system is not only capable of measuring the mean velocity components, including reverse flow, it can also provide accurate estimation of turbulence components.

INTRODUCTION

Research on vehicle aerodynamics requires a good understanding of airflow characteristics around the car body. Studies done to date on reduction of aerodynamic resistance of vehicles have usually been based on observations of general flow characteristics around the body. Such observations can be accomplished by using a balance method to measure the aerodynamic forces acting on a vehicle or by using smoke or tufts to visualize flow. New insights into the factors causing drag have also been obtained recently through the use of 5-hole pitot tubes to make quantitative measurements of flow in the wake region and of underbody flow.(1) These observations of general flow patterns and experimental data on macro flow properties around the body have provided a great deal of basic knowledge for reducing aerodynamic resistance and obtaining good crosswind stability. In addition to the attainment of improved aerodynamic vehicle designs, there has been a growing interest recently in studies of the flow details around the body, especially of turbulent flow phenomena.

Flow measurements have generally been made with pitot tubes and hot-wire anemometers because these devices are simple and easy to use. At the same time, these instruments also have certain drawbacks which require the exercise of care. Since flow measurements are obtained by placing the devices directly in the flow, they can affect flow characteristics to varying degrees. It is also very difficult to obtain accurate flow measurements when reverse flow is present. In addition, changes in temperature, density or other physical properties of the flow being measured can make it necessary to provide suitable calibration.

This paper describes a laser Doppler velocimeter system which employs an optical non-invasive measuring technique in place of these conventional methods. An overview of the system is presented along with examples of results obtained in scale model wind tunnel measurements.

LASER DOPPLER VELOCIMETER

A laser Doppler velocimeter (LDV) provides a non-intrusive method for measuring flow velocities by making use of the Doppler shift in the frequency of laser light scattered from particles in the fluid. While a fluid normally transmits light, tracer particles can be suspended in the fluid to scatter the light. Assuming that the tracer particles are completely following the fluid flow, the motion of the fluid can be quantified indirectly by measuring the velocity of the particles.

In terms of working principle involved, the LDV method provides the following advantages over conventional flow measurement techniques.

(1) Flow velocity measurements are obtained without inserting the device into the flow to be measured.

(2) Calibration is not necessary because there is a linear relationship between the Doppler frequency and the flow velocity.

(3) Spatial resolution is high. A measuring volume having a diameter of around one hundred microns and a length of nine hundred microns can be formed.

(4) Flow velocities can be measured over a wide range from 10^{-3} m/s to 10^3 m/s.

(5) Measurement of reverse flow is possible.

(6) Turbulence properties, such as Reynolds shear stress, can be measured.

The basic 2-beam LDV that was available initially has now been expanded into a wide variety of commercial devices, including a 2-channel monochrome 3-beam LDV, a 2-color 3-beam LDV, a 2-color 4-beam LDV and even a 3-channel 2-color 5-beam LDV and 3-color 6-beam LDV. While multichannel LDVs offer many different functions, they also have the disadvantage of being more difficult to use because of the vastly increased size of the optical system.

In recent years, optical fiber has come to be used increasingly in the field of communications. Since the early 1980s, there has been a succession of reports, such as those by Buchhave and Knutsen,[2] Durst and Krebs,[3] and Nguyen and Binh,[4] on efforts to apply optical fiber to LDVs in order to create smaller, easier-to-use devices. The optical fiber LDV prototypes that were built represent important pioneering research in this field. One aspect that remained to be improved, though, was the low quality of the output beam due to the presence of crosstalk, speckle or other undesirable phenomena.

The research groups of Maeda et al.[5] and Sato et al.[6] devised LDVs in which the probe was built with just a simple lens system. Their devices were achieved by applying a polarization-preserving type of optical fiber to the optical path of conventional LDV. The main feature of the optical system was that it allowed the use of a double Bragg cell to obtain the desired frequency shift. That was accomplished through innovations made to the portion where light was injected into the optical fiber, which eliminated the influence of changes in the optical axis caused by the frequency shifter employed for reverse flow measurement. As a result, it was no longer necessary to use an electronic circuit to perform a downmix operation on the signal, which was required with a conventional single Bragg cell. This had the effect of reducing the load on the signal processor, and it yielded a high-quality signal with little noise. In addition, optimization of the lens system made it possible to obtain virtually the same measuring volume size and fringe numbers as with the optical systems of conventional LDVs. Measurement accuracy was also equal to that of previous optical system.

Prototype LDVs have been built recently using laser diodes as the light source.[7] However, there are various problems with the light intensity, stability and difficulty of adjustment that are impeding practical application of these devices. Vigorous research work is also under way on PIVs[8] and PTVs,[9] which measure flow properties in a certain area, in contrast to the point-type measurement of conventional LDVs. At present, however, their use is limited to the laboratory because of the narrow range of measurement provided and the lengthy time required for processing, among other problems. In view of these various advantages and disadvantages, it would seem that an optical-fiber LDV is the most practical non-invasive method available at present.

While researchers at VW[10] have reported the use of an LDV in measuring airflow around the vehicle body in full-scale wind tunnel tests, the literature shows few examples of its use in scale model wind tunnel experiments.

SYSTEM DESCRIPTION

Figure 1 shows the devices making up the 2-channel LDV system used in scale model wind tunnel testing at the Nissan Research Center. The system consists mainly of six elements; an argon-ion laser, main optical unit, optical fibers, optical probe head, signal processor and micro computer that performs control and data processing tasks.

A 2-watt water-cooled argon-ion laser is used as the light source. The main optical unit incorporates a color separator, beam splitters, frequency shifters for measuring reverse flow, manipulators for injecting the laser beams into optical fibers and receiving optics for dividing the scattered light into two-channel electrical signals.

The optical fiber cable connecting the main optical unit to the probe head consists of four polarization-preserving single-mode fibers and one multi-mode fiber.

The probe was newly made for the system, as there was no existing product that could provide measurements over a distance of approximately 1,000 mm, which is one requirement for use in scale model wind tunnel testing at our Research Center.

An existing LDV probe (K-type LDV prove of KANOMAX Corp.) was rebuilt by removing the focusing lens, which had a small diameter and a short focal length (diameter = 31.5 mm and f = 200 mm), and replacing it with a beam expander and a large-diameter focusing lens having a long focal length (diameter = 120 mm and selectable settings of f = 750 or 1200 mm).

Optical-fiber LDV probes are often applied to systems for measuring complex shapes because of their ease to use, and they may be inserted directly into the flow when a large-scale flow is to be measured. In such applications, they are generally used with focusing lens having a relatively short focal length.[11] The LDV probe employed in this system has certain innovative features which have rarely been seen in other applications of these kinds of devices. As described above, the measurable distance has been extended through the combined use of a beam expander and a focusing lens with a long focal length.

The optical system is attached to an overhead 3-D traverser, which has been used in previous systems. As a result, this arrangement makes it possible to measure any arbitrary point in the wind tunnel by remote control without disturbing the flow.

The scattered light signals received from the measured point can be processed to yield velocity components in two directions. The scattered light signals are collected in backscatter, focused onto the multi-mode fiber at the probe head, and transmitted to the main optical unit, where they are converted to electrical signals by photomultipliers in the receiving optics. The output signals of the photomultipliers are input into the signal processor, which performs a digital autocorrelation function. The processor is connected directly to the host computer and its operation is controlled entirely by the software. After the processor converts the Doppler signals to frequency data by means a hardware operation, the results are sent to the 32-bit host microcomputer. Conversion of the frequency information to velocity data and all other statistical processing tasks are handled by the software run on the microcomputer.

CONSTRUCTION OF THE LDV PROBE

The laser beam is transmitted from the main optical unit to the LDV probe via polarization-preserving single-mode fibers. A schematic diagram showing the construction of the LDV probe is given in Fig. 2.

A micro rod lens and a beam collimator installed at the output end of the probe compensate for the divergence of the laser beam transmitted through the optical fiber and optimize the beam diameter, resulting in the output of two parallel beams. (In actuality, because the LDV has two channels, another pair of parallel beams are present at right angles to the central axis as can be seen in the A-A cross section, although the beams are not represented in the figure.)

The beam expander increases the beam diameter and spacing between the two (more precisely, four) parallel beams, and then the focusing lens functions to converge the beams so that they intersect at a forward point and form the desired measuring volume. Because of the conditions under which this system is used in wind tunnel testing, the focal length of the focusing lens can be selected to two setting of 750 mm and 1,200 mm with changing lenses.

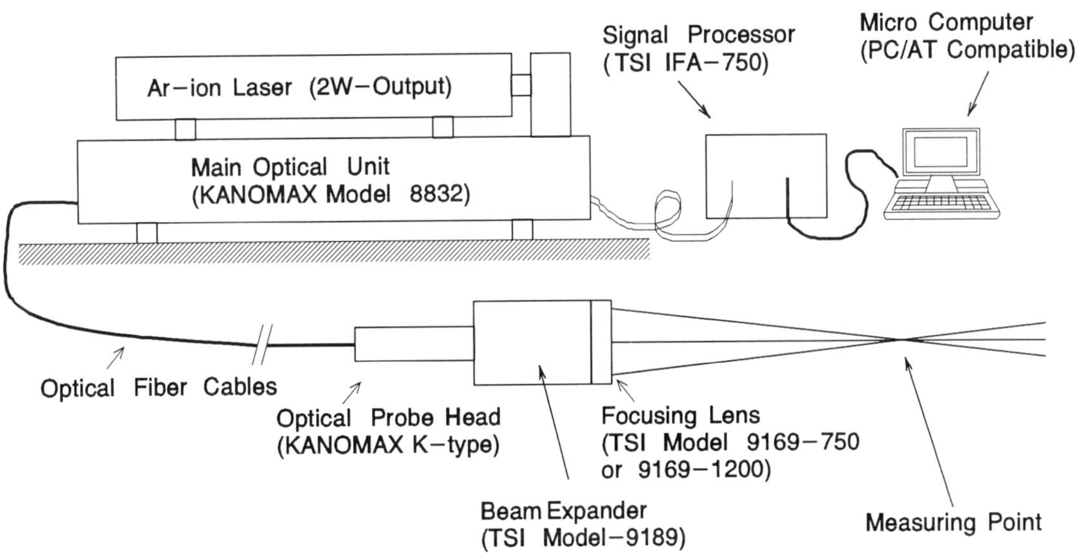

Fig. 1 Schematic Drawing of the LDV SYSTEM

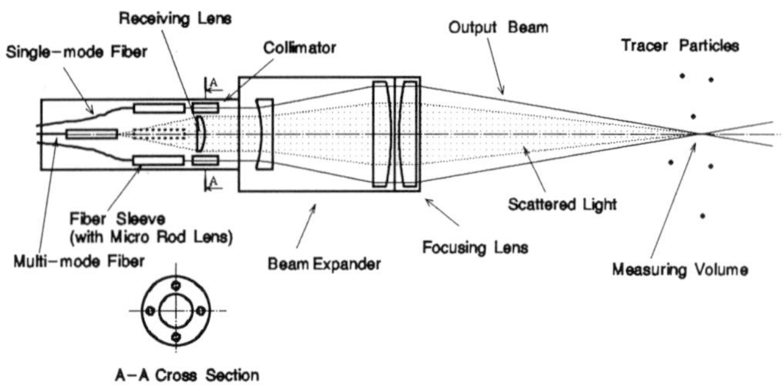

Fig. 2 Construction of the LDV Prove

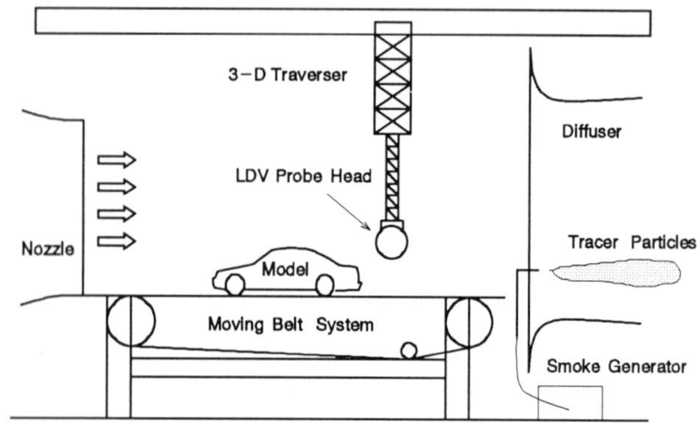

Fig. 4 Schematic Drawing of the Test Section

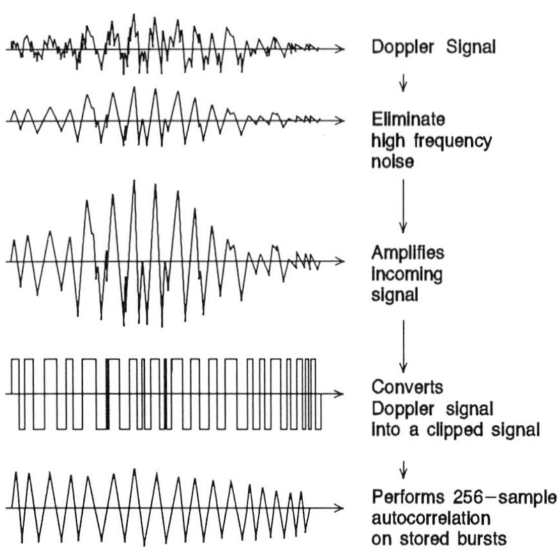

Fig. 3 Signal Processing Scheme

SIGNAL PROCESSING

HARDWARE-The photomultipliers output two channels of Doppler signals which are processed by the digital autocorrelation type of signal processor of TSI IFA-750 commercial device. Figure 3 outlines the signal processing scheme used in the IFA-750 processor.

Various types of high-frequency or pulse-like noise components, such as background noise and shot noise, are contained in the output of the photomultipliers. The signals input to the processor are first passed through a band-pass filter to eliminate the high-frequency noise and the low-frequency pedestal components. After that, they are amplified and fed into a circuit which converts the Doppler signal into a clipped signal.

The digitized signals are then input to a burst detector and a sampler. The burst detector provides approximate frequency and start- and end-of-burst signals. At the start-of-burst signal, the sampler initiates a sampling process. Upon conclusion of the burst signal, the sampled data stored in the buffer memory are forwarded to the autocorrelator. This device finds the autocorrelation using the center of the burst signal detected in the burst detector as the baseline criterion. Frequency information is then forwarded through the interface to the computer.

SOFTWARE-Signals processed in a hardware operation by the processor are sent to the micro computer as 16-bit time data and are converted to velocity data by the software. By means of statistical processing, the software calculates the mean velocity, turbulence intensities, Reynolds shear stress and correlation coefficient, among other values. Usually, data acquisition alone is carried out in advance and data are stored on a hard disk. Then, after a series of measurements has been completed, the stored data are processed. Subsequently, the calculated results are graphed with use of plotter.

APPLICATION TO FLOW MEASUREMENT IN THE WAKE OF A CAR

The LDV system described above was applied to flow measurements in the wake of scale car models in order to verify its practicality as a measurement system.

Flow measurements were made in a scale model wind tunnel equipped with a moving belt system, as shown in Fig. 4. The LDV probe head was attached to an overhead 3-D traverser, and laser beams were projected from the side of the wind path to obtain measurements of velocity components in two directions. The coordinate system consisted of the Y-axis in the flow direction, the Z-axis perpendicular to the flow and the X-axis along the span traveled by the probe head.

Used as the tracer particles were hot atomized particles of liquid paraffin, which were produced by a smoke generator of the type commonly employed for flow visualization. The particles were mixed into the flow from the wind tunnel diffuser. Since the scale model wind tunnel is built as a closed loop circuit, once the tracer particles are charged from the diffuser they flow along the wind path together with the air stream. Relatively large tracer particles that do not follow the flow very well tend to adhere to the walls or floor of the wind tunnel or the mesh. As a result, the air stream discharged from the nozzle contains only tracer particles having a small diameter (bellow 10 microns). For this reason, it can be assumed that the tracer particles are fully following the flow in the section where measurements are made.

One-fifth scale models of sedan passenger cars (PRIMERA or INFINITI G20) were used in the experiments. Schematic drawing of the test model is shown in Fig. 5. Two types of models were employed, one configured with normal body styling and the other equipped with a rear spoiler. The experiments were conducted at a main flow velocity of 25 m/s and without operating the moving belt. The Reynolds number based on the wheel base is set at Re = 843,000. As shown in Fig. 6, measurements were made at 50 mm intervals in Y-axis direction to distance of 400 mm from the rear bumper (Y=0) and 10 mm intervals in the Z-axis direction to a height of 300 mm from the ground. The number of samples taken at each measurement point was set at 2,000 on the basis of previous investigations.

Figure 7 shows the velocity vector profiles found for two models from the mean velocity components measured in two directions. The upper figure shows the results for the normal car body and the lower figure shows the results for the model equipped with a rear spoiler. Without the rear spoiler, the overall profile shows a pattern of strong down wash flow. In contrast, the profile with a rear spoiler shows a nearly parallel flow directly behind the car, except for the upward flow coming from beneath the rear of the car. It can also be seen that reverse flow developed immediately behind the car. The difference of the velocity vector profiles can be seen in the wake region. For the with-spoiler condition, the wake region is relatively larger. These results indicate that this LDV system is capable of providing measurements of flow including reverse flow, which could not be obtained with previous techniques such as pitot tubes or hot wire anemometry which cannot be apply to reverse flow region. Thus, the system makes it possible to analyze flow phenomena, including reverse flow, more accurately.

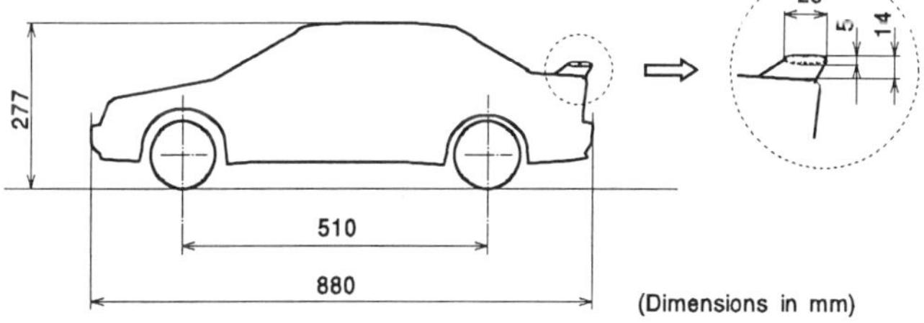

Fig. 5 Schematic Drawing of the Test Model

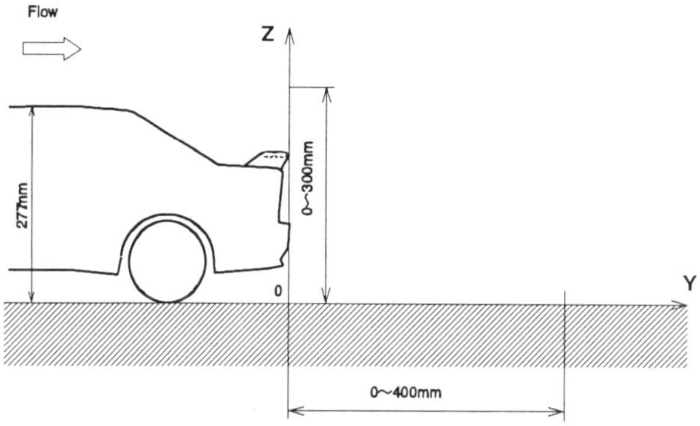

Fig. 6 Measuring Region

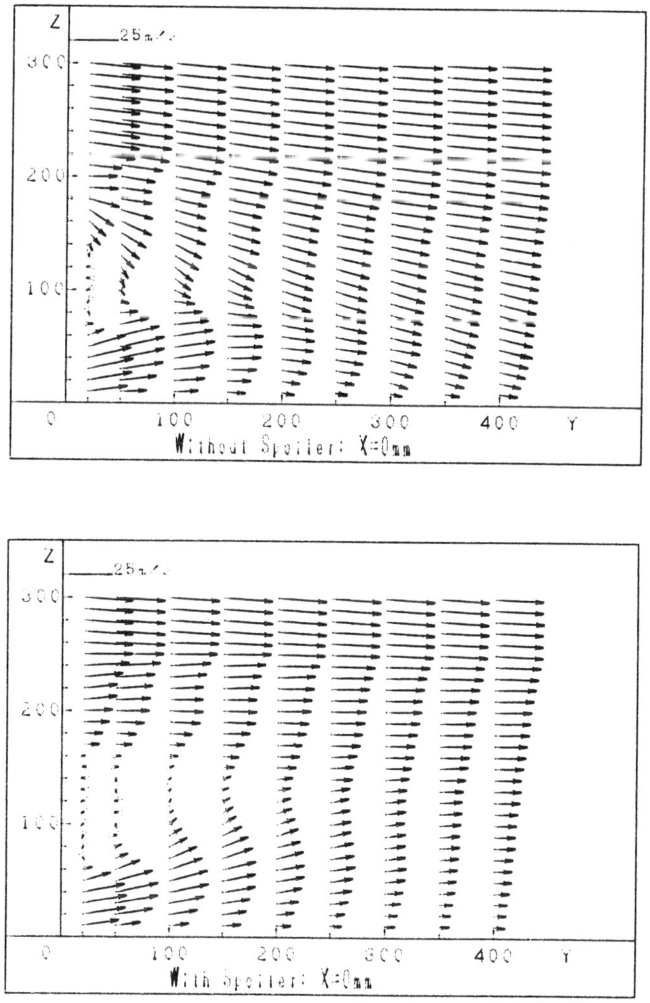

Fig. 7 Velocity Vector Profiles behind a Car

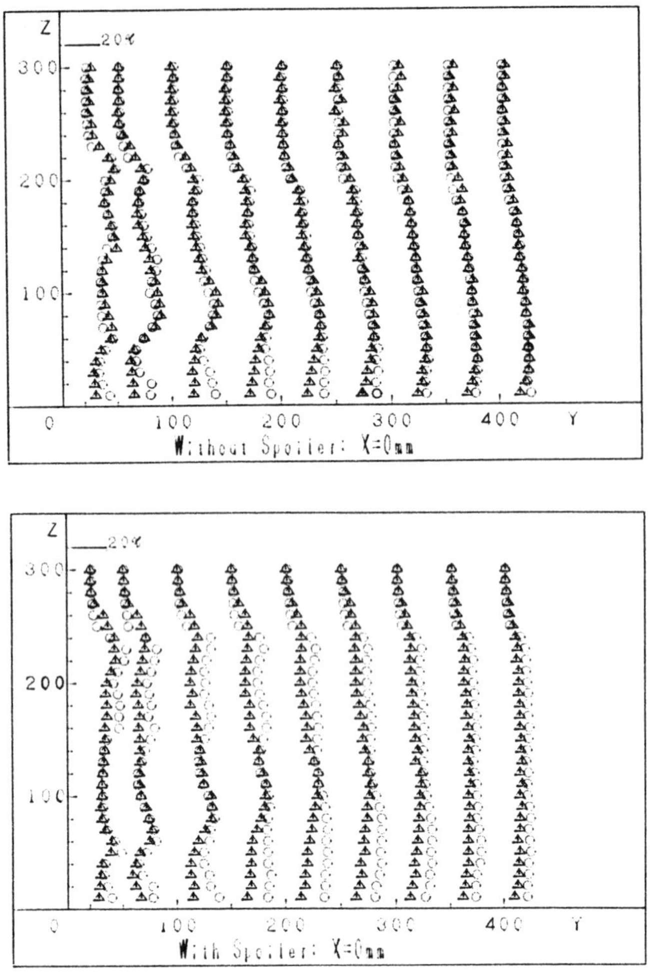

Fig. 8 Turbulence Intensity Profiles behind a Car (○:Y-component, △:Z-component)

Figure 8 shows the turbulence intensity profiles for the two models calculated from the variance of the velocity components. The open circles in the figures indicates the Y-axis turbulence component and the open triangles indicate the Z-axis turbulence component. Turbulence intensity is expressed as a percentage, found by dividing the RMS value by the main flow velocity.

Similar to the difference seen in the mean velocity vectors, the turbulence intensity profiles differ between the models with and without the rear spoiler. For both models, turbulence intensity peaks are observed in the vicinity of the shear layer that is formed between the main flow region and the wake. In the upper figure for the model without the spoiler, the turbulence intensity peaks have been moved downward toward the ground by the mean down wash flow and have also spread outward. Excluding the region near the ground, both the Y-axis and Z-axis components show virtually the same distribution pattern.

On the other hand, for the model with the rear spoiler (lower figure), the Y-axis component shows turbulence peaks around a height of 230 mm immediately behind the car, and the peaks stay at virtually the same height without being attenuated on almost all the measured line. In this case, the distribution pattern has an inflection point in the vicinity of Z=250 mm. A comparison of the Z-axis turbulence component (open triangles) and the Y-axis turbulence component (open circles) in the vicinity of Z = 250 mm shows that the former is smaller throughout the entire measurement region, indicating that the turbulence is anisotropic.

The foregoing results suggest that a rear spoiler changes not only the mean velocity profile but also the turbulence intensity distribution behind a car.

SUMMARY

A 2-channel optical-fiber LDV system has been implemented at the Nissan Research Center as a new general-purpose measurement system which can be used to measure various types of flow in scale model wind tunnel testing.

The optical system is built with an upgraded version of the ordinary fringe mode type of argon-ion 4-beam LDV, and the compact probe head has been separated from the main optical unit through the use of optical fibers. As a result, a major feature of the optical system is its vastly improved ease of use. Backscattering is used to obtain velocity measurements in two directions simultaneously, and reverse flow measurements are also possible. The optical system is mounted on an overhead 3-D traverser, which allows flow measurements to be obtained at virtually any desired point in the scale model wind tunnel test section.

This LDV system was used to measure flow velocities in the wake region behind one-fifth scale models of passenger cars. The results indicated that the system is not only capable of measuring the mean velocity components, including reverse flow, it can also provide accurate assessments of turbulence components. These results confirmed that the new system is effective in analyzing flow phenomena around car models in wind tunnel tests.

This application of the new LDV system has shown it to be an extremely useful flow measurement tool. By contributing to a better understanding of the mechanism causing various types of flow around the vehicle body the system is expected to be valuable in improving aerodynamics of vehicle.

ACKNOWLEDGMENT

The authors would like to acknowledge Mr. K. Umezaki for help in performing in experiments being described in this paper..

REFERENCES

(1) I. Yamaguchi, K. Nakatsuru, M. Takagi and Y. Simpo, "The effect of a Moving Belt on Automotive Underbody Airflow in Model Wind Tunnel Tests," JSAE Review, Vol. 12 (1991), No. 3, pp. 24-28.

(2) P. Buchhave and J. Knutsen, "Fiber-Optics Laser Anemometer Measurements," Selected Papers from 1st Int. Symp. on Application of Laser-Doppler Anemometry to Fluid Mechanics (1982), pp. 323-334.

(3) F. Durst and H. Krebs, "LDA-Optics Developments for Measurements in Internal Combustion Engines," Proc. of 2nd Int. Symp. on Applications of Laser Anemometry to Fluid Mechanics (1984), pp. 6.5.1-6.5.5.

(4) Than T. Nguyen and Le N. Binh, "A Fiber-Optic Laser-Doppler Anemometer," Appl. Phys. Lett., Vol. 45, No. 11 (1984), pp. 1163-1165.

(5) M. Maeda, K, Hishida, and M. Ono, "Development of Fibre LDA and Measurements of Turbulence in Boundary Layer on a Flat Plate," Bulletin of JSME, Vol. 28, No. 240 (1985), p. 1050.

(6) H. Sato, K. Hishida, and M. Maeda, "Development of Fibre LDA and Two-Component Velocity Measurements in Boundary Layer on an Inclined Heated Plate," Fluid Control and Measurement 1985 Tokyo, Vol. 2 (1985), pp. 1065-1070.

(7) S. Bopp, F. Durst and R. Muller, "Small Laser-Doppler Anemometers Using Semiconductor Lasers and Avalanche Photodiodes," Selected Papers from 4th Int. Symp. on Application of Laser-Doppler Anemometry to Fluid Mechanics (1988), pp. 315-337.

(8) K. Kobayashi, J. Sakakibara, K. Hishida and M. Maeda, "Time-Series Measurements of Turbulent Flow Field Using Image Processing System," Experimental and Numerical Flow Visualization, ASME FED-Vol. 128 (1991), pp. 155-162.

(9) T. Kobayashi, T. Saga, T. Haeno and N. Tsuda, "Development of a Real-Time Velocity Measurement System for High Reynolds Fluid Flow Using a Digital Image Processing Technique," Experimental and Numerical Flow Visualization, ASME FED-Vol. 128 (1991), pp. 9-14.

(10) R. Buchheim, M. A. Beeck, W. Hentschel, R. Piatek, D. Schwabe and F. Durst, "Advanced Experimental Techniques and Their Application to Automotive Aerodynamics, SAE Paper 870244 (1987).

(11) A. Cogotti and H. Berneburg, "Engine Compartment Airflow Investigations Using a Laser-Doppler-Velocimeter," SAE Paper 910308 (1991).

930294

Development and Use of LDV and Other Airflow Measurement Techniques as a Basis for the Improvement of Numerical Simulation of Engine Compartment Air Flows

Helmut Berneburg
GM Europe

Antonello Cogotti
Industrie Pininfarina S.p.A.

ABSTRACT

Modern vehicles require a low aerodynamic drag to minimize fuel consumption. A not negligible share of the overall C_D-value of a vehicle is produced by the engine compartment air flow. Therefore this share has also to be optimized. Furthermore, customer wishes for higher powered engines as well as for more safety and comfort result in more tightly packed engine compartments. Even the reduction of pass-by-noise required by legal reasons is often achieved with the help of underbody covers which in turn affect the engine compartment flow.

All these items may lead to rising underhood temperatures. To reduce the development time of new vehicles, numerical simulations of engine compartment air flow are more and more used to predict high temperature fields and to show ways to develop suitable remedies in the concept phase of the vehicle development. The experimental basis for such codes is provided by aerodynamic investigations in a wind tunnel. At least the three components of local velocities and the volumetric air flow at the inlet and the outlet plane have to be determined.

This paper first reports the development and the use of a modern aerodynamic "Test Radiator" for the investigation of velocity distribution and for the volumetric air flow measurements through a car cooling package. For this purpose air vanes of small diameter, able to detect even reverse flows, are used.

Furthermore, the development of various LDV-techniques, suitable for the investigation of complex flow fields, for instance behind a rotating fan are reported. And a newly developed 3-D miniature-probe is shown.

The influence of various cooling configurations, with/ without condenser in front of the radiator, and with stationary/rotating puller fan and with/without vehicle speed is also shown.

DURING THE LAST 20 YEARS the C_D-values of passenger cars have been reduced drastically by a third (about .45 to .30). For some vehicles as the OPEL Calibra (C_D = .26) even lower values have been reached. The reason for this new interest of car aerodynamics, which was already at a relatively high standard in the thirties, was the first oil crisis in the early seventies. The second oil crisis and the environmental problems in the eighties gave an additional impulse.

To face and overcome these new challenges the automotive industry engaged itself in a drastic reduction of the average fleet fuel consumption. (For instance OPEL aims a reduction of 25% in fuel consumption from the year 1979 to 1995, [1]). Wind tunnels for the improvement of C_D-values have been built all over the world in recent years. Existing wind tunnels have been equipped with modern and high-technology testing devices.

The reduction of C_D-values is one of the most effective way to achieve low fuel consumption because, when driving at constant speed, aerodynamic drag becomes the dominant parameter compared to rolling and other resistances.

One big share of the overall C_D-value is the engine compartment air flow drag, $\blacktriangle C_{Dc}$, which has been greatly reduced in the past from values of $\blacktriangle C_{Dc} = .050 \div .070$ to $\blacktriangle C_{Dc} = .010 \div .020$. This drag $\blacktriangle C_{Dc}$ is produced both by the air flow streaming onto and around parts within or underneath the engine compartment and by the changes in the outer flow around the car body, that may occur when the air flow through the front end openings is modified. And to reach these very low values of $\blacktriangle C_{Dc}$, both the position and the area of the front end openings, as well as the internal use of the cooling air flow, have to be optimized, in close cooperation with the cooling engineers.

Another important aspect of engine compartment air flow is that local hot areas within the engine compartment are to be avoided. Though the cooling of the engine can be in some cases not critical or at least borderline some "hot spots" can influence the functionality of the vehicle.

Wind tunnels equipped with high technology testing techniques are of great help in solving these problems, through the reconstruction and the analysis of the internal air flows, so as to find ways to improve the air flow distribution.

In the following paragraphs, the paper reports information about two of these test techniques, as well as some example of experimental results. The first technique is currently used to measure the flow volume through the cooling package and to map velocities through the radiator. It is based on a so called "Test Radiator" which is now very reliable and accurate, as it has been in operation for more than ten years, and in the meantime has been continuously upgraded.

THE TECHNIQUE FOR MEASURING AIR FLOW VELOCITIES AND VOLUME THROUGH THE COOLING PACKAGE FOR THE REDUCTION OF $\blacktriangle C_{Dc}$

To optimize the front end openings and to achieve a minimum engine compartment air flow drag, $\blacktriangle C_{Dc}$, that is to minimize and to better distribute the cooling air flow for a better utilization of the radiator core for heat exchanging purposes, an appropriate tool to measure this radiator air flow volume rate and its distribution over the radiator core is needed.

The importance of having such a device for optimizing the front end openings of a vehicle was already recognised 15 years ago. It was decided to build an "Aerodynamic Test Radiator" to measure the required velocity distribution over the radiator core and, by integrating these velocities over the radiator core, to get better information about the cooling air flow volume.

The first "Test Radiator" built by the OPEL aerodynamic group was made of:

- a frame having the same size as a real radiator and two screens, one at the front and one at the back side of the frame to reproduce the pressure loss of a real radiator,
- five small air vanes supported by a horizontal arm were fitted between the two screens,
- the horizontal arm on which the air vanes were mounted, was traversed vertically by a small electrical motor,
- five air flow velocity profiles were measured along vertical lines.

The "Test Radiator" replaced the real radiator at the same position. A sketch of this arrangement is shown in Fig. 1. The fan and shroud were also installed.

The main advantages of this system were that the fan(s) (either pushing or pulling) as well as any other part, such as a condenser etc., could remain installed as for the normal cooling package. Five spurs of velocity profiles were written by an ultraviolet analog paper recorder when the horizontal arm was moved upwards. Typical test recording time was about five minutes. But the data evaluation time was very time consuming, as every calculation had to be done "by hand". Therefore it was very complicated to use this procedure as an aerodynamic development tool.

In a close cooperation with PININFARINA the testing- and data processing system could then be simplified. The readings of the air vanes and the motion of the electrical motors were automated. The velocity profiles and the air flow volume were computed automatically by the wind tunnel computer. Testing time and especially evaluation time were drastically reduced, so that the "Test Radiator" could be used as a suitable aerodynamic development tool. At the end of a wind tunnel run, five velocity profiles and a number indicating the flow volume rate were available. To match various radiator dimensions some typical "Test Radiators" were built.

Later it was felt necessary to have better

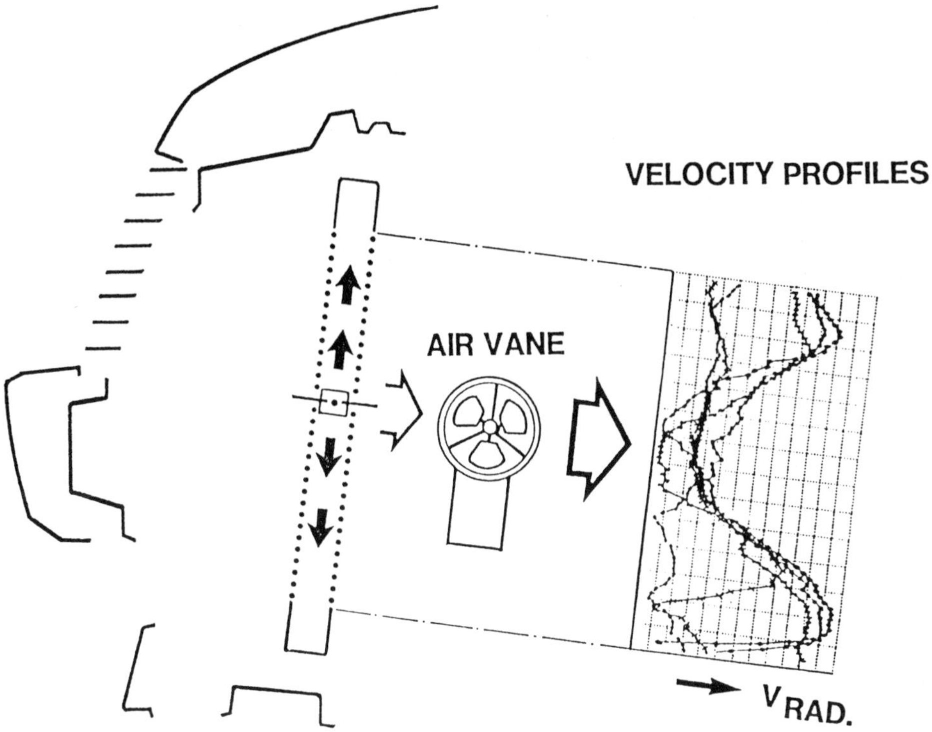

Fig. 1: Old "Test Radiator" and velocity profiles measured by 5 airvanes

Fig. 2: Pininfarina new "Test Radiator" with 11 airvanes

information of the cooling air flow distribution over the width of the radiator. Especially for wide radiators a reading of only five air vanes along five vertical lines reproduced the distribution of the overall flow in a very coarse way. Therefore it was decided to add further six air vanes to improve the accuracy of the flow volume rate and the information of the velocity distribution. The data acquisition system and the software were also further improved. The results could now be presented as "velocity maps", i.e. isotachs.

As this device proved itself to be quite helpful in the optimization of vehicle cooling systems, PININFARINA completely redesigned the "Test Radiator" system in 1991 by using in particular:

- new motors (stepper type) to control more accurately the actual position of the air vanes and their displacement in the time,

- 11 new air vanes which are capable of detecting reverse flows.

By a better arrangement of these air vanes within the "Test Radiator" the velocity distribution over almost the full height of the radiator could be measured.

The detection of reverse flows was found to be necessary, as more and more driving conditions, as for example stop-and-go or hot-idle traffic, occur on the road when the fans are operating at their full power, at low or even no vehicle speeds, i.e. without ram air. Under these conditions, a fan sucks the air from any region it can get it. So it even sucks the hot air back from the region around the engine through the openings in the radiator plane or through the open unshrouded radiator core areas, which means that the radiator works in these areas as a heater instead of a cooler.

This third generation of "Test Radiator", shown in Figure 2, now has been built in three different sizes to match typical radiator cores. Different pressure losses can be realized by a choice of various screens in front and behind the air vanes.

Once again the software was improved. Results are now presented in real time in black and white or in colour prints to easily identify and quantify radiator areas where velocities are comparably too high or too low or even reverse flow occurs. The flow rate is given as the total sum but it is also specified as positive and negative flow rates.

EXAMPLES OF VELOCITY DISTRIBUTION OVER THE RADIATOR CORE FOR VARIOUS TEST CONDITIONS

Fig. 1 shows vertical velocity profiles measured with the prototype "Test Radiator". This test configuration represents the typical cooling package of most European vehicles, i.e. without air conditioning condenser in front of the radiator. Typical high velocities are measured on the radiator core behind the grille- and the front-bumper openings. The area in the middle of the radiator is shielded by the front bumper and its wake. The velocities in that area are only 25% or less than the velocities in the upper or lower area of the core.

One can easily conclude that, with such a velocity distribution, the radiator core will not be used at its optimum. In this picture one can also see that one would probably not get the right information of the overall radiator cooling air flow. The arrangement of the air vanes in the prototype "Test Radiator" was not able to measure the velocity distribution all over the complete height of the radiator core.

Additionally one can see that - at least for this case without condenser in front of the radiator - only a small number of air vanes with a large diameter installed in front of the radiator would not give the best information about the velocity distribution and the overall cooling air flow volume. These large-diameter air vanes would give velocities which are integrated over a too big core area. In the case of an installed condenser which produces a higher air flow resistance and therefore acts as a flow straightener, these air vanes are probably more successful.

A better information of the cooling air flow distribution is seen in Figures 3 a-c. These velocities have been measured using the new PININFARINA "Test Radiator", a sketch of installation is shown in Figure 3d, and evaluated with the new software printing vertical velocity profiles as well as velocity maps. Results are shown in Figures 3 a-c. The test condition for these tests was:

- Test Radiator (without condenser)
- a puller fan with its shroud behind the radiator
- 1) wind only
- 2) no wind and running fan
- 3) wind and running fan

Approximately 1/5th of the radiator core at the left side (seen from the front side) of the radiator was not closed by the shroud so that

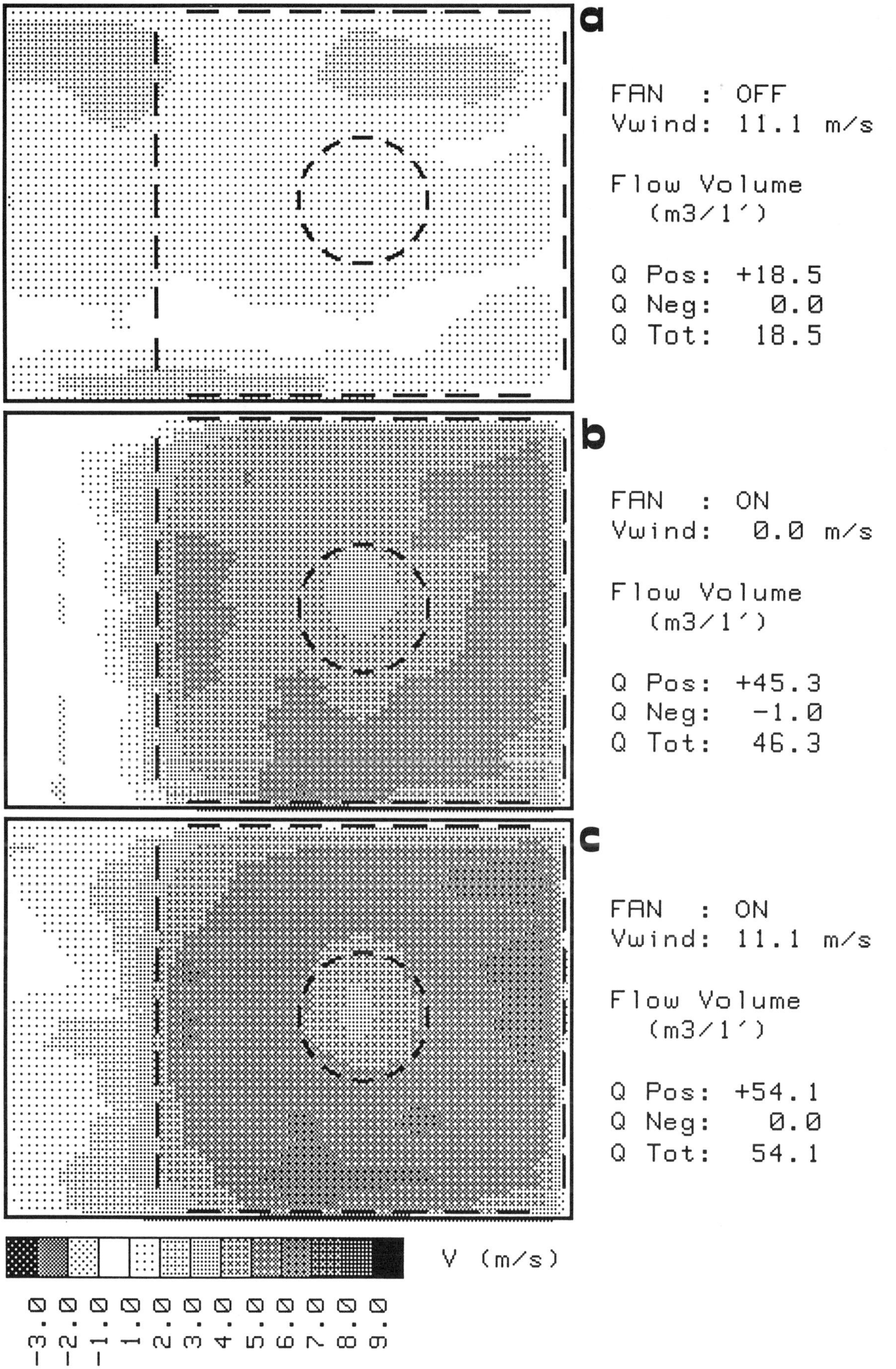

Fig. 3 a-c : New Test Radiator + Puller Fan
(seen from the front)
Velocity distribution trough the PF Test Radiator
at Vwind = 0 & 11.1 m/s, Puller Fan ON & OFF

the air flow may pass through this part of the core more easily.

Figure 3a, for the case of a stationary fan, shows, when compared to the velocity distribution in Figure 1, an already better distribution of flow all over the radiator core; although in the upper and the lower part higher velocities are still discovered. This is due to a better design of the upper and lower front end openings. The air flow rate amounts to 18.5 m^3/min at V_{wind} = 11.1 m/s.

With the fan running, Figure 3b, its influence on the velocity distribution at V_{wind} = 0, can be clearly seen. The fan alone increases the velocity by about 400%. Through the open unshrouded area some reverse flow can be seen.

The effect of the running fan can be also seen, when Figures 3a and 3c are compared. In both cases the wind velocity is 11.1 m/s. All over the area in front of the fan blades the velocity is increased by approximately 500%. The running fan increases the air flow rate by about 200%. In the case of running fan the increase of the air flow rate due to the wind velocity of 11.1 m/s is only 20%.

A very smooth velocity distribution all over the radiator core can be seen, when a condenser is installed in front of the "Test Radiator" (see Figure 4a). This device acts aerodynamically as a flow straightener. There is more even distribution of velocity all over the radiator core. The air flow volume is about 20% lower than without the condenser. In the case of no wind and running fan approximately the same total air flow volume is measured, yet a higher rate of reverse flow is shown. When the wind is blowing and the fan is running, a 10% lower flow volume is measured in comparison to the case without condenser. In this case a reverse flow through the open area of the radiator can be noticed.

In all cases with the running fan, it can clearly be seen that in the region in front of the fan hub a lower velocity is measured. This is due to the fact, that, because of packaging reasons, the distance between the fan hub and the radiator core has to be kept very small. In this area the radiator core is not used optimally as a heat exchanger. For better use, the distance between core and fan hub should be increased, but this would result in a longer front end of the vehicle, a situation which is often unacceptable.

To summarize these cooling air flow volume measurements, Figure 5 shows the dependency of the cases with/without running fan and with/without condenser as a function of vehicle speed. The measurements with/without condenser show the same tendency. The only difference is, that a lower flow volume rate is measured with the condenser because of the higher air flow resistance. In the case of no or low vehicle speed the fan gives most of the cooling air flow. With increasing wind speed ram air overtakes this task. This will be in the case of more than approximately 140 km/h (not shown in this diagram).

From the analysis of the various velocity distributions with/without running fan, with/without condenser it can be seen, that, for numerical methods, it is very important to know the real distribution of the cooling air flow all over the radiator core. (In one of the next paragraphs about LDV-measurements it will be shown, that it is even very important to know the local velocities in 3-D components.) So, although this "Test Radiator" was originally designed to replace the installed radiator and to get information about the cooling air flow distribution for the optimization of front end openings in respect to lower C_D-values, now this "Test Radiator" is also used to provide the first air flow data at the inlet plane for the calculation of underhood temperatures within the engine compartment.

TECHNIQUES FOR MEASURING AIR FLOW VELOCITIES WITHIN THE ENGINE COMPARTMENT

The wish of the customer for more powerful engines, more comfort, e.g. with power-steering and air conditioning systems, more safety, resulting in additional aggregates, e.g. ABS, means that engine compartments are more and more tightly packed. It is therefore becoming more and more difficult to reach aggregates or electronic parts with air of at least moderate temperatures [1].

Also the lower end of engine compartments are more and more enclosed by engine compartment covers for the fulfilment of reduced by-pass noise regulations. So the hot air within the engine compartment is hindered from escaping already at the lower front of the engine compartment. The lack of outlets due to the noise regulations, prevents the heated air flow to escape easily from the engine compartment.

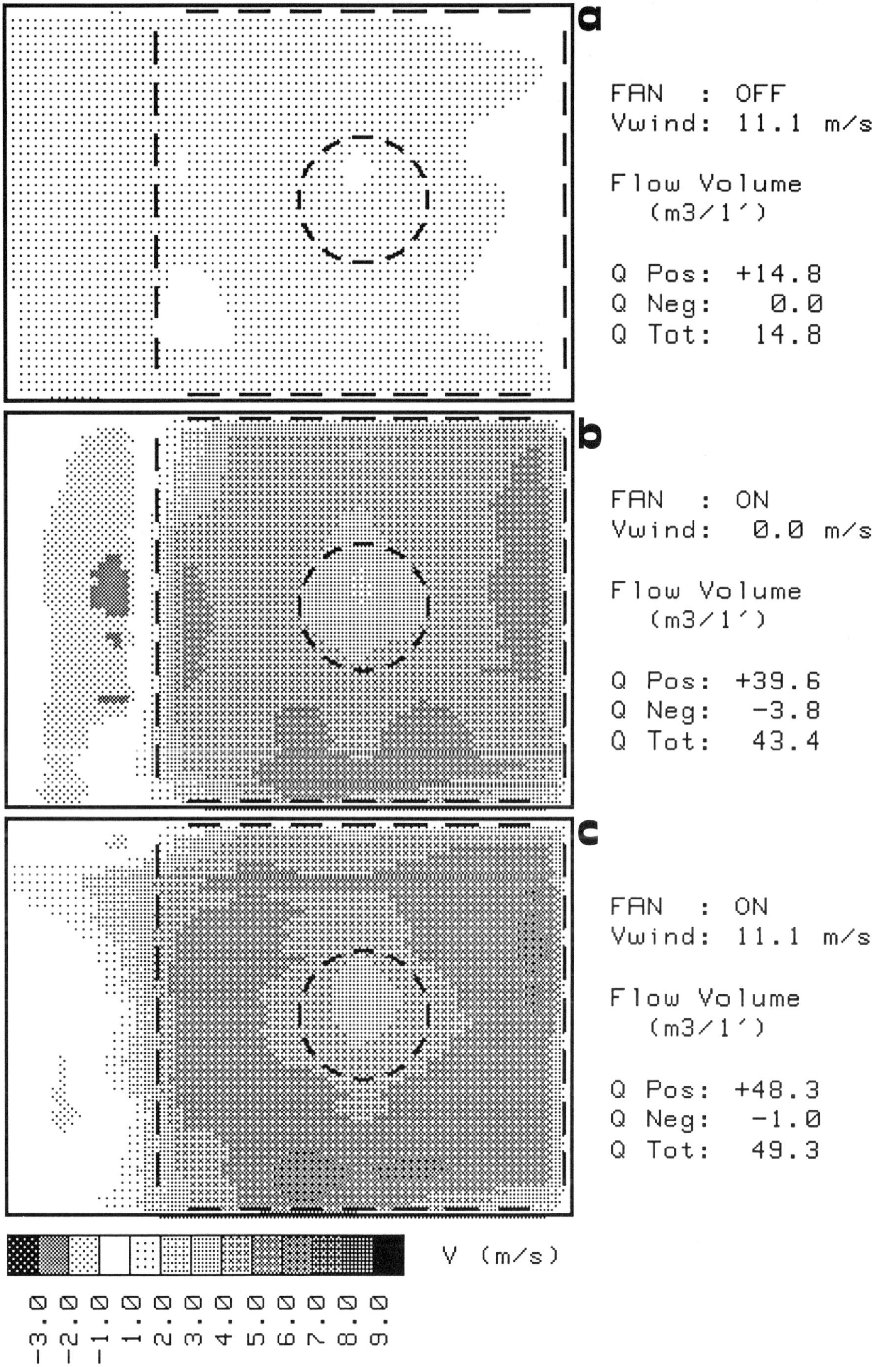

Fig. 4 a-c : Condenser + New Test Radiator + Puller Fan
(seen from the front)
Velocity distribution trough the PF Test Radiator
at Vwind = 0 & 11.1 m/s, Puller Fan ON & OFF

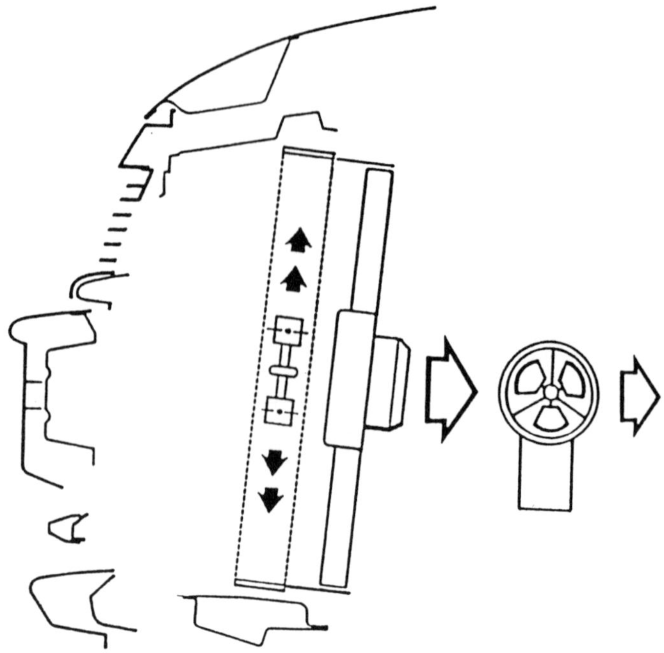

Fig. 3d: Simplified layout of the new "Test Radiator" + a Puller Fan.
The corresponding velocity distributions are shown in figs. 3a,b,c.

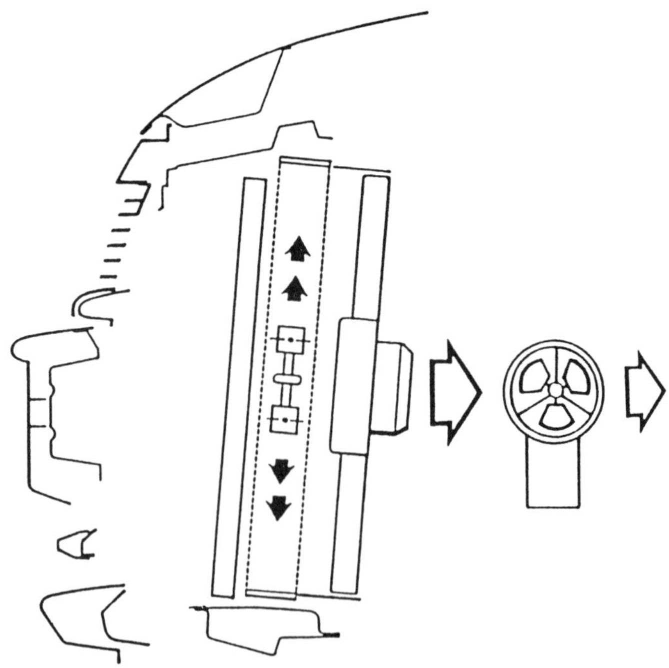

Fig. 4d: Simplified layout of a Condenser, the new "Test Radiator" and a Puller Fan.
The corresponding velocitv distributions are shown in figs. 4a,b,c.

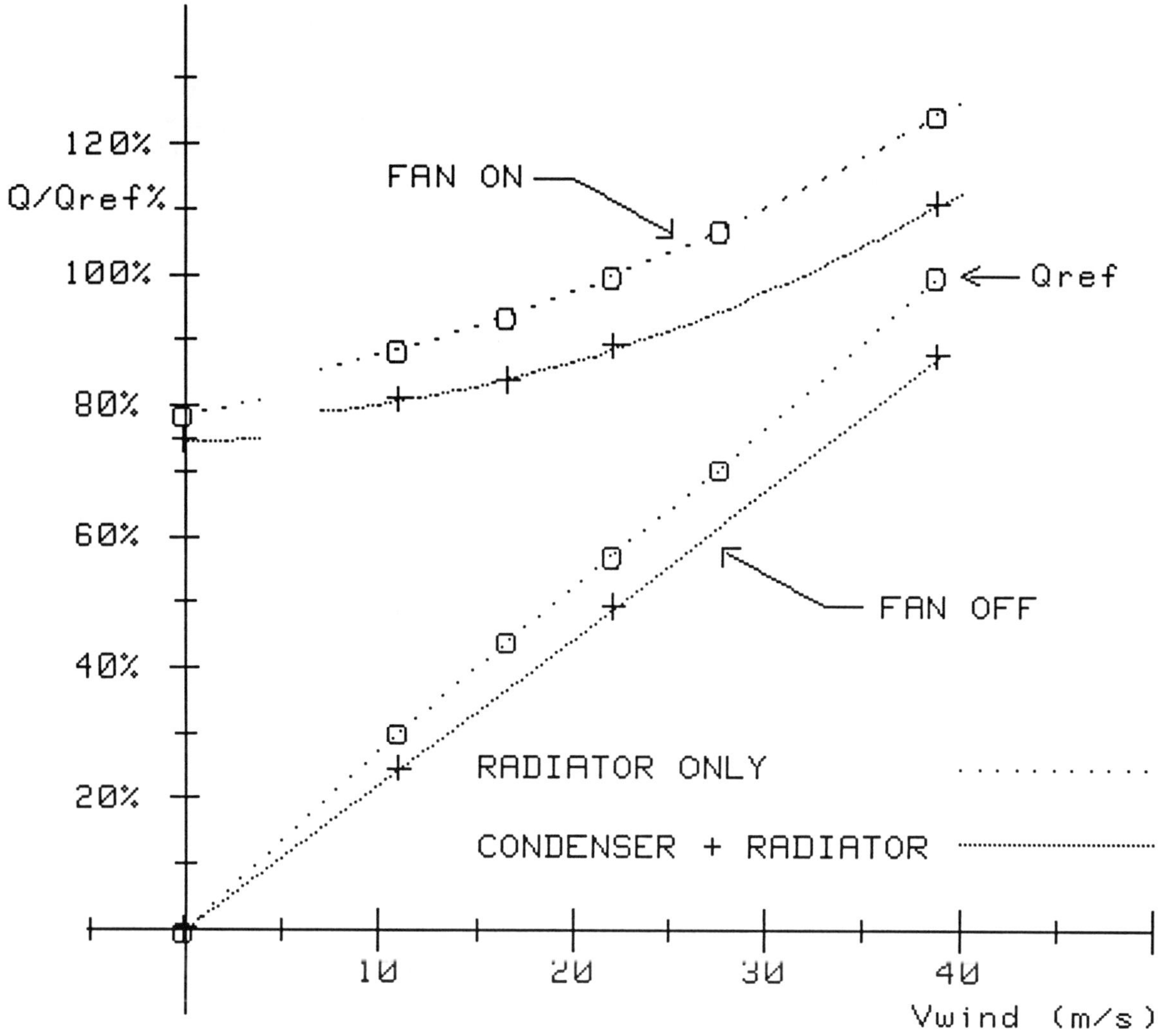

Fig. 5 : Flow volumes ,with Fan ON and Fan OFF , for
. Radiator only
. Condenser + Radiator
Qref = Q at Vw = 38.9 m/s (= 140 Km/h), Fan OFF

Furthermore nowadays hot-idle, stop-and-go--traffic and traffic at low speed occur often within cities and on motorways because of the increasing number of vehicles on the road.

All these reasons explain why, especially in recent years, driving conditions with higher temperatures within the engine compartments of modern vehicles are getting more and more common. Sometimes so called "hot spots" may lead to a failure of the vehicle or at least some components.

Attempts of air flow velocity investigations within the engine compartment were already done about ten years ago [1]. However, as the engine compartment air flow is, in general, a low velocity, highly-turbulent and often non stationary flow, the difficulty of measuring the three components of local velocities with a reasonable accuracy was evident.

The use of NTC-probes (**N**egative **T**emperature **C**oefficient) is very easy, but these probes can only detect whether there is a flow of a specific point and give an averaged velocity value. Information about the direction of the flow are precluded.

Multiwired hot-wire-probes might be able to give an indication about the flow direction and the three velocity components, however the reading of one or more wires can be in error due to the mutual interference of the wakes of the other wires. This is particularly true in the case of measurements inside the engine compartment where the flow direction is in principle unknown.

Therefore LDV-probes (LDV = **L**aser-**D**oppler-**V**elocimetry) which work without disturbing the flow at the measuring point and which are able to detect all three components of the flow velocity are, at present, the only reliable devices for the measurements of this type of flow.

THE LASER-DOPPLER-VELOCIMETRY SYSTEM

A two-component fibre optic LDV system was installed in the PININFARINA wind tunnel in 1987 [2].This system included a COHERENT Argon Ion Laser Generator, a main optical system, a two-component fibre optic link by TSI and two BSA (**B**urst-**S**pectrum **A**nalyzer) by DANTEC. It had been used for several years to investigate both external flows around car bodies and internal flows mainly in engine compartments [3], [4].

In general, this system worked rather well. However, it became apparent very soon that the measurement of the third velocity component was almost always needed. To get this third component, the only possibility was to repeat every measurement by rotating the arm of the traversing system, which carries the probe, so as to reach the same measuring point with a different angle.
In particular this meant:

- The wind tunnel time needed for a given investigation was usually doubled

- Sometimes, the measurement of the third component was not feasible at all. In fact, due to physical constraints, for instance inside the engine compartment, the rotation of the probe was impossible.

- When this rotation was possible, it was however almost impossible to check if the measuring point of the second measurement coincided exactly or not with the first one. This was mainly important in case of measurements in regions of high velocity gradients.

To overcome one of these limitations, that is the difficulty of rotating the probe supporting arm inside the engine compartment, an improvement to the LDV system was made in 1990, when the use of a "side looking mirror" attached to the probe, became common practice to measure the third component, Fig.7.

However, the two other limitations mentioned before, that is the doubling of the wind tunnel time as well as the uncertainty of the position of the measuring point during the second measurement, brought us to the decision to develop a new three-component miniature probe for the laser velocimeter.

This project was carried out by PININFARINA in close cooperation with DANTEC. A new 3-D probe of very compact size was designed and built. It is capable of being used both in the outer flow of the vehicle and inside the engine compartment. This new three-component fibre optic LDV system is now in operation since March 1992, (see Figures 8 and 9).

This new LDV-system consists of the following items:

- the same (5.0 W) Argon Ion Laser Generator as mentioned above,

- a new main optical system, named " Fibre Flow",

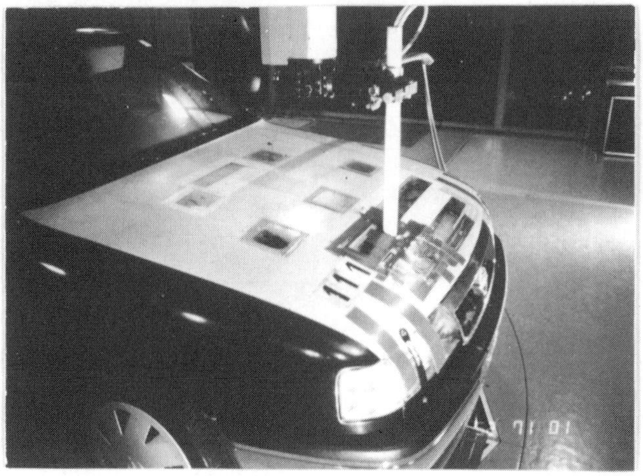

Fig. 6: The main traversing system and the 3D-LDV probe during the wind tunnel tests on an Opel Vectra

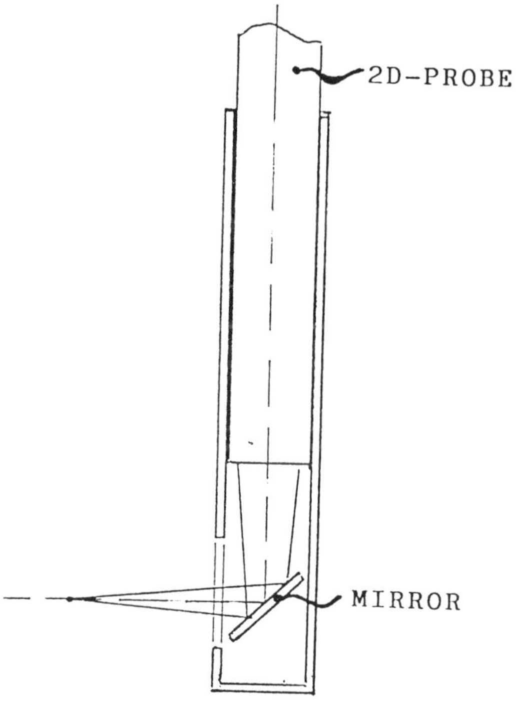

Fig. 7: Side looking mirror

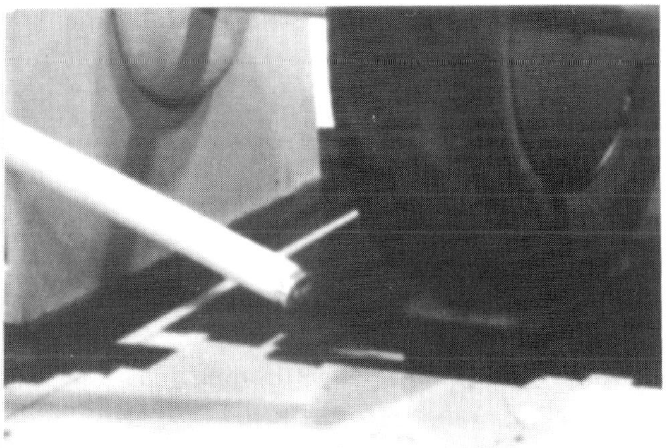

Fig. 8: The new 3D miniature LDV probe

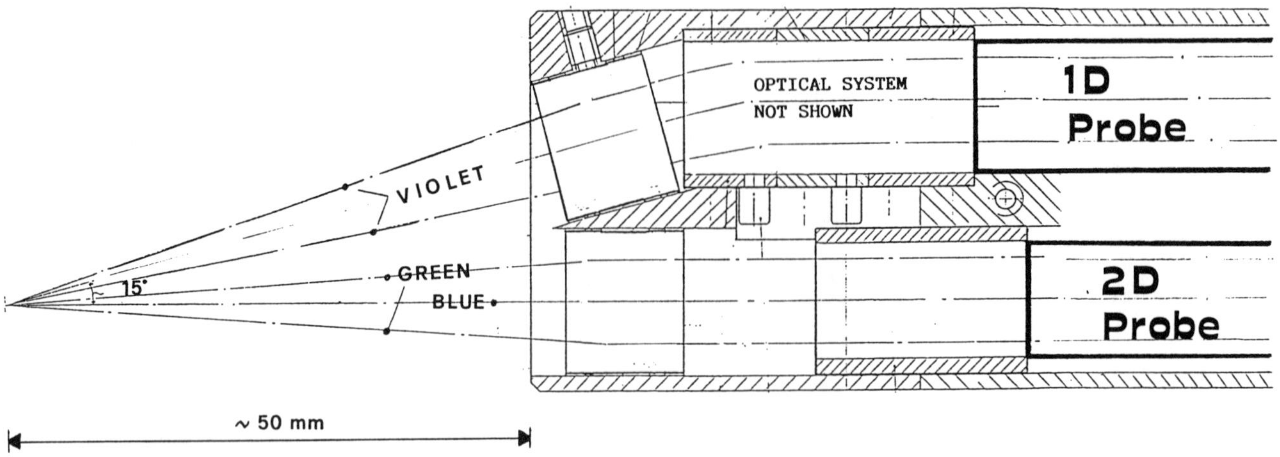

Fig. 9: Simplified drawing of the miniature 3D-probe

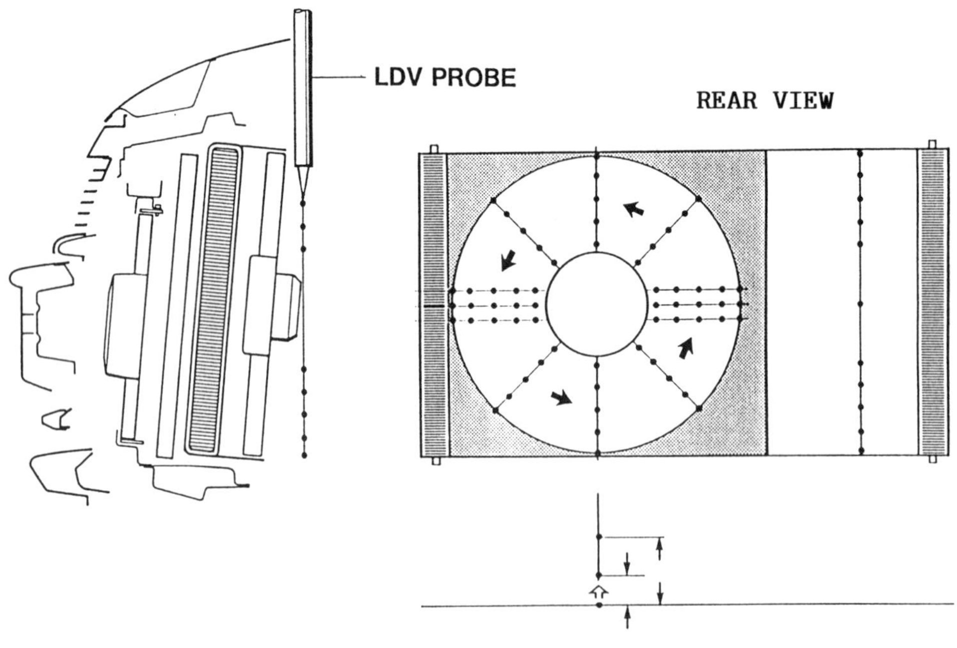

Fig. 10: Layout of the measuring points behind the rotating fan

and a fibre link 40 m long,

- an end-probe made of a standard 2-D probe and a 1-D probe, combined together in a compact 3-D probe. It uses the main three colour lines of the laser generator, that is to say, green, blue and violet.

- a third BSA, added to the existing two, in order to process the third component (violet).

The complete system is controlled by an HP 9000-380. The software, written by PININFARINA, controls both the three BSA and the traversing system which carries the probe. All that allows to greatly reduce the measuring time which is now typically 2 minutes per point including data acquisition,data processing, data storage and the movement of the probe to the next measuring point, when velocities along a straight line are investigated. An example of a 3-D velocity distribution measured using this LDV system is shown hereunder.

EXAMPLE OF VELOCITY DISTRIBUTION BEHIND A ROTATING SHROUDED FAN

The use and validation of numerical codes regarding the underhood thermal management, requires the knowledge of the three components of the velocities at the inlet plane of the engine compartment (as well as at the outlet plane). In this case, the inlet plane is located directly behind the shroud and fan assembly as well as the radiator and its surroundings.
It is the air coming through this plane that mainly contributes to the engine compartment air flow. And it depends mainly on the flow through the radiator which in turn depends greatly on the fan operation.
In fact, for modern vehicles other air flows coming, for instance, from leakages around the radiator or from underneath the front bumper, have been greatly reduced, the first by appropriate cooling air ductings, the second ones by the noise insulating underbody covers.

It is therefore of great importance to map the flow behind the rotating fan, as it mostly contributes to the overall flow volume entering the engine compartment and in the same time it defines the distribution of the local velocities at the inlet plane.
An impressive example of this velocity distribution at the inlet plane, that is the distribution of the longitudinal components V_x, the tangential components V_t, and the radial components V_r, measured behind a running puller fan, are shown in Figures 11 and 12. The single points at which the readings were taken are shown in Figure 10.

The tested vehicle was in fully dressed version, which means that there was a running pusher fan and a condenser installed in front of the radiator. Because of the backwards inclined radiator/shroud fan-assembly and of the standing heights of the vehicle, unfortunately it was not always possible to measure velocities at the same distance behind the fan blades.

Whereas most of the testing could be done from above by inserting the last probe through some openings in the hood, see Figure 6, some points had to be measured from below the vehicle. Fortunately, the PININFARINA wind tunnel has a well designed balance room underneath the vehicle where a special traversing system could be installed for this purpose.

The influence of a rotating fan on the velocity distribution at no vehicle speed (V_{wind} = 0 km/h), a condition which could be compared with the hot idle case, after quick motorway driving or stop-and-go driving in a city, is shown in Figures 11 a-c. Contrary to experience of commonly known investigations behind rotating fans, an inhomogeneous velocity distribution is measured when the fan is installed inside an engine compartment of a real vehicle with the engine just behind it and all its aggregates around and behind it. Normally when pressure flow rate characteristics of a fan are measured in a laboratory an equally distributed air flow over the area of the fan blades is measured, but there are no special objects in front of or behind the tested fan.

Whereas high positive longitudinal velocities (V_x = 8 m/s) are measured at the right hand side behind the fan, approximately 40% lower velocities are determined at the left side, see Figure 11a. There is also a relatively large area of reverse flow at the left and lower side of the surroundings of the electrical motor. The reverse flow in the direct proximity of the motor can be explained either by the influence of the ring of the shroud carrying the motor and/or by the fan blade roots which are probably not working at the optimum, i.e. that there may be stalled air flow.

In the region where Figure 11a shows a reverse flow, the tangential velocity components are directed clockwise and in Fig. 11c the radial components are directed inwards to the motor

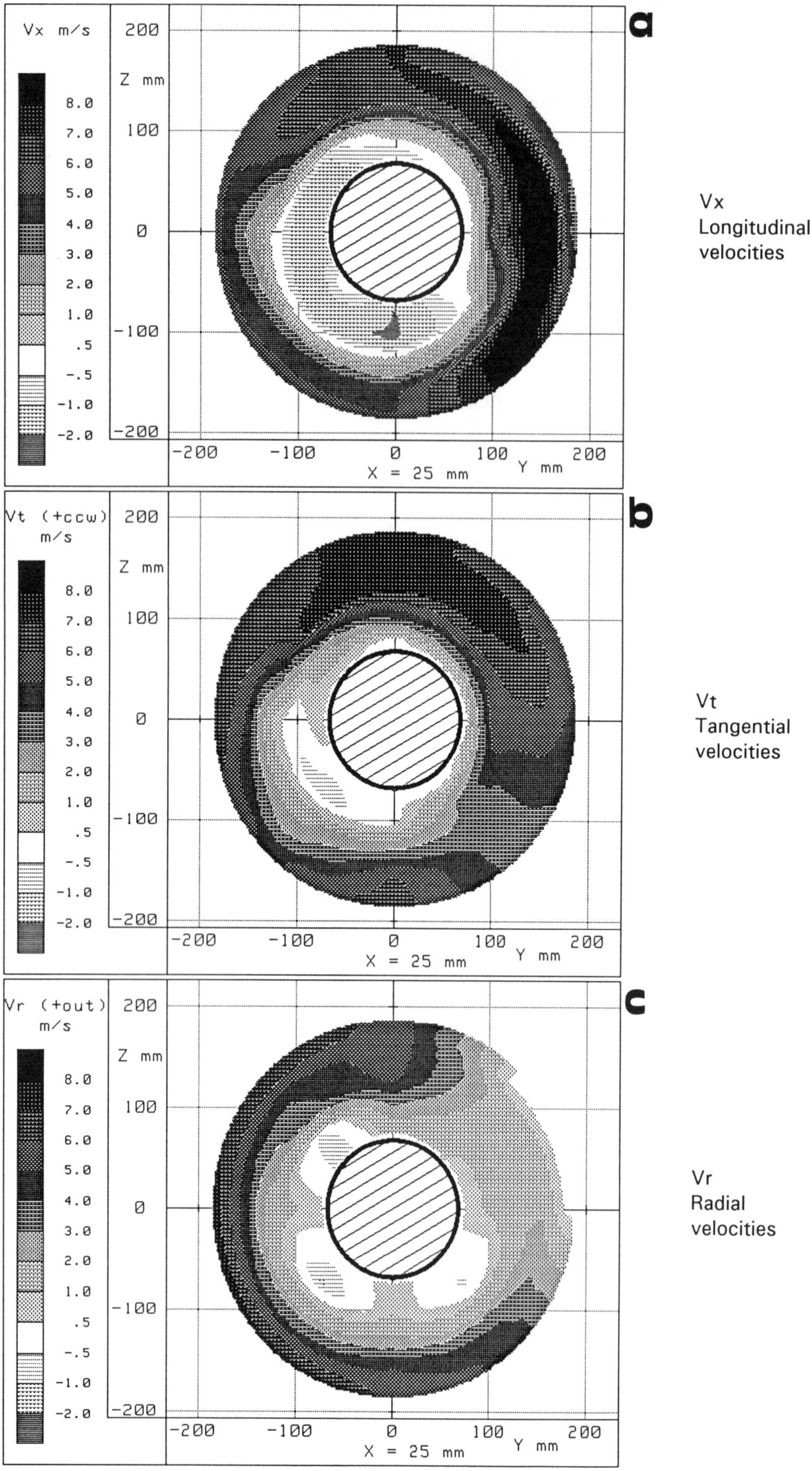

Fig. 11: Velocity maps behind a rotating fan at Vwind = 0.

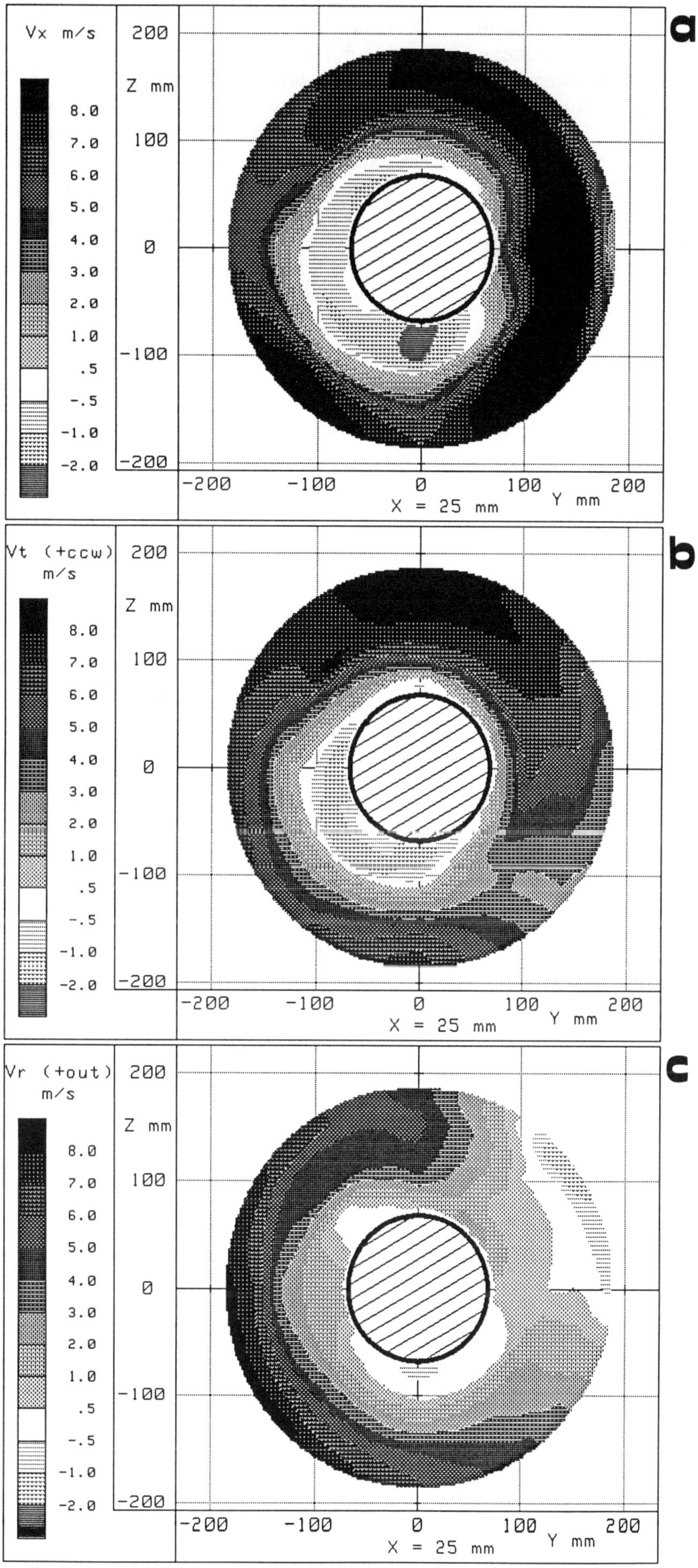

Vx
Longitudinal
velocities

Vt
Tangential
velocities

Vr
Radial
velocities

Fig. 12: Velocity maps behind a rotating fan at Vwind = 60 Km/h.

axis. This confirms the existence of a large area of stalled air flow. In addition to the above mentioned reasons, the air flow resistance of the radiator and all other parts in front of the fan may also be of influence.

High tangential velocities are measured in the region of the upper and lower part of the fan air flow. Highest radial velocities are exhibited in a "halfmoon-like" area along the top and the lower left side of the stream.

There is not a great difference in the behaviour of the flow, when wind is added. Figure 12 a-c show the velocity maps for the already known test condition, but now V_{wind} is 16.7 m/s. The area of high longitudinal velocities now spans approximately all the right half of the fan flow. The area of no or reverse flow in the upper, left and lower surroundings of the electrical motor still remains. The tangential velocities, Figure 12b, behave in the same way. Highest tangential velocities are measured in the upper part of the air flow. The region of clockwise flow direction in the left and lower vicinity of the motor did not change significantly. Only the area of inward directed velocity components around the electrical motor was reduced, Figure 12c. It is now mainly concentrated underneath the motor.

Results of these investigations show how important it is to know the exact distribution of the three velocity components at the inlet plane of the engine compartment. Only with a good representation of this flow a reliable basis for numerical codes can be achieved.

NUMERICAL METHODS FOR ENGINE COMPARTMENT AIR FLOWS

Numerical methods for engine compartment air flow have recently been developed for early prediction of underhood temperatures, i.e. for underhood thermal management, and to help in reducing vehicle development times. One of these methods, described in the next presentation of this conference by Ashmaway et al. [5], uses as the dynamic input, the flow conditions behind the radiator and fan/shroud plane, and the flow conditions underneath the front bumper as entrance conditions for the calculation. The outlet plane for the calculation is situated in the lower firewall area and wind tunnel floor. The accuracy of this calculation method and thereby the reliability of the predicted flow- and temperature values depend a great deal upon the accuracy of the measured in-flow and out-flow data.

In reality, in the course of the experimental work a number of investigations not only in the inlet plane but distributed also all over the engine compartment with/without engine compartment covers and in the outlet plane, were carried out in the PININFARINA wind tunnel. All these tests prove that only a wind tunnel which is equipped with high technology testing technology is able to deliver testing data which can be the basis for modern calculation codes.

CONCLUSIONS

The air flow can be shown as vertical velocity profiles or as velocity maps, i.e. as isotachs. The small air vanes within this newly improved "Test Radiator" are able not only to detect the air flow but also to distinguish between positive and negative (reverse) flow direction. The flow resistance of this special radiator can be easily matched to the resistance of the real installed radiator. For the first time it enables measurement of wind velocities in their three components at the same place, simultaneously. So, compared to conventional LDV measuring test procedures testing time is reduced and, more important, the third component now can even be measured in areas where it was impossible to obtain data because of physical space constrictions.

REFERENCES

[1] H.J.Emmelmann,H.Berneburg,"Aerodynamic Drag and Engine Cooling -Conflicting Goals?" FISITA paper 905128, May 1990.

[2] A. Cogotti,"A Two-component Fibre - optic LDV System for Automotive Aerodynamics Research", SAE paper 880252, February 1988.

[3] A. Cogotti, H. Berneburg,"Engine Compartment Air Flow Investigations Using a Laser-DopplerVelocimeter", SAE paper 910308, February 1991.

[4] B. Bonis, A.Cogotti, D. Vitali, " Numerical and Experimental Flow Field Survey of an Aerodynamically Efficient Passenger Car", ATA 1992

[5] M.Ashmawey et al., "A Numerical Evaluation of the Thermal Effects of the New V6 Engine on the Underhood Environment of the 1993 OPEL Vectra", presented at the Int'l SAE Congress, March 1993

950995

A New Multi-Component-Wind-Tunnel-Balance for Automotive Windtunnels

L. Polansky and W. Matich
Carl Schenck AG

J. T. Kutney Sr.
GE (Ret.)

ABSTRACT

An advanced technology 7-Component-Wind-Tunnel-Balance for the automotive industry has been developed by Carl Schenck AG, Germany and is now fully operational at the Volkswagen Wind Tunnel, Wolfsburg, Germany. The balance is based on the computer separated technology introduced by Schenck AG in the DNW wind tunnel in 1982. The seperation of the aerodynamic components inside the balance was performed -since the begining of wind tunnel testing- with mechanical separation systems. With the development of the computer technology it was possible to transfer this separation to the computer. This new computer separated balance has demonstrated outstanding performance and reliability.

Common practice in automotive wind tunnel testing is to measure the four lift forces on each wheel pad, the drag force and yawing moment. The side force acting on the front axle and rear axle are calculated with the data given by the measured yawing moment and the theoretical wheel base. The other two aerodynamic moments are calculated with the lift forces and the theoretical (or design) wheel base and wheel track. In practice wheel base and wheel track depend on tire pressure and mechanical fabrication tolerances. Therefore the calculation of the moments is subject to errors.

The Schenck AG advanced technology balance provides the ability to determine the actual wheel base and wheel track as influenced by the wind tunnel airflow. Each of the four wheels is located on four mechanically independent computerized 6-Component-Balances. The load signal of each balance combined with the load cell arrangement of each balance enables "ON-LINE" calculation of the actual wheel base and wheel track.

This new balance by Schenck AG will give to the automotive engineers more accurate test data and therefore can have a favourable result with better handling for safety and better ride quality for comfort.

1. BACKGROUND

The competitive economical situation in the world demands that the automotive industry improve not only the manufacturing capabilities, but also the time required to develop a new car.

Therefore the car companies have built in the past few years their own wind tunnels or modified existing wind tunnel facilities to improve the wind tunnel capabilities for faster and more accurate measurements. In wind tunnels, modifications on the car underbody must be done within the shortest possible time. It is important to allow the wind tunnel engineers to do this modification in-side the test section with the car in the measuring position and not outside.

The Volkswagen AG, Germany, which owns one of the largest climatic wind tunnel in the world decided for the mentioned reasons to replace the existing 25 year old wind tunnel balance by a new balance system which represents the state of the art and fulfills the following main requirements.

The measuring ranges for the various components were specified as follow:

Drag D	=	+ 1500 N
Side Force S	=	+/- 1500 N on each axle
Lift L	=	+/- 1500 on each wheel

The required accuracies to full scale for the components were:

Drag D	=	+/- 0,03 %
Side Force S	=	+/- 0,03 % on each axle
Lift L	=	+/- 0.03 % on each wheel

A high degree of flexibility was required to adjust - within 5 minutes each pad - different wheel bases and wheel tracks contiuously within the following ranges:

Wheel track	1000 mm - 2000 mm
Wheel base	1500 mm - 3400 mm

The balance and the turntable was required to yaw within a range of +/-180° with a variable speed of up to 2°/sec with a positioning accuracy of 0.05°.

The turntable was to be designed such that plasticine models with a total mass of 4 tons could move over the total area of the turntable including the adjustment area for wheel track and wheel base.

Furthermore, a system had to be incorporated into the balance to elevate the car 400 mm above the test section floor with a positioning accuracy of less than 2 mm in each position. The lift system had to be designed for operation over the balance yaw range of +/- 180°.

Depending on the wide range of cars with different tire sizes and thus different foot prints, different sets of wheel pads had to be provided adaptable from 100 mm to 250 mm in width, in steps of 50 mm, over a constant length of 180 mm. Alternatively a set of pads with the size of 250 mm by 180 mm could be provided. This solution required the pads to have a differential pressure measuring system which corrects automatically - that means "ON-LINE" - the lift forces measured on each wheel. This correction becomes necessary because of the occurring shunt lift force. This shunt force is produced by a small air pressure difference between the air pressure in the test section and the ambient pressure in the balance room acting on the rest of the wheel pad area which is not covered by the foot print of the tire.

In addition, the new wind tunnel balance had to fit into the existing space without any modification of temperature-controlled balance room and test section floor. Furthermore, the existing base frame had to be used to support the new balance.

2. COMPARISON OF THE DIFFERENT MEASURING SYSTEMS

The type of balance to be used was discussed with Volkswagen engineers during the tender period.

Two different types of external balance systems are used today in the automotive wind tunnels, i.e. the 7-Component-Balance and the 6-Component-Balance. Since load cells are mainly used to measure the acting forces and moments, only load cell balances are discussed in the following.

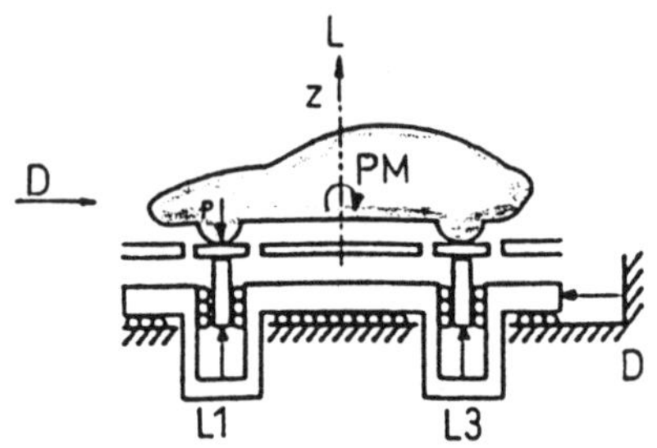

Fig. 1: 7-Component-Balance

Fig. 1 shows the principle of a conventional 7-Component-Balance. This system exclusively measures the acting lift forces, drag force and yawing moment. The car stands with its four wheels on four individual platforms (1) measuring the lift forces acting on each wheel via a force reduction system, i.e. lever system or hydraulic system. Furthermore, with a mechanic or hydraulic decoupling system (2), the lift measuring system is mounted in a so called swing frame (3). Also with aid of a force reduction system, the swing frame is connected to three load cells. The first load cell (4) measures the total acting drag force, the other two (5) are measure the yawing moment. The yawing moment is calculated by means of the load cell signal and the exact length between the load cells.

The side force acting on the front axle and rear axles are calculated by means of the theoretical wheel base. The two aerodynamic moments - pitching moment and rolling moment - are calculated by using the theoretical or design wheel base and wheel track.

The exact separation or decoupling of the three different components is necessary other wise each acting component would influence the others.

To guarantee the best possible separation of the different mechanical measuring platforms and frames, the separation system must be friction free and ensure that the three force vectors are exactly aligned in the axis of the aerodynamic coordinate system. Also the balance design must assure the long term reliability of the system and that no settling phenomenon of the foundation and the balance frames changes the calibration charateristic of the balance.

With this measuring method, fabrication tolerances of the car and load distribution in the tire footprints are not taken into account. These two facts lead to a significant falsification of the moment calculation which be will discussed later.

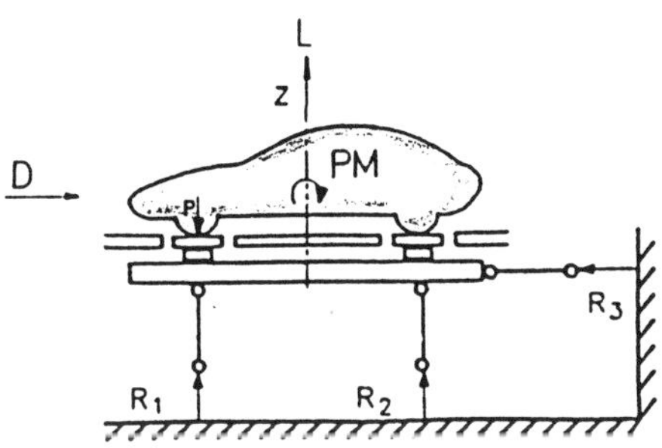

Fig. 2: 6-Component-Balance

Fig. 2 shows the principle of the computer separated 6-Component-Balance. The computer separated wind tunnel balance demonstrates the latest state of the art.

The car is standing with its four wheels on a single common platform. All six aerodynamic components are introduced into this platform. The platform is connected to six load cells by six so called connecting rods. A certain deflection of the platform is also necessary for the same reason as with a 7-Component-Balance. So for better flexibility the connecting rods are equipped with flexures on both ends which enables the platform to move in each direction repeatably and with a minimum on force.

To avoid faulty creeping of the mechanical structure of the balance system caused by temperature effects or movements of the foundation, the measuring platform should be kept static by using by a load cell arrangement such as shown in Fig. 3.

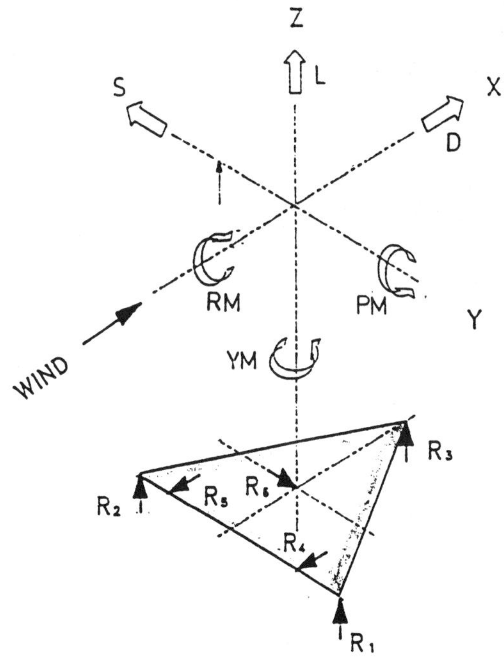

Fig. 3: Statically Determined Load Cell Arrangement

Under the assumption that the alignment of the connecting rods is made exactly in the axis of the areodynamic coordinate system and the balance system is absolutely rigid, the relation between the six aerodynamical components and the load cell signals are discribed with the following equation system.

$$D = -(R_4 + R_5)$$

$$S = -(R_6)$$

$$L = -(R_1 + R_2 + R_3)$$

$$RM = (R_1 - R_2) * a - R_6 * h$$

$$PM = (-R_1 - R_2) * b + R_3 * c + (-R_4 - R_5) * h$$

$$YM = (R_4 + R_5) * d$$

Based on the fact that a balance system is not rigid and the alignment of the connecting rods cannot be made with the required accuracy, the balance will be calibrated for all six aerodynamic components. All influences, as misalignment of mechanical parts, spring rates of flexures and mechanical parts, unknown sensitivity of load cells, are taken into account during calibration and form part of the measuring matrix.

This type of balance is characterized by fast measurement, variable integration time of the measuring signal, high reliability, freedom from maintenance and simple modular design.

3. DESCRIPTION OF THE NEW 7-COMPONENT-BALANCE

After Volkswagen had decided to use a 7-Component-Balance, Schenck developed during the tendering period a new type of 7-Component-Balance which combines the advantages of the 6-Component-Balance with the measuring principle of the 7-Component-Balance.

Fig. 4 shows the concept of the new 7-Component-Balance.

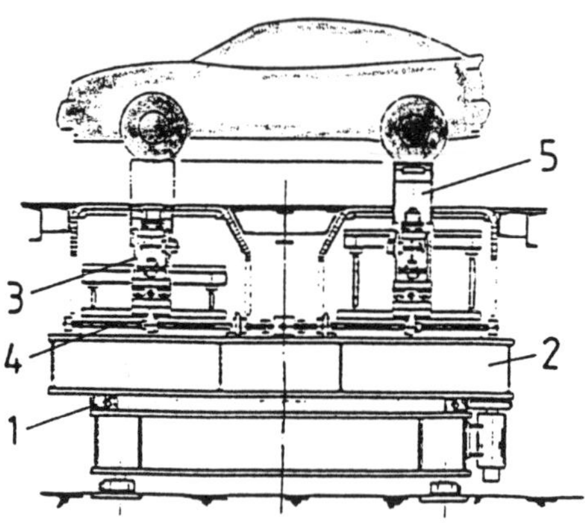

Fig. 4: New 7-Component-Balance

Its construction is quite simple. The new balance system needs no special frames to separate the aerodynamic components. It starts with a high precision bearing with gear rim (1) and the balance base frame (2). An electric drive yaws the balance in the desired position. On the frame, four crossing support systems (4) are mounted. On the crossing units four independent compact measuring units (3) are located. With one single electric drive it is possible to change the position of the measuring units to the required wheel bases. The different wheel tracks of the front and rear axles can also be changed independently by using two electric drives. In the balance, an electric motor-driven elevating system (5) is integrated allowing the car to be elevated up to 400 mm. The upper part of the balance is formed by the turntable. In the turntable, a so called F_x-Revolving-Shutter is installed which is mechanically connected to the wheel base adjusting electric drive. This permits synchronized adjustment of both systems -the measuring units and the revolving-shutter system-. For the wheel track, a F_y-Revolving-Shutter covers the turntable.

From the mechanical point of view the measuring units are based on the measuring principle of the 6-Component-Balance described above. Either of the four balances is able, by using the necessary electronical equipment, to measure all six areodynamic components. So in total 6 components can be measured on each wheel. But with the requirement of the Volkswagen engineers to measure only the seven components it was not necessary to measure each load cell signal separately.

Fig. 5 shows the electronical connection of all 24 load cells for the 7-Component-Balance Configuration. The vertical three load cells R_1 to R_3 of each measuring unit are interconnected to measure the four single lift forces. For measuring the two side forces acting on the rear and front axles, load cell R_6 of the respective measuring units are connected. Load cell R_4 and R_5 of all four measuring units are interconnected to measure the total drag force.

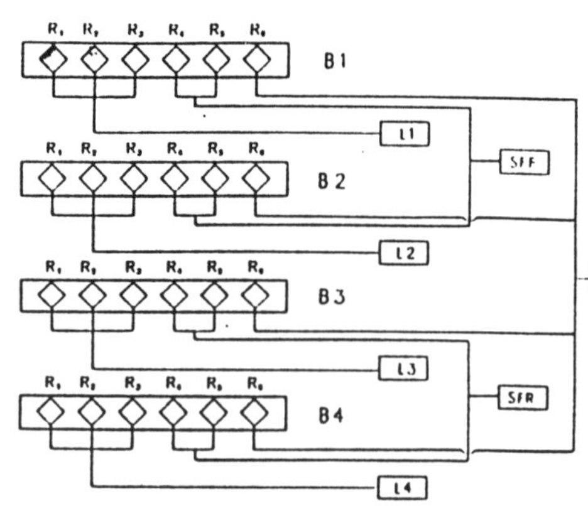

Fig. 5: Electronical Load Cell Connection

4. LOAD CELLS

Along with the development of the computerized wind tunnel balance a load cell was required which fulfills the requirements on best accuracy, high long-term stability, low creep, high resolution, high repeatability and the possibility to measure tension and compression forces. With the Schenck Load Cell a load cell is available which has demonstrated its outstanding capabilities over the past 30 years. For wind tunnel balances, only selected and specially prepared load cells are used to ensure highest accuracy and repeatabilty.

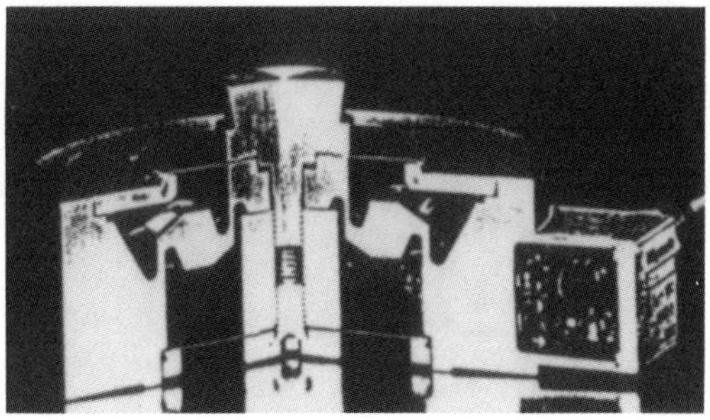

Fig. 6: Schenck Load Cell

Fig. 6 presents a vertical cut through the Schenck Load Cell. Special ring shaped strain gauges with a resistance of 4000 ohms permit up to 60 volts power supply to be used. With a sensitivity of 2.85 mV/V the output signal is 171 mV which enables a resolution of 1.7 million digits to be realized.

The Schenck Load Cell is also used for measuring tension and compression loads in a tension compression unit. Fig. 7 shows a tension-compression unit.

Fig. 7: Tension-Compression-Unit

A tension-compression-unit (TCU) comprises two load cells which preloaded with 50 % of the rated capacity. The mechanical preloading parts are going through the vertical axis of the load cell and therefor not to be seen in the picture. Depending on special design of the TCU it is possible to load the unit with the rated capacity of one single load cell and not only with 50 %.

The main features of the Schenck Load Cell are listed below:

Combined Error	:	< 0.01 %
Hysteresis	:	< 0.01 %
Repeatability	:	< 0.002%
Non-Linearity	:	< 0.01 %
Creep	:	< 0.01 %
Sensitivity	:	2.85 mV/V
Power Supply	:	max. 60 V

5. IMPROVEMENT OF WHEEL BASE AND WHEEL TRACK MEASUREMENT

The balance measurements include "unknown effects". "Unknown effects" are influences which can not be detected by the balance system as non-aerodynamic effects.

Depending on the balance measuring principle, two effects influence the 7-Component-Balance and the 6-Component-Balance in different ways.

1. The first "unknown effect" occurs by using a 7-Component-Balance. The car is standing on four independent measuring platforms and the calculation of the aerodynamic moments is based on the assumption that the actual wheel bases and wheel tracks are equal to the theoretical distances and do not change during measurement. However, with respect to the elasticity of car and tires, the car shifts with the air load. Since the car is standing on four independent measuring platforms, the movement of the center of gravity will not influence measurement. All the same, this movement is acting on the blocked tires of the car thus changing tire characteristics. As a result, foot print force vectors shift and cause the acting wheel base and wheel track to change.

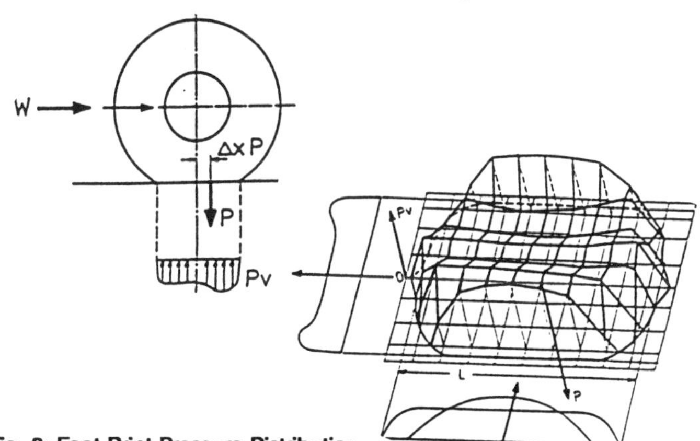

Fig. 8: Foot Print Pressure Distribution

Also fabrication tolerances and different tire pressure changes the foot print area and therefore the actual wheel base and wheel track. These effects together cause the actual wheel bases and wheel tracks to deviate from the theoretical values: assuming that the difference between theoretical and actual wheel base is 1 cm - this is a realistic value - and that a lift force of 1500 N is measured on each wheel, an additional pitching moment of 30 Nm will be realized (with two tires blocked).

2. The second "unknown effect" caused by using a 6-Component-Balance depends also on the elasticity of the car. The center of gravity moves in the three force directions depending on load. This displacement of the car mass produces, relative to the reference point of the balance, aerodynamic moments which cannot be separated from the actual aerodynamic moments by the balance. As a function of the different elasticities in the three directions X, Y and Z, the amount of the additional causing moments will be different. To give an example, the following calculation shows the amount of the additional pitching moment. A car with a mass of 1000 kg and a center of gravity movment of 0.5 cm will produce an additional pitching moment of 50 Nm.

To summarize the explanations, with a 7-Component-Balance, the shifting of the car mass has no influence on the moment calculations but, depending of the measuring principle the difference between theoretical and actual acting wheel base and wheel track lead to errors in pitching and rolling moment calculation. The side forces are also calculated with the theoretical wheel track and these are leading to same effect.

With a 6-Component-Balance only the shifting of the car mass influences the measured moments. All other effects depending on car suspensions, antivibration systems and tires are irrelevant and for the six aerodynamic component measurement it has no influence.

How the "unknown effects" can be eliminated will be described in the following.

For correction of the different effects, additional measuring devices were necessary in the past. For the 6-Component-Balance a measuring system for all six degrees of freedom is necessary. Fig. 9 shows how the six different movements can be measured.

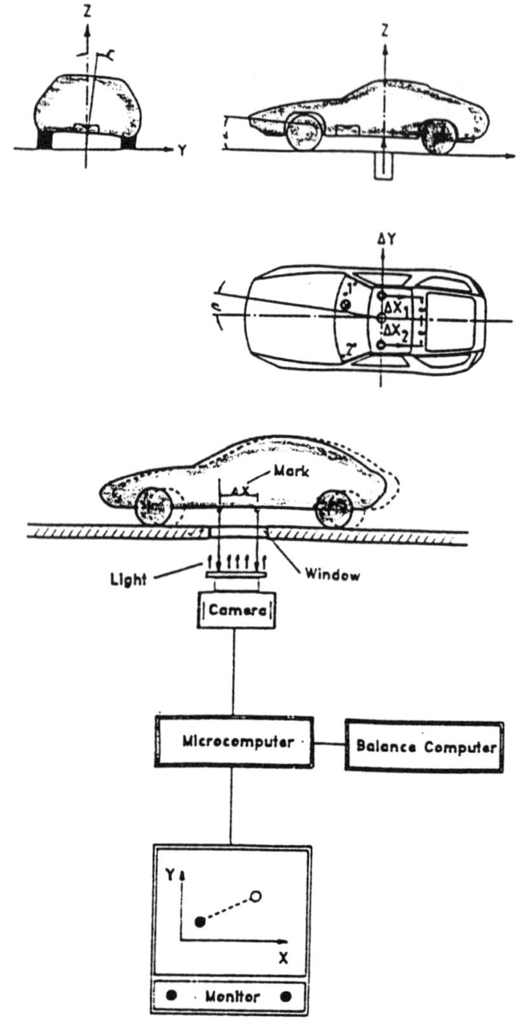

Fig. 9: Deflection Sensor Arrangement

A video camera monitors an identification mark located on the underbody of the car and a high resolution photo interpretation system measures the movements in the X-, Y- and Z-directions. The inclinations of pitching moment and rolling moment can be measured by two inclinometers. The movement caused by the yawing moment can be measured with a differential measuring system acting like a second camera. The measured mass and deflection of the car permit the actual yawing moment to be calculated.

With the new 7-Component-Balance system, no additional measuring devices are necessary. All relevant data are measured with the four single 6-Component-Balances.

The single mechanical balance geometry and the single load cell signal supply all information required to calculate the actual wheel base and wheel track, i.e. acting on the road.

To get all information all load cell signals have to

be measured separately and for this reason the number of digital voltmeters has to be extended from 12 to 24.

Fig. 10 shows the measuring principle.

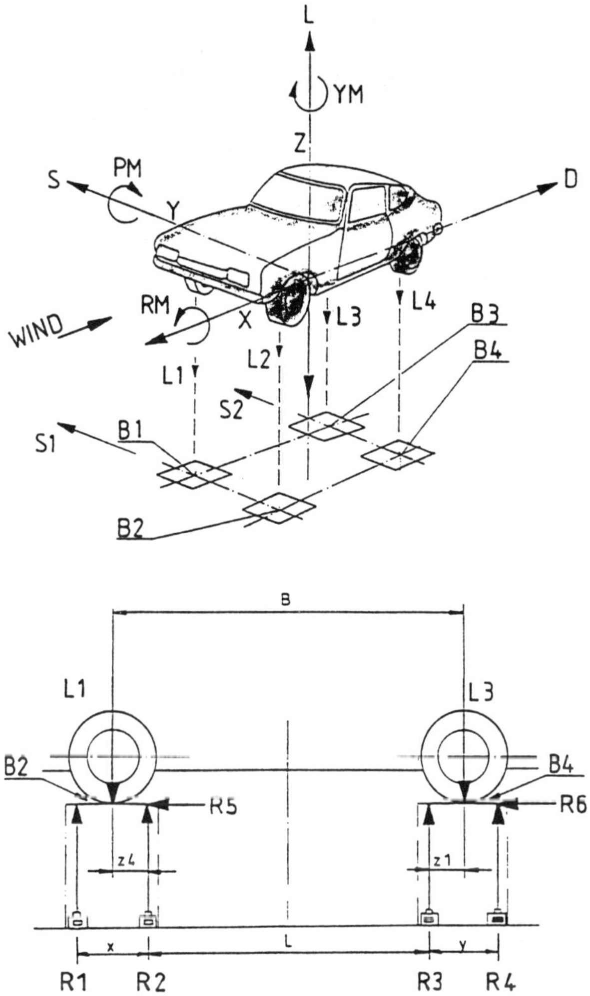

Fig. 10: Measuring Principle of the New 7-Component-Balance

6. IMPROVEMENT OF YAW MOMENT MEASUREMENT

With the conventional 7-Component-Balance the side forces are calculated with help of the theoretical wheel base and the calculated yaw moment. The yaw moment however is calculated with the measured forces in side force direction and the distance between the load cells. Then the side forces on the front axle and rear axle is re-calculated from the yaw moment with the following formula:

$$S_F = S_R = YM/2a$$

where:

YM	=	Yaw Moment
S_F	=	Side Force Front Axle
S_R	=	Side Force Rear Axle
a	=	1/2 Axle Base

It is obvious that this international agreement to calculate the side forces are based on the measuring principle of the conventional 7-Component-Balance. It is questionable if the way to calculate the side force and the yawing moment is comparable with the actual side force and yawing moment.

Depending on the non-symmetrical shape of the car and the non-symmetrical behaviour of tires and car shockabsorber system, the drag forces are producing a yawing moment because the measured single drag forces acting from the tire via the footprint on the balance differ by amount. However, the sum of all single drag forces equals the measured total drag force corresponding to the physical law "action equal reaction". Fig. 11 shows all yawing moment producing forces on a single platform.

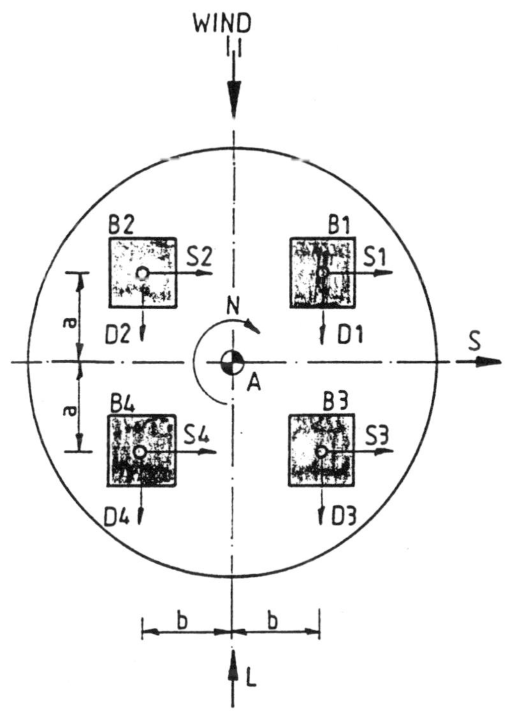

Fig. 11: Real Acting Horizontal Forces on Balances

The new 7-Component-Balance permits, in addition to the single lift force, all single drag forces and side forces acting on platforms A_1-A_4 to be measured, as shown in Fig. 11. This information makes it possible to calculate the actual yawing moment by using the following formula.

$$YM = (S_1+S_2+S_3+S_4) * a + (D_1+D_2+D_3+D_4) * b$$

where:

YM	=	Yaw Moment
S_{1-4}	=	Single Side Forces
D_{1-4}	=	Single Drag Forces
a	=	1/2 Axle Base
b	=	1/2 Wheel Base

7. OTHER FEATURES

1. As the new system with the four single balances requires less space, it was possible to install a motor-driven synchronized lifting system designed to elevate the car by up to 400 mm. The non-synchronization of the four elevators is less than 1 mm and is operational over the total yaw range of +/- 180° with a max. capacity of 1500 kg for each unit.

2. For the wheel pads, we selected the solution to use only 1 pad size. Each wheel pad is equipped with a pressure chamber. The top plate of the wheel pad has 1 mm holes in a grid of 10 mm. With a differential pressure system the mean pressure around the wheel foot print and the ambient pressure in the balance room is measured. With the differential area of the wheel pad and the foot print area and the measured differential pressure, the acting lift force is calculated ON-LINE and the aerodynamic lift force acting on each wheel is corrected.

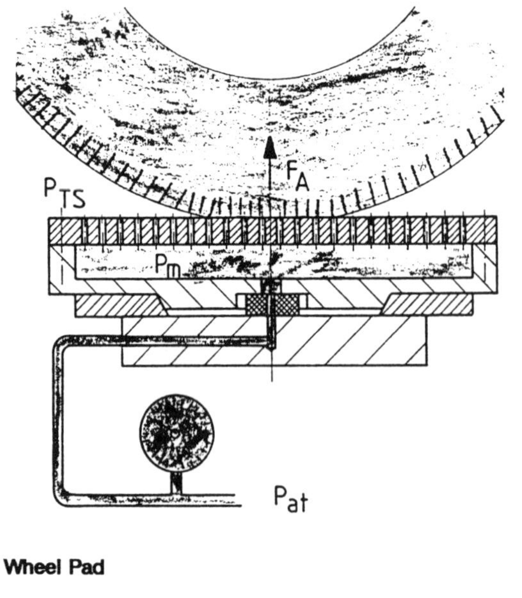

Fig. 13: Wheel Pad

8. CONCLUSION

With the new 7-Component-Balance concept, a considerable improvement of wind tunnel testing has been made. At this time, the Balance as delivered to Volkswagen is not able to measure the actually acting wheel bases and wheel tracks. However, if necessary, the actual forces used to calculate a better yawing moment can be measured. The Volkswagen balance has been in operation for more than one year and has proved perfectly reliable. Due to the fast data acquisition, wind tunnel testing time could be reduced and model preparation time be minimized - two factors which essentially contribute to increase wind tunnel productivity.

REFERENCES

(1) Wolf-Heinrich Hucho
Aerodynamik des Automobiles
1981

(2) P. Giesecke
Development Stages of External wind tunnel Balances,
La Dynamique des Véhicules Automobiles, Lyon, 1985

(3) P. Giesecke
Sechs-Komponenten-Plattform-Windkanalwaage mit "On-Line" Korrektur der Schwerpunktsverschiebung, ATZ, Vol. 87 Nr.1, S.31 pp., 1985

(5) Tim Preusser
Lubomir Polansky
Peter Giesecke
Advances in the Development of wind tunnel Balance Systems for Experimental Automotive Aerodynamics,
SAE Technical Paper Series, 1993

(6) Lubomir Polansky,
J.T. Kutney Sr.
A New and Working Automatic Calibration Machine for Internal Wind Tunnel Force Balances,
AIAA 93-2467 29th Joint Proposion Conference and Exhibit, Monterey, 1993

(7) SAEJI594 JUN87
Vehicle Aerodynamics Terminology, SAE Recommended Practice Issued June 1987

940419

Recent Developments of Experimental Techniques and Their Applications at NASA Langley Research Center

Scott O. Kjelgaard and William L. Sellers III
NASA Langley Research Center

ABSTRACT

The need for highly accurate measurements of velocity, temperature, pressure and density has required the development of new experimental techniques. While the majority of these development efforts at NASA Langley are focused toward applications for aeronautical programs such as the High-Speed Civil Transport, Advanced Subsonic Transport, and the National Aero-Space Plane, a number are applicable to other fields. The intent of this paper is to review recent instrumentation developments and applications at NASA Langley Research Center that may have applications in automotive testing. Five experimental techniques are described along with recent results obtained in NASA facilities.

INTRODUCTION

The earliest studies of fluid dynamics consisted of observations of nature. Fluid physics were observed through the study of bird flight and the flow water around obstacles. In later years, flow visualization was combined with quantitative measurements, to obtain a better understanding of the fluid flow. Flow visualization remained an important tool even when results were limited to qualitative observation, because it rendered certain properties of a flow field directly accessible to visual perception. When advances in both flow visualization and quantitative measurements were combined with mathematical models of flow, understanding grew further. Techniques such as the hot wire anemometry and laser Doppler velocimetry allowed time varying flows to be measured at one or several points. Application of Taylor's Hypothesis allowed the time varying measurements at a point to be related to the spatial details of flow structure. For highly complicated flows, better understanding could be obtained by combining time varying point measurements with flow visualization to generate a correlated picture of the flow field. In high speed flows, only planar images of qualitative data could be obtained using techniques like laser vapor screens, schlieren, or interferometry. This limited our ability to understand many of these flows.

Recent advances in computer speed and memory have allowed numerical solutions to many important fluid dynamic problems. Time-averaged solutions of the Navier-Stokes equations are becoming common for more practical configurations. Direct simulations of the Navier-Stokes equations are being performed for flows with simple geometries. Detailed experimental flow field data is required to assess the abilities of these methods to accurately predict the critical physics of the flow. In addition, detailed experimental data is needed to improve our physical understanding of many flows too complex for a numerical solution. In fact, experimental techniques have not generally taken full advantage of data acquisition and processing made possible by new computing capabilities and in some cases, experimental techniques are significantly lagging the computational methods.

The need for highly accurate measurements of velocity, temperature, pressure and density has required the development of new experimental techniques. While the majority of these development efforts at NASA Langley are focused toward applications for aeronautical programs such as the High-Speed Civil Transport, Advanced Subsonic Transport, and the National Aero-Space Plane, a number are applicable to other fields. The intent of this paper is to review recent instrumentation developments and applications at NASA Langley Research Center that may have applications in automotive testing. Five different techniques are described along with recent results obtained in NASA facilities.

TECHNIQUE DESCRIPTIONS

FOCUSING SCHLIEREN - Three methods have been widely used for the visualization of density gradients, shadowgraphy, schlieren, and interferometry. A major drawback to these methods is that they integrate the density variations over the entire optical path, often obscuring the flow phenomenon of interest. Focusing schlieren techniques that allowed examination of thin planes were first developed over 40 years ago by Kantrowitz and Trimpi (ref. 1) and Burton (ref. 2). While variations of focusing schlieren were used intermittently (refs. 3 - 5), these early focusing schlieren designs suffered from various limitations, such as low brightness, small field of view, large depth of focus and were difficult to use. Weinstein (ref. 6 and 7) has developed a focusing schlieren system that eliminates some of these limitations and provides a unique capability for imaging density variations in thin slices of a three-dimensional flow field. The ability of the focusing schlieren system to image a thin plane allows the use on non-optical quality windows. This fact alone can make focusing schlieren a significantly cheaper method than conventional schlieren. The focusing schlieren can be used for the detection of thermal gradients or thermal tracers in low-speed flows. This provides the ability to detect thermal convection in under-hood flows or heating and cooling systems.

Figure 1 shows the schematic of the large-field focusing schlieren system. The basic design is similar to the version described by Burton (ref. 3) with the primary differences being the use of a Fresnel field lens for the diffuser and a small, but extended light source. The Fresnel lens concentrates the light source at the imaging lens location. This results in an image over 100 times brighter than the diffuser version. The source grid can be either closely spaced small points or thin lines that fill the source area. For the current paper, only thin parallel-line grids are considered, but the results would be similar for other types of grids. The source grid has to be about twice the size of the test area since it is about twice as far from the lens. The cutoff grid is a photographic negative image of the source grid. Orientation of the grids (horizontal or vertical) depends on the desired direction of sensitivity. Different positions in the flow field can be brought into focus by moving the image plane location. As the source grid is made larger, the field of view gets larger, and different parts of the source grid are used to examine different parts of the flow field.

While focusing schlieren provides the ability to examine a thin plane of the flow fields of interest there are trade-offs between this technique and conventional schlieren systems. The sensitivity of a schlieren system is proportional to the image height and the distance between the lens and cutoff grid (knife-edge). In general, focusing schlieren systems have shorter distances between the lens and cutoff grid yielding lower sensitivities than conventional schlieren systems. The field of view of the focusing schlieren system is dependent on the size of the Fresnel lens and source grid used. Typical focusing schlieren systems require these elements to be twice the desired flow field diameter. The detailed design of the system requires trade-offs between resolution, depth of focus, and image uniformity and the details of the design process are described in reference 6.

Figure 2 presents an example of the ability of focusing schlieren to image a plane of interest. Focusing schlieren flow visualizations of two jets (M = 0.1) with separations of 4 and 8 jet diameters are presented with one jet out of focus. In both cases, the schlieren system is focused on the jet in the rear and the system is looking through the turbulence of the jet in front. A 10 nanosecond pulse from a Nd:Yag laser was used as the light source to provide detailed features of the turbulent jet. With a separation of four jet diameters, the blending of fine detail occurs, while in the case of 8 jet diameters the medium-sized features are blended. Large features may persist as brightness variations, but their features are clearly out of focus and generally easy to ignore. Figure 3 presents the image of the shock structure over the space shuttle configuration in the NASA Langley 0.3-m Transonic Cryogenic Tunnel. The use of conventional schlieren in this facility has been unsuccessful because of the large temperature gradients outside the test section were integrated into the optical path. The density gradients in the plenum totally obscured the density gradients in the test section. The use of focusing schlieren has provided this facility its first capability to visualize the shock structures over complex aircraft configurations.

LASER VELOCIMETRY TECHNIQUES - Laser velocimetry has been widely used for the last two decades for the study of flow fields. The early years focused on point-measurement laser velocimetry(LV), while the last decade has seen the arrival of planar measurement techniques (particle imaging velocimetry-PIV). Both fringe-type laser velocimetry and PIV are widely used and reaching mature status, therefore, the paper will focus on one experiment that has used both techniques to develop a unique data base for the validation of turbulence models.

Flow Over a Backward-Facing Step - The flow over a backward facing step has been used as a benchmark experiment for the validation of turbulence model for many years (references 7-9). Turbulence modeling data bases would ideally contain data describing the mean velocities and full Reynolds stress tensor and information for determining the spatial correlations of the turbulence. The use of both LV and PIV provide both these information types

LV System Description - The measurement of all nine terms in the Reynolds stress tensor required the use of a three-component laser velocimeter. Two-component laser velocimeter configurations using color separation have been increasingly used in the study of the flow over backward-facing steps, and their optical configurations are well established. Configurations for three-component laser velocimeter systems are not as well established. The various optical configurations for obtaining the third component range from reference beam systems to off-axis fringe systems. The measurement accuracy of these configurations was investigated by Meyers (ref. 10). A three-fold increase in the measured turbulence intensity, with corresponding increases in statistical uncertainties in the computed mean of the third component, can occur from these various optical configurations. The greatest accuracy occurs when the third component measures the transverse component directly, that is, orthogonal to the remaining components.

The three-component laser velocimeter system configuration used in studying the reattachment of the flow downstream of a backward facing step is described below. The need to measure a boundary layer profile from the floor to the ceiling of the duct required that the entire optical system be rotated 45° from the vertical. The 476.5 nm, 488.0 nm, 496.5 nm and 514.5 nm wavelengths from a 5-watt argon-ion laser were used to measure the longitudinal, free stream, rotated vertical (+45°), and rotated lateral (-45°) velocity components, respectively. The sample volume for this system is spherical and is approximately 100 microns in diameter. The scattered light from the seed particles are received at 135° backscatter through a single lens. Figure 4 presents a photograph of the three-component laser velocimeter used in the backward-facing step study.

A fourth wavelength of the argon-ion laser was used to measure the free stream velocity. A fiber-optic link was used to transmit the 488.0 nm wavelength from the moving traverse system to the fixed measurement point located 12 step heights (H), upstream of the step on the centerline of the inlet channel. The scattered light from the particles was received in a 180° backscatter mode. The laser velocimeter optics and laser move as a unit on a three-axis traverse system that provides 1 meter of travel, with 10 micron resolution in all axes.

LV Results - Figures 5 and 6 present the results obtained from the flow over the backward-facing step at a Reynolds number of 50,000 based on step height. Details of the LV system and a lower Reynolds number experiment can be found in reference 11. The higher Reynolds number data has not yet been formally published. Figure 5a shows a vector plot with streamlines superimposed. Fourteen profiles were obtained with a nominal spacing of 0.02H in the vertical direction and 0.5H in the longitudinal. Figure 5b presents the contours of the primary shear stress, $\overline{uw} / U_\infty^2$. It is important to provide measures of the uncertainties in any data set and figure 6 provides a sample of the uncertainty associated for the shear stress at x/H = 4.

PIV System Description - The PIV system, illustrated in figure 7, uses two frequency-doubled, 140-mJ Nd:YAG lasers to form a double-pulsed light sheet. Each laser had a pulse duration of 10 nsec, and were connected to fire in a sequence with an adjustable inter-laser pulse separation time. The output from the two lasers were optically coupled to be spatially collinear and used to form a light sheet approximately 100 mm wide and 1 mm thick. This light sheet, directed into the seeded flow, was positioned to bisect the tunnel along its centerline. To obtain time-resolved photographs, a 70-mm Hulcher sequence camera was synchronized with the laser system. Doubled-exposed photographs were acquired at a framing rate of 15 Hz, yielding a frame to frame time separation of 66.7 msec. A maximum of 450 cm (15 feet) of photographic roll film could be loaded into the camera, yielding 55 frames of data per run. The laser and camera systems were mounted on a movable traversing rig to allow sequences of photographs to be obtained at various centerline locations ranging from two step heights upstream to nine step heights downstream.

To resolve directionality in the flow, a polarization-based calcite crystal image shifter was attached to the camera system (references 12 and 13). The image shifter relied on controlling the polarization of the scattered light from the seed particles. The first light sheet pulse was designed to produce a polarization parallel to the principal plane of the calcite crystal, while the second pulse generated a polarization angled at 90 degrees with respect to the first pulse. This change in polarization caused the photographic image field to be shifted between laser pulses. The shift caused the most negative fluid velocity represented in the flow to produce a positive displacement of the second particle image with respect to the first.

Data acquisition consisted of exposing sequences of up to 55 photographs taken at 15 exposures / second. The particle image separations at the film plane were controlled by adjusting the inter laser pulse separation time. Each photograph in the sequence was analyzed by placing the negative in a holder attached to an X-Y translation stage with 1 micrometer positioning resolution. This level of resolution allowed accurate incremental scanning over the entire imaged flow field. The negative was interrogated to obtain velocity information by imaging adjacent square regions (with a side length of 1.5 mm) of the negative onto a CCD camera connected to a frame buffer. The frame buffer digitized and mapped these interrogation regions to a 512 x 512 pixel grid using 8 bit resolution. Details of the calibration process can be found in reference 14.

To determine the velocity vectors associated with each interrogation region, the region was first captured and stored in the frame buffer. Each image was enhanced using a thresholding and background noise subtraction process. The average particle image displacement vector was then obtained using two-dimensional spatial autocorrelation (ref. 15).

PIV Results - Figure 8 presents the results obtained from a study of the flow over the backward-facing step at a Reynolds number of 15,000. The data obtained from this study are being used to evaluate large-eddy simulations. Details of the PIV system and results can be found in reference 16. Figure 8a shows a vector plot for an instantaneous data set in the corner made between the vertical face of the step and the floor of the duct. The vortical structures present in this region are clearly evident. Figure 8b presents the average of 52 frames for the same location in the flow. In this case the smaller vortical structures have averaged into the large corner vortex noted in previous experimental investigations.

DOPPLER GLOBAL VELOCIMETRY - In the last three years a new technique referred to as Doppler global velocimetry (DGV) has been introduced. DGV like PIV also uses a laser light sheet to illuminate a particle field within the flow. The DGV measures the Doppler shift of light scattered from the illuminated particles. This approach improves on PIV by providing the capability to measure the full three-component velocity field without directional ambiguity. A short description of Doppler global velocimetry follows, but detailed system characteristics can be found in references 17-19.

Doppler global velocimetry uses the edge of an absorption line in molecular Iodine to serve as a frequency discriminator to directly measure the Doppler shift of the collected scattered light (ref. 20). An Argon-ion laser operating in single-line mode at 514.5 nm is tuned by tilting the intercavity etalon to an optical frequency corresponding to a point midway along the edge of an absorption line of an Iodine absorption line filter (ALF), figure 9. Scattered light collected from a stationary object or cloud of particles will be attenuated by 50 percent as it passes through the ALF. If the object or particle cloud is moving, the attenuation through the ALF will increase (or decrease, depending on the direction of movement) by an amount proportional to the Doppler shift. By using the ALF as a filter for a CCD camera, an entire laser light sheet can be viewed and the velocity field determined.

In practice, particle size distribution, number density, and the laser light sheet intensity profile are other factors influencing the amount of collected scattered light reaching the viewing camera. These influences can be minimized by viewing the same scene with a second camera without an ALF, figure 10, to provide a reference signal used to normalize the viewing camera output. The two cameras must be aligned with corresponding pixels in each camera viewing the same portions of the light sheet. Other influences from nonuniform optical elements and variations in pixel sensitivities in the CCD cameras are removed using pixel-by-pixel ratio calibrations. A prime advantage of this technique is the ability to map out multiple velocity components in a plane at essentially video frame rates (30 Hz).

DGV Results - While the DGV technique is relatively new, a few attempts have been made to apply the technique in large facilities. Figure 11 shows the velocity contours obtained over a delta wing at Mach = 2.8 in the Unitary Plan Tunnel at NASA Langley. Vapor condensation was used to provide particles for this test condition. The transverse component of velocity was measured and the vortical flow and shock structure can be seen. The techinique has been applied to flows with freetream velocities of 180 ft/sec. At these low speeds the accuracy of the technique is limited by two factors, the response function (sensitivity) of the atomic line filter and the quality of the CCD cameras. A test later in the year in the NASA Langley Basic Aerodynamics Research Tunnel will compare results obtained with both LV and DGV and should help define the accuracies and applicability of the DGV technique in low-speed flows.

PRESSURE-SENSITIVE PAINT - Pressure- and temperature-sensitive paints are another recent development in fluid dynamic measurements. Much of the pioneering work in this area has been performed by McDonnell Douglas Aerospace (MDA), the University of Washington, NASA Ames Research Center, and a Russian and Italian joint venture company called Inteco. References 21-24 describe recent results from these investigators.

Figure 12 shows a sketch of the measurement concept. Both pressure- and temperature-sensitive coatings contain photoluminescent probe molecules, when excited by the proper wavelength of light, luminesce with an intensity that can be related to the property of interest. In the case of pressure sensitive paints (PSP), the luminescent intensity is related to the partial pressure of oxygen through the Stern-Volmer equation. The probe molecule is raised to an excited state by absorbing a photon of appropriate energy. There are several mechanisms by which a molecule can lose the excess energy and return to the ground state and photoluminescence is only one. For pressure-sensitive coatings, the excess energy can also be absorbed by oxygen molecules in a process called dynamic quenching. The temperature-sensitive paints (TSP) under development at MDA are insensitive to pressure. To simplify integration with the pressure sensitive probe molecule, TSP's are being developed that are excited by the same illumination light source.

PSP data analysis usually involves the ratioing of a "wind-on" image and a "wind-off" or reference image. It is extremely important to maintain careful alignment of these two images before ratioing. The alignment process being developed at MDA and NASA Ames incorporates careful mapping of the image and the reference data to model coordinates, which accounts for aeroelastic deflections and distortions in the camera.

Basing the measurement on luminescent lifetimes provides an alternative to ratioing images. A modulated light source is used to excite the probe molecules and heterodyning techniques are used to evaluate the decay of the photoluminescence. One advantage to this technique is that careful alignment of data from a reference and test image is not required. A second advantage is that both temperature and pressure information can be extracted from the same image data by using separate modulation frequencies for the pressure and temperature probe molecules. It may also allow for measurements to be obtained under normal daylight conditions (rather than complete darkness).

Unfortunately, the method is considerably more sophisticated and requires customized instrumentation. It may also require long integration times, which would introduce error related to any model motion. Ironically, many of the image manipulation procedures required to align the two images of the previous analysis method are necessary to map the data from the lifetime technique to spatial coordinates of the model. The mapping scheme must still compensate for any model deflection and/or motion and is essential for the evaluation of model loads (which is a prime motivation for using these methods). Dividing the intensities from two images is a more immediate solution, and as a result, the content of this paper will be restricted to the first approach.

PSP Results - Most of the PSP data that has been published to date has been obtained in the transonic and supersonic speed range. Flow fields in these speed regimes usually contain shocks or large pressure gradients that are more easily detectable with pressure sensitive paint. Figures 13 and 14 shows pressure data for a 76°/40° swept delta wing in a low-speed wind tunnel. The data was obtained by MDA and represented a considerable challenge due, to the low pressure differentials in this particular flow. Figure 14 shows an excellent agreement between the PSP results and conventional pressure port data.

CONCLUDING REMARKS

While the experimental techniques described in this paper may be useful to solving fluid dynamic problems in automotive engineering , methods for improving communication between the aeronautical and automotive industries need to be discovered. Other techniques that are capable of measuring species, temperature, and density are under development which may also be of use in studying internal and combusting flows. It is hoped that papers such as this will help promote this interchange process.

REFERENCES

1. Kantrowitz, A.; and Trimpi, R. L.: "A Sharp-Focusing Schlieren System," *Journal of Aeronautical Science,* Vol. 17, No. 5, 1950, pp. 311-314.
2. Burton, R. A.: "A Modified Schlieren Apparatus for Large Areas of Field," *Journal of the Optical Society of America,* Vol. 39, No. 11, Nov. 1949, pp. 907-908.
3. Fish, R. W.; and Parnham, K.: "Focusing Schlieren Systems," Aeronautical Research Council, CP 54 (13.865), ARC Technical Rept., TN IAP 999, London, Nov. 1950.
4. Merzkirch, W.: *Flow Visualization,* Academic Press, New York, 1974, pp. 98-99.
5. Roedeker, L. R.: "Analysis and Construction of a Sharp Focusing Schlieren System," M. S. Thesis, Dept. of Aeronautics and Astronautics, Massachusetts Inst. of Technology, Cambridge, MA, 1959.
6. Weinstein, L. M.: "Large-Field High-Brightness Focusing Schlieren System," *AIAA Journal,* Vol. 31, No. 7, July 1993, pp. 1250-1255.
7. Eaton, J. K.; and Johnston, J. P.: "Turbulent Flow Reattachment: An Experimental Study of the Flow and Structure Behind a Backward-Facing Step," Stanford Report MD-39, 1980.
8. Kim, J.; Kline, S. J.; and Johnston, J. P.: "Investigation of Separation and Reattachment of a Turbulent Shear Layer: Flow Over A Backward-Facing Step," Stanford Report MD-37, 1978.
9. Pronchick, S. W.; and Kline, S. J.: "An Experimental Investigation of the Structure of a Turbulent Reattaching Flow Behind a Backward-Facing Step," Stanford Report MD-42, 1983.
10. Meyers, J. F.: "The Elusive Third Component," Symposium on Laser Anemometry, ASME 1985 Winter Annual Meeting, Miami, Florida, Nov. 17l-21, 1985.
11. Kjelgaard, S. O.: "Three-Component Laser Velocimeter Surveys of the Flow Over a Backward-Facing Step," Laser Anemometry - Advances and Applications, 1991, Vol. 2, pp. 529-541.
12. Adrian, R. J.: "Image Shifting Technique to Resolve Directional Ambiguity in Double Pulse Velocimetry," *Applied Optics,* Vol. 25, pp. 3855-3858, 1986.
13. Landreth, C. C.: and Adrian, R. J.: "Electrooptical Image Shifting for Particle Image Velocimetry," *Applied Optics,* Vol. 27, pp 4216-4220, 1988.
14. Humphreys, W. M.; Bartram, S. M.; and Blackshire, J. L.: "A Survey of Particle Image Velocimetry Application in Langley Aerospace Facilities,: AIAA Paper No. 93-0411, Reno, NV, 1993.
15. Adrian, R. J.: "PIV Processing Technique: Image Plane and Fourier Plane," von Kaman Institute for Fluid Dynamics Lecture Series 1988-06: Particle Image Velocimetry, Brussels, Belgium, March 21-25, 1988.
16. Humphreys, W. M.; Kjelgaard, S. O.; and Bartram, S. M.: "A Time-Resolved Particle Image Velocimeter for Unsteady Flow Field Studies," Manuscript submitted for Symposium on Laser Anemometry: Advances and Applications, Lake Tahoe, NV, June 19-23, 1994.
17. Meyers, J. F.: "A Three Dimensional View of Velocity Using Lasers," 10th International Invitational Symposium on Unification of Finite Element Methods in Theory and Test, Worcester, MA, July 18-19, 1991.
18. Meyers, J. F.; and Komine, H.: "Doppler Global Velocimetry, A New Way to Look at Velocity," ASME Fourth International Conference on Laser Anemometry, Cleveland, OH, August 23-27, 1991.
19. Meyers, J. M.; Lee. J. W.; and Cavone, A. A.: "Signal Processing Schemes for Doppler Global Velocimetry," IEEE 14th International Congress on Instrumentation in Aerospace Simulation Facilities, Rockville, MD, Oct. 27-31, 1991.
20. Komine, H.; Brosnan, S. J.; Litton, A. B..; and Stappaerts, E. A.: "Real-Time Doppler Global Velocimetry,: AIAA 28th Aerospace Sciences Meeting, Reno, NV, AIAA Paper No. 91-0337, Jan. 7-10, 1991.
21. Morris, M. J.; Benne, M. E.; Crites, R. C.; and Donovan, J. F.: "Aerodynamic Measurements Based on Photoluminescence," AIAA Paper No. 93-0175, Selected Technical Papers from the Aerodynamic Measurement Technology Committee, 31st AIAA Aerospace Sciences Meeting and Exhibit, Reno, NV, Jan. 11-14, 1993.
22. Donovan, J. F.; Morris, M. J.; Pal, A.; Benne, M. S.; and Crites, R. C.: "Data Analysis Techniques for Pressure- and Temperature-Sensitive Paint," AIAA Paper No. 93-0176, Selected Technical Papers from the Aerodynamic Measurement Technology Committee, 31st AIAA Aerospace Sciences Meeting and Exhibit, Reno, NV, Jan. 11-14, 1993.
23. McLachlan, B. G.; Bell, J. H.; Gallery, J.; Gouterman, J.; and Callis, J.: "Boundary Layer Transition Detection by Luminescence Imaging,: AIAA Paper No. 93-0177, Selected Technical Papers from the Aerodynamic Measurement Technology Committee, 31st AIAA Aerospace Sciences Meeting and Exhibit, Reno, NV, Jan. 11-14, 1993.
24. Bell, J. H.; and McLachlan, B. G.: "Image Registration for Luminescent Paint Applications," AIAA Paper No. 93-0178, Selected Technical Papers from the Aerodynamic Measurement Technology Committee, 31st AIAA Aerospace Sciences Meeting and Exhibit, Reno, NV, Jan. 11-14, 1993.

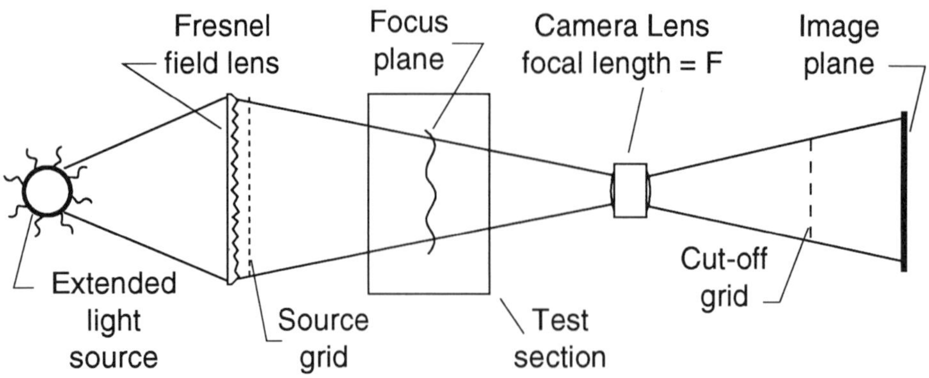

Figure 1. Sketch of Focusing Schlieren Optical Arrangement

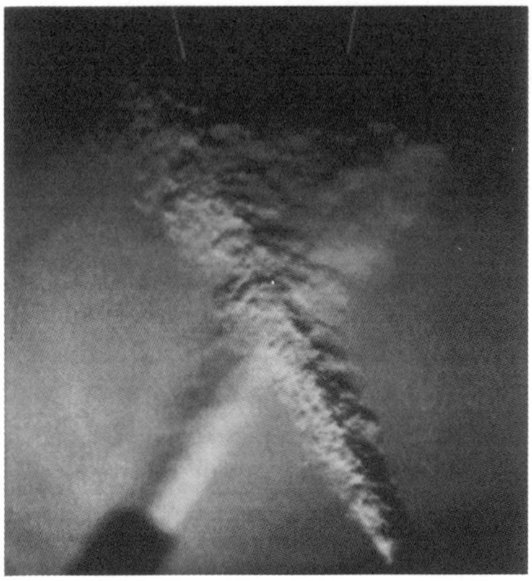

Figure 2a. Focusing schlieren flow visualization results for 2 turbulent jets with a 4 jet diameter separation.

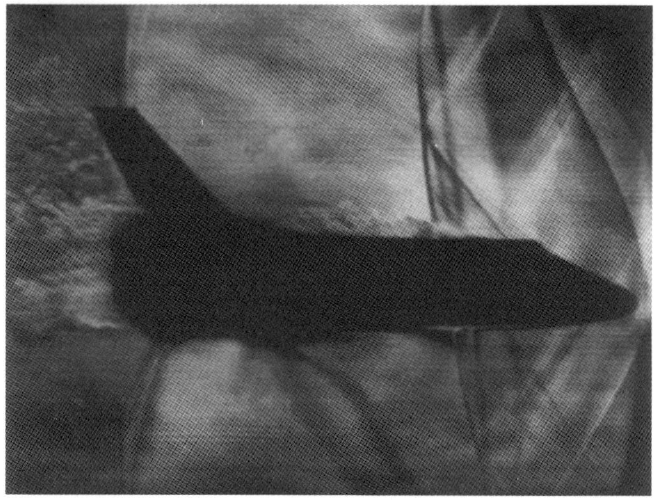

Figure 3. Flow over shuttle at M = 1.2 in 0.3m Transonic Cryogenic Tunnel.

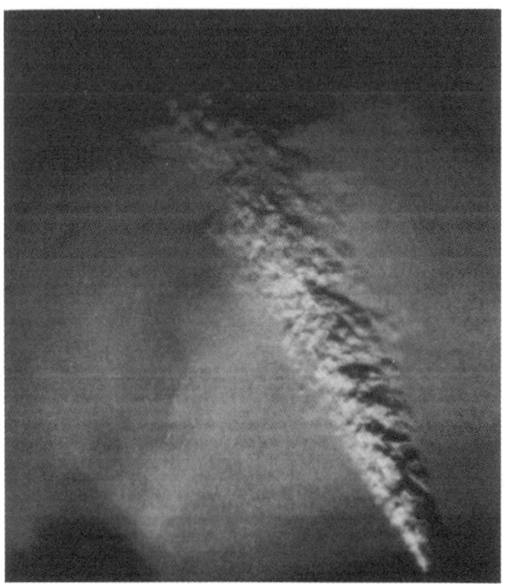

Figure 2b. Focusing schlieren flow visualization results for 2 turbulent jets with a 8 jet diameter separation.

Figure 4. Photograph of the three-component laser velocimeter used in the study of the flow over a backward-facing step.

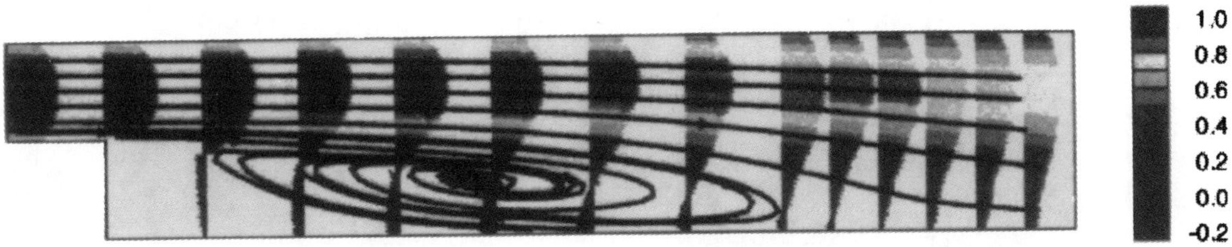

Figure 5 a. Vector plot with streamlines showing flow over the backward-facing step for a Reynolds number based on step height of 50,000.

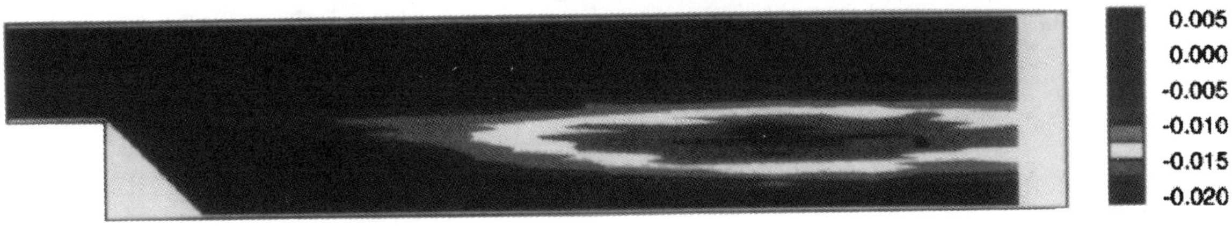

Figure 5 a. Shear stress contours, $\overline{uw} / U_\infty^2$, for flow over the backward-facing step (same conditions as figure 5a).

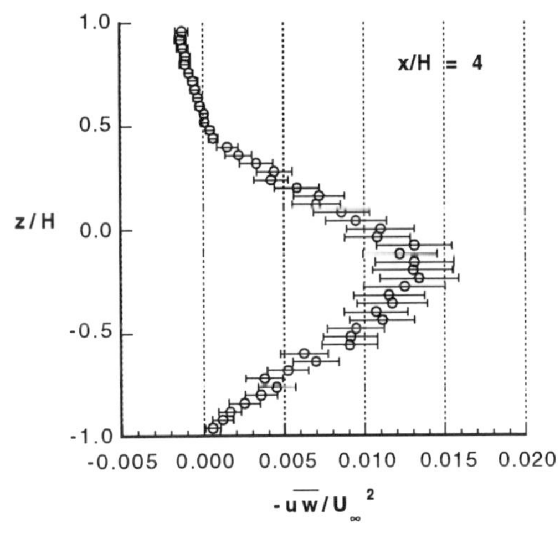

Figure 6. Uncertainty in shear stress measurements.

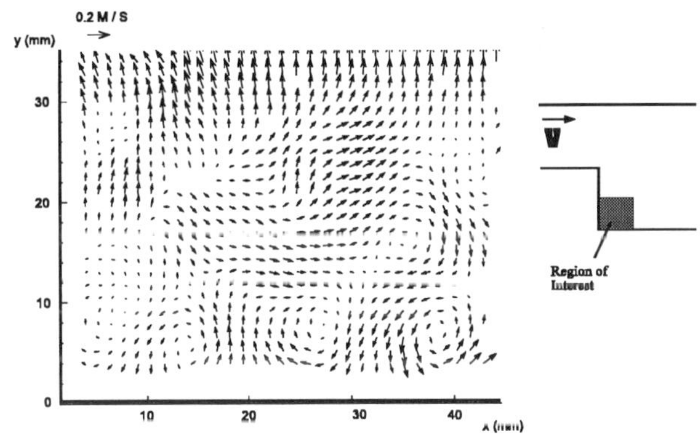

Figure 8a. Sample result of instantaneous velocity map in corner flow of the backward-facing step.

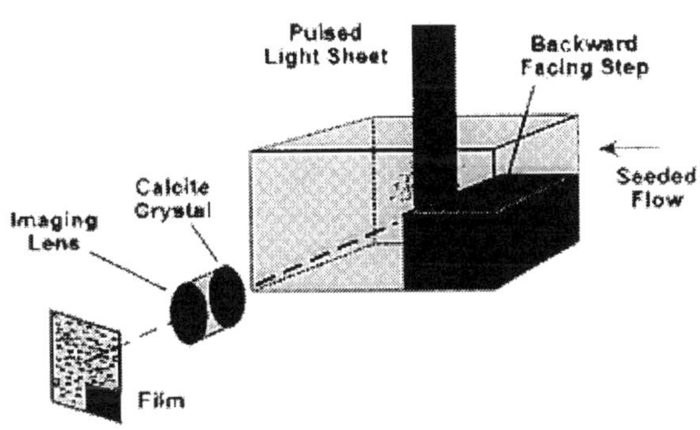

Figure 7. Sketch of PIV system used in backward-facing step experiment.

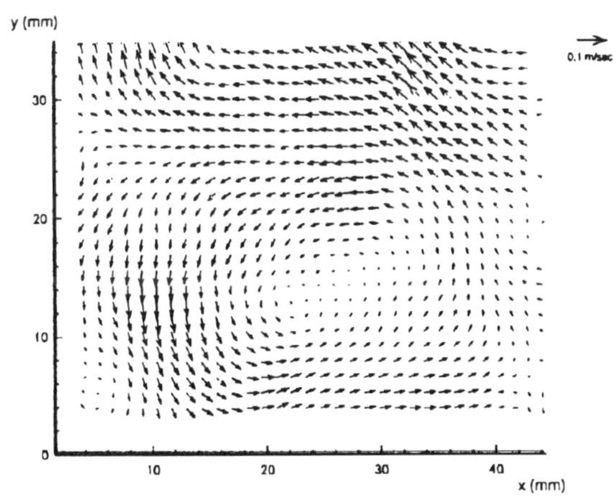

Figure 8b. Mean velocity map determined from 52 instantaneous measurements (same region as figure 8a).

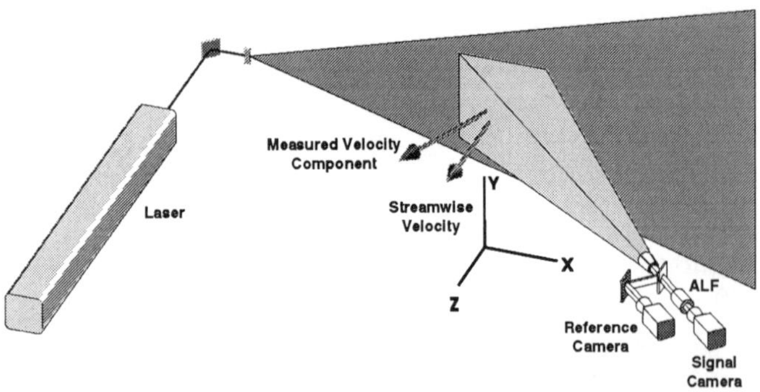

Figure 9. Sketch of Doppler Global Velocimeter optical arrangement

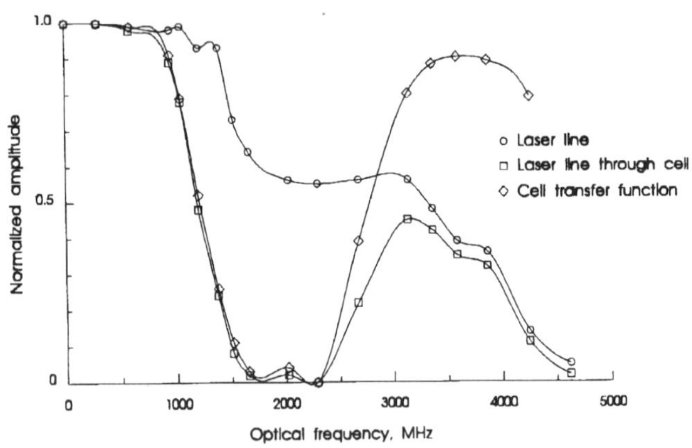

Figure 10. Transmission characteristics of iodine atomic line filter.

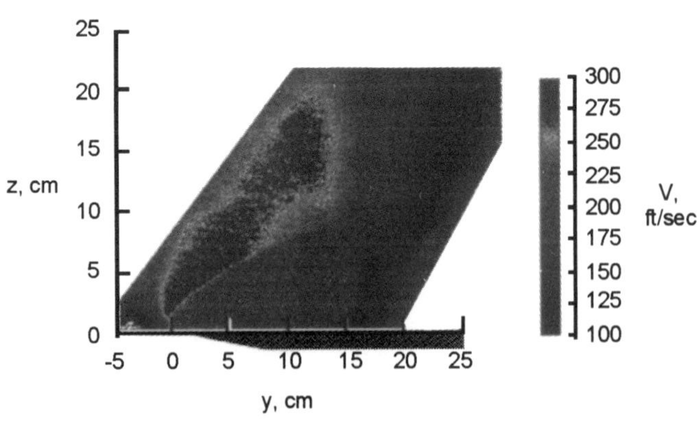

Figure 11. Transverse velocity contours obtained with one-component DGV for a delta wing at M = 2.8.

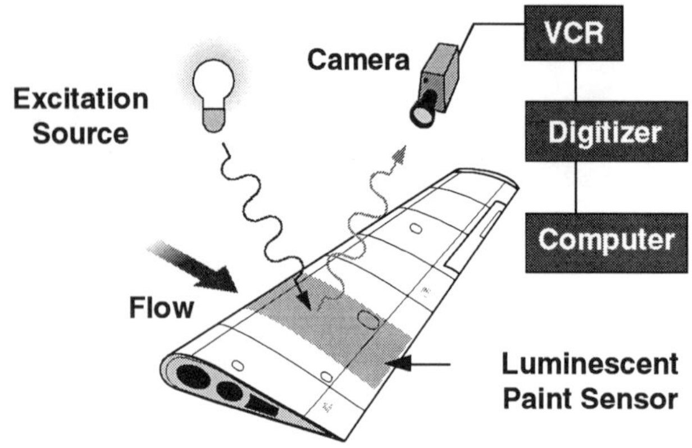

Figure 12.. Sketch of pressure-sensitive paint concept.

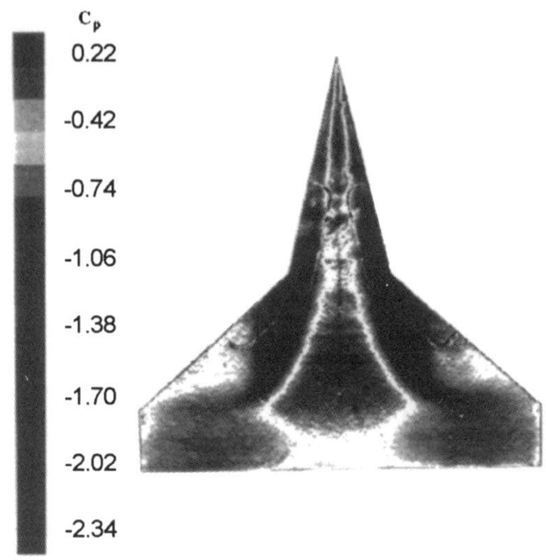

Figure 13. Pressure contours obtained by pressure-sensitive paint for a 76°/40° double-delta wing.

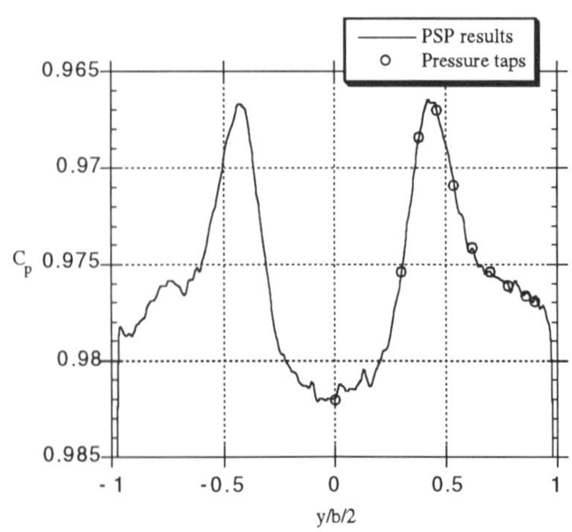

Figure 14. Detailed comparison of pressure-sensitive paint results versus pressure port results for 76/40 double delta.

Section III
On-Road Testing

Section III
On-Road Testing

On-road testing of vehicles for their aerodynamic properties brings the scene of investigation to a vehicle's natural operating environment. This dispenses with the need for expensive and elaborate facilities like wind tunnels. At the same time, on-road testing eliminates some compromises (for example, manner of ground simulation, far-field boundary condition, etc.) that the aerodynamicist is forced to endure while testing in a wind tunnel. So, there is much to recommend the apparent inherent simplicity of on-road testing versus wind tunnel testing. However, there is much more that should be considered before a choice is made. As may be expected, the choice is predicated by the objectives of a test program.

One purpose that on-road testing has been put to is the estimation of a vehicle's aerodynamic drag. Typically a coast-down method is adopted and the vehicle's dynamics are modeled through the equation of motion. The primary measured variable is the history of the vehicle's speed with time (while power delivery to the driven wheels is interrupted). This is used in the modeled equation to extract the aerodynamic drag. As with a model of any physical system, certain simplifying assumptions are inherent. Sources of error in the final drag estimate are therefore related to the approximations in the equation model, and to the selection and quality of the measured data. The increased capabilities of on-board data acquisition systems, together with better understanding of other loss (drag) mechanisms in a vehicle's drivetrain will no doubt increase the value of this methodology.

On-road tests have been most appropriate for studying special aerodynamic issues that arise from operational conditions. In certain driving environments, dirt or dust deposition on a vehicle's surfaces (leading to obstructed visibility for the driver) is a hazard and is influenced by the flow around the vehicle. Likewise, the spray generated by a vehicle when operated on a wet roadway can be detrimental to visibility for the driver in a following vehicle. These are examples of problems that are best studied in actual operating conditions. With suitable understanding of the manner and causes of these problems, the aerodynamics of the vehicle can be suitably modified to result in safer and better designs.

Another aspect of safe vehicle operation that is influenced by aerodynamics is the stability of a vehicle in cross-winds. While a vehicle's steady-state side-force coefficients may be measured in a wind tunnel, the same coefficients, in the presence of transient (gusty) side winds, may be significantly different. The transient conditions are more accurately simulated on special road segments where controlled gusts may be generated (with transverse-blowing fans) and the resulting dynamics of a vehicle assessed in conjunction with known wind data.

950626

ABCD - An Improved Coast Down Test and Analysis Method

Frank T. Buckley, Jr.
University of Maryland and Frank T. Buckley Associates, Inc.

ABSTRACT

An improved coast down method that incorporates the input of data measured with an onboard anemometer into the governing equation of motion is used to evaluate models representing the aerodynamic and mechanical drag characteristics of a wide variety of vehicles. Results from the application of the improved technique are presented to demonstrate the repeatability and precision of the analyzed aerodynamic drag coefficients, to address the merits of different wind tunnel blockage correction methods, to illustrate the method's usefulness in the study of tractor-trailer and light truck-travel trailer aerodynamic drag, to evaluate the influence of floor blowing in wind tunnel studies of the effects of various changes to the underbody of an automobile, and to compare track and dynamometer determinations of the variation of mechanical drag with speed on a vehicle equipped with a four-speed automatic transmission.

NOMENCLATURE

A, B, C, etc.	coefficients used in mechanical drag model
A	aerodynamic reference area
a_{TN}, b_{TN}, etc.	coefficients used in variation of tire rolling resistance with normal load model
a_i	coefficients used in aerodynamic drag coefficient model
$C_D(\psi)$	aerodynamic drag coefficient at yaw angle Ψ
$C_L(\psi)$	aerodynamic lift coefficient at yaw angle Ψ
D_{AERO}	aerodynamic drag
D_F	total drag due to inboard losses at front axles
D_{GRAV}	drag due to gravitational effects
D_{MECH}	total mechanical drag in the absence of aerodynamic lift effects
D_R	total drag due to inboard losses at rear axles
D_{TIRE}	tire drag in the absence of aerodynamic lift effects
f	$1 + (g/W)(I_F + I_R)/R^2$
g	acceleration of gravity
h	local track elevation
I_F	total rotational inertia at front axle
I_R	total rotational inertia at rear axle
k_1	measured yaw angle correction factor
k_2	measured relative airspeed correction factor at zero yaw
k_3	factor to correct k_2 for yaw effects
k_t	fractional decrease of mechanical drag with ambient temperature at T_{ref}
L_{AERO}	aerodynamic lift
N	normal load on tire
R	loaded radius of tire
s	distance along track
T_{ref}	ambient temperature at reference condition
T_{test}	ambient temperature during test
t	time
V	vehicle speed
V_r	relative airspeed
W	vehicle test weight
Ψ	yaw angle
Ψ_0	measured yaw angle offset correction
ρ	mass density of air
Subscripts:	
F	front axle
m	measured
R	rear axle
T	tire

INTRODUCTION

The coast down method is often used to estimate the drag forces that act on a full-scale vehicle when operating in its natural environment. The experimental technique, in which the vehicle is accelerated to a desired upper speed, declutched, and then allowed to decelerate under the action of the various drag forces that act on it, is remarkable in its simplicity. In contrast, the accuracies of the results that are produced in a coast down analysis are strongly dependent on the type and quality of the measured data, and on the realism of the analytical models used to describe the drag forces that were experienced during the test.

Numerous papers have been written on the subject of coast down testing. With a few exceptions, these studies fall into two broad categories. The first category consists of studies where the drag results are produced based on the analysis of but one parameter measured during the test, namely, the vehicle speed-time history. Examples of the latter studies are contained in the papers by Andersen et al (1), White and Korst (2), and Yasin (3). The second category consists of more recent studies wherein additional simultaneous measurements of the relative airspeed and yaw angle histories encountered by the vehicle during the test are incorporated into the analysis. Examples of the newer so-called anemometer-based methods are contained in the papers by Buckley, Walston and Marks (4, 5), Hamabe et al (6), and Passmore and Jenkins (7).

Though non-anemometer methods are simple in their application and analysis, the analyzed results can contain significant errors when tests are conducted in the presence of winds. And although corrections for wind effects have been developed based on the assumptions that the speed and direction of the wind are constant during the test, such conditions only exist in the rarely encountered calm wind condition. Because analysis errors associated with non-uniform wind effects can be significant, the application of non-anemometer methods is typically restricted to very low wind conditions. Even with such a restriction, however, non-anemometer methods generally employ additional modeling simplifications that introduce errors into the analyzed aerodynamic/mechanical drag breakdown and in the estimation of the variation of total vehicle drag with speed. To their credit, non-anemometer techniques have been widely used to provide very worthwhile estimates of the total drag acting on a vehicle at 80.5 km/hr, and thus have found wide application in vehicle federal certification activities (8).

In contrast to non-anemometer techniques, anemometer-based coast down methods can be employed in the analysis of data measured under almost any wind condition. Depending on data quality, they can also be used in the evaluation of drag models that more realistically describe the forces that act on the vehicle. Thus, anemometer-based test methods have the potential for providing significantly more accurate estimates of the aerodynamic and mechanical drag characteristics of a vehicle, and of the variation of its total drag with speed. The added complications of anemometer-based testing are associated with the facts that the anemometer contributes somewhat to the drag on the vehicle, and that the anemometer must be calibrated in the presence of the vehicle to account for the aerodynamic interference of the vehicle on the air flow at the anemometer's location.

The author's early work with anemometer-based coast down methods reported in references 4 and 5 led to his further development of the technique and to the production of a series of twelve proprietary technical reports over the following eleven year period for the Ford Motor Company. These reports detail the incremental improvement of anemometer-based coast down methods to provide answers to fundamental questions regarding the aerodynamic and mechanical drag characteristics of a number of automobiles, light trucks, and tractor-trailer trucks. As an outcome of these activities, an effort was undertaken to incorporate the matured testing technology into a tool that would facilitate the convenient use of anemometer-based coast down testing in the typical automotive test environment. This effort resulted in the development of the ABCD (Anemometer Based Coast or Drive) computer-assisted testing system, which was first acquired by the Ford Motor Company in 1988, and is currently utilized by all three United States automobile manufacturers.

This paper sets forth the equation of motion employed in the current version of ABCD, describes the method of analysis utilized in its evaluation, provides practical information regarding the use of the analytical tool, presents some historical results that were obtained during the development of the improved procedure, and gives examples of some results that have recently been obtained with the ABCD testing system. Although the ABCD methodology can and has been successfully utilized with drive tests as well as with coast down tests, the discussion will be limited to the coast down test situation only.

ABCD COAST DOWN ANALYSIS MODEL

The ABCD algorithm is based on the adoption of certain assumptions regarding the way the various contributions to the drag on a freely coasting vehicle can be represented. Some of these are more fundamental than others, but all give rise to modeling simplifications that appear justified in light of current knowledge. Since ABCD is a dynamic product, the algorithm is upgraded whenever an underlying assumption fails to conform with current experience.

Current Assumptions

1) The very small vehicle slip angles that occur in the presence of aerodynamic side forces and yawing moments can be ignored when formulating the equation of motion.

2) The tire slip angles developed in the presence of aerodynamic side forces and yawing moments have a negligible effect on tire rolling resistance.

3) The variation of the non-measured mechanical drag contribution with speed can be adequately modeled with no more than a fifth-order polynomial in speed.

4) The reduction in mechanical drag due to aerodynamic lift effects is very small and can be estimated in terms of the effect on tire rolling resistance.

5) The aerodynamic drag forces can be adequately modeled with a quasi-steady relation which is evaluated using instantaneous freestream relative airspeeds and yaw angles inferred from measurements provided by a boom-mounted anemometer situated ahead of the vehicle at its mid height.

6) The variation of the aerodynamic drag coefficient with vehicle speed is insignificant over the speed range of the typical test.

7) The sensitivity of the aerodynamic drag coefficient to the turbulence levels and non-uniformities which are introduced into the relative airstream when winds are present is negligible.

8) The variation of the aerodynamic drag coefficient with yaw angle can be adequately modeled with no more than an eighth-order polynomial in yaw angle.

Assumptions 1) and 2) are supported by the results of an analysis contained in an earlier proprietary study which considered the aerodynamic side force and yawing moment characteristics for a mid-size sedan, a mid-size station wagon, a full-size pickup truck, and a full-size van; and the rolling resistance and side force characteristics of two typical radial-ply tires as functions of slip angle and normal load. In that study, the tire rolling resistance in the presence of crosswinds was found to increase with the square of the product of the yaw angle and the square of the relative airspeed. The increase was less than 1% when the described product was less than about 3500 rad-$(km/hr)^2$ for the sedan, 2800 rad-$(km/hr)^2$ for the station wagon, 2100 rad-$(km/hr)^2$ for the pickup, and 1600 rad- $(km/hr)^2$ for the van.

The modeling in assumption 3 is based on a review of independent test results for the mechanical drag characteristics of a variety of tires, and driven and non-driven axles. Typically a lower-order polynomial (linear or quadratic) is adequate for modeling the rolling resistance of the tires, the drag due to hub losses inboard of the tires at the non-driven axles, and the drag due to hub and driveline losses inboard of the tires at the driven axles for vehicles equipped with manual transmissions or older three-speed automatic transmissions. A higher-order polynomial helps to capture the effects of the contribution of the driveline losses of newer four-speed automatic transmissions, whose drag characteristics can be quite irregular (see examples in Figure 1). Though one could argue that the latter characteristics might be better modeled with other than a polynomial, experience to date indicates that the level of the modeling error associated with the use of the higher-order polynomial has an insignificantly small effect on the analysis results.

Assumption 4 is well-supported by calculations utilizing typical wind-tunnel-determined aerodynamic lift characteristics and independent data for the variations of bearing drag and tire rolling resistance with normal load.

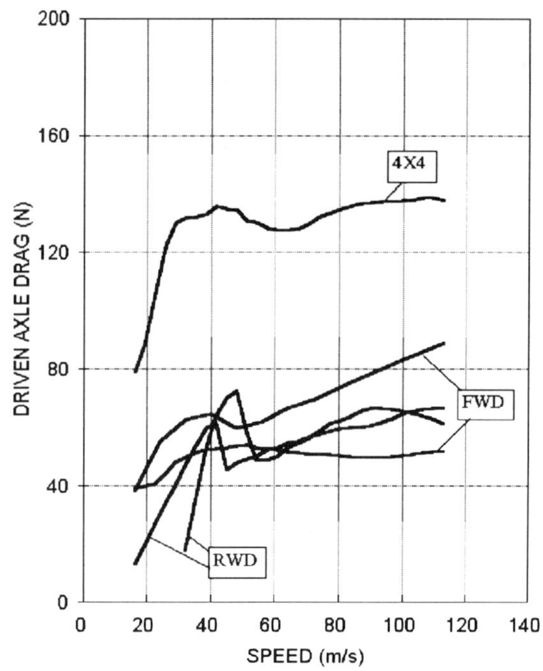

Figure 1 - Examples of Drag due to Combined Hub and Driveline Losses Inboard of the Tires on Vehicles Equipped with Four-Speed Automatic Transmissions

Assumption 5 is justified in the absence of winds, and, as concluded from the results of detailed computer simulations produced in an unreported study, is also appropriate in the presence of winds provided the dynamic responses of the anemometer and the vehicle are similar, and/or the test data are properly filtered. In either case, the evaluation of the vehicle drag coefficient versus yaw angle variation requires that the anemometer data be corrected to account for the modifications of the freestream relative airspeed and yaw angle produced by the interference of the vehicle flow field at the anemometer location.

Assumption 6 implies that the variation of drag coefficient with Reynolds number is negligibly small over the Reynolds number range associated with the usual test speed range (about 115 km/hr to 15 km/hr), which agrees with typical wind tunnel test results.

Assumption 7, while appearing to be supported by comparisons of results from wind-tunnel studies and coast down tests of cars and light trucks, requires further evaluation before its general validity for two-body vehicles such as tractor trailer trucks can be confirmed.

The level of modeling described in assumption 8 has been found appropriate to model the variation of drag coefficient with yaw angle for the most yaw sensitive vehicles (tractor-trailer trucks) over the range of yaw angles typically utilized in an analysis.

Equation of Motion

The general form of the equation of motion, which is equally applicable to coast down and drive tests, can be written as,

$$-fW/g(dV/dt) = D_{MECH} + D_{AERO} + D_{GRAV} - L_{AERO}(dD_{TIRE}/dN)\big|_W \qquad (1)$$

where

$$D_{MECH} = D_{TIRE} + D_F + D_R \qquad (2)$$
$$D_{AERO} = 1/2\ \rho\ C_D(\psi) A V_r^2 \qquad (3)$$
$$D_{GRAV} = W(dh/ds) \qquad (4)$$
$$L_{AERO} = 1/2\ \rho\ C_L(\psi) A V_r^2 \qquad (5)$$

and $(dD_{TIRE}/dN)\big|_W$ is the variation of tire rolling resistance with normal load N evaluated at the normal load produced under the vehicle test weight W.

Mechanical Drag Modeling

In the absence of the input of independent data for the losses inboard of the tires at the front and rear axles, the total mechanical drag is modeled as

$$D_{MECH} = A_{T+F+R} + B_{T+F+R}\ V + C_{T+F+R}\ V^2 + D_{T+F+R}\ V^3 + E_{T+F+R}\ V^4 + F_{T+F+R}\ V^5 \qquad (6a)$$

where A_{T+F+R}, B_{T+F+R}, ..., and F_{T+F+R} are constants that describe the combination of the drag contributions from the rolling resistance of the tires and the inboard losses at the front and rear axles, in the absence of aerodynamic lift effects and under the conditions of the test.

In cases where independent data are utilized to represent the losses inboard of the tires at the front and rear axles, the total mechanical drag is represented as the sum of the axle drags represented by the independent data and a modeled tire rolling resistance. Thus,

$$D_{MECH} = D_{TIRE} + D_F + D_R = A_T + B_T\ V + C_T\ V^2 + D_T\ V^3 + E_T\ V^4 + F_T\ V^5 + D_F + D_R \qquad (6b)$$

where A_T, B_T, ..., and F_T are the tire rolling resistance model constants in the absence of lift and under the conditions of the test; and D_F and D_R are the independently determined, speed and temperature dependent, axle drags. Because the variation of tire rolling resistance with vehicle speed is typically linear over the speed range utilized in the typical coast down test, the coefficients of the nonlinear terms are often constrained to zero when implementing the tire rolling resistance model.

The variation of tire rolling resistance with normal load utilized in the evaluation of the small correction for lift effects is modeled as

$$(dD_{TIRE}/dN)\big|_W = a_{TN} + b_{TN}\ V + c_{TN}\ V^2 + d_{TN}\ V^3 + e_{TN}\ V^4 + f_{TN}\ V^5 \qquad (7)$$

where $a_{TN} = (dA_T/dN)\big|_W$, $b_{TN} = (dB_T/dN)\big|_W$, ..., and $f_{TN} = (dF_T/dN)\big|_W$ are coefficients estimated from independent test data. Because the variation of tire drag with vehicle speed over the speed range of the typical coast down test is essentially linear, the coefficients of the non-linear terms are normally set to zero.

Aerodynamic Drag Modeling

The aerodynamic drag coefficient is modeled as

$$C_D(\psi) = \sum_{0}^{n} a_i \Psi^i \qquad (8)$$

where a_i are constants whose values are determined in the data analysis and n is an even number beyond which the uncertainty in the analyzed value of a_{n+2} exceeds the absolute value of a_{n+2}. The requirement that n be even is related to the fact that the typical $C_D(\psi)$ characteristic is essentially symmetrical about $\psi = 0$ so that the analyzed values of a_i for odd values of n are usually very small and less in absolute value than their uncertainty.

Solution Method

Combining Equations (1-8), rearranging and integrating over a time interval Δt, and rearranging once again obtains

$$-fW/g(\Delta V/\Delta t) - W\overline{(dh/ds)} + 1/2\ \rho A\ \overline{C_L(\psi)V_r^2\ (dD_{TIRE}/dN)\big|_W}$$
$$= A_{T+F+R} + B_{T+F+R}\ \overline{V} + C_{T+F+R}\ \overline{V^2} + D_{T+F+R}\ \overline{V^3} + E_{T+F+R}\ \overline{V^4} + F_{T+F+R}\ \overline{V^5}$$
$$+ 1/2\ \rho A\ (\ C_D(0)\overline{V_r^2} + \sum_{1}^{n} a_i\ \overline{V_r^2\ \Psi^i}\ ) \qquad (9)$$

where the overbars represent the time-average values of the respective quantities over time interval Δt and $C_D(0) = a_0$ is the aerodynamic drag coefficient at zero yaw angle.

Equation (9) forms the basis for the most common ABCD analysis wherein the evaluation of the mechanical drag is limited to a determination of the total mechanical drag variation with speed. A similar equation is employed when independent data for the losses inboard of the tires at the front and rear axles are incorporated in the analysis. In that case, the constants A, B, ..., and F describe the rolling resistance characteristic of the tires.

The quality of the results produced from an analysis of Equation (9) is strongly related to the quality of the data utilized in the analysis. In addition to minimizing instrument calibration errors, whose effect is easily predicted, it is also important that "local errors" produced by non-meaningful inputs be excluded, that the effects of random measurement errors be kept small, and that care be exercised to ensure that the measurements describe the motion of the vehicle.

Local errors are related to one or more isolated happenings that introduce some type of modeling inconsistency. For example, spurious outputs produced when a fifth wheel speed-measuring device bounces when passing over an irregularity in the track surface describe the effects of "forces" that are not actually acting on the vehicle. (Spurious outputs can also be found with use of some radar-based speed measuring systems.) As a contrasting example, the speed measurement can give evidence of the effects of forces that are not modeled in the equation of motion, such as could be produced if the vehicle passed over a major obstruction on the track surface which impulsively changed its speed. To reduce errors due to such local effects, a statistically-based data rejection algorithm, which incorporates an examination of the errors in a preliminary fit of the measured data, is employed to identify meaningless data for exclusion from the final analysis.

The reduction of random measurement errors is accomplished with use of a sequential combination of analog and digital filtering. Three-pole modified Butterworth low-pass filters with a 0.7 Hz corner frequency are utilized during the data acquisition process. Then, after application of the data-rejection algorithm, an FFT (Fast Fourier Transform) digital filtering algorithm is used to further smooth the data. The latter filter has a further use as described below.

Whereas the evaluation of the total drag and the modeling of the mechanical drag in terms of a vehicle speed measurement is straightforward, the description of the aerodynamic drag in terms of relative airspeed and yaw angle measurements requires some care. Since winds are unsteady, the relative airspeed and yaw angle encountered by a vehicle will also be unsteady. Because of its size, a vehicle will not respond to (the aerodynamic force will not correlate with) the highest frequencies in its aerodynamic environment. Thus, in order for the relative airspeed and yaw angle measurements to be predictive of the aerodynamic drag, the dynamic response of the anemometer should ideally match that of the vehicle. Practically, this can be achieved by utilizing an anemometer comprised of an airspeed sensor whose distance constant (length of air passage for 63% recovery) is of the order of the vehicle length and a yaw angle sensor with a similar dynamic response; and by filtering the data with a low-pass filter having a sufficiently low corner frequency. The latter is accomplished with use of the FFT filter described above.

It is also desirable, though not mandatory, that the anemometer output be insensitive to occasional vertical motions of the instrument which can occur when the vehicle passes over bumps on the track surface. Such insensitivity can be obtained by utilizing an airspeed sensor whose output varies with the cosine of the angle between the ground and the vector representing the speed of the air relative to the sensor. The influence of these motions can also be reduced by employing methods to identify and reject data subject to local measurement error effects such as the above-described data-rejection algorithm.

The acceptable, filtered data (together with required physical and geometric data, local test course slopes evaluated from track survey data, and approximate values of the aerodynamic lift coefficient and the tire rolling resistance variation with normal load to estimate the small reduction in mechanical drag due to lift) are then employed in writing Equation (9) for successive Δt's of one second during each coast. Since each coast produces about 120 such equations, a typical test involving 10 coasts produces about 1200 linear equations involving up to a maximum of 15 unknown values of the sought-after mechanical drag constants A, B, ..., and F, and the aerodynamic drag coefficient constants a_i . The solution for these constants is achieved through application of a least-squares, linear regression algorithm. Included in the algorithm is the ability to introduce various constraints in the solution, such as through the specification of the value(s) or the ratio(s) of the value(s) of certain of the unknown aerodynamic or mechanical constants.

ANEMOMETER DRAG AND CALIBRATION

In the usual installation, the anemometer is supported at the end of a boom that extends forward from the underside of the vehicle (see Figure 2). With careful design and installation practice (by utilizing components whose shapes and sizes represent practical compromises among function, strength, and low aerodynamic drag; by minimizing any increase to the projected frontal area of the vehicle; and by positioning any bluff components in regions of low-speed air flow), the drag increase produced by the boom-mounted anemometer can be made very small. Based on the results from wind-tunnel studies, the increase in C_DA at zero yaw will typically range from nil to about 0.01 m^2 .

Figure 2 - Typical Anemometer-Equipped Test Vehicle

In typical installations on automobiles and light trucks, the relative airspeed and yaw angle measurements are obtained at the mid-height of the vehicle and approximately 2m forward of its bumper. As indicated earlier, the flow past the vehicle affects the flow past the anemometer with the result that the relative airspeed is lower (10% to 15%) and the yaw angle higher (15% to 25%) than the freestream values. Because the aerodynamic drag coefficient is inversely related to the square of the relative airspeed, the percentage error in the analyzed drag coefficient is roughly equal to twice the error in the relative airspeed. And although errors in yaw angle have no affect on the evaluation of the zero yaw drag coefficient, they do affect the drag coefficient evaluations at non-zero yaw angles. Thus, proper anemometer calibration is extremely important.

Based on the results of wind tunnel studies, the freestream relative airspeed and yaw angle can be related to those measured at the anemometer location as follows,

$$\Psi = k_1 (\Psi_m - \Psi_0) \qquad (10)$$

$$V_r = k_2 V_{rm} (1 - k_3 \Psi^2) \qquad (11)$$

where subscript m denotes the value measured at the anemometer location and Ψ_0 is an offset correction to account for errors in the measurement at zero yaw angle.

The values of k_1, k_2, k_3, and Ψ_0 utilized in an ABCD analysis are obtained using an on-road, in-situ anemometer calibration technique. Although the full details of this technique are proprietary, the algorithm is an extension of the method reported in reference 5. The analysis rests on the assumption that the spatial averages of the wind speed and direction along a common section of test track are the same during two successive vehicle passes back and forth along the track, and that k_1 and k_3 are functions of k_2. Thus, the values of k_1, k_2, k_3, and Ψ_0 are evaluated from an analysis of pairs of data sets measured either during the vehicle warm-up phase that precedes the coast down test phase, or during the coast down test. A variation of the calibration technique facilitates an evaluation of the dynamic threshold of the anemometer which is utilized in the calculation of V_{rm}.

PRACTICAL CONSIDERATIONS AND ANALYSIS OPTIONS

Mechanical Drag Modeling - Although the equation of motion provides for the evaluation of the coefficients of up to a fifth-order polynomial mechanical drag model, it is useful to limit the order of the polynomial to the highest order beyond which the uncertainty in the analyzed value of the highest order coefficient exceeds the absolute value of the coefficient. The choice of this limit should also involve a consideration of the highest order that might be expected based on the results from independent tests of the types of mechanical elements that the mechanical drag model must represent.

Low-Wind Testing - Special care must be exercised in the analysis of data measured in very low wind conditions. An examination of Equation (9) reveals that, in the special case of zero wind ($V_r = V$; $\Psi = 0$), the aerodynamic drag term is perfectly correlated with the quadratic mechanical drag term. Thus, in situations where the mechanical configuration of the vehicle is such that the mechanical drag model must include a non-zero quadratic term, the analysis of a meaningful aerodynamic/mechanical drag breakdown can only be obtained by either assuring that the test is conducted in winds exceeding some minimum headwind/tailwind condition (approximately 3 to 4 km/hr), or, failing that, by constraining C or $C_D(0)$ in the analysis.

Constrained Analyses - Frequently, studies are conducted wherein many tests of the same vehicle are utilized in the evaluation of the effects of changes to its aerodynamic or mechanical configuration. In such cases, the precision of the analysis can be significantly improved through use of some type of analysis constraint.

For example, in the evaluation of the effects of aerodynamic shape changes, and on the assumption that the test-to-test variation of mechanical drag only reflects the effect of changes in ambient temperature, it is worthwhile to perform two sets of analyses. The first set would be performed with no constraints and a model of the form

$$A, B, C, \ldots \big|_{(T_{ref})} = A, B, C, \ldots \big|_{(T_{test})} * (1 + k_t (T_{test} - T_{ref})) \qquad (12)$$

would be used to compute a set of estimates of the mechanical drag coefficients for some reference ambient temperature T_{ref} from the coefficients analyzed at the various test ambient temperatures T_{test}. (Reference 8 recommends k_t = .0086/deg C; all three United States automobile manufacturers are currently using ABCD in a cooperative effort to confirm or reestablish the value of k_t for contemporary vehicles.) These estimates would then be averaged (using uncertainty weighting methods; see Young (9)), and the averages back-substituted into Equation (12) to obtain improved estimates of the mechanical drag coefficients at the various test conditions. The second set of analyses would utilize mechanical drag constraints based on use of the latter values. Although this constraint could be performed in an absolute sense, a weaker constraint that takes into consideration the possible imperfection of the temperature correction model is preferred. Thus, the analysis would employ constraints of the ratios B/A, C/A, etc., instead of constraints of the values of A, B, C, etc., in the final analysis for the sought-after aerodynamic drag results.

As another example, the same aerodynamic configuration could be used in a study to evaluate the effect of some change to the mechanical drag of the vehicle. Here again, two sets of analyses would be performed. The first set would be performed unconstrained and would provide a set of estimates of the aerodynamic drag coefficient versus yaw angle variation for the vehicle. An uncertainty-weighted average of these estimates could then be utilized to constrain either the zero yaw drag coefficient, or the entire drag versus yaw characteristic, in a second set of analyses for the desired mechanical drag result.

In situations where uncertainty-weighted averages are calculated from the results of unconstrained analyses, it is important that results containing non-random error be excluded from the data set utilized in the computation of the average. An example of results that should be excluded are those produced in the analysis of data measured in low-wind conditions for a vehicle that exhibits a quadratic mechanical drag characteristic.

Combined Analysis - The combined analysis is an alternative method for producing an average result from a set of tests of the same vehicle configuration. The method has the advantage that data measured in low-wind conditions for a vehicle that exhibits a quadratic mechanical drag characteristic can be incorporated into the analysis. The analysis rests on the assumptions that the test-to-test variation of mechanical drag only reflects the effect of changes in ambient temperature, and that the sensitivity to temperature is known. The combined analysis is implemented by expressing $A, B, C, .. |_{(T_{test})}$ in Equation (9) in terms of $A, B, C .. |_{(T_{ref})}$ with use of Equation (12). When this is done, Equation (9) involves unknowns $A, B, C, ... |_{(T_{ref})}$ and the unknown aerodynamic drag characteristic constants a_i. Since these values are fixed for a given vehicle configuration, the matrices that are formed for their solution from each set of test data can be combined to produce a single solution which incorporates the data from all tests of the configuration. Provided that some of the tests are carried out in winds exceeding the established minimum headwind/tailwind condition, the inclusion of low-wind data into the analysis should not preclude the evaluation of a meaningful aerodynamic/mechanical drag breakdown.

EXAMPLES OF SOME RESULTS FROM PAST STUDIES

Examination of Wind Tunnel Blockage Correction Models

In 1984, a study was undertaken to compare wind-tunnel-determined aerodynamic drag coefficients for baseline and aerodynamically-modified versions of three different sized European-manufactured sedans with similar results analyzed from coast down data.

All vehicles were equipped with manual transmissions and were coasted with their disc brake calipers removed, their drum brake shoes backed-off, and the drum brake self-adjusters disabled. The mechanical drag characteristics of the losses inboard of the axles at the driven axles were determined as functions of speed and temperature using axle coast-down test methods (4). The rolling resistance characteristics of the tires when operating in the thermal condition characteristic of that found during a coast down test were determined in flat belt tests at the Calspan Corporation's Tire Research Facility. Coast down tests were performed at tracks in Homestead, Florida and Romeo, Michigan. In addition to tests involving changes in their aerodynamic configuration, two of the vehicles were also tested with modified mechanical configurations. In these instances, two of the vehicles were tested with altered tire pressures; and one vehicle was tested with its brakes installed in their normal operating configuration.

The aerodynamic characteristics of the test vehicles were evaluated in the Lockheed-Georgia Low-Speed Wind Tunnel. The data were corrected for blockage effects using two different methods, namely, the pressure-signature method developed by Hackett and Wilsden (10) and the so-called continuity method.

The track data were analyzed using unconstrained and constrained analysis techniques to evaluate the effects of the various aerodynamic and mechanical changes on the drag of the vehicle. Because of the linearity of the mechanical drag characteristics found in the independent supporting tests, the values of the higher-order mechanical drag terms were set to zero, and the ratio of B to A was constrained based on the independent test data and on the incremental mechanical drag track test results.

Table 1 presents a summary of the zero yaw drag coefficients produced in the constrained analysis of each test (comprised of a set of 10 coasts performed alternately in opposite directions). The values in parentheses represent the drag coefficient uncertainties at the 95% confidence level. The last column presents the uncertainty-weighted averages of the individual test results for each aerodynamic configuration. An inspection of the results in Table 1 reveals that, when considered in terms of their uncertainties, the analyzed drag coefficients for each aerodynamic configuration were repeatable from test to test and from track to track, independent of mechanical configuration.

Figure 3 presents a typical variation of drag coefficient with yaw angle that was analyzed from the coast down data. The figure presents the variation from the two Florida runs of configuration F1 (heavy solid line) together with its associated uncertainty band (light solid lines) in comparison with wind-tunnel data corrected with the pressure-signature (WT-PS) and continuity (WT-CONT) methods. The results indicate that the drag coefficient variation with yaw angle analyzed from the track data was very similar to those found in the wind-tunnel, and, for this configuration, suggest better correlation with the continuity-corrected data than with the pressure-signature-corrected data.

Figure 4 presents a comparison between the track-determined, uncertainty-weighted-average zero yaw drag coefficients for the tested aerodynamic configurations (the height of each box representing the uncertainty of the average) and the corresponding wind-tunnel data. Figure 4 indicates that the track-determined results generally fell between the continuity-corrected and pressure-signature-

Configuration	Description	Florida C_{D0}	Michigan C_{D0}	Average C_{D0}
F1	Base	.425(.005)	.427(.005)	.426(.002)
		.422(.005)	.429(.006)	
F1(TP)	Base+Tire Pressure Reduction		.426(.006)	
			.432(.006)	
F2	Base+Rear Roof Spoiler	.490(.006)	.497(.004)	.492(.003)
		.487(.006)	.491(.005)	
S1	Base		.369(.006)	.369(.002)
			.371(.007)	
S1(TP)	Base+Tire Pressure Reduction		.372(.007)	
			.368(.008)	
S1(BRK)	Base+Brakes	.370(.006)	.375(.007)	
		.371(.006)	.359(.007)	
S2(BRK)	Base+Rear Deck Wing+Brakes	.363(.007)	.357(.008)	.356(.003)
		.357(.006)	.347(.007)	
S3(BRK)	Base+Spoiler Lip+Window	.366(.007)	.363(.010)	.366(.004)
	Strakes+Brakes	.364(.006)	.369(.007)	
E1	Base	.393(.006)		.393(.004)
		.393(.005)		
E2	Base+Rear Roof Spoiler	.443(.005)		.445(.004)
		.446(.005)		
E3	Base+Blocked Grille	.373(.007)		.372(.004)
		.372(.005)		

Table 1 - Summary of Aerodynamic Drag Results from 1984 Track Tests

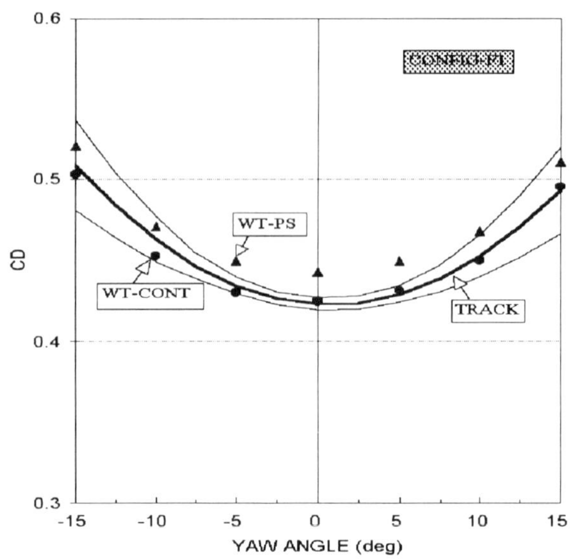

Figure 3 - Typical Aerodynamic Drag versus Yaw Characteristic from 1984 Automobile Test Program

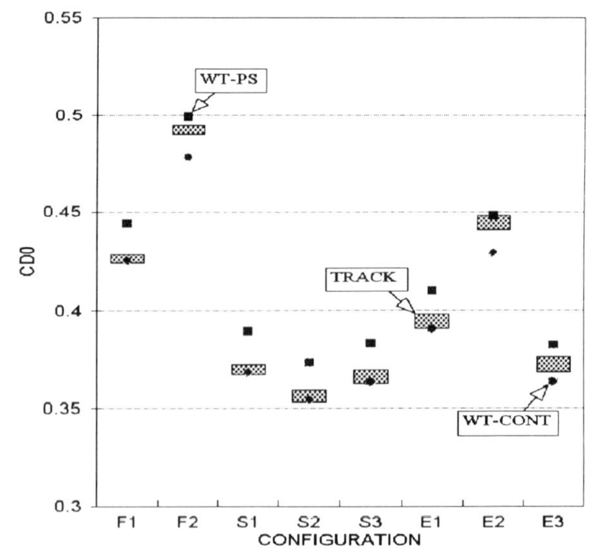

Figure 4 - Correlation of Wind-Tunnel and Track Results from 1984 Automobile Test Program

corrected data. In five of the eight cases studied, the continuity-corrected data exhibited better correlation with the track results. In the two cases that involved the addition of a drag-increasing spoiler that was transversely-mounted across the rearward roof edge (F2 and E2), the pressure-signature-corrected data displayed better correlation with the track results. In the only case which involved a blocked grille configuration (E3), the track result fell midway between the two corrected wind-tunnel data points.

The 1984 test results reported above are indicative of numerous results that have been obtained since that time. In all of these studies, great care was exercised to minimize the effects of random measurement error and any bias associated with the effects of inappropriate choices in the modeling of the mechanical drag term. The weight of these results indicates that, although the physics embodied in the pressure-signature method are clearly more appealing than those associated with the continuity method, drag coefficients analyzed from coast down data have almost consistently ranged somewhere between the drag coefficients produced with application of either correction, with neither method being the uniformly-better predictor of the track result. It should be recognized, of course, that any comparison between track and wind-tunnel results is clouded by the fact that the usual wind-tunnel simulations of the on-road operation of a vehicle do not reproduce the motion of the ground relative to the vehicle or the rotational motion of its wheels, both of which affect the flow past the vehicle to some extent. Thus, as the results above demonstrate, the use of an inferior method of blockage correction can often result in the calculation of drag coefficients that, fortuitously, more closely agree with track results than would be obtained with use of a more fundamentally-sound method of correction.

Heavy Truck Aerodynamic Drag Characteristics

Because of the size of the vehicle and the number and location of its ground contact points, the evaluation of the aerodynamic drag characteristics of a full-scale tractor trailer truck in a wind-tunnel presents a formidable task. Thus, drag data obtained from model tests of a pre-production vehicle in a wind-tunnel cannot be readily compared with similarly obtained results for the full-scale production vehicle. For the same reasons, the wind-tunnel cannot be conveniently used for evaluating the aerodynamic drag characteristics of competitive, full-scale tractor-trailer trucks.

A number of studies were carried out in the period from 1982 to 1987 which involved the application of anemometer-based coast down methods in the determination of the aerodynamic drag characteristics of a variety of full-scale tractor-trailer truck configurations. The anemometer was typically supported about 3 m in front of the tractor using a tripod-like frame that attached to the bumper and grille. Independent tests of tire rolling resistance and of the losses inboard of the tires at the driven and non-driven axles revealed that the drags of the mechanical components are essentially linear, and that any non-linear drag contributions are small and well-represented with a quadratic model. The final results in the studies were produced from combined data analyses utilizing B/A, and sometimes C/A, constraints based on the averages of the results obtained in unconstrained analyses or from use of the independent test data. In all cases, drag evaluations were based on the analysis of two tests (involving 10 coasts each) per configuration.

Figure 5 presents one example of the type of results that were obtained in the evaluation of heavy truck aerodynamic drag characteristics. In this instance, a cab-over-engine (COE) tractor was attached to a 4.11 m high, 2.44 m wide, 12.19 m long, smooth-sided trailer having 254 mm radii along its front vertical corners. The combination was tested in its baseline configuration (B) and following the attachment of devices designed to reduce its drag. The latter devices included a three-dimensional, roof-mounted fairing designed to reduce the impingement of air flow on the face of the trailer (F), trim tabs that attach to and extend outward and rearward from the fairing's vertical trailing edges (ST), and cab side extenders that attach to and extend rearward from the tractor cab's vertical trailing edges (CE). The side trim tabs and cab extenders work to reduce the drag due to cross-flow through the gap between the tractor and trailer at non-zero yaw angles. The side trim tabs also helped to reduce the flow impingement on the trailer face.

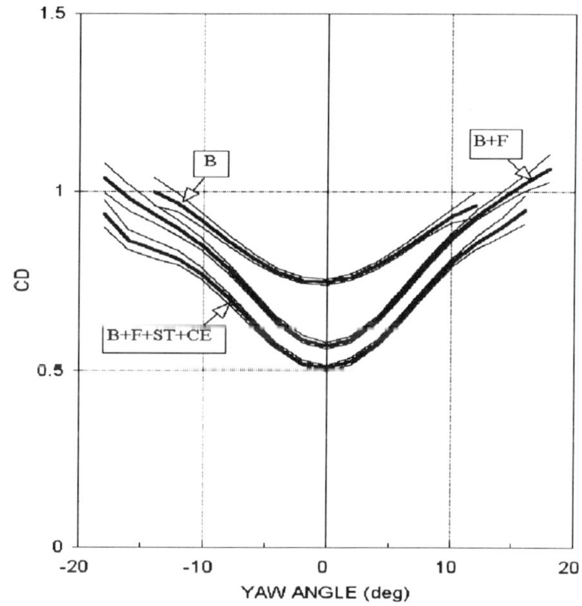

Figure 5 - Typical Aerodynamic Drag versus Yaw Characteristics for Base and Aerodynamically-Modified Tractor-Trailer Truck from 1984 Heavy Truck Test Program

In contrast with results for cars, light trucks, and vans, the drag characteristics for the baseline tractor-trailer truck combination demonstrate higher drag coefficient levels and greater sensitivity of the drag coefficient with yaw angle. These familiar features are associated with the behavior of the flow in the gap between the tractor and trailer. As can be seen, the addition of the fairing resulted in significant drag reductions at zero yaw which diminished as the yaw angle increased. The further addition of the side trim tabs and cab extenders provided additional drag reductions at zero yaw which were maintained as the yaw angle increased.

The interpretation of the benefit of such drag-reducing devices is aided with the calculation of the wind-averaged drag coefficient, which is used to predict an average aerodynamic drag for some specified set of operating conditions (11, 12, 13). Thus, assuming that, in long-term operations, the vehicle driving cycle can be represented with a constant operating speed of 88.5 km/hr, the wind is equally-likely to approach the vehicle from all directions to its motion, and the average drag can be adequately predicted using a constant wind speed of 11.3 km/hr, the percentage reductions in the wind-averaged drags produced by the fairing, and by the combination of the fairing, side trim tabs, and cab extenders, were 16.2(1.3) and 25.0(1.3), respectively. These percentage reductions are notably lower than the 23.9(1.3) and 32.2(1.3) percent reductions found at zero yaw.

Light Truck-Travel Trailer Aerodynamic Drag Characteristics

The determination of the full-scale aerodynamic drag characteristics of light truck-travel trailer combinations in a wind-tunnel presents the same difficulties as described for full-scale tractor-trailer truck combinations. Consequently, a series of coast down tests were performed in 1986 to assess the aerodynamic drag increase found in situations when such trailers are towed by various pickup trucks and vans. The tests involved eleven different vehicles and three different trailers. Figure 6 presents a typical example of the types of aerodynamic drag characteristics that were analyzed. In this case, a 4.9 m long utility trailer was towed behind a pickup truck having an approximate frontal area of 3.1 m^2 . The frontal area of the truck-trailer combination was

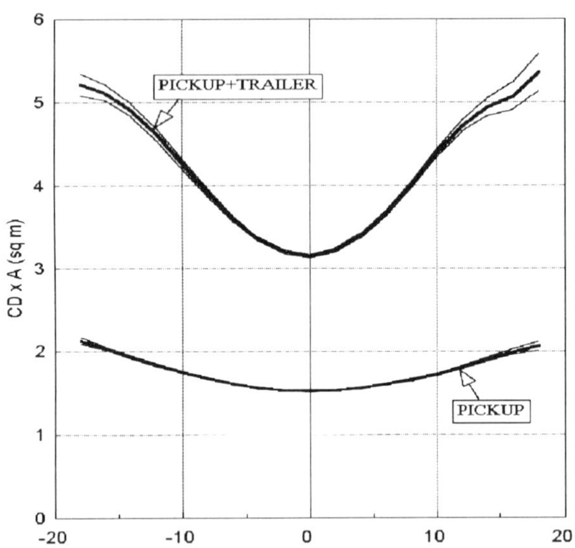

Figure 6 - Typical Aerodynamic Drag versus Yaw Characteristics from 1986 Light Truck-Travel Trailer Combination Test Program

approximately 5.6 m^2 . As can be seen, the addition of the trailer more than doubled the aerodynamic drag on the vehicle. Also, because of the formation of the gap between the two vehicles, the rate of drag increase with yaw angle for the combination was significantly higher than for the truck alone.

The results in Figure 6 are typical of those found for 10 of the 11 combinations that were tested. In the exceptional case, the vehicle with the highest zero yaw drag when operating alone was tested in combination with the largest trailer. The trailer design differed from those used with the other vehicles in that its front face was stepped so that its upper portion extended well forward over the pickup box and attached to a fifth wheel that was mounted in the bottom of the box. Thus, although the 5.9 m^2 approximate frontal area of this combination was the highest of all combinations tested, the percentage drag increase produced by the addition of the trailer was about one-half the increase found with the other trailers, and the total drag on the combination was among the lowest of all combinations involved in the study.

Tunnel-to-Road Correlation of Drag Changes Produced by Various Underbody Modifications

A recent study was undertaken to examine the correlation of wind tunnel evaluations of the drag reductions produced by various changes to the underbody of a typical sedan with similar evaluations of the drag reductions obtained from coast down tests.

Six underbody configurations (in addition to that of the baseline vehicle) were investigated in wind tunnel tests carried out in the Lockheed-Georgia Low-Speed Wind Tunnel. Results were obtained with use of a blowing system to energize the flow in the boundary layer along the tunnel floor (blowing-on) and in the usual test situation where no such energization takes place (blowing-off). All drag coefficient data were reduced using the continuity blockage correction method. Table 2 describes the underbody modifications that underwent testing.

Configuration	Description
1	Baseline (B)
2	B+Downturned flange at bottom of rear bumper (F)
3	B+F+Panel from front bumper to 1/3 underbody length (P1)
4	B+P1
5	B+P1+Panel beneath rear axle (P2)
6	B+P1+P2+Flat panel from rear axle to rear bumper (P3)
7	B+P1+P2+Diffuser panel from rear axle to rear bumper (P4)

Table 2 - Underbody Modification Configurations

Coast down tests were performed and analysis results produced using the ABCD test system. The results for each vehicle configuration were analyzed from data collected in three tests per configuration, with each test consisting of 16 coasts each. The analyses incorporated the input of independent data for the losses inboard of the tires at the driven and nondriven axles, and utilized B/A constraints established from uncertainty-weighted averages of the analyzed tire rolling resistance characteristic. The final drag coefficients for each configuration were calculated from the uncertainty-weighted averages of the results analyzed from the data measured in each of the three tests of the configuration.

The comparison of wind-tunnel and track underbody drag results is presented in Figure 7. The heights of the blocks used to mark the track results indicate the ranges of their uncertainty.

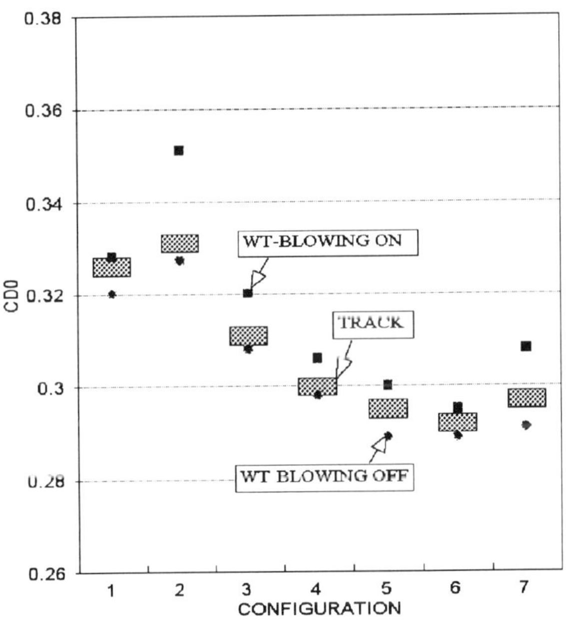

Figure 7 - Correlation of Wind-Tunnel and Track Results from Recent Automobile Underbody Test Program

An inspection of Figure 7 reveals that, in the absence of blowing, the wind-tunnel drag coefficients generally ranged somewhat below the drag coefficients analyzed from the coast down data. It is apparent, however, that the changes in drag produced by all of the underbody modifications in these two test situations were very similar .

Noting that floor blowing is intended to improve the wind-tunnel simulation of the on-road operating condition, the results in Figure 7 demonstrate that the wind-tunnel drag coefficients generally increased in the presence of blowing. It is also evident that, with the exceptions of Configurations 2 and 7, which had underbody modifications in the vicinity of the rear bumper, the drag changes produced by the underbody modifications in the presence of blowing were very similar to those produced in the absence of blowing.

A further examination of Figure 7 shows that, on average, the drag in the presence of blowing was higher than that found on the track. However, with the exceptions the results for Configurations 2 and 7, the drag changes produced by the underbody modifications in the presence of blowing were very similar to those produced on the track. In view of the trends of the data, and the differences between the track results and the wind-tunnel data in the absence and presence of blowing, it appears that the implementation of blowing in the present study resulted in a less consistent prediction of the influence of the various underbody modifications on the vehicle drag experienced on the track.

A Comparison of Dynamometer and Track Determined Mechanical Drags

A recent preliminary investigation was undertaken to compare the variation of total mechanical drag with speed determined from an analysis of the data measured in a coast down test with the sum of similar results obtained in dynamometer evaluations of the mechanical drags at the driven and non-driven axles in the laboratory. The vehicle utilized in the experiments consisted of a large, four-wheel drive sports utility vehicle equipped with a four-speed automatic transmission. All tests were carried out in the two-wheel drive operating mode.

A series of coast down tests were performed prior to the dynamometer experiments. All data were measured and analyzed with the ABCD test system. The results of the unconstrained analyses that were carried out in the aftermath of these tests were used in the evaluation of an uncertainty-weighted-average zero yaw aerodynamic drag coefficient for the vehicle. This value was 1% higher than the drag coefficient that had been established for an identical vehicle in the vehicle manufacturer's wind-tunnel. A final coast down test was performed on the day of the dynamometer tests. The analysis of the data measured in this test provided the mechanical drag characteristic for comparison with the dynamometer results. This characteristic was obtained by constraining the aerodynamic drag coefficient in the analysis to the uncertainty-weighted average value that had been established earlier, and by increasing the order of the polynomial utilized in the mechanical drag model to the highest order beyond which the uncertainty in the analyzed value of the highest order coefficient exceeded the absolute value of the coefficient. This resulted in the determination of a fourth-order polynomial representation of the mechanical drag variation with speed.

The dynamometer experiments were carried out on a newly-installed Horiba 122 cm single roll electric dynamometer contained in a test cell maintained at 24 deg C. The mechanical drags due to the combinations of tire rolling resistance and the losses inboard of the tires at the driven and non-driven axles were evaluated in separate tests. Prior to conducting these evaluations, the axles were warmed-up on the dynamometer. In the case of the driven axle, the axle was driven with the vehicle's engine at the same speed and under essentially the same road load as had existed during the warm-up period for the preceding coast down test. In the case of the non-driven axle, the axle was driven with the dynamometer at the same speed as had been utilized during the track warm-up period. Drag evaluations were carried out using dynamometer coast down and steady speed techniques. The same drag characteristics were found with either method. These drags were then corrected to estimate the drag at the ambient temperature under which the coast down tests were conducted using Equation (12).

The comparison of the dynamometer and coast down determined mechanical drag characteristics is presented in Figure 8. The lighter lines in this figure bound the

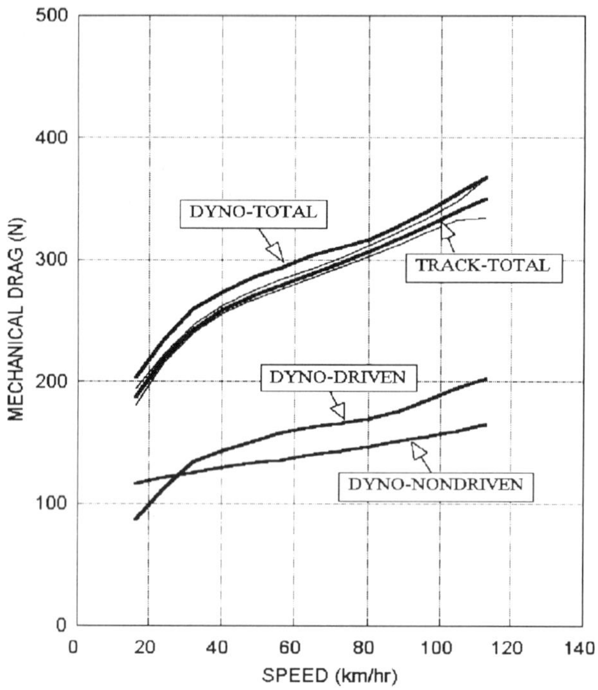

Figure 8 - Comparison Between Dynamometer and Track Mechanical Drag versus Speed Results for a 4x4 Truck Equipped with a Four-Speed Automatic Transmission

uncertainty range of the track results. With reference to the inboard loss characteristics presented in Figure 1, it is seen that the dynamometer result for the sum of the tire rolling resistance and the drag due to the losses inboard of the tires at the driven axle demonstrated a drag variation with speed typical of those found with use of four-speed automatic transmissions. It is also apparent that the shape of the variation of total mechanical drag with speed determined on the track was very similar to that found from the dynamometer total, but that the level of drag on the track was somewhat lower. The latter difference was not unexpected in view of the facts that the tire rolling resistance in the laboratory was established on a curved rather than a flat surface, which gives rise to higher drag, whereas, owing to the absence of the cooling airstream experienced on the track, the thermal condition of the tire and driveline on the dynamometer at the same ambient temperature is warmer than on the track, which acts to decrease drag. The difference between the smoothness of the track and dynamometer surfaces could also have contributed to the observed difference in drag levels.

SUMMARY AND CONCLUSIONS

1) An improved coast down method that incorporates the input of data from an on-board anemometer has been presented and its practical use discussed.

2) The equation of motion utilized in the improved procedure provides for the evaluation of models representing the aerodynamic and mechanical drag characteristics of a wide variety of road vehicles.

3) A comparison of the aerodynamic drag characteristics of eight different aerodynamic configurations and three different mechanical configurations determined in coast down tests of three different vehicles under arbitrary wind conditions and on two different test tracks confirm the ability of the procedure to produce repeatable and precise estimates of aerodynamic drag.

4) A comparison of wind-tunnel and coast-down-determined aerodynamic drag coefficients revealed that the track results fell between wind-tunnel data corrected with the continuity and pressure-signature blockage correction methods. In five of eight cases, the continuity blockage correction method provided a better prediction of the track-determined aerodynamic drag than did the pressure signature method. In two cases involving vehicle configurations that employed roof-mounted spoilers to increase the drag on the vehicle, the pressure-signature-corrected wind-tunnel data provided the better prediction.

5) The improved coast down method has been demonstrated to be a useful tool for evaluating the aerodynamic drag characteristics of full-scale vehicles that cannot be readily tested in a wind tunnel, such as tractor-trailer and light truck-travel trailer combinations.

6) The aerodynamic drag model employed in the improved method has been demonstrated to capture the highly yaw dependent drag characteristics of two-body vehicles containing a gap between the bodies.

7) A comparison of continuity-blockage-corrected wind-tunnel and coast-down-determined aerodynamic drag coefficients for various underbody configurations of a sedan revealed that although the track results fell between the wind-tunnel results obtained in the absence and presence of tunnel floor blowing, the most consistent prediction of the track-

determined aerodynamic drag was obtained from tests with tunnel floor blowing-off.

8) A comparison of the variation of mechanical drag with speed determined on a vehicle equipped with a four-speed automatic transmission in track coast down tests with like results obtained in laboratory tests on a 122 cm single roll electric dynamometer revealed that the shapes of the two characteristics were very similar, but that the overall drag level on the track was somewhat lower than that on the dynamometer.

ACKNOWLEDGEMENTS

The author would like to acknowledge the following individuals from the Ford Motor Company: Mr. Jim Chabot for his initiation of the programs that nurtured the refinement of the anemometer-based coast down method; Mssrs. Larry Socha and Bill Quinlan for supporting programs that allowed that refinement to continue; and Mr. Norm Peters for his many efforts to encourage the development and adoption of the ABCD test system. The author would also like to acknowledge Mr. Paul Karas of the Ford Motor Company, Mr. Dave Pruess of the Chrysler Corporation and Mr. Roger Leonard of the General Motors Corporation whose experiences with the use of ABCD have provided valuable feedback for its continued improvement.

REFERENCES

1. Anderson, J. W., Firey, J. C., Ford, P. W., and Kieling, W. C., "Truck Drag Measurements by Road Test Measurement", *SAE Transactions*, Vol. 73, Sec. 1, September 1965.

2. White, R. A. and Korst, H. H., "The Determination of Vehicle Drag Contribution from Coast-Down Test", SAE Paper 720099, 1972.

3. Yasin, T. P., "The Analytical Basis of Automobile Coastdown Testing", SAE Paper 780334, 1978.

4. Walston, W. H., Jr., Buckley, F. T., Jr. and Marks, C. H., "Test Procedures for the Evaluation of Aerodynamic Drag on Full-Scale Vehicles in Windy Environments", *SAE Transactions*, Vol. 85, Sec. 1, September 1976.

5. Buckley, F. T., Jr., Walston, W. H., Jr., and Marks, C. H., "Analysis of Coast-Down Data to Assess the Aerodynamic Drag Reduction on Full-Scale Tractor-Trailer Trucks in Windy Environments", *SAE Transactions*, Vol. 85, Sec. 3, September 1976.

6. Hamabe, K., Kitoh, K., Ogata, H., and Kobayashi, T., "An Estimation of Aerodynamic Drag Coefficient of a Passenger Car by Coast-Down Tests in Windy Environments", *JSAE Review*, p. 114, April 1985.

7. Passmore, M. A. and Jenkins, E. G., "A Comparison of the Coastdown and Steady State Torque Methods of Estimating Vehicle Drag Forces", SAE Paper 880475, 1988.

8. SAE Recommended Practice J1263, "Road Load Measurement and Dynamometer Simulation using Coastdown Techniques", issued June 1979, revised June 1991.

9. Young, H. D., Statistical Treatment of Experimental Data, McGraw-Hill, New York, NY, 1962, ISBN 07-072646-9.

10. Hackett, J. E. and Wilsden, D. J., "Estimation of Wind-Tunnel Blockage from Wall-Pressure Signatures", A Collection of Technical Papers, AIAA Tenth Annual Aerodynamic Testing Conference, San Diego, CA, April 1978.

11. Buckley, F. T., Jr. and Sekscienski, W. S., "Comparisons of Effectiveness of Commercially Available Devices for the Reduction of Aerodynamic Drag on Tractor Trailers", *SAE Transactions*, Vol. 84, Sec. 3, September 1975.

12. SAE Recommended Practice J1252, "Wind Tunnel Test Procedure for Trucks and Buses", August 1979.

13. Buckley, F. T., Jr., "Aspects of Over-the-Road Testing of Truck Drag Reducing Devices", *SAE Transactions*, Vol. 91, Sec. 4, September 1982.

950631

Reducing Splash and Spray of Trucks and Passenger Cars

Hans Goetz and Ronald Schoch
Mercedes-Benz AG

ABSTRACT

The problem of effectively reducing water spray formed by motor vehicles on wet roads remained up to now unsolved.Although numerous experimental investigations have been published, and comprehensive patent literature is available, the suggested solutions appear to be problematic under real-life conditions.
In this paper, a configuration applicable both to commercial vehicles and passenger cars is proposed, which is restricted exclusively to the wings, and - with today's design principles - does not require any special advance preparation. A grooved channel profile in the wing causes the water spray to be reduced considerably without affecting the vehicle's suitability for everday use.
An optoelectronic measuring instrument which is carried along the vehicle, makes it possible to conduct integral water spray measurements over time periods of various length.

INTRODUCTION

Vehicles which are driven on wet road surfaces, produce a spray mist consisting of fine drops of water. This mist impairs both the visibility of oncoming traffic as well as the visibility of vehicles driving behind and overtaking. The result is a reduction in road safety and an increase in the risk of accidents. The reduction of trails of spray as a result of suitable measures on the vehicle is becoming an increasing necessity, in particular because of the growing traffic density.
The obstruction to visibility resulting from trails of spray is influenced both by the size and number or density of single droplets as well as by the flow conditions. Whereas large drops do not follow the flow, or, do so only for a short time, and fall relatively rapidly because of their mass, the small, mist-like droplets of water follow the air flow and, dependent on the flow conditions, are the essential cause for obstruction of visibility.
Improvements to visibility now requires design measures to the vehicle which prevent or reduce the formation of spray mist and/or alter the flow conditions accordingly.
The primary objective is to increase road safety by improving visibility when the vehicle ist driven on wet roads.

MEASURING DEVICE

To measure the spray mist which vehicles produce on a wet road surface, a measuring method based on measuring the scattered light of the mist, has been developed and successfully applied for the first time in connection with this subject.Rapid signal detection and processing makes it possible to provide results immediately after the test. The results can then flow instantly into the next improvement measure. This is a compact test device which is suitable both for the laboratory as well as for field trials.
As an opto-electronic measuring system, it has the benefit of operating continuously, without contact and with scarcely any effect on aerodynamics. The prinicipal task of the method, though, is to register and assess the spray mist in a defined manner over a large area.

THE MEASURING SYSTEM

MEASURING PRINCIPLE - The trail of spray which exists from the wheel arch, is illuminated in a defined two dimensional way by a powerful source of light which is attached to the vehicle. The source of light produces a "light curtain" at right angles to the direction of travel (Figure 1). When droplets of mist flow through the light curtain, they reflect the light. The more water that leaves the wheel arch, and the denser the mist, the more light is reflected and registered by the detector. The detector consists of a reception lens and a light-sensitive sensor.

LIGHT SYSTEM - Requirements in respect of the light source:

- resistant to moisture;
- well-defined illuminated area;
- high luminous intensity.

The light source consists of 4 halogen bulbs each with an output of 150 watts, a fibre-optic cross-section converter and an auxiliary lens.

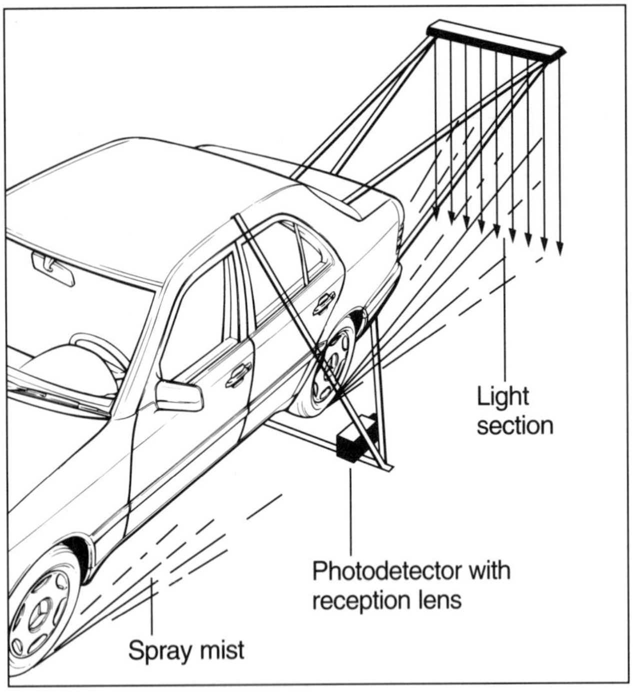

Figure 1: Principle of scattered light measuring method

The heart of the lamp is based on the idea that a cylindrical light inlet is converted into a very small rectangular one (Figure 2). Four fibre bundles which consist of 150 light-conducting fibres each, are coupled to a halogen lamp. Each of these fibres has a diameter of 1 mm. At the point where light is emitted, the light-conducting fibres are arranged one next to the other so that a total of 4 times 150 light conductors, i.e. 600 fibres, are located in a row.
Upstream of the fibres a cylindrical lens is located, which bundles the light in such a way that at a certain distance from the lens the gap is accurately focused. The entire configuration is located in a water-proof aluminium housing. Figure 2 shows the complete cross-section converter. Thus, a light curtain with a width of 10 mm on the road surface and an area of $1m^2$ can be produced.

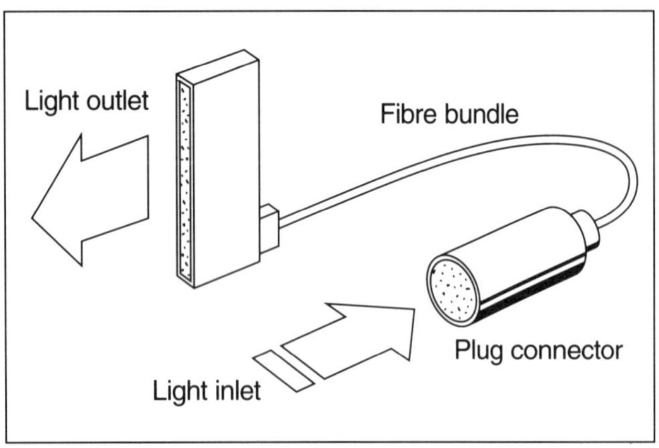

Figure 2: Principle of light system for producing a "light curtain"

DETECTOR - Linearity and sensitivity were the critical factors in choosing a suitable sensor element. A silicon photoelement operating in the short-circuit mode produced the best results.

If light strikes the transitional range between the Positive (P) doped and Negative (N) doped range of the photoelement, pairs of electron holes are produced, which generate a current if the element is short-circuited. This current is proportional to the luminous intensity. The current output by an element is very small (a few microamperes).For this reason, four elements are connected in parallel, which results in an addition of the individual currents.

After the current has passed through the current/voltage converter and downstream voltage amplifier, signals in the volt range are available at the output. These are 10 V in the case of 100 lux. Luminous intensities in excess of this can also be processed by choosing different lens apertures. An automatic offset adjustment and an overshoot display are likewise integrated in the amplifier unit (Figure 3).
The photo-sensitive sensor is housed in a 35 mm camera, precisely in the film plane. The advantage of this is that it is possible to produce a sharply focused image of the light section plane, on the sensor in a particularly simple way by setting the range of the lens. In addition, it is possible to elegantly monitor the measuring field in the spray mist.
The watertight housing has an orifice with which the measuring field is defined.

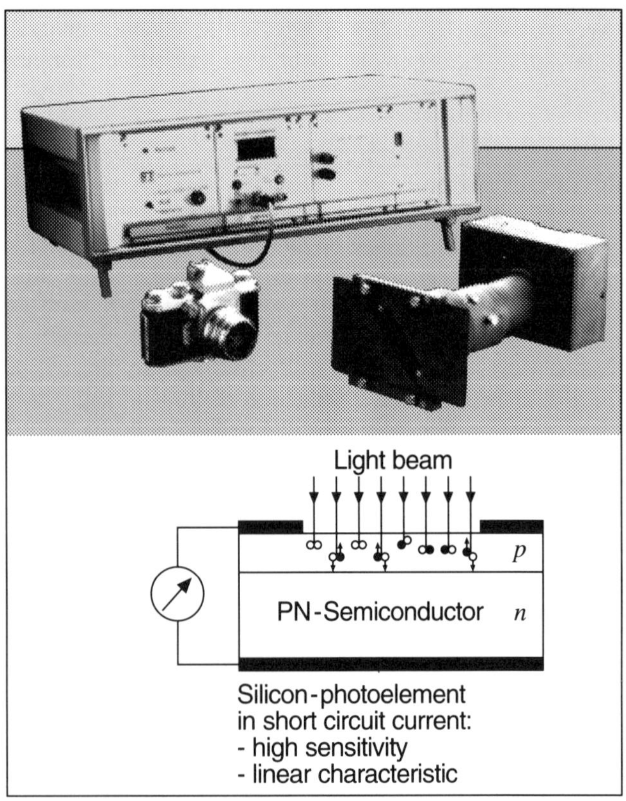

Figure 3: Components of the spray measuring system

FIELD MEASUREMENTS - A watered stretch of asphalt was available for the field trials. The trials were conducted only in darkness in order to avoid any influence of other light sources on the signal obtained. Figure 4 shows the test setup. The fibre-optical light source was mounted on the side of the vehicle to enable it to register the spray mist exiting from the front wheel arch. An outrigger was fabricated for attaching the sensor.

A mobile power unit was used for supplying the current for the light source. The lens of the detector was focused on the light plane. Figure 5 shows a diagram of the initial state and following modifications to the vehicle. The rapid rise in the signal when the spray commences can be clearly seen. The signal level fluctuation provides more detailed information regarding the portion of mist in the spray pattern of the particular vehicle version.

Figure 4: Test setup on vehicle

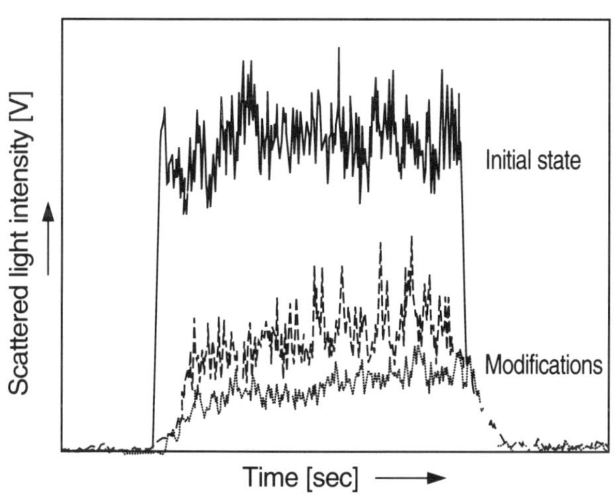

Figure 5: Signal pattern of spray mist

CONCLUSION - The presently applied laser process has the advantage that, owing to the spectral purity of the light and the great light intensity, interference from other light sources is virtually ruled out. For this reason it is possible to make the measurements also in dim daylight.

The basic drawback is that the measurement characteristics are restricted to one light spot, i.e. measurements are taken only at a specific height line. In most cases the standard "eye height" is taken.

But what would be the "correct height" for measuring the obstruction of the driver's field of vision?

One way out of this situation could be vertical scanning of the water spray, which, however, would have to take place in a very short time, because the vehicle passes the measurement set-up at a relatively high speed.

Another disadvantage of this process is the fact that no time sequences can be assessed, i.e. that dynamic characteristics of the water spray cannot be measured. The measurement set-up is stationary and the vehicle passes by. Thus it is possible only to make a snapshot measurement of the spray conditions.

An analysis of the new measuring method reveals that it is not the quantity of water which is of decisive significance for the obstruction to visibility resulting from spray mist, but the optical density and distribution of the mist.

The major advantage of this measuring method is the fact that the impression which a road user receives of the trail of spray of a motor vehicle, can be analysed and stored in meteorological terms.

This method has its strengths when it comes to comparing variants, in other words for assessing different measures on the vehicle. The result is available as an electrical measured value immediately after the test whereas film or video recordings permit a certain freedom in respect of the subjective interpretation.

The development status of the measuring method applied up to now was characterized by the fact that an integral signal was obtained from an observation area of about 1 m^2. It was possible to conduct comparisons between individual variants in respect of mist density and mist fluctuations or by means of vortex frequencies. A local resolution within the observation area is now possible.

This is achievable by replacing the total area of the sensor by nine individual sensors in the focal plane of the detector. An area segment in the light curtain is then assigned to each individual sensor. This will make it possible to also determine changes in the position of the spray trails between the individual test variants.

STATE OF THE ART AND REGULATIONS

COMMERCIAL VEHICLES - It is important to equip all heavy commercial vehicles having a certain design-related minimum speed with splash water protection systems for improving road safety. The shape of the wheel cover presented in Figure 6 corresponds to the national regulation as specified in § 36a STVZO (German Construction and Use Regulation/FMVSS).

The present standard wheel cover of a modern truck is shown in Figure 7.

EEC Directive 91/226 aims at laying down a standard function test. For EEC type testing, two types of spray protection devices which have already been commercialized, were taken into account, namely the water absorber, Figure 8, and the air/water separator, Figure 9.

The water absorber may be part of the wheel cover and/or the mud flap and/or the skirt which absorbs the energy of the water which is sprayed up and thus reduces the spraying of the water mist.

The air/water separator may be part of the skirt and/or of the mud flap which allows the air to flow through while retaining part of the water sprayed up (spray water).

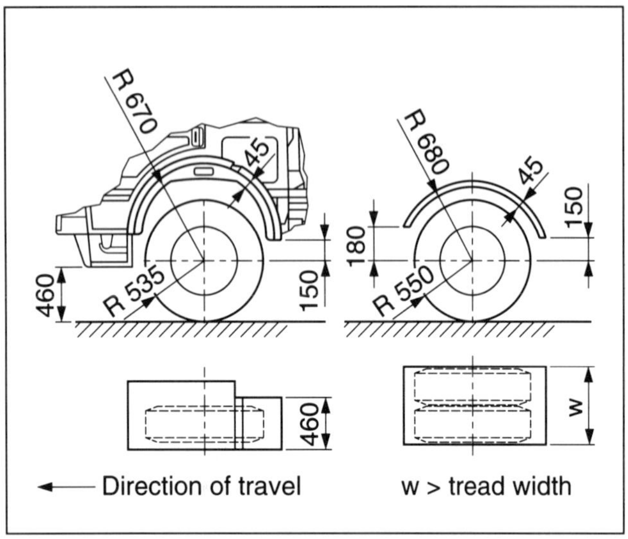

Figure 6: Fender skirt according to national regulation

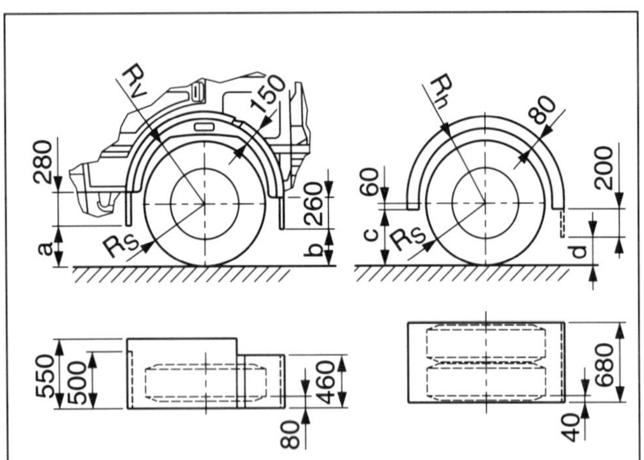

	normal	high
Rs	550	–
Rv	650	730
a	400	480
b	350	430

	normal	high
Rs	550	–
Rh	620	720
c	450	550
d	250	350

Figure 7: Serial fender skirt of a modern truck

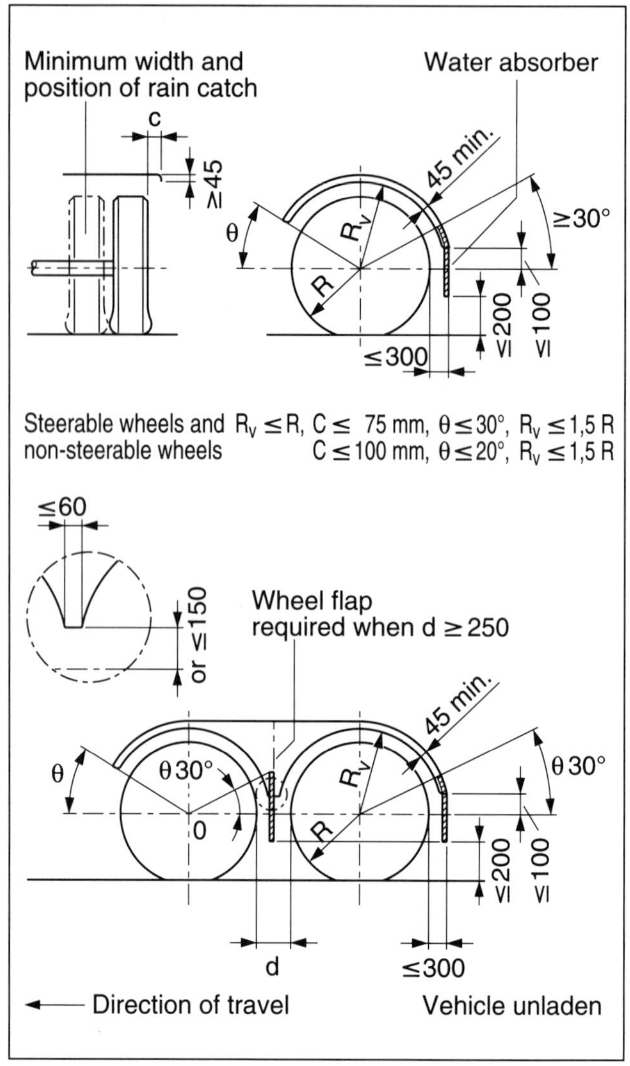

Figure 8: Wheel cover (mudguard) with water absorber

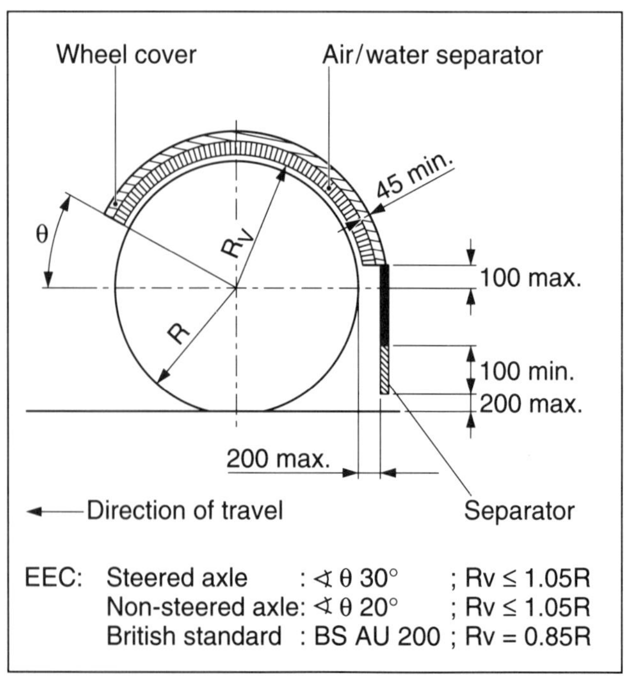

Figure 9: Splash protection system with air/water separator

TEST IN ACCORDANCE WITH EEC DIRECTIVE

WATER ABSORBER - The EEC directive on spray suppression devices for commercial vehicles suggests a procedure with which the "ability of a device to retain water, which is sprayed directly on it from a number of spray nozzles" should be determined. "With a flow rate of 0.675 litre/sec., a minimum of 90 litres of water must be sprayed horizontally onto the test piece from a distance of 500 mm". The test procedure is shown in Figure 10.

Then the percentage ratio between the amounts of water collected and deflected are determined. The water absorber will have passed the test, if a minimum of 70% of water has been collected.

With the test following the EEC directive it is possible to determine whether or not a specific material is suitable to absorb water spray.

The classification is independent of the vehicle model, because only laboratory tests in accordance with the above-mentioned directive are required. Thus, the test procedure is a compromise which - although accurate results can be obtained and comparisons can be made - does not consider the overall vehicle system in which a large number of parameters are responsible for water spray formation.

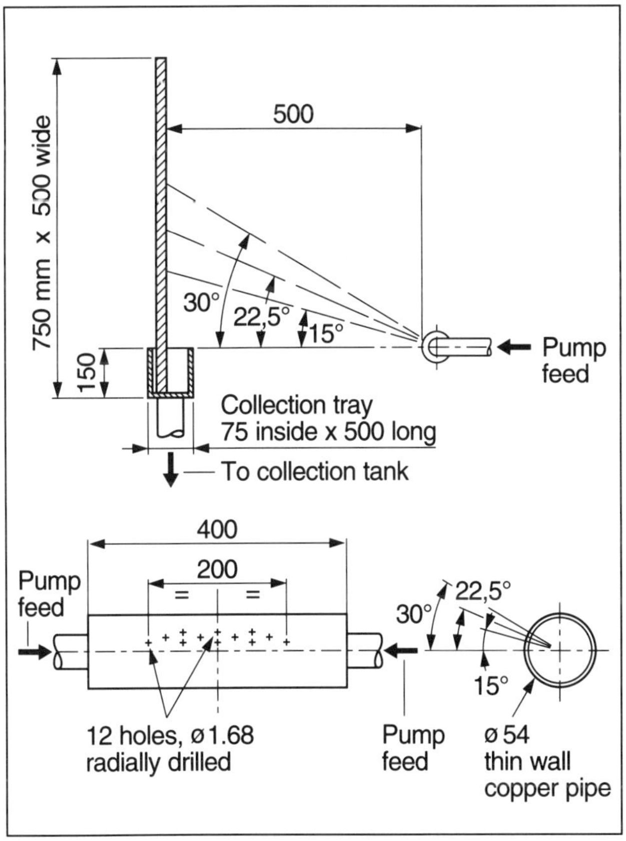

Figure 10: Apparatur for testing a suppresion material or device

AIR/WATER SEPARATOR - The relevant test equipment is shown in Figure 11. This test serves the purpose of showing to which extent a material is suitable to retain water sprayed on it from an air/water mechanical atomiser. With a flow rate of 1litre/minute and a pressure of 5 bar, the test piece is subjected to spraying from a distance of 200 mm in circles with a diameter of 50 mm. The diameter of the spray nozzle is 5 mm. The amount of water collected must not be less than 85 % of the water sprayed on the test equipment.

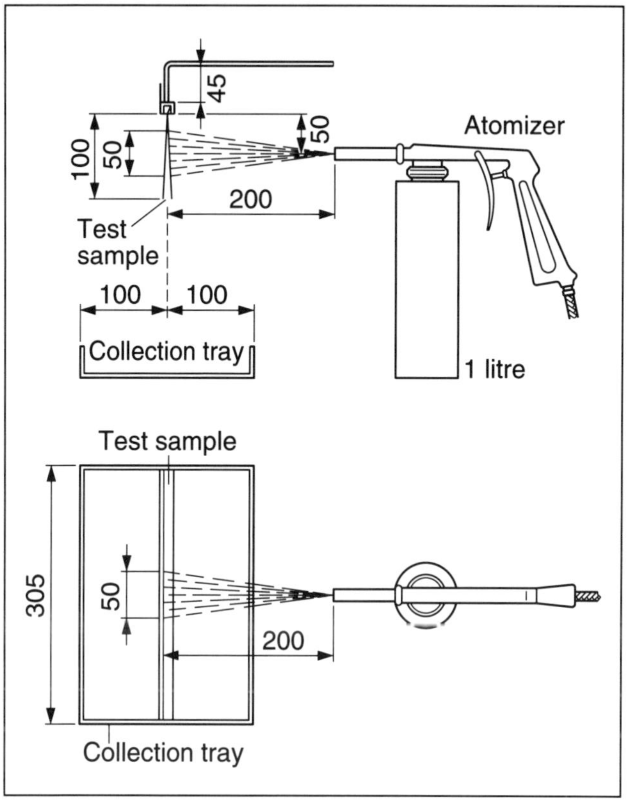

Figure 11: Procedure of a splash protection system with air/water separator

NEW APPROACH TO SOLUTION

Although the subject of spray mist has been presented on numerous occasions in patent literature, the majority of patents have failed because of the demands for day-to-day driving capability and on the non-integral approach at the vehicle.
The new approach is based on the idea of not allowing the spray water to atomize, but to concentrate it so that it can be channeled off in the form of numerous small streamlets. The grooved sections running longitudinally in the mudguard enable the water to collect without atomizing, regardless of the spray angle, and, as a result of the curvature of the mudguard, allow the water to run off by itself to the front and rear ends of the mudguard, Figure 12.
The self-cleaning effect and suitability for everyday motoring have been demonstrated in road tests.

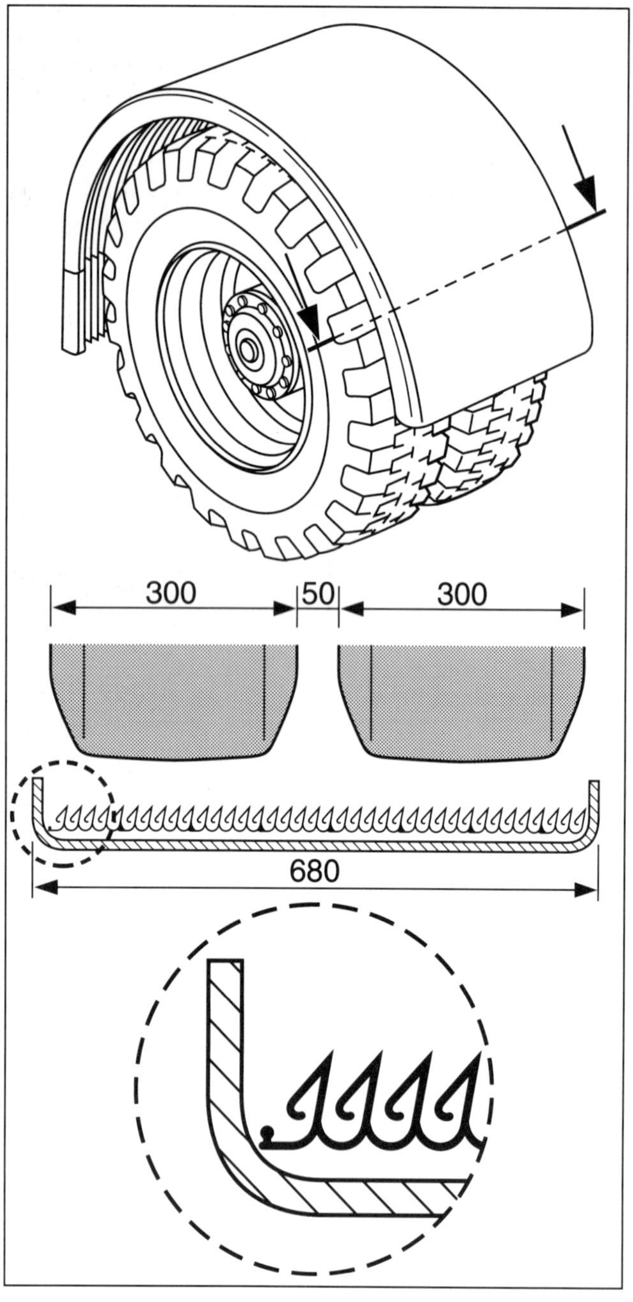

Figure 12: Cross-section of mudguard with grooved section

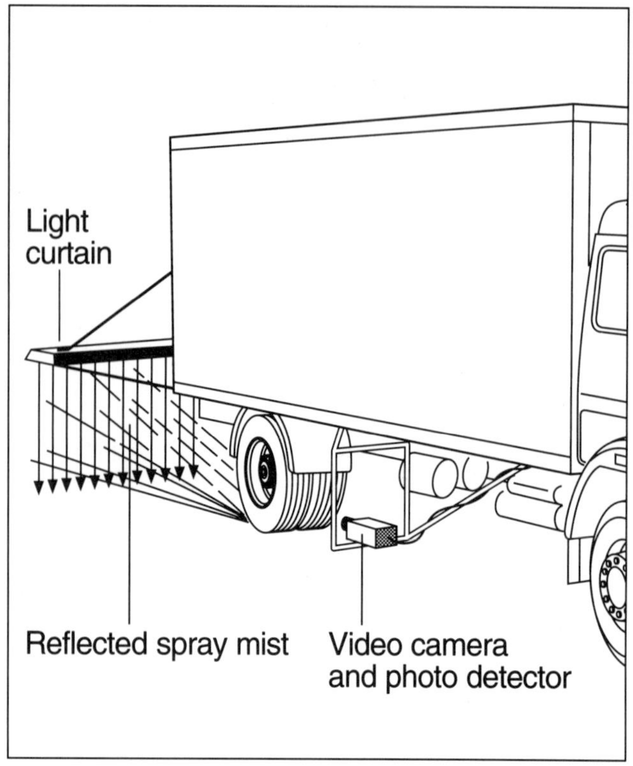

Figure 13: Test setup on commercial vehicle

COMPARATIVE MEASUREMENTS AND ASSESSMENT OF DIFFERENT SYSTEMS

The principle of mist scattered light measurement was applied to determine the quantity of spray mist produced by the vehicle. The test setup, Figure 13, also took into account the water which flows out of the wheel arches to the side. Further assessment criteria were the impairment of the vision of a car driver behind or overtaking, the subjective observations on the passing truck and the optical size of the spray mist trail formed in the light curtain, Figure 14.

DIFFERENT SPRAY PROTECTION CONFIGURATIONS - For testing, the vehicle has been equipped completely with the relevant variants at all wheels. I.e., the measurement results are the mean value from six measurement runs, the water stretch being crossed from both directions. The subjective observations confirm the measurement results; Table 1.

VEHICLE SPEED - As the results of the mist scatter light measurement show the formation of spray mist increases significantly as vehicle speed rises. The formation of spray mist is more than doubled when speed increases from 60 to 80 km/h. On the basis of the present standard state significant improvements are possible depending on vehicle speed (80 and 60 km/h, respectively).

GROOVED SECTION - The fairing of the mudguards with the grooved section reduces spray mist by 36% and 50% respectively. For the most part, it is the splash water which is swirled within the wheel arch and then flows out to the side, which is collected and specifically channeled toward the center of the vehicle.
Extending the grooved section with a splash flap to 50mm above the road surface additionally keeps the water wedge, which is ejected to the rear by the wheel, flat and reduces the spray mist in total by 60% and 73% respectively, compared to the standard state.

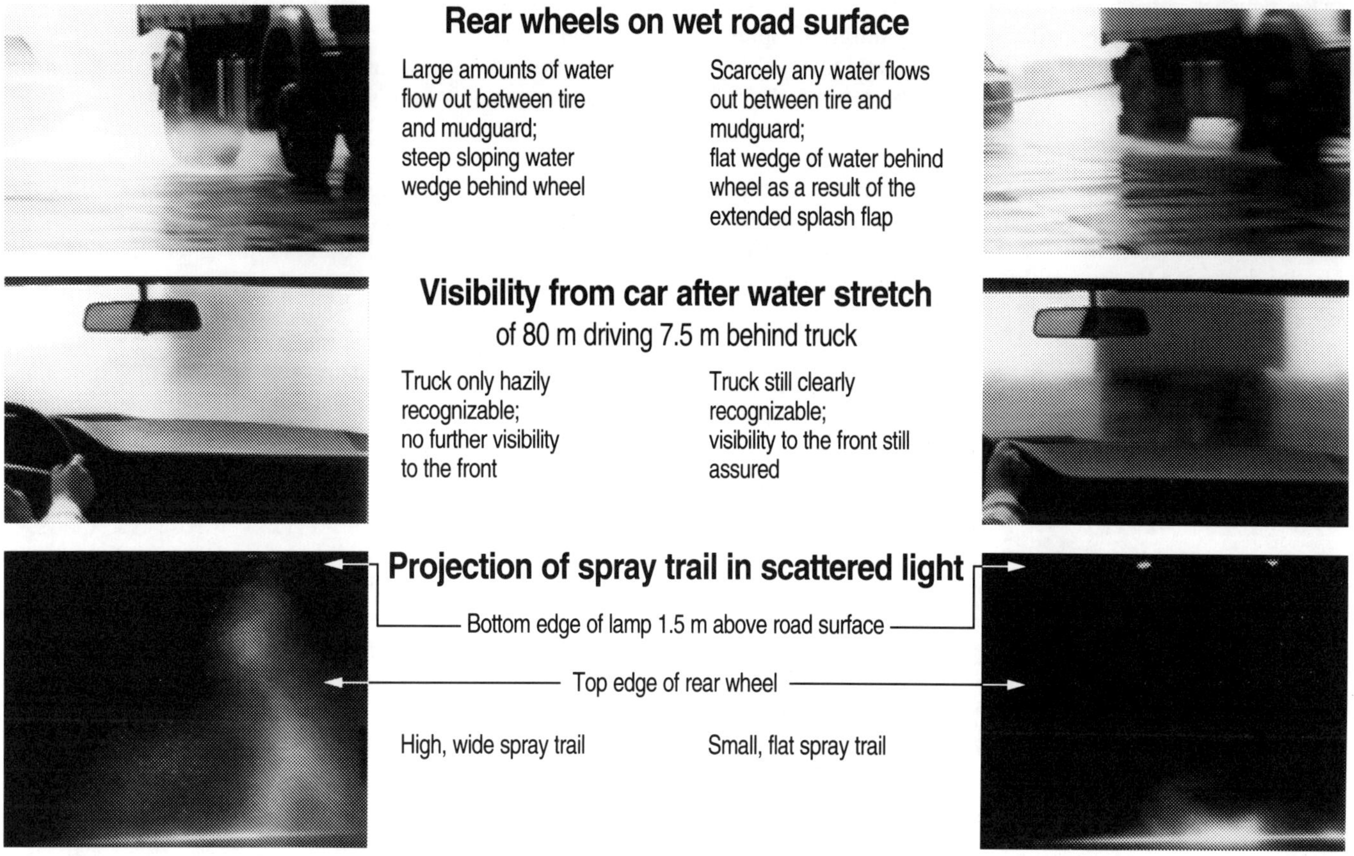

Figure 14: Comparison of standard mudguard (left) and grooved section version (right)

Variant	Scattered light intensity at			
	60 km/h absolute (V)	relative (%)	80 km/h absolute (V)	relative (%)
Standard mudguard without mud flap at rear axle	0.30	100	0.52	100
Standard mudguard with newly developed grooved channel profile	0.15	50	0.33	64
Standard grooved channel mudguard with longer mud flap (up to 50 mm above road surface)	0.10	27	0.21	40
EEC mudguard with water absorber	0.32	107	0.70	135
Standard mudguard with brushes, air/water separator	0.13	43	0.39	75
Full fairing	0.07	23	0.12	23

Table 1: Test Results of Variants with the Scattered Light Intensity Method

SEALING OF THE CLEAR SPACE - Sealing of the clear space between wheel and wheel arch likewise makes it possible to significantly reduce the formation of spray mist. The brushes fitted radially to the mudguard skirt produced an improvement of 57% and 25% respectively. The relatively slight effect achieved at 80 km/h is attributable to the fact that the brushes are greatly fanned open by the air stream as vehicle speed increases and thus allow more splash water to again flow out at the sides. In addition, it was found that the effectiveness of the brushes rapidly decreased the higher the mileage, and that they did not prove satisfactory in the road endurance test.

Condition after driving in muddy terrain

Condition after driving on wet road surface afterwards over a distance of 1 km

Figure 15: Wheelhouse fairing with grooved section; demonstration of self-cleaning effect

FULL-FAIRING - As was to be expected, the best results were achieved with the full fairing. With this variant, even if somewhat extreme, it was possible to reduce the formation of spray mist by 77%, respectively, to a minimum even at high speeds.

EEC MUDGUARDS - The EEC mudguards were significantly worse than the present standard state as the large clearance between wheel and wheel arch allows significantly more splash water to flow out at the side.
The measurement revealed a deterioration by 7-10% at 60km/h and 33-42% at 80km/h. Practical trials did not demonstrate any, and measurements scarcely any, reduction in spray mist resulting from the so called "absorber mats" as splash flaps of the same shape and size with a smooth inner surface produced practically identical results as obtained with absorber mats.

DETAILS IN PHOTOGRAPHY - Figure 15 shows a demonstration of self-cleaning effect. Figure 16 shows the rear mudguard design. Figure 17 represents the system of air/water separator. Brushes are fitted radially to the mudguard skirt. Figure 18 shows the full fairing. It represents the maximum in splash and spray suppression which is achievable.

Figure 16: Rear mudguard: The fairing also performs the function of the splash flap. Clearance from road surface to bottom edge of channel: 160 mm (vehicle unladen).

Figure 17: Wheel covers with air/water separator

Figure 18: Truck with full fairing

PREVENTING SPRAY TRAILS ON CARS

The grooved channel mudguard fairing proved just as satisfactory when fitted to a car as to a truck. Excellent improvements are achieved if the wheel fairing is divided into two, in such a way that the area behind the wheel (splash water catch unit) can be adjusted vertically and extends down to about 30 mm from the road surface. This necessitates a control to instantly detect and compensate for spring compression and rebound, body roll and driving over obstacles and potholes.

Sensors measure the moisture or wetness on the road, the vehicle speed, the temperature of the outside air and the compression and rebound of the running gear.
On roads with very irregular surfaces, the distance to the road surface is increased to approx. 50 mm.
The splash water catch unit behind the wheel is retracted and therefore not visible in the following states: dry road surface, reversing, driving forward at $\leq$ 30 km/h and air temperatures below -3° C.

Figure 19: New fender design inside

Figure 20: Pneumatic elements to move the rear part up and down close the road surface

Figure 19 shows the inside of the right front fender with grooved section. The rear part is movable up and down with pneumatic elements as shown in Figure 20. Figure 21 shows the standard version compared with the variant fitted with a grooved channel profile. The intensity of the water spray is approx. 5 times lower. The local division of the light curtain into nine fields with the same number of sensors at the level of the camera film, allows an accurate analysis and pin-pointed development of the water spray procedures to be carried out. Figure 22 is a comparison of mist vortices behind a car. Above is standard and below is modification. Figure 23 shows the overall view of modified car.

SUMMARY

All tests which were carried out at completely calm conditions, took place on an airfield which is no longer in use, without the influence of other vehicles. The level airfield ground features a slight downhill gradient of 3% at right angles to the direction of travel. By spraying water on the ground over a length of 80metres, a constant, irregular water film with a thickness between 0.5 and 1.5mm was formed; the irregularities were caused by the rough surfacing of the airfield ground.

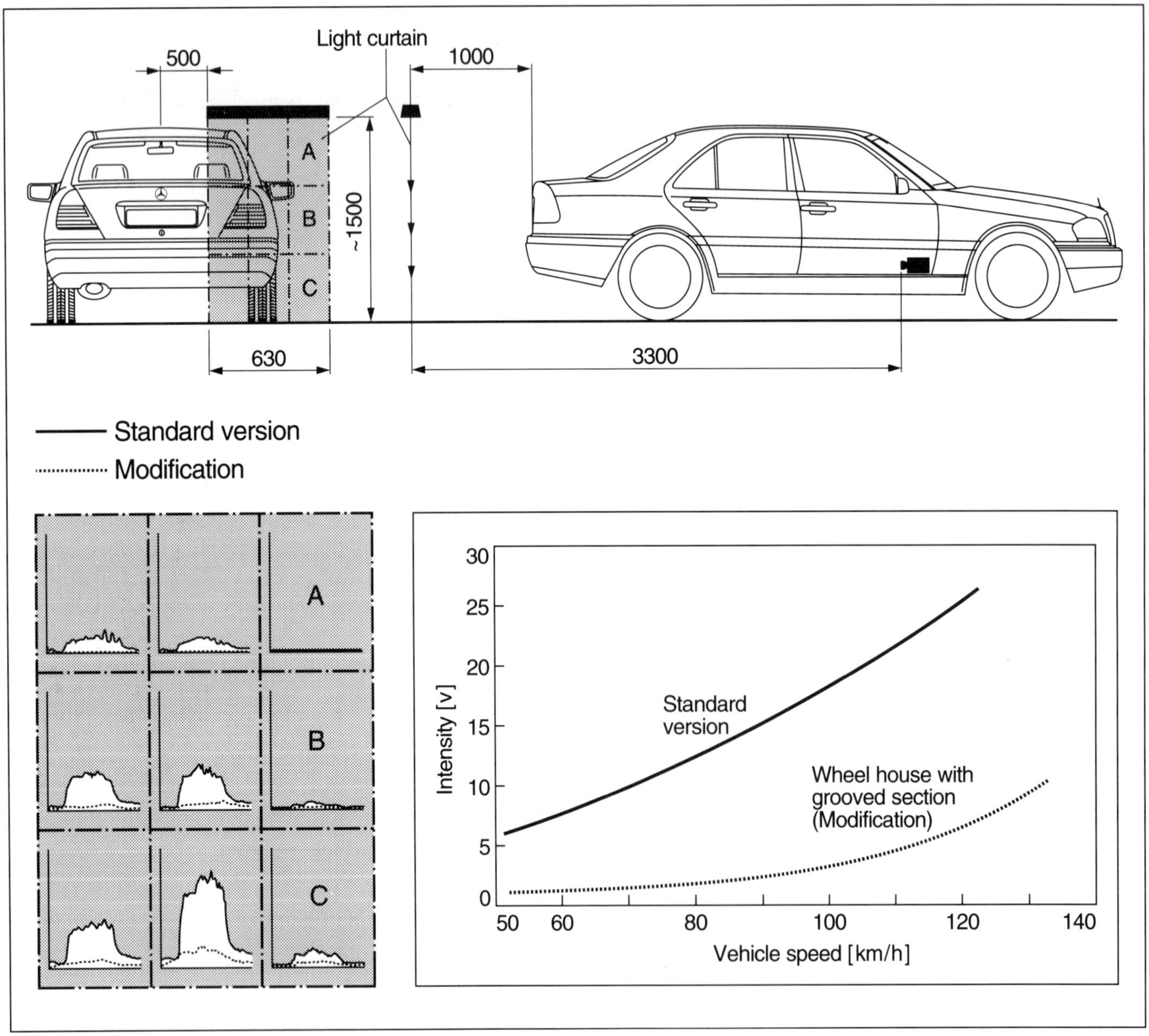

Figure 21: Spray trail intensity of a passenger car

The vehicles used for the tests were an 18-ton truck with platform and tarpaulin and a C-class sedan. The tire make and vehicle body remained unchanged through all the tests. Improvements in the area of the wheelhouse were investigated exclusively.

The subjective obversations confirm the measurement results and lead to the following CONCLUSIONS FOR TRUCKS - The current state of production can be considerably improved. The grooved channel profile extremely reduces the water spray. Sealing the free space between wheel and wheelhouse, e.g. with brushes has good effects at speeds of up to 60km/h. At a vehicle speed of 80km/h the brushes fan out owing to the head-on wind, and an increased amount of water spray emanates. Wheel covers meeting the minimum requirements of the EEC directive have a poorer effect than the current production state. Absorber mats hardly have any water spray reducing effect in practical application.

CONCLUSION FOR CARS - Reducing the water spray on cars increases safety since the car can be recognized earlier by other traffic participants. In addition it has the pleasant side effect that the car gets less dirty.

The grooved channel profile proves its excellent efficiency also on cars.

Figure 22: Mist vortices behind a car

The water spray, however, is extremely reduced if a vertically moveable grooved channel profile, which looks like a commercially available anti-spray mud flap in extended condition, can be kept at the smallest possible distance from the road surface.

Extended grooved mudguard flap

Retracted mudguard flap

Figure 23: Modified car

REFERENCES

1. Patent document DE 41 35 930 C1

2. Hucho, W.H.: Aerodynamics of Road Vehicles, chapter 8, Butterworths & Co. Ltd, 1987

3. Hucho, W.H.: Aerodynamik des Automobils, Kapitel 8, VDI-Verlag, 1994

Crosswind Sensitivity of Passenger Cars and the Influence of Chassis and Aerodynamic Properties on Driver Preferences

C. C. MacAdam and M. W. Sayers
The University of Michigan, Transportation Research Institute

J. D. Pointer and M. Gleason
The Chrysler Motors Corporation Aerodynamics Department

Reproduced, with permission, from Swets & Zeitlinger Publishers
publication: Vehicle Systems Dynamics (1990)

SUMMARY

Results of vehicle crosswind research involving both full-scale driver-vehicle tests and associated analyses are presented. The paper focuses on experimental crosswind testing of several different vehicle configurations and a group of seven drivers. A test procedure, which utilized wind-generating fans arranged in alternating directions to provide a crosswind "gauntlet", is introduced and described. Driver preferences for certain basic chassis and aerodynamic properties are demonstrated and linked to elementary system responses measured during the crosswind gauntlet tests. Based on these experimental findings and confirming analytical results, a two-stage vehicle design process is then recommended for predicting and analyzing the crosswind sensitivity of a particular vehicle or new design.

1. INTRODUCTION

This paper is based on recent findings from a vehicle aerodynamics research project [1] sponsored by the Chrysler Motors Corporation at the University of Michigan Transportation Research Institute. The general thrust of that research was directed at the crosswind sensitivity of passenger cars, and specifically, the influence and interaction of chassis characteristics and aerodynamic properties on driver preferences. The key elements considered in that study are outlined in Figure 1 and included: (A) vehicle dynamics considerations (e..g., weight distribution, tire, and suspension characteristics), (B) vehicle aerodynamic properties, (C) steering system characteristics (most notably steering system compliance, friction and torque assists), and (D) driver closed-loop steering behavior and preferences obtained from experimental crosswind tests. This paper focuses on the experimental crosswind testing conducted during the project using several different vehicle configurations and a group of seven drivers. It reports on driver preferences for certain basic chassis and aerodynamic properties and demonstrates a linkage to elementary system responses measured during those crosswind tests.

The paper begins with an examination of previous research findings related to crosswind sensitivity of passenger cars. A computer model, developed under this

* The University of Michigan, Transportation Research Institute and
** The Chrysler Motors Corporation Aerodynamics Department

research program, is then briefly described. The use of that model at different stages of the research helped to identify and probe certain vehicle-related mechanisms identified as possibly significant contributors to the issue of crosswind sensitivity. As will be seen, the ability of the model to predict basic dynamic behavior patterns, observed in experimental measurements of driver-vehicle systems operating in crosswinds, is an important factor for recommending it as a tool within the vehicle design process. The basis of the conclusions of this paper, however, rest upon experimental measurements and evaluations of a group of seven test drivers operating seven distinctly different vehicle configurations during nearly identical crosswind conditions. The crosswind tests were conducted using a set of eight fans (arranged in an alternating direction over a length of several hundred feet) to approximate a random-like crosswind "gauntlet" driven repeatedly by each driver. On-board measurements of the driver-vehicle responses, combined with the subjective evaluations of the test drivers during these tests, form the basis of the paper's conclusions.

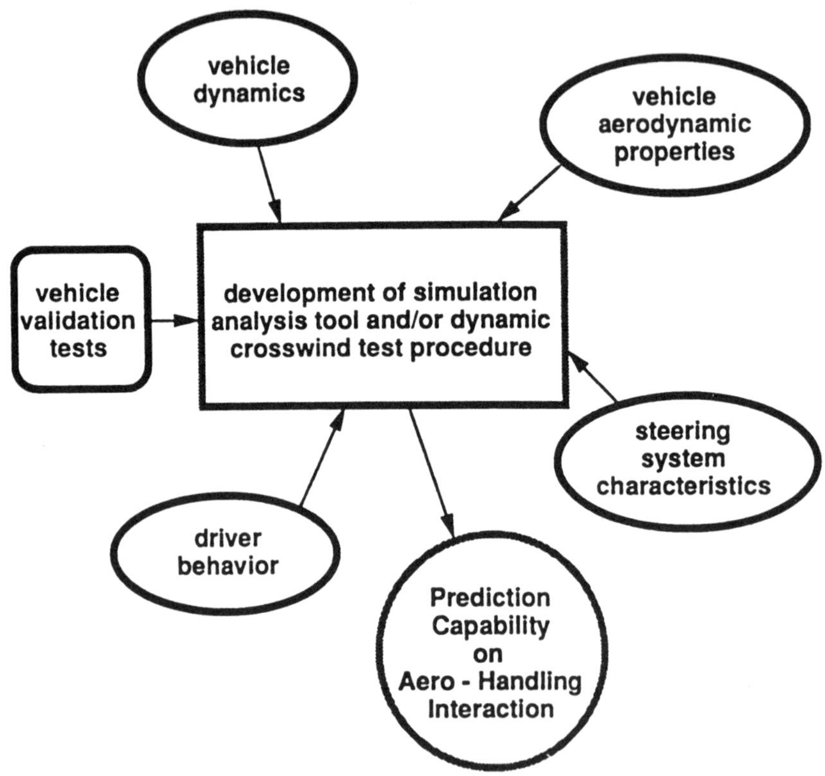

Fig. 1. Key Elements in Vehicle Crosswind Stability.

2. PREVIOUS RESEARCH

The trends during the last decade or so to lighter, more fuel efficient cars in response to changing energy policies, combined with more recent trends toward higher performance passenger cars, have led to increased interest in aerodynamic styling as a means for providing low drag configurations and for mitigating any high-speed crosswind sensitivities. In many cases, attempts at streamlining passenger cars for minimizing drag have led to unwanted increases in crosswind sensitivity. As noted in such comprehensive texts such as Hucho [2] and Scibor-Rylski [3], this tradeoff was observed as early as 1933 by Kamm [4] out of which arose the well-known truncated rear-end design ("Kamm-back") which helped to offset much of the the crosswind susceptibility introduced from streamlining. More recent observations, such as Kohri and Kataoka [5] or Hucho [2], have contributed to improved understandings on the subtle influences relating to A- and C-pillar styling designs and their importance in affecting crosswind sensitivity of passenger cars. Studies such as Noguchi [6] have also noted the importance of certain suspension properties (e.g., roll steer and lateral force compliances) as elements not to be discounted when considering a vehicle's crosswind sensitivity. Other studies relating to crosswind sensitivity of passenger cars have been compiled in such documents as Kobayashi and Kitoh [7], which was primarily concerned with literature related to the crosswind sensitivity of light weight cars.

Analytical Work

Numerous formulations aimed at simplified identifications of the crosswind sensitivity of passenger cars have been offered in the technical literature. In 1974 at the FISITA Congress, Watari et al [8] offered a crosswind sensitivity formulation based upon the steady-state lateral acceleration response of a vehicle to a constant aerodynamic side force. This paper is an example of many studies that have noted the importance of the neutral steer point and its relationship to the mass center and aerodynamic center of pressure in the question of crosswind sensitivity. (Very often in the literature, researchers have puzzlingly identified the mass center as the key point of reference in crosswind sensitivity considerations, suggesting that aerodynamic side forces acting at the mass center, instead of the neutral steer point, would produce zero steady-state turning responses.) In 1985, Bucheim et al [9], proposed a formula for predicting the crosswind sensitivity based upon the initial transient response of a vehicle to a step input of crosswind and its subsequent steady-state turning response. However, that formulation is largely dominated by the steady-state response of the vehicle and, in effect, is similar in many applications to what others have proposed based upon steady turning analyses. The static turning analysis method offered by MacAdam [10] also identified the importance of the relative locations of the neutral steer point, the mass center, and the aerodynamic center of pressure in addressing the crosswind sensitivity problem. (That method is further extended in this paper to explicitly include vehicle aerodynamic properties within a formulation that relates a vehicle's static turning response to a constant crosswind input.)

More comprehensive analyses of vehicle crosswind sensitivity have also been undertaken which utilized computer simulation to analyze the time history responses of driver-vehicle systems operating under natural, or random-like, crosswind conditions . For example, in 1985 Uffelmann [11] argued for the need to include driver steering behavior under natural crosswind conditions to more accurately evaluate the "active" crosswind sensitivity of the combined driver-vehicle system. Uffelmann outlined an example of how chassis-related and aerodynamic-related properties might be traded off to help maintain a similar level of crosswind sensitivity for the total driver-vehicle system. In 1989, MacAdam [12] conducted a simulation-based study involving systematic parameter variations of a validated computer model by Sayers et al [13] that included detailed representations of the vehicle dynamics, aerodynamic properties, steering system characteristics, and driver steering behavior. That study utilized step-wind inputs to characterize the "passive" wind sensitivity of the vehicle alone, and then, contrasted those results with "active" crosswind sensitivities obtained from driver-regulated maneuvers during simulated random crosswind conditions. A crosswind handling index, involving root mean square values of vehicle yaw rate and driver steering power, was proposed as a candidate method for more precisely identifying crosswind sensitivities of driver-vehicle systems during dynamic side wind conditions.

Full-Scale Testing

Numerous full-scale test programs have also been conducted which attempted to identify and illustrate influences of chassis and aerodynamic properties on vehicle crosswind sensitivity. Many of these full-scale studies have employed crosswind fan generator facilities, most of which are located in Europe and Japan. An example of a simple test track survey of eleven European passenger cars appeared in *Auto Motor and Sport* in 1985. Each vehicle was driven past a German fan facility (with a fixed steering wheel) while lateral path and yaw rate responses deriving from the side wind were recorded to illustrate the degree of variation in the selected sample of cars. In 1975, Barter [14] conducted crosswind tests with fans at the MIRA facility in England as part of a doctoral thesis, primarily to obtain disturbance inputs to support the system identification work that was the main thrust of his thesis. While Barter noted that the short duration and quality of the crosswind tests were largely insufficient for his system identification purposes, it was observed that closed-loop regulation tests did result in a "rather more violent" system response than the more conventional fixed steering wheel drive-bys also conducted. Such amplifications for the closed-loop driver-vehicle system were presumably associated with driver steering behavior and driver lags.

Bundorf et al [15] in 1963 reported on a small hydrogen peroxide rocket engine system used to study the passive response of a station wagon to lateral force inputs at various fore-aft locations. The lateral disturbance input from the rocket engine was intended to approximate an aerodynamic disturbance force.

Different vehicle responses were observed with the the rocket motor located at several different longitudinal positions. The authors noted that minimal steady-state yaw disturbance was imparted to the vehicle when the disturbance was applied at the vehicle's neutral steer point, and, that minimal lateral displacement (immediately following the force application) was imparted to the vehicle when the disturbance force was applied slightly aft of the neutral steer point.

In 1980, Klein and Hogue [16] reported on findings of a full-scale test program involving five distinctly different U.S. vehicles. Crosswind tests were conducted with a newly-developed crosswind fan facility [17]. (These same fans, and the only ones available in the U.S., were also utilized in the test work appearing in this paper.) During that study, open-loop (fixed steering wheel) and closed-loop (driver regulated) tests of each vehicle were conducted at forward speeds of 50 mph. Crosswind fan speeds were nominally 35 mph. Based upon these test conditions, Klein and Hogue concluded that "Passenger cars, pickup trucks, and most vans do not have a directional control crosswind sensitivity problem." However, as part of that test program, a strong correlation was observed between increased yaw rate levels and driver steering wheel angles and poorer subjective ratings by the test drivers during the crosswind tests. Since many of these tests were conducted at fairly large aerodynamic sideslip angles, the authors also noted the importance of accounting for the nonlinear aerodynamic influences under such sideslip conditions that usually result in a rearward shifting of the center of pressure, and an in turn, an eventual reduction in the crosswind yaw disturbance.

In 1987 van den Hemel et al [18] conducted similar tests with a group of 15 drivers at vehicle speeds of 100 km/h. A European crosswind fan facility was utilized which provided aerodynamic slip angles of 20 degrees at such test speeds. The authors noted difficulty in discriminating between vehicle configurations because of the short crosswind duration provided by the fans, the magnitude of crosswinds, and the relatively modest level of differences in aerodynamic properties of the vehicles tested. However, van den Hemel et al did note that yaw rate and lateral acceleration appeared to be the most useful measurements as discriminators of different vehicle configurations during these tests.

Lastly, Willumeit et al [19], also in 1987, reported on a sequence of cross wind driver-vehicle tests conducted under natural crosswind conditions along a North Sea coast motorway in 1985. Five drivers and five test vehicles were used. (Initial closed-loop tests were also conducted at a European crosswind fan facility, but later proved to be unsatisfactory because of the short duration of crosswind input and driver anticipatory behavior to the fan disturbance.) Willumeit's study indicated that the passive (non-driver, fixed steering wheel) vehicle response to crosswinds does not "fully correlate with driver's impressions of side wind sensitivity." The study carefully analyzed a number of driver-vehicle responses recorded during normal driving under random crosswind conditions and observed that yaw velocity, lateral acceleration, and driver steering wheel angular displacement were vehicle/driver responses most perceived and utilized by drivers in evaluating the crosswind sensitivity of vehicles. Also observed as important were vehicle yaw angle, steering wheel velocity, vehicle roll angle, and driver steering wheel input power. Smaller levels of all of these responses correlated with better subjective ratings. Vehicle lateral velocity and displacement were observed to be of secondary importance. Opposite and less significant correlations were identified for vehicle yaw acceleration.

Wind Measurement Technology

In addition to the studies cited above, a number of technological advancements in on-board wind measurement capabilities have also occurred in the last two decades which have helped to promote further understanding of the nature of crosswinds acting on vehicles in open areas, as well as in the vicinity of fixed roadway objects and other moving vehicles. Smith in 1972 [20] described a MIRA wind measurement device (utilizing anemometers) and its use in measuring a number of different crosswind profiles. More recently, Tran [21] presented a pressure transducer system for measuring the wind forces and moments acting on a vehicle by recording the pressures at approximately 10 points along the circumference of a vehicle and then combining this information with a uniform flow model. A wind transducer developed by the the Chrysler Corporation aerodynamics department utilizing a strain-gauged sphere in combination with an inertial mass compensating accelerometer, is described by Pointer in [22]. This latter device was also used during the crosswind testing described subsequently in this paper.

A Driving Simulator Study

Lastly, a study conducted by Alexandridis et al [23], utilizing a moving-base simulator, is noteworthy because of the inherent ability of such devices to carefully control test conditions and systematically vary vehicle parameters without the normal accompaniment of test track background noise. Using both step gusts and random crosswind inputs to the simulator (for 12 test subjects and 12 vehicle configurations), the authors observed that driver subjective ratings favoring rearward aerodynamic center-of-pressure locations were statistically significant. At a simulator vehicle speed of 60 mph, lane deviations decreased as the aerodynamic center of pressure was moved rearward, or, as steering control sensitivity was increased. For one very forward center-of-pressure location (at the front axle), lane deviations were observed to decrease with an increased level of understeer. Similar results were not observed for more rearward center-of-pressure locations and increased understeer. Steering wheel activity was seen to be reduced for increased understeer and control sensitivity and also for rearward center-of-pressure locations. In general, Alexandridis et al noted that the aerodynamic center-of-pressure location (ranging from the front axle to as much as 37% behind the front axle) was the most dominant factor when compared with the level of variations in understeer (3 and 5 deg/g) and control sensitivity (0.8 and 1.2 g/100 deg) examined.

3. A DRIVER-VEHICLE CROSSWIND MODEL

A computer model for predicting the interaction between vehicle and driver during crosswind conditions is introduced and outlined briefly in this section. Its use as an advanced tool, that can be used in place of, or in conjunction with, certain types of dynamic random crosswind test procedures, is subsequently being recommended as part of a total crosswind sensitivity design process. The model was developed during this research and was used to help identify and separate different mechanisms of the driver-vehicle system contributing significantly to the crosswind sensitivity of the system. An example of its use in understanding and interpreting the experimental findings from this study will be demonstrated subsequently during the discussion of findings in Section 5.

Figure 2 shows a block diagram outlining the principal components of the crosswind driver/vehicle model. The technical details of the computer model can be found in Sayers et al [13]. The four primary components of the model are described briefly in the following.

Vehicle Model

The vehicle model is characterized by five degrees of freedom for the sprung mass, constant forward speed, and massless suspension/wheel assemblies. Tire and suspension compliances are also included. Tire lateral force is treated as largely linear except for cornering stiffness dependency on vertical load. The basic dynamics of the vehicle are very similar to that developed by Segel [24]. High speed test data, collected during the course of the aerodynamic crosswind stability research program at UMTRI [1], were used to validate the baseline model behavior through direct comparisons with model predictions. Test track handling measurements and aerodynamic wind tunnel measurements of a baseline passenger car, in several different aerodynamic configurations, were conducted at nominal speeds of 50 and 100 mph to validate the model. As part of the model validation process, a stable platform and a variety of transducers were used to measure all body motions. Steering wheel displacement and torque measurements, front wheel rotations, and other steering system functions (power boost pressures, pitman arm motion, and tie rod forces) were also recorded and utilized in the model validation.

Aerodynamics

A linear-regime aerodynamics model is employed and includes the three conventional forces (drag, side force, and lift) and the three corresponding moments acting about each body axis. Each force and moment is characterized by a constant coefficient and a linear or quadratic variation with respect to aerodynamic yaw angle. The aerodynamics used in the computer model for examining different vehicle configurations were based upon aerodynamic measurements performed by Chrysler at the Lockheed wind tunnel in Atlanta, Georgia. The nature of wind

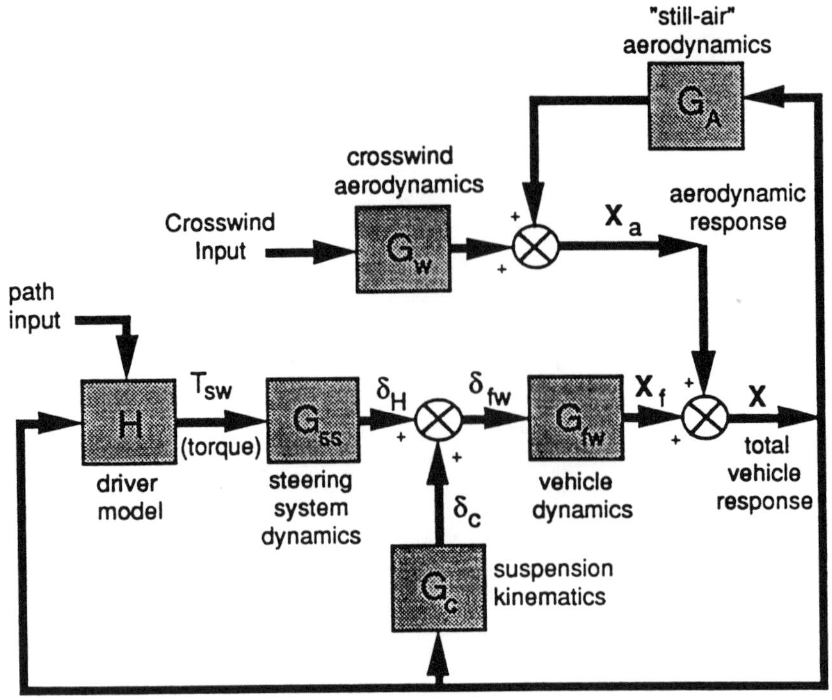

δ_H front wheel angle contribution from steering command and compliance

δ_c front wheel angle contribution from suspension kinematics

δ_{fw} total front wheel steer angle

X_f non-aerodynamic vehicle response

Fig. 2. Chrysler/UMTRI Crosswind Vehicle Model.

tunnel data collected under this work is analogous to traditional, previously published passenger car aerodynamic data, such as seen in Hogue [25].

Driver Characteristics

An existing driver steering control model by MacAdam [26, 27] was extended within this study to provide for generation of steering wheel torque, T_{sw}, seen in Figure 2 as the output from the driver model block. Steering wheel torque produced by the driver model served as the primary control input to the dynamic steering system model during most of the simulated closed-loop driving scenarios studied. It also (in combination with steering wheel angular displacement) provided a means for calculating energy expenditures for simulated drivers during either discrete steering maneuvers or simple path regulation tasks as demonstrated in [12].

Steering System

A simplified second order dynamic steering system model, as seen in Figure 3, augments the aforementioned sprung mass dynamics and relates steering wheel torque and displacement to front wheel motion. Properties of the steering system include: compliance, friction, steering wheel inertia, lash, and power steering boost characteristics. The basic steering system model is patterned after that originally proposed by Segel [24, 28]. However, the high frequency dynamics of the front wheels included by Segel were replaced here by a quasi-static treatment that included rotational motion of the front wheels through a lash element. A power steering boost torque approximation was also added. The power steering boost was treated as a driver-assisting torque, linearly proportional to the sum of the front tire aligning torques.

Figure 4, from [12], shows an example comparison for the steering system model and test data collected at 100 mph during which low frequency sinusoidal-like steering torque inputs were applied by a test driver for purposes of validating the baseline vehicle model and its steering system properties. This type of steering input is similar to that proposed by Norman [29] as part of an on-center handling test procedure (normally conducted at 100 kmh) for evaluating basic properties of the steering system and associated vehicle responses. Other example uses of the model for predicting total system responses during dynamic

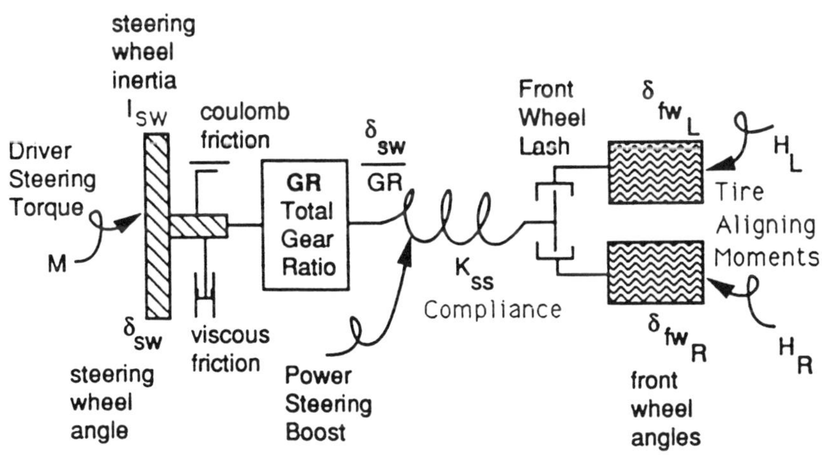

Fig. 3. Steering System Model.

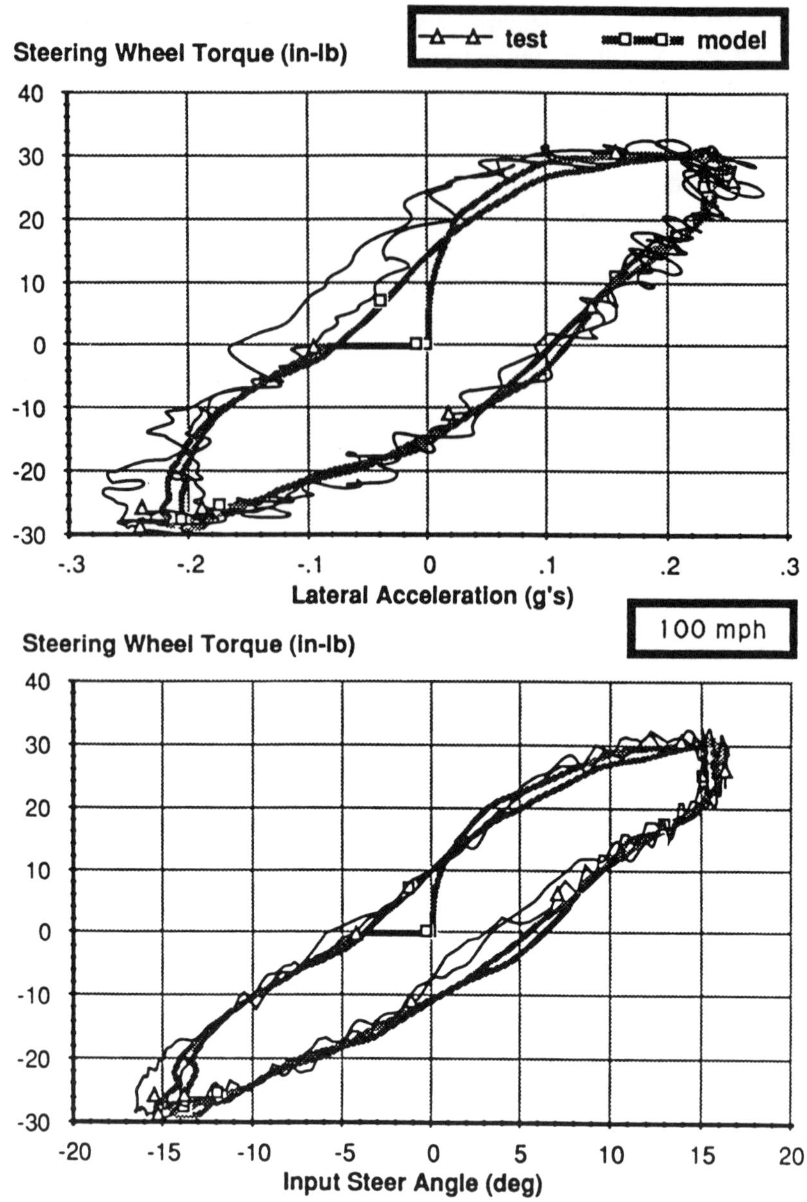

Fig. 4. Steering System Measurements and Model Predictons.

crosswind conditions appear in [12]. Its use as a tool in explaining certain findings from the crosswind tests is also illustrated by the example calculation appearing in Section 5.

4. EXPERIMENTAL CROSSWIND MEASUREMENTS

The results of experimental crosswind tests of driver-vehicle systems conducted during the research program are presented in this section. All full-scale testing was performed on the vehicle dynamics facility at the Chrysler Proving Grounds in Chelsea, Michigan. The vehicle crosswind testing results were obtained with the use of eight U.S. Government-owned wind generating fans described in detail by Klein and Jex in [17].

The use of fan-generated crosswinds in this research program was based upon several needs. One need was to obtain crosswinds of sufficient magnitude that the test drivers would be given a definite subjective impression of the different vehicle configurations being driven, and, that would also produce a system response that could be readily measured by on-board instrumentation. Test speeds of 90-100 mph were selected so that the aerodynamic inputs provided by sizeable crosswinds (25 mph or so) would still permit a linear-regime characterization of the vehicle aerodynamics. A second need was to have available a generally repeatable set of crosswind conditions so that different drivers could be exposed to more or less the same crosswind experiences at widely varying times during the test program. And finally, there was the practical need to be able to schedule test drivers on a regular basis and be guaranteed that sufficient crosswind test conditions would be available.

Against this back-drop of perceived needs was the clear observation, well reported in the literature, that fan-generated crosswinds were generally insufficient for obtaining reliable subjective evaluations from test drivers. Since nearly all of the previous uses of fans for subjective evaluations were for short duration drive-by scenarios, in which drivers attempted to regulate the lateral path in response to a short-term pulse of crosswind, a different approach was considered for this research program. It was decided that the traditional, closely-grouped fan arrangement, which provides a short-duration pulse of crosswind, would be used only to evaluate the passive (non-driver) vehicle response. A new arrangement of the fans, in the form of a crosswind "gauntlet" course, would be used, instead, to expose the test drivers to the fan-generated crosswinds over a longer period of time for collecting their subjective evaluations.

Vehicle Configurations

A total of seven different vehicle configurations were used. Of the seven configurations, one served as the nominal baseline vehicle configuration. Two additional configurations provided variations of aerodynamic properties, and two other configurations were used to vary certain chassis properties. The last two configurations consisted of simultaneous combinations of aerodynamic and chassis variations. Table 1 lists the particular vehicle configurations studied.

As noted in Table 1, Configuration #1 corresponds to the baseline vehicle. Configurations #2 and #3 represent significant fore-aft modifications of the vehicle aerodynamic center of pressure, CP. Configuration #2 moves the CP forward

Table 1. Vehicle Configurations Examined in Test Program.

Configuration	Description
1	Baseline Vehicle (see Figure 5 for specific information)
2	Aerodynamic CP Moved Forward 10 inches from Baseline
3	Aerodynamic CP Moved Rearward 30 inches from Baseline
4	320 lb Ballast Load Added to Rear Trunk of Baseline Vehicle
5	Baseline Vehicle with Roll Stiffness Reduced 35%
6	Configuration #2 with Roll Stiffness Reduced 35%
7	Configuration #2, with 320 lb Ballast, and Reduced Roll Stiffness

10 inches from the baseline position, while Configuration #3 moves the CI rearward approximately 30 inches from the baseline configuration. Configuratior #4 corresponds to the baseline vehicle with 320 lb of ballast weight added to the vehicle trunk resulting in a 10 inch rearward shift of the vehicle's mass center. I also produced an estimated 22% increase in the vehicle's yaw and pitch moment: of inertia and 11% increase in weight. Configuration #5, due to disconnection o: the front suspension roll bar, corresponds to a roll stiffness reduction from the baseline vehicle of approximately 35%. The last two configurations involvec simultaneous aerodynamic and chassis modifications. Configuration #6 represents the combined effects of the forward CP location (configuration #2) and the reduced roll stiffness variation (configuration #5). Lastly, configuration # 7 corresponds to a worst case scenario, combining the forward CP location (#2), with a rearward mass center (#4), and a reduced roll stiffness (#5). Numerical values for several of the key parameters are seen in Figure 5 which furthei describes the baseline vehicle and each of the alternate configurations.

All of the aerodynamic modifications were accomplished through the use of small vertical fins designed by the Chrysler aerodynamics department and mounted on the vehicle with light-weight tubular framing at distances of approximately 5 feet fore or aft of the vehicle ends. Full-scale wind tunnel measurements of the fin-equipped vehicle configurations were also performed in the Lockheed wind tunnel to identify their exact aerodynamic properties. The primary intent of the fins was to move the aerodynamic CP of the baseline vehicle (no fins) forward and rearward by significant distances by altering the aerodynamic yaw moment, without also significantly affecting the remaining aerodynamic properties (such as lateral force) of the baseline vehicle. The subsequent wind tunnel measurements verified the intended effects. Table 2 list the aerodynamic force and moment coefficients obtained from the Lockheed wind tunnel measurements. The data shown here apply for yaw angles as large as 20 degrees. The reference area for these measurements was 21 ft^2. All measurements are with respect to a point located on the vehicle center-line, at the mid-wheelbase position, and in the ground-plane. (It should be noted that the data seen here correspond to a research test vehicle and are not necessarily indicative of production vehicles manufactured by the Chrysler Corporation or other automotive manufacturers.)

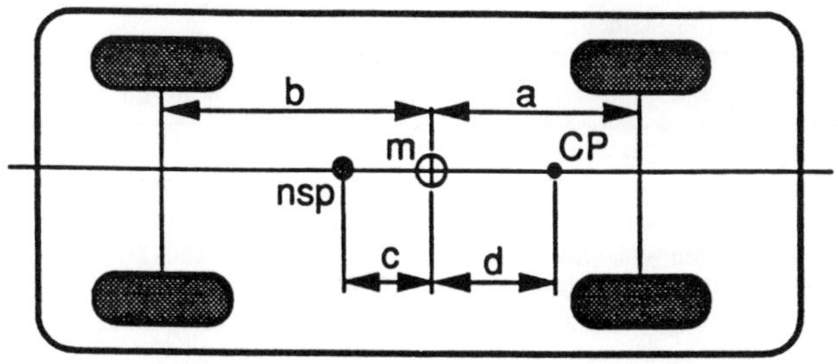

a distance from mass center to front axle
b distance from mass center to rear axle
nsp neutral steer point ("center of tire forces")
m mass center
CP aerodynamic center of pressure
c distance from nsp forward to mass center
d distance from mass center forward to CP

Config #	a	b	c	d	Weight	Roll Stiffness
	(in)	(in)	(in)	(in)	(lb)	
1	37.4	59.6	18.7	19.8	3160	baseline value
2	37.4	59.6	18.7	29.8	3160	baseline value
3	37.4	59.6	18.7	-9.7	3160	baseline value
4	47.4	49.6	14.0	29.8	3480	baseline value
5	37.4	59.6	22.2	19.8	3160	65% baseline
6	37.4	59.6	22.2	29.8	3160	65% baseline
7	47.4	49.6	18.2	39.8	3480	65% baseline

Fig. 5. Description of Test Vehicle Configurations.

At speeds of 100 mph, the aerodynamic lift force and pitch moment data seen in Table 2 indicate a relatively small lifting force (e.g., 135 lb at zero yaw) as a percentage of vertical load that acts almost entirely at the front axle. Consequently, most of the aerodynamic inputs affecting the test vehicle with regard to crosswind sensitivity issues, are being generated by yaw-plane forces and moments.

Table 2. Aerodynamic Wind Tunnel Data.

Aerodynamic Coefficients	Config #1	Config #2	Config #3
(variations are per degree or degree^2)			
Linear side force variation with yaw	0.0326	0.0353	0.0408
Linear yaw moment variation with yaw	00104	0.0149	0.0006
Linear roll moment variation with yaw	0.0087	0.0092	0.0117
Constant drag force coefficient at zero yaw	0.359	0.359	0.362
Quadratic drag force variation with yaw	0.000302	0.000309	0.000295
Constant lift coefficient at zero yaw	0.253	0.249	0.252
Quadratic lift force variation with yaw	0.000655	0.000670	0.000687
Constant pitch moment coefficient at 0 yaw	0.0939	0.102	0.106
Quadratic pitch moment variation with yaw	0.000106	0.000092	0.000096

Steady turning tests at 50 mph were conducted for both the baseline vehicle and the rear ballast configuration to measure their respective understeer gradients. Interestingly, even though configuration #4 represents a significant 10 inch rearward shift of the mass center, the understeer gradient of the loaded test vehicle did not change significantly from the baseline condition. This was a result of the test vehicle's tires having a significant sensitivity of cornering stiffness to vertical load. These tests indicated approximate understeer gradients of 3.2 deg/g for the baseline vehicle and 2.7 deg/g for the loaded configuration. The net effect of the rear loading, therefore, was to produce a rearward shift of not only the vehicle's mass center, but its neutral steer point as well. The amount of rearward shift for the neutral steer point varied, depending upon the loading and roll stiffness (as noted previously in Figure 5).

Test Procedures

Two basic test maneuvers were used to evaluate the response of the various driver-vehicle configurations. The first maneuver was a fixed steering wheel drive-by of a constant pulse of fan-generated crosswind. It was used to primarily characterize and verify differences in the passive crosswind behavior of the different vehicle configurations. The second maneuver, serving to evaluate the active (driver and vehicle) crosswind system behavior, employed the same fans spread out over a longer distance but arranged in alternating directions. Active steering for path regulation by each driver was required during this latter test. All tests were conducted for vehicle speeds of 90-100 mph. Further descriptions of the these two test procedures follow.

• *Crosswind Pulse*. This maneuver was conduced with the eight fan units grouped together and facing perpendicular to the test track as seen in Figure 6. Fixed steering wheel drive-by tests were then performed for each vehicle configuration to evaluate their respective passive (no regulatory driver steering) wind behav-

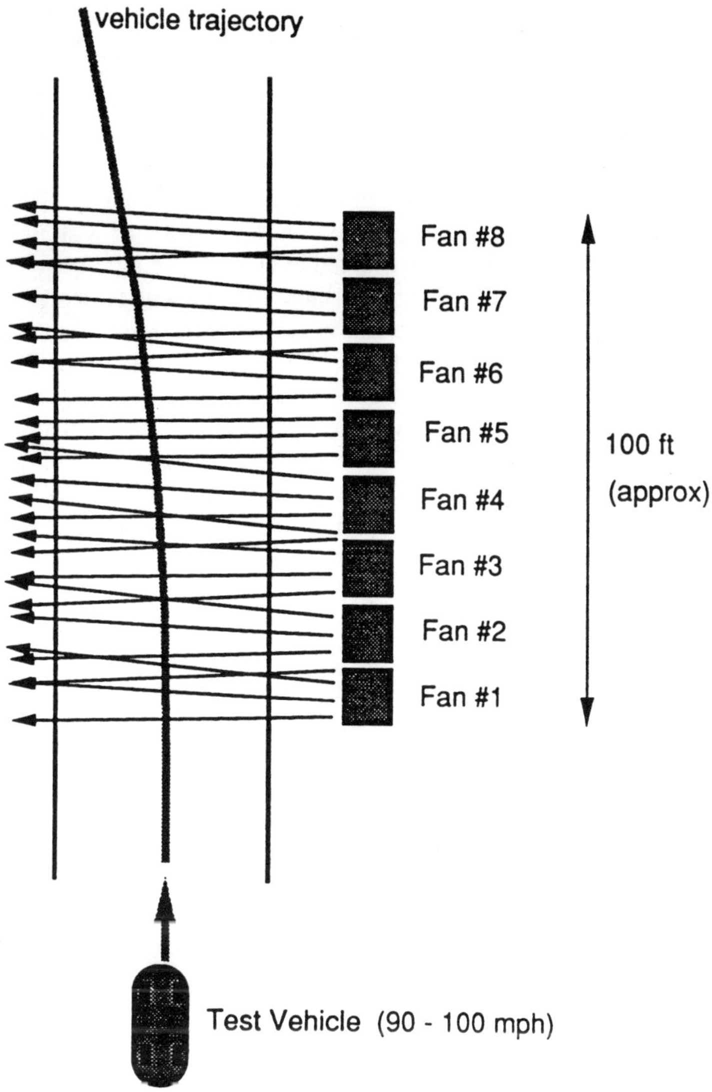

Fig. 6. Crosswind Pulse, Fixed Steering Wheel Test.

ior. Each test vehicle was driven in a straight line at 100 mph from the ambient environment past the fans whereupon it encountered an approximately constant 25 mph crosswind for a period of nearly 0.7 seconds. A pulse-like vehicle response due to entering and then exiting the crosswind stream was recorded by on-board instrumentation. The resulting peak yaw rate and lateral acceleration responses observed for each vehicle configuration were then used to confirm that significant and distinctly measurable differences in the vehicle aerodynamic and chassis configurations were present in the passive crosswind behavior of each vehicle. (The short duration of the crosswind pulse at these speeds did not permit the vehicle to fully establish itself in a steady-state turning condition.)

• *Crosswind "Gauntlet" Maneuver.* For this set of tests, the eight fan units were located, in an alternating manner, along opposite sides of a straight-line test course (single lane width) and distributed over a longitudinal travel distance of approximately 350 feet. See Figure 7. This arrangement presented the driver-vehicle system with a series of fluctuating pulses of crosswind from one side and then the other in a repeating sequence. The spacing between the fans was approximately 52 feet. Wind output from each fan was set at an approximate level of 25 mph and the test course was driven at speeds of 90 to 100 mph. An inner lane width of approximately 8 to 9 feet was defined by a series of traffic cones along the center of the course in order to require each driver to regulate the vehicle path (without undue demand) within those bounds during traversal of the crosswind course. All of the subjective evaluations collected during the test program were obtained from this crosswind test procedure.

Impressions of several drivers who drove past the fans in both arrangements (Crosswind Pulse versus the Crosswind Gauntlet) noted significant differences. The closely grouped pulse arrangement had a very small effect on a driver's subjective and objective response as the crosswind pulse was encountered. The primary inputs to the drivers were reported to be sound (from wind noise) and a mild change in direction as the fans were passed. Drivers also commented that the experience was too brief. In contrast, the driving experience through the crosswind gauntlet generally made a much stronger impression on the test drivers. This was most likely due to the longer length of time of crosswind exposure provided by the gauntlet course and significantly increased driver-vehicle system responses during this test. For the same levels of fan wind output, the gauntlet course produced noticeably amplified system responses compared to those obtained from simple drive-bys of the closely-spaced pulse arrangement. The alternating pulses and their input frequency during the gauntlet test produced a more resonating dynamic response that further contributed to subjective driver impressions. The on-board measurements, as well as simple visual observation of the different vehicle configurations traversing the gauntlet, confirmed the amplifying qualities of the gauntlet test procedure. As an example, in a few of the worst-case vehicle configurations, peak lateral accelerations greater than 0.6 g were recorded at the driver head position and more than 0.45 g's on the stabilized platform. Levels approximately half these would have been recorded in the crosswind pulse test for the same fan output.

Drivers

Seven different drivers were utilized in the test program to provide subjective evaluations of each of the vehicle configurations during the crosswind gauntlet

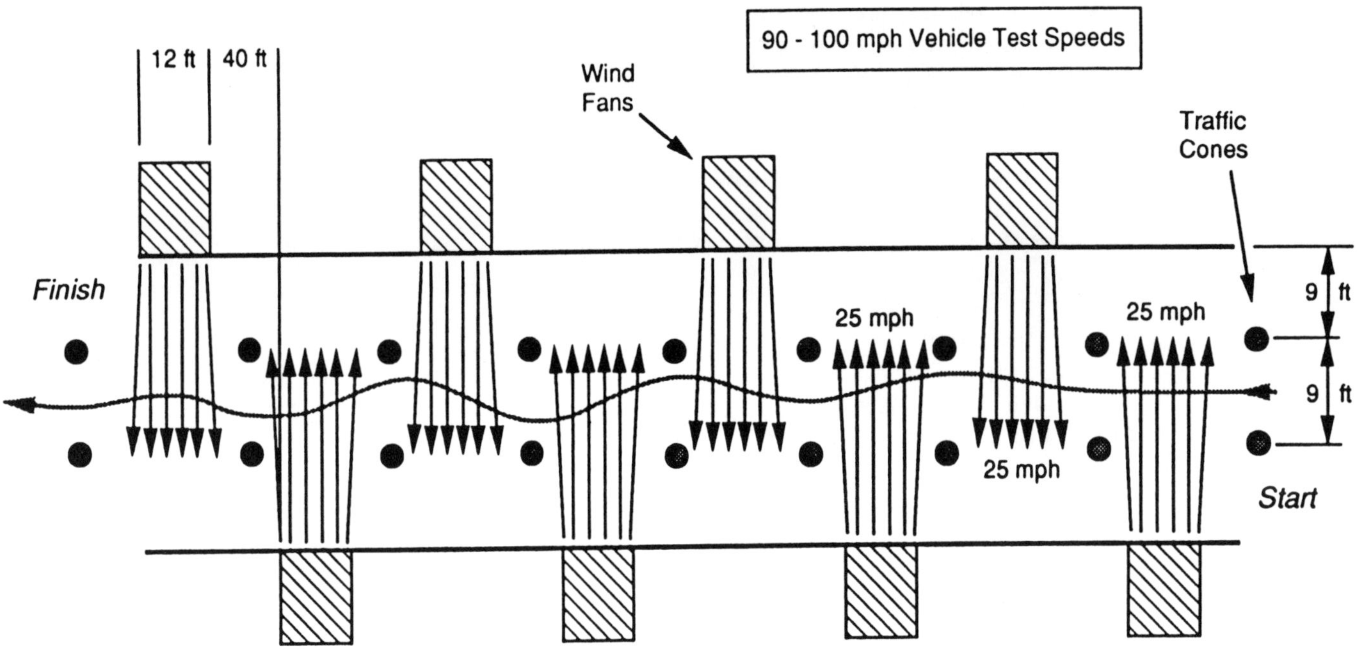

Fig. 7. Crosswind Gauntlet Test Course Used in Study.

tests. Objective measurements of the total driver-vehicle system responses were also collected for four of these test drivers to obtain representative system responses during the gauntlet tests. All of the drivers were males with backgrounds as engineers or technicians ranging in age from 25 to 55. All drivers, but one, were associated to varying degree with the crosswind research program. (Other drivers also participated intermittently during the testing but were not included in these results because of not having driven each of the different vehicle configurations, or, because their chosen speeds fell significantly outside the nominal 90-100 mph range.)

Instrumentation

A stable platform and a variety of transducers were used to measure primary driver-vehicle responses during the gauntlet tests and included: lateral and longitudinal acceleration, yaw rate, roll rate, forward speed, steering wheel displacement and torque. In addition, an accelerometer was mounted laterally in the driver's head-rest to measure accelerations, deriving primarily from lateral vehicle motion and body roll, in the vicinity of the driver's head position. A Chrysler-developed wind transducer [22], utilizing a strain-gauged sphere in combination with an inertial mass compensating accelerometer, was also employed on top of the test vehicle to record the crosswind profiles experienced during the tests. Figure 8 shows example wind profile measurements of aerodynamic sideslip [30] from both test maneuvers at 80 mph using this transducer. The zero offsets seen for each of these examples reflect ambient wind conditions prevailing at the time of these tests.

Crosswind Pulse Measurements

An example test result (fixed steering wheel) illustrating the baseline vehicle response (configuration #1) and the degree of variation provided by the two fore/aft CP variations (configurations #2 and #3) are seen in Figure 9. Peak lateral accelerations, ranging from about 0.11 g's to 0.24 g's, are seen for each of these aerodynamic configurations. Figure 10 shows a similar comparison between the baseline vehicle, the rear ballast configuration (#4), and the worst case configuration (#7, rear ballast and forward CP). The peak lateral accelerations seen here range from about 0.15 g's to 0.25 g's. (The data seen in Figures 9 and 10 were collected on two different days, with different ambient wind conditions.)

In general, the results from the crosswind pulse tests indicated that distinct and measurable degradations to the baseline vehicle (as reflected by peak lateral acceleration and yaw rate responses) were associated with forward movement of the CP and for rearward shifts in the mass center/neutral steer point due to the rear ballast. In contrast, rearward movement of the CP produced a noted improvement in the vehicle's passive crosswind sensitivity. The effects of reduced roll stiffness were not evaluated with the crosswind pulse test procedure. However, the reduced roll stiffness was estimated to produce a small increase in the vehicle's understeer gradient, and correspondingly, a small decrement in its pas-

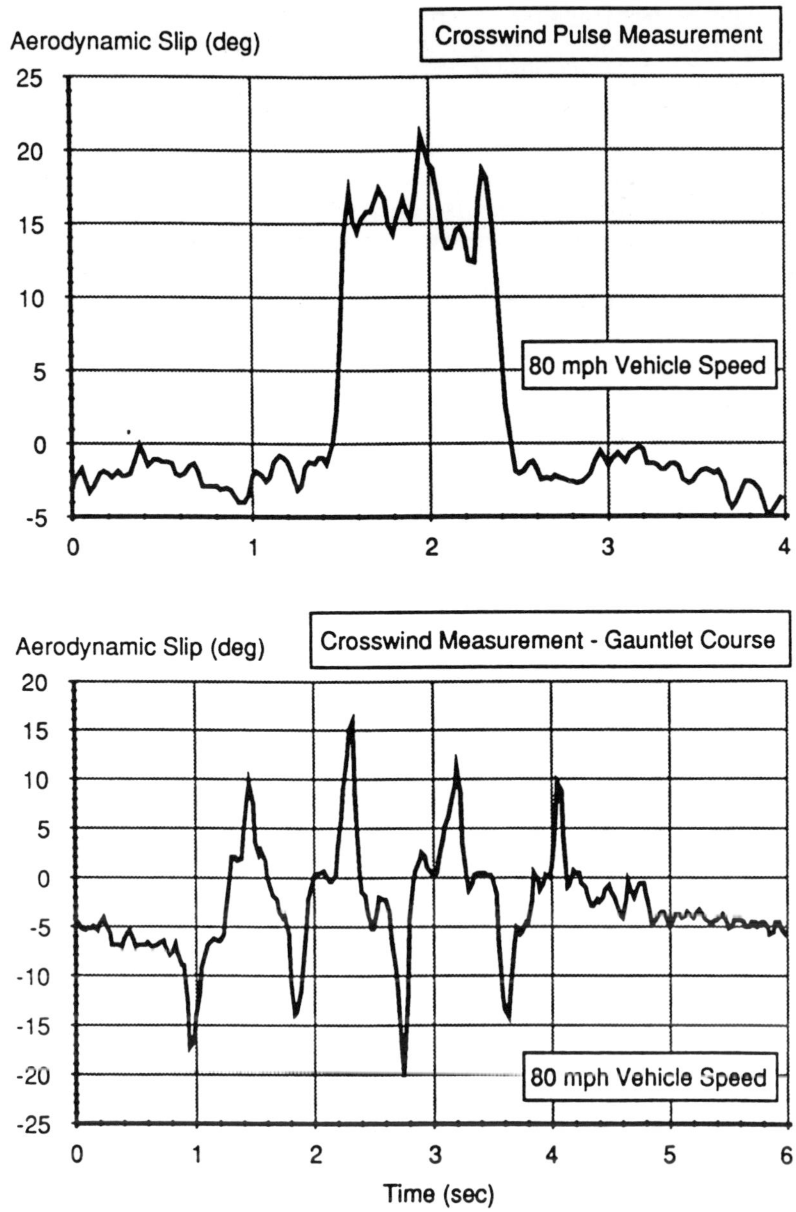

Fig. 8. Example On-Board Measurements of Crosswind Profiles Using the Chrysler Wind Transducer (22).

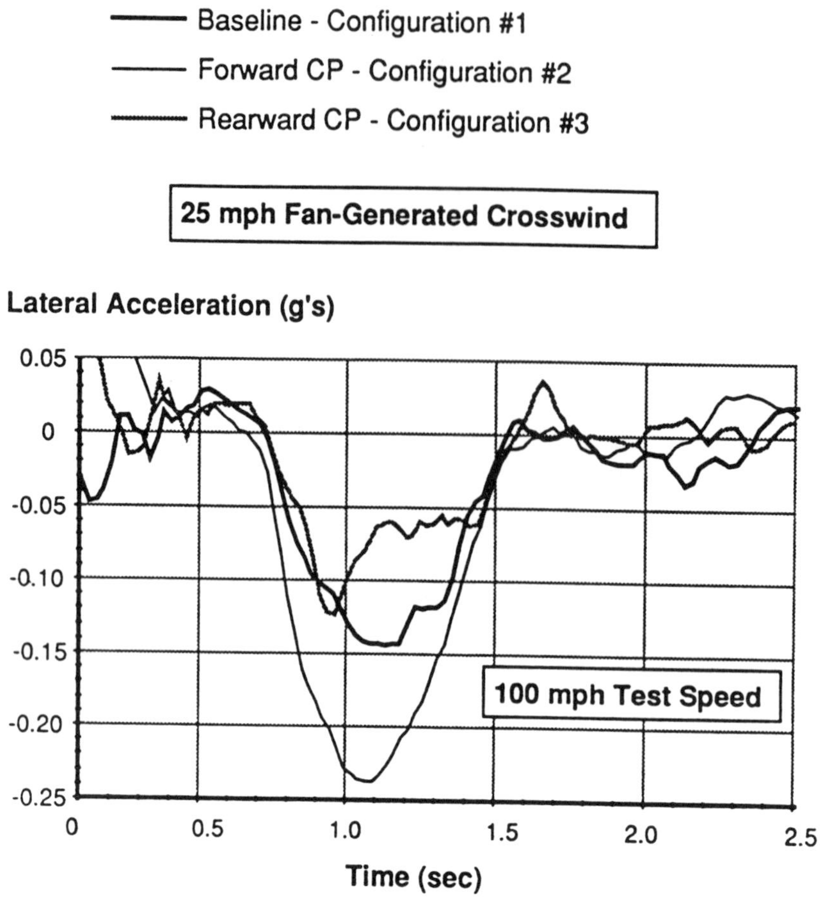

Fig. 9. Example Measurements of Vehicle Response Due to Pulse of Fan-Generated Crosswind. CP Influence.

sive crosswind sensitivity. This particular variation was primarily introduced for the subsequent closed-loop crosswind gauntlet tests in order to obtain driver subjective reactions to increased levels of roll compliance.

Having confirmed that known and measurable differences in passive crosswind sensitivity were being introduced by these vehicle modifications (aerodynamic and loading), the primary test procedure, the crosswind gauntlet, was then undertaken utilizing a variety of drivers.

Crosswind Gauntlet Measurements

To reiterate, the primary purpose of the crosswind gauntlet was to collect driver reactions (in the form of subjective evaluations and representative system response measurements) to these same vehicle modifications during a closed-loop

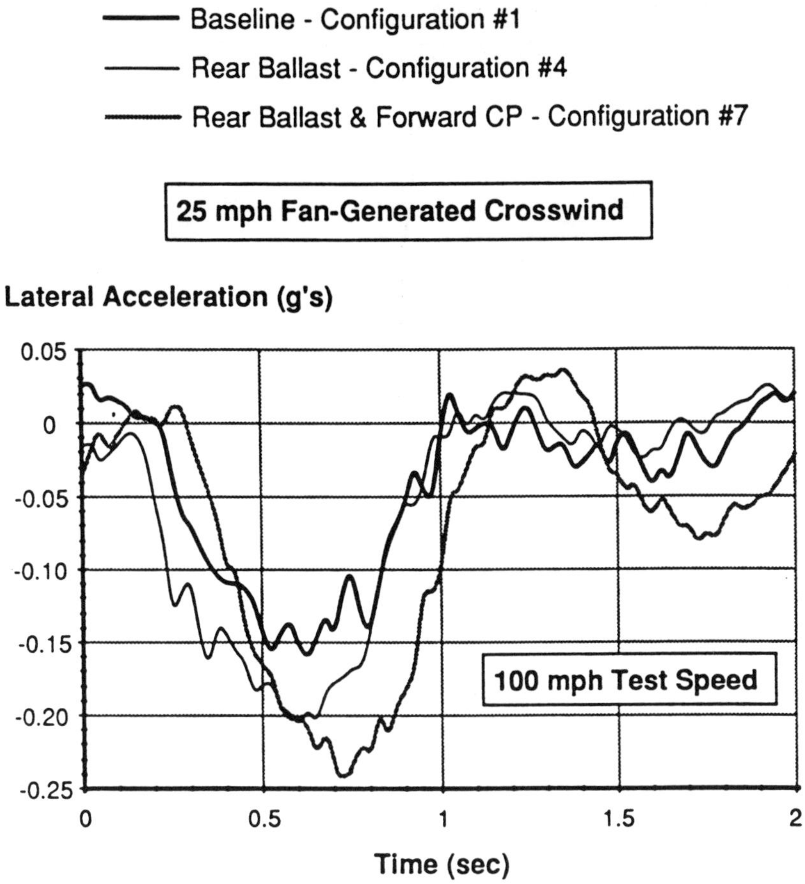

Fig. 10. Example Measurements of Vehicle Response Due to Pulse of Fan-Generated Crosswind. Load Influence.

regulation task in the presence of dynamic, and presumably more realistic, crosswind conditions. Figure 11 shows a representative set of time histories corresponding to the baseline vehicle traversing the gauntlet test course at a nominal speed of 100 mph. Seen here are driver steering wheel displacement, yaw rate, and lateral acceleration time histories from the same test. The sinusoidal-like, qualitative nature of these time histories were similar for each of the drivers. The magnitudes, however, would vary depending upon the degree of steering control elected by each driver in regulating the vehicle's lateral and yaw motion through the gauntlet course.

The primary difference in driver behavior seemed to derive from how tightly a driver elected to control the vehicle. While some steering regulation was re-

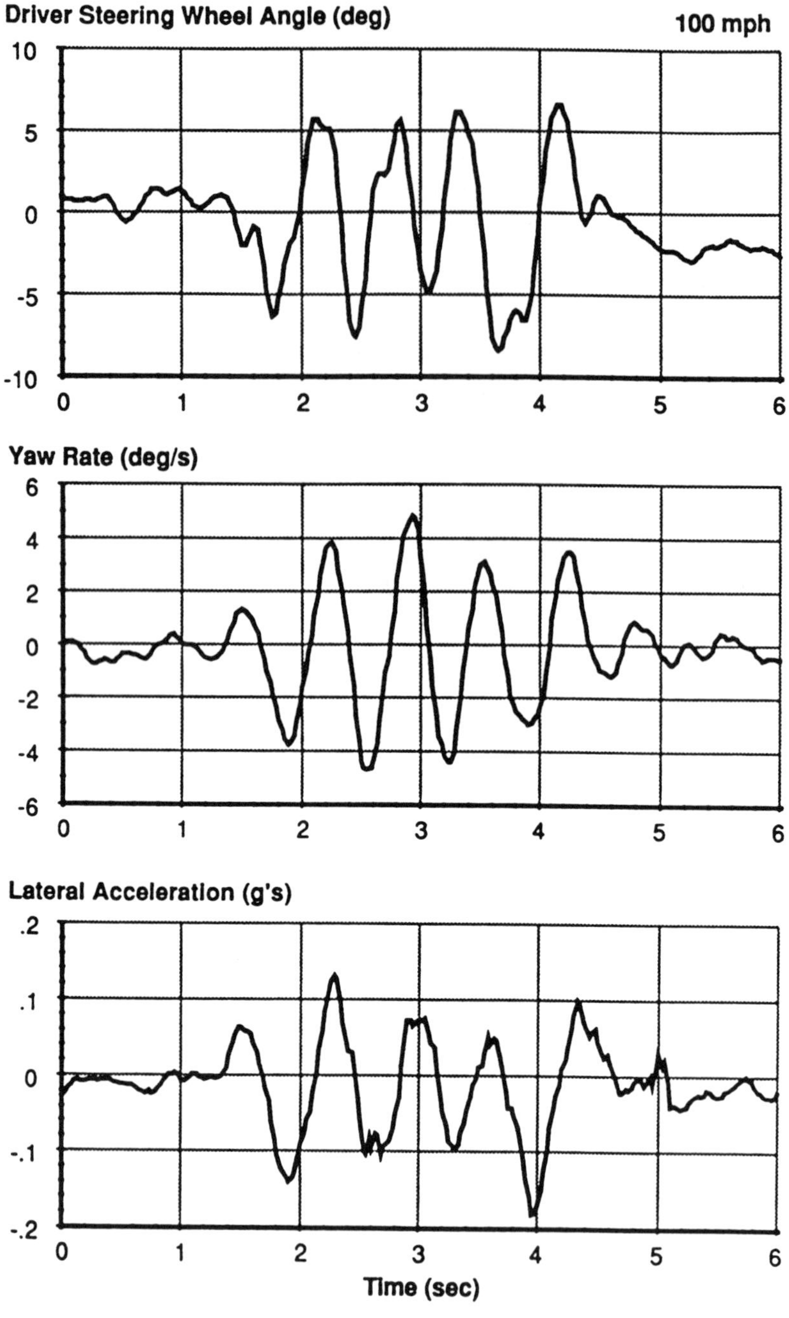

Fig. 11. Example Driver-Vehicle Response Measurement from the Crosswind Gauntlet Course/Baseline Vehicle.

quired to maintain the gauntlet course and avoid cones strikes, driver behavior seemed to separate into two basic types: a) those who chose to provide minimal steering through the course, and b) those who tried to closely regulate the lateral path deviations caused by the crosswind. It was observed that poorer lateral performance and amplified system responses were exhibited by drivers who attempted to closely regulate the vehicle during the crosswind course. This is likely explained by the inherent time lags of most drivers while attempting to respond to relatively high frequency disturbance inputs (as provided here by the crosswind pulses). Those drivers who more or less "went along for the ride" provided smaller corrective steering inputs and focused more on controlling any significant path deviations. Despite such differences in observed driver steering behavior patterns during the crosswind gauntlet, the subjective evaluations of most drivers still generally agreed with one another regarding preferences for different vehicle configurations.

In Figure 12, an overlay of three yaw rate responses (not synchronized in time) are seen to illustrate the degree of variation routinely observed from the

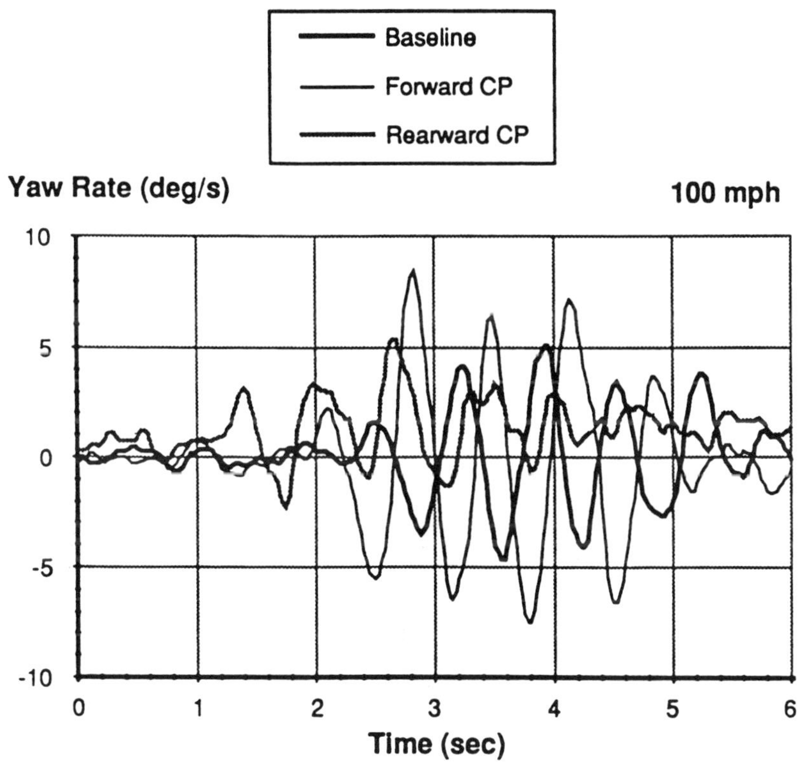

Fig. 12. Example Yaw Rate Measurements from the Crosswind Gauntlet Course.

three different aerodynamic configurations. The three runs seen in Figure 12 are for the same test driver. Again, the absolute magnitudes of system responses such as these may have varied from driver to driver, but the relative scaling from vehicle configuration to configuration tended to remain about the same. For example, if driver A utilized 25% more steering activity than driver B through the gauntlet course to control vehicle configuration #1, it was likely that driver A would also use about 25% more steering control than driver B in controlling vehicle configuration #2.

Testing was conducted for seven drivers, across all seven vehicle configurations, and for three repeats each driver/vehicle configuration. Subjective ratings were obtained from each driver regarding general impressions of controllability and preferences. A 10-point rating scale was used to rank each vehicle configuration with the extremes of the scale implying such descriptions as "uncontrollable" for a value of 10, to "ideal - no noticeable crosswind effect" corresponding to a value of 1. All ratings fell in the range from 3 to 8.5, with 4.8 being the average rating for the baseline configuration in these tests. Figure 13 shows a plot of the driver subjective ratings for the gauntlet tests versus vehicle configuration averaged over all seven drivers. The degree of scatter about each average value is also indicated by the small vertical bars seen in Figure 13.

Objective measurements of driver-vehicle responses noted above were also collected for four of these same tests drivers with an on-board data acquisition system. RMS values of each of these responses were calculated over a time period starting approximately 1 second prior to commencement of the crosswind gauntlet and ending 1 second after the gauntlet traversal. These RMS values were then normalized by the baseline configuration values so that a value of 1 repre-

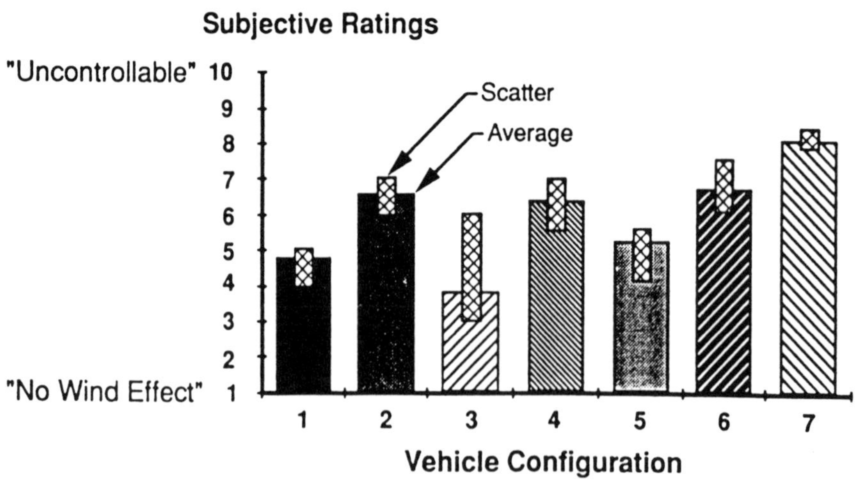

Fig. 13. Subjective Ratings from 7 Drivers - Crosswind Gauntlet Course.

Table 3. Normalized RMS Responses from the Crosswind Gauntlet Tests.

	Vehicle Configuration						
RMS Driver/Vehicle Response	1	2	3	4	5	6	7
Yaw Rate (2.29 deg/s)	1.0	1.56	0.63	1.52	1.13	1.63	2.41
Lateral Acceleration (0.069 g)	1.0	1.52	0.99	1.28	1.07	1.51	2.03
Driver Steering Angle (2.9 deg)	1.0	1.39	0.85	1.28	1.07	1.69	1.76
Driver Steering Torque (6.7 in-lb)	1.0	1.31	1.07	1.15	1.0	1.03	1.81
Head-Rest Acceleration (0.113 g)	1.0	1.57	0.90	1.38	1.11	1.59	2.27

sents the baseline response, values larger than 1 correspond to amplified responses relative to the baseline response, and values less than 1 correspond to responses less than the baseline. Table 3 lists the average RMS values (across 4 drivers and 3 repeats each) of several different driver-vehicle responses and all seven vehicle configurations.

The average, non-normalized baseline responses are also listed in parentheses in Table 3 to gauge the absolute response levels being experienced. In general, *peak* values of these responses are approximately twice the RMS values listed here. For example, an estimate of the average peak value of head-rest acceleration for configuration #7 would be approximately: 0.113 g (Baseline value) x 2.27 (Configuration #7 normalized ratio) x 2 (peak/RMS approximate ratio) = 0.51 g's.

5. FINDINGS

Direct observation of the test results of Section 4 indicate certain driver preferences regarding vehicle aerodynamic properties, weight distribution, and roll compliance as represented by the different vehicle configurations tested. Further analysis of those preferences will be shown to be related to certain basic driver-vehicle responses, which in turn, can be predicted (to first order) by simple static analyses, and more completely, by use of dynamic simulation.

The results of Figure 13 indicate clear driver preferences for more rearward locations of the aerodynamic center-of-pressure (Configurations #1, #2, and #3). Similarly, a clear preference is seen for the more forward mass center location (Configuration #1 versus #4). A less pronounced, but definite preference is also indicated for less roll compliance in Figure 13 (Configurations #1 and #2, versus #5 and 6). Finally, Configuration #7, which combined the detrimental qualities of a forward center-of-pressure, rearward mass center, and increased roll compliance, is given the lowest subjective rating of all vehicle configurations tested.

If a linear regression analysis is performed between the subjective evaluations appearing in Figure 13 and each of the driver-vehicle normalized responses appearing in Table 3, some interesting results are observed. For example, Figure 14 shows the linear regression between subjective ratings and normalized RMS vehicle yaw rate measurements. (The y(x) regression equation and its r^2 correlation is seen in the inset box appearing in Figure 14.) The high degree of correlation between subjective ratings and the measured yaw rate responses is surprisingly strong, suggesting that nearly all of the variation in subjective ratings can be accounted for by the RMS levels of yaw rate alone.

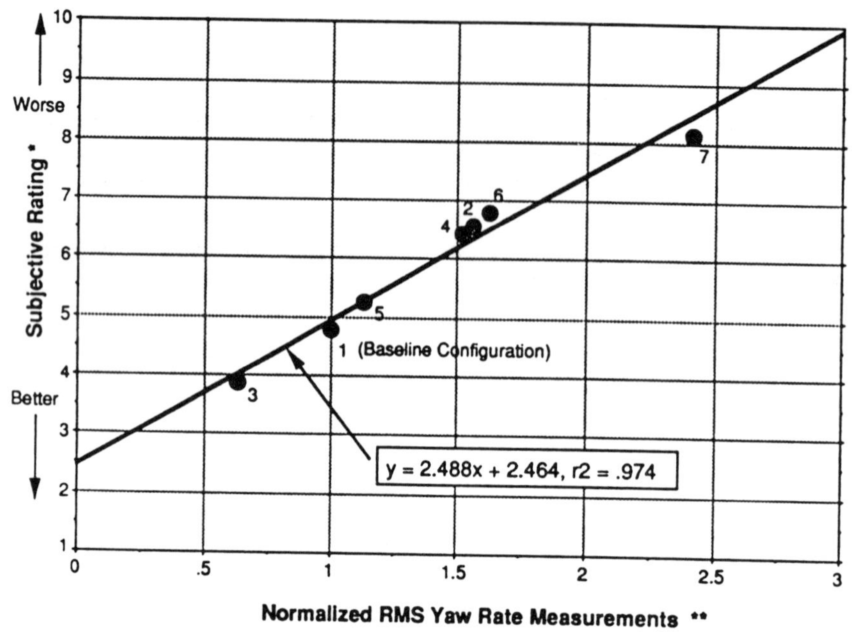

Vehicle Configurations:

1 Baseline
2 Forward CP
3 Rearward CP
4 320 lb Rear Ballast
5 Baseline & Reduced Roll Stiffness
6 Forward CP & Reduced Roll Stiffness
7 Forward CP, Rear Ballast, & Reduced Roll Stiffness

* Average of All 7 Drivers
** Average of 4 Instrumented Drivers

Fig. 14. Correlation between Subjective Ratings and Normalized RMS Yaw Rate Responses during Crosswind Fan Test.

If similar regression analyses are performed, one at a time, on each of the remaining response variables appearing in Table 3, the r^2 correlation coefficients rank the responses in the order listed in Table 4. Also included in Table 4 is a numeric, S, defined as the ratio of the ensemble average for the configuration with the highest RMS level to that configuration with the lowest RMS level. This numeric provides an approximate description of the "spread" in the data and identifies those system responses having greater discriminatory ability or sensitivity to changes in vehicle configuration. It also happens that the numeric, S, ranks the response variables here in the same order as the r^2 correlation coefficient.

The results appearing in Table 4 rank vehicle yaw rate first, with head-rest acceleration (which includes body roll motion effects), driver steering wheel displacement, and vehicle lateral acceleration (stable platform measurement; no direct body roll influence) also as strong correlating responses. Driver steering wheel torque showed relatively poor correlation with the subjective ratings. Since a strong correlation normally exists between most of these same responses in typical steering maneuvers, further interpretation of these results without additional test data and statistical analysis is probably not warranted here. However, it is clear from these results (yaw rate, in particular, and the other measurements representing various types of driver-vehicle responses) that *increased* levels of these motion responses during crosswind conditions, are clearly associated with *poorer* subjective ratings by the test drivers. On the basis of the results seen in Table 4, the indicated rankings would suggest which response variables best predict the subjective ratings of drivers during these specific crosswind gauntlet tests.

Whether, these test results and analyses can be extended to more natural crosswind conditions and other vehicles is a legitimate question. Evidence supporting the generalization of these findings is found in several studies appearing in the technical literature [7, 16, 18, 19] – all of which are in basic agreement on the importance of the indicated responses as key variables related to driver subjective evaluations of crosswind sensitivity. Furthermore, the one study conducted during purely natural crosswind conditions [19] also supports the same basic observation noted here regarding the special significance of yaw rate, lateral acceleration, and driver steering wheel displacement as responses significantly correlated with driver subjective ratings.

Table 4. Rankings of the Response Variables by Degree of Correlation with Driver Subjective Ratings.

Response Variable	Raking	r^2Correlation	S
Yaw Rate	1	0.974	3.83
Head-Rest Acceleration	2	0.925	2.52
Driver Steering Wheel Displacement	3	0.909	2.07
Lateral Acceleration	4	0.890	2.05
Driver Steering Wheel Torque	5	0.561	1.81

Since the RMS yaw rate measurements from the gauntlet tests correlate so well with the driver subjective evaluations of crosswind sensitivity, a pertinent question is to ask how well the steady-state yaw rate responses (due to a constant crosswind and fixed steering wheel) might correlate with the same subjective ratings. Using the dynamic crosswind model [13], steady turning responses, induced by a constant crosswind with a fixed steering wheel, were calculated for each of the test vehicle configurations. A regression analysis, similar to above, was then performed to obtain the relationship between the subjective ratings from the gauntlet tests, and the calculated steady-state yaw rate responses (normalized by the baseline vehicle response), for each of the test vehicle configurations. Figure 15 shows the result of this regression. (The yaw rate response of the baseline vehicle has a gain of 0.122 deg/sec per degree of crosswind). As seen, the calculated steady-state yaw rate responses provide a reasonably good correlation. This would suggest, that even in the absence of test measurements, a *first-order* prediction of a vehicle's subjective rating may be provided by its passive (non-driver, fixed steering) turning response to a constant crosswind (also referred to as a vehicle's "passive crosswind sensitivity").

Static Analysis

Further insight regarding a vehicle's passive wind sensitivity and the basic vehicle parameters that affect it can be seen in the diagram and equation of Figure 16. The contents of this figure are based upon reference [10]. The equation seen in Figure 16, is shown here as Equation (1),

$$r / Fa = [1 / m U] (c + d) / (c + \zeta) \qquad (1)$$

and relates a vehicle's yaw rate response, r, to an applied aerodynamic side force, Fa, in terms of certain fundamental vehicle properties. These fundamental properties are simply the relative locations of a vehicle's mass center, its neutral steer point, the aerodynamic center of pressure, and the tire yaw damping moment arm — all defined by the parameters c, d, and ζ in Figure 16. If the aerodynamic side force, Fa, is expressed in terms of the conventional aerodynamic properties,

$$Fa = q\, C_S\, A\, \alpha \qquad (2)$$

where,

- q is the dynamic pressure = 0.5 x air density x V^2
- V is the magnitude of the relative wind vector
- C_S is the aerodynamic side force coefficient
- A is the cross-sectional reference area

and,

- α is the aerodynamic slip angle (corresponding to a constant crosswind, αV, and forward vehicle speed, U)

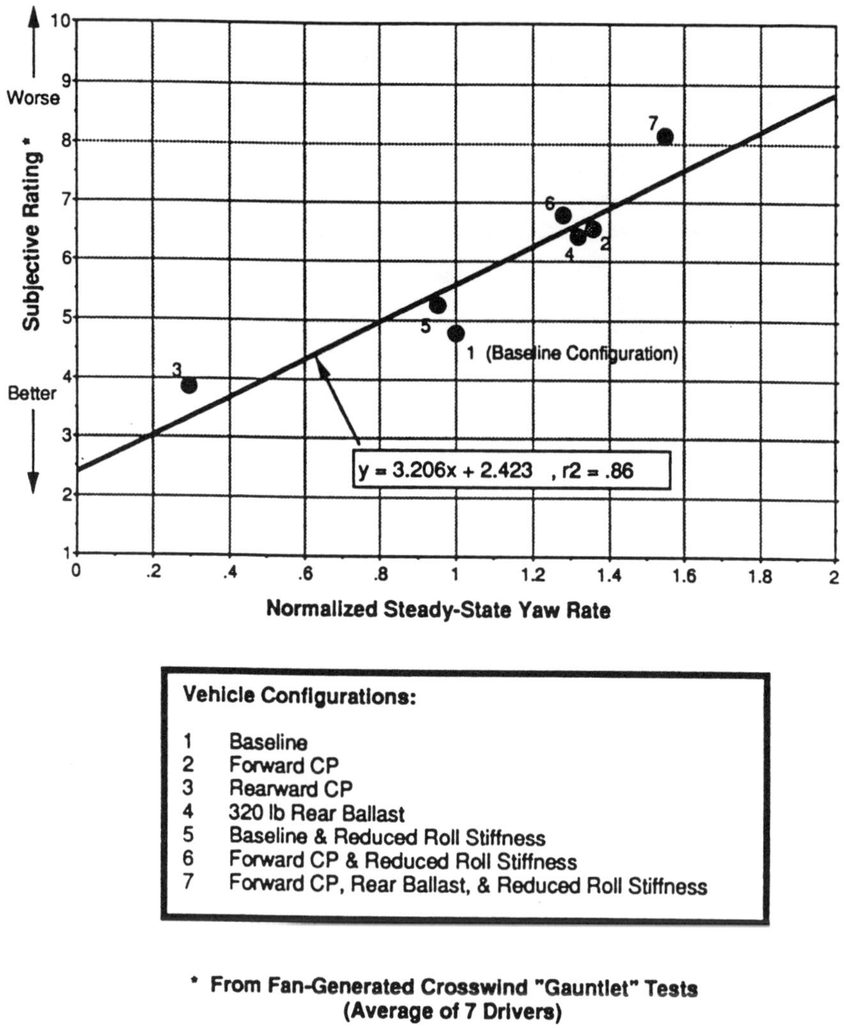

Fig. 15. Correlation Between Subjective Ratings and the Steady-State Yaw Rate Response due to a Constant Crosswind.

Equation (1) can be re-arranged to express the passive wind sensitivity of a vehicle in terms of its yaw rate response *per angle of aerodynamic sideslip*, as given by Equation (3):

$$r / \alpha = [q\, C_S\, A / m\, U]\, (c + d) / (c + \zeta) \qquad (3)$$

Equation (3) predicts a vehicle's static yaw rate response per angle of aerodynamic sideslip (as produced by a constant crosswind of magnitude αV), when travelling at a speed, U. By noting that d is given by $(a - L / 2 + L\, C_{mz} / C_S)$,

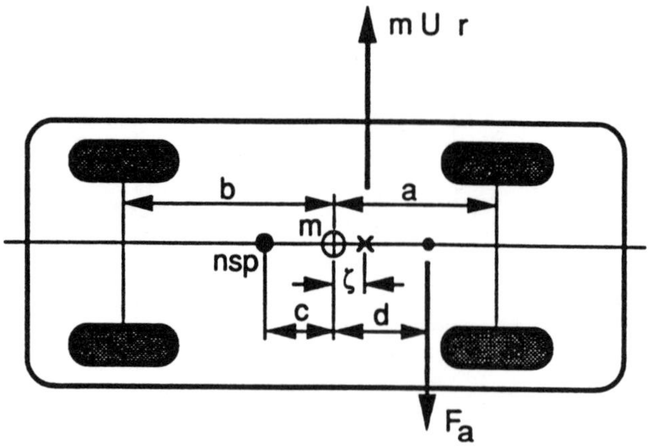

$$\frac{r}{Fa} = \frac{1}{mU}\,\frac{(c+d)}{(c+\zeta)}$$

where,

a	distance from mass center to front axle
b	distance from mass center to rear axle
nsp	neutral steer point ("center of tire forces")
m	vehicle mass
CP	aerodynamic center of pressure
c	distance from nsp forward to mass center $= (b\,C_r - a\,C_f)\ /\ (C_f + C_r)$
d	distance from mass center forward to CP
U	vehicle speed
r	vehicle steady turning yaw rate response
Fa	constant, aerodynamic side force
ζ	moment arm proportional to the tire force yaw damping moment about the nsp $= (a+b)^2\,C_f\,C_r\ /\ [\,(C_f + C_r)\,m\,U^2]$ *
C_f	effective total tire cornering stiffness of front axle
C_r	effective total tire cornering stiffness of rear axle

* e.g., at 100 mph, $\zeta = 3.5$ inches for baseline vehicle configuration with $C_f = 300$ lb/deg and $C_r = 400$ lb/deg

Fig. 16. Static Turning Response of a Vehicle Subject to a Constant Aerodynamic Crosswind Force.

where,

$L = a + b$ is the vehicle wheelbase

and

C_{mz} is the aerodynamic yaw moment coefficient

Equation (3) may be further expanded to express a vehicle's passive crosswind sensitivity completely in terms of its aerodynamic properties,

$$r / \alpha = [q\, C_S\, A / m\, U]\, [\, c + (a - L / 2 + L\, C_{mz} / C_S)\,] / (\, c + \zeta\,) \qquad (4)$$

For either Equation (1), (3), or (4), a principal factor determining the yaw rate response to a crosswind is the ratio of moment arms, $(c + d) / (c + \zeta)$. The distance c locates the neutral steer point and is affected by a vehicle's tire force properties, steering system compliance, suspension kinematics, and mass distribution. The distance, d, between the aerodynamic CP and mass center, is clearly determined by the vehicle aerodynamics and weight distribution. And finally, ζ, the tire yaw damping moment arm, is influenced by tire and weight properties, but primarily by forward speed. Consequently, the opportunity to alter a vehicle's passive crosswind sensitivity can be affected by any number of vehicle and aerodynamic modifications as well as forward speed. Fortunately, the relatively simple formulation provided by Equations (1) or (4) permits one to anticipate and predict, rather easily, how proposed chassis or aerodynamic modifications are likely to influence the passive crosswind sensitivity of a vehicle. For example, Equation (3) or (4) can be used with the numerical values contained in Table 2 and Figures 5 and 16, to calculate the normalized steady turning yaw rate responses appearing in Figure 15. (It should be noted that such analytical calculations normally need to represent the effects of steering system compliance and suspension properties, bounce-camber and bounce-steer, by means of *equivalent* tire cornering stiffnesses which, in turn, are reflected in the moment arms c and ζ.)

Dynamic Analysis

While the static turning analyses noted above are useful for understanding and estimating a vehicle's passive wind sensitivity, the reliance upon such properties *alone* for predicting driver subjective evaluations can be misleading in certain cases. For example, configurations #5 and #6, corresponding to increased levels of front suspension roll compliance relative to those seen in configurations #1 and #2, will normally introduce more understeer, and thereby produce a decreased crosswind turning response under these *static* conditions. This would then result in a prediction of improved subjective ratings under this scheme. However, under *dynamic* wind conditions, increased roll compliance was seen in the test data to produce a poorer driver subjective rating due to amplified motions caused by the increased roll compliance. Consequently, the use of the static turning analysis for predicting driver crosswind subjective ratings on the basis of steady-state yaw rate responses is viewed here only as a first-stage screening method.

More careful analyses should subsequently be conducted with dynamic simulation models, or vehicle crosswind testing, to further refine any static analysis. That is, simulations or vehicle tests can be conducted, under dynamic crosswind conditions, to examine effects of other vehicle properties (e.g., steering system characteristics and suspension properties) that are related to crosswind sensitivity in a more subtle manner.

To illustrate this latter point, computer simulations of the baseline vehicle and its more roll-compliant counterpart (with no front suspension roll bar) were conducted for a gauntlet-like crosswind input seen in Figure 17. The crosswind model described in Section 3 and reference [13] was utilized for these calculations. The closed-loop driver model [26] is activated to steer and regulate the vehicle along a straight-line course in response to the dynamic crosswind buffeting. Time histories of driver steering wheel angle, vehicle yaw rate, and lateral acceleration are seen in Figure 18 for both vehicle configurations. Simple examination of these results show that, if these time histories were processed to obtain RMS values, the vehicle configuration with the increased roll compliance would exhibit higher RMS levels. The increased RMS value of yaw rate for the more roll compliant vehicle would imply a poorer subjective rating, as was previously observed for the crosswind gauntlet test data. Thus, the use of a more comprehensive dynamic analysis (through simulation or testing) is seen as a recommended means for fine-tuning and extending the basic crosswind sensitivity predictions provided by the simplified static turning analysis, to more realistic dynamic crosswind conditions. Similar examples of subtle conflicts that can arise between predictions of crosswind sensitivity obtained from static turning analyses, versus crosswind sensitivity predictions obtained from RMS system responses and dynamic analyses, are also provided in reference [12] using this same approach.

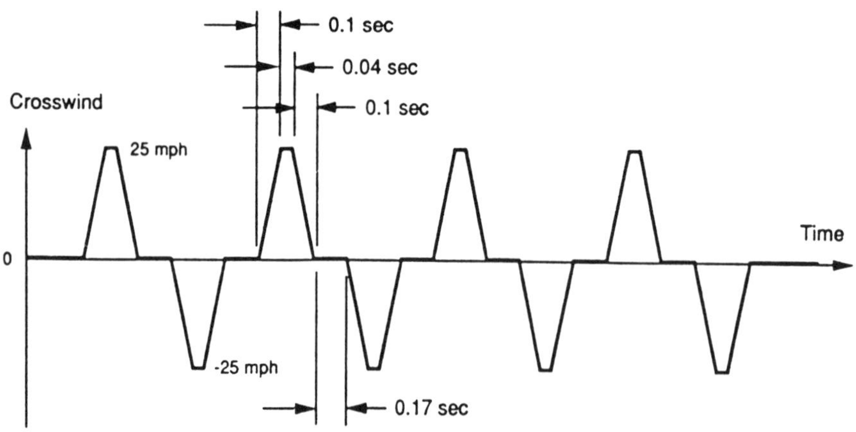

Fig. 17. Crosswind Profile Used in Example Calculations of Figure 18.

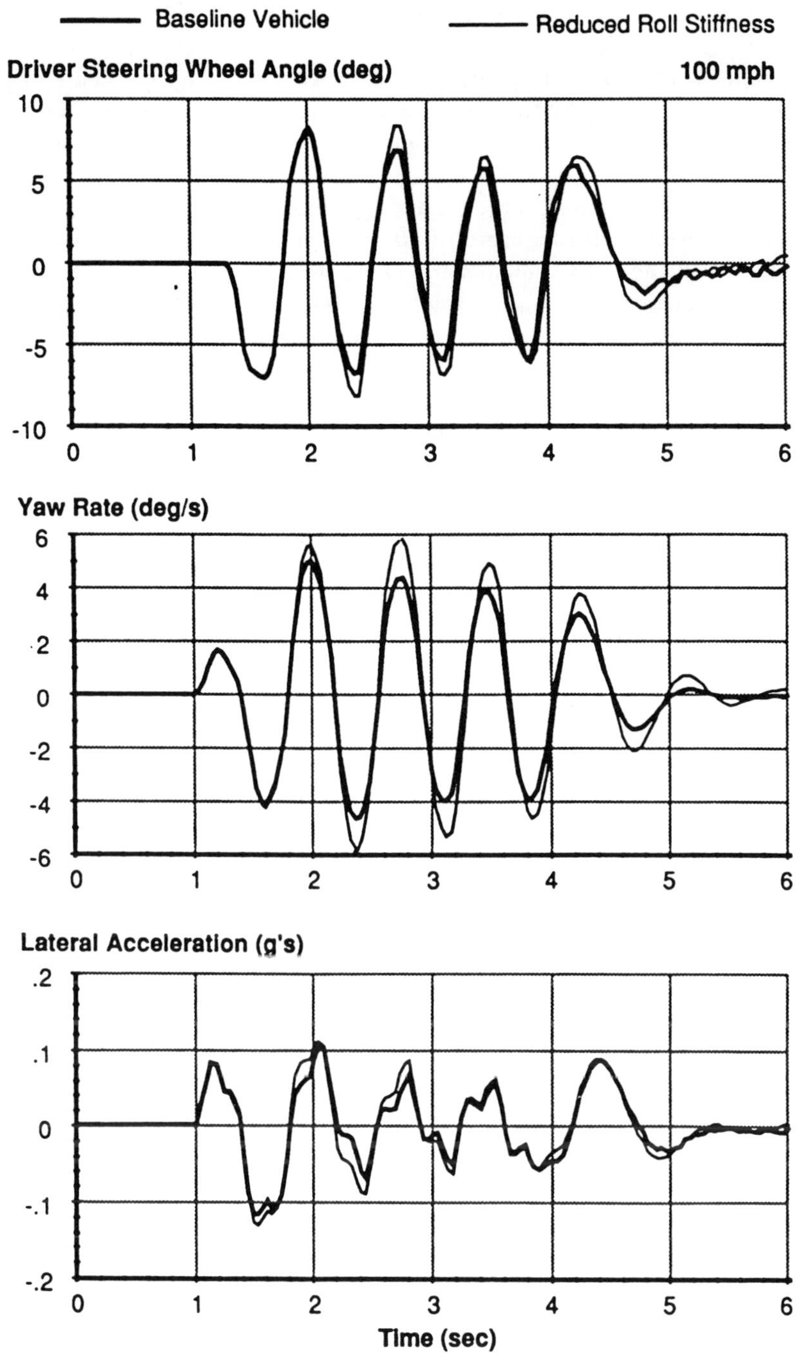

Fig. 18. Simulated Driver-Vehicle Response Using the Dynamic Crosswind Model – Gauntlet Course Example.

6. CONCLUSIONS

The results of the full-scale driver-vehicle crosswind tests presented above, in combination with static turning analyses of vehicles in constant crosswinds and more complete dynamic crosswind simulations, suggest the following conclusions:

- A vehicle's static turning response due to a constant crosswind input and fixed steering wheel angle is a useful, *first-stage* predictor of driver's likely subjective evaluation of a vehicle's crosswind sensitivity.
- That same static turning measure will also frequently predict a vehicle's likely ranking of RMS responses obtained during dynamic crosswind maneuvers.
- A more reliable and accurate method for predicting subjective evaluations of vehicle crosswind sensitivity is with RMS yaw rate values obtained from full-scale testing or comprehensive dynamic simulation of driver-vehicle systems during dynamic, random-like or natural crosswind conditions.
- Increased roll motion due to decreased suspension roll stiffness was associated with lower driver subjective evaluations of vehicle crosswind sensitivity.
- At vehicle speeds of 90-100 mph, variations in fore-aft weight distribution played as important a role as comparable variations in aerodynamic center-of-pressure location in influencing both subjective and objective evaluations of vehicle crosswind sensitivity.
- A two-stage vehicle design process is recommended for analyzing the crosswind sensitivity of a potential vehicle or new design: (1) A preliminary screening of candidate vehicle designs for crosswind sensitivity, based upon the simplified statics analysis of Equations (3) or (4), should first be conducted to screen out ineligible candidate designs having unsuitable vehicle properties (e.g., relative locations of mass center, neutral steer point, and aerodynamic CP that promote high passive crosswind sensitivity). (2) Conduct a more in-depth and comprehensive analysis of the "final round" candidate designs using dynamic analysis (such as the crosswind model described above). The dynamic model should employ random-like, natural crosswind inputs to examine likely driver-vehicle responses to systematic variations of vehicle chassis properties (suspension, steering system, and weight distributions particularly) and different aerodynamic designs. RMS values of system responses (e.g., yaw rate) can be used to evaluate the influence of alternate designs.
- The dynamic crosswind model developed under this work can be used to further explain and analyze the crosswind sensitivity of driver-vehicle systems in a dynamic context, particularly for driving scenarios involving active driver steering control during representative random crosswind conditions. Its use as a tool to systematically examine the influences of vehicle sub-components on crosswind stability is especially useful.
- Further man-machine research into likely driver preferences regarding body roll motion and driver-centered motion experiences deriving from aerodynamic crosswind forces and moments is recommended.

- Re-examination by other parties of the crosswind gauntlet test procedure (or similar procedures) utilizing wind-generating fans is also recommended. Based on the experiences reported here, this type of crosswind test procedure appears to offer significant promise for collecting driver-based subjective data of vehicle crosswind sensitivities. Whether such procedures can effectively replace natural crosswind testing as a reliable method for collecting driver subjective information, remains to be seen.

REFERENCES

1. Vehicle Crosswind Stability. Chrysler Challenge Fund Project – #2000533, UMTRI, Sponsored by the Chrysler Motors Corporation, 1987-1990.
2. W. Hucho, *Aerodynamics of Road Vehicles*. Butterworths, 1987.
3. A. J. Scibor-Rylski, *Road Vehicle Aerodynamics*. Revised by D.M. Sykes ed. Wiley, New York, 1984.
4. W. Kamm, Anforderungen an Kraftwagen bei Dauerfahrten. ZVDI, vol 77, 1933.
5. I. Kohri and T. Kataoka, Research on Improvement of Cross-Wind Properties of Passenger Cars. *JSAE Review*, Vol. 10, No. 3, 1989. pp. 46-51.
6. H. Noguchi, Crosswind Stability and Suspension Properties. MIRA Translation by G. Wood, Rept No. Translation 2/87, 1987.
7. T. Kobayashi and K. Kitoh, Cross-Wind Effects and the Dynamics of Light Cars. *Int. J. of Vehicle Design*, Vol. No. Special Publication SP3, 1983. pp. 142-157.
8. A. Watari, S. Tsuchiya and H. Iwase, On the Cross-Wind Sensitivity of the Automobile. *Proceedings of the 15th FISITA Congress*, Paris, 1974.
9. R. Buchheim, J. Maretzke and R. Piatek, The Control of Aerodynamic Parameters Influencing Vehicle Dynamics. SAE Paper no. 850279, 1985.
10. C. C. MacAdam, Static Turning Analysis of Vehicles Subject to Externally Applied Forces – A Moment Arm Ratio Formulation. *Vehicle System Dynamics*, Vol. 18, No. 6, 1989. pp. 345-357.
11. F. Uffelmann, Influence of Aerodynamics and Suspension on the Cross-Wind Behaviour of Passenger Cars – Theoretical Investigation under Consideration of the Driver's Response. *Proceedings of the 9th IAVSD Symposium of the Dynamics of Vehicles on Roads and Tracks*, Ed. Nordstrom. Linkoping, Sweden, Swets & Zeitlinger B.V. - Lisse, 1985.
12. C. C. MacAdam, The Interaction of Aerodynamic Properties and Steering System Characteristics on Passenger Car Handling. *Proceedings of the 11th IAVSD Symposium of the Dynamics of Vehicles on Roads and Tracks*, Ed. Anderson. Kingston, Canada, Swets & Zeitlinger B.V. - Lisse, 1989.
13. M. Sayers, C. C. MacAdam and Y. Guy, Chrysler/UMTRI Wind-Steer Vehicle Simulation. User's Manual, Version 1.0, vols I and II, Report No. UMTRI-89-8/1-2, 1989.
14. N. F. Barter, *The Transient Response of Automobiles to Steering and Wind Gust Inputs*. Doctoral Thesis, The Loughborough University of Technology, 1975.
15. R. T. Bundorf, D. E. Pollack and M. C. Hardin, Vehicle Handling Response to Aerodynamic Inputs. SAE, 1963.
16. R. H. Klein and J. R. Hogue, Effects of Crosswinds on Vehicle Response – Full-Scale Tests and Analytical Predictions. SAE Paper no. 800848, 1980.
17. R. H. Klein and H. R. Jex, Development and Calibration of an Aerodynamic Disturbance Test Facility. SAE Paper No. 800143, 1980.
18. H. van den Hemel et al., The Cross-Wind Stability of Passenger Cars: Development of an Objective Measuring Method. Fourth IAVD Congress, 1987.
19. H. P. Willumeit et al., Method to Correlate Vehicular Behaviour and Driver's Judgement Under Side Wind Disturbances. *Proceedings of the 10th IAVSD Symposium of the Dynamics of Vehicles on Roads and Tracks*, Prague, Cz, 1987.
20. J. P. Smith, Wind Gusts Measured on High-Speed Roads. The Motor Industry Research Association, Rept No. 1972/7, May, 1972.

21. V. T. Tran, Determining the Wind Forces and Moments Acting on Vehicles by Means of Pressure Transducers. SAE Paper No. 900313, 1990.
22. J. D. Pointer, On-Road Calibration of the Dynamic Pressure Vector System. 10/17/88 Memorandum to G. Romberg, Chrysler Motors Corporation, 1988.
23. A. Alexandridis, B. Repa and W. Wierwille, The Influence of Vehicle Aerodynamic and Control Response Characteristics on Driver-Vehicle Performance. SAE Paper no. 790385, 1979.
24. L. Segel, On the Lateral Stability and Control of the Automobile as Influenced by the Dynamics of the Steering System. *Journal of Engineering for Industry*, Vol. 66, No. 8, 1966. pp.
25. J. R. Hogue, Aerodynamics of Six Passenger Vehicles Obtained from Full Scale Wind Tunnel Tests. SAE Paper no. 800142, 1980.
26. C. C. MacAdam, Application of an Optimal Preview Control for Simulation of Closed-Loop Automobile Driving. *IEEE Transactions on Systems, Man, and Cybernetics*, Vol. SMC-11, No. 6, 1981. pp. 393-399.
27. C. C. MacAdam, Development of Driver/Vehicle Steering Interaction Models for Dynamic Analysis. Final Technical Report, UMTRI-88-53. TACOM Contract DAAE07-85-C-R069, 1988.
28. L. Segel and C. C. MacAdam, The Influence of the Steering System on the Directional Response to Steering. *Proceedings of the 10th IAVSD Symposium of the Dynamics of Vehicles on Roads and Tracks*, Prague, Cz, 1987.
29. K. D. Norman, Objective Evaluation of On-Center Handling Performance. SAE Automotive Engineering Congress, Paper no. 840069, 1984.
30. Society of Automotive Engineers, SAE Recommended Practice, Vehicle Dynamics Terminology, SAE J670e. 1978.

Section IV
Computational Fluid Dynamics

Section IV
Computational Fluid Dynamics

The last decade has witnessed the ascendancy of Computational Fluid Dynamics (CFD) in the analysis of vehicle aerodynamics, and its role can be expected to grow in the coming years. Fundamentally, the appropriate governing equations of fluid motion are used to construct a computational model which is used to simulate desired flow situations. The pursuit of these simulations is motivated by the desire for fuller, faster or even easier understanding of aerodynamic behavior.

The evolution of CFD application in vehicle aerodynamics has gone through several phases. Drawing on much of the CFD technology developed within the aerospace industry, many initial applications focused on the use of CFD codes to compute the flowfield around a vehicle and to estimate its drag coefficient. However, a key distinguishing feature of road vehicle aerodynamics is the presence of a large separated wake directly behind vehicles, which is a consequence of typical vehicle shapes (classified as *bluff-bodies* in aerodynamic parlance). Such separated flow regions are usually absent, and indeed avoided, in most aeronautical applications. Complete understanding of these inherently unsteady flow regions remains elusive, and our ability to describe them through modeling is limited. In light of this handicap, it is easy to understand why it is difficult to accurately calculate the pressure over the rear surface of a test model and, therefore, its overall drag.

These limitations have led CFD analysts to focus on assessing the ability of computational methods to predict specific flow situations. Through detailed comparison with measurements of essentially the same situation, different efforts in the literature have described code *validation*, code *verification* and code *calibration* exercises. While the terminology remains confusing and the differences subtle, some conclusions are easy to comprehend and less contentious. Starting with the formulation of the governing equations, construction of the computational model and mathematical definition of the test problem, numerous approximations and simplifications are introduced perforce. During the subsequent application of the simulation, one must be cognizant of these limitations. CFD simulations are most effective when the objectives of the simulation are within the true capability of the simulation.

Many different CFD methodologies have found application in vehicle aerodynamics. In instances where an economy of computational effort is desired, relatively simple boundary integral methods (such as the Panel method) have been applied with success. However, where separated flow regions exist and are ill-defined, Reynolds-averaged Navier-Stokes (RANS) formulations are *de rigueur.* Although many turbulence models find use among different codes, the k-ε model is the most popular, in spite of its limited accuracy in dealing with massively separated flow regions. Considering that separated flows are inherently unsteady, some computations of these flows employ Large Eddy Simulations (LES), or different formulations of Direct or Quasi-Direct Simulations. However, for the latter methods, the lack of sufficient grid resolution due to current computer limitations, and our incomplete understanding of the full consequences of numerical dissipation are problems.

In recent years, CFD simulations have been successfully applied to study the aerodynamics of components (such as spoilers) and vehicle sub-systems (such as engine-cooling airflow). Another role that has seen effective contributions from CFD is the ranking of vehicle shapes on the basis of their aerodynamics. In this case, the utility of CFD simulations is their ability to provide useful information rather quickly and without recourse to the expensive and time-consuming construction of test models. This capability is expected to receive more recognition as a valuable companion to wind tunnel testing. The biggest challenge confronting aerodynamicists in coming years could well be the effective integration of CFD into the design and development process of future vehicles.

To reach this stage, many specific hurdles need to be addressed. The entire issue of handling complex geometry information prior to creating a computational mesh, as well as the manner and quality of mesh generation, requires further development. The errors inherently introduced by many numerical schemes currently in use also demand attention. Likewise, the physical models (conventional and sub-grid scale turbulence models) have proved to be inadequate in some important applications. On the other hand, the search for a universally applicable turbulence model is unlikely to provide an answer for some time to come, if ever. Finally, the goal of integrating aerodynamic simulations into the framework of a multi-disciplinary design optimization methodology will continue to challenge our current capabilities and, in turn, spur future development.

921092

Steady and Unsteady Computational Aerodynamics Simulations of the Corvette ZR-1

J. Michael Summa
Analytical Methods, Inc.

ABSTRACT

Two integral CFD methods for the simulation of steady-state and transient vehicle aerodynamics, respectively, are described in this paper. Both codes couple potential flow and boundary layer schemes with simple, thin-shear-layer wake models which ensure relatively fast computational times. Calculations of the steady-state yaw effects for the Corvette ZR-1 compare favorably with experimental pressure data located on the hood. Furthermore, the code's rapid computational speed makes multiple design simulations even on today's low-end workstations quite practical. Finally, time-stepping calculations of an unsteady pitching motion of the Corvette ZR-1 demonstrates one example of the current capability of the unsteady computational method.

INTRODUCTION

In the last decade, road vehicle aerodynamics has become an important design consideration. Originally, the design of low-drag shapes was given high priority due to worldwide fuel shortages that occurred in the mid-seventies. More recently, however, as vehicle weight has been reduced in order to improve fuel economy, there has been increased interest in the role aerodynamics plays in vehicle stability and passenger safety. Consequently, transient aerodynamics and also the aerodynamics of road vehicles in yaw have become important issues at the design stage.

While there has been tremendous progress in Navier-Stokes methodology in the last few years, the physics of bluff-body aerodynamics is still very difficult to model correctly. Moreover, the computational effort to carry out the Navier-Stokes simulations from the geometric stage to complete flow solutions requires too much time to effectively impact the design cycle. As a result, in the short run, simpler methods must be used for such complicated problems.

In this paper, two integral CFD methods are briefly described for the simulation of steady-state and transient vehicle aerodynamics. For steady-state simulations, program VSAERO (1), which couples integral methods for potential and boundary layer flows, is used to compute the external aerodynamics of the Corvette ZR-1 for several yaw conditions. Calculated surface pressures are compared with available experimental data located on the hood. These results, in addition to those already published (2), (3), illustrate that the forebody aerodynamics of such vehicles can be accurately computed with this method. For transient simulations, program USAERO (4), which couples unsteady integral methods for potential and boundary layer flows, is used to compute the aerodynamics of the Corvette ZR-1 undergoing a longitudinal pitching motion. While no experimental data are available for this case, it serves to illustrate an automobile transient simulation in addition to applications of this method already published (5), (6).

COMPUTATIONAL METHODS

The mathematical model that is solved by program VSAERO is an idealization of the real flow past arbitrarily shaped bodies. The flow is assumed to be steady. At small distances from the body surface or any separated wake region, the flow is inviscid and irrotational due to the high Reynolds number characterizing automobiles; hence, the potential flow representation coupled with a boundary layer method should be a good approximation of this region of the flow. Open separations and longitudinal vortex formations are computed using vortex line filaments and vorticity panels shed from the body. Special procedures are also included in the program for computing the local flow trajectory of such vortex lines in the presence of the body. Massive separations at the rear of automobiles are modeled by thin shear layers whose strengths are coupled with the vorticity carried by the body near the separation. Obviously, this simplification of massive separations is the most difficult approximation of the method and continues to be an area of research. Engine inlet and exhaust effects can also be simulated by allowing the flow to pass through the appropriately identified inlet and exhaust areas. Finally, the ground plane can be simulated by simple imaging or by computing the effect of a tunnel floor and its boundary layer characteristics.

The coupled viscous/potential flow is modeled in VSAERO by an iterative solution. The scheme starts with the steady-state potential flow problem. The potential flow solution is found by solving the integral equation for the velocity potential on the body surface. Numerically, the body surface is represented by a number of flat quadrilateral panels over each of which doublet and source distributions are assumed constant. The source strengths of each panel are determined by the flow tangency boundary condition and any wake surfaces are represented by triangular panels with a linear doublet distribution which is in turn related to the body doublet strengths by wake auxiliary conditions. If there are N panels on the body, then N simultaneous equations are obtained in N unknown doublet (or perturbation velocity potential) values. Currrently, solutions have been obtained for up to 12,000 unknowns.

Following the solution of the surface potential, the surface velocity vectors are obtained from the gradient of the potential using a second-order distribution through the panel control point values. Forces and moments are obtained by pressure integration.

For the initial solution, the wake geometry is prescribed starting at an assumed separation line geometry. At this stage, iterative methods for relaxing the wake sheets to force-free locations are available in the code, and the potential flow solution is updated during each relaxed wake cycle.

Once the relaxed wake cycle is completed and the surface potential is known, a family of *external* surface streamlines is calculated and boundary layer calculations are then performed along each surface streamline. The boundary layer codes are based on locally axisymmetric conditions with surface curvature and streamline convergence/divergence effects included.

The calculated boundary layer displacement values are then used to reset the panel source values. Source values for panels not crossed by streamlines are obtained by interpolation. If forebody separations are computed and are expected to reattach, then streamlines with turbulent boundary layer properties can be restarted near the reattachment line. Otherwise, separated wakes must be attached near the computed separation line. At present, this must be accomplished manually along panel edges. The potential flow solution is then repeated in the presence of the boundary layer sources. Usually, for the simpler cases involving fixed separation locations, only three viscous/potential flow iterations are required.

The flow idealizations required by USAERO are the same as those for VSAERO, except that the flow can now be unsteady. This program calculates the transient aerodynamic characteristics by utilizing a time-stepping procedure which allows relative motions of components of the configuration. The potential flow equation is now solved for each time step according to the transient motion, and new wake panels are generated at each time step while old wake panels are convected downstream. Of course, surface pressures now include the unsteady apparent mass term. Additionally, the boundary layer method includes unsteady terms and is computed for each time step if viscous effects are required. Obviously, USAERO requires much more in the way of computer resources than VSAERO, but is still much less demanding than time-accurate, Navier-Stokes calculations.

COMPUTATIONAL RESULTS

The surface definition of the Corvette ZR-1 was kindly provided by General Motors Corporation (7). This CAD data was utilized to generate the surface panel model that is illustrated in Figure 1. The wheel well openings and

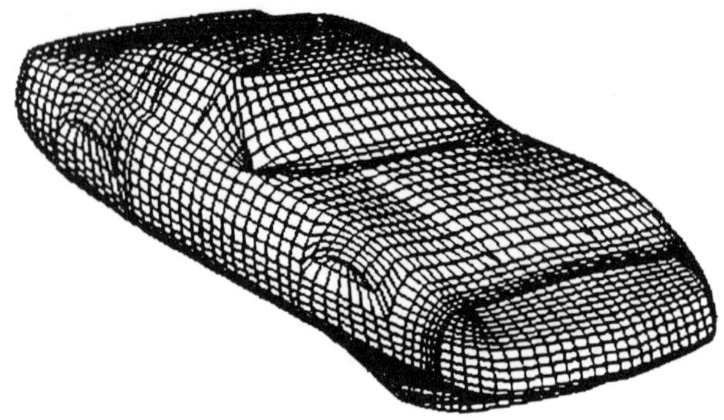

Fig. 1. Corvette ZR-1 Surface Panel Model.

body underside were simply paneled in order to obtain a closed surface; hence, tire blockage, wheelhouse and detailed underbody effects are not included in the simulations. The original geometry was developed to compute the effects of a rear wing design for increased downforce in braking and high speed maneuvers, and the VSAERO simulation of this model was reported orally in 1990 (8). Since that time, wind tunnel experimental pressure measurements (7) on the hood for several yaw angles became available for comparison. While the surface panel density could be increased on the hood near the fender lip and near the windshield for possible improved comparisons, the current panel model was felt to be adequate. Furthermore, all current VSAERO calculations were to be made on a Silicon Graphics 4D/35TG Personal Iris workstation and the model already has 2,659 panels (unknowns) on the half-body for zero yaw calculations. For the non-zero yaw calculations, therefore, the model includes 5,318 panels. In all cases, the wind tunnel floor is simulated by the image technique, and the model front spoiler lip was set to 0.14 meters above the tunnel floor. Finally, all computations correspond to a speed of approximately 100 kmh, and the heat exchanger intake flow under the front bumper is simulated by flow transpiration.

STEADY-STATE CALCULATIONS - VSAERO calculations for the Corvette ZR-1 are started with a base separation wake geometry that trails straight back from the rearmost part of the car. While this wake simulation is not necessary for computing front-end aerodynamics, the wake simulation will provide an initial base separated flow effect. In each of the yaw simulations, the wake geometry is then relaxed for 3 iterations and two viscous iterations are computed by a program restart. In all cases, the boundary layer computations are started fully turbulent in order to more closely match wind tunnel testing conditions.

A comparison of the computed ZR-1 centerline pressure distribution (yaw = 0) for the inviscid solution and the final viscous iteration is shown in Figure 2. The most notable feature in the comparison is the reduction in pressure due to the predicted forebody separation at the rear of the hood and the reattachment on the windshield. This separation bubble is modeled in VSAERO by a constant pressure approximation with reattachment boundary layer characteristics obtained from values just upstream of the predicted separation location. The cowl vent pressure coefficient at the centerline is reduced from a potential flow value of 0.60 to a final predicted value of 0.25. Viscous effects are also noticeable in reducing the peak suctions at the beginning and end of the roof. The reduction in pressure evident near the lower portion of the backlight and the rear deck is due to the combined boundary layer and wake relaxation effects. The boundary layer has, in fact, indicated flow separation near the lower part of the backlight as shown in the figure. At this stage, no attempt has been made to model this separation which will be required in order to improve the pressure calculations on the rear of the car.

Figure 3 illustrates the computed surface streamlines, separation zones, and wake shapes for the ZR-1 at 0°, 6°, and 12° yaw. The yaw effects on the streamlines appear qualitatively correct and the forebody separation moves to the windward side while the separation near the rear deck moves to the leeward side. This rear separation zone is most likely another bubble that reattaches to the rear deck although VSAERO does not yet include the methodology to predict such events.

The computed yaw effects on the surface pressure distributions for the ZR-1 are shown in Figure 4. Again, these variations appear to be qualitatively correct in general with pressures increasing on the windward side while decreasing on the leeward side. The three-dimensional change in pressure levels near the cowl vent location due to the forebody separation is also quite plausible; however, the pressure distributions on the rear deck are not realistic solutions, since the rear separation is not included in these computations.

Wind tunnel measurements of the surface pressures on the hood of the ZR-1 were taken along 6 buttline cuts. The pressure taps were located at lateral y-stations at 0.120, 0.220, 0.375, 0.475, 0.575, and 0.665 meters from the centerline. These data cuts are illustrated in Figure 5. It is noted that the data for y = 0.665 meters is at the top edge of the fender. Wind tunnel data was provided for zero yaw and also for positive and negative yaw angles of 6° and 12° in order to obtain windward and leeward hood pressures.

The quantitative comparisons of the computed and experimental pressure data are shown in Figure 6. Only the most lateral 3 data cuts at y = 0.475, 0.575, and 0.665 meters are illustrated, since the comparisons at the inboard data cuts show no plottable difference between the calculated and experimental values. Pressure distributions are displayed only for yaw = 12°, 0°, and -12° for visual clarity. The yaw = 6° case shows similar trends. Finally, the geometric contour of the data cut is also shown in Figure 6. The comparisons with the experimental data are re-

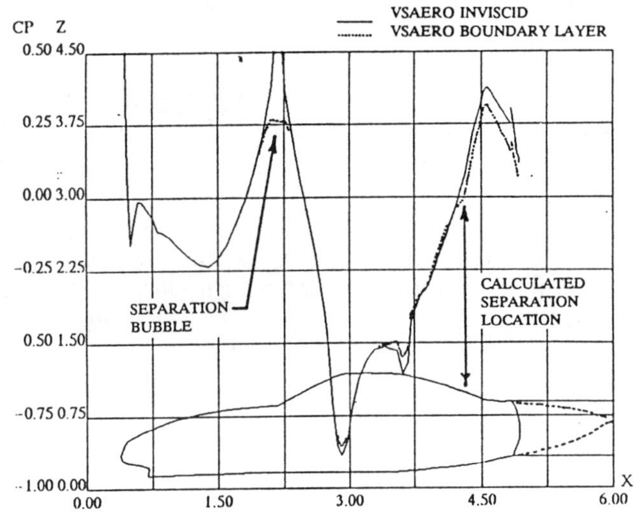

Fig. 2. Calculated Centerline Pressures for Corvette ZR-1 at Yaw = 0.

markably good with the exception of the leeward pressure at the very edge of the fender (y = 0.665). Even here, the computed pressure is only about 0.03 to 0.05 less than the experimental data. Most likely, a leeside vortex separation occurs at the fender edge for the 12° case that VSAERO does not correctly predict for the current panel model.

Although again there is not any experimental data for comparison, the change in lift due to yaw for the front surface body panels of the Corvette ZR-1 is illustrated in Figure 7. The panel pressures that are integrated to compose these lift changes are also displayed in the figure. It is noted that this is not the change in forebody lift due to yaw since wheelhouse and underbody effects are ignored. However, the trend and increase in uplift by about 30% for 12° yaw would seem to be plausible.

Finally, on the SGI Personal Iris (or equivalently, on the Indigo workstation) these computations required about 3/4 hour for the zero-yaw case and about 1 3/4 hours for the each of the lateral yaw cases. Moreover, only 12 minutes of CPU time is required for the zero-yaw simulation with no wake relaxation and 27 minutes for each lateral yaw case with no wake relaxation. Hence, forebody aerodynamics simulations that still include boundary layer separation predictions for several different body designs can be completed within 1 day on today's low-end workstations.

UNSTEADY CALCULATION - A single USAERO simulation of the Corvette ZR-1 undergoing a pitch oscillation has been computed. No experimental data is available for this case and, in fact, the transient motion was chosen purely to demonstrate the application of the calculation method for an example problem. While the transient effects of a cross wind or the passing of another vehicle are simulation capabilities of USAERO, these problems would require the full panel geometry (5,318 unknowns) of the ZR-1 and were rejected for the present demonstration. Currently, the program is being validated for complete aircraft transient motions and also for trains passing in a tunnel.

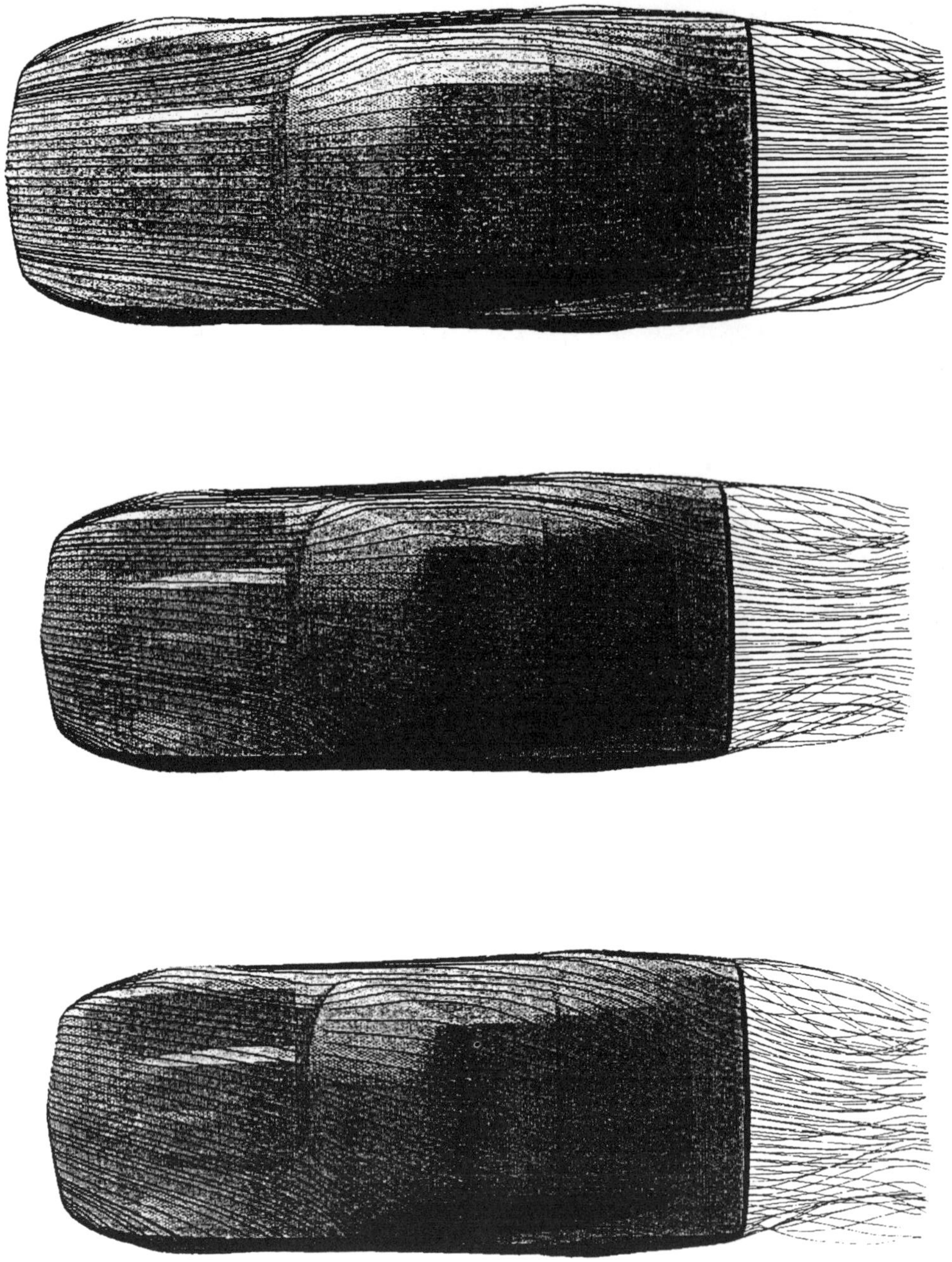

(a) Yaw = 0°

(b) Yaw = 6°

(c) Yaw = 12°

Fig. 3. Computed Yaw Effects on Streamlines, Separation Zones, and Wake Shape for Corvette ZR-1.

The pitch axis of oscillation was located at 0.14 meters above the ground plane and just ahead of the rear wheel opening. The amplitude was set to plus and minus 1 degree which means that the lower front spoiler lip will periodically move downward and upward about 0.035 meters from its nominal position (0.14 meters above the ground plane). The frequency is set such that during one full cycle of oscillation, the car moves a distance of 4 meters or just under one car length.

(a) Yaw = 0°

(b) Yaw = 6°

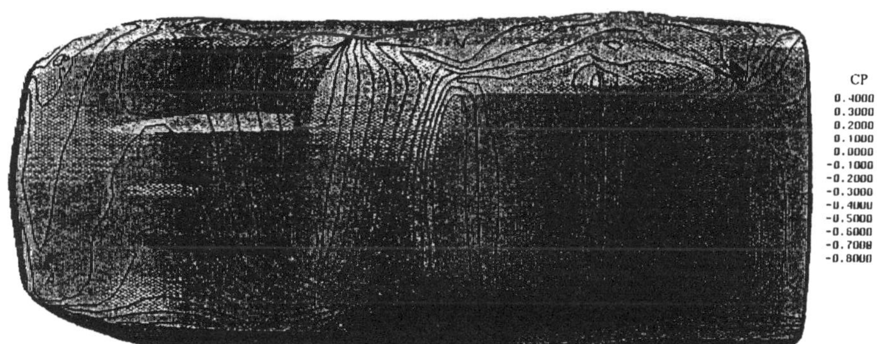

(c) Yaw = 12°

Fig. 4. Computed Yaw Effects on Surface Pressure Distribution for the Corvette ZR-1.

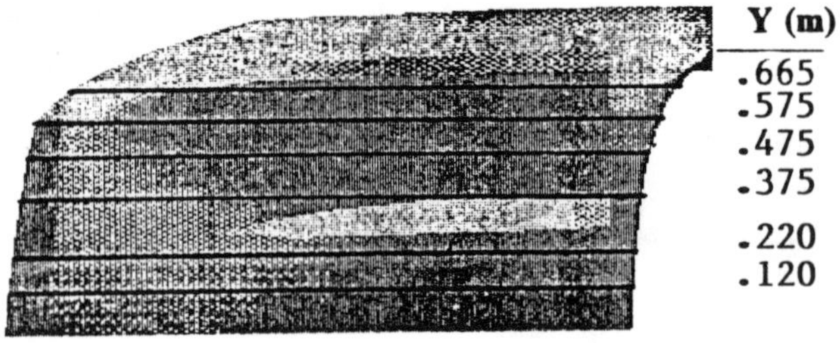

Fig. 5. Experimental Data Cut Locations on Corvette ZR-1 Hood.

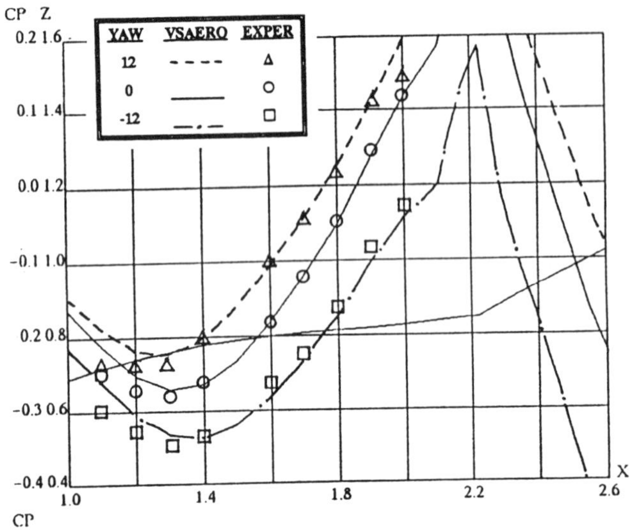

(a) Y = 0.475 m

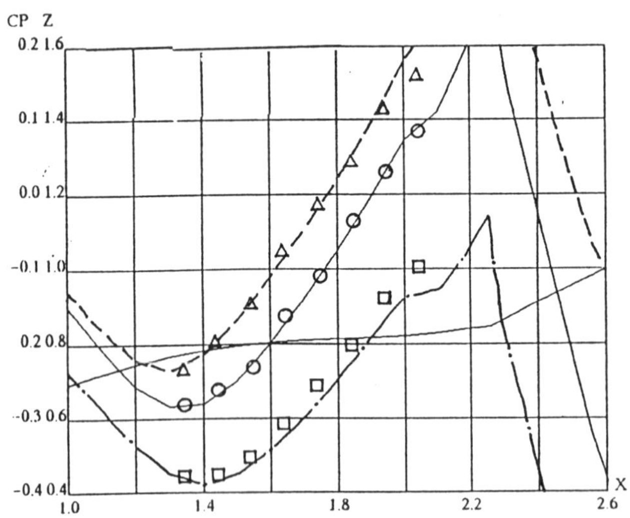

(b) Y = 0.575 m

Fig. 6. Comparison of Calculated and Exsperimental Hood Pressure Yaw Variations for the Corvette ZR-1.

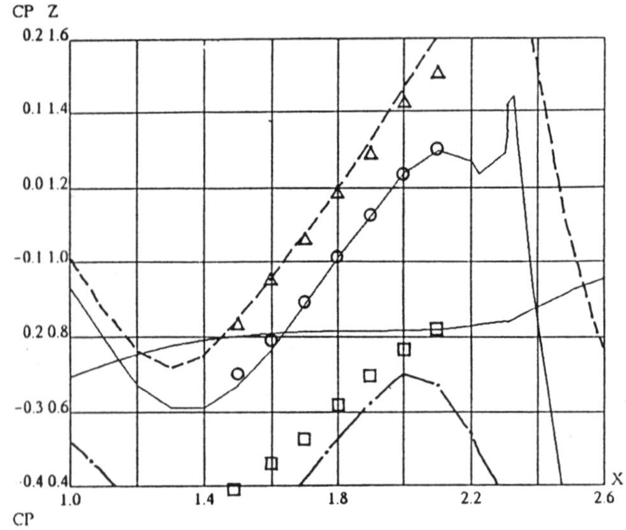

(c) Y = 0.665 m

Fig. 6. Concluded.

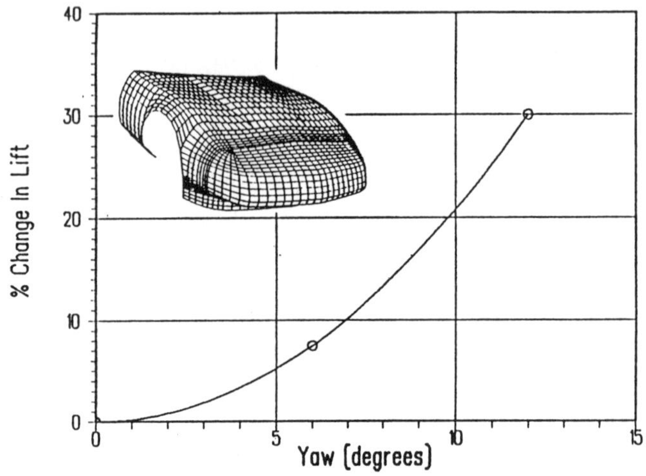

Fig. 7. Computed Change in Lift Due to Yaw for Front Surface Body Panels of Corvette ZR-1.

Figure 8 illustrates the computed unsteady centerline pressure distributions for two instants of time which correspond to the minimum and maximum pitch angles. Although boundary layer calculations are performed for each time step, these effects are not depicted in this figure. The calculation clearly shows the increase in pressure on the upper-surface at the downstroke and the decrease in pressure on the upper surface at the upstroke. The general pressure distribution trend also compares favorably with the VSAERO prediction as one would expect.

The three-dimensional pressure contours are compared in Figure 9 for three time instants that correspond to the -1°, 0°, and +1° pitch. As these do not include boundary layer effects or the forebody separation modeling that is predicted in the unsteady calculation, the contours represent potential flow results. Still, the transient motion of the pressure contours with pitch attitude is at least qualitatively correct.

Figure 10 shows the time history of the computed lift and drag coefficient contributions of the forebody surface panels. Of course, these are periodic with a phase shift in the drag coefficient indicated.

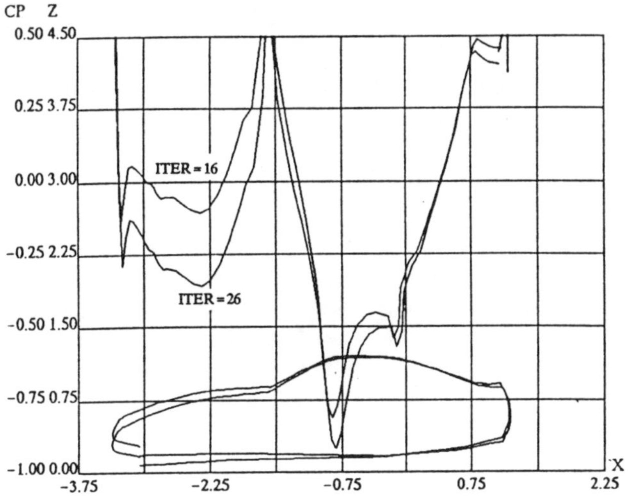

As discussed earlier, the purpose of the calculation is merely to demonstrate the current capability of USAERO for a particular automobile transient simulation. The calculation was performed on a CRAY XMP-48 and required slightly over 2 hours of CPU time.

Fig. 8. Computed Centerline Unsteady Pressures for the Corvette ZR-1 in Pitching Motion.

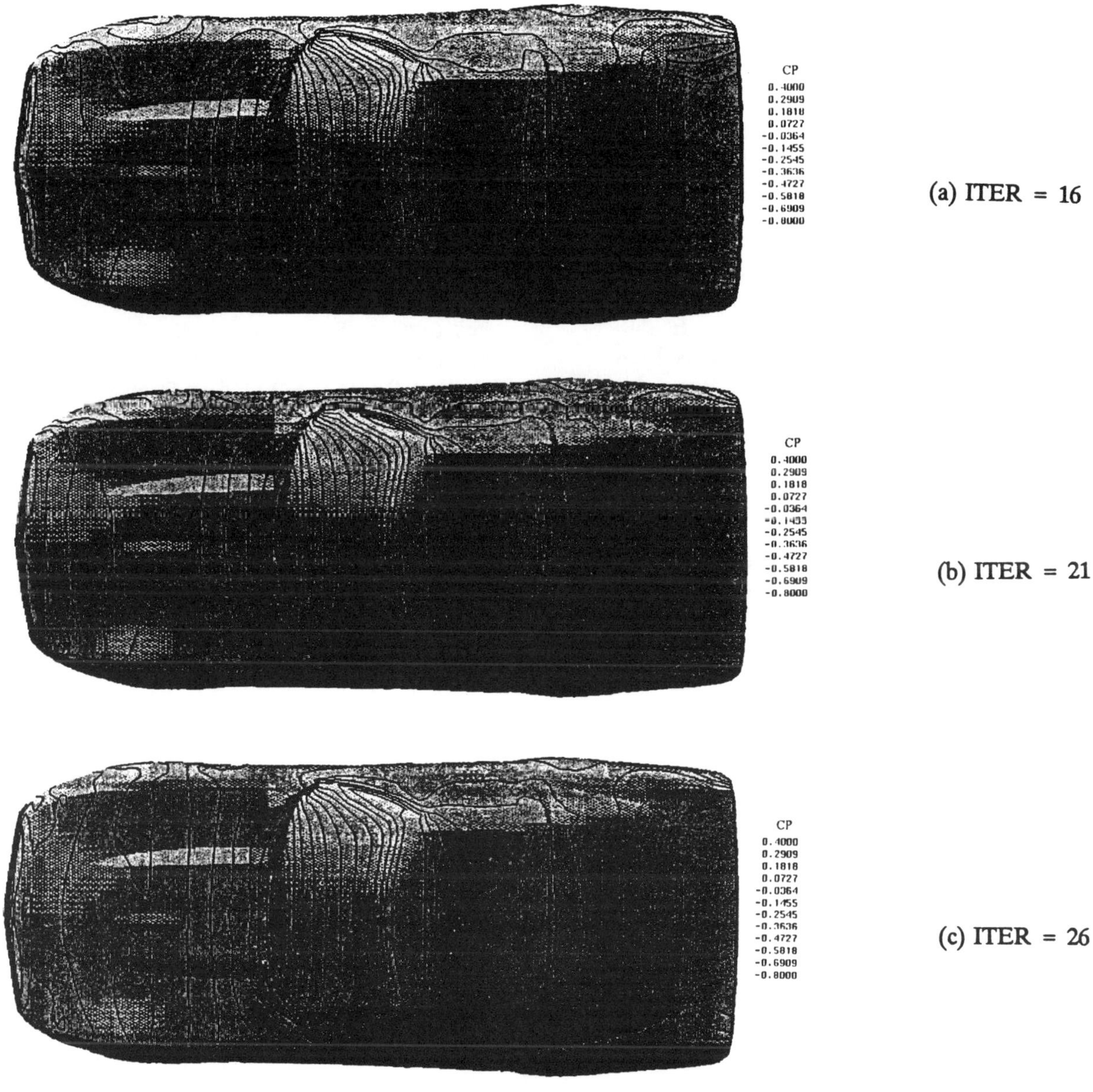

Fig. 9. Computed Unsteady Pressure Distributions for the Corvette ZR-1 in Pitching Motion.

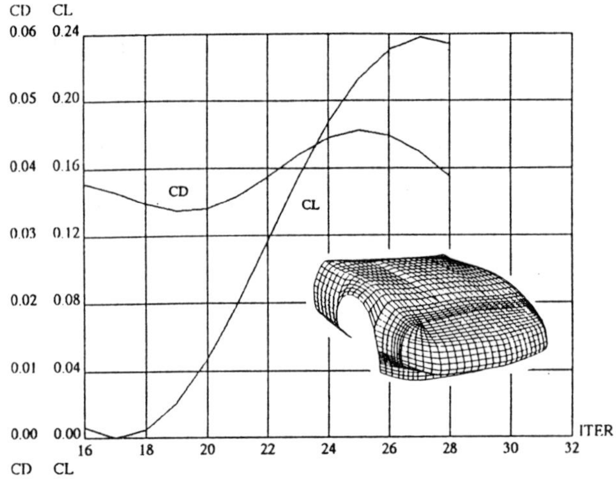

Fig. 10. Computed Unsteady Lift and Drag History for the Front of the Corvette ZR-1 in Pitching Motion.

CONCLUSIONS

Aerodynamics continues to play an important role in the design of new automobiles. Wind tunnel simulations have traditionally supplied the required aerodynamic data for decision making, but current computational methods can provide valuable companion information at a pace compatible with the design cycle. Additionally, numerical solutions of some vehicle aerodynamic problems such as transient behavior can be obtained while experimental measurements may not be possible or practical. In this paper, two integral CFD methods are briefly described for the simulation of steady-state and transient vehicle aerodynamics. Since both methods couple integral potential and boundary flow schemes, they are each probably the simplest techniques that can be of practical use in auto aerodynamics calculations. For the steady-state simulations, program VSAERO is briefly described and applied to the computation of yaw effects for the Corvette ZR-1. While overall force and moment experimental data are not available and would not be accurately represented in these calculations, the following conclusions are supported in the paper:

1. Comparison with wind tunnel data shows that yaw effects on the pressure distribution on the hood of the ZR-1 can be accurately predicted by the code.
2. The calculation shows a 30% increase in the lift at the 12° yaw case compared to the zero yaw case for the front-end body surface.
3. Forebody aerodynamics simulations for the Corvette ZR-1 that include boundary layer separation predictions can be completed in about 12 minutes for zero yaw and 27 minutes for non-zero yaw conditions on the SGI Personal Iris workstations; consequently, multiple design studies are quite practical on today's low-end workstations.

For unsteady simulations, program USAERO is briefly described and demonstrated by applying the code to the calculation of an example pitch oscillation of the Corvette ZR-1.

ACKNOWLEDGEMENTS

The author would like to thank David Lednicer for his excellent work in the geometric modeling of the Corvette ZR-1, Jahan Saghafi and Dan Almosnino for running the USAERO program, and Cray Research for providing the computer time for the development of the USAERO simulation. Finally, a very special thanks is due Robert Niemiec of the Chevrolet-Pontiac-Canada Group, General Motors Corporation, for supporting the original Corvette ZR-1 project and supplying the experimental data.

REFERENCES

1. *Program VSAERO, A Computer Program for Calculating the Nonlinear Aerodynamic Characteristics of Arbitrary Configurations, Users' Manual*, Analytical Methods, Inc., Redmond, WA, U.S.A., 1991.
2. Summa, J.M. and Maskew, B., "Predicting Automobile Aerodynamic Characteristics Using an Iterative Viscous/Potential Flow Technique", *SAE Technical Paper Series*, No. 830303, 1983.
3. Summa, J.M. and Dvorak, F.A., "Computing Automobile Aerodynamics by an Integral Method", *Proc. Internat'l Conf. on Supercomputer Applications in the Automotive Industry*, Zurich, Switzerland, October 1986.
4. Maskew, B., *USAERO, A Time-Stepping Analysis Method for the Flow about Multiple Bodies in General Motion, Users' Manual*, Analytical Methods, Inc., Redmond, WA, May 1991.
5. Clark, D.R. and Maskew, B., "A Re-examination of the Aerodynamics of Hovering Rotors Including the Presence of the Fuselage", Presented at Internat'l Technical Specialists' Meeting on Rotorcraft Basic Research, Atlanta, GA, March 1991; also at 47th National Forum of the Am. Hel. Soc., Phoenix, AZ, May 1991.
6. Maskew, B., "USAERO/FSP: A Time-Domain Approach to Complex Free-surface Problems", Symp. High-Speed Marine Vehicles, Naples, Italy, 1991.
7. Niemiec, R., Private Communication, General Motors Corporation, Chevrolet-Pontiac-Canada Group, 1990.
8. Summa, J.M., "Road Vehicle Aerodynamics and Computational Simulations", AIAA-90-3054; Invited Paper, Applied Aerodynamics Conf., Portland, OR, August 1990.

930293

Computational Analysis of Flow Around a Simplified Vehicle-Like Body

Akiyoshi Yamada and Shingo Ito
Toyota Motor Corp.

ABSTRACT

The flow around a simplified vehicle-like body was computed. The aerodynamic characteristics of this body depended on the afterbody geometry, especially the rear slant angle. In order to examine reliability of computation, the computations were performed for various rear slant angles. Regarding the computational method, two kinds of methods were applied : a Navier-Stokes solver that employs a k-ε turbulence model and a quasi-direct simulation with a third-order upwind difference scheme.

Comparing the results with a wind tunnel test data of the flow fields and aerodynamic forces, it was found that, the k-ε model had potential for prediction of flow field and the quasi-direct simulation for prediction of aerodynamic forces.

1. INTRODUCTION

Recently, the progress in super computers and computational fluid dynamics (CFD) has motivated the CFD technology to be applied actively to the flow around a vehicle body [1-7] . Although vehicle aerodynamics are improved by wind tunnel tests in general, CFD is expected to be one of the effective tools to realize aerodynamic improvement of high quality.

In applying the CFD technology to the vehicle aerodynamics, computation reliability is an important factor to be considered. Therefore , it is necessary to examine in detail, whether the variation of flow field and aerodynamic forces with the vehicle shape could be simulated by the computation or not.

On the other hand, of the computational methods presently applied, two kinds of methods, i.e. k-ε turbulence model and third-order upwind difference scheme without turbulence model especially, have shown promise. Each method has often been applied and their accuracies and efficiencies have often been studied. However, these methods have seldom been compared for the same computational object [8].

In this study, flows around several types of bodies were computed by the two above mentioned methods and the reliability of each method was considered by detailed comparison with the wind tunnel test.

2. COMPUTATIONAL OBJECT

A bluff body , shown in Figure 1, was selected as the computational object.

This bluff body is a simplified model which was designed to study the wake structure of the vehicle basically and is similar to

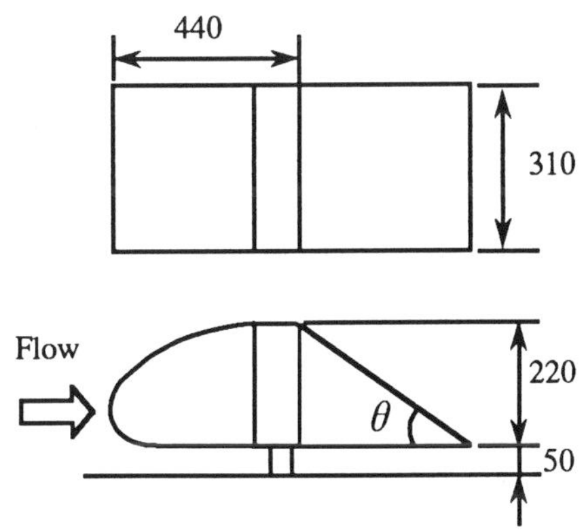

Fig.1 Bluff body (Dimensions in mm)

Ahmed's body [9]. The aerodynamic characteristics of this model depend on the afterbody geometry, especially the rear slant angle. For example, when the rear slant angle θ is increased, the drag coefficient C_D is found to peak at $\theta = 35°$ and thereafter, to drop rapidly to $\theta = 40°$ (see Fig.2). (Figure 2 shows the experimental data. The experimental condition will be mentioned later.)

In order to examine whether these characteristics are computable or not, the following rear slant angles were investigated : 25° , 35° , 40° , 45° ,and 55° .

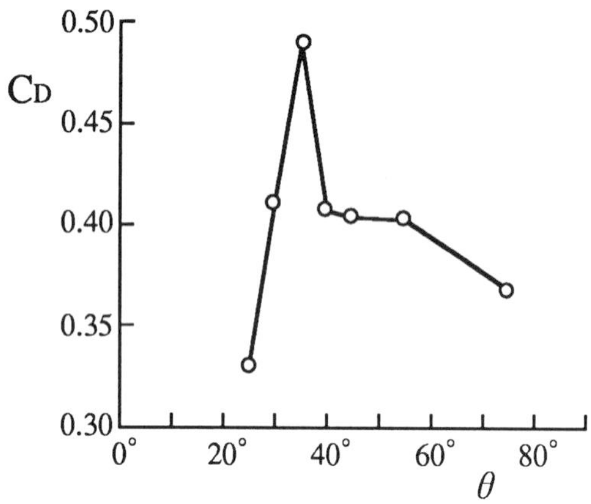

Fig.2 Variation of drag coefficient with rear slant angle

3. METHOD OF COMPUTATIONAL ANALYSIS

3.1 BASIC EQUATIONS AND COMPUTATIONAL METHOD - The flow around vehicle body can be assumed to be incompressible. Therefore , the basic equations can be described in dimensionless form as follows :

Equation of continuity

$$\nabla \cdot \mathbf{v} = 0 \tag{1}$$

Navier-Stokes equation

$$\frac{\partial \mathbf{v}}{\partial t} + (\mathbf{v} \cdot \nabla)\, \mathbf{v} = -\nabla p + \frac{1}{Re} \Delta \mathbf{v} \tag{2}$$

where $\mathbf{v}$ and p are the velocity vector and pressure, respectively, and Re the Reynolds number.

The above equations were solved by the CFD code "SCRYU" (developed by Software Cradle Co.,Ltd.). SCRYU is based on a body fitted coordinate system and a finite volume method. One of the features of this code is to be able to select a number of various computational methods by combining a turbulence treatment and difference scheme for the convection terms of EQ(2) [3]. As mentioned before, two kinds of computational methods were applied in this study and their distinctions are shown in Table 1. In method A, to achieve a good computation stability , compared with the third-order upwind scheme , the first-order upwind scheme is preferable for the convection terms and therefore has been adopted in this study. For simplicity reasons , method A and B are hereafter called the k-ε method and quasi-direct simulation, respectively.

In both methods, the first-order implicit Euler method is used for the time integration and the second-order central difference scheme for the spatial derivatives except convection terms.

Table 1 Computational methods

	Method A	Method B
Turbulence treatment	Standard k-ε model	No turbulence model
Difference scheme for convection terms of EQ(2)	First-order upwind scheme	Third-order upwind scheme [10]

3.2 COMPUTATIONAL GRID - Figure 3 shows the computational grid around the bluff body ($\theta = 35°$) . To exactly simulate the experimental condition (mentioned later) , the body supporting strut is also considered. The total number of grid points is 101×81×87. The same grid was used for both computational methods, and the same number of grid points in all rear slant angles were used.

Fig.3 Computational grid

The computational grid was generated by operating the grid generation system which was developed for a general vehicle. Its procedure can be outlined as follows :

1) The body surface (in this case, the bluff body) is drawn as a CAD data.
2) The surface grid of the vehicle body is generated by the transfinite interpolation method and projected on to the body surface.
3) The computational domain is defined and divided into multi-blocks.
4) The computational grid is generated in each block and smoothed throughout the computational domain.

3.3 INITIAL AND BOUNDARY CONDITIONS - In applying the k-ε method , each computation was started from the static state. Moreover , in applying the quasi-direct simulation, the solution of the k-ε method was utilized for the initial condition to save on CPU time.

The boundary conditions were the following :

1) At the inflow boundary , the constant velocity condition was imposed.
2) At the outflow boundary , the constant pressure condition was imposed.
3) At the wall boundaries (the body surface and ground plane), two kinds of conditions were employed separately according to the computational methods. In the case of k-ε method, the standard log-law wall functions were adopted. In the case of quasi-direct simulation, the no-slip conditions were adopted.
4) At the outer boundaries , the free-slip conditions were imposed.

As regards the quasi-direct simulation, since it is essential for the unsteady flow to be simulated, more sophisticated boundary conditions should be applied. However, because of extreme definition difficulties , the above mentioned simple conditions were chosen.

3.4 COMPUTER SYSTEM - Figure 4 shows the composition of computer system. This computer system is connected by a local area network (LAN). The computational process from the grid generation (pre-process) to the display of computed results (post-process) was achieved by suitable hardware for each process.

4. EXPERIMENTS

The experiments for the verification were performed in a wind tunnel with an open test section of 1m×2m×2m. The bluff body was supported by a strut having a tear-drop shaped cross-section (10mm wide at its thickest point) to achieve a smooth as possible flow.

The experimental items and methods are shown in Table 2 . The velocity vectors and pressure distribution around the body were measured on the grid points of 10mm×10mm in three representative sections (see Fig.5). In all experiments, except the streamlines, wind speed was 30m/s and the Reynolds number, based on the body height, was 4.4×10^5.

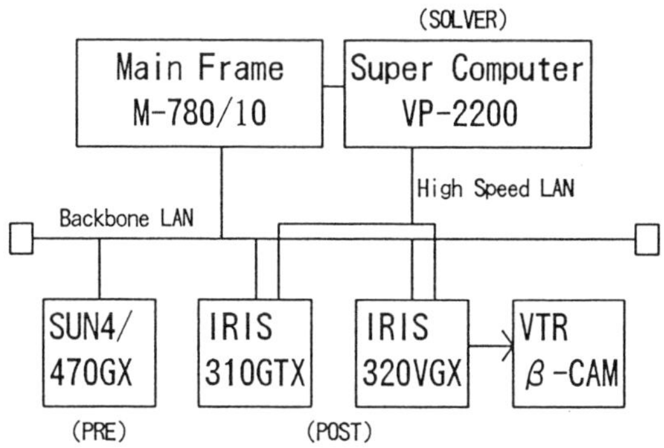

Fig.4 Computer system

Table 2 Experimental items

Item	Method
Streamlines	Visualization by smoke
Aerodynamic forces	Measurement by balance
Surface pressure	Measurement by sensors via tubes connected to surface holes
Velocity vectors & pressure distr. around the body	Measurement by 5-hole Pitot tube [11]

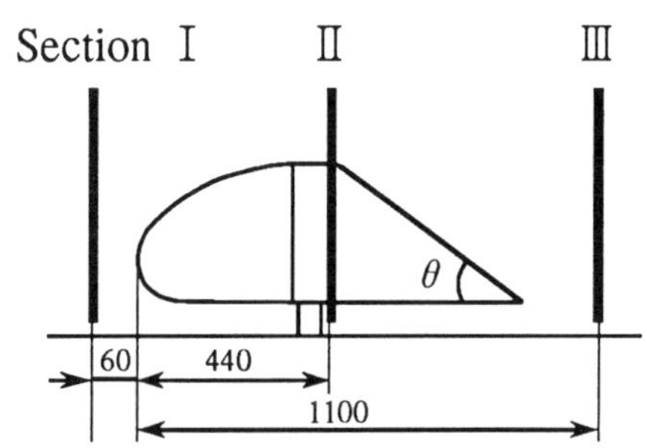

Fig.5 Measurement sections

5. RESULTS AND DISCUSSION

The flow at the Reynolds number of 4.4×10^5 was computed to correspond to experimental condition. In the case of the k-ε method, the computation was carried out until the solution became a steady state. In the quasi-direct simulation, a lengthy computation was executed to achieve a best possible mean flow. The simulation solution was time-averaged and compared with other results, i.e. with experimental data and solution of k-ε method. All computations were processed by Fujitsu VP2200 and the CPU time for one computation was about 40 hours.

In this section, discussion is centered on the rear slant angles of 25° , 35° , and 55° .

5.1 COMPARISON OF FLOW FIELD - Figure 6 shows a comparison of the experimental and computational streamlines in the symmetry plane. For θ = 25° configuration, although the flow runs almost along the upper body surface in the experiment and k-ε method, in quasi-direct simulation, the flow separates from the body surface at the roof end. For θ = 35° configuration, compared with the ex-

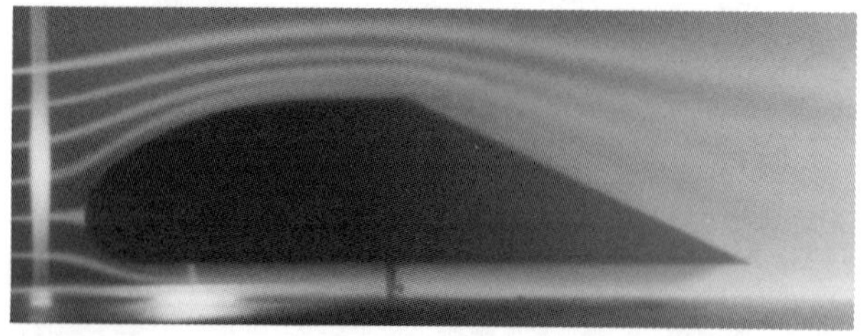

Experiment

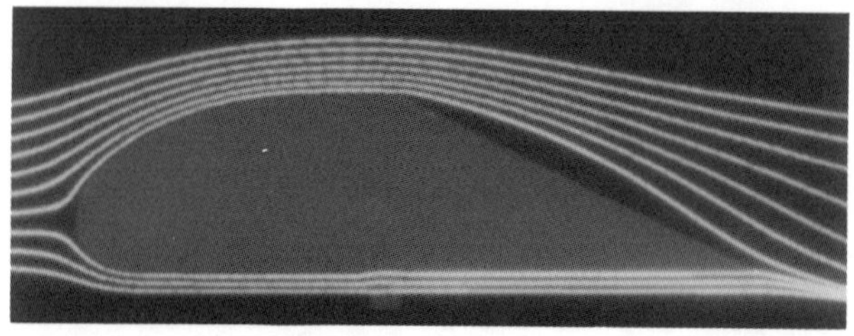

k-ε method

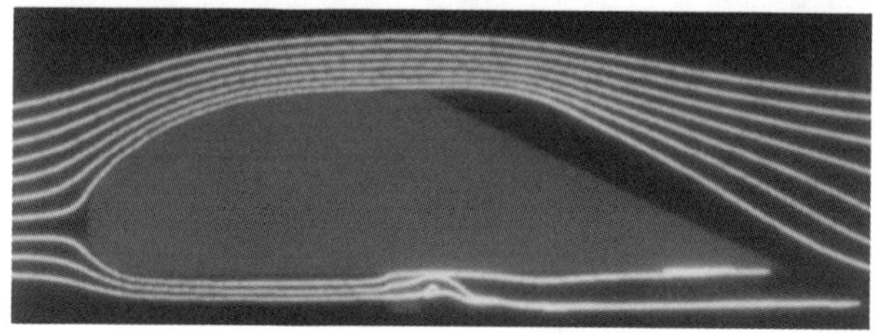

Quasi-direct simulation

(a) θ = 25°

Experiment

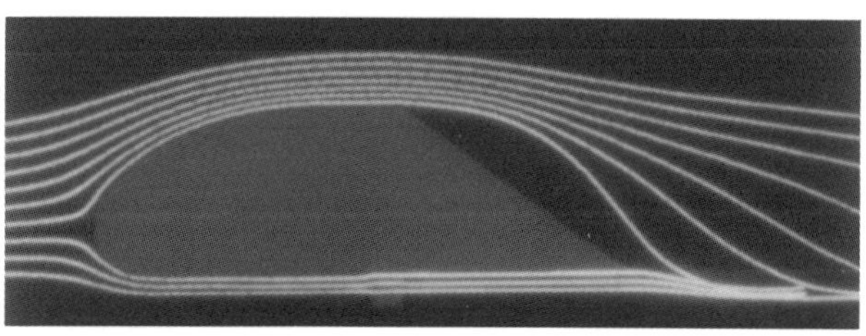

k-ε method

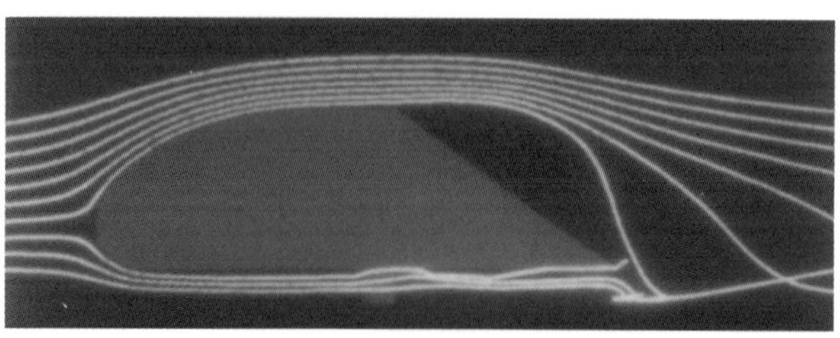

Quasi-direct simulation

(b) θ = 35°

Experiment

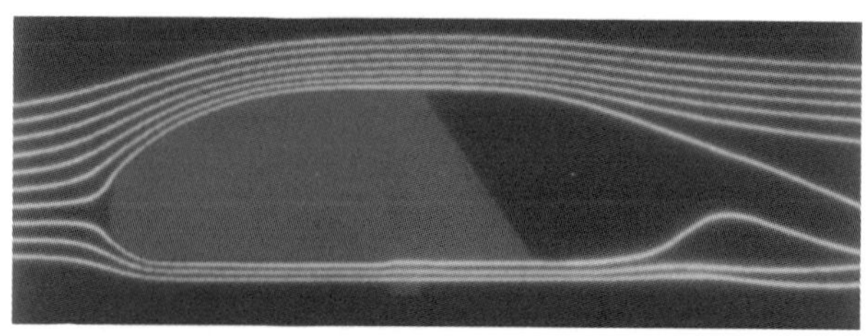

k-ε method

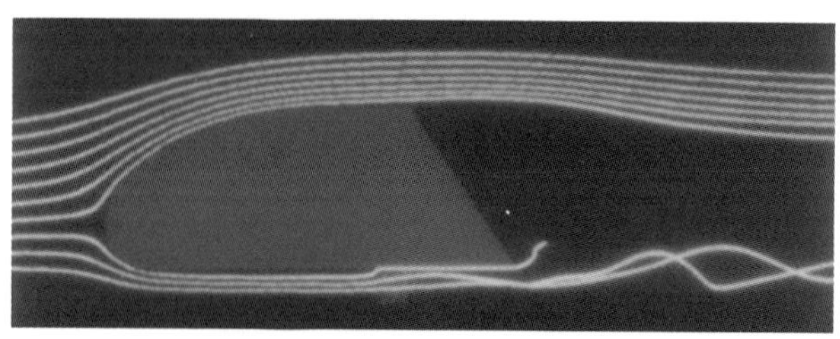

Quasi-direct simulation

(c) θ = 55°

Fig.6 Streamlines in the symmetry plane

perimental separation area at the back of body, that of k-ε method is smaller, but, conversely, that resulting from quasi-direct simulation is larger . For $\theta = 55°$ configuration, although there is a little dissimilarity in the separation area size between k-ε method and quasi-direct simulation, the experimental and computational flow patterns are very similar. Generally, the k-ε method predicts the experimental flow pattern better than quasi-direct simulation which is inclined to predict a larger separation at the back of body.

Figure 7 shows the surface pressure distributions in the symmetry plane. For $\theta = 25°$, the solution of k-ε method, rather than quasi-direct simulation, is in good agreement with experimental data. The disagreement between the solution of quasi-direct simulation and experimental data is due to the difference of flow patterns, shown in Fig.6(a). For $\theta = 35°$, the pressure profile of quasi-direct simulation is closer to experimental data, and the k-ε method can not well predict the experimental peak position of negative pressure on the rear slant. This unsuccessful prediction of k-ε method substantiates the fact that the flow, which separates from the upper body surface at the roof end, lacks curvature (see Fig.6(b)). For $\theta = 55°$, both computations are in good agreement experimentally.

Moreover,to explicitly examine the three-dimensional correspondence between computations and experiments, the flow fields are not only compared in the symmetry but also transverse plane.

Figure 8 shows a comparison of velocity vectors in section II (shown in Fig.5). The computed results of k-ε method are in fairly good agreement with the experimental data for all rear slant angles. On the other hand, in the case of quasi-direct simulation, from experiments, the following differences can be found: the computed velocity vectors just over the center of upper surface(in this case, roof end) are oriented upwards, and the computed longitudinal vortices at the roof and floor corners, which are considered to be generated at the front side corners, are slightly more intensive.

Figure 9 shows the velocity vectors in section III. Although the variation of trailing vortices, in the wake of the body, with rear slant angle is well predicted by both computations, the following characteristics of each method can be seen : in the case of the k-ε method, the computed trailing vortices are circular and calm in general, and in quasi-direct simulation, the computed center positions are either low or outward.

Figure 10 shows the total pressure distributions in section III. The trailing vortices and the after images of longitudinal

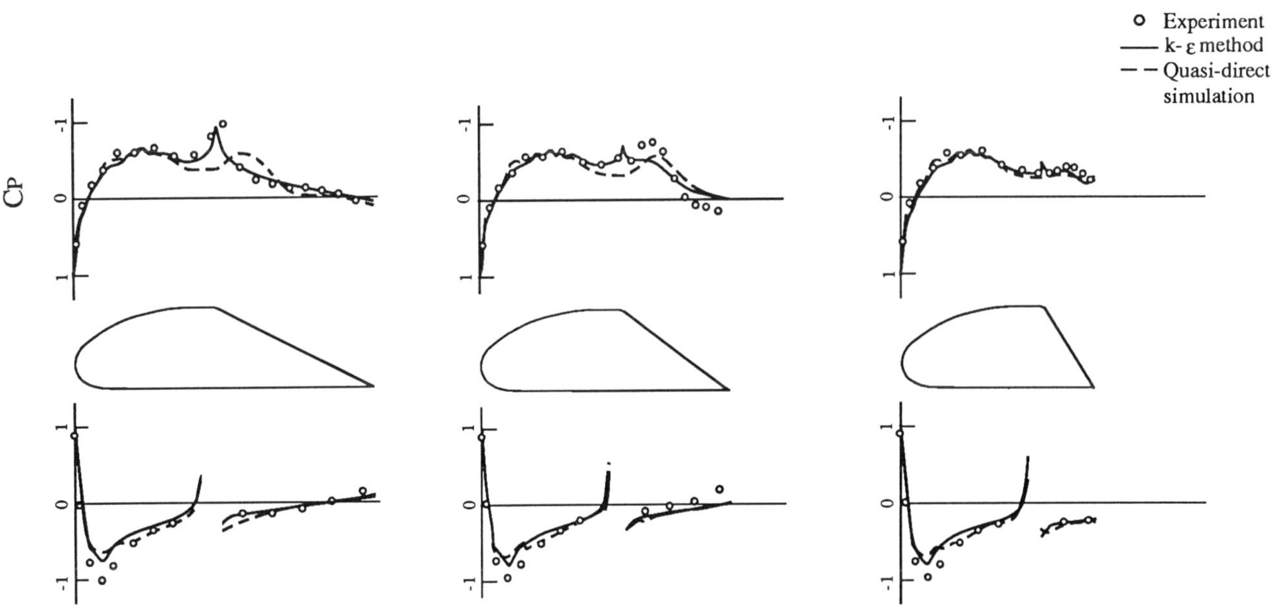

(a) $\theta = 25°$ (b) $\theta = 35°$ (c) $\theta = 55°$

Fig.7 Surface pressure distributions in the symmetry plane

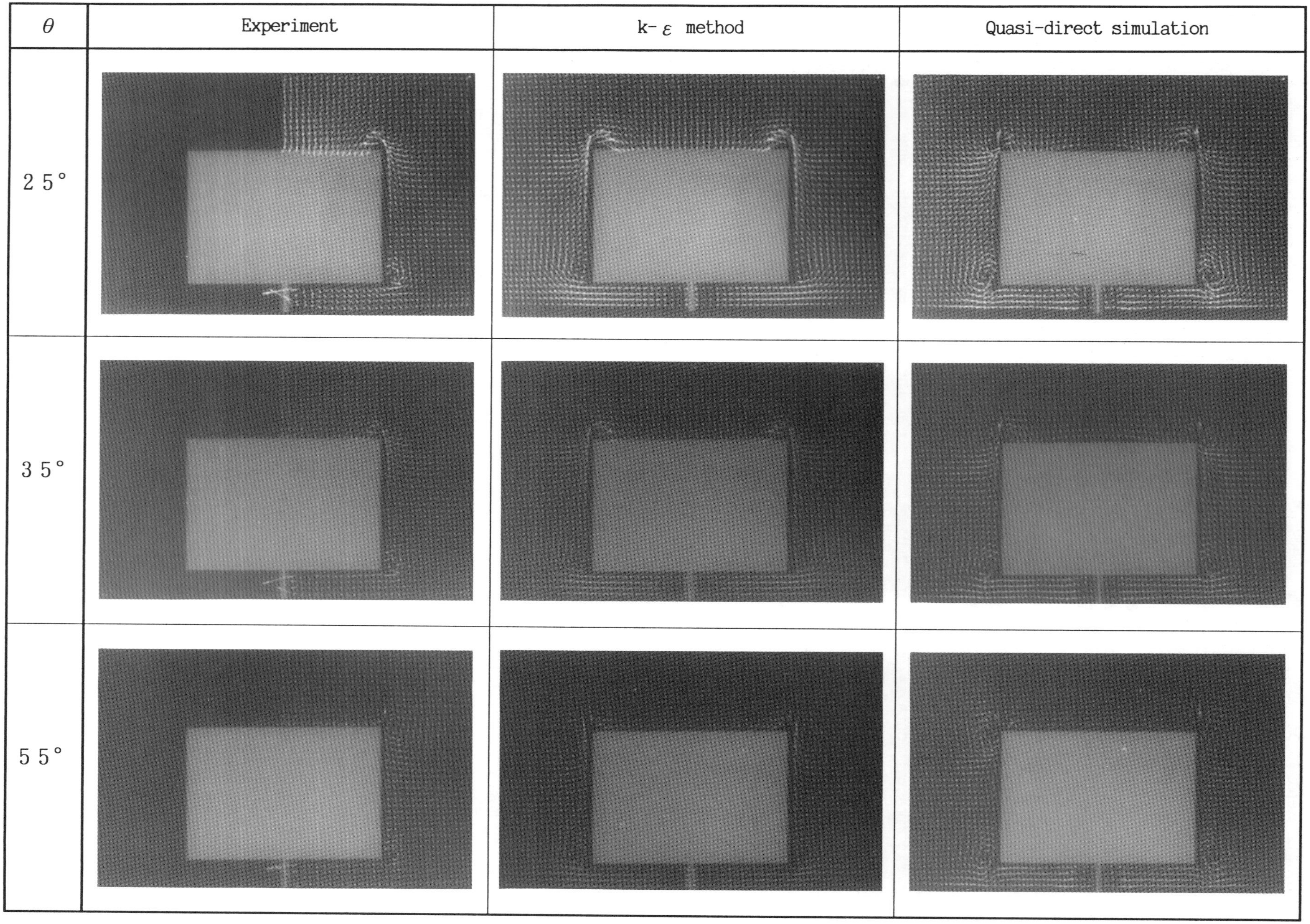

Fig.8 Velocity vectors in section II

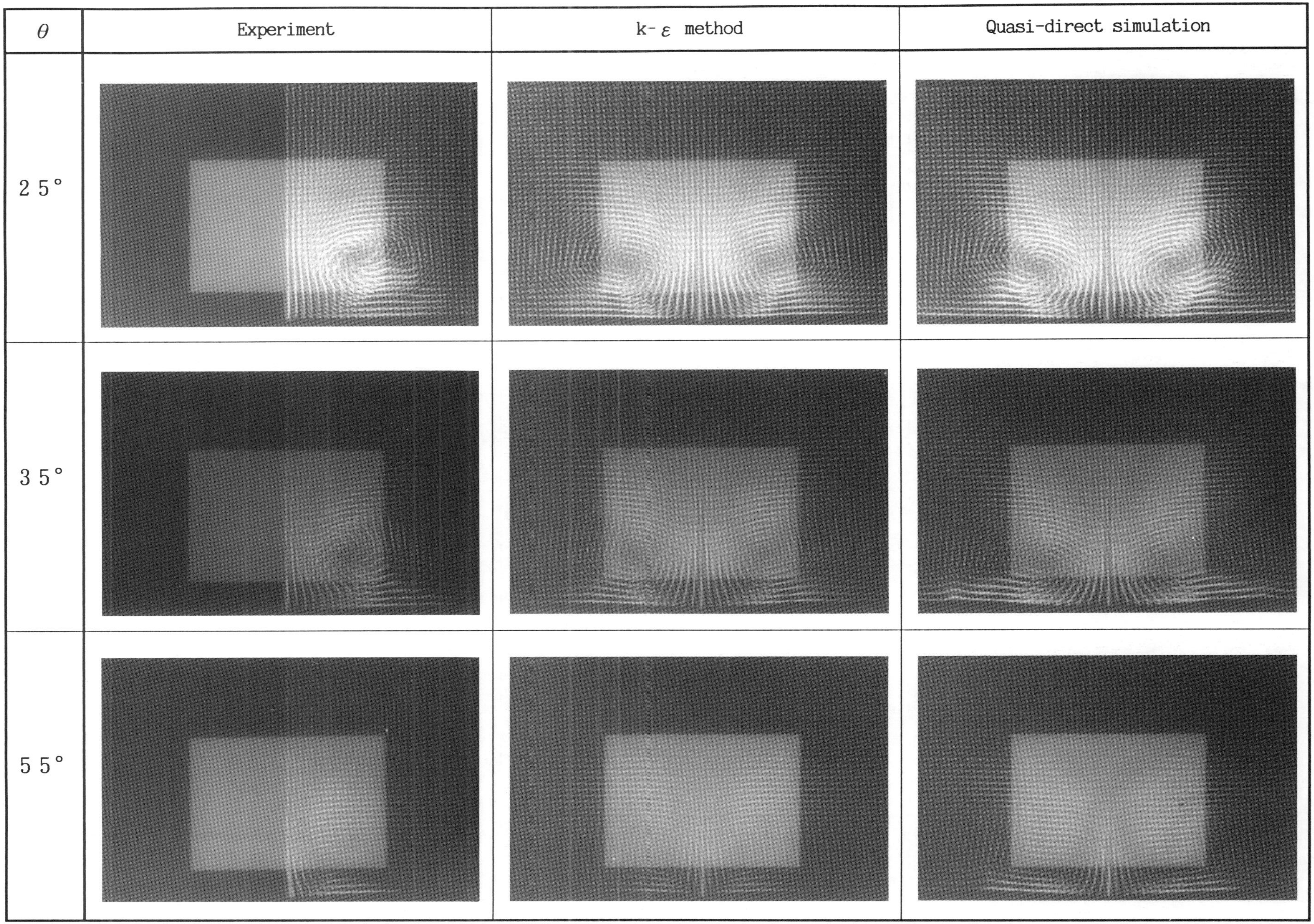

Fig.9 Velocity vectors in section Ⅲ

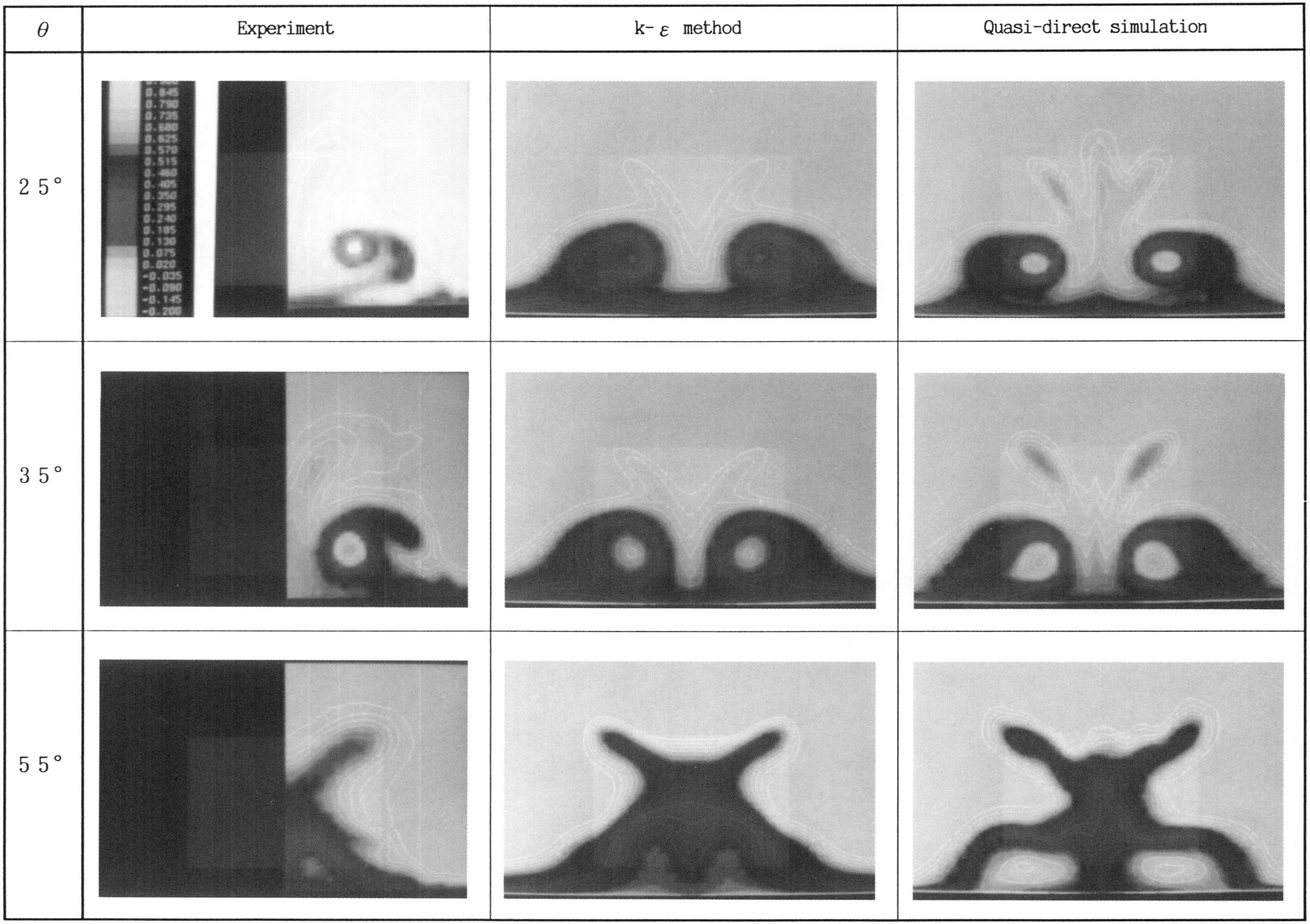

Fig.10 Total pressure distributions in section III

vortices, mentioned in Fig.8, can be confirmed in all cases. Judging from the intensities of trailing vortices, the solutions of quasi-direct simulation, compared with k-ε method, are closer to experimental data. However, judging from locations of trailing vortices, the above conclusion of quasi-direct simulation and k-ε method is reversed.

The correlations between computed results and experiments of velocity vector components (u,v,w) and pressure (p) for θ= 35° configuration, are shown in Fig.11. The dots representing (u, v, w, and p) values consist of all measurement grid points in the three sections, shown in Fig.5. As shown in Fig.11, the k-ε method seems to be, to some extent, superior in the velocity components, but on the contrary, the quasi-direct simulation, to some extent, superior in the pressure.

To grasp the quantitative correspondence between computations and experiments , the correlation coefficients are calculated. The results are shown in Table 3. In this table, the larger or equal corresponding value of each method is marked with a circle or triangle, respectively. From Table 3, the

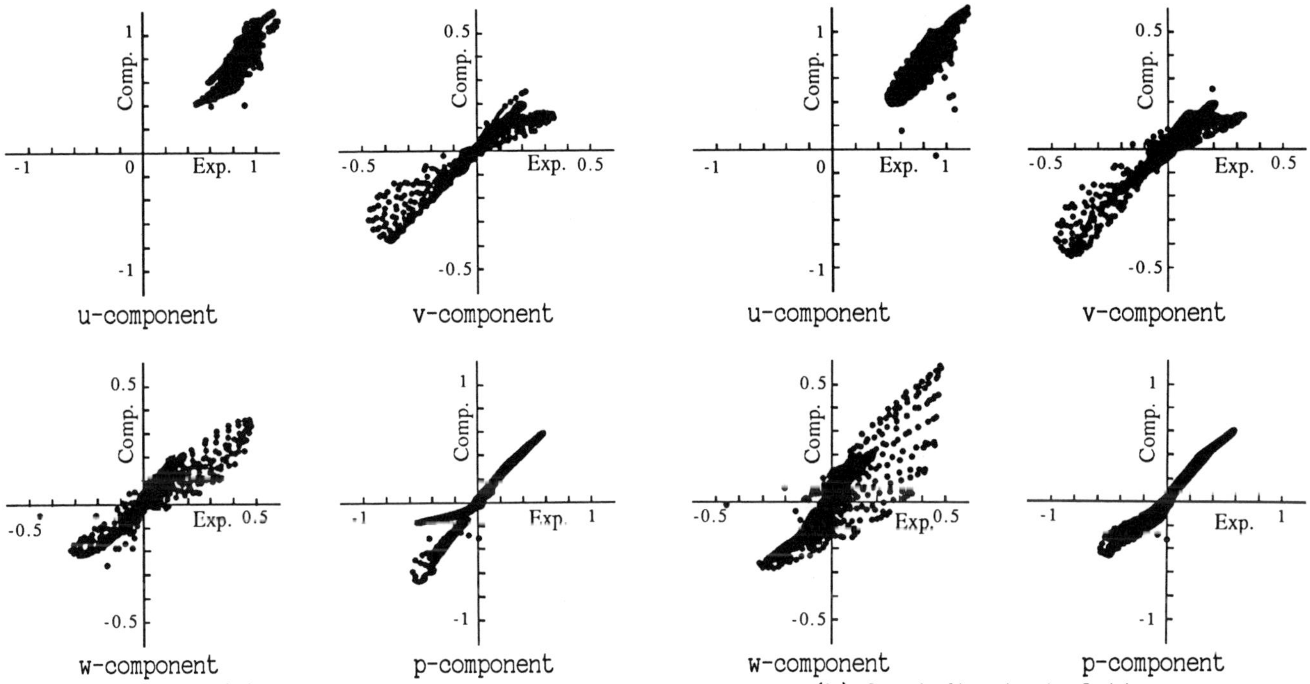

(a) k-ε method (b) Quasi-direct simulation

Fig.11 Correlations between experimental data and computed results for θ= 35° configuration

Table 3 Correlation coefficients between experimental data and computed results

θ	k-ε method					Quasi-direct simulation				
	u	v	w	p	total p	u	v	w	p	total p
25°	0.880	◯0.973	◯0.939	◯0.976	◯0.853	◯0.897	0.959	0.928	0.959	0.837
35°	◯0.915	◯0.960	◯0.926	0.965	0.874	0.900	0.952	0.910	◯0.974	◯0.876
40°	0.844	◯0.910	△0.929	0.979	0.788	◯0.890	0.909	△0.929	◯0.980	◯0.866
45°	0.890	◯0.958	◯0.951	△0.987	0.856	◯0.919	0.952	0.948	△0.987	◯0.906
55°	◯0.932	◯0.977	◯0.938	◯0.985	◯0.911	0.885	0.931	0.937	0.980	0.864

following is found : for the rear slant angles of 35° , 40° ,and 45° near the critical angle at which the drag breakdown occurs (see Fig.2), the static and total pressure distributions are predicted better by the quasi-direct simulation than the k-ε method. In other respects, the solutions of k-ε method are generally in better agreement with experimental data. The aerodynamic condition around the bluff body, whose rear slant angle is near the above mentioned critical angle, is highly fluctuant. Consequently, such a condition may not be within the application limit of k-ε method which is based on the time-averaging. Still, since the velocity field and pressure field are strongly related, a slight inconsistency can be found. As Table 3 demonstrates, the quasi-direct simulation can not relatively well predict the velocity components (v and w), except the main-flow velocity (u), for all rear slant angles. This simulation inaccuracy may be the paradoxical reason that the computational method with better prediction accuracy in the velocity field is different from that of the pressure field.

5.2 COMPARISON OF AERODYNAMIC FORCES - The influence of the skin friction on aerodynamic forces is thought not to be as significant as the pressure in the flow at a high Reynolds number, which is discussed in this paper. Therefore, in calculating the aerodynamic forces from the computed flow, only the effect of pressure is considered.

Comparison of the experimental and computational drag coefficients C_D are shown in Fig.12. The phenomenon, that the drag coefficient increases initially and then decreases with increasing rear slant angle, is roughly predicted by both methods. However, the remarkable drag breakdown which occurs from θ= 35° to 40° is not clearly captured by k-ε method. On the other hand, the same breakdown is well captured by the quasi-direct simulation.

The lift coefficients C_L are shown in Fig.13. As shown in this figure, the rapid lift reduction, similar to drag breakdown, can be observed in the experiment, and is also comparatively well predicted by quasi-direct simulation.

The circumstance,that the rapid variation of aerodynamic forces is relatively well computable by quasi-direct simulation, is due to the fact that the quasi-direct simulation

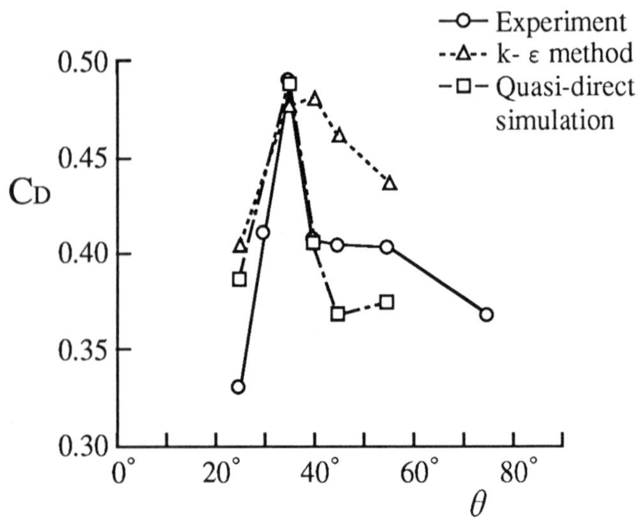

Fig.12 Drag coefficients

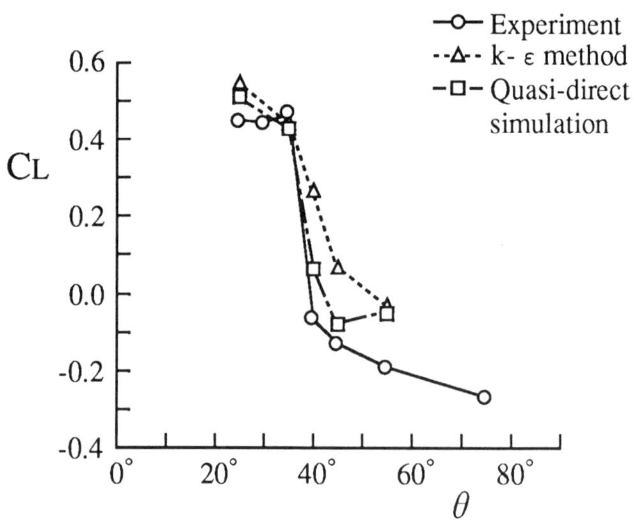

Fig.13 Lift coefficients

is superior to k-ε method in the correlation coefficients of static and total pressure for the rear slant angles where the forces change rapidly (see Table 3). On the other hand, the aerodynamic forces for θ= 25° and 55° are also well predicted by quasi-direct simulation, even if the influence of the skin friction is taken into account. However, this simulation method is not necessarily superior to k-ε method as long as the surface pressure distributions are examined (see, for example, Fig.7). This phenomenon results from the fact that the aerodynamic forces are the integral values.

6. CONCLUSIONS

Flows around the bluff bodies, with various rear slant angles, were computed by two representative computational methods. Comparing the computed results with wind tunnel test data, the following was found :

1) Concerning the flow field, the k-ε method is superior to the quasi-direct simulation, except for severely changing aerodynamic conditions.
2) Concerning the aerodynamic forces, the quasi-direct simulation is superior to k-ε method, particularly for rapid variation of drag and lift.

In the k-ε method, applied in this paper, the standard k-ε model and first-order upwind difference scheme were adopted for the treatment of turbulence and convection terms, respectively. It is known that the standard k-ε model can not well simulate the separated flow because of the isotropic representation ; moreover, the first-order upwind scheme is extremely diffusive. Therefore, the employment of either the anisotropic k-ε model or higher-order difference scheme is generally expected to promote the improvement of the overall computation accuracy.

ACKNOWLEDGEMENT

The authors would like to thank Dr. V. Sumantran of General Motors Research Laboratories for reviewing the manuscript and making a number of helpful suggestions.

REFERENCES

[1] Shaw, C.T. and Simcox, S., The Numerical Prediction of the Flow around a Simplified Vehicle Geometry, Proceedings of the Second International Conference on Supercomputing Applications in the Automotive Industry (1988), 219, A Cray Research,Inc. book.

[2] Hutchings, B.J. and Pien, W., Computation of Three-Dimensional Vehicle Aerodynamics Using FLUENT/BFC, Proceedings of the Second International Conference on Supercomputing Applications in the Automotive Industry (1988), 233, A Cray Research,Inc. book.

[3] Kuriyama,T.,Numerical Simulation on Three-Dimensional Aerodynamics Using "SCRYU", Proceedings of the Second International Conference on Supercomputing Applications in the Automotive Industry (1988), 273, A Cray Research,Inc. book.

[4] Han, T., Computational Analysis of Three-Dimensional Turbulent Flow around a Bluff Body in Ground Proximity, AIAA Journal, Vol.27, No.9(1989), 1213.

[5] Hashiguchi,M., et al., Computational Study of the Wake Structure of a Simplified Ground-Vehicle Shape with Base Slant, SAE Paper 890597(1989).

[6] Himeno, R., et al., Numerical Analysis of the Airflow around Automobiles Using Multi-Block Structured Grids, SAE Paper 900319 (1990).

[7] Kataoka, T., et al., Numerical Simulation of Road Vehicle Aerodynamics and Effect of Aerodynamic Devices,SAE Paper 910597(1991).

[8] Kobayashi,T. and Kitoh,K., A Review of CFD Methods and Their Application to Automobile Aerodynamics, SAE Paper 920338(1992).

[9] Ahmed,S.R., et al., Some Salient Features of the Time-Averaged Ground Vehicle Wake, SAE Paper 840300(1984).

[10] Kawamura,T. and Kuwahara,K., Computation of High Reynolds Number Flow around a Circular Cylinder with Surface Roughness, AIAA Paper 84-0340(1984).

[11] Maeda, K., et al., Analysis of Air Flow Behavior around a Vehicle to Improve Vehicle Aerodynamics, JSAE Review, Vol.11, No.1(1990),18.

Optimization of Bluff-Body Rear-End Shape for Minimum Drag in Ground Proximity

T. Han, D. C. Hammond, Jr., and C. J. Sagi
Engineering Mechanics Dept., General Motors Research Lab.

Abstract

Three-dimensional Navier-Stokes analysis was applied to optimize the rear-end shape of a vehicle-like body in ground proximity. The flow analysis was coupled with an optimization program to find values of the shape parameters: backlight angle, boat-tail angle, and ramp angle which produce a minimum aerodynamic drag coefficient. The approach of this method is to create a localized quadratic approximation to the objective function, in this case drag coefficient, in terms of the three shape parameters. Values of the objective function are calculated using Navier-Stokes analysis of proposed optimum geometries, the localized objective function approximation is updated and used to select a new geometry, and the iteration process is repeated until the objective function converges to a minimum value. The optimum design geometry (17.8-degree backlight, 18.9-degree boat-tail, 9.2-degree ramp) was obtained after fifteen Navier-Stokes analyses. The predicted drag coefficient reduction was 0.10 referenced to the constant-cross-section afterbody. It was observed that the optimum afterbody shape minimized the trailing vortices in the wake and also produced near zero afterbody lift force. The experimentally determined optimum, for a similar but not identical body, was very "flat" but fell in the range: 15- to 18-degree backlight, 15- to 22-degree boat-tail, 9- to 14-degree ramp. The measured drag coefficient reduction was 0.13.

Introduction

Reducing the drag on bluff bodies in ground proximity has important practical applications to ground vehicles. In contrast to the situation in aeronautics, where flow separation from body surfaces is largely avoided, ground vehicles inherently exhibit significant regions of flow separation of both quasi-two-dimensional and fully three-dimensional character. Most of the drag on bluff bodies is pressure drag, directly due to flow separation from the body surface. Therefore, minimizing the drag of bluff bodies is an exercise in controlling the location and scale of unavoidable flow separations.

It is often observed that the drag coefficient is not sensitive to the forebody shape if the flow over that region is free from large-scale separations.[1] (Traditionally, on automobiles the forebody is that portion ahead of and the afterbody that portion behind the transverse plane of maximum cross-sectional area. On bluff bodies like the one considered here the exact division is unclear and unimportant.) By careful body shaping, flow separation on the forebody can be easily minimized for the turbulent flows usually experienced.

In contrast, flow separation on the afterbody of ground vehicles is unavoidable, and much less controllable. Separation on the afterbody is the major source of aerodynamic drag for bluff bodies (with separation-free forebodies). Therefore, reducing the aerodynamic drag produced by the afterbody becomes one of the most important subjects in ground vehicle aerodynamics.

Boat-tailing is a well known method of drag reduction and has been applied to two-dimensional and axisymmetric bodies[2] as well as three-dimensional vehicle-like bluff bodies. By tapering the afterbody surface, boat-tailing permits pressure recovery in the flow along the afterbody surface, thereby reducing the pressure drag. Excessive boat-tailing produces flow separation and can even increase drag above that of the un-boat-tailed geometry. Three independent boat-tailing angles (upper, side, and lower surfaces) are required to specify the geometry of three-dimensional bluff bodies with rectangular cross sections. These three angles are called the:

* Member AIAA

backlight angle, boat-tail angle, and ramp angle and are shown in Fig. 1. In addition to the quasi-two-dimensional separation which can occur on two-dimensional boat-tailed bodies, three-dimensional bodies can exhibit three-dimensional separation, i.e. vortex shedding from longitudinal edges. This results from flow interaction among upper, side, and lower surfaces that is complex and highly non-linear.

Heretofore, afterbody shape optimization has relied heavily on wind tunnel testing.[3] This paper reports a computational technique for optimizing afterbody shape to minimize drag.

Previous Computational Study of Ahmed's Body

Computations were previously performed for three-dimensional turbulent flows around Ahmed's vehicle-like body.[4] The detailed computational method and results were reported for various afterbody backlight angles for a pure fastback afterbody at a Reynolds number of 4.3 million based on body length.[5] A grid refinement study was presented which established a mesh of 97 streamwise, 51 vertical, and 31 horizontal nodes as sufficient for producing grid-independent solutions. This previous computational study predicted most of the essential measured features of the flow field around Ahmed's body in ground proximity, such as the formation of trailing vortices and flow separation with recirculation at the vertical base of the afterbody. It also established that the incremental changes in the drag coefficient produced by changes in backlight angle were predicted with reasonable accuracy for the backlight angles between 0 and 25 degrees. This quantitative validation prompted exploring use of the analysis for aerodynamic optimization.

Solution Procedure

The contribution of skin friction drag to the total drag is small for bluff bodies. In Ahmed's experiment, about 80 percent of the total drag is pressure drag. For Ahmed's body with its blunt-but-rounded front end, the forebody pressure drag is only about 6 percent of the total drag. Therefore, more than 70 percent of the total drag comes from the afterbody pressure drag. In the present study, we will minimize the afterbody pressure drag by altering the three afterbody angles, which will approximately minimize the total drag.

The previously validated three-dimensional Navier-Stokes analysis, VINE3D[5], was manually coupled with a parametric optimization analysis, COPES[6], and used to find the optimum afterbody shape design variables: backlight, boat-tail, and ramp angle, which produce a minimum aerodynamic drag coefficient. The Navier-Stokes analysis used has been fully described by Han[5]; in the interest of brevity, that description will not be repeated here. A brief summary of the optimization analysis follows.

The approximate optimization technique, COPES[6], is suitable for problems with less than ten design variables and where the computational cost of each analysis is very high. The method used is to create an analytic approximation to the objective function, the drag coefficient in the present study, in terms of the three design parameters (backlight, boat-tail, and ramp angles).[7] We then perform the optimization using these approximations and search for a proposed optimum design. The objective function for the drag coefficient is then updated by including the results of a Navier-Stokes analysis of this proposed optimum design, and the process is continued until convergence to a precise optimum design geometry has been achieved.

The approximate function for the drag coefficient can be obtained by a simple Taylor series expansion.

$$C_D(\mathbf{X}) = C_D(\mathbf{X}^o) + \Delta C_D(\mathbf{X}^o)\,\delta\mathbf{X} + 1/2\,\delta\mathbf{X}\,\mathbf{H}(\mathbf{X}^o)\,\delta\mathbf{X} \tag{1}$$

where $\mathbf{X}$ is the vector of design variables, three afterbody angles in the present study, and superscript "o" denotes the nominal design values. $C_D(\mathbf{X})$ is the objective function, drag coefficient. $\Delta C_D(\mathbf{X})$ is the gradient, and $\mathbf{H}(\mathbf{X})$ is the Hessian matrix. The perturbation of design variables $\delta\mathbf{X}$ is defined as

$$\delta\mathbf{X} = \mathbf{X} - \mathbf{X}^o \tag{2}$$

In order to have full quadratic approximation of the objective function, we have to determine $m = (n+3)n/2$ unknowns where n is the number of design parameters ($n=3$ and $m=9$; in the present study) in addition to a nominal design $\mathbf{X}^o$.

In the present study, our main objective is to minimize the drag coefficient subject to simple linear constraints of the design variables, i.e.

$$\text{minimize} \quad C_D(\alpha,\beta,\gamma)$$

subject to

$$0 < \alpha < 30 \text{ degrees}$$
$$0 < \beta < 30 \text{ degrees}$$
$$0 < \gamma < 20 \text{ degrees}$$

where α, β, and γ are backlight, boat-tail, and ramp angles respectively. In the present study, the above constraints are specified simply to maintain a positive rear vertical base area. These constraints can be subsequently modified based on specific styling or packaging constraints. $C_D(\alpha,\beta,\gamma)$ is the pressure drag of the afterbody having angles, α, β, and γ and is evaluated numerically by solving the Navier-Stokes equations.

The overall accuracy and stability of the optimization depends on how the objective function is approximated and, also, how the objective function spans the design space to include a global optimum. In order to cover a wide range of the design space the initial objective function was approximated from an initial distribution of the design variables based on Taguchi orthogonal arrays.[8] The Taguchi orthogonal arrays for three levels and three design parameters are shown in Table 1. The three level design values were specified to cover a wide range of design space, i.e. 0, 10, and 25 degrees for both backlight and boat-tail angles and 0, 7, and 15 degrees for ramp angles.

Present Computational Study

Ahmed's body has an overall length of 1.04 m and length:width: height ratios of 3.36:1.37:1. In the present study, the afterbody shape is defined by the three afterbody angles and the length of the afterbody is specified the same as the height of the body ($l_b/H = 1.0$). It consists, therefore, arbitrarily of a forebody, a long mid-section and an afterbody. Again all computations were based on a free stream velocity of 60 m/s, which corresponds to a Reynolds number of 4.3×10^6.

Boundary Conditions

Due to the relatively long mid-section, the flow interaction between the rear and the front is very weak. It was observed from both experiment and computation that there are no noticeable differences in the surface pressures on the forebody even with large changes in the afterbody. In the present study, the numerical computations were performed after the mid-section($x/L = -0.5$) of Ahmed's body with various afterbody shapes in order to reduce overall computational cost. Therefore, the inlet of the computational domain was located at the mid-section($x/L = -0.5$) of the body and boundary conditions were specified from the previously reported full domain calculations.[5] The three velocity components and the turbulent kinetic energy and its dissipation rate were specified at the inlet and at the outer boundaries from the full domain calculations. The exit plane was located at roughly two body lengths downstream of the body and the constant pressure boundary condition was imposed. Symmetry boundary conditions were enforced at the symmetry plane. At the body surface and the ground plane, no-slip boundary conditions were imposed. Near the wall, the standard log-law wall functions were used for the velocity distribution.[9]

Computational Grid

A simple sheared grid system was generated by applying algebraic stretching with a hyperbolic tangent function. In the present work, the computations were performed only after the mid-section of the body by skipping the first 36 grid planes in the (97x51x31) mesh; therefore, a (61x51x31) mesh was utilized to simulate the flow around the afterbody and the downstream wake.

Computational Performance

It takes about 200 flow-solver iterations to converge the aerodynamic drag coefficient within an accuracy of 5×10^{-4}. It is estimated that the CPU time per point and iteration (on the CRAY YMP), including data input/output is about 6×10^{-5} seconds. Therefore, it takes roughly 30 minutes of CPU time for one complete calculation of an afterbody with a (61x51x31) mesh.

Results and Discussion

The search for the optimum afterbody shape was terminated when the differences of drag coefficient between consecutive iterations were less than 0.001. Among the nine orthogonal arrays, Case 9 ($\alpha=25$, $\beta=25$, $\gamma=7$) produced a minimum pressure drag coefficient of 0.124. Optimization started with this afterbody shape as a nominal design after the initial nine function evaluations to establish the quadratic approximation of the drag coefficient. Six additional Navier-Stokes analyses were required to reach an optimum afterbody ($\alpha=17.8$, $\beta=18.9$, $\gamma=9.2$), which reduced the drag coefficient by 0.014. The overall convergence pattern is shown in Fig. 2. The body shape is shown as a surface grid in Fig. 3. The predicted drag coefficient reduction was 0.115 from Case 3 ($\alpha=0$, $\beta=25$, $\gamma=15$). The experimentally determined optimum for a similar but not identical body, indicated a very wide range: 15 to 18 degrees backlight, 15 to 22 degrees boat-tail, 9 to 14 degrees ramp. The experimentally measured drag coefficient reduction was 0.13.

Drag and Lift Variations

Contributions to pressure drag from backlight C_α, boat-tail C_β, ramp C_γ, and vertical rear end base C_b were calculated by integration of the axial component of the calculated pressure over the individual surfaces. The afterbody lift coefficient was calculated by integrating the normal component of surface pressure on the slanted surfaces. The drag breakdown for various afterbody shapes is listed in Table 2 along with the afterbody lift coefficient, the base pressure coefficient, and the base area ratio. Major contribution to the total pressure drag comes from the slanted surfaces and vertical base surface of the rear end. For Case 1, (so called squareback), rear end pressure drag is wholly due to the flat base. The vertical base area decreases with increasing slant angles, but the base pressure increases as do the relative contributions to the drag of the various afterbody surfaces. The exact level of the base pressure is controlled by both the shedding of longitudinal vortices from the body surfaces and the reverse flow separation near the base.

The present analysis predicts that the optimum afterbody shape corresponds to maximizing the product of base pressure and base area as well as minimizing the afterbody lift.

Wake Flow Structure

The structure of the wake is a consequence of the flow around the body and can offer insight into drag-producing mechanisms. The computed flow will be summarized by presenting results in the plane of symmetry and and in a transverse plane at one body length downstream. There were dramatic differences in the predicted wake flows for the fifteen cases computed, and four will be discussed here.

Case 1 (0, 0, 0): The wake behind the body is characterized by a large separation zone as shown in Figs. 4a and b. At the symmetry plane, two recirculating vortices are clearly visible and the separation bubble closes at roughly a half-body-length downstream of the body. As listed in Table 2, a relatively low base pressure coefficient, -0.209, was predicted due to the large flow separation at the base. The total pressure contours of Fig. 4c show the turbulent wake at one body length downstream; also shown are the thickened turbulent boundary layers formed on the stationary ground plane. The shape of the wake boundaries were defined as the position of the 0.9 total pressure contour. The height of the low energy wake region (total pressure contour 0.3) decreases downstream due to the downwash induced by the weak longitudinal vortices generated downstream of the wake closure as shown in Fig. 4c. The pair of weak longitudinal vortices was not generated directly from the body surface, but was formed as the result of a slight vorticity imbalance between the upper and the lower vortices shown in Fig. 4a, which are part of a toroidal vortex structure.

Case 3 (0, 25, 15): This afterbody shape produced the highest total pressure drag among the 15 afterbody shapes evaluated in the present study, higher even than Case 1 which has a constant cross-section afterbody. It is interesting to observe the behavior of the wake at the symmetry plane as shown in Figs. 5a and b. The height of the wake boundary increases rapidly downstream. A pair of strong longitudinal vortices is generated along the upper longitudinal edges and is clearly visible in Fig. 5c. No vortices were formed along the lower longitudinal edges due to a proper match of ramp and boat-tail angles. Due to the strong inflow towards the separation bubble at the rear base, the reverse flow region is relatively short (as shown in Fig. 5a) and narrow (as shown in Fig. 5c).

As shown in Fig. 5c, a sharply defined pair of longitudinal vortices move up and towards each other downstream of the body. This predicted behavior is consistent with the potential theory for vortices in ground proximity.[10] The strong upwash associated with the formation of counter rotating longitudinal vortices is responsible for the spread of the wake boundaries downstream of the separation bubble. As shown in Figs. 5a and b, the height of the wake boundary was extended to roughly two body heights at X/L = 1.0. This large wake disturbance is responsible for the high drag coefficient of this afterbody.

Case 9 (25, 25, 7): This case produced a minimum base area and also produced a minimum pressure drag among the first nine afterbody shapes which were based on the orthogonal arrays. As shown in Figs. 6a-b, the effect of backlight angle tends to dominate the flow pattern in the separation bubble behind the base, and a relatively large upper vortex was generated. No apparent longitudinal vortices were found along the upper edges. However, a pair of relatively weak vortices was generated along the lower longitudinal edges as shown in Fig. 6c. These vortices deform the total pressure contours, Fig. 6c; the total pressure contours wrap around the vortex core.

Case 15 (17.8, 18.9, 9.2): Six additional iterations were performed to obtain this afterbody which minimized the afterbody pressure drag. This optimum afterbody reduced the overall pressure drag coefficient by 0.115 from Case 3 ($\alpha=0$, $\beta=25$, $\gamma=15$), and by 0.1 from the Case 1 with the constant afterbody cross section. This body further produced a 0.014 reduction of pressure drag coefficient from Case 9 even with a 50% larger base area. The resulting flow field is characterized by well balanced upper and the lower recirculating vortices in the separation bubble as shown in Figs. 7a and b. This afterbody also minimizes the longitudinal vortices downstream, as shown in Fig. 7c. Previous studies have similarly concluded that minimizing the trailing vortices in the wake will reduce drag.[11,12] The total pressure contours, as shown in Fig. 7c, bulge out slightly along the four corners. This is probably due to the interaction of two turbulent boundary layers which were developed along the slanted surfaces. In order to reduce this boundary layer interaction, the sharp longitudinal edges should be rounded.

Validation of the Optimization

Data from an independent experimental study can be used to show the qualitative validity of the present computational optimization study. The overall length of the experimental models is 1.44m and the length:width:height ratios are 3.79:1.38:1.00. The afterbody lengths (l_b/H) of the experimental models are 0.80 and 1.0 body heights. Thus, the experimental model with $l_b/H=0.80$ has a slightly shorter afterbody and a slightly longer mid-section than the computational model. Also, the experimental models were equipped with wheels that were mounted with clearance into wheel wells; the afterbody started just downstream of the rear wheel wells. The experiments were carried out at a free stream velocity of 60 m/s, which corresponds to the same unit Reynolds number as the computations. The backlight, boat-tail, and ramp angles, (α,β,γ), complete the specification of the afterbody geometry.

It was observed from computation and experiment that the drag coefficient is very "flat" near the optimum design variables. A wide range of design parameters(backlight, boat-tail, and ramp angles) near the optimum produces a relatively low afterbody drag coefficient. Therefore, various combinations of these three afterbody angles will be equally acceptable. In a practical engineering sense, it is not sensible to search for a precise optimum because the amount of further improvement of the drag coefficient will be small and the computational cost large. It is more practical to identify the range of design variables which produce a relatively low drag coefficient rather than to obtain the precise optimum.

A quadratic polynomial approximation of the objective function for drag coefficient was obtained from a weighted least-squares fit in which the design information near the optimum are weighted most heavily. Altogether information from fifteen afterbody shapes is utilized to establish the analytic function for the drag coefficient in terms of the three afterbody angles, α, β, and γ. This analytic expression for the drag coefficient provides the overall behavior of the drag coefficient near the optimum afterbody shape.

Consider first the comparison of the predicted optimum afterbody geometry to the experimentally determined optimum. Figs. 8a, b, and c show the interactions among three afterbody angles near the optimum afterbody shape. The experimental data is too sparse and the experimental uncertainty too high to produce contours of drag differences to compare directly to the predicted values. However, the experimentally determined optimum geometries (15,15,9), (18,18,9) and (21,18,12), have been plotted on Figs. 8a, b, and c since all three geometries were found to attain the minimum drag of 0.220. Note that the points with 18-degree boat-tail angle were obtained from cross plots of the actual data points; these geometries were not tested. The experimentally determined optimums all fall close to the predicted optimum (17.8,18.9,9.2). The level of agreement is surprisingly good when the experimental uncertainty of the data (±0.0035) and the differences in the model geometries are considered. The range of the experimentally determined optimum geometries also substantiates the predicted "flat" nature of the minimum.

The interaction between backlight and ramp angles, as shown in Fig. 8a, make it necessary to change the backlight angle, α, and the ramp angle, β, simultaneously to maintain a relatively low drag coefficient. This is probably because the pair of longitudinal vortices generated from the backlight and the ramp oppose each other and tend to be weakened in strength if the angles are "balanced." Fig. 8a also shows that a relatively small ramp angle is required to compensate for a relatively large backlight angle. This is due to the stronger vortices formed from the ramp than from the backlight for the same slant angles, α and γ.

Fig. 8b shows the interaction between the backlight angle, α, and the boat-tail angle, β, at the optimum ramp angle of 9.2 degrees. As shown in Fig. 8b, the relative change of the drag coefficient due to the change of the backlight angle and the boat-tail angle are nearly the same.

Fig. 8c shows the interaction between the boat-tail angle, β and the ramp angle γ, near the optimum backlight angle of 17.8 degrees. The drag coefficient is more sensitive to changes in the boat-tail angle than to changes in the ramp angle near the optimum afterbody shape. However, this may not be generally true, particularly when the strength of the longitudinal vortices becomes a major source of the pressure drag. Away from the optimum, It was observed that the drag is the most sensitive to the ramp angle if the other two angles are fixed at zero.

The computational and experimental studies are also compared in Figs. 9a-c where the incremental changes of afterbody pressure drag coefficient are presented as functions of each of the afterbody parameters, using the optimum afterbody value as a reference. As previously discussed, the approximation function for the drag coefficient was obtained from a weighted least-squares fit of the computed drag coefficients of all fifteen afterbodies. Since those afterbodies near the optimum were weighted more heavily, the approximation for drag coefficient increments will be most accurate near the optimum afterbody. The experimental data are made up of composites of the available drag data closest to the predicted optimum afterbody geometry. In order to make direct comparisons, both the computational and experimental data were adjusted to give the same value of afterbody pressure drag for the optimum afterbody, Case 15. In each comparison there is good agreement in the shape of the curves, the location of minima, and the drag change from the optimum afterbody (17.8,18.9,9.2).

As shown in Fig. 9a, the drag coefficient increases monotonically with the ramp angle for a squareback (α = 0, β = 0), shown as a dotted line, which is substantiated by the experimental data. However, for the optimum afterbody (α = 17.8, β = 18.9), shown as a solid line, the drag coefficient is much less sensitive to the change of the ramp angle. The drag coefficient initially decreases slightly with the ramp angle and then increases gradually after roughly nine degrees of ramp angle. For the optimum afterbody, the weak dependence of the pressure drag on ramp angle was predicted very well, and it agrees with the experimental trend for ramp angles greater than five degrees.

Fig. 9b shows the effects of the backlight angle on the drag coefficients for both squareback (β=0, γ=0) and the optimum afterbody shapes (β=18.9, γ=9.2). For the optimum afterbody, as shown as a solid line, the drag coefficient initially decreases with increasing backlight angle reaching a minimum near eighteen degrees corresponding to 0.08 drop in drag coefficient from the value at zero backlight angle. This is in good agreement with the experimental data, Fig. 9b, which shows approximately a 0.088 drop in drag coefficient from the value at zero backlight angle. For the squareback, as shown as a dotted line, the minimum drag coefficient occurs near thirteen degrees of backlight angle. As the strength of the longitudinal vortices increases, the drag coefficient starts to increase after thirteen degrees of backlight angle. The overall trends are consistent with the experimental data, as shown in Fig. 9b.

Fig. 9c shows the effects of the boat-tail angle on the drag coefficient. Both the squareback (α=0, γ=0) and the optimum afterbody (α=17.8, γ=9.2) produced a minimum drag coefficient near nineteen degrees of boat-tail angle, and the drag coefficient starts to increase for greater angles.

The variation of the drag coefficient with boat-tail angle for the optimum afterbody is very well predicted for angles less than twenty degrees. However, experimental values increase rapidly beyond twenty degrees of boat-tail angle (indicated as filled symbols); this behavior is likely due to flow separation on the boat-tail surfaces, which is thought to occur at approximately twenty-five degrees of boat-tail angle. At this critical angle, the near-wake flow changes abruptly from one dominated by vortices formed along the longitudinal edges to one dominated by a separation bubble. The present computation predicted attached flow on the boat-tail surfaces up to thirty degrees of boat-tail angle, and thus did not capture the observed wake transition.

The performance of the k-ϵ turbulence model in the adverse pressure gradients encountered on the boat-tail surfaces is suspect. It overpredicts skin friction and the increase in length scale normal to the wall, both of which tend to produce attached flow.

For the optimum afterbody, the experimental data, Fig. 9c, show approximately a 0.078 drop in drag coefficient from the value at zero boat-tail angle, and the computations show a 0.075 drop in the afterbody drag coefficient. For the optimum afterbody, as shown in Fig. 9c, the change of drag coefficient due to the change of the boat-tail angle and the backlight angle is almost identical near the optimum, which is also consistent with the result shown in Fig. 8b.

As a summary, the drag data shows that the computations correctly predict not only the optimum geometry but also the flat behavior of the drag surface in the vicinity of the minimum. The effects of the backlight angle and the boat-tail angle on the drag coefficient are similar. However, the ramp angle affects the drag coefficient differently. For the ramp angles, shown in Fig. 9a, the strong longitudinal vortices overpower the boat-tailing effects and the drag coefficient increases monotonically for the squareback afterbody. It has been observed that the strength of the ramp vortices are intensified by the ground plane, which is a unique feature associated with ground vehicles.

Conclusions

1. This optimization method and the underlying Navier-Stokes analysis can find optimum designs in a space containing dramatically different wake-flow structures. However, the performance in predicting the onset of pressure-gradient-driven flow separation at boat-tail angles exceeding twenty-five degrees needs to be improved.

2. Minimizing drag coefficient yields an afterbody geometry which: minimizes afterbody lift coefficient, minimizes longitudinal vorticity in the wake, and maximizes the product of base pressure and base area.

3. The drag coefficient hypersurface is very "flat" near the optimum geometry, permitting a number of different afterbodies having near-optimum drag.

4. A local quadratic approximation to the drag coefficient is adequate both for optimization and characterizing the interactions among the design variables.

5. Boat-tail angle can be optimized independently of backlight and ramp angles. However, backlight and ramp angles must be increased or decreased together to minimize drag.

6. These optimization calculations are relatively expensive, in this case consuming almost 8 CPU hours on a Cray YMP. However, the reader must judge the cost relative to an experimental optimization.

References

1. Morel, T., "Theoretical lower limits of forebody drag," Aeronautical Journal, Paper No. 639, Jan. 1979.

2. Mair, W. A., "Reduction of Base Drag by Boat-Tailed Afterbodies in Low Speed Flow," Aeronautical Quarterly, Vol. 20, 1969, pp. 307-320.

3. Kohri, L. and Numata, N., "Effects of rear-end configuration of vehicles on aerodynamic drag," Int. J. of Vehicle Design, Vol. 9, No. 2, 1988.

4. Ahmed, S. R., Ramm, G., and Faltin, G., "Some salient features of the time-averaged ground vehicle wake," SAE Paper 840300, 1984.

5. Han, T., "Computational analysis of three-dimensional turbulent flow around a bluff body in ground proximity," AIAA Journal, Vol. 27, No. 9, September 1989.

6. Madsen, L. E. and Vanderplaats, G. N., COPES - A Fortran Control Program for Engineering Synthesis, NPS69-81-003, March 1982.

7. Vanderplaats, G. N., Numerical Optimization Techniques for Engineering Design, McGraw-Hill, New York, 1984, pp. 211-215.

8. Taguchi, G., Introduction to Quality Engineering, Asian Productivity Organization, pp. 181-188, 1986.

9. Launder, B. E. and Spaulding, D. B., "The Numerical Calculation of Turbulent Flows," Computational Methods in Applied Mechanics and Engineering, Vol. 3, March 1974, pp. 269-289.

10. Teitjens, O. G., Fundamentals of Hydro- and Aeromechanics, based on Lectures of L. Prandtl, Dover, New York, 1957, pp. 201-213.

11. Morelli, A., "Aerodynamic basic bodies suitable for automobile applications," Proceedings of the International Association for Vehicle Design: Technological Advances in Vehicle Design; SP3, editors M. A. Dorgham and U. L. Businaro, 1983.

12. Bearman, P. W., "Some observation on road vehicle wakes," SAE Paper 830922, Feb. 1983.

Table 1 Taguchi Orthogonal Arrays and Design Variables

Design Variables

Level	Backlight α	Boat-tail β	Ramp γ
1	0 deg.	0 deg.	0 deg.
2	10 deg.	10 deg.	7 deg.
3	25 deg.	25 deg.	15 deg.

Taguchi Orthogonal Arrays

Array	Backlight α	Boat-tail β	Ramp γ
1	1	1	1
2	1	2	2
3	1	3	3
4	2	1	2
5	2	2	3
6	2	3	1
7	3	1	3
8	3	2	1
9	3	3	2

Table 2 Drag breakdown and lift coefficient for various afterbodies

Case Runs	Back-light α	Boat-tail β	Ramp γ	Cd Total Drag Coeff.	Cl After-body Lift Coeff.	Cα Back-light Drag Coeff.	Cβ Boat-tail Drag Coeff.	Cγ Ramp Drag Coeff.	Cb Base Drag Coeff.	Cp Base Press. Coeff.	Base Area Ratio
1	0	0	0	.209	-.024	0	0	0	.209	-.209	1.0
2	0	10	7	.199	-.120	0	.05	.027	.122	-.186	.657
3	0	25	15	.225	-.079	0	.142	.041	.042	-.146	.287
4	10	0	7	.204	-.095	.028	0	.035	.141	-.200	.704
5	10	10	15	.151	-.075	.026	.039	.054	.032	-.075	.425
6	10	25	0	.133	.011	.024	.122	0	-.013	.041	.320
7	25	0	15	.183	.005	.112	0	.071	0	.0	.319
8	25	10	0	.159	.070	.084	.039	0	.036	-.083	.433
9	25	25	7	.124	.009	.053	.078	.014	-.021	.113	.186
15	17.8	18.9	9.2	.110	.004	.041	.063	.021	-.015	.053	.283

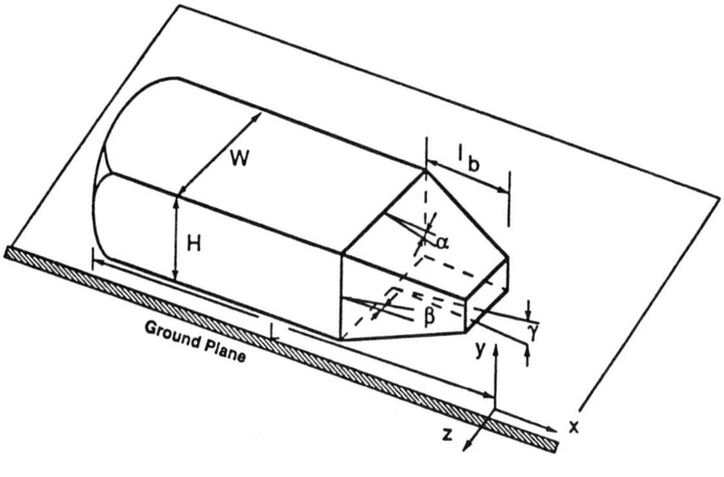

Fig. 1. Schematic of the vehicle-like body with three boat-tail angles in ground proximity.

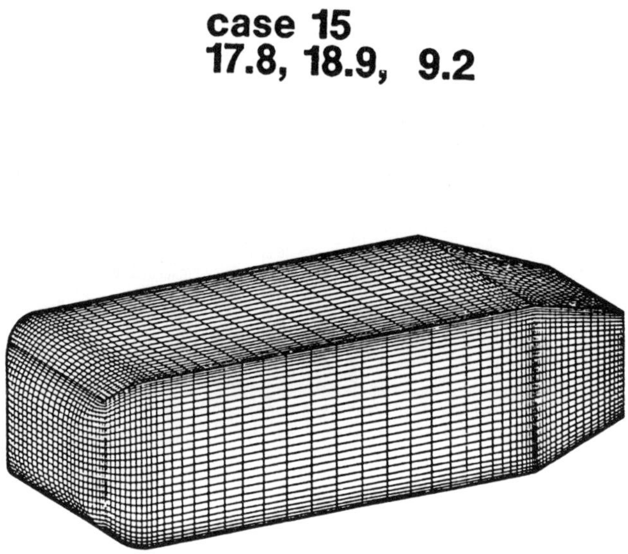

Fig. 3. Surface grid distribution (97x51x31) on Ahmed's body with three boat-tail angles ($\alpha = 17.8$, $\beta = 18.9$, $\gamma = 9.2$).

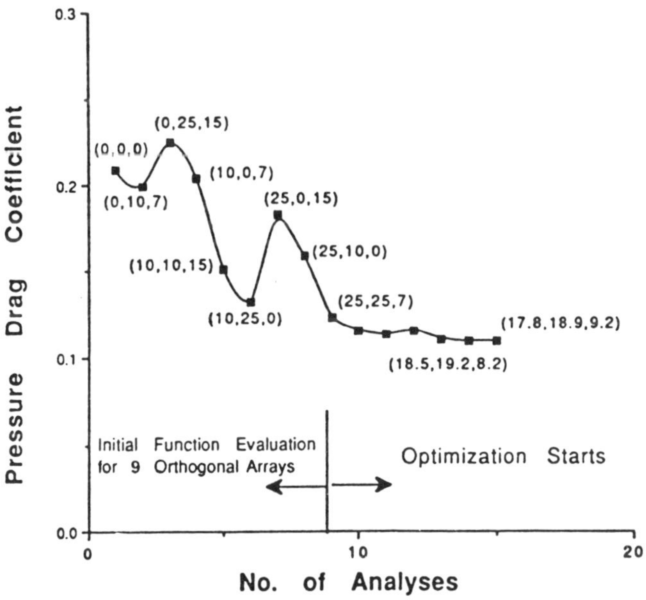

Fig. 2. Convergence history of the pressure drag coefficient with the number of Navier-Stokes analyses (The numbers within the parentheses are the three boat-tail angles, α, β, and γ)

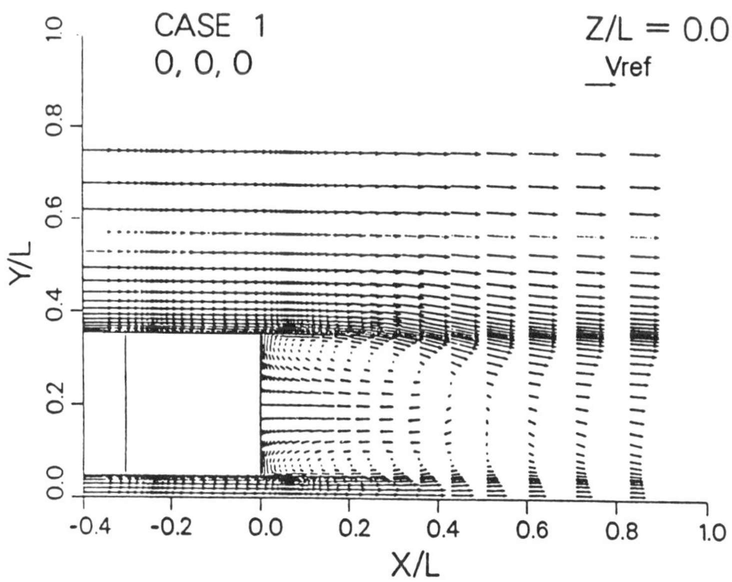

Fig. 4a. Velocity vectors in the plane of symmetry for Case 1 ($\alpha = 0$, $\beta = 0$, $\gamma = 0$).

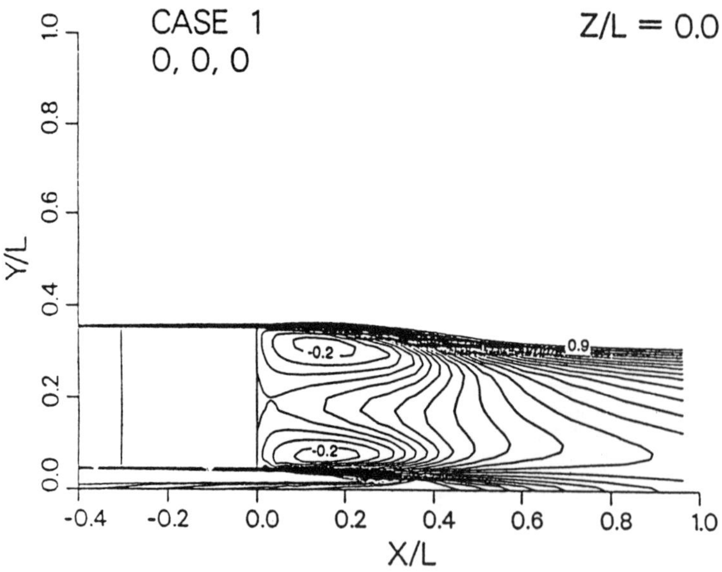

Fig. 4b. Total pressure contours in the plane of symmetry for Case 1 ($\alpha=0$, $\beta=0$, $\gamma=0$).

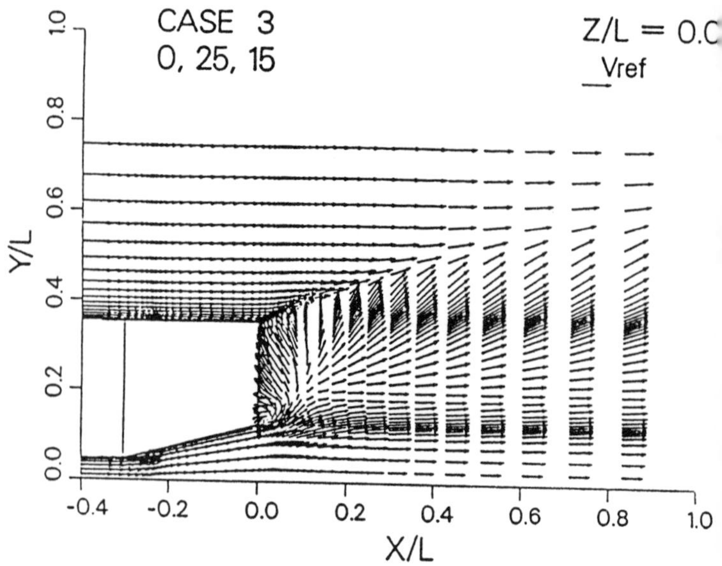

Fig. 5a. Velocity vectors in the plane of symmetry for Case 3 ($\alpha=0$, $\beta=25$, $\gamma=15$).

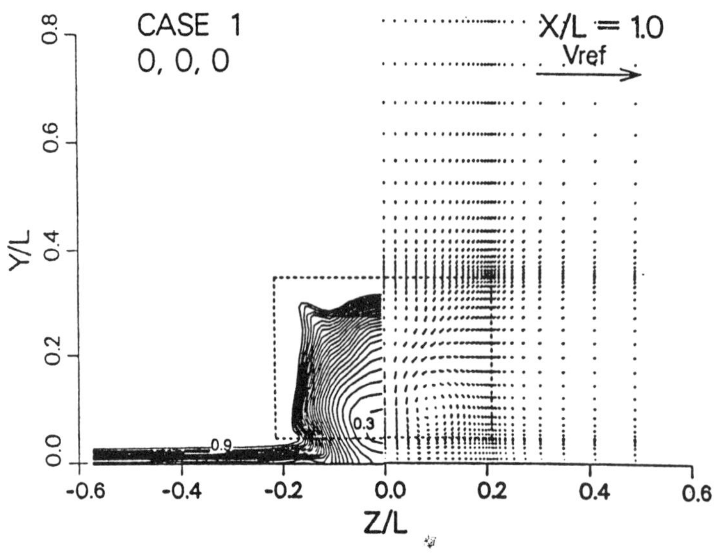

Fig. 4c. Total pressure contours and cross-flow velocity vectors in the transverse plane at x/L = 1 for Case 1 ($\alpha=0$, $\beta=0$, $\gamma=0$). (Dotted lines indicates the body cross section and the base and the total pressure contours greater than 0.9 are not shown).

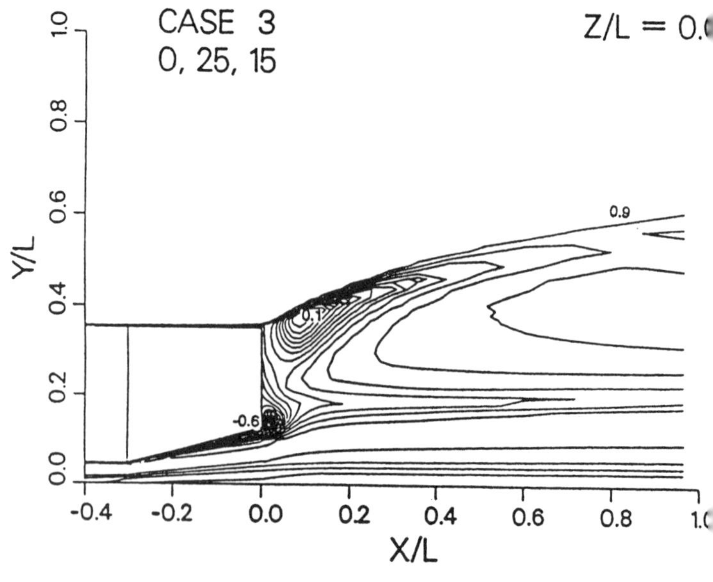

Fig. 5b. Total pressure contours in the plane of symmetry for Case 3 ($\alpha=0$, $\beta=25$, $\gamma=15$).

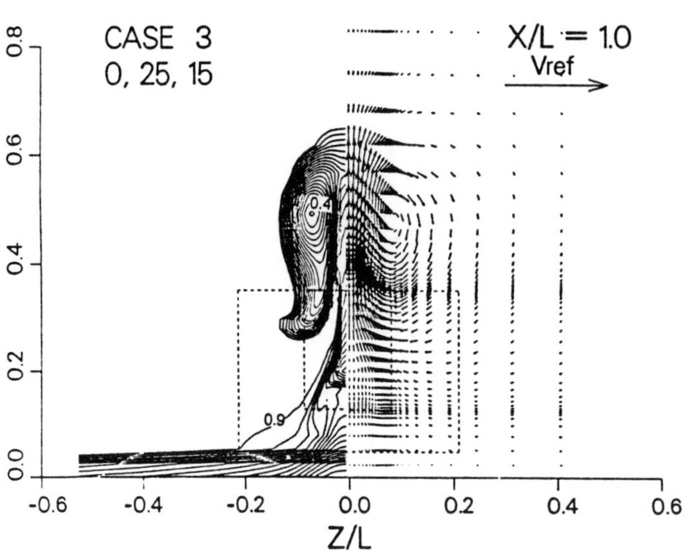

Fig. 5c. Total pressure contours and cross-flow velocity vectors in the transverse plane at x/L = 1 for Case 3 ($\alpha=0$, $\beta=25$, $\gamma=15$). (Dotted lines indicates the body cross section and the base and the total pressure contours greater than 0.9 are not shown).

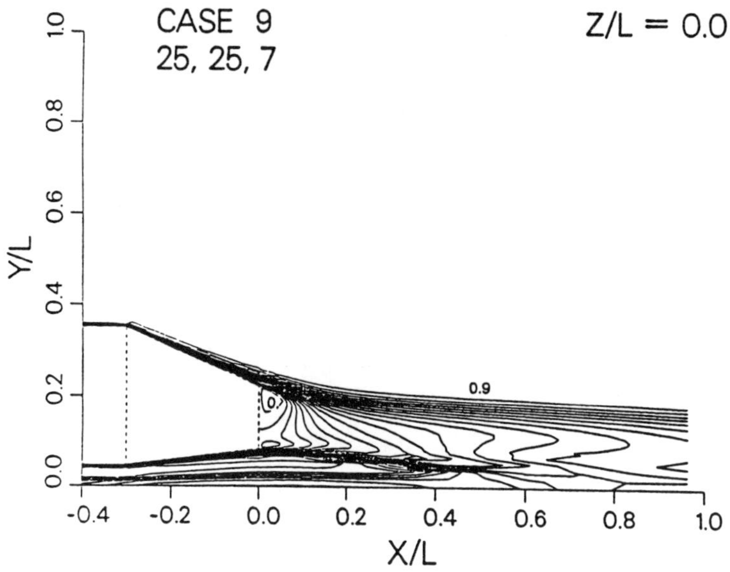

Fig. 6b. Total pressure contours in the plane of symmetry for Case 9 ($\alpha=25$, $\beta=25$, $\gamma=7$).

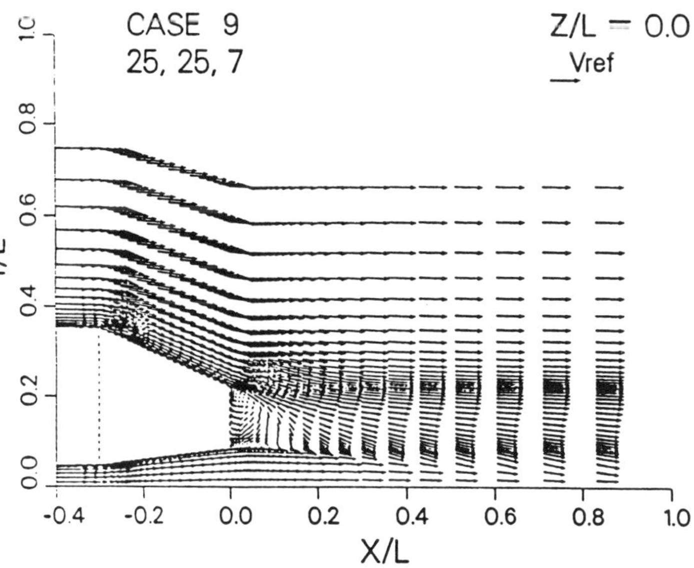

Fig. 6a. Velocity vectors in the plane of symmetry for Case 9 ($\alpha=25$, $\beta=25$, $\gamma=7$).

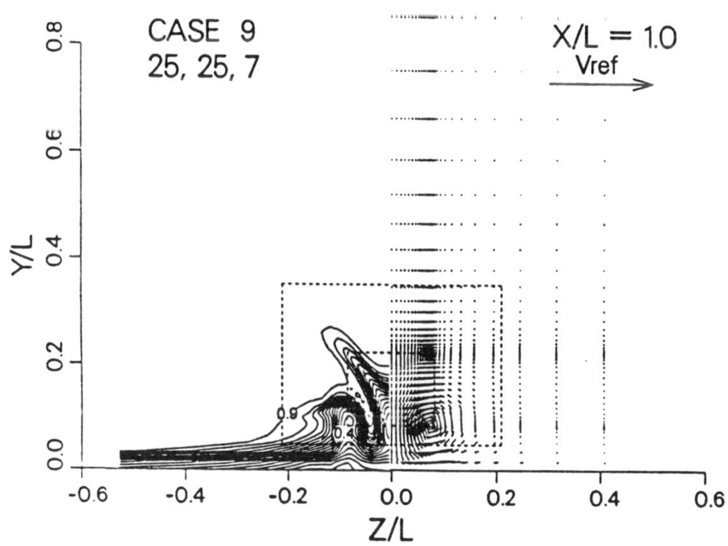

Fig. 6c. Total pressure contours and cross-flow velocity vectors in the transverse plane at x/L = 1 for Case 9 ($\alpha=25$, $\beta=25$, $\gamma=7$). (Dotted lines indicates the body cross section and the base and the total pressure contours greater than 0.9 are not shown).

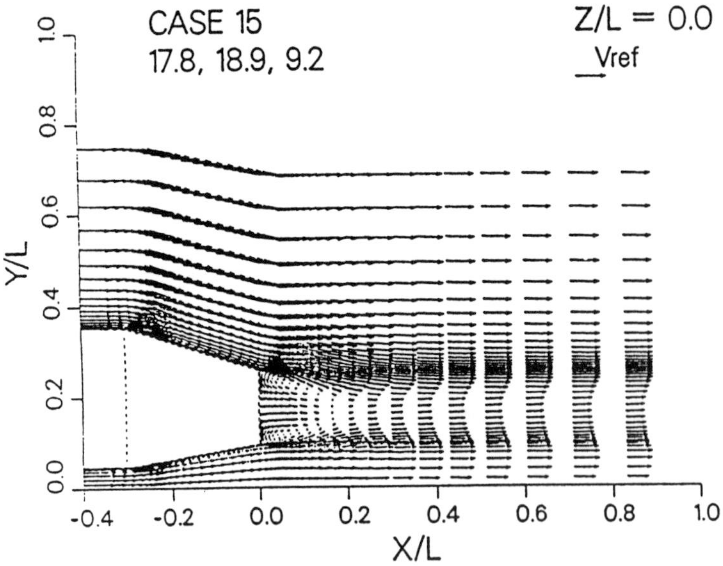

Fig. 7a. Velocity vectors in the plane of symmetry for Case 15 ($\alpha = 17.8$, $\beta = 18.9$, $\gamma = 9.2$).

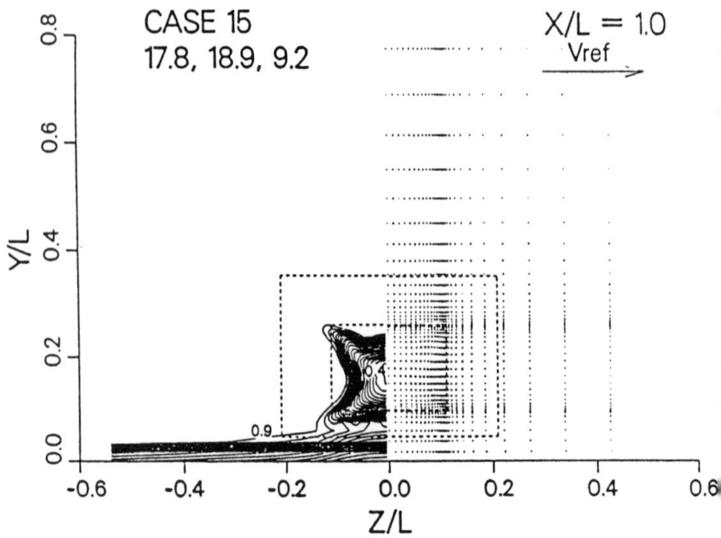

Fig. 7c. Total pressure contours and cross-flow velocity vectors in the transverse plane at x/L = 1 for Case 15 ($\alpha = 17.8$, $\beta = 18.9$, $\gamma = 9.2$). (Dotted lines indicates the body cross section and the base and the total pressure contours greater than 0.9 are not shown).

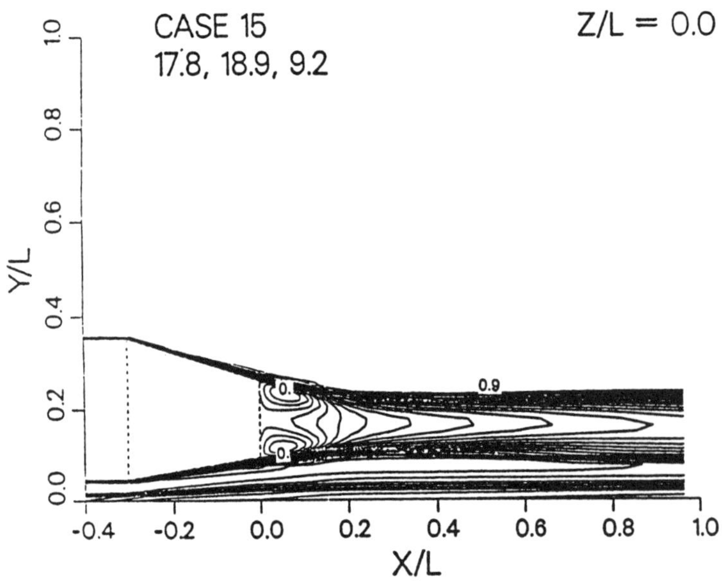

Fig. 7b. Total pressure contours in the plane of symmetry for Case 15 ($\alpha = 17.8$, $\beta = 18.9$, $\gamma = 9.2$).

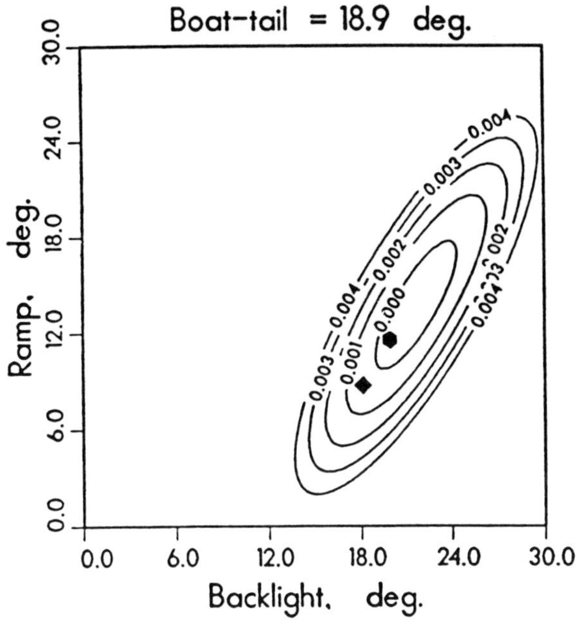

Fig. 8a. Iso-contours of the computed afterbody drag coefficient increment near the optimum shape parameters with a fixed boat-tail angle of 18.9 degrees and selected experimental data (symbols) for comparison.

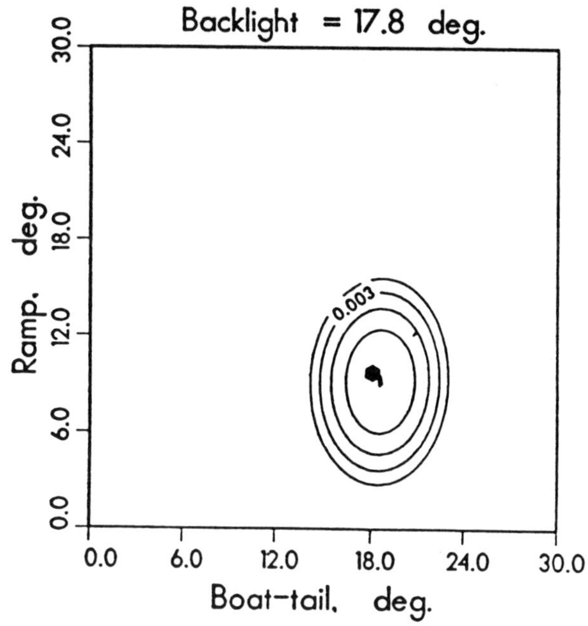

Fig. 8c. Iso-contours of the computed afterbody drag coefficient increment near the optimum shape parameters with the fixed backlight angle of 17.8 degrees and selected experimental data (symbols) for comparison.

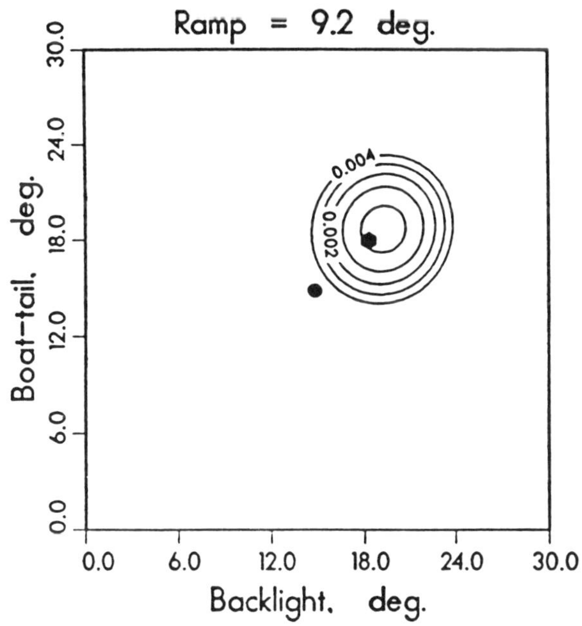

Fig. 8b. Iso-contours of the computed afterbody drag coefficient increment near the optimum shape parameters with a fixed ramp angle of 9.2 degrees and selected experimental data (symbols) for comparison.

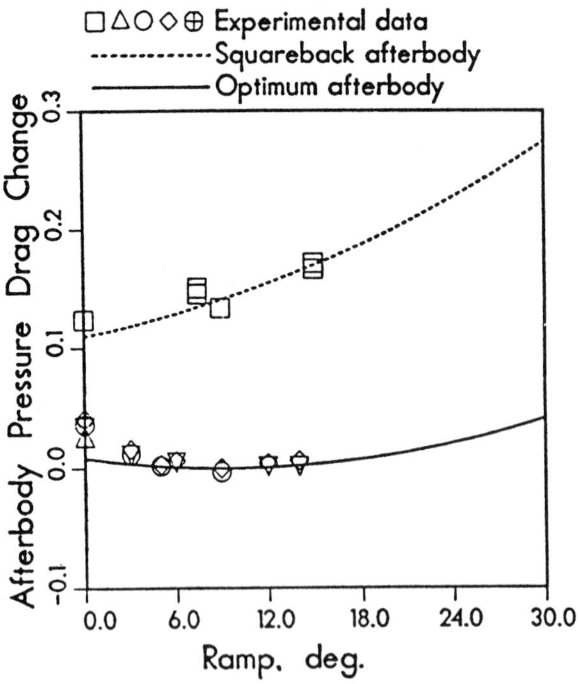

Fig. 9a. Computed (lines) and measured (symbols) afterbody drag coefficient change with ramp angle for the optimum ($\alpha = 17.8$, $\beta = 18.9$) and squareback ($\alpha = 0$, $\beta = 0$) afterbodies.

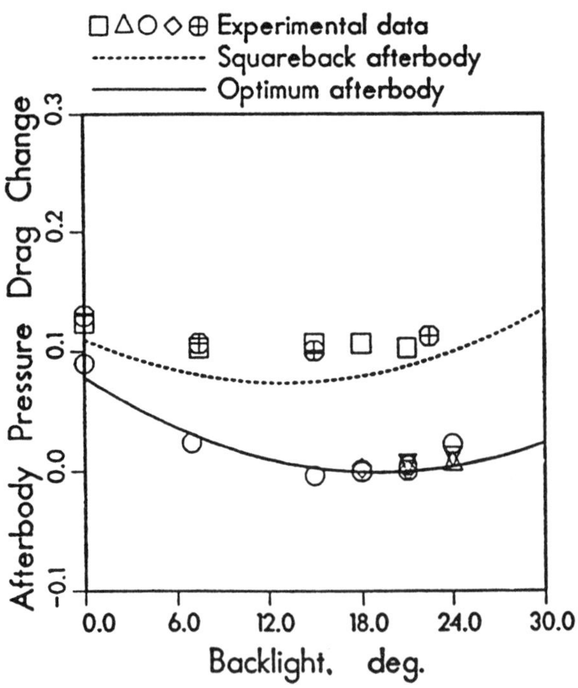

Fig. 9b. Computed (lines) and measured (symbols) afterbody drag coefficient change with backlight angle for the optimum ($\beta = 18.9$, $\gamma = 9.2$) and squareback ($\beta = 0$, $\gamma = 0$) afterbodies.

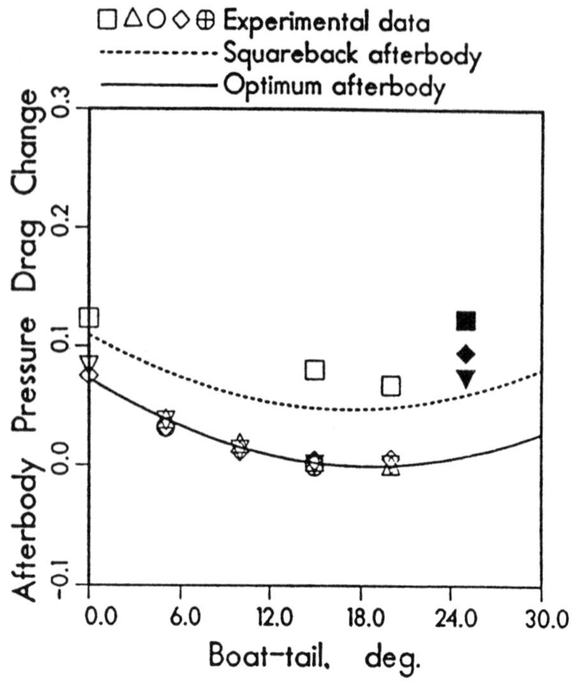

Fig. 9c. Computed (lines) and measured (symbols) afterbody drag coefficient change with boat-tail angle for the optimum ($\alpha = 17.8$, $\gamma = 9.2$) and squareback ($\alpha = 0$, $\gamma = 0$) afterbodies.

940323

A Calibration Study of CFD for Automotive Shapes and CD

J. Williams and W. J. Quinlan
Ford Motor Co.

J. E. Hackett and S. A. Thompson
Lockheed Aeronautical Systems Co.

T. Marinaccio and A. Robertson
Adapco

ABSTRACT

An extensive calibration study has been initiated to assess the predictive ability of CFD (Computational Fluid Dynamics) for the aerodynamic design of automotive shapes. Several codes are being checked against a set of detailed wind tunnel measurements on ten car-like shapes. The objective is to assess the ability of numerical analysis to predict the C_D (drag coefficient) influence of the rear end configuration. The study also provides a significant base of information for investigating discrepancies between predicted and measured flow fields and for assessing new numerical techniques.

This technical report compares STAR-CD predictions to the wind tunnel measurements. The initial results are quite encouraging. Calculated centerline pressure distributions on the front end, underbody and floor compare well for all ten shapes. Wake flow structures are in reasonable agreement for many of the configurations. Drag, lift, and pitching moment trends follow the experimental measurements. The predicted C_D levels, however, are higher than the measurements and, despite the fine computational grid, modeling of the rear end flow is incorrect for some configurations.

INTRODUCTION

In today's competitive environment there is a need to shorten product development time and bring new designs to the marketplace more quickly (1, 2, 3)*. The automotive designers need aerodynamic guidance on the many shapes under consideration, early in the design process and before the final theme selection. Computational Fluid Dynamics (CFD) is one possible answer to this need. To be successful, however, the CFD process must provide the engineer and designers with high confidence engineering answers equivalent to wind tunnel measurements, in a shorter time, and at a lower cost (1, 4).

Several technical publications and a variety of simulation methods are available in the literature for assessing the usefulness of CFD for road vehicle aerodynamics — References 1, 2 and 3 are constructive reviews of the current technology. Many of the publications, however, are research-oriented studies of critical shapes that can have significant unsteady effects (5 - 9). These shapes can cause large, sudden changes in the size and structure of the wake, and large corresponding changes in drag (1, 8). A time-accurate CFD model is necessary for these shapes. But, is a time accurate model necessary for more car-like shapes with non-critical features?

Time-averaged drag and lift values are needed for product design and are reported by some investigators (10 - 16). DNS (Direct Numerical Simulation) computations in (17) and (18) report drag and lift forces that compare quite well with experimental measurements. However, the computational and personnel resources for these calculations appear to be large, and CFD results cannot be accepted solely on the basis of C_D (drag coefficient) (1). A correct representation of the wake is essential (19).

Early exploratory CFD studies at Ford on a typical sedan shape raised other interesting questions. The simulation did not show the expected separation bubble over the backlight, and the vorticity in the wake suggested the body was generating downforce instead of lift. Because these results were not consistent with testing experience, there was an initial temptation to discount the CFD process. However, the simplified model did not include many of the geometric details of a production car, such as underbody detail, glass offsets, moldings, engine cooling openings, and mirrors. Since test data on the simplified shape was not available for comparison, the computational results may actually have been correct. A critical assessment of the CFD solution could not be made.

Against this background, a calibration study has been initiated to address many of these issues. By carefully modeling and testing the same configurations, maintaining controlled and coordinated experimental and computational processes, an assessment of the predictive ability of CFD can be made. The objective is to assess the ability of numerical analysis to rank the shapes aerodynamically by predicting the C_D (drag coefficient) influence of the rear end configuration. The study also provides a significant base of information for investigating discrepancies between predicted and measured flow fields, and for assessing new numerical techniques.

---

* Numbers in parentheses refer to the references at the end of the paper.

Overall, the results of the numerical simulations discussed in this work-in-progress report are quite encouraging. The wake flow structures are in reasonable agreement for many of the shapes. Drag, lift and pitching moment trends follow the wind tunnel measurements. The C_D values are higher than the measurements, however, and despite the fine computational grid, the modeling of the separation and reattachment patterns over the backlight region are incorrect for some configurations.

CALIBRATION CONCEPT

The overall philosophy of the calibration study is to create a synergism between the EFD (Experimental Fluid Dynamics) and CFD processes. The diagram in Figure 1, changed from reference (20) for this report, illustrates the calibration process. The EFD experiments will provide information to improve the modeling of the various flow structures and reveal the code's ability to predict specific design parameters. The CFD calculations from the pilot code will allow a complete interrogation of the flow field. It will allow examination of flow structures that are very difficult to measure. This synergism can only improve both technologies and the calibration results.

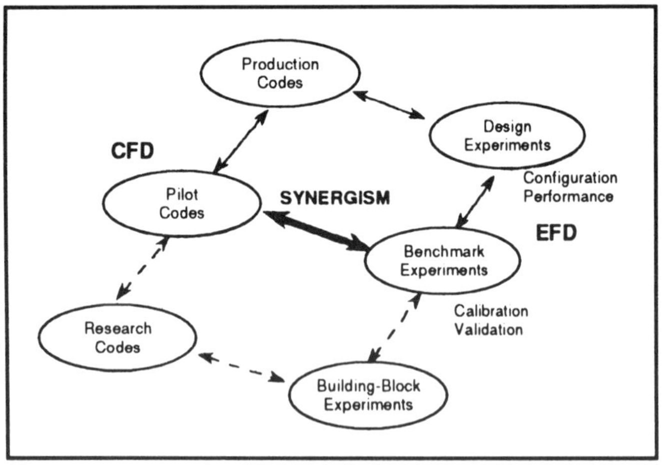

Figure 1
CFD CALIBRATION PROCESS CHART

Close coupling of benchmark experiments with pilot code simulations is the key to creating this synergism. Compared to a research code, a pilot code is at a more mature stage of development. The code documentation and building-block test cases are complete. Analysts not involved in software development create the mesh and conduct the simulations. They understand the empirical input parameters and convergence criteria, and have experience with grid resolution and distribution effects. In other words, the code is ready to be applied to configurations of engineering interest.

The word "validation" is frequently used to describe the process of verifying a CFD simulation. It is a misleading descriptor of this report. "Validation" implies a strong endorsement of the analytical process. High confidence simulations over a wide range of configurations are possible, because the validated code has a strong physical foundation and an application process that is under control. Since the code reproduces the flow details quite well, frequent comparisons to EFD experiments are not necessary. The validated CFD process can extrapolate with confidence, and model flow structures not seen before.

With today's technology, however, validating a CFD process is not possible because models of the flow details essential for validation are not available (1, 3). In addition, RANS (Reynolds Averaged Navier Stokes) codes contain empirical input parameters, truncation errors, mesh dependent approximations, numerical approximations to the convection derivatives and explicit turbulence models for closure. So, frequent comparisons to EFD experiments are essential to calibrate the process and check the results.

Although a reasonable representation of the essential physics is obviously necessary, a calibrated code may not have to be completely correct everywhere in the space and time domains. It is possible, for example, that car-like shapes may not require a time-accurate model of the unsteady effects. A steady state CFD interpolation between tested configurations may be able to provide useful design information.

This is the essence of the calibration study — can a calibrated CFD process provide useful, quantitative, product design information? And, if so, what is the range of applicability? How should critical flow phenomena, local or global, be identified and handled?

Assuming success, the next steps should involve more detailed experiments to understand the more difficult flow structures and to confirm that all the important flow features are modeled correctly. These rigorous investigations would lead to a verification or validation of the physics in the numerical simulation. They would also provide a stronger endorsement of the application process and make the code applicable to a wider range of configurations. The robust production CFD process would then be ready for extrapolation and design applications.

EXPERIMENTAL PROCESS

Careful and detailed wind tunnel measurements were collected for ten car-like shapes, using a 1/2-scale model of a simplified sedan, strut-supported from the floor of the Lockheed Low Speed Wind Tunnel (Figure 2). The model scale of 1/2 provided a low blockage environment (area blockage ratio of 1.4%) in a long test section (about two model lengths ahead of and three model lengths behind the model center). The setup removed any measurable interaction with the tunnel speed setting and diffuser.

The model was attached to the under-floor force balance and supported by struts above the test section floor (Figure 2). Internal screw jacks were used to position the model parallel to the floor in roll and pitch. The 127 mm (5.0 inch) ground clearance and strut supports, instead of wheels, significantly reduced viscous interactions with the floor boundary layer. Viscous side lobes in the wake, seen in earlier fixed and moving ground treatment studies (21), were eliminated. The tunnel speed was 176 kph (110 mph).

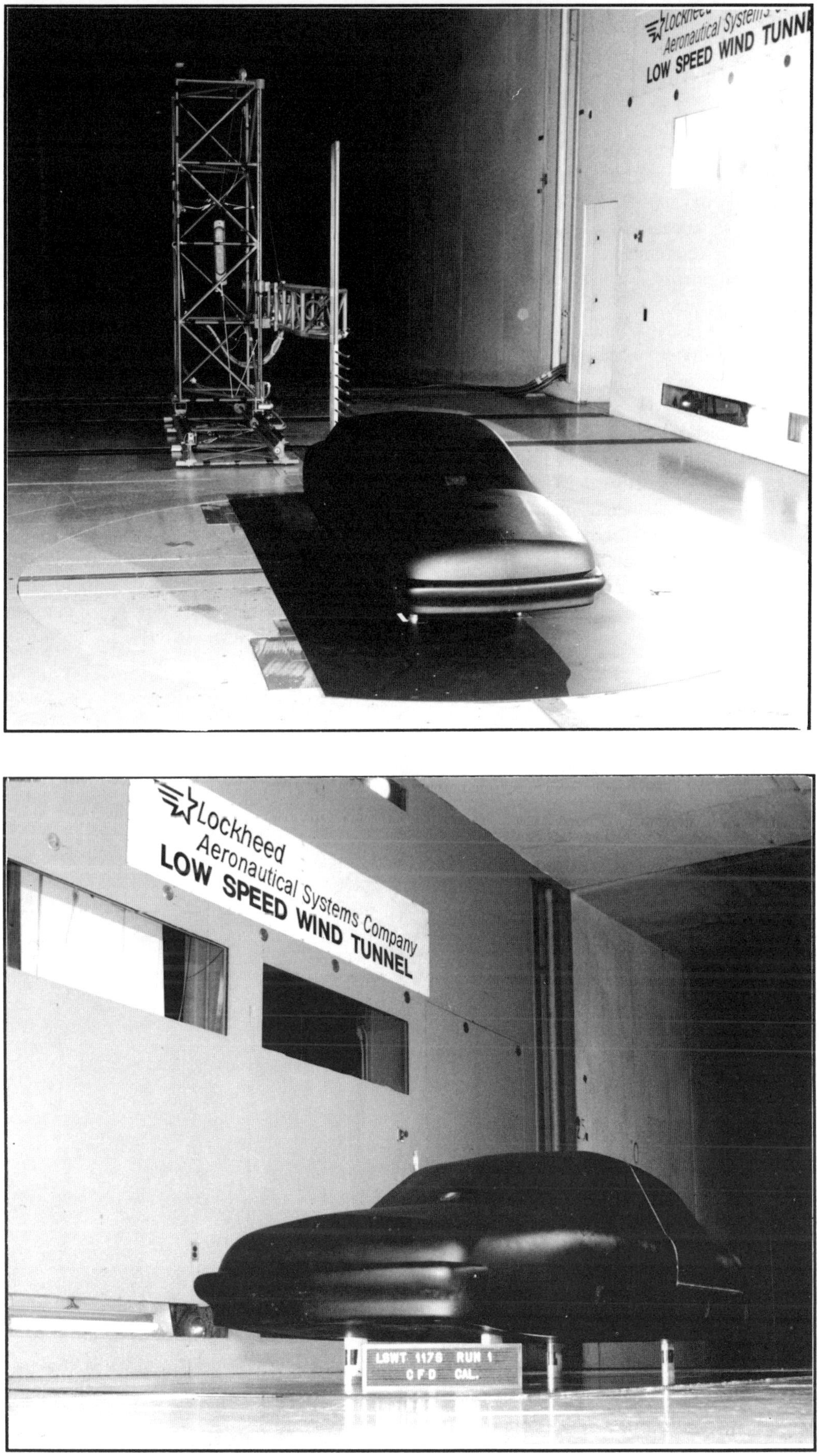

Figure 2
1/2-SCALE MODEL IN THE LOCKHEED WIND TUNNEL

Balance, surface pressure (model and floor), and wake flow data (cross flow and centerplane) were collected and averaged over five seconds for each configuration. The wake system was removed for the balance and pressure measurements. Separation and reattachment features of the flow were documented with surface flow visualization (Figure 3). Two model installations in the wind tunnel, four months apart, provide some insight into the repeatability of the experimental measurements. Qualitative comparisons of balance, surface pressures, wake measurements, and surface flow topology showed very good agreement.

On-line display of the wake flow measurements from the 5-hole-probe automatic survey system (Figure 2), confirmed the quality and completeness of the data. Deliberately locating the cross flow surveys near the bubble closure behind the model, increased the vortical information without exceeding the probe angular calibration limits. After the test, however, a small region of reverse flow was noted in the center of the surveys and has been deleted from the cross flow velocity plots. In the total-pressure-deficit and longitudinal-vorticity plots, this small area has been numerically filled-in to avoid distorting the valid contours outside the region. The dotted line in the plots shown later in this paper indicates the filled-in region.

Because of the plan to use a steady-state CFD simulation and a symmetry condition at the model centerplane, the possibility of unsteady flow across the centerplane was a concern in the early discussions. Supporting flow visualization studies at the Texas Tech University tow tank provided some insight. The shedding fluctuations were vertical with very little sideways activity at the centerplane. The present wake flow measurements also indicated good symmetry with very little lateral activity in the mean flow.

CAR-LIKE SHAPES

The ten 1/2-scale, car-like shapes share a common front end with various rear ends to provide a diverse mix of Studio themes as well as aerodynamic modeling difficulty (Figure 4). The configurations include a station wagon (SW), four fastbacks (FB1, FB2, FB3, FB4), three notchbacks (F1, C1, C2) and two sculptured fastbacks (B1, B2). The sculptured-fastbacks have a slope change between the rear window and deck, and an increasing slope to each side of the deck centerline. Some of the shapes have sharp trailing edges to fix the flow separation location.

The surfaces were developed mathematically on a CAD workstation, rather than use the more time-consuming clay

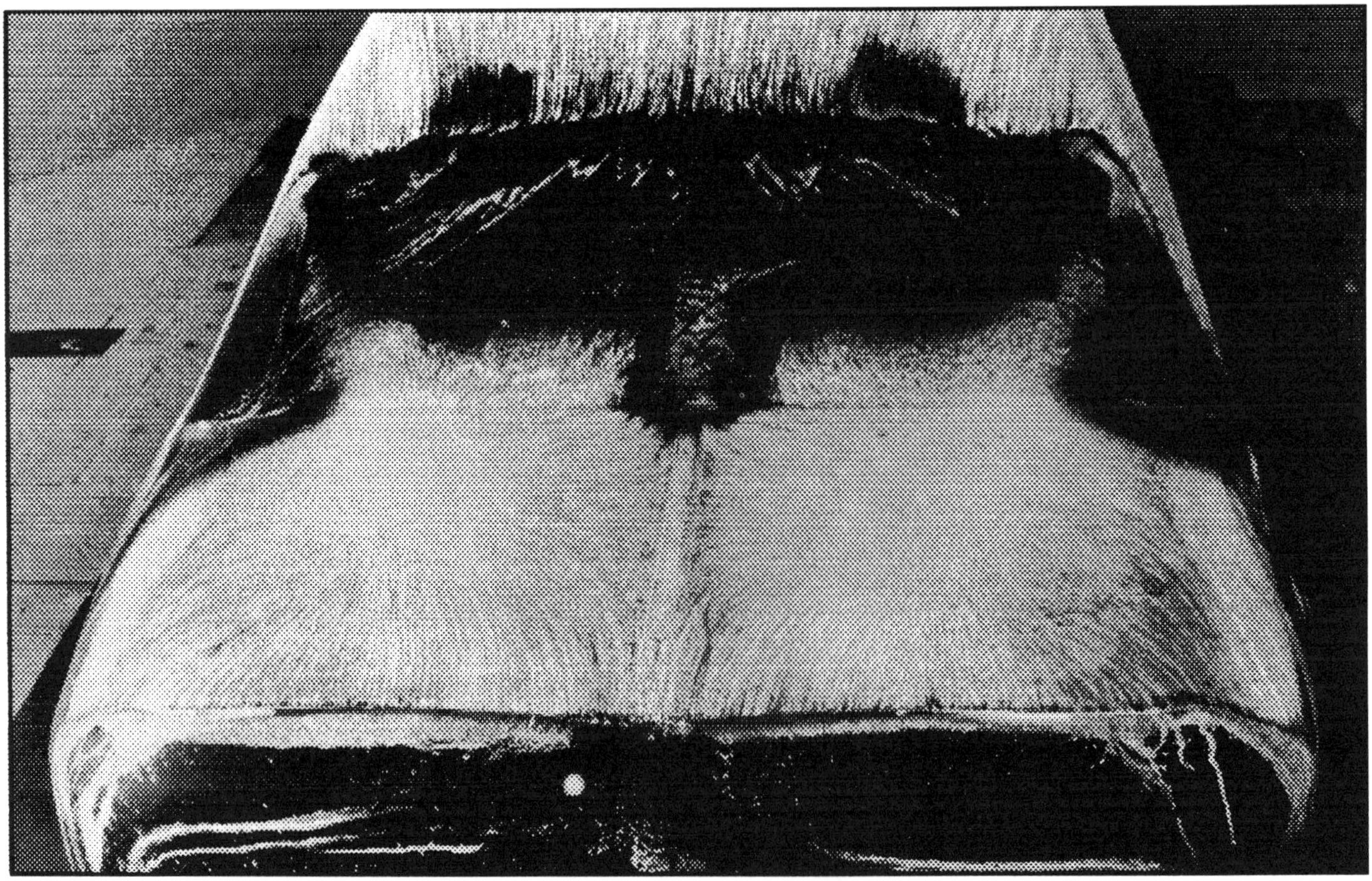

Figure 3
SURFACE FLOW VISUALIZATION OF THE F1 NOTCHBACK

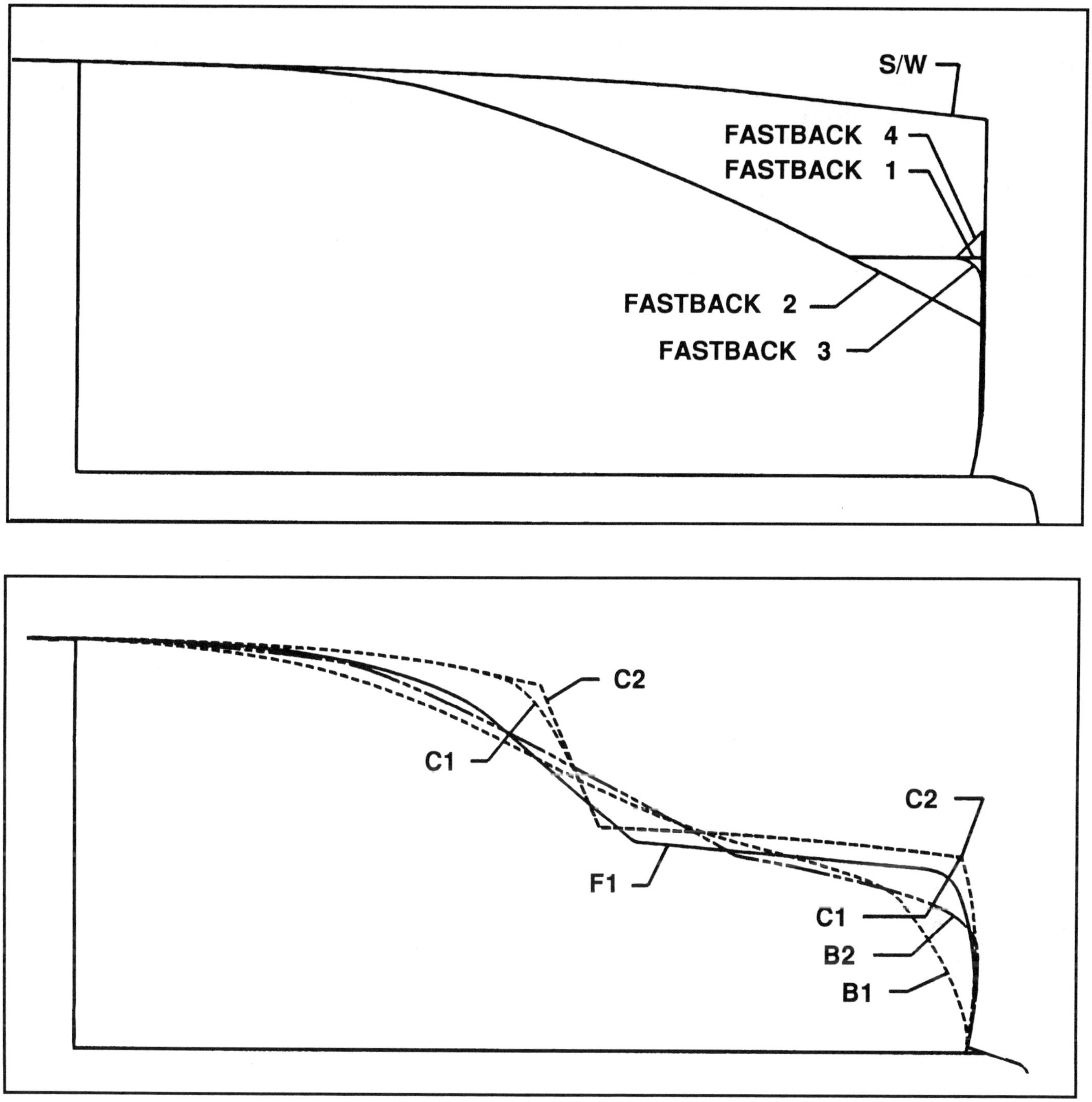

Figure 4
SIDE-VIEW PROFILE OF THE TEN REAR END SHAPES

and fiberglass approach for model build. The electronic surface data, which was used to generate the milling cutter paths for the physical model, was also used to create the computational mesh. This made the experimental and computational properties essentially identical. A scan of the models confirmed the lateral symmetry, the accuracy of the critical sections, and the location of the surface pressure taps.

Table I lists some significant characteristics of four of the models — two notchbacks (F1 and C1), the station wagon (SW), and a sculptured-fastback (B1). Because they are representative of the observed flow structures and forces, the comparisons later in this report will concentrate on these shapes.

NUMERICAL ANALYSIS

The numerical simulations with the commercial CFD package STAR-CD carefully paralleled the wind tunnel tests. Reference (22) contains a full description of the equations in the Code, its numerical methodology, and capabilities.

Great care was taken to accurately model the test geometry and wind tunnel. As mentioned earlier, the vehicle shapes were obtained from the same electronic surface data used by the experimental model builders. The computational domain extends from a symmetry condition at the model centerplane to the walls of the tunnel. The model includes

Table I
Selected Models and Flow Structures for Comparison

Baseline Notchback **F1**	**Notchback** **C1**	**Station Wagon** **SW**	**Sculptured Fastback** **B1**
Separation at rear header, reattachment to deck and separation from deck edge	Separation at rear header, reattachment to deck and separation from deck edge	Fixed separation at the rear header and upper sides	No classical separation bubble on the rear window and deck
Natural unsteadiness due to deck reattachment	Natural unsteadiness due to deck reattachment	Very stable flow pattern	Unsteady separation from large radius deck edge
		High Drag	Low Drag
		High Downforce	Hightest Lift and Vortex Drag
		Largest Wake	Smallest Wake

the airfoil struts and the same ground clearance with the ground boundary layer represented.

An unstructured grid was assembled and fitted to the surface of the model. The mesh closest to the vehicle surface is an "O" grid of six cell layers that coats the vehicle and provides a fine inner grid, which is essentially normal to the surface. About 14,000 faces define the vehicle surface. A "D" grid surrounds the inner "O" grid and extends out from the body and to the tunnel floor. Small cells are non-uniformly distributed next to the vehicle and floor, as shown in Figure 5 for the F1 notchback. A relatively fine grid extends from the rear to capture the wake.

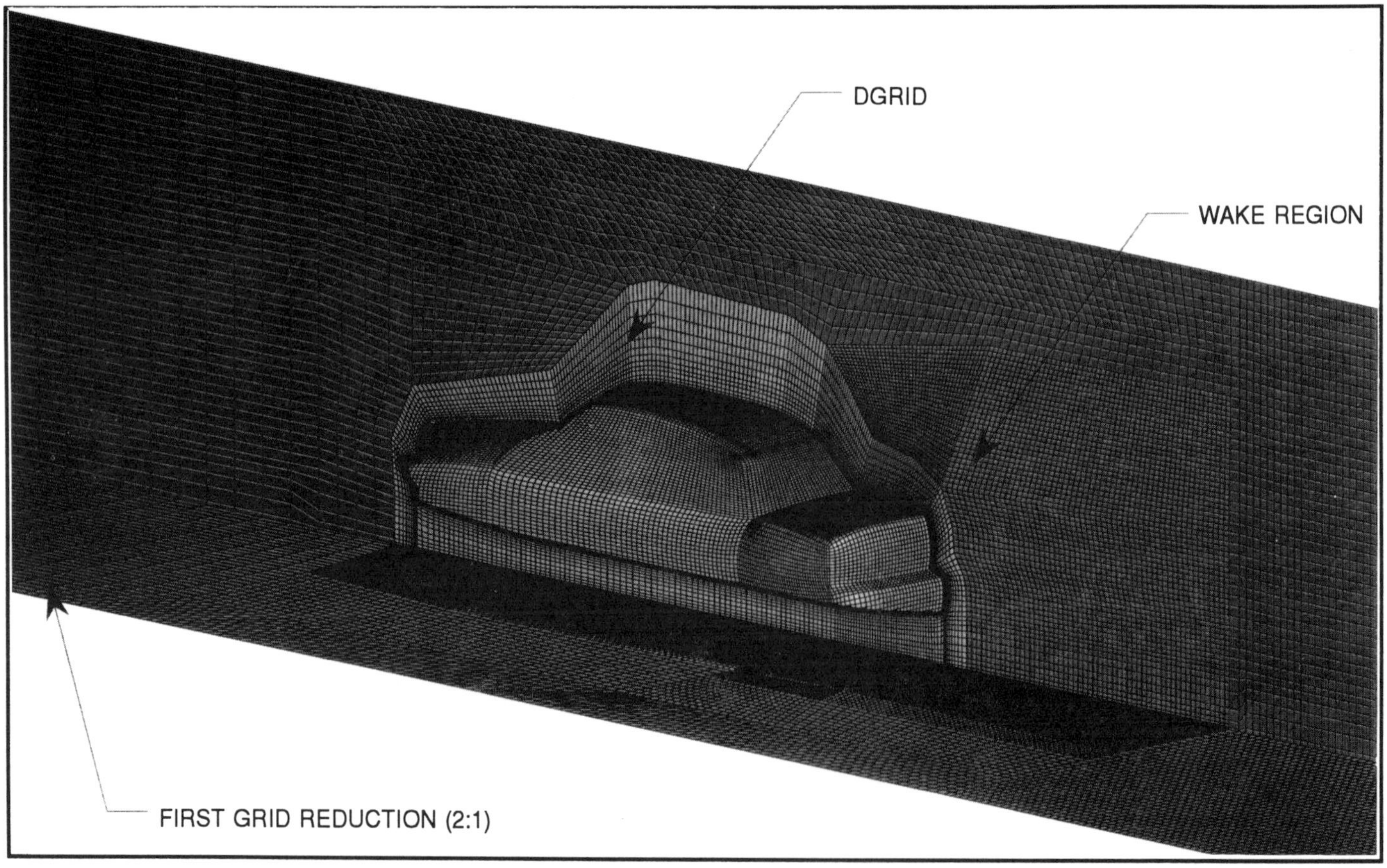

Figure 5
INNER-GRID REGION, F1 MODEL

The Code allows the grid to contain discontinuities of cell density with certain restrictions. The remainder of the domain uses this feature, called "embedded grids", to reduce cell density away from the vehicle and wake. As shown in Figure 6, a box with the first 2:1 grid reduction surrounds the "D" grid and wake cells. The second and third grid reductions reduce the cell density away from the model even further. A (2:1) grid reduction, for example, means that four smaller faces have one larger neighboring face. There are about 800,000 hexahedral cells in the computational domain.

The numerical analysis used mass and momentum conservation equations in three-dimensions for steady state, compressible flow. A second-order accurate linear upwind differencing scheme (LUD) was used for modeling the convective terms of the momentum equations. For the present study a k-ε turbulence model with logarithmic turbulent wall functions was employed, and no attempts were made to tailor or adjust the coefficients based on the computed results.

Measured floor boundary layer and free stream velocities were used for the upstream boundary conditions. A no-slip boundary condition was applied to the tunnel floor and walls. Force and moment values were calculated from surface pressure and skin friction integrations. For reference, the skin friction around another smooth, bluff body in ground proximity was about 20% of the total drag (5). The IBM RS6000/560 workstation calculations required about 290 Megabytes of core memory and between 150 - 200 CPU hours for each simulation.

To ensure a consistent comparison to the experimental measurements, wake data was extracted using post-processing nodes or "sensors". These sensors were placed in the computational domain in the same locations as the experimental probes, and field values were interpolated from adjacent cell data. To create a "computational measurement data set", the information was reformatted into data arrays, following the same pattern as the experiment. This data set was then manipulated and post-processed in the same manner as the experimental data.

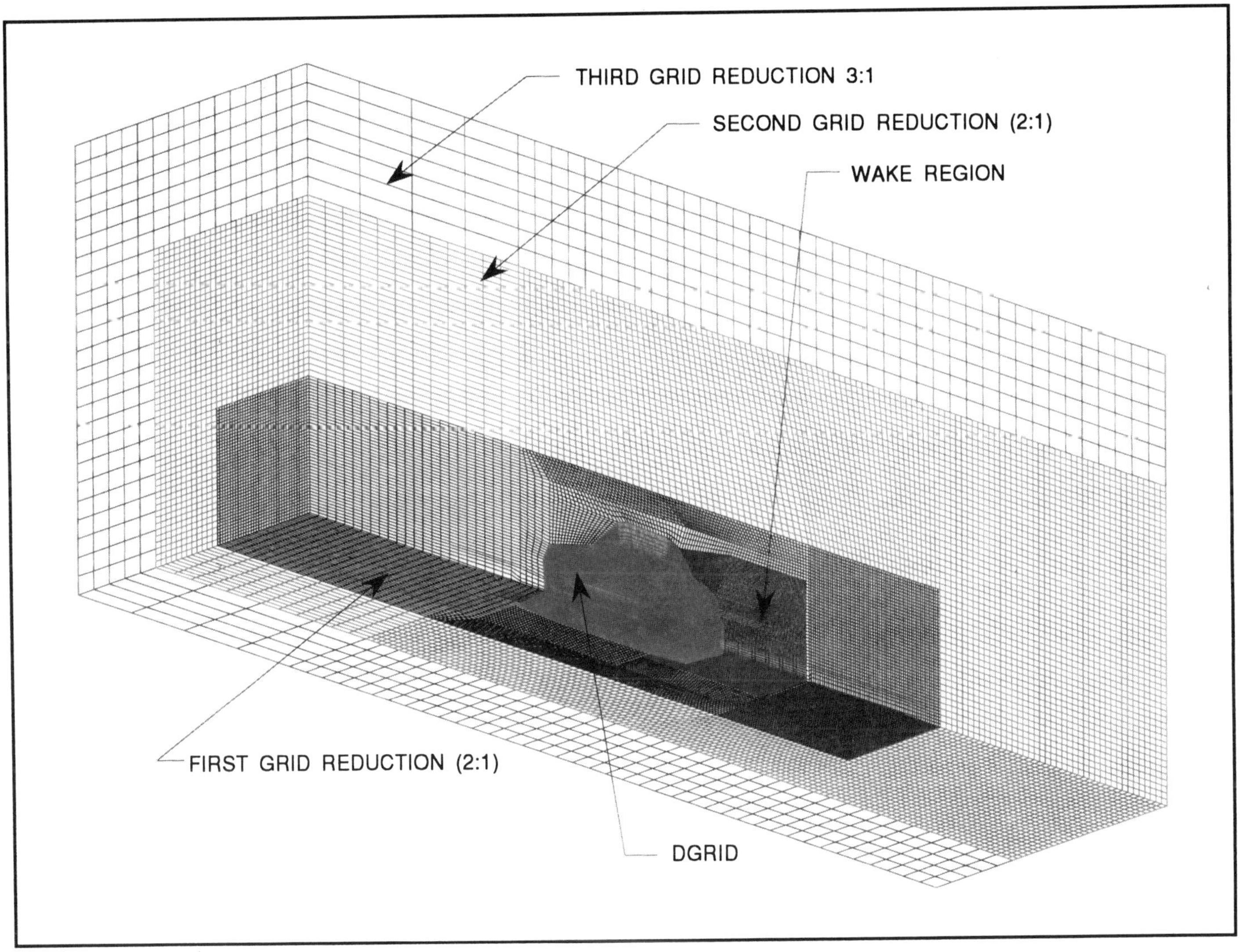

Figure 6
COMPLETE MODEL DOMAIN, F1 MODEL

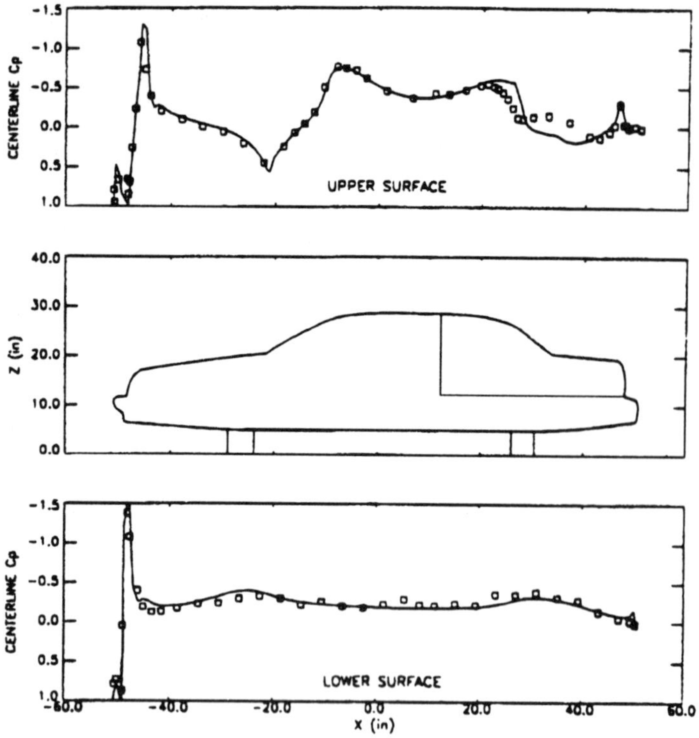

COMPARISON OF RESULTS

Overall, the numerical simulations of the FB1, FB2, FB3, B1, B2, and SW shapes were very encouraging. The F1, C1, C2, and FB4 simulations, however, will need more work because of the unsuccessful calculation of the backlight separation and deck reattachment patterns. As mentioned earlier, only four of the shapes — two notchbacks (F1 and C1), the station wagon (SW), and a sculptured fastback (B1) — will be used to compare the CFD simulations to the measurements.

Front End, Underbody and Floor Centerline Pressure Distributions — Generally, the calculated centerline pressure distributions on the front end, underbody and floor are in good agreement with the measurements for all the shapes. The F1 notchback comparisons in Figure 7 are typical. The data points are from the test and the line is from the calculation.

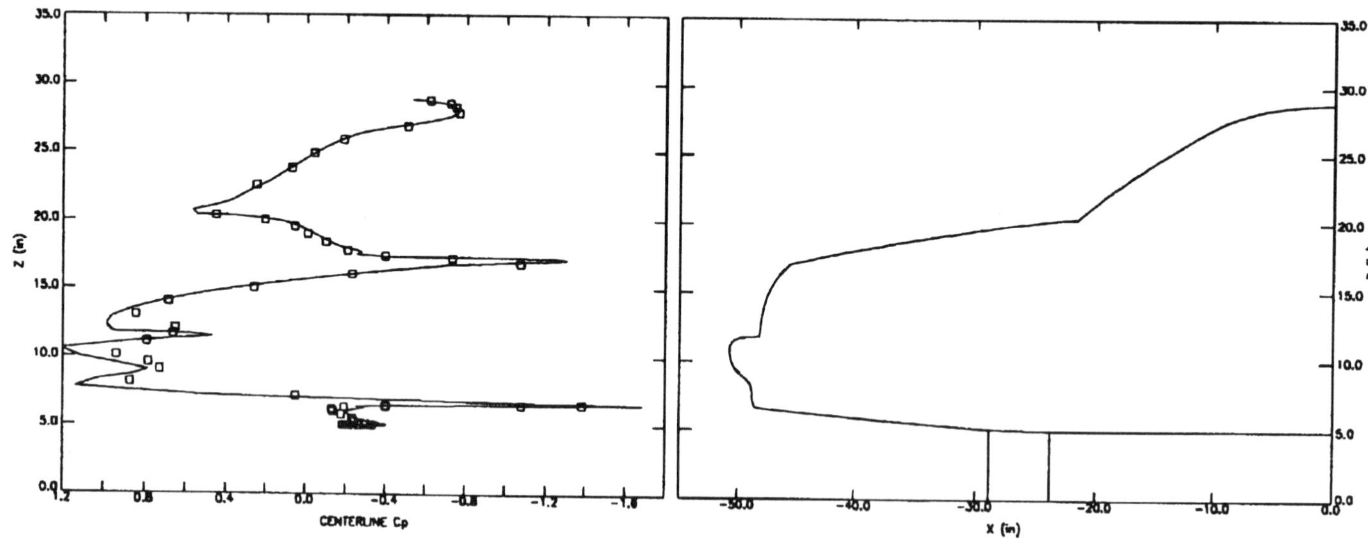

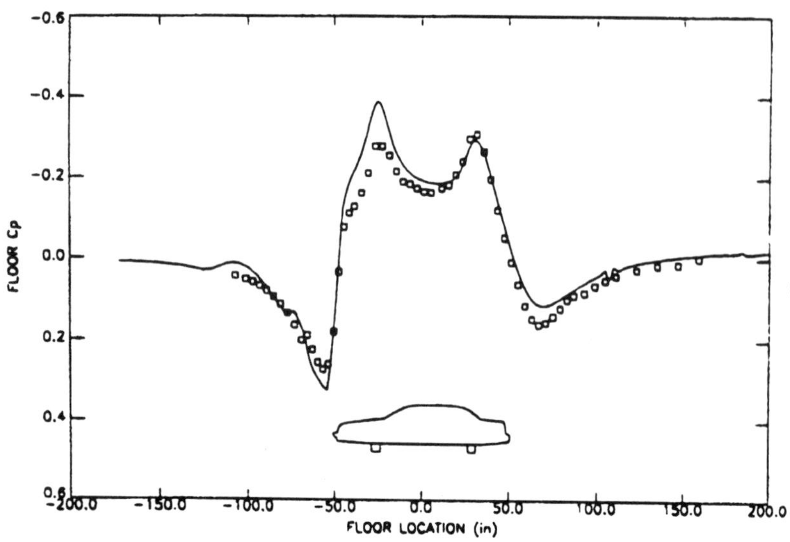

Figure 7
CENTERLINE PRESSURE DISTRIBUTION FOR THE F1 MODEL

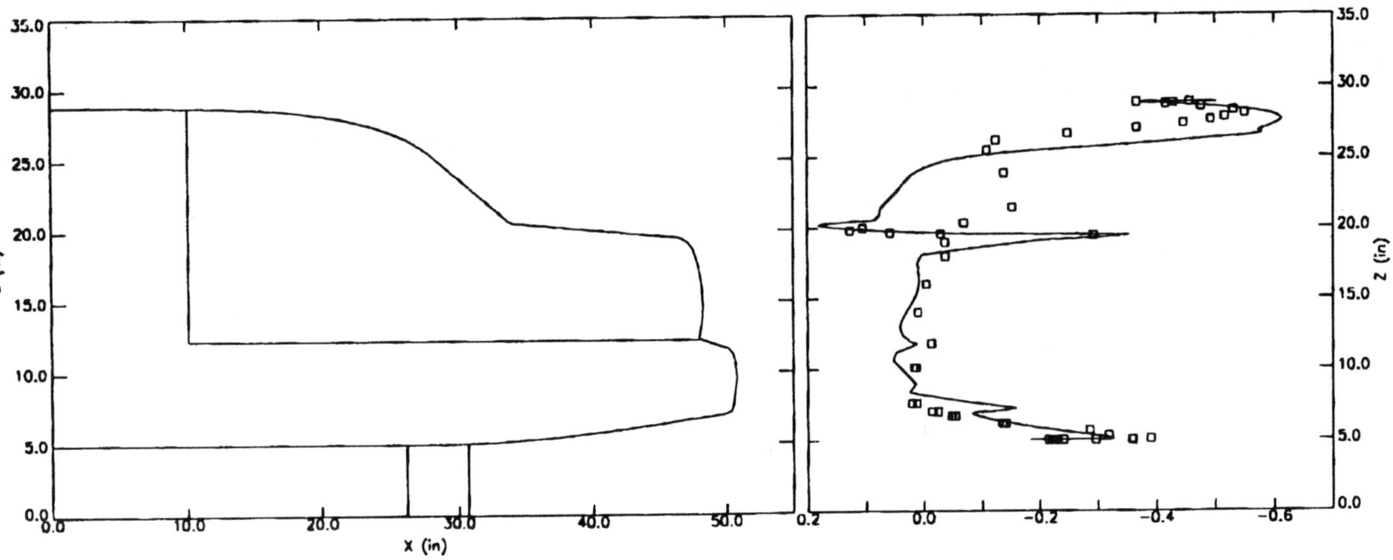

Figure 8
F1 NOTCHBACK DRAG PLOT OF REAR END PRESSURE DISTRIBUTION

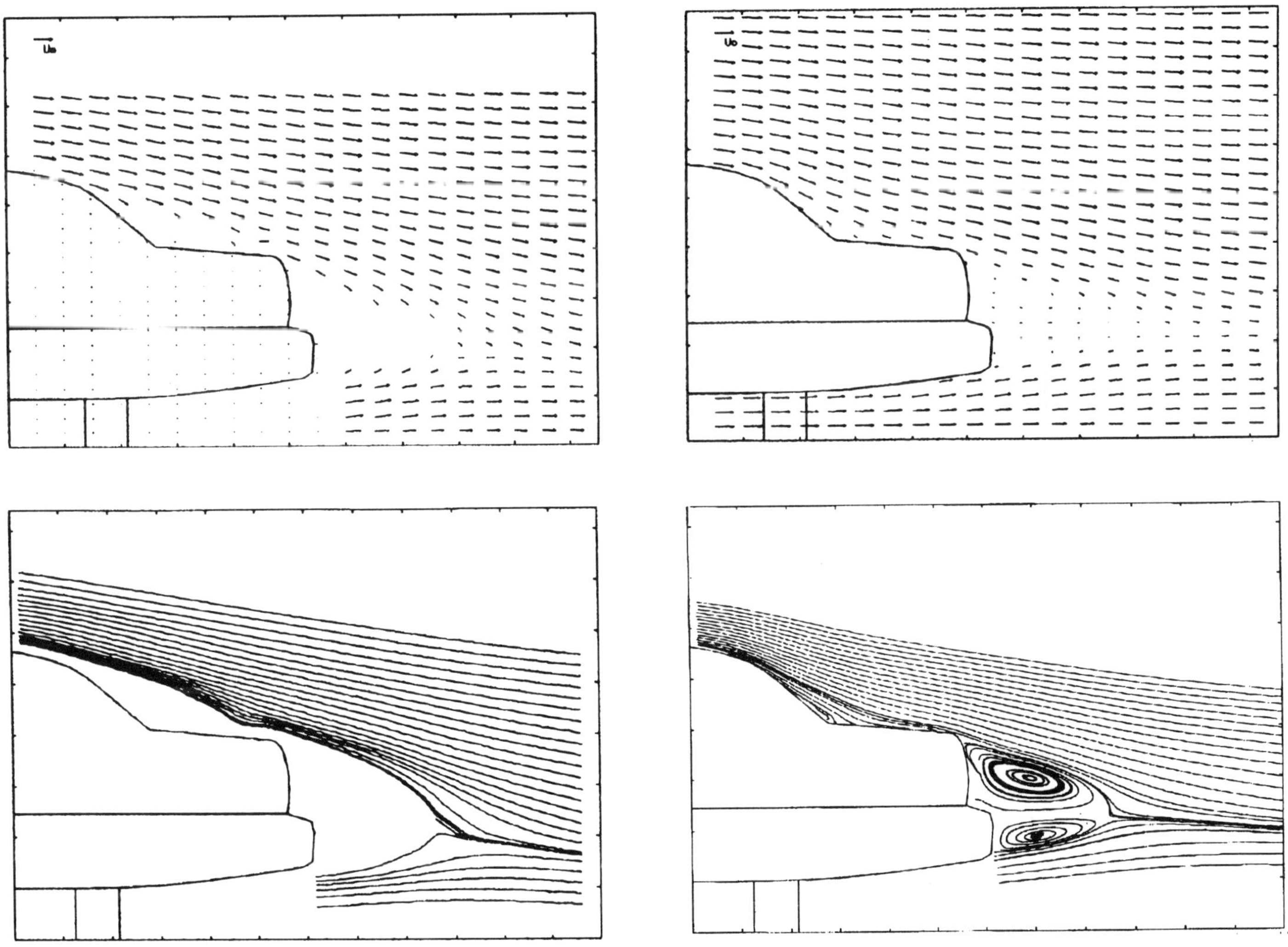

Figure 9
F1 NOTCHBACK CENTERPLANE STREAMLINES

As shown in the "drag" plot of the pressures, (Figure 7, middle), the calculated front end stagnation pressures are a little high. The reasons are presently unclear. The location of the stagnation line, however, agreed well with the measurements, and the expected separation and trapped vortex on the top of the bumper shelf were evident in the results. During the tests, the pressure distribution under the bumper (around the valance corner) was noted to be sensitive to Reynolds number, and surface flow visualization indicated a small laminar separation and reattachment downstream. Since the CFD code cannot recognize laminar boundary layer stability and transition, the solution did not contain this local effect.

Notchback Rear End Flow Structures—As was expected from the beginning, the viscous flow patterns over the notchback rear ends were difficult to model. The simulation overpredicted the centerplane acceleration and turning around the backlight (or back window) header; see the upper surface pressure distribution for the F1 notchback in Figure 7. This delayed separation contributed to the higher pressure recovery and the missed reattachment pattern on the deck.

The drag plot in Figure 8 is another comparison of the notchback rear end pressures. The area between the measurements (data points) and the calculation (line) shows the drag effect of the centerline differences. The lower pressures at the rear header, causing higher drag, are partially canceled by the higher pressures on the backlight. These off-setting drag and thrust forces will mislead C_D comparisons, and care must be taken. Reasonable agreement with the measured C_D does not mean that the numerical simulation is acceptable. It is a necessary, but not sufficient condition for a CFD calibration (1). Inspection of the flow structure is essential.

Streamline tracking of the measured and computed centerplane velocities, based on a fourth-order Runge Kutta time integration of the velocities, is another illustration of the difficulty with the notchback flow (Figure 9). The data presentation format here is typical of the remainder of the report: the measurements (left) are compared to the CFD simulation (right). Measured reverse flow regions are not shown because the multi-hole probes were out-of-range. But, the outline of the backlight separation area can be seen and, although the bubble size is under-predicted, the simulation does suggest some interesting flow structures that would be difficult to measure. The simulation does not show the centerplane separation on the F1 notchback.

The off-centerplane deck-surface flow topology was also not well predicted for the notchbacks. Surface flow visualization on the C1 notchback uncovered indications of sources and saddles on the deck (Figure 10) that were not fully evident in the simulation (Figure 11). Computed skin friction lines would be a better illustration of the reattachment patterns because the near-surface velocity vectors may not represent the actual on-surface flow structure.

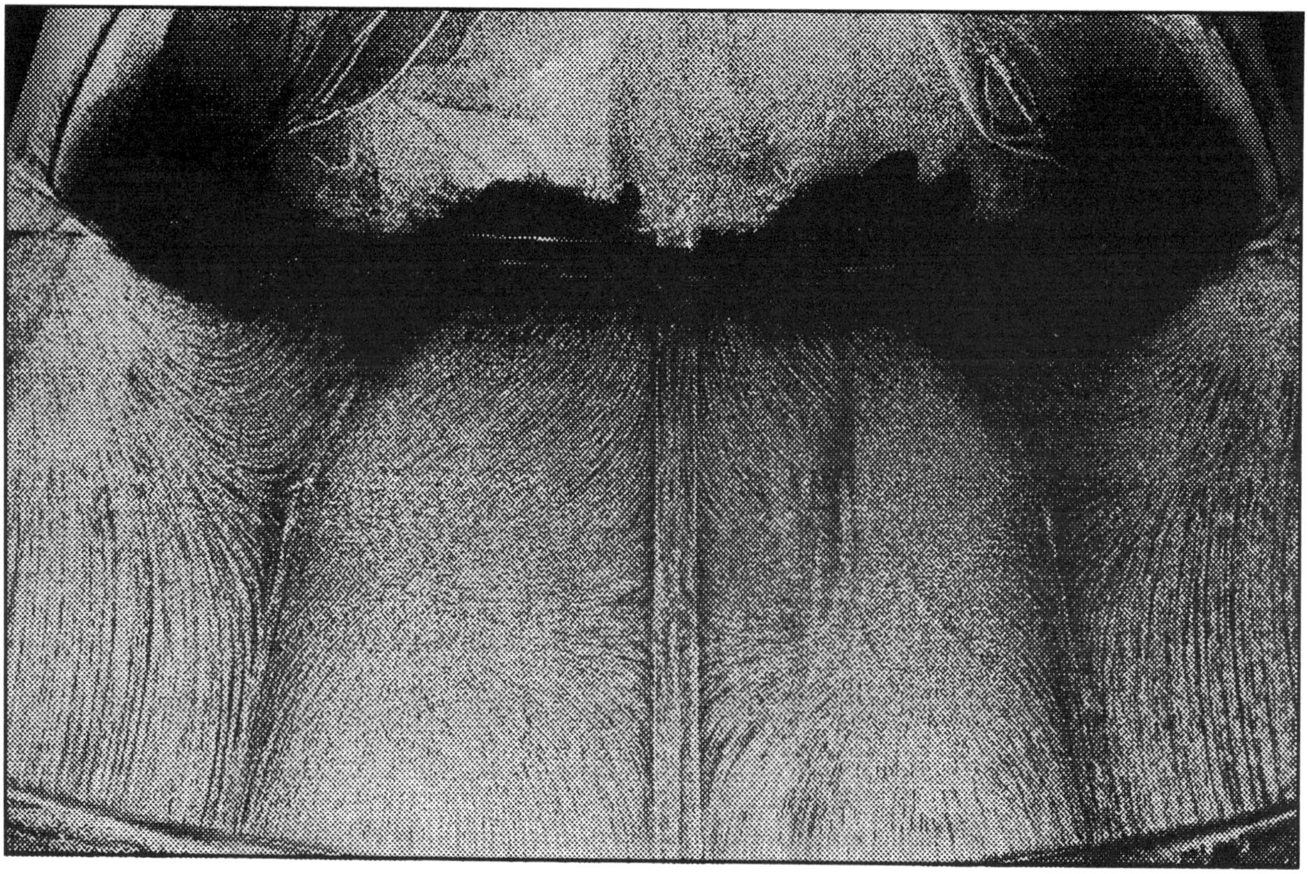

Figure 10
C1 NOTCHBACK DECK SURFACE FLOW TOPOLOGY
(Continued on next page)

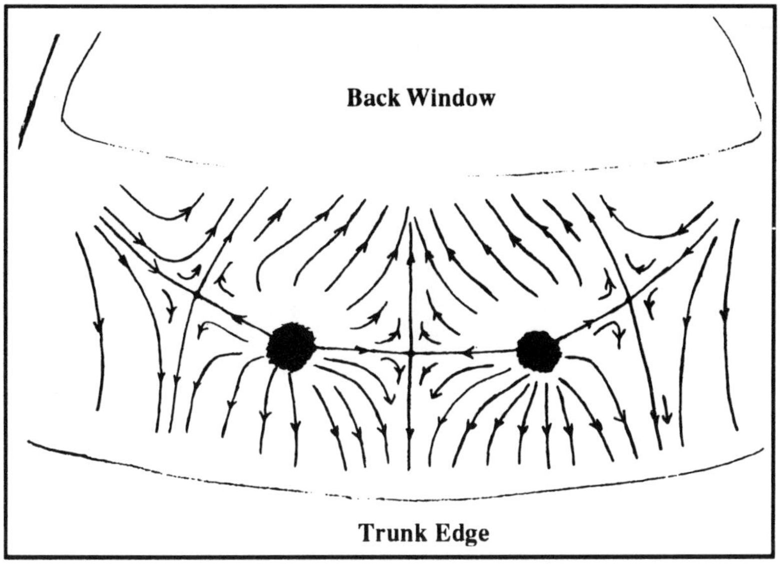

Figure 10
(Continued)
C1 NOTCHBACK DECK SURFACE FLOW TOPOLOGY

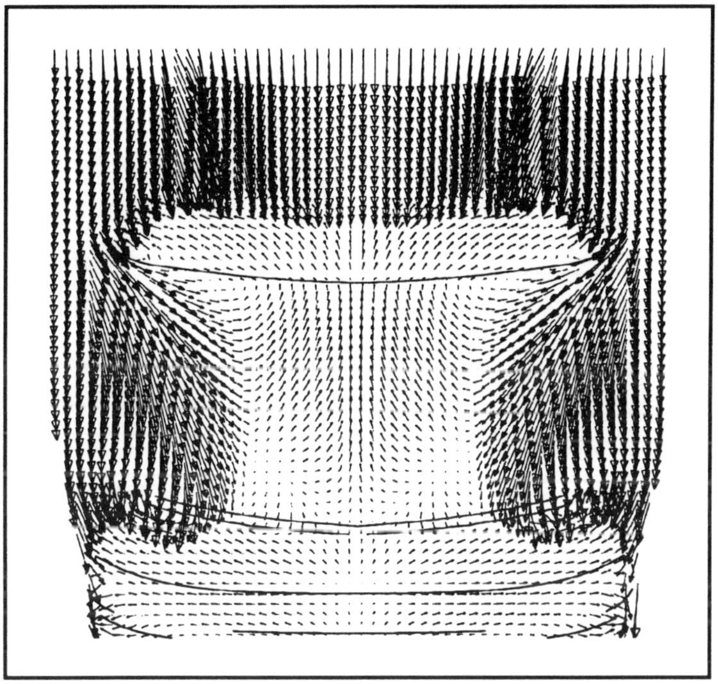

Figure 11
C1 NOTCHBACK CALCULATED DECK VELOCITY VECTORS

Lateral Wake Flow Patterns — The time-averaged, cross flow experimental measurements were very well behaved and symmetrical. Regions of high longitudinal vorticity were accompanied by high total-pressure deficits. High vortex drag was associated with high model lift measurements, positive or negative. Probe flow angle and cross-flow kinetic energy measurements verified lateral symmetry. There was very little mean-flow activity across the centerplane and symmetry was good.

Figures 12, 13 and 14 compare the F1 notchback, station wagon, and B1 sculptured fastback wake flow measurements to the corresponding CFD simulations. The measurement plane locations are about 0.62, 0.62 and 0.77 model lengths behind the center of the F1, B1 and station wagon, respectively. Cross plane velocity vectors (upper), and longitudinal vorticity (middle) and total-pressure-deficit (lower) contour plots are shown in the Figures. The left plot in each pair shows the experimental measurement while the right plot is the simulation result. In the vorticity plots, a solid line represents clockwise rotation while a dashed line is counterclockwise. As mentioned earlier, the dotted line in the contour plots indicates the filled-in region.

The off-centerplane wake simulations in these Figures show some promising results. The pressure and vorticity patterns are in reasonable agreement with the measurements. Regions of high vorticity are accompanied by high total-pressure deficits and the vorticity distribution is good. However, the near-centerplane simulations, particularly for the notchbacks, are not quite as good. In Figure 12 for example, the difficulty in predicting the separation and deck reattachment patterns of the F1 notchback is evident in the wake. The middle total-pressure-deficit peak and near-centerplane vorticity are missing.

The simulation is much better for the Station Wagon shown in Figure 13. Because of the sharp rear edge over the top and upper sides, the flow separation characteristics are fixed. This, in part, helps explain the good simulation results. Although not shown, the centerline base pressures were also in excellent agreement.

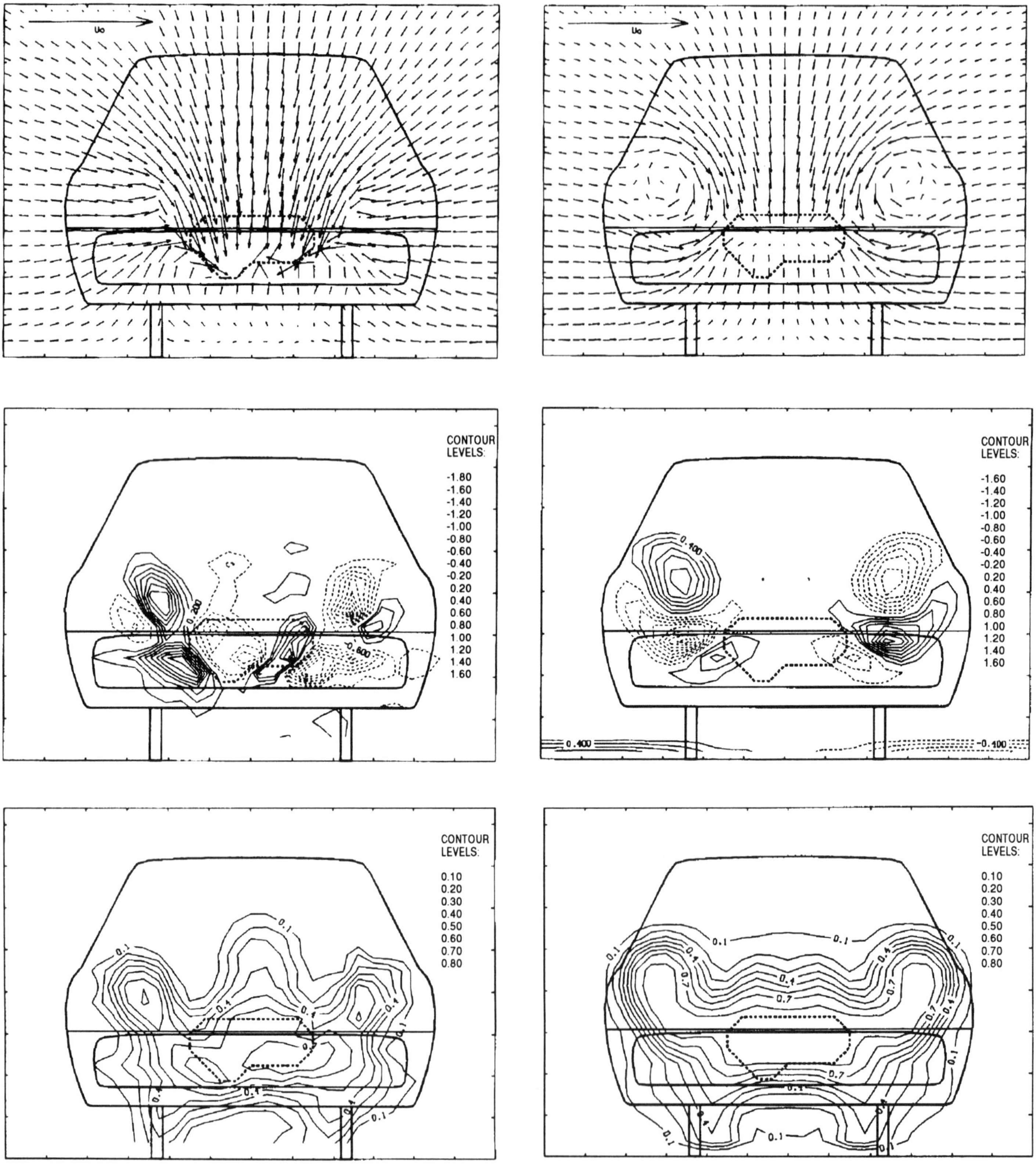

Figure 12
WAKE COMPARISONS OF CROSS-FLOW VELOCITIES, TOTAL-PRESSURE DEFICIT, LONGITUDINAL VORTICITY FOR F1 NOTCHBACK

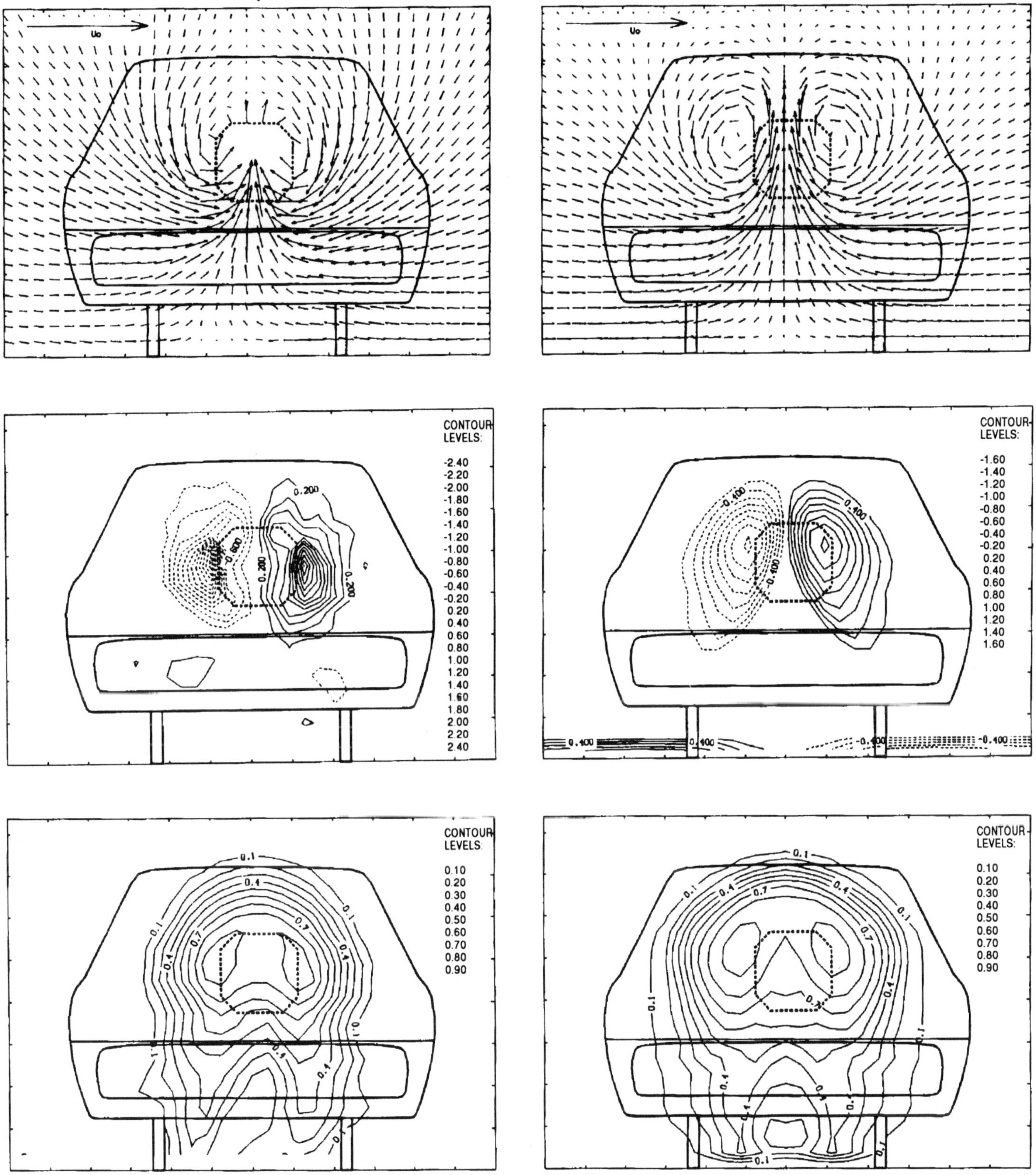

Figure 13
WAKE COMPARISONS OF CROSS-FLOW VELOCITIES, TOTAL-PRESSURE DEFICIT, LONGITUDINAL VORTICITY FOR STATION WAGON

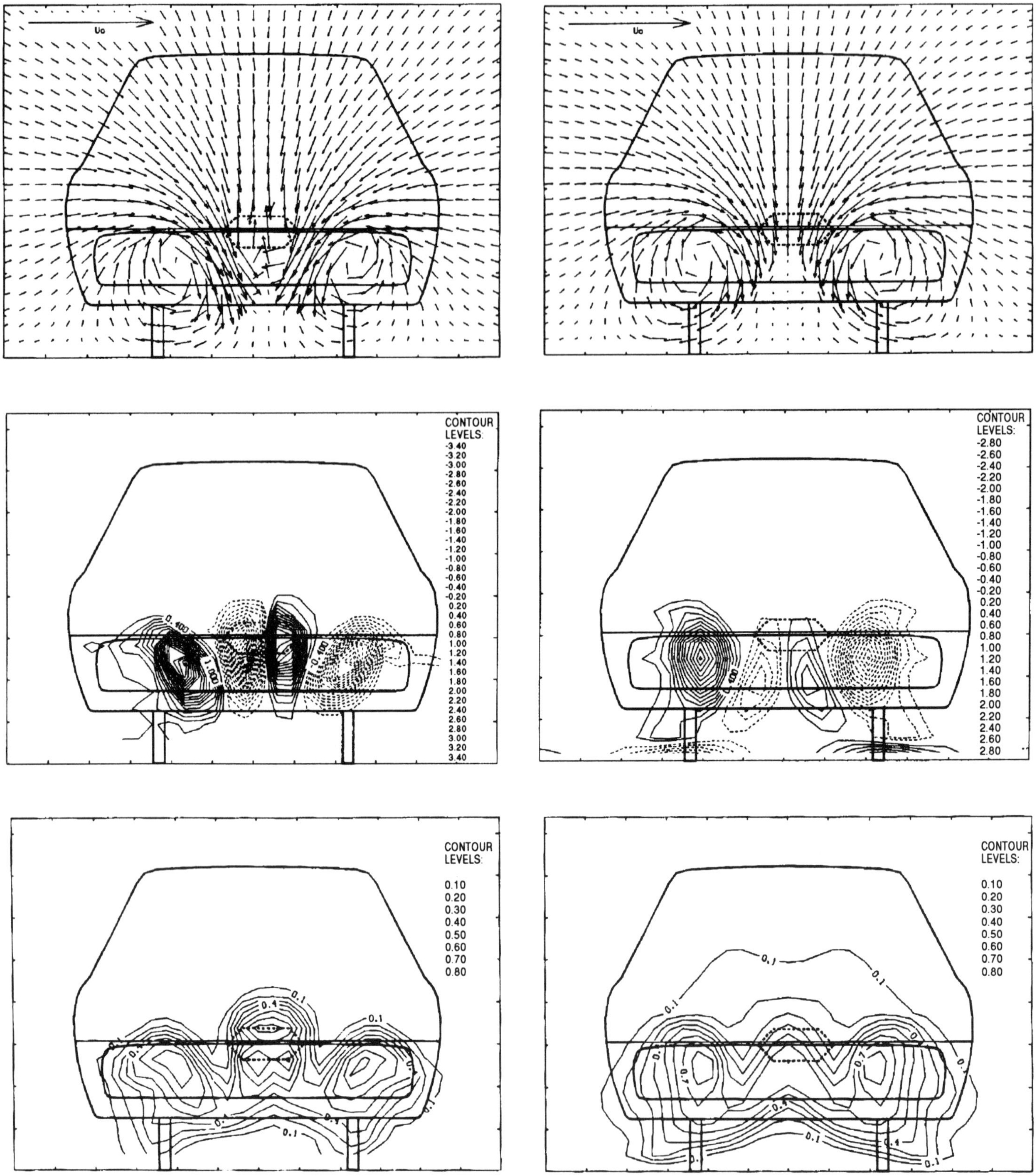

Figure 14
WAKE COMPARISONS OF CROSS-FLOW VELOCITIES, TOTAL-PRESSURE DEFICIT, LONGITUDINAL VORTICITY FOR B1 SCULPTURED FASTBACK

The B1 comparisons in Figures 14 and 15 are also good. The CFD pressure-deficit and vorticity plots show several areas of agreement. However, as with the F1 notchback, modeling the near-centerplane flow (deck edge flow and upflow-producing vortex pair) is presenting some difficulty.

The B1 aft-end geometry is a sculptured fastback with a slope that increases to each side of the centerline. The flow over the surface comprises a strong outer primary vortex pair (downflow producing) (Figure 14, middle left) and an almost-as-strong inner secondary pair (upflow producing) near the centerplane.

As on a slender delta wing, there is a separation line down the spine of the model. The lateral surface flow towards this line is pulled up into the inner vortex pair, concentrating the boundary layer fluid into a central hump (Figure 14, lower left). In addition, the upflow from the separation line moves the experimental streamlines away from the surface, even though there is no separation bubble in the classical sense.

The primary vortex pair is present in the CFD solution (Figure 14, middle right), and it is in reasonable agreement with the experiment. The secondary vortex pair is also present, but it is much weaker than for the experiment. Consequently, the total pressure hump is much reduced and the stand-off of the centerplane streamlines (Figure 15) becomes negligible. This is also reflected in the rear end pressure distribution plot, where there is an over-prediction of the turning around the large-radius deck edge.

Though the CFD solution has captured the primary pair quite well, it has largely failed in predicting the secondary vortex pair.

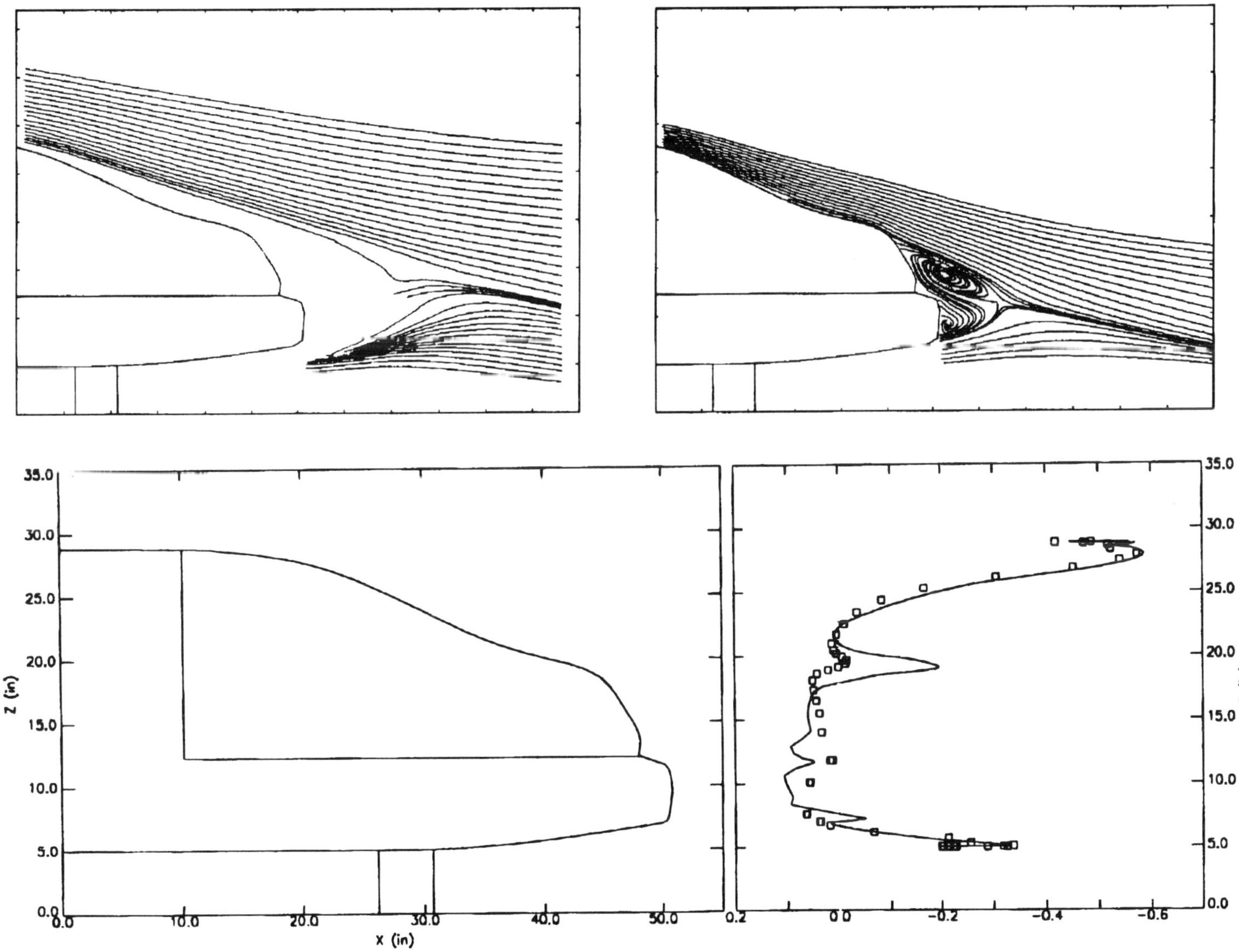

Figure 15
CENTERPLANE STREAMLINES AND REAR END PRESSURE DISTRIBUTION FOR B1 SCULPTURED FASTBACK

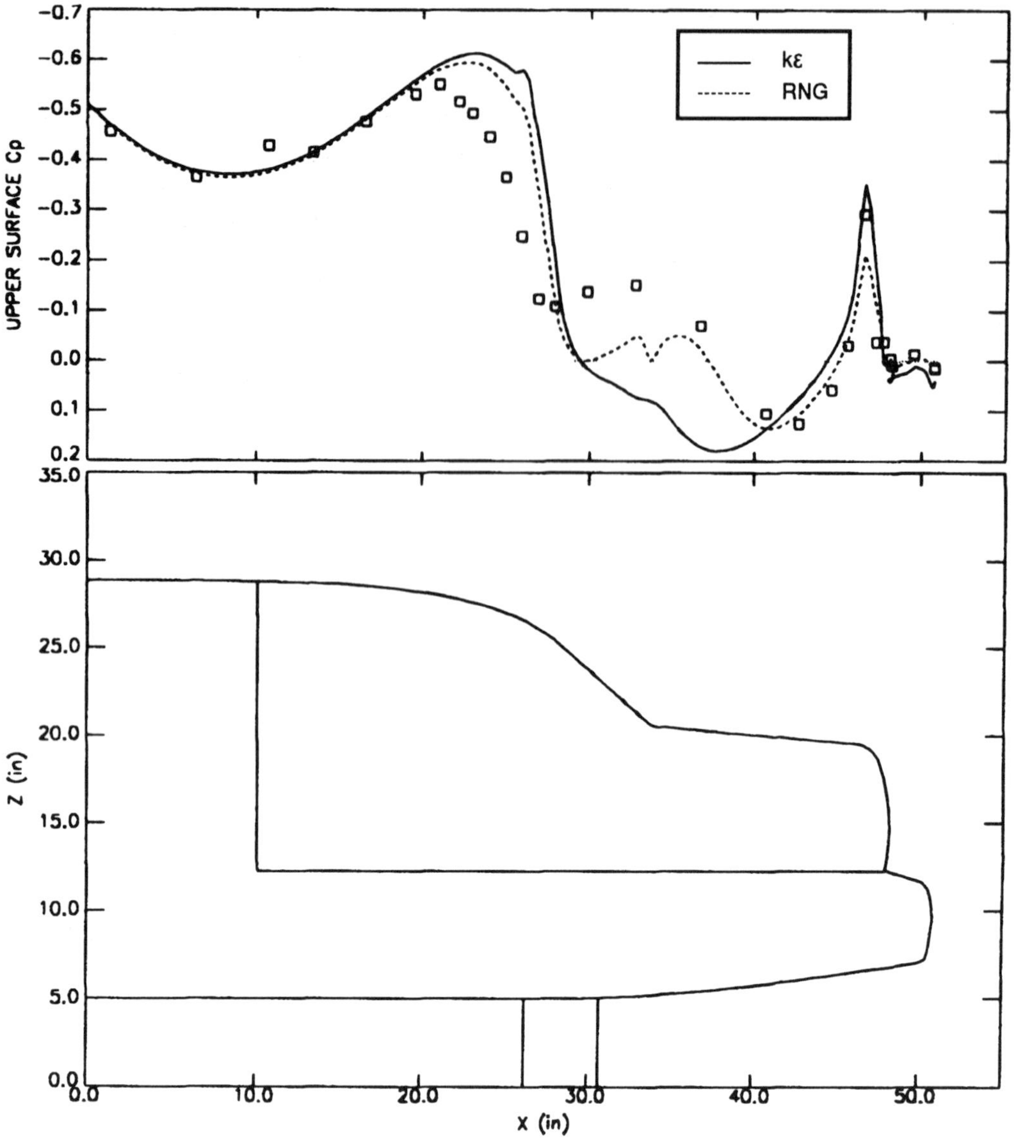

Figure 16
REAR END PRESSURE DISTRIBUTION
FOR ALTERNATIVE TURBULENCE MODEL FOR F1 NOTCHBACK

Alternative Turbulence Model – Notchback — It is well known that the k-ε turbulence model was not designed for some of these flow problems, so an alternative formulation based on Renormalization Group (RNG) methods was investigated (23). Calculations on the notchbacks initially looked very promising. The centerline rear end pressure distributions for the F1 notchback began to look better (Figure 16), and the lateral wake flow patterns started to improve (compare Figure 17 to Figure 12). There were indications of an upflow producing vortex pair near the centerplane (Figure 17), which was absent from the earlier calculation (Figure 12). The middle hump in the pressure-deficit contours was better defined and the tighter contour spacing suggested a better model of the wake gradients. Interestingly, the viscous drag represented by the pressure-deficit contours actually increased.

Similar calculations on another notchback shape also looked encouraging. Figure 18 shows the streamline tracking of the C1 centerplane flow: measurements (upper), k-ε model simulation (middle), and alternative turbulence model (lower). The limiting streamline between the above-and-below-model flow comes closer to the ground and there are suggestions of some organization in the bubble. Deck flow reattachment on the centerplane, however, is not present.

As the centerplane flow appeared to improve, the cross-flow wake comparisons deteriorated. Figure 19 compares the measurements (upper), k-ε model simulation (middle) and alternative turbulence model (lower). Increased vorticity above the deck suddenly appeared (lower picture), the directional sense of the vorticity remained incorrect, and the pressure-deficit envelope did not improve. In fact, considering the region outside the dotted boundary, neither solution correctly modeled the primary vortex pair. Although an improved turbulence model is certainly part of the final solution for a calibrated CFD design tool, this particular formulation does not appear to be the answer. Since investigation of flow around simple bluff bodies (24) indicates the existence of regions of negative eddy-viscosity, use of second-order closure models may be more successful.

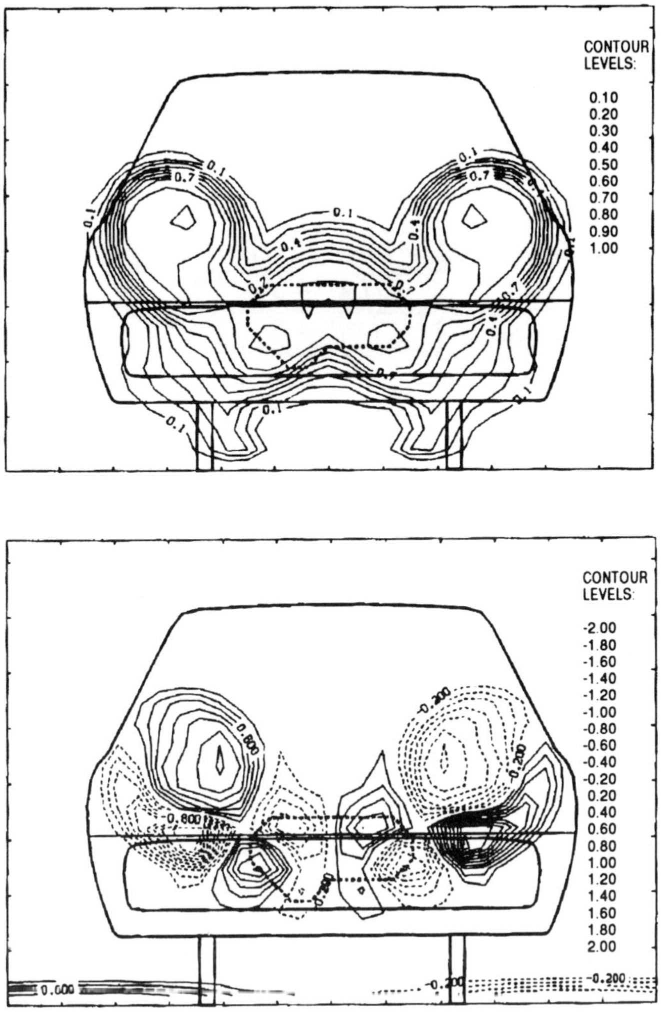

Figure 17
WAKE TOTAL-PRESSURE DEFICIT AND LONGITUDINAL VORTICITY FOR ALTERNATIVE TURBULENCE MODEL FOR F1 NOTCHBACK

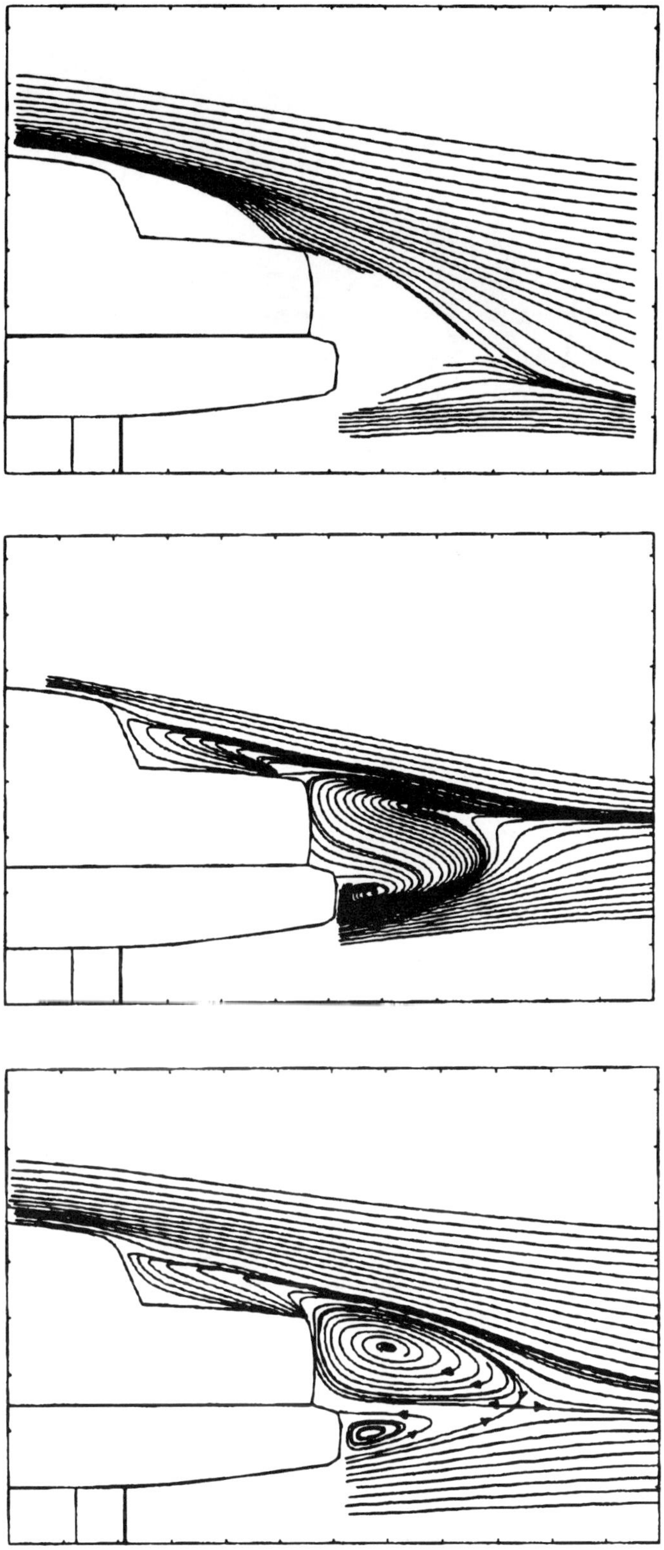

Figure 18
CENTERPLANE STREAMLINE COMPARISONS FOR C1 NOTCHBACK: TWO TURBULENCE MODELS AND EXPERIMENT

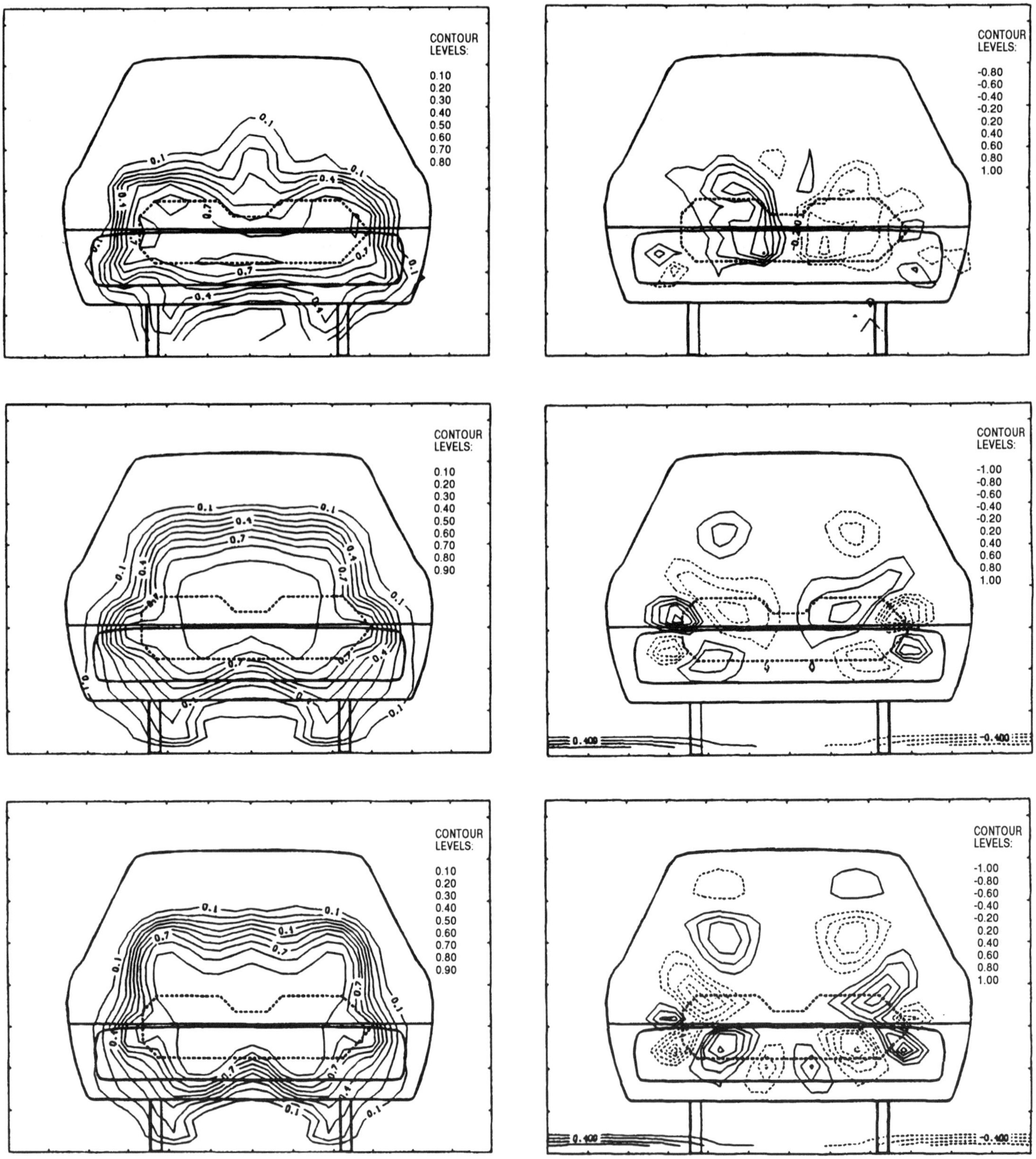

Figure 19
WAKE COMPARISONS OF TOTAL-PRESSURE DEFICIT AND LONGITUDINAL VORTICITY FOR TWO TURBULENCE MODELS FOR C1 NOTCHBACK

C_D, C_L and C_{PM} Comparisons — Figure 20 shows a comparison of the calculated drag (C_D), lift (C_L) and pitching moment (C_{PM}) coefficients to the measurements. The regression line is provided for visual reference. Except for the notchbacks, the trends are quite encouraging relative to the primary objective of ranking the shapes aerodynamically. Despite the fine computational grid, however, the drag coefficient values are higher than the measurements. The C_L and C_{PM} trends are better, partly because they are less dependent on viscous effects. It should be noted that the model underbody is smooth and that the viscous effects of normal roughness may alter the generality of this statement.

The difficulty in modeling the notchback separation and reattachment regions is clear in the drag comparisons. The F1 and C1 notchbacks are further from the trend lines. Their flow structures require more work.

Just before publication, a critical inspection of the turbulence level imposed at the upstream boundary revealed a higher than expected effective viscosity, effectively lowering the Reynolds number of the simulations. Preliminary calculations indicated reduced drag levels and modified wake flow structures. The overall implications are unknown at present but will obviously be examined further as the calibration study continues.

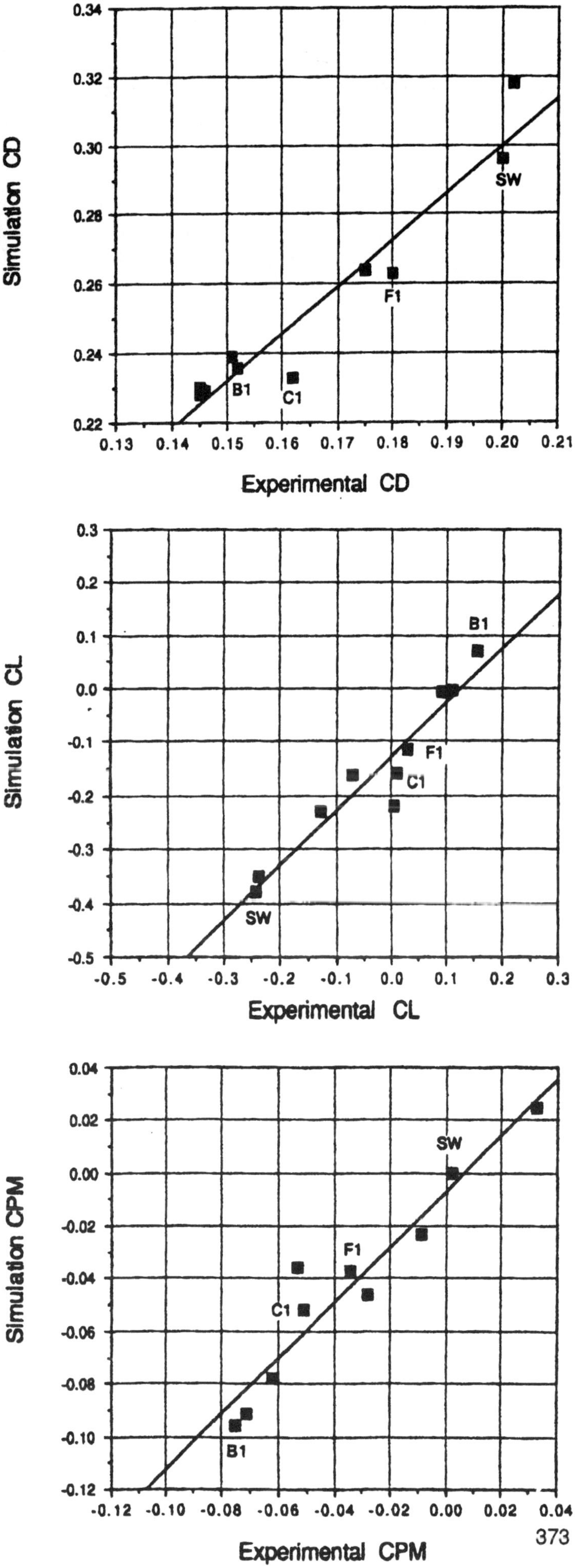

Figure 20
COMPARISON OF DRAG, LIFT AND PITCHING MOMENT CALCULATIONS TO MEASUREMENTS

SUMMARY

A framework and synergistic approach for the assessment and calibration of CFD processes has been initiated and presented in this work-in-progress report. The approach emphasizes close collaboration between the EFD and CFD technologies. The significant data base of experimental information on ten car-like configurations can be used to assess current and new numerical procedures, and investigate discrepancies between predicted and measured flow fields.

The experimental wake measurements are very well behaved and symmetrical about the centerplane. As expected, regions of high longitudinal vorticity are accompanied by high total-pressure deficits.

The numerical simulation results are quite encouraging. The calculated centerline pressure distributions on the front end, underbody and floor compare well with the measurements for all the shapes. The wake flow structures are also in reasonable agreement for many of the shapes.

However, as expected from the beginning, modeling of the viscous flow patterns over notchback rear ends is difficult. Prediction of the backlight separation line and deck reattachment pattern remains the Achilles heel of the CFD simulations. The notchback simulations remain a challenge, and more work is needed to improve the modeling of these flow structures.

For most of the shapes the CFD analysis shows the correct C_D (drag coefficient), lift and pitching moment trends. More work is needed to understand the C_D offset from the experimental measurements.

ACKNOWLEDGMENTS

The authors appreciate the efforts of the many supporting personnel —

Rob Lietz of Ford for help in the overall project, definition of the shapes and model surface development; Randy Glimpse and Bill Hayden of Ford for their expert help in the design and build of the model; Al Curtis and the Ford test crew for the precision tunnel installation of the model and general test support; Derek Wilsden of Lockheed for his generous help in the model design, pressure measurements, wind tunnel testing and surface flow visualization; Joe Patrick of Lockheed for wake flow data acquisition and reduction; and Paul Hancock of Adapco for his significant efforts on the computational side.

REFERENCES

1 Hucho, W. and G. Sovran. Aerodynamics of Road Vehicles. Annual Review of Fluid Mechanics, 1993, 25:485-537.

2 Kobayashi, T. and K. Kitoh. A Review of CFD Methods and Their Application to Automobile Aerodynamics. SAE 920338, February 1992.

3 Ahmed, S. R. A Critical Review of the CFD-Methods Available to Compute Flow Around Automobiles. Innovation and Reliability in Automotive Design and Testing, Vol. 3, 1437-1456, April 1992.

4 Bangert, L. H., C. E. Johnson and M. J. Schoop. CFD Applications in F-22 Design. AIAA 93-3055, July 1993.

5 Han, T. Y. A Navier-Stokes Analysis of Three-Dimensional Turbulent Flows Around a Bluff Body in Ground Proximity. AIAA 88-3376, July 1988.

6 Hashiguchi, M., et al. Computational Study of the Wake Structure of a Simplified Ground-Vehicle Shape with Base Slant. SAE 890597, February 1989.

7 Matsunaga, K., et al. Finite-Difference Simulation of 3D Vortical Flows Past Road Vehicles. SAE 920339, February 1992.

8 Nouzawa, T., et al. Unsteady-Wake Analysis of the Aerodynamic Drag of a Notchback Model with Critical Afterbody Geometry. SAE 920202, February 1992.

9 Yamada, A. and S. Ito. Computational Analysis of Flow Around a Simplified Vehicle-Like Body. SAE 930293, March 1993.

10 Hutchings, B. J. and W. Pien. Computation of Three-Dimensional Vehicle Aerodynamics Using FLUENT/BFC. Proc. 2nd International Conf. on Supercomputing Applications in the Automotive Industry, Seville, p. 233-255, Cray Research, 1988.

11 Adey, P. C. and J. R. A. Greaves. The Application of a 3-D Aerodynamics Model to the Sterling 825 Body Shape and Comparison with Experimental Data. SAE 880456, February 1988.

12 Shaw, C. T. Predicting Vehicle Aerodynamics Using Computational Fluid Dynamics — A User's Perspective. SAE 880455, February 1988.

13 Malone, J. B., et al. Three-Dimensional Navier-Stokes Solution for a Car-Like Body. AIAA 88-3769, July 1988.

14 Kawaguchi, K., et al. Computation Study of the Aerodynamic Behavior of a Three-Dimensional Car Configuration. SAE 890598, February 1989

15 Himeno, R., et al. Numerical Analysis of the Airflow around Automobiles Using Multi-Block Structured Grids. SAE 900319, February 1990.

16 Ono, K., et al. Simultaneous Computation of the External Flow around a Car Body and the Internal Flow Through its Engine Compartment. SAE 920342, February 1992.

17 Kataoka, T., et al. Numerical simulations of Road Vehicle Aerodynamics and Effect of Aerodynamic Devices. SAE 910597, February 1991.

18 Shiozawa, H., et al. Numerical Analysis of the Effect of a Rear Spoiler on Flow Fields Around a Car. Innovation and Reliability in Automotive Design and Testing, Vol. 3, 1479-1485, April 1992.

19 Bearman, P. W. Some Observations on Road Vehicle Wakes. SAE 840301, February 1984.

20 Marvin, J. G. Accuracy Requirements and Benchmark Experiments for CFD Validation. NASA-TM-100087, May 1988.

21 Hackett, J. E., et al. On the Influence of Ground Movement and Wheel Rotation in Tests on Modern Car Shapes. SAE 870245, February 1987.

22 "STAR-CD Version 2.1 Manuals". Computational Dynamics Limited, 1991. (Available from Adapco in Melville, NY)

23 Yakhot, V., S. A. Orsag, et al. Development of Turbulence Models for Shear Flows by a Double Expansion Technique. Physics of Fluids, A 4(7), p. 1510, 1992.

24 Franke, R., W. Rodi and B. Schoenung. Analysis of Experimental Vortex Shedding Data with Respect to Turbulence Modeling. Seventh Symp. on Turbulent Shear Flows, 24-4, 1989.

920342

Simultaneous Computation of the External Flow Around a Car Body and the Internal Flow Through its Engine Compartment

Kenji Ono and Ryutaro Himeno
Nissan Research Center, Nissan Motor Co., Ltd.

Katsuro Fujitani
Engineering Systems Center, Nissan Motor Co., Ltd.

Yutaka Uematsu
Body Experiment Dept., Nissan Motor Co., Ltd.

ABSTRACT

Both the external and internal flows of cars are simulated simultaneously. A third-order upwind-difference scheme is used in these simulations. Computational grids are generated by a multi-block transformation and a trans-finite method. Engine compartments are modeled by grid systems but the heat exchanger is simulated as a pressure loss proportional to the dynamic pressure of the flow passing through it.

First, the flow for a very simple test model with no wheels and nothing in its engine compartment is simulated and compared with experimental results in order to validate a simulation method for the engine compartment. Pressure distributions on the inner surfaces agree very well with measured values, while pressure distributions on the external surfaces show reasonable agreement except for the roof end and the leading edge of the floor. The predicted drag coefficient is 7% larger than the experimental value.

This method is next applied to a prototype car. A grid system is generated including an engine block, a transmission, an air-cleaner and suspension arms. Both the drag coefficient and the airflow rate through the radiator are calculated. These values are compared with experimental results and the agreement is good. We conclude that the present method is quite useful for predicting and optimizing both the drag coefficient and the airflow rate through the radiator.

INTRODUCTION

Computational Fluid Dynamics (CFD) has developed greatly, due to the progress in super-computers and computational schemes. Various 3-dimensional analyses of the airflow around an automobile are carried out nowadays[1-3]. A third-order upwind-difference scheme was developed by Kawamura et al. It was successfully applied to the analysis of the drag crisis of a two-dimensional circular cylinder and the turbulent inner flow between parallel flat plates without using any turbulence model, even in the high Reynolds-number region[4,5]. The present authors have simulated flows related to automobiles using the third-order upwind-difference scheme and reported that the scheme has enough accuracy to predict their aerodynamic characteristics[6-8]. It has become possible to predict the drag coefficient of a production car with an error of less than 2 percent[8].

Simulation of cooling flow is another important application of CFD. The shape of inlets is important from both the styling and engineering points of view. It is quite useful to be able to predict the amount of cooling airflow through heat exchangers in the early design stage. For this analysis, it is essential to calculate the external and internal flows simultaneously. There are several papers which simulate the engine cooling flow of cars[9-11]. However, almost all of them use rectangular grid systems which are not suitable for calculating external flows along smooth surfaces. Now we report on the application of a third-order-upwind scheme combined with a curvilinear coordinate system for the simultaneous prediction of both the aerodynamic forces and the cooling airflow of a car.

BASIC EQUATIONS AND COMPUTATION SCHEMES

The airflow around automobiles is incompressible. The continuity and Navier-Stokes equations can be described as follows with a coordinate system fixed on the vehicle moving with velocity $\mathbf{v}_0$ in a stationary fluid.

*Equation of continuity

$$\nabla \cdot \mathbf{v} = 0, \tag{1}$$

*Navier-Stokes equation

$$\frac{\partial \mathbf{v}}{\partial t} + \{ (\mathbf{v} - \mathbf{v}_0) \cdot \nabla \} \mathbf{v} = -\nabla p + \frac{1}{Re} \nabla^2 \mathbf{v}, \tag{2}$$

where $\mathbf{v}$ and p are, respectively, the non-dimensional velocity vector based on $\mathbf{v}_0$ and the non-dimensional pressure based on twice the dynamic pressure of $\mathbf{v}_0$, and Re is the Reynolds number. The Poisson equation with respect to pressure of a high-Reynolds-number flows is as follow,

$$\nabla^2 p = -\nabla \cdot \{ (\mathbf{v} - \mathbf{v}_0) \cdot \nabla \} \mathbf{v} - \frac{\partial D}{\partial t}, \tag{3}$$

where $D = \nabla \cdot \mathbf{v}$. Then the second term of the right hand side of this equation is expressed by the finite-difference approximation, $(D^{n+1} - D^n) / \Delta t$, where superscripts n and n+1 denote the present and the next time step respectively, and Δt denotes a time increment. Although D should be 0 due to the equation of continuity, Dn is not omitted so that any error in satisfying the equation of continuity at the present time step n can be compensated. Finally, the pressure equation to be solved becomes as follows:

$$\nabla^2 p = -\nabla \cdot \{ (\mathbf{v} - \mathbf{v}_0) \cdot \nabla \} \mathbf{v} + \frac{D^n}{\Delta t}, \tag{4}$$

The equation of continuity is not directly satisfied, but is implicitly satisfied. This derivation is based on a MAC method[12]. Next, eqs. (2) and (4) are transformed into a generalized coordinate system and discretized. A first-order implicit Euler method is used for time derivatives. Convective terms are discretized using a third-order upwind-difference scheme[4,5]. A second-order central-difference scheme is used for the other spatial-derivative terms.

RADIATOR MODEL

In order to reduce the number of grid points, the flow through the radiator is not directly calculated but is modeled as a pressure loss. The pressure-loss characteristics of the radiator used in the prototype car are shown in figure 1, where the horizontal and vertical axis represent, respectively, non-dimensional velocity based on 15 m/s and non-dimensional pressure loss based on twice of the dynamic pressure of the flow at 15 m/s. This way of normalization is the same as the one used in the basic equations, (1) and (2). Here, the pressure loss is assumed to be

$$\Delta p = cv^2, \tag{5}$$

where c is a constant coefficient and v is a non-dimensional velocity component normal to the radiator. When this coefficient c has a value of 8.0, the pressure-loss characteristics are as shown in figure 1, which corresponds well to the characteristics of the production model.

The length of the simplified model is 0.5 m, about 1/9 of the full-size car length, and the radiator is specially made to simulate the one used in the prototype. The characteristics are shown in the same figure 1. They do not agree with each other in the velocity region higher than 0.3, but they show reasonable agreement in the lower-velocity region, which contains the range of velocities at which the air passes through the radiator in the experiment.

COMPUTATIONAL GRID SYSTEM

A multi-block-transformation is used to generate the grid systems. This method is more suitable than a single-block approach for the analysis of the flow around objects with complicated geometry[13].

For this grid generation, the grid distribution on the boundary surfaces of the computational domain is determined first. The surface data for the vehicle are taken from CAD data. The grid in each block is generated by a transfinite interpolation[14] using the grid distributions on the boundaries. In the next step, all blocks are combined into one block. Finally, the grid is modified to be smooth through block boundaries and orthogonal to the body surfaces, as these are important requirements for an accurate numerical solution. These procedures are described in more detail in reference 15.

SIMPLIFIED TEST MODEL - The test model has an inlet on its front end, an outlet beneath the floor and a slant slope in the rear as shown in figure 2(a). The model has two cavities inside the body, connected with a radiator in order to simulate a cooling flow through the engine compartment of a car. (See figure 2(b).)

Figure 3 shows a close-up view of grid distributions on the body surface, the ground and the symmetry plane. This grid system has 138x105x92 grid points. The dimensions of the computational space are 2 times the body length upstream, 10 times downstream, 5 times the width sideward and 5 times the height upward, respectively. The minimum spacing of the grid is set as 0.0005 based on the body length.

PROTOTYPE CAR - A grid system is first generated for a prototype car of the front-wheel-drive three-box-sedan and shown in figures 4 and 5. Principal parts in the engine compartment, such as lower suspension arms, a powertrain, a radiator and its support, are modeled in the grid system as shown in figure 5. The size of this grid system is 189x131x80. The minimum spacing is about 0.00023 based on the car length. The grid system has a flat floor, while the prototype car has roughness on it. It took about 5 weeks for one person to generate this grid system.

Then, the grid system is modified to have a license plate and two holes near both edges of the radiator support, without increasing the number of grid points (see figure 6). The holes are modeled to represent not only the head-light holes but also all of the small holes on the radiator support and all of the gaps between the radiator support and the body or the radiator. So, the total area of the two holes is set equal to the summation over those small holes and gaps. This modification took one week.

COMPUTER SYSTEM AND PROGRAMS

The computer hardware used for this study is shown in figure 7. A mini-supercomputer Convex C240 is used for the testing of the grid systems, the computation of particle tracing and also as a file server. Graphics work stations IRIS210GTX and TITAN3000 are used for grid generation and the display of computed results. Video tape recorders, two sets of Sony BETACAM, and an optical disc system are connected to graphic workstations through a controller for taking pictures frame by frame to make animation films.

The program for computing flows is written in FORTRAN77 and is vectorized 99 percent and parallelized over 98 percent on a CRAY YMP8. It requires about 60 Mwords of its memory space and achieves about 1.2GFLOPs on a CRAY YMP8/664. Programs for flow visualization are written in C-language. The program for particle tracing is vectorized and parallelized on a CONVEX C240. All programs, the grid generation system, the Navier-Stokes solver and the flow visualization system, have been developed by the authors.

RESULTS AND DISCUSSION

In the calculation of both the simplified test model and the prototype car, the model is moving with velocity v_0 in a stationary fluid, and the boundary condition of the ground is set as stationary with velocity 0. Aerodynamic forces are calculated by integrating pressure over whole surfaces. The friction forces are neglected in the calculation.

SIMPLIFIED TEST MODEL - Reynolds number based on the length of the model is 588,000 in both the calculation and the experiment. The experiment is performed in a scale wind tunnel with a moving belt, making the boundary condition for the ground the same as the one in the calculation.

Flow Field - The computed flow field is compared with experimental visualization in a smoke wind tunnel, although the Reynolds number in the experiment is only about 1/3 of that in the computation. Figure 8 (a) is a photograph of the flow around the simplified model visualized by smoke, and figure 8 (b) shows the streamlines in a time-averaged computed flow field. Comparing these two figures, a computed separation in the rear is slightly larger than that in the experiment, but they agree well qualitatively. A pair of trailing vortices in the wake is shown in figure 9. Flow near the body sides comes across the side edges of the slanted surface and is vented toward the center, due to the strong vortices.

To investigate an internal flow field, the time-averaged pressure field on the symmetry plane is shown in figure 10. While pressure varies a great deal in the first cavity in front of the radiator, the pressure difference in the second cavity after the radiator is very small and the pressure is almost equal to zero, i.e. the same as the free stream value. It is obvious that the second cavity has little effect on the aerodynamic coefficients of this model, while the first cavity has considerable influence on the drag force.

Pressure Distribution on the Symmetry Plane - Calculated pressure distributions on the outer surfaces are first compared with experimental measurements in figure 11. The measured data indicates sharp negative-pressure peaks at the leading edges of the roof and the floor, and also a negative peak at the joint of the roof and the slanted surface where the flow is accelerated. These are the regions where the peaks are due to strong streamline curvature. Simulation results agree well with measurements, especially at the front end, the trailing edge of the roof, and beneath the rear half of the floor. However, the simulation does not predict well the peaks of negative pressure at the beginning of the slant and at the leading edge of the floor. This is probably due to coarse grid distributions there.

Next, pressure distributions on inner surfaces are compared in figure 12. The agreement is excellent.

Amount of Internal Airflow - The amount of internal flow is calculated and compared with experimental results in table 1. In the experiment, the velocity is measured at 50 points in front of the inlet using a Laser-Doppler anemometer. The agreement is also excellent. Thus, the computational model for the radiator has been validated.

Drag and Lift Coefficients - The predicted drag coefficient using numerical simulation is about 7 percent higher than that of the experiment, as shown in table 2. This is primarily due to the over-estimation of the negative pressure in the rear (see figure 11). The calculated lift coefficient does not agree well with the measured value and is about 2/3 of the experimental measurement.

Next, the calculated drag force that shows reasonable agreement is decomposed into three parts according to figure 13 using a portion of the drag coefficient C_D^* defined as follows:

$$C_D^* = \frac{f_D^*}{1/2\rho v^2 S} = \frac{f_D^*}{f_D} C_D , \qquad (6)$$

where f_D^* and f_D are the drag force acting on a specific part and the total drag force acting on the whole model respectively, and ρ , v and S are the density of the air, the free-stream velocity and the frontal area of the model respectively. Two of the three parts in figure 13 are the frontal half and rear half of the outer surfaces. The other is the inner surfaces. Table 3 shows that the internal flow causes 22 percent of the total drag force.

PROTOTYPE CAR - The external flow around a prototype of a front-wheel-drive car and the internal flow through its engine compartment are calculated simultaneously to estimate the amount of the engine-cooling flow and the aerodynamic coefficients for two cases. The possibility of predicting those values for an actual production car is also investigated. The Reynolds number in two cases is 4.3×10^6 based on the length of the car and the free stream velocity of 15 m/s, and it is the same as that of the experiments using the actual-size prototype car in a full size wind tunnel[16].

Unsteady flow fields in two cases are calculated for non-dimensional time 6.3 on a CRAY YMP8/664 and the total CPU time for each case is about 180 hours. In the case of the car with a license plate, flow fields are computed for another 2.3 non-dimensional time for time-averaging on a CONVEX C240. The Strouhal number is about 1.7 in both cases and the 2.3 non-dimension time contains 4 period of the fluctuation.

Airflow Rate - First, a time-averaged airflow rate is calculated in the case of the prototype car without a license plate and holes on the radiator support, and compared with experimental measurements as shown in table 4. In the experiment, pressure differences between the two sides of the radiator are measured at 40 points. Then, the corresponding velocities are obtained from a table giving the pressure loss as a function of velocity. The flow rate is defined here on the basis of the averaged velocity through the heat exchanger as a percentage of the velocity of the car, v_0 (i.e. 15 m/s in this case). Experiments are performed using the same configuration as in the grid system in each case, except for the roughness beneath the floor. The calculated flow rate of 15.1 % in the case of the car without a license plate agrees excellently with the measured value of 14.2 % using the prototype car with the same configuration, but it is more than twice as much as the airflow rate of 5.8 % for a fully-equipped car that can run legally on a road. This discrepancy means that CFD is not able to predict the amount of cooling flow practically in the case of a production car. Next, we try to improve it.

Our approach is to include the influence of a license plate and bypass flows through holes on the radiator support and gaps between the radiator support and the body or the radiator. Comparing figures 14 and 15, it is obvious that the license plate disturbs the airflow coming into the engine compartment. Including a license plate can improve the accuracy considerably, as shown in table 4, but the disagreement still exists. This is caused primarily by neglecting several parts in the engine compartment, such as a shroud or a cooling fan, which may have much influence on the cooling flow. The discrepancy between the simulation and the experiment with the same configuration is probably due to the air escaping through the holes. Table 5 shows that the amount of airflow passing through the holes is about 15% of the amount going through the radiator. The area may be under-estimated, because it is very hard to measure all holes and gaps. The position of holes in the simulation may affect the accuracy, as a hole near the radiator should have a greater effect than a more distant one. The coarseness of the grid system may also have some influence on the accuracy.

Time-averaged Flow Fields and Aerodynamic Coefficients - The time-averaged flow field is calculated only in the case of the model with a license plate. Figure 16 (a) shows streamlines around the car and they visualize the A-pillar vortex and trailing vortices. Figure 17 shows pressure distributions on the surface of the car. It is observed that the pressure decreases to have minus-peaks on the leading edge of the hood and roof, and on the side window. The pressure distribution on the body surface in the symmetry plane is shown and compared with experimental results in figure 18. They agree very well except for the roof.

Figure 19 (a) shows the internal flow in the engine compartment. The license plate disturbs the airflow coming into the engine compartment. Part of the flow through the inlet passes through holes near both edges of the radiator support, as shown in figure 19 (b). The airflow through the radiator flows over the engine and transmission, and goes out beneath the floor.

The estimated drag coefficient is 0.29. As the drag coefficient of a production model of this car is 0.32 without door mirrors, the simulation underestimates the drag by about 10 percent (see table 6). This discrepancy is acceptable, considering the difference between the simulation model and the production model. The grid system has a flat floor while the production model has roughness on it.

CONCLUSION

Both internal and external flows are calculated simultaneously for two kinds of models using a third-order upwind-difference scheme combined with a curvilinear coordinate system. We found that the present method can simulate flows well, and that the accuracy of predicted aerodynamic coefficients, airflow rates, and pressure distributions is reasonable. In the prototype car, the license plate and the holes on the radiator support significantly influence the airflow rate through the radiator. A multi-block transformation is powerful enough to generate complicated grid systems for these computations.

REFERENCES

[1] Shaw, C.T. and Simcox, S., The Numerical Prediction of the Flow Around a Simplified Vehicle Geometry, Proceedings of the Second International Conference on Supercomputing Applications in the Automotive Industry, Seville, Spain, October 1988, A Cray Research, Inc. book, pp.219-232, 1988.

[2] Hutchings, B.J. and Pien, W.S., Computation of Three Dimensional Vehicle Aerodynamics Using FLUENT/BFC, Proceedings of the Second International Conference on Supercomputing Applications in the Automotive Industry, Seville, Spain, October 1988, A Cray Research, Inc. book, pp.233-256, 1988.

[3] Kawaguchi, K., Hashiguchi, M., Yamasaki, R. and Kuwahara, K., Computational Study of the Aerodynamic Behavior of a Three-Dimensional Car Configuration, SAE paper, 890598, 1989.

[4] Kawamura, T., Takami, H. and Kuwahara, K., Computation of High Reynolds Number Flow around a Circular Cylinder with Surface Roughness, Fluid Dynamic Research 1, 1986, pp.145-162.

[5] Kawamura, T. and Kuwahara, K., Direct Simulation of a Turbulent Inner Flow by Finite-Difference Method, AIAA paper, AIAA-85-0376, 1985.

[6] Himeno, R., Fujitani, K. and Kuwahara, K., Finite-Difference Analysis of Two-Dimensional Flow around an Automobile, Transactions of the JSME, Vol. 486, Series B, pp.356-362, No.86-0407A, 1986.

[7] Fujitani, K., Himeno, R. and Takagi, M., Unsteady 3-D Calculation of Flow around a Car and its Visualization, Automotive Simulation: Proceedings of Second European Cars/Trucks Symposium, Schliersee, FRG, May 1989, Moshe R. Heller (Editor), Springer-Verlag Berlin, Heidelberg, 1989.

[8] Himeno, R., Takagi, M., Fujitani, K. and Tanaka, H., Numerical Analysis of the Airflow around Automobiles Using Multi-block Structured Grids, SAE Paper 900319, 1990.

[9] Kuriyama, T., Numerical Simulation on Three Dimensional Flow and Heat Transfer in the Engine Compartment Using "STREAM", Proceedings of the Second International Conference on Supercomputing Applications in the Automotive Industry, Seville, Spain, October 1988, A Cray Research, Inc. book, pp.283-292, 1988.

[10] Shibata, S., Hosaka, S., Fujitani, K. and Himeno, R., A Numerical Analysis Method for Optimizing Inter-cooler Design in the Vehicle Development Process, SAE Paper 900080, 1990.

[11] Aoki, K., Numerical Simulation of Three Dimensional Engine Compartment Air Flow in FWD Vehicles, SAE Paper 900086, 1990.

[12] Roache, P. J., Computational Fluid Dynamics, Hermora Publishers, 1976.

[13] Sawada, K. and Takanashi, S., A Numerical Investigation on Wing/Nacelle Interfaces of USB Configuration, AIAA paper, AIAA-87-0455, 1987.

[14] Thompson, J. F., Warsi, Z. A. U. and Mastin, C. W., Numerical Grid Generation : Foundations and Applications, North-Holland, 1985.

[15] Fujitani, K. and Himeno, R., A CAD data oriented grid generation system and its application to automotive aerodynamics, Numerical Grid Generation in Computational Fluid Dynamics and Related Fields, Proceedings of the Third International Conference on Numerical Grid Generation in Computational Fluid Dynamics and Related Fields, Barcelona, Spain, 1991, A.S.-Arcilla, J.Hauser, P.R.Eiseman and J.F.Thompson (Editors), North-Holland, Amsterdam, 1991.

[16] Ogata, N., Iida, N. and Fujii, Y., Nissan's Low-Noise Full-Scale Wind Tunnel, SAE Paper 870250, 1987.

Table 1. Comparison of flow quantity for the simplified test model.

	Cal.	Exp.
Flow quantity (m^3/s)	2.33×10^2	2.28×10^2
Mean velocity (m/s)	6.65	6.50

Table 2. Comparison of aerodynamic coefficients.

	Cal.	Exp.
C_D	0.356	0.327
C_L	0.192	0.300

Table 3. Partial drag coefficients of the simplified test model in computational result.

	Front	Inner	Rear
C_D	0.084	0.081	0.184

Table 4. Comparison of airflow rate for prototype car.

Difference		Airflow rate (%)	
License plate	Bypass hole	Cal.	Exp.
Exist	Exist	9.0	6.9
Non	Non	15.1	14.2

Table 5. Comparison of flow quantity for the prototype car in computational result.

Measurement position	Radiator	Bypass hole
Sectional area (m^2)	0.208	0.059
Flow quantity (m^3/s)	0.293	0.043

Table 6. Comparison of aerodynamic coefficients.

Prototype car	C_D	C_L
Calculation	0.29	0.44
Experiment	0.32	0.24

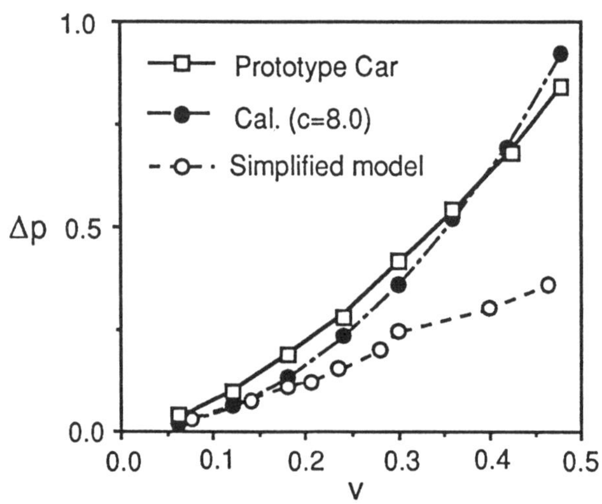

Figure 1. Non-dimensional pressure loss characteristics for simulation models.

(a) Shaded image of outer view.

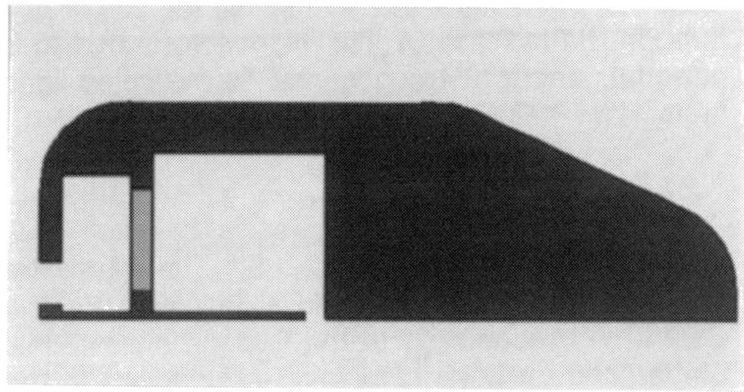

(b) Section of symmetric plane.

Figure 2. Configuration of the simplified test model.

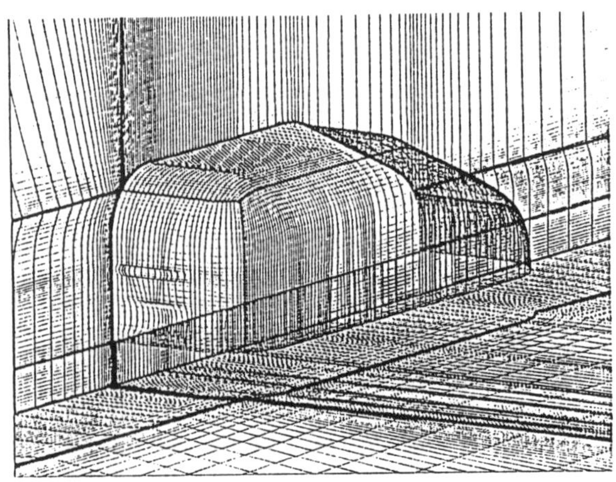

Figure 3. Grid system for the simplified test model. (138 x 105 x 92)

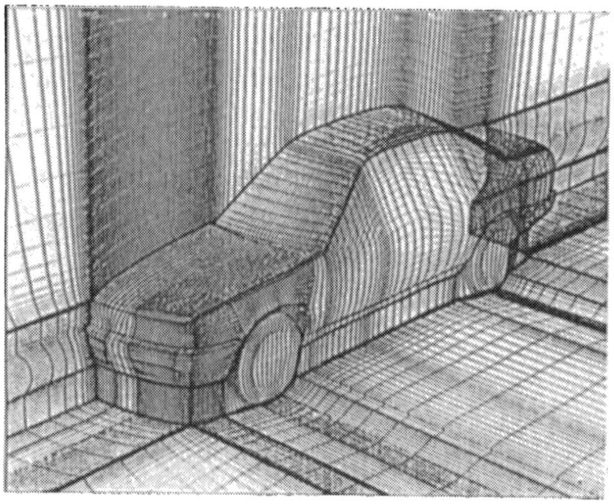

FIgure 4. Grid system around the prototype car. (189 x 131 x 80)

Figure 5. Surface grid distributions in the engine compartment.

Figure 6. Bypass holes located on the radiator support.

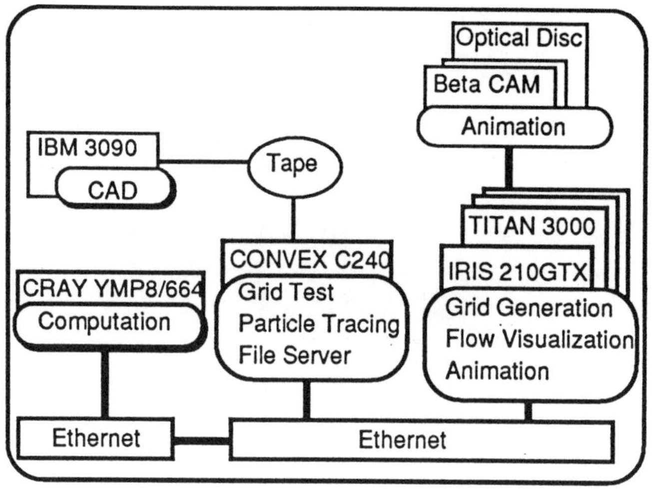

Figure 7. Computer system.

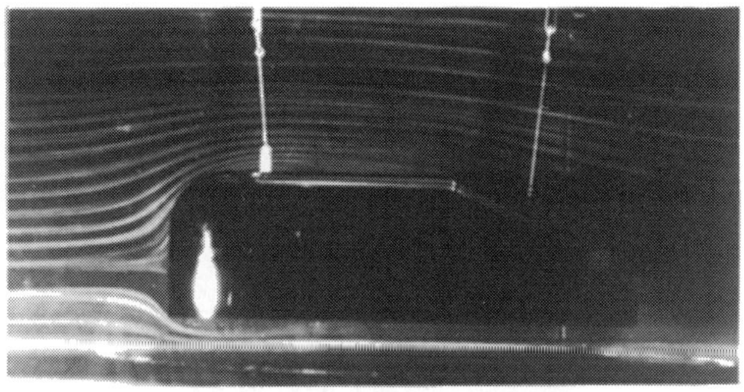

(a) Experimental results. (Re=176,000)

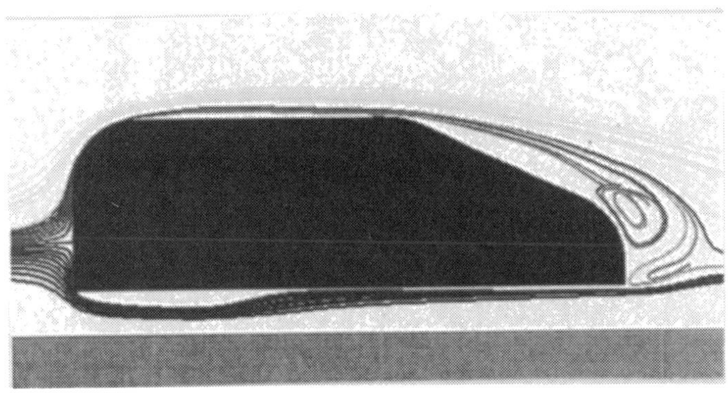

(b) Computational results. (Re=588,000)

Figure 8. Flow around the simplified test model.

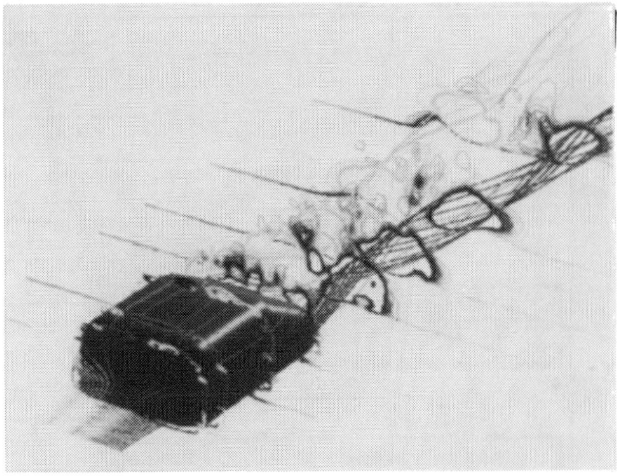

Figure 9. Streamlines and contours of the total pressure in the time-averaged flow field.

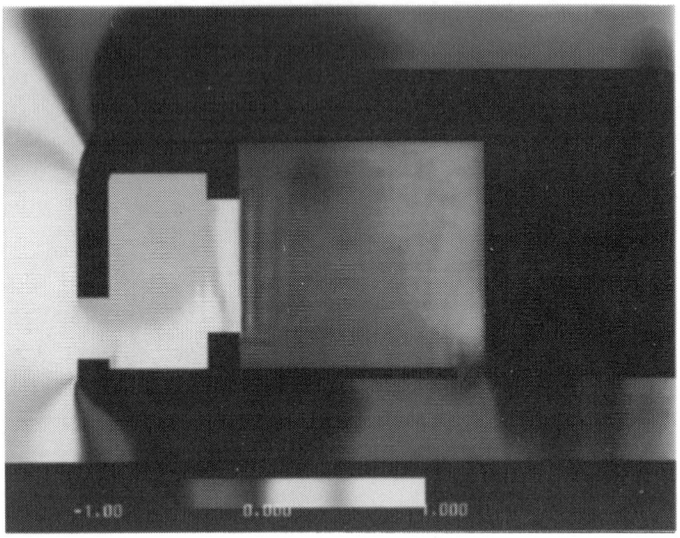

Figure 10. Time -averaged pressure field on the symmetry plane.

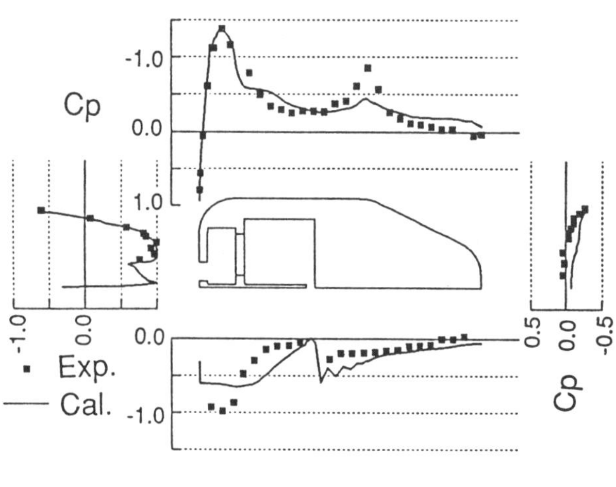

Figure 11. Comparison of pressure distributions on outer surfaces on the symmetry plane.

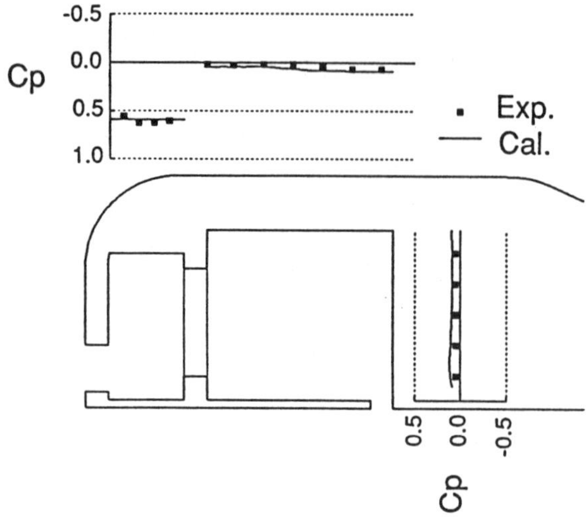

Figure 12. Comparison of pressure distributions on inner surfaces on the symmetry plane.

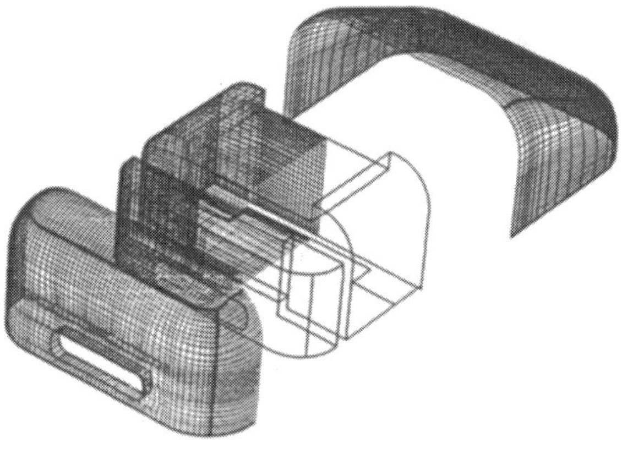

Figure 13. Definition of parts for the simplified test model.

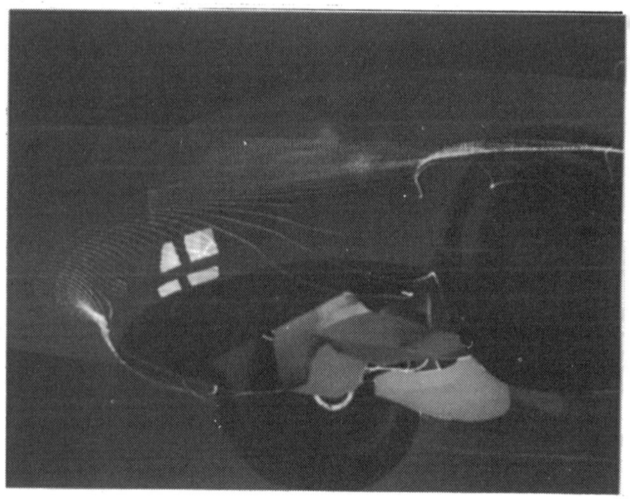

Figure 14. Instantaneous streamlines around air-inlet without a license plate.

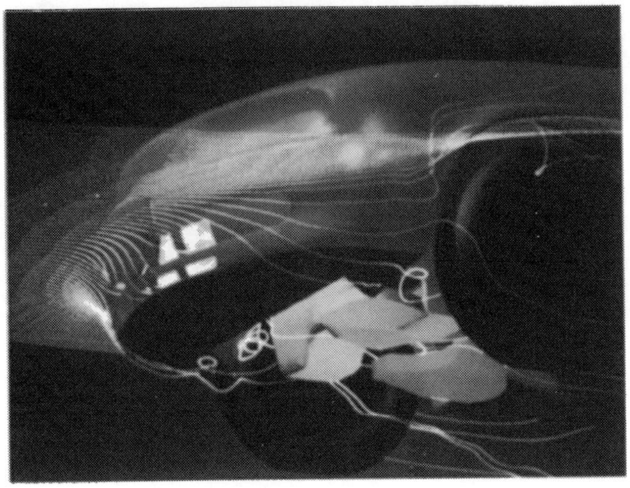

Figure 15. Instantaneous streamlines around air-inlet with a license plate.

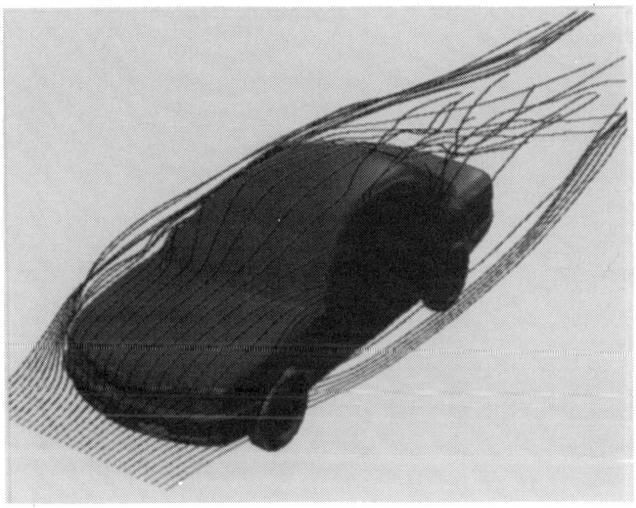

Figure 16. Time-averaged Streamlines around the car .

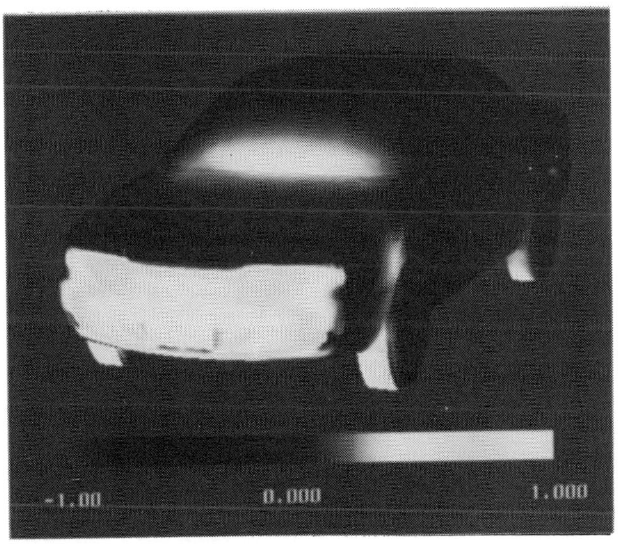

Figure 17. Pressure distributions in the time-averaged flow field.

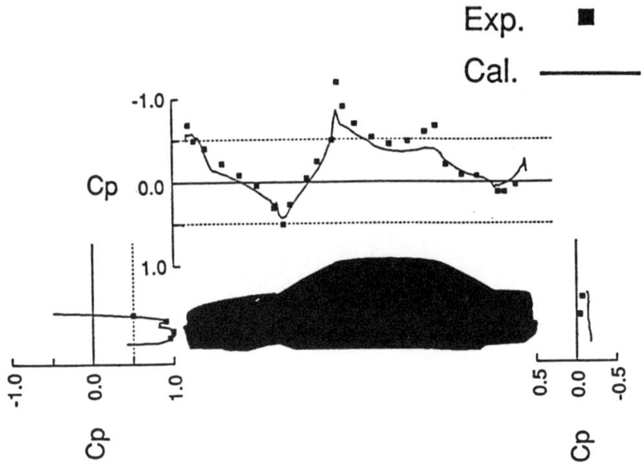

Figure 18. Comparison of pressure distributions on the symmetry plane.

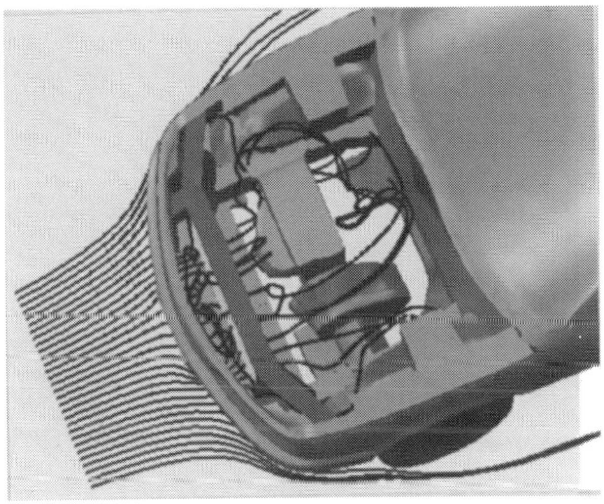

(a) Streamlines in the engine compartment.

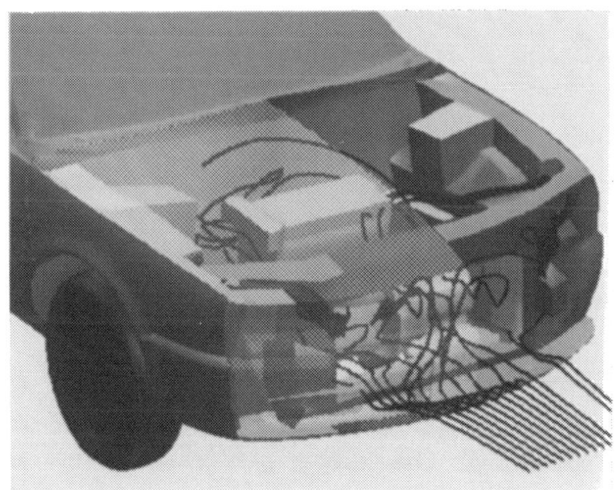

(b) Streamlines through bypass holes.

Figure 19 . Flow in the engine compartment.

950629

Practical Aerodynamic Simulations (C_D, C_L, C_{YM}) Using a Turbulence Model and 3rd-Order Upwind Scheme

Kenji Okumura and Toshihiko Kuriyama
Daihatsu Motor Co., Ltd.

ABSTRACT

Conventional aerodynamic simulations have been carried out by using Supercomputer and over a hundred thousand grid points. It takes a long CPU time to get a result. So this method has been mainly used as a research and demonstration tool.

In this paper a practical aerodynamic simulation for early phases of the automobile design is presented. This new method basically uses the conventional (wall function used) k-ε turbulent model which is regarded as the most promising in engineering and industrial fields.

Firstly, the third-order upwind scheme is introduced to the convective term to improve flow field and pressure distribution. Secondly, the modified turbulent energy production method (M.P.Method) which was proposed by B.E.Launder[3] in 1993, is introduced to the k-ε turbulent model to reduce the excessive generation of turbulent kinetic energy.

It takes 20 hours to generate a computational mesh from design CAD data by a new semi-automatic mesh generator and 10 hours to get the steady flow result by using EWS (IRIS Indigo2 R4400,1CPU).

It takes only 4 days to get an aerodynamic solution within an error of 5% for C_D (Drag Coefficient) and C_L (Lift Coefficient). This new method has been applied to the aerodynamic simulation system, a designer has been able to carry out the practical simulation on early designing phase of an automobile.

We also present a C_{YM} (Yawing Moment Coefficient) simulation between 0 and 45 degree and the influence of a outside rear view door mirror on Two-Box automobile.

INTRODUCTION

CFD(Computational Fluid Dymamics) has been able to apply to industrial fields in these days. High performance EWS and Supercomputer of next generation are going to be appeared, so that CPU time is shortened and more fine-mesh model can be calculated. Recently aerodynamic simulations have been carried out by using Supercomputer and over hundreds thousands grid points to get higher accuracy.

Although for practical use of CFD in industrial fields there is a little need to calculate in high accuracy as well as a research CFD that will do for a experiment. Because a practical CFD must be a tool for a designer and a engineer to give some guidance for efficiency of development. This tool is more effective if it is used on early phase of the automobile design.

So in this paper the practical aerodynamic simulations are presented the following

COMPUTATIONAL METHOD

It is very important in industrial fields to decrease computing cost to keep in good accuracy.

Firstly the low cost hardware is necessary. So high performance EWS (Engineering Work Station) is developed in low price and low maintenance cost recently.

Secondary to shorten the analysis time is necessary. So the pre-processing time is shortened by interactive semi-auto pre-processor "PROSNER". The CPU time is shortened by the steady state solver code "SCRYU" which presents a good accuracy with relatively rough mesh.[1] Both are made by software CRADLE Co.,Ltd. in JAPAN, are sold in JAPAN. This solver is basically used the conventional k-ε turbulent model, and the third-order upwind scheme MUSCL[2] is introduced to the convective term to improve flow field and pressure distribution. Moreover the modified turbulent energy production method (M.P.Method) which was proposed by B.E.Launder[3] is applied in this computational method.

The post-processing time is shortened by interactive software such as "FIELDVIEW" or "Atrac". The former is created in USA by Intelligent Light,the latter is created in JAPAN by software CRADLE.

Table 1 shows the computational conditions for

Table1.Calculational Conditions

	Backward Facing Step	Automobile(C D,C L)	Cross Wind
Turbulence model	Conventional k-ε model		
Modified Production	Yes		No
Convective Term	3rd order upwind MUSCL		QUICK
Coordinate System	Unstructure,BFC (SCRYU)		BFC
Computer	IRIS Indigo2 (R4400)		CRAY YMP
Calculational Method	Steady		Unsteady
Wall Function	Logarithmic Law		
Outlet Conditions	Pressure Boundary (zero dp/dx)		
Inlet Conditions	U,k,ε are based on LES data	Velocity = 27.8 m/s k, ε = $1x10^{-4}$	Diagonal Velocity
Re Number	$4.6x10^4$	$2x10^6$	
Element Number	$5.1x10^3$	$7.4x10^4$,$1.6x10^5$	$1.1x10^5$
CPU Time	0.7 hours	10 hours,40 hours	30 hours

Backward Facing Step and Automobile Models. The coodinate system can be chosen BFC (Boundary Fitting Coodinate) or boundary fittied Unstructured (hexahedral) Mesh in SCRYU. The used computer is mainly IRIS Indigo 2 (R4400). Only in cross wind simulation CRAY Y-MP is used. Because it needs full model and accuracy even in the oblique flow.

In this paper the backward-facing step simulations are investigated to verify the accuracy of the solver. These investigation are carried on two dimensional meshes ($5.1x10^3$ elements).

Then automobiles investigation are presented, those are C_D,C_L simulations using half model meshes of real scale($7.4x10^4$ elements). Moreover C_{YM} simulation is investigated using full model meshes of real scale($1.2x10^5$ elements) at any angle,and the influence of door mirror is investigated using half model of real scale($1.6x10^5$ elements).

The CPU time of backward-facing step is 0.7 hours, and that of C_D,C_L simulation in steady flow is only 10 hours by EWS in case of Unstructured simulations.

For the wall function, the logarithmic-law boundary conditions are applied in all cases. The smallest Y^+ is controlled to be from 40 to 500 by modifying the mesh aspect adjacent to a wall. But the Y^+ near the separational region such as the corner of automobile is adapted about 40.

Inlet conditions (U:velocity and k:kinetic energy and ε:dissipation rate distribution) of backward-facing step are used the reported Large Eddy Simulation data[4].

Inlet velocity of automobile cases are given the constant velocity of 27.8m/s (100 km/h). Outlet conditions are pressure boundary (zero dp/dx) in all cases.
In the case of cross wind simulations, inlet velocity is given at windward side diagonally. On the other hand outlet condition of pressure boundary is given at another leeward side.

Re Number of backward-facing step is $4.6x10^4$,that is based on step height, and Re Number of automobile cases is $2x10^6$,that is based on overall height of automobile.

BACKWARD-FACING STEP

The calculated reattachment point of the backward-facing step is often underpredicted in case of the conventional k-ε model[4]. So in order to improve its deficiency, a lot of works have been carried on.

COMPUTATIONAL MODELS

Fig1,Fig2 show the two dimensional grids in this computation. The flow field is a channel with a backward-facing step having an expansion ratio of 1.5.

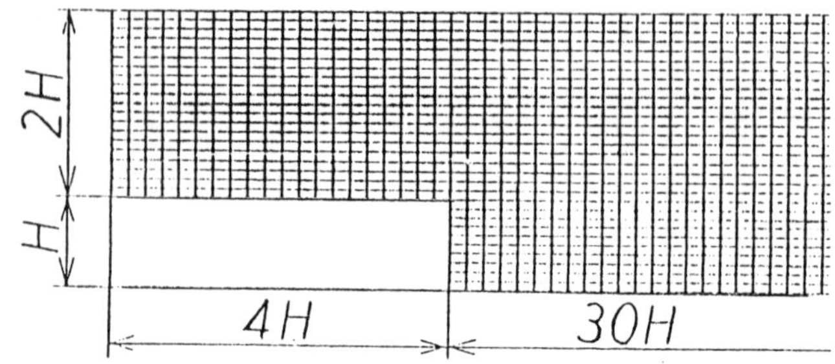

Fig1.Uniform Mesh of 171x31

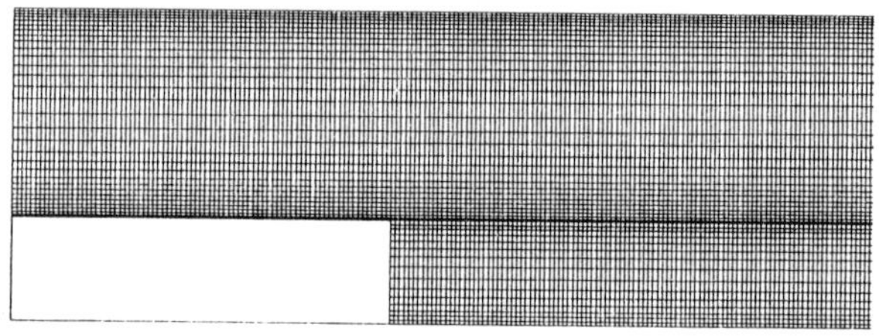

Fig2.Unstructure Mesh of 681x46

RESULTS AND DISCUSSIONS

Table2 shows the calculated results of reattachment point. The experimental value is reported as 7.2 X/H at Re=46,000[5] which is based on step height(H). Fig3 shows the stream lines and the reattachment point of CASE1 and CASE6. In these results, when the third order upwind scheme and M.P.Method are used (CASE6), the calculated reattachment point is fairly improved in spite of rough grids. QUICK Scheme gives more acuurate than MUSCL Scheme from the results

Table2.Calculated Results of Reattachment Point

	Convective Term	Element Number	Re Number	Reattachment Point
CASE1	1st-order	170 x 30	46000	**5.2** X/H
CASE2	QUICK	170 x 30	46000	**5.8**
CASE3	1st-order	680 x 45	46000	**6.0**
CASE4	QUICK	680 x 45	46000	**6.3**
CASE5	MUSCL	170 x 30	46000	**5.5**
CASE6	MUSCL and Modified Production	170 x 30	46000	**6.7**
CASE7	1st-order	170 x 30	46000	**5.0**
CASE8	1st-order	680 x 45	46000	**5.9**

(CASE7,8 are Star-CD results) Experiment 7.2 X/H

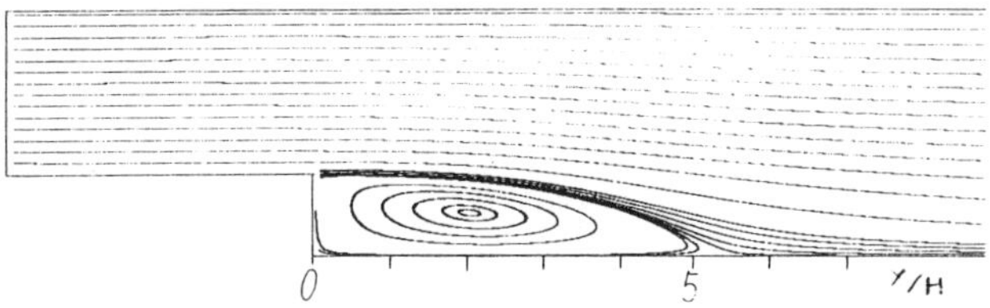

CASE1

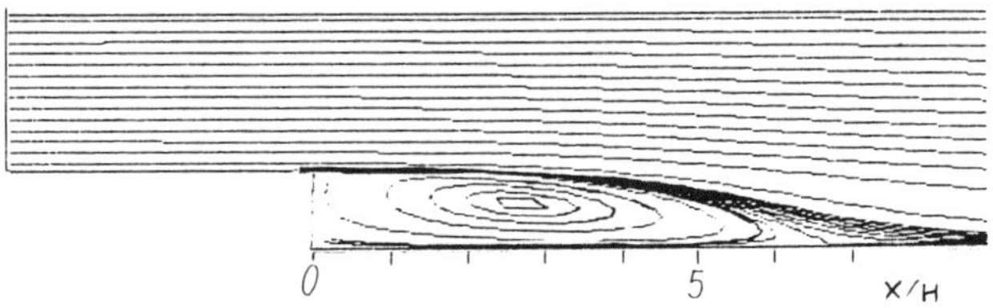

CASE6

Fig3.Comparison of Stream Lines

of CASE2 and CASE5. But QUICK Scheme is not applied to unstructued steady solver and M.P.Method now,it is unsteady time accurate solver and its CPU time is about ten times(7 hours) to steady solver(0.7 hours). Because QUICK Scheme is used only finite difference method and it is difficult to make a cooding for steady flow simulation of unstructured mesh. This steady solver is adopted only colocated element in SCRYU. Therefore we usually use steady state solver with MUSCL and M.P.Method to shorten computing time except cross wind simulation which needs time accurate solver to get a fluctuate phenomemon around automobile.

Consequently CASE6 (MUSCL and M.P.Method used) gives the best result in those investigation to calculate in high accuracy and a short term.

THE AERODYNAMIC SIMULATION SYSTEM

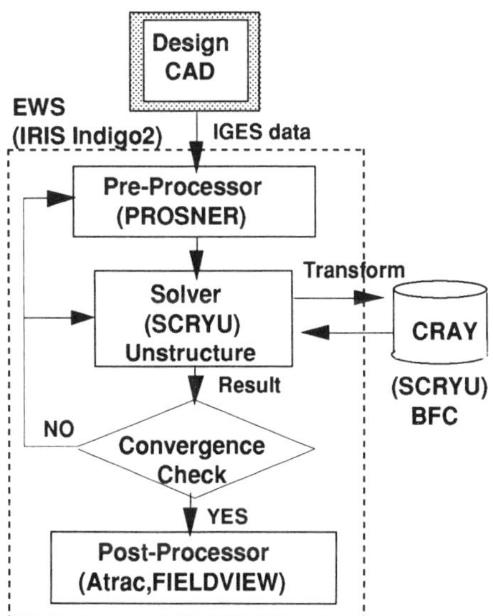

Fig4.Aerodynamic Simulation System

Fig4 shows the flow chart of our aerodynamic simulation system. This system is used the CAD data by transforming the IGES format.

At first the pre-processor "PROSNER" brings operaters to generate computing models easily by interactive semi-automatic operation. Fig5 shows its example of door mirror model.

To shorten mesh generating time, the BFC Mesh is applied in Unstructured Mesh topology. Because it is difficult to make a fine Unstructured Mesh around automobile body, on the other hand the BFC Mesh is easily generated by projecting body surface mesh.

After all, the pre-processing time is shortened from 10 days to 3 days by using BFC Mesh in unstructured steady state solver.

Secondly, the steady flow simulations (C_D,C_L) are usually carried on by using MUSCL Scheme and M.P.Method of unstructured solver. If more accuracy is required such as cross wind unsteady simulation, Supercomputer "CRAY" is appllied in double precision by using QUICK Scheme of BFC Mesh.

Thirdly, after the convergence check, the interactive post-processor are used. Fig6 shows the example of Atrac, and Fig7 shows the example of FIELDVIEW. The former is used for the convergence check and making technical data. The later is used for making visualization data such as volume rendering and real time animation.

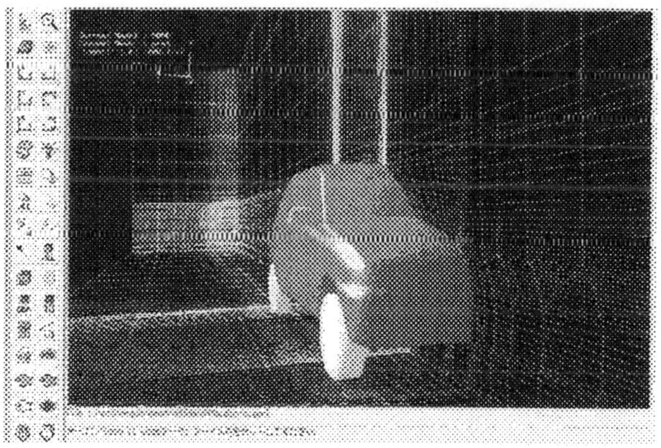

Fig5.Example View of Pre-Processor (PROSNER)

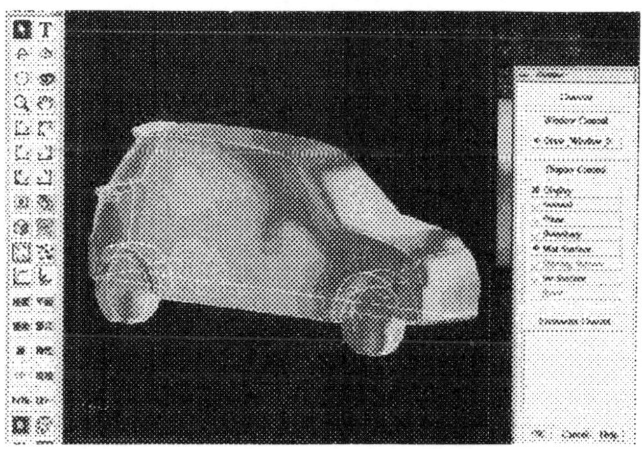

Fig6.Example View of Post-Processor (Atrac)

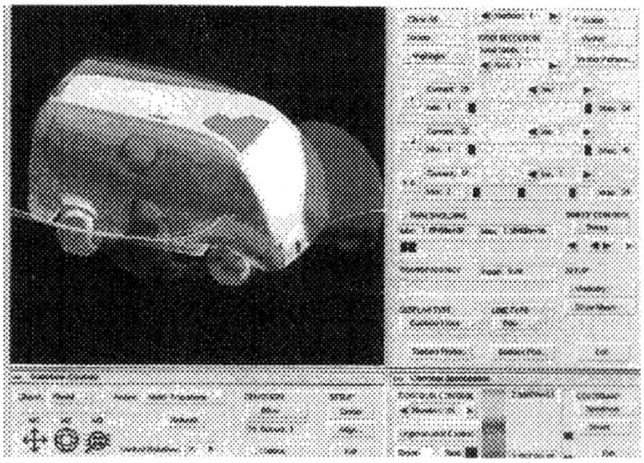

Fig7.Example View of Post-Processor (FIELDVIEW)

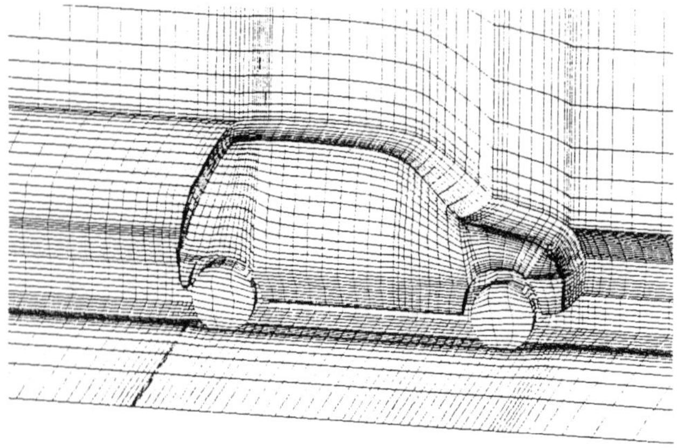

Fig9.Model B Mesh

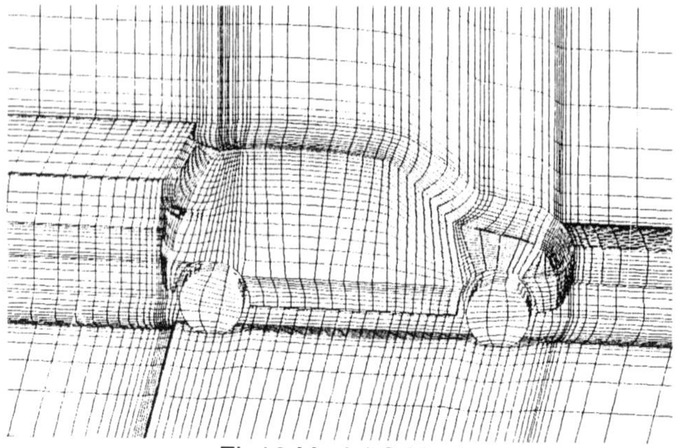

Fig10.Model C Mesh

AERODYNAMIC SIMULATIONS OF AUTOMOBILES

The various shapes such as One-Box,Two-Box automobiles are investigated,and the follows.

COMPUTATIONAL MODELS

Table3 shows the type of models in this simulation. Model A and Model B are Two-Box automobiles. Model C is with aero-spoiler of Model B. Model D is with side view door-mirror of Model B. Model E is One-Box automobile. Model F is with the aerokick shape[6] of front bumper of Model E. Model G is with the cross wind front spoiler of Model E. Fig8-Fig13 are shown some of them. To shorten the mesh generating time the computational mesh topology are basically made of BFC Mesh construction.

Table3.Calculational Models

	Two Box Automobile				One Box Automobile		
	Shape 1	Shape 2			Shape 3		
Model	A	B	C	D	E	F	G
Type	Normal	Normal	Aero-spoiler	Door-mirror	Normal	Aerokick	Cross Wind Spoiler

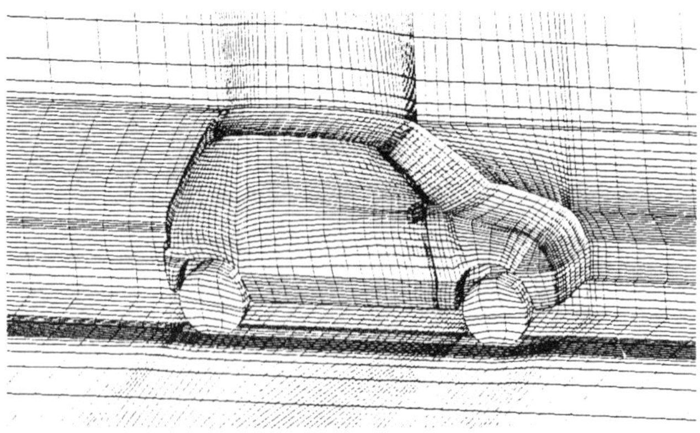

Fig11.Model D Mesh

Fig8.Model A Mesh

Fig12.Model E Mesh

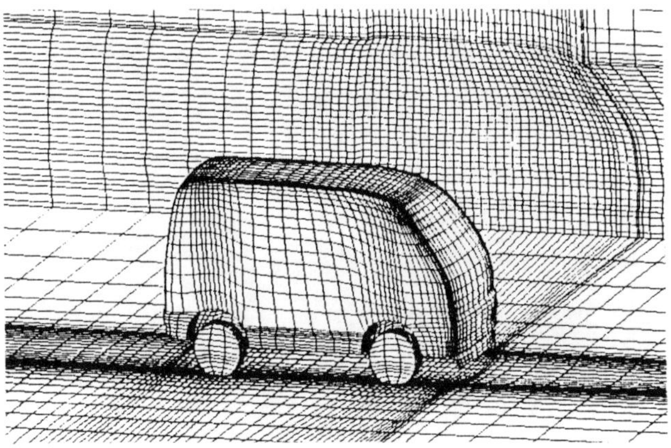

Fig13.Model G Mesh

FLOW VERIFICATION

Tested experimental models are clay models and have a smooth underbody. Wheels are stopped. The cooling slots are closed. The calculational models are the same as experimental models.

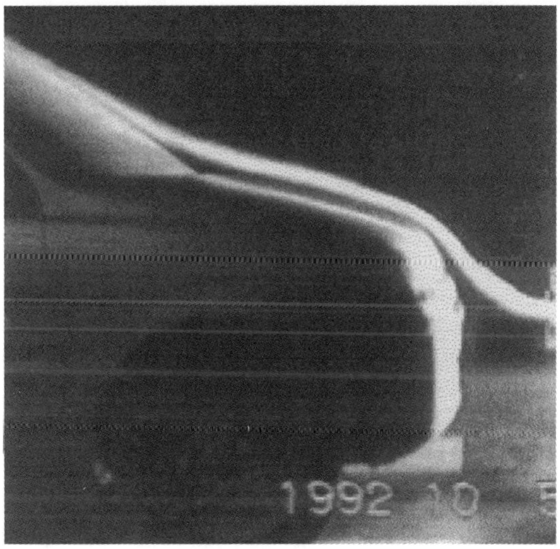

Experiment

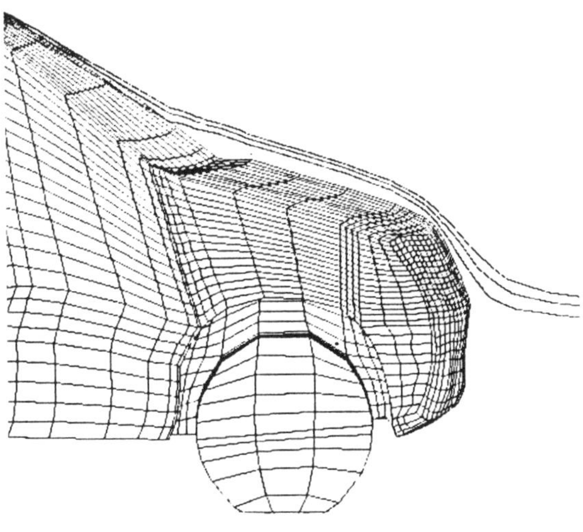

Calculation
Fig14.Stream Lines on hood (Model B)

Fig14,15 show the stream lines of Model B between experiment and computation at 100 km/h speed in wind tunnel. Calculated stream lines on the hood and stagnation points are good agreement with experiment. If the calculated stagnation point would not agree with experiment, the pressure distribution around automobile might differ to the experiment. Because the stagnation point is the highest pressure all around automobile, its balance effects the backward flow of automobile.

Furthermore it is ideal that the calculated maximam pressure is equal to the dynamic pressure.Because we presented the pressure force occupies 98% of all drag at high Re number[1] and the pressure distribution is important to calculate drag. In this computation the error of maximam pressure is carried out within 3%. Coventional k-ε model using 1st order upwund) is over estimated it.

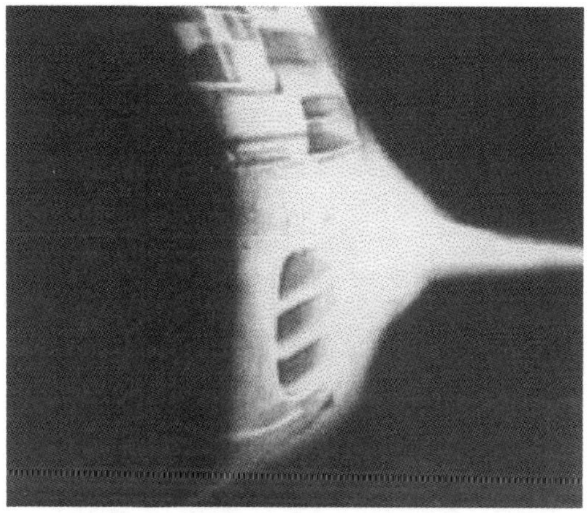

Experiment (cooling slots are closed)

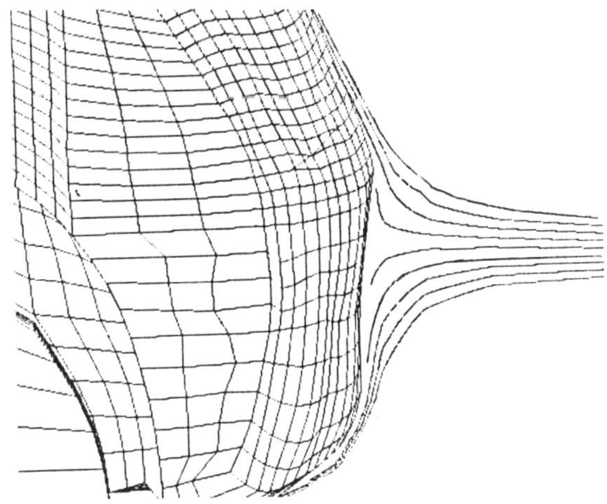

Calculation
Fig15.Stream Lines arround Stagnation Point (Model B)

DRAG COEFFICIENT (C_D) VERIFICATION

C_D is calculated as the steady flow condition using unstructured solver. The error of C_D and CPU time by EWS are shown in Fig16. All the error between experiment and computation seem to calculate a little larger. But those are within 5%. All the CPU time are within one day computation. Fig17 shows the comparison of C_D between experiment and calculation on various research models. Also good agreements are shown in these results.

	Two Box			One Box
	Model A	Model B	Model C	Model E
C_D Error % (0–5)	+1.5%	+2.4%	+1.1%	+3.5%
Element Number	7.4x10^4	7.4x10^4	7.4x10^4	1.2x10^5
CPU Time (IRIS Indigo2)	10H	10H	10H	16H

Fig16.C_D Error and CPU Time

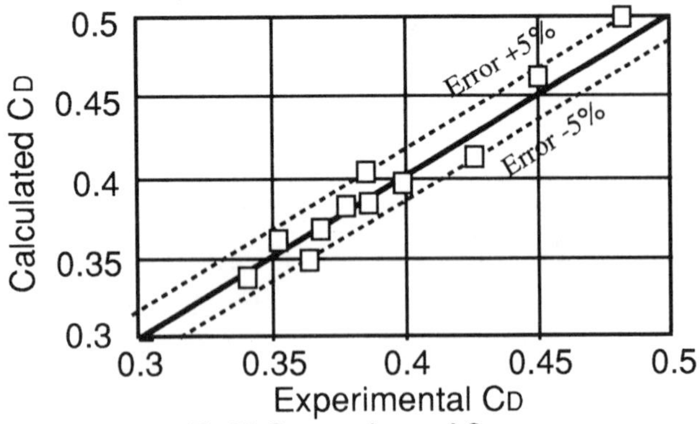

Fig17.Comparison of C_D

LIFT COEFFICIENT (C_L) VERIFICATION

Fig18 shows the backward flow of Model B. This flow is so turbulent just behind the back window.

Experiment

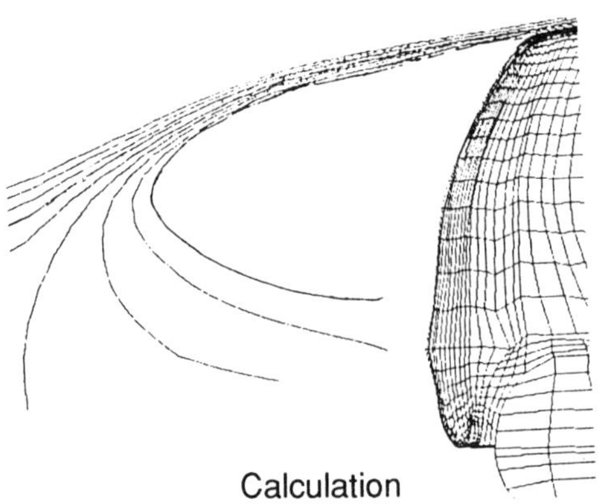
Calculation

Fig18.Comparison of Backward Flow (ModelB)

Fig19 shows the flow in case of aerospoiler model of Fig18. The flow is separated from the back window clearly.

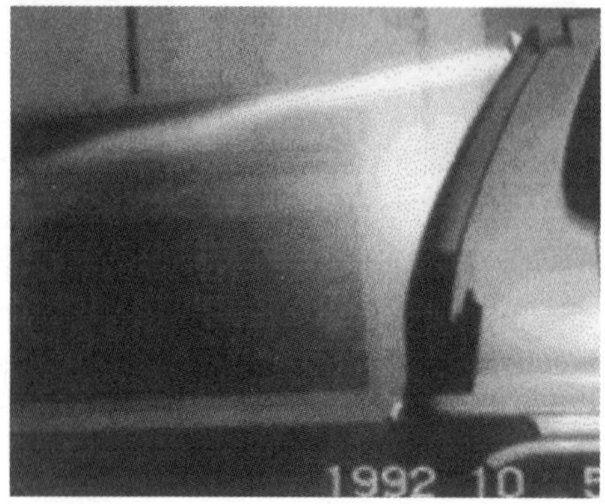

Experiment

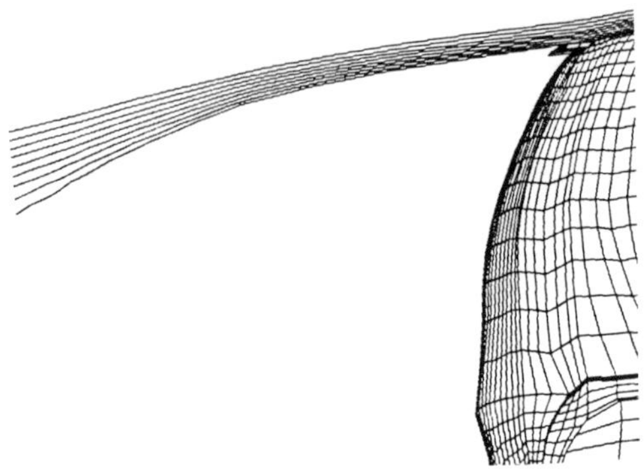
Calculation

Fig19.Effect of Aerospoiler (ModelB)

The comparison of C_L between experiment and calculation on various research models are shown in Fig20.

Calculated C_L are little small,but the tendency of almost results are shown in the same low level.

From these results the calculated results are good agreement with experiments.

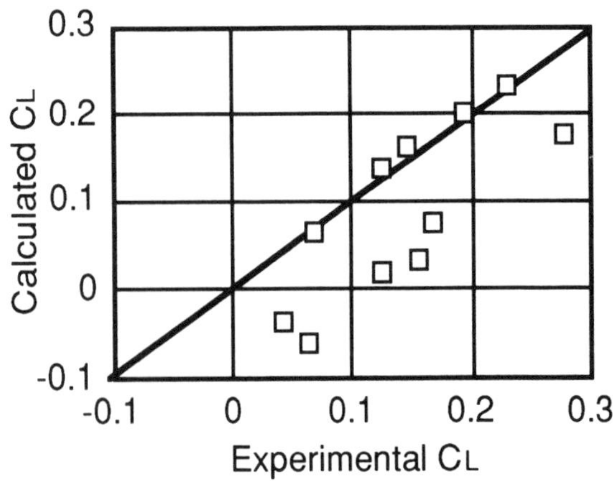

Fig20.Comparison of C_L

Fig21 shows the effect of spoiler based on Model A by calculated turbulent kinetic energy. Type A is normal (no spoiler). Type B is only with a rear spoiler. Type C is with aerospoiler (front and rear spoiler). Calculated turbulent kinetic energy shows the balance of backward twin vortices. Calculated C_D of Type B is increased +5.3% (exp. is +4.7%) to Type A (normal), and Type C is decreased -3.0% (exp. is -2.2%). Calculated ΔC_L between Type A and Type B is -0.038 (exp. is -0.048). Consequently, Type C has a good balance, and gives the best improvement for C_D and C_L.

This result indicates that the adjustment of spoiler must be done carefully to balance backward twin vortices. Otherwise the ability of aerospoiler would not present effectively. Although this CFD method teaches how to adjust, so designer and engineer can use it as practical simulation.

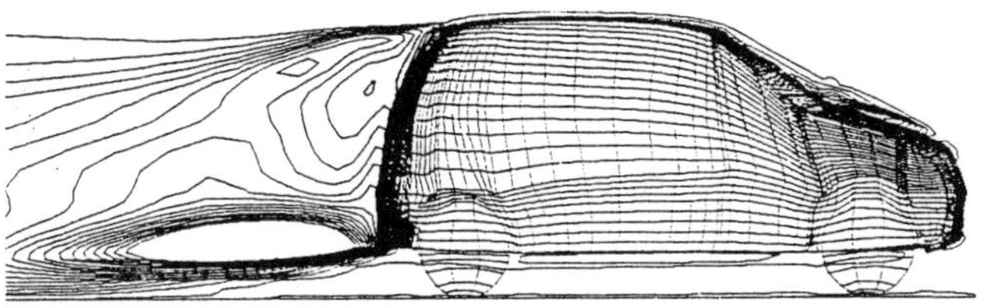

Type A (No Spoiler)

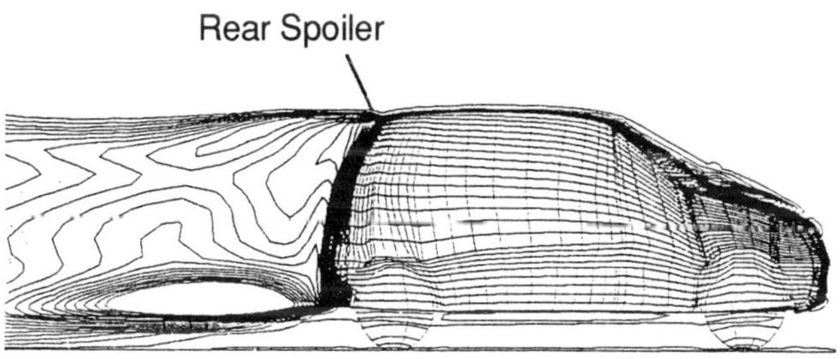

Type B (only Rear Spoiler)

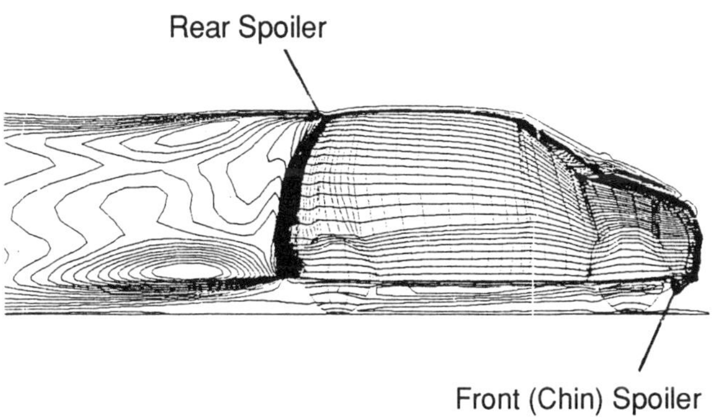

Type C (Rear and Front Spoiler)

Fig21.Influence of Spoilers on Turbulence Kinetic Energy

INFLUENCE OF REAR VIEW DOOR MIRROR

Fig 22 shows the models of this investigation. No door mirror model is Model B. Shape A is Model D. Shape B is filled up the cutting groove of shape A.

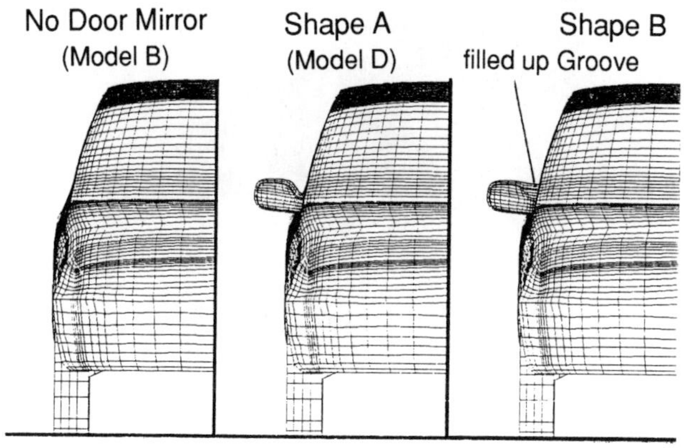

Fig22.Investigated Models

The flow on the side window saface between experiment (Tuft Method) and calculated velocity vectors are shown in Fig23,Fig24,Fig25.

Fig23 (Model B) is no door mirror, Fig24 (Model D) is with a door mirror shape A. Fig25 is with a door mirror shape B which is filled up the cutting groove of shape A. The reattachment line (that is indicated in each figure as line) from front pillar are good agreement between experiments and computations. Moreover the calculated twin vortices backward door mirror and its reattachment area (that is indicated in Fig24 and Fig25 as area) are good agreement with experiment. But in case of no

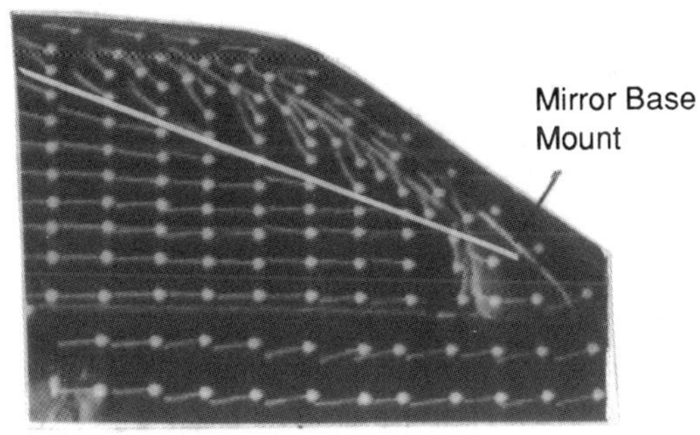

Experiment (Tuft Method)

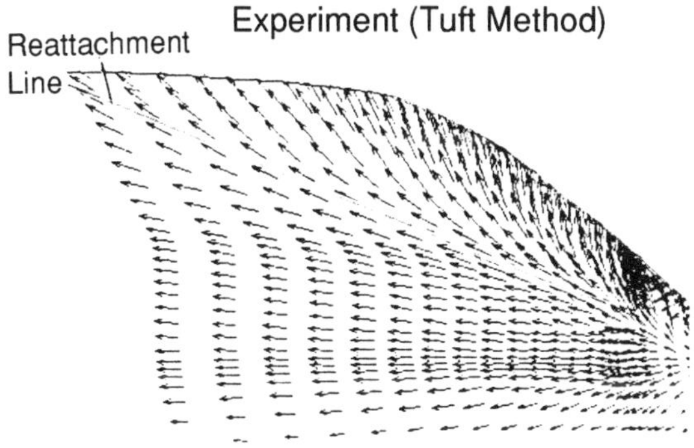

Calculation (Surface Vector)

Fig23.Side Wind Surface Flow (no door mirror)

door mirror (Fig23),it seems a little different flow backward the mirror base mount becase there is a little distance between the mirror base mount and window surface. The calculatinal models have not the mirror base mount.

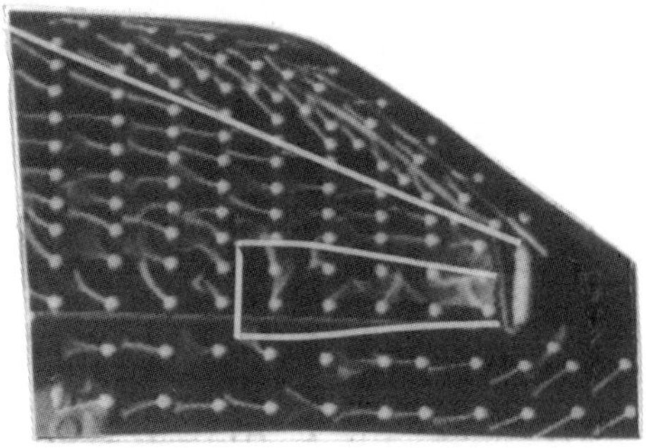

Experiment (Tuft Method)

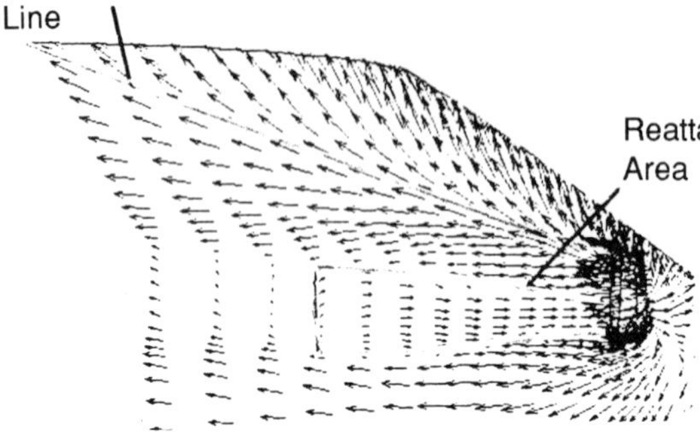

Calculation (Surface Vector)

Fig24.Side Wind Surface Flow (door mirror shape A)

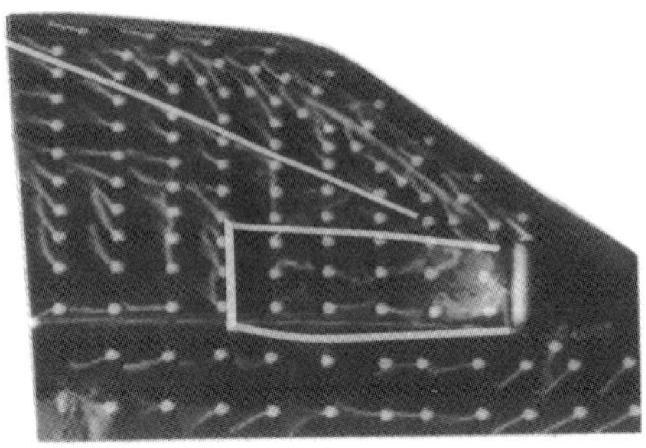

Experiment (Tuft Method)

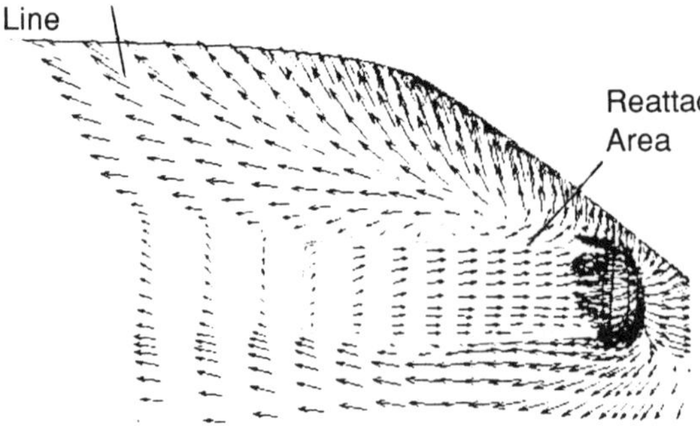

Calculation (Surface Vector)

Fig25.Side Wind Surface Flow (door mirror shape B)

STREAM LINES AND PRESSURE ON SIDE WINDOW

Fig26 shows the comparison of stream lines,Fig27 shows the comparison of pressure distribution on the side window. There is a intensive spilal flow around front pillar in no door mirror, and the negative pressure level is the lowest.

In case of door mirror shape A, the spiral flow is weakened and spreaded, because the door mirror obstructs the flow and the reflected flow interferes the spiral flow of front pillar in Fig26. Also the maximam negative pressure center around the front pillar is shfted upper side and weakened from -1250 Pa to -1000 Pa in Fig27.

In case of door mirror shape B, the upper side of twin vortices backward door mirror is strengthened and the spiral flow is more weakened in Fig26, because its gap is filled up and the reflected flow of door mirror is shifted more upper side. Also the maximam negative pressure center around the front pillar is shfted more.upper side and weakened from -1000 Pa to -950 Pa.in Fig27 too.

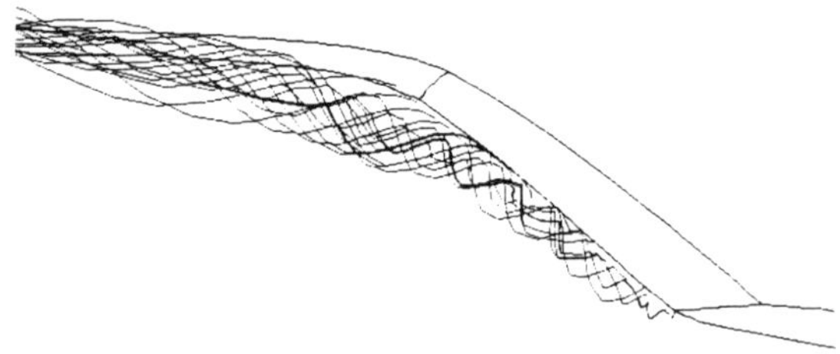

no door mirror (Model B)

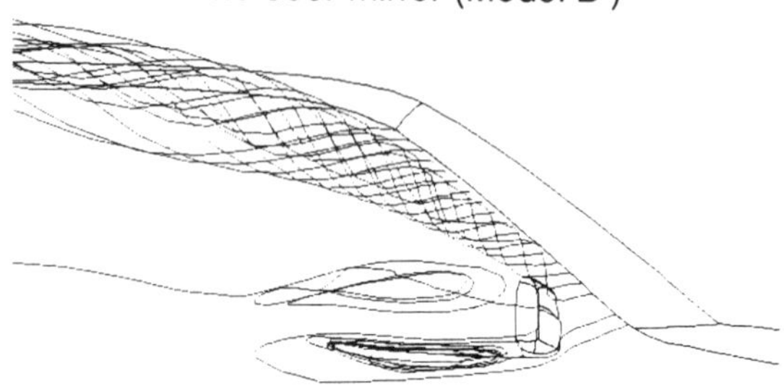

door mirror shape A (Model D)

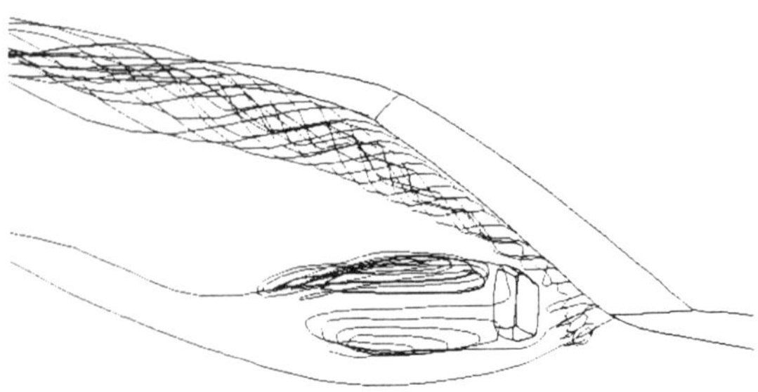

door mirror shape B

Fig26.Comparison of Stream Lines

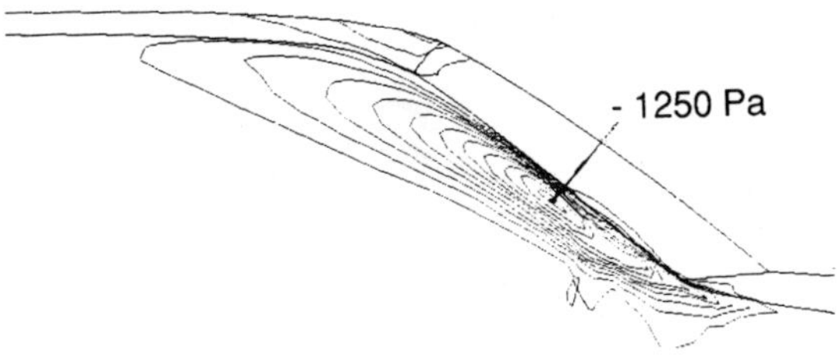

no door mirror (Model B)

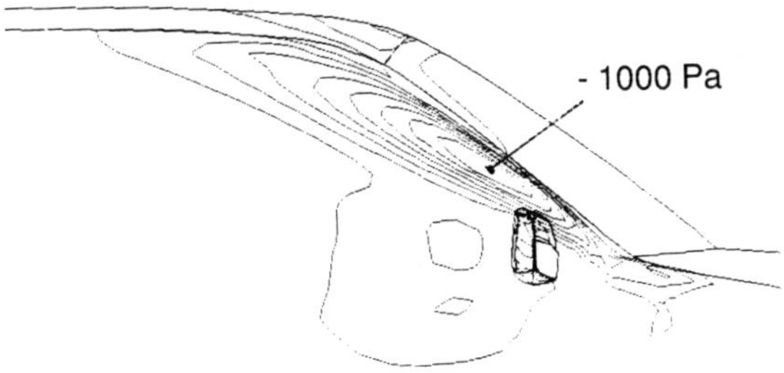

door mirror shape A (Model D)

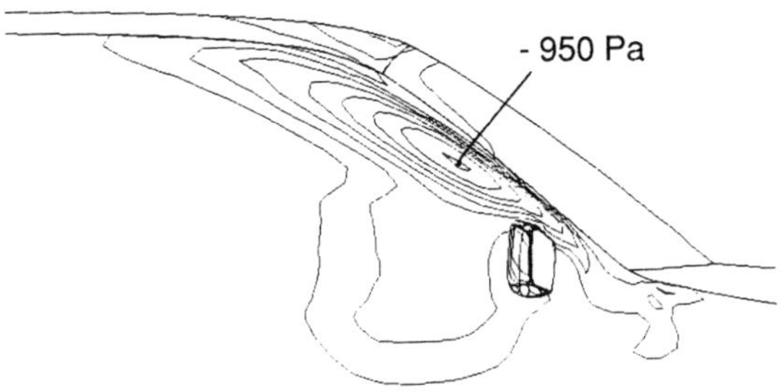

door mirror shape B

Fig27.Comparison of Pressure Distribution

TURBULENCE KINETIC ENERGY AND SOUND PRESSURE LEVEL

Recently the mechanical sound noise of automobile has been reduced. So the wind noise has been remarkable, and it must be decrease especially for comfortable driving on high speed. It is known door mirrors generate the external sound and the influence of outside rear view mirror investigated[7].

This section is investigated the relationship between the calculated turbulent kinetic energy (k) and external sound pressure level (SPL) around the front pillar and door mirror.

Fig28 show the comparison of k distribution, and Fig29 is the comparison of over all SPL distribution.

In case of no door mirror (Model B), k is concentrated on the upper side of front pillar (Fig28), also the measured strong SPL is shown in the same place (Fig29).

When door mirror shape A is installed (Model D), k is generated around twin vortices behind the mirror in Fig28. The concentration of k on the upper side of front pillar is weakened. The SPL of door mirror shape A

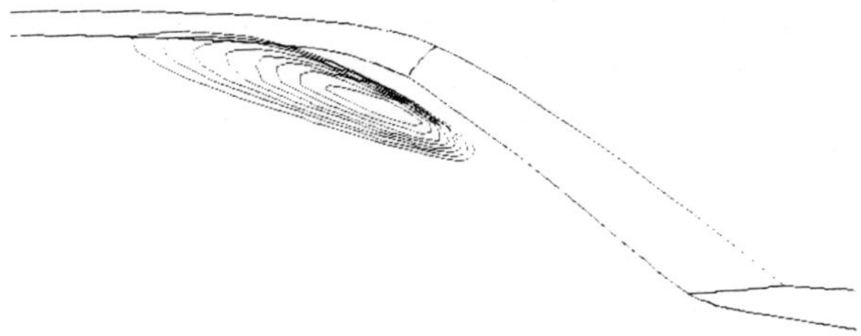

no door mirror (Model B)

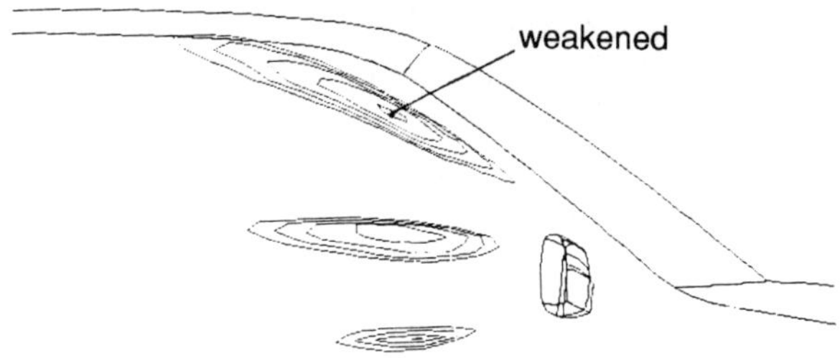

door mirror shape A (Model D)

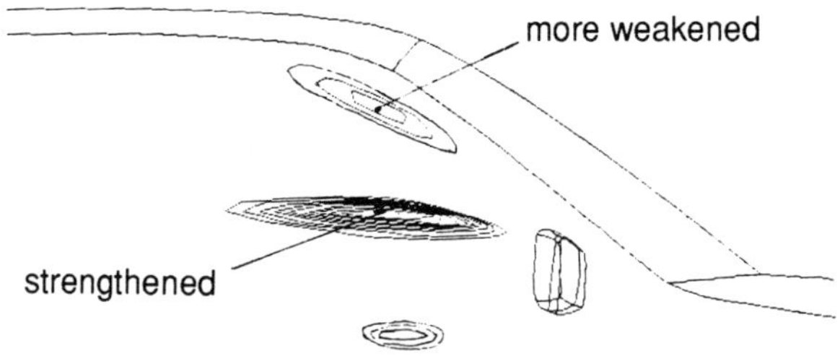

door mirror shape B

Fig28.Comparison of Turbulence Kinetic Energy (k)

no door mirror (Model B)

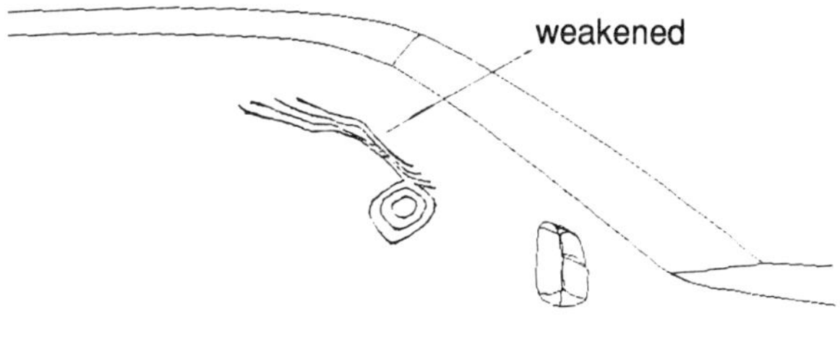

door mirror shape A (Model D)

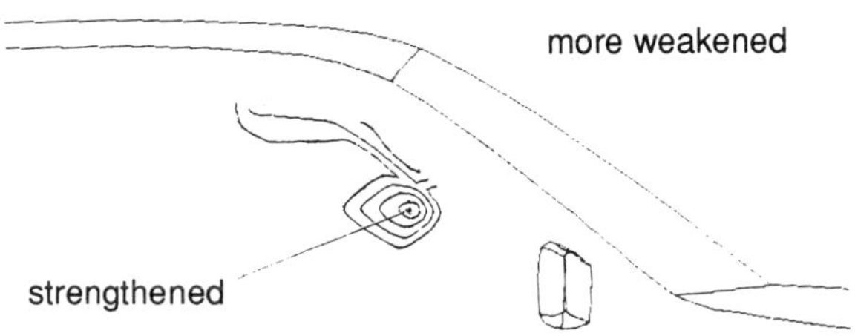

door mirror shape B

Fig29.Comparison of Sound Pressure Level

seems the same tendency in Fig29. Also the sound pressure is generated from twin vortices. (the under vortex behind the mirror is out of measuring region. Therefoe it is not indicated in Fig29)

In case of door mirror shape B, the concentration of k is more weakened and the upper side behind door mirror is more strengthened because of the influence of no cutting groove (Fig28). As for the SPL the same tendency seems in Fig29

From these results there is a relationship with k and SPL, so that the influence of door mirror on sound will be able to estimate roughly by this method.

Although theoretical proof have not been carried out at present.

YAWING MOMENT (CYM) VERIFICATION

This section describes the generating mechanism of the yawing moment on automobile in cross wind by using CFD simulation.

After several investigations, it is found that the calculated accuracy of the cross wind is worse than the zero yaw angle such as CD,CL simulation. Because the separational region is unsymmetric and lager, also the separational points move as the yaw angle, moreover the inlet velocity vector flow across the computational mesh diagonally as shown in Fig30.

Therefore it is difficult to calculate CYM at each yaw angle correctly by using the same mesh and the steady state simulation. We use the unsteady flow simulation of double precision with the k-ε turbulence model and 3rd order upwind QUICK Scheme, and also use Supercomputer CRAY Y-MP (1CPU).

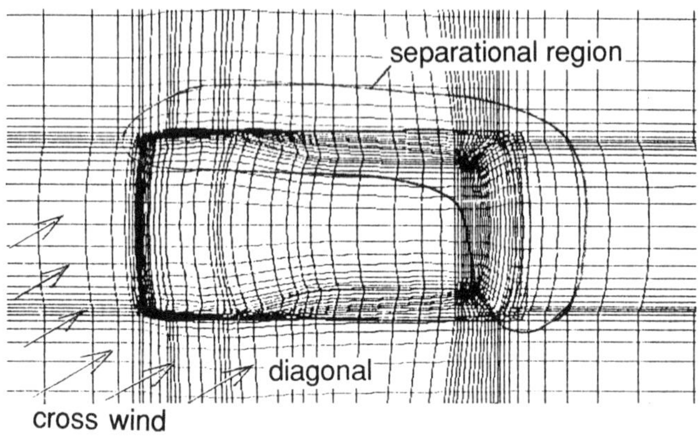

Fig30.Mesh Construction of CYM calculation

Yaw characteristics of Two Box automobile - Fig31 shows the comparison between calculated and experimental CYM variations. A good agreement is obtained. Both have a peak about 40 degree. We investigated the factors of this phenomenon by using cross wind simulations.

The pressure iso-surface of Model A with the yaw angle from 10 degree to 45 degree are shown in Fig32.

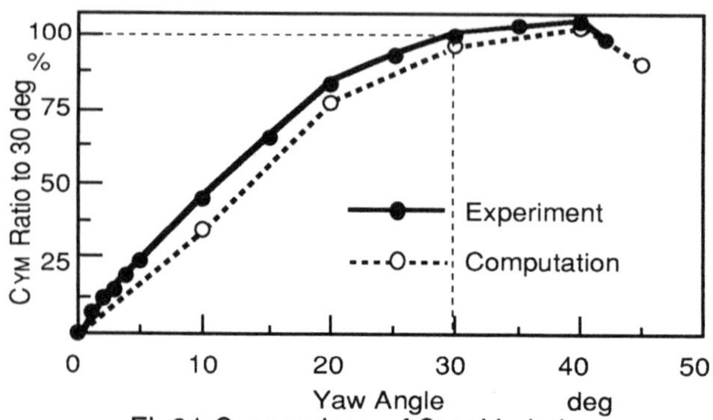

Fig31.Comparison of CYM Variations

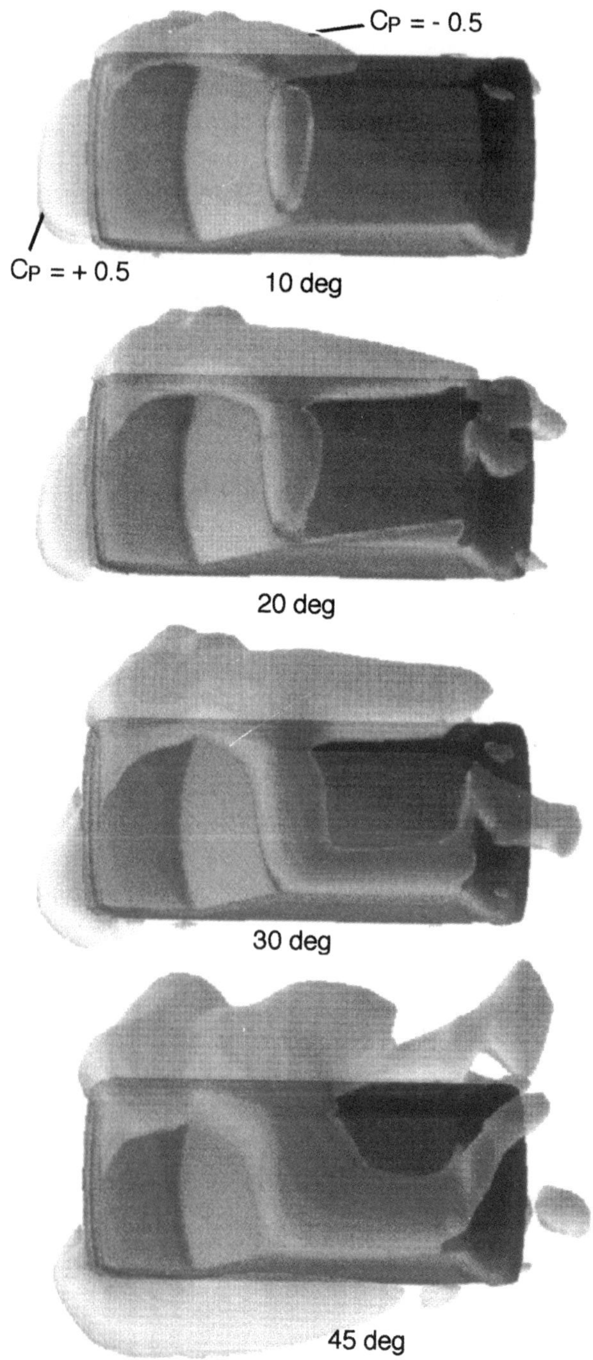

Fig32.Pressure Iso-surface of Model A

The positve pressure iso-surface (CP=+0.5) concentrates to windward front corner (A) as the yaw angle is close to 30 deg.

At last when it reaches 45 deg, the positive pressuer iso-surface spreads to the rear, and the negative pressure iso-surface (CP=-0.5) which concentrates to leeward front corner (B) is spreaded to rear too.

The reverse force generate on the windward corner (D) by the positive pressure and on the leeward corner (C) by the negative pressure,so that the yawing moment is weakened at 45 degree.

That is the reason why there is a peak of CYM variations (Fig31) between 30 degree and 45 deg.

Yaw characteristics of One Box automobile - The stability of One Box automobile on highway is often influenced by cross wind and a large truck. These phenomenon cause the unexpected behavior of automobile. So almost driver of One Box automobile require to improve its aerodynamic stability after fuel consumption.

In this section the effect of aerokick which is proposed to reduce yawing moment is investigated.

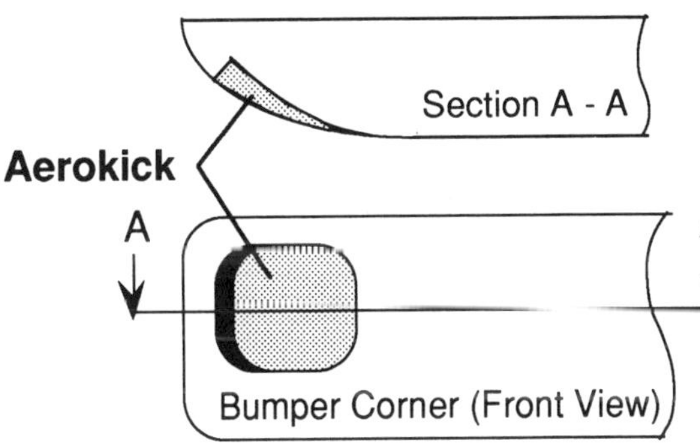

Fig33.Concept of Aerokick Bumper

Aerokick front bumper concept is shown in Fig33. The cross wind is flicked by the dent of bumper corner. The flow is separated from bumper corner, and the turbulency around bumper corner is weakened[6].

Fig34,Fig35 show the verification of the stream lines between normal bumper (Model E) and aerokick bumper (Model F). The experimental conditions are that velocity is 50 km/h (14m/s), Yaw angle is 30 degree, clay model is used. The calculational model is adopted in the same conditions as experiment.

The leeward side flow of bumper corner is so turbulent in normal bumper.The experiment shows the so turblent flow from the bumper corner by smoke in Fig34. The smoke is not able to see backward because of its intensive turblency. In case of calculation, it can be captured the same tendency as experiment.

On the other hand,the experimental flow of aerokick bumper is separated clearly in Fig35. Also calculated result are good agreement with experiment.

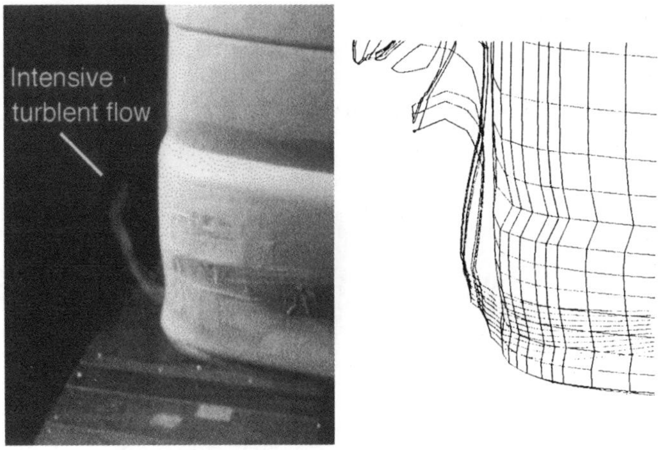

Experiment Calculation

Fig34.Stream Lines from Normal Bumper Corner

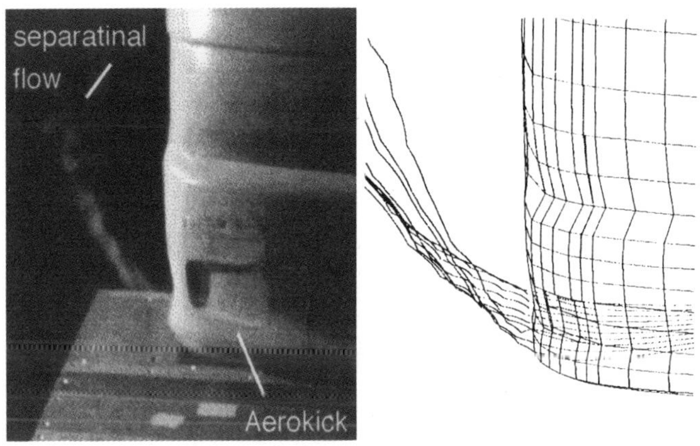

Experiment Calculation

Fig35.Stream Lines from Normal Bumper Corner

Fig 36 show the calculated results of stream lines, pressure iso-surface and turbulent kinetic distribution from the top view of those above results. Comparing between normal bumper and aerokick bumper, the stream lines are separated clearly by aerokick bumper. The negative pressure iso-surface concentrate near the front bumper corner, and retreats backward. After all the turbulent kinetic energy is weakened and the negative pressure is weakened too in case of aerokick bumper by separatinal effect. So that the moment which is generated intensively on bumper corner is reduced.

Fig37 shows CD,CYM errors of Model E and Model F with the yaw angle 0 degree and 30 deg. Calculated results are good agreement with experiments. Although the error at 30 degree are worth than the zero yaw angle because of the more complex wake region and the diagonal flow for mesh. That is the same as Two Box automobile.

After all the experimental effect of aerokick is 35% down compared with the normal bumper CYM, and the computation is 26% down.

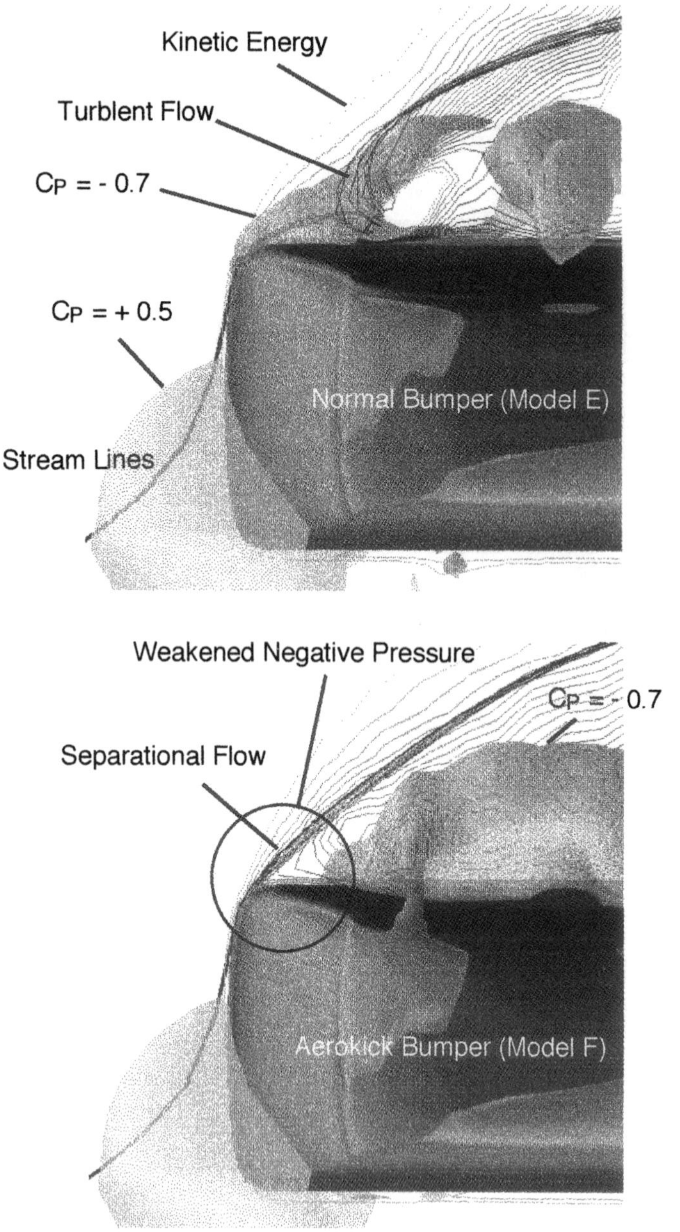

Fig36.Effect of Aerokic Bumper

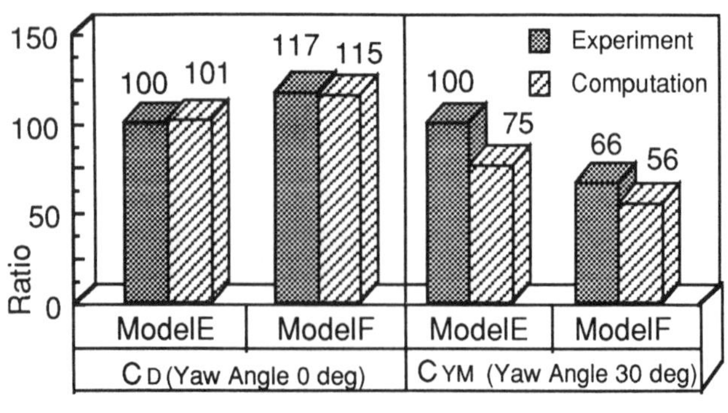

Fig37.CD,CYM Errors of Model E and Model F

Moreover the comparison of the stream lines on bumper corner at 30 deg yaw angle are shown in Fig38. Model E is with normal bumper. Model F is with aerokick bumper. Model G is with the cross wind front spoiler which was developed by the investigation of bumper corner flow. The cross wind spoiler figure is basically front airdam, which is lager than conventional chin spoiler. This spoiler cuts off the cross wind flow not to blow up from under-floor to body side and front wheel-houses. In case of Model E, there is turbulent flow which spirals up from the bumper corner because of the intensive negative pressure region around it. Oppositionally in Model F and Model G, that spiral flow is disappeared.

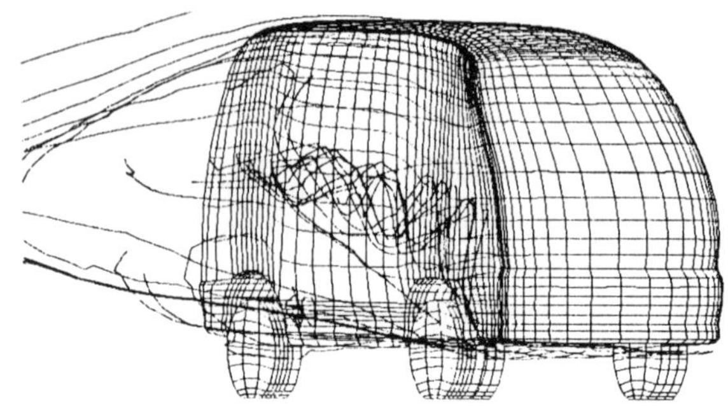

Normal Bumper (Model E)

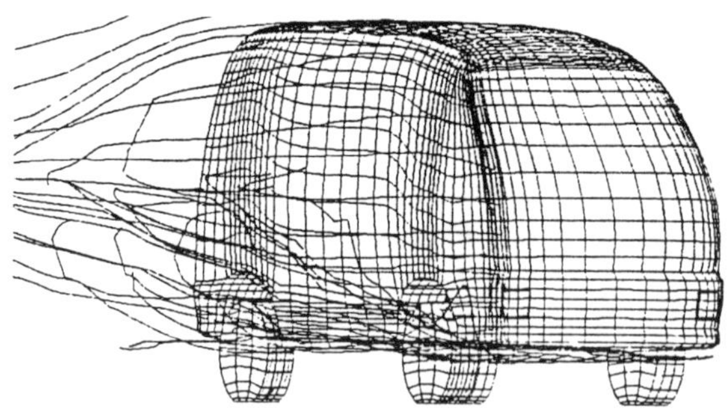

Aerokick Bumper (Model F)

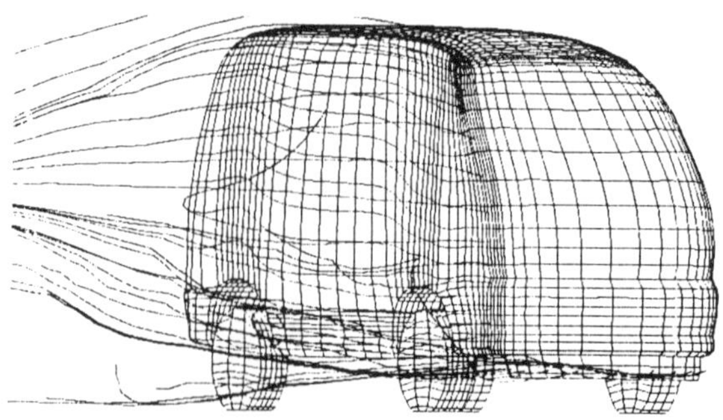

Cross Wind Spoiler (Model G)

Fig38.Comparison of Stream Lines

Fig39 shows the devided block in four corners of One Box automobile and shows the comparison of calculated yawing moment on those blocks. The moment mainly generates on block A and B from these results.

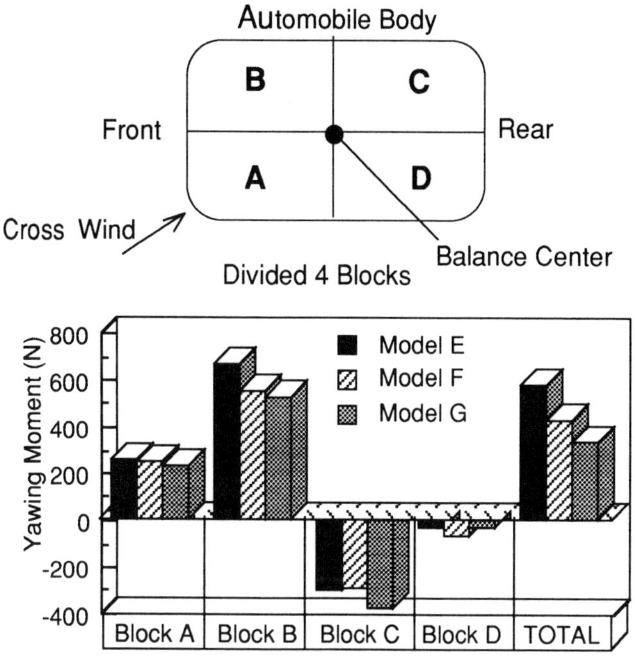

Comparison of Yawing Moment at 30 deg

Fig39.Computed Yawing Moment on divided Blocks

The total moment of normal bumper (Model E) is the biggest one. That of aerokick bumper (Model F) or cross wind spoiler (Model G) is smaller,especially in block B. That is the front bumper corner of leeward side.

The reducing effect rate of yawing moment are shown next. The calculated result is 26% by aerokick bumper (Model F), 43% by cross wind spoiler (Model G). The experiment is 34% by aerokick bumper, 53% by cross wind spoiler.

Moreover the comparison of experimental C_D, C_L, C_{YM} are shown in Fig40. Those percentage are based on nomal bumper. The cross wind spoiler is better at any character. Additionally Fig41 shows the evaluation of cross wind spoiler. The improvement of movement is reached to 25%. This spoiler will promise the good aerodynamic stability for safty on highway by the effective reduction for yawing moment and the increasing down force.

	C_D Ratio	C_{YM} Ratio	C_L Ratio	
Aerokick	+17%	-34%	-5%	Good ○
Cross Wind Spoiler	+7%	-53%	-30%	Superior ◎

Fig40.Effect of Aerokick Bumper and Cross Wind Spoiler

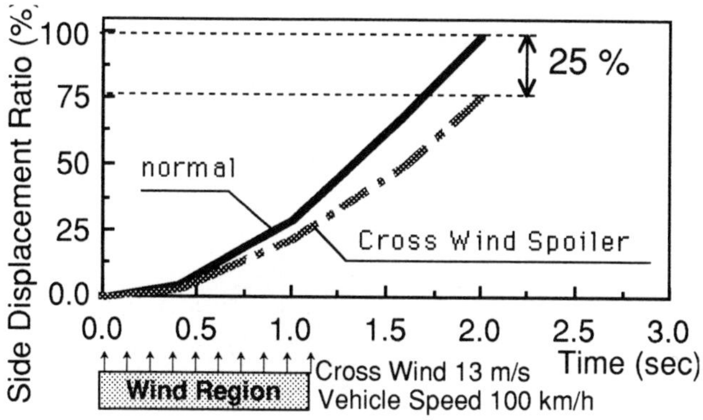

Fig41.Evaluation of Cross Wind Spoiler

CONCLUDING REMARKS

The practical aerodynamic simulation by using the k-ε turbulent model with M.P.Method and the 3rd-order upwind scheme are investigated and verified.
Conclusions are the following.

(1) The reattachment point of backward-facing step is well improved even with a relativery coarse mesh.

(2) C_D for various models are calculated within an error of 5%, and C_L,C_{YM} tendency comparatively show good agreements with experiments.

(3) CPU time of C_D simulations by using EWS (IRIS Indigo2: R4400) is only 15 hours in steady flow.

(4) The total analysis time is shortened from 3 weeks to 4 days by using interactive pre,post processor.

REFERENCES

[1] K.Okumura,T.Kuriyama:Three-Dimensional Aerodynamic Simulations of Two-Box Automobile using Turbulence Model and QUICK Scheme,26th ISATA,Super Computer Applications in the Automobile Industries,45-52(1993-9)
[2] van Leer,B.:Toward the Ultimate Convection Difference Scheme.4,A New Approach to Numerical Convection,Journal of Computational Physics,23,276-299(1977)
[3] M.Kato,BELaunder:The Modelling of Turbulent Flow around Stationary and vibrating Square Cylinders,NINTH Symposium on TURBULENT FLOWS,16-18(1993-8)
[4] Kobayashi,Kito:"A Review of CFD Methods and Their Application to Aerodynamics",SAE,920338
[5] N.Kasagi et al.:"Turbulence mesurement in a separated and reattaching flow over a backward-step with the aid of three-dimensional particle tracking velocimetry",3rd Joint Symposium on Fluid Control and Mesurement,Tokyo(1991)
[6] K.Sumitani et al.:Development "Aero Slit" Improvement of Aerodynamic Yaw Characteristics for Commercial Vehicles,SAE,890372
[7] Robert J. Chapman et al.:Wind Noise and Aerodynamic Drag,Optimization of Outside Rear View Mirror,SAE,931292

940325

The Aerodynamic Development of a Small Specialty Car

Kazuo Yanagimoto, Kunio Nakagawa, Hiroshi China, Tomio Kimura, Minoru Yamamoto, and Takashi Sumi
Mitsubishi Motors Corp.

Hiroki Iwamoto
Nimura Design Inc.

ABSTRACT

Aerodynamic drag reduction is one of the most important aspects of enhancing overall vehicle performance. Many car manufacturers have been working to establish drag reduction techniques. This paper describes the development process of a new small speciality car which achieved coefficient of drag(C_D) of 0.25. A description of the test facilities and the systems used for developing the aerodynamic aspect of the car are also introduced briefly.

INTRODUCTION

Recently in car development, drag reduction has been one of the most important targets as one of energy saving means. It is widely known that the reduction of aerodynamic drag coefficient(C_D) contributes more effectively as the cruising speed becomes higher. That is to say, it provides less fuel consumption and higher maximum cruising speed. We obtained examples of these effects by tests, which are shown in Fig.1. So we made a goal of reducing C_D rate for a newly developed small speciality car. The target of the rate is over 10% less in the C_D value than that of the current model.

We describe the development process of the car in this paper. The car has always been the aerodynamically leading model since its original model. To attain the goal, we have employed every technique for C_D reduction. Fig.2 indicates the tendency of C_D value. These data are based on officially released information excluding our products.

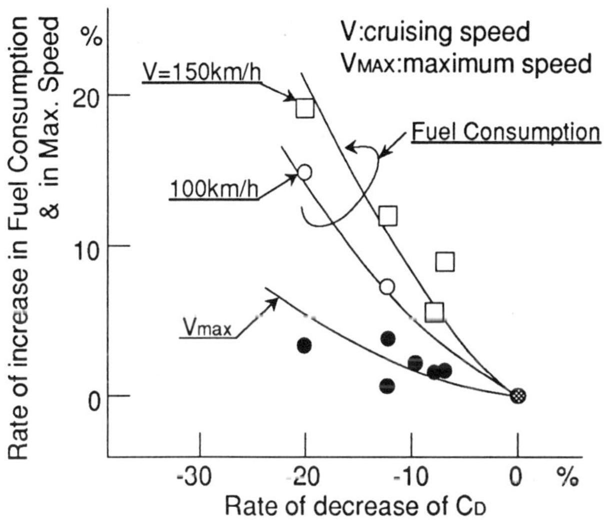

Fig.1 C_D reduction Effects on Fuel Consumption and Maximum Speed

During the past ten years, C_D values have reduced by about 30%, and the same can be said with our products. The C_D value of our new model was measured to be 0.25. This C_D value is now the lowest for a mass production car which we have ever developed. The value is about 15% less than that of the current model, and consequently the fuel consumption and the maximum speed are much improved as is expected. These effects are confirmed by tests and shown in Table 1. Here the computational analysis, flow visualization techniques and wind tunnel tests are performed to evaluate the aerodynamic characteristics of the newly developed car.

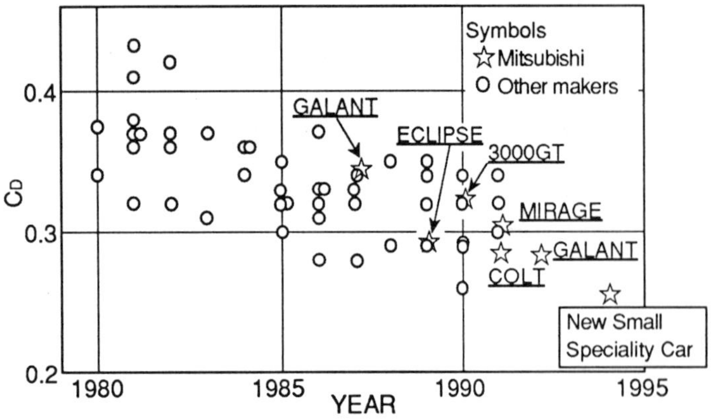

Fig.2 Recent Trend in Drag Coefficient of Passenger Cars

Generally aerodynamic engineers and styling designers usually are in close cooperation to determine key areas on vehicle models from their initial stages of development. Thus, the final configuration of the new small speciality car is developed through many discussions and tests mentioned above.

In this paper, test facilities and tests performed by using the flow visualization systems are introduced. Second, we show the results of applying the Computational Fluid Dynamic(CFD) techniques to the car. Third, the wind tunnel test results of a small scale model are described. Finally, we mention the results of the prototype model tests.

Table 1 Comparison of Fuel Consumption and Maximum Cruising Speed

Item	Fuel Consumption (MPG)			Maximum Speed (km/h)
	City	Highway	Combi-nation	
New Model	24.3	39.8	29.5	207
Current Model	24.1	36.8	28.5	195

TEST FACILITIES

WIND TUNNELS - Two types of wind tunnels were available for the development. One is a full size tunnel, and the other is a small scale size, where we usually use a 1/5 scale model. They were constructed in 1981. The outlines and the specifications of these facilities are shown in Fig.3 and in Table 2 (detailed in Ref.[1]). The full scale tunnel has several technical features as shown in Table 2. It is usually used as a closed circuit type with a closed test section. In the flow visualization tests by smoke, it is changed to an open-circuit type. A large contraction ratio and a boundary layer suction system enable us to make a high quality air flow in a test section and the silencers make the wind tunnel run in suppressed noise level. Thus this wind tunnel enables us not only to test to improve aerodynamic characteristics, but also to reduce wind noise and to measure engine cooling air flow rate. For a blockage correction, we use an area ratio method. The blockage ratio of a full size vehicle in the test section is about from 7 to 9%. As far as aerodynamic characteristics are concerned, we have compared data measured in our wind tunnel with published data measured in other known wind tunnels to confirm the accuracy of our wind tunnel.

On the other hand, the shapes of 1/5 scale concept models are aerodynamically refined in the small scale wind tunnel at the initial stage of the car development. This small scale wind tunnel has almost same performance and system compared with the full size tunnel excluding its sizes. Specifications are also indicated in Table 2.

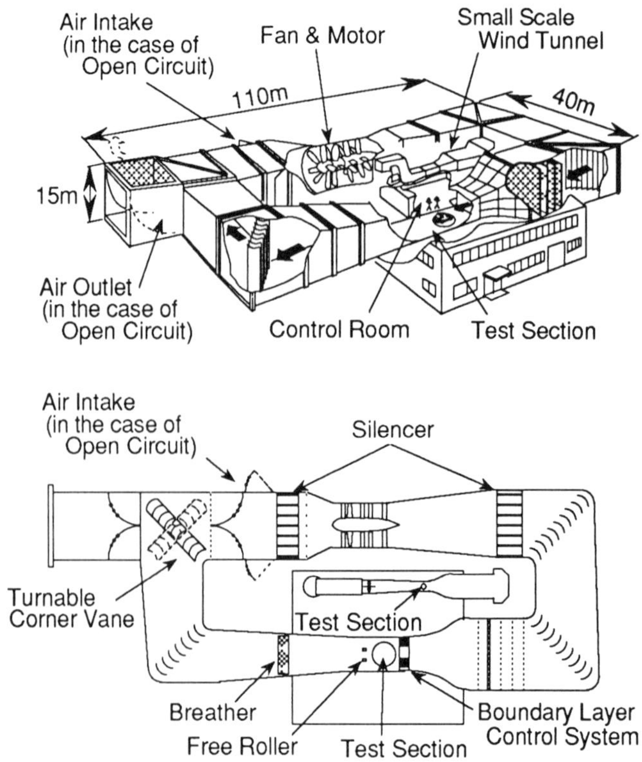

Fig.3 Wind Tunnel Laboratory

Table 2 Main Specifications of Wind-tunnels

	Full Scale Wind Tunnel	Small Scale Wind Tunnel
Type	Open/Closed Circuit	Open Circuit
Test Section Type	Closed, Solid Wall	Closed, Solid Wall
Contraction Ratio	5	5
Max. Speed (m/s)	60	60
Test Section (m)	$6^W \times 4^H \times 12^L$	$1.4^W \times 0.9^H \times 2.5^L$
Motor Power (kW)	2350	100
Fan Diameter (m)	8	2
Appurtenances	Aerodynamic Balance Boundary Layer Control System On-Line Data Acquisition & Processing system Silencers Breather for Air Cooling (Full Scale) Free Rollers for Wheel Rotation (Full Scale)	

Table 3 Specifications of Smoke Wind Tunnels

		2-Dimensional Smoke Wind Tunnel	3-Dimensional Smoke Wind Tunnel
Specifications	Wind Speed Range	4～20m/s	
	Visualization Wind Speed	4～15m/s	4～12m/s
	Test Section Size (m)	$0.2^W \times 2.0^H \times 2.9^L$	$1.4^W \times 0.9^H \times 2.5^L$
	Contraction Ratio	15 : 1	16 : 1
	Smoke Nozzles	ϕ 2mm 80 lines	Vertical ϕ 2mm 40 lines Horizontal ϕ 2mm 40 lines
	Smoke Generator	Heating Type	Heating Type Smoke Wire Type
	Blower	15 kW (in Common)	

In addition, the flow visualization tests were often held to verify the flow around the 1/5 scale model of the car. Specifications of the flow visualization test facilities are shown as following paragraph.

Flow visualization equipments were constructed as one of the major facilities in 1989. Their outlines are shown in Fig.4. There are a 2-dimensional and a 3-dimensional smoke wind tunnels and a water tunnel. They are mainly used for various kinds of flow visualization tests, such as those around a vehicle body, in an engine compartment, in a cabin and so on. Two smoke wind tunnels are driven by one blower. Their specifications are described in Table 3. As for the flow visualization method, heated oil mist are usually used as a tracer. On the other hand, complicated flow patterns like a vortex structure in a wake of a vehicle are often investigated in detail in the water tunnel.

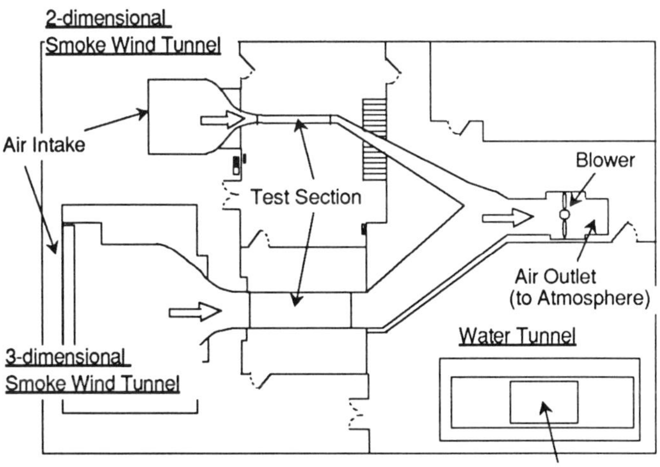

Fig.4 Flow Visualization Facilities(Plane View)

Table4 Specifications of Water Tunnel

Specifications	Water Speed Range	Close 0 ～ 1.5 m/s Open 0 ～ 1.2 m/s
	Uniformity of the Velocity Distribution	Within 3 % (at 1m/s)
	Turbulence Level	Within 2 %
	Test Section Size (m)	$0.8^W \times 1.0^H \times 3.0^L$
	Contraction Ratio	1.5 : 1
	Impeller	5.5 kW × 2

Because in the water, Reynolds Number is easily adjusted as same level as that of a full size vehicle's actual condition. Consequently, vortex structures and separation conditions are more exactly searched by using a water tunnel than by using a smoke wind tunnel. The water tunnel's specifications are described in Table 4. The main tracer particles used for visualization are spheres of nylon12 whose diameter is about 1mm.

COMPUTATIONAL FLUID DYNAMICS (CFD)

On account of recent remarkable development of computer hardware and calculation techniques, the computational analysis has come to be widely used for many research fields. In the field of aeronautical engineering, CFD techniques have been applied to evaluate aerodynamic characteristics of aircraft since the 1970's. On the other hand, in vehicle aerodynamics, this method had been applied for only limited cases because of complicated air flow around a vehicle.

Our original CFD system which was constructed in 1990[2] has been used to estimate aerodynamic characteristics of new cars and their aero-parts. In this system, one of the flow solver which is based on finite difference method is used. The convection terms in Navier-Stokes equation are discretized into third order upwind difference. The grid generation is based on a transfinite method. Block diagram of CFD system is illustrated in Fig.5. In the development of our new small speciality car, we evaluated initial aerodynamic characteristics by using this CFD system. Besides this system is used to evaluate some areas that we have to refine for creating good aerodynamic performance. Fig.6 shows the comparison between the calculation results and experiment in wind tunnel for C_D. The calculated results correspond well to the experimental ones.

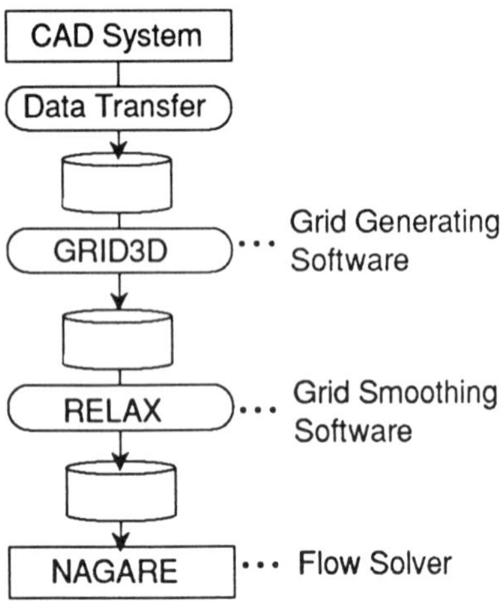

Fig.5 Block Diagram of CFD System

APPLICATIONS OF CFD METHODS - Air flow phenomena around vehicles are simulated by using CFD techniques as mentioned above. It has the advantage to depict not only the flow around a vehicle but also the pressure distribution on the vehicle surface. Fig.7 shows the grid distribution of our new small speciality car. The calculated streamlines around its body and the pressure distribution on its surface are shown in Fig. 8 and 9. For calculation, we constructed the model of the underside virtually same as the actual one. It is shown in Fig.10. Through the CFD simulation, we estimate its C_D value and find the items which we must improve for C_D reduction.

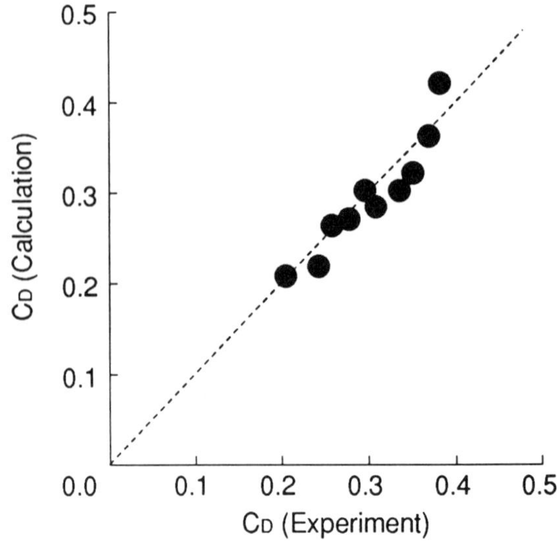

Fig.6 Comparison of C_D between Calculation and Experiment

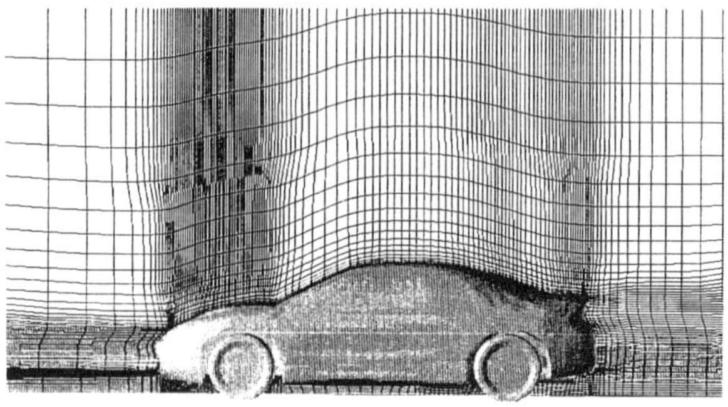

Fig.7 Grid Distribution

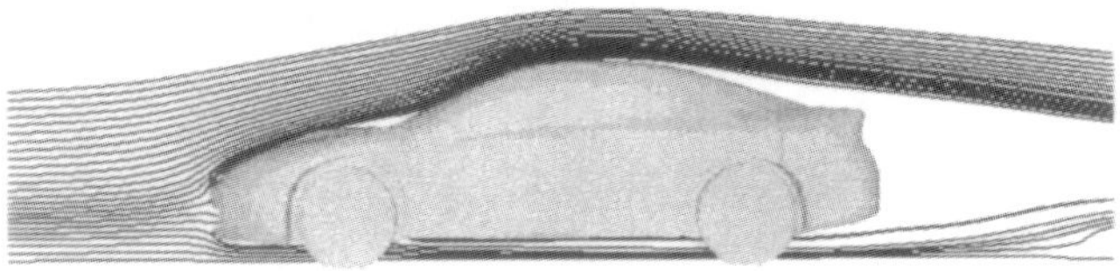

Fig.8 Calculated Result of External Flow (Center Plane)

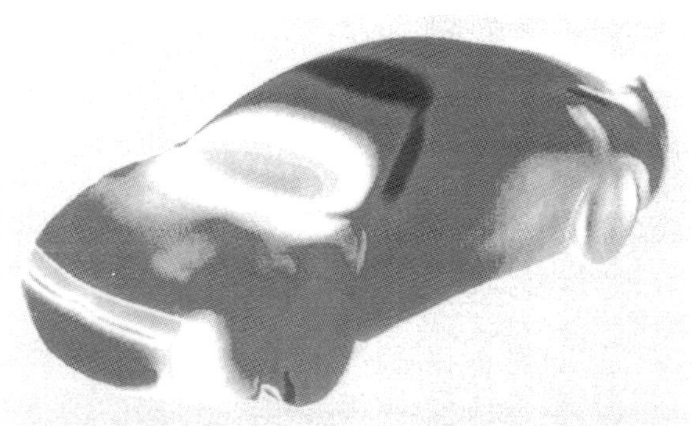

Fig.9 Pressure Distribution on Surface

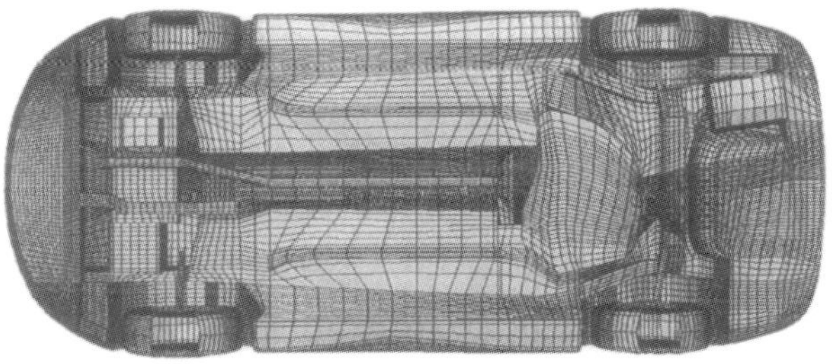

Fig.10 Underside of CFD Model

SCALE MODEL INVESTIGATIONS

As the design evolves, it becomes more necessary to analyze the aerodynamic influences of each sections in detail. At this stage, we make the 1/5 scale model for evaluating the vehicle's aerodynamic characteristics and improving the details of parts at the small scale wind tunnel. Fig.11 shows the 1/5 scale model of the car. Power trains and suspension systems are precisely fabricated like those in the full scale vehicle.

Fig.11 1/5 Scale Model

In order to investigate the lowest drag silhouette, we have established guidelines which can be used for styling designers to mold the aerodynamic shapes. This advantage was that the body exterior was almost completed aerodynamically at the initial design stage. It is noted that the CFD results were also reflected on the concept model.

Initial examination and optimization work were started from flow visualization tests. Fig.12 shows the flow around the body by using the 3-dimensional smoke wind tunnel. It is recognized that the smoother streamlines are formed from the front nose to the roof end and the wake region is well controlled. Fig.13 shows the visualization results of the wake formation compared with the current body. It is obvious that the wake region in the new model is smaller than that of the current one. Moreover, the flow patterns over the body surface are shown in Fig.14. This is oil flow test results. By using this visualization technique, separation points and some other flow phenomena are recognized in detail. From those results, we could grasp the key flow characteristics around the car.

After the visualization tests mentioned above, we began to optimize the shape of the car aerodynamically.

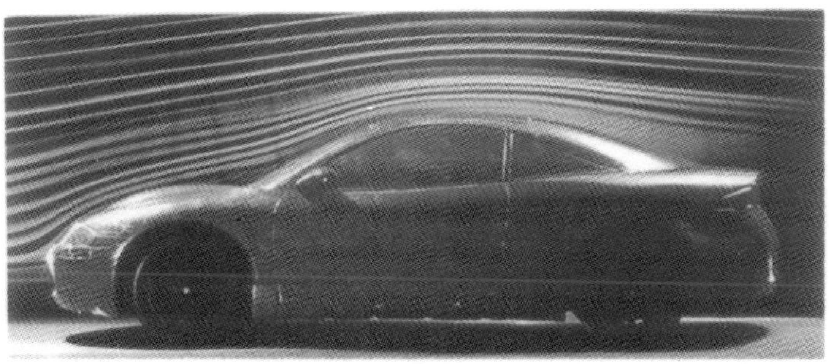

Fig.12 Flow Visualization result in Smoke Wind Tunnel

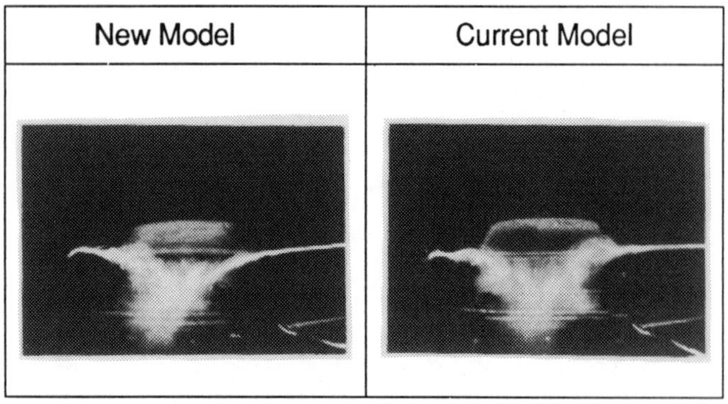

Fig.13 Comparison of Wake Regions
(Rear View in Smoke Wind Tunnel)

Fig.14 Oil Flow Pattern

FRONT END SHAPE - To reduce aerodynamic drag, careful attention must be paid to the shapes of vehicle's front corners. On our new small speciality car, the investigations of the front corner shapes, namely the front nose, the bumper side corners, the front pillars and so on, were carried out to prevent the flow separating from the body surface. To restraint the flow separation from occurring at those corners, we have developed a simple method to determine corner curvature (detailed in Ref.[3]). We proposed and adopted the following formula for the geometry of the curved surface of a newly developed car.

$$(x/a)^m + (y/b)^n = 1 .$$

We selected the optimum numbers of a, b, m and n to design each front corners. So the air flow is created good feature without separating from each front corners. From Fig.14, it is obviously seen that the flow over new vehicle's front corners is very smooth.

REAR END SHAPE - In order to apply the boat-tailing shape to a newly developed car, we investigated some tapering shapes. Fig.16 shows the C_D variation reference to the ratio of the rear tapering versus the car width. They are the experimental results from the 1/5 scale model of the new car. We decided the new car's tapering ratio about 20% from these results.

The rear corner's roundness is also important as well as the tapering. The small curvature will make a flow separated early and create strong vortices into the vehicle wake. And it makes a strong negative pressure region behind the vehicle body. From an aerodynamics viewpoint, rear end corners have to be made sharp to prevent strong vortices being created in the vehicle wake.

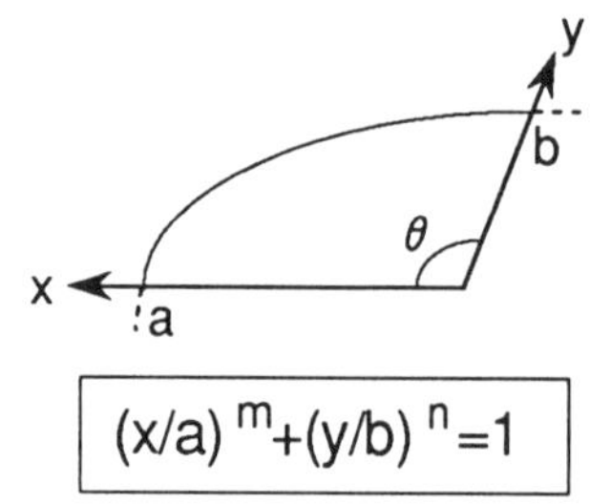

Fig.15 Formula to Determine Corner Curvature

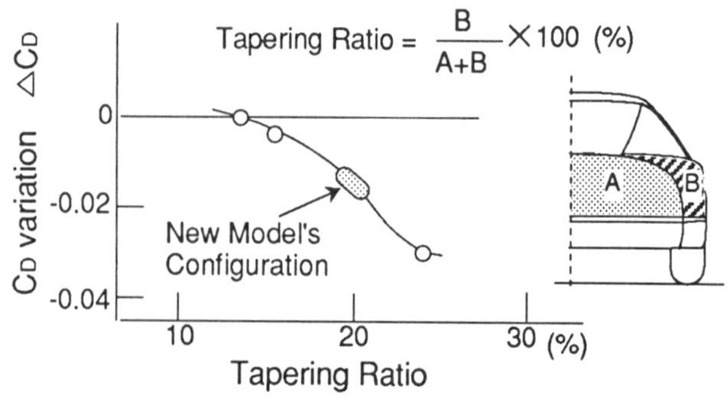

Fig.16 Tapering Ratio Effect on C_D variation

On the model, we proposed a new shape for the rear combination lamp for avoiding vortex formation. The proposed shape is shown in Fig.17. The sharp edge is formed on side surface. The effect of C_D reduction due to this sharpened shape is shown in Fig.18 with the other additional parts' effects.

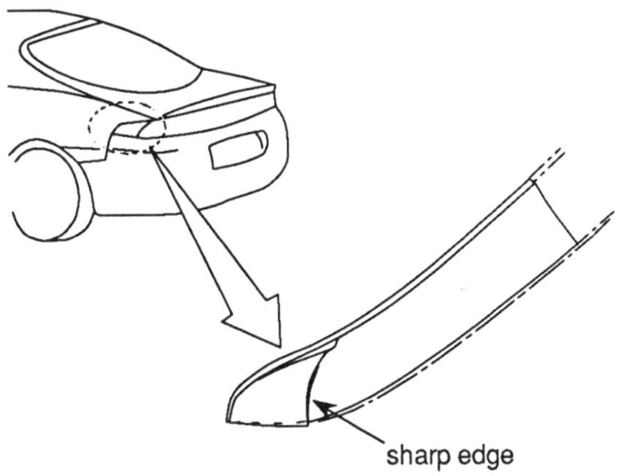

Fig.17 Shape of Rear Combination Lamp

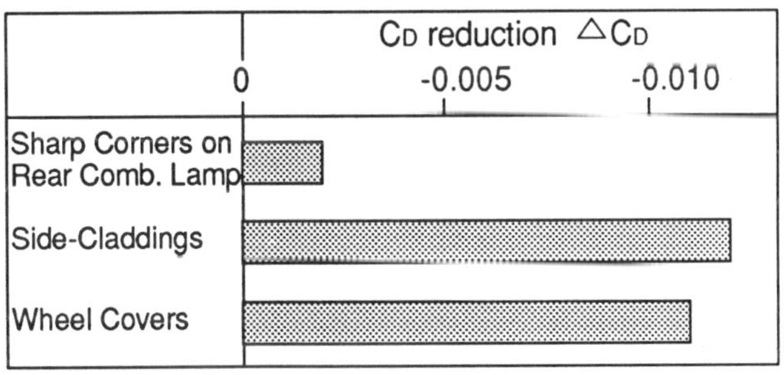

Fig.18 C_D Reduction Effects on 1/5 Scale Model

ESTABLISHMENT OF SIDE-CLADDINGS - A side-cladding is an aero-part attached to body side between front and rear wheel. It restrains flow from flowing in and out of the underside of the vehicle body and prevent air flow from striking against rear wheels [4]. As the result, it makes smoother streamlines along the vehicle body sides. Fig.19 illustrates the pressure distribution around the rear wheel with the side-cladding as compared to the one without it. It is obvious that the positive pressure region becomes smaller by attaching the side-cladding. In case of the model, the side-claddings reduce C_D to about 0.01. It is also indicated in Fig.18.

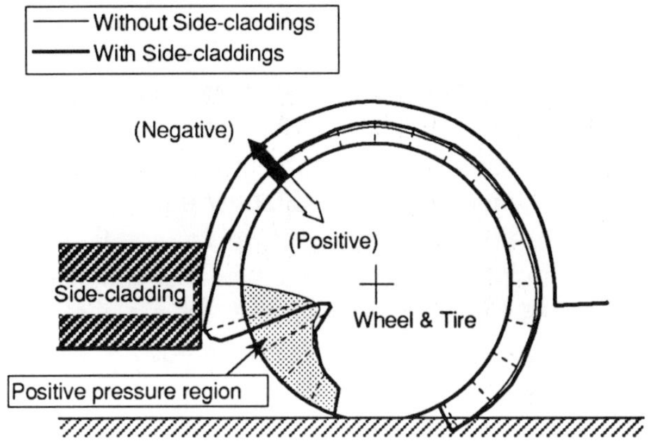

Fig.19 Pressure Distribution over Rear Wheel (experimental data)

FLAT SURFACE WHEEL COVERS - To make smoother streamlines on the side of vehicle body, air flow along the wheel is not to be neglected. We proposed to design a new flat surface wheel cover for a new small speciality car. The difference of the body side flow between two cases are shown in Fig.20. It is obvious that the streamlines on the body side surface become smoother and the velocity increases.

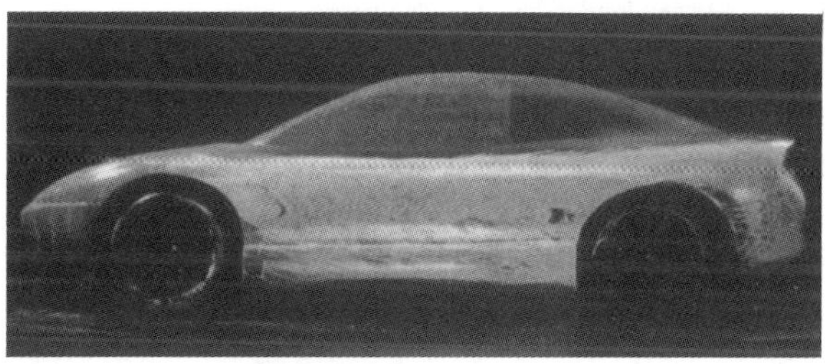

(a)Without Flat Wheel Covers

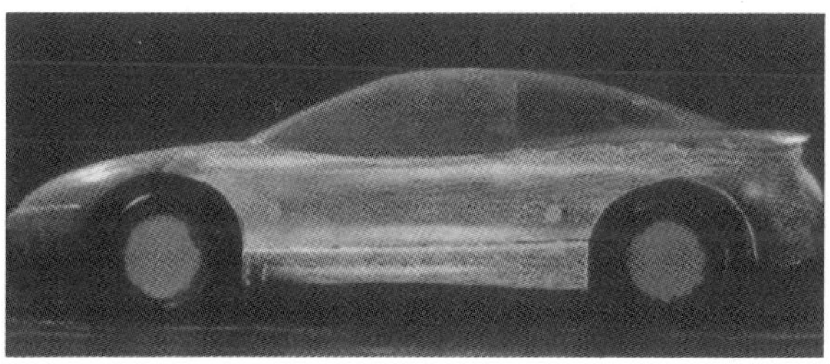

(b)With Flat Wheel Covers

Fig.20 Oil Flow Patterns on Body Side

To realize the non turbulent air flow on the outside surface of wheels, the surface of wheel covers must be extremely flat. So we equipped the outlets of the brake cooling air flow near the edge of the cover. Fig.21 shows its structure. The step height between the tire outside surface and the cover surface is very low. The effect of C_D reduction by equipping these flat wheel covers is also shown in Fig.18, where measurement is made with the wheel covers fixed to the wheels which do not rotate. We have confirmed that the effect of non-rotating flat cover can be extended to the rotating condition.

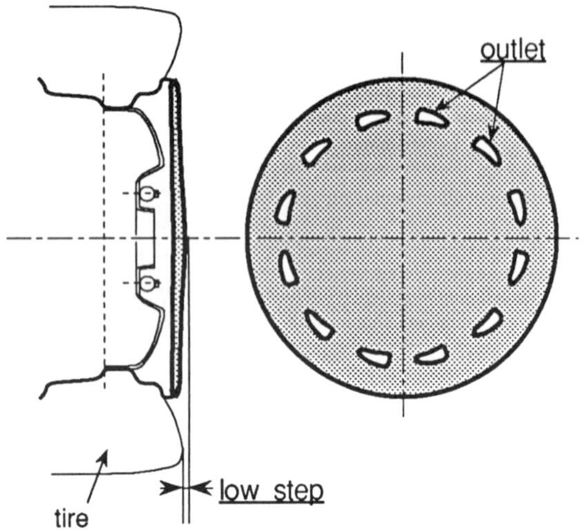

Fig.21 Structure of Wheel Cover

UNDERSIDE - The contribution of the vehicle's underside is as important as that of the upper body. On the underside of the vehicle, many obstacles are located against the air flow, namely transmission, suspensions, cross-members and so on. Up to now, it is said that the reduction of C_D value achieved by complete paneling of the underside is about from 0.04 to 0.05. That is to say, the rate of drag caused by a vehicle's underfloor ranges from 10 to 20% of the total C_D value of a vehicle [5]. Accordingly the aerodynamical improvement of the underside had also become the key target for reducing C_D value.

Fig.22 illustrates the underside fairing parts. Front under cover and other covers set behind the front floor panel reduced C_D value to about -0.015. To determine the aerodynamically optimum shape of the front under cover, we used the CFD technique. Fig.23 shows calculated results of the flow on the final shape of the front under cover. This shows that the center flow is smoother and the air-dam shapes which were located in front of wheels prevent the air flow from striking against the front wheels effectively as is expected.

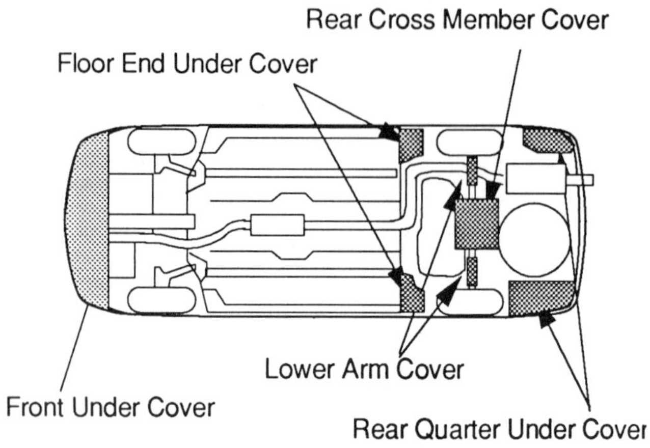

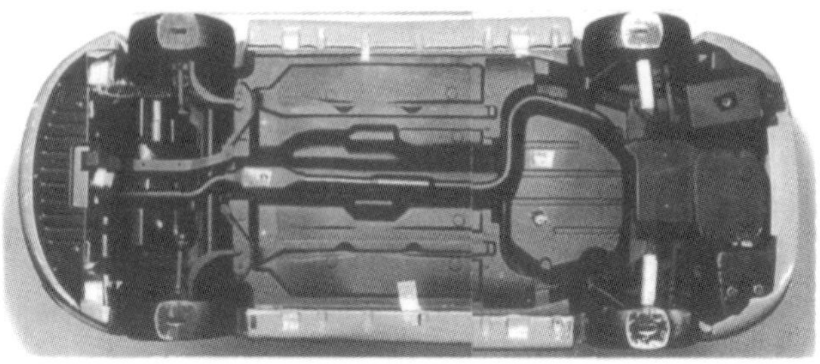

Fig.22 Aero-parts on Underside

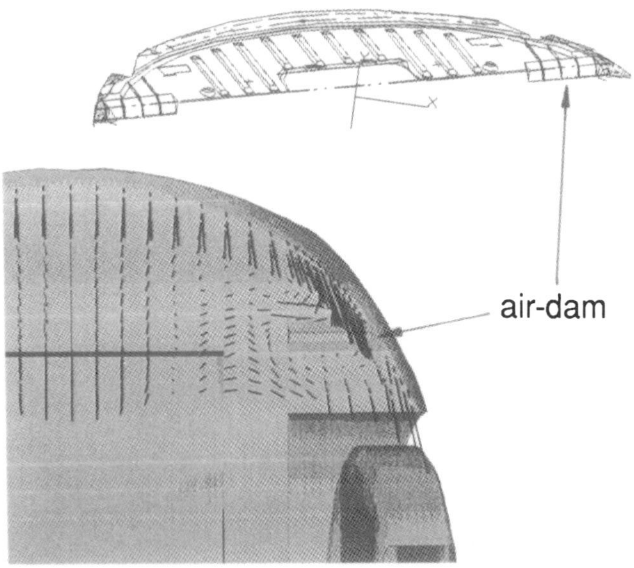

Fig.23 Flow Pattern on Front Under Cover (Calculation)

COOLING AIR FLOW

The flow around a vehicle is classified into two flows. One is the external flow which we described above, and the other is internal flow, namely the engine cooling air flow and the ventilation in the compartment. The engine cooling air flow especially has a great influence on vehicle's drag. So it usually becomes one of the most important areas for C_D reduction. To get air flow into a engine compartment smoothly, we examined duct shapes from the front grill to the heat exchanger by the CFD technique and tested by using the 1/1 scale mock-up front nose shape model. Fig.24 shows the test results of the cooling air flow rate and C_D increment with respect to the air intake opening area of the front grill. Fig.25 illustrates the flow patterns obtained by CFD. There is the small separated region in the duct, but it is recognized that the almost smooth flow is realized and it spread to the radiator without any loss. In this case, the engine shape is simplified in the calculated model.

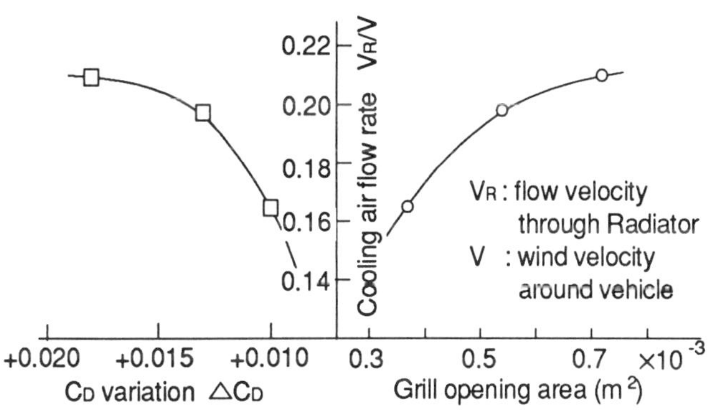

Fig.24 C_D and Cooling Air Flow Variation reference to Grill Opening Area

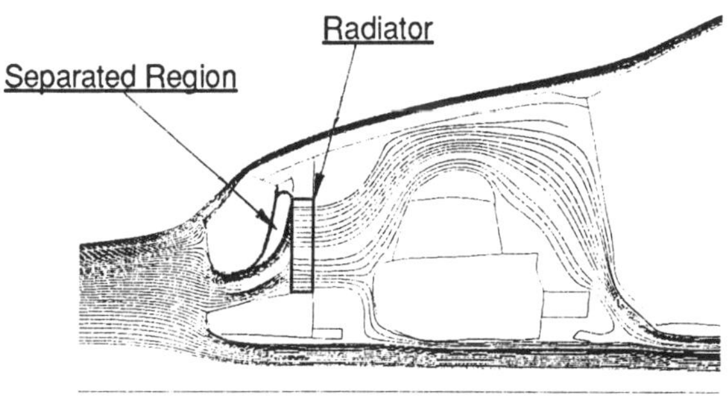

Fig.25 Flow in Engine Compartment (Calculation)

PROTOTYPE MODEL TESTING

In the full scale wind tunnel, we tested the prototype vehicle model in order to confirm aerodynamic characteristics of the new model. The influences of aerodynamical additional parts were also verified. At this stage, further aerodynamic refinement to the prototype model were achieved, although each portion of the prototype had been aerodynamically improved by incorporating the results from the 1/5 scale model test stage. This full size model has a 2.0 liter DOHC engine and front wheel drive (shown in Fig.26). Its major dimensions are shown in Table 5.

Fig.26 Full size vehicle model tested in the Wind tunnel

Table 5 Major Dimensions

ITEMS	SIZES
Overall Length (m)	4.375
Overall Width (m)	1.735
Overall Height (m)	1.300
Wheelbase (m)	2.510
Front Tread (m)	1.515
Rear Tread (m)	1.500
Tire Size	215/55 R16
Frontal Area (m^2)	1.90

And at this final stage, we have to confirm the flow on the vehicle body surface, especially the areas where lies some possibility that the flow could separate. The flow on the rear bumper side of the base shape is compared with that on the improved shape in Fig.27. Calculation and experimental results are illustrated side by side. On the calculated directions and velocities are indicated with vectors. It is recognized that the flow on the base bumper is separated from its surface just behind the rear wheel. By widening the bumper width, we improved the surface flow. The flow over the refined bumper surface flows smoothly. This improvement reduces C_D value by 1%. This is a typical example of the body optimization by using the flow visualization and CFD technique.

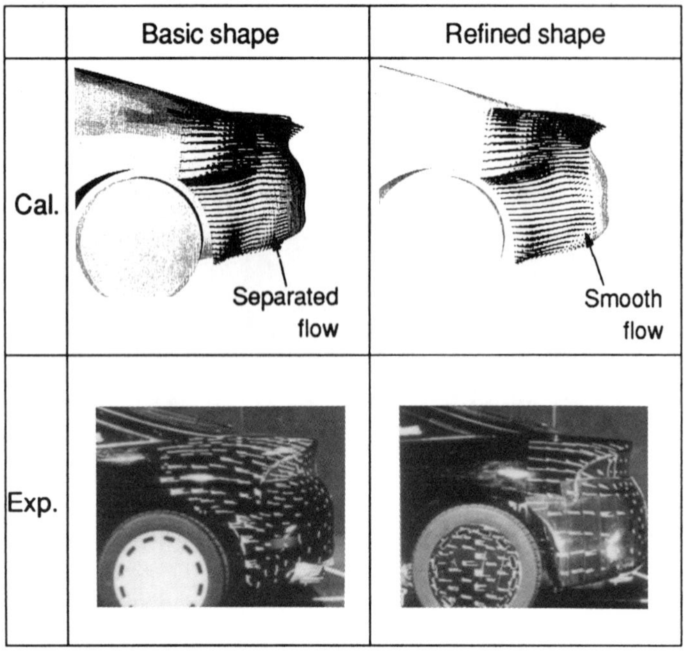

Fig.27 Flow on Rear Bumper Side (Calculation and Experimental Results)

AERO-PARTS ON UNDERSIDE - In Fig.28, C_D reduction effects of underside aero-parts on the prototype model are shown in comparison with the results of the 1/5 scale model. Taking the construction accuracy of the scale model into consideration, the differences of C_D reduction effects of the aero-parts between on the scale model and on the prototype are small and their tendencies of effects correspond well. The air flow under the vehicle is well controlled by these parts as we expected from 1/5 scale model tests. The panels attached on the underside have definite functions, which are the prevention of the flow striking wheels and chassis, the restraint of the flow from entering into the underfloor, etc. By making the flow under the vehicle smooth and fast, the drag produced by the flow attaching to the obstacles on the underside is much reduced.

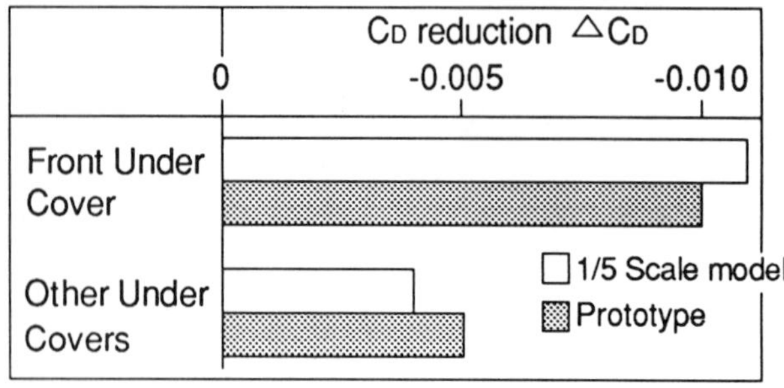

Fig.28 C_D Reduction Effects of Underside Aero-parts

C_D reduction effect of the front under cover is realized in Fig.29. This figure shows the total pressure loss distribution behind the vehicle. It is the examined result by using the other product which has similar underside geometry to the new small speciality car. It shows that the total pressure loss becomes smaller than the normal underside by adding the front under cover. A total pressure loss is closely concerned with a loss of momentum energy. As the aerodynamic drag of vehicle is mostly caused by momentum energy loss due to flow separations, vortex shedding, and lots of other turbulence in the flow over its surface, it is evident that the reduction of the total pressure loss shown in Fig.29 is directly connected with C_D reduction.

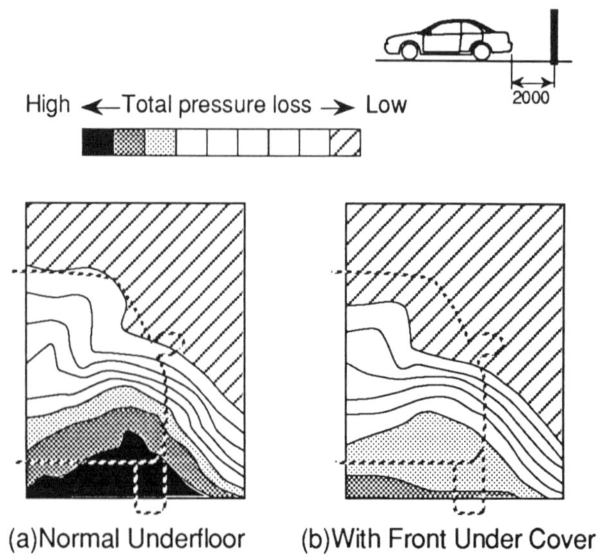

Fig.29 Total Pressure Loss Distribution behind Vehicle

INLET OF ENGINE COOLING AIR - Since the duct from the front grill to the heat exchanger restrains the flow separating and leaking outside, it enables us to get the adequate cooling air flow into the heat exchanger. And this smooth flow enables us to reduce the drag caused by the air flow taken in the engine compartment. The C_D increment caused by the engine cooling air flow of the recent mass production cars are shown in Fig.30 with reference to the engine displacement. The C_D value increment of our small speciality car is under 0.015. It is much smaller than average number.

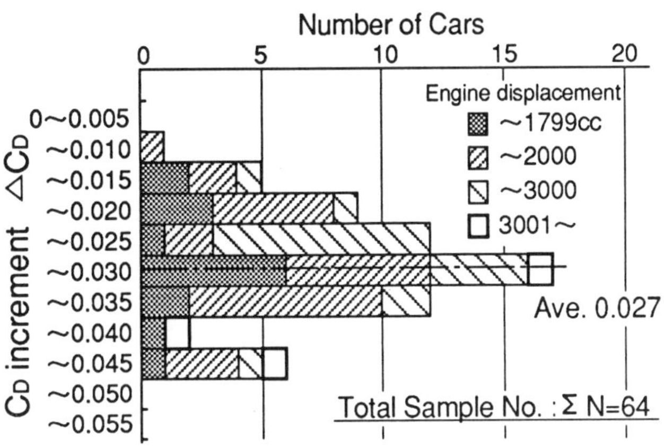

Fig.30 C_D increment reference to Engine Displacement

OUTSIDE MIRRORS - Much attention is also paid to the shape of the outside mirror case from the aerodynamics viewpoint. To make its wake region smaller and to prevent the flow from separating, we adopted a bullet like shape keeping an adequate view area as a production car. Fig.31 indicates the increase in C_D value caused by attaching outside mirrors. The C_D increment 0.005, which is our case, lies apparently within satisfactory.

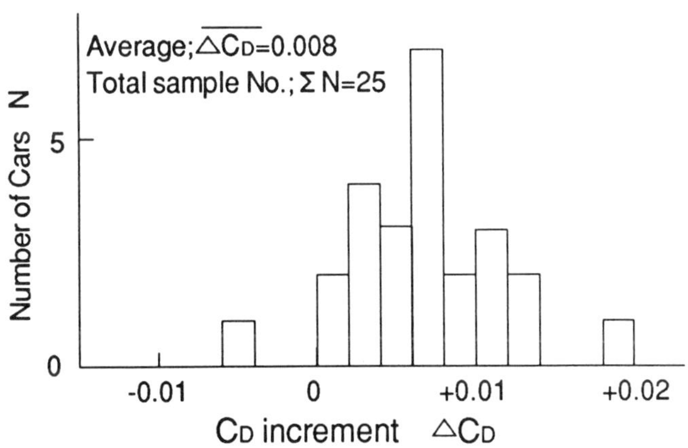

Fig.31 C_D Increment due to Outside Mirrors

CONCLUSIONS

In this paper we introduced the development process of a low drag car. The points of the process are summarized as follows.

(1)By using the CFD techniques, we confirmed aerodynamic characteristics of an initial design concept model.

(2)In the 1/5 scale model stage, we estimated the flow around the vehicle body by flow visualization techniques. Next we optimized the body shape aerodynamically by small scale wind tunnel tests.

(3)We also optimized the engine cooling air flow by using a mock-up model in the full scale wind tunnel. CFD techniques were independently implemented for body shape improvement.

(4)At the final developing stage, we confirmed the aerodynamic characteristics of the car and its additional parts in the full scale wind tunnel.

REFERENCES

[1] T.Shibata, et al. : A Study on Aerodynamic Characteristics of Passenger Cars, IPC, 1983

[2] T.Kataoka, et al. : Numerical Simulation of Road Vehicle Aerodynamics and Effect of Aerodynamic Devices, SAE paper No.910597

[3] I. Kohri, et al. : Aerodynamic Development of an Experimental Car, SAE paper No.890373

[4] H. Ohno, et al. : Improvement of Aerodynamic Characteristics of Passenger Car by Side-Airdams, JSAE Review Vol.12, No.3, 1991

[5] W. H. Hucho, et al. : 'Aerodynamic of Road Vehicle', Butterworths, England, 1987

[6] Hans-Joachim Emmelmann, et al. : The Aerodynamic Development of the Opel Calibra, SAE paper No.900317

950628

Numerical Investigation of Vehicle Aerodynamics with Overlaid Grid System

Nariaki Horinouchi, Yoshihiro Kato,
Sayoko Shinano, and Tsuguo Kondoh
Toyota Central R&D Labs, Inc.

Yuichi Tagayashi
Toyota Motor Corp.

ABSTRACT

The drag reduction mechanism in newly developed low aerodynamic drag model car is investigated through numerical simulation. In order to deal with the computational domain around a three-dimensional complicated vehicle body, the method of overlaid grid system is employed.

The results of computational case study on the body shape demonstrate that the lateral tapering near the rear end and the spats around the wheels bring better flow properties for drag reduction, such as the pressure recovery in the wake.

INTRODUCTION

The numerical analysis of vehicle aerodynamics with Computational Fluid Dynamics (CFD) is becoming more and more practical, owing to the recent progress in computational techniques and computer hardware performance. But in order to obtain reliable results, it is important to adopt a computational grid system with good properties. In the finite-difference method, that is employed in our numerical simulation, the grid system should satisfy the following requirements as well as possible:

- The grid lines cross orthogonally.
- The aspect ratio of a grid cell is nearly equal to 1.
- The spacing between the sequential grid points along each grid line varies smoothly.
- Grid points concentrate near the boundary.

However, it is very difficult to cover the whole computational domain including a three-dimensional complicated vehicle body with a single block-structured grid because of the following reasons:

- It takes a large amount of human effort.
- It requires a grid generation skill of a very high degree and even so sometimes it is not possible to generate a single block-structured grid.
- The total number of grid points tends to become very large.

Thus the method of "Overlaid Grid System [1] [2] [3] [4]" is employed in the present study, where the whole computational domain is covered with multiple grid blocks overlapping each other. Each grid block (individual block constructing overlaid grid system) is generated around the local boundary shape to satisfy the above requirements.

In this study, we investigated the mechanism of significant drag reduction which was achieved in a newly developed model car (shown in Figure 1 and hereafter called a "model car") with the following aerodynamics refinement:

- Lateral (Planform) tapering of rear-end body shape. (Boat tailing)
- Completely flat under-floor panel.
- Covering of the rear wheels.
- Body fairing around wheels (called "Spats").

Among these items, we evaluated, through the numerical simulation, the effects of the first one (planform tapering at the rear, Figure 2) and the fourth one which is achieved by adding attachments under the wheelhousing (Figure 3).

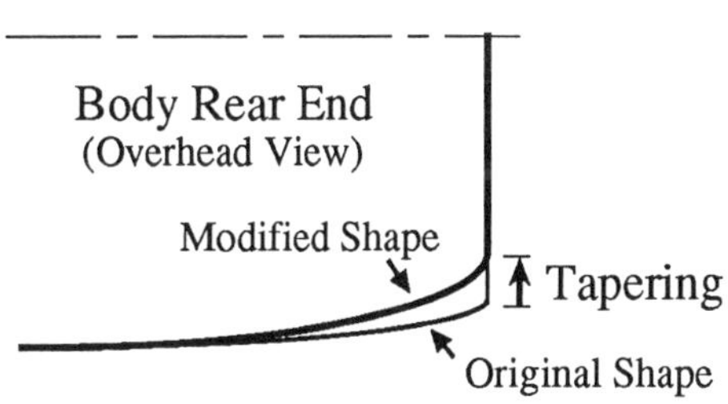

Figure 2: Lateral Tapering of Rear End

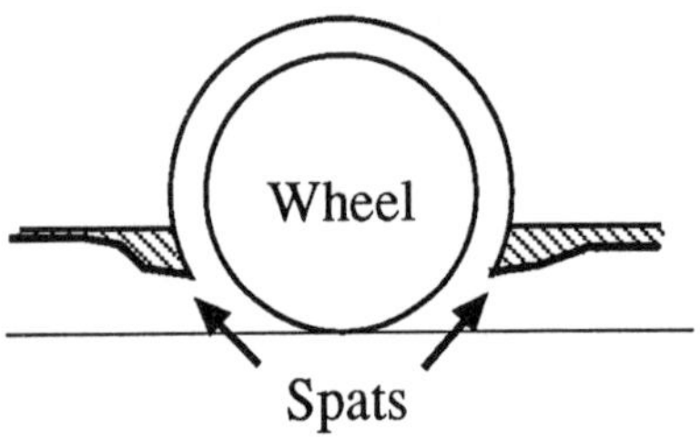

Figure 3: Spats

NUMERICAL METHOD

GOVERNING EQUATION - The flow around a road vehicle is unsteady, incompressible and viscous at a high Reynolds number. This is mathematically described by the following equations, i.e., the equation of continuity,

$$\nabla \cdot \vec{u} = 0, \qquad (1)$$

and the Navier-Stokes equation,

$$\frac{\partial \vec{u}}{\partial t} + (\vec{u} \cdot \nabla)\vec{u} = -\nabla p + \frac{1}{Re}\Delta \vec{u}, \qquad (2)$$

where Re is the Reynolds number. Unfortunately, we have difficulty in obtaining pressure values using these two equations. Therefore, we derive the following Poisson equation for pressure from these two equations, where $D = \nabla \cdot \vec{u}$.

$$\Delta p = -\nabla \cdot ((\vec{u} \cdot \nabla)\vec{u}) - \frac{\partial D}{\partial t}. \qquad (3)$$

In actual computation, EQs (2) and (3) are solved by the finite-difference method. Though the second term of the right-hand side of EQ (3) should mathematically be equal to zero, it is kept here to correct the mass-conservation error involved in the computation.

EQs (2) and (3) are transformed into the generalized coordinate system and then discretized by the finite-difference method.

FINITE-DIFFERENCE APPROXIMATION - In the present study, the second-order-accurate Crank-Nicolson scheme was used for the time integration of the Navier-Stokes equation (2). The convection terms of EQ (2) were discretized by the third-order-accurate upwind scheme [5] and the other spatial derivatives were evaluated by the central-difference scheme.

As a result of the discretization of EQs (2) and (3), a set of linear equation systems are obtained for each component of the velocity vector and pressure. In these matrix equations, the coefficient matrices are of very large scale and sparse. Therefore, they were solved by the Successive Over Relaxation (SOR) method in this study.

Our computational codes introduced various optimization techniques including the memory access method called "Multi-Color Ordering [6] [7]" for the SOR method. As a result, extremely high computing speed was achieved on the supercomputer NEC SX-3/14R.

OVERLAID GRID METHOD - An example of the overlaid grid system, which is used to simulate the flow around a two-dimensional circular cylinder, is shown in Figure 4. In this example, the grid system consists of two grid blocks which overlap each other. One of them covers the whole computational domain and the other, which is the so-called "O-type" grid, covers the local region around the cylinder. On the grid points near the outer rim of each grid block in the overlapping regions, the physical quantities are determined by interpolation on the other grid block.

In this example, a bi-linear interpolation is used to interpolate the physical quantities. As illustrated in Figure 5(a), suppose that a point to be interpolated (hereafter called P) is involved in a cell enclosed by four vertices, where $\vec{x}_P$ is the coordinate of the point P, i, j are the indices of the grid points of the vertices, and $\vec{x}_{i,j}, \ldots, \vec{x}_{i+1,j+1}$ are the coordinates of the vertices. If the location of P is expressed in terms of the normalized local coordinates ξ', η' as shown in Figure 5(b), the interpolated value f_P of an arbitrary physical quantity f at P is determined by the following bi-linear interpolation:

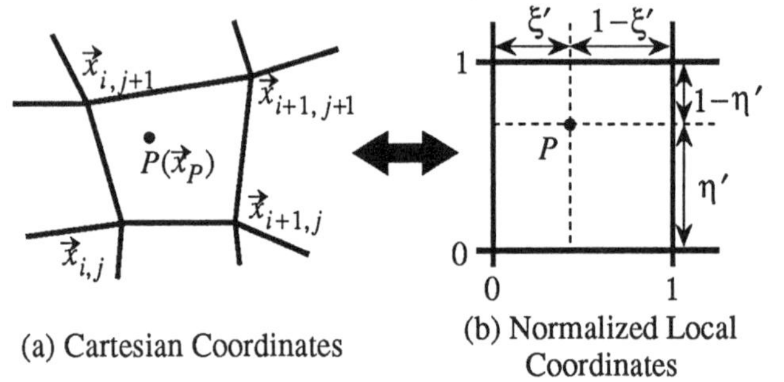

Figure 5: Bi-linear Interpolation

$$\begin{aligned} f_P &= (1-\xi')(1-\eta')f_{i,j} + \xi'(1-\eta')f_{i+1,j} \\ &+ (1-\xi')\eta' f_{i,j+1} + \xi'\eta' f_{i+1,j+1}, \end{aligned} \qquad (4)$$

where $f_{i,j}, \ldots, f_{i+1,j+1}$ are the values of f at the cell vertices. The normalized local coordinates ξ', η' are given by solving the following system of non-linear equations:

$$\begin{aligned} \vec{x}_P &= (1-\xi')(1-\eta')\vec{x}_{i,j} + \xi'(1-\eta')\vec{x}_{i+1,j} \\ &+ (1-\xi')\eta'\vec{x}_{i,j+1} + \xi'\eta'\vec{x}_{i+1,j+1}. \end{aligned} \qquad (5)$$

In the similar way, the tri-linear interpolation, which is defined as EQ (6), is employed for the three-dimensional computation. In EQ (6), ξ', η', ζ' are the normalized local coordinates and i, j, k are the indices of cell vertices.

$$
\begin{aligned}
f_P &= (1-\xi')(1-\eta')(1-\zeta')f_{i,j,k} \\
&+ \xi'(1-\eta')(1-\zeta')f_{i+1,j,k} \\
&+ (1-\xi')\eta'(1-\zeta')f_{i,j+1,k} \\
&+ \xi'\eta'(1-\zeta')f_{i+1,j+1,k} \\
&+ (1-\xi')(1-\eta')\zeta' f_{i,j,k+1} \\
&+ \xi'(1-\eta')\zeta' f_{i+1,j,k+1} \\
&+ (1-\xi')\eta'\zeta' f_{i,j+1,k+1} \\
&+ \xi'\eta'\zeta' f_{i+1,j+1,k+1}.
\end{aligned} \tag{6}
$$

This method has the following advantages over the conventional methods:

- It is easier to generate a grid block in each local computational region and assemble them to cover the whole computational domain than to generate a single grid system in the whole computational domain.
- It is easy to make the grid fitted to the complicated boundary shape.
- It is very easy to deal with the change in the local boundary shape because the modification of the grid is limited to a local and small amount.

These advantages are very effective in reducing the efforts in grid generation and are also very useful in promoting practical applications. Moreover, they contribute to improving the computational accuracy through the use of a grid with better properties.

COMPUTATIONAL RESULTS

COMPUTATIONAL CASES - The following four cases of calculations were performed where the lateral tapering of the rear end and spats over the front wheel were changed.

Table 1: Computational Cases

	Tapering	Front Spats
Case 1	0mm	yes
Case 2	80mm	yes
Case 3	160mm	yes
Case 4	80mm	no

Among these cases, case 2 corresponds to the model car.

COMPUTATIONAL CONDITIONS - In the present study, the above case studies were performed under the following conditions.

- Boundary Conditions: The boundary conditions used in the present computation are shown in Table 2. The derivative $\partial/\partial n$, which appears as the Neumann boundary condition, indicates the gradient in the direction normal to the boundary surface. $\vec{u'}$ is the velocity vector on the rotating wheel surface.

Table 2: Boundary Conditions

	Velocity	Pressure
Inlet	$u=1, v=w=0$	$\partial p/\partial n=0$
Outlet	$\partial\vec{u}/\partial n=0$	$p=0$
Body Surface	$\vec{u}=0$	$\partial p/\partial n=0$
Wheel Surface	$\vec{u}=\vec{u'}$	$\partial p/\partial n=0$
Others	$u=1, v=w=0$	$\partial p/\partial n=0$

In the followings, the computational results will be compared with the experimental data. Since the measurements were performed in the wind-tunnel, the ground was stationary. However, the experimental data have been converted to the data corresponding to the actual driving condition by an empirical formula. Therefore, the moving ground condition was adopted in the present calculation.

The computational domain was the half region bounded by the vehicle central plane (the right-half region viewed from the front), which is illustrated in Figure 6.

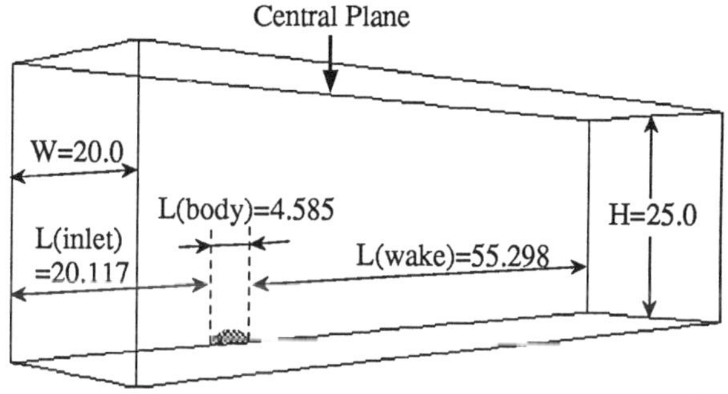

Figure 6: Computational Domain in Half Region (in meter)

- Reynolds Number: The model in the wind-tunnel testings was full-scale and the velocity of air flow was 30 m/sec. The Reynolds number based on this velocity and the unit length of 1 meter is 2.0×10^6. This is common to the computation and the experiment.
- Initial Conditions:
  - The velocity field was given as the potential flow. This is obtained by solving the Laplace equation for the velocity potential ϕ:

$$\Delta\phi = 0, \tag{7}$$

    and applying the gradient operator to ϕ,

$$\vec{u} = \nabla\phi. \tag{8}$$

  - The pressure field was given by applying the theorem of Bernoulli.

COMPUTATIONAL GRID SYSTEM - Nine grid blocks, among which seven were used around the vehicle body and wheels, were overlapped in the whole computational domain. The total number of grid points was about 1,270,000. The construction of the total grid system is shown in Figure 7.

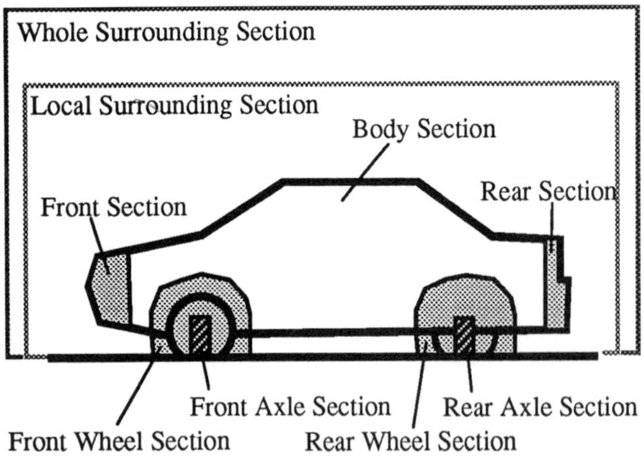

Figure 7: Configuration of Overlaid Grid System

Figure 8 shows the computational grid system on the body surface for case 2. In cases 1, 3 and 4, the computational grid was constructed by changing only the partial grid blocks for case 2.

EFFECT OF LATERAL TAPERING - Figure 9 shows the computational grids on the body surface viewed from top, in which green, red and blue lines correspond to cases 1, 2 and 3, respectively.

Figure 10 shows the change in drag coefficient C_D with the rear-end tapering. It is found that C_D decreases substantially with tapering width. For both the absolute value (C_D) and the rate of reduction (ΔC_D) agreement between calculations and experiments is very good. For example, the difference in the value of C_D is about 5%.

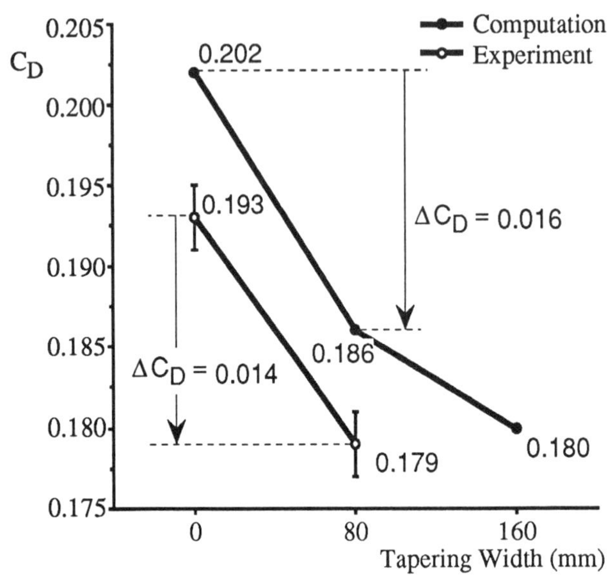

Figure 10: Change in C_D with Tapering

The pressure distribution is shown as a color map of the pressure coefficient C_p in Figure 11, where the pressure changes according to the color legend (blue: low pressure ↔ red: high pressure). Figure 11 reveals that with increasing tapering width, the pressure on the rear end surface becomes higher. This is because the velocity becomes smaller as the flow area increases due to the tapering. The drag reduction is typically attributed to better pressure recovery at the base due to the closure of wake.

However, to obtain this benefit it is necessary that the flow never separates until it reaches the rear end. Figure 12 illustrates velocity vectors on the cross-section through the center of the front and the rear wheels in case 1. Around the front wheel (Figure 12(a)), which is rotating as mentioned before, the vortices due to the secondary flow appear in the both lateral sides, and a complicated flow field is generated there. On the other hand, around the rear wheel (Figure 12(b)), which is also rotating, the secondary flow exists only near the ground; it is observed that the flow is moving in the main-stream direction along the wheel cover. The streamlines are shown in Figure 13. In any case, no significant flow separation occurs until the rear end.

EFFECT OF SPATS - "The spats," already illustrated in Figure 3, are attachments put under the wheelhousing upstream and downstream of the wheel. These devices are introduced to make the flow more smooth around the wheel.

In order to investigate the effects of the spats, the computational results for case 2 and case 4 are compared. While the spats are attached in case 2, the spats are removed from the front wheelhousing in case 4. As mentioned before, C_D in the computation is 0.186 for case 2. On the other hand, C_D becomes 0.192 as a result of the calculation in case 4. It is demonstrated that C_D is reduced by attaching the spats.

C_p distributions along the body surface on two longitudinal sections A-A' and B-B', which are illustrated in Figure 14, are plotted in Figure 15. The results on the front and upper surface, where the pressure distributions are almost the same in both cases, are omitted. In Figure 15, solid lines and dotted lines correspond to the computational results for case 2 and case 4, respectively. The difference of C_p is definitely recognized at the lower surface and the base on both sections. At the front part of the lower surface, C_p is greater in case 4. On the contrary, at the rear part of lower surface and the base, C_p is higher in case 2. If we consider the direction of body surface where the difference of C_p is observed, it is derived that C_D must be smaller in the case with the spats.

Figures 16 illustrates the streamlines around the front wheel. Figure 17 shows the streamlines near the

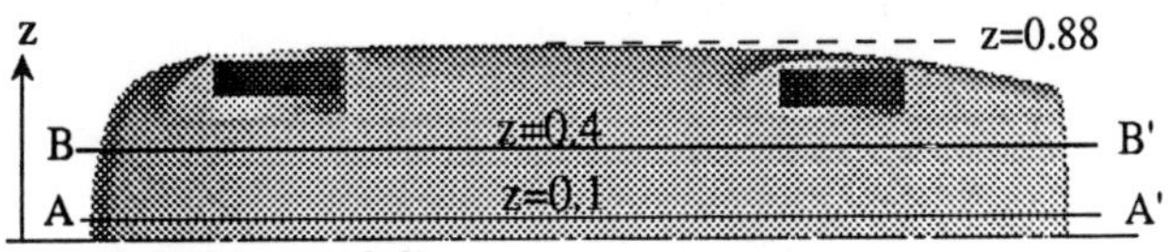

(Central Plane: z=0.0)

Figure 14: Longitudinal Sections (Underside View)

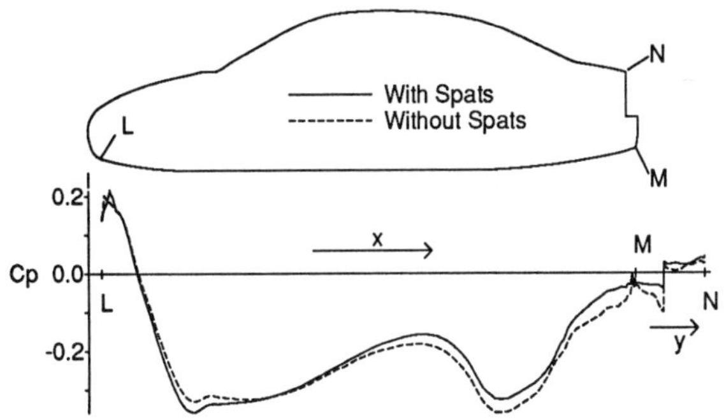

(a) C_p Distribution on A-A' Section ($z = 0.1$)

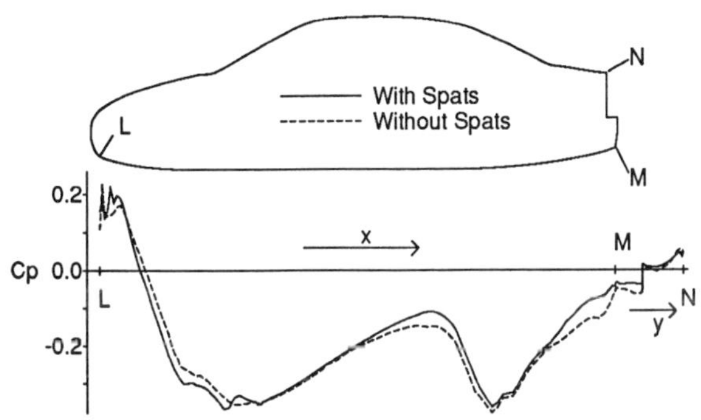

(b) C_p Distribution on B-B' Section ($z = 0.4$)

Figure 15: Change in C_p Distribution with Spats

lower surface viewed from the bottom. It is observed that the streamlines are considerably different depending on whether the spats are attached or not.

CONCLUSIONS

In this paper, an aerodynamic analysis using the overlaid grid system for a newly developed model car has been conducted.

The effects of the main geometrical optimization of this low aerodynamic drag model, i.e., "the lateral tapering of the rear end" and "the spats," were numerically simulated and evaluated. The important features of the flow field, which brought about a significant C_D reduction, are discussed. From the case study on the lateral tapering of the rear end, it was found that the larger the tapering width, the greater the pressure recovery on the rear end surface for the tested range of the tapering, and accordingly, C_D decreased. The difference between C_D's, calculated and measured, is as small as 5 %, and the rate of C_D reduction was also in good agreement. As for the effects of the spats, it was found that the spats brought better pressure distribution on the body surface, especially on the lower surface and the base. Consequently, C_D was reduced by attaching the spats.

In the present study, with the advantage of "the overlaid grid system," only a small effort was required to modify the grid system for the case study once the overlaid grid system was established in the whole computational domain for the first case. Each grid block could be made to have a very good property by taking care that the skewness of the grid lines intersection was decreased as far as possible. In addition, the total number of grid points was saved as compared with the single grid system.

REFERENCES

[1] Benek, J.A., Buning, P.G. and Steger, J.L., "A 3-D Chimera Grid Embedding Technique," AIAA Paper, 85-1523-cp, 1985

[2] Suga, K., Aoki, H. and Shinagawa, T., "Numerical Analysis on Two-Dimensional Flow and Heat Transfer of Louvered Fins Using Overlaid Grids," JSME International Journal, Series II, Vol.33, No.1, 1990

[3] Chessire, G. and Henshaw, W.D., "Composite Overlapping Meshes for the Solution of Partial Differential Equations," J. of Comp. Phys., 90, 1990

[4] Petersson, N.A. and Malmliden, J.F., 'Computing the Flow around a Submerged Body Using Composite Grids," J. of Comp. Phys. 105, 1993

[5] Kawamura, T. and Kuwahara, K., "Computation of high Reynolds number flow around a circular cylinder with surface roughness," AIAA 22nd Aerospace Sciences Meeting, AIAA-84-0340, 1984

[6] Poole, E.L. and Ortega, J.M., "Multicolor ICCG Method for Vector Computers," SIAM J. Numer. Anal., Vol. 24, No. 6, pp.1394-1418, 1987

[7] Doi, S. and Hoshi, A., "Large-numbered multicolor MILU preconditioning on SX-3/14," Intern. J. Computer. Math., Vol. 44, pp.143-152, 1992

[8] Thompson, J.F., Warsi, Z.U.A. and Mastin, C.W., Numerical Grid Generation, Foundations and Applications, North-Holland, 1985

[9] Hucho, W.-H.(Ed.), Aerodynamics of Road Vehicles, Butterworths, 1987

[10] Himeno, R., Takagi, M., Fujitani, K. and Tanaka, H., "Numerical Analysis of the Airflow around Automobiles Using Multi-block Structured Grids," SAE Paper, 900319, 1990

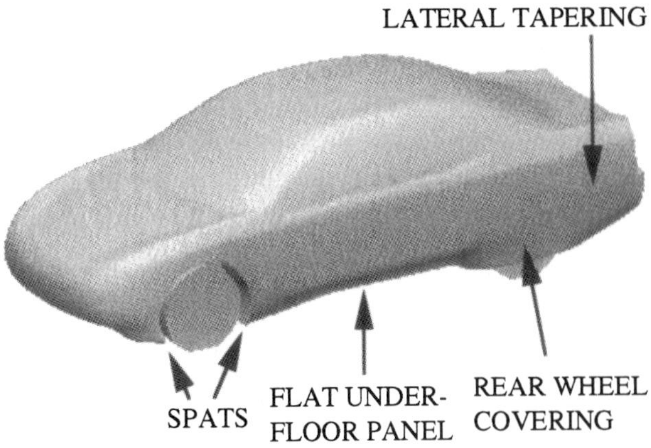

Figure 1: Newly Developed Model Car

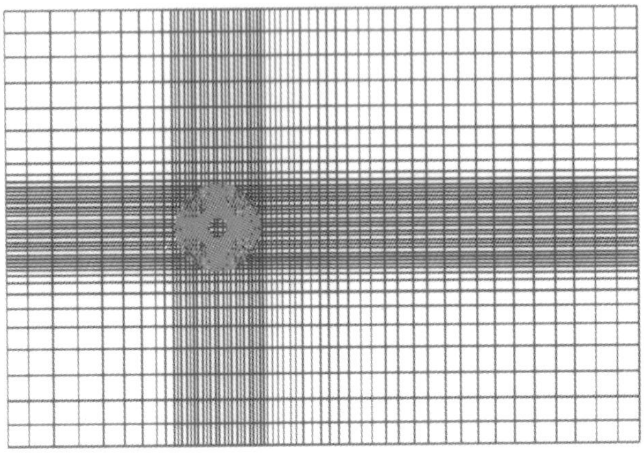

(a) Whole View

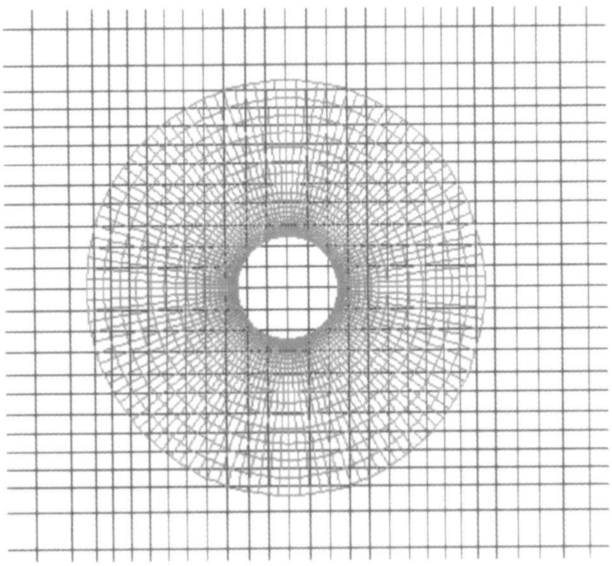

(b) Near View

Figure 4: Example of Overlaid Grid System
(For Simulation of Flow around Circular Cylinder)

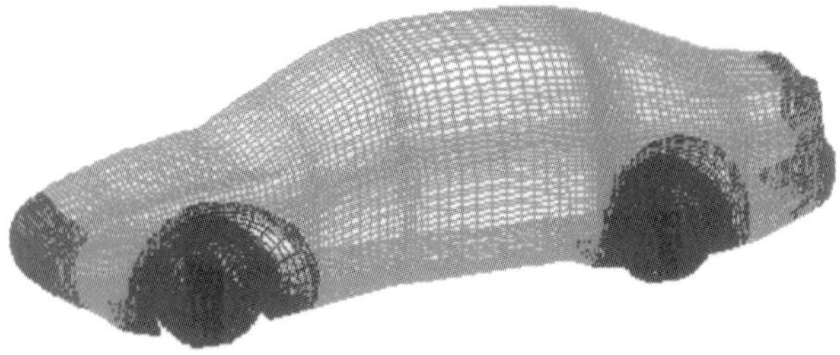

(a) Whole View

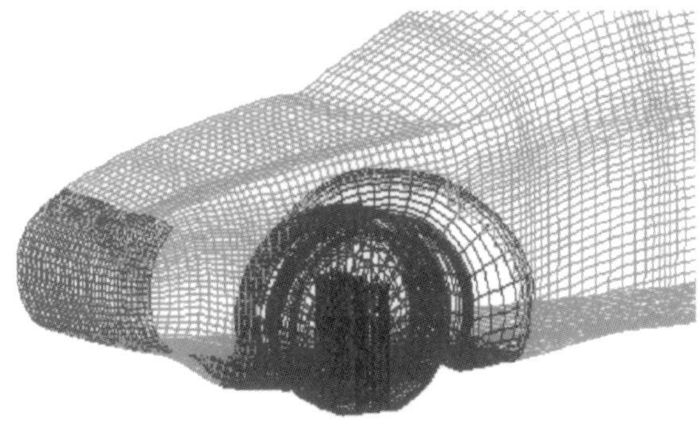

(b) Front View

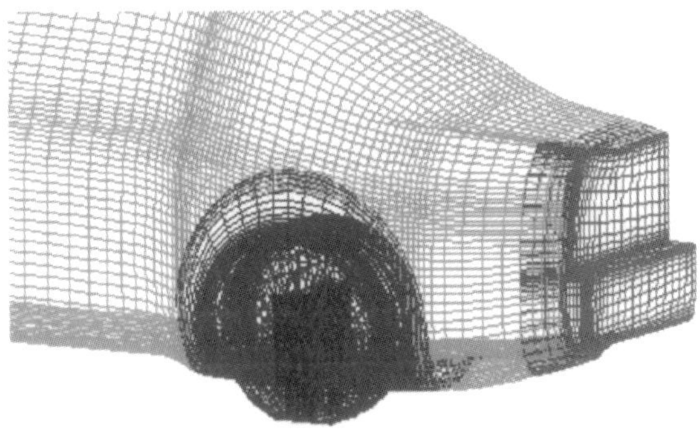

(c) Rear View

Figure 8: Overlaid Computational Grid on Body Surface
(In Half Region)

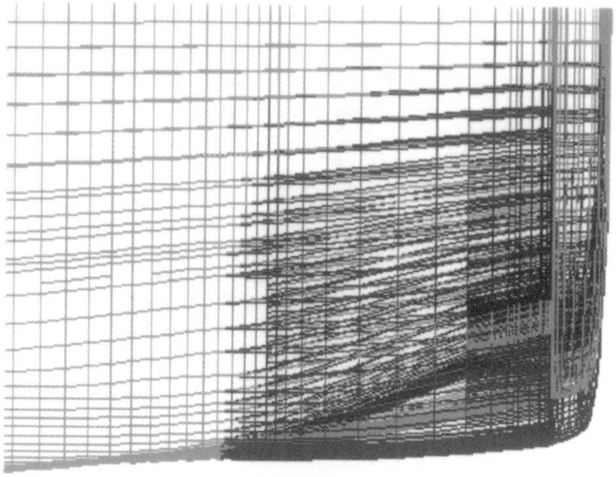

Figure 9: Computational Grids
on Body Surface in Rear End

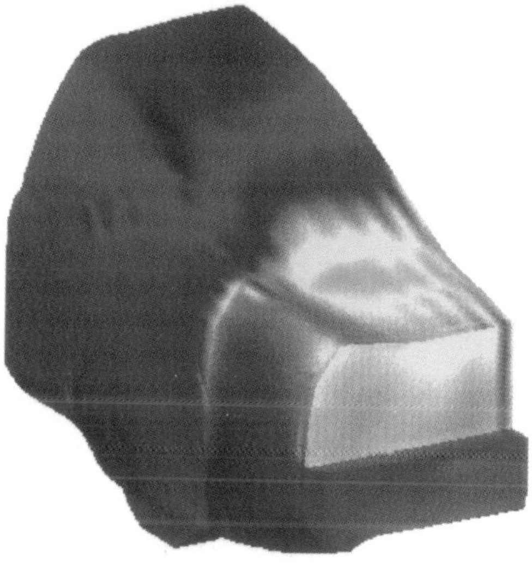

(a) Tapering Width: 0mm

(b) Tapering Width: 80mm

Figure 11: Comparison of Pressure Distribution
(In Half Region)

(c) Tapering Width: 160mm

Figure 11: Comparison of Pressure Distribution
(In Half Region, Continued)

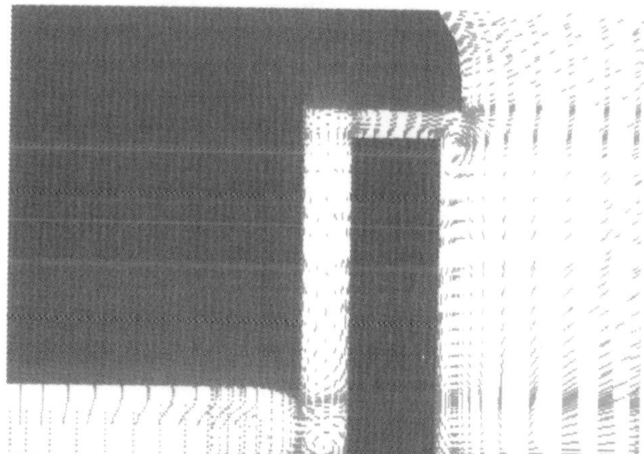

(a) around Front Wheel

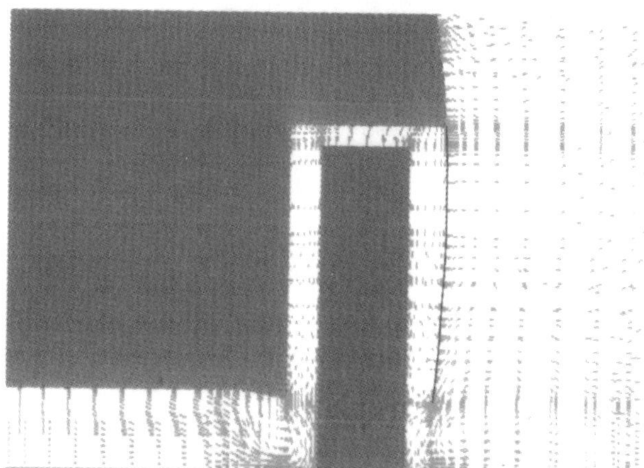

(b) around Rear Wheel

Figure 12: Velocity Vector around Wheel

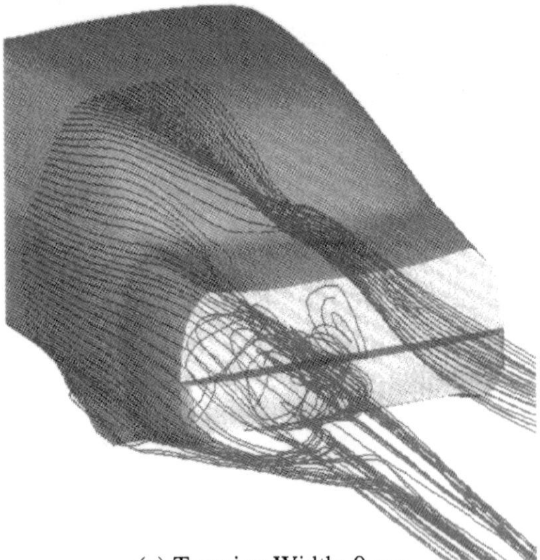

(a) Tapering Width: 0mm

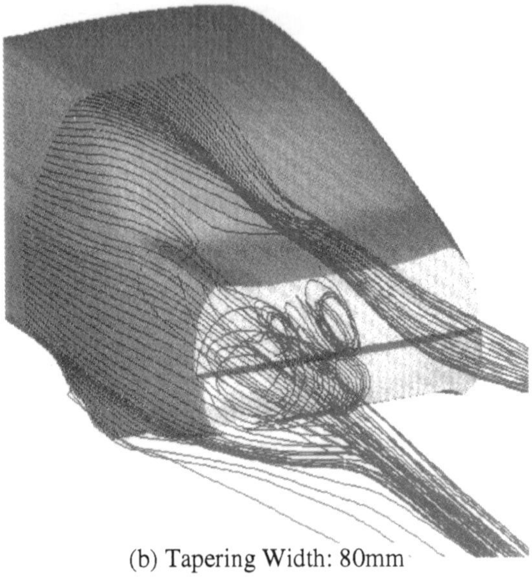

(b) Tapering Width: 80mm

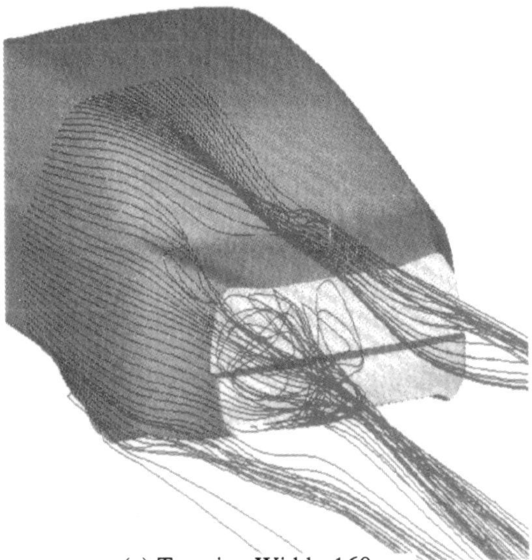

(c) Tapering Width: 160mm

Figure 13: Streamlines in Rear End

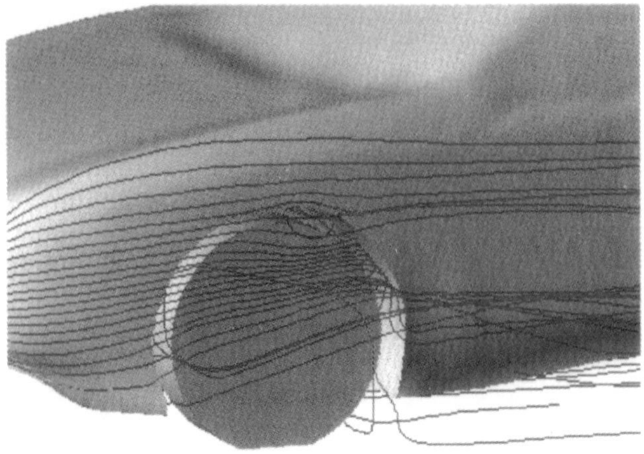

(a) With Spats

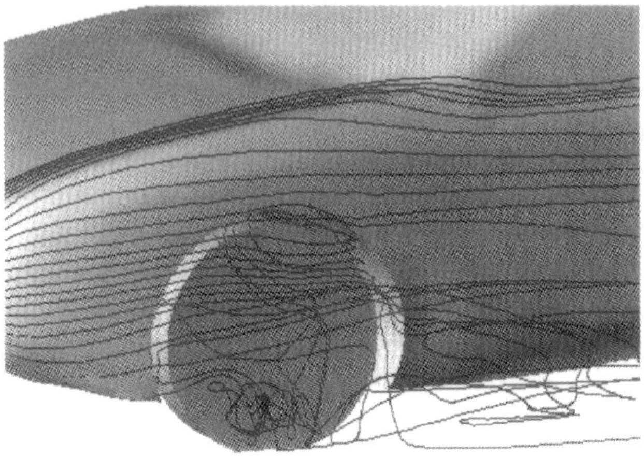

(b) Without Spats

Figure 16: Streamlines around Front Wheel (Side View)

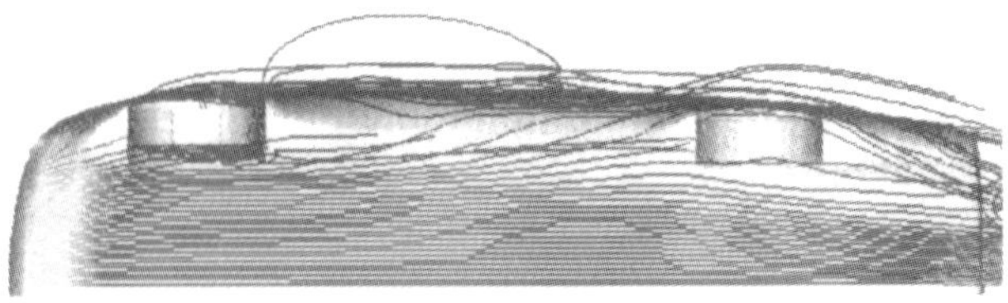

(a) With Spats

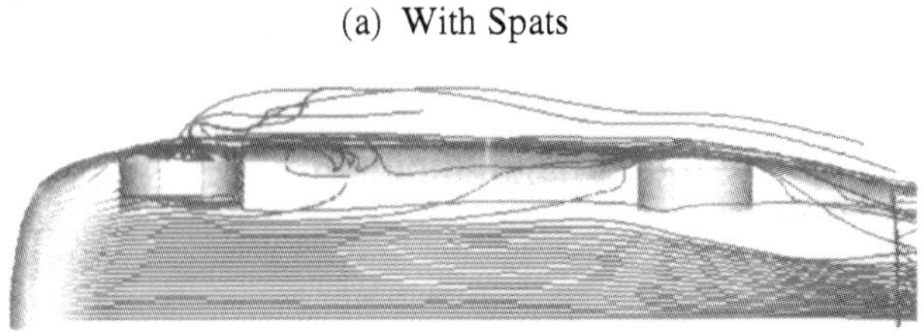

(b) Without Spats

Figure 17: Streamlines around Lower Surface (Bottom View)

931896

Finite-Volume Simulation of 3-D Vortical Flow-Fields About Road Vehicles with Various After-Body Configuration

Kiyohira Aoki and Toshihazu Ohbayashi
Isuzu Motors, Inc.

Ming Zhu and Hideaki Miyata
University of Tokyo

ABSTRACT

The flow structures about a car-like bluff body with a variety of after-body configurations are numerically studied based on a finite-volume solution method with the sub-grid scale turbulence model. The numerical calculations are compared with the experiments in both the water basin and wind tunnel. Different detachment features are demonstrated in the case of different after-body configuration by the numerical analysis and the computer-graphics visualization. Some critical features such as high and low drag states are revealed in the numerical simulation in the case of critical-angle configuration and their mechanism are discussed.

It is well known that there are critical after-body configurations for the road vehicle where the drag force shows a discontinuous change due to the critical angle of back-face slant. This means one could obtain more than two states at the critical angle, high drag and low drag, such as those of a circular cylinder during the transient Reynolds numbers. This kind of phenomenon is firstly found in the pioneer work of Hucho [1]. In the development of a hatch-back car, he found that there are drastic changes of drag force when varying the angle of back-face slant. When the back-face slant angle becomes about 30 degrees, i.e., the critical angle of back-face slant, the drag force suddenly changes its value. Morel [2] studied in detail on this phenomenon in his experiments using a bluff body. He found that there is critical geometry for the after-body of bluff body and showed that when the back-face slant angle is critical, there is two states in the measurement, i.e., high drag and low drag state. Bearman [3] recognized the existence of the critical geometry, and investigated the longitudinal vortex structure in the wake measurement. He showed that the drag force can be estimated by means of wake survey. He also estimated the drag component due to vortex motion and noted the correlation between the drag force and the vortex structure. Ahmed et al. [4] at first time measured both the time-averaged velocity and surface pressure about the so called "Ahmed model" in series test by changing the angle of back-face slant. He demonstrated the different vortex structure on the back-face with respect to the different drag state at the critical angle. Nouzawa et al. [5] defined a slant angle of effective back-face which is in a plane including the roof end and the trunk end for the notchback type bluff bodies. They showed that for the notchback shape ground vehicles there are also critical angles of back-face slant about 25°. They reported that there is arch type vortex shedding from the rear window in the experimental observation. At this time, all of experimental studies are based on the time-averaged measurements except the experimental visualization. Therefore, because of the difficulties and uncertainty of the experiment for the unsteady and three-dimensional flow-fields, the mechanism of producing the critical state have not yet been clarified. This is really complex phenomenon of flow detachment, in which interactions play an important role among the shed unsteady vortices and those between the vortex and mean flow.

Computational Fluid Dynamics (CFD) based on the Navier-Stokes solution has been introduced into the vehicle aerodynamics for almost one decade. Although many techniques have been developed and the color computer graphics show the possibility of understanding the complex flow feature [9][10], however, still very few is devoted to the elucidation of the mechanism of flow structure. Han [7] calculated the three-dimensional turbulent flow about the Ahmed model using k-ε two equation turbulence model. Recently, Han and his colleagues developed a design method based on his numerical prediction method [8]. However, it seems that his calculation does not fully succeed in revealing the high state of drag force at the critical angle in case of Ahmed model. On the other hand, Nouzawa et al. [11] showed a time-dependent numerical simulation without turbulence model to study the unsteady wake about a notchback model. They showed that the high drag state takes place only once after the acceleration of the flow about the vehicle and demonstrated that the arch type vortex is shed from the rear window. Hashiguchi et al. reported similar results in their numerical simulation for the Ahmed model [6]. However, it is questionable that the high drag state appeared only once just after the acceleration, where the reverberations of pressure wave due to the acceleration are possible to still exist. Being different from the flow about the front face of the road vehicle, the flow about the rear window is composed of the fully developed turbulent boundary layer beyond the roof and on the side body, and the breakdown of these boundary layer, i.e., the three-dimensional flow detachment. Regarding the degree of grid resolution in their calculation, it seems that such kind of unsteady turbulent phenomena are not well predicted. In fact, the predicted drag forces versus the angle of rear window slant in their calculations show a different features from the experiment.

The authors developed a finite-volume solution method based on the Navier-Stokes equation called WISDAM-V method [12],[14] for the unsteady, incompressible viscous flow past a road vehicle with practical body configuration. In the numerical method, the Navier-Stokes equations and the continuity equation are discretized using the finite-volume technique in the curvilinear coordinate system and solved with a time-marching solution method. A sub-grid scale turbulence model is incorporated and validated in the case of a viscous flow about a modified spheroid of revolution investigated by Patel et al. [13]. Series of numerical tests on both notchback and hatchback type car models are performed and excellent agreement is obtained for the drag force in almost all cases with the experimental results in both water basin and wind tunnel. It is noted that the similar fluctuation of the lift force is present in the experiment. Especially, the present method seems to be successful in revealing the high and low drag state in time for the critical geometry and demonstrated the differences of flow structure about the rear window.

NUMERICAL METHOD AND ANALYSIS METHOD

NUMERICAL METHOD - The numerical method is called WISDAM-V developed by Zhu [12]. In this method, the Navier-Stokes equations and the continuity equation for the unsteady, incompressible fluids are discretized using finite-volume techniques in the curvilinear coordinates.

A curvilinear coordinate system is employed. The so-called O-type grids are used for the transverse section and H-type for the longitudinal planes. That is curvilinear coordinates (ξ,η,ζ) are chosen so that ξ directs in the longitudinal direction, η in the radial, and ζ in the circumferential.

The metrics of the three-dimensional curvilinear coordinates are defined using area vectors and are used for the formulation of the governing equations. Cartesian coordinates are chosen as the basic coordinate system of the physical region and the governing equations are formulated in a partial transformation approach. Suppose that the position vector of a coordinate system is

$$\mathbf{r} = \mathbf{r}\left(x^1,x^2,x^3\right) \quad (1)$$

then the transformation of the coordinate system between the physical region and the transformed region is

$$\text{grad}_\xi \mathbf{r} = \left(\frac{\partial \mathbf{r}}{\partial \xi^1}, \frac{\partial \mathbf{r}}{\partial \xi^2}, \frac{\partial \mathbf{r}}{\partial \xi^3}\right)^T \quad (2)$$

Hence the Jacobian of (2), which corresponds to the local volume about that definition point, becomes

$$J = \det\left(\text{grad}_\xi \mathbf{r}\right) \quad (3)$$

The element of the inverse matrix of (2) is

$$\frac{\partial \xi^l}{\partial x^p} = \frac{1}{J}\left[\frac{\partial x^s \partial x^t}{\partial \xi^m \partial \xi^n} - \frac{\partial x^t}{\partial \xi^m}\frac{\partial x^s}{\partial \xi^n}\right] \quad (4)$$

where (l, m, n) and (p, s, t) are in cyclic order. In the ordinary finite-difference method, equations (2) and (4) are adopted as a pair of transforming metric coefficients. In the present paper, in order to employ a finite-volume approach for the formulation of the governing equations, the area vector is adopted instead of equation (4):

$$S^l_p = J\frac{\partial \xi^l}{\partial x^p} = \left[\frac{\partial x^s}{\partial \xi^m}\frac{\partial x^t}{\partial \xi^n} - \frac{\partial x^t \partial x^s}{\partial \xi^m \partial \xi^n}\right] \quad (5)$$

where the superscript of the area vector expresses its respective contravariant direction, while the subscript means its Cartesian components; that is, an area vector is defined for the respective contravariant direction which is normal to the surface of the local volume element, and its value is the area of the surface of the local volume. The vector form for the area vector $\mathbf{S}^m$ in Cartesian coordinates is

$$\mathbf{S}^m = \left(S^m_1, S^m_2, S^m_3\right) \quad (6)$$

Therefore, the Jacobian J and the area vectors $\mathbf{S}^m$ are defined as the geometric coefficients in the WISDAM-V.

The governing equations are made dimensionless with respect to a constant reference velocity U_0 and the vehicle length L, and the Reynolds Number is defined as

$$R_e = \frac{U_0 L}{\nu} \quad (7)$$

For the incompressible viscous flow, Navier-Stokes equation can be written in the following conservation form:

$$\frac{\partial \mathbf{u}}{\partial t} + \text{div}\,\mathbf{T} = \mathbf{F} \quad (8)$$

where $\mathbf{F}$ is the external force and is neglected in this study. $\mathbf{T}$ is defined as

$$\mathbf{T} = p\mathbf{I} + \mathbf{uu} - \frac{1}{R_e}\,\text{def}\,\mathbf{u} + \overline{\mathbf{u}'\mathbf{u}'} \quad (9)$$

where $\mathbf{I}$ is the unit tensor, p is the pressure divided by fluid density, and the fluid deformation is defined as

$$\text{def}\,\mathbf{u} = \text{grad}\,\mathbf{u} + \left(\text{grad}\,\mathbf{u}\right)^T \quad (10)$$

Therefore the stress tensor $\mathbf{T}$ is composed of the normal stress of the pressure, nonlinear stress, viscous stress, and sub-grid scale Reynolds stress.

By separating the pressure term from others by the MAC-type solution method, and assuming that external force is negligible, equation (8) becomes

$$\frac{\partial \mathbf{u}}{\partial t} = -\,\text{grad}\,p + f \quad (11)$$

and

$$f = \text{div}\,\overline{\tau} \quad (12)$$

Then the stress in equation (12) are

$$\overline{\tau} = -\,\mathbf{uu} + \frac{1}{R_e}\,\text{def}\,\mathbf{u} - \overline{\mathbf{u}'\mathbf{u}'} \quad (13)$$

By use of the Jacobian of equation (3) and the area vector of equation (5), equation (11) and (12) are described in tensor form as

$$\frac{\partial u^i}{\partial t} = -\frac{1}{J}\frac{\partial}{\partial \xi^j}\left(S^j_k p \delta^{ik}\right) + f^i \tag{14}$$

and

$$f^i = \frac{1}{J}\frac{\partial}{\partial \xi^j}\left(S^j_k \tau^{ik}\right) \tag{15}$$

where the stress tensor of equation (15) becomes

$$\tau^{ij} = -u^i u^j + \frac{1}{R_e}\left(2e^{ij}\right) - \overline{u'^i u'^j} \tag{16}$$

and the strain tensor is

$$e^{ij} = \frac{1}{2}\left(\frac{\partial u^i}{\partial x^m}\delta^{mj} + \frac{\partial u^j}{\partial x^n}\delta^{in}\right) \tag{17}$$

Suppose that the sub-grid scale Reynolds stress of equation (9) is defined by analogy to an eddy viscosity concept:

$$\overline{u'^i u'^j} = \frac{2}{3} k\delta^{ij} - 2\nu_s e^{ij} \tag{18}$$

where ν_s is the subgrid-scale eddy viscosity, which will be defined in the following section, k is the turbulent kinematic energy, and e^{ij} is the strain tensor defined by equation (17). Therefore equation (14) to (16) can be rewritten as

$$\frac{\partial u^i}{\partial t} = -\frac{1}{J}\frac{\partial}{\partial \xi^j}\left(S^j_k H \delta^{ik}\right) + \frac{1}{J}\frac{\partial}{\partial \xi^j}\left(S^j_k \tau^{ik}\right) \tag{19}$$

in which turbulent kinematic energy is included into the pressure term, and the sub-grid scale eddy viscosity term of equation (18) into the viscous term,

$$H = p + \frac{2}{3} k \tag{20}$$

and

$$\tau^{ij} = -u^i u^j + \left(\frac{1}{R_e} + \nu_s\right)\left(2e^{ij}\right) \tag{21}$$

Therefore in the present study, H is solved instead of the pressure p.

The continuity equation for the incompressible flow is

$$\mathrm{div}\,(\mathbf{u}) = 0 \tag{22}$$

and its vector form is written as

$$\frac{1}{J}\frac{\partial}{\partial \xi^j}\left(S^j_k u^k\right) = 0 \tag{23}$$

Pressure and velocity components are defined in the staggered manner as shown in Fig.1. As shown in Fig.2, the momentum conservation of equation (19) to (21) is expressed as the divergence of momentum fluxes $(S^i \cdot \overline{\tau})$ evaluated at the respective velocity point, while the mass conservation of equation (23) is expressed as the zero-divergence of mass fluxes $(S^i \cdot \mathbf{u})$ estimated at the pressure point.

The MAC-type algorithm is used. Velocity is updated by semi-implicit time-marching method in which the time derivative is differenced explicitly for the convection term while semi-implicitly for the diffusion term in a manner of the fractional-step method. The pressure and the velocity are solved by a simultaneous iterative method.

Suppose that the velocity field is determined for the n-th time level, and the convection term **C** and the diffusion

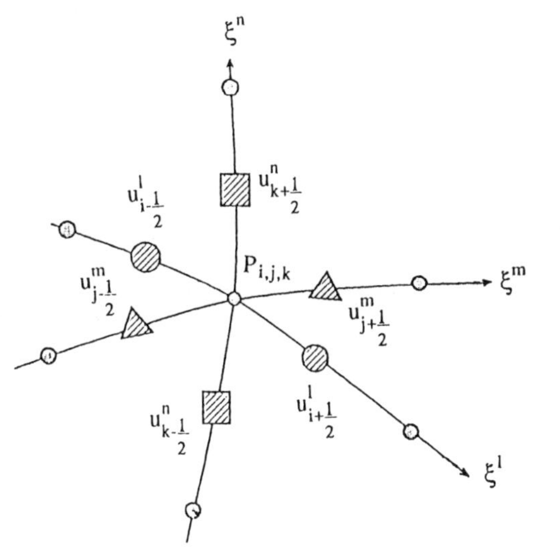

Fig.1 Definition Sketch of velocity and pressure points.

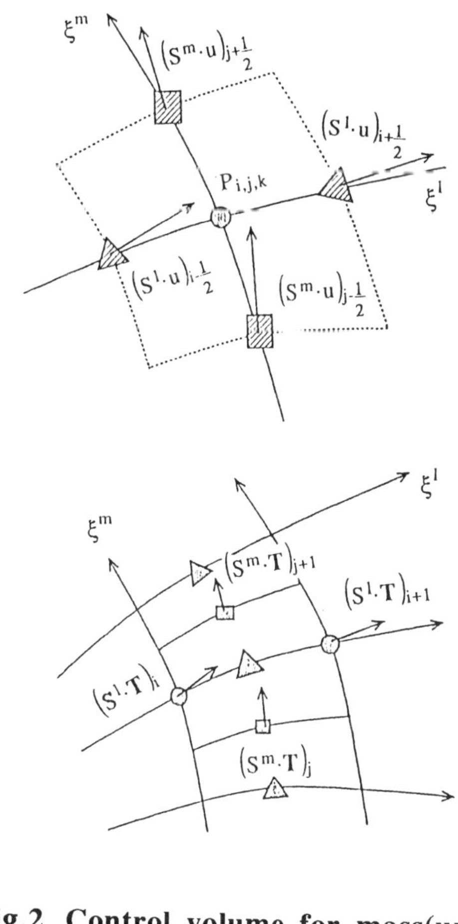

Fig.2 Control volume for mass(upper) and momentum(lower) conservation.

term **D** are calculated in equation (21). Then the semi-implicit time-marching formulation can be factorized approximately by a factor λ as

$$\frac{\mathbf{u}^{(n+1)} - \mathbf{u}^{(n)}}{\Delta t} = -\,\mathrm{grad}\,(H) + \mathbf{C}^{(n)} + \mathbf{D}^{(n)}_c + (1-\lambda)\mathbf{D}^{(n)}_d + \lambda \mathbf{D}^{(n+1)}_d \tag{24}$$

where the superscript denotes the time level and subscripts (d) and (c) denotes the diagonal and cross elements of the diffusion term, respectively. When λ is set at 1, it corresponds to the fully implicit time-marching formulation. Then the velocity changes in the right-hand-side of the equation (24) except the pressure term becomes

$$\delta\tilde{\mathbf{u}} = \Delta t\left[\mathbf{C}^{(n)} + \mathbf{D}^{(n)}_c + (1-\lambda)\mathbf{D}^{(n)}_d + \lambda \mathbf{D}^{(n+1)}_d\right] \tag{25}$$

Then the equation of updating velocity is written as

$$\mathbf{u}^{(n+1)} = \mathbf{u}^{(n)} - \Delta t\ \mathrm{grad}\,(H) + \delta\tilde{\mathbf{u}} \tag{26}$$

In the present study, the diagonal element of diffusion term with the eddy viscosity at the (n+1) time level $\mathbf{D}^{(n+1)}_d$ is defined as

$$\mathbf{D}^{(n+1)}_d = \left(\frac{1}{R_e} + \nu_s\right) g^{ii} \frac{\partial^2 \mathbf{u}^{(n+1)}}{\partial \xi^i\, \partial \xi^i} \tag{27}$$

where g^{ij} is the contravariant metric tensor, which is defined as

$$g^{ij} = \delta^{mn} \frac{\partial \xi^i}{\partial x^m} \frac{\partial \xi^j}{\partial x^n} \tag{28}$$

By setting $\lambda = 1$, from the equation (26) and (27), equation (25) can be written into the following " delta form " as

$$\delta\tilde{\mathbf{u}} = \Delta t\, \mathbf{f}^{(n)} + \Delta t\left(\frac{1}{R_e} + \nu_s\right) g^{ii} \frac{\partial^2 \delta\tilde{\mathbf{u}}}{\partial \xi^i \partial \xi^i} \tag{29}$$

where $\mathbf{f}^{(n)}$ is the divergence of the stresses defined in equation (21) at the n-th time level. Then the velocity change $\delta\tilde{\mathbf{u}}$ is solved in the following equation by the implicit approximate factorization method:

$$\left(1 - \Delta t\left(\frac{1}{R_e} + \nu_s\right) g^{ii} \frac{\partial^2}{\partial \xi^i \partial \xi^i}\right)\delta\tilde{\mathbf{u}} = \Delta t\, \mathbf{f}^{(n)} \tag{30}$$

Then the velocity and the pressure are solved by a simultaneous iterative method in the following way:

$$\hat{\mathbf{u}} = \mathbf{u}^{(n)} - \Delta t\ \mathrm{grad}\left(H^m\right) + \delta\tilde{\mathbf{u}} \tag{31}$$

$$H^{(m+1)} = H^{(m)} - \frac{\omega}{|\nabla^2|} \frac{\mathrm{div}\,\hat{\mathbf{u}}}{\Delta t} \tag{32}$$

where the superscript (m) indicates the iteration level, ω is the relaxation factor, **u** is the velocity predictor, and $|\nabla^2|$ is the magnitude of normalized Laplacian operator of the Poisson equation. When the pressure is converged, the predicted velocity will be the velocity at the (n+1) time level. In the present study, ω is set at 0.8, and the pressure is assumed to be converged at every time step when

$$\left|H^{(m+1)} - H^{(m)}\right| \le 5\times 10^{-5} \tag{33}$$

This algorithm has the advantage of fast convergence. Each time level normally requires less than ten iterations.

As for the boundary conditions, no-slip condition is imposed on the body surface and no wall function is used. Zero normal gradient condition is imposed on the outer boundary except the ground boundary. Uniform velocity is given on the inflow boundary, and zero normal gradient is given on the outflow boundary. On the ground boundary, the uniform velocity or the vehicle's advancing speed is imposed. This corresponds to the condition of vehicle advancing steadily in the still air and on the stationary ground. As for the initial condition, the air is stationary and then gradually accelerated in 1000 time steps.

ANALYSIS METHOD - Result of computation consists of basic variables, i.e. pressure and velocity vector for each time step of the computation. Aerodynamic force is calculated from the integration of tangential and normal stresses over the body surface. Therefore frictional drag can also be calculated.

In order to obtain the understanding of the unsteady vortical flow, helicity (34) and Lamb-vector (35) are introduced

$L_s = \omega \cdot \mathbf{u}$ (34) .... Helicity or Lamb-scalar

$L_v = \omega \times \mathbf{u}$ (35) .... Lamb-vector

Helicity or Lamb-scalar represents the vorticity component in the direction of the velocity while the Lamb-vector represents the vorticity component perpendicular to the velocity vector.

TURBULENCE MODEL

SGS TURBULENCE MODEL - It is beyond the present computer capacity to treat all the eddies for the practical high Reynolds number concerning to flow field around vehicle directly by fine mesh size. Therefore turbulence models are necessary. The k-ε 2-equation model has been most popular and useful for the practical analysis. Relatively light load on computation of the model even enables to treat problems of design optimization.[8] However many works report that it would be difficult to treat separated flow by the k-ε model. In the computation for the Ahmed model series with k-ε model, T. Han [7] reports that the discrepancy of Cd between computation and experiment increases as the back-face slant angle increases. Moreover, the Reynolds-averaged models such as the k-ε model might have the limitation due to the time-averaged concept. Since the main purpose of the present paper is to resolve the time-dependent motion of large scaled vortices behind a vehicle, the time-averaged model will not fit to the purpose. Some of recent works in Japan[9][10] use a kind of direct method, where they introduce a higher-order upwind differentiation of the advection term of the Navier-Stokes equations instead of introducing turbulence model. However the validation of the method is left to be carried out.

Another model called sub-grid scale (SGS) model which is used in the large eddy simulations seems to have more versatile properties, having simple equations and the least number of empirical constants. The previous studies of Zhu [12] and of Miyata [14] show the advantages of the use of SGS model for the wake flow.

The SGS model used in the present study is described following Smagorinsky et al. [15] as:

$$\nu_s = L_s^2\left(2\overline{e^{ij}}\ \overline{e_{ij}}\right)^{1/2} \tag{36}$$

where ν_s is eddy viscosity appeared in (18), and L_s is the length scale. L_s is set as

$$L_s = \min\left(0.1V^{1/3},\ 0.5\Delta x^1,\ 0.5\Delta x^2,\ 0.5\Delta x^3\right) \quad (37)$$

As shown by the above formula, since the length scale is dependent on the mesh size or mesh division, and of course on the Reynolds number (Re) of the phenomenon, the empirical constants are to be tuned with the help of experimental results. The Van-Driest type damping function (38) is introduced near the wall.

$$D_f = \left[1 - \exp\left(- y^+/A^+\right)\right] \quad (38)$$

Here, y^+ is the viscous unit length from the body surface and A^+ is set at 26.0. Then (37) becomes as:

$$L_s = D_f \min\left(0.1V^{1/3},\ 0.5\Delta x^1,\ 0.5\Delta x^2,\ 0.5\Delta x^3\right) \quad (39)$$

TUNING OF SGS MODEL - For the tuning and the validation of the turbulence model and the present computational method, several test computations are carried out varying the grid spacing and other factors, such as the length scale of SGS model, for the flow around the modified spheroid and computed results are compared with the experimental results of Patel et al. [13].

The modified spheroid is a six-to-one prolate spheroid with conical tail piece. The Reynolds number based on the body length and uniform flow velocity in the experiment is 2.75×10^6 and is common with the computation. Grid system is shown in Fig.3. Number of grid points in ξ, η, ζ direction is 141, 26,and 37 respectively. The minimum mesh spacing of the innermost velocity point from the body surface is 5×10^{-5}, and time increment is set at 1×10^{-4}. Computation is carried out until the dimensionless time T reaches 2.5.

Fig.4 shows the radial distribution of the computed axial velocity at the longitudinal position of 90%, 93%, and 96% of the body length downstream from the fore end, and longitudinal distribution of the surface pressure is presented in Fig.5.

Introduction of the Van-Driest damping function into the length scale much improves both the velocity and pressure distribution. Although pressure distribution shows fairly good coincidence with the experiment, the velocity distribution needs to be more improved near the body surface for the better prediction of frictional drag. However, because pressure drag is dominant in the case of the bluff body as shown by Ahmed [4] , the present method will be effective for the prediction of drag force of the vehicle-like bluff body.

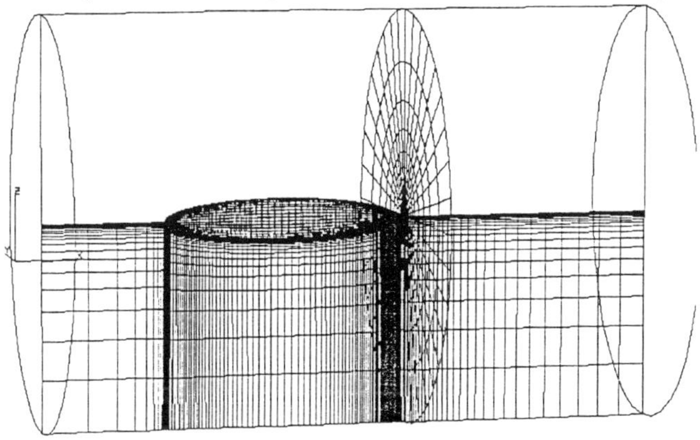

Fig.3 Computational grid for modified spheroid

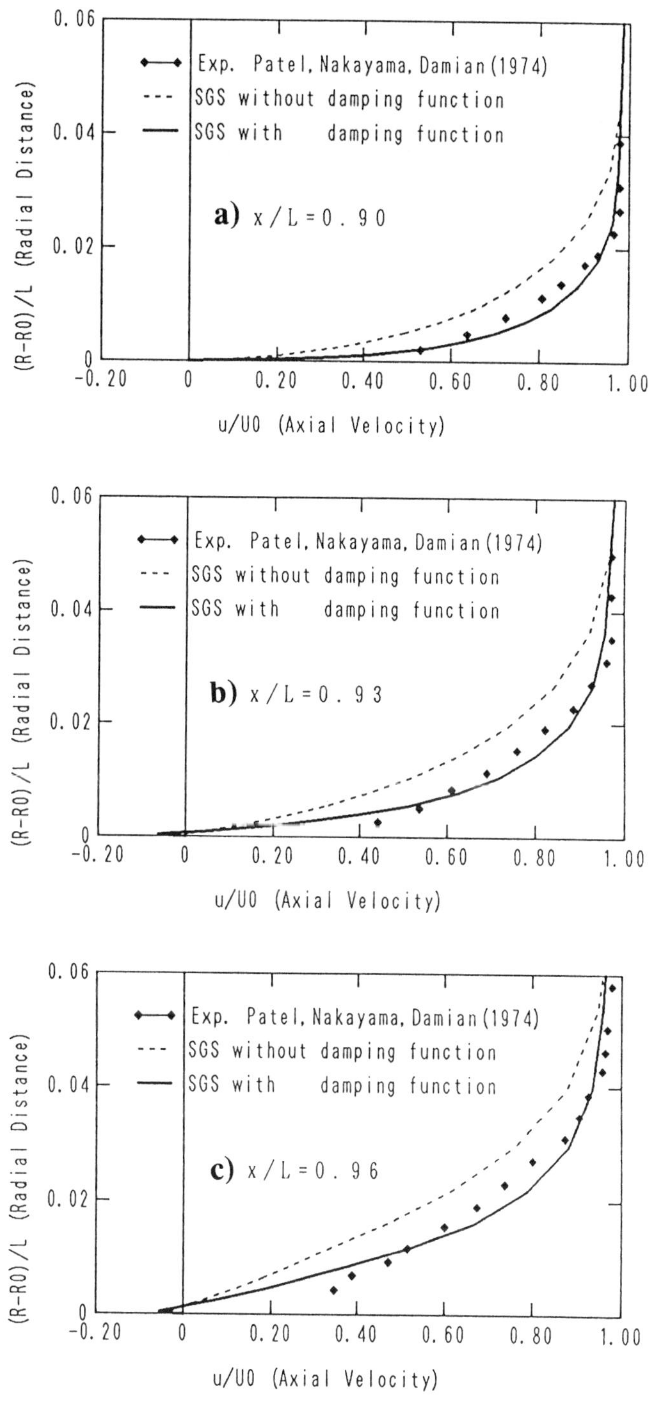

Fig.4 Comparison of radial distribution of the axial velocity for modified spheroid. a).Longitudinal position at 90% of the body length from the fore end. b). same at 93%. c). same at 96%

SIMULATION AND ANALYSIS

CONDITION OF SIMULATION - Model configurations for both numerical and experimental tests are shown in Fig.6, in which the series of the tested model are listed. The model has a simplified car-like body configuration with no wheels, with no wheel openings, and with flat floor. Openings for air-intake and other miscellaneous details are omitted. The fore half body configuration is chosen from a contemporary production car which is aerodynamically refined and accordingly generates small separation and vortices. This is convenient to make clear the effects of the after-body configurations on the aerodynamic force. Although after-bodies are composed of flat planes and straight lines, practicability is fully taken into consideration. The roof end position is determined from the head point of the passenger in the rear seat and the trunk deck height is limited by the rearward visibility and cargo space. After-body configurations are characterized by the two parameters α and β following Carr [16]. Parameter α is a rear window inclination angle and β is an angle between the horizontal and the line connecting the roof end and the rear deck end point. In the case of notch-back (N/B) model, β indicates the height of trunk deck, and $\alpha = \beta$ in the case of hatch-back(in this paper, it is called as slant-back and noted as S/B). All the S/B models and three N/B models with different trunk deck height are provided for computation. The road clearance is 0.208h for the N/B models and 0.381h for the S/B models, where h is the body height(distance between floor of the body and roof top).

In the curvilinear coordinate system (ξ, η, ζ), 111×26×73 grid points are employed. Fig.7 is an example of computational grid for model No.4. The Reynolds number based on body length is set at 5×10^5 and the minimum grid spacing in the η direction is 5×10^{-4}. Computations are carried out until the dimensionless time based on the model length and the velocity of the uniform flow reaches T=11.6 in the case of N/B and T=12.0 in the case of S/B. In the case of S/B model No.8 which is a critical geometry with 30° slant, the computation is carried out until T=20.0.

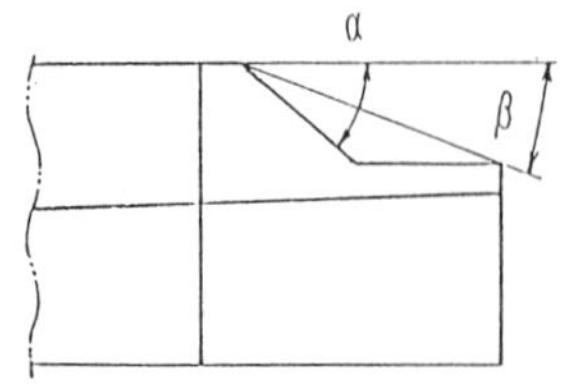

	α (°) 0	5	10	15	20	25	30	35	40	45	50
β (°) 0	1										
5											
10											
15									2		
20					3		4		5		6
25									7		
30							8				
35											
40									9		
45											
50											

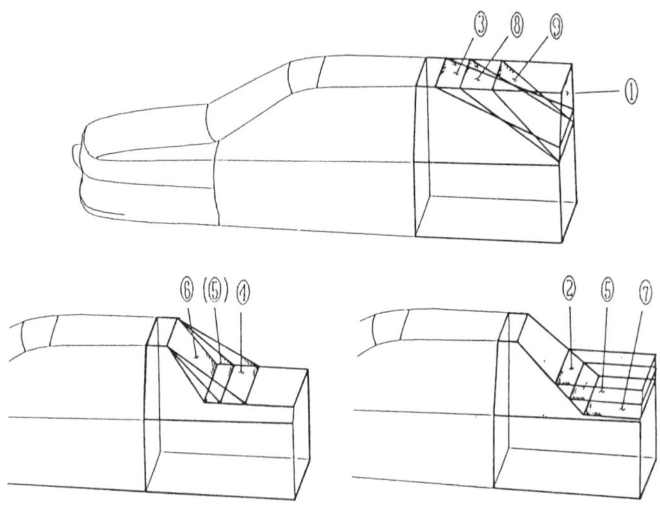

Fig.6 Series test cases for car model

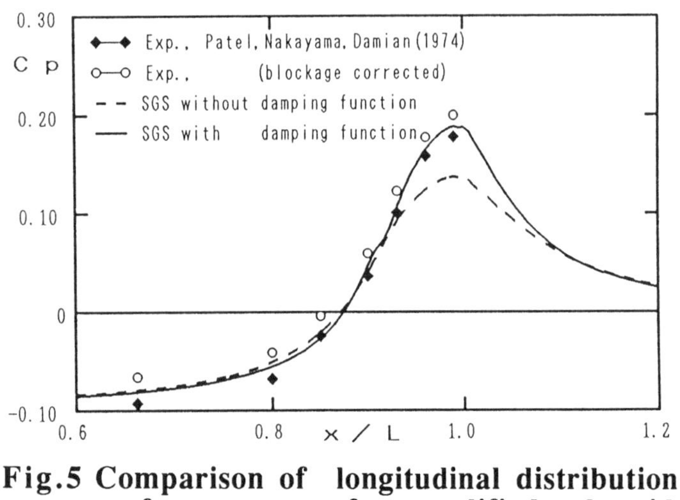

Fig.5 Comparison of longitudinal distribution of the surface pressure for modified spheroid.

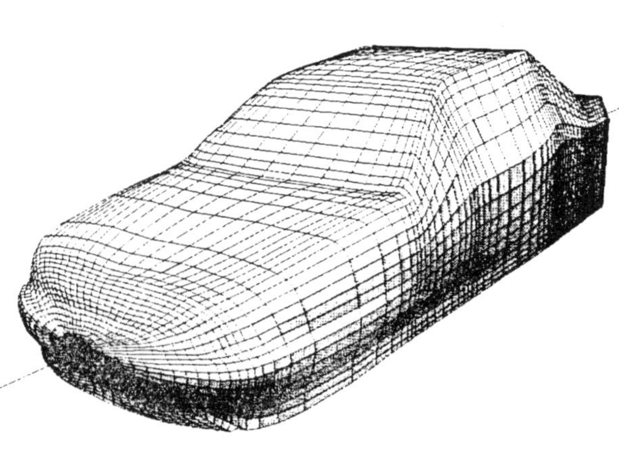

Fig.7 Grid system for car model.

COMPARISON WITH EXPERIMENT - For the validation of the computation, water basin test and wind tunnel test with a moving belt are carried out using 1/5 scale models.

Water Basin and Wind Tunnel Test - Water basin test method [17] is developed for the simulation of moving ground effect. The model is fixed to the towing carriage and towed in the still water on the bottom wall of the basin which represents the ground. Because of the existence of the free surface, wave making is brought about. Therefore, in order to remove the influences of wave making from the measurements, the conditions for towing speed and submergence are important.

The detail of the method is present in the author's previous paper[17], and the test conditions in this paper are the same as those shown in the previous paper. That is, towing speed is set at 0.7m/sec. and the submergence(distance between roof and the free surface) is at 0.7m. The dimensions of the 1/5 scale model are 0.85 m in length, 0.34 m in width, and 0.242 m in height. Then the Reynolds number in the water basin test is 5×10^5 and the Froude number based on the submergence is 0.24.

Use is made of the wind tunnel of the Göttingen type with open test section. Model is supported by the strut with wing-like section. Moving belt is used for the realization of the moving ground effect. The Reynolds number is at 5×10^5 and at the wind-speed 8.7m/sec. The detail of the test method is also present in the Ref.[17].

Comparison with Experiment - Fig.8 shows the computed time history of the Cd(drag coefficient) of the N/B models. When T<3, the influence of acceleration still remains. Therefore the averaged Cd and Cl are calculated for T>3.

Fig.9 shows the comparison of computed Cd with the measured one and Fig.10 shows that of Cl. As for model No.8 (critical geometry), Cd and Cl has two states in the experiment and they are encountered at random. In Fig.9 and Fig.10, averaged Cd and Cl are shown.

In Figs.9 and 10, computed Cd shows good coincidence with the measurements both in the water-basin and in the wind tunnel. On the other hand, Cl is not so well predicted as Cd, showing smaller value than that of experiment and the discrepancies are large in the case of notchback type(model No.2, No.4, & No.7). Although some discrepancies exist in lift, the over-all tendencies of the variations of the forces due to the differences of configurations are correctly reproduced in the computation. Therefore it is expected that the differences of the flow field due to the large scale vortical motions are correctly reproduced to some extent in the computation. In the following sections, the unsteady flow field is investigated through the visualizations of the computed results.

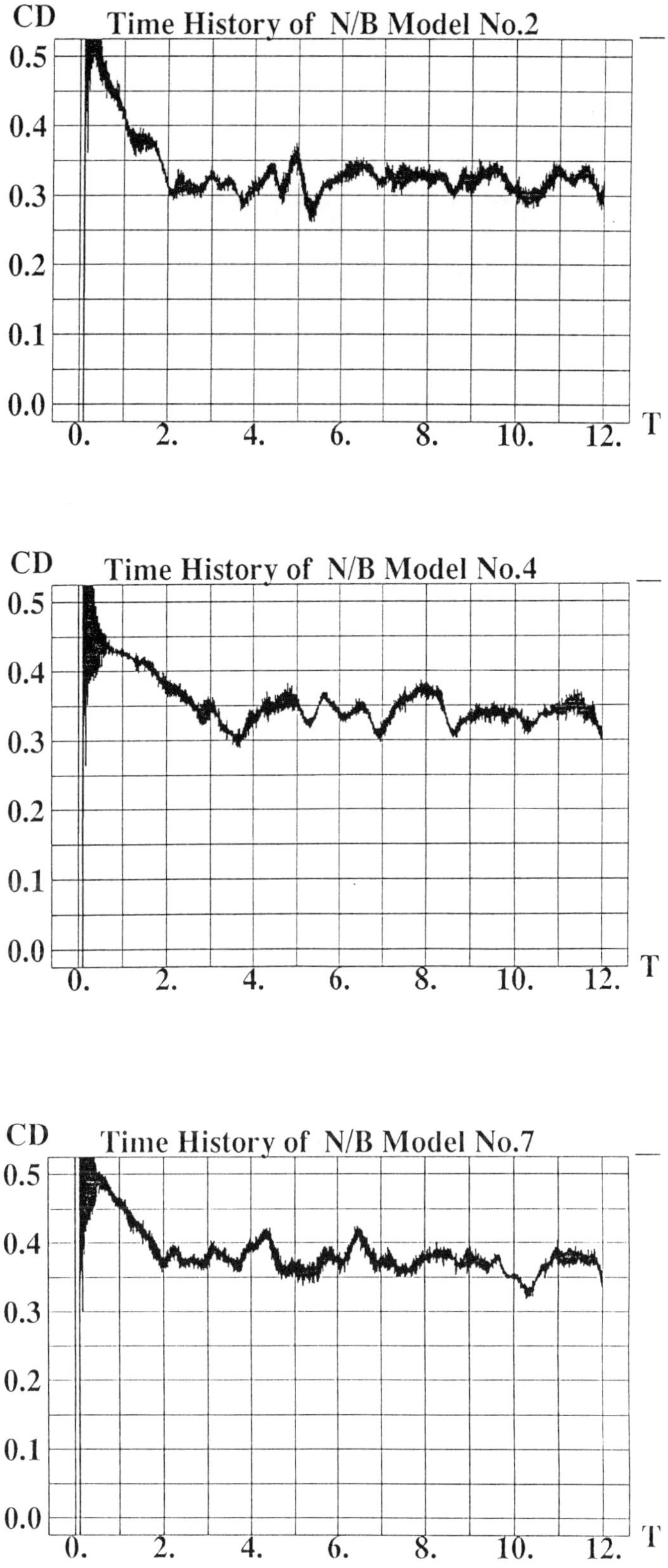

Fig.8 Time history of drag.

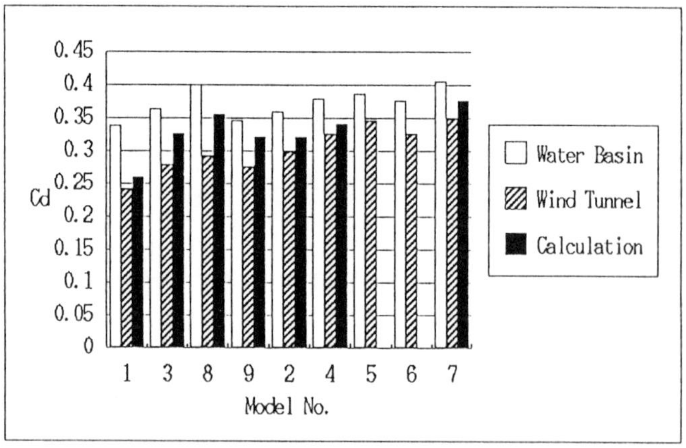

Fig.9 Comparison of computed and measured drag.

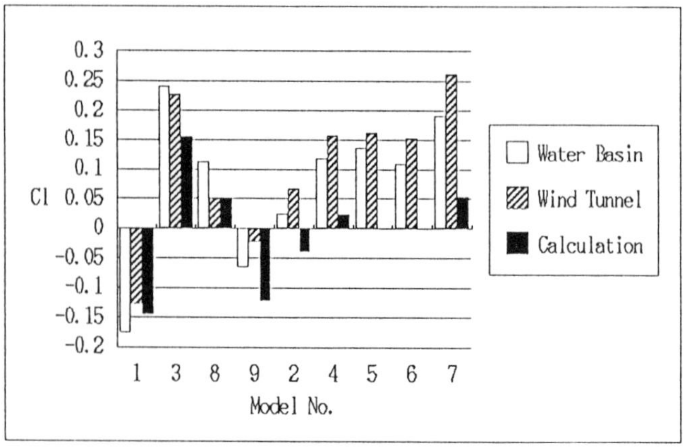

Fig.10 Comparison of computed and measured lift.

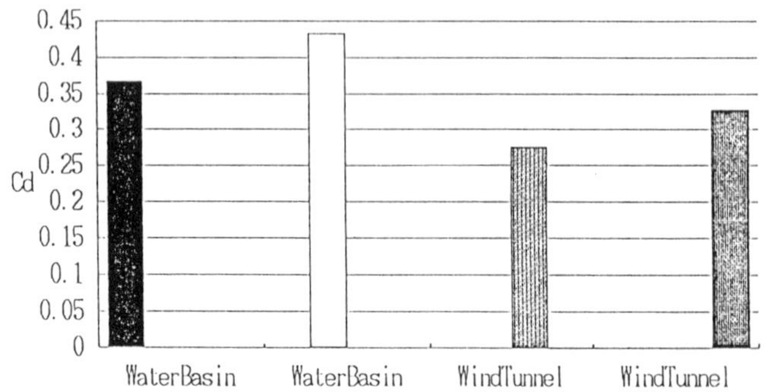

Fig.11 Measured Cd both in the high-drag state and the low-drag state for the critical geometry with back-face slant of 30°. (model No.8)

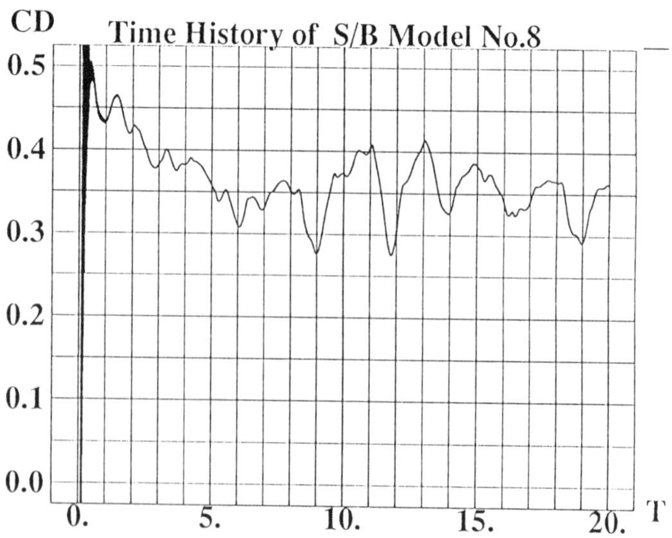

Fig.12 Computed time history of Cd of the critical geometry with back-face slant angle of 30°. (model No.8),T=0.0~20.

CRITICAL GEOMETRY OF HATCHBACK MODEL - As is well known, bluff body with back-face slant angle of 30 ° is a kind of critical geometry and it has two states of the flow; i.e., the high Cd state and the low Cd state. Model No.8 is a kind of critical geometry. Actually in the experiment, shown in the Fig.11, model No.8 shows two states of Cd.

Fig.12 shows the computed time history of Cd. The amplitude of the Cd curve is outstandingly large.compared to other cases. Therefore, it might be said that the computation reproduces the alternative high Cd state and low Cd state. Periodical variation is observed although the interval is varying.

Fig.13 shows the stream streak for T=9.0, 11.0, and 11.6 of the low, high, and again low Cd state respectively. For the high Cd state, the stream streak is smooth and attached flow is observed. On the other hand, the streak is tortuous and vortex whose axis is normal to the uniform flow is observed in the case of low Cd.

In Fig.14, the iso-helicity surface at the value +4 and -4 is shown for the high Cd and the low Cd state. In the low Cd state, the breakdown of vortex is clearly observed on the slanted back surface, whereas the the shape of the iso-helicity surface is calm and flat at the high Cd state.

Fig.15 shows the velocity vectors in the center plane. As the helicity suggests, when the Cd is low, the flow from the upstream changes its direction upward around the middle of the slanted surface and collides with the reverse flow from down stream which is induced by the shed vortex behind the body. This generates region of relatively higher pressure on the slanted surface, which results in the low Cd state. On the other hand, at the high Cd state, the flow above the slant back-face flows nearly along the surface, but the direction of the flow near the body is reverse.

Fig.16 shows the surface pressure contours. When the Cd is low, the contour-lines are nearly parallel on the upper half of the slant back face, whereas they forms an "eye" on the lower portion. On the contrary, the contours form two "eyes" on the upper portion of the slanted surface, which are nearly symmetry about the center plane, and forms low pressure region when the Cd is high.

The above observations are summarized as follows.

1) When the breakdown of vortex formed on the slant back face, the low Cd state is brought about. The flow from upstream collides with the reverse flow from downstream, and as the consequence, the high pressure zone is prevailing over the slant back face, which results in the low Cd state.
2) After the vortex breakdown, high Cd state appears. At this time, the flow is along the slant back-face and stream streaks are smooth.
3) Processes 1) and 2) are repeated, and the high Cd and the low Cd state appear alternatively.

The computed results coincide with the facts observed in the experiment of Ahmed [4] and of Morel[2], and it is safe to say that the present computation succeeds in capturing the unsteady features of the phenomenon which is difficult to capture in the experiment and has never been reproduced in the previous computational studies.

QUASI-CRITICAL GEOMETRY OF NOTCHBACK MODEL - It is supposed that a flow field of N/B type after-body is more complicated than that of S/B [11], and the detailed investigation is on the way. In this section, computed results for three kinds of N/B type after-bodies are presented and discussed.

Three types of N/B after-bodies are chosen for the computation. They are Model No.2 (α=40°, β=15°.), No.4 (α=30°., β=20°.), and No.7 (α=40°., β=25°.). The differences between the three models are mainly in the trunk deck heights. The model No.2 has the highest trunk deck height and the model No.7 has the lowest.

The stream streaks and the iso-helicity contours (the values are +4 and -4) of the three N/B models for the typical higher-Cd-state and for the lower-Cd-state are shown in Figs.17 to 19.

The stream streaks show the common features with that observed in the case of critical geometry of S/B (model No.8); that is, the stream streak is smooth when the Cd is higher, whereas it is tortuous and is forming vortices when Cd is lower. This is most evident in the case of No.7 which showed the highest Cd of the three. Difference of flow field between the higher-Cd-state and the lower-Cd-state is also observed in the iso-helicity-contours. The burst of vortex is clearly observed in the lower-Cd-state, whereas it is not so evident in the higher-Cd-state in the case of Model No.4 and No.7. In the case of No.2 which showed the lowest drag among the three models, the difference of the pattern of helicity is not so clear between the lower-Cd-state and higher ones. These results show that the formation of vortices and its breakdown has close relation with the fluctuation of the drag in time. It is noted that the difference of magnitude of the vortex burst (seen in the iso-helicity contour) is of the order of the magnitude of drag force, i.e., the smaller magnitude of the vortex burst results in the lower drag force.

Nouzawa et al. [11] reported that (quasi-) critical geometry for N/B type after-body exists around β=25°. in their experiment when the rear window inclination is greater than 25°, and that the arch type vortex becomes strong at the(quasi-) critical geometry. Model No.7 in the present study is supposed to be the critical geometry. Thus the arch type vortex is investigated using Lamb-vector. Fig.20 shows the iso-Lamb-vector contour of the model No.7 at several time levels. At T=6.4, when the Cd is higher, the contour develops on the roof end, and then succeedingly (as is not indicated in the figure) rises up around the center plane showing the symptom of the arch vortex. At T=6.7, the arch vortex is clearly observed. The arch vortex is carried downstream and shed in the wake and breakdown of the vortex occurs(T=7.4, lower-Cd-state), and then dissipates. Nouzawa et al. [11] reported in their numerical simulation that breakdown of the arch vortex is connected with the decrease of Cd. Contrary to their result, the formation and shedding of the arch vortex are periodically repeated in the same way with the experiments in the present computation. This seems to demonstrate the effectiveness of the present computational method for the separated flow.

The influence of the after-body configuration on the formation of arch vortex is investigated. Fig 21 shows the

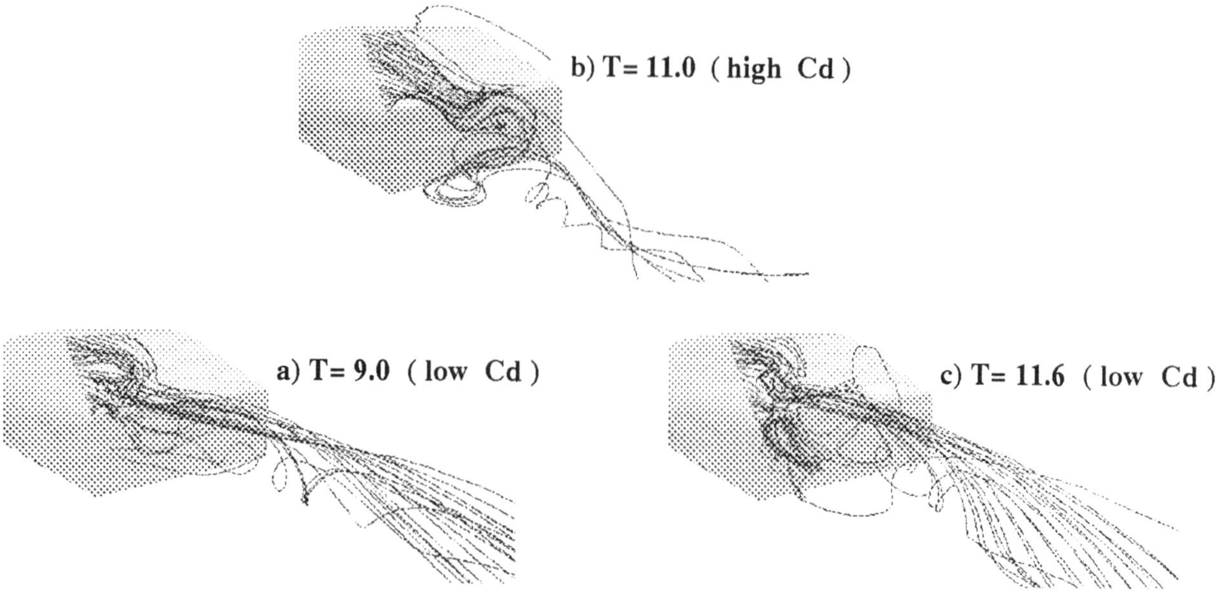

Fig.13 Stream streak of the critical geometry with back-face slant angle of 30°. (model No.8) a) T=9.0 (low Cd), b) T=11.0 (high Cd), c) T=11.6 (low Cd)

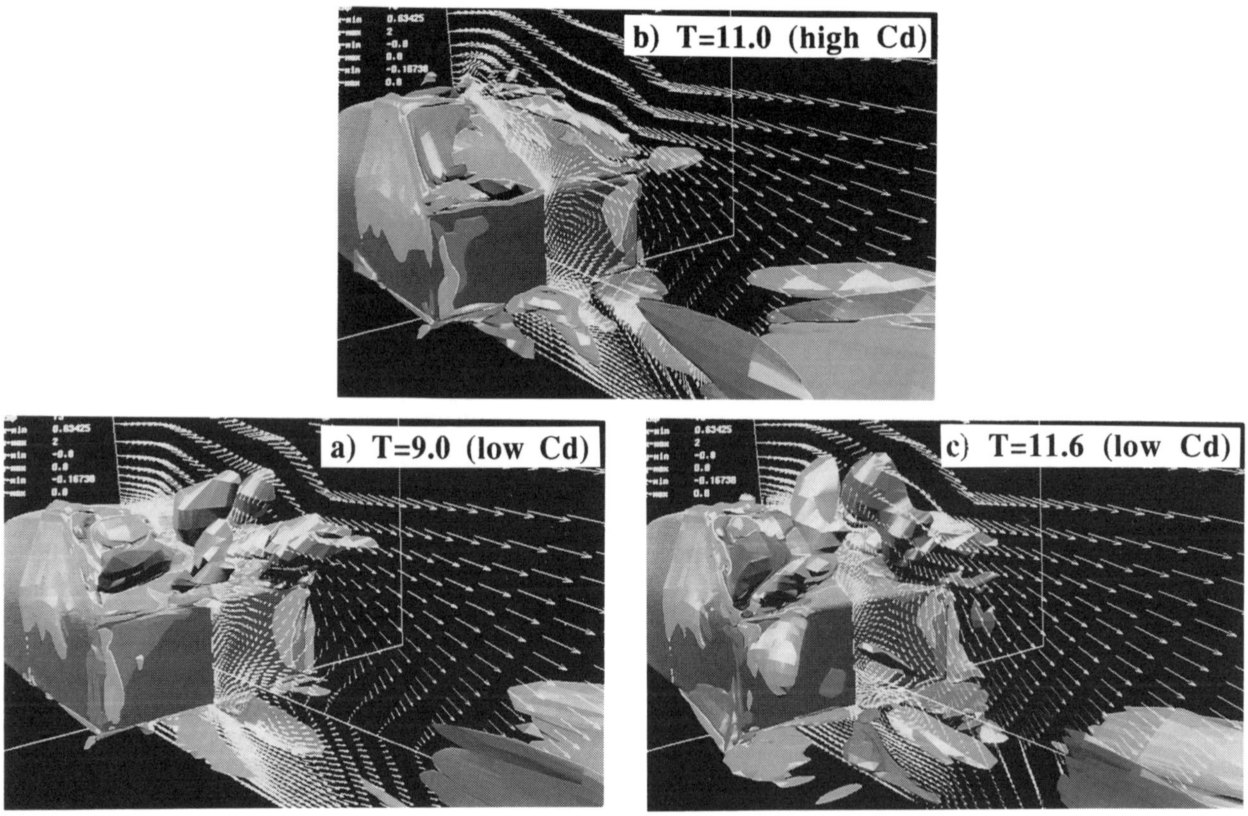

Fig.14 Iso-helicity surface of the critical geometry with back-face slant angle of 30°. (model No.8) a) T=9.0 (low Cd), b) T=11.0 (high Cd), c) T=11.6 (low Cd)

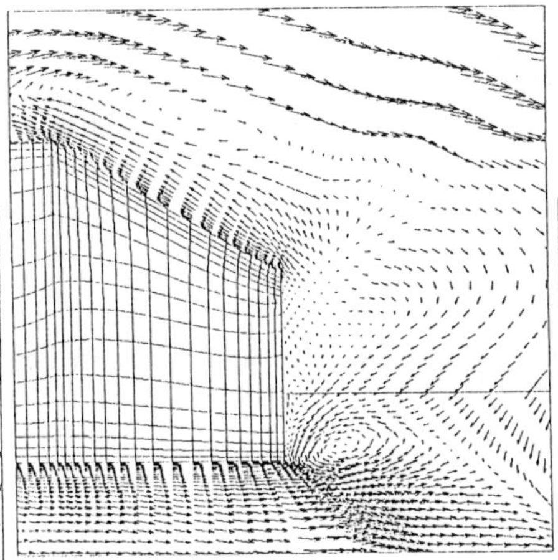

b) T=11.0 (high Cd)

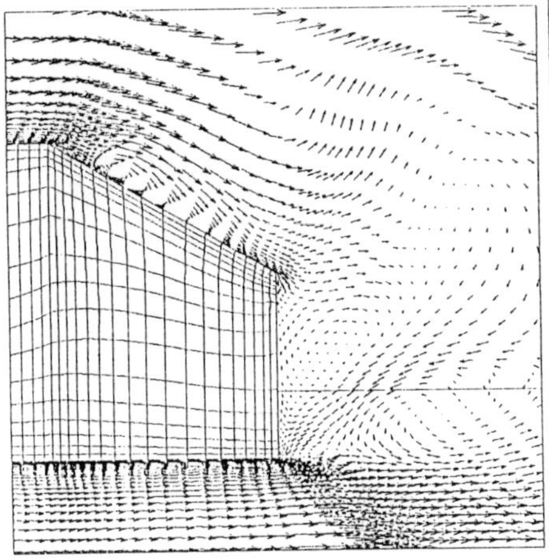

a) T=9.0 (low Cd)

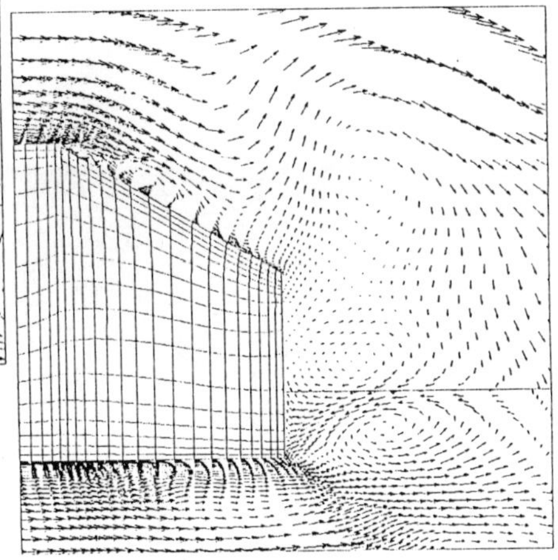

c) T=11.6 (low Cd)

Fig.15 Velocity vector in the center plane of the critical geometry with back-face slant angle of 30°. (model No.8)
a) T=9.0 (low Cd), b) T=11.0 (high Cd), c) T=11.6 (low Cd)

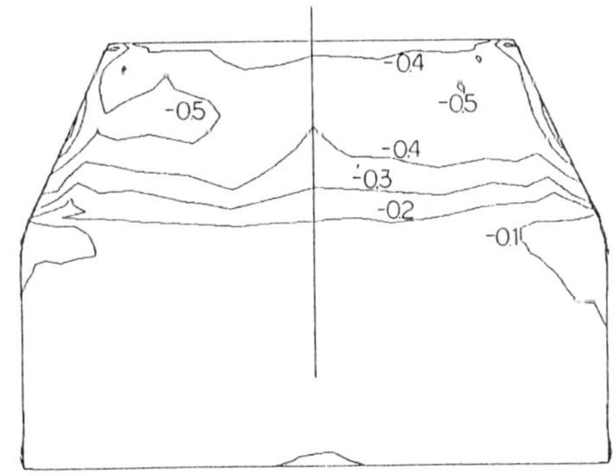

b) T=11.0 (high Cd)

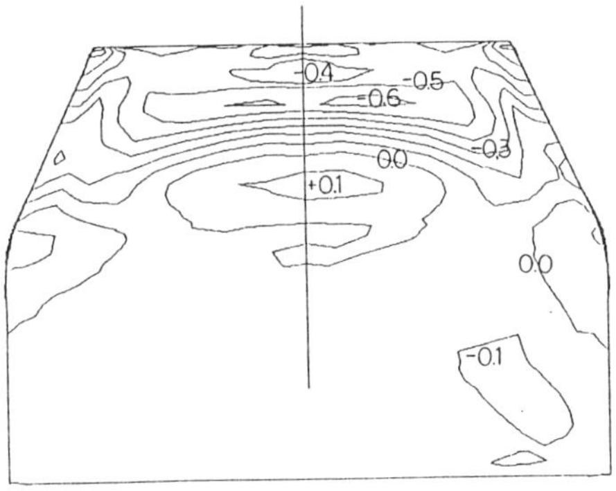

a) T=9.0 (low Cd)

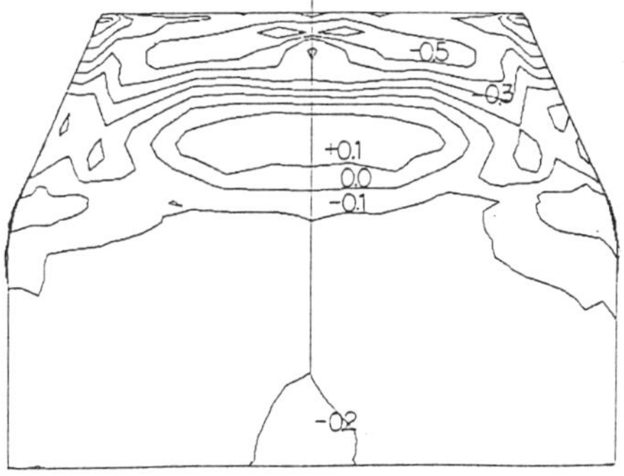

c) T=11.6 (low Cd)

Fig.16 Surface pressure distribution of the critical geometry with back-face slant angle of 30°. (model No.8) a) T=9.0 (low Cd), b) T=11.0 (high Cd), c) T=11.6 (low Cd)

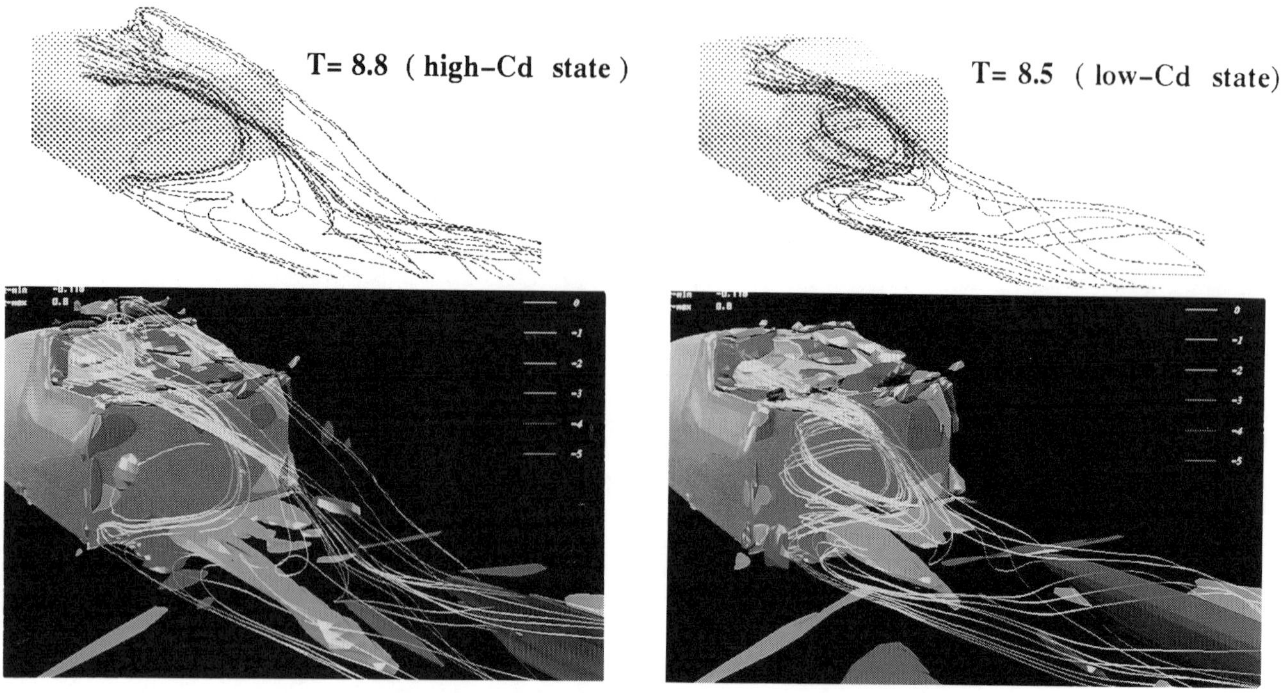

Fig.17 Stream streak(upper) and iso-helicity surface(lower)
of N/B model No.2
high-Cd state: (T=8.8) left-hand-side
low-Cd state : (T=8.5) right-hand-side

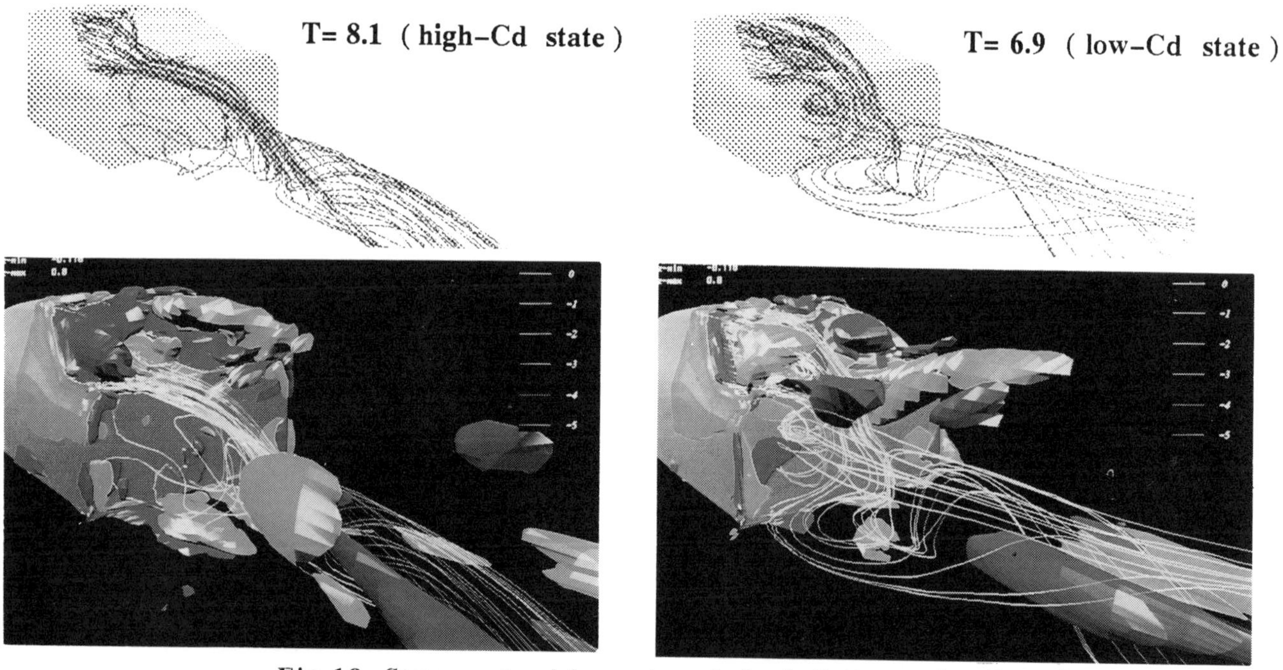

Fig.18 Stream streak(upper) and iso-helicity surface(lower)
of N/B model No.4
high-Cd state: (T=8.1) left-hand-side
low-Cd state : (T=6.9) right-hand-side

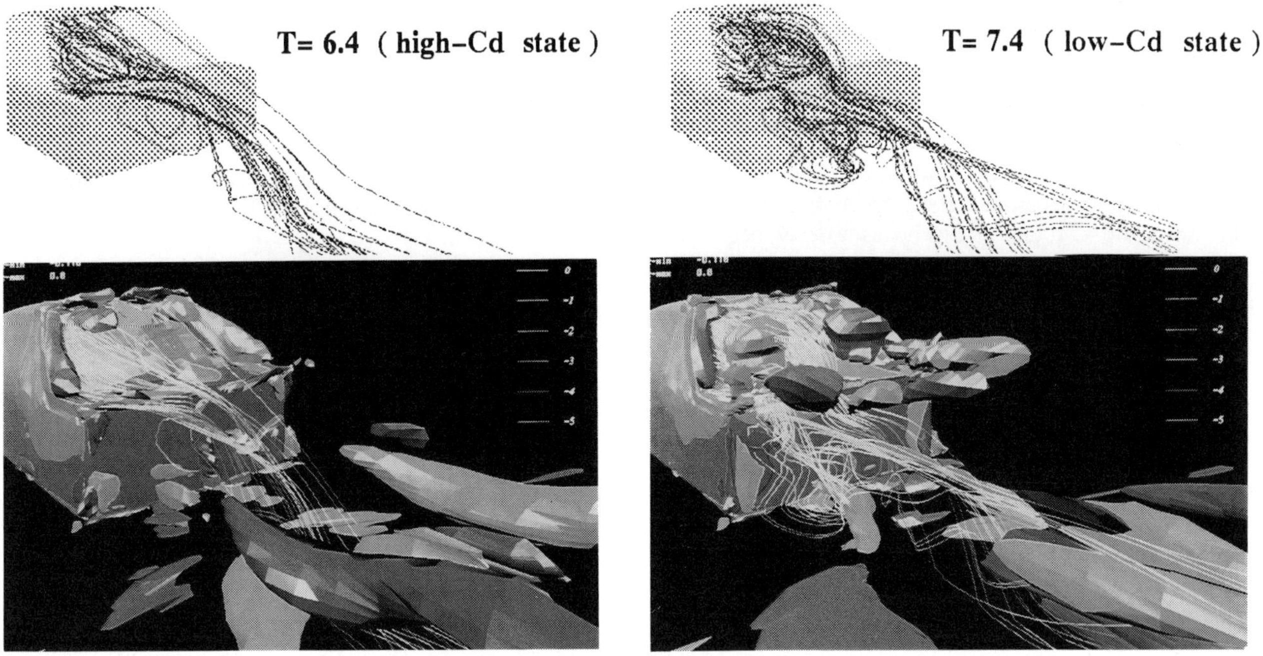

Fig.19 Stream streak(upper) and iso-helicity surface(lower) of N/B model No.7
high-Cd state: (T=6.4) left-hand-side
low-Cd state : (T=7.4) right-hand-side

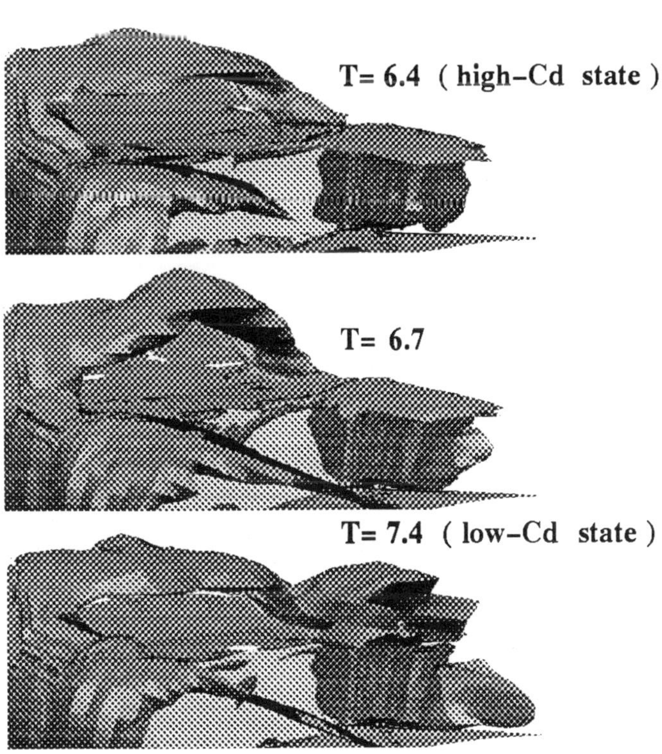

Fig.20 Time sequence of iso-Lamb-vector surface of N/B model No.7.
T=6.4 (high-Cd state), T=6.7, and T=7.4 (low-Cd state)

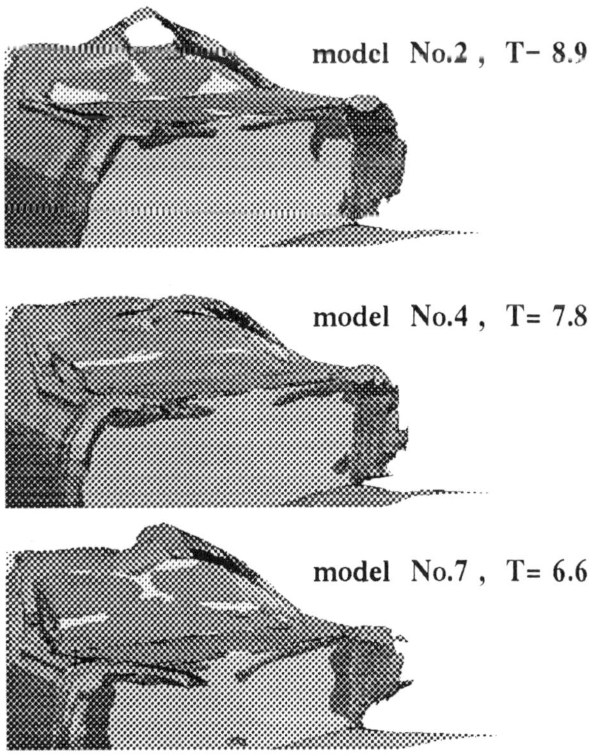

Fig.21 Comparison of arch vortex (iso-Lamb-vector surface) between N/B model No.2, No.4, and No.7

comparison of iso-Lamb-vector contours (at the value 18) for the three N/B models. The differences among the three are evident; high-drag-generating configuration accompanies wider and stronger spread of the arch vortex.

CONCLUDING REMARKS

The principal conclusions are as follows:

1. Numerical simulation method for the prediction of the 3-D vortical flow around road-vehicles is developed in the framework of a finite-volume method in a curvilinear, boundary-fitted coordinate system.
2. The sub-grid scale turbulence model is introduced and tuned with the help of experimental results.
3. Series test computations for the vehicle-like bluff body with various after-body configuration are carried out. The results are validated by the experimental results both in the water basin and in the wind tunnel. The agreement of the computed drag force with the experimental ones is satisfactory.
4. Some important features such as high and low drag states are revealed in the numerical simulations of the case of critical geometry and their vortical mechanisms are elucidated.
5. It is proved that the present method is useful for the aerodynamic design of road vehicles.

ACKNOWLEDGEMENT

The computations were executed by CRAY-YMP super computer system of ISUZU Motors Ltd. and the CPU time was about 13 hours per unit dimensionless time for each case. The pre- and post-processings were executed by an engineering workstation SGI IRIS-340GTX-B.

The authors would like to thank Mr. T. Iguchi and Mr. K. Okamoto of the Engineering Systems Department of ISUZU Motors Ltd. for their collaboration in the computations.

REFERENCES

[1] W. H. Hucho, L. J. Janssen, and H. J. Emmellmann, "The Optimization of Body Details - A Method for Reducing the Aerodynamic Drag of Road Vehicle", SAE paper 760185, 1976

[2] T. Morel, "Aerodynamic Drag of Bluff Body Shapes Characteristic of Hatch-Back Cars", SAE paper No.780267, 1978.

[3] P. W. Bearman, "Some Observations on Road Vehicle wakes", SAE paper No.840301, 1984.

[4] S. R. Ahmed, G. Ramm, and G. Falin, "Some Salient Features of he Time-Averaged Ground Vehicle Wake", SAE paper No.840300, 1984.

[5] T. Nouzawa, et al., "Analysis of Wake Pattern for Reducing Aerodynamic Drag of Notchback Model", SAE paper No.900318, 1990.

[6] M. Hashiguchi, et al., "Computational Study of the Wake Structure of a Simplified Ground-Vehicle Shapes with Base Slant", SAE paper No.890597, 1989

[7] T. Han, "Computational Analysis of Three-Dimensional Turbulent Flow Around a Bluff Body in Ground Proximity", AIAA Journal, Vol.27, No.9, pp.1213-1219, Sept., 1989

[8] T. Han, D. C. Hammond Jr., and C. J. Sagi, "Optimization of Bluff Body for Minimizing Drag in Ground Proximity", AIAA Journal , Vol.30, No.4, April, 1992

[9] R. Himeno, M. Takagi, K. Fujitani, and H. Tanaka, "Numerical Analysis of the Airflow around Automobiles Using Multi-block Structured Grids", SAE paper No.900319, 1990

[10] T. Kataoka, H. China, K. Nakagawa, K. Yanagimoto, M. Yoshida, "Numerical Simulation of Road Vehicle Aerodynamic and Effect of Aerodynamic Devices", SAE paper No.910597, 1991

[11] T. Nouzawa, K. Hiasa, T. Nakamura, A. Kawamoto, and H. Sato, "Unsteady-Wake Analysis of the Aerodynamic Drag of a Notchback Model with Critical Afterbody Geometry", SAE paper No.920202, 1992.

[12] M. Zhu, H. Miyata, and H. Kajitani, "Finite-Volume Method for the Unsteady Flow about a Ship in Generalized Coordinate Systems", Journal of The Society of Naval Architects of Japan, Vol.167, pp.9-16, 1990

[13] V. C. Patel, A. Nakayama, and R. Damian, "Measurements in the thick axisymmetric turbulent boundary layer near the tail of a body of revolution", J. Fluid Mech., vol.63, part 2, pp.345-367, 1974

[14] H. Miyata, M. Zhu, and O. Watanabe, "Numerical Study on a Viscous Flow with Free-Surface Waves About a Ship in Steady Straight Course by a Finite-Volume Method", Journal of Ship Research, Vol. 36, No.4, Dec. 1992, pp.332-345

[15] Smagorinsky, J., Manabe, S., and Holloway, J. L. "Numerical results from a nine-level general circulation model of the atmosphere", Monthly Weather Review, 93, Dec. 1965, pp. 727-768.

[16] G. W. Carr, "Influence of Rear Body Shape on the Aerodynamic Characteristics of Saloon Cars", MIRA Rep. No.1974/2, 1974

[17] K. Aoki, H. Miyata, M. Kanai, Y. Hanaoka, and M. Zhu, "A Water-Basin Test Technique for the Aerodynamic Design of Road Vehicles", SAE paper No.920348, 1992

Section V
Vehicle Aerodynamic Development and Drag Reduction

Section V
Vehicle Aerodynamic Development and Drag Reduction

The objective of understanding vehicle aerodynamics is to apply this understanding to the design and development of vehicles. For both passenger and commercial vehicles, the design and development process must encompass diverse operational, regulatory, manufacturing and market *requirements*. Beyond these requirements, manufacturers seek to endow their products with *attributes* that establish some level of superiority over competing products. The *aerodynamic* development of these vehicles must therefore involve comprehension of not only the basic requirements, but also the relative importance (weight) of each of the attributes. As an example of the latter, styling has remained a major factor in determining the exterior shape of a passenger car, whereas this attribute commands much less emphasis for a commercial vehicle.

An aerodynamically efficient vehicle shape, possessing a very low drag coefficient, can help significantly in lowering fuel consumption. This is particularly true in cases where sustained operation at high speeds is envisaged or where significant improvements have been made to other vehicle systems such as the powerplant and drivetrain. The worldwide trend, over the last three decades, of improved overall level of fuel efficiency for vehicles has been assisted by a trend to lower drag coefficients for both passenger and commercial vehicles.

For most road-vehicle aerodynamicists, the pursuit of drag reduction remains one of the principal objectives. When viewed in totality, the aerodynamic efficiency of a vehicle arises not only from the basic shape of the vehicle's envelope, but also from its myriad detailed geometries that interact with the exterior flow, such as the cooling air inlet, underbody components, etc. As progress continues to be made in understanding and refining the basic shapes of vehicles, the importance of understanding and improving the aerodynamic effects of these details grows.

In some cases, drag reduction is pursued through the addition of aerodynamic devices to existing vehicles. For many commercial vehicles, where the shape of the load-carrying volume may not be altered, special components may be added to the exterior of the body to influence the airflow around the vehicle and reduce its drag coefficient.

Understandably, for specialized vehicles such as racing cars or other competition vehicles the role of aerodynamics is very different. The design and development of these vehicles demands special attention to aerodynamics so that specific performance targets are attainable. Downforce, for better traction during cornering, takes on a role that is equal to or greater than that of drag for many of these situations.

900317

The Aerodynamic Development of the Opel Calibra

Hans-Joachim Emmelmann, Helmut Berneburg and Josef Schulze
Technical Development Centre of GM Europe

ABSTRACT.

The Opel Calibra Coupe is an example of a newly developed series production vehicle which has been conceived with a view to low drag and good fuel consumption. A great deal of world-wide aerodynamic expertise went into its development, including newly developed measurement techniques and investigations to provide inputs for computational codes. The aerodynamic development program for the vehicle, from the 1:5 scale model stage to the full size detail development stage is outlined as well as specific examples of some interesting individual investigations.

INTRODUCTION.

In the last ten years there have been intensive world-wide efforts to reduce fuel consumption by reducing the aerodynamic drag of motor vehicles. It is therefore not surprising that most of the fuel consumption reduction which has been made has been achieved by the reduction of aerodynamic drag. This fuel consumption reduction is illustrated in the Opel fleet consumption chart(Fig.1) which also includes vehicles produced by Vauxhall in England. Fig.2 shows the corresponding development of drag coefficients for the Opel model range.

The Opel Calibra is now the last link in a chain of vehicles which Opel has presented with good aerodynamics during recent years. At their introduction each of these vehicles has been amongst the best in their class in terms of aerodynamic excellence.

In contrast to the majority of recent publications on the subject of aerodynamics, which have presented windtunnel technology and detail aspects of local areas or flow calculations, this publication describes features of the aerodynamic development of a real vehicle, referring to the application of modern windtunnel measurement technology and inputs for the discussion about the applicability of numerical simulations.

The Opel Calibra, with a drag coefficient of c_D=.26, currently the lowest value for a series production car, is particularly well suited for use as an example of an

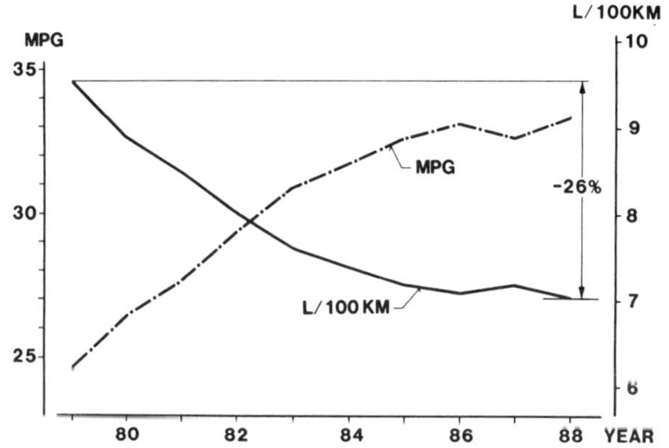

Fig.1: Opel fleet average fuel economy (Euromix) over the last ten years.(Includes Vauxhall production)

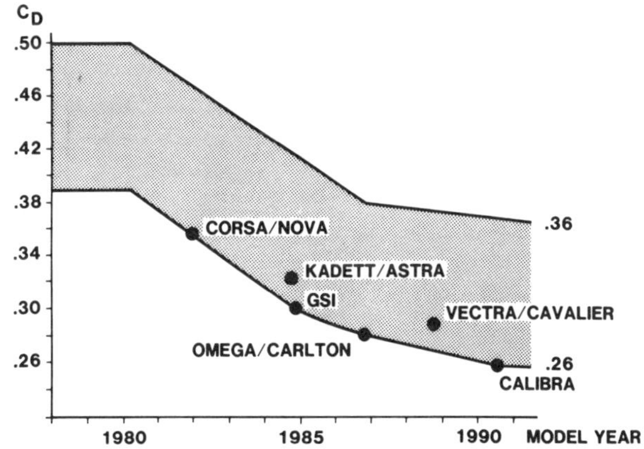

Fig.2: The development of the drag coefficient of Opel/Vauxhall vehicles since 1978.

aerodynamic development program.

It cannot be over-emphasised that such a low drag coefficient is only achieveable with the application of all the world-wide available aerodynamic "know-how" and the subsequent adoption of all the aerodynamic

requirements into the styling and construction concept of the vehicle.
In this respect aerodynamic development at Opel TDC (Technical Development Centre of GM Europe) benefits from the position of the aerodynamics department within the organisation and the aerodynamic development process itself.

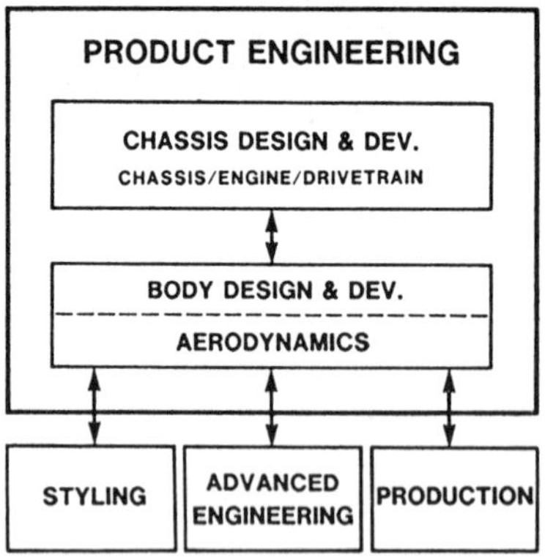

Fig.3: The placement of the aerodynamics group within the organisation.

The aerodynamics department is located in the body design and development area,(Fig.3). We are of the opinion that this placement ensures an optimum follow-up of the whole development phase, starting with the first styling drawings and continuing up to production. On the contrary most other manufacturers have integrated the aerodynamic group into advanced engineering, styling or research, which means, that the priority of aerodynamics is limited to the development stages during which the individual departments are strongly involved in the development process.

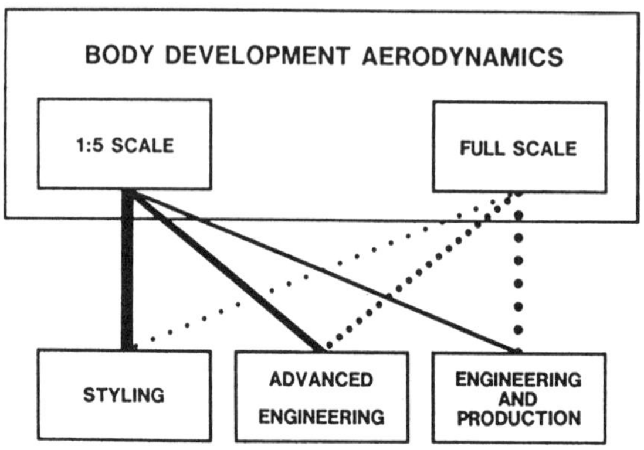

Fig.4: Schematic illustration of the interaction between the aerodynamics group and the rest of the organisation.

The aerodynamics department is set up in such a way that the entire vehicle development process can be followed (Fig.4). The 1:5 scale group works on the optimisation of 1:5 scale models in conjunction with design, some of these investigations being conducted prior to the commencement of the styling activity. The full scale group is mainly concerned with the optimisation of full scale wind-tunnel models, prototypes and pilot vehicles, right up to production.
In a similar manner to modern "just-in-time" production logic, in order to be successful the aerodynamicists must make aerodynamic data available to styling, advance development and design departments, as soon as it is required. In order to achieve this it is necessary to work strictly to the aerodynamic development plan, (Fig.5).

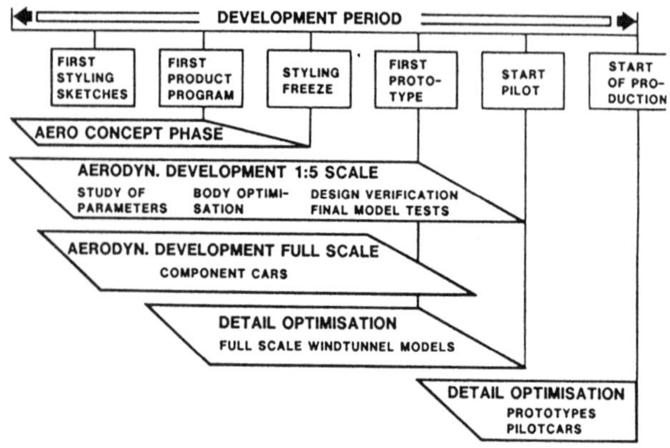

Fig.5: An aerodynamic timing chart.

Already prior to the styling phase it is necessary to optimize the basic body shape. This work is conducted on 1:5 scale models. The actual 1:5 scale optimization phase runs concurrent with the styling activity, during which the stylists and the aerodynamicists must work closely together. The result of this development is a 1:5 scale model which exactly represents the full-size styling model, including all the aerodynamic optimization, with the exception of the fine detail optimization.
However, prior to the completion of the styling model, the optimization of so-called component vehicles is already begun. These vehicles, which are basically production vehicles, are modified according to test requirements, e.g. using fibre glass parts in local areas to represent the styling model. With these vehicles it is possible, at a very early stage, to begin combined aerodynamic and cooling tests. These tests are then continued on full-scale wind tunnel models, which are plastic bodied running vehicles, and later on prototypes. By wind tunnel tests on prototypes and pilot vehicles it is then possible to check that the obtained optimisation features have actually been incorporated and that they are effective.
This timing program and the organisational measures are a pre-requisite for optimum aerodynamic development.

TEST FACILITIES.

For the aerodynamic development of the Opel Calibra various wind tunnels throughout Europe were used (Fig.6).

While the majority of testing in both 1:5 scale and full scale was conducted in the Stuttgart University wind tunnels, other wind tunnels were used due to their specific high technology test facilities.
For example both the Pininfarina Wind Tunnel and the DNW tunnel offer specific test facilities of this nature:

Pininfarina has a system for Flow Pattern Investigation [1]* and a Laser-Doppler Anemometer System [2] which, with the help of fibre optic tubes, allows measurements at almost any position outside, or inside, the vehicle. DNW (German Dutch Wind Tunnel) has equipment which can support a full size vehicle above a rolling road(moving belt), with the balance being contained within the test object [3].

MODEL SCALE	WIND TUNNEL	TYPE OF TEST
1 : 5	- FKFS, STUTTGART UNIVERSITY - PININFARINA, ITALY	FORCE AND FLOW MEASUREMENTS ON • SCALE MODELS
1 : 1	- FKFS, STUTTGART UNIVERSITY - GERMAN DUTCH WINDTUNNEL (DNW),NETHERLANDS - PININFARINA , ITALY - GMAL, USA	FORCE AND FLOW MEASUREMENTS ON • WIND TUNNEL MODELS • PROTOTYPE TESTS • PILOT AND PRE - PRODUCTION VEHICLES
	- EIDGENÖSSISCHES FLUGZEUGWERK (EFW), SWITZERLAND - FKFS, STUTTGART UNIVERSITY	CONTAMINATION TESTS

Fig.6: Wind tunnels used for the Calibra aerodynamic development.

REDUCED SCALE INVESTIGATIONS.

Aerodynamic testing using typical Opel 1:5 scale models, as previously mentioned, was conducted prior to and during the styling phase of the Calibra. These models typically demonstrate very high quality external surfaces (as shown in Fig. 7) and the engine compartment and underbody are very accurately represented.
As the Calibra is a derivative of the Opel Vectra 2000 a model underbody was already available. Using this, the first model was constructed according to stylists sketches (Fig. 8). Test work then proceeded with the integration of data from the full scale styling models into the 1:5 scale model. At the end of the development the 1:5 scale model exactly represented the finally released raw data plan.

Fig.7: A 1:5 scale wind tunnel model.

* Numbers in parentheses designate references at the end of the paper.

Fig.8: An early styling sketch.

As can be seen from the sketch in Fig.8, the basic essentials for a low drag coefficient are already evident, according to the aerodynamic concept. They consist of a front end with smooth rounded transitions, a droplet shaped plan view, sharp flow separation edges around the rear and considerable tapering of the 'C' pillar and rear end.

FRONT SPOILER LIP.- Although it is now a well known fact that low drag coefficients are mainly achieved by optimizing the rear end of a vehicle [4] a basic requirement is still separation free airflow around the front end. Naturally this is only valid for the hood transition and the flow around the front corners to the vehicle sides. However, in the lower area it is more effective to work with a front spoiler which is designed for optimized separation. By this technique the impingement of high speed airflow onto underbody parts is considerably reduced. In the Calibra development the height of the front spoiler above the ground plays an important part in drag reduction (Fig. 9). From a starting point of 210 mm clearance between front spoiler and ground, i.e. 42 mm for the 1:5 scale model tests, the front spoiler was progressively lowered. As a result of this work the optimum spoiler height was found to be 150 mm, but for ground clearance reasons it was eventually only lowered to 180 mm. The front spoiler area was addressed again during the Calibra development, but this will be outlined in the section dealing with the full scale development.

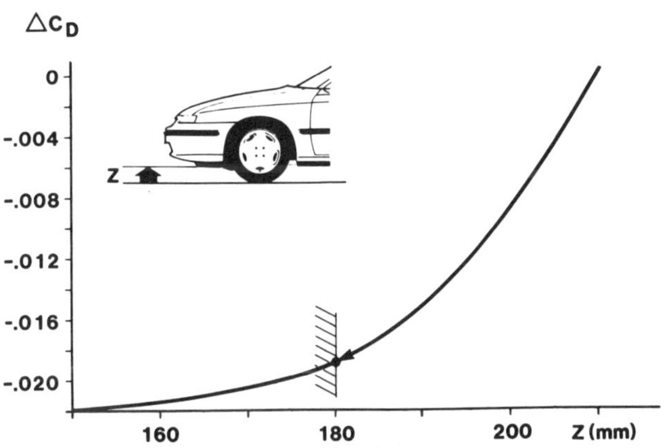

Fig.9: The influence of front spoiler ground clearance on the drag coefficient.

REAR END OPTIMISATION.- Rear end optimization is especially important during the 1:5 scale phase. The rear end is a typical example involving large panels which are difficult and time consuming to change, therefore the design must already be aerodynamically acceptable at the end of the 1:5 scale work. Late changes become continually more difficult to incorporate as they can jeopardise the complete timing plan.

The graph in Fig.10 shows the improvement in drag coefficient which was achieved by tapering the rear sides of the vehicle, including the parallel displacement of the 'C' pillar, but not further tapering of it. These results were measured using the previously optimised rear end height of 940 mm.This rear end height was established from the curve shown in Fig.11. As the results of these investigations are interdependent, a considerable number of parameter variation tests were necessary. The results shown in Fig.10 are only a small part of the total investigation.

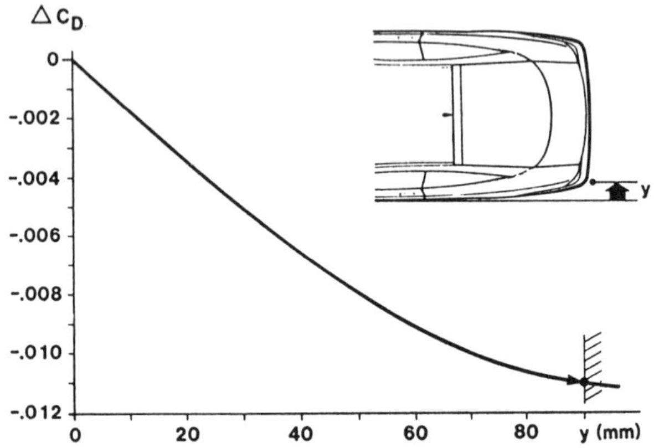

Fig.10: The influence of rear end taper on the drag coefficient.

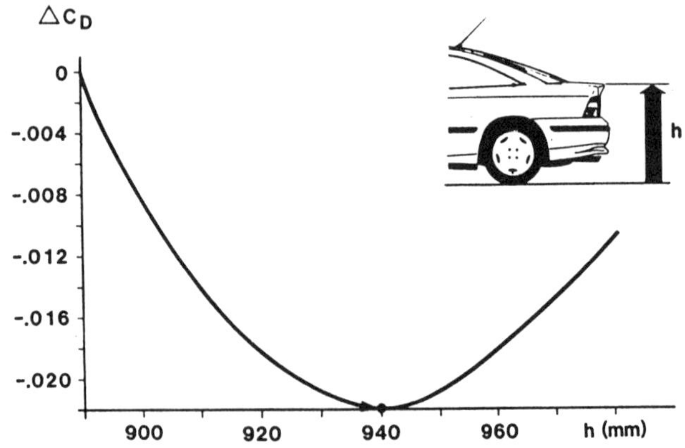

Fig.11: The influence of decklid height on the drag coefficient.

Due to the fact that the following parameter variations give important information for the applicability of CFD development when used for real vehicles, these results are shown in their entirety.

SPECIAL STANDING HEIGHT/UNDERBODY ROUGHNESS INVESTIGATIONS.- A series of tests were run with a range of vehicle standing heights and different vehicle configurations (Fig.12).

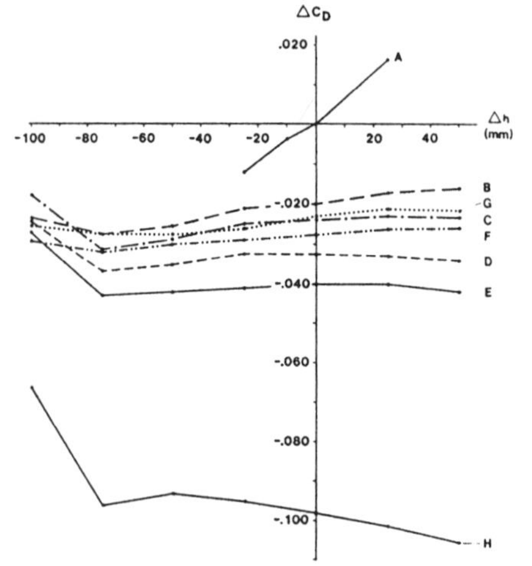

	A	B	C	D	E	F	G	H
BASIS	X	X	X	X	X	X	X	X
WITHOUT WHEELS		X	X	X	X	X		
WITH LOWER ENGINE COMPARTMENT COVER			X	X	X			
WITH PARTIAL UNDERBODY COVERS				X	X			
WITH WHEELHOUSE COVER PANELS					X	X		
WITH PARTIAL WHEEL SIMULATION							X	
SMOOTH UNDERBODY, WHEELHOUSE FILLED TO BODY AND UNDERBODY CONTOUR								X

Fig.12: The influence of vehicle standing height and body roughness on the drag coefficient.

The curve for the basis configuration"A", a standard Calibra, shows the well known dependency of the drag coefficient on standing height variation: with increasing ground clearance the drag coefficient increases relative to the base configuration and with decreasing ground clearance it decreases. Following the base measurements step by step changes were made to change the body into a smooth surfaced vehicle, similar to that which is typically used in numerical calculations (Fig.13).

Configuration "B" (Fig.12) shows the results for the vehicle without the wheels, thin supports being used to fix the model to the wind tunnel balance.

Configuration "C" involved the addition of an engine compartment lower cover, and

configuration "D" included covers over particularly aerodynamically "rough" areas of the underbody.

For configuration "E" the wheelhouses were closed flush to the body on the outside, but the underside was still open.

Configuration "F" was the same as configuration "E" except that the engine compartment lower cover and the underbody covers were removed.
For configuration "G" the lower parts of the wheels, which protruded outside the body surface were cut off, but the gaps around the wheels remained. Configurations "F" and "G" were only conducted to complete the investigation.
Finally configuration "H" represents a completely smooth model.
As expected, the drag coefficient is seen to reduce as the vehicle body becomes more smooth. But the slope of the curve, dependant on the vehicle attitude, is also seen to change. Without wheels (Config.B) the slope of the curve is considerably flatter, with underbody covers (Config.D) it already shows an opposite trend to the basis configuration.
This trend is further reinforced by the completely smooth configuration (Config.H). These results were also obtained from full scale tests, in a windtunnel with both fixed and moving ground planes.
If numerical codes are able to accurately represent the physics of the airflow around smooth bodies they will produce results similar to those obtained by configuration "H". However this means that trend-statements made on the basis of numerical code analyses of simplified body shapes cannot be expected to be valid for real vehicles which are represented by configuration "A".

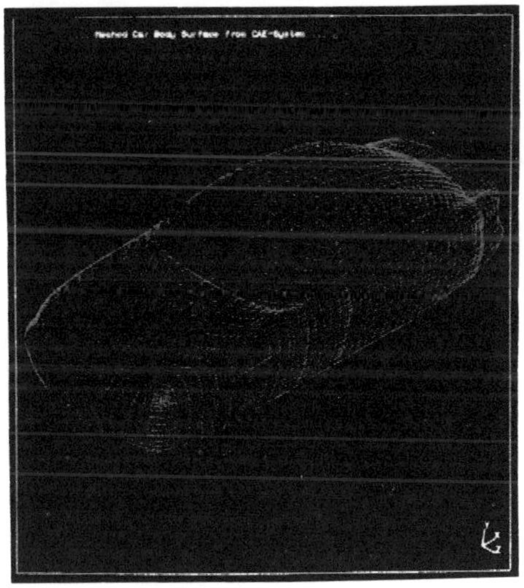

Fig.13: A Calibra CFD model.

FULL SCALE WIND TUNNEL MODEL.

When most of the results of the 1:5 scale testing had been incorporated into the full scale styling model a fibre glass moulding of the complete outer body shape was made. With the help of a tubular frame construction it was mounted onto the running underbody of a Vectra 2000, and Plexiglas windows were installed.
With this full size fibre-glass bodied running vehicle, which was available approximately one year ahead of the first prototype, it was possible to conduct not only further aerodynamic optimization, but also thermodynamic investigations, contamination tests, dynamometer tests and driving tests at the proving ground.

Fig.14: The Calibra full size wind tunnel model, shown mounted by a sting above the rolling road in the DNW tunnel.

Fig.14 shows this vehicle in the DNW tunnel, as previously mentioned, mounted on the sting attachment which suspends the vehicle above the rolling road.
Along with the engine cooling group it was possible to establish the grill and spoiler openings for various engine, gearbox and load variations. The engine compartment lower cover for the 16 valve engined vehicle and its cooling air inlets were also developed on this vehicle.
During the development of vehicles which are especially intended for the European market it is necessary to overcome higher windnoise levels and closing forces on functional components. It was therefore especially advantageous that it was possible to conduct functional tests on the spoiler sun-roof at an early stage. Due to unlimited top speeds on many german "autobahns" it is possible for driving speeds to be reached which are more than twice as high as those allowed on freeways in U.S.A. and many other countries. Aerodynamic forces more than four times higher than normal are associated with these high speeds.

CHOSEN EXAMPLES OF FULL SCALE DEVELOPMENT.

FRONT SPOILER/UNDERBODY OPTIMISATION.- As previously mentioned, the front spoiler, which was optimized at the 1:5 scale stage, was to be further optimized with the help of the full size wind tunnel model. This optimization investigation illustrates a typical example of the application of modern wind tunnel measurement and evaluation methods. It began with the use of the Pininfarina flow evaluation techniques and continued with force measurements in the Stuttgart University wind tunnel. The results were then verified in the DNW tunnel with the help of the rolling road, all of which influenced the integration of aerodynamic underbody measures.

Although front spoiler optimization is being discussed here, the origin of this work came from flow investigations behind the vehicle.

The wake behind a vehicle was measured and evaluated in the Pininfarina wind tunnel using a 14 hole Kiel probe.[1]. The results were produced as diagrams which showed velocity distribution, total pressure distribution(Fig.15), turbulence distribution and flow direction. From the representative singular results a so-called Micro-drag diagram was produced.

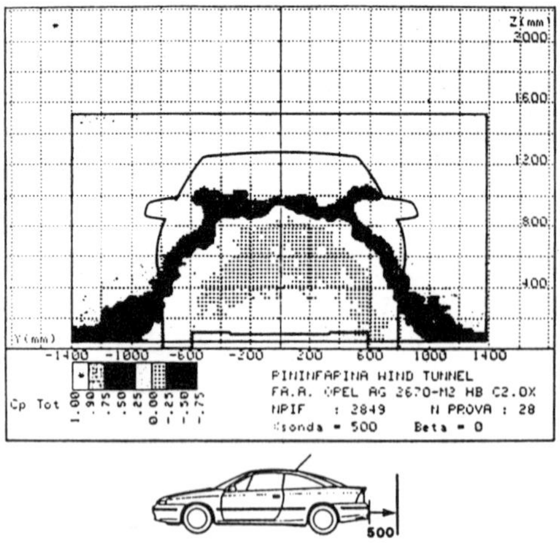

Fig.15: A black and white wake survey (total pressure map) from the full scale wind tunnel model of the Calibra.

Fig.15 shows a black and white diagram of the total pressure on a plane 500mm behind the vehicle, which is obtained immediately at the completion of the test. Colour diagrams, which make the results easier to read, follow in a short time. The wake form shown here is typical for a fully optimised fastback model with no noticeable "tip" vortices. It does however exhibit "ears" which result from vortices caused by the "A" Pillar and outside rear view mirrors.

In comparison to previous tests on other vehicles it is noticeable that the wake was relatively wide slightly above ground level. In order to establish the reason for this, further measurement planes at the sides of the vehicle were investigated until it was discovered that the wide wake was being generated by the front wheels (Fig.16).

As a result, flow visualisation in front of the front wheels was conducted revealing that the flow ahead of the front wheels approaches at an angle of approximately 30° from the centre of the vehicle in plan view. In spite of the already low c_D value of .28 which had been achieved it was thought as a result of these findings that a further potential existed if the flow angle to the front wheels could be reduced.

The strategy to make use of this potential was to make the flow more parallel by reducing the pressure differential between the flow under the vehicle and the flow around the sides of the vehicle.

As the flow around the sides of the vehicle, close to the ground, (below 200 mm) is barely influenced by the body contours it was only possible to consider changes which would affect the underbody flow. In order to reduce the pressure in this area it was necessary to increase the underbody flow velocity.

To achieve this the 40 mm front-spoiler lip was cut away in the centre, as shown in Fig. 17, but a 10 mm flange was maintained. The improvement was first noted at 600 mm cut out width but the maximum improvement was achieved with 965 mm

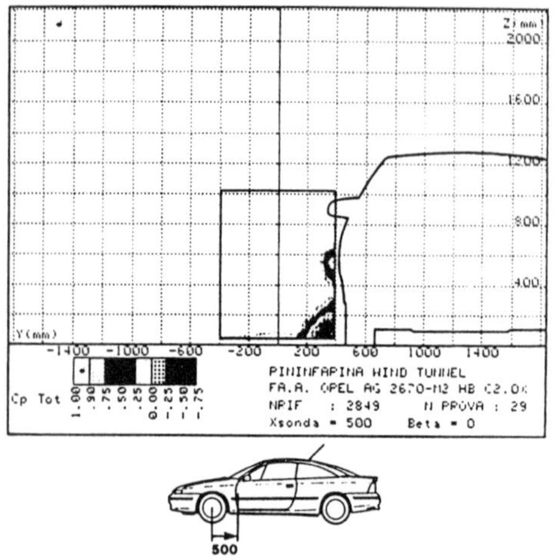

Fig.16: Additional wake planes were investigated at the sides of the vehicle. A total pressure map 500mm behind the front axle is shown.

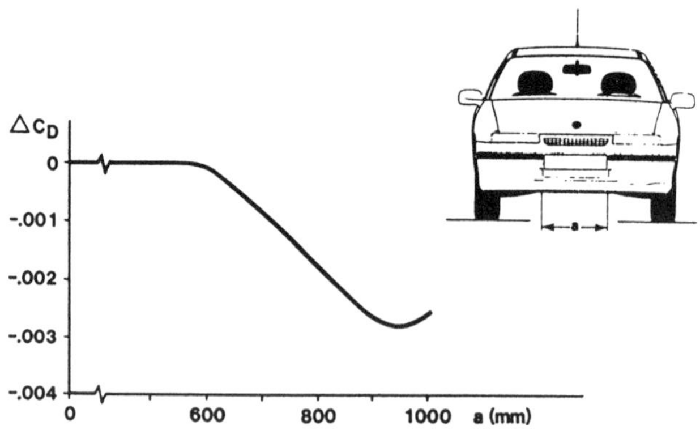

Fig.17: The influence of the width of the front spoiler cut-out on the drag coefficient.

In order to reinforce the effect of this measure further variations of the front spoiler corners in height and position were investigated.

By moving the sides of the spoiler 60mm outboard, to cover more of the front wheels, a c_D reduction of .006 was obtained. This change, implemented as a body, is later referred to as spoiler corners. Subsequent variation of the height resulted in the curve shown in Fig.18. In total a drag coefficient improvement of .009 was achieved by changes to the front spoiler corners. All the measures in the front spoiler area including the spoiler cut-out were found to give a drag coefficient reduction of .012.

Fig.19 shows an overview of the shape development of the spoiler area up to the stylistically developed final configuration. As there is a danger that the spoiler will contact curb-stones the spoiler was designed in a flexible material, but it had to be stiff enough to resist deformation by air forces.

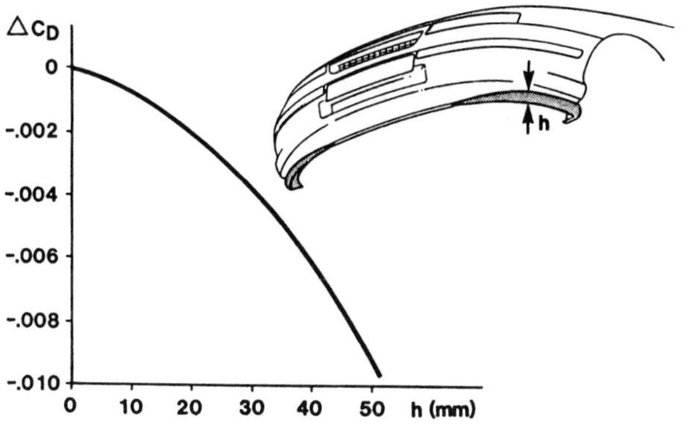

Fig.18: The influence of spoiler corner height variation on the drag coefficient.

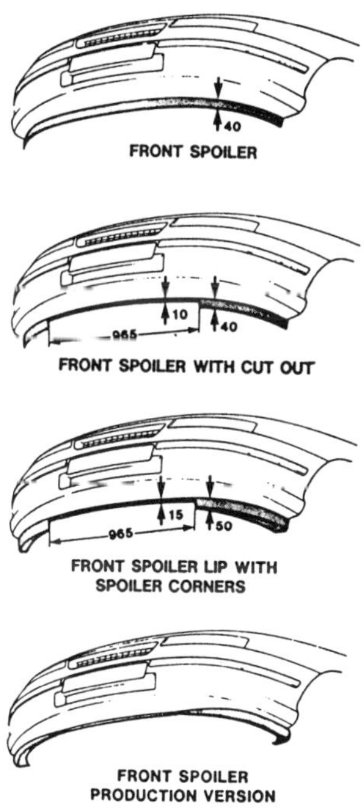

Fig.19: The development of the front spoiler and spoiler corner shape.

When the spoiler area had been modified to suit the styling and constructional requirements, the above mentioned measures still brought a c_D reduction of .010, which is still a very acceptable improvement to the final product.
When the optimisation of the front spoiler was being conducted the measures on the underbody, which resulted from the wake flow tests, were not yet finalised. It was assumed that the influence of the spoiler cut-out, as shown in Fig.17, would be reduced due to the fact that the associated higher flow velocities under the vehicle would lead to increased losses in the underbody area. The assumption was that increased pressure under the rear part of the vehicle would lead to a sideways airfow between the front and rear wheels, which in turn would present angled flow to the rear wheels.
In order to reduce this effect the rocker was lowered between the tyres to the minimum height which was acceptable from a ground clearance standpoint.

Naturally to lower the steel rocker panel itself was impossible, therefore the lowering was realised by the addition of plastic rocker mouldings. From a styling standpoint it was possible to integrate the rocker mouldings into the design as a continuation of the front spoiler corners(Fig.20). The detail development of the final production rocker mouldings will not be dealt with here, but it can be reported that their addition reduces the Calibra's c_D by .005.

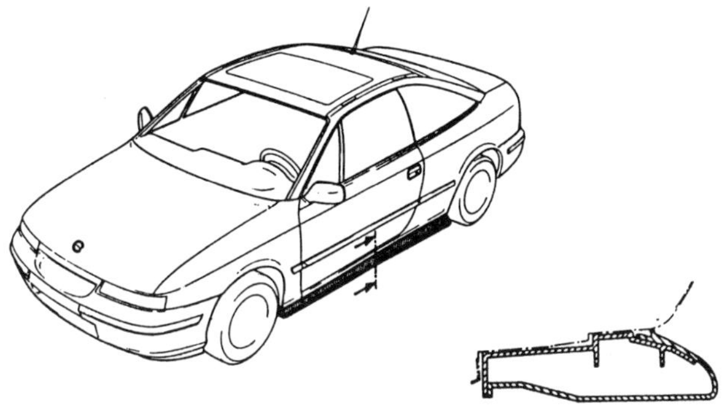

Fig.20: Rocker mouldings, standard on the Calibra, improve aerodynamics and appearance, whilst reducing body contamination.

Various measures in the underbody area were investigated, but most of them were rejected due to either their low potential, or the inability to integrate them into the production concept.
Only one aerodynamic feature remains on the underbody. The plastic tank spoiler, between the fuel tank and the rear bumper, (Fig.21), along with a c_D reduction of .003, also reduces rear end contamination.

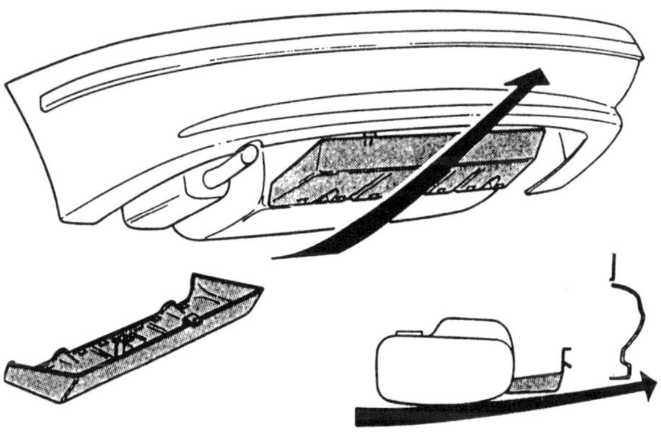

Fig.21: The tank spoiler improves the c_D and reduces rear end contamination.

After it had been established that the underbody measures (spoiler corners, rocker mouldings and tank spoiler) effectively helped to achieve a drag coefficient of c_D = .26 in conventional wind tunnels, it was necessary to verify that they were fully effective during driving on the road, where no boundary layer is present. As coast-down tests are not sensible, due to their inaccuracy and poor reproduceability, wind tunnel tests were once again used, however this time boundary layer influencing measures such as tangential blowing and moving ground were adopted.
These tests were conducted in the German-Dutch Wind Tunnel (DNW) using the publicised equipment outlined in the reference [3], which will be explained later.
Individual results of these tests will be shown in the respective sections, but as a summary it can be said that the achieved improvements measured with moving ground tunnel conditions and other methods for simulating road conditions were better than those measured in conventional wind tunnels.
As the starting point for the spoiler/underbody optimisation which has been presented, was a wake measurement in the full scale Pininfarina wind tunnel, it is appropriate that for comparison the same measurement on a prototype vehicle with all the above measures incorporated should be shown at the end of this development section.
From Fig. 22 and Fig. 23. it can be clearly seen that the sides of the wake near to the ground are considerably smaller. This confirms the strategy that it was necessary to increase the airflow velocity in the underbody area at the centre of the vehicle (Fig. 22) to reduce the wake size and thereby lower the vehicle drag.

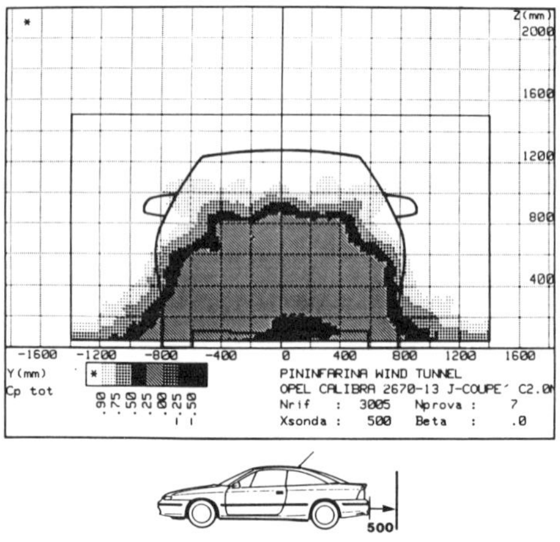

Fig.22: Wake survey output (total pressure) for a prototype Calibra in the Pininfarina Wind tunnel.

BODY CONTAMINATION.

In addition to the aerodynamic development of the Calibra other aspects of the flow around the vehicle were taken into consideration:

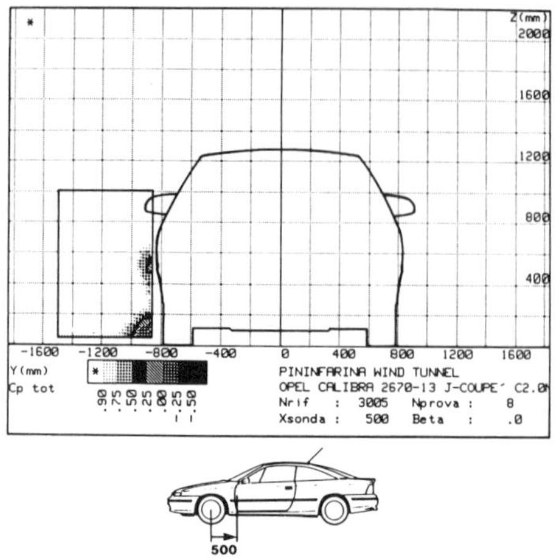

Fig.23: A total pressure map at the side of a prototype Calibra. The map 500 mm behind the front axle is shown.

ANTI-CONTAMINATION LIPS - The lips integrated into the rubber windshield moulding serve the purpose of directing dirty water from the windshield up the 'A' pillar and onto the roof thereby preventing the side windows becoming contaminated in the area of the rear view mirror (Fig.24). Also the outside rear view mirrors have a lip to release dirty water from the mirror housing in such a way that its deposit onto the side windows and mirror glass is prevented.

Fig.24: Anti-contamination lips are integrated into the rubber windshield moulding and the outside rear view mirrors.

THE TANK SPOILER.- The lower surface of the tank, the tank spoiler and the lower surface of the rear bumper have been tuned to induce smooth separation of the airflow which passes below the vehicle. This airflow carries dirty water which is picked up by the tyres from the road, therefore it is important that it is directed away from the wake of the vehicle to keep the rear end clean.

THE ROCKER MOULDINGS.- The rocker mouldings, which are standard on all Calibra versions, not only improve the aerodynamic performance of the vehicle, but they also reduce stone chipping and contamination of the rocker panels and lower door areas.

COOLING AIR INLETS.

The optimum positioning of the front cooling air inlets on a vehicle plays an important part in achieving a low c_D value along with good engine cooling characteristics. From an aerodynamic standpoint the optimum front would be completely closed, as airflow into the engine compartment causes drag. On the Calibra, it was necessary to find a compromise which would allow sufficient airflow for engine and under-hood component cooling whilst minimizing drag.

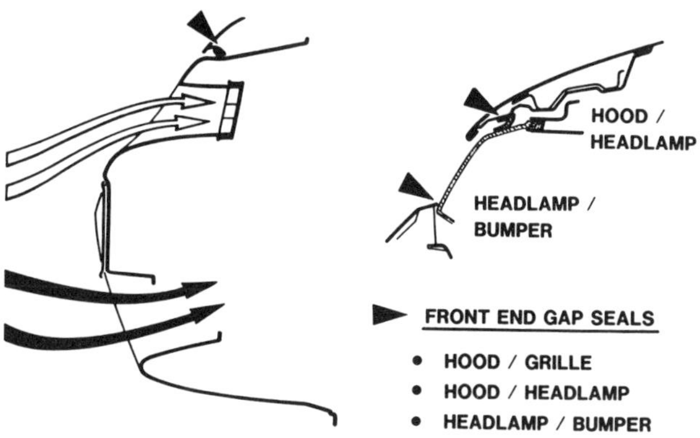

Fig.25: A centre line section through the front of a base Calibra showing cooling air inlets (upper inlet closed) and gap seals.

The front openings (Fig.25) were therefore tuned to provide the correct amount of cooling airflow. On the base vehicle ** it was possible to close the grill inlet above the bumper. When the vehicle is used for towing or automatic transmission is installed it is necessary to open this upper grill, whereby the plastic closing panel will not be installed.
Drag inducing airflow into the engine compartment is minimized by the use of flexible mouldings between the leading edge of the hood and the headlamps and grill and also between the headlamp assemblies and the bumper. A further seal between the bumper and the radiator crossmember is achieved by fastening the bumper facia direct to the crossmember (not shown). It prevents cooling airflow which enters the lower inlet, through the bumper, from passing ineffectively below the radiator.
With the help of this front end opening and front end sealing system it was possible to reduce the aerodynamic losses caused by airflow passing through front end openings and gaps to a creditable $\blacktriangle c_D$=.014, which is very low for such a modern high performance vehicle.

** The base Calibra has a 2.0 litre fuel injection two valve per cylinder engine with catalytic converter and front wheel drive. A five speed manual transmission is installed.

On the sixteen valve engine version the engine compartment is closed from below with a cover (Fig. 26). The front of this cover has air inlets to improve component cooling in the engine compartment. When four wheel drive is installed in combination with this engine, improved gearbox cooling is achieved by shortening the front spoiler lip to improve the airflow into these openings.
From an aerodynamic standpoint there are a number of requirements relative to the front bumper and its air inlets:
- The integrated cooling air inlets must be so positioned and designed that the air which enters passes through the radiator optimally, i.e. if possible with even velocity distribution over the whole surface.
- Air must not be allowed to flow uncontrolled between the bumper and the body because this airflow does not contribute to cooling but it does increase the aerodynamic drag.
- The lower edge of the front bumper must be so designed that it is optimised from an aerodynamic standpoint and also allows temperature sensitive components to be adequately cooled.

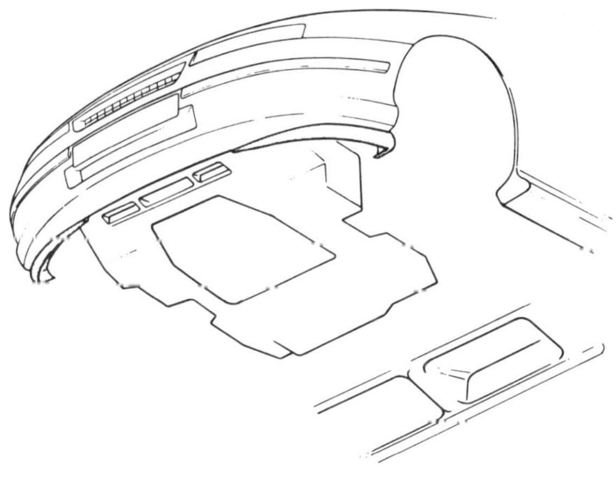

Fig.26: Engine compartment lower cover required for the high performance version, to meet noise legislation.

AERODYNAMIC PERFORMANCE UNDER SIDEWIND CONDITIONS.

During the aerodynamic optimisation of the Opel Calibra considerable attention to vehicle behaviour under sidewind conditions was given.
The optimisation target was a flat curve of drag coefficient over the yaw angle, low lift coefficients on the front and rear axles and low side forces and yawing moments.
Figs.27-31 show the results which were achieved: The drag coefficient curve is relatively flat, even when compared to competition vehicles which have much higher zero degree drag. The same is valid for lift coefficients. These results are even more noteworthy as they have been achieved not with the use of an add-on adjustable rear spoiler, but by the shape of the body exterior surfaces themselves. Side forces and yawing

moment coefficients are comparable with those of modern vehicles exhibiting drag coefficients at or above c_D=.30. The more than ten percent lower drag coefficient of the Calibra, at c_D=.26 has not been achieved at the expense of side-wind sensitivity.
This fact proves that contrary to the statement made in reference [5], which implies that all vehicles with low aerodynamic drag are automatically more side-wind sensitive, enough possibilities exist in the frame of aerodynamic development to avoid the negative aspects of side-wind sensitivity.

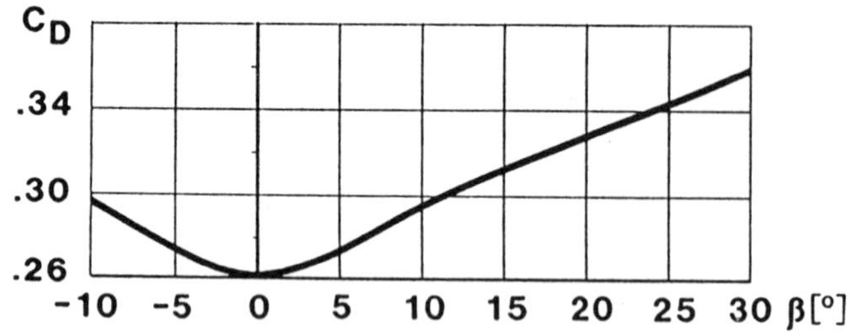

Fig.27: The drag coefficient curve is relatively flat over the yaw range -10° to +30°.

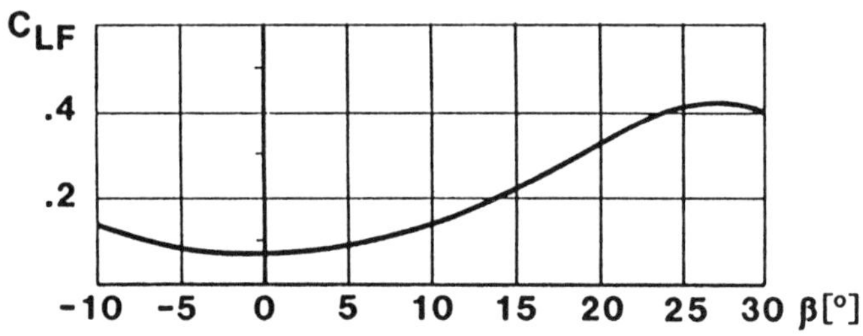

Fig.28: The front lift coefficient over the yaw range.

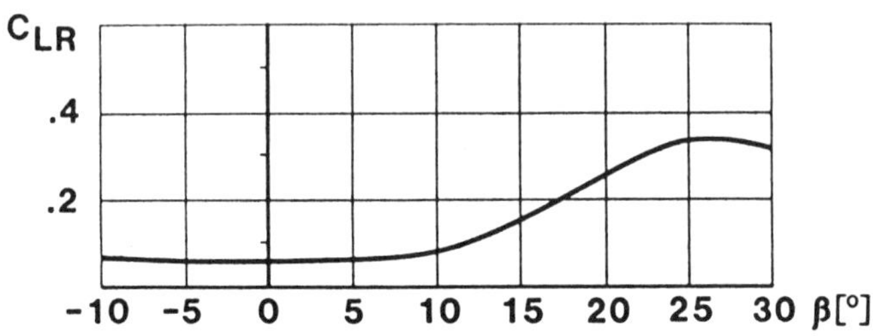

Fig.29: The rear lift coefficient over the yaw range.

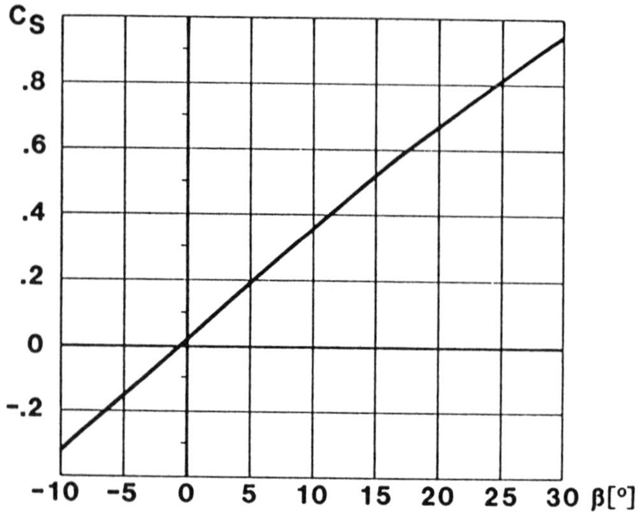

Fig.30: Side force coefficient variation over the yaw range.

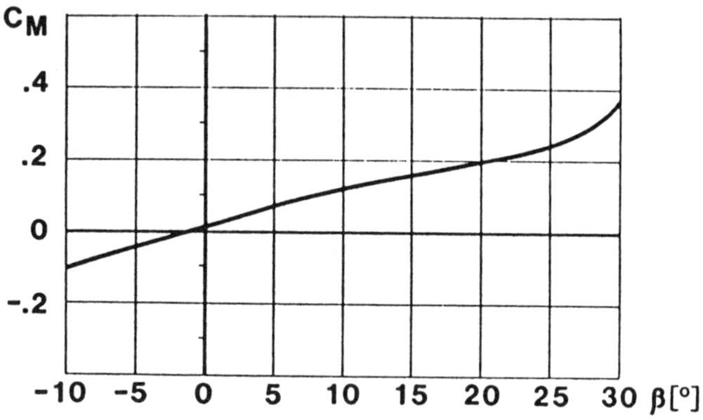

Fig.31: Yawing moment variation over the yaw range.

SUMMARY.

In recent years there have been intensive world-wide efforts to reduce fuel consumption by reducing the aerodynamic drag of motor vehicles in order to contribute to resource savings and exhaust emission reductions.
The aerodynamic development of the Opel Calibra with its c_D of .26, currently the best value for a production vehicle, has shown, with the help of examples, that it is only possible with the use of modern measurement and simulation methods to achieve such results. The fact that an optimum organisation which allows the aerodynamic requirements to be integrated into the design and construction principle must exist is also shown.
At the same time, during the development of the Calibra, aspects relative to future numerical code models were confirmed. Cavities such as wheelhouses, underbody roughness and obstacles such as mirrors need to be fully represented in such CFD models.

ACKNOWLEDGEMENTS.

The authors would like to thank the operators of the various wind tunnels for their support during the tests and the discussion of the results which followed and Mr. C. Makepeace for the translation into the english language.

ATTACHMENT.

INVESTIGATIONS WITH BOUNDARY LAYER INFLUENCING MEASURES.

As previously mentioned, actual road conditions can only be appropriately simulated in the wind tunnel by using a wind tunnel floor with a moving belt. However, this technique requires that the test vehicle be fitted with an internal balance if aerodynamic forces are to be measured.
The vehicle itself must be supported by a so called "sting" above the moving ground. Due to the large proportions of the internal balance, which, along with the aerodynamic forces, must also support the vehicles' weight, it was necessary to extensively modify the inside of the vehicle. This modification work makes the technique extremely time consuming and complicated. It is therefore worthwhile to attempt to find other techniques which require less outlay than the moving belt system, but could deliver the same accuracy in order to achieve equivalent results for development work.
The DNW tunnel has such a technique available, the so called "Tangential Blowing" Technique (Fig.32), [3]. It is true by this technique that the kinematic boundary conditions of a moving vehicle are not met, but by feeding the boundary layer on the wind tunnel floor it is almost possible to fulfil the mechanical flow conditions which occur on the road.

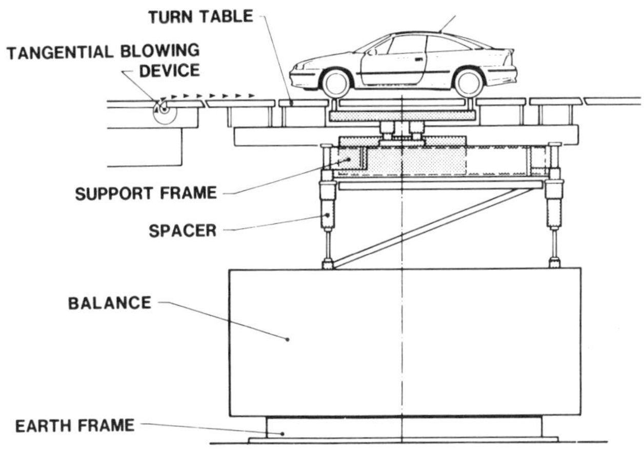

Fig.32: Schematic showing the installation of the tangential blowing set-up.

During the aerodynamic investigation of the Opel Calibra this tangential blowing technique was used, also with regard to future development tasks.

In the following sections results will be shown which refer to tests which were conducted with unmodified boundary layer, with tangential blowing and with moving ground simulation. In addition some tests with a narrower moving belt were conducted. The aim of these tests was to establish whether the results differed when a moving belt between the wheels was used instead of a full width belt, which was wider than the vehicle. The advantage of the narrow belt is that the tyres of the vehicle could stand on a conventional under-floor balance. The narrow belt technique could then be implemented as a routine procedure without the necessity for major modifications to the test vehicle.

INSTALLATION OF THE VEHICLE WITH AN INTERNAL BALANCE.- During the tests with moving belts, 4.8m and 1.2m wide respectively, the belt speed was always matched to the wind speed and the vehicle was supported on a rear sting (Fig.33). The clearance between the tyre patch on the simulated hard foam tyres and the ground was 10mm.

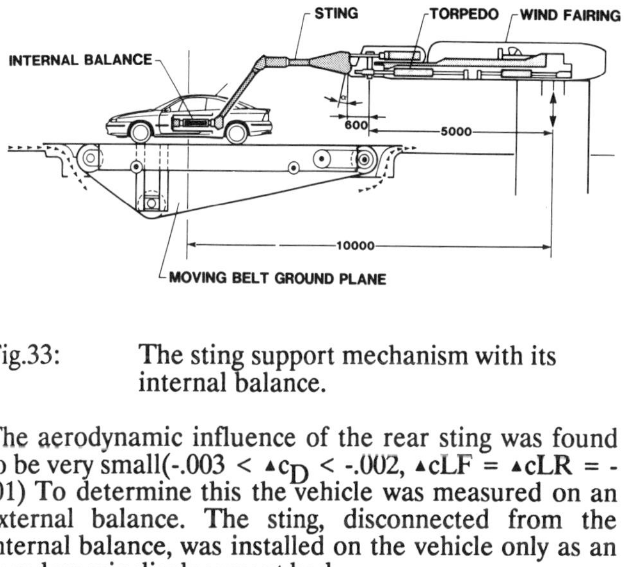

Fig.33: The sting support mechanism with its internal balance.

The aerodynamic influence of the rear sting was found to be very small($-.003 < \Delta c_D < -.002$, $\Delta cLF = \Delta cLR = -.01$) To determine this the vehicle was measured on an external balance. The sting, disconnected from the internal balance, was installed on the vehicle only as an aerodynamic displacement body
A detailed description of the various systems and the flow properties has been presented in [3].

ENGINE COMPARTMENT FLOW LOSSES WITH SMOOTH AND ROUGH UNDERBODIES.- For the interpretation of the influence of the various road surface simulation methods it is important to know the effect of open as well as closed fronts and at the same time rough or production realistic smooth underbodies. Fig.34 shows the effect of these parameters in comparison to a vehicle with a closed front and a rough underbody, when measured in a conventional wind tunnel (CONV).
When compared to the conventional measurements it can be seen that the c_D values for the closed front version with the boundary layer simulation methods tangential blowing (TB) and wide moving belt (MBC) are approximately .010 higher. With a smooth underbody these two methods, TB and MBC, bring a c_D

reduction equivalent to twice that measured with the conventional method.

Also with open front the c_D values for TB and MBC are approximately .010 higher than by convential measurements(CONV), but both values are at the same level.

Contrary to the closed front measurements, a smooth underbody caused a higher c_D value on the open front version. The worsening by conventional measurements was relatively small, Δc_D = + .003, but with TB and MBC it was twice as high.

The share of the drag coefficient losses caused by the engine compartment through flow increased with the smooth underbody configuration. By conventional measurement (CONV) the increase is Δc_D=.009, for TB Δc_D=.008 for MBC Δc_D=.019.

These tests show that it is very important to correctly simulate the engine compartment flow if the effect of underbody measures is to be investigated.

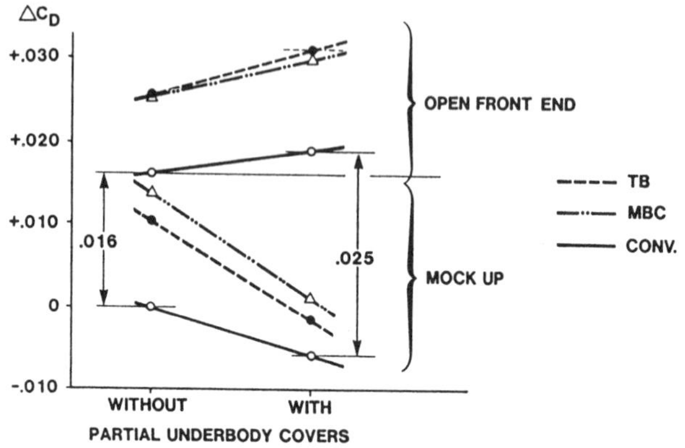

Fig.34: The effect of the ground simulation technique on the aerodynamic losses due to engine compartment airflow.

INVESTIGATION OF VARIOUS FRONT SPOILER LIP LENGTHS.- As previously explained, the optimisation of the front spoiler lip on the Opel Calibra was an important part of the concept. The development of the spoiler corners and the spoiler lip between them has already been presented, therefore only the tests concerning various spoiler lengths will be reported now.

Spoiler lengths down to 50 mm above ground were tested, which is of course unrealistically low for series production vehicles.

Fig. 35 shows the variation of the drag coefficient depending on the spoiler length (constant height over vehicle width). No underbody covers were fitted for these tests and the vehicle front was open. The reference value for the comparison is the drag coefficient measured without spoiler with a clearance of 200 mm to ground (CONV).

Initially it was observed that by reducing the spoiler to ground clearance ranges which were realistically achieveable the drag coefficient was improved. The curve is very similar to that shown in Fig.9. Possibly the flow picture changes very rapidly in this area such that no connection to the curve in the lower ground clearance region was found. This phenomenon will be

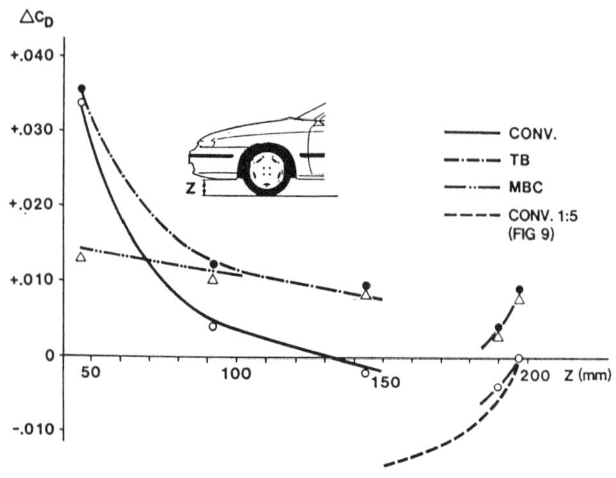

Fig.35: The effect of various front spoiler ground clearances on the drag coefficient.(With various road simulation methods.)

more extensively investigated at a later date. In close ground proximity (z=50mm) there were however large differences between MBC and TB, making the use of TB under these extreme conditions undesireable. In order to explain the rapid increase in drag coefficient for TB and CONV it will also be necessary to conduct further investigations

VARIATION OF THE ROCKER MOULDING HEIGHTS.- The evaluation of the rocker moulding height is shown in Fig. 36. The reference results for this diagram are those measured with conventional boundary layer (CONV) and no rocker mouldings.

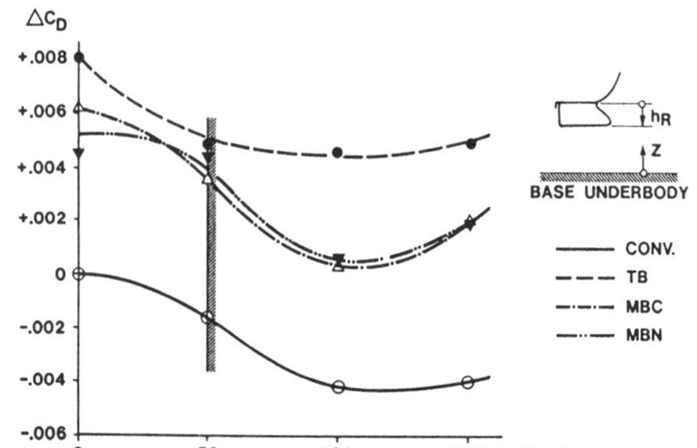

Fig.36: The effect of the rocker moulding heights with various simulation methods.

In addition to the results for TB, those for the 4.8m wide moving belt (MBC) and the reduced width, 1.2m wide moving belt (MBN) are shown. As almost the same values were measured for both moving belt methods it can be assumed that for the purposes of rocker moulding development a narrow moving belt is sufficient.

In contrast, the results with tangential blowing (TB) are all higher than those measured with other methods. Possibly the blowing rate for the tangential blowing was set too high.
All three methods indicated that the lowest drag coefficient was achieveable with a rocker moulding height of 100 - 120 mm (80 - 100 mm ground clearance), but for functional ground clearance reasons this optimum height could not be realised on a series production vehicle. As a result the final spoiler height on the Calibra is 50 mm.

STANDING HEIGHT VARIATIONS.- The influence of various vehicle standing heights on the drag coefficient is shown in Fig. 37. The vehicle was measured with an open front. The tests were only conducted with moving belt road simulation (MBC), therefore a comparison with the other simulation methods is not possible. The reference result for these tests was that measured without underbody covers with design standing heights. In general it was found, as with conventional measurements, that with ground clearance reduction the drag coefficient reduced. The same was established with underbody covers but the c_D level was approximately .005 higher.

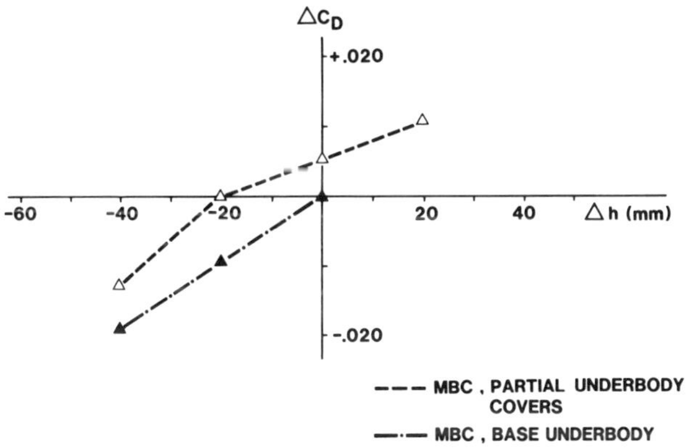

Fig.37: The influence of standing heights with various simulation methods.

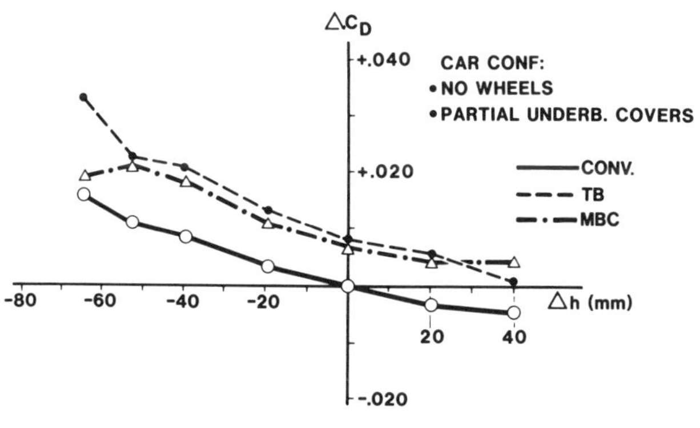

Fig.38: The influence of standing heights with various simulation methods.

Fig. 38 shows the vehicle with underbody covers but no wheels, a configuration which can be compared to configuration 'D' in Fig. 12, except with an open front. Here again the influence of standing heights on the drag coefficient is shown.
In contrast to Fig. 37, the complete vehicle, the observed tendency was reversed. As standing heights were lowered the drag coefficient increased. The same results were observed as previously reported on a 1:5 scale model when measured in a conventional wind tunnel.
From the results in Fig. 38 it is evident that good correlation once again can be seen between moving belt and tangential blowing simulation methods. Only in the lowest vehicle position, was a c_D variation of -.010 evident. However during that test the minimum ground clearance was only 35 mm. Without further testing it is difficult to say whether this ground clearance is the limit of use for the tangential blowing road simulation method, or whether due to the small clearance between the underbody and the moving belt and the associated suctional forces the belt lifted, resulting in falsification of the test results.

INVESTIGATION OF THE LIFT BEHAVIOUR.- As previously stated, an important development goal for the Opel Calibra was a low drag coefficient along with good driving stability and comfort and low front and rear axle lift coefficients.
Fig. 39 shows the front lift coefficient, cLF, of the basic body without wheels with various standing heights.

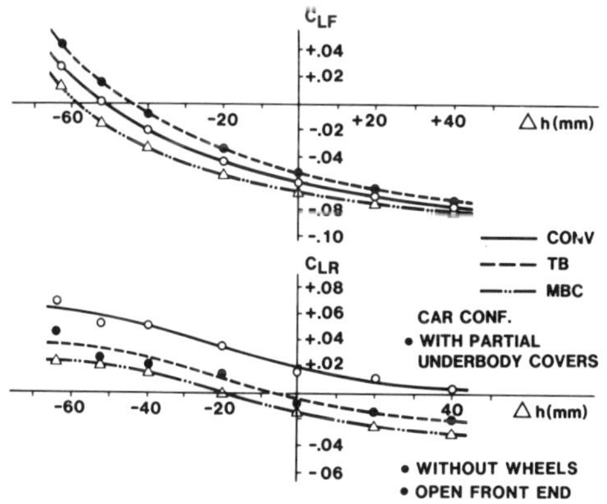

Fig.39: The influence of the standing heights of the basic body shape (without wheels) on the lift coefficients.

It was found that for large ground clearances the three road simulation methods gave identical results. This is plausible since the effect of the boundary layer is reduced as the vehicle's ground clearance increases. When compared to the results with moving belt continually increasing variation was observed as ground clearance was reduced.
It was not possible without further testing to explain why the values measured with tangential blowing (TB) were higher than the values from conventional tests (CONV).

In contrast it can be seen that good correlation of the rear axle lift results exists between the moving belt (MBC) and tangential blowing (TB) methods (Fig. 39). Clearly higher lift coefficients were found with conventional measuring technology. These results are also plausible, since the implementation of ground simulation techniques, as detailed in [6], increases the airflow mass which passes under the vehicle resulting in higher suctional forces which in turn have the effect of reducing lift values.
Similar tendencies were also observed with the complete vehicle, with wheels. Fig. 40 shows these results for the vehicle with and without partial underbody covers. This range of tests with standing height variations was only run with the moving belt simulation.

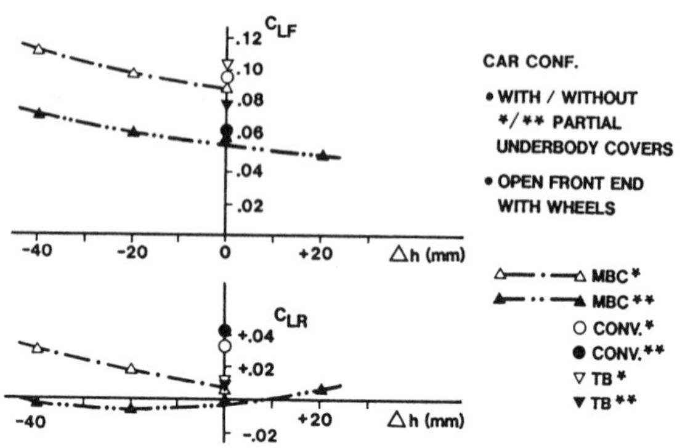

Fig.40: The influence of the standing heights of the complete vehicle on the lift coefficients.

Therefore, for comparison also the values for TB and CONV for the vehicle in the design position have been indicated. As these measurements were made with a gap between the fixed (simulated) wheels and the moving belt it can be assumed that the front lift coefficients for the tests with MBC were measured too low.(See the respective hints in [7]). As detailed in [3], the influence of the gaps is reduced at the rear wheels and very good correlation between the results with moving belt and tangential blowing was achieved.

CONCLUSIONS.

When a road simulation method is used (tangential blowing or moving belt) the results obtained are different to those obtained from conventional measurements. In comparison the drag coefficients with simulation methods are higher.
However there is good correlation between results measured with tangential blowing and moving belt techniques. Only variations between the measured front lift coefficients need clarification. The lift measured with the moving belt system is too low due to the gaps between the simulated tyres and the belt, and the resultant through-flow. This influence is greater on the front axle. Correlation of lift coefficient at the rear axle is good.

Underbody covers heavily influence the share of the drag coefficient which is caused by airflow through the engine compartment. It is therefore very important that the flow through the engine compartment is accurately simulated for the evaluation of underbody covers.

References:

[1] A. Cogotti, "A strategy for optimum surveys of passenger car flow fields", SAE paper 890374, Detroit, February 1989

[2] A. Cogotti, "A two-component fiber-optic LDV system for automative aerodynamic research", SAE paper 880252, Detroit, February 1988

[3] E. Mercker, H. W. Knape, "Ground simulation with moving belt and tangential blowing for full-scale automotive testing in a wind tunnel", SAE paper 890367, Detroit, February 1989

[4] W.-H. Hucho et. al., "Aerodynamic of road vehicles", Butterworths, 1986, ISBN 0-408-01422-9

[5] U. R. Sorgatz, R. Buchheim, "Untersuchungen zum Seitenwindverhalten zukünftiger Fahrzeuge", ATZ 84 (1982) 1. (Investigations into the side-wind behaviour of future vehicles.)

[6] A. Berndtsson et al. The Effect of Groundplane Boundary Layer Control on Automotive Testing in a Wind Tunnel SAE paper 880248, Detroit, Feb. 1988.

[7] A. Cogotti Aerodynamic Characteristics of Car Wheels - Impact of Aerodynamics on Vehicle Design. Int. Journal of Vehicle Design SP3, London 1983

912125

Aerodynamic Performance of a Drag Reduction Device on a Full-Scale Tractor/Trailer

Wendy R. Lanser and James C. Ross
NASA Ames Research Center

Andrew E. Kaufman
Continuum Dynamics, Inc.

ABSTRACT

The effectiveness of an aerodynamic boattail on a tractor/trailer road vehicle was measured in the NASA Ames Research Center 80- by 120- Foot Wind Tunnel. Results are examined for the tractor/trailer with and without the drag reduction device. Pressure measurements and flow visualization show that the aerodynamic boattail traps a vortex or eddy in the corner formed between the device and the rear corner of the trailer. This recirculating flow turns the flow inward as it separates from the edges of the base of the trailer. This modified flow behavior increases the pressure acting over the base area of the truck, thereby reducing the net aerodynamic drag of the vehicle. Drag measurements and pressure distributions in the region of the boattail device are presented for selected configurations. The optimum configuration reduces the overall drag of the tractor/trailer combination by about 10 % at a zero yaw angle. Unsteady pressure measurements do not indicate strong vortex shedding, although the addition of the boattail plates increases high frequency content of the fluctuating pressure.

INTRODUCTION

Following the 1972 fuel crisis, increased gasoline and diesel fuel prices and the uncertainty of future fuel supplies generated immense interest in road vehicle efficiency. Because the aerodynamic drag of road vehicles is a major contributor to the fuel consumption at highway speeds, the Dryden Research Center conducted drag reduction studies on automobiles and trucks (1)*. The aerodynamic drag of modified box-shaped ground vehicles such as vans, trucks, motor homes, and low-drag tractor-trailer configurations were studied by measuring the drag using the coast-down method. Using the coast-down method it was found that the aerodynamic drag coefficients of the test van, at highway speeds were 0.302 and 0.307 respectively, for the full boattail and the truncated boattail configurations. However, road-test techniques proved too time consuming for configuration development, and the repeatability of the test conditions was difficult. In the past In order to overcome these difficulties, most manufacturers have performed small-scale wind-tunnel tests of prototype vehicles providing for more efficient configuration development and greater accuracy in the drag measurements. Full-scale wind-tunnel testing is not common in tractor/trailer aerodynamic studies due to the scarcity of appropriate facilities. Correlation of the small-scale wind-tunnel results with careful road fuel consumption tests has allowed accurate extrapolation of the small-scale test results to full-scale tractor/trailer vehicle configurations.

In 1988, a full-scale test program was conducted in the 80- by 120- Foot Wind Tunnel of the National Full-Scale Aerodynamic Complex

* Numbers in parentheses designate references at end of paper.

(NFAC) at NASA Ames Research Center to measure the effectiveness of a base drag reduction device on a tractor/trailer road vehicle. The device studied is referred to as an aerodynamic boattail and consists of four panels that are set inboard from the edges of the trailer. Figure 1 illustrates the boattail concept for a rigid boattail and the boattail plates. The air flow around the rigid boattail stays attached and smoothly flows over the aft-end of the vehicle

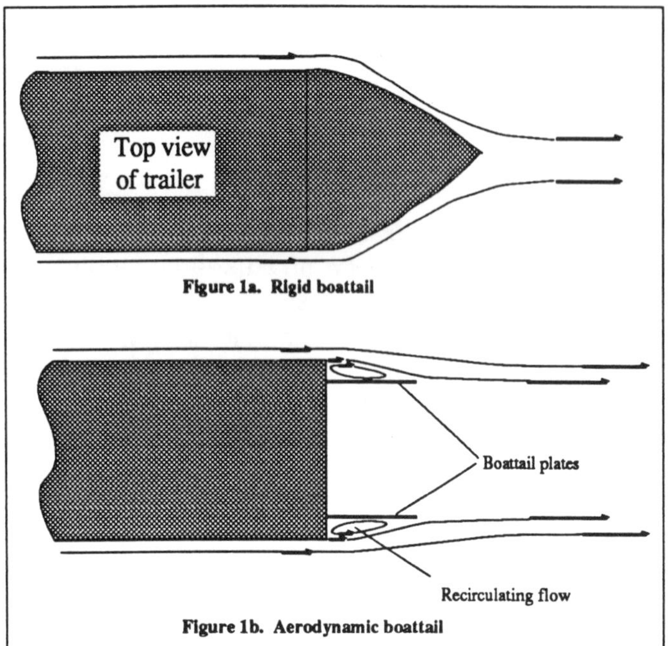

Figure 1. Aerodynamic boattail concepts.

(Figure 1a) so that the base pressure is increased and the drag is reduced (2). Figure 1b shows a sketch of the aerodynamic boattail plate that approximates the rigid boattail effect by trapping a recirculating flow in the corner formed between the device and the rear corner of the trailer. The result is to turn the flow inward slightly as it separates from the rear edge of the trailer: this effectively reduces the base area of the truck. By this process, the pressure acting on the base of the trailer is increased, and the overall drag is reduced.

The full-scale tests were considered significant since there was no information available on the Reynolds number sensitivity of the aerodynamic boattail concept. Additionally, the thickness of the boundary layer on the sides and top of the trailer can have a significant effect on the drag reduction potential; this makes small-scale test results unreliable. This paper presents a description of the full-scale wind tunnel tests along with a summary of the aerodynamic data collected. Included are the drag characteristics of the tractor/trailer as a function of the yaw angle both with and without the boattail plates. The effect of the plates on the pressure distributions over the rear-door panel of the trailer is also presented along with the time varying pressure data on the rear door. Lastly, the results of a limited optimization effort will be discussed.

MODEL DESCRIPTION

A Navistar International Transportation Corporation 9700 tractor and a Fruehauf Corporation 48 foot trailer were used in the wind-tunnel tests: the photograph in Figure 2 shows the tractor and trailer mounted in the 80-by 120- Foot Wind-Tunnel test section.

Figure 2. Photograph of the tractor/trailer in the 80-by 120-Foot Wind Tunnel.

In order to mount side skirts along the length of the trailer, trailer modification was necessary. The rear door panel was also modified in order to mount the boattail panels.

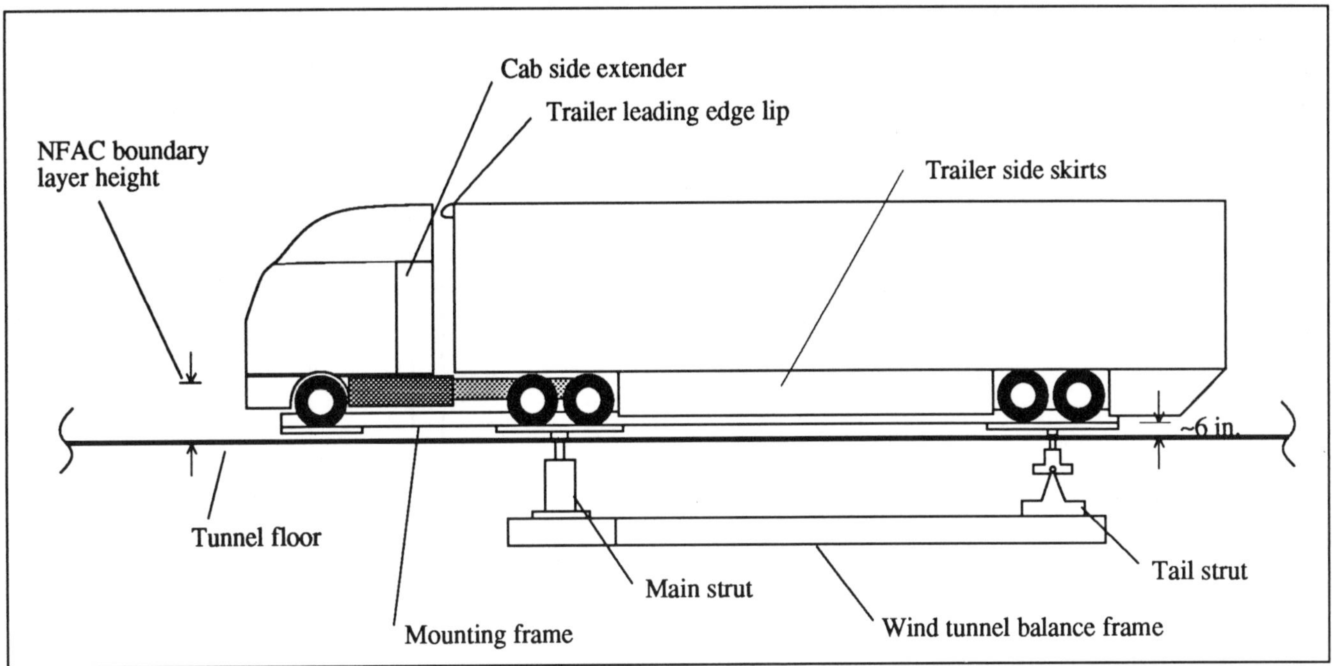

Figure 3. Sketch of truck mounting (side view) connected to the wind-tunnel balance system.

The tractor and the trailer (henceforth referred to as the model) were secured to a support frame. The vehicle tires rested on this support frame (shown in Figure 3), which was connected to the wind-tunnel scale system. This mounting arrangement placed the model approximately 6 inches above the test section floor. The boundary layer thickness at the front of the model was approximately 30 inches; this measurement corresponds to a displacement thickness of about 10 inches (assuming a flat plate turbulent boundary layer in zero pressure gradient). While at the rear of the model, the boundary layer is approximately 44 inches thick with a displacement thickness of about 15 inches. No attempt was made to remove or to reduce the height of the boundary layer. The floor boundary layer and the mounting frame had a small effect on the overall drag measurement. Since changes in the drag due to the boattail plate were the primary interest of the experiment, the mounting frame and boundary layer interference had a small effect on the overall test objectives.

EXPERIMENT DESCRIPTION

The baseline configuration used in the study included the basic tractor/trailer combination with a cab fairing, cab-side extenders (forming about a 17 inch gap between the cab and the trailer), a leading edge lip on the upper forward edge of the trailer, and trailer side skirts, which had an approximate 1 foot gap between their lower edge and the wind-tunnel floor. The drag coefficient of this configuration at 0° yaw was the reference drag coefficient used in the data analysis.

Three independent methods were used to measure the base drag reduction that resulted from the boattail device: 1) the wind-tunnel scale system; 2) load cells that supported the rear door panel; and, 3) pressure distribution on the base panel. Since the wind tunnel scale system measured the total drag on the model and the mounting frame, the reference drag mentioned previously was used to determine changes in base-drag. The load cells and the pressure distributions indicated the base drag directly. All three techniques gave base drag increments that agreed to within ± 10 lbs.

A total of 241 pressure taps and transducers were used to measure the pressure distribution on the rear door of the trailer. The pressure range of each transducer was 10 inches of water, and the majority of the pressure taps were located on the right half of the base panel. In order to check the symmetry at the zero yaw condition, a single row of pressure taps spanning the entire width of the door was installed. Figure 4 shows the location of pressure taps. At 0° yaw, the base drag was estimated by doubling the integrated pressure force acting on the right half of the base panel.

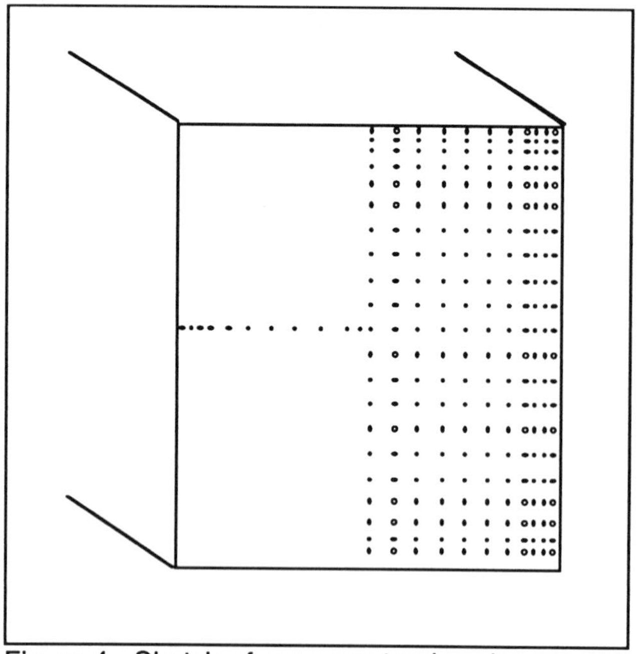

Figure 4. Sketch of pressure tap locations on rear of trailer.

The majority of the test was conducted at a dynamic pressure of 8.4 lbs/ ft^2 (approximately 58 miles per hour): a speed comparable to highway speeds. Additional data were also taken at 24 lbs/ ft^2 (approximately 99 miles per hour) to obtain better resolution on the pressure distribution measurements, and measurements were taken over a yaw range of 0° to ± 15°.

Figure 5 shows a sketch of the boattail device mounted on the trailer. The effects of the plate length, horizontal offset, and vertical offset were examined during the test in order to define an optimum configuration. The geometrical configurations studied are listed in Table 1: L_p ranges from 0 to 0.44; D_v ranges from 0 to 0.15; and D_h ranges from 0 to 0.15.

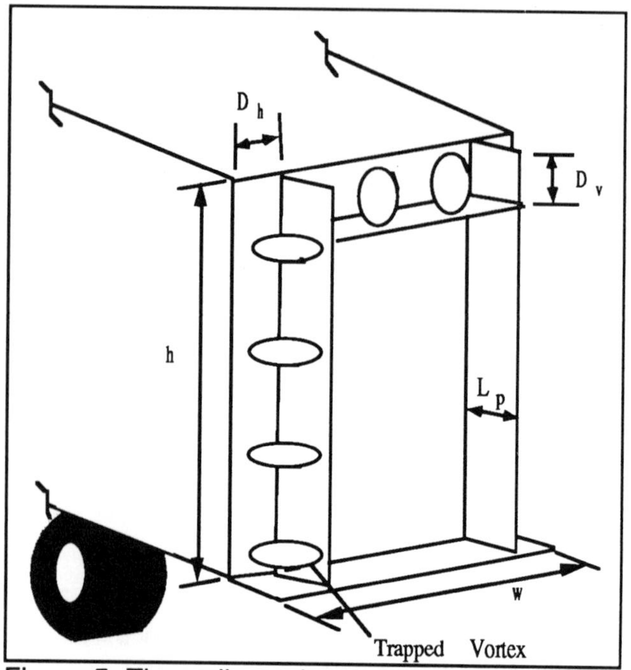

Figure 5. Three dimensional view of the aerodynamic boattail plates mounted on rear of trailer. All plates have the same extension distance from the trailer base, but the inset distance may differ.

Table 1. Geometrical configuratuions tested. All lengths and distances are normalized by the trailer width.

L_p	D_v	D_h
0.0	0.0	0.0
0.24	0.04	0.04
0.24	0.06	0.06
0.24	0.12	0.12
0.30	0.04	0.04
0.30	0.06	0.06
0.30	0.09	0.09
0.30	0.04	0.06
0.30	0.06	0.09
0.36	0.04	0.04
0.36	0.06	0.06
0.36	0.12	0.12
0.36	0.15	0.15
0.36	0.04	0.06
0.44	0.04	0.04
0.44	0.06	0.06
0.44	0.09	0.09

RESULTS AND DISCUSSION

The principle governing the effectiveness of the boattail plate is an increase in the pressure on the rear surface of the trailer that decreases the overall drag on the vehicle. Figure 6 presents the horizontal pressure distribution across the rear-door panel for both the baseline and the optimum boattail-plate configuration at 58 miles per hour and at 0° yaw. Distributions are given for nondimensional distances of z/h (z/h = 0.1, 0.5, and 0.9) from the top of the trailer . Figure 6b shows substantially higher pressure over the area inside the perimeter of the boattail plates than for the no plate case. The pressure is reduced, however, between the boattail plates and the edge of the trailer, and the pressure distribution across the door panel is noted to be nearly symmetric about the vertical centerline of the door panel for zero yaw conditions.

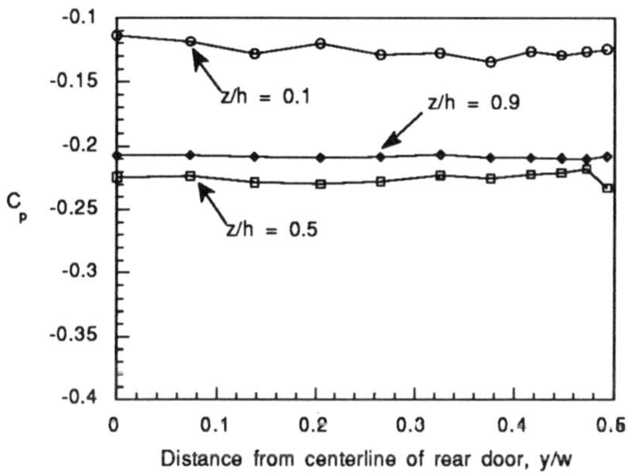

a). Baseline configuration.

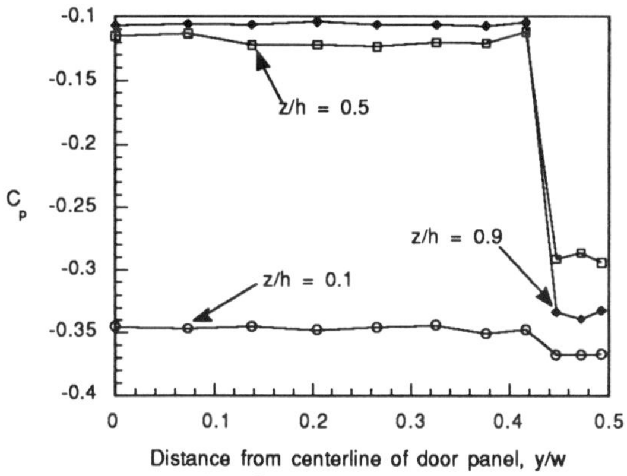

b). Optimum configuration.

Figure 6. Pressure variation across the rear door; yaw = 0°, V = 58 mph.

Figure 7 presents the vertical pressure distribution from the top to the bottom of the rear-door panel for both the baseline and the optimum boattail-plate configuration at three lateral locations across the door panel. In the baseline configuration, the vertical pressure distributions are nearly identical across the width of the door (see Figure 7a). The addition of the boattail plates (Figure 7b) results in a pressure increase over the area enclosed by the boattail plates as seen at y/w of 0.1 and 0.35. A decrease in pressure occurs over the area between the plate and the trailer edge at y/w of 0.45.

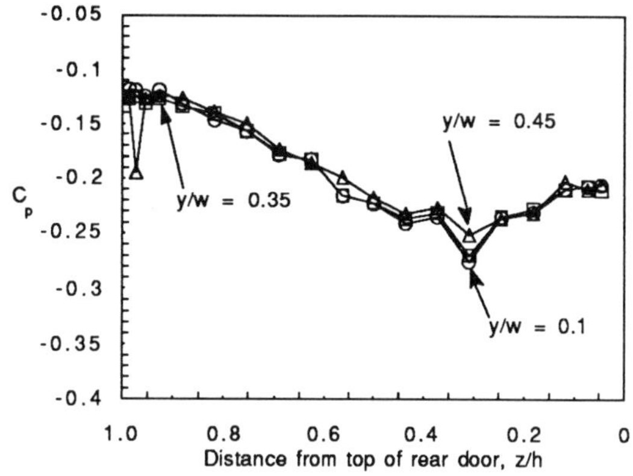

a). Baseline configuration

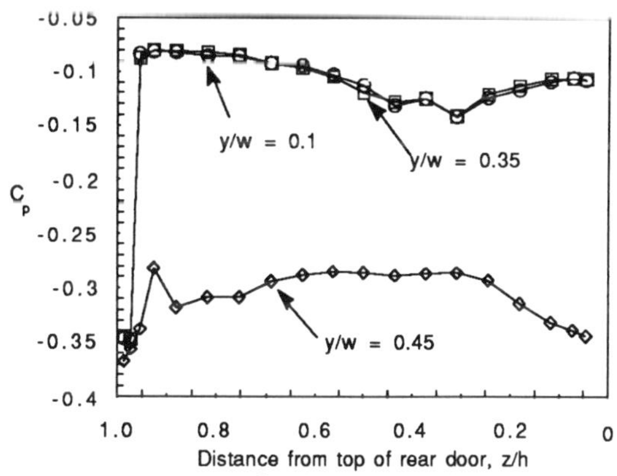

b). Optimum configuration

Figure 7. Pressure variation from top to bottom of the rear door; yaw = 0°, V = 58 mph.

The pressure coefficient data illustrate the same characteristics when the tractor/trailer is yawed at ± 6°; Figure 8. The pressure is fairly uniform across the rear of the baseline configuration at all three yaw angles, but is lower for ± 6° than at 0° yaw. Similar results were obtained for the optimum configuration (Figure 8b). For the yawed conditions, the minimum pressure is located on the leeward side of the base plate, outboard of the plates.

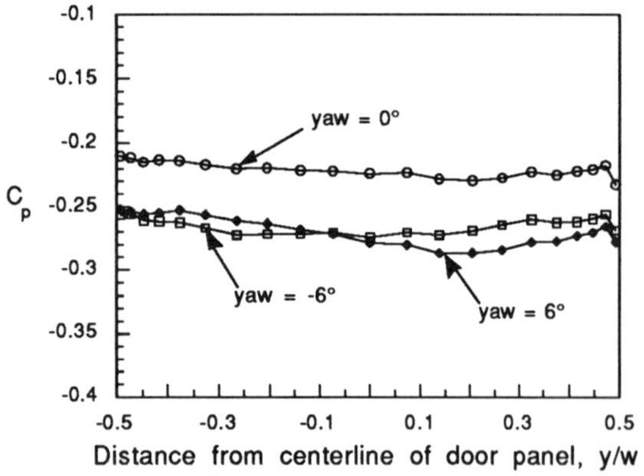

a). Baseline configuration.

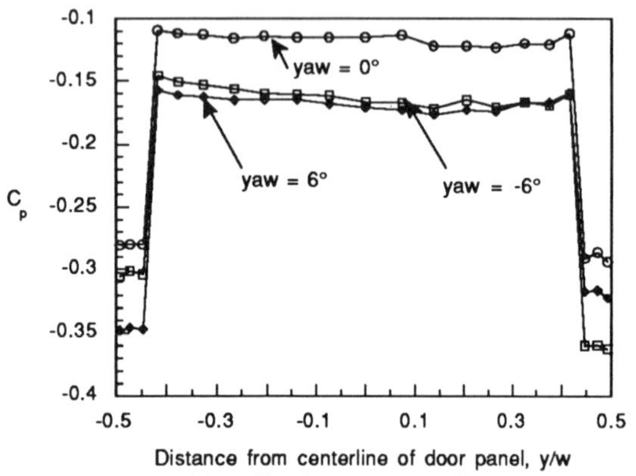

b). Optimum configuration

Figure 8. Pressure variation across the center of the rear door, V = 58 mph, z/h = 0.5.

As mentioned previously, measurements made with the wind tunnel balance system were used to determine the increments in total drag between configurations. The geometric configurations examined consisted of variations in the length of the boattail plates in addition to the horizontal and vertical offsets of the plates from the trailer edges. Figure 9 shows the incremental changes in the normalized drag coefficient as a function of yaw angle for a series of boattail-plate configurations. The data were obtained at a velocity of 58 miles per hour and yaw angles of 0° and ± 9° to determine the geometrical configuration that produced the largest decrease in drag. This optimum geometrical configuration was then examined over the entire yaw range of ± 15°. Data screened at yaw angles of 0°, and ± 9° show the greatest decrease in the drag coefficient at a horizontal and vertical offset of 0.06w and a nondimensional plate length, L_p = 0.36 (Figure 9c). For plate lengths greater than 0.30w, the drag at 9° yaw is more strongly affected by the offset distance than at 0° yaw.

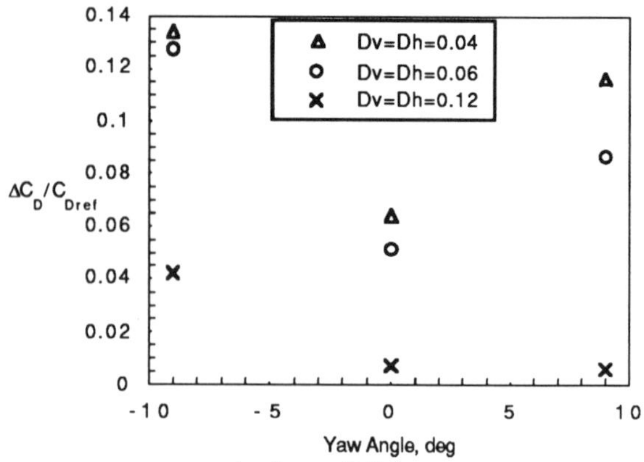

a). $L_p = 0.24$

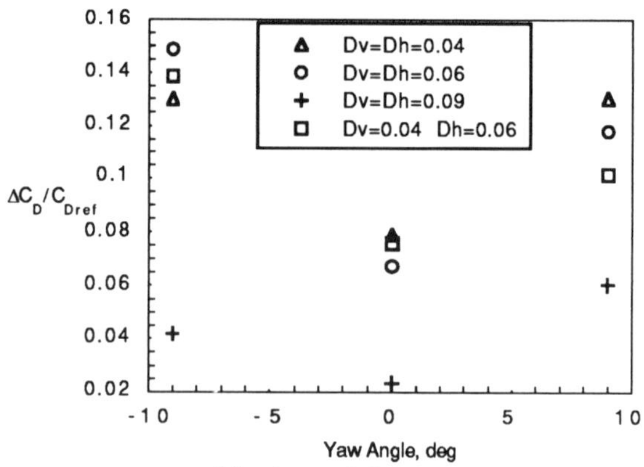

b). $L_p = 0.30$

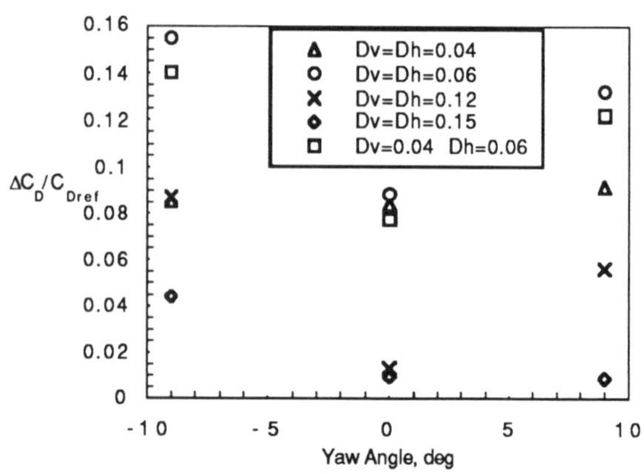

c). $L_p = 0.36$

10 percent over the full yaw range used in the test program.

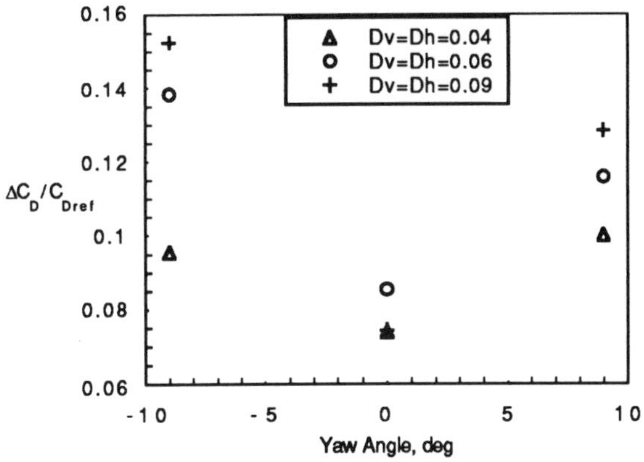

d). $L_p = 0.44$

Figure 9. Effect of yaw angle on base drag for various boattail offset distances and plate lengths.

Figure 10 presents the normalized base drag coefficient as a function of the normalized boattail-plate length for several vertical and horizontal offset positions at 0° yaw. The largest reduction in drag occurs a L_p = 0.36 and $D_v = D_h$ = 0.06.

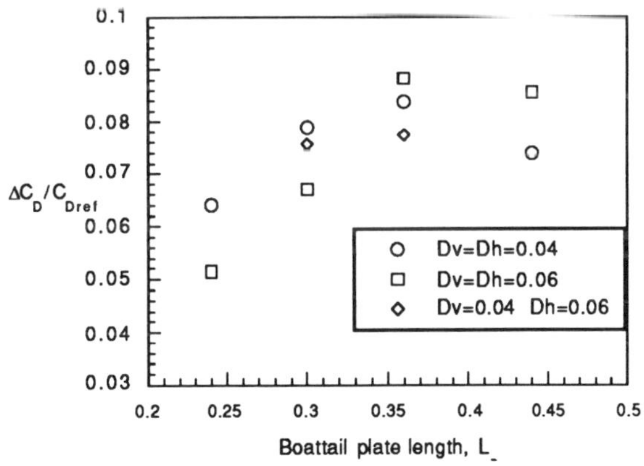

Figure 10. Effect of boattail plate length and offset on base drag, yaw = 0°.

Figure 11 presents the change in drag from the baseline configuration to the optimum configuration over the entire range of yaw angles tested. The change in drag is greater for the yawed conditions than at zero yaw indicating that the drag reduction is sensitive to yaw angle. The base drag reduction device consistently showed a reduction in the normalized drag coefficient of

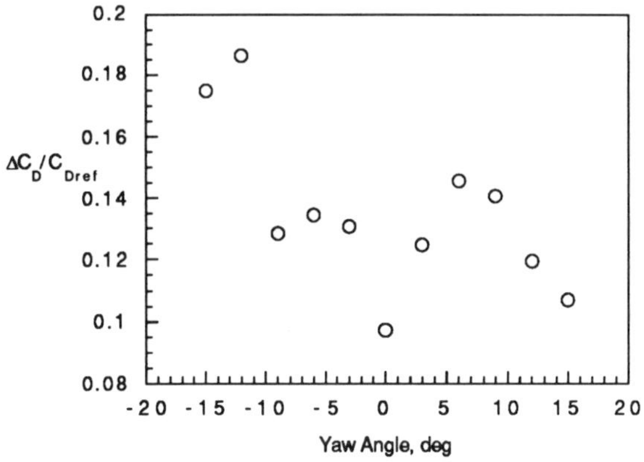

Figure 11. Change in base drag coefficient from the baseline to optimum configuration as a function of yaw angle.

Since the drag is a strong function of the yaw angle, a consistent way to include the effect of yaw in a single number is of interest. A commonly used technique is to compute a wind-averaged drag value using the SAE 7-55 rule (3). This is a weighted integration of the drag data taken at yaw angles up to 15 degrees. The integration is based on a truck speed of 55 miles per hour and an average wind velocity of 7 miles per hour which has equal probability in all directions. By use of such a method the optimum boattail configuration resulted in a wind-averaged drag reduction of approximately 9.8 percent. This wind-averaged drag computation agrees well with the measured results for the yaw range tested.

Unsteady pressure signals from taps located at the center of the trailer rear panel are shown in Figure 12. The signal in Figure 12a is for the baseline configuration while Figure 12b shows the data for the optimum boattail plate configuration. The data was collected at 200 Hz over a period of 10 seconds and was high-pass filtered digitally at 0.6 Hz. Filtering was applied because the trailer and mounting frame had a small amplitude rocking motion when the wind tunnel was running at a frequency of approximately 0.5 Hz.

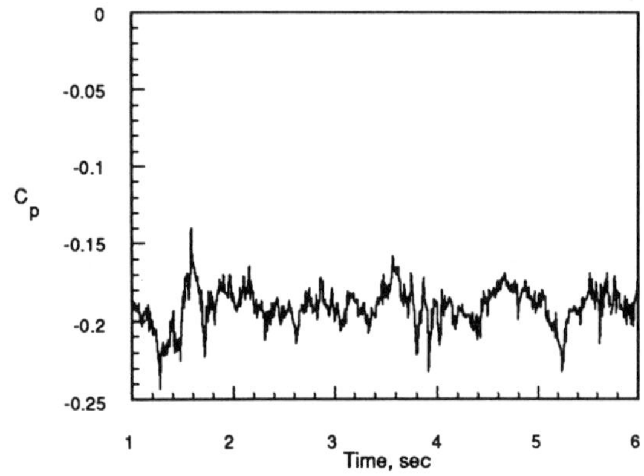

a). Baseline configuration.

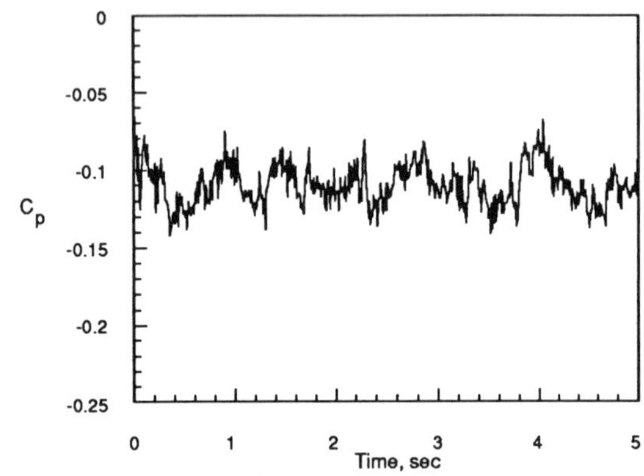

b). Optimum configuration.

Figure 12. Time history at the centerpoint of the rear door; yaw = 0°, V = 99 mph.

The character of the pressure fluctuations is not obviously influenced by the presence of the plates. Figures 13a and 13b present this data as power spectral densities (PSD) for the baseline and optimum configurations respectively. A large peak appears in both cases at 0.8 Hz which may be a remnant of the mechanical oscillations. The only significant and repeatable difference caused by the plates is an overall increase in the high frequency content of the fluctuating pressures. There is a consistent peak between 1.8 and 2.0 Hz which corresponds to a Strouhal number of 0.12 based on the trailer width. This could well be a relatively organized vortex shedding. Smoke and laser light sheet flow visualization showed evidence of discreet vortices being shed into the wake but it was not possible to determine a shedding frequency from the recorded video data.

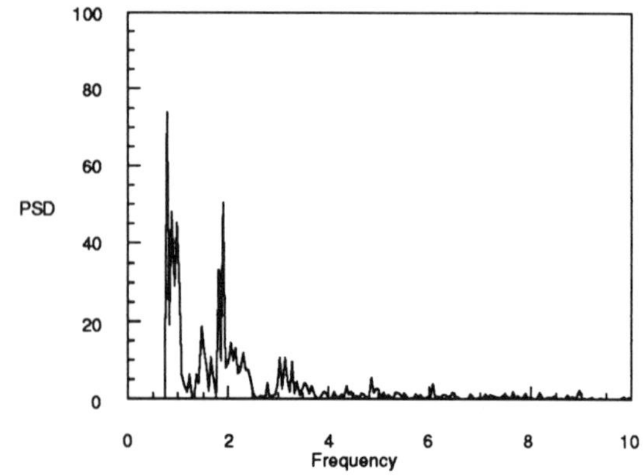

a). Baseline configuration

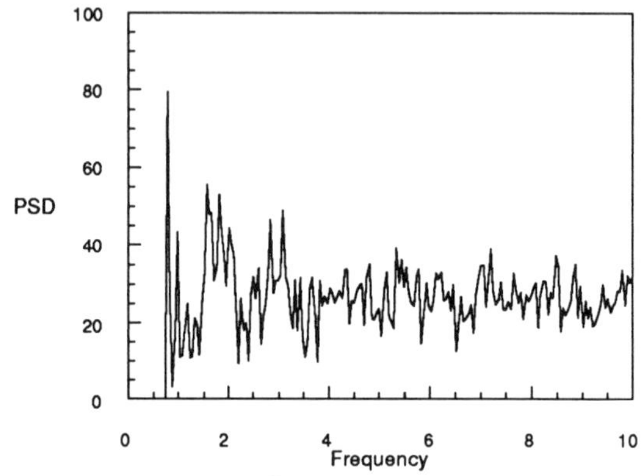

b). Optimum configuration.

Figure 13. PSD at the centerpoint of the rear door; yaw = 0°, V = 99 mph.

CONCLUSION

The results of the present full-scale test indicate that, the aerodynamic boattail is effective for reducing the wind-averaged drag of a tractor/trailer by about 9.8 % over a yaw range of ± 15°. Pressure measurements made on the rear door of the trailer showed that the boattail plates increase the pressure on the area enclosed by the plates. Unsteady pressure measurements on the rear door indicated that the strong vortex shedding by the baseline configuration changed when the boattail plates were installed so that the high frequency activity was increased.

NOMENCLATURE

C_D	drag coefficient = $\frac{D}{qS}$
C_{Dbase}	drag coefficient of base panel = $\frac{Dbase}{qS}$
C_{Dref}	drag coefficient of baseline configuration
C_p	coefficient of pressure
D	drag force, lbs
D_{base}	drag force of base door panel, lbs
D_v	vertical offset of boattail plate divided by the width of trailer
D_h	horizontal offset of boattail plate divided by the width of trailer
h	height of trailer, ft
L_p	boattail plate length divided by the width of trailer
q	dynamic pressure = $\frac{1}{2}\rho V^2$
S	model reference area, 110 ft^2
V	wind-tunnel air velocity
w	width of trailer, ft
y	lateral position referenced to center of rear of trailer
z	vertical position referenced to the top of rear of trailer
ρ	air density

ACKNOWLEDGEMENT

This study was performed under the sponsorship of the National Science Foundation (4). The aerodynamic boattail device is proprietary to Continuum Dynamics, Inc. under Patent # 4682808.

REFERENCES

1. Saltzman, E.J., "A Summary of NASA Dryden's Truck Aerodynamic Research", SAE Paper 821284, Truck & Bus Meeting & Exposition, Indianapolis, Indiana, November 1982.

2. Peterson, R.L., "Drag Reduction Obtained by the Addition of a Boattail to a Box Shaped Vehicle", NASA CR-163113, August 1981.

3. Schaub, U.W., Olson, M.E., and Raimondo, S., "Correction of Wind Tunnel Force Data for Yawed Full-and Half-Scale Truck Models Using a Modified Pressure Signature Method", SAE Paper 90-0187, 1990 SAE International Congress and Exposition, February 1990.

4. Bilanin, A.J., Williamson, G.G., and Geyer, R.G., "Reducing U.S. Import Oil Dependence by One Billion Gallons of Diesel Fuel Per Year", C.D.I. Report No. 85-06, July 1985.

901868

The Aerodynamic Development of the University of Maryland's "Pride of Maryland" Solar Powered Vehicle

Dr. Jewel Barlow, Stephen Brady, Laurence Long, and Timothy Long
Dept. of Aerospace Engineering, University of Maryland

TABLE OF CONTENTS

ABSTRACT

This paper details the aerodynamic development of the University of Maryland's *Pride of Maryland* solar powered race car. This vehicle was built for the General Motors Sunrayce U.S.A., held in July of 1990. Two 3/8 scale models were built and tested in the university's wind tunnel. In addition, computational fluid dynamics and full scale testing were used to verify the design. Race experience further proved the worthiness of the design. Conclusions were drawn between the 3/8 scale model test results and the full scale results.

1.0 INTRODUCTION

1.1 General Introduction

In 1987, the General Motors experimental vehicle "Sunraycer" won the inaugural World Solar Challenge race. This vehicle won by two days over its nearest competition at an average speed of 67 kph. To encourage even more practical research in the field of solar vehicles, General Motors, the U.S. Department of Energy, and the Society of Automotive Engineers sponsored a solar car race in 1990.

GM Sunrayce USA was a trans-national U.S., solar-powered car race for North American university teams. On July 9 the cars left from Walt Disney World's Epcot Center in Lake Buena

Vista, Florida, and traveled 2630 kilometers north on secondary state and county roads. The finish was eleven days later in Warren Michigan. General Motors awarded the top three teams sponsorship for the 1990 World Solar Challenge in Australia.

This paper outlines the aerodynamic development of the *Pride of Maryland*, shown in Figure 1, which began in June of 1989.

1.2 Rules of the Race

Although there were many rules for the G.M. Sunrayce, only those which affect the aerodynamic design of the vehicle will be discussed. The primary factors which effected the body shape and ride height were solar array size, overall dimensions, and aerodynamic stability. The solar array had to be constructed so as to fit inside a right rectangular parallelepiped 4 meters by 2 meters by 1.6 meters. The maximum size of the vehicle while in motion was limited to 6 meters long by 2 meters wide by 1.6 meters high. The vehicle had to have a minimum height of 1 meter. Because of the difficulty in determining the aerodynamic stability of the many different entries, stability was observed during the race. Vehicles demonstrating any aerodynamic instability were to be disqualified from the race. These rules, along with other practical considerations, dictated the rough shape the car would take.

2.0 PRELIMINARY DESIGN

2.1 Design Objectives

Early performance analysis centered on identifying important parameters in maximizing speed on level ground and up constant grades. To aid in this analysis, a steady state simulation code was developed and various test cases were run. Initial values were estimated from previously designed solar cars (Kyle, 1988). Variations about these values were run through the simulation to identify the most important parameters on which the design effort would focus. The variables examined were as follows: weight, aerodynamic drag coefficient, frontal area, rolling resistance coefficient, and power system throughput.

The results of this steady state model coupled with previous competition experience led the team to develop the following list ranking performance factors in decreasing order of importance:

(1) Reliability/Safety,
(2) Power System Efficiency,
(3) Weight,
(4) Aerodynamics
(5) Rolling Resistance
(6) Cost/Ease of Manufacturing.

Due to the relatively few proven solar race car designs, it was decided to use the G.M. Sunraycer as the standard to beat. The final configuration would have to meet or exceed all of the Sunraycer's performance specifications. For example, the Sunraycer's CdA was given as 0.147 m^2 (Hibbs, 1989). Only limited data was available for the Sunraycer's lift performance in yaw (Hibbs, 1989). This information was utilized more in a qualitative sense as a goal to improve upon the stability in cross winds. The consensus was that any car with performance characteristics less than the Sunraycer would not be competitive in the Sunrayce.

Due to uncertainty of the methods and reporting of the G.M. Sunraycer scale results, the CdA of the full scale vehicle was used as the goal for drag area. In studying the available data, it was determined that 0.147 m^2 was not a realistic CdA, but one with modifications made that were not used on the vehicle in race configuration. From the limited information available it was concluded that a realistic CdA for the for the G.M. vehicle was 0.162 m^2, and this was set as the design goal.

2.2 Design Constraints

The problem of aerodynamic development of a solar car was complicated by the constraints imposed by solar array geometry, time and resource availability. In order to have high electrical system efficiency the operating voltage must be in the 100 volt range. This requires a large number of cells to be wired in series. To achieve the highest efficiency possible each of the cells connected in series must be oriented at the same angle to the sun. This electrical concern significantly constrains the aerodynamic design by prohibiting a large amount of curvature in the portion of the body containing the solar array.

Another constraint which became evident as the design progressed is one of resource availability. Wind tunnel time, for exanple, was available only when paying clients were not using the facility. The fall of 1989 was a particularly busy period since two major American car manufacturers were using the facility for scale model development.

The greatest constraint on design options was the severly compressed development schedule. Aerodynamic development was begun in early July of 1989 and essentially complete by early October of 1989. Construction of the full scale molds was scheduled to begin in November of 1990. Many ideas were not pursued because insufficient time existed for resolving complicated interactions.

2.3 Configuration Selection

The major areas of consideration when choosing between competing designs were drag, solar array layout, weight and stability. Since a solar car has a small power potential, it is imperative to minimize the drag. The lowest cost of fabrication for the solar array was for a single flat panel. As the geometric complexity of the solar array increases, so does the price of fabrication. Therefore, it was desired to keep the array shape limited to a flat panel or a minimal variation from a flat panel, for example, a twisted flat panel. Weight had to be kept low so that the rolling resistance would be low. In addition, increased weight requires increased energy for hill climbing. Safety also played a primary roll in the selection process. The rules of the Sunrayce had indicated that the roads would not be closed to other traffic, thereby exposing the solar cars to cross winds from automobiles and trucks. The design chosen had be stable in yaw and not have a tendency to lift up or be forced off the road.

Based upon the available information we considered four basic configurations: pod and table top, pod and tilting table top, pod and fixed table top, and super streamliner (Hibbs, 1989). Figure 2 illustrates these four designs. The three table top designs all had difficulties with stability in crosswinds due to their large projected side area. Furthermore, the addition of a table top increases the wetted area, and hence the drag. The super streamliner avoids this drag penalty by blending the array into the body shape, which also helps to lower the projected side area and improves the stability. Although the fixed tilted table top design has the largest array area and therefore has the best solar gain, it has more weight than the other designs. The super streamliner design has a lower solar gain than the tilting table top or fixed tilted table top; however, this is offset by its excellent stability, drag and weight characteristics. Based on this evaluation, it was decided to attempt a super streamliner design with a flat panel for ease of fabrication.

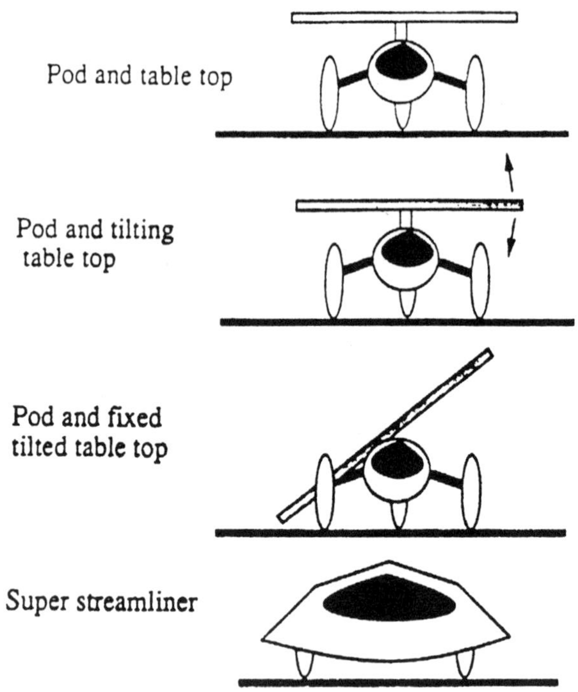

Figure 2

3.0 SCALE MODEL WORK

3.1 Model #1

Since there was not time to do an independent shape development, for low drag, airfoil sections were considered for the basis for choosing the vehicle shape. The first airfoil considered was a NASA LS(1)-0417; however, it soon became apparent that this airfoil was not easily modified to provide a flat array area. Following the LS(1)-0417, it was decided to use a Clark Y-14, since this airfoil already had a flat underside. It was known that interaction with the ground would strongly influence the aerodynamics. No useful data on an inverted airfoil in close proximity to the ground was available to the team. A wind tunnel test was chosen as the quickest way to obtain reliable information. The frontal area was dictated by the volume necessary to hold the driver and components.

In August of 1989, the configuration was tested in the Glenn L. Martin Wind Tunnel at the University of Maryland. This design proved to have a CdA which we deemed uncompetitive (see Figure 3). The vehicle developed a large downforce, which increased the rolling resistance and induced drag. This is similar to ground effect cars used in many racing series. The design was also unstable, having excessive yawing and other moments. The large pitching, rolling and yawing moments in crosswinds would cause vehicle handling problems. If the design had shown more promise, it might have been modified to obtain the desired performance characteristics. However, since time was running short, it was decided that a more reliable approach would be to choose a new baseline shape more closely resembling the G.M.

Sunraycer and accepting the increased cost of a non-planar solar array.

CdA's of Various Configurations

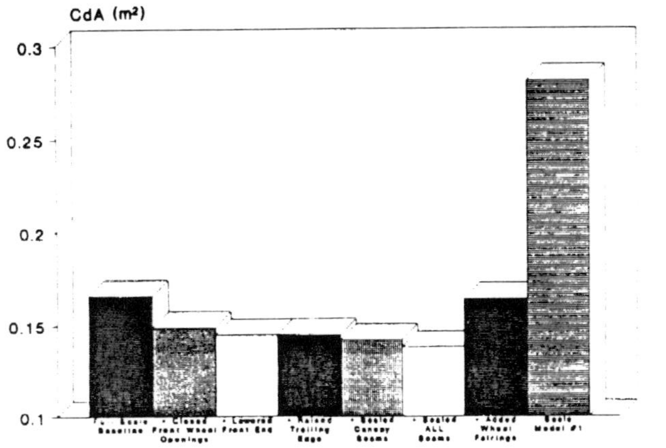

Figure 3

3.2 Model #2

3.2.1 Design

During the testing of model #1 the design team was notified that another three days of wind tunnel time would become available starting September 25, 1989. This, coupled with the poor results for model #1, forced a quick revision of the aerodynamic design. Protracted discussions were held with the electrical portion of the design team. The aerodynamics team argued that the aerodynamic advantages of a twisted panel out weighed the expected loss in array power. When all the dust had settled four days remained in which to construct a new wind tunnel model.

Three concepts more closely resembling the Sunraycer type were developed each with varying degrees of panel twist (see Figure 4). From these drawings a decision was made to construct the model concept with the twisted panel on the basis of frontal area. It was felt that this would provide the best performance of the three, but would result in the greatest increase in panel cost. The wing section profile chosen was a NACA 63-012, due to its low drag characteristics and the fact that its shape is designed to encourage laminar flow. To this wing section, the mean camber formulas were applied, as outlined in Abbot and Von Doenhoff, 1959. This was done to adjust the height of the lateral edge to the chosen value of 1/3 the thickness. After the airfoil camber had been adjusted, further modifications were made to eliminate the cusp along the bottom of the airfoil section. This was done in an attempt to eliminate some of the expected lift. General Motors, using VS AERO, a numerical flow field simulation code, had shown that this change would not greatly effect the other characteristics of the airfoil. The back of the top surface of the profile section was also flattened to accommodate the solar array.

The resulting airfoil section geometry generated by the above process was then three dimensionalized to obtain a triangular frontal projection for the upper

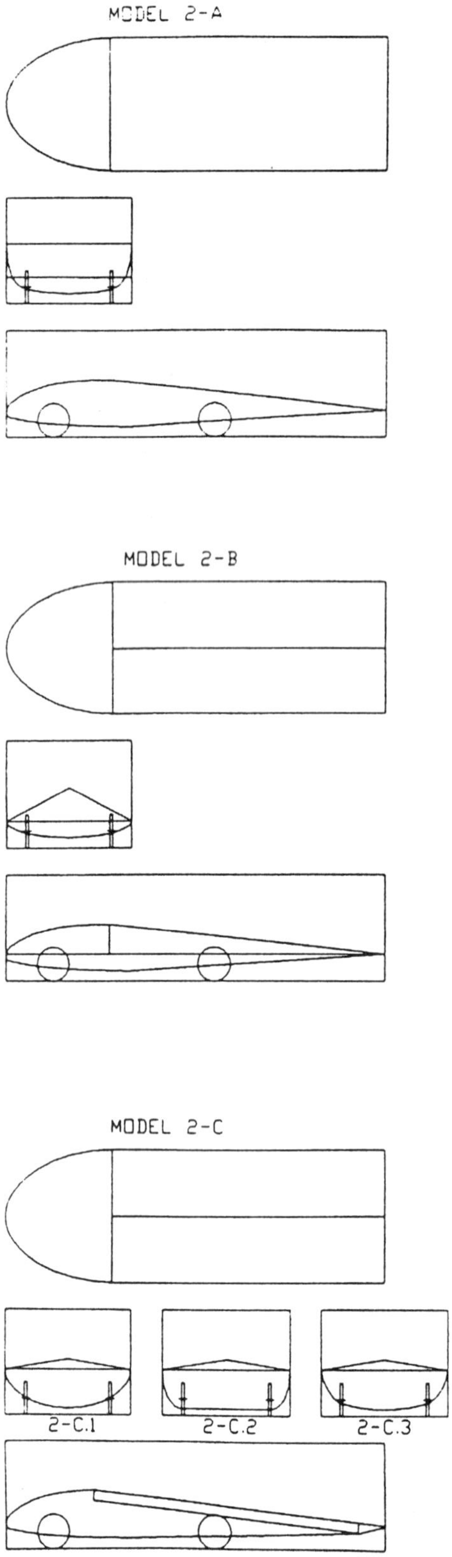

Figure 4

portion of the car and an elliptical front projection for the lower portion of the car. This combination was selected on the basis of need to minimize frontal area while obtaining sufficient volume for all components plus driver, and limiting array distortion from a plane surface to a simple twist. The latter constraint produces the triangular form for the upper part of the car.

The output of the solar array of the finished car exceeded that of any other single seat car in the Sunrayce competition. This provided verification of the electrical part of the design goal of maintaining high array output while striving for good aerodynamics.

3.2.2 Model #2 Results

The second wind tunnel model was tested in the Glenn L. Martin Wind Tunnel. Data collected included drag, lift, side force, yawing moment, pitching moment and rolling moment (see Figures 5-10). Ride height and attitude were varied to find optimum values yielding the least drag. The use of wheel fairings to reduce drag was also explored.

The vehicle height and attitude were chosen on the basis of minimum achieved drag with unfaired wheels. Figures 5 through 10 show forces and moments as a function of yaw for the optimum height and attitude. Results are shown for faired and unfaired wheels from model #2 along with full scale (to be

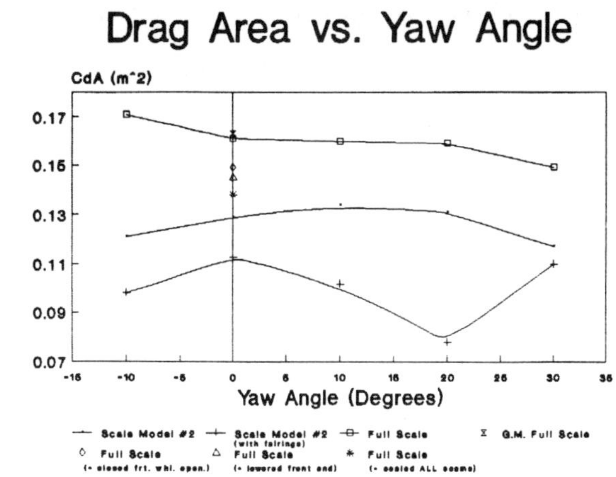

Figure 5

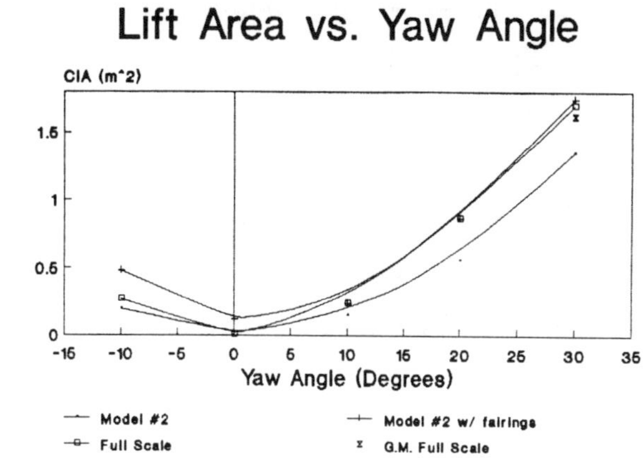

Figure 6

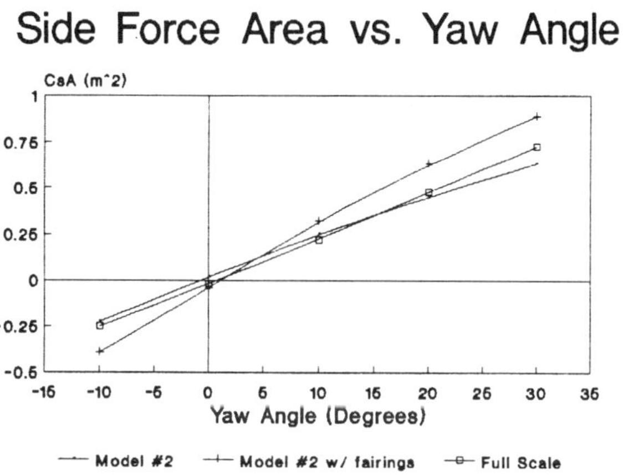

Figure 7

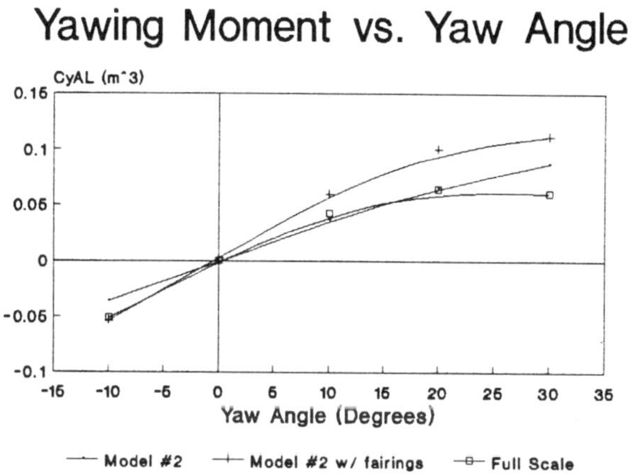

Figure 8

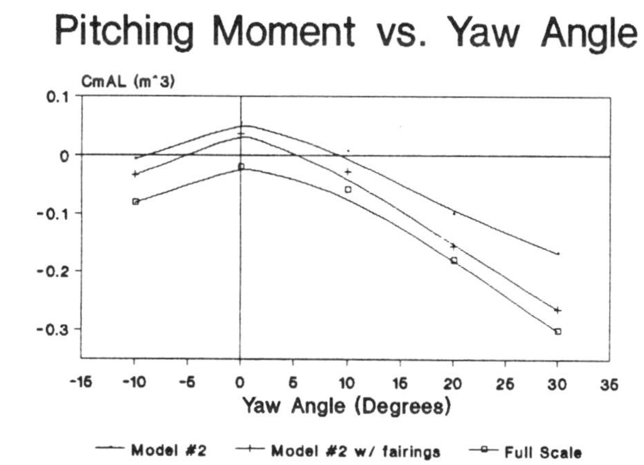

Figure 9

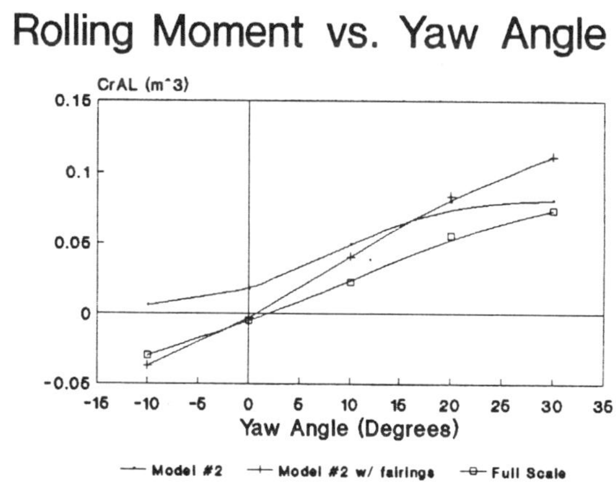

Figure 10

discussed later) and available data from the G. M. Sunraycer (Hibbs 1989). In Figure 5, the drag area of model #2 is shown to be 0.125 m^2 which is below the operational full scale G.M. target of 0.162 m^2, however adding wheel fairings decreased the drag area to 0.112 m^2. The drag curve for the faired wheels took on an interesting shape in that it has a dip in the curve at around 20 degrees yaw. This effect was repeated for all test runs with fairings added. Figure 6 represents the lift area of the model as a function of yaw angle. A yaw of 30 degrees represents a strong cross wind. Using this information the effect of a 50 kph cross wind was calculated to impose 53.74 kg of lift on the vehicle at a road speed of 80 kph. The vehicles weight was projected at 260 kg, so that a lift of 53.74 kg was judged acceptable for stability. The addition of wheel fairings increased the lift on the vehicle, but it was still at an acceptable level. Figure 7 displays the increasing side force with increasing yaw angle. The vehicle was determined to be stable because the crosswind component of 50 kph as above, imposed only a 30.41 kg side loading on the vehicle. This is only a 0.12g lateral load and the vehicle was designed to exceed 0.7g lateral loads. The addition of wheel fairings generated higher side forces, but again at acceptable levels. Figure 8 shows the effect of yaw angle on yawing moment. The side forces exerted on the front wheels due to the 50 kph crosswind were calculated at 25 kg and the rear wheels 5 kg. Some concern for the yaw stability of the vehicle was caused by the fact that the G. M. Sunraycer which is a similar configuration, had fins added to the rear of the vehicle to increase its yaw stability. Fins were constructed to install on the model but lack of time prevented testing. Experience with the full scale vehicle later proved them to be unnecessary. The addition of wheel fairings increased the yawing moment on the vehicle. In Figure 9 the pitching moment characteristics are displayed. Using this information, the calculations for the 50 kph crosswind creates 12.101 kg downforce on the front wheels which is advantageous. The rear wheels however experience a lifting force of 53 kg. The rear wheels were expected to carry a load of 130 kg so it was determined that the lift would not present any control problems. The addition of wheel fairings increased the negative pitching moment. Figure 10 shows the effect of yawing on the rolling moment of the vehicle. The previously described 50 kph crosswind exerts only about 25 kg of lift on the upwind wheels of the vehicle which is much less than it would take to overturn the vehicle.

4.0 CFD ANALYSIS

4.1 Test Facilities

The *Pride of Maryland* design team was provided with an opportunity to apply a VS AERO solution to study the configuration aerodynamics late in the design process. Earlier thought had been given to using a panel method code for the initial aerodynamic design but the codes available on campus at that time had just been installed and were not fully operational; furthermore, there was no simple way of entering the geometry, making quick changes, and viewing the results. After the testing of model #2, a contact was made with Dr. Nils Salvesen at Science Applications International Corporation (SAIC) in Annapolis, Maryland. SAIC has developed IDEAS (Interactive Design Evaluation and Design Analysis System). This this system as outlined in Figure 11 permits the use of complex general purpose flow codes in a systematic, iterative manner.

Work was started at SAIC as construction began on the full scale plug. It was obvious that no substantial changes to the car would be made; but it was decided to proceed with the CFD work in order to gain experience with the IDEAS system and provide a foundation of CFD work on which to build for any future competition. This also allowed a chance to evaluate IDEAS with an eye towards its possible use as a teaching tool. The experience with IDEAS was positive and the system is now being acquired for use at the University of Maryland.

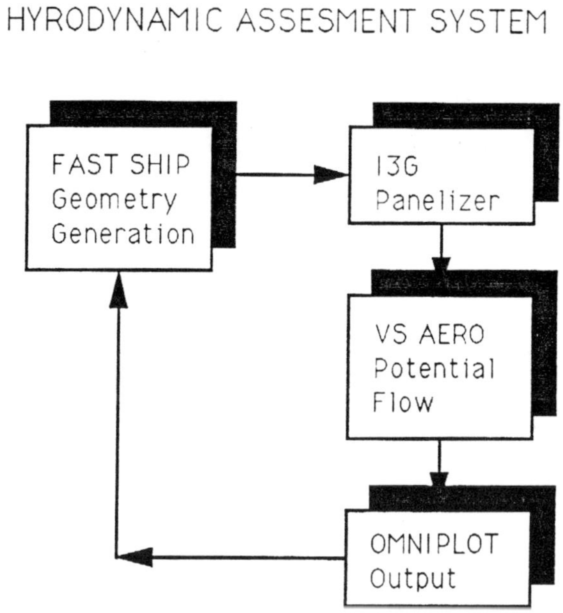

Figure 11

IDEAS actually consists of four separate programs. The first program used in starting the design analysis process is Fast Ship, which facilitates geometry input and subsequent manipulation through on screen interactions. Surfaces were defined by a net of spline control points. By moving the control points of the spline net large scale changes could be implemented quickly. The model two geometry rendered in Fast Ship is shown in figure

12. The I3G program was used as an interactive panelizer to prepare the geometry for the flow code The panelization is shown in Figure 13. The flow code utilized was VS AERO from Analytical Methods, Inc. It used a three-dimensional panel method. After the car's surface had been panelized, a doublet was placed on each panel and adjusted by the program until there was no flow through the body. A wake of trailing vortices could then be attached to the trailing edge to model the wake and allow the computation of lift and induced drag (Maskew, 1983). The output of VS AERO is displayed using OmniPlot, a graphics post processing code.

4.2 CFD Results

The results of a single VS AERO run were voluminous. One way to present this data was through the use of color surface plots. OmniPlot could also render 2-D section cuts along any plane. Figure 14 shows the coefficient of pressure plotted over the surface of the body, while Figure 15 shows the coefficient of friction plotted over the surface of the body. The iterated wake is also shown. Figure 16 shows a plot of surface streamlines. It reveals that the flow on the surface of the car is well behaved. The smoke studies conducted on both the scale models and the completed car also demonstrated similar qualities.

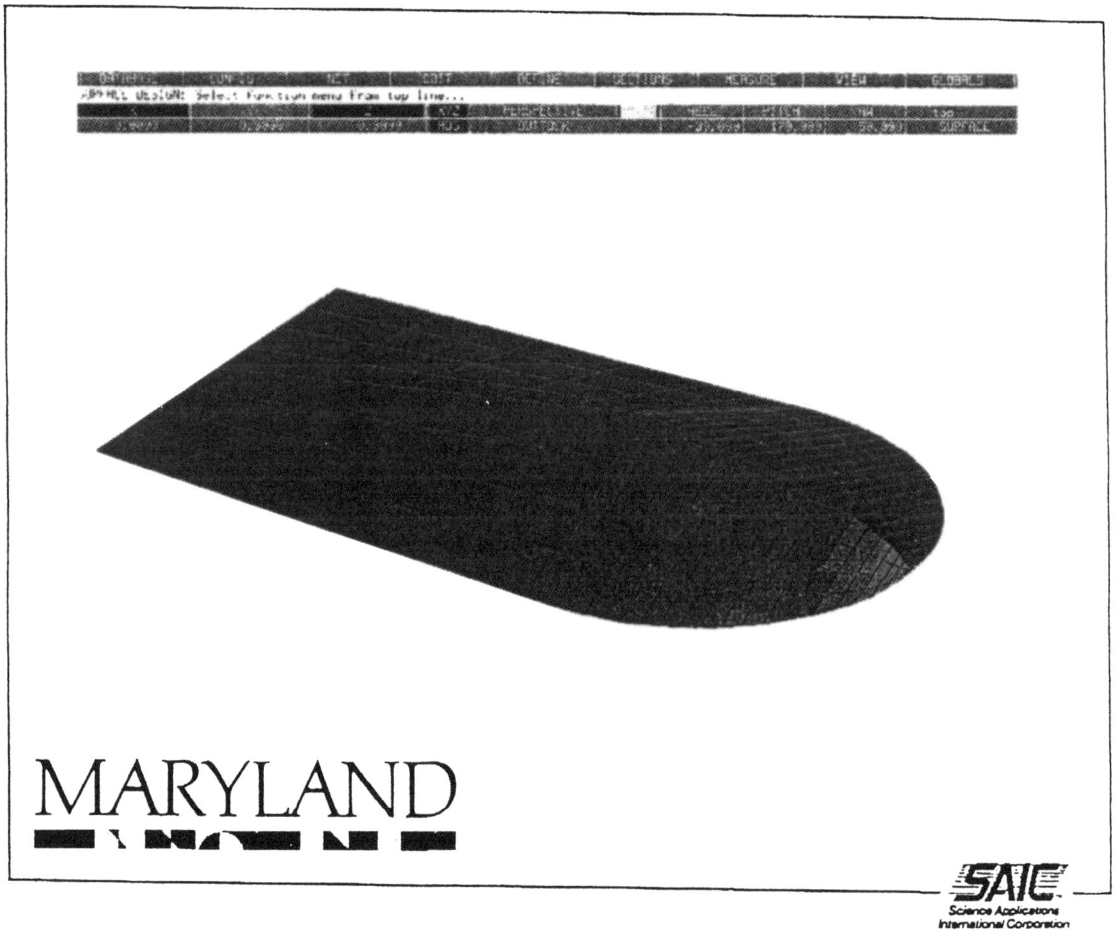

Figure 12

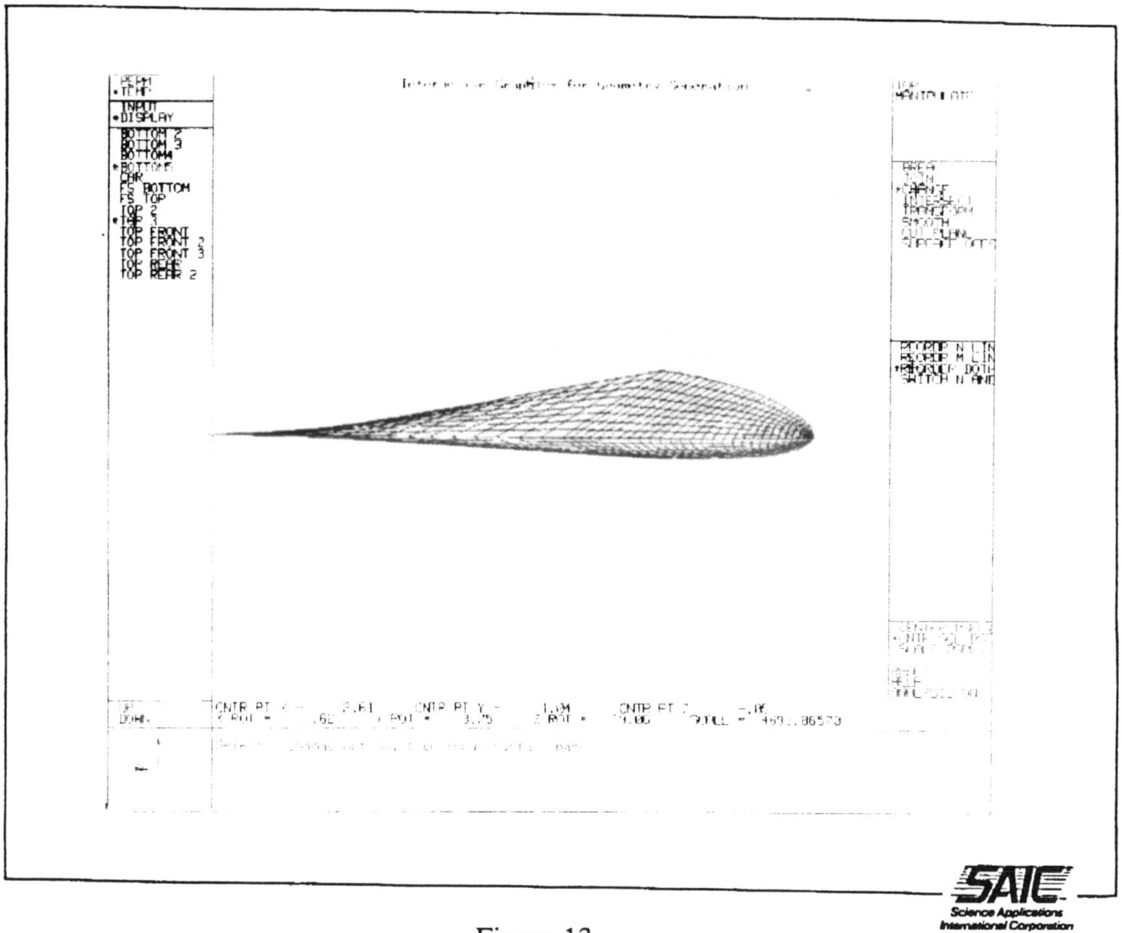

Figure 13

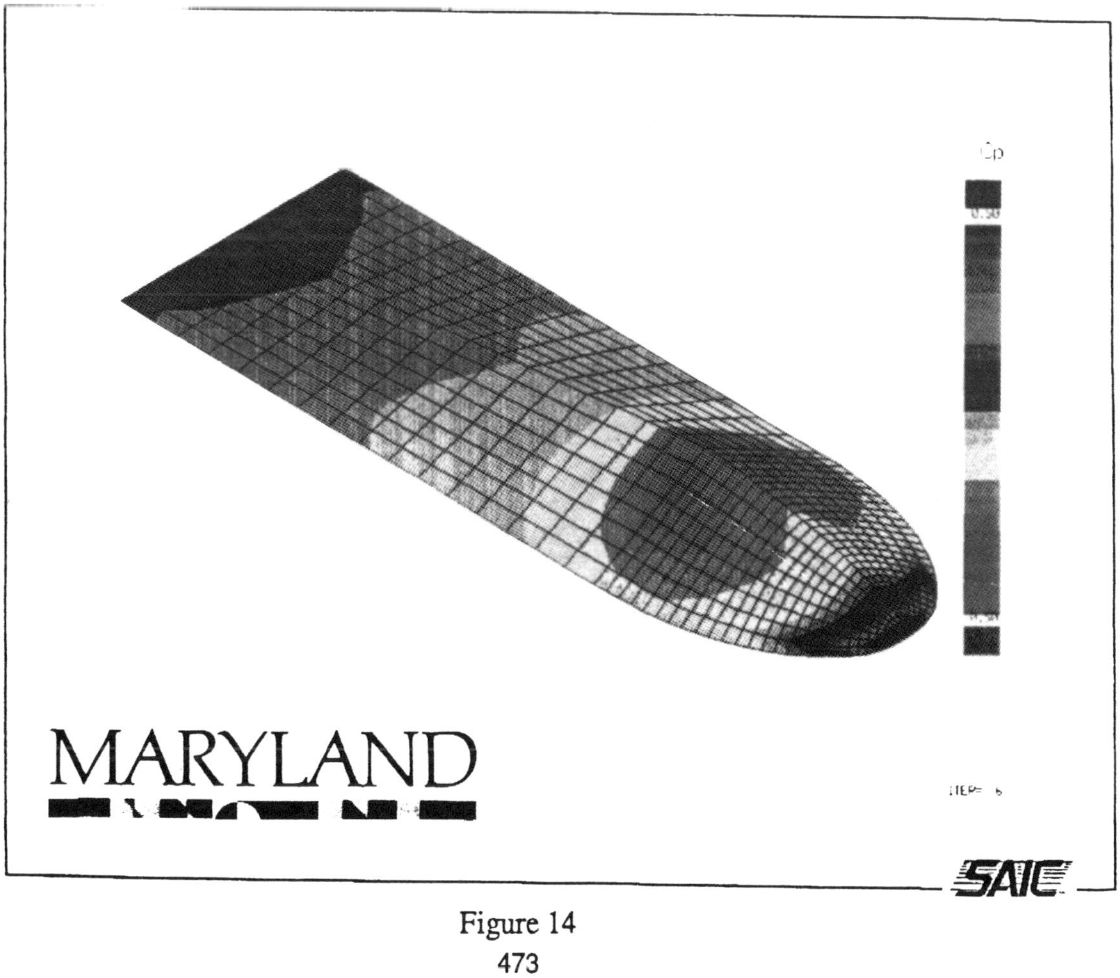

Figure 14

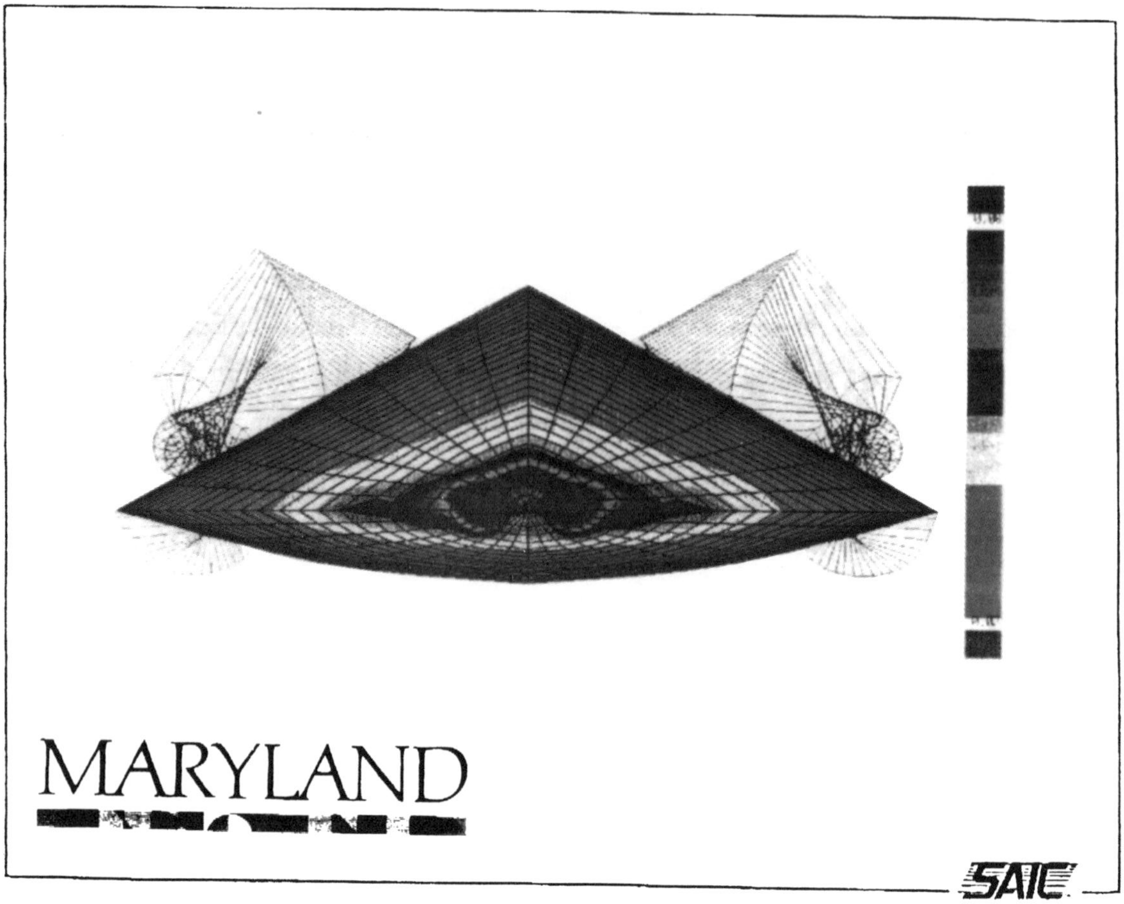

Figure 15

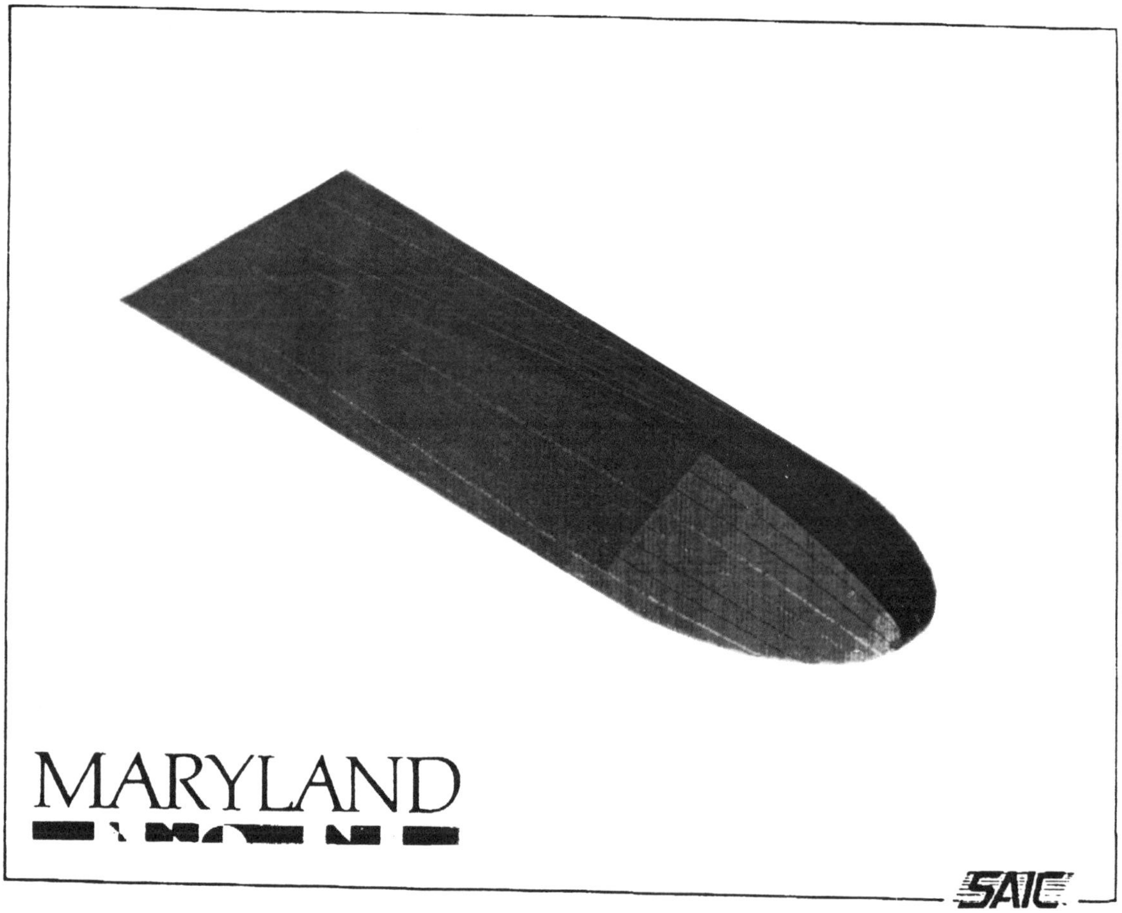

Figure 16

Figure 17 shows a plot of coefficient of pressure over the two center surface streamlines, one over the top surface the other over the bottom. The large pressure gradient seen around the peak on the upper surface can also be seen in the Cp surface plot (Figure 14). This result is due to the fact that the geometry entered in the computer had a discontinuity in slope about this point . On the scale model and the actual car this discontinuity was removed in the manufacturing process.

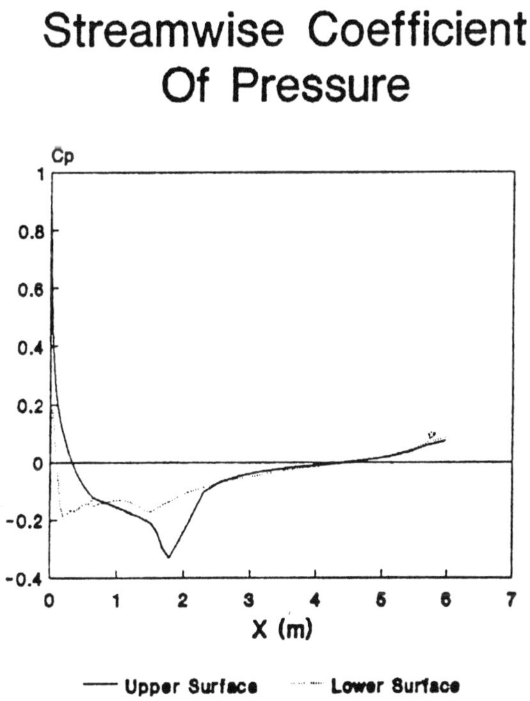

Figure 17

5.0 FULL SCALE RESULTS

Several design modifications were required due to suspension clearance and canopy optical properties. It was discovered that insufficient clearance for wheel travel existed in the optimum low-drag configuration. Furthermore, it was learned that it would be impossible to form a distortion free canopy to a triangular cross-section.

The wheel travel problem was solved by narrowing the track and wheelbase, this brought the wheels closer to the maximum thickness of the body. This change was permitted by a late rule change which relaxed the tip-over stability requirement. To solve the canopy problem, it was decided to make the nose elliptical in cross-section and smoothly blend the new cross-section to the triangular array cross-section. Coincidentally, this modification further alleviated thc suspension travel problems by providing a larger clearance for the front wheels.

These changes were incorporated into the computer code which provided the coordinates for the surface of the body. The data was output in a format so that full scale cross-sections could be plotted for plug construction. It was decided that although these geometry changes were significant, they did not warrant building a third 3/8 scale model. In addition, the race date was rapidly approaching and it was doubtful if the tests could be completed in time for the results to be useful.

5.1 Race Experience

The G.M. Sunrayce was conducted on two-lane public highways and secondary state roads open to all traffic. The car exhibited excellent aerodynamic stability during the race at speeds up to 100 kph. At those speeds, the lift forces did not effect the vehicle's handling. On one day, the car encountered constant cross winds greater than 24 kph; however, no steering compensation was required. Large trucks were anticipated to cause the most stability problems when either approaching head on or overtaking the vehicle in passing. Initially approaching trucks were announced to the driver over the radio; however, after observing the car's stability, this practice was soon abandoned. Even though the trucks overtaking the vehicle appeared to have a greater effect than those approaching head on, no corrective action was required by the driver. The overall stability of the solar car surpassed that of the lead van and chase van.

5.2 Test Facilities

In August of 1990, after the G.M. Sunrayce competition, the team was offered a chance to test the car in the General Motors full scale wind tunnel. The G.M. tunnel is equiped with a six component balance accurate to one part in 10^5. The test section measures 6.1 meters by 6.1 meters. Although the tunnel has a top speed of 250 kph, the car was only tested up to 129 kph. Flow visualization was accomplished using smoke generated by boiling propylene glycol. In addition to the aerodynamic testing, the G.M. tunnel provided facilities to measure the frontal area using a laser measurement system.

5.3 Test Results

The objectives for the day in the full scale tunnel included: a) checking the optimum attitude; b) checking effects of details of fit and finish; and c) evaluating wheel and suspension fairings. The suspension was modified to allow the attitude of the car to be adjusted over a wide range. Body modifications included sealing body seams and wheel openings and adjusting trailing edge geometry.

In Figures 5 through 10 the full scale results are shown for comparison to the scale model data and available G.M. Sunraycer full scale data. The baseline full scale data is for the configuration under which the vehicle competed in the G.M. Sunrayce. The frontal area of the full scale vehicle is larger than the scaled up model area due to shape changes necessitated by additional room required to fit components of the suspension. As a result the forces and moments are greater for the full scale vehicle (0.932

m^2) than for the model (0.826 m^2). The forces and moments of the full scale vehicle generally correspond to the predictions of the scale model, except for the drag area. Modifications were made in an attempt to reduce drag. It was decided after the first yaw run that the rest of the runs would be made at 0 degrees yaw. This was due to the large amount of time necessary to conduct the yaw runs and the limited amount of time available for testing. Figure 3 shows the cumulative drag areas at 0 degrees yaw for the modifications tested. Closing the front wheel openings resulted in a reduction of CdA by 0.017 m^2. The optimum drag attitude lowered the CdA by an additional 0.004 m^2. The optimal drag attitude had the nose lower on the full scale vehicle than on the scale model. This is due to more of the brakes, wheels and suspension system being contained inside the body. This effect was not demonstrated by the scale model because the suspension and braking systems could not be accurately modeled at the time of the wind tunnel test. Sealing all the body seams reduced the CdA by 0.0075 m^2 (Figure 3). Wheel fairings were also added to the full scale vehicle, but proved to increase CdA by 0.026 m^2 (Figure 3), in contrast to the scale model, again a result of not accurately modeling the braking and suspension systems when fairings were added to the scale model. It is felt that significant drag reduction could be realized if more time could be devoted to developing wheel fairings. In another modification the trailing edge of the vehicle was raised by 4 degrees to reduce lift on the rear wheels. This resulted in the rear lift coefficient being reduced from 0.059 to -0.03 with no increase in drag (Figure 3).

6.0 CONCLUSIONS

The final vehicle generally met the aerodynamic objectives set out early in the project. It is important to keep in mind that CdA, not Cd, is the proper measure of performance. The full scale wind tunnel measurements show the importance of fit and finish on the achieved drag performance. The full scale best drag result was 0.138 m^2 compared to the best result projected from the model of 0.128 m^2, and the design goal of 0.162 m^2. However, during the development, changes were necessary which produced a full scale frontal area of 0.932 m^2 compared to 0.826 m^2 for the scaled up model shape. Comparing these numbers indicates a lower drag coefficient for the full scale test than for the model. It must be kept in mind that the shapes are different, and the protrusions of wheels, brakes, etc. are different so that this is not a comparison of a scaled shape.

Our experience with the panel code computations showed good qualitative comparisons with the flow observed in the wind tunnel. The methods can be used to explore design options if sufficient time is available. One significant result of the panel code was the indication of a sharp peak in pressure at the maximum thickness point. The actual model was smoother than than the numerical geometry and may not have had that peak in pressure distribution. Our experience in this project was that the wind tunnel provided a faster and more reliable result, especially for such necessary details as wheels, protruding suspension components, etc.

The wheels and associated suspension components represent a sizeable contribution to total drag. Our model tests indicated about a 25% increment due to wheels. This represents a potential for improvement, but presents many difficulties in implementation on the operational car. The CFD results indicate that smoothing the peak may also be an area for further drag reduction.

APPENDIX A - REFERENCES

Abbot and Von Doenhoff, Theory of Wing Sections,Dover Publications, 1959

Hibbs, B., "Sunraycer Aerodynamic Development", G.M. Sunraycer Case History, AeroVironment Inc., 1989.

Kyle, C., "How Design Factors Affect Solar Car Performance", SAE 880728 - International Congress and Exposition, Feb. 29 - Mar. 4, 1988.

La Bozzetta, W. F., Interactive Graphics for Geometry Generation Users Manual AFWAL-TR-87-3064, Lockheed Aeronautical Systems, 1987

Maskew,B., VSAERO User's Manual, Analytical Methods Inc., 1983.

Software Users Manual for Fast Ship Design Systems Services, 1989

Section VI
Aerodynamics of Engine Cooling

Section VI

Aerodynamics of Engine Cooling

One aspect of vehicle design that is significantly influenced by aerodynamics is engine cooling. Typically, cooling of the engine and its components occurs not only at dedicated heat-exchangers (radiators, transmission coolers, etc.) but also by heat transfer directly to the air within the engine compartment (underhood). The airflow that performs both of these functions is entrained from the external flowfield (typically at the front of a vehicle) and subsequently returned to it. This process impacts the external aerodynamics of the vehicle to some degree.

The design of efficient engine-cooling systems is impacted by many factors. With the increasing focus on fuel efficiency and *low-drag designs*, engine-cooling airflow must be carefully controlled to minimize adverse effect on aerodynamic drag. *Styling* and packaging requirements dictate the overall size and shape of the engine compartment, which in turn influences the ability to control engine-cooling airflow. Modern designs frequently employ small and low-to-ground front faces, together with small (and sometimes hidden) air openings. Consumer preference influences not only the level of vehicle performance but also passenger comfort. The former frequently leads to larger engines which have increased engine heat rejection. To improve the overall level of comfort, air-conditioning systems are increasingly used and this demands the presence of another heat-exchanger (condenser) and increases the demand for cooling airflow. Finally, the growing trend towards quieter vehicles has meant that engine compartments are becoming encapsulated for noise isolation, and this causes additional restriction to engine-cooling airflow.

The need to accommodate these multiple constraints when determining a "best compromise" for engine-cooling design underscores the fact that this is not merely a technical challenge but also a challenge for the design and development process itself. To aid this process, numerous methodologies have evolved to provide better understanding of the behavior of engine-cooling airflow and its influence on external aerodynamics, as well as more accurate assessments of cooling performance.

Wind tunnels continue to play the primary role in measuring and understanding engine-cooling airflow performance. While methods to measure the airflow through heat-exchangers have been fine-tuned over several decades, an ability to study the flow within the engine compartment is limited due to the complexity of the region and its limited access. Traditional diagnostic techniques such as flow visualization and flow surveys (with different types of measurement probes) are not easily adaptable to the heat-exchanger and underhood environment.

The complexity of the underhood geometry has also presented a tremendous challenge to math-based simulations of this environment. The need to deal with numerous and complex geometric entities such as the engine block, exhaust system, engine-cooling fan and alternator imposes a severe burden on the mathematical representation of the necessary geometry.

In spite of these challenges, numerous efforts (both experimental and computational) have succeeded in obtaining ever-increasing understanding of the flow in this domain. These advances have led to improved vehicle development processes whereby low-drag designs that permit good engine-cooling performance are feasible.

90512

Aerodynamic Drag and Engine Cooling-Conflicting Goals?

H. J. Emmelmann and H. Bernebur
Adam Opel AG

ABSTRACT

The two oil crises, increased interest in environmental consciousness and good fuel economy have made lower drag coefficients more important than ever before. Modern aerodynamic development work now includes c_D reduction by minimizing cooling flow due to front end and front opening optimization. Further restricted air flow through engine compartments results from modern noise reduction laws which have been met by using engine compartment covers. Direct airflow onto heat sensitive components is limited by more and more packed engine compartments. As a result it is now necessary for aerodynamicists and cooling engineers to work closely together to deal with the conflict of interest which results when designing new vehicles. This paper outlines ways to overcome the resulting engine and engine component cooling problems.

AERODYNAMIC DEVELOPMENT ACTIVITIES WITH THE OBJECTIVE OF IMPROVING FUEL ECONOMY

To better understand the connection between aerodynamic drag and engine cooling it is important to delve somewhat deeper into aerodynamic development activities and their necessity in order to improve fuel economy.

This necessity became obvious at the beginning and at the end of the 70s when we faced the first and the second oil crises. The Club of Rome, which had first met in 1968, made it clear that the raw materials for fossil fuels are only available to a limited extent. Additionally, environmental consciousness had grown considerably during recent years. These facts imposed the obligation upon the automotive industry, to significantly improve the fuel economy of vehicles. Low drag coefficients of motor vehicles became more important than ever before.

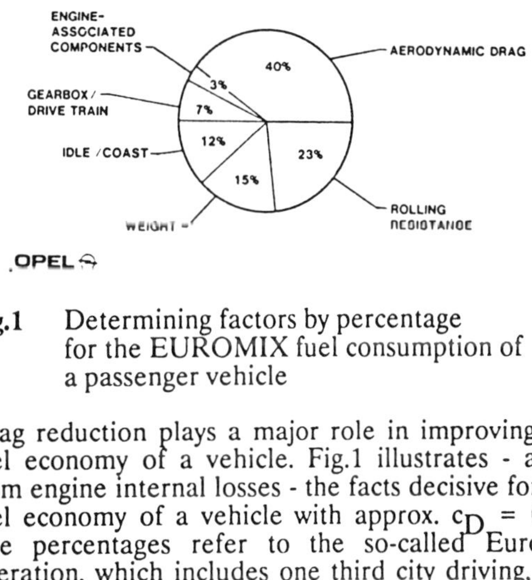

Fig.1 Determining factors by percentage for the EUROMIX fuel consumption of a passenger vehicle

Drag reduction plays a major role in improving the fuel economy of a vehicle. Fig.1 illustrates - apart from engine internal losses - the facts decisive for the fuel economy of a vehicle with approx. c_D = 0.40. The percentages refer to the so-called Euromix operation, which includes one third city driving, one third constant speed driving at 90 km/h, and one third constant speed driving at 120 km/h. The largest share in fuel consumption results from overcoming aerodynamic drag, it amounts to 40 %. Rolling resistance takes a share of 23 %. The remaining 37 % may be split up into 15 % for vehicle weight, 12 % for idle and coast during the city cycle, 7 % for gearbox and drive train, and 3 % for engine associated components.

The two most important driving resistance parameters influencing the fuel economy of a vehicle are clearly therefore aerodynamic drag and rolling resistance, adding up to a total of 63 %. At higher speeds the influence of aerodynamic drag dominates. It is known that the aerodynamic drag rises as the second power of driving speed. The rolling resistance rises to a minor extent in the speed range up to approx. 100 km/h. Above this speed it rises more rapidly.

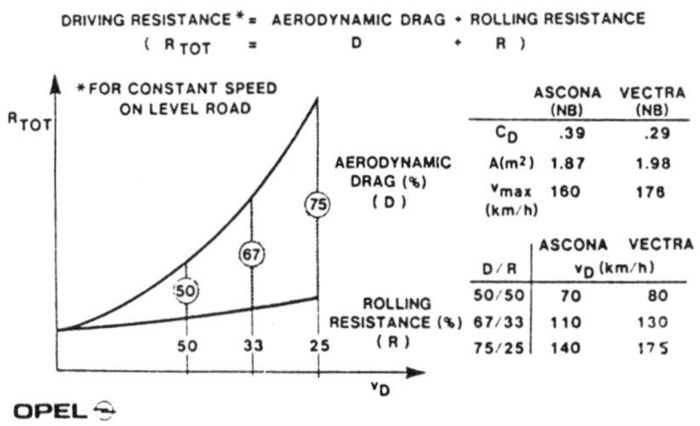

Fig.2 The relationsship between aerodynamic drag and rolling resistance over the speed range.

When comparing a 1982 Ascona C and a 1989 Vectra both with the same engine and gearbox, the influence of aerodynamic improvement becomes obvious (fig. 2). The selected ratios D/R = 50/50, 67/33 and 75/25 are related to 70, 110 and 140 km/h for the Ascona. Due to a more favourable c_D value the same ratios are shifted to 80 and 130 km/h for the Vectra. The third ratio 75/25 is first reached at 175 km/h shortly before the top speed. Or, in other words, if the selected speeds remain constant the ratios will change because drag reduces and as a result also fuel consumption. The intensified aerodynamic development of the Vectra made a significant contribution to the considerable improvement of fuel economy.This improvement will have dramatic effect on Opels sales weighted fuel consumption as the Vectra is a high volume selling vehicle.

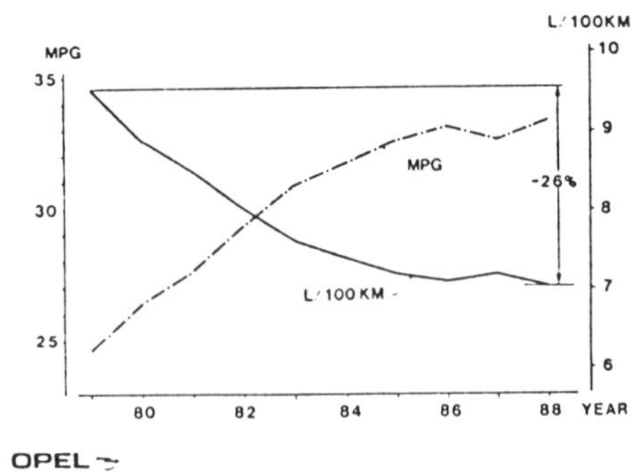

Fig.3 An improvement of fuel consumption is indicated by the sales weighted average fuel economy chart for Opel vehicles

Fig. 3 shows the development of the sales-weighted average fuel economy for Opel vehicles between 1979 and 1988. It can be seen that by the end of 1988 a fuel consumption reduction of 26 % was achieved since 1979. In future it will be more difficult to achieve such a big improvement. Increased sales of high displacement engines is one of the reasons for this difficulty. A decreasing diesel engine share also had its effect.
Approximately 20 years ago knowledge about the positive influence of low c_D values on fuel economy was revived and special attention has again been paid to aerodynamic development. Good examples are the Omega, the Vectra and the new Calibra. During this period it has been possible to lower the c_D values of many cars by more than 40 % from above 0.50 to 0.30 and below. An example of such an improvement is the 1975 Kadett City with c_D = 0.51 and the 1985 Kadett GSi with c_D = 0.30, a c_D reduction of 41 %.

By reducing aerodynamic drag to improve fuel economy aerodynamic development has to overcome a number of legal and functional conflicts.

ENGINE COOLING PROBLEMS OF MODERN VEHICLES

One big obstacle in aerodynamic development is engine cooling.
On modern passenger cars the front end design has a decisive influence on the heat development in the passenger and engine compartment and therefore on passenger comfort and driveability. Front design is partially dependent upon the air required for engine cooling, the relevant temperatures in the engine compartment and the components installed within it.

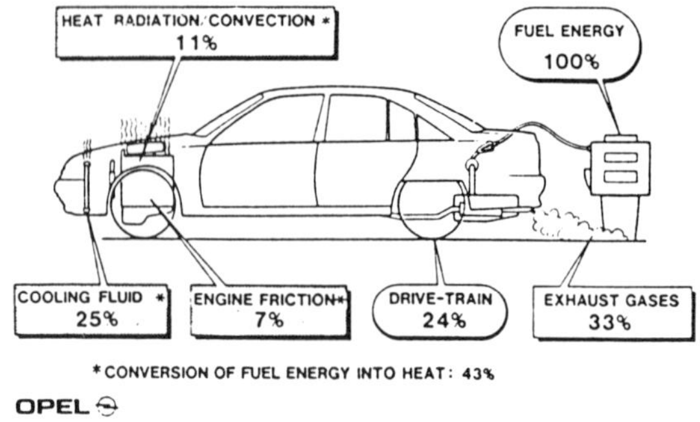

Fig.4 The energy balance of a motor car

The energy balance of a motor car (fig. 4), shows that 36 % of the energy contained in fuel is directly converted to heat 11 % is given off as radiated and convected heat, 25 % goes to engine coolant and a further 7 % is dissipated indirectly through engine friction. This means that 43 % of the energy consumed is converted into heat all of which must be dispersed.

For vehicles with low drag coefficients this seems to be a problem at first sight. The previously mentioned reduction of the drag coefficient by more than 40 % during the past 15 years was partially achieved by optimizing and thus minimizing the air flow through the engine compartment. During the past 5 years the consequence for many vehicles was that the share of the engine compartment air flow Δc_{Dec} of the overall aerodynamic drag was reduced by approx. 80 %, from values above 0.05 on aerodynamically unfavourable vehicles, to 0.01 on vehicles which have undergone excellent aerodynamic development. The Audi 100 already reached this value in 1982.

In view of such low engine flow rates, it is obvious that further aerodynamic gains by use of a controllable radiator grille, which would almost allow the aerodynamics engineer to reach the optimum of no cooling losses, Δc_{Dec} = 0, are hardly worthwhile (i.e. this solution is not cost effective).

THE DEMANDS OF LEGAL REQUIREMENTS -- Further new problems for engine cooling and the temperatures in the engine compartment have arisen during recent years due to additional legal requirements concerning exterior noise and the constituents of exhaust emissions of motor vehicles.

Development engineers achieved the step by step reduction of pass-by noise by such techniques as engine encapsulation and engine compartment covers. In this way the air flow and the temperature levels in the engine compartment have been influenced significantly.

To reduce pollutants in the exhaust emmissions of engines, the catalytic converter concept is increasingly being used. The effective functioning of the catalyst involves high operational temperature levels, such that suitable heat protection measures become necessary.

CUSTOMER REQUIREMENTS AND STYLING TRENDS ARE ALSO FORCING CHANGE -- The tendency towards more engine output has the effect that bigger and more powerful engines are being installed into engine compartments of vehicles which were not designed for such engines. This applies not only to gasoline engines, but diesel engines as well. Furthermore there is an increased usage of turbochargers.

Another factor is the continuing drive towards the improved operating comfort of vehicles. This results in an increase in the number of auxiliary components in the engine compartment e.g. power steering pumps, ABS, carburettors and injection systems with voluminous air inlets, large volume air cleaners and resonators, air conditioning with condensors, additional fans and compressors, and transmissions with larger numbers of gears and bigger housings. In this manner the free space in the engine compartment is continually being reduced. Very often, these auxiliary components themselves are additional heat sources. This trend can be seen by comparing a 1985 Kadett E GSi 16V and a 1982 Kadett A. The engine compartment of the Kadett GSi 16 V shows a very high package density. In comparison the engine compartment of a Kadett A is a good example of an earlier and simpler vehicle with much more free space around the engine (figs. 5 and 6)

Fig.5 Kadett E GSi engine compartment packed with components, leaving very little space for cooling air circulation

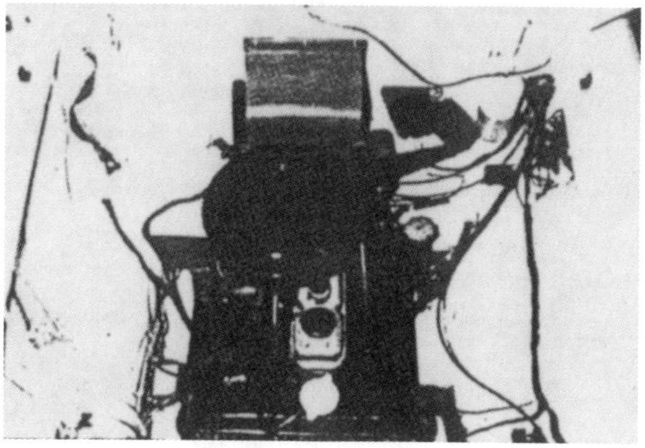

Fig.6 Kadett A engine compartment with plenty of space for free circulation of cooling air

Last but not least, the design of modern vehicles influences the air flow into the engine compartment to a large extent. A low front end and a high rear end changes the pressure field around the whole body. Positive-pressure areas from which it is necessary to supply air to the engine compartment become continuously smaller. An enlargement of the front openings into negative-pressure areas involves the risk that the air required for cooling will be extracted before it flows through the radiator.

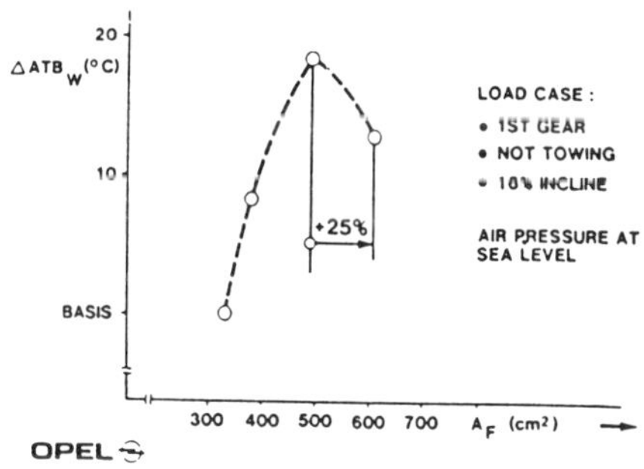

Fig.7 Effect of the front inlet area on the engine cooling results

An example from the Vectra development program (fig. 7) shows the relationship between an increasing front opening area A_F and the cooling capacity. Cooling capacity is expressed as a critical limit temperature ATB_W, i.e., the ambient temperature condition under which the boiling point of the cooling system is reached under specified load conditions. A very specific optimum front opening area was found to give a particular, high ATB_W temperature. After this temperature was reached, it would drop by 25 % even if the front opening area was further increased. This meant that a further improvement of the ATB_W by simply increasing the front opening area was not possible.

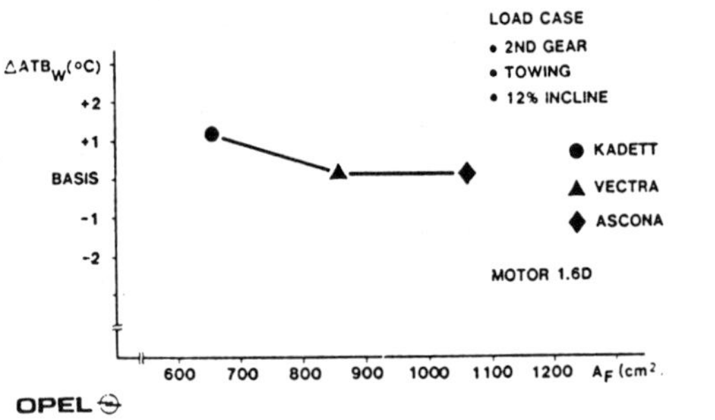

Fig.8 A comparison of the effect on engine cooling by differently located front inlets

Positioning the front openings at the best possible place is the more promising approach, even though the size of these openings may be below the maximum size which can be realized. A comparison between a Vectra, it's predecessor the Ascona, and a 1985 Kadett E (all models equipped with a 1.6 D engine) confirms this theory (fig. 8). The ATB_W temperature of the Vectra is just as high as that of the Ascona even though its cooling air intake area of 850 cm^2 is 200 cm^2 or 20 % smaller. On the Kadett E, the conditions are even more favorable. Its ATB_W temperature is 1° higher with a front opening area of only 650 cm^2, which is even 20 % smaller than the Vectra. This effect results from a particularly favourable positioning of the openings in the front stagnation area. A much more uniform wind speed results for the flow to the radiator core.

In addition to the ATB_W temperature for the cooling system, temperature limits for components and parts within the engine compartment and on the underbody which are particularly temperature sensitive must be considered. These include fuel lines and recirculation lines, fuel injection valves, cooling fluid hoses, the power braking system, ABS hydraulic components and insulation on the hood, dash panel and lower engine compartment cover. The ignition system, battery case, starter relay, fuel injection system wiring, fuel control unit, flow meter and oil pressure sensors are also equally temperature sensitive. For driveability and durability reasons the maximum acceptable temperatures for these parts and materials must not be exceeded.

It is important to know how such temperatures are caused. The type of heat, heat source and heat and the type of heat dissipation, whether radiation or convection, must be determined.

By convection, i. e., heat dissipation by air, only a very small share of the heat is carried off. Particular attention must be paid to this when stopping suddenly after extended periods of driving at high engine loads, which interrupts the ram air flow, or during stop-and-go traffic in the city or on highways. Some radiative heat sources are exhaust manifolds turbochargers, exhaust pipes, and catalysts. Heat is for instance predominantly conducted from the cylinder head to the spark plugs.

SOLUTIONS TO THE PROBLEM OF HIGH COOLING SYSTEM AND MATERIAL TEMPERATURES

When solving high cooling system and material temperature problems in the engine compartment and on the underbody, cooling engineers and aerodynamicists work closely together. Different types of stringent tests in the wind tunnel, on test stands, and on the road are carried out.

In order to evaluate the performace of the cooling system relative to the proposed package, cooling system component cars and wind tunnel models are built. These fully driveable full scale reinforced fibre glass cars are hand built and available at a very early stage in the development programm, long before prototypes exist.

RECORDING OF THE AIR FLOW IN THE ENGINE COMPARTMENT -- Entry of the air into the engine compartment and flow through it can be investigated in the wind tunnel. But, since the flow shows heavy turbulence, it is hard to record it by simple means. Results are also dependent upon the probe location and wind speed, as the recordings of two different test points in the engine compartment of an Omega with an open front show (fig. 9).

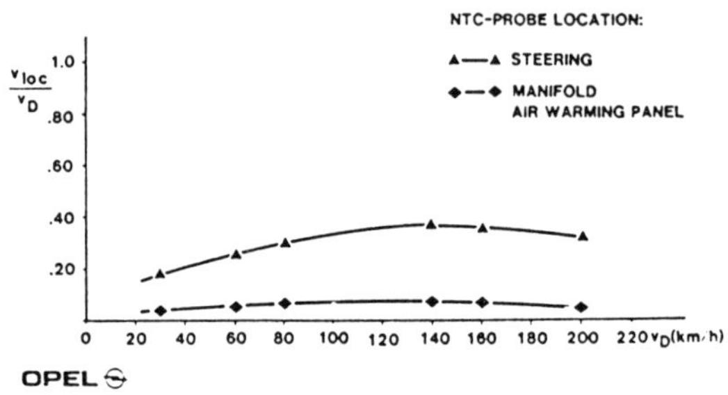

Fig.9 The influence of driving speed and probe location on the airflow velocities in the engine compartment of a Opel Omega

Flow within the engine compartment is furthermore extremely influenced by cooling fan operation. Flow patterns can be completely changed by operating fans.

Among the test methods used we distinguish between qualitative and quantitative procedures, the first category makes the air flow visible and the second includes measurement and recording techniques.

METHODS TO MAKE AIR FLOW VISIBLE -- The best known method to make air flow visible is without doubt the use of a smoke wand. The smoke discharged from a probe enters the front openings. With a plexiglass hood installed, the underhood flow can be watched and recorded by videocamera. In this way the structure of the flow can be roughly evaluated. An integrating view from the top to the bottom of the engine compartment can be made visible. The flow directly under the hood can be clearly defined. A recording of smoke layers is desireable, but it is difficult to accomplish, particularly in the area of components susceptible to high temperatures, such as turbo-chargers and manifolds, which are awkwardly positioned.

A method using helium bubbles has been optimized in the Pininfarina wind tunnel. It is similar to the smoke method, but has certain advantages over it, because the bubbles are in an almost weightless state and can enter the airflow almost impulse-free. The air flow may be observed in more detail than with smoke.

MEASURING TECHNIQUES -- NTC probes, (thermistors) may be used to sense mean local speeds. The test values in fig. 9 were measured with such probes. Even though only mean speeds are indicated and the accuracy does not exceed ± 10 %, first indications about local wind speeds and their distribution can be made.

Hot wire probes may also be used to measure time - averaged speed. With so-called X-wire probes, consisting of two wires forming an angle of 45°, it is possible to determine the flow rate in all three planes. However, due to their very high susceptibility to mechanical damage, installation in the engine compartment, which is often very cramped, is very difficult.

OPEL made wind speed measurements using Laser Doppler method in the engine compartment in cooperation with Pininfarina. The results turned out to be very encouraging.

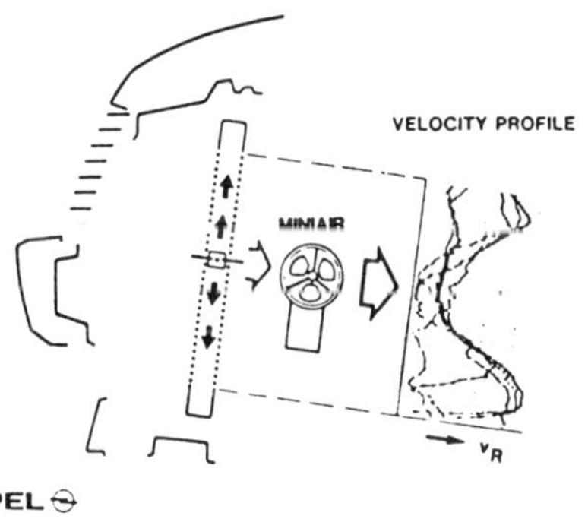

Fig.10 Air flow measurement radiator

A specially developed test radiator is used to measure the cooling airflow into the engine compartment (fig. 10). This special radiator replaces the actual radiator. Its flow-through resistance is adjusted to that of the proposed radiator by perforated screens. A bridge which moves between the screens is used to carry eleven small impeller anemometers vertically and to record the local wind speeds. The integration of the velocity profiles over the radiator core surface, gives the volume flow rate. All data is collected and processed automatically by computer. In this way it is possible to determine the effect of different front openings and their placement relative to wind speed distribution and cooling air flow volume is calculable. To assure optimum use of the radiator core a uniform cooling air and wind speed profile is most desireable.

TEMPERATURE MEASUREMENTS IN THE ENGINE COMPARTMENT, ON THE UNDERBODY AND IN THE COOLANT CURCUIT -- With the above described methods, tools are on hand to measure the air flow into and within the engine compartment and on the underbody, and to determine if and how air reaches the components which are susceptible to high temperatures.

The question whether or not the parts concerned will be sufficiently cooled can only be answered with the help of temperature measurements under load. These tests can be driving tests on test tracks or simulations with a towing dynamometer, simulation using a dynamometer in a climatic wind tunnel, or on a cooling test stand. Presently it is difficult possible to verify the test results without road tests.

REMEDIAL MEASURES TO AVOID HIGH COOLANT AND MATERIAL TEMPERATURES -- A number of special measures can be applied to avoid high coolant and component temperatures. These measures fall into two different categories, the first type may be introduced on current models, the second on new vehicle concepts only.

MEASURES INTRODUCEABLE ON A SHORT TERM BASIS I. E. ON CURRENT MODELS -- The following three features can be implemented relatively quickly on current models.

OPTIMUM POSITIONING OF THE FRONT OPENINGS -- The effective use of the available airflow which passes into the engine compartment is the first priority.

By careful matching of the front openings to the cooling components the available airflow can be optimized for best use. To achieve this the front inlets should be located in a position that allows an even flow distribution across the whole radiator surface. Experience has shown that front inlets located in the stagnation area of the front end with the adoption of a flow-through-bumper concept, as on the 1985 Kadett E, are not only the most effective from a cooling standpoint but also from an aerodynamic standpoint.

The changes to the grille and front bumper of a vehicle to implement such an improved cooling concept can already be accomplished during a minor facelift on a relatively short term basis. Full cooling air ducting between the front inlets and the radiator ensures 100 % usage of the air which flows into the front by avoiding leakage around the radiator, below the vehicle and into the wheelhouses.

The radiator itself must be tuned to the requirements. The core area, thickness, fin and tube spacing all plays a role. The coolant flow pattern, whether down, cross, U or Z can also affect the heat dissipation performance of the cooling system.

The design and location of the fans must also be matched to the radiator location, the flow resistance of the front openings and the radiator. Various kinds of fans can be used, varying from single or multi-speed electric fans to belt driven fans having fixed or viscous coupled drive mechanisms. Shrouds with flap valves to avoid backflow can also be used.

VENTILATION INLETS AND OUTLETS TO THE ENGINE COMPARTMENT -- By special openings in the outer body panels or in the engine compartment cover, better air flow to and air extraction from the engine compartment can be assured. For this purpose advantage may be taken of differential pressure levels in the air flow around the body to support the air flow into and out of the engine compartment.

We have for example used air extraction openings in the hood on the OPEL GT, the Ascona 400, and the Kadett GSi. In such cases care must be taken to ensure that the hot air from the engine compartment is not taken in by the passenger compartment ventilation inlet which in general is located at the lower edge of the windshield.

Hood outlets must therefore be located away from the airflow path to the passenger compartment fresh-air ventilation inlet, i. e. at the outer edges of the hood. Under small yawing angles airflow passing over this area which picks up the departing under-hood warm air, flows steadily outboard to pass around the A pillars onto the side of the vehicle. Of course the behaviour of the flow under sidewind conditions (larger yaw angles) must always be considered and investigated.

Another effective location for warm air outlets is the front fender directly behind the transition radius from the headlamp/blinker area to the fender (ahead of the front wheels). This area normally exhibits lower pressure than the engine compartment. Therefore it is feasible that warm air which has passed through the radiator could be dispelled from outlets located there.

Another favourable area is the front fender between the front wheelhouse and the door opening line. Examples of vehicles which have used this location include the famous Mercedes 300 SL (ca. 1954) and the new Mercedes 450 SL.

Air intakes according to the "NACA" principle in aircraft construction are an efficient way to ventilate the engine compartment. An opposing pair of vortices which produce a strong flow of the close-to-body air into the engine compartment, is produced by the curved trailing edges.

The aerodynamic function of this special double vortex system can be more powerful than local surface negative pressures which may exist on the body. For instance tests with "NACA" intakes have previously shown that when installed in some areas of negative preasure i. e. behind the large radius of the hood air did not flow outwards, but instead, due to this vortex system resulting from the intakes, cool air from outside was forced into the engine compartment. This effect could naturally further be improved by raising the inlet surface of the duct relative to the hood panel.

In addition to the ventilation openings described, thermostatically controlled doors which open automatically under critical temperature conditions particularly in the engine compartment cover may be used. In this way best possible aerodynamics and reduced noise radiation will be achieved when driving conditions are not critical for engine cooling.

To improve flow into the encapsulated engine compartment large radii should be used on the available inlets. The inlets also should not be located downstream from spoiler lips which may be necessary for aerodynamic reasons. In some special cases it may even be necessary to delete such lips in favour of improved cooling, but optimization of such measures is always necessary from both cooling and aerodynamic standpoints.

In order to provide sufficient cooling airflow for the Calibra 4WD when the 16 V engine is installed it was necessary to do just that. This version has an engine compartment lower cover with a cooling flow inlet. The inlet, by necessity, is located directly behind a small central spoiler lip at the front of the vehicle. By rounding the corner of the inlet and by cutting away the lip on this version with a slight aerodynamic penalty, an oil temperature reduction of 30°C in the add-on gearbox was achieved.

Naturally all the technical measures mentioned can only contribute to lower engine compartment temperatures during driving i. e. when the vehicle sees active flow.

Under "idle" conditions, with the vehicle stationary, aerodynamics cannot help maintain low engine compartment temperatures. For this condition the use of additional fans and ducting is necessary. With correct installation, again the result of considerable development activity, such installations can draw air from "cold" regions, (e. g. from the wheelhouses) to reduce temperatures in areas of high temperature sensitivity.

FURTHER REMEDIES -- Further possible measures against high engine compartment temperatures are shields and dividing panels to separate hot and cold regions. When the more recently developed sandwich construction for such panels is used heat insulation properties can also be improved.

Air deflectors can also be used, for instance on the underbody, to deflect airflow onto temperature sensitive components. The use of such deflectors can of course worsen aerodynamic performance.

MEASURES INTRODUCEABLE ON A LONG TERM BASIS -- In the long run a different placement of the components in the engine compartment relative to the air flowing in or through is less costly than any other secondary measures such as, openings, shields, or additional fans. The objective is the creation of "hot" and "cold" zones, which would allow the placement of components susceptible to high temperatures on the cold engine side. Direct passage of cooling air to very hot components on the warm side must be maintaned.

Naturally consideration must be given to the temperature of the so-called cooling air which is already in the engine compartment. Normally in an optimized aerodynamic/cooling/noise package this cooling air has already passed through the radiator and been warmed, therefore it's potential for cooling heat sensitive areas is severely reduced.

By increasing the combustion efficiency, the exhaust gas temperature on the engine side can be lowered, with the consequence that high temperatures are lowered at source. Modern engines with high compression ratios and multi-valve technology as a result are already showing this trend. By use of this technique it has been possible to reduce the exhaust gas temperature by approx. 100°C at the manifold outlet on Opel's new 3.0 liter 24 valve engine.

At the same time it is necessary to ensure that exhaust gas temperatures do not sink below such a level that they no longer activate catalytic converters, a fact which is especially true when catalytic converters have been in operation for a considerable time.

PROSPECTS FOR THE FUTURE

Various parameters, such as legal specifications, good fuel economy, customer requirements with regard to improved noise and temperature control, product and production costs all contribute to high cooling system and component temperatures. These requirements reduce the latitude for technical innovation.

It is the task of aerodynamicists in close cooperation with design and development engineers, to create new body shapes which provide good c_D values and sufficient cooling openings which are located in the right places on the body. Correctly placed additional openings in body panels for ventilation or air extraction can make a further contribution to the reduction of high temperatures in the engine compartment. In places particularly susceptible to high temperatures, however, shielding measures and the use of additional fans, especially for stop and go driving, will still be required.

In particular engine designers are encouraged to develop engines which produce less combustion heat and thereby reduce cooling problems. This is also desireable from an environmental point of view, when one considers that from the energy balance shown previously 36 % of consumed energy during car operation is transformed into heat and subsequently lost, wasted into the environment without further use as a source of power.

Improvement of the energy balance by reduction of this loss would result in less fuel consumption, a saving of oil resources and a cleaner environment.

At the beginning of the project phase, conflicting points of view of design engineers, aerodynamicists and development engineers are often unavoidable. However experience shows that good solutions can be realized if modern tools are used sensibly.

Symbols, Indices, Abbreviations

Aerodynamic measures

A	= frontal area (m^2)
A_F	= front opening area (cm^2)
c_D	= Aerodynamic drag coefficient (-)
c_P	= Pressure coefficient (-)
D	= Aerodynamic drag
R	= Rolling resistance
R_{TOT}	= Driving resistance
V_D	= Driving speed (km/h)
V_{Lec}	= Local velocity (m/s)
v_{max}	= Maximum speed (km/h)
v_R	= Velocity within the airflow (m/s) measurement radiator

Other Abbreviations

ABS	= Antilock Brake System
ATB_W	= Ambient temperature conditions (°) under which the boiling point of the cooling system is reached under specific load conditions (Air Temperature to Boil Water)
HB	= Hatchback Vehicle
NB	= Notchback Vehicle
MPG	= Miles per gallon
NACA	= National Advisory Committee for Aeronautics
NTC-Probe	= Negative Temperature Coefficient-Probe (Thermistor)
4WD	= Four wheel drive

910309

Ram-Recovery Coefficient Correlations for Automotive Cooling Airflows

J. W. Oler, C. M. Roseberry, and D. P. Jordan
Texas Tech University

J. E. Williams
Ford Motor Co.

ABSTRACT

The design of more efficient automotive cooling systems requires better understanding of the losses in fluid mechanical energy through conversion to thermal energy for the air flow through the front end cooling openings. A wind tunnel study with simplified models has been conducted to investigate the factors affecting the cooling opening mechanical energy losses. The data from this study have been sucessfully corrolated using newly defined dimensionless paramoters. The influence of the controlled geometrical factors are clearly identifiable with the data presented in dimensionless form. An algorithm has been devised which accurately correlates the behavior of a combination of cooling openings given the behavior of each of the openings operating individually.

NOMENCLATURE

A_∞, A_1	Streamtube cross sectional area
A_{model}	Cross sectional area of model
A_{open}	Area of opening
c_p	Specific heat
h_∞, h_1	Enthalpy
$\dot{Irr}$	Irreversibility rate
K_{loss}	Front end loss coefficient
K_{RAM}	Ram-recovery coefficient
$\dot{m}$	Mass flow rate
P_∞, P_1	Static pressure
$\Delta P_{T,fe}$	Front end total pressure loss
Q_{model}	Volumetric flow rate through model
R	Gas constant
s_∞, s_1	Entropy
T_∞, T_1	Temperature
V_∞, V_1	Velocity
V_{model}	Average velocity inside model
V_{open}	Average velocity through opening
ρ	Density

INTRODUCTION

Efficiency and reliability have become increasingly important issues in automotive design. Unfortunately, the task of optimally incorporating the cooling system into the overall design often brings these two concerns into conflict. Adequate cooling is essential to provide reliable operation; however, the sizable power requirements to provide convective heat transfer must be minimized in the interests of fuel economy.

The energy input required to induce an airflow through the cooling system typically comes from two immediate sources - the radiator fan and "ram pressure" resulting from the forward motion of the vehicle. Accordingly, to optimize the cooling system both the external and internal flow conditions must be considered. Moreover, the external and internal flow fields interact to some degree. The interfaces between the two flow fields are the external cooling openings of the vehicle and the underside of the engine bay. The mechanical to thermal energy exchange across the front end cooling openings (which is the focus of this study) is the first in a series of energy exchanges through the cooling airflow circuit. The other energy exchanges are due to the effects of the radiator, condenser, fan and shroud, and engine blockage.

While it is recognized that energy is conserved and never "lost," the transformation of fluid mechanical energy to thermal energy is in effect a loss since it results in a lost capacity or increased requirement for flow work in generating airflow through the cooling system. Consequently, the term "energy loss" is used within the current context to mean the loss of fluid mechanical energy through transformation to thermal energy.

In addition to being functionally important, the cooling air inlets are often a distinctive styling feature of a particular automobile. Consequently, there is great variation in the geometry and location of frontal cooling openings. At one extreme, are the large vertical openings with grille inserts located above the bumper which are typical of many of today's trucks and luxury cars and the majority of past car designs. In contrast are the below bumper inlets with little or no projected frontal area which are no longer only common to sports cars.

Since the mid 1970s, the design of more efficient cooling systems has been addressed by several automotive researchers. Olson [1]* recognized the need to introduce scientific methods to replace trial and error methodology in optimizing vehicle cooling systems. By performing full scale wind tunnel tests with traversing anemometers mounted behind the radiator, Olson quantified the energy losses for flow through the front end openings and identified the effects of various front end configurations. Hawes [2] outlined many possible improvements to cooling systems and recognized that an overly large frontal intake will not necessarily increase cooling airflow.

Schaub and Charles [3] addressed the interaction between the ram airflow and the radiator fan. The streamtube concept was used to present a key argument in the study concerning the energy losses through the frontal openings. The researchers asserted that the entrance area to the cooling airflow streamtube is infinitely large when the vehicle is stationary with the fan running; whereas, at high speeds this entrance area can be quite small (less than one square foot). Schaub and Charles offered this extreme streamtube variation as an explanation for the large changes in the total pressure losses through the frontal openings. Like Olson, Schaub and Charles used traversing anemometers mounted behind the radiator which measured velocities and flow rates during both road tests and full scale wind tunnel tests. To investigate the influence of the fan in the wind tunnel tests, the engine bay backpressure was independently controlled by a catchment sealed to the engine bay underside. This catchment was in turn connected by a duct to a separate fan. This approach provided a way to measure the front grille total pressure loss over a wide range of cooling flow rates at a constant wind tunnel speed.

Williams [4] showed that grille open area (the amount of the open grille area that can be frontally projected onto the radiator) is not a reliable parameter for predicting cooling drag and ram airflow. By testing in both aerodynamic and environmental wind tunnels, Williams also determined that over 95% of the physical variation in measured cooling performance (compared to radiator test stand data) could be attributed to factors involved in the cooling airflow.

Renn and Gilhaus [5] placed emphasis on the importance of providing ducting between the cooling openings and the radiator and condenser. This feature increases the effectiveness of the cooling system by preventing the air from bypassing the heat exchangers and by reducing hot air recirculation. These researchers noted that, based on cooling drag data, most cooling systems are not closely optimized. After qualitatively outlining other front end effects, they concluded that "aerodynamic improvements need not necessarily interfere with cooling requirements."

A primary difficulty in optimizing automotive cooling systems is the inability to accurately predict the cooling performance in a new vehicle before a production prototype is constructed and tested. Because many produc-

---

* Numbers in parenthesis designate references at the end of the paper.

tion tooling decisions have already been made at this point in the vehicle development, cooling system changes can only be made at great expense. For this reason, cooling systems are frequently designed conservatively.

The purpose of this study is to provide a better understanding of the general fluid mechanical energy loss behavior of various configurations of frontal cooling openings over a wide range of flow rates and vehicle speeds. The findings of this study are intended to be relevant to design efforts involving frontal cooling openings.

THEORETICAL BASIS

The loss of mechanical energy through conversion to thermal energy across the front end cooling openings is the first in a series of energy exchanges through the cooling airflow circuit. The other energy exchanges are due to the viscous and heat transfer effects for flow through the radiator, condenser, fan and shroud, and engine bay.

Consider a streamtube which extends from far ahead of the vehicle (indicated with a subscript '∞') through a frontal opening to the front face of the first heat exchanger in the cooling system (indicated with a subscript '1'). It is assumed that in the thermodynamic process for the flow through the streamtube, the air can be treated as a perfect gas with a constant specific heat. It is also assumed that there is no heat transfer to or from the air as it flows from the environment through the frontal cooling openings to the face of the first heat exchanger. Finally, it is assumed that the air speeds during this variable area, adiabatic process are low relative to the sonic speed. Subject to these assumptions, it can be shown [6] that the air density is approximately constant.

The Conservation of Mass principle requires that

$$\rho V_\infty A_\infty = \rho V_1 A_1$$

and the Conservation of Energy Principle for this process is

$$h_\infty - h_1 = C_p(T_\infty - T_1) = \frac{1}{2}(V_1^2 - V_\infty^2)$$

The irreversibility rate for the process can be shown [7] to be

$$\dot{Irr} = \dot{m}T_\infty(s_1 - s_\infty)$$

where the entropy change is

$$s_1 - s_\infty = c_p \ell n\frac{T_1}{T_\infty} - R\,\ell n\frac{P_1}{P_\infty}$$

$$s_1 - s_\infty \approx c_p\frac{T_1 - T_\infty}{T_\infty} - R\frac{P_1 - P_\infty}{P_\infty}\ .$$

Combining this expression and the Conservation of Energy result yields the following for the irreversibility rate,

$$\dot{Irr} = \dot{m}\left[\left(\frac{P_\infty}{\rho} + \frac{1}{2}V_\infty^2\right) - \left(\frac{P_1}{\rho} + \frac{1}{2}V_1^2\right)\right].$$

Hence, it is observed that, subject to the assumptions described above, the irreversibility per unit mass is equal to the change in fluid mechanical energy or, equivalently, the irreversibility per unit volume is equal to the change in total pressure.

We may define a loss coefficient, K_{loss}, on the basis of the freestream kinetic energy as

$$K_{loss}\left(\frac{1}{2}V_\infty^2\right) = \left(\frac{P_\infty}{\rho} + \frac{1}{2}V_\infty^2\right) - \left(\frac{P_1}{\rho} + \frac{1}{2}V_1^2\right)$$

such that the irreversibility rate is

$$\dot{Irr} = K_{loss}\dot{m}\frac{1}{2}V_\infty^2$$

and the static pressure at the heat exchanger face is

$$P_1 = \left(P_\infty + \frac{1}{2}\rho V_\infty^2\right) - \frac{1}{2}\rho V_1^2 - K_{loss}\left(\frac{1}{2}\rho V_\infty^2\right).$$

The choice of the freestream kinetic energy as the scaling parameter for the energy losses is appropriate at high freestream or vehicle velocities where the cooling flow is driven primarily by the freestream ram or dynamic pressure and the fan has little effect and is often unpowered. In this situation, the limiting values on K_{loss} are zero and approximately one.

The lower limit on K_{loss} implies zero losses and an ideal dynamic-to-static pressure exchange as the flow is decelerated from freestream conditions to the front face of the first heat exchanger. This ideal situation can actually be approached when the cooling flow rate and internal velocities are small relative to the freestream velocity due to downstream flow restrictions. The resulting static pressure at the heat exchanger face is approximately equal to the freestream stagnation pressure.

The upper limit on K_{loss} occurs when the losses are approximately equal to the freestream dynamic pressure and the static pressure at the heat exchanger face is close to the freestream static pressure. The cooling flow rate in this situation is nearly zero depending on the engine bay back pressure to the heat exchangers.

At zero vehicle speed, with fan induced flow through the frontal openings, the preceding definition of the loss coefficient leads to an infinite result since the freestream kinetic energy is zero. For that situation, it is more appropriate to define a loss coefficient on the basis of a kinetic energy derived from the cooling flow rate. This could be the kinetic energy of the flow at the face of the heat exchanger or in the frontal openings themselves. At intermediate vehicle speeds, a complete correlation of the losses should include the effects of both the freestream kinetic energy and an "internal" kinetic energy derived from the cooling volumetric flow rate. The determination of the form of this correlation was the objective for the study reported herein.

It is convenient to define a ram-recovery coefficient as

$$K_{RAM} = 1 - K_{loss}$$

such that an alternate expression for the static pressure at the heat exchanger face is

$$P_1 = P_\infty + K_{RAM}\tfrac{1}{2}\rho V_\infty^2 - \tfrac{1}{2}\rho V_1^2 .$$

The ram-recovery coefficient reflects the ram or dynamic pressure recovery and corresponding static pressure rise at the front face of the heat exchanger. Based on the preceding discussion of K_{loss}, the ram-recovery coefficient can be expected to fall between one (at high vehicle speeds and low flow rates) and minus infinity (at zero vehicle speed with fan induced flow).

Figure 1 Bluff and streamlined front end models with supporting duct and ground plane

EXPERIMENTAL PROCEDURE

The primary objective of this study was to determine the relationships between the front end losses, cooling air flow rate, vehicle speed, and the size and location of the openings. To more clearly reveal these relationships, the design of the wind tunnel models represented only the front portion of an automobile. This approach was taken to isolate the front end cooling openings from any unique aerodynamic effects associated with the other geometrical features of a real automobile and to make the results of the study as generic as possible.

The basic apparatus consisted of two simply shaped automotive front ends of approximately one-fifth scale which could be mounted interchangably above a ground plane in a 3x4 ft open jet wind tunnel (Figure 1). The *bluff model* illustrated in Figure 2 was simply a rectangular prism with slightly rounded edges. In contrast, the *streamline model* illustrated in Figure 3 was formed from elliptical sections. This variation in streamlining somewhat reflects the extremes observed in automotive design. To compare the performance of different cooling opening configurations, both models featured frontal openings, bottom openings, and air dams that could be interchanged and an adjustable ground clearance. Each of the adjustable geometric elements was tested at a "nominal" size or position and with a "larger" and "smaller" perturbation (Table 1) to determine the corresponding trends associated with the geometric feature.

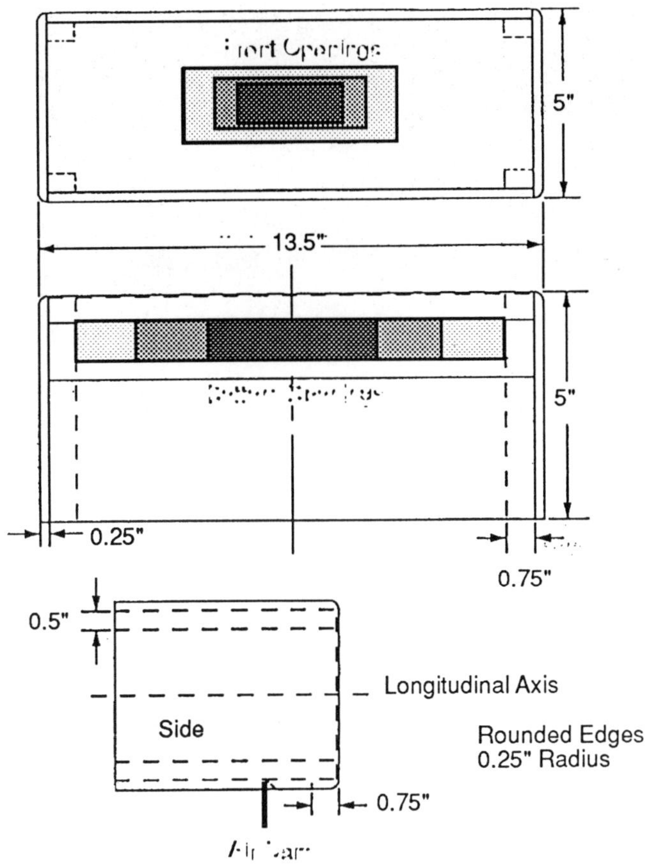

Figure 2 Bluff front end model

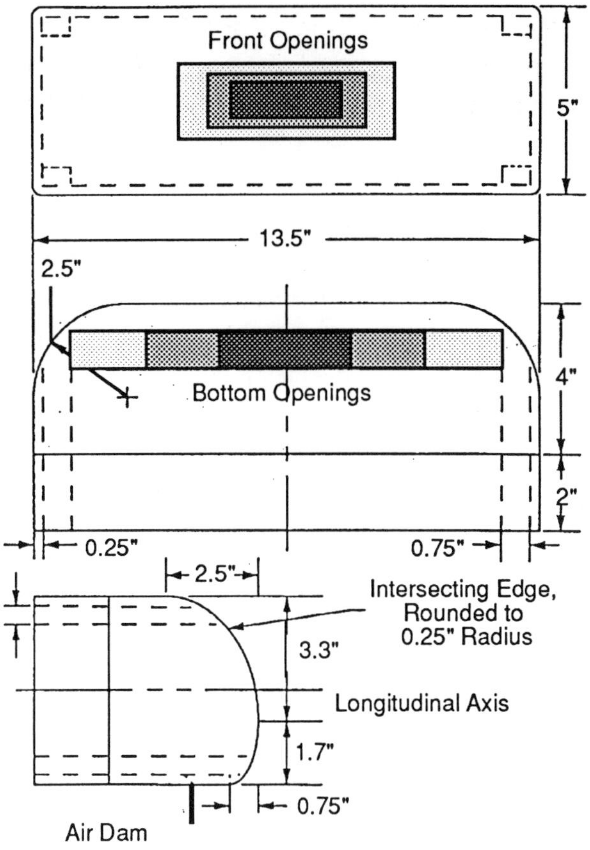

Figure 3 Streamlined front end model

The simplified front end models were fitted individually to a smooth elongated duct of the same cross-section (Figure 4). The duct was, in turn, connected by a hose to the suction side of a high pressure centrifugal blower which was also fitted with a laminar flow element for flow rate measurements.

In the preceding section, it was shown that a loss coefficient based on the freestream dynamic pressure could be calculated from the change in the total pressure between freestream conditions and the front face of the first heat exchanger. A direct measurement of the total pressure is straightforward in the freestream; however, in the region downstream of the frontal openings, the measurement is more difficult to the highly three-dimensional turbulent nature of the flow in that area. The approach taken for the present study was to determine the static and dynamic pressure in the duct separately and then to calculate the corresponding total pressure from their sum.

The total volumetric flow rate through the cooling open-

Table 1
Geometrical Parameters

	Small	Nominal	Large
Frontal Opening			
height (in.)	1.03	1.41	2.03
width (in.)	2.83	4.08	5.78
% of Frontal Area	5.1	10.0	20.3
Bottom Opening			
length (in.)	0.64	0.64	0.64
width (in.)	3.50	7.50	11.50
% of Frontal Area	3.9	8.3	12.8
Air dam			
height (in.)	0.38	0.64	0.97
width (in.)	11.50	11.50	11.50
Groundplane Clearance (in.)	0.93	1.85	2.78

ings and the duct was measured directly with the laminar flow element. The average velocity and dynamic pressure in the duct could then be easily determined.

To obtain meaningful static pressure measurements in the duct, the turbulence generated by the flow through front openings was dampened with honeycomb flow straightener material (0.25 in. cells, 2 in. thick) and finely meshed screen (36 mesh, 59% open area) which were placed 2 in. aft of the removable front ends or 8 in. aft of the front of the assembled models. A ring of pressure taps was mounted on the interior duct surface, 8 in. downstream of the screen and manifolded together such that an average static pressure in the duct could be obtained. The pressure drop versus flow rate relationship for the honeycomb and screen combination was obtained in a separate test using similarly mounted up- and downstream pressure taps in a long duct which provided near uniform flow throughout. It is assumed that the highly three-dimensional flow aft of the frontal openings in the wind tunnel models caused only minor changes in the pressure drop versus flow rate relationship. Unfortunately, this assumption could not be directly verified. The static pressure aft of the frontal openings and upstream of honeycomb was calculated by adding the known pressure drop associated with the honeycomb and screen to the measured duct pressure.

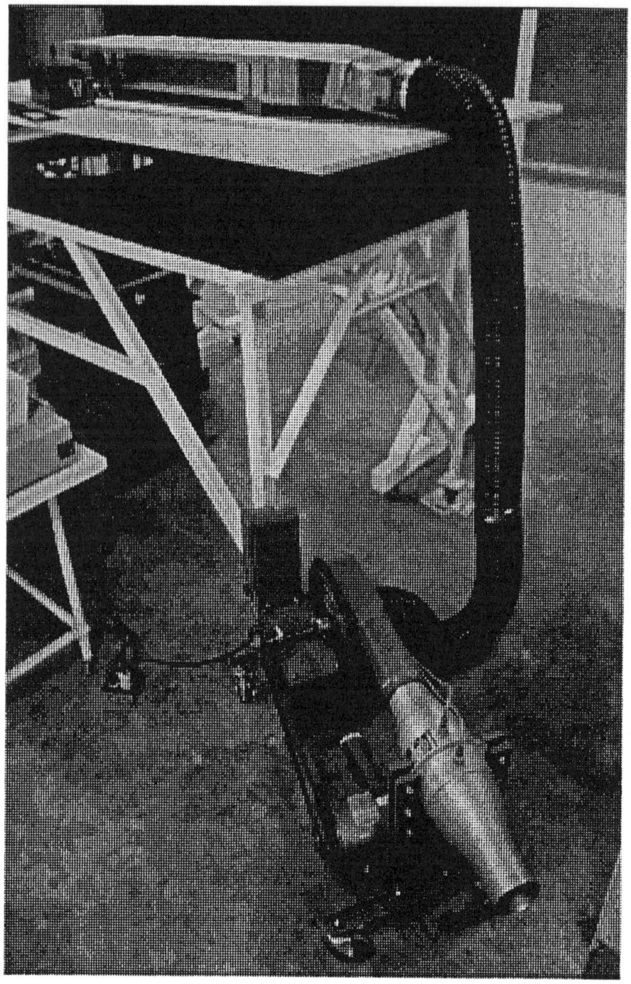

Figure 4 Front end models, duct, blower, and laminar flow element

The actual connections of the instrumentation for the measurements are illustrated in Figure 5. A differential pressure transducer was used to obtain the difference between the freestream total pressure and the static pressure in the duct downstream of the honeycomb and screen, i.e.,

$$\Delta P_{meas} = \left(P_\infty + \frac{1}{2}\rho V_\infty^2\right) - P_{model}$$

The internal dynamic pressure and honeycomb and screen pressure drop obtained from the measured flow rate were subtracted from this difference to obtain the desired change in total pressure,

$$\Delta P_{T,fe} = \Delta P_{meas} - \Delta P_{honeycomb\,\&\,screen} - \frac{1}{2}\rho V_{model}^2$$

$$\Delta P_{T,fe} = \left(P_\infty + \frac{1}{2}\rho V_\infty^2\right) - \left(P_1 + \frac{1}{2}\rho V_{model}^2\right)$$

from which K_{loss} and K_{RAM} could be obtained.

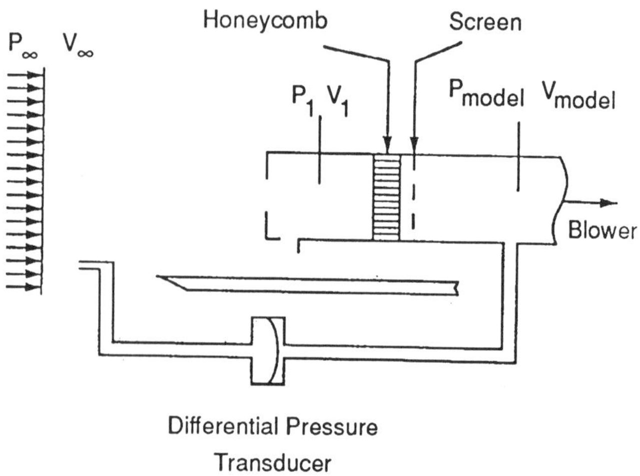

Figure 5 Instrumentation Layout

EXPERIMENTAL RESULTS

In the analysis of experimental data, the use of dimensionless parameters offers great advantage in formulating correlations with a reduced number of parameters, identifying trends, and revealing underlying relationships between the experimental parameters. In this section, measurement results from a more or less realistic combination of frontal and bottom openings, air dam, and ground clearance on the the streamlined model will be used to demonstrate the basic form of the correlations for front end cooling opening energy losses. The correlations for data from separate measurements with a single, variable sized frontal opening and a single, variable sized bottom opening will then be presented to illustrate modified forms of the correlations which can account for variable opening size. This will be followed by the presentation of an algorithm with which the performance of a combination of frontal and bottom openings can be predicted based on measurements for each of the openings tested individually Finally, the applicability of the correlations to real automobiles will be demonstrated by applying the data reduction techniques to previously published data by Schaub and Charles [3] for a 1976 Ford Granada.

Figure 6 is a dimensional plot of the loss in total pressure across the frontal openings versus flow rate for the streamlined model with *nominal* (Table 1) frontal opening, bottom opening, air dam, and groundplane clearance. These data are typical in that each freestream velocity results in a separate curve with increasing freestream velocities leading to higher total pressure losses and irreversibility rates. This indicates that, apparently, the front end fluid mechanical energy losses scale with the freestream kinetic energy.

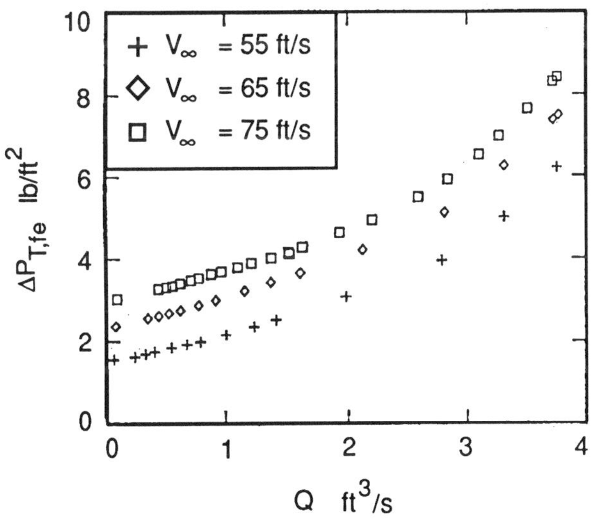

Figure 6 Total pressure drop versus volumetric flowrate for the streamlined front end with nominal front and bottom openings and air dam

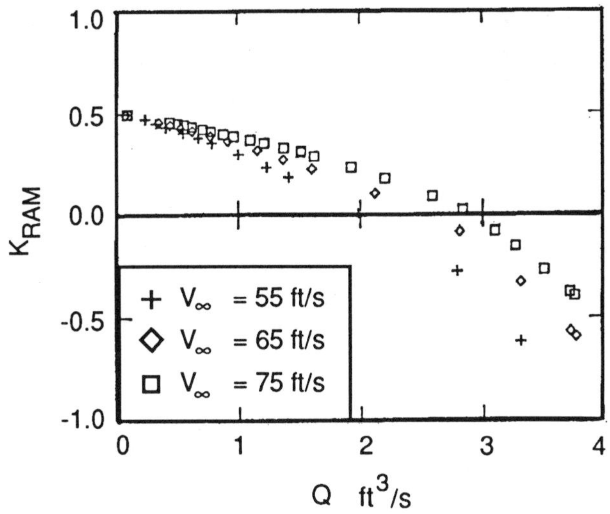

Figure 7 Ram recovery coefficient versus volumetric flowrate for the streamlined front end with nominal front and bottom openings and air dam

If the front end total pressure losses shown in Figure 6 are used to calculate dimensionless ram-recovery coefficients, K_{RAM}, (scaled with the freestream kinetic energy), Figure 7 is obtained. In the limit, as the flow rate goes to zero, the ram-recovery coefficient for all three freestream velocities is 0.5. This indicates that half of the freestream dynamic pressure has been recovered as static pressure within the model. A ram-recovery coefficient of 1.0 with zero flow rate would indicate an internal static pressure equal to the freestream total or stagnation pressure. For the geometry leading to the data presented in Figure 7, the maximium pressure rise in the duct is limited to the maximum pressure rise that can be achieved at the bottom opening, ahead of the air dam. For this particular geometry, there is flow into the front opening and out the bottom opening when the flow rate through the duct is zero.

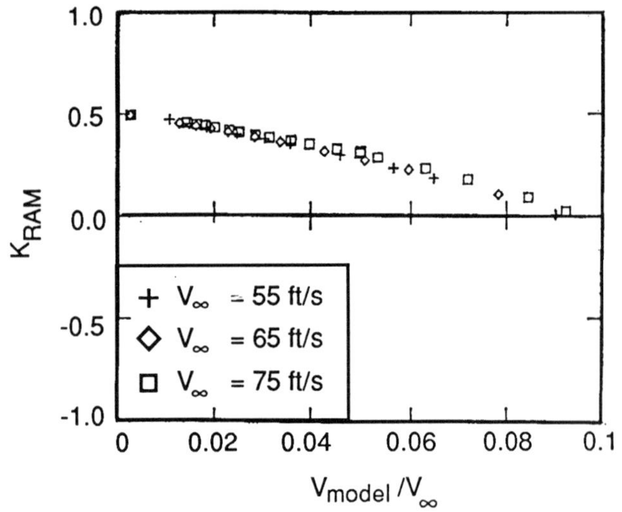

Figure 8 Ram-recovery coefficient versus velocity ratio for the streamlined front end with nominal front and bottom openings and air dam

For the data in Figure 7 at higher flow rates, the ram-recovery coefficient decreases or equivalently, the energy loss increases. At very high flow rates, relative to the freestream velocity, the ram-recovery factor becomes negative. This indicates that the high fan induced flow rates result in energy losses through the frontal openings that exceed the freestream kinetic energy. Schaub and Charles [3] offer an explanation for this variation in terms of the exterior flow streamline pattern and hypothetical streamtubes leading to the frontal openings. They argued that the entrance area to the cooling airflow streamtube is infinitely large when the vehicle is stationary with the fan running; whereas, at high speeds this entrance area can be quite small. This extreme streamtube variation is responsible for the large changes in the total pressure losses through the frontal openings.

It may also be noted from Figure 7 that the rate of decrease of the ram-recovery coefficient with flow rate is dependent on the freestream velocity. This indicates that the freestream kinetic energy alone is not sufficient to correlate the total pressure losses as a function of flow rate. Following Schaub and Charles, it is suggested that kinematically similar streamtube geometries and streamline patterns should produce dynamically similar energy losses. Furthermore, kinematically or geometrically similar streamline patterns should exist between flow fields at different freestream velocities if the ratios of a suitably defined internal velocity to freestream velocity are the same. This premise is tested by plotting the ram-recovery coefficient variation against the ratio of the average internal velocity to the freestream velocity with the results presented in Figure 8. As can be seen, all data collapse onto a single curve, thereby indicating a successful correlation of the three parameters, fluid mechanical energy loss, flow rate, and freestream velocity, in terms of two dimensionless parameters, ram-recovery coefficient and velocity ratio. This collapse occured for all test configurations when plotted with the above dimensionless parameters.

Figure 9 shows the energy loss versus flow rate results, in dimensionless format, for both front ends for the three sizes of frontal openings without bottom openings. Altough they are not differentiated in the plot, the data for each opening size include a range of flow rates at three freestream velocities. With decreasing opening area, the losses clearly increase more rapidly with velocity ratio or flow rate. In the absence of a bottom opening, the limiting value of the ram-recovery coefficient can approach unity as the internal pressure is near the freestream stagnation pressure. The streamlined front end is less effective in this regard due to sensitivity to the relative positions of the effective stagnation point on the front of the model and the

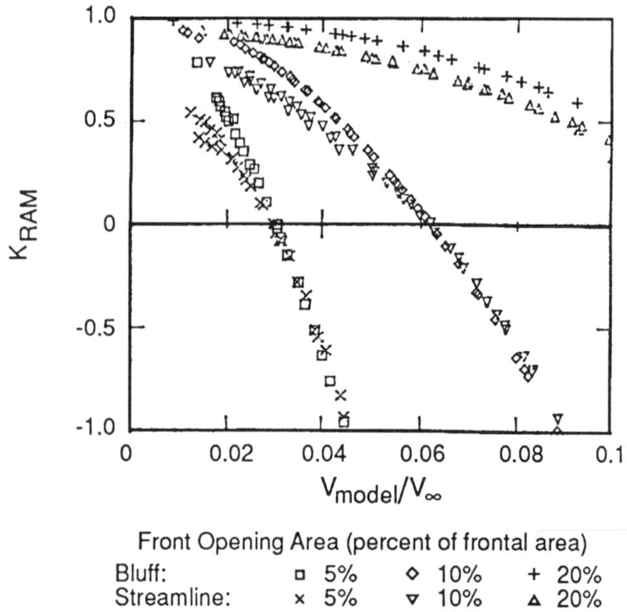

Figure 9 Ram-recovery coefficient for isolated front openings; velocity ratio is computed from the average velocity in the duct

location of the opening.

The selection of the area used in calculating a representative internal velocity from the flow rate through the model is an arbitrary consideration. If the area of the opening is used to define the velocity ($V_{open} = Q_{model}/A_{open}$) and V_{open} is used instead of V_{model} in the velocity ratio, Figure 10 is obtained from the same data presented in Figure 9. The collapse of the data indicates that in spite of moderate changes in the geometry of the openings, the external flow fields and corresponding fluid mechanical energy losses at the openings are similar if the ratios of velocity at the opening to freestream velocity are the same.

The effect of varying the bottom opening area is shown in Figure 11. For these tests, the transverse dimension of the bottom opening was varied while the longitudinal dimension of the opening and the air dam dimensions were held constant. The data for the different bottom opening areas are also reasonably well correlated when put into dimensionless format.

Extensive data were collected for configurations that combined frontal and bottom openings. An algorithm was devised to predict the ram-recovery coefficient for a particular combination of openings using the data from the individual openings tested separately. The primary assumption is that the drop in total pressure is the same for both the frontal and bottom opening flow paths although the flow rates may be different. Accordingly, the ram-recovery coefficients for the two openings must match since the upstream conditions are identical for both openings.

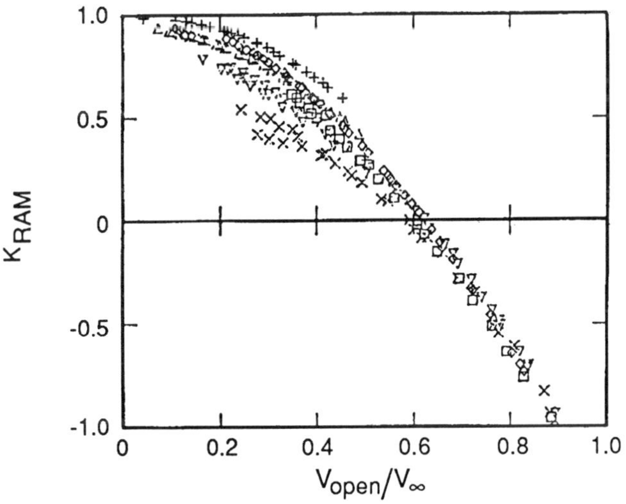

Front Opening Area (percent of frontal area)
Bluff: □ 5% ◇ 10% + 20%
Streamline: × 5% ▽ 10% △ 20%

Figure 10 Ram-recovery coefficient for isolated front openings; velocity ratio is computed from the average velocity in the inlet opening

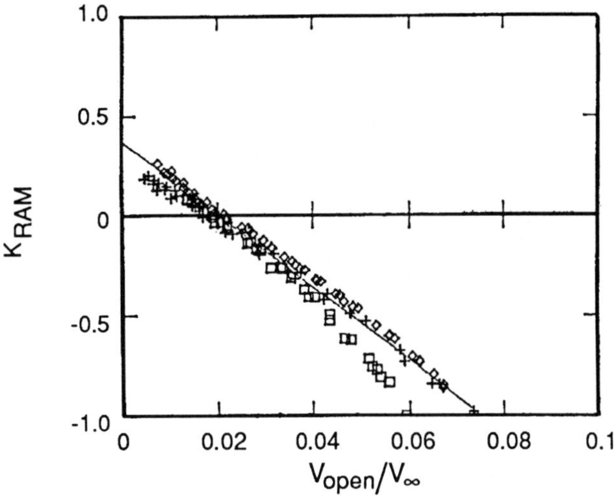

Bottom Opening Area (percent of frontal area)
□ 3.9% ◇ 8.3% + 12.8%

Figure 11 Ram-recovery coefficient for isolated bottom openings; velocity ratio is computed from the average velocity in the inlet opening

Treating the ram-recovery coefficient as the independent variable, the corresponding velocity ratios (V_{open}/V_{∞}) are found from correlations for each opening, multiplied by the ratio of the opening area to the cross-sectional area and added together to obtain the velocity ratio (V_{model}/V_{∞}) for the combination at a particular ram-recovery coefficient,

$$\left[\frac{V_{model}}{V_{\infty}}\right]_{combination} = \left[\frac{V_{open}}{V_{\infty}}\right]_{frontal}\left[\frac{A_{open}}{A_{model}}\right]_{frontal} + \left[\frac{V_{open}}{V_{\infty}}\right]_{bottom}\left[\frac{A_{open}}{A_{model}}\right]_{bottom}$$

Figure 12 shows data from combined opening configurations and predictions made from the data for separately tested openings.

As noted previously, Schuab and Charles [3] also investigated the front end pressure loss. Figure 13, a dimensional plot of front end pressure loss versus flow rate, is an excerpt from their paper describing wind tunnel tests of a 1976 Ford Granada. If the data taken from this figure are presented in the dimensionless format described above, Figure 14 is obtained. The use of the dimensionless parameters, $V_{radiator} / V_{\infty}$ and ram-recovery coefficient, seems to effectively collapse the full scale data. This result provides some assurance that the observations for the simplified front end models are also applicable to full scale automobiles.

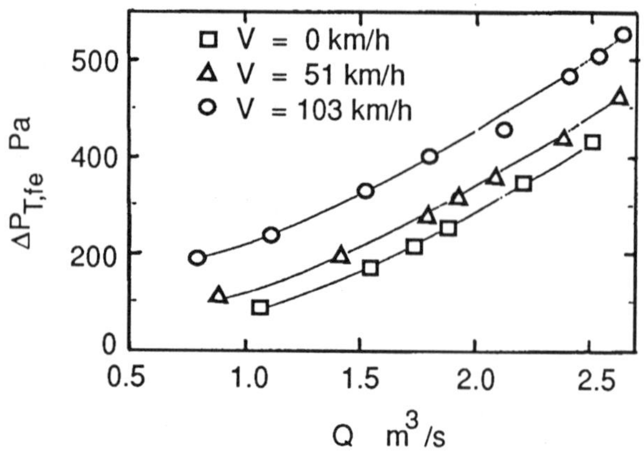

Figure 13 Front end total pressure loss results from Schaub and Charles [3]

CONCLUSIONS

The dimensionless parameters defined in this study have helped in understanding the behavior of front end cooling openings by revealing the relationships between the mechanical energy losses, freestream velocity, and cooling system air flow rate. It was shown that the variations in energy losses can be interpretted in terms of variations in the external streamline patterns as suggested by Schaub and Charles [3]. The dimensionless data reduction technique also reduces the number of experimental parameters from three (pressure loss, freestream velocity, and flow rate) to two (ram-recovery coefficient and velocity ratio). This is useful in reducing the number of experimental data points which must be taken to evaluate a cooling system and in extrapolating from a given data set. Since the effects of variation of cooling opening size on the energy losses are included, the dimensionless correlations may be useful in selecting the appropriate cooling opening for particular cooling system requirements. In addition, an algorithm has been presented that may be used to predict the ram-recovery coefficient for a combination of openings given K_{RAM} data for the individual openings.

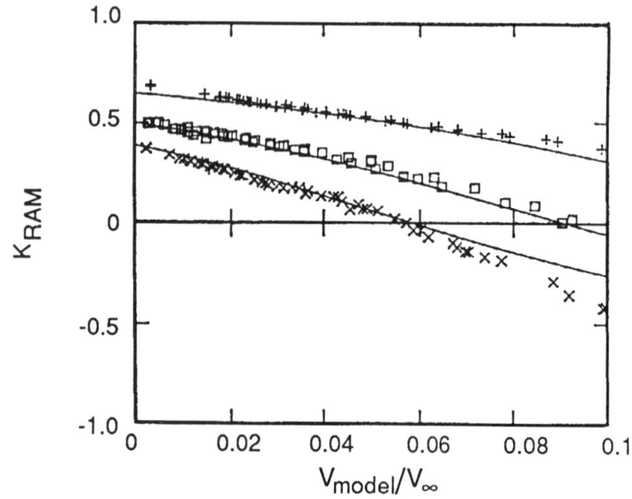

Front Opening Area (percent of frontal area)
× 5% □ 10% + 20%
——— Calculated Ram-Recovery Coefficient

Figure 12 Calculated and measured ram-recovery coefficients for combinations of front and bottom openings

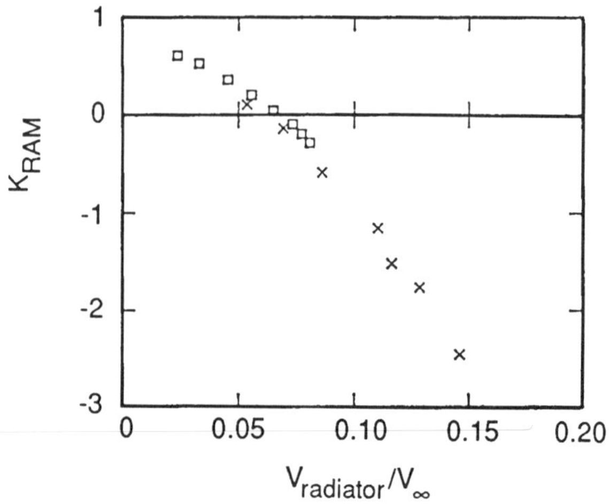

Figure 14 Ram-recovery coefficient versus velocity ratio for the data presented in Figure 13

REFERENCES

1. Olsen, M.E., "Aerodynamic Effects of Front End Design on Automobile Engine Cooling Systems," SAE Paper 760188, February, 1976.

2. Hawes, S.P., "Improved Passenger Car Cooling Systems," SAE Paper 760112, February, 1976.

3. Schaub, U.W. and Charles, H.N., "Ram Air Effects on the Air Side Cooling System Performance of a Typical North American Passenger Car," SAE Paper 800032, February, 1980.

4. Williams, J., "An Automotive Front-End Design Approach for Improved Aerodynamics and Cooling," SAE Paper 850281, 1985.

5. Renn, V. and Gilhaus A., "Aerodynamics of Vehicle Cooling Systems," **Journal of Wind Engineering and Industrial Aero-dynamics**, Vol 22:339-346, 1986.

6. Shapiro, A.H., **The Dynamics and Thermodynamics of Compressible Fluid Flow, Vol. 1**, Ronald Press Co., 1953.

7. Van Wylen, G.J., and Sonntag, R.E., **Funda-mentals of Classical Thermodynamics**, Third Edition, John Wiley & Sons, New York, 1986.

8. Roseberry, C.M., "Automotive Cooling Airflow Correlations," M.S. Thesis, Texas Tech University, 1990.

932889

Prediction of the Airflow Through Automotive Radiators Using 3-Dimensional Analysis

Tsuyoshi Yasuki, Takao Nishimura, Takayuki Abe, and Jirou Takamitsu
Toyota Motor Corp.

ABSTRACT

Air flow around a car front end configuration and through the radiator and condenser was computed simultaneously. Although the engine compartment was simplified to reduce computational cost, comparison of experimental data with the analysis showed excellent prediction of the air flow through the radiator and condenser.

INTRODUCTION

The purpose of this paper is to outline numerical approach for predicting the air flow through radiator and the condenser.

Recent vehicle designs are emphasizing low hood lines associated with low external aerodynamic drag. This design trend can produce more needs of predicting the cooling air flow through the radiators. An availability of a flow-through-radiator analysis tool could greatly reduce the development time and experimental cost by providing approximate result before building hardware.

On the other hand, flow analysis through the radiator and condenser is complicated due to complex boundary conditions, irregular geometries, and non-linear governing equations. For ease in handling such problems, the engine compartment was simplified, since the dominant factor controlling flow through the radiator and condenser is the central plane shape. Such simplified engine compartment model provided reasonable effect on the prediction of the cooling airflow.

MODEL FORMULATION

GOVERNING EQUATION - The governing equations for mean velocity can be derived from Reynolds-Averaged Navier-Stokes equations in the Cartesian coordinates system as follows:

Equation of continuity

$$\frac{\partial ui}{\partial xi} = 0 \tag{1}$$

Equation of momentum

$$\frac{\partial ui}{\partial t} + \frac{\partial ui\,uj}{\partial xi} + \frac{\partial \overline{ui'uj'}}{\partial xi} = -\frac{1}{\rho}\frac{\partial p}{\partial xi} + \frac{\partial}{\partial xj}\,\nu\left(\frac{\partial ui}{\partial xj} + \frac{\partial uj}{\partial xi}\right) \tag{2}$$

where t is time, density, p static pressure, u mean velocity, u' velocity fluctuation, and ν molecular viscosity. Turbulent stresses in the momentum equations are related to mean rate of deformation through eddy viscosity .

$$\overline{ui'uj'} = \nu t\left(\frac{\partial ui}{\partial xj} + \frac{\partial uj}{\partial xi}\right) - \frac{2}{3}\delta ij\ k \tag{3}$$

The eddy viscosity νt is related to the turbulent kinetic energy k and its dissipation rate ε.

$$\nu t = C\mu\frac{k^2}{\varepsilon} \tag{4}$$

$$k = \frac{1}{2}\overline{ui'ui'} \tag{5}$$

$$\varepsilon = \nu \overline{\frac{\partial u_i'}{\partial x_j} \frac{\partial u_j'}{\partial x_j}} \qquad (6)$$

where $C\mu$ is a model constant and k and ε are obtained from a two-equation turbulence model[1].

In order to implement a finite volume solution and treat boundary conditions for general curved boundaries, the independent space variables (x,y,z) in the above equations are transformed into a boundary-fitted coordinate system(ξ, η , ζ). Through the transformation, the governing equations can be expressed in the following form:

$$\frac{\partial \vec{q}}{\partial t} + \frac{\partial \vec{E}}{\partial \xi} + \frac{\partial \vec{F}}{\partial \eta} + \frac{\partial \vec{G}}{\partial \zeta} = \vec{S} \qquad (7)$$

$$\vec{q} = J^{-1} \begin{bmatrix} 1 \\ u \\ v \\ w \end{bmatrix} \qquad (8)$$

$$\vec{E} = J^{-1} \begin{bmatrix} U \\ Uu + \xi^x P - \xi^x \tau_{xx} - \xi^y \tau_{xy} - \xi^z \tau_{xz} \\ Uv + \xi^y P - \xi^x \tau_{yx} - \xi^y \tau_{yy} - \xi^z \tau_{yz} \\ Uw + \xi^z P - \xi^x \tau_{zx} - \xi^y \tau_{zy} - \xi^z \tau_{zz} \end{bmatrix} \qquad (9)$$

$$\vec{F} = J^{-1} \begin{bmatrix} V \\ Vu + \eta^x P - \eta^x \tau_{xx} - \eta^y \tau_{xy} - \eta^z \tau_{xz} \\ Vv + \eta^y P - \eta^x \tau_{yx} - \eta^y \tau_{yy} - \eta^z \tau_{yz} \\ Vw + \eta^z P - \eta^x \tau_{zx} - \eta^y \tau_{zy} - \eta^z \tau_{zz} \end{bmatrix} \qquad (10)$$

$$\vec{G} = J^{-1} \begin{bmatrix} W \\ Wu + \zeta^x P - \zeta^x \tau_{xx} - \zeta^y \tau_{xy} - \zeta^z \tau_{xz} \\ Wv + \zeta^y P - \zeta^x \tau_{yx} - \zeta^y \tau_{yy} - \zeta^z \tau_{yz} \\ Ww + \zeta^z P - \zeta^x \tau_{zx} - \zeta^y \tau_{zy} - \zeta^z \tau_{zz} \end{bmatrix} \qquad (11)$$

$$\vec{S} = J^{-1} \begin{bmatrix} 0 \\ gx \\ gy \\ gz \end{bmatrix} \qquad (12)$$

where J is the Jacobian, U, V, and W contravariant velocities, gx, gy, and gz gravity vectors, and τ_{xx} etc. are the partial differential.

NUMERICAL PROCEDURE - The governing differential equations were discretized by using a finite volume approach. In this approach, the governing differential equations were integrated over an elemental control volume. The solution of mean transport variables u,v,w,k, and ε are obtained by SIMPLE algorithm[2]. The system of algebraic equations formed by the Poisson equation for the static pressure field is solved by an iterative conjugate gradient method. SCRYU (version 1.3) flow analysis code by Software Cradle Co. Ltd. was used to solve these equations.

RADIATOR AND CONDENSER MODEL - To simulate front end air flow and engine compartment flow simultaneously, the flow model for both radiator and condenser was developed, and the capability of SCRYU was enhanced in the process. Flow through the radiator and condenser can be simulated by specifying the pressure drop characteristics, which depend on the type of radiator and the Reynolds number. The pressure drop versus cooling air flow rate characteristics curves can be obtained by testing the radiator and condenser. In this study, the relationship between pressure drop and cooling air flow rate within the radiator and condenser was assumed to be as follows.

$$\delta p = \rho \, (au + bu^2) \qquad (13)$$

where a and b are the empirical coefficients dependent on the type of radiator and condenser. These empirical coefficients were determined using the least square fit of the experimental data for the pressure drop versus cooling air flow rate.

VEHICLE MODEL - A computational vehicle model had a simpler shape than an actual vehicle since generating grids for the external shape of an actual vehicle are very time consuming. Longitudinal length of the car's front end configuration to be calculated in the simulation was 1000 mm as shown in Fig. 1. Experimental data of the front end configuration was obtained from TOYOTA MOTOR CORPORATION wind tunnel testing for validating the cooling air flow rate through a radiator and condenser. The pressure distribution at the front configuration showed excellent agreement with the pressure distribution of an actual vehicle as shown in Fig. 2.

Experimental data on pressure drop through the under-hood packaging was also obtained from TOYOTA MOTOR CORPORATION wind tunnel testing. As shown in Fig. 3, pressure drop through the engine compartment is negligible in the flow through the under hood. It is assumed that the flow is quasi three dimensional near the rectangular radiator inlet, therefore a slice of the engine compartment center plane was extrapolated to the width of the radiator. An orifice was added under the dash board to adjust the pressure drop characteristics around the engine block. Therefore the engine block was eliminated as shown in Fig. 4. The flow rate through the radiator and condenser of car front end configuration with such simplified engine compartment showed excellent agreement with that of an actual vehicle, as shown in Fig. 5.

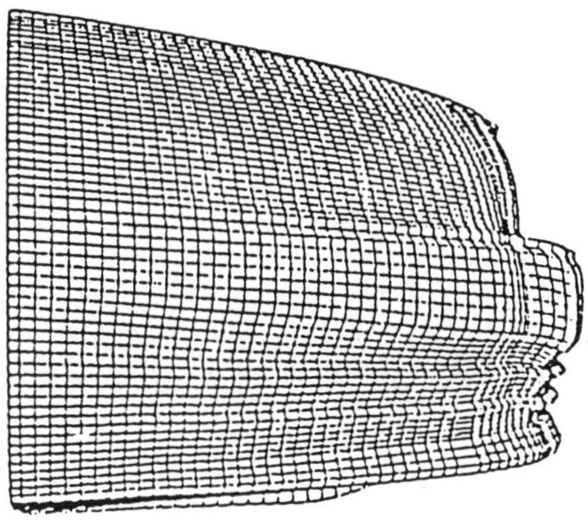

Fig. 1 Car front end configuration

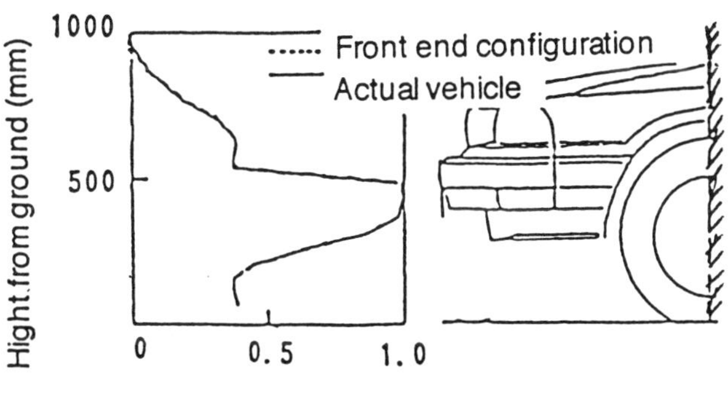

Static pressure coefficient

Fig. 2 Comparison of pressure distribution in front of the vehicle

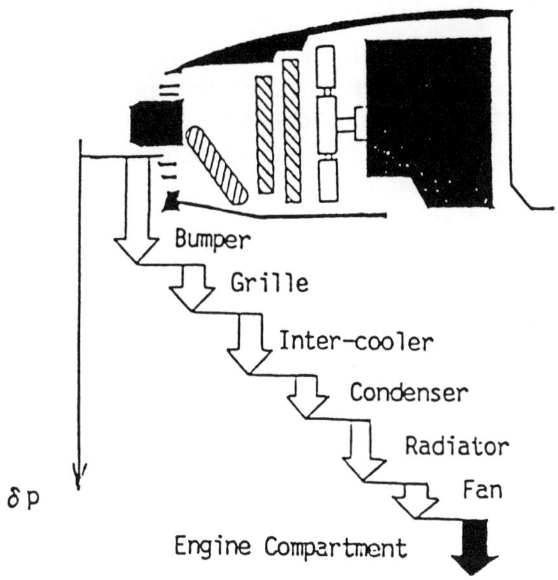

Fig. 3 Schematics of pressure drop through under hood packaging

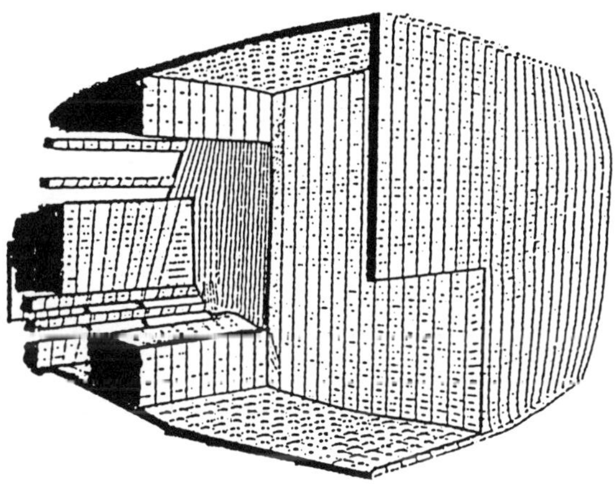

Fig. 4 Simplified engine compartment

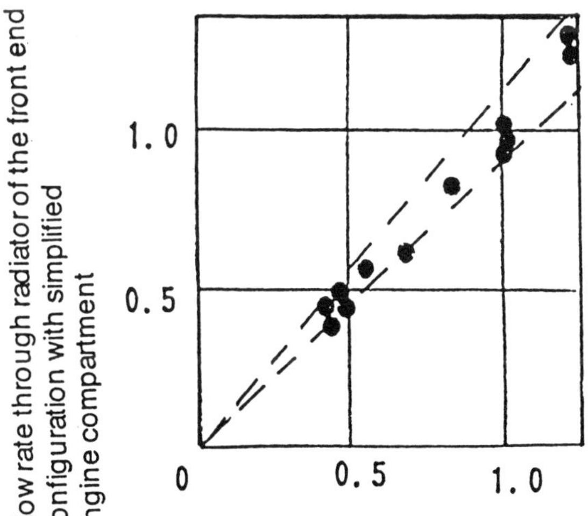

Flow rate through radiator of the actual vehicles

Fig. 5 Comparison of flow through radiator between analysis and experiment

GRID GENERATION - The grid distribution around the vehicle's front end configuration is shown in Fig. 6 . A total of (74×35×85) grids were used in the simulation. The computational grids were generated by an in-house grid generation code as described below:

1)Solid shape is defined as three-dimensional surface data.

2)Grids on solid surfaces are generated by a trans-finite interpolation method and projected onto the solid surfaces.

3)Computational spatial domain is divided into several blocks.

4)Spatial grids for each block were generated by trans-finite interpolation method and smoothed over the whole computational spatial domain.

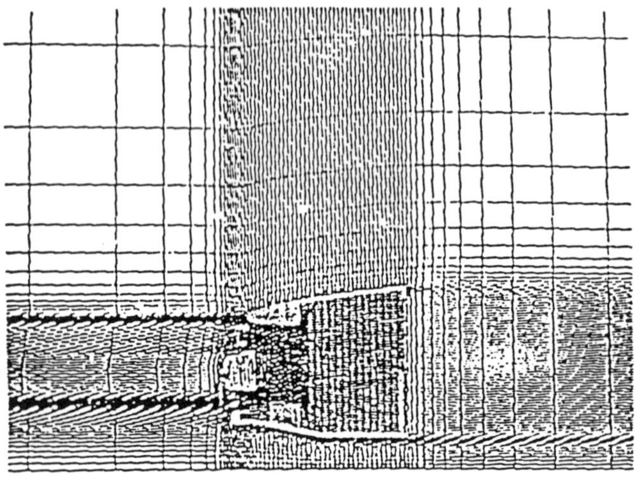

Fig. 6 Grid distribution at the center plane of a car front end configuration

BOUNDARY AND INITIAL CONDITIONS - The inlet plane is located approximately two car-body lengths upstream of the front end of the vehicle,where free stream velocity specified. The turbulent kinetic energy and its dissipation are specified at the inlet assuming constant turbulence. The exit plane is located approximately three car body lengths down stream from the rear end of the vehicle, where constant pressure boundary conditions imposed. On the solid surfaces, no-slip boundary conditions are imposed with turbulent wall function treatment for the momentum equations. The ground plane is specified as moving with the vehicle speed of 120 km/h. Flow through the radiators and condenser is simulated directly from the pressure drop characteristics.

At all computational grids excluding locations where no-zero boundary conditions are imposed, u,v,w,k, and ε were set to zero at the initial time step.

COMPUTATIONAL RESULTS

In this section, the computational results for a sedan type vehicle with a simplified engine compartment will be discussed. The main purpose of this study was to improve front end flow of a car for enhanced cooling of the radiator using computational fluid dynamics. This requires validating the capability of the Navier-Stokes code for predicting the front-end flow distribution and the cooling air flow rate through a radiator.

INITIAL CONFIGURATION - As shown Fig. 7, there are two inlets for cooling air. As shown in Fig. 8,the stream lines at the center plane stagnate near the license plate. Relative location of the license plate is crucial for maximizing the cooling air flow. SCRYU predicted a cooling air flow rate of 101.0m^3/min. for the initial configuration ,which agrees well with the wind tunnel test data of an actual vehicle of 111.6m^3 /min. The air flow at the center plane diverges mostly into the opening above the bumper. As shown in Table 1, the cooling air flow distribution over the two openings diverges mostly into the opening above the bumper.

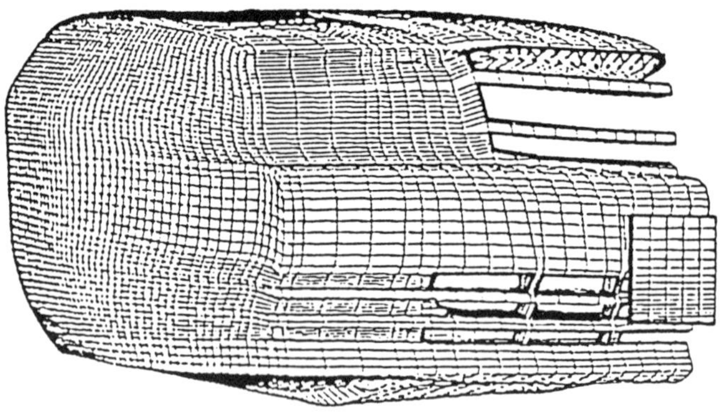

Fig. 7 Openings of front end configuration

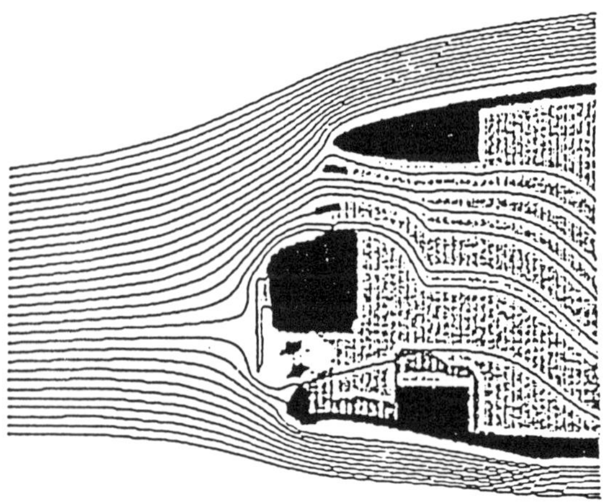

Fig. 8 Stream lines at the vehicle center plane for the initial configuration

Air flow rate [m^3/min.]	Total	Opening above bumper	Opennings below bumper
Base Case	100.1	67.3	32.8
License Plate (-35mm down)	97.8	69.9	27.9
License Plate (30mm up)	101.4	65.5	35.9
License Plate (60mm up)	102.4	64.4	38.0
License Plate (23mm forward)	96.7	65.3	31.4

Table 1 Cooling air flow distribution for the initial configuration and various design modification

LICENSE PLATE EFFECTS - To increase the air flow rate through radiators,we examined the simple method of shifting the position of the license plate. Fig. 9,10,11, and 12 show the effects of shifting the position of the license plate.

By shifting the license plate position to a 35 mm lower location, the air flow at the vehicle center plane diverges into the opening above the bumper, as shown in Fig. 9, and the cooling air flow rate was reduced to 97.8 m^3 /min..This is a 2.3 percent decrease from the initial configuration, as shown in Table 1. Although the total cooling air flow rate was reduced by 2.3m^3/min., the air flow rate through the opening above the bumper was increased by 2.6 m^3 /min. Such shifting of the license plate into a lower location reduces vehicle cooling performance.

By shifting the license plate to a 30 mm or 60 mm higher location, the air flow at the vehicle center plane diverges not only into the openings above the bumper but also into the openings below the bumper as well, shown in Figs. 10 and 11, respectively. The cooling air flow rates were increased to 101.4 m^3/min. or 102.4 m^3/min. which is a 1.2 percent or 2.3 percent increase from the initial configuration, respectively as shown in Table 1. Although the air flow through the openings above the bumper was decreased, the total cooling air flow rate was increased. Such shifting the license plate to a higher location increases vehicle cooling performance.

By shifting the license plate to a 23 mm forward location, the air flow rate at the vehicle center plane diverges into the openings above the bumper, as shown in Fig. 12. The cooling air flow rates were reduced to 96.7 m^3/min. ,which is a 3.4 percent decrease from the initial configuration, as shown in Table 1. Since recent vehicle designs emphasizing low hood lines tend to shift the license plate to a forward location,further design modifications are necessary to raise the cooling air flow rate through the radiator and condenser. This will be discussed in the next section.

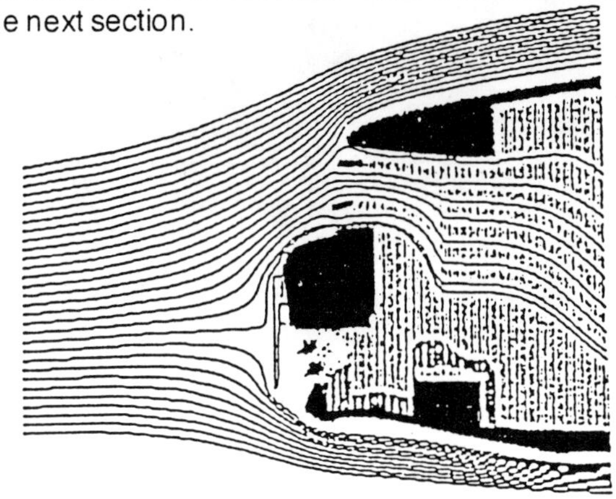

Fig. 9 Stream lines at the vehicle center plane for the license plate 35 mm down

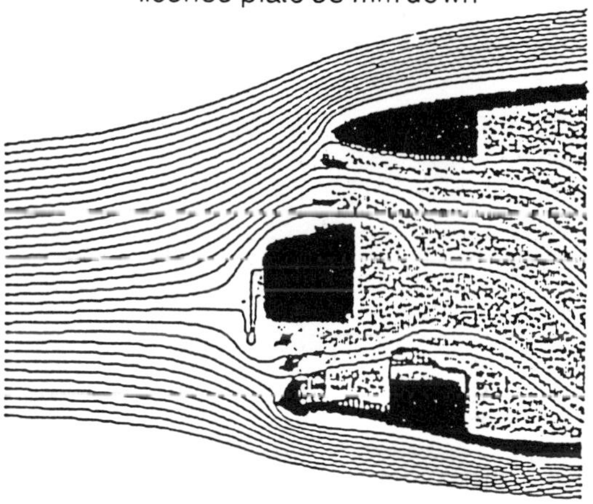

Fig. 10 Stream lines at the vehicle center plane for the license plate 30 mm up

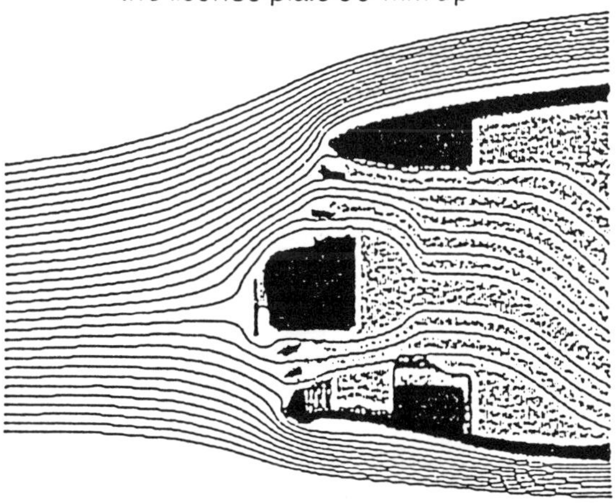

Fig. 11 Stream lines at the vehicle center plane for the license plate 60 mm up

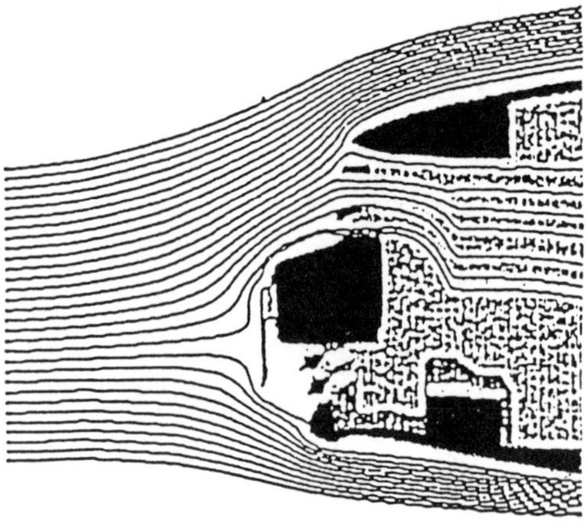

Fig. 12 Stream lines at the vehicle center plane for the license plate 23 mm forward

SMOOTH BUMPER EFFECTS - To increase the flow rate through the radiator and condenser , we proposed smoothing the bumper shape. As shown in Fig. 13, a sharp edge at the front upper corner of the bumper license plate in the 23 mm forward position was eliminated. This modification produced a smooth air flow from the stagnation point to the opening above the bumper at the vehicle center plane and increased the air flow rate to 98.1 m^3/min., as shown in Table 2. This is a 1 percent increase over the license plate with 23 mm forward position. Cooling air flow through the openings below the bumper increased as a result of shifting the stagnation point to a slightly lower location. As shown in Fig. 14, a sharp edge at the front lower corner of the bumper license plate in the 23 mm forward position was eliminated. This modification did not produce a smooth air flow at the vehicle center, but increased the air flow rate to 99.2 m^3/min., as shown in Table 2. This is a 3 percent increase from the license plate in the 23 mm forward position because of increased cooling air flow through the openings below the bumper of 2.5 m^3/min.

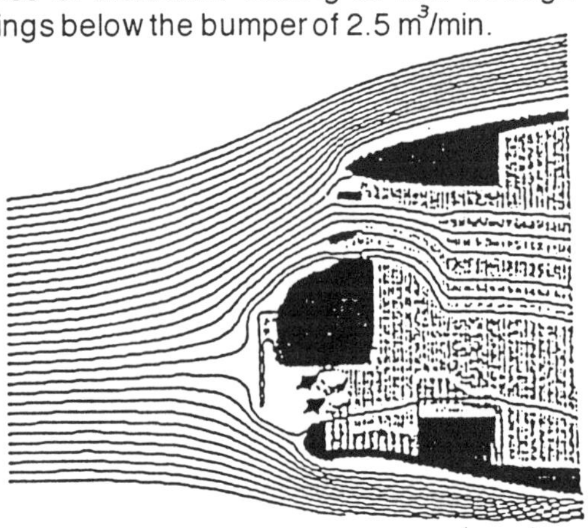

Fig. 13 Stream lines at the vehicle center plane for the license the plate 23 mm forward with bumper front upper corner smoothed

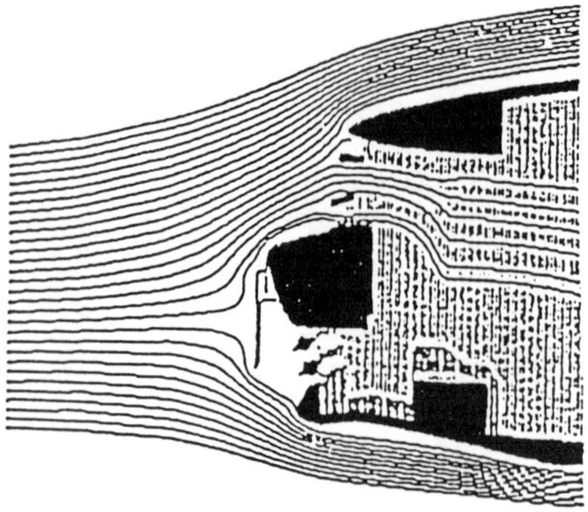

Fig. 14 Steam lines at the vehicle center plane for the license plate 23 mm forward with bumper lower corner smoothed

Air flow rate [m^3/min.]	Total	Opening above bumper	Opennings below bumper
Base Case	100.1	67.3	32.8
License Plate (23mm forward)	96.7	65.3	31.4
bumper front upper smoothed	98.1	65.4	32.7
bumper front lower smoothed	99.2	65.3	33.9

Table 2 Cooling air flow distribution for the initial configuration and various design modification.

CONCLUSION

A finite-volume method for predicting cooling air flow through radiator and condenser has been developed and confirmed. The agreement between experimental and numerical cooling air flow rates was excellent.

By utilizing computational fluid dynamics, relationship between the cooling air flow rate and the relative location of the license plate on the bumper was investigated.

This flow analysis tool reduces development time and experimental costs by providing approximate results before building hardware.

REFERENCES

1. Launder,B.E., and Spalding,D.B.,"The numerical Calculation of Turbulent Flows",Computational Method in Applied Mechanics and Engineering,vol.3 1974,pp 269-289

2. Patankar,S.V.,"Numerical Heat Transfer and Fluid Flow", 1980,McGraw-Hill

910307

Water Flow Simulation of Automotive Underhood Airflow Phenomena

J. E. Williams
Ford Motor Co.

J. W. Oler
Texas Tech University

J. E. Hackett
Lockheed Aerospace Co.

L. Hammar
SSPA Maritime Consulting AB

ABSTRACT

An experimental investigation of the underhood cooling airflow of a production passenger car has been carried out with a 3/8-scale model in a water filled tow tank at Texas Tech University and with a production vehicle in the Maritime Dynamics Laboratory tow basin of SSPA Maritime Consulting AB in Gothenburg, Sweden. The primary objectives for both the 3/8 and production vehicle investigations were to obtain a better understanding of the cooling airflow behavior within the engine compartment of an automobile and to identify the major factors influencing the flow field. The tests consisted of running a fan on a stationary and moving vehicle with pressure measurements and extensive flow visualizations within the engine bay.

NOMENCLATURE

C_p	Coefficient of pressure
g	gravitational acceleration (m/s^2)
G_r	Grashof number
L, L_c	Characteristic vehicle length scale (m)
N	Fan rotation speed (rps)
P	Fluid pressure (Pa)
P_{sp}	Static pressure sensed by pitot static probe (Pa)
$P_{s\infty}$	Freestream static pressure (Pa)
q_p	Dynamic pressure sensed by pitot static probe (Pa)
q_∞	Freestream dynamic pressure (Pa)
q_{shroud}	Dynamic pressure of shroud flow (Pa)
R	Fan radius (m)
R_e	Reynolds number
T_s	Engine block surface temperature (°K)
T, T_∞	Fluid temperature (°K)
U	Fluid velocity (m/s)
β	Coefficient of thermal expansion (1/°K)
V	Vehicle velocity (m/s)
ρ	fluid density (kg/m^3)
μ	fluid viscosity coefficient (Pa-s/m^2)
ν	fluid kinematic viscosity (m^2/s)

INTRODUCTION

Recent trends in automotive styling and design such as lower hood profiles, smaller grille openings, transverse mounted front-wheel drivetrains, elevated engine operating temperatures, and emphasis on low drag exterior flows have made the effective management of underhood thermal environments increasingly difficult. The automotive engineer must be concerned with providing adequate cooling airflow through the radiator and condenser, allowing sufficient space behind the radiator for efficient fan operation, protecting temperature sensitive underhood components, and limiting excessive underbody heat rejection leading to fuel tank heating and possible recirculation through the cooling system under idle conditions.

In light of these concerns, the objectives of the study

Figure 1 Texas Tech tow tank and 3/8 scale Taurus model

described in this paper were to develop a better understanding of what happens to heated radiator airflow after it enters the engine bay and to identify the major factors which influence underhood airflow. Armed with this background information, it is possible to isolate the causes of difficulties known to be associated with underhood airflow and to identify techniques to improve the underhood thermal environment. The basic approach utilized in the research program was water flow visualization and pressure measurements within the underhood area of reduced scale and full size vehicles.

The underhood airflow for a rear wheel drive vehicle has been investigated by Clemens [1] and [2]. The tests were conducted in an environmental wind tunnel and utilized flow visualization and velocity, flowrate, and temperature measurements. Clemens noted that under idle conditions, the majority of the fan discharge air dumps down to the front of the engine and out underneath the vehicle with approximately 25% of the flow recirculated back to the front of the radiator. Very little of the fan discharge reached the rear of the engine compartment. The flow distribution was improved by tilting the fan and radiator forward and applying recirculation seals. These modifications result in a reduction of the average underhood air temperature by 5 °F and temperature reductions for selected underhood components of 10-20 °F.

Figure 2 The 3/8-scale 1989 Taurus test vehicle

EXPERIMENTAL TECHNIQUES

Scale Model Testing at Texas Tech

The Texas Tech University tow tank has been utilized for classical aerodynamic studies for the past twelve years. Recent research projects include investigations of unsteady flows and dynamic stall on two dimensional airfloils, parachute opening dynamics, Darrieus wind turbines, and the exterior and underhood aerodynamics of automobiles.

The overall dimensions of the Texas Tech tow tank are 5 m x 10 m with a water depth of 1.2 m. Models may be propelled through the tank at speeds up to 1 m/s with a cable towing system or by a sting extending into the water from an above-water beam and carriage traversing system running the length of the tank as shown in Figure 1.

A 3/8-scale model of a 1989 3.0L Ford Taurus (illustrated in Figure 2) was tested in the Texas Tech University tow tank facility (Oler and Fuller, [3] or Williams, [4]). The model featured a detailed engine and engine compartment

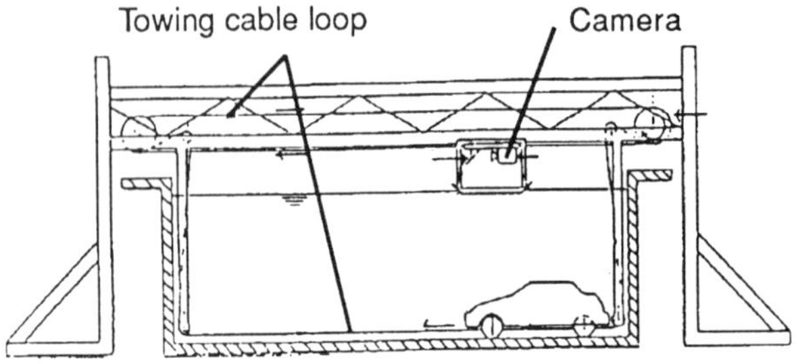

Figure 3 Texas Tech vehicle towing system

with a clear plexiglass hood through which the interior flow could be observed. The model included a porous screen representation of the radiator and condenser and a scale fan that was driven by a shaft extending through the model engine from a variable speed DC motor in a water tight compartment behind the dash panel. A coupled cable and carriage towing system was used to provide synchronized movement of the model and a plexiglass bottom "boat." Video records were obtained from a camera which was mounted with the plexiglass bottom boat to the above-water beam and carriage (Figure 3).

The 3/8-scale tests demonstrated the viability of water tank testing for underhood flow visualization and provided the initial characterization of the basic flow patterns. The relatively small tank and model provided a convenient and low cost testbed for the development of the experimental procedures and flow visualization techniques which were later utilized in the full scale tests that are the subject of the current paper.

The disadvantages of the 3/8-scale testing were that it was not possible to achieve the equivalent full scale Reynolds numbers (less by a factor of two) or absolute geometrical accuracy in the scale model The principal modelling discrepancies were the omission of most of the tubing and wiring in the underhood area and an incorrectly positioned front structural cross-member.

In spite of these differences, the observed underhood flowfield was essentially identical to the full scale observations which are described later in the current paper. This indicates that the bluff body dominated underhood region is not critically sensitive to the Reynolds number. However, there was a significant discrepancy in the degree of idle recirculation of the underbody flow which was traced to the incorrectly positioned cross-member.

Figure 4 The SSPA Maritime Dynamics Laboratory Tow Basin

Production Vehicle Testing at SSPA

The Maritime Dynamics Laboratory at SSPA Maritime Consulting is a facility for research and development in areas such as the behavior of ships in a seaway, offshore structures, automobiles, etc. The computer-controlled multi-motion carriage which spans the basin (Figure 4) can be programmed to perform different types of ship maneuvering tests. A model may be connected to the carriage with a multi-component force balance, thus enabling measurements of the hydrodynamic forces and moments in all degrees of freedom. Seakeeping tests can also be performed in which the carriage follows a free-running model.

The applicability of the SSPA water basin, towing carriage, and measuring eqiupment to automotive aerodynamic testing was demonstrated in a previous study (Larsson, et. al., [5]) in which detailed comparisons of wind tunnel and towing basin aerodynamic force measurements were made. It was found that water basin testing yields results with similar accuracy and provides significant advantages when the presence of a moving ground plane is important, e.g., when optimizing the underbody shape of an automobile. Although aerodynamic force measurements were not a part of the tests described herein, the prior study serves as validation of the technical approach of water basin testing in automotive aerodynamics.

The overall dimensions of the SSPA tow basin are 88 m x 39 m. A water depth of up to 3.2 m is possible; however, the depth was set at 2 m for the underhood flow visualization tests to allow the observers to be as close to the vehicle as possible. Although the multimotion carriage is capable of speeds up to 3.5 m/s, the maximum speed required for the tests reported herein was 1.22 m/s.

The production vehicle tested in the SSPA tow basin

(Hammar, [6]) was a production 1989 3.0L Ford Taurus equipped with air conditioning, power steering and brakes, and automatic transmission (Figure 5). The original hood was replaced by a double surfaced, vacuum formed clear plexiglass hood which replicated both interior and exterior surface contours. The fan was driven by a flexible cable extending forward through the radiator and condenser to a 90 degree gearbox hidden within the production bumper. A second flexible cable extended from the gearbox, through the interior of the bumper to exit slightly ahead of the driver-side wheelhouse. The cable was driven by a variable speed motor (with speed control to reduce sensitivity to loading variations) located above the water surface. No attempt was made to drive the engine accessories (alternator, etc.) or the associated belts. Although it is felt that this is not a critical omission, the assumption has not been verified at the current time.

Figure 5 The production 1989 Taurus test vehicle

Flow passageways between the engine compartment and front fenders were established by cutting ventilation openings at the upper left and right hand corners of the dash panel into the outer portions of the cowl plenum and at adjacent locations in the interior body shotgun structure as illustrated in Figure 6. Removable covers were fitted to the dash panel openings so that the original underhood configuration could be restored for baseline testing. Four symmetrically located vents were cut in the left and right side fenders as illustrated in Figure 7. These openings were also equipped with removeable covers.

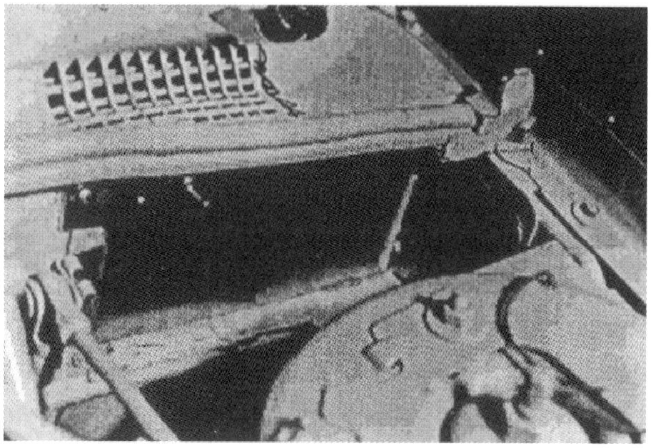

Figure 6 Dash panel openings to fender vents

The fender vents offer the potential of providing a means of ventilating selected regions of the underhood area, reducing the total heat rejection beneath the car, and increasing the total cooling flow rate.

The interior of the vehicle was stripped for testing in water and to facilitate attachment to the SSPA towing equipment. The roof of the passenger compartment was opened to allow a streamline strut to be mounted to the reinforced floor as illustrated in Figure 8. The ground clearance was set by suspending the vehicle beneath the towing apparatus and allowing the water filled tires to rest on the bottom of the basin. For stationary testing a wall was positioned in front of the vehicle to eliminate the influence of any secondary currents on forward recirculation.

Figure 7 Fender vents

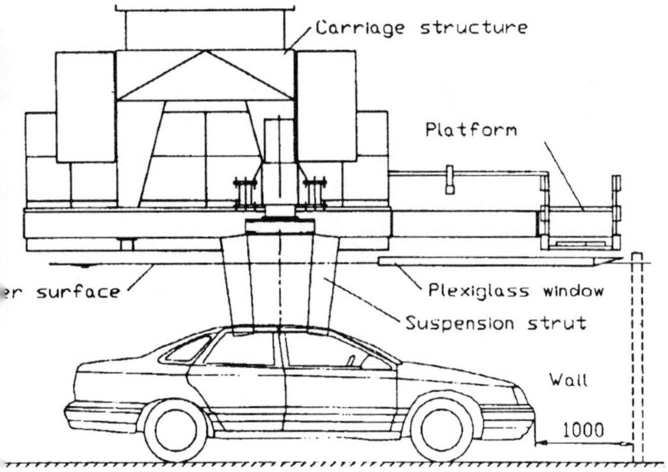

Figure 8 The SSPA towing system

Photographic records of the experiments consisted of video images obtained from two synchronized cameras - one located above the car and directed down through the plexiglass bottomed tray and hood and the other positioned underwater to observe the flow through the hood and fender vents. Multiple 35 mm still frame photographs were also obtained during each test run from an above water camera. In addition, sketches and handwritten notes were formed independently by the authors and later used to compare and contrast observations.

To enhance visualization of the flow, the entire underhood areas in the test vehicles were painted a uniform nonreflective light grey color. Lighting was provided by a combination of above and below water spotlights. In addition, 23 small lamps were located within the engine compartment itself to illuminate dark recesses.

The flow patterns were made visible by observing the motion of short tufts of yarn attached to surfaces throughout the underhood area. These tufts were attached perpendicular to the respective surfaces, thereby avoiding a bias in the tuft direction. In addition, controlled amounts of red dye were injected at preselected locations distributed within the engine bay. The dye was made from organic food coloring and alcohol with the ratio chosen such that the mixture was neutrally buoyant in water.

Thermal Buoyancy

One potential problem in water testing of the underhood environment is the resulting isothermal flow and the implicit neglect of buoyancy effects. As described by Maxwell and Sullivan [7], convective heat transfer may be categorized according to the nature of the flow. If the flow is driven by buoyancy forces, the heat transfer is referred to as free or natural convection. The buoyancy forces are a consequence of density variations induced by temperature gradients. The magnitude of the buoyancy force relative to the viscous force acting in the fluid is characterized by the Grashof number

$$G_r = \frac{g\rho^2\beta(T_s - T_\infty)}{\mu^2}$$

where the symbols are defined in the Nomenclature at the front of the report. If, instead, the flow is driven by some external source such as a fan or the movement of an automobile, inertial forces are important and the heat transfer is referred to as forced convection. The magnitude of the inertial force relative to the viscous force acting in the fluid is characterized by the Reynolds number

$$R_e = \frac{\rho U L_c}{\mu}$$

In a driven, nonisothermal flowfield such as automotive underhood flow there is a possibility that both the inertial and buoyancy forces are significant. It can be shown [8] that the relative magnitude of the buoyancy and inertial forces can be characterized by the combination of Grashof and Reynolds numbers

$$\frac{G_r}{R_e^2} = \frac{g(T_s - T_\infty)L_c}{U^2 T}$$

If $G_r / R_e^2 << 1$, then the buoyancy forces may be neglected. If $G_r / R_e^2 >> 1$, then the inertial forces may be neglected. If $G_r / R_e^2 \cong 1$, then both the buoyancy and inertial forces must be considered.

It may be noted that the buoyancy forces will be large relative to the inertial forces when the temperature differences within the flow are large and the fluid velocities are small. A global estimate of G_r / R_e^2 for an automotive underhood flow may be made by setting the fluid

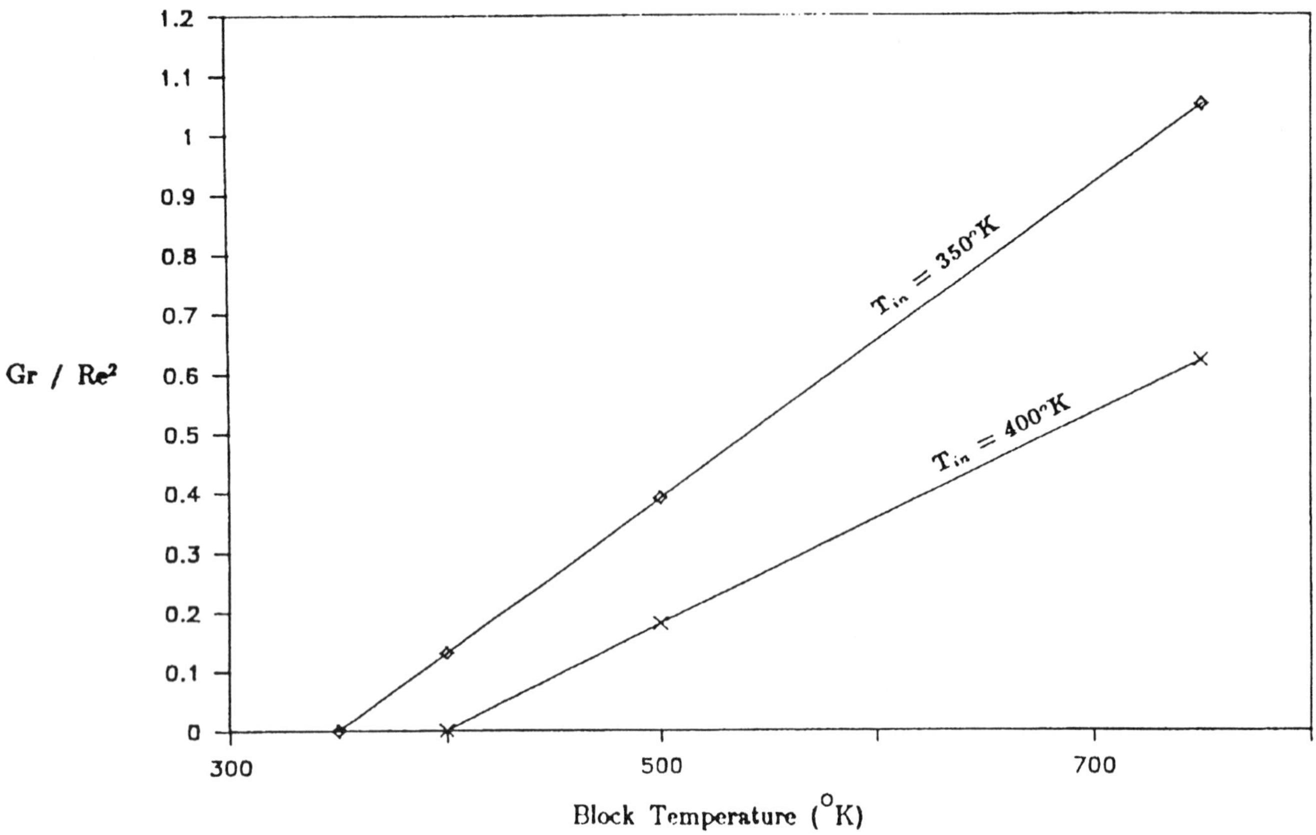

Figure 9 Global estimate of the significance of buoayancy in an underhood flow (Maxwell and Sullivan, [7])

temperature to the radiator-exit temperature T_{in}, the surface temperature to a representative engine block temperature and the fluid velocity to a nominal 3 m/s. The height of the engine block is used as the characteristic dimenstion. The results of this analysis (taken from [7]) are presented in Figure 9. From the plot, it may be noted that for an inlet temperature of 373K (100C) and block temperatures less than 423K (150C), that G_r / R_e^2 is much less than 1.0. Thermal buoyancy is even less important in regions away from the exhaust manifold and in areas of higher fluid velocity.

Maxwell and Sullivan tested this hypothesis further by comparing the results of two dimensional CFD analyses of a simplified underhood geometry with isothermal and nonisothermal flows. The nonisothermal flow calculation utilized an inlet temperature and velocity of 400 °K and 1 m/s. The temperature of the external surfaces was set to 350 °K and the engine block temperature was set to 750 °K. Velocity vector plots of the resulting flow fields are presented in Figures 10 and 11. Since there are no significant buoyancy effects evident it appears reasonable to neglect the buoyancy forces and to perform isothermal tests such as the current water flow experiments.

Reynolds Number Similarity

The principal fluid mechanical similarity parameters of interest in this investigation were the Reynolds numbers for the car motion and fan speed. If the Reynolds numbers for the water tests match those in air, then all other kinematic and dynamic properties of the flow can be expected to match as well. The primary advantage of testing in water is that velocities in water for equivalent Reynolds numbers are lower making flow visualization easier.

For the Reynolds number to be the same in air and water, the following velocity relation must be satisfied:

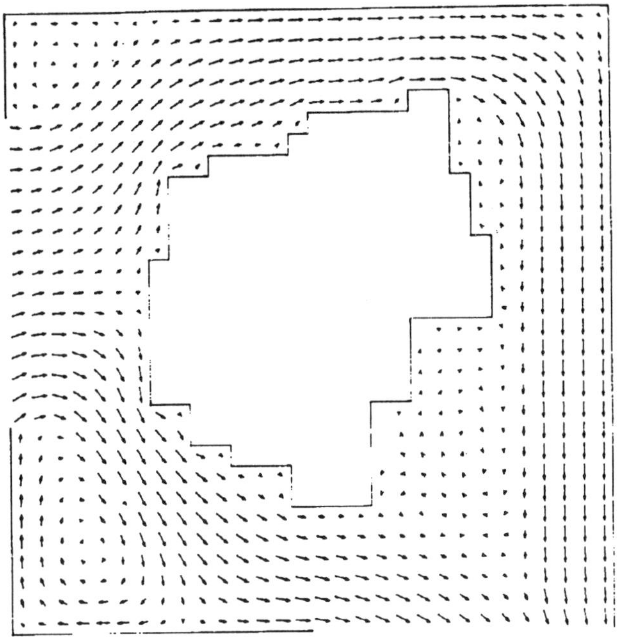

Figure 10 Isothermal velocity field for simplified engine compartment (Maxwell and Sullivan, [7])

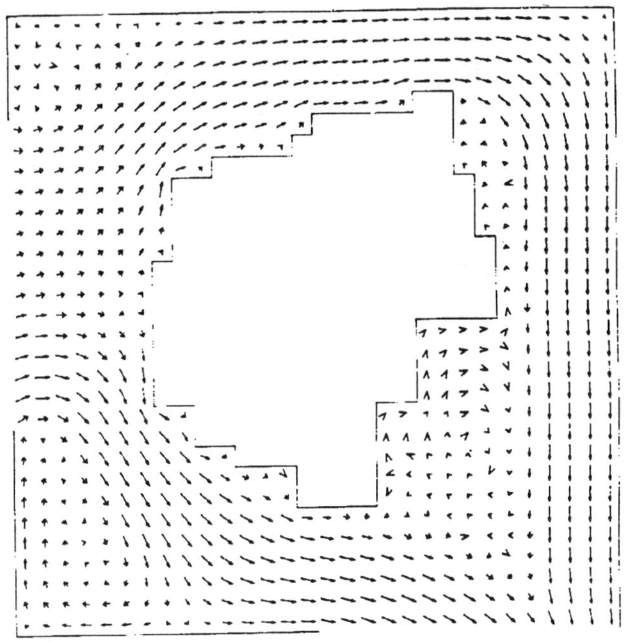

Figure 11 Nonisothermal velocity field for simplified engine compartment (Maxwell and Sullivan, [7])

$$\frac{\rho_{water} V_{water} L_{water}}{\mu_{water}} = \frac{\rho_{air} V_{air} L_{air}}{\mu_{air}}$$

or

$$\frac{V_{water}}{V_{air}} = \frac{\nu_{water}}{\nu_{air}} \frac{L_{air}}{L_{water}}$$

For ambient air during a vehicle cooling test at 45°C, $\nu = 17.3 \times 10^{-6}$ m²/s, while for water at 18°C, $\nu = 1.05 \times 10^{-6}$ m²/s. Therefore, the towing speed in water for the production vehicle is 6.1%.of the equivalent speed in air.

The fan speed should also be adjusted in accord with Reynolds number considerations, i.e.,

$$\frac{\rho_{water} 2\pi N_{water} R^2_{water}}{\mu_{water}} = \frac{\rho_{air} 2\pi N_{air} R^2_{air}}{\mu_{air}}$$

or

$$\frac{N_{water}}{N_{air}} = \frac{\nu_{water}}{\nu_{air}} \left(\frac{R_{air}}{R_{water}} \right)^2$$

Assuming no heat addition from the radiator, the fan speed in water is 6.1% of the equivalent fan speed in air. This would be 4.6% if the hotter, radiator-exit air temperature was used to establish the similarity relationship. In order to avoid having two scaling relationships in the flow field, 6.1% was used for both the fan and the towing speeds.

The principle simulated test conditions for the flow visualization experiments were idle and 47.8 km/h (30 mph) with the fan turning at 2700 rpm. This translates to a fan speed of 164 rpm and a towing speed of 0 and 0.81 m/sec in water.

Pressure Measurements

The ratio of corresponding pressures in air and water is equal to the ratio of the dynamic pressures,

$$\frac{P_{water}}{P_{air}} = \frac{\frac{1}{2} \rho_{water} V^2_{water}}{\frac{1}{2} \rho_{air} V^2_{air}}$$

and at equivalent Reynolds numbers,

$$\frac{P_{water}}{P_{air}} = \frac{\rho_{water}}{\rho_{air}} \left(\frac{\nu_{water}}{\nu_{air}} \frac{L_{air}}{L_{water}} \right)^2.$$

Consequently, for full scale tests, it can be expected that the measured pressures in water are higher than those in air by a factor of 3.32.

The vehicle tested at SSPA was fitted with surface mounted

TABLE 1 PRESSURE TAP LOCATIONS	
Engine Compartment Pressure Taps	
Tap	**Position**
1.	Transmission, underside
2.	Oil pan, underside
3.	Coolant Reservoir, front face
4.	Battery, front face
5.	Shock tower, left
6.	Shock tower, right
7.	Brake booster, lower right
8.	Dash panel, lower left
9.	Dash panel, upper center
Hood Pressure Taps	
Tap	**Position**
10.	Front right, driver's side
11.	Front middle, right position
12.	Front middle, left position
13.	Front left, passenger's side
14.	Right side, middle position
15.	Left side, middle position
16.	Rear right, driver's side
17.	Rear middle, right position
18.	Rear center
19.	Rear middle, left position
20.	Rear left, passengers side

static pressure taps at nine locations in the underhood area (listed in Table 1) and 11 positions on the inside of a clear plastic hood (illustrated in Figure 12) (Hammar, [6]). Four low range pressure transducers, sampled at a rate of 25 kHz, were utilized for the pressure measurements. A switching panel with valves was utilized to connect the 20 pressure taps in five groups to the four transducers. All pressure tubing, manifolding, and transducers were water-filled with the air removal accomplished by means of a pump connected to the highest points in the tubing systems.

The static pressure from a pitot-static probe in front of the car was used as a reference pressure. It can be assumed that the influence of any long wavelength surface waves on this reference pressure source is equivalent to that on the pressures to be measured in the car. In this way, the effects of the waves on the differential pressure measurements were eliminated.

Two corrections to the pressure measurements were made during the data reduction. The first had to do with the disturbance at the pitot-static probe due to the presence of the car. To obtain an undisturbed reference pressure, the static pressure at the probe was corrected for this effect.

The total pressure sensed by the pitot-static probe is the same asthat which would result without the car's disturbance by Bernoulli's equation, i.e.,

$$P_{s_p} + q_p = P_{s_\infty} + q_\infty$$

where q_p is measured directly and q_∞ is calculated from the car velocity. The deviation of the static pressure sensed at the pitot-static probe due to the car's disturbance can be found from

$$DP_s = P_{s_p} - P_{s_\infty} = q_\infty - q_p$$

This correction was applied in the reduction of the static pressure measurements.

A second correction was applied to compensate for residual currents in the basin which caused the relative velocity between the fluid and car to be different from the car velocity. If it is assumed that the static pressure measurements on the car vary in proportion to the dynamic pressure sensed at the pitot-static probe, then a corrected pressure measurement may be found from

$$P_{corrected} = P_{measured} \frac{q_\infty}{q_p}$$

This correction results in a measured value which is "normalized" to the nominal speed of the car.

The measuring chain including transducer, amplifier and A/D converter allowed an accuracy of ±5 Pa; however, the repeatability can be below this value, as was found from the measurements. The repeatability was as high as ±2 Pa with the car stationary. With the car rolling, the

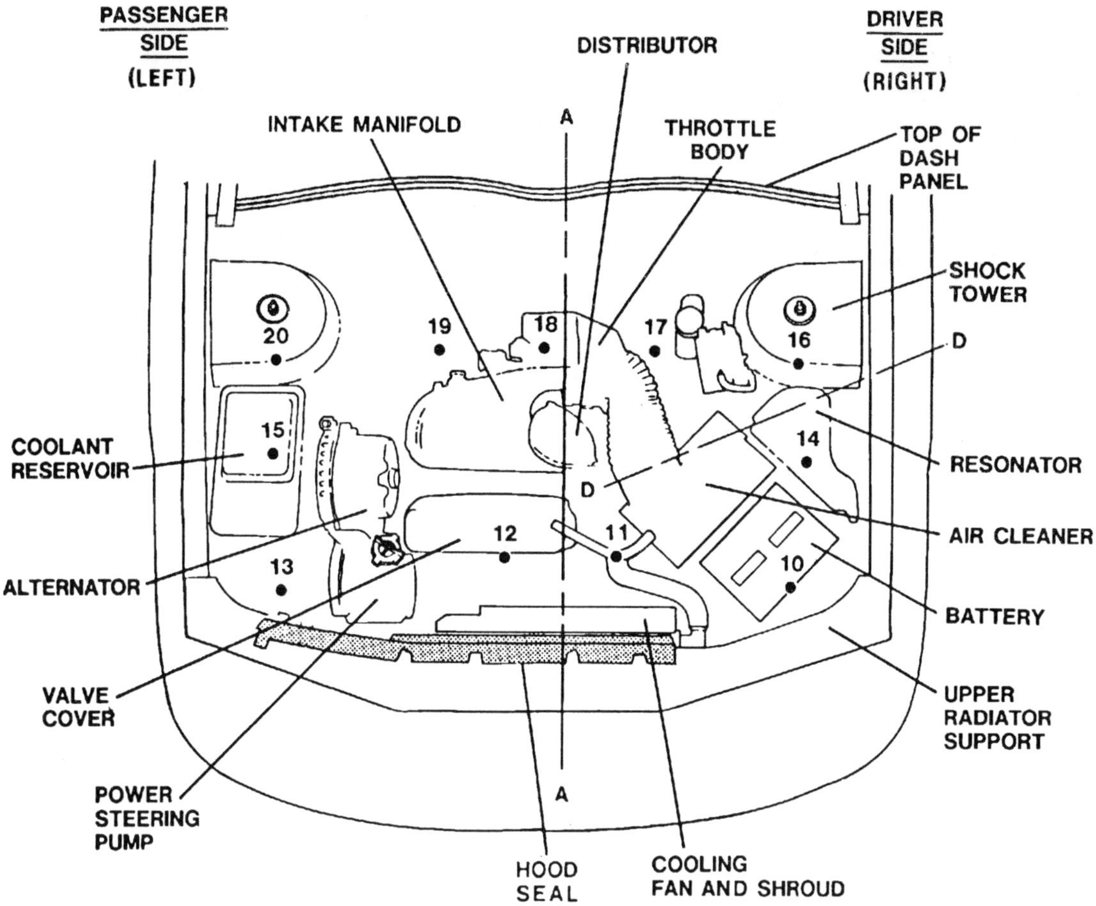

Figure 12 Underhood components and pressure tap locations

repeatability was within ±4 Pa for 0.81 m/s and ±5 Pa for 1.22 m/s.

There are two dynamic pressures that may be defined for normalizing the pressure measurements into coefficient form: the freestream and shroud flow dynamic pressures. The most obvious of these is the freestream dynamic pressure which can be expected to scale the underbody pressure and the ram-induced pressure rise at the front of the vehicle. When the vehicle velocity is low, the freestream dynamic pressure is negligible and the fan induced flow through the shroud provides a more appropriate dynamic pressure reference. By alternately scaling measured pressures by the freestream and shroud flow dynamic pressures, q_∞ and q_{shroud} , respectively, it is possible to determine which effect provides the dominant controlling influence at particular points in the underhood region under various operating conditions.

The pressure measurements were made at towing speeds of 0, 0.81, and 1.22 m/s simulating 0, 47.8 (30 mph), and 72 km/h (45 mph) in air. For presentation purposes in this report, the pressures are normalized by the shroud flow dynamic pressure, and plotted as Cp versus tap number (Figures 17 and 18). Where appropriate, the effects of normalizing by the freestream dynamic pressure will be described in the text.

Although the fluid velocities through the shroud were not measured directly, they can be estimated on the basis of conventional radiator flow measurements in air. The resulting freestream and shroud dynamic pressures are summarized in Table 2 for the car in water.

FLOW VISUALIZATION RESULTS

In this section, a relatively detailed description will be

TABLE 2 NORMALIZING DYNAMIC PRESSURES		
Towing Speed	q_{∞}	q_{shroud}
0.0 m/sec	0 Pa	152 Pa
0.81 m/sec	328 Pa	182 Pa
1.22 m/sec	743 Pa	227 Pa

given of the flowfield for the standard Taurus configuration while stationary with the fan running. The descriptions of the flowfields for other configurations and operating conditions which were tested will be given in terms of their differences with respect to this flow. The major underhood components, which will be referenced in the following discussion, are identified in Figure 12. The engine on the front wheel drive Taurus is offset to the passenger's side, while the transmission is located on the driver's side underneath the air cleaner and ducting to the intake manifold. Because of the orientation of the observers during the testing, the convention in this report will be to refer to the driver's side as the right side of the car .

Underhood Flowfield When the Vehicle Is Stationary

As might be expected, the flow is highly three dimensional, quite complicated, and multibranched. The flow coming into the engine bay from the fan shroud is split into three principle branches or paths after impacting the engine block (Hackett, [9]) and several related secondary branches. The nature of the flow split is directly controlled by the presence of the exhaust manifold, at approximately the fan midplane, and the lateral offset of the transverse mounted engine to the passenger's side of the Taurus engine compartment. The three branches are: 1) flow up and over the engine, 2) a midlevel counterclockwise circulatory flow in a tilted, horizontal plane, and 3) flow down the front of the engine block, toward the ground. The secondary paths consist of several low speed recirculating flows at the front of the engine compartment and a separation bubble over the air cleaner. The essential features of these flows were also noted in the 3/8-scale tests at Texas Tech University (Oler and Fuller [3]).

Stream 1)

The first principle branch is the upper part of the fan exit flow, which contacts the engine block above the exhaust manifold. It runs up the front of the engine block and separates to form a quasi-two dimensional jet (E in Figure 13). This flow over the exhaust manifold would probably carry a significant amount of heat with the engine running. Next, the upgoing flow hits the underside of the hood and forms a transverse attachment line. This appears as stagnation point F in the cross section of Figure 13 and an attachment line in Figure 14. The flow then splits a second time and forms forward-moving and aft-moving wall jets on the underside of the hood. Where the forward moving jet ends up is not clear; however, the aft-moving jet slows as it spreads laterally, thickens, and attaches to the intake manifold, which protrudes into it (note stagnation point G in Figure 13). This part of the spreading flow also washes around the top of the distributor (Figure 14). On the driver's side, the flow heads for the dash panel and joins with a down going stream that exits below the car.

Stream 2)

Due to the lateral asymmetry of the obstruction presented by the Taurus engine block to the fan flow, a second major flow path consists of a counterclockwise circulation around the engine, spiralling upwards as illustrated in Figure 15. This flow, which probably carries the heaviest heat load from the engine and exhaust system, is diverted to the driver's side of the engine bay, over the transmission and below the flexible duct connecting the air cleaner and intake manifold. It continues to the dash panel where it is turned laterally along the back of the engine with a general upward drift (J). There is also a significant fraction of the flow against the dash panel that is diverted downward to the ground. Next, the circulatory flow turns forward and passes between the alternator and the coolant reservoir, with some fluid moving directly beneath the alternator. Still moving upwards, the flow passes over the top of the fan efflux and starts a second circuit at a higher level in the engine compartment. This is identified as H in Figures 15, 14 & 13. The swirl in the top part of the fan efflux is consistent with the upper level flow but is not believed to be a major contributor to it. The upper part of this stream

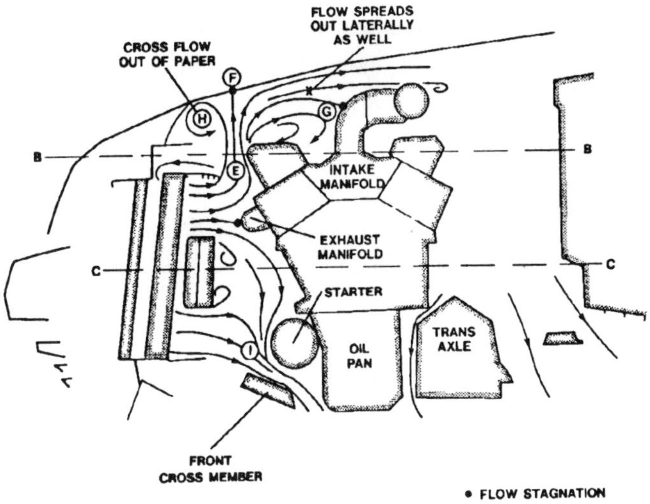

Figure 13 Vertical section at fan centerline (Section A-A from Figure 12)

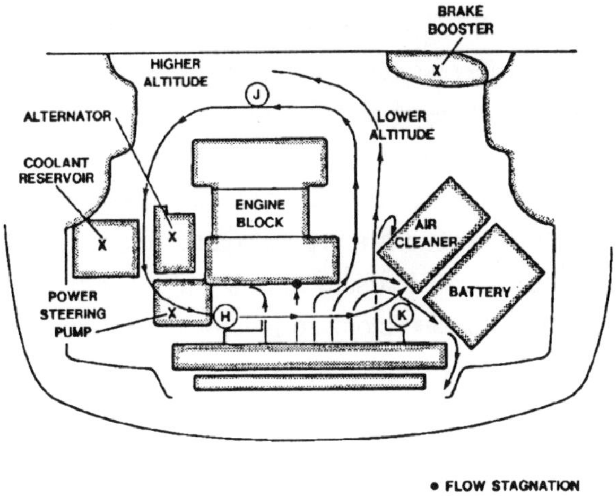

Figure 15 Midlevel circulatory flow in horizontal planes (Sections B-B and C-C from Figure 13)

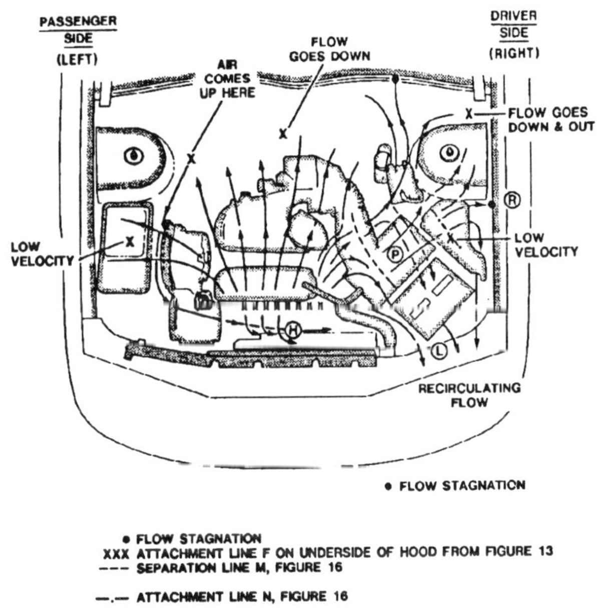

Figure 14 Flowfield in the region just below the hood surface

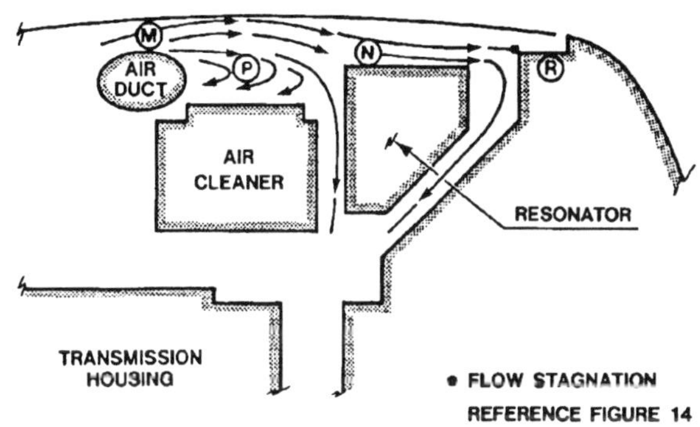

Figure 16 Detail of separation bubble over air cleaner (Section D-D from Figure 12)

does not complete the circuit but drifts upward and impacts the air cleaner (discussed below and in Figure 16). It is speculated that the lower part of this stream makes a second circuit. If this is true, there may be implications to cooling.

The circulatory flow just described develops primarily because the engine block is offset relative to the fan. More of the fan efflux passes to the driver's side of the engine and a circulatory motion is set up. The strength of the circulation is determined by the offset and also by the clearance between the block and the air cleaner. Increasing the clearance by repositioning the air cleaner or by removing it's forward corner would increase the midlevel circulation.

Stream 3)

In general, the fan efflux which impacts the engine block below the exhaust manifold on the passenger's side of the engine is diverted toward the ground to form the third major underhood flow path. Virtually all of this flow is channeled between the engine block and a chassis cross-member (I in Figure 13). On the drivers side, the contoured surface on the top of the cross-member diverts the downward flow smoothly across the bottom of the transmission housing and toward the rear of the car. At the center of the engine compartment, the downward flow is channeled through an opening bounded by the cross-member on the bottom, and the oil pan, starter motor, and transmission bell housing flange on the top. (The transmission housing is located underneath the air cleaner.) The vertical face of the oil pan directs the flow to a stagnation point on the ground. Since the flow direction is inclined slightly, the majority of the flow continues toward the rear along the ground with some lateral dispersion. Due to minor modeling inconsistencies on the 3/8-scale model, the downward flow direction was more vertical and resulted in a significant upstream dispersion and recirculation of the flow from the stagnation point on the ground. This is a important demonstration of the attention to detail which is necessary, not only for model building, but for car design itself.

Other Flows

In addition to the three principle flow paths, there are several secondary flows which have a significant effect on the thermal environment in the underhood area. Each of these secondary flows results in the presence of low speed air in the front corners of the engine compartment. This is significant since, due to the action of the fan, the pressure in the engine bay is higher than the region between the bumper assembly and the front face of the condenser. Consequently, low speed air in the front corners of the engine bay is drawn forward, below and through gaps in the radiator support, over the top of the radiator (Figure 13), through the portion of the radiator not covered by the fan shroud, and through the gap in the hood seal on the driver's side of the radiator support (L in Figure 14). These flows recirculate directly back through the condenser, radiator, and fan, thereby raising the average air inlet temperature to those components.

Another significant secondary flow path results from the lateral deflection of the fan flow off the engine block to the driver's side of the engine (Figure 15). This flow is split by the front corner of the air cleaner box with part going rearward over the transmission and the remainder directed forward along the front face of the air cleaner and continuing to the battery (K). Since this flow path comes directly from the radiator and across engine block and exhaust manifold, it has the effect of raising the average temperâture of the battery in addition to providing a recirculation path to the front face of the condenser.

The separation bubble over the air cleaner is illustrated by section D-D (from Figure 12) shown in Figure 16. The flow against the hood, that is dispersed on the driver's side of the underhood area, goes over the flexible duct connecting the air cleaner and intake manifold. A separation line forms over the relatively sharp corner of the air cleaner cover (M) thereby creating a low energy vortex bubble (P) over the top of the air cleaner itself. The resonator, protruding upwards, closes the downstream end of this bubble. The separated fluid from the air cleaner duct impacts the side of the resonator (N) and passes over its top. In Figure 14, the forward part then exits down the side of the nearby fender while the aft part washes over the top of the shock tower and exits downwards next to the dash panel. A stagnation point on the side of the inner fender separates these two flows (R). An important feature of the vortex bubble over the air cleaner is that flow along its axis is directed forward, toward the battery. After flowing over the battery, this fluid leaves the underhood area at the front corner through the gap in the hood seal (L in Figure 14).

Underhood Flowfield at 47.8 km/h

The primary additional influence on the underhood flowfield due to vehicle movement is the ram-induced pressure rise on the front of the vehicle. At velocities as low as 40-50 km/h, this effect is sufficient to raise the pressure at the front face of the condenser above that of the engine bay. In addition, as flow is channeled beneath the vehicle, the average fluid velocity increases and the pressure decreases. This effectively lowers the backpressure to the exit paths for the underhood airflow and increases the pressure drop from the front face of the condenser to the engine bay.

The major consequence of the change in direction of the

pressure gradient is to prevent the recirculation of heated air from the front corners of the engine bay to the front face of the condenser and through the unshrouded portion of the radiator. Instead, the dynamic pressure rise forces fresh air through these openings into the engine bay, particularly through the unshrouded area of the radiator and past the gap in the hood seal on the driver's side. Consequently, the battery and accessory package are provided with a limited cooling flow at speeds above 40-50 km/h.

At the 47.8 km/h test condition, the underhood flow exhibits the same three principle flow paths that appear at idle with the exception of the midlevel circulatory flow, which exits along the base of the dash panel. The fan exit flow, diverted below the exhaust manifold and down the front face of the engine (I in Figure 13), exits and mixes with the flow beneath the car with no tendency for forward recirculation.

The flow against the engine block and above the exhaust manifold is dispersed against the inner hood surface with a greater proportion of the flow going towards the rear of the engine compartment and less going laterally toward the coolant reservoir on the passenger's side and the battery on the driver's side; however, the accessory package remains heavily washed by the flow up and over the engine. (Reference dye pictures in Figure 18)

The counterclockwise underhood circulation induced by the lateral flow around the engine and over the transmission is similar to the idle flow pattern with the exception that the flow behind the engine has a much stronger downward velocity component and exits beneath the car more readily. There is no tendency for the flow behind the engine to recirculate forward below the accessory package.

The most significant secondary flow which persists from the stationary to the 47.8 km/h test conditions is the lateral flow off of the engine block, which is diverted forward along the front faces of the air cleaner and battery (K in Figure 15). Consequently, it can be expected that the battery may still exhibit elevated temperatures at the 47.8 km/h test condition.

Overhead Flow Visualization Pictures

The still frame and video records of the dye injection tests captured the dynamic behavior and turbulent nature of the unsteady flow and were very instrumental in formulating an understanding of the flow field. The photographs in Figures 17 - 19 illustrate some of the visualization results described above.

The behavior of dye released from a location at the front exhaust manifold illustrates the convection effects of fluid near the front of the engine. At idle, this heat is quickly transported around the whole compartment (Figure 17). The first picture shows the initial release of dye into the water and the second picture shows the pattern after about 10 seconds (0.6 seconds in air) The dye has been convected by the lateral flow over to the battery and air cleaner (reference K in Figure 15), and has started to recirculate forward over the top of the radiator support (L in Figure 14). The counter clockwise circulation over the transmission has carried the heat under the air cleaner duct to the rear of the engine. The unsteady nature of the flow has also allowed some of the dye to be entrained by the flow going up and over the top of the engine (E in Figure 13). Although it can't be seen in these pictures, the influence of the fan causes some of the dye to recirculate forward through the unshrouded portion of the radiator.

The ram-induced pressure rise at the front of the vehicle stops the recirculating flows at the front of the engine compartment and alters the mid-level circulatory flow. At the 47.8 km/h test condition, shown in Figure 18, nearly all of the dye from the same exhaust manifold location is convected up and over the engine and accessories, to the rear of the compartment on the passenger side. The time scale here is about 5 seconds (0.3 seconds in air) suggesting that the fliud velocities are higher than at idle. It exits behind the engine and shows no tendency to circulate forward. The ram-induced flow through the unshrouded part of the radiator has also changed the lateral velocities so that this heat is not transported to the battery, but rather to the alternator on the other side of the compartment. The tufts on the coolant reservoir, radiator support, battery and resonator have all changed direction from the idle condition, indicating that the forward

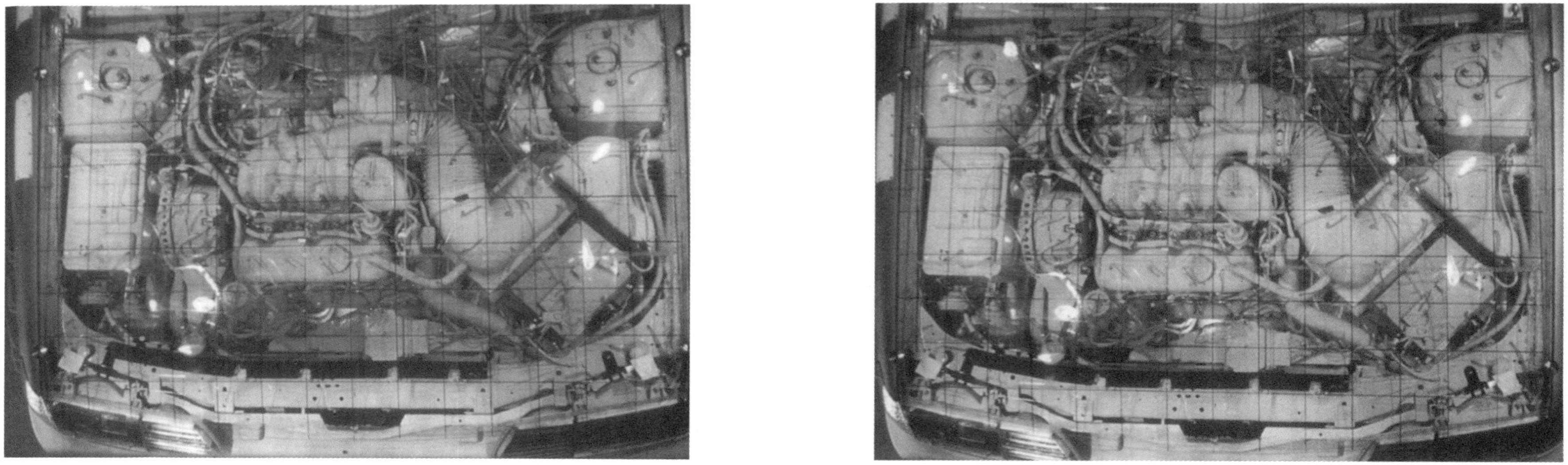

Fig. 17 — Dye Flow Pattern from the Exhaust Manifold at Idle

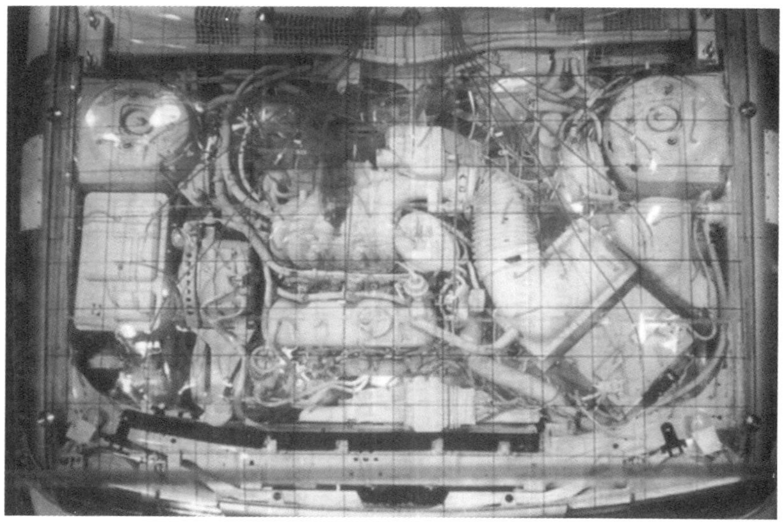

Fig. 19 — Dye Flow Pattern over the Intake Manifold at 47.8 km/h

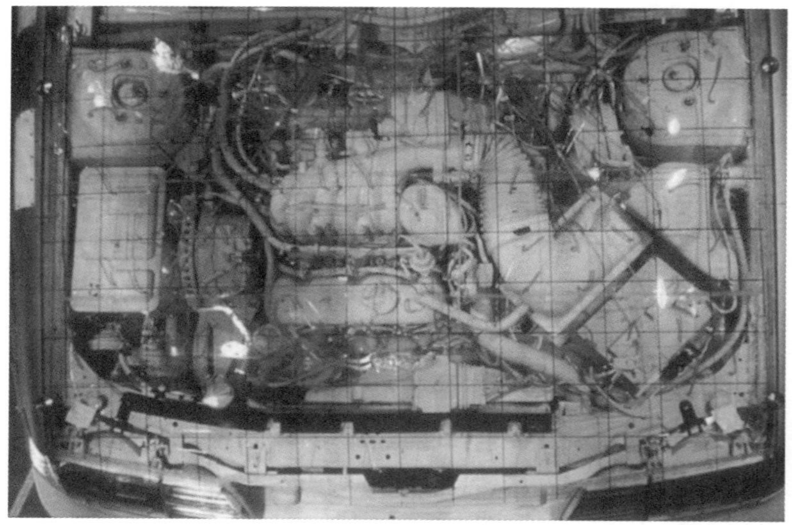

Fig. 18 — Dye Flow Pattern from the Exhaust Manifold at 47.8 km/h

recirculating flows have been reversed.

The fluid velocities over the intake manifold at the 47.8 km/h test condition are higher yet and lateral jet spreading is minimal (Figure 19). The reduced hood clearance at this location increases the local velocities and accelerates the fluid directly to the rear of the compartment. The time scale here is about 3 seconds (0.2 seconds in air).

Effect of Fender and Wheelhouse Vents at Idle Conditions

The fender and wheelhouse vents provide alternate escape paths for the underhood flow through dash panel openings in the top left and right corners. The dash panel openings are partially shielded by the shock tower housings and are not directly in line with any of the three major underhood flow paths. Consequently, for these alternate paths to be effective at idle conditions, the fan must provide a pressure differential between the underhood region and the ambient exterior to drive the fluid to the dash panel openings. In the case of the Taurus, the standard underbody exit paths have the effect of pressure equalization between the engine bay and outside fluid. Consequently, although a small amount of dye does leave the fender vents, there are no significant detectable differences in the flowfield at idle, with and without the dash panel-to-fender openings.

Effect of Fender and Wheelhouse Vents at 47.8 km/h

When the vehicle is in motion, the exterior pressure distribution is altered, with generally positive increments on the front and decrements over the sides, underbody, and rear of the vehicle with respect to the ambient pressure. Therefore, the required pressure differential for flow from the underhood region to the fender and wheelhouse vents can be generated by the exterior flow. For the Taurus at 47.8 km/h, it was found that the lower two fender vents behind the wheelwell were most effective in this regard (vents #3 and 4 in Figure 7).

The presence of the dash panel openings at 47.8 km/h has a negligible effect on the flow down the front of the engine and around the right side of the engine over the transmission. The flow following these paths exits beneath the vehicle as in the standard configuration. While interesting from a fluid mechanical point of view, the alterations to the front end structure, which would be required for the fender vents, may make them impractical with regard to crash safety, corrosion protection, and manufacturing costs.

Mechanisms for Manipulating Underhood Flows

Some general comments may be made concerning how to manipulate underhood flows effectively (Hackett, [9]). Stagnation points and attachment lines divide the flow into connected regions that contain fluid which either recirculates or exits. Minor geometric alterations that change stagnation point locations could redistribute the balances among these regions significantly. In other words, fluid amplifier effects might be found if changes are made with care. One example is to change the direction of a free flow slightly before it attaches to a surface. This changes the flow split upon attachment. For example, if the direction of the upward flow (E) on the valve cover was changed slightly as it separates, the flow split under the hood could be changed significantly. Similarly, a slight change in the flow direction upon impingement with the ground is apparently the distinguishing feature between a recirculating ground flow as observed in the Texas Tech experiments and the relatively benign rearward flow along the ground noted in the SSPA study. Deliberately separating a flow, using an angled plate for example, might also be effective. For maximum effect, the key is to make the changes near sensitive locations, such as stagnation points, attachment lines, separation lines and corners, rather than attempt to manipulate the major flows directly.

PRESSURE MEASUREMENT RESULTS

The pressures in the engine compartment are generated either by the fan (stationary car), by the freestream dynamic pressure (ram-induced effect of car moving at high speed), or by both the fan and the freestream dynamic pressure (car moving at moderate speed). To determine which influence dominates at a particular point in the underhood region it is instructive to plot the variation of the pressure coefficient based on either the freestream or shroud flow dynamic pressure, q_∞ and q_{shroud},

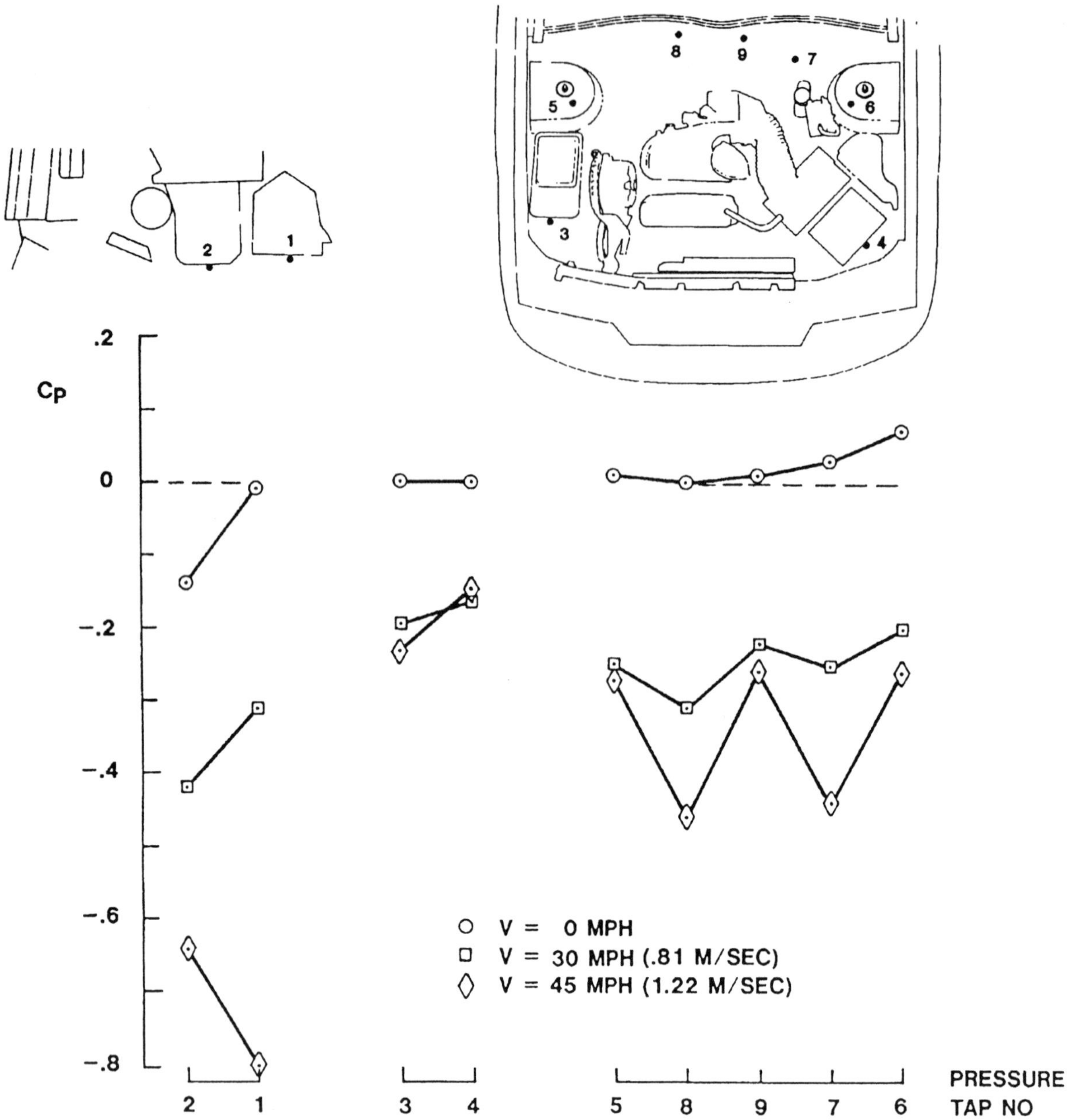

Figure 20 Pressures within the engine bay

respectively, as the vehicle speed increases. If the freestream dynamic pressure provides the dominant influence, the corresponding pressure coefficient would be expected to be approximately constant with respect to vehicle speed. Conversely, the pressure coefficient based on the shroud flow dynamic pressure would remain constant if the fan provides the dominant effect.

For the plots, pressure taps located in the same general area of the engine compartment are assembled into labeled groups and the measurements within each group are connected with lines for clearness. On the horizontal axis, the pressure taps are shown with their respective position numbers. Separate plots showing the pressures in the engine compartment and on the underside of the hood are presented for the various operating conditions.

The engine compartment and underhood pressure coefficients, based on the values of q_{shroud} given in Table 2 for the baseline car at 0, 0.81 and 1.22 m/s (0, 47.8 and

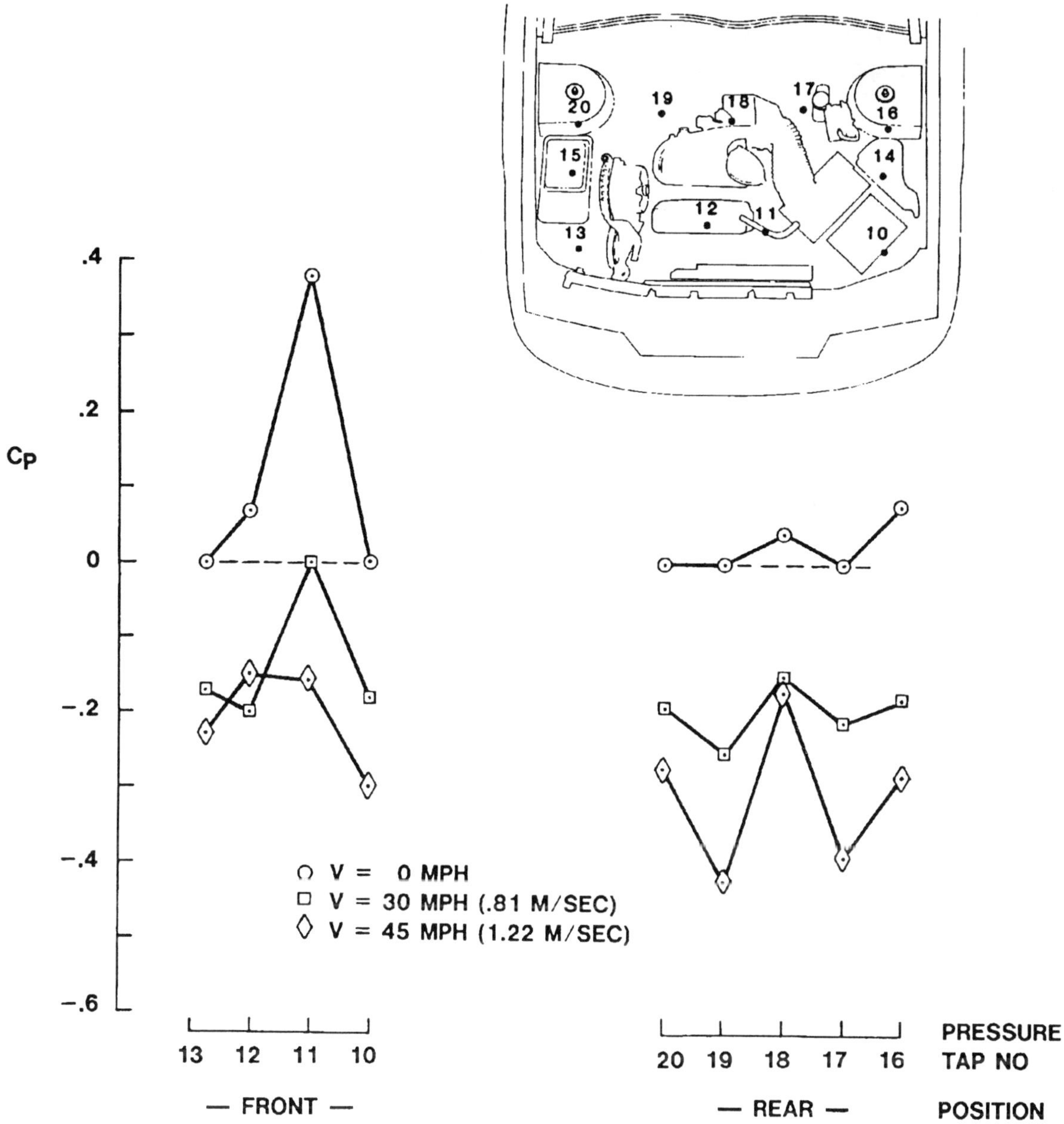

Figure 21 Pressures on the inner hood surface

72 km/h), are presented in Figure 20. As expected, the pressures in the engine compartment generally decrease with increasing speed due to the reduced pressure associated with the accelerated flow beneath the car.

Upon examining pressure tap Nos 3 and 4 located in the front corners of the engine bay, it is apparent that q_{shroud} is appropriate for scaling the measured pressures. This is probably a consequence of the deflection of the shroud flow off of the front face of the engine directly into these regions. This interpretation is consistent with the flow visualization results, K in Figure 15. Similar observations may be made concerning the pressures at tap Nos 5, 9, and 6. These pressure taps are located high in the engine bay at positions where they experience dynamic pressure effects from the portion of the shroud flow which is deflected up the front face of the engine and along the underside of the hood, E in Figure 13.

In contrast, the pressures measured beneath the car at

tap Nos 1 and 2 on the transaxle and engine oil pans do not scale with q_{shroud}. If q_∞ is used instead, the pressure coefficients for the taps are approximately -0.20 for both the 0.81 and 1.22 m/s test conditions. Similarly, the pressure coefficients at tap Nos 7 and 8 located low on the dash panel are relatively constant at -0.13 when formed from q_∞.

The measured pressures on the inside surface of the hood normalized by q_{shroud} are presented in Figure 21. Pressure tap Nos 12 and 13 located at the front of the hood apparently scale satisfactorily with the shroud flow. The same would probably be true for tap Nos 10 and 11 if the hood seal at the top of the radiator support were extended to the driver's side fender. Instead, the ram-induced flow through the gap (which was observed in the flow visualization experiments) dominates the pressures. The q_∞ based coefficients for tap 10 are -0.1 and -0.09 at 0.81 and 1.22 m/s, respectively. The pressure coefficients at tap 11 are less consistent. It is speculated that the flow through the gap in the hood seal and other leakage paths cause a rearward movement of the transverse reattachment line on the undersurface of the hood away from tap 11 (attachment line F in Figure 14 & 13). Consequently, pressure tap 11 experiences a diminished shroud dynamic pressure effect.

Pressure tap Nos 16, 17, 19, and 20 along the back of the hood are all influenced by the the freesteam dynamic pressure. Pressure tap 18 is an exception in that the pressure variation scales with q_{shroud}. This is because it is located at a point over the throttle body where the hood clearance is almost zero and consequently presents an obstruction to the shroud flow over the engine.

In summarizing the results of the pressure measurements, it can be concluded that when the car is moving at speeds as low as 47.8 km/h (30 mph), the underhood pressures are controlled by the underbody pressure and consequently, scale with the freestream dynamic pressure. The exception to this general statement occurs at points near the front of the engine compartment which are exposed to the direct or deflected shroud flow. It may be noted that these observations are consistent with a previous windtunnel investigation by Schenkel [10] of the underbody and underhood pressures in a simplified 3/8-scale sedan. That model included a 50% porous screen to represent the radiator and condenser flow resistance but did not have a fan.

CONCLUSIONS

Using water for the underhood flow visualization gave excellent viewing with very low fluid velocities. With the set-up in the laboratories, very good observation conditions were obtained. These provisions enabled most of the very complicated flow patterns, to be correctly simulated, observed in detail, recorded and mapped.

The underhood flowfield at idle was noted to have three primary flow branches resulting from the impingement of the shroud flow on the front face of the engine: flow downwards toward the ground and turning toward the rear of the vehicle, flow upwards to an attachment line on the inner hood surface with subsequent rearward and lateral dispersion, and a counterclockwise circulatory flow resulting from the asymmetry of the engine block obstruction to the shroud flow. An erroneous indication of an underbody recirculatory flow was found in the 3/8-scale tests and attributed to an incorrect positioning of the front cross-member.

The results from the pressure measurements have indicated that the freestream dynamic pressure provides the dominant influence for most of the underhood region with the exception of the front where the direct fan influence is experienced.

The production tests have provided a validation of the 3/8-scale tests thereby indicating that precise Reynolds number matching is not required for the bluff body dominated underhood flowfield, and that moderate levels of modeling detail are sufficient for representing the primary underhood flow obstructions. The scale model tests offer the advantage of relatively low test facility costs and operational simplicity and convenience. The low testing costs are somewhat offset by the high initial model fabrication costs. The use of production vehicles for full scale testing guarantees geometric accuracy. Although the vehicles are relatively inexpensive and easy to prepare for water testing, the actual facility cost for the tests is higher.

REFERENCES

1. Clemens, W.J., (1974). **Improved Underhood Airflow - Baseline Studies**, Ford Motor Co. Car Research Technical Memorandum, CRM 74-1.
2. Clemens, W.J., (1974). **Improved Underhood Airflow - Airflow Development Studies**, Ford Motor Co. Car Research Technical Memorandum, CRM 74-1.
3. Oler, J.W., and Fuller, D.R., (1988). **An Experimental Investigation of Automotive Underhood Airflow**, Texas Tech University, Department of Mechanical Engineering Technical Report TR-FMC-88-3.
4. Williams, J.E., (1988). **Aerodynamics of Automotive Cooling Flow**, oral presentation, Aerotech '88 - SAE Aerospace Technology Conference & Exposition, Anaheim, CA.
5. Larsson, L., Hammar, L., Nilsson, L.U., Berndtsson, A., Knutson, K., and Danielson, H., (1989). **A Study of Ground Simulation - Correlation Between Wind-Tunnel Water-Basin Tests of a Full-Scale Car**, SAE Technical Paper 890368, presented at the 1989 SAE International Congress and Exposition, Detroit, MI.
6. Hammar, L., (1990). **Investigation of the Underhood Cooling Flow of a Production Passenger Car**, SSPA Maritime Consulting Technical Report 5567-1.
7. Maxwell, T.T., and Sullivan, P.T., (1988). **Buoyancy Effects on Engine Compartment Airflow**, Texas Tech University, Department of Mechanical Engineering Technical Report TR-FMC-88-1.
8. Incropera, F.P., and DeWitt, D.P., (1985). **Fundamentals of Heat and Mass Transfer**, Second edition, John Wiley and Sons, Inc.
9. Hackett, J.E., (1989). **Underhood Flow for a 3.0L 1989 Taurus Under Idle Conditions**, Lockheed Contract Report.
10. Schenkel, F.K., (1977). **The Origins of Drag and Lift Reductions on Automobiles with Front and Rear Spoilers**, SAE Technical Paper 770389, presented at the 1977 SAE International Congress and

9

A Numerical Evaluation of the Thermal Effects of the New V6 Engine on the Underhood Environment of the 1993 Opel Vectra

Mohamed Ashmawey
Electronic Data Systems Corp.

Helmut Berneburg, Winfried Hartung, and Frank Werner
Adam Opel AG

ABSTRACT

High performance engines required in contemporary vehicles are causing underhood components to operate under hostile temperature environments. Aerodynamic styling and the addition of new components to the engine compartment further add to the problem by decreasing the volume of underhood cooling air flow. The addition of engine compartment coverings required to meet environmental noise reduction standards further restrict and debilitate air flow cooling.

The above conditions demand that the analysis of air flow patterns and heat transfer phenomena under the hood be an essential part of early systems design of new engines, engine compartment components, and underhood component packaging. A numerical approach to calculate cooling air flow velocity and temperature distribution of the air and engine compartment components is utilized.

Air flow is calculated using a finite volume Computational Fluid Dynamics code on a 220,000 cell representation of the flow domain. A three dimensional Navier-Stokes Equation solver with a "k-epsilon" turbulence model is used to calculate the flow field and convective heat transfer coefficients. Inlet boundary conditions for the code are provided using Laser Doppler Velocimetry measurements performed in the Pininfarina wind tunnel. Results of the code are mapped to a super block lumped mass representation of the underhood environment. The three modes of heat transfer (convection, conduction, and radiation) are modeled using a lumped parameter resistor-capacitor solver.

The application of the above method to the 1993 Opel Vectra equipped with a newly-developed V6 engine is presented. Areas of restricted air flow and high underhood component temperature were identified. Trade-off studies of several insulation and/or shielding strategies were performed to create a cooler environment for the operation of thermally sensitive components under severe operating conditions. The tool can be used to optimize operating parameters of new engines and packaging strategies under the hood to reduce potential thermally induced problems.

1. INTRODUCTION :

High underhood operating temperatures may be blamed for more and more failures of thermally sensitive engine compartment components. More heat sources are being added in the engine compartment due to high performance expectations from contemporary engines, fuel economy expectations, and new emission standards. Environmental noise reduction standards are forcing motor vehicle manufacturers to seal the engine compartment, thus decreasing the opportunity to cool over-heated components.

General Motors Corporation (GM) developed a systems-oriented methodology to simulate the thermal behavior of engine compartment components. Two main objectives were established for this simulation; initially to identify regions of relatively high and low temperatures within the engine compartment and thereby aid in trade-off packaging strategies of thermally vulnerable components, and secondly to examine the impact of system modifications on lowering local underhood temperatures.

To assess the utility of this modeling technique for studying the thermal environment in a vehicle engine compartment, this enhanced "Underhood Thermal Modeling" (UTM) method was applied to several GM vehicle platforms already in production.

This paper discusses the application of the UTM methodology to the 1993 Opel Vectra equipped with a newly-developed V6 engine. In this case, most of the modeling activities took place before the vehicle was in production. Thus, measurement data required to validate model predictions was not available. However, this project represents the first application of the method in which the results obtained may be used to help in the packaging and shielding strategies early in the design phase of the engine compartment.

A brief description of the UTM method is given in the next section. Temperature predictions at 60 km/h, road load driving condition will be presented. The impact of the addition of the engine underbody plastic cover designed to reduce engine noise will be discussed. Trade-off studies involving the effect of adding an alternator duct specially designed to deliver sufficient ambient air to cool the alternator will be presented. Finally, trade-off studies of different shielding and/or insulation strategies applied to the exhaust system will be discussed.

2. OVERVIEW OF THE UNDERHOOD THERMAL MODELING PROCESS:

The Underhood Thermal Modeling process uses a lumped-parameter approach to represent the thermal system of interest. In this approach, the thermal system is modelled using an equivalent resistor-capacitor network. Each component in the engine compartment is represented by one or more lumps of finite size, called super elements. The air surrounding the components is also divided into super elements.

The thermal mass of each super element is then concentrated into a node located at its centroid, with air nodes having zero thermal mass. The temperature at a given node is assumed to be uniform over the entire volume of the represented super element. Thus, for example, the alternator can be represented by one or two nodes, the battery may be represented by six nodes, and so on. Thermal interactions between super elements take place along one-dimensional resistors which connect their respective nodes.

To calculate the temperature map of the engine compartment, the UTM process may be divided into three major steps. First, the geometry of the model is developed, usually by extracting geometric information from a crash worthiness model of the vehicle. The engine compartment is then divided into super elements. Thermal properties, like emissivity and conductivity, are then associated with each super element. Second, the thermal network which represents the model is created. Several programs are required in this step to calculate the thermal interactions between the nodes due to the three modes of heat transfer, conduction, convection, and radiation. Steady state temperature of each node in the network is then calculated by satisfying the energy equation at that node.

A simplified flow chart for the modeling process is presented in figure (1). Details of the three major steps of the process are discussed in the following sections.

2.1. Geometry Development:

In the present study, the basic engine compartment geometry was built by simplifying a detailed crashworthiness model provided by the Vehicle Computations Department at Adam Opel AG, figure (2). Detailed front and rear views of the engine compartment, after simplification, are presented in figures (3) and (4) respectively. In these figures, elements which represent the hood, the front surface and the air are removed for clarity. A three-dimensional cartesian coordinate system is mapped on the model so that the x-direction is along the length of the car, i.e. in the direction of air flow from front to back. The y-direction is vertical, from ground surface to the hood, and the z-direction is across the car from passenger side to driver side.

In the spirit of lumped-parameter methodology, the engine compartment was divided into 12 layers of lumped masses (or super elements) in the x-direction, 10 layers in the y-direction, and 17 layers in the z-direction. In figures (3) and (4) model super elements are shrunk for better representation of component details.

Beside the new V6 engine in the Vectra, two new components are also worth mentioning. First, a plastic underbody cover is used to meet environmental noise reduction requirements. The cover completely seals the bottom of the engine compartment, except for front and rear openings. A special computational layer of super elements was included in the model to accommodate the cover. With this layer, the model consists of 2040 hexagonal super elements.

The second new component introduced in this model year is the alternator cool-air duct which is added to ensure that sufficient fresh air is available to aid in cooling the alternator internal electronic components. Figure (4) shows the alternator cool-air duct routing from a rear view.

2.2. Thermal Network Creation:

In this step, three versions of the engine compartment super element model developed in the above section are created. The first version contains all the super elements representing the solid surfaces and the air surrounding them. This version is used to calculate the air flow connectors and convection connectors between the air and solid surfaces. The second and third versions contain solid surfaces only. These two versions will be used to calculate the conduction and radiation connectors. Connectors, or resistors, which represent all three modes of heat transfer between each node and the remainder of the thermal system are then assembled in the thermal network.

2.2.1. Flow Model:

A computational fluid dynamics code, VINE3D-H[1], developed by GM Research Laboratories, is used to predict the air flow field under the hood. VINE3D-H solves the Navier Stokes and energy equations using a finite volume technique applied on a three-dimensional structured mesh. Turbulence is modeled using the high Reynolds number k-ϵ model.

Standard wall functions are used to model different variables near rigid walls. VINE3D-H provides computed velocity field at each air node, and convective heat transfer coefficient data between each solid surface and neighboring air nodes. Details of VINE3D-H can be found in [1].

Super elements representing the air are divided into smaller finite volume elements. Each super element is divided into approximately 220 elements. This allows VINE3D-H to capture and model the small scale structures which occur in such complex turbulent flow fields with reasonable accuracy. Figure (5) shows a vertical cross-section of the engine compartment . In figure (5,a), super elements of the air and solid components are presented. Figure (5,b) shows the computational fluid dynamics mesh of the same cross-section.

Using special programs, the air flow rate between air super elements is calculated by integrating the flow velocity field over the interface between the elements, and is then represented in the model as a one-way resistor. Convective heat transfer resistors between each air super element and touching solid surfaces are also calculated and added to the thermal network as two-way resistors.

Computational fluid dynamics calculations were performed on a 190,000 finite volume mesh, for the 60 km/h model without the underbody cover, and 220,000 finite volume mesh, for all the other cases. Flow field calculations required three CPU hours on a Cray Y-MP super computer. Radiator inlet boundary conditions for the flow calculation were measured by Opel engineers in Pininfarina, Italy [2] using Laser Doppler Anemometry techniques.

Computational fluid dynamics calculations using VINE3D-H were performed for the following operating conditions:

o 60 km/h, road load, without underbody cover.
o 60 km/h, road load, with underbody cover.

The cover was represented in the model by thin horizontal surface elements separating the engine compartment from the air flow. In the above two models, the inlet axial and circumferential (swirl) velocity components were obtained from "mini-air" anemometers and a Laser-Doppler Anemometry system respectively. Details of these measurements which were taken by Opel in the Pininfarina wind tunnel can be found in [2].

2.2.2. Conduction Model:

Conduction connectors between adjacent solid surfaces are calculated using a special pre-processor conduction translator program. This program's output file includes all the nodes representing the volume and surface solid elements as well as the conduction resistors.

Brackets which hold several components to the inner surface of the fenders, the dash and the engine are then represented in the conduction model. Connectors representing the thermal conductivity of these brackets are calculated manually and added to the output of the conduction translator pre-processor.

2.2.3. Radiation Model:

A Hughes Aircraft developed radiation pre-processor, CVIEW, converts the geometry file of the radiation model into a TRASYS input file. TRASYS [3], (Thermal Radiation Analyzer SYStem), is a program which computes the shape factor between each two surface elements, and then calculates the radiation heat transfer resistor between these two elements.

Due to the large number of connectors produced in this step, only the connectors which have values above a certain cut-off factor are actually used in the calculations. Engineering judgement and experience help determine the cut-off factor. In the present model 43600 radiation connectors were used for the calculations without underbody cover, and 31900 connectors for the calculations with underbody cover.

2.2.4. Subsystem Models:

As described above, conduction and radiation calculations are performed on a fairly coarse mesh. Even though fluid flow calculations are performed on a smaller sized mesh, results are averaged over the super element size before they are added to the network. Using the above method, the average temperature of each super element in the model can be calculated. But the method will not provide temperatures of internal component parts.

Thus, for example, in the case of an alternator, the above method will calculate the temperature of the whole alternator as one lump, but will not give any information about the temperature of some of its thermally vulnerable internal electronic components. However, experience tells us that, while the alternator average temperature may be within safe operating limits, the regulator, for example, might be operating at very high and unsafe conditions.

To circumvent this draw-back, models simulating the thermal behavior of selected engine compartment components have been developed. These models provide predictions of heat dissipation from these components to the rest of the thermal network. They also predict the component's thermal interactions. Thus, the use of these sub-system models reduce dependency on experimental data to measure the heat dissipation from thermally active components. These models also predict the temperature of sensitive internal parts of modelled components.

2.2.5. Assembly and Execution:

The complete network provided by the above calculations can now be assembled. Super elements are now the nodes of the network. Output from the flow, conduction, and radiation connector calculations for each driving condition under consideration are the resistors which connect appropriate nodes. Submodel internal networks (nodes and connectors) are also added to the network. The thermal network is now a representation of the thermal interactions which take place in the engine compartment for a specific driving condition.

The assembled network is solved using CINDA[3], a resistor-capacitor network analyzer developed by Chrysler Corporation and enhanced by Hughes Aircraft. CINDA calculates the temperature of each node in the network, including submodel internal nodes. CINDA also provides information about the amount of heat exchanged between any pair of nodes in the network. The following discussion shows that this information helps in the solution of over-heating problems of thermally sensitive components.

Nodal temperatures are post processed using a graphical finite element post-processor package. In this program, the temperature field is mapped on the model geometry to identify locations of high and low temperatures.

3. Discussion of Results:

The underhood thermal environment of the 1993 Opel vectra was investigated using the thermal network developed in the above sections. Two driving conditions were chosen for the investigation, 60 km/h road load with underbody cover, and 60 km/h road load without underbody cover.

Base-line calculations were performed for a vehicle with no underbody cover, a single wall exhaust system, and no cool-air duct for the alternator. Predictions obtained for this vehicle are compared to those obtained for the same vehicle with the underbody cover installed. Results obtained for both cases will be presented in the sections which follow.

3.1. Base-Line Design:

Figure (6) shows the front view of the engine compartment normalized temperature distribution for the 60 km/h road load driving condition. In this paper, temperature scales are normalized using a reference temperature. Model predictions of underhood component temperatures agree with engineering judgement and experimental data of similar underhood geometry. The passenger side of the engine compartment experiences higher temperatures, on average, than the driver side. Thus, the driver side is a better place to locate thermally sensitive underhood components.

With no shielding or insulation strategies applied to the exhaust system, parts of the secondary dash in close proximity to the rear exhaust pipe experience temperatures higher than those measured in earlier model year versions of this vehicle. Alternator submodel calculations provide temperatures for internal electronic components. Submodel temperature predictions were comparable to those seen in earlier underhood models of this vehicle under the same driving conditions.

3.2. Addition of Underbody Cover:

Figure (7) shows a front view of the engine compartment temperature map for the 60 km/h, road load driving condition after the addition of the underbody cover. On average, underhood temperatures are higher when the underbody cover is installed compared to the base-line design. In particular, the plastic cover on top of the engine block, which is used for safety and cosmetic reasons, is 100°C warmer than its temperature at base-line operating conditions.

One reason for the observed temperature rise is the increase in radiative heat transfer from the high temperature exhaust system to the rest of the engine compartment. In this case, heat which used to be dissipated from the exhaust to the ground by radiation is now reflected within the engine compartment by the underbody cover. Several design strategies will be applied to the exhaust system to in a later section to decrease the heat loss from the system to engine compartment components.

The presence of the underbody cover also prevents the cooling effect of the underbody air flow. Figure (8) shows a comparison between the temperatures of selected components with and without the underbody cover. The figure reveals that components located in the lower half of the engine compartment are more affected by the presence of the underbody cover than higher components. Underhood packaging engineers rely on the mixing which takes place between the ambient underbody air and the warmer air under the hood to cool such components. The presence of the cover prevents this mixing.

CINDA output reveals that the temperature of the air at the bottom of the engine compartment increase by about 65°C when the underbody cover is installed. Also, since air velocity is lower in this case, convective heat transfer coefficients are almost half their value compared to base-line case. The combined effect of these two factors when the underbody cover is installed causes lower component to experience larger temperature increases than higher components. For example, the air-conditioning (A/C) compressor is 66°C warmer, while the A/C accumulator, which is located near the hood is only 8°C warmer than their temperature at base-line conditions.

Figure (9) confirms the above discussions. Figure (9,a) shows the variation in the axial component of air velocity in a horizontal plane through the engine compartment for the base-line case. In figure (9,b), the variation of the axial component of air velocity at the same location in the engine compartment after the underbody cover is installed is presented. The figures show that, in the two cases examined in this paper, air velocities, and thus cooling effects by convection heat transfer, are higher in the driver side compared to those in the passenger side.

Comparison of the two figures also reveal the adverse effect of the underbody cover on the amount of radiator air flow. Wind tunnel measurements [2] show that the installation of the underbody cover reduced radiator air flow by around 6%. This is due to pressure increase downstream of the radiator when the underbody cover is present.

4. Trade-Off Studies:

To study the effect of system modifications on underhood temperatures, several trade-off studies were performed. Another objective of the trade-off studies is to identify the optimum packaging and shielding strategies for the Vectra to avoid any thermally induced failures. In this paper, the addition of the alternator cool-air duct, and different exhaust system design strategies will be discussed.

4.1. Addition of Alternator Cool-Air Duct:

Due to the location of the alternator (recall figure (4)), surrounding air temperature is around 90°C. At this temperature the cooling effect of the air drawn inside the alternator by its fan is minimal. To decrease the temperature of surrounding cooling air, a special duct was designed to allow enough ambient air to reach the alternator.

Front and rear views of the duct are shown in figures(3) and (4). As figure (10) reveals, the use of the cool-air duct decreased the operating temperature of all the internal electronic components of the alternator. For example, the temperature of the negative rectifier diode dropped 42°C compared to the base-line operating temperature.

4.2. Exhaust Design Strategies:

One way to decrease the average underhood temperature when the underbody cover is installed is to shield the entire exhaust system which represents the largest radiation heat source under the hood. Shielding the entire exhaust system, however, will retain more heat in the exhaust gas. This will elevate the exhaust gas temperature throughout the system and, in particular, at the system exit to the catalytic convertor. This, in turn, may adversely affect performance of the catalytic convertor.

Several insulation and shielding options were investigated to identifying the optimum design strategy for the exhaust system. This optimum design is expected to decrease the heat dissipated from the exhaust system to over-heated components, like the engine plastic cover, on the one hand. On the other hand, this design is expected to dissipate enough heat from the exhaust gas to the cooling air so that the gas temperature at the catalytic convertor would be within safe limits.

Since the above two expectations of the optimum exhaust system design contradict each other, the required design is a balance between the two extreme designs. Figure (11) shows a schematic representation of the exhaust system. In figure(11,a) the exhaust system is divided into several pipes and runners to satisfy the requirements of the exhaust submodel. In figure (11,b) a cross section of a typical exhaust pipe is shown. This figure shows the different exhaust system design strategies that are used in contemporary vehicles. To decrease the heat dissipated from the exhaust system to the environment, several methods are suggested by the figure. These methods include insulating the inner or outer surface of the exhaust pipes, using a double wall with an air gap, or shielding the outer surface of the exhaust pipe. Combinations of these methods can also be used in extreme situations.

To satisfy the first expectation of the optimum design, three design strategies were studied, namely a double wall with an air gap, outer insulation, and shielding of the entire system. Figures (12,13,&14) show the temperature map of the engine compartment when applying the three designs. The three methods help, to different degrees, in relieving the overheated components. However, insulation has the highest impact on decreasing the amount of heat dissipated from the exhaust system to the underhood. As expected, however, the gas temperature at the catalytic convertor in this case was unacceptably high.

CINDA output provides information about the quantity of heat exchanged between any pair of nodes in the network, thus, exhaust system pipes which dissipate heat to a specific underhood component can be distinguished. Figure (15) shows exhaust system pipes which radiate heat to the engine plastic cover. Shielding only these pipes would relieve the plastic cover, yet leaving enough unshielded surfaces on the exhaust system to dissipate heat to the cooling air. Figure (16) shows the effect of this strategy on the underhood. The temperature of the engine plastic cover dropped 75°C compared to its temperature when a single-walled pipe design was applied to the exhaust system. At the same time, the exhaust gas temperature at the outlet of the system was 30°C cooler than its temperature when a full shielding design was implemented to the exhaust system.

5. Conclusions:

Results obtained in this paper confirm the usefulness of this lumped-parameter approach as a tool that can be used early in the design phase of new vehicles to give valuable insight to the underhood thermal environment. The tool can also be used to study the effect of trade-off packaging alternatives on the temperature of the underhood system. The following conclusions can be drawn from the calculations performed in this paper to simulate the thermal behavior of the 1993 Opel Vectra:

1. On average, the passenger side is warmer than the driver side. This is due, in part, to the larger quantities of cooling air flow in the driver side compared to those in the passenger side.

2. The addition of the underbody cover increases the temperature of the underhood environment. Three reasons were suggested for that increase. First, the underbody cover prevents radiation heat transfer from the exhaust system to the ground. Instead, the cover reflects this heat back in the engine compartment. Second, the underbody cover prevents the mixing of underbody ambient air with warmer air under the hood, thus adversely affecting its cooling effect. This was especially noticeable in components located in the lower half of the engine compartment. And third, the underbody cover causes the pressure downstream of the radiator to increase. This reduces the radiator air flow.

3. The addition of a cool-air duct to deliver sufficient ambient air to cool internal alternator components decreased the temperature of these components 40°C, on average.

4. Partial shielding of the exhaust system represents the optimum design to decrease the temperature of the underhood environment without adversely affecting the catalytic convertor.

6. References:

[1] Han, T., "A Navier-Stokes Analysis of Three-Dimensional Turbulent Flows around a Bluff Body in Ground Proximity," AIAA Paper 88-3766,1988.

[2] Berneburg, H., and Cogotti, A., "Development and Use of LDV and Other Airflow Measurement Techniques as a Basis for the Improvement of Numerical Simulation of Engine Compartment Air Flows," Accepted for publication, SAE Conference, Detroit, Michigan, USA, March 1993.

[3] Ingersoll, J. G., et al, "Automotive Passenger Compartment Thermal Model for General Motors," Hughes Ref. No. 722620/1514, pp.86-95, Jan. 1990.

7. Acknowledgements:

The authors would like to express deep thanks to EDS, GM and Adam Opel AG management for the support, advice and help they provided for this work. Numerous discussions with Dr. G. Mandrusiak enriched the project. Mr. J. Mesojedec helped in the final draft of this paper.

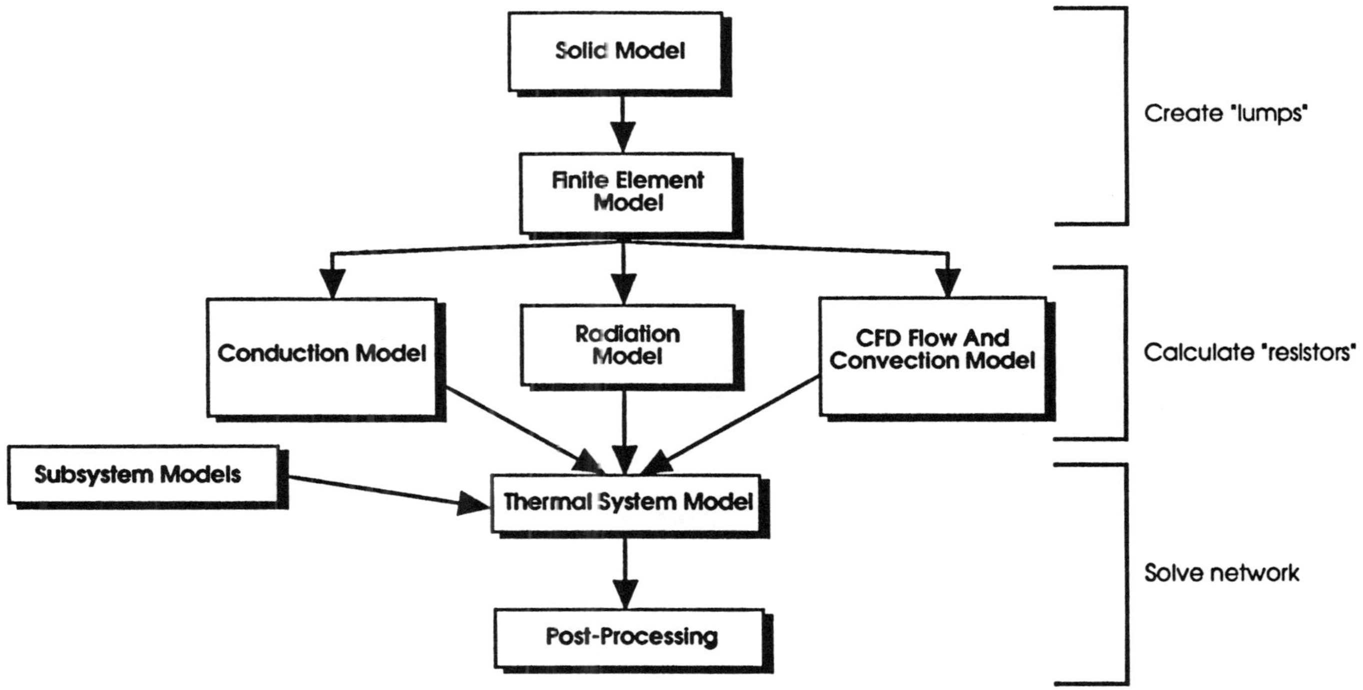

Figure 1 - Simplified underhood model building flowchart

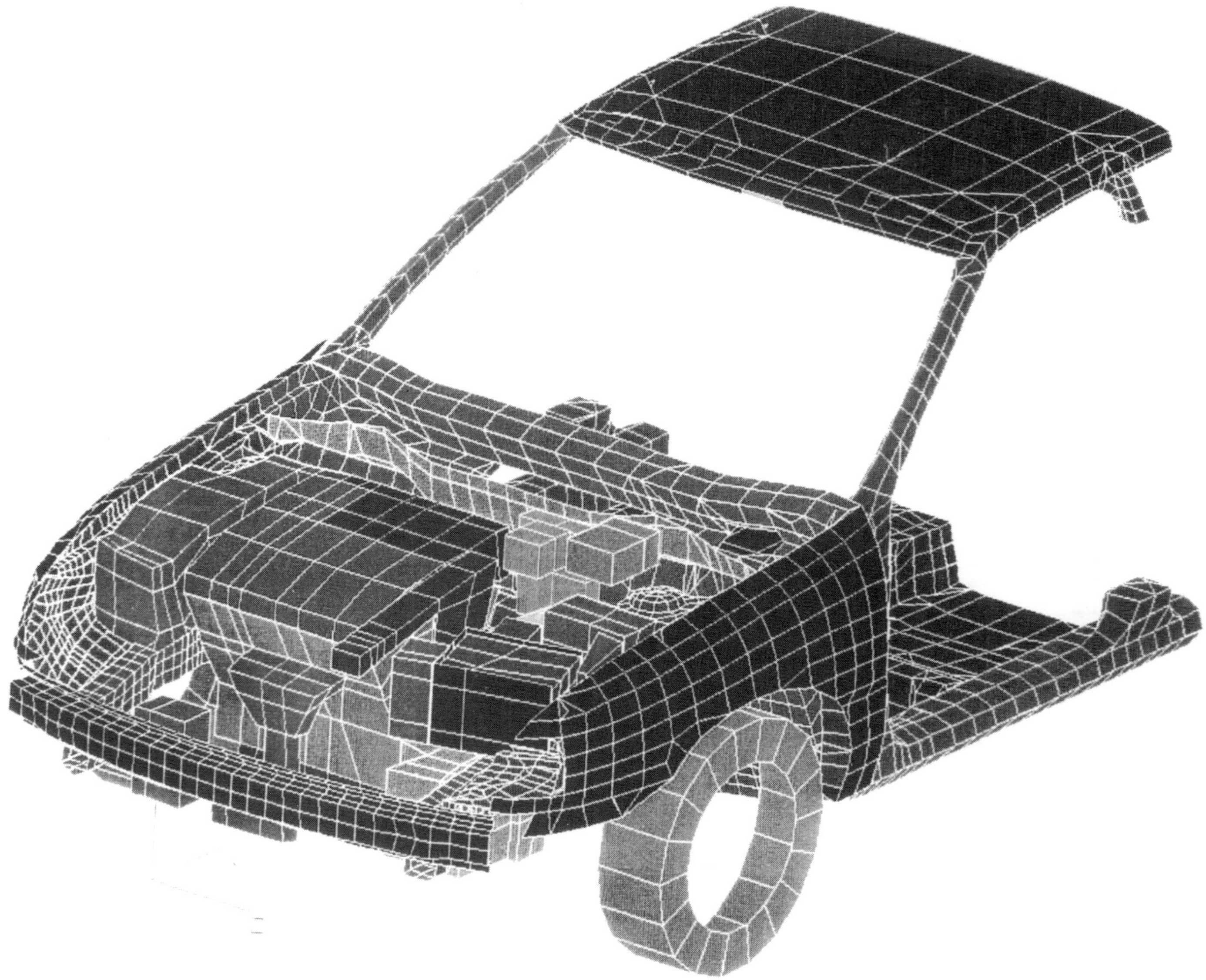

Figure 2 - Basic engine compartment geometry of the Opel Vectra in a more detailed crashworthiness model of the vehicle

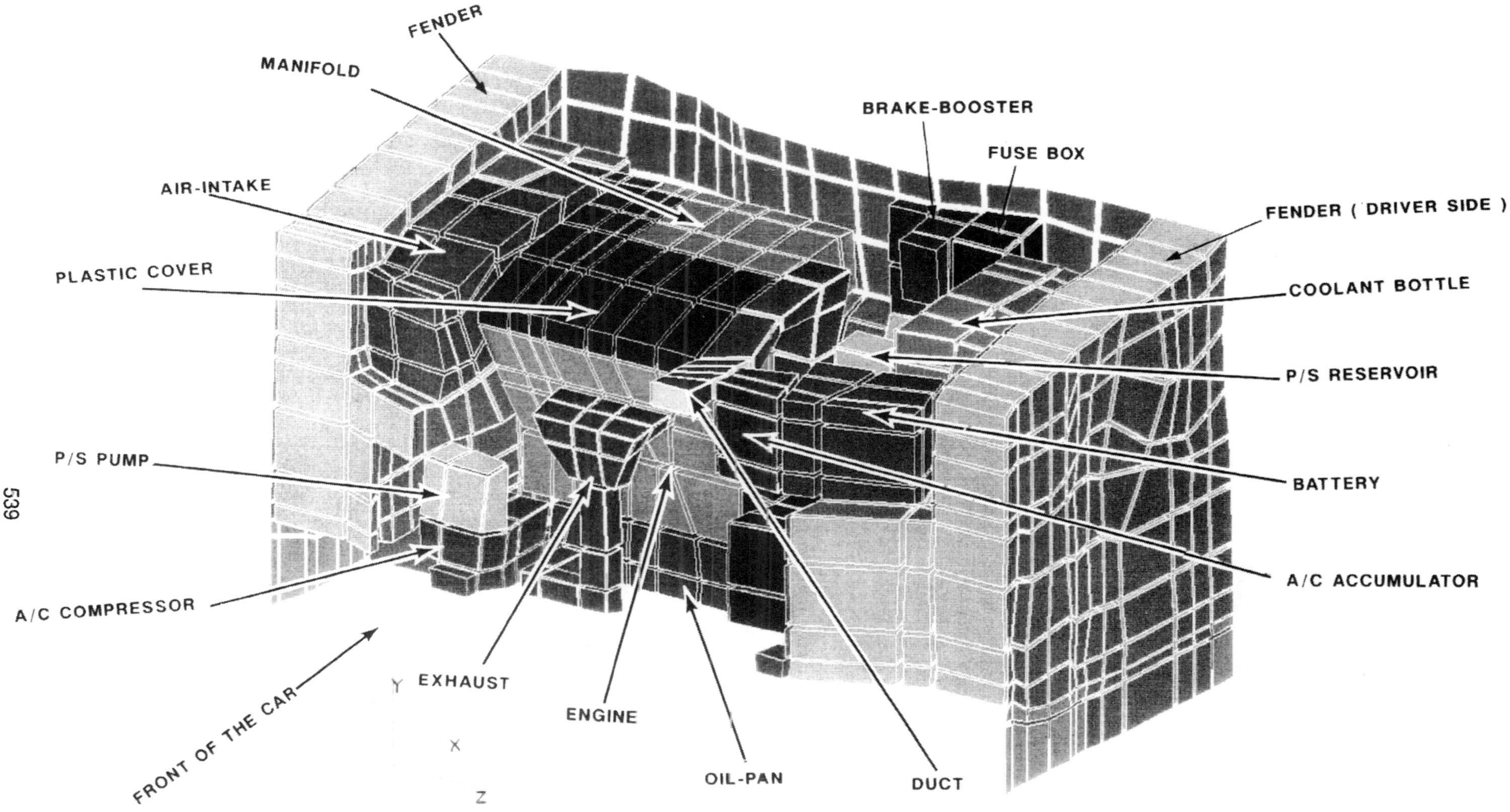

Figure 3 - Detailed super element representation of the engine compartment - Front view

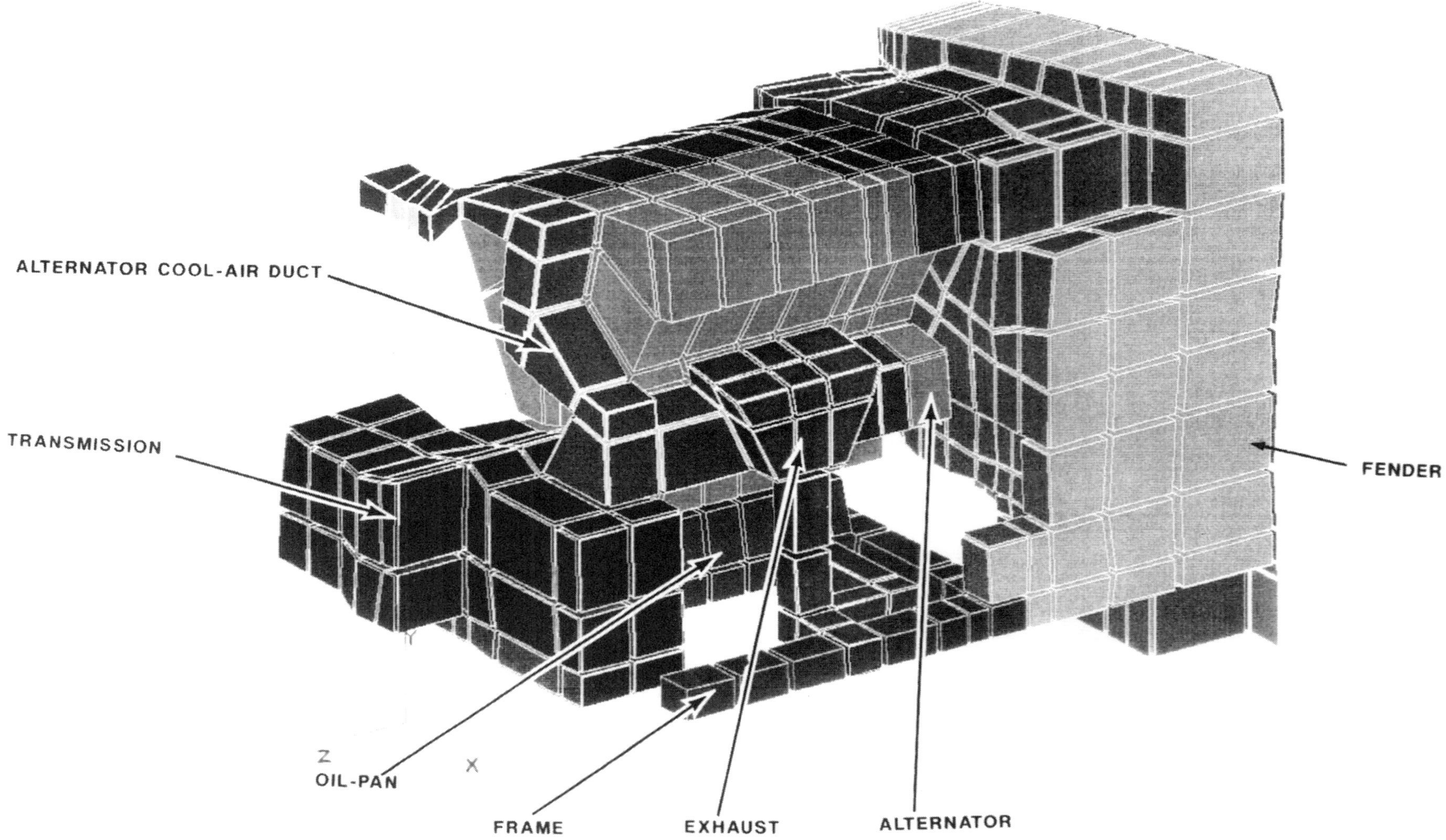

Figure 4 - Detailed super element representation of the engine compartment - Back view

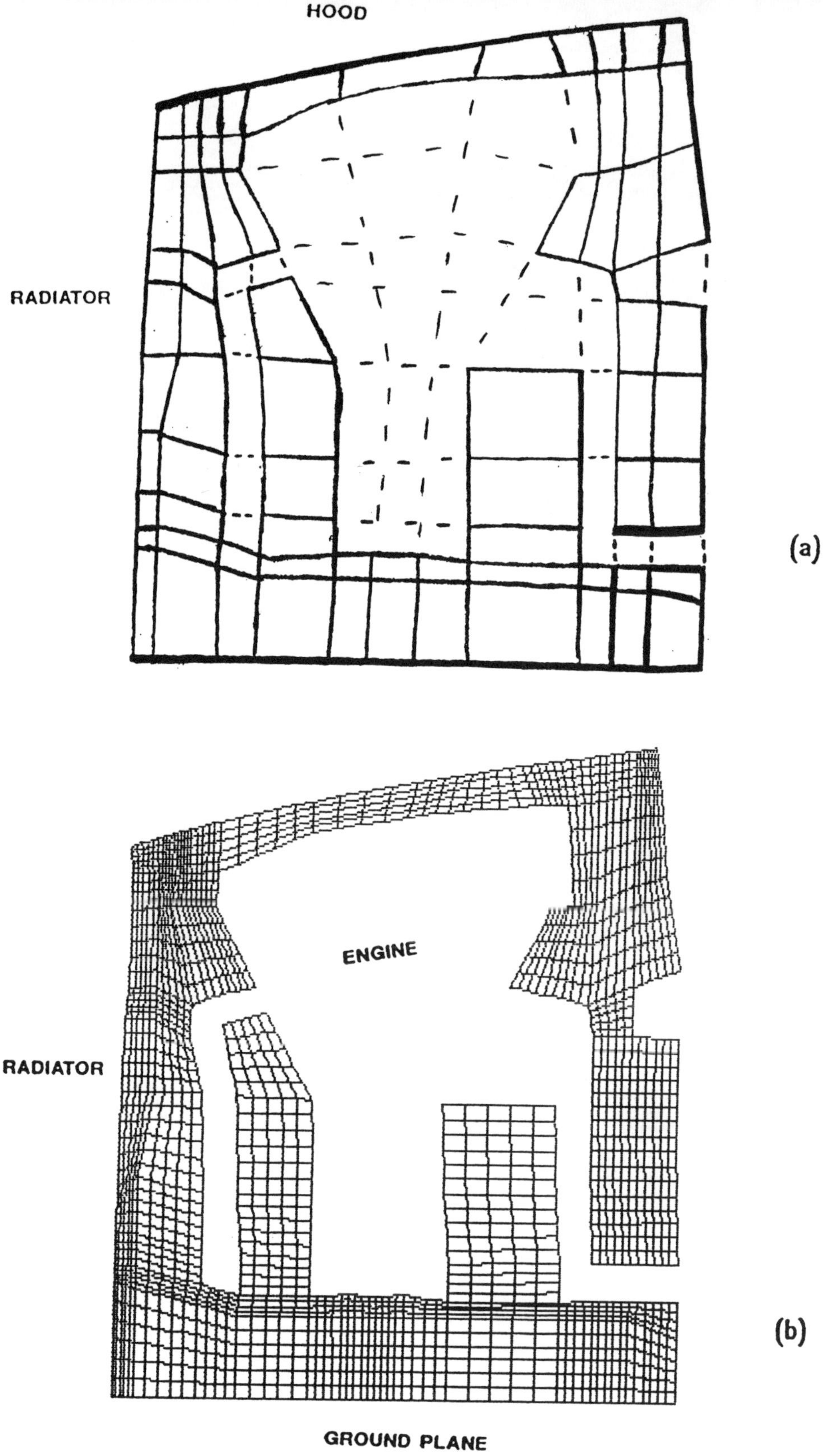

Figure 5 - Vertical cross-section of the engine compartment: (a) super elements representation of air and solid components; (b) computational fluid dynamics mesh representation

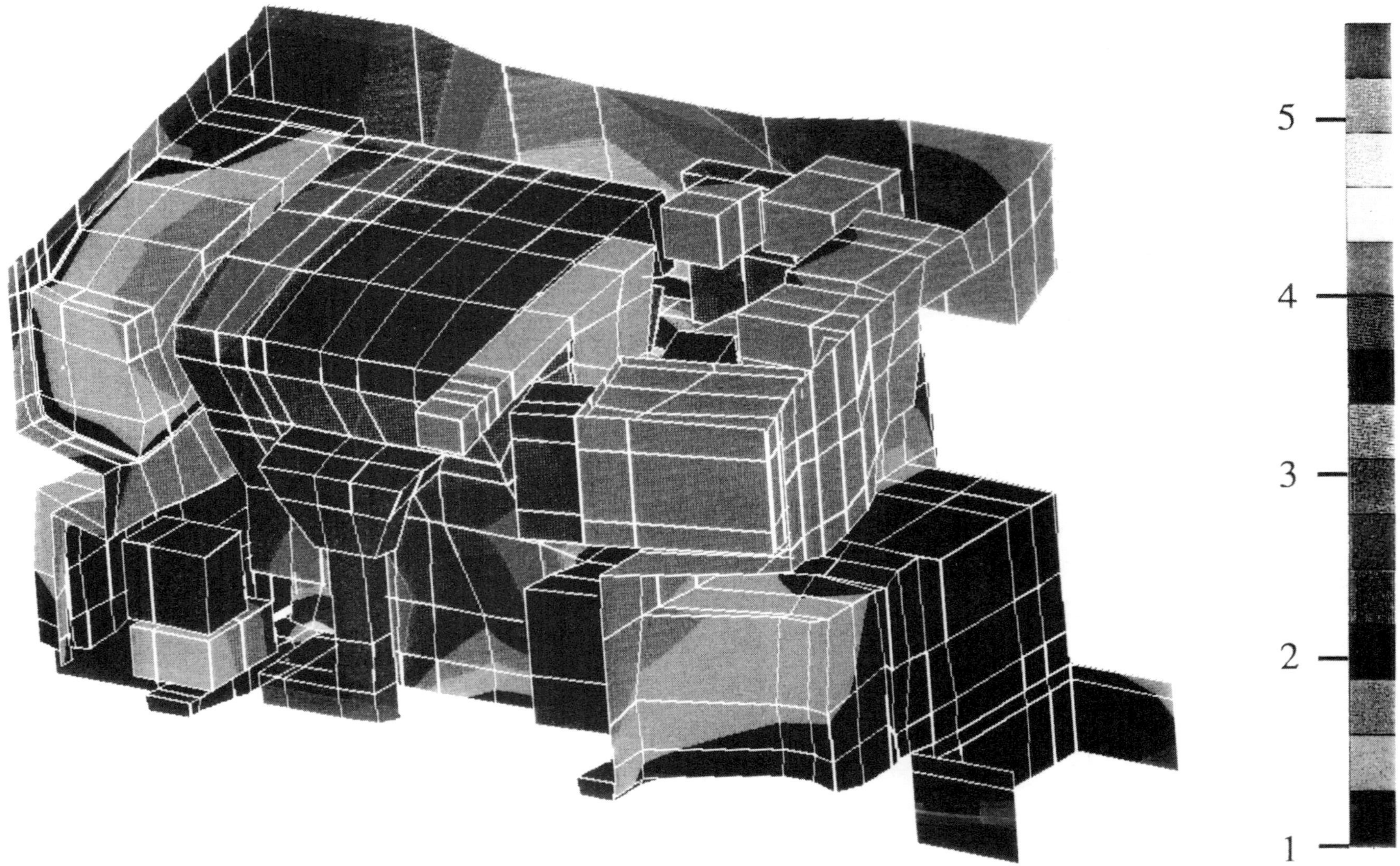

Figure 6 - Normalized temperature distribution of engine compartment surfaces for 60 km/h, road load driving condition

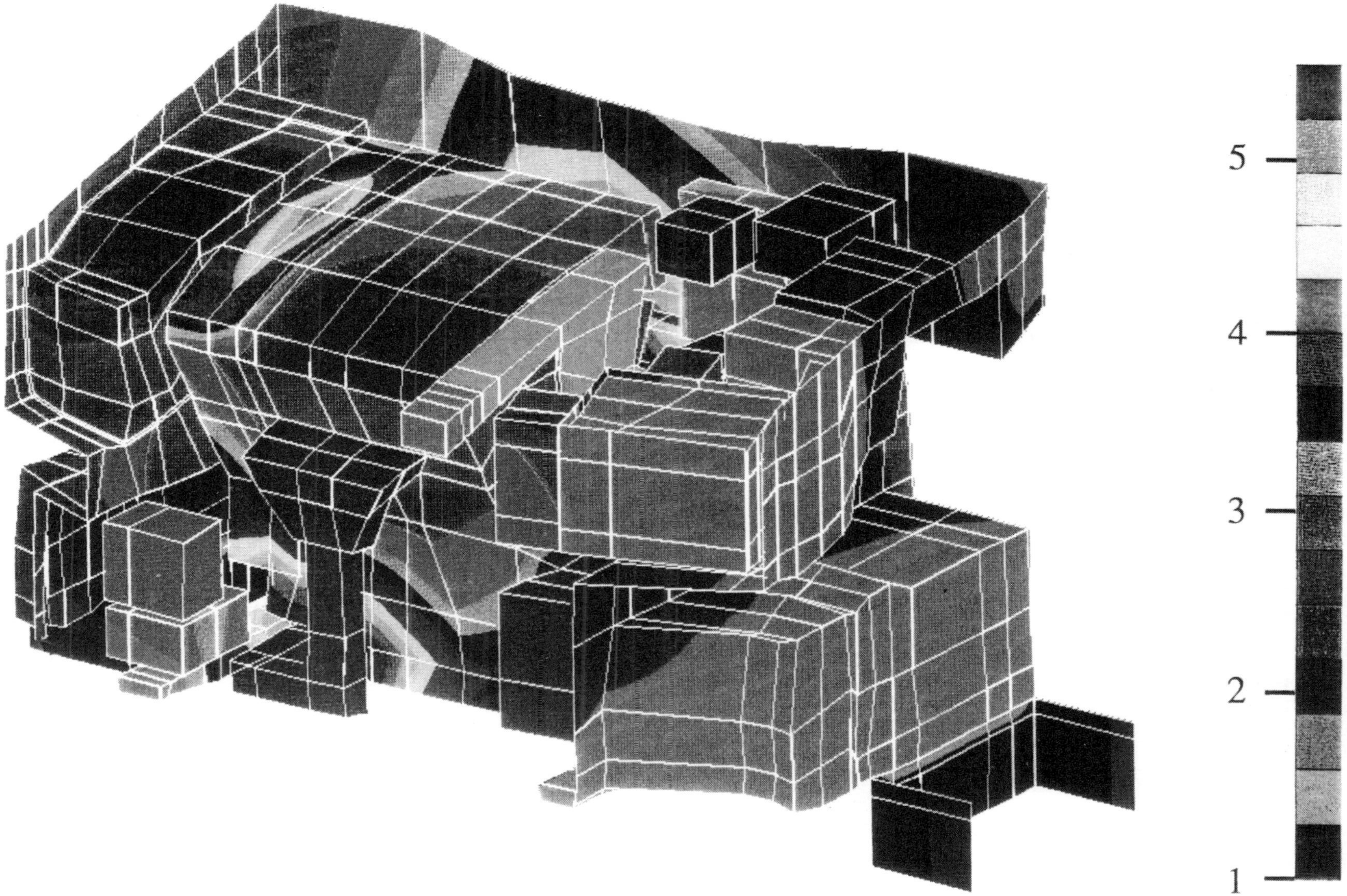

Figure 7 - Normalized temperature distribution of engine compartment surfaces for 60 km/h, after the addition of the underbody cover

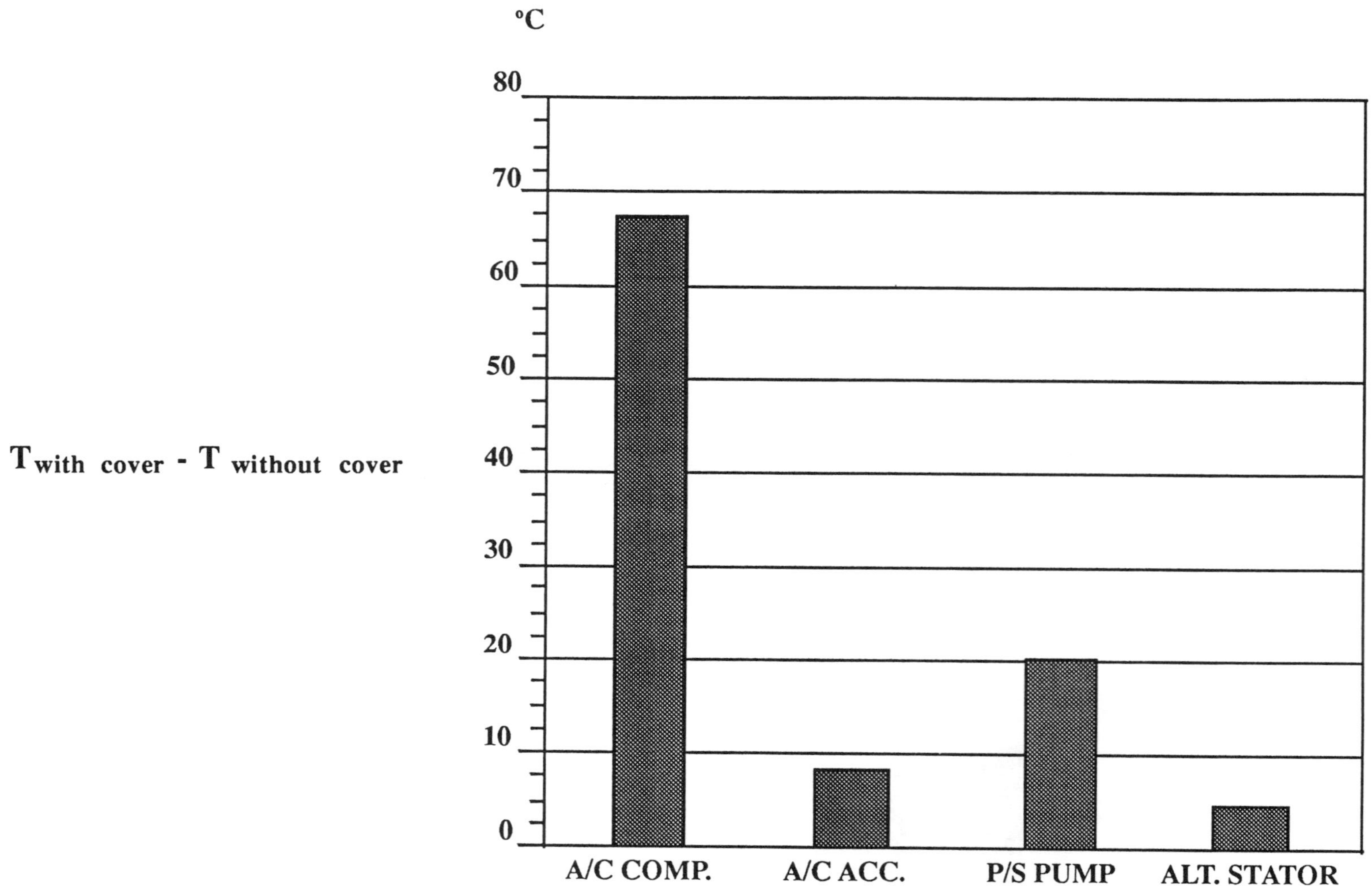

Figure 8 - Effect of adding the underbody cover on the temperature of selected components

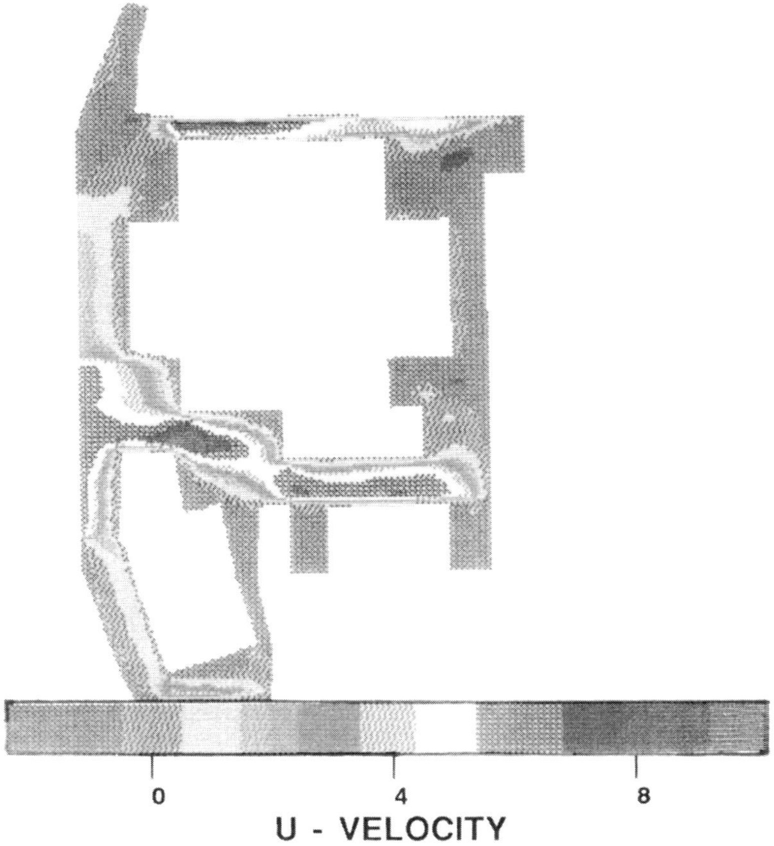

a) 60 Km/h - no cover

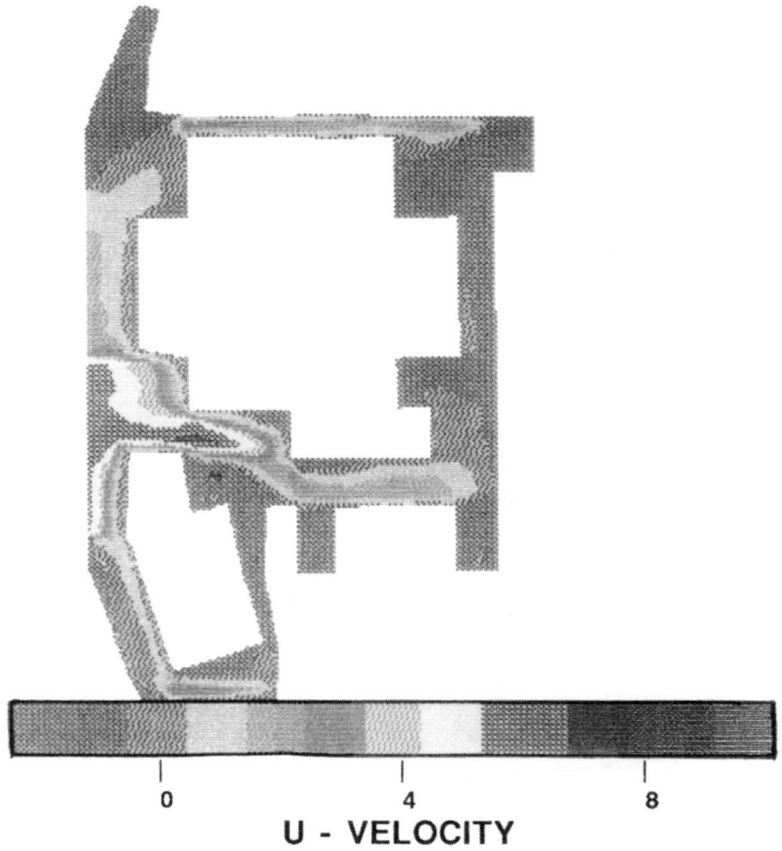

b) 60 Km/h - with cover

Figure 9 - Variations in the axial component of air velocity in a horizontal plane through the engine compartment: (a) base-line case; (b) with the underbody cover

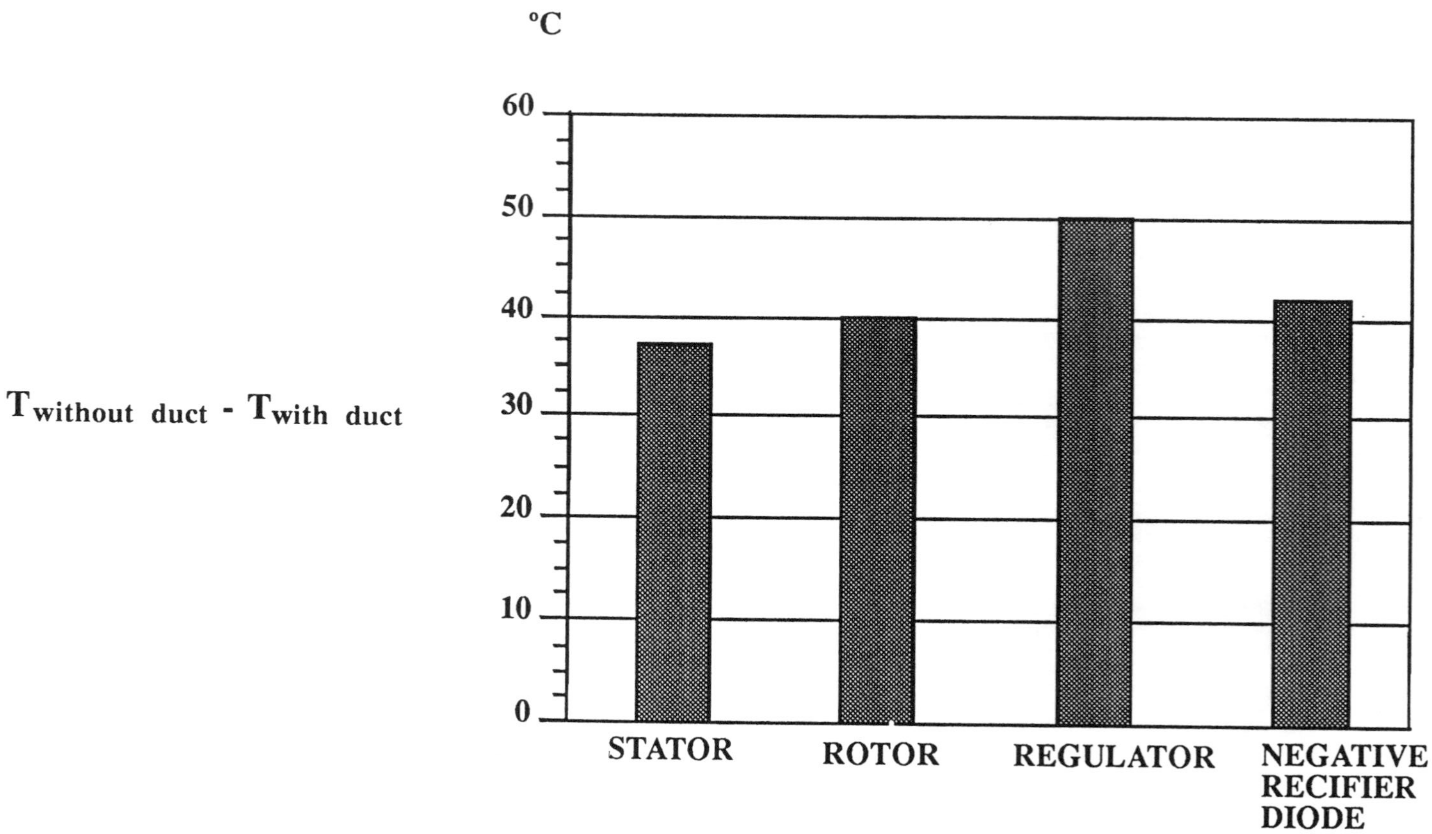

Figure 10 - Effect of using a cool-air duct on the temperature of selected alternator internal components

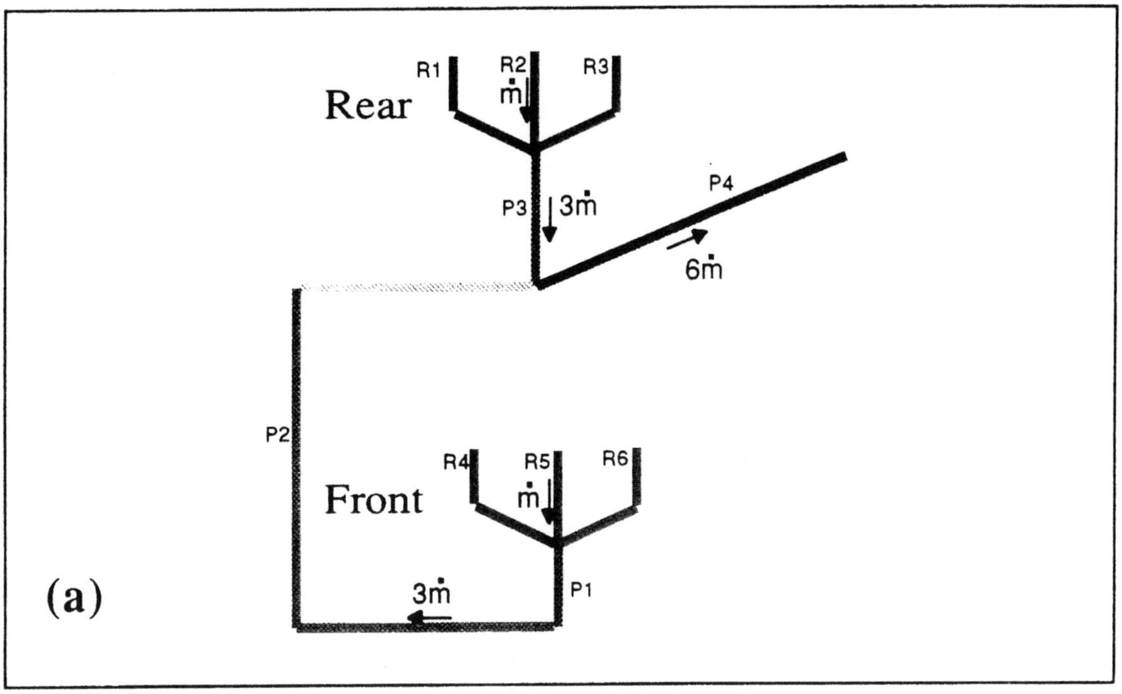

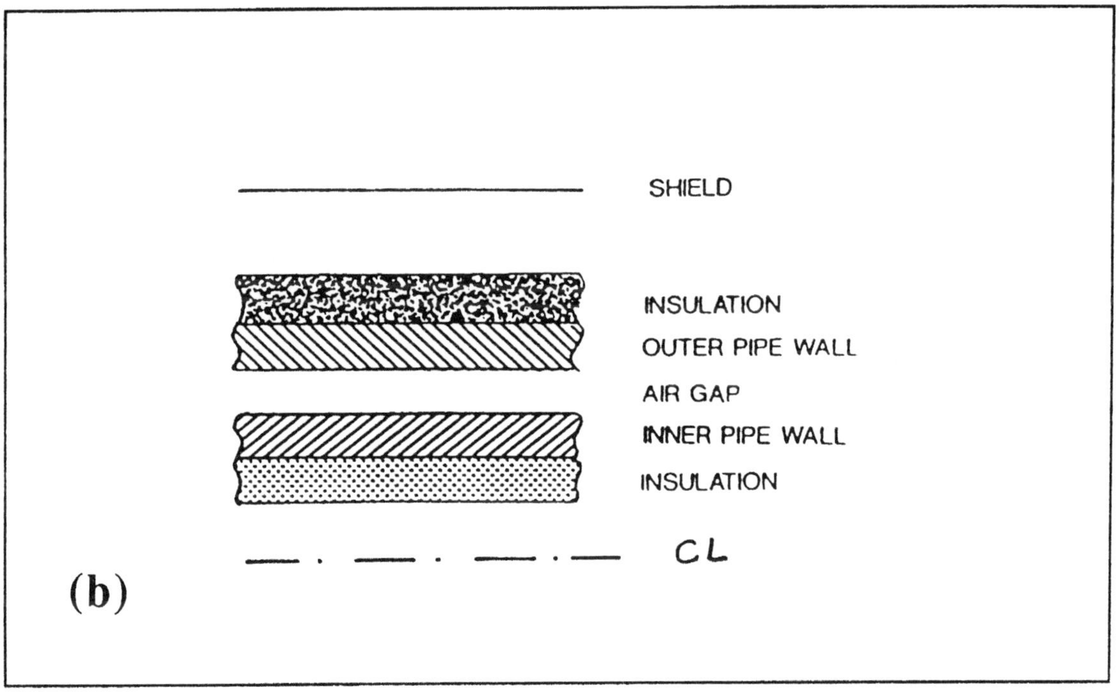

Figure 11 - Schematic representation of the exhaust system: (a) exhaust system submodel pipes and runners; (b) cross-section of a typical exhaust pipe

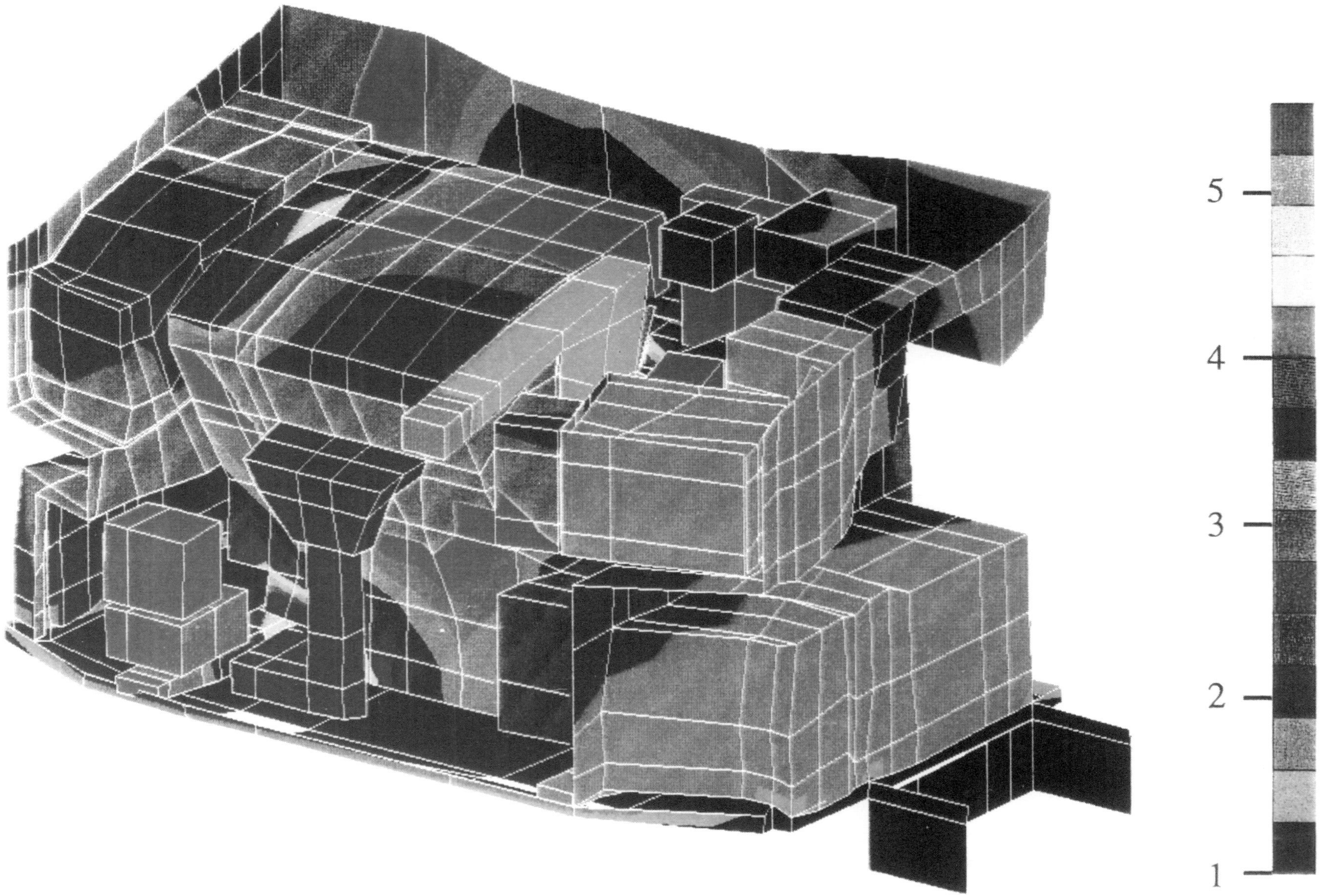

Figure 12 - Normalized temperature distribution of engine compartment surfaces for 60 km/h, with cover, double-walled exhaust system

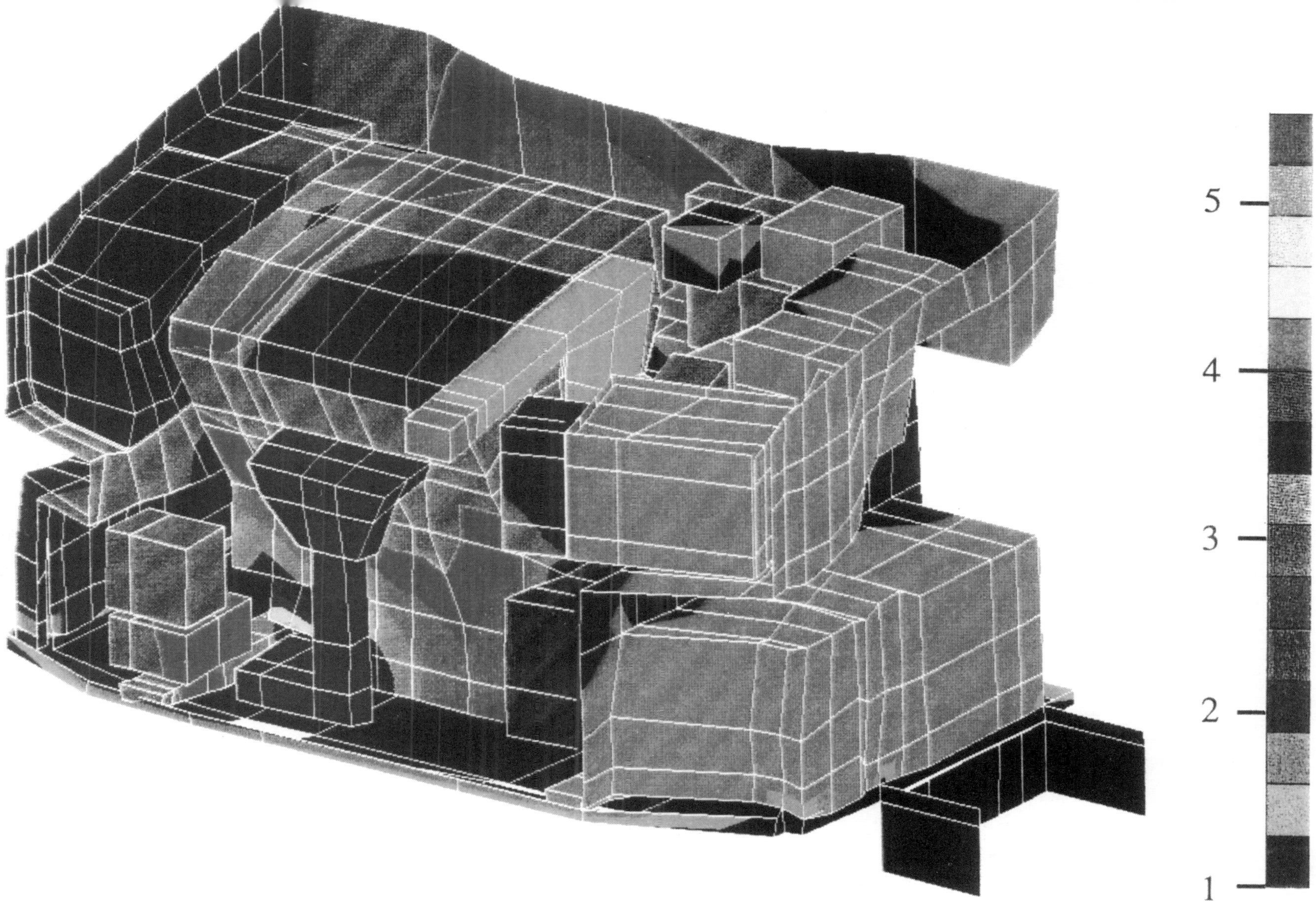

Figure 13 - Normalized temperature distribution of engine compartment surfaces for 60 km/h, with cover, fully insulated exhaust system

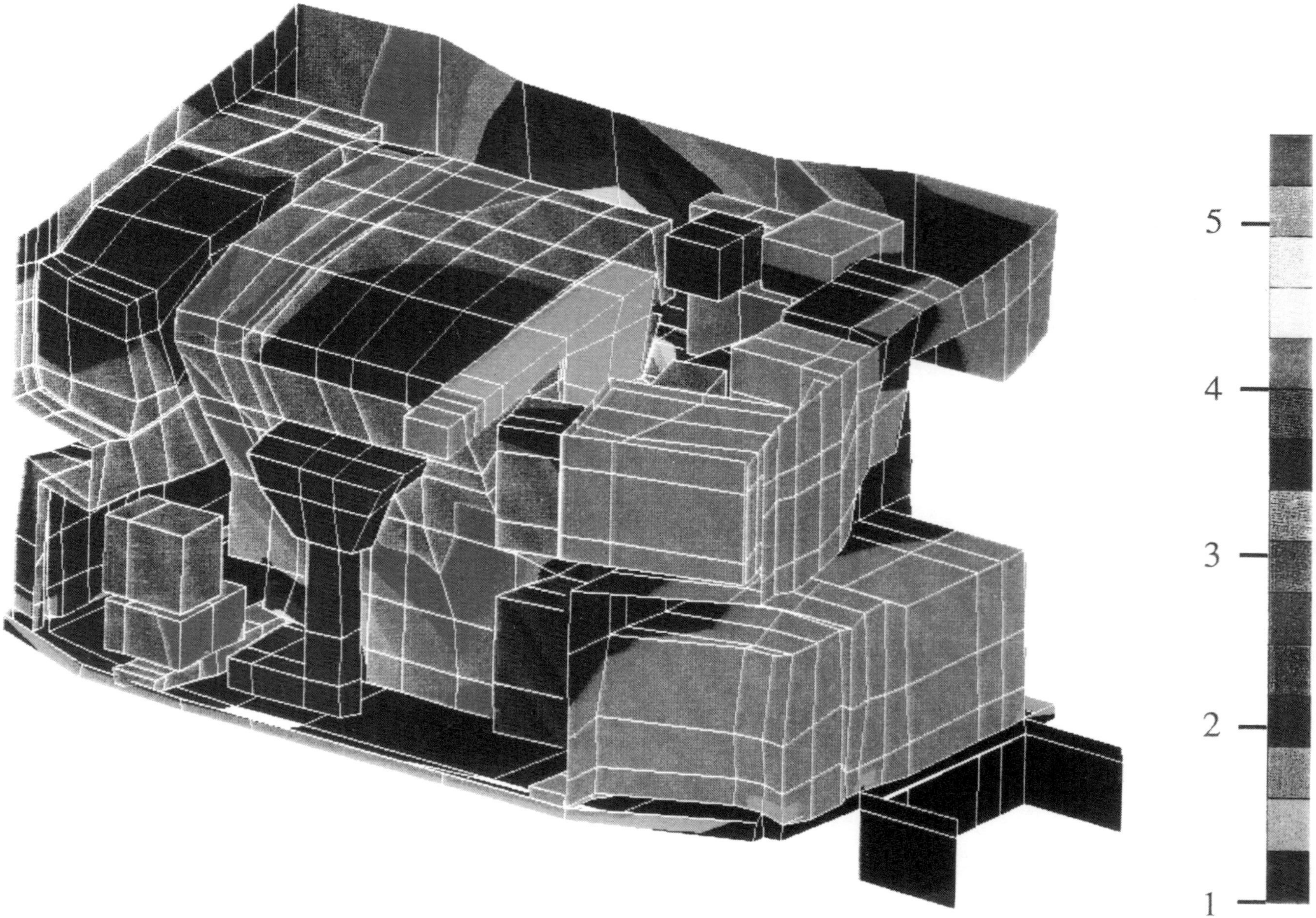

Figure 14 - Normalized temperature distribution of engine compartment surfaces for 60 km/h, with cover, fully shielded exhaust system

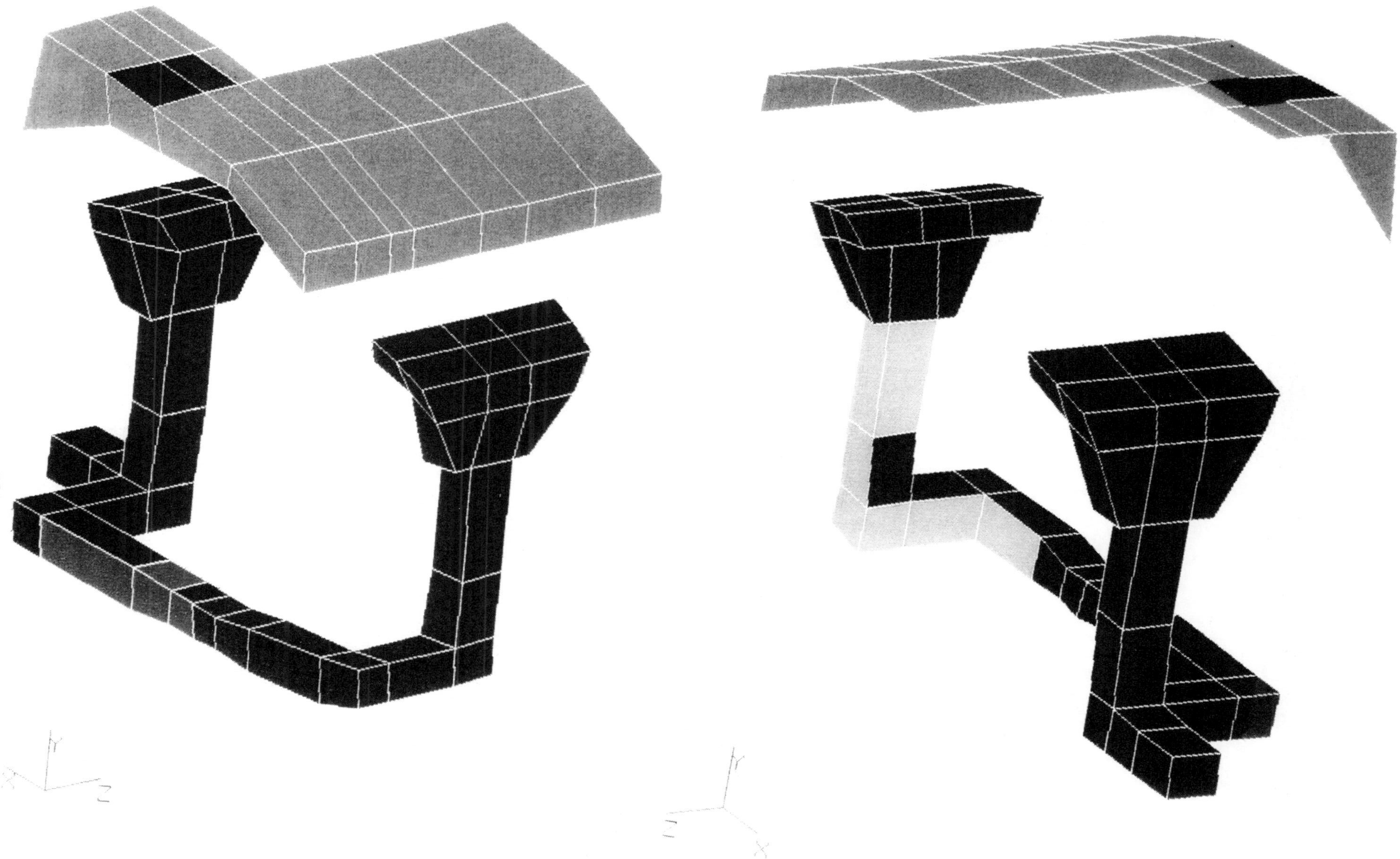

Figure 15 - Exhaust system pipes radiating most heat to the plastic cover

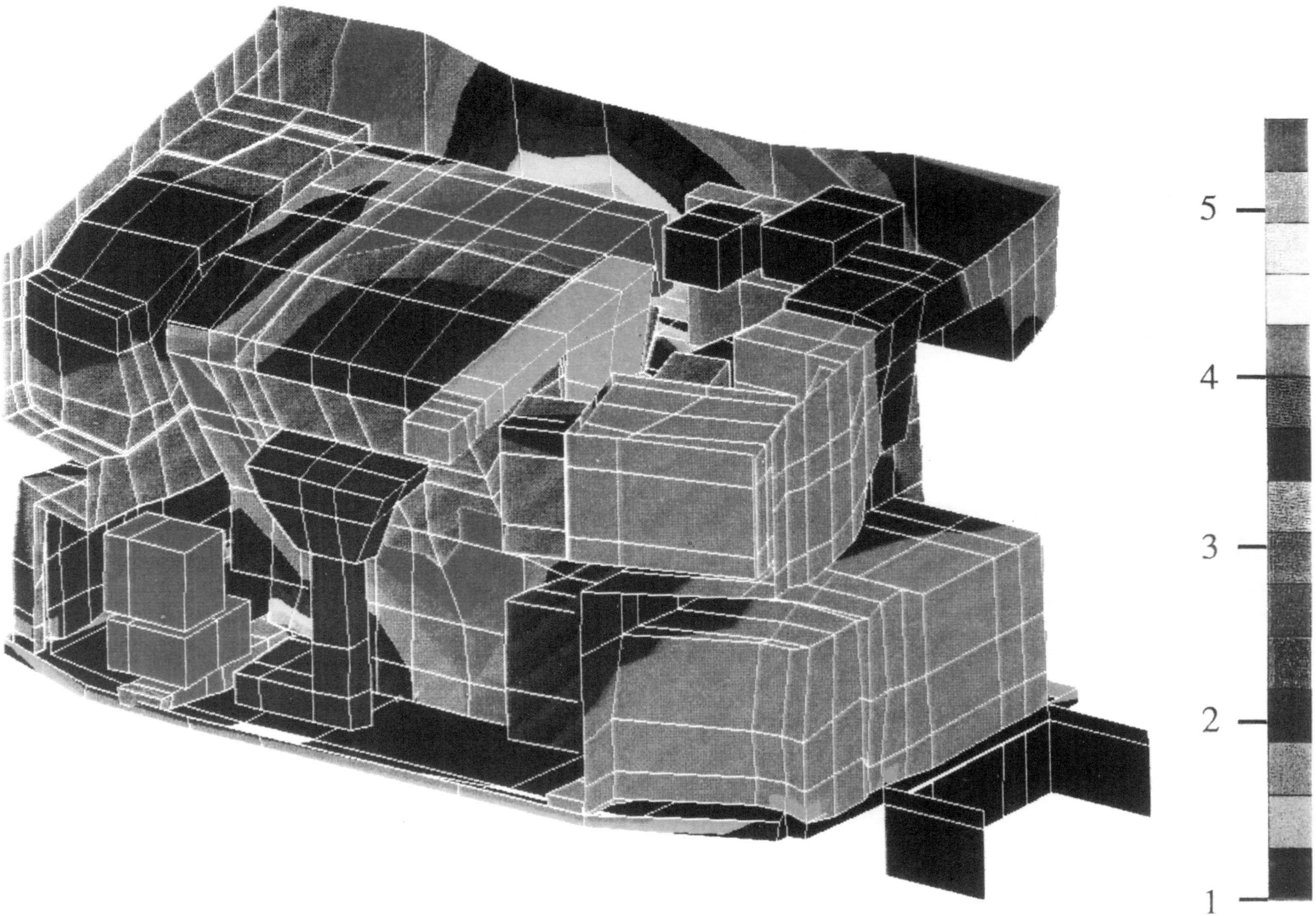

Figure 16 - Normalized temperature distribution of engine compartment surfaces for 60 km/h, with cover, partially shielded exhaust system

Section VII
Vehicle Aerodynamics and Aeroacoustics

Section VII
Vehicle Aerodynamics and Aeroacoustics

It is perhaps not surprising that the topic of aeroacoustics has commanded increased attention from vehicle aerodynamicists and acousticians in recent years.

There are two basic dimensions to the automobile noise issue. One is the noise field experienced by the occupants of a vehicle and the possible degradation of comfort and perception of vehicle quality it may cause. The second is the far-field noise (transmitted from a vehicle) that is increasingly subject to legislation as communities act to curb the degree of noise pollution in the vicinity of highways. The last two decades have seen societies around the world make strides in slowing different forms of pollution and degradation of our environment. Noise pollution and its effect on our environment has naturally come under scrutiny. Historically, powertrain and road/tire noise sources have commanded the most attention. As these areas see continuous improvement, aeroacoustic (flow-induced noise) sources are being studied in more detail.

The source of wind noise in vehicles is unsteady aerodynamics. The severe buffeting that occupants experience when driving at high speed with an open window can create large-amplitude pressure fluctuations within the passenger compartment. This class of problem (which includes open sun-roofs) is typically treated separately from the one where the vehicle's interior is "sealed" from the exterior. In the latter case, aerodynamic noise results from turbulent boundary layers and regions of separated flow over the vehicle. The vehicle's surface in these regions is subjected to unsteady pressures that can be perceived as aerodynamic noise by the occupants.

Elimination of all flow separation around a practical vehicle is not possible, and therefore much effort is dedicated to understanding aerodynamic-noise generation and transmission so that they may be minimized. A factor that contributes to the complexity of the problem is the large influence of small scales. While small geometric details typically give rise to small (local) hydrodynamic pressure disturbances, the accompanying acoustic disturbances can be large. Furthermore, the levels of (acoustic) sound pressure are several orders of magnitude smaller than the hydrodynamic pressure disturbances, making the study of noise-generation mechanisms very difficult.

The increased attention devoted to aeroacoustics has led to significant improvements in acoustic measurement capability in wind tunnels. While the flow stream in a wind tunnel test section is inherently noisy, it is important that this background flow noise be significantly smaller than the level of noise generated by a test vehicle. This has led to the construction of several aeroacoustic or "quiet" wind tunnels, as well as the conversion of several older wind tunnels to aeroacoustic quality. These efforts demand very significant resources and commitment from the parent organizations, and underscore the importance of aerodynamic noise to the course of a vehicle's design and development.

This emphasis has also given birth to the discipline of Computational Aeroacoustics (CAA), where developments in the field of CFD are coupled to calculations of the acoustic pressure field. Most of the work in this area is in its infancy. Apart from the difficulty in computing unsteady flow, numerous challenges remain with respect to correlating the hydrodynamic computations to computations of the acoustic field. The small scales of fluid motion that are involved (linked to the desired range of frequencies of interest) as well as the time-dependent nature of the simulations demand extraordinarily large computational effort. In addition, appropriate treatment of boundary conditions for the simulations remains unclear and continues to be the subject of research.

900315

Automobile Aerodynamic Noise

Albert R. George
Sibley School of Mechanical and Aerospace Engineering

ABSTRACT

As engine, tire, and other automobile noise is reduced and as driving speeds increase, aerodynamic noise sources on ground vehicles are becoming relatively more important. They often dominate at cruise speeds above 60 mph. Aspiration and leak noise are strong sources but generally can be controlled by known methods. Turbulent pressure fluctuations due to separated and vortical flows are also strong sources. Much interior noise is caused by transmission of these external pressure fluctuations through windows and other surfaces. The paper presents the variety of aeroacoustic sources on automobiles and reviews the state of experimental data, of analysis methods, and noise reduction principles. A new correlation method for predicting external fluctuating pressures in separated regions is presented.

INTRODUCTION

When automobiles were first developed the engine was the primary cause of automobile noise, but over the years engines and drive trains have become quieter due to many man-years of engineering attention. As engines and drive trains became quieter, tires began to be recognized as an important noise source on automobiles, especially, of course, snow or traction type tires. Again, as time has gone on, the noise reaching the passenger compartment from conventional tires has been significantly reduced until now tire noise is significantly lower than engine and drive train noise in most cars at speeds below, say, 100 mph.

As engine, power train, and tire noise have been reduced, wind noise has become relatively more important. In fact, in 1972, Buchheim et al. showed that, for a "typical" car of that date, wind noise became as important as any of the other noise sources at speeds of about 100 mph.(1) But since 1972, development has further quieted engine and drive train noise, especially in luxury cars, and now many cars' interior noise levels are significantly affected by wind noise at speeds on the order of 60 mph, again under cruising conditions. Under accelerating, hill climbing, or other high power conditions, many cars are still dominated by engine noise, which is from engine vibration (including that transmitted through the automobile structure), intake, exhaust, and accessories such as fans. But under high speed cruise conditions, the aerodynamic noise becomes important. This relative increase in aerodynamic noise has implications for both the external noise of ground vehicles and the passenger compartment noise. For example, many of the high speed highways and rail lines traveling through populated areas in Europe have embankments or walls built along the sides of them to reduce annoyance from the traffic or rail noise.

It has been predicted by King and Beckert that the aerodynamic noise of the planned future high-speed trains in Europe will become even more important than wheel/rail noise.(2) The reason for this is that wheel/rail noise, although dominant at present train speeds, only increases as approximately velocity cubed (V^3), whereas aerodynamic noise increases approximately as velocity to the sixth power (V^6), and at about 200 mph, it becomes as important as wheel/rail noise. This means that even if a silent rail/wheel system, magnetic levitation, or other silent suspension is successfully introduced, aerodynamic noise at speeds above 200 mph will be considerably worse than the conventional wheel/rail noise expected at such speeds.

The inside noise in an automobile is important to its sales, particularly for luxury cars. However, even with sporty cars, people are sensitive to the fatigue factor of excessive noise. Even a level of 70 to 80 dBA is tiring on a long trip, and designing cars with lower noise levels is a high priority. Let us consider the magnitude of the problem of aerodynamic noise. In tests of about 15 different production cars reported in 1972, Buchheim et al. found that at 70 mph the aerodynamic noise in cars ranged from 62 to 78 dBA, whereas at 110 mph, a common cruising speed in Germany, the aerodynamic noise ranged from 72 to 87 dBA.(1) Clearly 87 dBA is excessive, it is even greater than the 85 dBA OSHA industrial workplace limitations.

As we will see in this paper, certain aerodynamic external pressure fluctuations can be reduced but some aspects of noise transmission into the inside of a car are extremely difficult to control. Also, there are somewhat fewer intrinsic possibilities for aerodynamic noise reduction through increased sound insulation and absorption than there are for the corresponding noises from tires and engines and drive trains.

AUTOMOBILE AEROACOUSTIC SOURCES, PATHS, AND RECEIVER - In looking at any noise problem, it is always beneficial to look at the sources, the paths, and the receiver. In automobile aeroacoustics the sources are unsteady aerodynamic pressures generated on various parts of the car. This paper is concerned with the noise sources originating in the unsteady aerodynamic pressures generated by turbulent flow on the outside and underneath automobiles and other ground vehicles. Another important source can be the fluctuating pressures of the engine cooling fan, fans used to circulate air in the heating and air conditioning systems, etc. Fan noise is a separate topic (see for example references 3, 4, 5, 6) and will not be treated in this paper.

External unsteady aerodynamic pressures can generate external noise, which propagates away from the vehicle and is annoying to people in the surrounding environment. These unsteady aerodynamic pressures can also propagate through various paths into the inside of the vehicle generating interior noise. One path by which noise can get inside the car is directly through openings or leaks in the basic structure of the car. The other path by which noise can get inside the car is through panels, windows, and seals, because their transmission losses are less than 100%.

Once the sound energy reaches the receiver, the inside of the car, it is eventually absorbed. High frequencies are easy to absorb with carpeting and upholstery, reducing reverberant sound pressure buildup in the interior of the car. However, low frequencies are more difficult to absorb and one needs to use Helmholz resonator type damping, damped panels, or, of present interest, active damping.

AEROACOUSTIC APPROACHES

In principle, fluctuating pressures and sound can be found as a part of a full solution of the turbulent aerodynamics of the flow around an actual car shape. Again, in principle, the propagation through various panels of the car can be solved using methods of solid mechanics, vibrations, and acoustics. However, there are extremely few aerodynamic noise problems that we can solve analytically or even computationally in any exact fashion. Thus we are forced in most situations to break up a real problem into various idealized models with varying degrees of approximation, and to couple these models together to get an estimate of the noise generated and propagated to the receiver. The models can range from fairly elaborate ones making an attempt to solve for the full flow, perhaps numerically; to those where the near field flow is solved, perhaps numerically, and a Kirchhoff method is used to find the far field[7]; to perhaps what is the most common approach of all, approximating the near field flow without sound, and then using some variation on the acoustic analogy to find the radiated sound. This is the primary point of view taken in this paper.

Approximate models for the radiation are usually employed; these are often monopole, dipole or quadrupole models. The monopole source effect from an unsteady introduction of volume into the surrounding fluid, is the most efficient generator of sound at low Mach numbers. The next most efficient generator of sound at low Mach numbers is the dipole effect of unsteady forces applied to the fluid. (These pressures are, of course, just the opposite of the unsteady aerodynamic pressures applied to the body by the flow.) And then finally, the least efficient sound source at low Mach numbers is that due to internal stresses or turbulence or quadrupoles in the interior of the flow.

The most notable volume addition or monopole source of noise for automobiles is the exhaust from an engine, where, due to the combustion process and the nature of the usual piston engine design, pulses of exhaust gas are released periodically from the engine through the exhaust system into the surrounding atmosphere, generating significant monopole exhaust sound, depending on the efficiency of the muffler system. With modern muffler technology, this source of sound can be reduced to be almost negligible. Similarly the intake noise due to the unsteady induction of air into the engine, can also be quieted through proper design of the air cleaner intake silencer assembly.

The unsteady surface forces of primary interest in this paper are found on the vehicle exterior. These unsteady forces are primarily due to turbulence in the flow, especially in separated and vortical portions of the flow. Automobiles and other road vehicles are, of necessity, bluff body shapes in order to have room for passengers and yet be within the size constraints of garages, roads, etc. Thus there are large regions of external separated flow on the rear of any automobile. In addition to the probably necessary separation at the rear of automobiles, there are many other separated flow regions associated with the wheels, the flow in the engine compartment, and the flow underneath the car. Not all of these are strictly necessary. However, present designs have large portions of separated flow associated with all these parts of the car.

Let us consider the scaling of acoustic sources for automotive type applications, for the case of the source size small compared to a wavelength. For a monopole source, the pressure is given by

$$p = \frac{\rho}{4\pi r}\frac{\partial \dot{Q}}{\partial t}.$$

If a typical velocity is V and a length L, then the volume flow $\dot{Q}$ is proportional to VL^2. Approximating the derivative by $\frac{\partial}{\partial t} \sim \frac{V}{L}$, and using the pressure to find the intensity, we find

$$I_{monopole} \sim \frac{\rho}{r^2 c} L^2 V^4.$$

On the other hand, for a dipole source, we have:

$$p = \frac{1}{4\pi\rho r c}\frac{\partial F}{\partial t}\cos\theta.$$

The force on a surface is proportional to $\rho U^2 L^2$, and then we can find for the intensity:

$$I_{dipole} \sim \frac{\rho}{r^2 c^3} L^2 V^6.$$

And then finally for a quadrupole source, which can be thought of as equal and opposite dipole sources, the intensity can be shown to be:

$$I_{quadrupole} \sim \frac{\rho}{r^2 c^5} L^2 V^8.$$

If we divide these to find the ratio of the strengths of the various sources, we find that the dipole strength divided by the monopole strength is proportional to the Mach number squared. Similarly the quadrupole strength divided by the dipole strength is again proportional to Mach number squared. This shows that at low Mach numbers, if there is a monopole source, it will be the primary noise source, while if there is not, the dipole will be the primary noise source. Similarly, at low Mach numbers the quadrupole effects are important only if the monopole and dipole effects are negligible.

By looking at some of the scale lengths and velocities in automotive flows we can get an idea of which flows contribute to which frequency ranges of noise. For example, considering the full car as a whole, with a velocity on the order of 30 meters/sec and a length on the order of 5 meters, we find that the frequencies associated with the large scale separation on the outside of the car, say at its rear and wake region, are about 6 Hz. On the other hand if we consider separation and turbulence on a smaller scale as associated with accessories and particular details of the body shape ranging from, say, 1 cm to 10 cm in size, but again with a freestream velocity on the order of 30 meters/sec, we find that the frequencies are in the range of 300 to 3000 Hz. If we look at the heating and air conditioning inside flow in the car, the velocities are more on the order of 3 meters/sec, the scale lengths are on the order of 1 cm for the parts of the heating and air conditioning system and grills, giving us a frequency in the range of about 300 Hz. And then finally if we consider the various fans used for engine cooling or heating and air conditioning, blade tip speeds are typically, say, 40 meters/sec and length scales range from 1 cm to 10 cm, giving frequencies in the range of 400 to 4000 Hz.

Putting these ideas together, we get the following approach to finding the aeroacoustic noise of automobiles, or for that matter, of anything. First, we must understand the aerodynamics of the vehicle and identify which are the important noise sources. Second, we must determine the unsteady aspects of the aerodynamics of the vehicle and any turbulence that might be generated by otherwise steady flows. Third, from the unsteady aerodynamics and turbulence, we need to find the unsteady pressures on the surfaces of the vehicle. Then these unsteady pressures can: (i) lead to unsteady volume flows into the inside of the vehicle, (ii) directly radiate sound to the outside of the vehicle, or (iii) excite structural motion which will then generate sound inside of the vehicle. Of course, it is easier to outline this approach than to carry it out.

AUTOMOBILE AEROACOUSTIC SOURCES - In order to get an idea of the origins of aerodynamic noise on cars, consider figure 1.[8] In this figure we see in the center a sketch of the flow around an automobile, and surrounding it, a number of details of the flow. Starting from the upper left we find flow through grills, radiators and fans, and possible separation on the hood of the car. At the base of the windshield there is a separation where the air inlets for most cars are located. In the area of the A-pillar (which is the structural member between the windshield and the side windows) there is a strong vortex flow generated. As we will see later, this vortex flow is strongly turbulent and is a major noise source on automobiles. Then on the side of the car around the windows there are a number of quasi-two-dimensional separations behind various body parts. There are extremely complicated separated flows around the tires and under the car in the vicinity of the engine compartment.

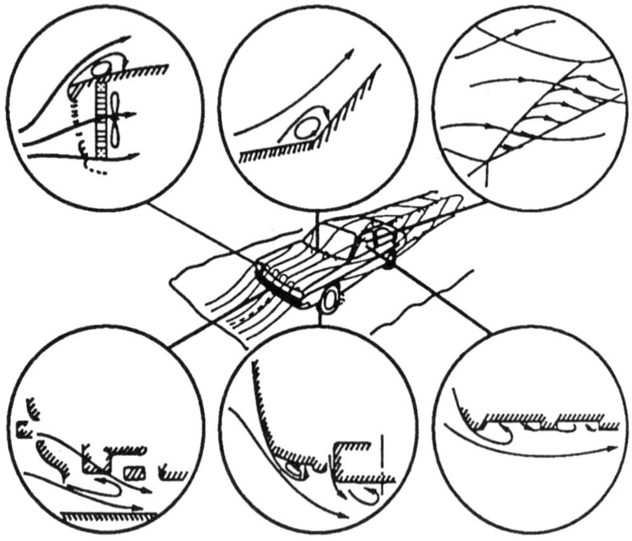

Figure 1. Flow around automobile and some details of noise source regions. (After reference 8.)

If we consider these sources from an aeroacoustic point of view, we find examples of most of the sources of interest in aeroacoustics. For example, the antenna of a car can support a vortex shedding type of flow, causing pure tones. Open windows or a sun roof can generate Helmholtz resonance or organ pipe type resonances. Joints between the doors and the body panels, if open, can also form resonant cavities. Leaks where air can come in or out of the car through cracks around the doors and windows are very important sources of noise. Various edges on the outside of the car, such as around doors, rain gutters, etc., can have high local fluctuating pressures on them due to turbulence in the upstream boundary layer, or due to local separation and reattachments with strong turbulence. Extraneous parts of the car, such as outside rear view mirrors or hood ornaments, grills, etc., can generate turbulent flows which then impinge upon other parts of the vehicle to generate noise. The underbody of a car has this happening to an extreme.

There are still many aspects of aircraft aeroacoustics which are not sufficiently understood, and this situation is even more pronounced in automotive aeroacoustics. If we refer to figure 2, showing schematically the various stages of development of experiment and analysis, we find that automotive aeroacoustics is still primarily operating in stages 1 and 2. People are doing exploratory experiments to pin down noise sources, and looking at very idealized models of these sources, and in the more advanced cases doing phenomenological experiments on particular identified generic source types and looking at the first concepts and analyses for predicting and reducing the noise from them. We are still far from having design concept verification experiments or specific design ideas and analyses used, even on a development basis. We have quite a long way to go before we will be doing developmental tests having standard design methods and analyses available for these problems.

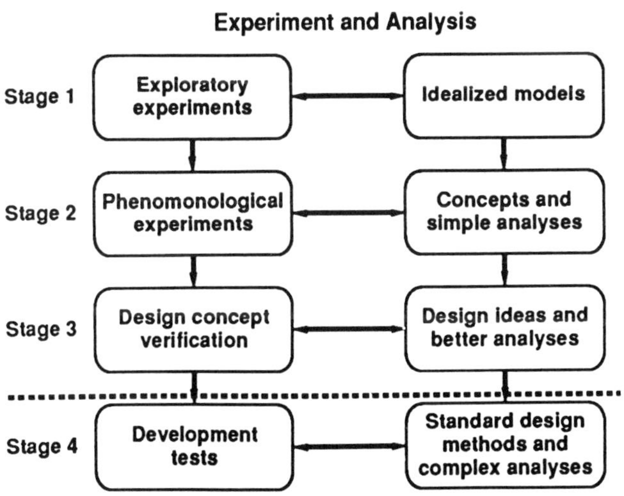

Figure 2. Stages of experiment and analysis.

EXPERIMENTAL METHODS - A number of methods are used to make measurements to determine the sources and magnitudes of aeroacoustic noise on automobiles. The most fundamental method is to test either idealized models or vehicle components in a quiet flow facility, in other words, in a facility where there is a flow of air with extremely low background noise enabling one to isolate a particular noise source by using an idealized model or component in the facility. Several such facilities exist, although they are primarily used for aeroacoustics of aircraft (see, for example, reference 9). Another fundamental way of measuring aeroacoustic effects on automobiles is in a full-scale quiet wind tunnel. There are several semi-anechoic wind tunnels used for automobile testing. BMW has one in Germany and Nissan has one in Japan(10). The DNW tunnel in the Netherlands has both a fully anechoic chamber and the possibility of operating a ground plane for testing car type models in a semi-anechoic situation. In these large quiet wind tunnels, a full scale car can be tested. Testing a full car gives a complete measurement of the total sound generated by the car. However, in order to determine what the sound sources are, more subtle techniques such as a "buildup test" have to be carried out. For example, a car can be completely covered with smoothed outer body work, all openings closed, all seams taped, etc., and sound level measured both inside and outside. Then various real aspects of the car are opened or uncovered and exposed to the air flow, one at a time. A few tests of this sort have been run and the results published, which will be mentioned later in this paper.

A method of testing for aerodynamic noise on automobiles without a wind tunnel is to carry out driving and coast-down tests on full cars. In this case with the engine off, the car is coasted at high speed past microphones or alternatively with onboard microphones. The difficulty with this kind of test is that it of necessity includes the noise due to tires and the underbody and the flow around the tires. If one is interested in the overall sound generated, of course, this is no problem. However, if one is interested in finding the noise of various components, or contributions to noise of various components, then this method poses difficulties. One way of partially eliminating tire and underbody noise is to drive the car next to a low sound barrier, which will then shield the higher-frequency components around the tires and underneath the car from the microphone, which is placed in the barrier's shadow with respect to the tires.(11) Other ways of separating out the aerodynamic noise include using the known velocity and power dependence of the noise components to separate them,(12,13) the insertion of a special acoustic-measuring enclosure behind a window noise transmission area,(14) or the direct measurement of pressure fluctuations at known noise sources.(15) However, these are at best makeshift ways of making such measurements. A possibly more satisfactory way of determining the noise effects of different components of a full car is to use intensity measurements or directional microphones to image the important sources. It is difficult to use intensity microphones in a flow, and it is also difficult to find a microphone system which is directional enough to do fine scale imaging. But the imaging method has the potential of being a useful technique in the future.

Finally, if a true anechoic wind tunnel is not available, an interesting technique has been developed at Fiat which can be used with reasonable success in certain situations.(16) The idea is to first measure the interior noise in the car when there is no external flow, but where an external noise spectrum is generated with loudspeakers which is a representation of what is measured in the wind tunnel flow without any car present. The idea is that this baseline measurement tells how much of the background sound pressure level in the wind tunnel is transmitted through the body of the car into its interior. Then when the car is placed in the wind tunnel flow any increase in interior sound level must be due to additional noise generated by the flow itself. Thus, in any frequency band or in any location inside the car, if the measured sound level is higher than what was measured with no flow but the outside noise, one can calculate how much extra sound pressure is due to the flow alone. In the situations where the interior sound with flow is louder than the reference case, this can be a useful technique. Some results of using this technique are shown later in this paper.

AEROACOUSTIC NOISE ANALYSES

ASPIRATION OR LEAK NOISE - The first of the aeroacoustic sources we will consider is that which is sometimes called aspiration, or perhaps more simply stated, leak noise.(11,16,17) Aspiration or leaks can lead to increased sound level inside the car via two mechanisms, the first of which is the direct generation of sound by air flowing through a leak either into or outside of the car. The second mechanism is the leak allowing sound to propagate from the outside of the car to the inside of the car without being attenuated by the car's structure. In other words, a leak can allow a short circuit of the sound around the panel or glass that would normally introduce a substantial transmission loss between the outside pressure levels and the inside sound levels.

If an actual leak flow occurs in a car, this generally dominates other aerodynamic sound sources. Figure 3, for example, shows the influence of a bad vent window seal on interior wind noise at 110 kph, compared to an automobile with a properly sealed vent window.(1) It is quite clear that above approximately 1000 Hz, we have an increase in sound spectrum level in the car on the order of 10 dB or more, contributing greatly to the dBA levels inside the car. Thus the most important factor in reducing aeroacoustic noise inside the car is the sealing of any leaks which allow air to flow around windows, doors, or through openings in the body work.

A leak air flow generates sound several ways as shown in figure 4. First, if the leak connects to a portion of the outside of the car which supports unsteady pressures due to turbulence or separation, then the leak will cause an unsteady mass flow, generating monopole sound inside the car. Figure 4 shows schematically how a leak around a seal can generate monopole sound due to a fluctuating external pressure at location 1 affecting the mass flow through location 2. On the other hand, even if the leak connects to a steady pressure source and a steady flow of air goes through the opening of the leak, this flow generally separates in a turbulent fashion after the minimum area at location 2 thus generating local fluctuating pressures on various portions of the door jamb, etc. These pressures give force or dipole type sound, which is transmitted into the car. As we have seen, if the monopole sound source exists, it will be the strongest. If it does not exist, but the pressure fluctuations acting on surfaces do exist, the dipole noise will be the strongest. Finally the free turbulence in the air flow will give quadrupole noise but it will be quite small for the Mach numbers which are attained in automotive type leak flows.

The second type of aspiration or leak noise occurs where an opening allows sound that exists outside the car to transmit in as acoustic waves. An example in figure 5 shows the reduction in sound level inside of the car by acoustically treating the ductwork for the ventilation system which connects the inside of the car to the windshield/hood area and allows aerodynamic noise (and perhaps engine noise) from that area to propagate in through the duct.[10]

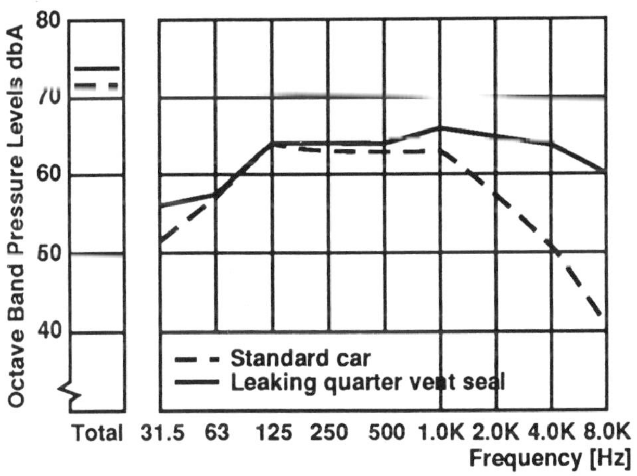

Figure 3. Effect of leaking vent window seal. (After reference 1.)

Once the car is sealed and any intentional ventilation air paths are acoustically treated, we consider some of the fluctuating pressure sources on the outside of the car. One important source is the influence of joints or gaps in the outside body work. For example, where the doors close, where the doors come up next to the door jambs, or between body panels, there are often gaps. The air flow over a gap can generate both broadband and sometimes discrete frequency noise due to the turbulent flow over the gap and unsteady flow involving the cavity and its resonances. In figure 6 we see the interior noise measured in a car at 180 kilometers per hour with the door joints in their standard condition, and then after all the door joints have been taped with a cloth tape to prevent air-flow-generated additional fluctuating pressures in the door joint areas.[1] It should be noted that the tape is not massive or stiff enough to introduce an additional transmission loss for the propagation path through the door joint. What it does is merely prevent the flow from generating additional acoustic pressures from flowing over an open cavity.

The second part of figure 4 shows how sound can be transmitted through a seal, even if there are no leaks. If there is an external pressure $p_1(t)$, this can move the seal back and forth slightly and generate sound on the inside. This will be resisted by the stiffness, damping, or mass of the seal. In common with the question of sound propagation through walls, the best way to reduce sound propagation through the seal area is to have multiple seals rather than to try to increase the mass or stiffness of any one seal.

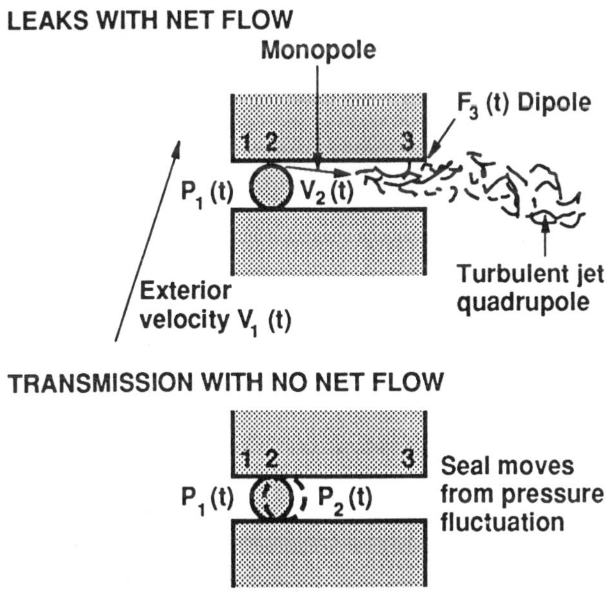

Figure 4. Types of aspiration/leak noise mechanisms.

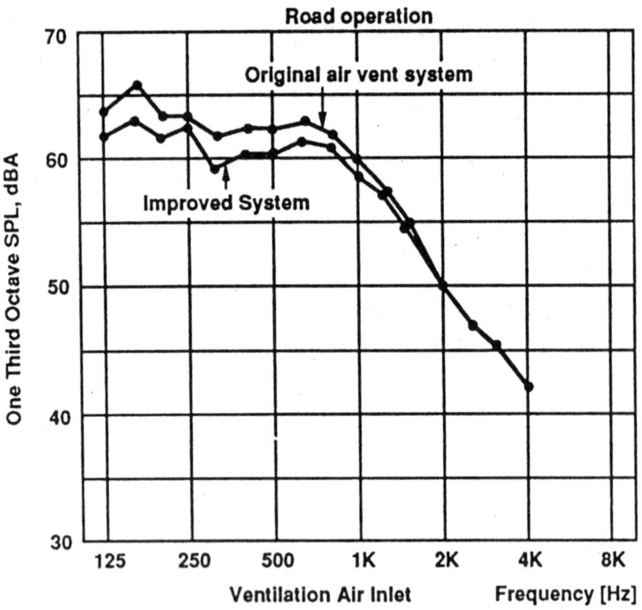

Figure 5. Effects of attenuation in ventilation ducting path. (After reference 10.)

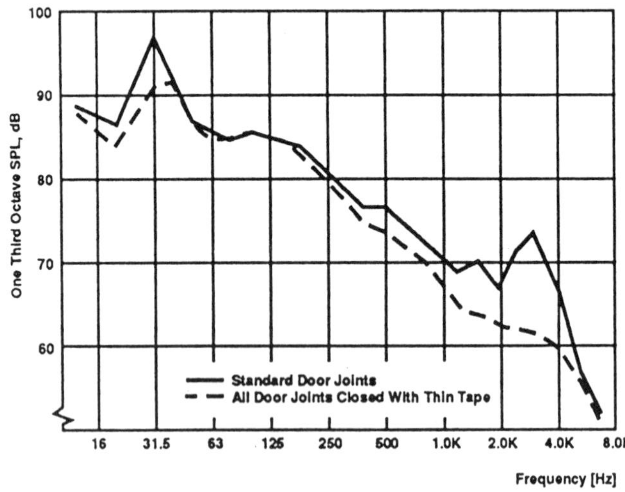

Figure 6. Reduction of noise due to flow over door joint gap openings by covering with low transmission loss tape. (After reference 1.)

Doubling the stiffness or mass will only give a 3 dB increase in attenuation, whereas using two seals, if the seal assembly is damped and will not support transmission or resonances between the seals, can double the attenuation decibel rating.

Figure 7, taken from reference 16, shows some of the numerous locations where aspiration noise sources can occur on a car body. Figure 8 shows some values of dBA contributions obtained by the subtraction technique, both before and after treatment of these sources. In this particular case the two righthand noise sources on the bar graph, the rear door seals and the door top drip area, seem to be the largest aspiration noise sources and the most difficult to treat; even after treatment, these two dominate the other noise sources of the car. Figure 9 shows the modified design for the seal on a drip rail at the top of the door which gave the approximately 5 dB noise reduction as shown in the previous figure. Here we see that the modified drip rail serves two functions: first, it bridges the gap openings so that the aerodynamics cannot generate any extra pressure from the flow over the gap, and it also introduces an additional seal between the outside fluctuating pressures and the inside of the car. Figure 10 shows a one-third octave band distribution of the reduction in interior noise with the improved seal showing 5 dB reductions, especially in the range of 800 to 3000 Hz. The use of such externally-flush seals is an increasingly used aerodynamic noise reduction measure.

PARASITE NOISE SOURCES - In the same way that the drag of a vehicle is affected by the parasite drag added by small extra components, automobile noise is affected by some of the extra components such as the wipers, mirrors, door handles, rain gutters, etc. For example, measurements made near wiper blades on automobiles show as much as a 6 dB increase in the local fluctuating pressures outside the car, whereas near the mirrors, the local fluctuating pressures can be as much as 20 dB higher than with the mirror not present.

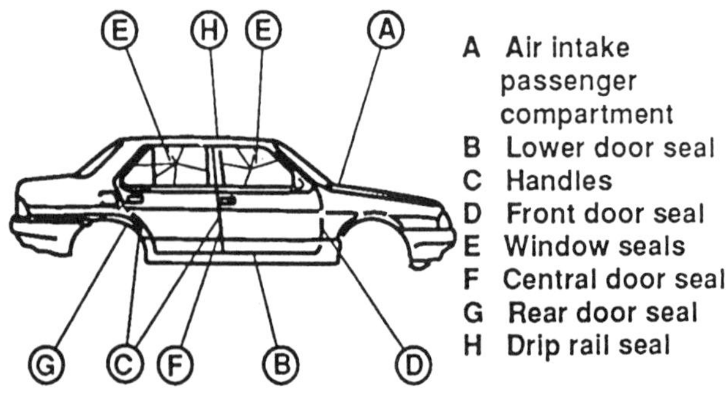

Figure 7. Possible aspiration/leak noise source locations. (After reference 16.)

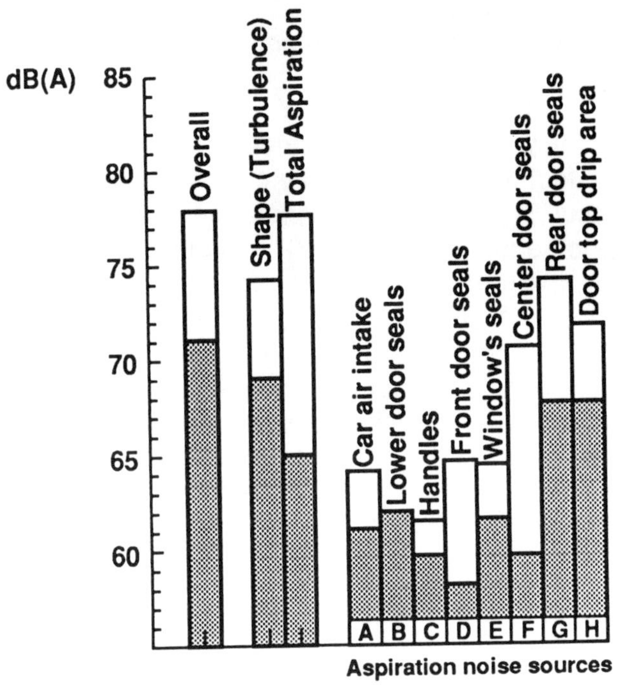

Figure 8. Reduction of internal noise by treatment of aspiration and leak noise sources. (After reference 16.)

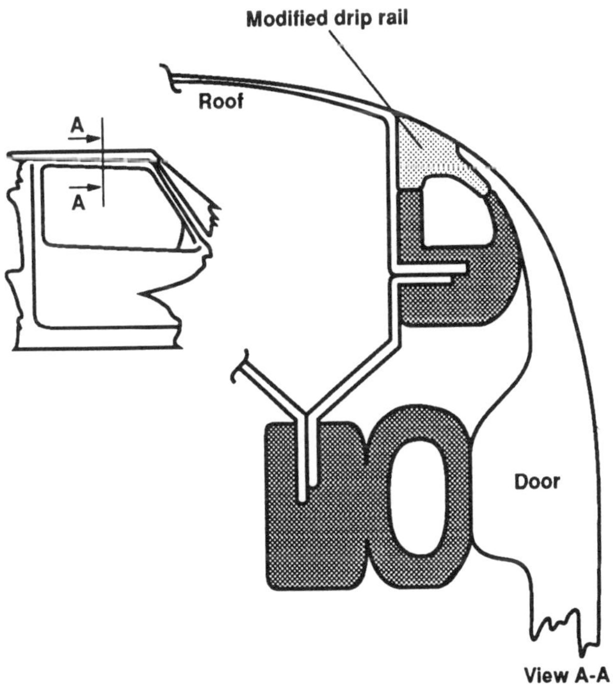

Figure 9. Modification to drip rail to reduce aspiration noise. (After reference 16.)

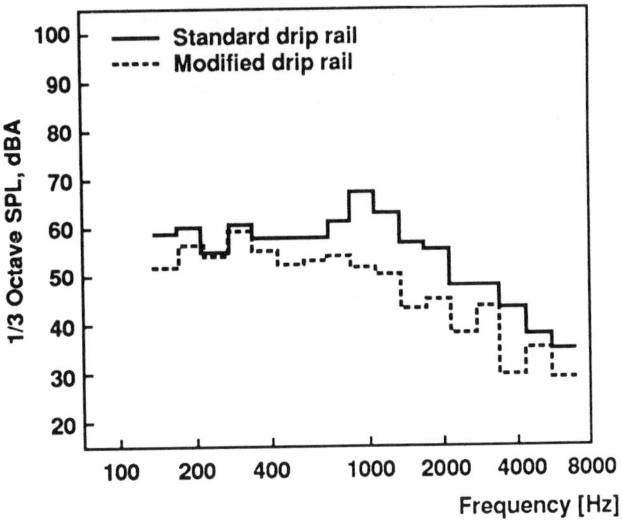

Figure 10. Internal noise spectrum with original and modified drip rail. (After reference 16.)

Similarly, the antenna on a car can generate a vortex street. This vortex street, in turn, generates sound directly from the antenna, which then propagates through the air to the car and can then be transmitted through the car's structure to the interior of the car. But what is generally of even more importance is that the vortex street generates a vibration of the antenna which is then transmitted down to the car structure and this vibration then excites sound inside the car. An example of the inside sound from an antenna due to von Karman vortex street is shown in figure 11.[10] Here the addition of the antenna to an automobile introduces a substantial peak on the order of 5 or 6 dB in the frequency range of about 1 kHz, which was the von Karman vortex shedding frequency for this particular case.

SEPARATED FLOW REGIONS - An example of external noise effects of a separated region and its increased fluctuating pressures are shown in figure 12. The upper curve is for an automobile configuration which had a separated flow right behind the front edge of the hood, whereas in the lower curve, the edge of the front contour of the hood has been rounded off enough so that the flow does not separate there and does not cause additional fluctuating pressures[10]. More effects of separation on internal and external noise will be discussed later in this paper.

UNDERBODY AND WHEELS - Due to the complex shapes and the various structures exposed to the unsteady local flow, the wheel and underbody sections of the car also generate very high local unsteady pressures, which can then be transmitted directly through the floor of the car or diffracted around the side of the car and enter

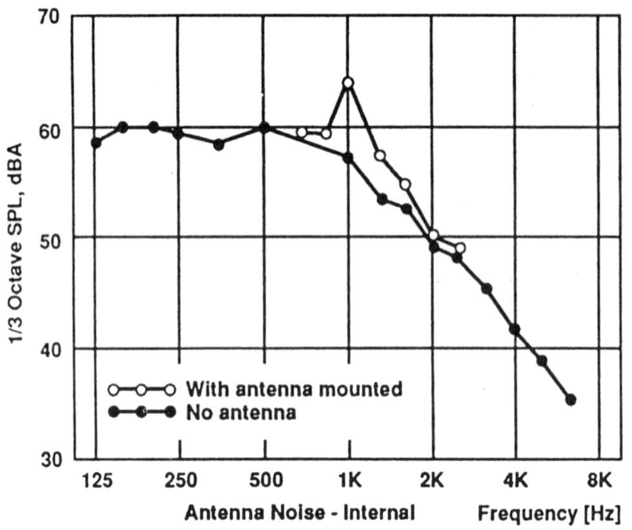

Figure 11. Noise due to antenna. (After reference 10.)

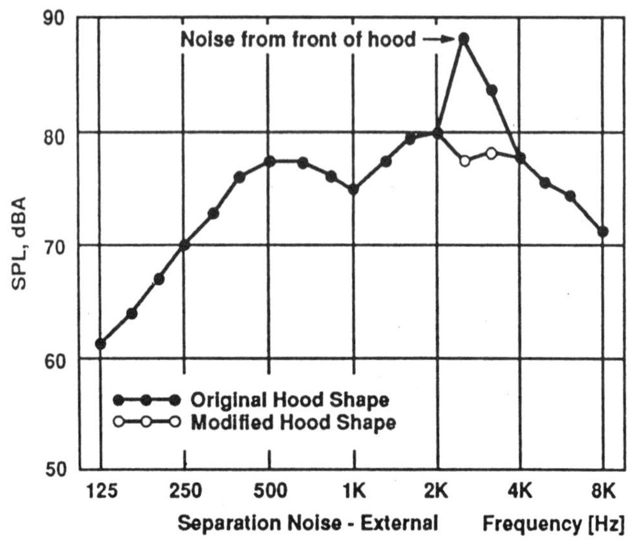

Figure 12. External noise from separation at front of hood. (After reference 10.)

through the other panels. Due to the extreme flow complexity these sources have not been investigated to date.

LOCAL VELOCITY EFFECTS - An extremely important but little appreciated effect is the effect of the local velocity on fluctuating sound pressures generated by separation or other flow phenomena. Since sound intensity generated from pressure fluctuations on a surface is proportional to the sixth power of the velocity, there is a very large increase in sound pressure with velocity. Alternatively, instead of considering velocity, one can consider pressure coefficients. Since the pressure coefficient on the surface of a body is usually controlled by the inviscid flow outside of the boundary layer or outside any thin separated region, it can be used to estimate the inviscid outer velocity V. For the outer flow one can show that:

$$(V/V_\infty)^6 = (1 - C_p)^3 .$$

Then this V will be the characteristic velocity driving any local flows and fluctuating pressures. For example, this external V will determine the order of magnitude of the turbulent velocity fluctuations in a local separation between the surface and the external flow. Thus, for a given turbulent penomena such as associated with a mirror, edge, or vortex, if the undisturbed flow pressure coefficient is minus one, the radiated sound is increased by 9 dB, according to the sixth power law compared to the case where the pressure coefficient is zero. If the pressure coefficient is minus two, which is not unusual in certain parts of a car, as will be shown later, there can be a 14 dB relative increase in radiated sound. Conversely, if the pressure coefficient is positive, as plus one, for example, at the stagnation point, there will be no velocity and therefore essentially no sound generated. If the pressure coefficient is even as high as plus one-half, then there will be a 9 dB relative decrease is sound radiated from that area.

In a similar manner, consider sound pressures generated if an object is placed in a stream which has been slowed by irreversible phenomena upstream such as separation. For example, in the flow underneath a car, near the rear of the car the air has been slowed down by passing over many projections, cavities, and objects underneath the car. This slowed-down air will generate less sound. Even a reduction of the velocity to three quarters of the free-stream value, following the sixth power law, gives a 7.5 dB reduction in radiated sound for a component in the flow. Or if the velocity has been slowed to a half by irreversible phenomena, there is an 18 dB decrease in radiated sound.

Thus it is extremely important in designing the location of items which will cause separation and unsteady pressure fluctuations. They should be placed in locations which have a low velocity either due to a high pressure coefficient in an inviscid flow, or due to air having been slowed irreversibly upstream. Many outside rearview mirrors are placed in poor locations.

CAVITY NOISE - In addition to the cavity noise discussed previously, where flow over small gaps such as joints around doors, etc., can generate external pressure fluctuations which are eventually transmitted into the car, there are also an extremely large noise source associated with actual openings such as open windows or open sunroofs. As shown in figure 13, there are two general types of sound generated by flow over open cavities. The first is broadband noise, which can be looked at as being broken down into several sources. If the boundary layer is turbulent upstream of the cavity, then when this turbulence passes over the front edge so-called "trailing edge noise" is generated. This has been analyzed extensively in conjunction with rotor flows in helicopter noise.[18] After the flow leaves the front or "trailing edge" of the cavity, a turbulent free shear layer develops which then impinges on the rear of the cavity generating "leading edge noise". Here the exact shape of the edge would be expected to have some influence on the sound radiated. In many cases, in addition to the broadband noise, a tone noise can be generated. In this case the turbulent shear layer becomes unstable and generates coherent structures, which when they interact with the rear of the cavity, generate strong tones. In addition, this interaction can generate disturbances which propagate forward through the cavity and influence the shear layer in the region where it is separating from the front of the cavity, giving a feedback and resonance type phenomenon. Tones can be excited which are related to the geometry of the cavity. This feedback can be acoustic or it can be convective, and it can involve the ordinary acoustic modes of a cavity or Helmholtz type resonances of the cavity. Reviews of cavity noise are given in references 19, 20, 21, 22. In automotive applications, noise control of sunroofs either tries to assure as smoothly matched flow as possible where the boundary layer reattaches to the rear edge of the sunroof opening, or in some cases, uses deflectors in order to assure that the shear layer will not impinge directly on this rear edge. Resonances of side window flows have also been investigated[23,24] but have generally not been of primary interest in design.

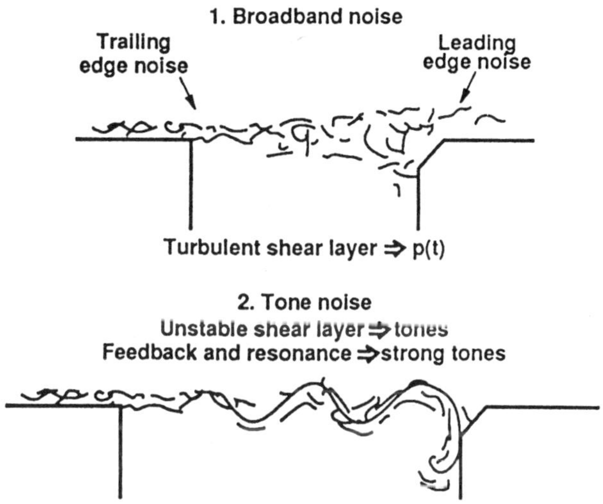

Figure 13. Noise mechanisms for flow over large open cavities.

PREDICTION OF NOISE

The prediction of noise from turbulent and unsteady pressures on the outer surface of an automobile can be separated into two separate problems: interior noise and exterior noise.

INTERIOR NOISE PREDICTION - First consider the prediction of interior noise due to external fluctuating pressures. The problem divides into two parts. The first is the aerodynamic problem of finding the external fluctuating pressures on the outside surface of the body. Then, once that is known, the second part is determining how much of this pressure fluctuation is transmitted into the interior of the car through the panels making up the outer surface of the car. This is a panel transmission problem.

External Pressures - In order to find the external fluctuating pressures, we must understand the aerodynamics well enough to be able to predict the fluctuating pressures as a function of position and time over the body. Both the amplitude and spectral characteristics of the fluctuating pressures will be of interest as well as the correlation lengths of the pressures over the body surface. In the case of excitation of structural waves in the body, we may have to know the correlation lengths defined for motion at the wave number of the bending waves in the panel.(11,25) Turbulent fluctuating pressures are generated everywhere on the car due to the basically turbulent boundary layer existing over the overwhelming majority of the surface of a car at normal operating speeds. However, by far the strongest fluctuating pressures are generated in separated flow regions, and in particular, in regions where vortex separation occurs. Figure 14 shows a flow pattern picture taken from reference 26 showing the visualization of a flow pattern on a front side window of an automobile. Here we see that the flow has separated at the A-pillar (which is the structural member dividing the windshield from the side window), and then has reattached along a diagonal line further back on the window. We also see a secondary stagnation line of the reattached flow indicating a more complicated vortex structure. Figure 15 shows a schematic representation of the flow field found in this region, including the reattachment line and the stagnation line where the vortex flow separates on the upstream side.(26)

The pressure distribution measured on a front side window in such a flow is shown in figure 16.(26) Here we see pressure coefficients near the A-pillar in the region marked AA approaching minus two. Almost all of the separated flow region has a pressure coefficient of at least minus one. Recalling the earlier analysis, a pressure coefficient of minus two gives a relative increase of fluctuating pressures of about 14 dB, while a pressure coefficient of minus one gives a relative increase of fluctuating pressures of about 9 dB. Thus it is not surprising to find that these vortex flows on the side windows are major sound sources within a car. As we will see later, because the glass is basically an undamped single layer, there is a limitation on the improvement in transmission loss that one can expect for this glass area. Thus the side glass is a strong contributor to the interior noise field.

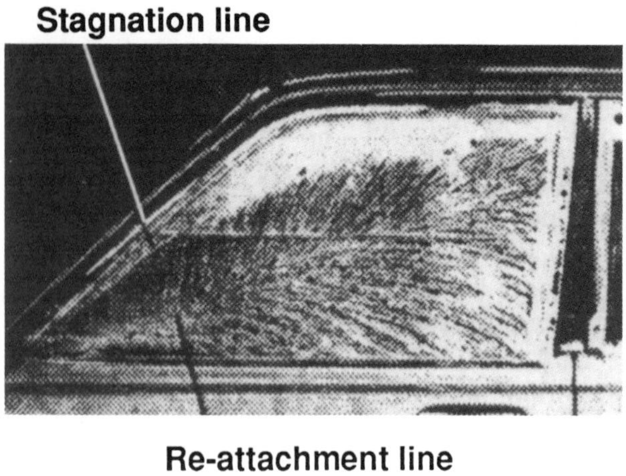

Figure 14. Flow pattern behind A-pillar on front side window. (After reference 26.)

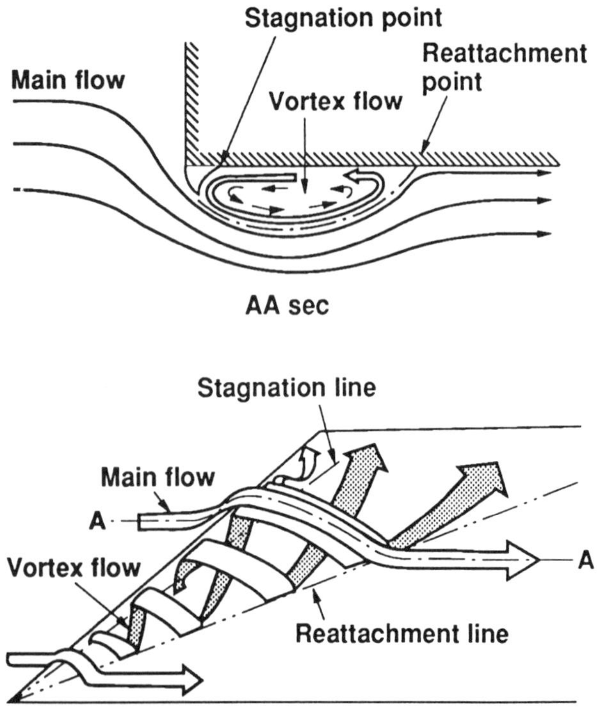

Figure 15. Schematic of flow field behind A-pillar on side window. (After reference 26.)

In order to better understand this kind of flow and noise generation, Stapleford and Carr[27] investigated the flow patterns for square-edged boxes at various yaw angles to the flow. Figure 17 shows the separated flow around a square-edged box at zero yaw angle. The flow separates at the upstream edge, and then reattaches about two-thirds of the way back along the side of the body as shown. The reattached flow moves forward from the reattachment point then separates at a point about a tenth of the way back from the front edge. In this study a number of pressure transducers were placed in the body surface to measure the fluctuating pressures. Figure 18 from the same study shows the square edged block at a 30 degree yaw angle, at which angle it supports vortex flows very similar to those occurring around the A-pillar and side window on an automobile.[27] Here too, pressures were measured with pressure transducers placed on the surface.

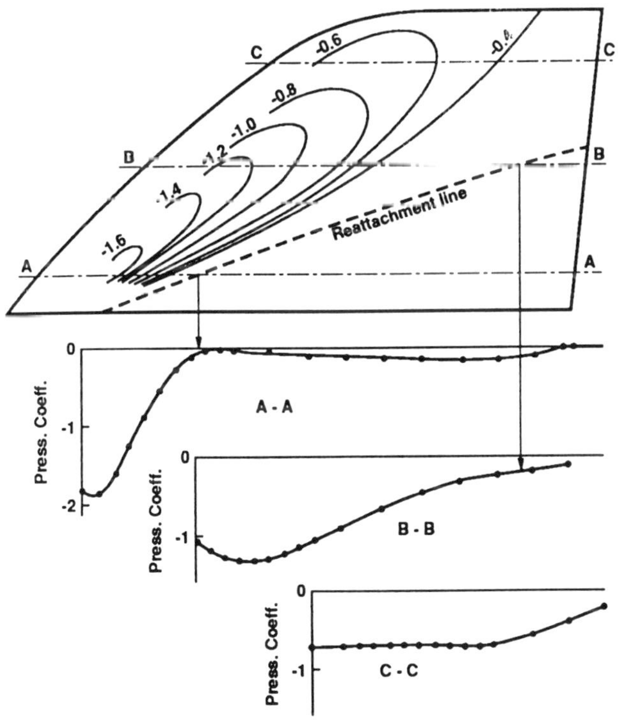

Figure 16. Pressure distributions on side window behind A-pillar. (After reference 26.)

Figure 19 shows some measured external noise spectra for different flows from reference 27. In this particular experiment the tunnel ambient background noise was not completely negligible. It is shown as the lower curve in the figure. The most interesting curves are the upper ones. The highest one, which is in underneath the vortex separation, shows very large fluctuating pressure levels, in this case almost 20 dB above the background noise level; these pressure fluctuations are generated due to the turbulent separated flow of the vortex "core". The turbulent pressures are this large because the vortex increases the local velocity and shear, which then leads to larger fluctuating pressures in its turbulent core.

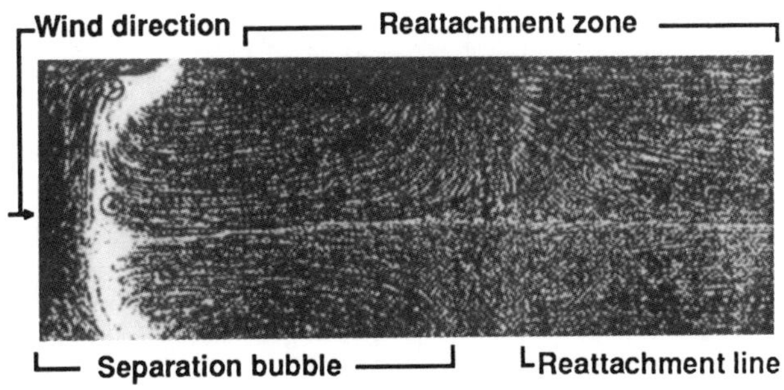

Figure 17. Flow pattern over sharp edged rectangular shape at zero yaw angle. (After reference 27.)

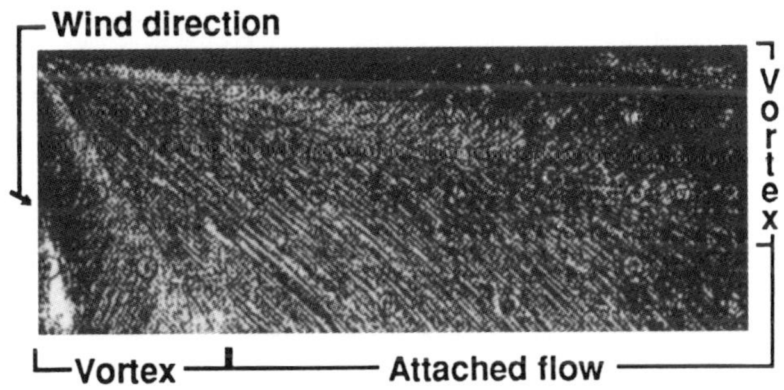

Figure 18. Flow pattern over sharp edged rectangular shape at 30° yaw angle. (After reference 27.)

The next lower curve shows the high fluctuating pressures at the reattachment point. Here the pressures are large because the turbulence is very strong in the shear layer between the external flow and the separated flow, and this turbulence impinges directly on the body at the reattachment point. The next curve down is from a pressure transducer in the quasi-two-

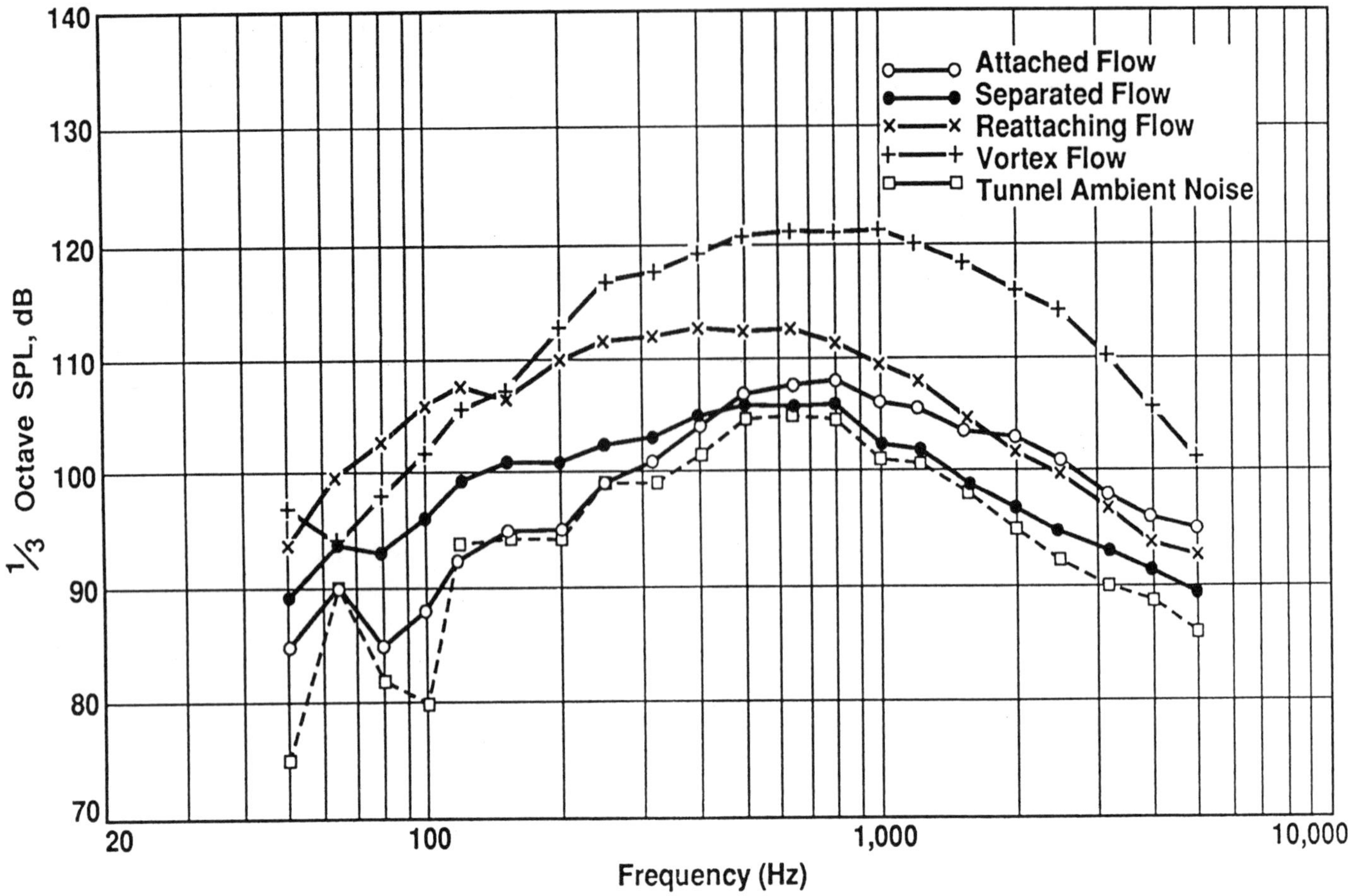

Figure 19. Measured pressure spectra for flows of figures 17 and 18. (After reference 27.)

dimensional separated flow region. Here the pressures are fairly large at low frequencies, due to the large scale turbulent structures in the separated region, although at the very high frequencies, because of the weak small scale velocity gradients in the separated region, they are slightly lower than even the attached flow case. In the attached flow case the small scale turbulence is on the order of the boundary layer thickness, and can be stronger than the corresponding small scale turbulence in the separated region, which has low average velocities and gradients.

Figure 20 shows a comparison of surface pressure fluctuations on a production car as measured by Dobrzynski et al. and compared to similar measurements made on a low-drag car.[1,11,25] Starting from position 1 on the front hood we see only minor differences between the fluctuating surface pressures for the production car, the low drag car, and an undisturbed wall boundary layer. However, in position 2, under the A-pillar vortex, there is a nearly 20 dB increase in surface pressure fluctuations due to the turbulent vortex. In this case the low drag car has almost the same values, showing that the A-pillar vortex separation had not been controlled in the low drag car. This was because the A-pillar vortex was not as important to the drag of the car as it was to the acoustics. On the other hand, Dobrzynski et al. later modified the low drag car somewhat to reduce the A-pillar separation vortex producing the lower curve as shown. The measured surface pressure fluctuations behind the A-pillar were reduced to become comparable to an undisturbed wall boundary layer.

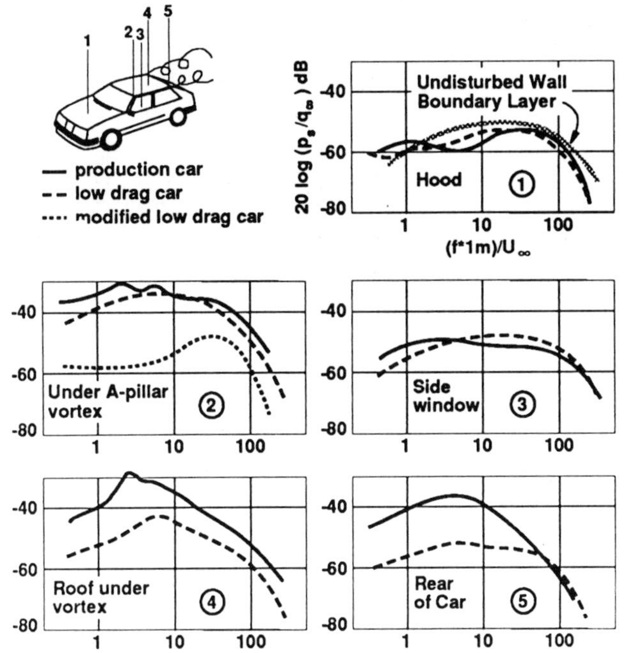

Figure 20. Some measured pressure spectra on automobiles. (After reference 1.)

Position 3 is well back along the window, away from the A-pillar vortex, and only a moderate increase in surface pressure fluctuations is observed at that position, whereas position 4 on the roof is under the location of the continuation of the A-pillar vortex. Here, the production car has a much higher value than the low drag car due to reduction of the vortex strength in the low drag car, presumably from a smoother roof-side window junction. At the rear of the car there is a large increase of fluctuating pressures, particularly at low frequencies, due to the large scale separation at the rear of the car. Here the design of the low drag car, which paid particular attention to reducing the separation at the rear of the car and its wake, gives a very large reduction in surface pressure fluctuations.

Prediction of External Fluctuating Pressures - Until now, researchers have not tried to predict these fluctuating pressures on automobile bodies due to the complexities of the flow field. However, some progress has recently been made in aeroacoustic research on noise due to tip vortices on rotor blades[4,28]. Methods have been developed for satisfactorily predicting surface pressure fluctuations in separated flow regions. The methods are based on the similarity of the fluctuating pressures for flows with similar turbulent flow structures. A separated turbulent flow will be determined by the flow geometry, the scale of the flow separation L, the maximum flow velocity V_m driving the shear flow, and the associated maximum dynamic pressure q_m. Based on these parameters, the measured pressure spectra in similar flows can be normalized based not just on dimensional reasoning but by also accounting for the physics of the turbulence. These methods will be

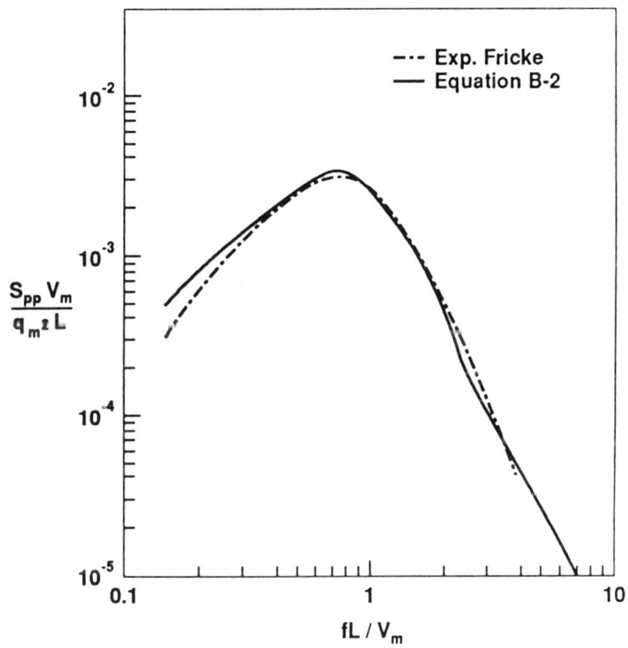

Figure 21. Fit of normalized spectrum to pressure fluctuations under two-dimensional separation. (After reference 4.)

reviewed and then applied to predicting the fluctuating pressure behind the A-pillar.

A measured fluctuating pressure spectrum is a function of frequency. The pertinent nondimensional frequency is defined as

$$\hat{f} = \frac{f\,L}{V_m}$$

with the Strouhal scaling based on the scale of the separation L, and the maximum external velocity V_m, which for a recirculating flow near a wall determines the time scale of the largest turbulent eddies. The pressure spectrum is then normalized as

$$\hat{S}_{pp} = \frac{S_{pp} V_m}{q_m^2 L}$$

by noting that the pressure fluctuations will be expected to scale with the maximum dynamic pressure $q_m = \frac{1}{2} \rho V_m^2$ and the frequency with the Strouhal scaling discussed above.

Figure 21 shows some results of references 4 and 28, which analyzed the pressure fluctuations produced by the formation of turbulent tip vortices on rotor blade or wing tips. A properly normalized spectrum is fit to the pressure fluctuations behind a two-dimensional separated step flow, as for example might occur behind a rain gutter with a flow normal to it. The curve fit shows an excellent collapse of the aerodynamically-related data in this case. These results could be applied directly to such separated regions.

As most of the separated flow regions on the side surfaces of the car are of the vortex type (for example around the A-pillar), the results of the previous two-dimensional normalization of a spectrum must be modified to fit to a cross flow/vortex flow case or data on three-dimensional vortex flow pressure fluctuations must be normalized. Figure 22 shows how the two-dimensional data was initially adapted to the tip vortex case, where the maximum velocity over the separation bubble was used as the normalizing velocity in the dimensionless curve fit. On the other hand, data for separated vortex flows was available from early tests (from when the Concorde supersonic transport was being developed) giving the pressure fluctuations beneath the rolled up vortices on a delta wing at angle of attack. Figure 23 shows a curve fit to normalized spectra from a variety of data for such rolled up turbulent vortices.[4] Here again the maximum velocity, in this case the three-dimensional maximum velocity magnitude around the vortex, is used to carry out the normalization with reasonably good results.

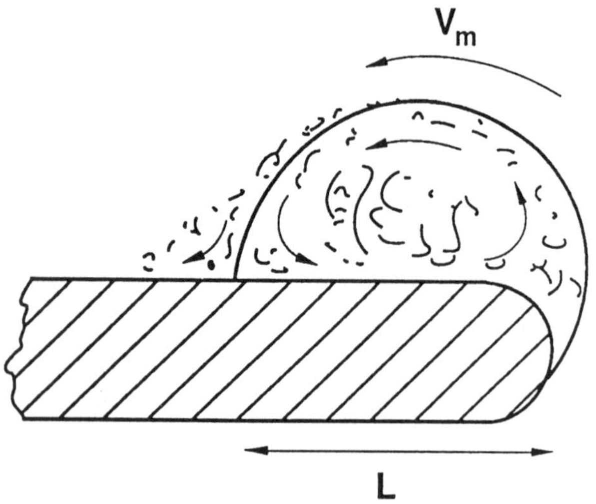

Figure 22. Cross flow plane model relating vortex flow separation to two-dimensional separation.

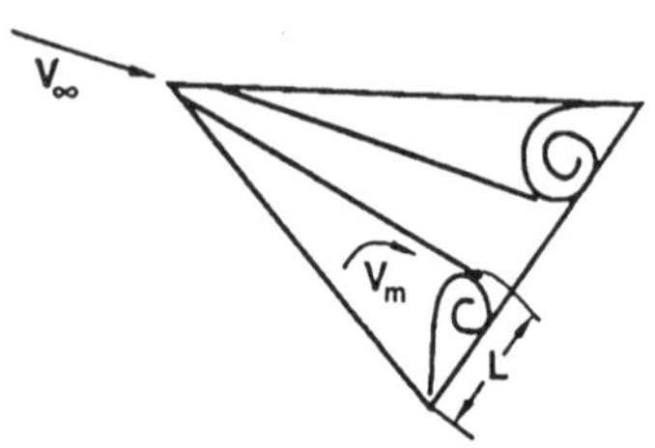

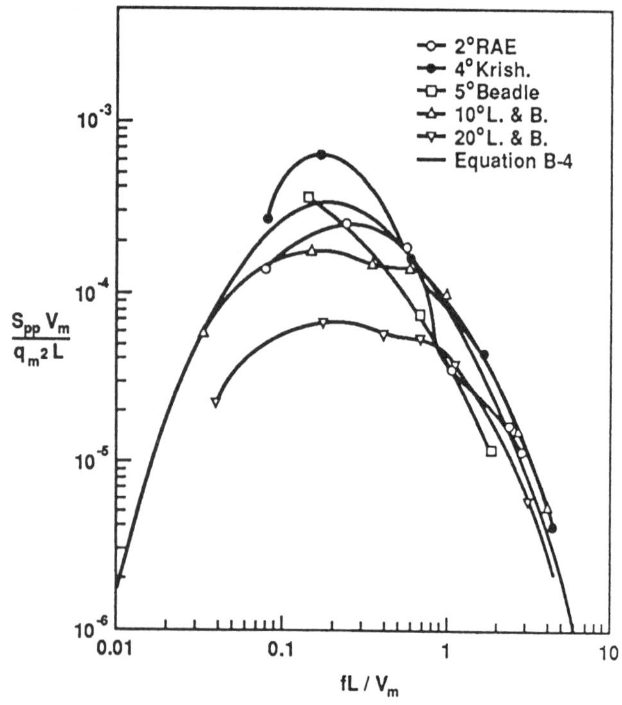

Figure 23. Fit of normalized spectrum to pressure fluctuations under vortices on delta wings. (After reference 4.)

This pressure fluctuation prediction method was applied to the current study of automobile aeroacoustics for the case of the separated vortex flow behind the A pillar, which is a primary aerodynamic noise source region. The results of the following analysis, figure 24, show good agreement qualitatively and quantitatively with the experimental normalized pressure spectra of reference 11. The analysis and normalization procedure were carried out in the following manner:

Experimental static pressure data from Watanabe, et al., reference 26 (see figure 16) was used to characterize the maximum velocities in

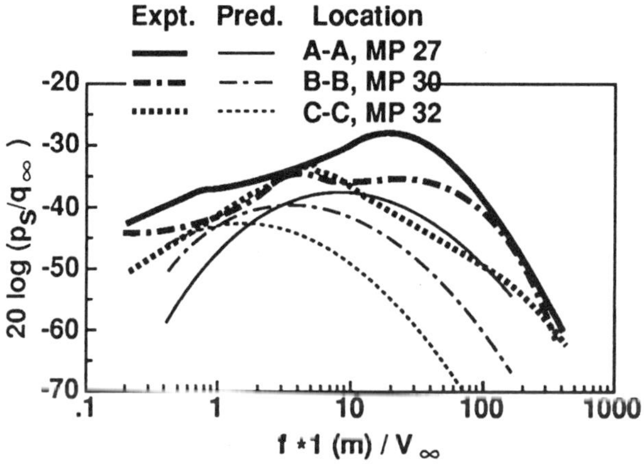

Figure. 24. Comparison of predicted dimensionless alternating pressure spectra with experiments of reference 11.

the vortex flow behind the A pillar. Since the external part of the vortex flow will have only small losses, the pressure coefficient is related to the velocity by Bernoulli's equation giving $Cp = (1/2)\rho[1 - (V/V_\infty)^2]$. Thus the minimum (most negative) Cp will determine V_m. The maximum velocity over the vortex separation bubble was determined from figure 16 for three positions corresponding to the measuring points of the fluctuating pressures on the different car of Dobrzynski in reference 11. The maximum velocities were calculated along the three sections: A-A, B-B, and C-C. These velocities correspond to microphone positions 27, 30 and 32 of Dobrzynski (see figure 25). The length scales, L, were approximated by comparing the vortex width of the three sections to window dimensions of a typical car which was presumably similar to the cars of Wanatabe and of Dobrzynski . A sensitivity analysis showed a ± 30% variation of the length scale to be relatively unimportant as the normalization process and the dB units of acoustics tend to mitigate any error in L.

Equation B-4 of reference 4 was used to predict a normalized spectrum of pressure fluctuations under the A pillar vortex. It is:

$\log_{10} \hat{S}_{pp} = -3.475 - 1.654 (\log_{10} \hat{f} + 0.82)^2$;
for $\log_{10} \hat{f} < -0.82$

$\log_{10} \hat{S}_{pp} = -3.475 - 0.984 (\log_{10} \hat{f} + 0.82)^2$; f
for $\log_{10} \hat{f} > -0.82$

S_{pp} was calculated with these equations for frequencies corresponding to the preferred 1/3 octave band center frequencies. In order to compare to the form of the experimental pressure data given by Dobrzynski,[11] the third octave fluctuating pressure magnitude p_s was calculated from:

$$p_s = \sqrt{S_{pp} \cdot \Delta f}$$

where Δf is the bandwidth corresponding to the 1/3 octave center frequency. These calculations were carried out for each of the three velocity and length scales corresponding to the different measurement location. The resulting pressure versus frequency curves were normalized to correspond to how Dobrzynski presented his experimental data, and plotted in figure 24.

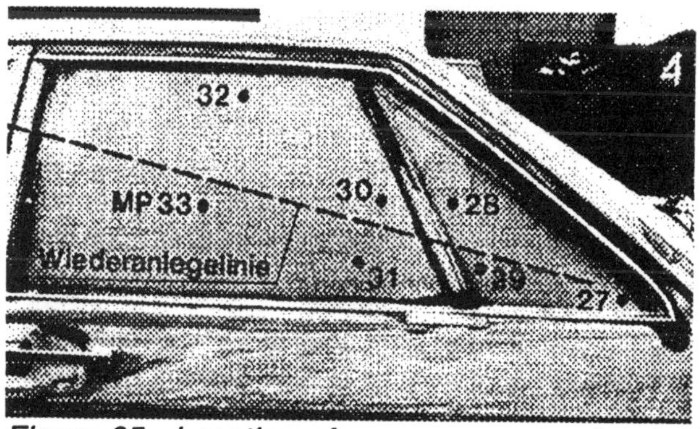

Figure 25. Location of measurement points in the A-pillar vortex regtion, showing delineation of vortex flow region (from reference 11).

A comparison of the predicted spectra with the experimental spectra shows good qualitative agreement, accurately predicting the shape of the pressure spectra and their variation with location in the A-pillar vortex. The predictions show how the pressure fluctuations decrease in magnitude and frequency as the vortex travels upward and rearward along the window. Quantitatively the predictions are also reasonably good, especially considering that the static and fluctuating pressure data were taken from different automobile shapes.

These methods show promise for future development for predicting the pressure fluctuations on other separated and vortex flow regions on automobiles including outside mirror wakes, etc. The prediction of the fluctuating pressures in base flow wake regions of an automobile still poses a difficult problem. There can be some coherence as well as statistical fluctuations in turbulent wakes,(29) but there is a long way to go before predictions of unsteady base flow pressures can be made. These large scale low frequency pressure fluctuations can be important to ride smoothness at high speeds.

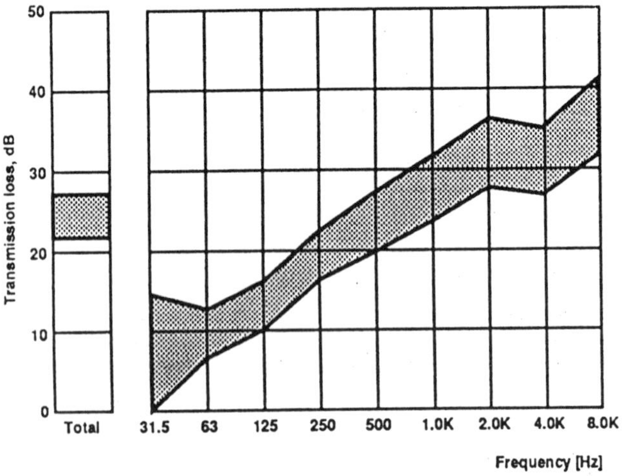

Figure 26. Approximate range of average transmission loss for automobiles. (After reference 1.)

Transmission Into the Interior - Once the pressures on the outer surface of the car are known, the problem remains to find how much of these pressure fluctuations are transmitted through the panel into the car interior. This problem has two aspects, first the transmission loss throught the panel, and secondly, the fairly simple problem of the reverberation or excitation

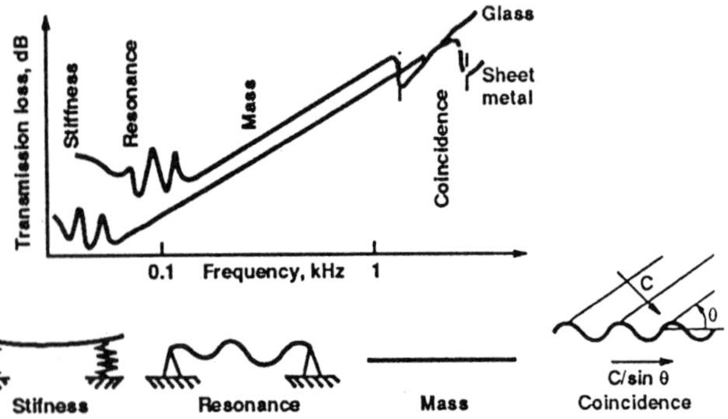

Figure 27. Panel transmission loss mechanisms.

of modes inside the car cavity. A range of transmission losses for present day production cars given by Buchheim et al.(1) is shown in figure 26. Here we see that transmission loss at high frequencies is about 30 dB, whereas at low frequencies it drops to perhaps 10 dB, indicating the difficulty in finding a structure which can give a high transmission loss at low frequencies. The reason for this difficulty can be seen by looking at figure 27, which shows the transmission loss characteristics as a function of frequency for single thin panels. As indicated on the figure, there are four regions of different physical mechanisms for transmission loss by thin panels. First of all, at the very lowest frequencies, the panel motion response to the pressure fluctuations is that due its stiffness alone. In this region, if the panel can be made structurally stiffer or its mountings can be made stiffer, the transmission loss can be improved. However, there is a limitation on the stiffness available for ordinary materials and structures for car bodies. The mass region of transmission loss is one where the mounting of the panel is not important, but it is the mass per unit area of the panel which determines its motion and response to an external pressure fluctuation. In this case doubling the mass can give a 3 dB reduction in transmitted sound. However, this adds substantial weight to the car structure for a small dB gain; it is much more efficient to use multiple panels with sound absorption between them to improve transmission loss. Between the stiffness and mass regions is the resonant region where the mass and stiffness combine to allow resonances or modal vibrations of the panel structure. Here, damping of these resonances is of primary importance. Finally, at higher frequency, the coincidence phenomena can occur, where if a wave is incident on a panel surface with the geometry such as to generate a trace velocity and a trace wavelength on the panel which coincide with the propagation of structural bending waves in the panel, we have a large dip in the transmission loss. In order to increase the frequency at which the coincidence dip occurs it is desirable to have massive but limp panels.

There is one published overall analysis of the transmission of sound through automotive structures into the interior; it has been carried out by Dobrzynski.[11] His analysis is based upon Cremer and Heckl's analysis of radiation from bending waves in a panel.[30] The basic assumptions made are (1) point loading, where the correlation lengths are small compared to the size of the panel; (2) negligible loading of the vibrating plate by the air; and (3) the plate is free to rotate at its supported edges. Once the plate vibration is known, a Green's function solution can be used to find the radiated acoustic waves. The radiation analysis considers two mechanisms shown in figure 28: that of the effect of a point load on an infinite plate, and that involving reflections of waves from supported edges. In the first mechanism, the point load on an infinite plate (which is modeling the small correlation area loading by fluctuating pressures) introduces a local singularity in the subsonic bending waves which would otherwise not radiate sound into the surrounding fluid. The point load is given by

$$F_s^2 = p'^2 A \ell_x \ell_y$$

where A is the panel area and ℓ_x and ℓ_y are correlation lengths.[11] Then the radiated power can be shown to be[30]

$$W = \rho F_s^2 / (2\pi c \sigma^2)$$

where σ is the panel mass/area. In the second mechanism, the subsonic bending waves, which would not radiate on an infinite plate, are incident upon supported edges of total length S, at which point they are reflected. This discontinuity in their propagation at the edge leads to radiated power given by

$$W = \rho F_s^2 S / (4\pi c \sigma^2 A k \eta),$$

where k is the wave number and η a loss factor[30]. Thus if the value of the effective point load magnitude, F_s, is known, one can evaluate the radiated sound from these two mechanisms. Dobrzynski evaluates F_s in terms of the fluctuating pressure magnitude and the correlation lengths of the pressure fluctuations, accounting for their resonance with lightly damped plate bending modes. Then the two mechanisms can be added, and then finally, assuming a reverberant sound field inside the automobile, the interior sound pressure level is given by

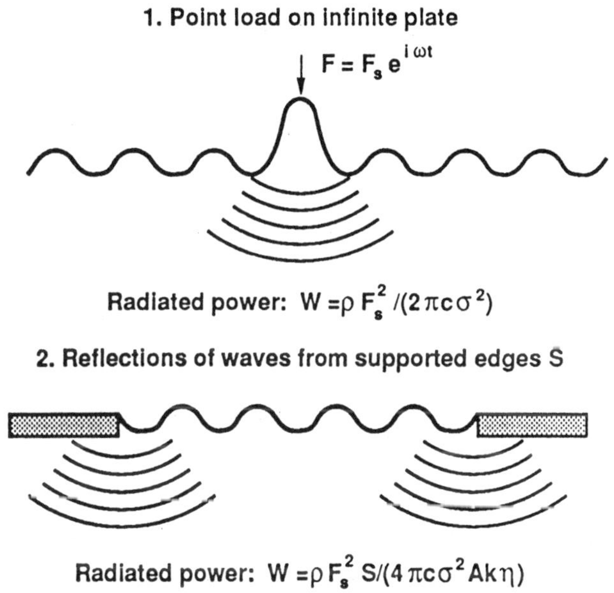

Figure 28. Mechanisms for radiation of sound from elastic panels excited by point loads. (After references 11 and 30.)

$$p^2 = 4\rho c W_{total} / (\bar{\alpha} A_{int})$$

where $\bar{\alpha}$ and A_{int} are the interior's average absorption coefficient and total absorbing area respectively. Dobrzynski finds that most of the reduced power is due to the supported edge effect, implying that few stiffeners should be used.[30]

The analysis makes a number of assumptions, and the evaluations of the correlation lengths, loss factor, and other parameters are not without approximation. For example, a damping factor, D, has to be estimated. Figure 29 shows a comparison between overall measured and calculated A-weighted interior noise level spectra for two values of the damping factor, D.[11] It can be seen that even with this wide range of D, the agreement is less than ideal. Figure 30 shows calculated interior noise contributions from different sections of the car in order to demonstrate which pressure fluctuations-area combinations dominate the interior sound field of an automobile.[11] Here we see that the primary contribution to the interior sound field is from area A_4, which is window directly behind the A-pillar, followed by the sound from the next rear side window, area A_5. The reasons that these areas dominate are two-fold. First, the pressure fluctuations on the outside of the car are the strongest on those side windows due to the A-pillar vortex. Secondly, there is only a single layer of glass with fairly low damping possible in the window area. By contrast, the other portions

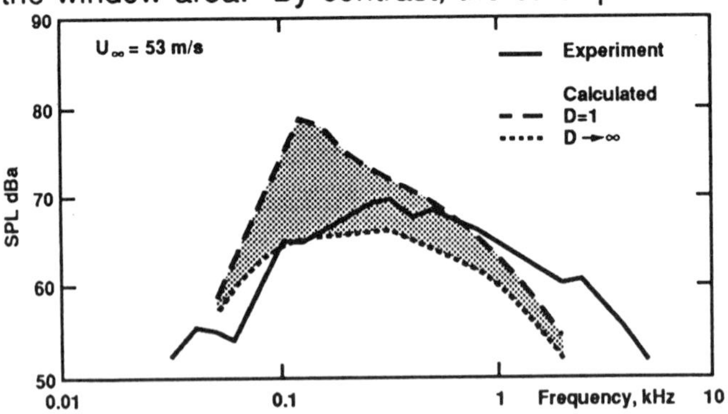

Figure 29. Comparision of measured interior noise spectra and those calculated by method of reference 11 for different values of damping factor D.

of the structure of a car are usually made of several layers, often with some kind of absorption between the layers, giving us a double wall, rather than a single wall as we have at the windows. One might ask, why not use double windows here. This indeed would be more beneficial than using thicker windows, but because of the optical and structural requirements of windows, it is difficult to decouple the two windows, possibly evacuating the cavity between them. Even in the evacuated cavity case, there would be a transmission of bending waves at the edges of the window which might partially defeat the evacuation. It is much more practical to minimize the fluctuating pressure source, the A-pillar vortex and separations.

Another result of interest from figure 30 is that at the very low frequencies, say 100 Hz or below, most of the acoustic energy in the car comes from the rear window, which is exposed to the large scale low frequency fluctuations in the wake of the rear of the car. In fact, much of the low frequency vibration of automobiles at high speeds is probably due to unsteady pressures associated with the overall wake and underbody flow. Figure 31 shows a comparison between measured surface pressures on the outside of the car body underneath the car body floor, compared to the side window area. Here we see that, again, at the very low frequencies, the large scale separations associated with the flow underneath the car are the strongest contributors, whereas at the higher frequencies the pressure fluctuations associated with the vortex flow on the side windows is the strongest contributor. While the floor of the car can be designed with carpeting, etc., to have quite substantial transmission losses at moderate to high frequencies, it is still difficult to obtain a high transmission loss at low frequencies. Thus much of the low frequency noise transmission problem inside of automobile bodies is unsolved. At low frequencies the modeling of the internal pressure field must be modal in nature, as opposed to the reverberant or diffuse sound approximation. Some analyses of these low frequency cavity modes of automobiles are given in references 31 and 32. Much of this work has been motivated in the past by vibration noise from engine and drive train, but as shown here, aeroacoustic excitation can also be important.

EXTERIOR NOISE PREDICTION - The first part of the prediction of external noise is finding the aerodynamic flow and the resulting fluctuating pressures on the external surface of the vehicle. This was discussed above along with internal noise prediction. For the second part we need to predict external noise from these fluctuating surface pressures. Once these

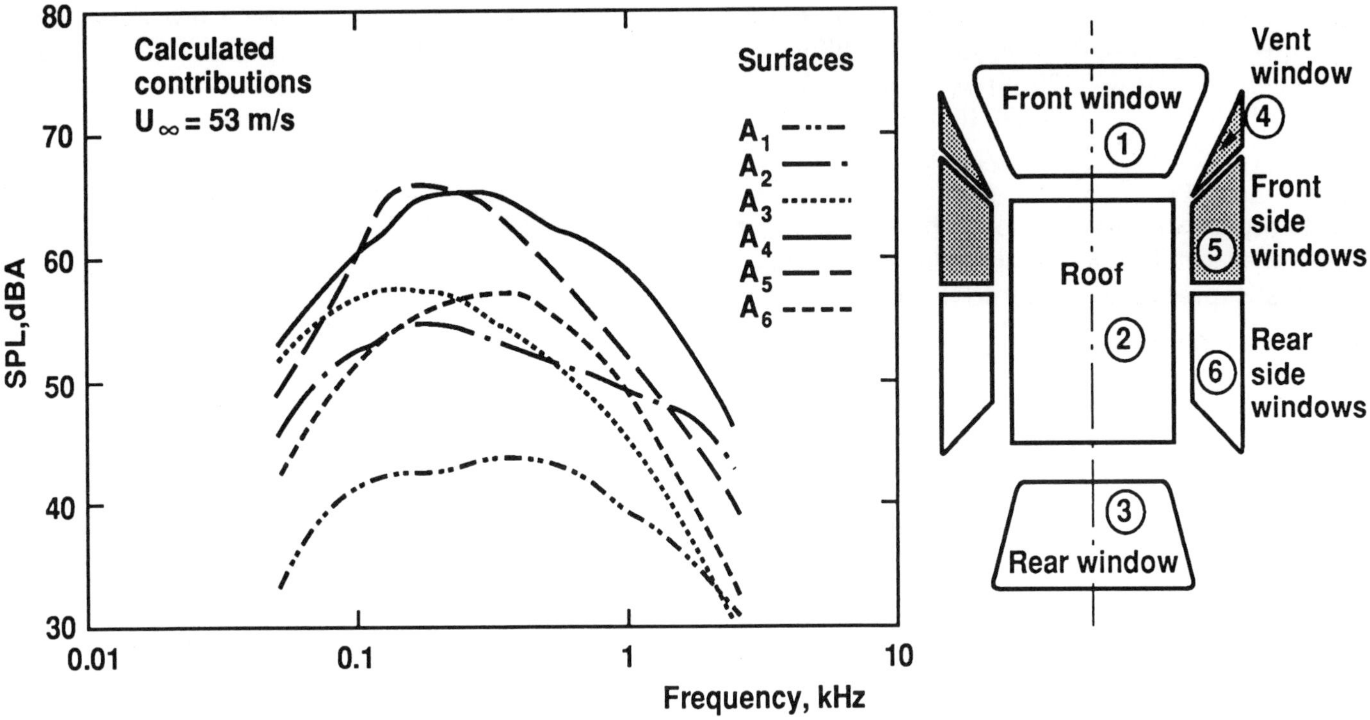

Figure 30. Calculated interior noise contributions from different parts of automobile body. (After reference 25.)

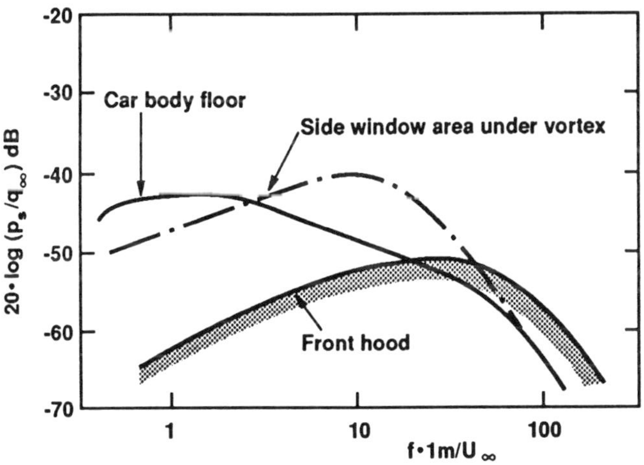

Figure 31. Measured pressure spectra under car floor and on side window under vortex separation compared to undisturbed boundary layer on front hood. (After reference 25.)

pressures are known, they radiate directly to the external flow essentially independently of the stiffness, rigidity, or transmission loss of the surfaces themselves. As far as the external flow is concerned, these surfaces can be considered infinitely rigid.

For the primary frequencies of interest in automobile noise, one can consider the correlation length, ℓ, of the unsteady pressures, which is on the order of an eddy size, to be small compared to the wavelength of the sound, λ. This means that one can assume that each correlation area of the surface of the automobile is an independent source, and we can then find the intensity radiated from each area independently and add the intensity from the various correlation areas. For low frequencies, where the correlation area is larger than the body size, we only have one coherent source radiating, and the acoustic pressure can be given as

$$p_{tot} = \frac{1}{4\pi\rho cr} \frac{\partial F_{total}}{\partial t} \cos\theta \propto p'L^2$$

where the total force is the integrated force over the whole surface of the body. Then the acoustic pressure squared is $p_{tot}^2 \propto p'^2L^4$. On the other hand, for the more important high frequencies where the correlation length is small compared to the body size, then we have a number of independent correlation lengths radiating. The pressure from each one can be given by

$$p_i = \frac{1}{4\pi\rho c r}\frac{\partial F_i}{\partial t}\cos\theta \propto p'\ell^2$$

but if we add the contributions up, we find that the mean squared acoustic pressure is the sum of acoustic pressures squared from each of the contributions, which is $p_{tot}^2 = \Sigma\, p_i^2 \approx Np'^2\ell^4$

where N is the number of correlation areas on the body. N can be approximated by body area divided by the correlation length area, giving the result

$$p_{tot}^2 \propto \frac{L^2}{\ell^2}p'^2\ell^4 \propto p'^2L^2\ell^2 .$$

This equation is important because it shows that with larger correlation areas, ℓ^2, as for large scale separations, more noise is generated.

CONCLUSIONS

Due to the reduction of other noise sources, aerodynamic noise is becoming important for automobiles and other ground vehicles, especially at higher speeds. The aeroacoustics of automobiles involve many interesting phenomena. The fluctuating pressures generally originate in turbulent flows. Strongly turbulent flows are associated with large and small scale separations on the vehicle body, with vortices, with large and small cavity flows, with flow in ventilation and engine cooling ducts and systems, and with fans. In most of these situations, dipole type sound due to pressure fluctuations acting on the surfaces is the most important. Several aeroacoustic sources and their relative importance still need to be analytically and experimentally identified and quantified. The basic approach to be taken in automobile aeroacoustics is first to understand the aerodynamics of the overall flow, including separated regions; second, from this aerodynamic information to find the fluctuating pressures generated on the surfaces of the vehicle; and then third, from these pressures, to find the external sound radiation, the aspiration noise transmitted through openings in the body and finally, to find how these pressures excite structural vibration and generate internal noise. As shown above, there are some approximate calculation methods which are in early stages of development for both the estimation of important fluctuating pressures and their transmission through the automobile body.

Noise can be reduced by eliminating air and sound leaks at door and window seals. Open gaps between body sections causing cavity noise can be avoided by using flush seal designs. The external shape of the vehicle should have rounded edges and flush windows in order to avoid two-dimensional and vortex type separations, especially the A-pillar type vortex and other separations on glass areas, which have limited transmission loss. The underbody, engine, and wheel housing areas should also be designed for reduced unsteady separated flows in the future. The transmission loss of body panels can be increased by designing for large panels with few stiffeners and using damped and double wall construction with high coincidence frequencies.

Improvements are needed in both the source and path modeling. Some progress may come from carrying over work from the aerospace field to the automotive field. As experimental and analytical understanding develops, it will naturally lead to improved capability for lower noise designs and further concepts for reducing aeroacoustic noise on given designs.

ACKNOWLEDGEMENT

The author would like to acknowledge many interesting discussions with colleagues at BMW AG, West Germany; NASA-Ames; NASA-Langley; and Cornell. In particular, the contributions of Charles D. Coffen to the fluctuating pressure calculations are acknowledged.

REFERENCES

1 Buchheim, R., W. Dobrzynski, H. Mankau, and D. Schwabe, "Vehicle Interior Noise Related to External Aerodynamics," *Int. J. Vehicle Design*, Special Publication SP3, 1983, pp. 197-209.

2 King, W. F., and D. Beckert, "Aerodynamic Noise Generated by High-Speed Trains," *Noise Control Engineering*, Vol. 13, No. 1, July-August 1979, pp. 13-23.

3 Chanaud, R. C., and D. Muster, "Aerodynamic Noise from Motor Vehicles," *J. Acoustic Soc. Am.*, Vol. 58, No. 1, July 1975, pp. 31-38.

4 George, A. R., and S.-T. Chou, "Broadband Rotor Noise Analyses," NASA Technical Report, NASA CR-3797, April 1984.

5 Chou, S.-T., and A. R. George, "An Analytical Parametric Study of the Broadband Noise from Axial-Flow Fans," Proceedings, Noise-Con 87, Noise Control Foundation, New York, 1987, pp. 83-88.

6 Heckl, M., and H. A. Müller, *Taschenbuch der Technischen Akustik*, Springer-Verlag, Berlin, 1975, Chapter 9.

7 Lyrintzis, A. S. and A. R. George, "The Use of the Kirchhoff Method in Acoustics," to appear in *AIAA Journal*, 1989.

8 Hucho, W.-H., "The Aerodynamic Drag of Cars," in *Aerodynamic Drag Mechanisms of Bluff Bodies and Road Vehicles*, Sovran, G., Morel, T., and Mason, W. T. (eds.), Plenum, New York, 1978, pp. 1-44.

9 Glover, B. M., and B. N. Shivashankara, "Aeroacoustic Testing in Wind Tunnels," AIAA 10th Aeroacoustics Conference, Seattle, 1986.

10 Ogata, N., N. Iida, and Y. Fukii, "Nissan's Low-Noise Full-Scale Wind Tunnel," SAE Paper 870250, 1987.

11 Dobrzynski, W., "Zur Bedeutung von Strömungswechseldrücken auf der karosserie-oberfläche für den Innenlärm von Personenkraftfahrzeugen," Forschungsbericht FB 83-28, DFVLR, Germany, 1983.

12 Romberg, G. F., and R. G. Lajoie, "An Objective Method of Estimating Car Interior Aerodynamic Noise," SAE Paper 770393, International Automotive Engineering Congress and Exposition, Detroit, 1977.

13 Oswald, L. J., "Exterior-Radiated Aerodynamic Noise of Vehicles at Highway Speeds," Inter-Noise 78, San Francisco, 1978, pp. 711-714.

14 Oswald, L. J., and D. A. Dolby, "A Technique for Measuring Interior Wind Rush Noise at the Clay Model Stage of Vehicle Design," SAE Paper 770394, International Automotive Engineering Congress and Exposition, Detroit, 1977.

15 Van der Brulle, P. J. and Ulrich, A., "Determination of the Sound Pressure Distribution along the Doorslits of a Passenger Car," Proceedings of Inter-Noise 81, Amsterdam, 1981, Vol. 2, pp. 719-722.

16 Lorea, A., V. Castelluccio, A. Costelli, and M. Masoero, "A Wind-Tunnel Method for Evaluating the Aerodynamic Noise of Cars," SAE Paper 860215, 1986.

17 Thomson, J. R., "Wind Noise - A Practical Approach," SAE Paper 840B, Detroit, Michigan, 1964.

18 George. A. R., and Y. N. Kim, "Trailing-Edge Noise from Hovering Rotors," *AIAA Journal*, Vol. 20, Sept. 1982, pp. 1167-1174.

19 Rockwell, D., and E. Naudascher, "Review—Self-Sustaining Oscillations of Flow Past Cavities," *Transactions of the ASME*, Vol. 100, June 1978, pp. 152-165.

20 Rockwell, D., and E. Naudascher, "Self-Sustained Oscillations of Impinging Free Shear Layers," *Ann. Rev. Fluid Mech.*, 1979, pp. 67-94.

21 Maull, D. J., and L. F. East, "Three-dimensional flow in cavities," *J. Fluid Mech*, Vol. 16, August 1963, pp. 620-632.

22 Panton, R. L., and J. M. Miller, "Excitation of a Helmholtz Resonator by a Turbulent Boundary Layer," *J. Acoust. Soc. Am.*, Vol. 58, No. 4, October 1975, pp. 800-806.

23 Bodger, W. K., and C. M. Jones, "Aerodynamic Wind Throb in Passenger

Cars," Paper 840A, SAE Automobile Week, March 1964.

24 Aspinall, D. T., "An Empirical Investigation of Low Frequency Wind Noise in Motor Cars," MIRA Report No. 1966/2, January 1966

25 Dobrzynski, W., "Wind Induced Interior and Farfield Radiated Exterior Noise from Automobiles," in Vehicle Aerodynamics, Von Karman Institute for Fluid Dynamics Lecture Series, 1986-05, March 17-21, 1986.

26 Watanabe, M., M. Harita, and E. Hayashi, "The Effect of Body Shapes on Wind Noise," SAE Paper 780266, 1978.

27 Stapleford, W. R., and G. W. Carr, "Aerodynamic Noise in Road Vehicles, Part 1: The Relationship Between Aerodynamic Noise and the Nature of Airflow," The Motor Industry Research Association, Report No. 1971/2, November 1970.

28 George, A. R., F. E. Najjar, and Y. N. Kim, "Noise Due to Tip Vortex Formation on Lifting Rotors," AIAA Paper 80-1010, June 1980.

29 Perry, A. E., and J. H. Watmuff, "The Phase-Averaged Large-Scale Structures in Three-Dimensional Turbulent Wakes," *J. Fluid Mech.*, (1981), Vol. 103, pp. 33-51.

30 Cremer, L., and Heckl, M., *Körperschall, Physikalisch Grundlagen und Technische Anwendungen*, Springer-Verlag, Berlin, 1967, Sections VI.6-VI.7. Also translated as: Structure-borne Sound (E. E. Ungar, translator), Springer-Verlag, Berlin, 1988.

31 Jha, S. K., and Cheilas, N., "Acoustic Characteristics of a Car Cavity and Estimation of Interior Sound Field Produced by Vibrating Panel," ASME Paper 76-WA/DE-1, 1976.

32 Succi, G. P., "The Interior Acoustic Field of an Automobile Cabin," *J. Acoust. Soc. Am.* 81(6), June 1987, pp. 1688-07.

950623

A Survey of Automobile Aeroacoustic Activities in Germany

Syed R. Ahmed
DLR

ABSTRACT

This survey reports on the salient features of the approaches being pursued in Germany by automobile manufacturers, research organizations etc. to control and inhibit the noise emission from automobiles. The results reviewed are those presented by a number of experts from industry, research organization and universities during a Symposium held in February 1994 in Germany. Topics of concern were among others: sources of noise in automobiles, noise emission measurement techniques, aeroacoustic optimisation of car body and air conditioning equipment, design of and operational experience with new aeroacoustic wind tunnels in Germany.

INTRODUCTION

This paper describes in brief the material presented at the Symposium 'Akustik und Aerodynamik des Kraftfahrzeugs' (Acoustics and Aerodynamics of Powered Vehicles) at the Haus der Technik, Essen, Germany in February 1994. The aim of this survey is to make the main results available to non-german speaking automobile engineering community and report about the current state-of-the-art practices pursued in the industry, universities, research organizations and government agencies to control, inhibit and regulate the noise emission from road vehicles and road traffic in Germany.

With the number of vehicles on the roads continually on the rise, they are recognized as a major source of noise in the densely populated urban areas of Germany. According to a recent poll conducted for a government agency, over 70% of the population in Germany felt subjected to serious annoyance through traffic noise. For the continued acceptance of automobiles and road traffic in the future, it would be essential to drastically reduce the external and interior noise generated by automobiles.

The noise from automobiles - exterior and interior - is generated essentially by the powertrain (engine, transmission, clutch etc.), by the tires (tire/road interaction) and by the flowfield around the vehicle (airflow/vehicle body interaction). The past 20 years have seen intensive efforts to reduce the powertrain and tire noise leading to significant reductions in engine and transmission noise. Less successful were the efforts to reduce the tire noise.

As engine and transmission noise were lowered, airflow/vehicle body interaction noise - more simply 'wind noise' - became more audible and important. This source of automobile noise is becoming an important subject of detailed research. Up to cruise speeds of 90 km/h (54 mph), powertrain and tires are the dominant noise generators. Beyond this speed wind noise becomes the major noise source, increasing approximately with the velocity to the power of six. The noise of high speed vehicles is thus closely related to their aerodynamics.

A survey of the literature on automobile aeroacoustics reveals a number of interesting contributions, many of them related to specific vehicle configurations where a myriad of configuration details influence the aerodynamics and aeroacoustics. These investigations, conducted mainly by research groups of automobile manufacturers are a must to solve the on-the-spot problems arising during the development process of a new vehicle. Unfortunately the results often do not lend to deduction of generally valid conclusions. This makes the quantification of gained experience difficult and often all the steps of an investigation have to be repeated for a new vehicle model.

Most of the aeroacoustics work on automobiles performed currently is of an experimental nature. Measurement techniques, test strategies and experimental facilities are the topics of main concern for the automobile aeroacoustician. This is reflected in the bulk of the themes covered in this survey. An almost total lack of theoretical approaches indicates the extreme challenge faced by theoretical methods to describe the complex unsteady aerodynamics and consequently the aeroacoustic phenomena associated with the flow in and around the automobiles. Only the noise generation mechanisms around isolated vehicle components - in an abstract and simplified manner - appear amenable to a theoretical treatment [1].

SOURCES OF NOISE IN AUTOMOBILES

The noise radiated from automobiles - farfield as well as inside the vehicle - is generated mainly by four sources: the engine, powertrain, interaction tire/road and interaction vehicle body/surrounding air. Due to the different generation mechanisms at work, the variation in the A-weighted noise levels from the above named sources with the vehicle speed is also different. Generally, the engine and powertrain noise dominates at low speeds, whereas for normal cruising speed the tire/road interaction noise is the major contributor to the overall noise. Depending upon the vehicle body shape, the wind noise completely dominates the other noise sources for speeds in excess of 100 km/h (62 mph) especially with regard to the far field radiation. The noise transmission into the car interior is dependent upon construction details of the body and a general statement as to which of the above three noise sources becomes a major contributor at low to normal cruise speeds is difficult to make. However, for high cruise speeds the aeroacoustic body/flow field interaction noise clearly becomes the major source of noise.

Fig. 1 gives an overview of the typical noise generating sources on an automobile body, [2]. It is worthwhile to differentiate here between noise which is generated through interaction between air and rigid body components and such which is generated when air enclosed in hollow cavities is excited into resonance by the airflow outside. A typical example of the latter type is the low frequency noise generated due to open side windows or a sun roof inside the car. In a similar manner the air flow over the gaps between door and window periphery and main body can excite resonance in the cavities lying adjacent to them. The amount of this noise radiated into the far field is, however, extremely small. The transmission of this noise inside the car is dependent upon construction details, for example, the sealing between door/window periphery and passenger compartment.

Well known examples of noise generated due to interaction between air flow and rigid vehicle body protrusions is the 'whistling' produced by antennas, roof racks or other cylindrical or prismatic components exposed to the free flow. The periodic vortex separations in the wake of such components induce fluctuating pressures on the vehicle body surface. In a similar manner, the fluctuating velocities in the turbulent boundary layer induce fluctuating pressures on the vehicle body surface. These fluctuations are more intense downstream of areas where the flow separates. These pressure fluctuations generate noise which is radiated out into the far field.

It is worthwhile to observe here that the generated noise caused by the fluctuating pressure-induced structural oscillations of the vehicle body plays an insignificant role in the noise radiated into the far field. Of main importance here is the variation of the gradient of the fluctuating pressure amplitude over the body surface. In contrast, the transmission of this noise into the vehicle interior is - as indicated earlier - dependent upon the (body structure-dependent) transfer function.

The discussion above reveals that the transmission of noise into the vehicle interior can be controlled through suitable body design measures, the fluctuating pressure-generated noise is more difficult to control and needs a careful aerodynamic/acoustic analysis.

SOME CHARACTERISTICS OF THE SEPARATED FLOW-INDUCED FLUCTUATING SURFACE PRESSURES - As an illustration of the characteristics of the fluctuating pressures generated on the front side window pane of a car, in the region of the A-pillar vortex, experimental results are shown in Fig. 2, [2]. Considering the pressures at the measuring point No. 22 (MP 22), the basic version of the car - with a steep windshield and recessed side window pane - shows a high broadband fluctuating pressure level. This is caused by the A-pillar vortex and the flow separation at the rearward facing step created by the recessed window pane. A less steeper windshield (configuration A), but especially a smoother transition from the windshield to the side window (configuration B) significantly reduces the level of the fluctuating pressures. In Fig. 2, P_s denotes the rms value of the fluctuating pressures and f the frequency.

Even if the car body is designed to be free of protrusions that generate pockets of separated flow, separation can still occur. On the average there is always an ambient wind. All ambient winds except those blowing along the roadway cause the relative wind to approach a vehicle at some angle of yaw, a typical range being ±10° to the vehicle's longitudinal axis. On the leeside, the flow separates causing highly unsteady flow conditions here leading to a dramatic increase in the level of the fluctuating surface pressure spectra. This is illustrated in Fig. 3, [2] where the spectral variation for the indicated measuring point on the front side window pane of a car is plotted as a function of the yaw angles. Negative value of yaw angle β means that the measurement point lies on the leeward side of the vehicle.

Fluctuating direction of the ambient wind causes variation in the yaw angle β and change in the magnitude of the fluctuating pressures. This leads to fluctuating interior noise levels which is most annoying. Vehicle body shapes optimised to have little or no separations on most of the front body surface for head-on wind conditions are prone to the above-mentioned phenomena during side wind situations.

INTERIOR NOISE CHARACTERISTICS - The interior noise in a vehicle is generated at the body panels such as doors, windows, floor etc. Due to multiple reflections the noise is diffused, its level rises from a point in the interior of the passenger compartment towards the body panels. An experimental result to demonstrate this is shown in Fig. 4 taken from [2]. The A-weighted airflow induced sound pressure level rises from the middle of the vehicle towards the side window, in the result shown by an amount of 7 dB. Typically, the left ear of the driver is exposed to the noise near field. This position is used as a standard location to measure the car interior acoustics as it is sensitive to the airflow induced interior noise levels.

FAR FIELD NOISE RADIATION - Noise generation on the surface of the vehicle body is due to the fluctuating pressures generated there and this noise is transmitted to the surrounding air. Typically, downstream of a separation region the fluctuating pressure gradient is positive and decays, after reaching a maximum, slowly to the initial value. It appears that this pressure rise is the significant factor influencing the far field noise radiation. As mentioned earlier, there are a number of locations on the vehicle body surface, such as the A-post, recessed window panels etc., where air flow separation occurs.

The effect of configuration changes to avoid flow separation and its effect on the far field noise has been investigated in [3], a sample result is shown in Fig. 5. The A-weighted sound pressure level measured at a ground level location of 15 m laterally from the passing location is shown for various configuration changes made on the vehicle. Up to a the driving speed of about 130 km/h (~ 81 mph), the effect of configuration changes (for the test vehicle considered) on the radiated noise is negligible. Beyond that speed, the basic configuration exhibits a sound pressure level rise proportional to velocity to the power of seven, whereas the aerodynamically treated, flow separation inhibited configurations show the levels rising with velocity to the power of four.

NOISE MEASUREMENT TECHNIQUES

Road vehicles are complicated structures whose noise emission characteristics are currently difficult or impossible to predict by purely theoretical means. The noise measurement techniques pursue the objective of analysing the generation, mutual interaction and propagation of noise in the structure and the surroundings and develop methods to inhibit the noise propagation. Investigations in laboratory and test stands complemented by road tests are important elements of the test technique strategies.

NOISE EMISSION MEASUREMENTS IN REVERBERATION CHAMBER - The total sound energy levels of a noise source can be determined, in principle, by measurement at a single location in a reverberation chamber provided isotropic energy distribution conditions prevail. This is rarely the case in the practice so that the spatial distribution of acoustic pressure has to be determined. There are two possibilities of doing this: either the microphone is traversed during the measurement, Fig. 6a, [4], or a number of microphones at fixed locations are arranged to acquire the spatial values, Fig. 6b. The disadvantage of the former approach is that the test conditions (for example the engine rpm) can only be changed in steps, while a continuous variation of the test conditions is possible with the latter method.

TRANSFER FUNCTION ANALYSIS

The technique of transfer function analysis can be employed to ascertain the effect of a force, acting at a point on the automobile body, and the generated sound pressure field, for example, in the car interior. Knowledge of the spatial distribution of the sound pressure level can help in devising ways and means to improve the acoustic quality of the car interior by way of changing the interior room geometry.

Fig. 7 (taken from [4]) illustrates the basic procedure of this approach. The car body is struck at an appropriate point with a hammer which is instrumented with a sensor to register the hammer-stroke force versus time history. The phase- and amplitude spectrum of this signal (as a function of the frequency) is evaluated with a Fast Fourier Transformation (FFT). Simultaneously the sound pressure level time history at a number of points in the car interior is registered with the help of microphones. Each sound pressure time history signal is put through an FFT analysis to obtain the corresponding phase- and amplitude spectrum. The division of the complex force spectrum by the corresponding complex sound spectrum leads to a transfer function, schematically indicated in the lower part of Fig. 7, for each individual microphone signal.

Instead of a hammer stroke, the force application can also be effected by a computer controlled oscillator. More details of this technique are given in [4].

AEROACOUSTIC OPTIMISATION OF AUTOMOTIVE COMPONENTS TO REDUCE INTERIOR NOISE

The success in reducing the noise generated by a number of sources has increased the passenger sensitivity to interior noise in automobiles. This is especially important in the medium to luxury category of automobiles. The noise transmitted into the inside of the car can be, among others, due to engine, powertrain, wind noise generated on the vehicle body and noise generated by the air conditioning equipment transmitted to the car interior from the exit ports of the air conduits. A feeling for the interior noise measured at a cruising speed of 160 km/h (99 mph) can be gained from the result of Fig. 8 taken from [5]. Shown is the A-weighted sound pressure level measured inside the car in an acoustic wind tunnel and the total noise measured at the same speed on the road. The major contribution of wind noise to the interior noise is clearly visible.

The subjective feeling of comfort felt by the car passengers is dependent not only on the absolute level of the sound pressure. Annoyance is felt when some frequencies stand out and a discontinuity in the frequency spectrum is present. This can happen, for example, when the door or window sealing system loses its integrity at high speeds.

EXAMPLES OF BODY COMPONENT OPTIMISATION TO REDUCE INTERIOR NOISE

Rear View Mirror - The flow around the rear view mirror contains a region of separated flow which reattaches downstream but generates a highly turbulent wake which may persist almost all along the length of the car. While the noise generated by this turbulent flow is transmitted into the car interior mainly as structure-borne noise, it can also be radiated into the inside through the sealing system or body leaks.

With suitable optimisation of the rear view mirror contour - with the goal of reducing the size of the wake and its direction downstream - a significant reduction of the noise level inside the car can be achieved. An example of this type of optimisation, performed on the Ford 'Mondeo' (see [5]) is illustrated in Fig. 9.

Vehicle Underfloor - The frequency analysis of interior noise in the car exhibits typically high sound pressure levels in the high frequency range. A subjective analysis of this shows a similarity to the noise generated by the road/tire interaction. This noise contribution is thus difficult to isolate from the road/tire noise measured during road tests and can be identified only in an acoustic wind tunnel measurement. Wind tunnel tests with the flow blocked underneath the car have shown (see [5]) that the high frequency noise contribution inside the car is due to the separated flow between the underfloor and the road which imposes strong fluctuating pressures on the underfloor panel. Fig. 10 taken from [5], illustrates the fluctuating pressure contours at a speed of 170 km/h (~ 106 mph), 70 mm above the road in the gap between underfloor and the road. High values are noticed in the wake of engine and front tires. Beside noise, these pressure fluctuations excite vibrations of the underfloor panel which are transmitted on to the passenger seats. An improvement can be achieved by an underfloor covering in the engine compartment or also through an air dam mounted below the front bumper.

Cavity Isolation - Cavities in the body structure are sources of noise transmission and excitation and can significantly influence the noise perceived in the car interior. A study in the BMW acoustic wind tunnel, shown in Fig. 11 ([6]), shows the noise spectrum measured at different locations of the car body (inside the cavity) before and after corrective measures were taken.

NOISE OPTIMISATION OF AIR CONDITIONING EQUIPMENT

A variety of components of the climatic control equipment contributes to the interior noise of the automobile. Fig. 12 taken from [7], gives an overview of the important components which generate air-borne and structure-borne noise.

Noise is generated in the air conditioning equipment due to fluiddynamic phenomena arising while air flows through the components. The significant reasons and the nature of noise contribution is depicted in Fig. 13a, b [7].

Air condition equipment noise is propagated mainly into the passenger compartment. It considerably influences the feeling of comfort and consequently the driving safety. The subjective feeling of comfort is governed not only by the noise level but also by the tone 'quality'; the objective of an optimisation is thus a reduction of noise level and an improvement of the tone quality.

Optimisation measures on air conditioning equipment are feasible under the constraints of low cost and satisfactory function in the available room for installation which is becoming increasingly smaller.

Noise reduction measures have to be, due to the physical nature of noise generation mechanisms, always considered in context of the complete automobile. This means that a true optimisation of the air conditioning equipment can be performed only by installation in the final vehicle prototype. Unfortunately this procedure is not practical and the developers of air conditioning equipment have to start early and be concurrent with design development of the vehicle prototype.

At the beginning of every optimisation, a detailed noise emission evaluation of the basic version of the proposed air conditioning unit is performed in a suitable noise chamber. The test set up should closely correspond to the real situation. At this point normally the air conduits and their layout is not known and the components are not available so that the evaluation of only the air conditioning unit is possible. For this purpose the air conditioning unit is connected on the suction side with a throttle valve, appropriately treated with a sound absorbing covering,

Fig. 14 (taken from [7]). This simulates the pressure drop in the real vehicle before the air enters the conditioning unit.

For the noise measurements either an anechoic or a reverberation chamber can be utilised. The advantage of the latter is that only one microphone is needed to measure the noise irrespective of the direction of radiation. This preliminary noise emission test reveals in most cases the predominant noise sources which can then be targeted for improvement. If this is not possible, more in depth investigations, such as the flow situation inside the conditioning unit has to be investigated. Very often this flow can be simulated in a water tunnel or basin. Also theoretical methods to simulate the flow may be applied. In both cases regions of high velocity gradients or flow separation have to be localised and eliminated. Also locations in the equipment structure where resonance may occur have to be identified and corresponding changes made.

SOME EXAMPLES OF NOISE OPTIMISATION OF AIR CONDITIONING EQUIPMENT - The blowers of air conditioning units operate over a wide range of operating conditions. Usually they are of the centrifugal type, drawing air at the center and blowing it radially outwards through channels formed by the blades. The leading edges of the blades are thereby not always optimally orientated to the air flow over the complete operational range. This leads to flow separation at the blade leading edge and increased noise generation. A remedy is the circular thickening of the blade leading edge with the result that the flow remains attached over a wide range of operating conditions and consequently a lower noise level is achieved. Fig. 15 taken from [7], shows the blade leading edge modification (insert b) and the noise reduction obtained over the basic version, (insert a).

A significant reduction of the blower noise can be obtained if the blower housing is made of foam material. The noise absorption of foam increases with the frequency and remains after reaching the maximum, at a high absorption level. This is impressively demonstrated by the result of Fig. 16 taken from [7], for a blower housing with a foam material thickness of 25 - 30 mm. Frequencies above 800 Hertz are reduced compared to the conventional blower. The main disadvantages are higher cost and bigger installation volume of the blower.

The elimination of flow separation in air conduits as an effective measure to suppress noise is demonstrated in Fig. 17a ([7]). The version 'a' shows regions of separated flow around the three guide vanes at the exit and an uneven velocity profile at the conduit outlet. Improved contouring of the conduit and proper positioning of a single guide vane (Fig. 17b) drastically reduced the separated flow regions and improved the velocity profile at the exit. The reduction of sound pressure level achieved with this optimisation is clearly visible in Fig. 17c.

AUTOMOBILE AEROACOUSTIC WIND TUNNELS IN EUROPE

Many of the older wind tunnels for automotive investigations in Europe were designed primarily for aerodynamic tests and exhibit a level of background noise which is considered to be high for acoustic tests. Some of the existing tunnels such as the IVK-tunnel in Stuttgart and the Pininfarina-tunnel in Italy have recently undergone extensive upgrading to enable aeroacoustic tests of full scale automobiles while the newer tunnels, for example, the DNW-tunnel in the Netherlands - which is used also for full scale automobile tests -, the BMW-Technik tunnel in Germany and the Volvo tunnel in Sweden are from the beginning designed for aeroacoustic investigations.

As automobile aeroacoustics is gaining importance, it is expected that more older tunnels will be upgraded to allow acoustic tests to be performed. This has to be done without sacrificing the aerodynamic qualities of the facility.

SOME CHARACTERISTICS OF WIND TUNNEL TEST SECTIONS - Automotive wind tunnels are commonly equipped with either a 3/4 open jet, closed or a slotted wall test section. The 3/4 ***open jet*** configuration seems to be the preferred choice for acoustic investigations because of the ease of access as well as the efficient sound absorption treatment which is possible in the surrounding plenum chamber where out-of flow and far field measurements can be performed. The main disadvantage of the open jet test section is the development of a shear layer around the jet perimeter which may introduce instabilities in the test section flow, and initiate oscillations [8], which have contamination effects on the acoustic measurements. Also the entrainment of air from the plenum chamber into the shear layer generates a flow recirculation which may also contribute to the contamination effects mentioned above.

Some of the newer automotive wind tunnels in Europe such as the one at Porsche in Germany and Volvo in Sweden operate with ***slotted-wall*** test sections. Intuitively one may expect a higher in-the-flow noise level in such a test section relative to a 3/4 open test section. The result of a comparative test [8], done in the Porsche tunnel which can alternatively be equipped with a slotted wall or 3/4 open jet test section revealed some interesting results.

It is to be noted that the plenum chamber of the Porsche wind tunnel, as well as other components are not provided with an acoustic treatment.

Fig. 18 (taken from [8]) shows the difference of the 1/12 octave in-the-flow band spectrum of the sound pressure level between the slotted wall and open jet test section, measured at the tunnel centerline and 1 m above the ground for the same wind speed. Over the ***whole*** frequency range, the slotted wall test section exhibits a lower background noise level (1 - 2 dB) than the 3/4 open test section. In the low frequency range, this is due to the stabilising effect of the test section walls on the flow. High frequency sound leaking through the wall slots is probably shielded by the solid wall segments from entering again the test section after reflection in the plenum chamber, thereby reducing the high frequency content of the in-the-flow spectrum. This same trend is exhibited also by the out-of-flow acoustic measurements. Also of interest was the result of interior noise measurements in a car, which revealed no change in the measured spectrum over the whole frequency range for slotted wall and 3/4 open jet test section configurations.

Except for the DNW-tunnel, which offers a closed test section for automotive tests, none of the European automotive wind tunnels have presently this test section configuration. The main advantage of a ***closed test section*** is the flow stability since no mixing layer is formed at the perimeter of the jet as in an open test section. No collector is needed to direct the air back into the air circuit, thus avoiding further aerodynamic interference effects which unfavourably affect the aeroacoustic measurements.

However, there are significant disadvantages which go with a closed wall test section since without acoustic treatment, the test section along with the air circuit (diffuser, settling chamber, nozzle) behave like a half open reverberation chamber. On the test section walls, as well as on the walls of the rest of the air circuit a boundary layer develops which generates high frequency noise. A further limitation is the restricted accessibility and space which do not allow the realisation of far field conditions.

AN EXAMPLE OF THE UPGRADING OF AN AUTOMOTIVE FULL SCALE WIND TUNNEL FOR AEROACOUSTIC INVESTIGATIONS

The most recent example of upgrading an existing full scale automotive wind tunnel for aeroacoustic investigations in Germany is that of the tunnel at the University of Stuttgart, Institute of Internal Combustion Engines (IVK). The acoustic treatment of this tunnel, which went into operation in 1988, was completed in a six month-period in 1993. Some salient features of the many innovative measures detailed in [9] are presented in this section.

For a reliable measurement of the noise generated by an automobile, it is necessary to maintain a difference of at least 10 dB between the generated noise and the background noise levels of the empty tunnel over a frequency range of 60 Hertz to 8000 Hertz. The current experience shows this difference to be just sufficient also for interior noise measurements.

The noise generated by a wind tunnel consists of contributions from the ventilator, the nozzle chamber components (sieves, flow straighteners, heat exchanger), mixing layer around the free jet, boundary layer on the floor, flow interferences with the collector and other fluiddynamic phenomena in the tunnel air circuit.

The main source of noise is the axial ventilator which in practice allows little room for acoustic optimisation under the constraints of economy. The noise reduction measures for the ventilator are thus applied in the IVK tunnel on the ventilator upstream and downstream sections of the air circuit. This is indicated in Fig. 19 (taken from [9]) for the downstream side:

1. The air circuit cross section is partitioned into unequal subsections with same acoustic damping and pressure loss

2. Walls acoustically treated by covering with membrane absorber elements

3. Vertical partitions formed with back-to-back mounted membrane absorber elements

4. Corner vanes coated with sound absorbing foam

The patented membrane absorbers are ideally suited for damping low frequency (< 300 Hertz) noise. Above these frequency levels the membrane absorbers need to be complemented by an additional covering of fiber free material such as open pore foam mats which exhibit efficient sound absorbing properties. A combination of frequency tuned membrane sound absorber and foam mat panels (a development of Fraunhofer Institut für Bauphysik (IBP) in Stuttgart) gives a highly efficient sound absorbing arrangement for the low to high frequency range as shown in Fig. 20 taken from [9].

In addition to the acoustic treatment of the air circuit and corner vanes, the plenum chamber walls enclosing the open test section were also covered with broadband sound absorbing arrangement consisting of foam mat panels and membrane absorbers explained above. The construction details are shown in Fig. 21, [9]. The foam mat panels are enclosed in perforated plate cassettes and arranged with an air gap before the membrane absorbers. A steel structure, connected to the air circuit's concrete walls and provided with elastic joints, carries the foam cassettes and membrane absorbers.

The final result of all these measures to upgrade the acoustic quality of the tunnel is depicted in Fig. 22 where the A-weighted sound pressure level in- and out-of-flow of the IVK wind tunnel before and after upgrading is shown as a function of wind speed. For comparison, the results for some other European wind tunnels are also included, [9].

For the in-the-flow measurements, a reduction of almost 20 dB compared to the original untreated tunnel values (curve nos. 1 and 6), was obtained over the speed range of 80 to 180 km/h (~ 50 to 112 mph). A remarkable result is also the small difference (~ 3 dB) between the in-the-flow and out-of-flow result for the IVK tunnel (after upgrading).

The comparison of the results for the IVK tunnel with other existing European wind tunnels shows the drastic reduction in the in-the-flow sound pressure level obtained after the careful implementation of the various acoustic treatment measures. An A-weighted in-the-flow sound pressure level of less than 80 dB over the speed range of 80 to 180 km/h is indeed a very good value.

REFERENCES

[1] George, A. R.: "Automobile Aeroacoustics", AIAA 12th Aeroacoustics Conference, San Antonio, Texas, Paper No. AIAA-89-1067, (1989)

[2] Dobrzynski, W.: "Windgeräuschquellen am Kraftfahrzeug", Proceedings of the Symposium 'Akustik und Aerodynamik des Kraftfahrzeugs', Haus der Technik, Essen, Germany, (1994)

[3] Dobrzynski, W.: "Zur Bedeutung von Strömungswechseldrücken auf der Karosserieoberfläche für den Innenlärm von Personenkraftwagen", DFVLR-FB 83-28, (1983)

[4] Klingenberg, H.: "Automobilmeßtechnik, Band A: Akustik", 2nd Edition, Springer-Verlag, (1991)

[5] Hoffmann, R.; Gilhaus, A.; Renn, V.: "Aeroakustische Optimierung von Fahrzeugkomponenten", Proceedings of the Symposium 'Akustik und Aerodynamik des Kraftfahrzeugs', Haus der Technik, Essen, Germany, (1994)

[6] Lindener, N.; Geib, W.: "Aeroakustische Effekte an Fahrzeugen analysiert im BMW-Akustikwindkanal", Proceedings of the Symposium 'Akustik und Aerodynamik des Kraftfahrzeugs', Haus der Technik, Essen, Germany, (1994)

[7] Hipp-Kalthoff, Ch.: "Geräuschoptimierung an Klimaanlagen von Kraftfahrzeugen", Proceedings of the Symposium 'Akustik und Aerodynamik des Kraftfahrzeugs', Haus der Technik, Essen, Germany, (1994)

[8] Mercker, E.: "Kfz-Windkanäle für akustische Untersuchungen", Proceedings of the Symposium 'Akustik und Aerodynamik des Kraftfahrzeugs', Haus der Technik, Essen, Germany, (1994)

[9] Potthoff, J.; Essers, U.; Eckoldt, D.; Fuchs, H. V.: "Der Aeroakustik-Fahrzeugwindkanal der Universität Stuttgart am Institut für Verbrennungsmotoren und Kraftfahrwesen", Proceedings of the Symposium 'Akustik und Aerodynamik des Kraftfahrzeugs', Haus der Technik, Essen, Germany, (1994)

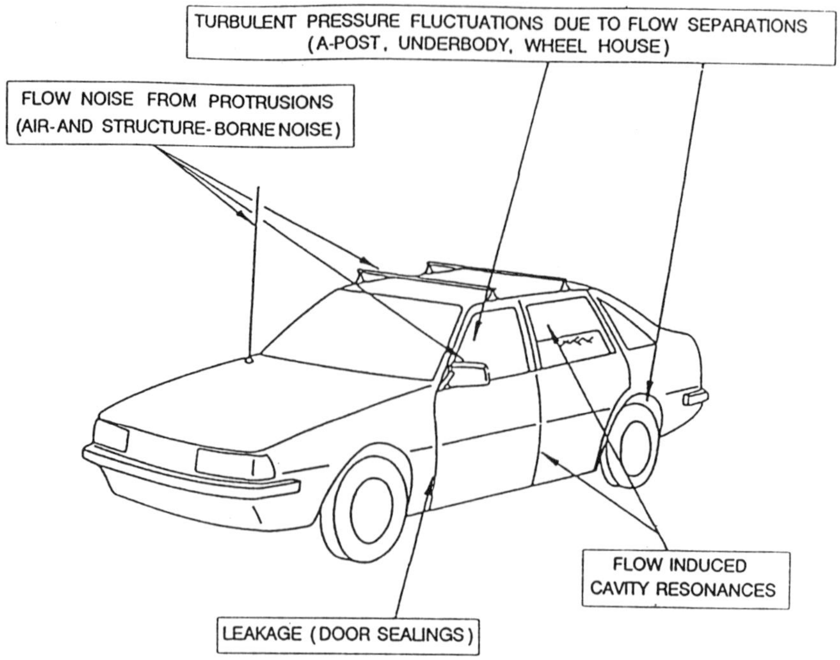

Fig. 1 Sources of aeroacoustic noise on an automobile, [1]

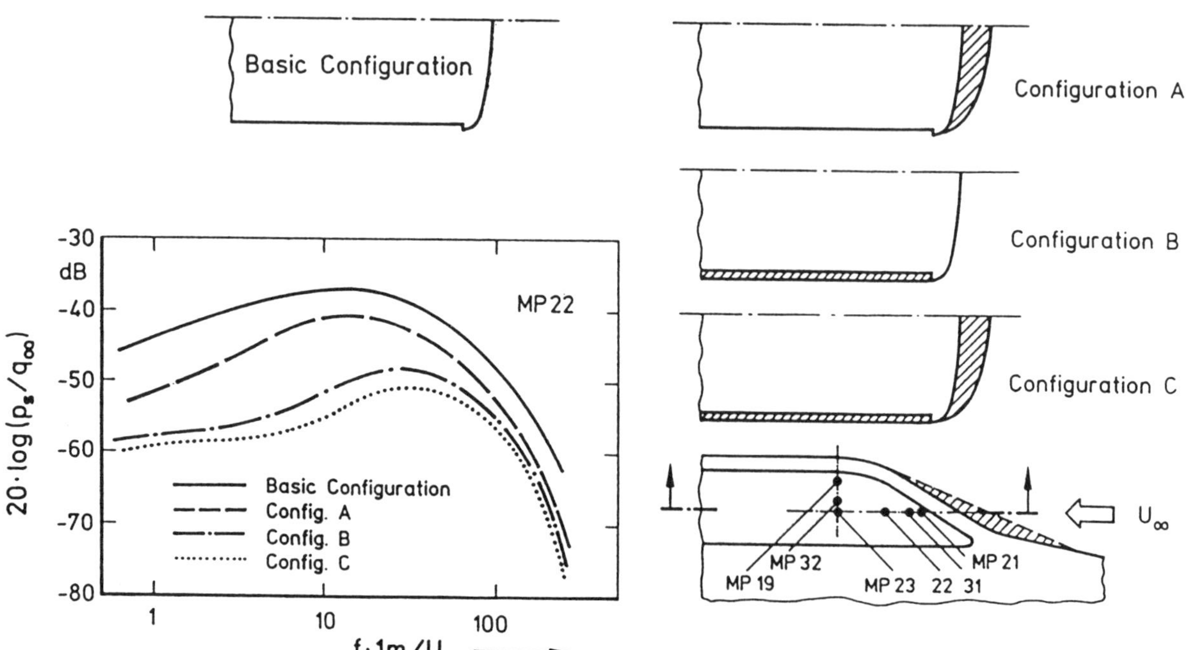

Fig. 2 Effect of body geometry changes on fluctuating pressure level at front side-window, [2]

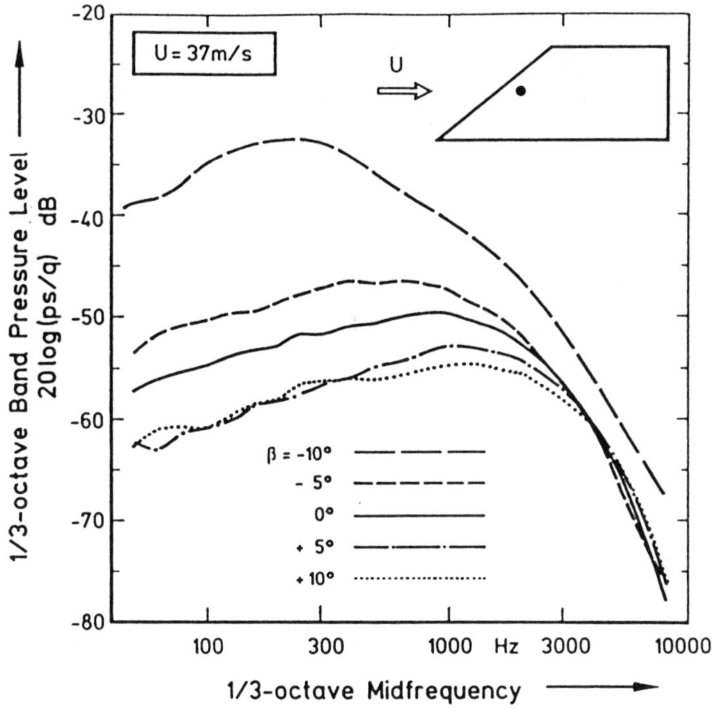

Fig. 3 Variation of fluctuating pressure level with yaw angle β, [2]

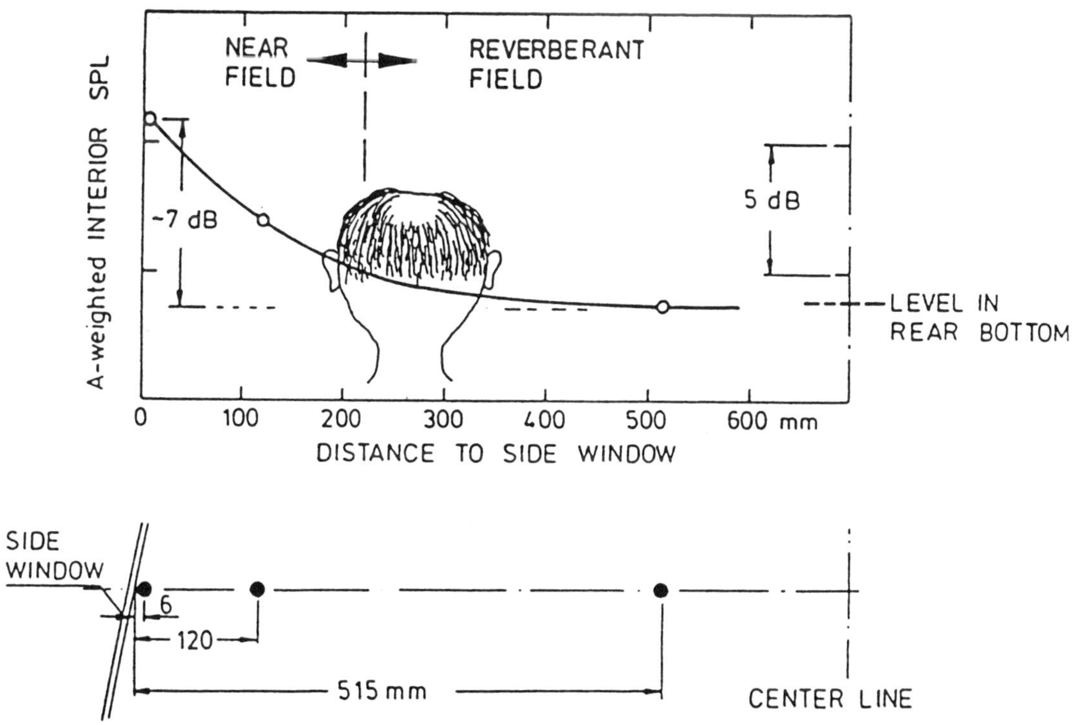

Fig. 4 Typical lateral distribution of A-weighted total sound pressure level in the car interior, [2]

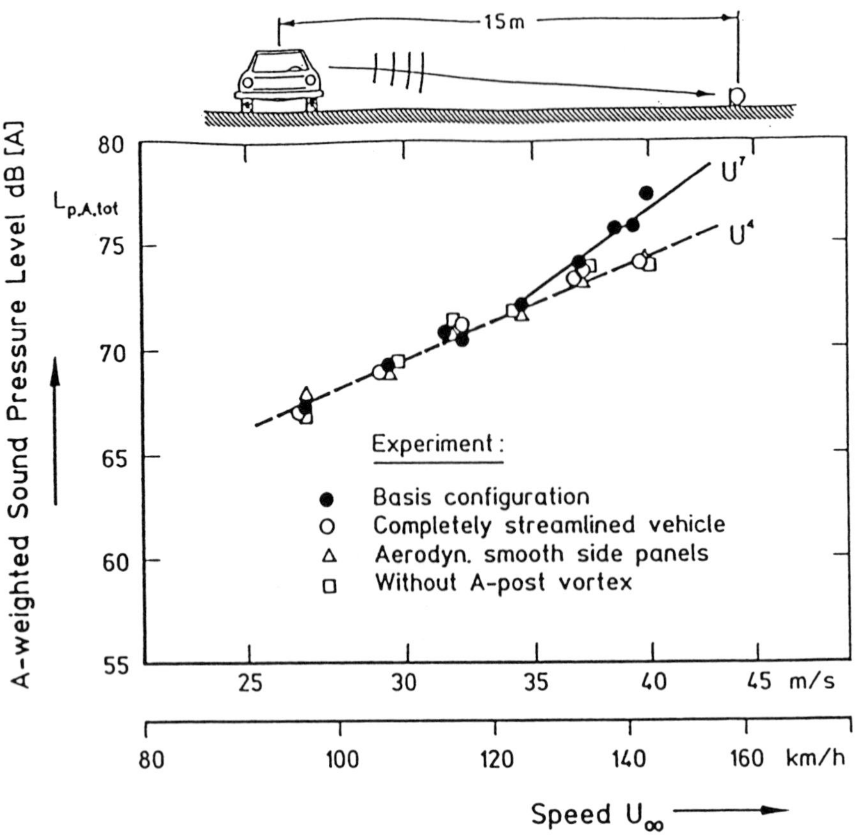

Fig. 5 Maximum A-weighted noise level during passing (with engine turned off), [3]

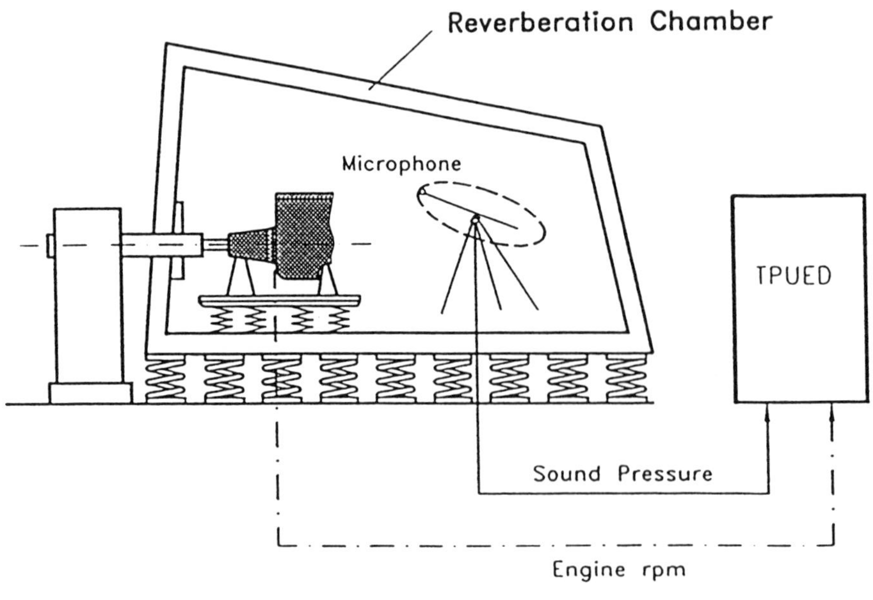

Fig. 6a Sound pressure measurement in a reverberation chamber with a traversing microphone, [4]

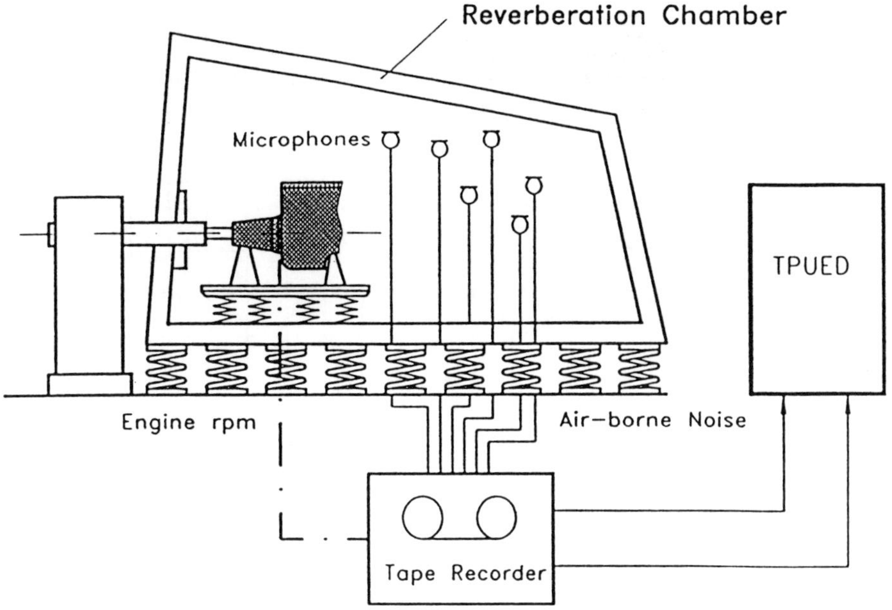

Fig. 6b Sound pressure measurement in a reverberation chamber with a fixed microphone array, [4]

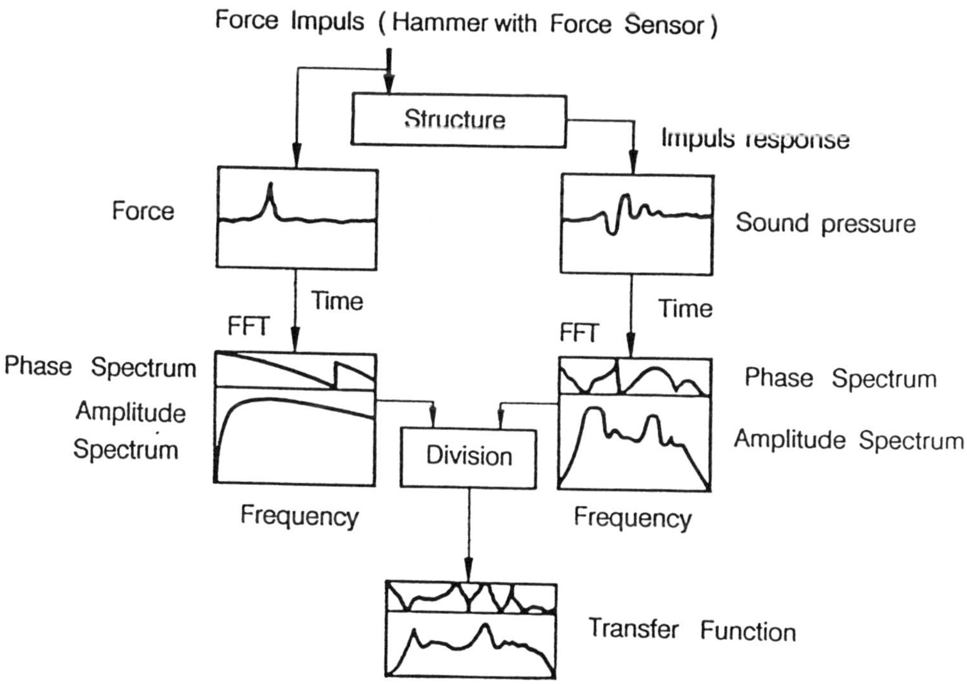

Fig. 7 Procedure to determine the transfer function for the force applied at a point on the car body and the sound pressure measured at a location in the car interior, [4]

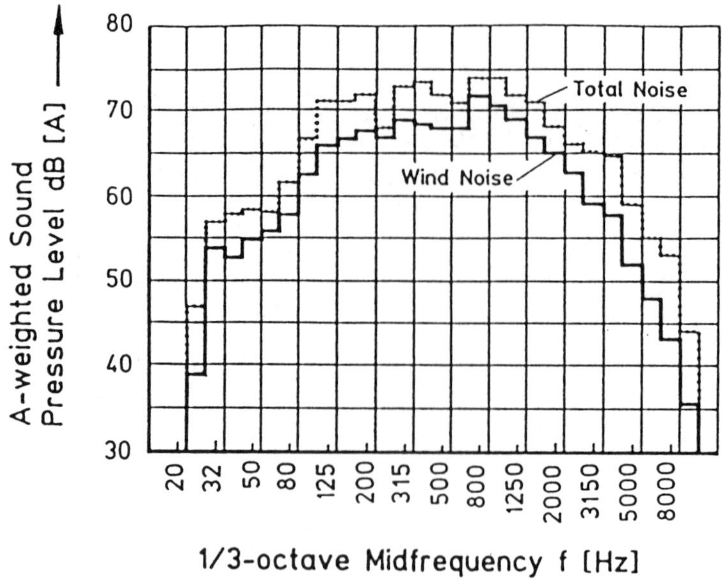

Fig. 8 Frequency analysis of the interior noise in a car travelling at 160 km/h (99 mph), [5]

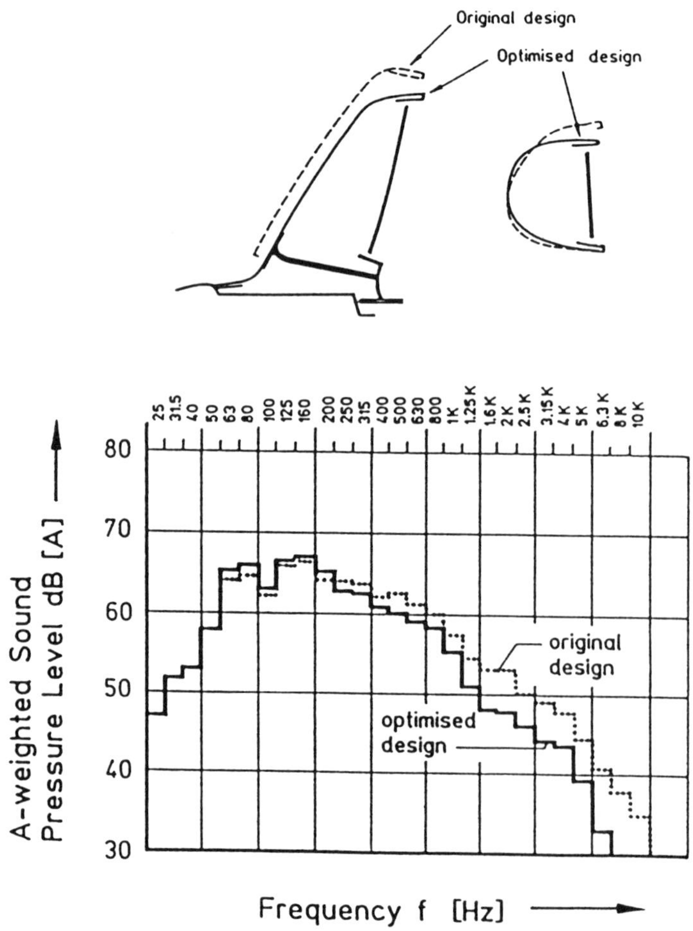

Fig. 9 Effect of rear view mirror optimisation on sound pressure level in the car interior, [5]

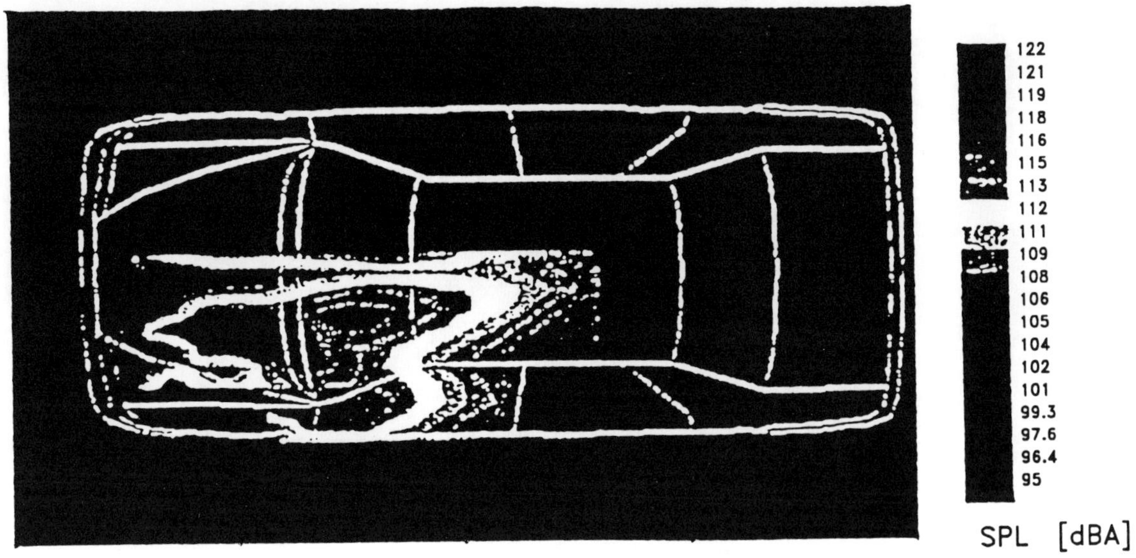

Fig. 10 Fluctuating pressure isobars beneath a car travelling at 130 km/h (81 mph), [5]

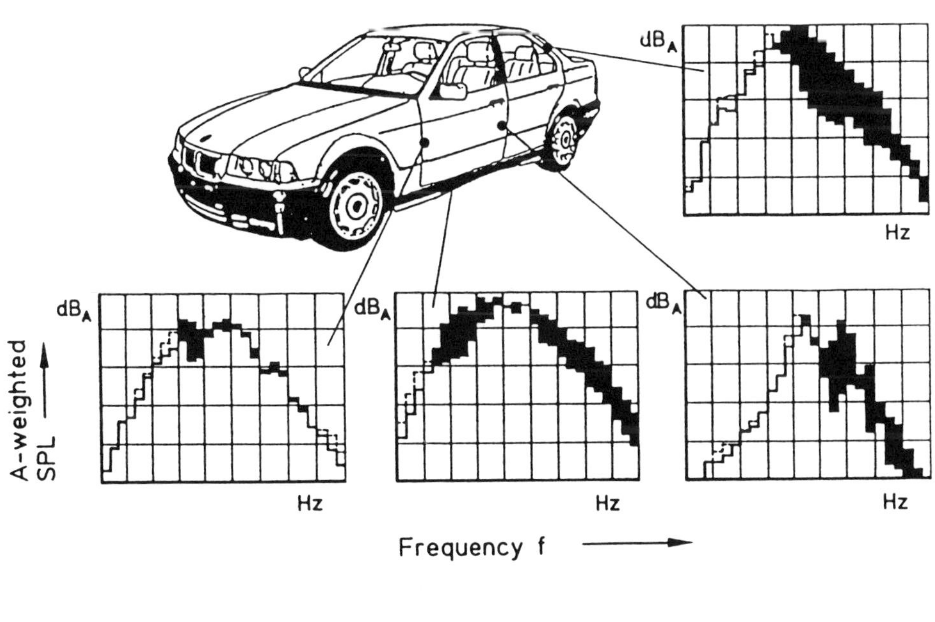

Fig. 11 Effect of isolation measures on the noise in car body cavities, [6]

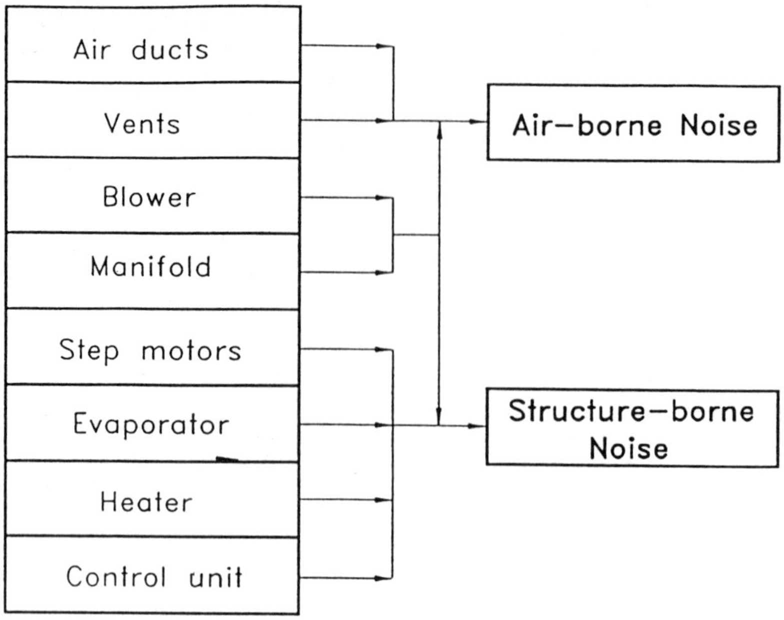

Fig. 12 Sources of noise in air conditioning equipment, [7]

Phenomenon	Contribution to Noise Spectrum
Spatially disturbed inlet flow to blower	discrete
Interference between blower rotor and guide vanes	discrete
Vortex separation flow due to fluid dynamic instability	narrowband
Turbulence in boundary layer	broadband
Free turbulence	broadband

Fig. 13a Sources of air-borne noise in air conditioning equipment, [7]

Phenomenon	Contribution to Noise Spectrum
Excitation of structure by pressure fluctuations in the flow	narrow-, broadband
Rotary angle dependent torque fluctuations of the electric motor	discrete
Vibrating brushes of the electric motor	narrowband
Imbalance of blower components	discrete
Stepmotors in operation	discrete
Air bubbles in heater	intermittent

Fig. 13b Sources of structure-borne noise in air conditioning equipment, [7]

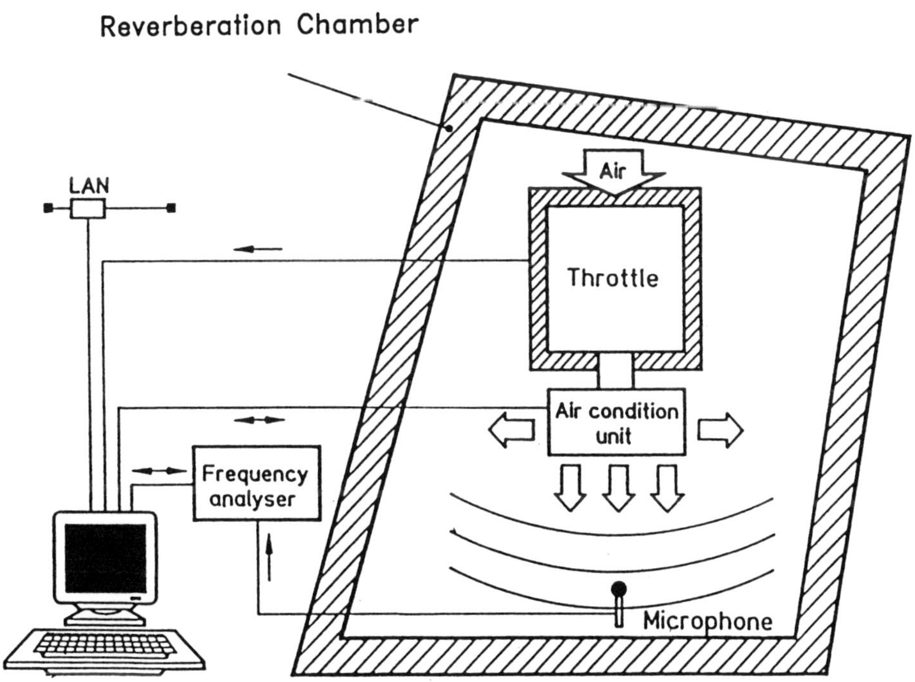

Fig. 14 Test set-up for noise measurements of air conditioning unit in a reverberation chamber, [7]

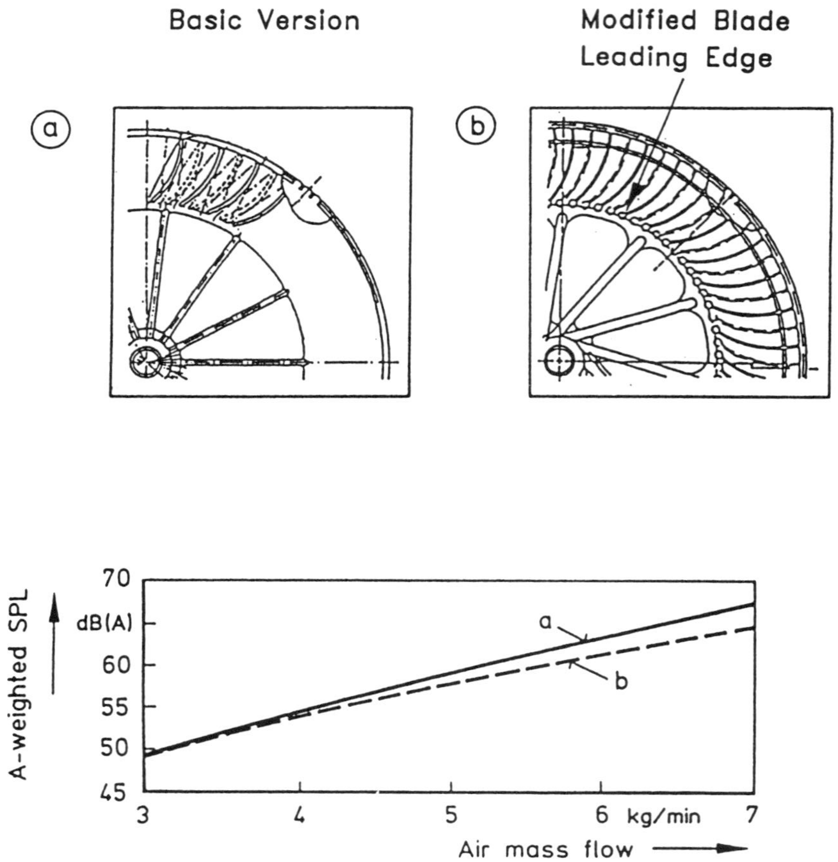

Fig. 15 Effect of blade leading edge modification on the sound pressure level, [7]

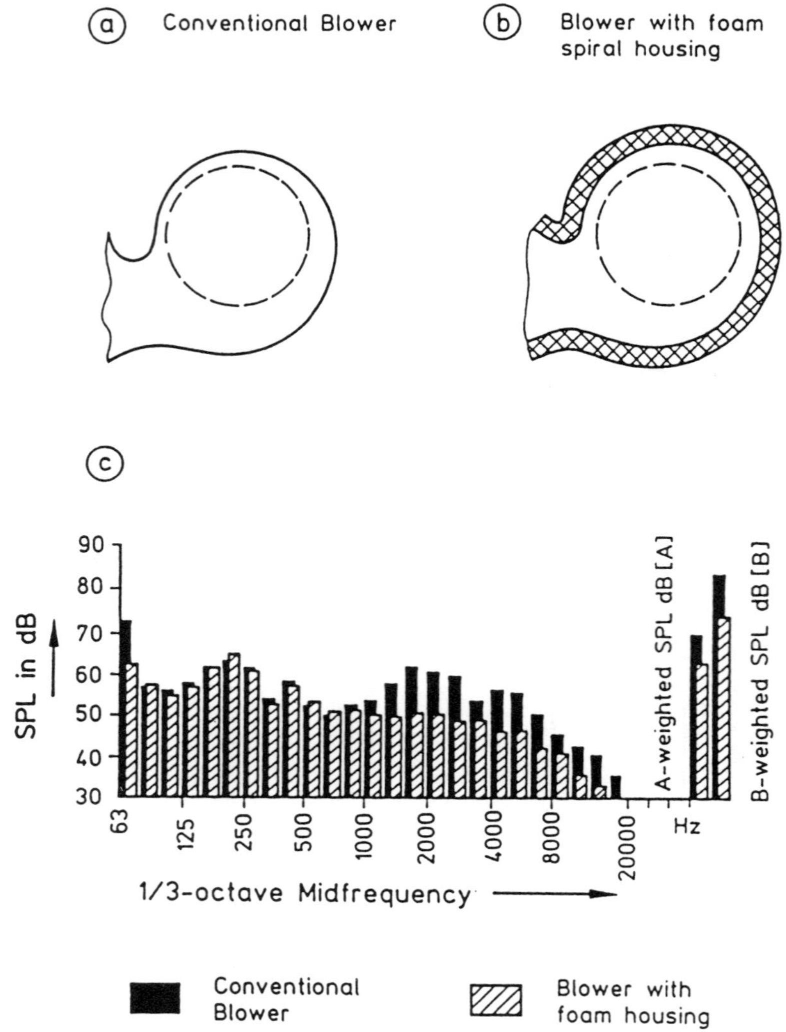

Fig. 16 Noise levels of blower with conventional and foam material housing, [7]

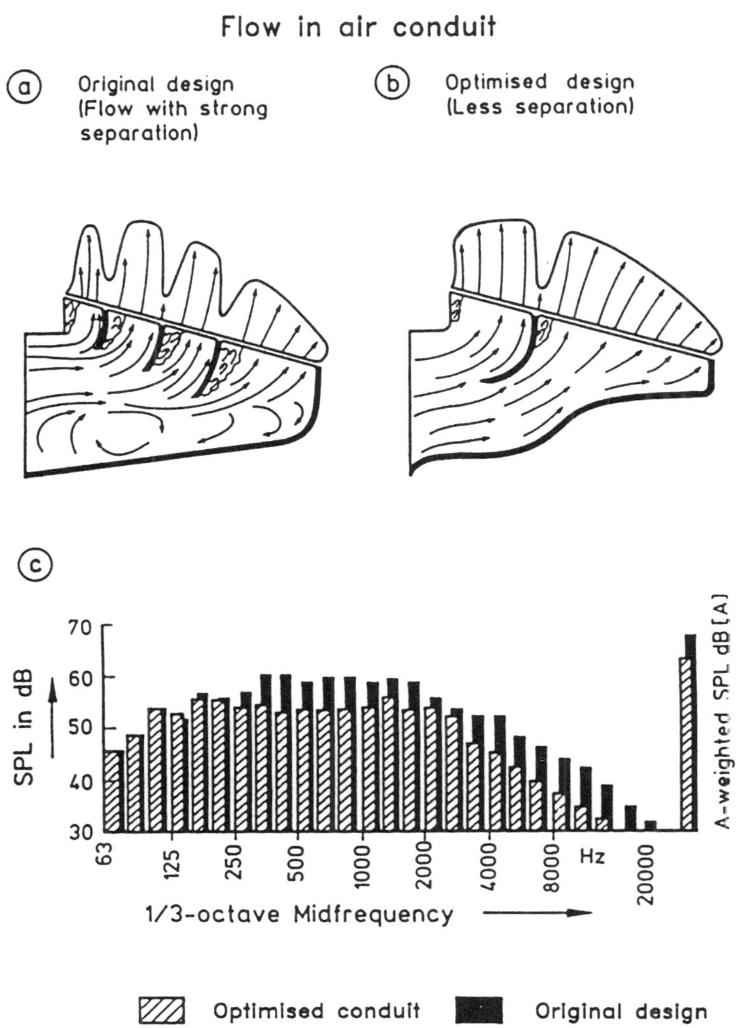

Fig. 17a, b Flow conditions in air conduit before and after optimisation
c: Spectrum and SPL before and after optimisation, [7]

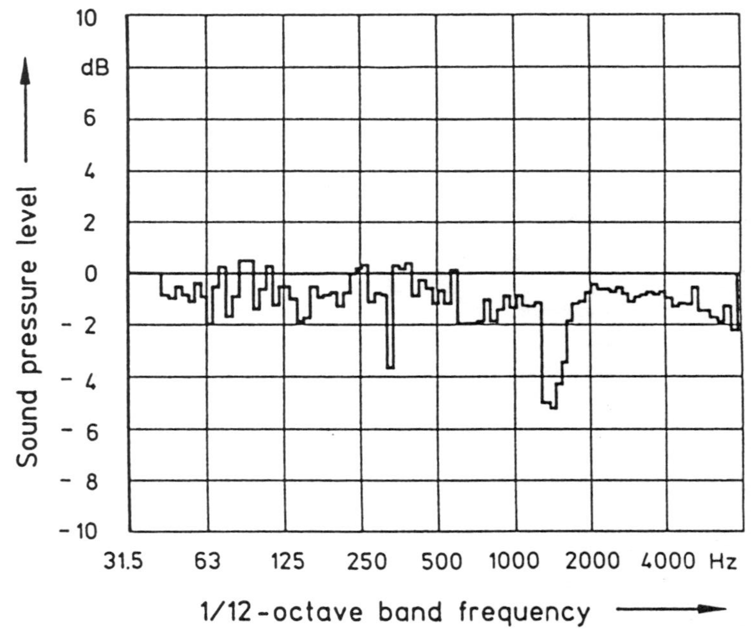

Fig. 18 In-the-flow difference spectrum of slotted wall minus open jet test section; Porsche wind tunnel test, [8]

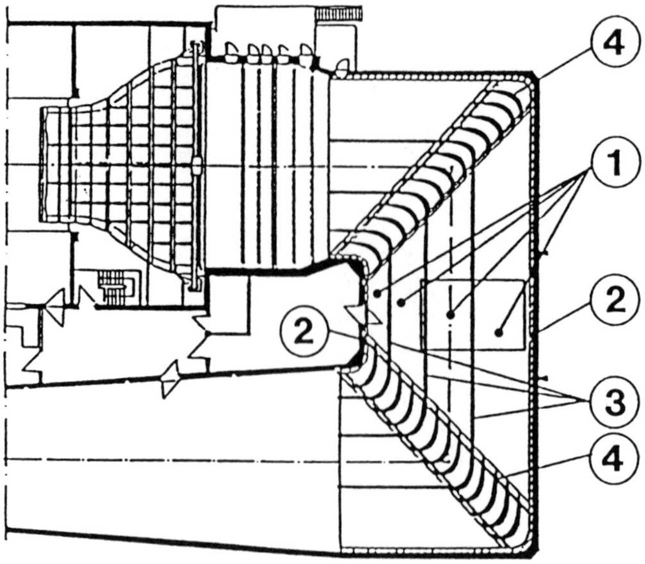

Fig. 19 IVK-Wind tunnel acoustic upgrading measures.
1: Partitioning of air circuit in subsections of equal pressure loss and damping;
2: Wall covered with membrane absorbers;
3: Partitions with back-to-back mounted membrane absorbers;
4: Foam coated corner vanes.

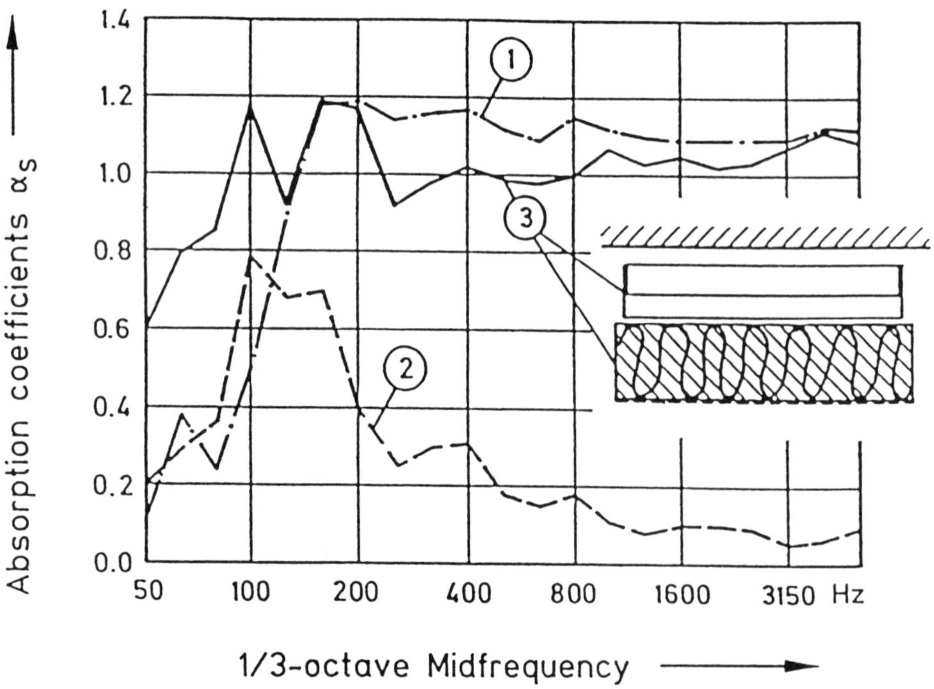

Fig. 20 Sound absorption of different wall covering.
1: 140 mm thick foam mat panels;
2: 100 mm thick low frequency membrane absorber;
3: Combination of 1 and 2 with 10 mm air gap, [9].

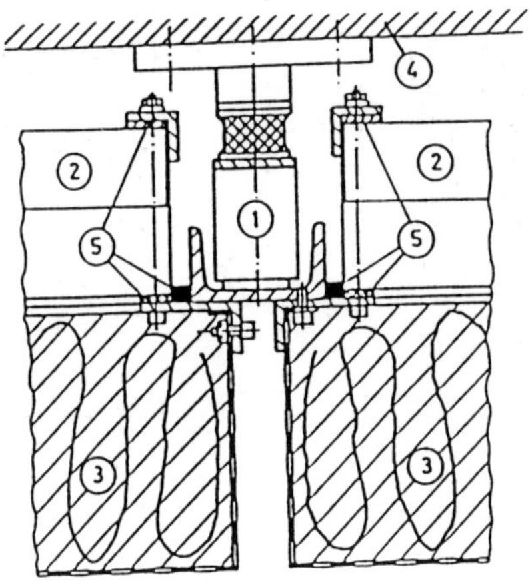

Fig. 21 Plenum chamber wall covering.
1: Steel structure elastically secured to wall;
2: 100 mm thick membrane absorber;
3: Foam mats encased in perforated plate cassettes;
4: Plenum chamber wall;
5: Foam rubber inserts, [9].

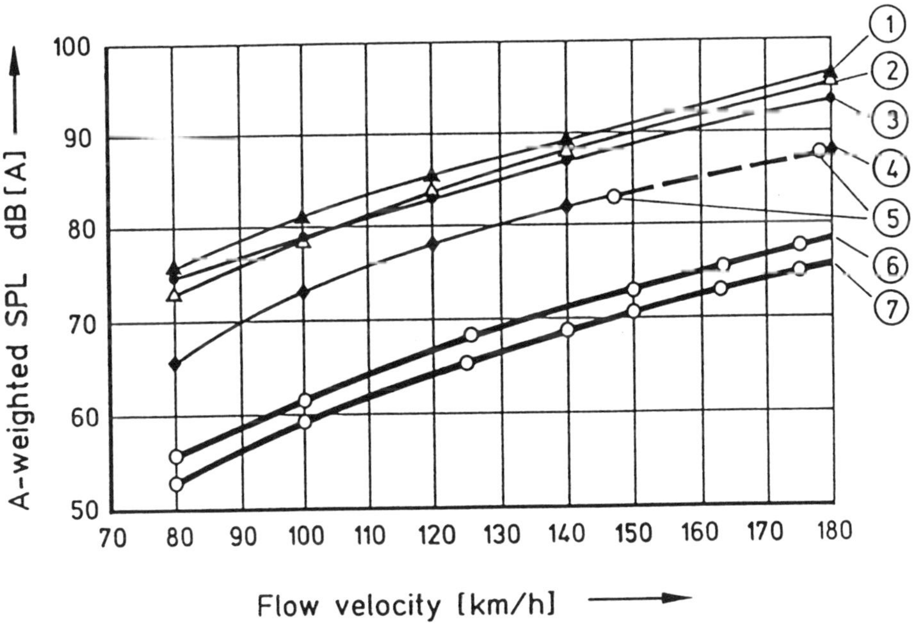

Fig. 22 In-the-flow sound pressure level of various European wind tunnels.
1: IVK-Tunnel before aeroacoustic upgrading;
2: Volvo-Tunnel with slotted wall test section;
3: DNW, 8 x 6 m closed wall test section;
4: DNW, open test section with anechoic plenum chamber;
5: BMW, 10 m² aeroacoustic wind tunnel;
6: IVK-Tunnel after aeroacoustic upgrading;
7: IVK-out-of-the-flow SPL after aeroacoustic upgrading, [9].

940415

On the Induced Noise of Test Sections in Different Wind Tunnels and in the Cabin of a Passenger Car

E. Mercker and K. Pengel
German-Dutch Wind Tunnel (DNW)

ABSTRACT

An acoustic comparison test of eleven European and two American full-scale automotive wind tunnels was carried out with a passenger car. The scope of investigation comprised in-flow and out-of-flow background noise measurements of the empty test sections as well as interior noise measurements inside the passenger compartment of a standard type Volkswagen Passat Variant CL. For this purpose seven different vehicle configurations were investigated in the various wind tunnels.

The aim of this experimental investigation was to establish physical criteria in order to carry out a suitable acoustic comparison between wind tunnels with different types of test sections and to establish upper limits of wind tunnel background noise levels where acoustic measurements inside a passenger compartment are questionable.

GENERAL DESCRIPTION AND CHARACTERISTICS OF TYPES OF TEST SECTIONS

For automotive testing in a wind tunnel, three different types of test section are commonly used: The 3/4 open jet test section, the closed wall test section and the slotted wall test section.

With respect to acoustic testing of cars, the 3/4 open jet is widely accepted to be the most suitable test section configuration, because the surrounding plenum chamber can be equipped with acoustic absorption material, thus simulating a quasi interference-free acoustic far field enabling acoustic out-of-flow measurements which might be important for noise source detection.

With a closed wall test section such measurements are more difficult to verify because the receiving microphone always has to be placed inside the flow and thus self-generated noise at the microphone is of concern even with the use of nose cones to protect the diaphragm. Furthermore, unless specially treated, the solid walls reflect the emitted noise created by the test vehicle. Thus, the test environment may be transcribed as a half open reverberation room which makes noise source detection more difficult, if not impossible.

In principle, the slotted wall test section permits acoustic out-of-flow measurements in the surrounding plenum chamber. However, there is a good chance that due to the solid parts of the test section boundaries, sound will be reflected. In addition, similar to the closed wall test section, some sound intensification may take place due to reverberation.

Another significant factor in different types of wind tunnels is the determination of air speed seen by the vehicle (q-determination). Depending on the procedure used, different "true" air speeds may be attained. Since aerodynamic noise is highly depending on air speed (i.e. U^6 typically), this aspect is of considerable importance.

THE OPEN JET TEST SECTION - Although it is very practical to use an open jet with an anechoic enclosure for acoustic testing of cars, the boundaries of the open jet may introduce some inherent drawbacks on the quality and stability of the flow. It is a fact that an open jet wind tunnel generates a shear layer between the potential core of the jet and the quiescent air outside the test section in the plenum chamber. Thus, at least for low wind speeds, noise generated by a vehicle placed into the test section is able to propagate out of the flow with little reflection. Consequently, an acoustic free field can be established and sampled in or out of the flow. However, due to turbulence and vortex flow generated in the shear layer the potential core flow will be disturbed to a certain degree and the jet may interact with the flow collector and the nozzle. Depending on the design of the different wind tunnel components, the entire jet may tend to oscillate and may couple with room acoustic modes of the plenum chamber. If additional resonance with the surrounding walls occurs, structural failure of the facility can follow. As already described by Soderman and Olson [1], the mechanism causing this oscillation can be explained as follows: Analytically it can be shown that a free jet collector should capture the jet core mass equal to that emitted from the nozzle to avoid a longitudinal static pressure gradient in the test section. Entrained flow in the shear layer would be cut off. In practice, however, this would require a collector immersed in the shear layer. Such a collector would experience strong unsteady pressure fluctuations which radiate upstream acoustically, and trigger vortex shedding from the nozzle lip. This feedback mechanism can generate flow oscillation at resonance conditions which can be described as an edge-tone type resonance. The frequency of this resonance is dependent on the vortex convection speed in the shear layer, U_c and the length of the open jet, ℓ:

$$f = (n + 1)\, U_c/\ell \qquad \text{where } n = 1, 2, 3 \ldots$$

As a consequence, the leading edge of a collector should not intrude too far into the shear layer. On the other hand, a collector that is too large will result in unnecessary high flow losses and a horizontal static pressure gradient which may alter the flow about the test vehicle. To delay the onset of flow oscillation, two methods are available. The first method is to install vortex generators on the nozzle lip. The second method is to optimize the collector shape. The vortex generators, however, can generate higher frequency noise and it seems to be advisable to use special foam material and special streamlining to avoid these effects. Also triangular vanes installed inside the nozzle exit have been tested successfully. A side effect of these vanes is that they reduce the area of the uniform flow in the test section.

Beside the dynamic effects described above, the shear layer also implies other direct aeroacoustic effects. Due to the vortex flow in the shear layer, fluid from the plenum is entrained into the shear layer causing a significant recirculation flow in the surrounding acoustic hall. Depending on the propagation angle of the noise source, the flow Mach number and the acoustic frequency, the shear layer may also distort, scatter and refract the transmitted sound.

In order to reconstruct the original acoustic radiation without shear layer, Amiet [2] has successfully developed a correction method. However, the situation is rather complex with a large noise source like an automobile and a nearby observer, because the observer cannot be sure where the sound originates as the ray/shear layer intercept angle is ambiguous. To eliminate this problem it would be necessary to enlarge the open jet.

Furthermore, some methods have also been proposed to deal with spectral broadening, amplitude fluctuations, and phase fluctuations due to shear layer interferences (Ross [3], Guedel [4]). Finally, if the noise source reveals an impulsive signature (exhaust pipe noise), the signal can be badly distorted in time and phase by the shear layer. It is therefore sometimes preferable to place the microphone inside the open jet although limitations might be imposed by the near acoustic field. Generally speaking, the theory of the corrections to be applied is complicated, inaccurate and, to most people, uninteresting. Unfortunately, it is often essential to make these corrections and every wind tunnel user must therefore understand them sufficiently well to apply them.

As mentioned before, the shear layer of a jet spreads in downstream direction. For a rectangular jet this spread angle can be several degrees less than that of a round jet. Beside that, the shear layer also influences the flow outside the mixing zone in the potential core region because of the unsteady velocities induced by vortices moving in the shear layer. The region of low turbulence is only about 40 to 50% of the core which is commonly defined as the region between the shear layers where the mean velocities are at least 99% of the centerline velocity. Outside this core region, the turbulence increases rapidly. Furthermore, due to the entrained flow into the shear layer, static pressure fluctuations will occur in the potential core. These facts are a fundamental problem with open jets. Depending on the specific model or vehicle size and test requirements, this may or may not be important for potential users of such facilities. If acoustic in-flow measurements are carried out with a microphone, the potential flow fluctuations will add to the pressure fluctuation generated by an acoustic source. By nature, the microphone cannot distinguish between different sources of pressure fluctuations and thus a sound pressure level will be put on which is enhanced by the level of the pseudo-noise.

THE SLOTTED WALL TEST SECTION - Slotted walls wind tunnels have been in use for automotive testing for several years. With regard to noise levels measured in the test sections it is generally assumed that a slotted wall wind tunnel may be louder than an open jet wind tunnel because of the flow disturbances generated at the slots. To answer this question an aeroacoustic comparison test between slotted walls and open jet test section was executed in the full-scale wind tunnel of Porsche AG in Weissach. This test facility permits the unique possibility to completely remove the slotted walls which exhibit an open area ratio of 30%, and change to a 3/4 open jet configuration. The plenum chamber surrounding the slotted walls or the open jet test section is big enough not to interfere with the jet shear layer and the flow collector can be adjusted adequately to attain a good horizontal static pressure distribution over most of the workable length of the test section. It is noteworthy, that the Porsche wind tunnel has been treated acoustically with a porous bulk-absorber protected by perforated sheet metal. The absorber is installed in the first cross leg of the air circuit. Furthermore, due to a modern design of the different wind tunnel components the background noise level can be considered as being very low.

From an acoustic point of view, the most striking advantage of having two interchangeable test section configurations is the possibility to study some basic noise source mechanism generated in the test section area of a wind tunnel with basically the same air circuit.

Figure 1 displays the difference of two 1/12 octave band spectra measured in-flow at the centerline (one meter above the ground) and at the same tunnel speed for two different test section configurations: the slotted wall test section and the 3/4 open jet test section. It can be seen, that over the whole frequency band (31.5 Hz - 9 kHz) the slotted wall test section yields a slightly lower background noise spectrum (1 - 2 dB). For the low frequency regime this result was not surprising because the flow will be stabilized by the slotted walls and the potential flow fluctuations, normally induced by the shear layer of the jet, were largely reduced. However, the lower noise levels in the higher frequency regime measured with slotted walls may be explained by sound radiated partly through the slots into the plenum chamber where it is damped out. In contrast, in the open jet configuration the sound will be reflected back into the jet at the untreated hard walls of the plenum chamber. Apparently, high frequency boundary layer noise generated at the solid parts of the slotted walls does not play a major role.

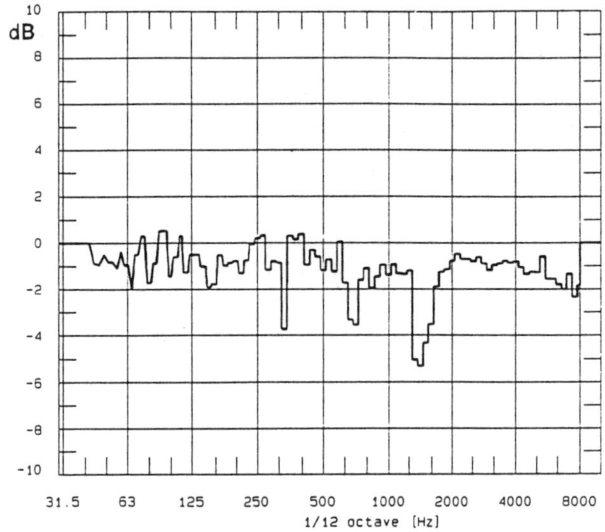

Fig. 1 In-flow-difference spectra between slotted wall (sw) and open jet (oj) test section for the Porsche wind tunnel (V = 140 km/h, sw-oj)

Furthermore, two frequency peaks at around 0.3 and 1.3 kHz are noticeable. Whereas the frequency peak at about 0.3 kHz can be attributed to some heterodyne noise originating from the thyristor-controlled fan motor, the frequency peak at 1.3 kHz can presumably be put down to collector noise or lip noise of the nozzle which is only present in the open jet mode.

The same trend as described above can be recognized, and emphasized more for out-of-flow measurements. Figure 2 shows the difference of two spectra from slotted wall and open jet test section measurements at the same tunnel speed.

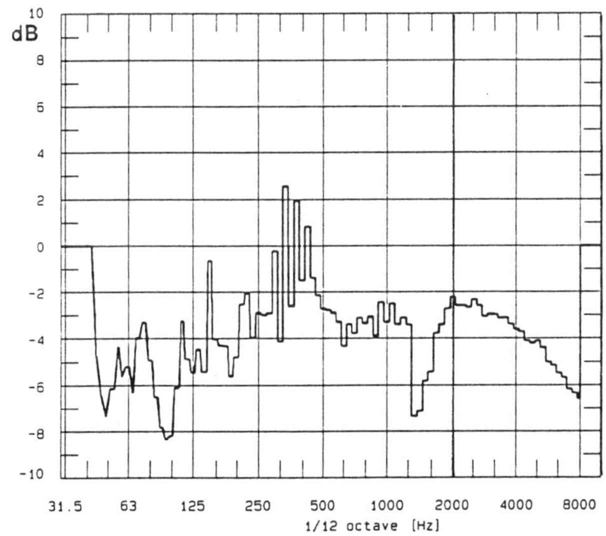

Fig. 2 Out-of-flow difference spectra between slotted wall (sw) and open jet (oj) test section for the Porsche wind tunnel, (V = 140 km/h, sw-oj)

In this case, the microphone was placed outside the jet in the plenum chamber. With the microphone located in the plenum chamber the recirculating entrainment flow generates low frequency components in the noise spectrum. During the in-flow measurements these components were largely reduced to potential flow fluctuations, as mentioned above. Consequently, judging the noise level of a wind tunnel at the model position inside the jet by out-of-flow measurements may be misleading. The considerably lower noise level in the high frequency regime for the out-of-flow measurements in the slotted wall test section can be attributed to a shelter effect of the solid part of the test section. Due to larger wave lengths in the low frequency range this effect was much reduced ($\lambda > 0.4$ m for $f < 800$ Hz) at the lower end of the scale.

It can be argued that the acoustic conditions may be altered for the slotted wall wind tunnels when a car is placed inside the test section, because the vehicle will displace fluid and thus some flow will escape from the test section through the slots. It is suspected that especially the low frequency portion of the noise floor inside the test section will then be increased. However, interior noise measurement inside a passenger car revealed that over the whole frequency regime almost identical spectra were measured in both test sections. Details about this comparison will be given in a following chapter.

THE CLOSED TEST SECTION - Regarding flow stability, the closed test section provides probably the most sufficient boundary conditions. Therefore, this type of test section can be used for very high tunnel velocities. Furthermore, due to the rigid walls, the flow cannot expand and thus, the horizontal static pressure gradient of the empty test section will be very low whilst the increasing thickness of the tunnel wall boundary layers is usually compensated by slightly diverging side walls. Additionally, the closed wall test section does not have a flow collector. Consequently, no additional noise can be generated by the downstream condition.

However, closed wall test sections do have some inherent drawbacks from an acoustical standpoint. The most obvious disadvantage is the presence of hard walls and ceiling, where sound reflection may take place. The test section together with the nozzle and the settling chamber in upstream direction and the diffuser in the downstream direction, can be looked at as a half open reverberation room. It is, however, imaginable to integrate some sound absorbing material into the rigid walls and ceiling. Only recently Fuchs et al. [5] introduced a novel type of panel sound absorber, which is able to withstand high velocity flow and also reveals a very high absorption coefficient over a large frequency band. Although the installation height of such absorbers is only about 240 mm, a retro-fit installation into a closed test section seems not to be feasible without changing the contour of the contraction and the diffuser.

Instead of developing a shear layer like an open jet, the closed test section develops wall boundary layers. They tend to generate noise in the higher frequency regime (> 3 kHz) which, however, might be of less importance for car testing due to two reasons:

1) Acoustic car testing is usually carried out with full-scale vehicles. Thus, no use of acoustic scaling laws has to be made (in scale model testing the spectrum shifts to higher frequencies).

2) The dominating frequency of the exterior flow noise of a vehicle as measured inside a vehicle, can be established at around 0.5 kHz. Thus, boundary layer noise may contribute less to the overall sound pressure level (OSPL) inside the cabin.

Whereas in a 3/4 open jet out-of-flow far field measurements are easily accessible, measurements in a closed test section have to be restricted to in-flow measurements.

SURVEY OF EXISTING WIND TUNNELS

Most of the full-scale wind tunnels used currently for automotive testing have been designed purely from an aerodynamic point of view. Thus, the background noise level inside the working section is often considered to be too high to execute acoustic measurements. Nevertheless, the absolute background noise level at which acoustic testing is still meaningful is an open question. Table 1 displays a list of all major wind tunnels which are known to be used for full-scale automotive testing. The list also publicizes those wind tunnels in which aeroacoustic testing was envisaged already during the specification and design phase of the facility. In the true sense of a word they can be called aeroacoustic wind tunnels and it is not surprising that these wind tunnels reveal the lowest background noise level, which will be shown below. At the time of writing this technical paper there were known to be two wind tunnels (IVK, Pininfarina) which have been upgraded from an acoustic point of view in order to lower the background noise level and to include acoustic

Table 1 Full-Scale Automotive Wind Tunnels

Wind Tunnel	T.S. Type	Nozzle Area (m^2)	Contr. Ratio	T.S. Length (m)	Acou. Treat.	Type
BMW AG	3/4 open	20.02	3.66	10.02	no	CC
BMW-Technik*	3/4 open	10.00	3.00	9.83	yes	OC
Mercedes-Benz	3/4 open	32.64	3.60	10.00	no	CC
DNW*	closed	36.00 48.00 90.00	12.00 9.00 4.80	16.00 20.00 20.00	yes	CC
FIAT	3/4 open	30.00	4.00	10.50	no	CC
Ford	3/4 open	23.75	4.00	10.50	no	CC
GM	closed	56.30	5.00	21.30	no	CC
Honda*	3/4 open				yes	CC
IVK**	3/4 open	22.45	4.41	9.50	yes	CC
JARI	closed	12.00	4.06	10.00	no	OC
Lockheed	closed	35.10	7.00	13.10	no	CC
Mazda	closed	24.00	6.00	12.00	no	CC
MIRA	closed	35.00	1.50	15.20	no	OC
Mitsubishi	closed	24.00		12.00	no	CC
NAMI	slotted wall	27.00	5.85	18.00	no	CC
Nippon Soken	3/4 open	17.50/12.00	3.60	12.50/8.50	no	CC
NRC	closed	81.90	6.00	22.90	no	CC
Nissan*	3/4 open	28/15	6.40/12.00	12.00	yes	CC
Pininfarina**	3/4 open	11.75	6.75	8.00	yes	OC
Porsche	3/4 open	22.30	0.00	13.50	no	CC
Porsche	slotted wall	22.30	6.06	13.50	yes	CC
St. Cyr S4	slotted wall	15.00	5.00	10.00	yes	CC
St. Cyr S10	slotted wall	15.00	7.70	10.00	no	CC
Toyota	closed	17.50	3.66	8.00	no	CC
Volkswagen	3/4 open	37.50	4.00	10.00	no	CC
Volvo*	slotted wall	27.06	6.06	15.82	yes	CC

Type: OC = open circuit (Eifel type)
CC = closed circuit (Göttingen type)

* Aeroacoustic wind tunnels
** Wind tunnel upgraded for acoustic testing

testing of road vehicles into the standard test routine. As acoustic aspects of passenger cars have become an important item for modern car development, many other wind tunnels will presumably follow and reduce their background noise level by installing suitable acoustic damping and absorption treatments. Naturally, this should be verified without sacrificing the aerodynamic quality to a large extent. However, the costs for such upgrading measures can be considerable. The main noise contribution in a wind tunnel is normally provided by the fan and its drive unit. It is a fact that beside broadband noise, tonal noise originating from the interaction of the rotating fan blades and the fixed stator vanes may also be generated revealing total sound pressure level of 120 dB and more in the fan area at a test section speed of 140 km/h. Considering an unweighted overall sound pressure level (OSPL) of 80 dB in the working section to be very acceptable for a good aeroacoustic wind tunnel, it becomes apparent that attenuating the background noise level sufficiently can be very expensive.

When it comes to comparing the background noise level of different open jet wind tunnels, normally reference is made to an A-weighted, out-of-flow, overall sound pressure level of the empty test section measured at some distance from the jet shear layer in the surrounding plenum chamber. However, this kind of measurement as a basis for comparison gives rise to some criticism:

a. The measuring position in the plenum chamber is often not specified.
b. Some of the noise generated by the wind tunnel is directionally sensitive and thus noise will be underestimated when measured out-of-flow.
c. A-weightening underestimates the effect of low frequency noise and hydrodynamic pressure fluctuations in the wind tunnel which is significant for cavity noise or booming noise of open sunroofs, for instance.
d. A comparison of out-of-flow data with in-flow data measured in closed or slotted wall test sections is questionable.
e. The shear layer may distort, scatter and refract the transmitted sound.

It therefore seems to be more sensible and meaningful to refer to acoustic in-flow measurements. The microphone then has to be supported by an aerodynamically shaped holder which tends to radiate noise mainly sidewise and from there it will be acoustically insignificant. In addition, the protection grid of the microphone diaphragm has to be replaced by a nose cone in order to prevent self-generated noise at the microphone. Furthermore, it is proposed to use a measuring position at the centre of the turntable (one meter above the tunnel floor seems to be sufficient). Accordingly, the background noise measurements of the empty test section are made at a position which will be occupied by the vehicle at a later stage, and hence the actual background noise level being relevant for the car can be evaluated.

A comparison between in-flow and out-of-flow background noise measurements can be taken from Figure 3 displaying results measured in the 3/4 open wind tunnel of BMW-Technik GmbH, which is surrounded by an anechoic plenum chamber. In the A-weighted out-of-flow spectrum the only noticeable broadband enhancement can be seen between 2 and 3 kHz, which has been identified during the calibration phase of the wind tunnel to originate from the direction of the flow collector and its integrated sound absorber. The in-flow spectrum, which is also included in Figure 3, reveals much higher pressure levels over the entire frequency range and the measured out-of-flow enhancement is completely masked by broadband noise. The reason for this effect may be seen in the contribution of some directivity noise, plus hydrodynamic pressure fluctuations in the jet. The directivity noise may have radiated from the nozzle or from the collector by-passing the out-of-flow measurements due to its directional wave pattern. These considerations are based upon the facts, that the jet itself did not radiate any significant sound due to the low Mach number and that the self-generated noise of the

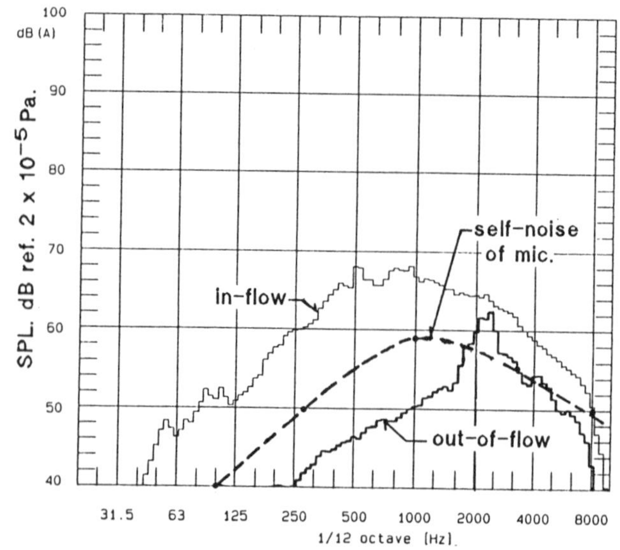

Fig. 3 In-flow and out-of-flow background noise of the BMW-Technik wind tunnel and self-noise of a microphone attached to a glider V = 140 km/h

microphone and its holder is negligible. To verify the latter, reference is made to some acoustic in-flow measurements carried out with the aid of a sailplane [6]. It has always been a major problem to establish the self-noise level of a microphone in an open jet wind tunnel, as the potential flow fluctuations (pseudo-noise) will accordingly alter the measured sound pressure level, see Fuchs [7]. The glider was equipped with several microphones, which were attached to the wing. Although it cannot be excluded, that also during these measurements airframe and turbulence noise was present, the results indicated an upper level of flow induced noise of a ¼" microphone, equipped with a nose cone. The results of these experiments are also included in Figure 3, revealing a considerable lower sound pressure level of self-induced noise of the microphone when compared with in-flow measurements in the wind tunnel. Therefore, the self-noise of the microphone can indeed be considered to be insignificant. The result, displayed in Figure 3, distinctly shows how effective the anechoic plenum chamber of the BMW-Technik wind tunnel has been designed. But it also shows, how misleading out-of-flow measurements would be for determining the background noise level which, after all, serves as a base for judging the noise floor generated by a particular test car. (Presumably, a different spectrum would have been measured if the microphone position is changed in the plenum chamber). On the other hand, in-flow noise spectra measured in open jets always contain some pseudo-noise contributions as mentioned above. Since frequency and convection speed determine the wavelength of these pressure fluctuations, the structure of a vehicle would respond differently when being exposed to an acoustic source. For a test speed of 140 km/h the acoustic wavelength is about 10 to 12 times longer than the wavelength of potential flow fluctuations at the same frequency. Due to its larger correlation length, the acoustic wave is more effective in excitating the outer panel of a car. Therefore, judging the background pressure level of a wind tunnel merely from a frequency domain may be misleading. In order to determine whether the pressure fluctuations are dominated by acoustic noise or by pseudo-noise, a cross-correlation between two microphones can be carried out. With the aid of the phase-angle measured with the two microphones the convection velocity can be evaluated and associated with the pressure fluctuation.

Finally, it shall be mentioned, that when a car is installed inside the working section of a 3/4 open wind tunnel, the air of the jet is displaced due to the solid body of the vehicle and its wake. Consequently, the flow towards the flow collector will be disturbed. For large blockage ratios in the test section the disturbances can be of such magnitude that additional noise will be generated at the outer edge of the collector, which may add to the background noise level of the wind tunnel. However, the enhanced noise floor in the test section is a consequence of the car and can only indirectly be assigned to the wind tunnel.

Figure 4 shows another example for the spectral difference between in-flow and out-of-flow measurements. The measurements were carried out in the full-scale wind tunnel of IVK of University of Stuttgart, Germany. Before this wind tunnel went through a complete acoustic upgrade program, the

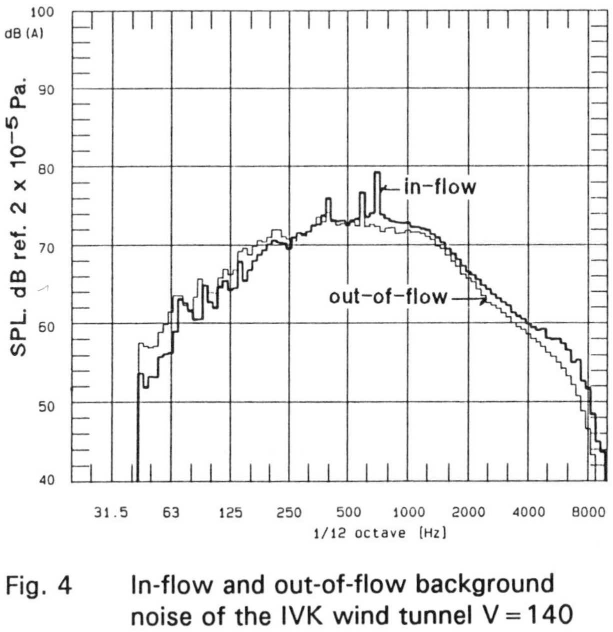

Fig. 4 In-flow and out-of-flow background noise of the IVK wind tunnel V = 140 km/h

plenum chamber existed of hard concrete walls yielding the characteristics of a more or less half open reverberation room. Apparently, the differences between in-flow and out-of-flow measurements have almost vanished. Presumably, the directivity noise is reflected now at the hard walls and is not absorbed in an anechoic room. Furtermore, it is noteworthy that this wind tunnel has not been treated acoustically at all in the remaining air circuit at the time of the measurements, and consequently the acoustic background noise may have masked the hydrodynamic pressure fluctuations of the jet.

It is noteworthy, that in the low frequency regime, the out-of-flow spectrum now reveals somewhat higher levels. They may have been generated due to recirculating secondary flow in the plenum chamber. In the high frequency regime the in-flow spectrum exhibits slightly higher sound pressure levels. The reason for this may be seen in the fact, that sound waves may escape through the opening of the nozzle and the collector, thus diminishing the effect of reverberation.

It is interesting to note that sometimes in 3/4 open jet wind tunnels which reveal somewhat higher background noise levels, acoustic testing of cabin noise is executed by the so-called "subtraction method". In order to compensate for the magnitude of the background noise of the wind tunnel on the actual acoustic measurements inside the passenger compartment, the test vehicle is first placed outside the jet in the plenum chamber with the tunnel running at the envisaged speed. After recording the cabin noise spectrum at this position, the car is rearranged inside the jet in a second step. The measurements are then repeated at the same wind speed. By off-line processing, the background noise spectrum can be subtracted from the final measurements inside the jet. The result will be the noise contribution due to the flow around the test vehicle. From Figure 4 it becomes apparent, that such kind of measurements is only sensible if the out-of-flow frequency spectrum and the in-flow frequency spectrum coincide as nicely as in the IVK wind tunnel. Consequently, it would not make sense to improve the acoustic quality of a wind tunnel by just improving the anechoic quality of the plenum chamber alone without lowering the in-flow background - and pseudo - noise floor adequately. A too low background noise level will thus be feigned and the induced flow noise inside the cabin will be overestimated.

Only for completeness, it shall also be mentioned that, in principle, due to reverberating sound waves, a test section with hard reflecting slotted or closed walls always yields a higher A-weighted overall sound pressure level than a test section exhibiting an open jet which is surrounded by an anechoic plenum chamber. Depending on the design of the flow collector and the stability of the jet, the situation may, however, be changed when unweighted levels are considered.

A comparison of two A-weighted in-flow background noise measurements is shown in Figure 5 for the aeroacoustic wind tunnels of BMW-Technik exhibiting an open test section with an anechoic plenum chamber and DNW where a closed test section with hard walls is used when it comes to car testing. Both wind tunnels reveal almost identical spectra for the low- and the high frequency regime. Yet, in the important audible range between 250 Hz and 3 kHz the closed test section wind tunnel yields about 5 dB higher sound pressure level.

For the sake of simplicity, the background noise level of a wind tunnel is often referred to an A-weighted overall sound pressure level in the literature. Figure 6 gives an example of such levels versus tunnel speed for a number of automotive wind tunnels. The results of such measurements are somewhat misleading, because it is a fact that due to weighting and averaging important information

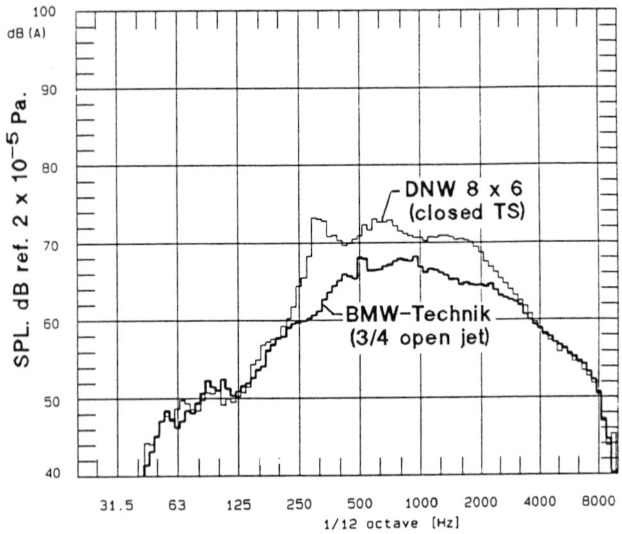

Fig. 5 In-flow background noise for the BMW-Technik wind tunnel and DNW 8x6, V = 140 km/h

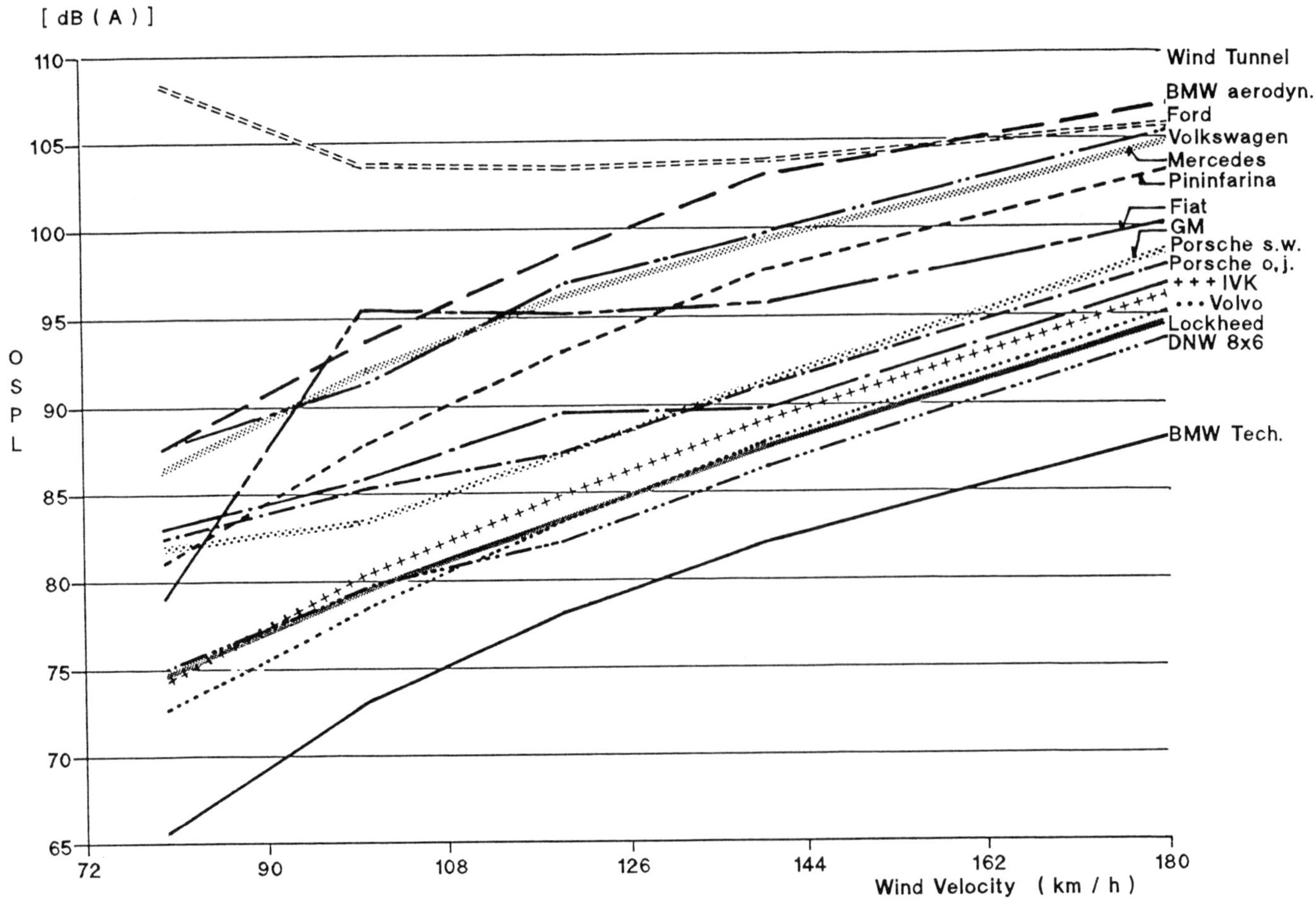

Fig. 6 A-weighted OSPL versus tunnel speed (in-flow measurements)

may be concealed. For example, Figure 7 displays a comparison of A-weighted, in-flow background noise levels of two wind tunnels with closed test sections on a spectral basis: The Lockheed 7x5 m^2 and the DNW 8x6 m^2 wind tunnel. Although the A-weighted overall sound pressure levels only differ by a fraction of 1 dB at 140 km/h, it is apparent that especially in the low frequency regime a considerable difference of up to 15 dB higher levels were measured in the Lockheed tunnel. Consequently, in the Lockheed wind tunnel this fact would make noise sources, originating from the test vehicle and radiating in this particular frequency regime, more difficult to detect by far field measurements.

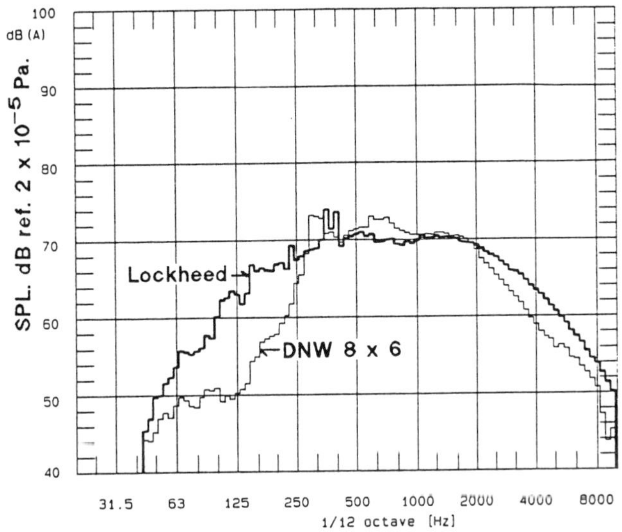

Fig. 7 In-flow background noise DNW 8x6 and Lockheed, V = 140 km/h
Note: the 500 Hz spike at Lockheed was generated by the microphone strut.

The background noise level inside the test section is very important and should be as low as possible because the noise radiating from the vehicle may otherwise be completely masked by the other noise created by the wind tunnel. A rule of thumb states that the signal-to-noise ratio should be at least 10 dB. How difficult it is to fulfill this requirement shall be illustrated by Figure 8, showing the out-of-flow noise spectra of an empty test section and a test section occupied by our test car. The measurements were taken in the wind tunnel of BMW-Technik, which is considered to be one of the best low noise automotive wind tunnels in the world. From Figure 8 it is evident that in the range between 80 Hz and 1.4 kHz, which is the most important audible range, a signal-to-noise ratio of more than 10 dB is measured. Above and below this regime the signal-to-noise ratio decreases 2 to 5 dB in the 1/12 octave band representation.

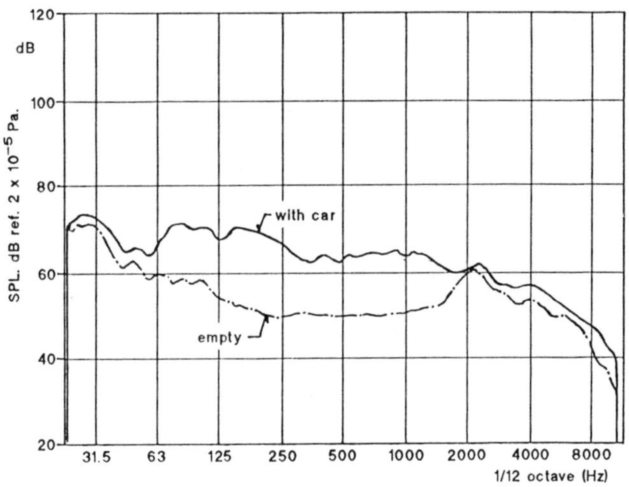

Fig. 8 Out-of-flow background noise empty test section and occupied by the car (BMW-Technik wind tunnel), V = 140 km/h

CABIN NOISE MEASUREMENTS IN A CROSS WIND FLOW

When it comes to cabin noise measurements inside the passenger compartment, the situation may be examined from a different point of view. Flow induced noise generated at the surface or at the extremities of the test vehicle may be transmitted into the cabin through structural vibrations of sheet metal or window glass, for instance. If the driver's ear is located in the proximity of such panels the noise will be experienced as very irritating. The excitation of the outer shell of a vehicle can take place in a very efficient way because panels, made from steel or glass, may be located in the hydrodynamic near-field of the source. Depending on the detailed aerodynamic shape of the vehicle, several kW of power due to fluctuating flow components may excite the outer shell at a certain test speed. In opposition to that, the power content provided by the noise floor of the wind tunnel consists of some Watts only. Thus, the ratio of excitation forces to background noise is sufficiently high enough to allow measurements inside the car. Consequently, regarding cabin noise, wind tunnels yielding a relative high background noise level may well be suited for acoustic optimization of a car to a certain degree.

An example elucidating this fact is in Figure 9, showing the relative overall sound pressure level measured at the outside ear of a rear passenger seated on the right hand side of the car. The vehicle was tested at different cross flow conditions in 14 different test sections of various wind tunnels.

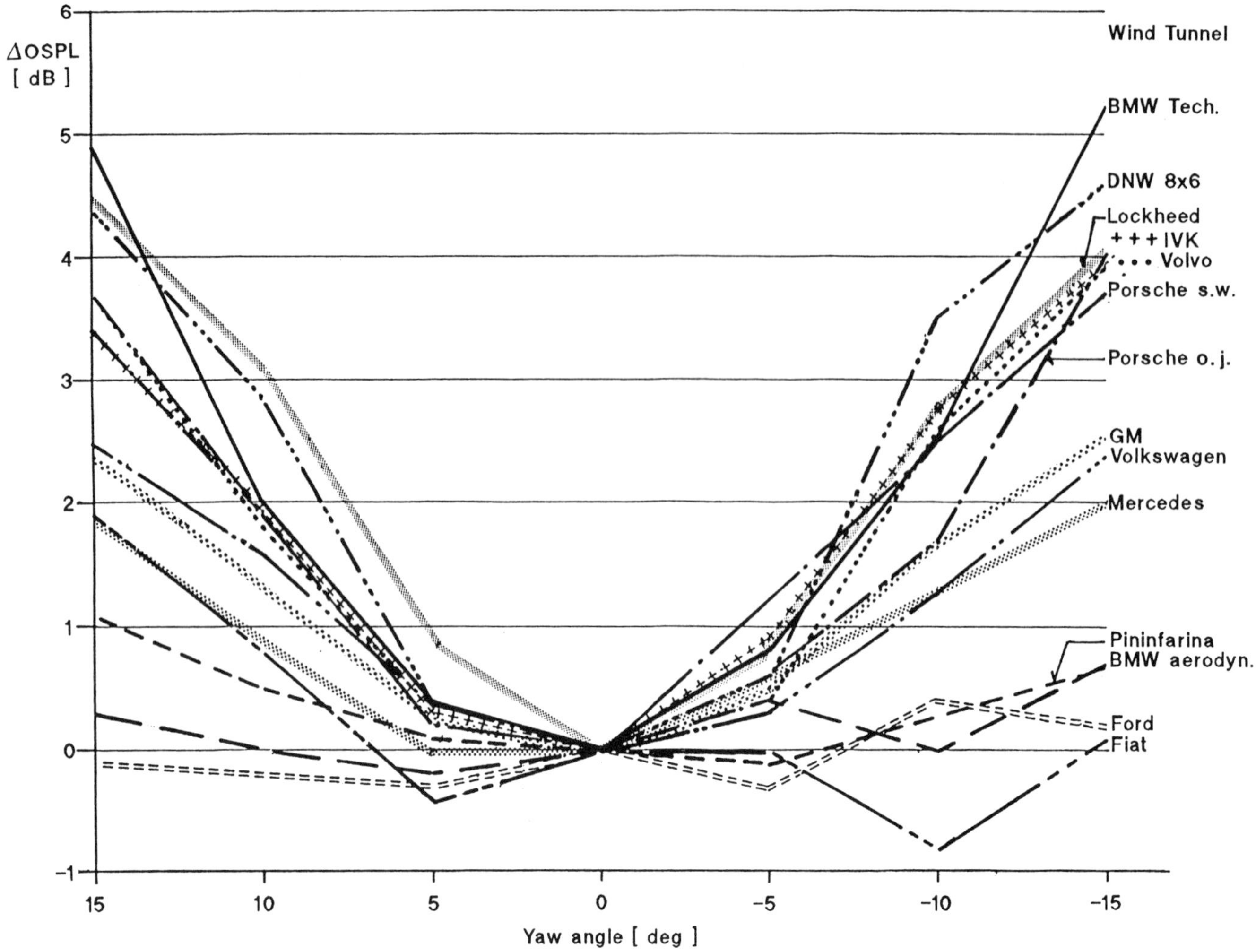

Fig. 9 Cabin noise at V = 140 km/h (ref. OSPL at 0° of yaw); Note: for negative yaw angles the microphone was at windward side

Reference is made to the overall sound pressure level measured at 0° of yaw in each particular test section.

Considering for the present, the results measured at a cross wind angle of +15°, the different wind tunnels can be classified into five groups.

Table 2 Acoustic classification of different wind tunnels

1. BMW-Technik, DNW, Lockheed
   In these wind tunnels approximately the same noise enhancement due to side wind was measured.

2. Volvo, IVK, Porsche slotted and 3/4 open jet test section
   In these wind tunnels merely 75% of the noise enhancement measured in wind tunnels of group 1 was revealed.

3. Volkswagen, GM, Fiat, Mercedes
   In these wind tunnels merely 45% of the noise enhancement measured in wind tunnels of group 1 was yielded.

4. Pininfarina
   This wind tunnel showed only 25% of the noise enhancement of the wind tunnels of group 1.

5. Ford, BMW-Aerodynamik
   In these two wind tunnels no noise enhancement due to a cross wind flow could be measured in the passenger compartment with our standard measuring equipment and data reduction system.

For a side wind of -15° basically the same classification can be proposed. Yet, the wind tunnel of Lockheed should then be grouped in 2., and the wind tunnels of Pininfarina and Fiat should be arranged in group 5 (thus, four groups will remain). The reason for this slightly different trend cannot be explained easily without further experiments. It is noteworthy that these three wind tunnels were the only facilities where the noise level inside the cabin decreased when the direction of the cross wind was changed from a positive angle to a negative angle of 15°. Finally, it shall be mentioned that yaw comparison can be misleading to a certain degree due to wind tunnel blockage effects. To do a proper comparison, care must be taken that the air speed the vehicle sees is the same between tunnels. Varying the yaw angle from 0° to 15° for instance, will result in an increased speed about the vehicle. Applying Mercker's [8] blockage correction method, a velocity increase of 1.2% can be calculated when our test car was placed in DNW's 48 m^2 closed test section and yawed to 15°. If the same vehicle is tested in an open jet wind tunnel of, say, 15 m^2 nozzle area a difference of approximately 1 dB in sound pressure level has to be considered due to blockage effects. Thereby it is assumed that the blockage effects in an open jet are about half the magnitude and of opposite sign than in an equivalent closed test section. Due to the fact that blockage corrections for open jet wind tunnels are still under discussion, we decided not to correct our data.

The frequency spectra of the cabin noise for the vehicle measured at 0° and at -15° of yaw are displayed in Figure 10. The measurements were carried out in the wind tunnel of BMW-Technik and were used as a benchmark because this wind tunnel revealed the lowest background noise level. Apparently, by yawing the car the noise level over the entire frequency range up to 3 kHz is increased.

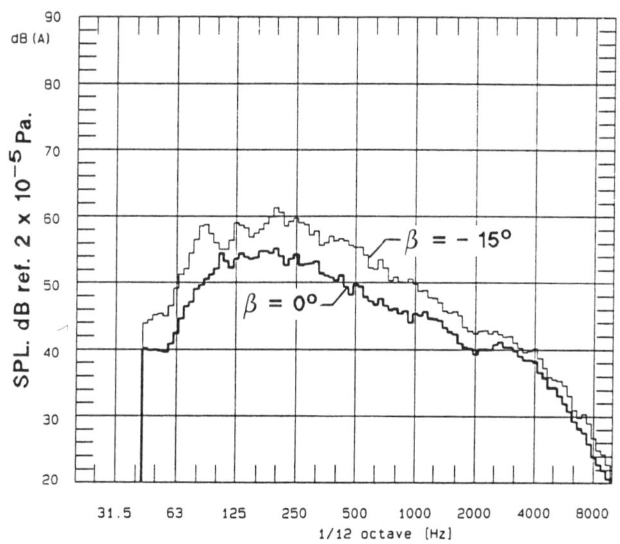

Fig. 10 Cabin noise at 0° and -15° of yaw measured in the BMW-Technik wind tunnel; V = 140 km/h
Note: microphone at windward side

The maximum enhancement takes place in the low frequency regime up to 500 Hz. Thus, those wind tunnels revealing comparatively high noise levels in the low frequency regime will be less suitable for such kind of measurements. Again, judging the acoustic capabilities of a wind tunnel merely from the A-weighted, overall sound pressure level may lead to bias conclusions. For example, from Figure 6 it can be taken that the wind tunnels of Porsche and of GM reveal almost identical OSPL at 140 km/h. However, in the latter wind tunnel only about 45% of the acoustic enhancement due to cross wind flow was measured relative to group 1, see Figure 9, whereas in the Porsche wind tunnel 75% of the noise enhancement was yielded. Notwithstanding, both wind tunnels reveal a much higher noise level in the low frequency regime when compared to DNW. From Figure 11, it can be taken that at 150 Hz a 20 dB higher background noise level was measured in the GM wind tunnel whereas in the Porsche wind tunnel this level was 15 dB higher. For frequencies in excess of 500 Hz the background noise level in both wind tunnels coincide quite nicely (the tonal components in both spectra do not play an important role for this consideration, because they cancel out in the relative representation of Figure 9). Apparently, a difference in the background noise level of only some dB in the low frequency band was decisive for the much better utility of the Porsche wind tunnel for this kind of measurement carried out with our test vehicle (in its best acoustic configuration our car yielded an OSPL of about 69 dB(A) inside the cabin).

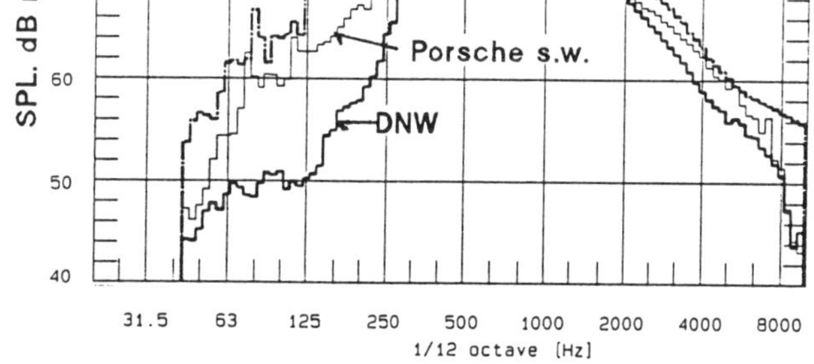

Fig. 11 In-flow background noise in DNW 8x6, GM and Porsche slotted wall; V = 140 km/h

Concluding this comparison, reference is made again to Figure 7, showing the background noise levels of the Lockheed wind tunnel and DNW. It has been shown above, that both wind tunnels reveal approximately the same utility when it comes to cross wind flow acoustic measurements, although in the Lockheed wind tunnel much higher sound pressure levels were measured in the low frequency regime. Up to 500 Hz these levels do agree quite well with those measured in the Porsche wind tunnel. But in the frequency band between 500 Hz and 2 kHz the Porsche wind tunnel yields about 5 dB higher level, possibly confining its utility in comparison to wind tunnels with a lower background noise.

It is interesting to note that for passenger cars of today the low frequencies of the spectrum contribute most to the noise level inside the driver's compartment. An example of such measurements have already been displayed in Figure 10. Evidently, the maximum A-weighted noise level for our car was measured typically at about 250 Hz illustrating how important the low frequency allotment of the background noise floor of a wind tunnel may be. Yet, it should be mentioned that also the tyres and the motor of today's cars radiate most of their acoustic power in the low frequency regime. Thus, aerodynamic noise may be partly buried and difficult to detect during driving on the road at a certain speed. Therefore, it may be argued that the classification suggested in Table 2 is not relevant because the used overall sound pressure level is putting too much emphasis on the low frequency part of the spectrum. Due to this reason, we recalculated the relative overall sound pressure levels for the frequency band of 0.8 to 1.25 kHz. However, non of the wind tunnels taking part in our comparison, revealed a different trend than our initial approach. During our experiments we were operating with omnidirectional free-field microphones. There is a good chance that for those wind tunnels classified in group 3, 4, and 5, and radiating their dominant background noise mainly from one direction, a diferent result may have been achieved if also the microphone revealed a directional characteristic and were oriented at right angles to the direction of sound propagation of the background noise. Under these circumstances some useful dedicated tests may indeed be possible.

It should be mentioned that beside the excitation of the outer shell of a vehicle, noise may enter the passenger compartment through gaps or leakages of rubber seals of doors and windows. In this case it may even be profitable to work in a wind tunnel with a higher background noise floor because noise may enter the cabin with little insertion loss and thus can easily be determined. Of course, it cannot be expected that the correct aerodynamic on-road level of the sound pressure in the cabin will be measured. Notwithstanding, for the optimization of certain vehicle components this may not be necessary. Only the verification of the final vehicle configuration may take place in an especially designed aeroacoustic wind tunnel. The absolute noise level inside the cabin of the combined influences of aerodynamic noise, motor noise and tyre noise may have to be verified on the road, anyhow.

CABIN NOISE MEASUREMENTS WITH DIFFERENT VEHICLE CONFIGURATIONS

Beside cabin noise measurements with the test car exposed to a cross wind flow, some additional experiments were carried out with different vehicle configurations. As mentioned before, the same car was then tested with the same measuring equipment in 13 different wind tunnels all over Europe and the USA (see Figure 6). Beside measurements with standard free field microphones, noise recording took place also with the aid of an "Acoustic Head" inside the passenger compartment. These experiments were carried out in most of the above mentioned wind tunnels in order to evaluate the sensation of psycho-acoustic effects. The results of this investigation may be published at a later stage by our colleagues Lindener and Zacharias from BMW-Technik who were also members of our test team.

The test car was a standard type Volkswagen Passat Variant CL and was supplied to us by the Volkswagen company. Altogether seven different vehicle configurations were investigated. During a check-out test program in the wind tunnel of BMW-Technik the different configurations were selected under the following criterion:

The configuration changes should

- exhibit only small contributions to the sound pressure level inside the cabin (2 to 4 dB in 1/12 octave bands),
- be audible in various frequency bands (between 60 Hz and 10 kHz).

Table 3 gives an overview and a brief description of the different vehicle configurations tested. Furthermore, the frequency bands in which the configuration changes were audible are indicated.

In order to provide the reader of this technical paper with a convenient survey of the results, it was decided to generate 1/12 octave band difference spectra of the cabin noise measured with different configurations. Thus, the following plots show the spectral difference of two configurations, respectively. Negative results indicating an increase

Table 3 Description of vehicle configurations tested

Conf. #	Modification	Frequency band	Effect on OSPL
7	Base configuration	-	-
6	Blocked cooling air	60 - 250 Hz	Increase
5	Sharp edged step at A-pillar (3 mm)	4 - 8 kHz	Decrease
4	Blocked triangular opening at outside mirror footing	250 - 750 Hz	Increase
3	Circular cross bar (12 mm ∅) fixed at roof rails	Strouhal frequency 650 Hz	Increase
2	10 cm leakage at rear door rubber seal	750 Hz - 2 kHz	Increase
1	Taped rear door gap (low noise configuration)	750 Hz - 2 kHz	Decrease

of the noise level and positive results reveal an improvement or decrease of the noise floor inside the cabin.

Figure 12 displays the difference spectra of all configurations as measured in DNW. Configuration 7 (base configuration) was used as a reference and the measurements were carried out at a tunnel speed of 140 km/h. One microphone was located at the driver's left ear and a second microphone was placed at the right ear of a rear seated passenger.

The comparison between data from the wind tunnel of BMW-Technik and DNW are shown in Figure 13 which displays the results for the configuration with taped rear door gap (configuration 1) and blocked cooling air flow (configuration 6). The wind tunnel of BMW-Technik is used as a reference tunnel because it yielded the lowest background noise level. It is unambiguous that the results in both wind tunnels agree very well. The same is true also for the other configurations investigated, which are omitted here. Apparently, the somewhat higher noise floor in the closed test section of DNW does not play a major role for the determination of these differences and thus it may be concluded, that the DNW wind tunnel can be used unrestricted for such kind of measurements.

The same statement can be put forward, more or less, for all those wind tunnels which have been classified into group 1 and group 2 of Table 2. Figure 14a shows good agreement between configuration 7 (base configuration) and configuration 6 (blocked cooling air flow) for the Volvo wind tunnel and DNW. It can be seen that the agreement between both wind tunnels is exceptionally good, although in the Volvo wind tunnel a slightly higher background noise level is measured in the low and the high frequency regime than in DNW. This fact can be taken from Figure 14b. Another example is displayed in Figure 15, showing a comparison between DNW and the Porsche wind tunnel with slotted walls. The test shows the noise difference between configuration 7 (base configuration) and configuration 2 (leakage at rubber seal). Except for the frequency band of 315 Hz the agreement between both tunnels is very good. The reason for the divergence at 315 Hz can be attributed to a possible beat interference of some tonal components in the frequency spectrum. As the displayed difference plots of the two configurations are not real time spectra, noise enhancement may have been feigned.

In those wind tunnels, which were classified in group 3 of Table 2, the magnitude of the noise decrease or increase is gradually diminished with increasing background noise floor. Therefore it might be difficult for the development engineer to decide in these wind tunnels, whether a further optimization of a vehicle is reasonable or not, because his development potentials may have been hidden. An example is given in Figure 16, where configuration 1 and configuration 4 are compared

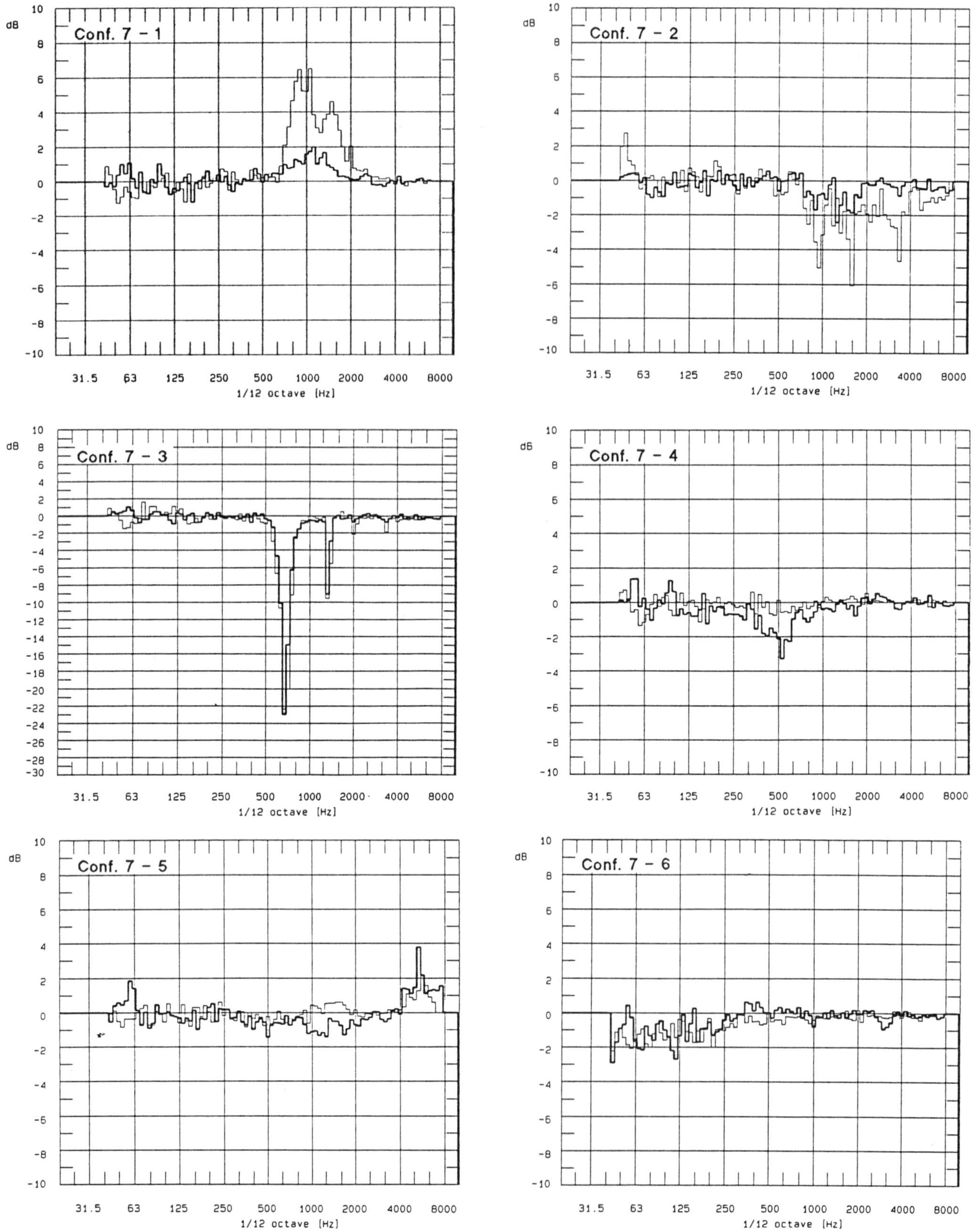

Fig. 12 DNW difference spectra for different configurations, V = 140 km/h
(Conf. 1 Taped rear door; Conf. 2 Seal leakage; Conf. 3 Cross bar, Conf. 4 Outside mirror change; Conf. 5 A-pillar step; Conf. 6 Mock-up; Conf. 7 Base configuration)
——— microphone front ——— microphone rear

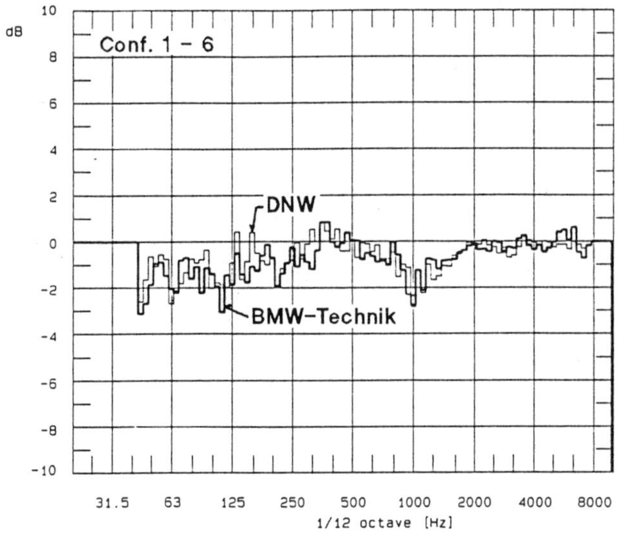

Fig. 13 Difference spectra of configuration 1 and 6 measured at BMW-Technik and DNW, V = 140 km/h

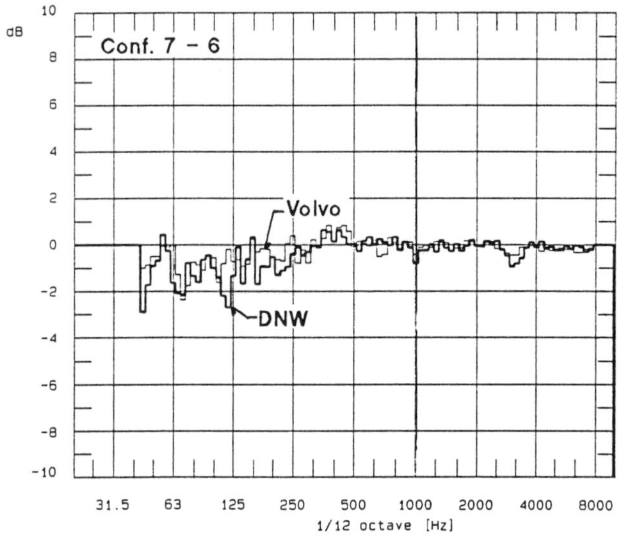

Fig. 14a Difference spectra of configuration 7 and 6 measured at DNW and Volvo, V = 140 km/h

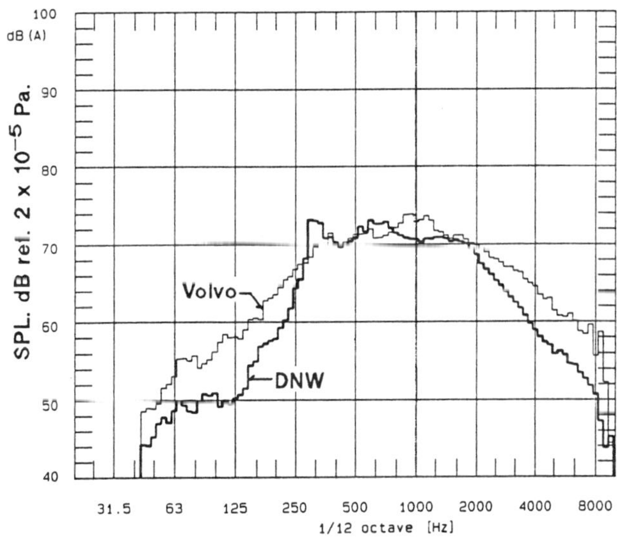

Fig. 14b In-flow background noise spectra at DNW and Volvo (empty test section), V = 140 km/h

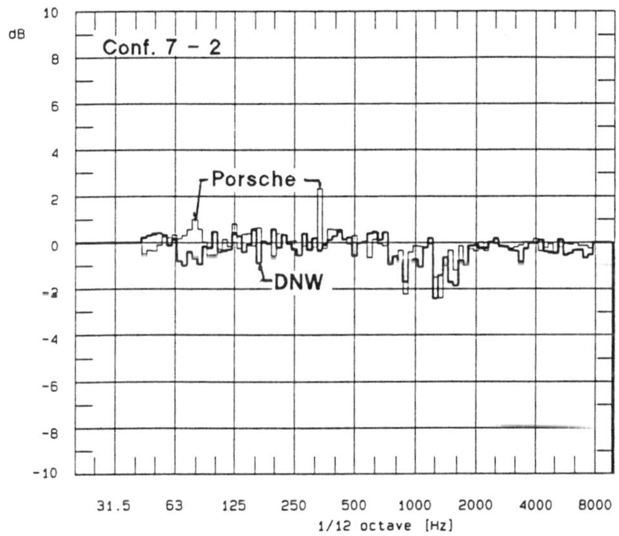

Fig. 15 Difference spectra of configuration 7 and 2 measured at DNW and Porsche (slotted wall), V = 140 km/h

with configuration 7 (base configuration) in the wind tunnels of Volkswagen and DNW. Whereas the influence of the taped rear door (configuration 1) is still perceivable on a reduced basis in the VW tunnel, the noise increase due to a modified outside mirror (configuration 4) is hardly discernible.

The rating of different wind tunnels as suggested above may give rise to the impression that an acoustical improvement of a test facility may not be necessary, whilst the A-weighted noise level is established somewhere around those wind tunnels classified in group 1 or 2 of Table 2. Unfortunately, this may only be true if, over the complete audible frequency range the A-weighted spectrum reveals a kind of smooth curvature. A good example for such a spectrum can be measured in the wind tunnel of Volvo, showing no disturbing tonal or broadband noise enhancement between 31.5 Hz and 10 kHz, see Figure 14b. An example for such tonal components, can for instance be found in the background noise spectrum of the Porsche wind tunnel, see Figure 11. These tones originate from the thyristor controlled fan motor and are generated by structural vibration of the pole shoes of the motor. Usually, it is extremely difficult to damp this structure noise. At DNW, where similar tonal

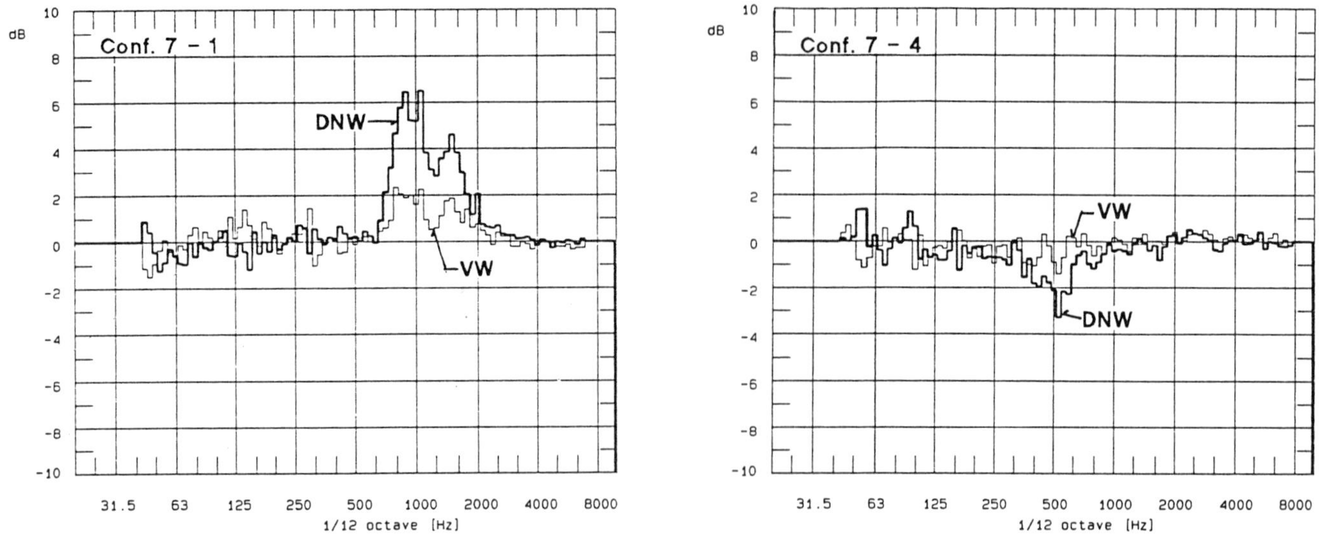

Fig. 16 Difference spectra measured at DNW and Volkswagen, V = 140 km/h

disturbances occurred, costly expenditures had to be taken to reduce the levels. Nevertheless, as long as the frequency bands of the disturbances are known, they can be considered during data interpretations.

SUMMARY AND CONCLUSION

An acoustic comparison test of eleven European and two American full-scale automotive wind tunnels was carried out with a passenger car. The scope of the investigation comprised in-flow and out-of-flow background noise measurements of the empty test sections as well as interior noise measurements inside the passenger compartment. For this purpose seven different vehicle configurations were investigated at the various wind tunnels.

In order to evaluate the results, we first tried to elucidate some principle drawbacks of different types of test sections commonly used for aeroacoustic investigations in the automotive industry. In this context, results show that the background noise level of a wind tunnel is best measured with an in-flow microphone at the exact position of the vehicle when installed at a later stage. Out-of-flow background noise measurement in an open jet surrounded by an anechoic chamber will feign a noise floor which is too low. The reason for this can be seen in the hydrodynamic pressure fluctuation generated by the shear layer, which will add to the acoustic pressure fluctuations. Thus, employing for instance the "subtraction method" to determine the aerodynamic interior noise of a car may lead to false conclusions.

Judging the utility of a wind tunnel for acoustic measurements, from an A-weighted overall sound pressure level of the empty test section, has to be done with considerable care. Today's passenger cars reveal aerodynamic interior noise spectra, yielding their most dominant content between 150 Hz and 500 Hz. If a wind tunnel reveals a high background noise level within this frequency band, A-weighting underrates the effect of low frequency noise, and the utility of the facility may be limited. Thus, a spectral representation of the background noise seems to be unavoidable when the noise level of two wind tunnels is being compared. As this sometimes might be inconvenient to verify, it is suggested to split the audible frequency range into three or more bandwidths. A good compromise seems to be taking an unweighted in-flow spectrum and calculating the overall fluctuating pressure level of a bandwidth of 63 Hz to 800 Hz, of 800 Hz to 3.6 kHz and of 3.6 kHz to 10 kHz. Thus, the low, the medium and the high frequency ranges are equally taken into account for a description of the acoustic characteristics of a wind tunnel. Furthermore, in order to take pseudo-noise into consideration, it is suggested to include the overall pressure level of the ⅓ octave band between 6.3 and 10 Hz.

For far field exterior noise measurements, the background noise level of a wind tunnel should be as low as possible. When a car is placed inside a test section, the signal-to-noise ratio should be at least 10 dB for the determination of the sound pressure level. However, even for the best acoustic wind tunnels it is difficult to meet this requirement in the low and high frequency regime. Therefore, more elaborate sound intensity measurements and correlation techniques have to be applied instead.

Regarding interior noise measurements, the background noise floor of the test section is not as decisive as for exterior noise measurements. The outer shell of a vehicle is usually located in the hydrodynamic near field of the noise source. Thus, the excitation of the outer panels can take place in a very efficient way and noise is transmitted into the cabin by structural vibration.

By means of interior noise measurements of a car exposed to a cross wind flow, a classification of different wind tunnels has been suggested. It was shown that in wind tunnels revealing an A-weighted background noise level of up to 10 dB higher than the most silent wind tunnel, 75% of the noise enhancement due to cross wind flow could successfully be measured.

Additionally, interior noise measurements with different vehicle configurations were executed. The configurations were chosen from the necessity to produce only minor changes in the interior noise level at different frequency bands. With some deduction regarding tonal noise effects, the following wind tunnels proved to be fully suitable for acoustic measurements: BMW-Technik, DNW, Lockheed, Volvo, IVK, Porsche.

REFERENCES

[1] P.T. Soderman, L.E. Olson: The design of test section inserts for higher speed aeroacoustic testing in the Ames 80 by 120 foot wind tunnel. DGLR/AIAA 14th Aeroacoustics Conference, May 1992.

[2] R.K. Amiet: Correction of open jet wind tunnel measurements for shear layer refraction. AIAA paper 75-532, 1975.

[3] R. Ross: Spectral broadening effects in open wind tunnels in relation to noise assessment. AIAA Journal, Vol. 19, No. 5, May 1981.

[4] A. Guedel: Scattering of an acoustic field by a free jet shear layer. Journal of Sound and Vibration, Vol. 100, No. 2, 1985.

[5] H.V. Fuchs, D. Eckoldt, U. Essers, J. Potthoff: New design concepts for silencing aeroacoustic wind tunnels. DGLR/AIAA 14th Aeroacoustics Conference, May 1992.

[6] B. Gelhar, B. Fuhrken: Vergleichsmessungen zum Eigengeräusch von Mikrofonsonden in Strömungen. IB 128-81/4, DLR-Abteilung Technische Akustik, Braunschweig, 1981.

[7] H.V. Fuchs: Measurements of pressure fluctuations within subsonic turbulent flow, Journal of Sound and Vibration, Vol. 22, No. 3, 10972.

[8] E. Mercker: A blockage correction for automotive testing in a wind tunnel with closed test section. Journal of Wind Engineering and Industrial Aerodynamics No. 22, pp 149-167, 1986.

930300

Audi Aero-Acoustic Wind Tunnel

Jochen Wiedemann and Gerhard Wickern
AUDI AG

Bernd Ewald and Christof Mattern
Technische Hochschule Darmstadt

ABSTRACT

The present paper reveals the design concept as well as results of experimental investigations, which were conducted in the early design stage of the planned AUDI Aero-Acoustic Wind Tunnel. This low-noise open-jet facility, featuring a nozzle exit area of 11 m^2 and a top speed of approximately 60 m/s, enables aerodynamic as well as acoustic testing of both, full-scale and model-scale ground vehicles. Ground simulation is provided by means of a moving-belt rig. The surrounding plenum is designed as a semi-anechoic chamber to simulate acoustic free-field conditions around the vehicle. Fan noise will be attenuated below the noise level of the open jet.

The work reported herein, comprises 1/8-scale pilot tunnel experiments of aerodynamic and acoustic configurations which were carried out at the University of Darmstadt. One of the primary goals was minimisation of collector induced broad-band noise and avoidance of discrete frequency open-jet excitations due to shear layer impingement on the collector. Furthermore, the pressure distribution along the empty test section and its pressure-loss had to be optimized. To match these requirements, a relatively long test section and a large collector appeared to be especially advantageous. Extensive experimental investigations were conducted to optimize the collector's aerodynamic as well as acoustic performance. The results suggested a test-section length of 10 m (= 3.1 hydraulic nozzle diameters) and a funnel-type collector featuring a collector-to-nozzle area of about 3.7. Tests with models in the test section showed that even with a blockage ratio of 18 %, reliable aerodynamic data can be obtained. It was found that pressure distributions along comparison models, taken in a large tunnel at low blockage (1.9 %), could be reproduced satisfactorily, if the model was positioned at about 40 % of the test-section length.

INTRODUCTION

In the development of new high performance cars there is a steady tendency towards increasing safety, economy and comfort. Naturally, this affects the scope of automotive aerodynamics significantly. Realisation of excellent overall aerodynamic and aero-acoustic performance has been one of AUDI's primary goals throughout recent years. Therefore, there is a growing demand for wind-tunnel testing capacity. This has been discussed e.g. in [1].

Against the background of these requirements, there have been on-going discussions at AUDI, as to wether the construction of an in-house wind-tunnel facility would be beneficial from a technical and economical point of view. After many iteration loops which started in 1984 from a pure 1:4-scale-model tunnel concept [2,3,4] and fairly extensive pre-investigations, both theoretical and experimental [5-11], in 1991 the advisory board gave way to the realisation of the Audi Aero-Acoustic Wind Tunnel for testing full-scale cars as well as 1:2.5/1:4-scale models. The facility will be located in the AUDI Technical-Development Centre in Ingolstadt, Bavaria. It will serve the entire Volkswagen Group, comprising AUDI, SEAT, SKODA and VW, as a reference and check-off tool for all aero-acoustic tests requiring free field conditions and low background noise level. Furthermore, due to the moving belt, it will provide unique potential for underbody optimizations of both, full-scale cars and scale models. The wind tunnel will also be available to customers from outside the Volkswagen Group.

Right from the start of the project, the AUDI Aerodynamics Department intended not only to control, but also to conduct much of the engineering work associated with the tunnel, and furthermore, to make use of specialists' expertise to solve certain problems, e.g. in the field of acoustics, mechanical engineering and civil engineering. This methodology was preferred to a turn-key solution.

THE WIND-TUNNEL CONCEPT

To achieve good flow quality in the test section, a Göttingen type closed-circuit air duct was considered desirable. Furthermore, this concept allows for minimisation of both, noise emission from the tunnel to the surroundings, as well as noise penetration from the environment into the testing hall. Contrary to a closed test section, an open jet in a large plenum allows for anechoic conditions, desirable for assessing noise radiation from the car to the environment. From an acoustic point of view, jet dimensions should be as small as possible to provide easy access to the vehicle, satisfactory spatial resolution of the acoustic field and sufficient signal-to- (shear-layer-) noise ratio.

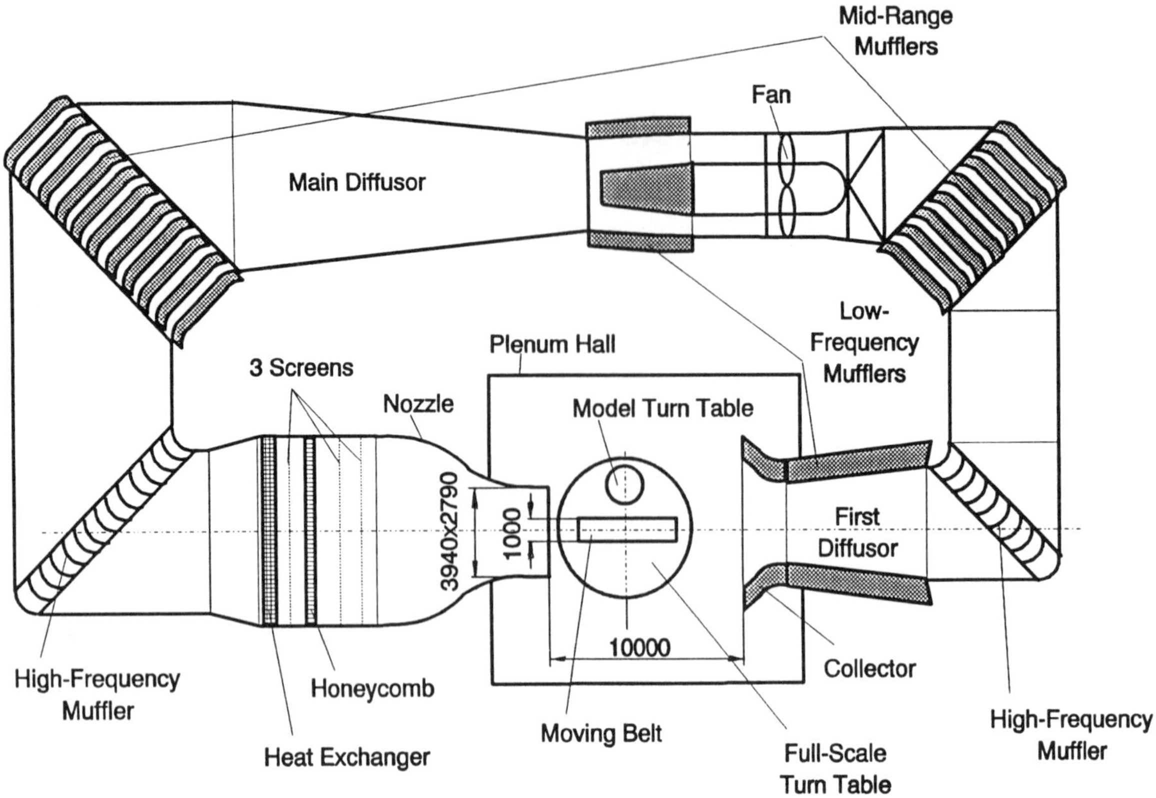

Fig. 1: Wind Tunnel Lay-Out.

Due to different Mach-number exponents of open-jet noise ($L_w \propto Ma^6$) and the noise of a vehicle ($L_w \propto Ma^{4.5}$), aero-acoustic test should be preferably performed at lower Mach-numbers, i.e. lower free stream velocities, to increase the signal-to-noise ratio of the car under investigation. On the other hand, for aerodynamic testing, especially with scale models, high airspeeds are desirable. Therefore, the tunnel is designed for a top speed of about 60 m/s. However, acoustic tests will be generally performed at velocities below 45 m/s.

Most of the air duct, except nozzle, collector and fan section will be made out of concrete. Therefore, only the aforementioned components will be housed inside a building. It also comprises workshops and laboratories for both full-scale cars and scale models. Furthermore, there will be offices and meeting rooms for the AUDI Aerodynamic Department as well as for external clients, who will be able to use the facilities separately under protection of their individual privacy.

Fig. 1 displays a sketch of the circuit lay-out the features of which will be discussed in the following sections.

AERODYNAMIC CONSIDERATIONS

In recent times it has been revealed that some phenomena, formerly termed "blockage effects", can be attributed to the interaction of the model with the nozzle and the collector [14-18]. The way of interaction is dependent on test-section length, collector size, breather size and location, model position and drag blockage. From the results given in [14,17,18], even test-section tuning appears possible, that is to say, collector and nozzle influence may, at least partially, compensate blockage effects emanating from finite jet dimensions. The nozzle area was chosen to be 11m^2, since this appears to be close to a lower limit for reliable measurements of full-scale cars.

The rectangular nozzle with a contraction ratio of 5.5 was designed at the Institute for Aerodynamics and Measurement Techniques of the Technische Hochschule Darmstadt (THD) according to the principles developed by Börger [19] and Schilawa [20]. 3d-computations of the potential flow field were conducted, using the NLR nozzle-design program [21], which has already been successfully applied during the development of the German-Dutch Wind Tunnel. Design goal for the nozzle shape was a velocity homogeneity better than ±0.2 % in the nozzle exit area, as well as avoidance of boundary-layer separation and relaminarization. According to the theoretical pressure distributions the development of the boundary layer along the nozzle walls was computed, both at THD with a 2d-integral method [22,23], and at AUDI using a 2d-finite-difference method [24].

According to the findings presented in [14,15,16] a long test section with $L_T/D_h \approx 3$ is desirable in order to minimise collector effects. However, the vehicle length has to be considered as well, i.e. the test-section length has to be an appropriate multiple of the car length. Therefore, L_T was initially chosen to be 10 m, which is a good average of existing automotive tunnels. This value was varied during the investigations. Experiments discussed in [14,17,18] revealed a considerable influence of model position along the test-section axis on the aerodynamic coefficients of the car. Furthermore, as the potential core is diminishing in a long open jet, the flow along the test object may be more turbulent than admissible [17]. In correspondence with these considerations and the present experimental findings discussed below, the model centre-of-wheelbase was fixed at 40 % of the test-section length. In the case of scale-models to be tested on a stationary ground plane, however, a more forward position is desirable to minimise the effect of the ground-plane boundary layer. Therefore, a separate scale-model turn-table, holding an external six-component model balance, is embedded in the main turn-table for full-scale cars (Fig. 1). By a 90°-rotation of the latter, the model turn-table is positioned close to the nozzle.

In recent publications the advantages of a moving ground plane and wheel rotation, even for tests with production cars, have been widely demonstrated, e.g. [11,25-33]. Therefore, apart from the conventional ground plane a high-speed moving belt (60 m/s) will be available for both full-scale and model testing. The belt length and width will be 5 m and 1 m, respectively. As can be seen from Fig. 1, it is integrated in the full-scale turn-table. In the case of full-scale cars, the belt runs within the cars' track width. Therefore, no sting support is required, and force measurements can be performed easily and in a time-efficient way by mounting the car on conventional wheel pads of the external balance. With the help of two perforated areas ahead of the belt, where suction is applied, the thickness of the approach boundary layer can be carefully controlled, i.e. according to potential-flow theory, the suction rate has to be adjusted such that the suction-induced negative flow angle remains tolerable. The result of such computations is published in [11]. To allow for wheel rotation, each wheel pad can be equipped with a small driven roller. When testing scale models on the moving belt, a sting support in conjunction with an internal balance or even the external model balance can be used. The above mentioned features make the tunnel very valuable for underbody optimizations of production cars, as well as ground-effect racing cars.

To allow for constant temperature in the test section, the tunnel will be equipped with a heat exchanger located in the settling chamber. The temperature will be kept below 25° C, which is especially valuable when testing clay models. Furthermore, screens and a honeycomb are positioned in the settling chamber.

ACOUSTIC CHARACTERISTICS

The design and construction of aero-acoustic automotive wind tunnels still requires some pioneering-spirit, though a limited number of publications on the subject has been published in recent times [34,35,36]. However, more literature is available on non-automotive facilities, see e.g. [37-40].

From [34,35,36] it can be deduced that in acoustic automotive tunnels the overall sound-pressure level (OASPL) at a free stream velocity of U=150 km/h ranges from about 72 dB(A) to 77 dB(A), whereas conventional facilities feature between 20 dB(A) and 30 dB(A) more than that. These figures apply for out-of-flow measurements taken along the jet's sideline at 5.2 m off axis and 5.5 m off axis, respectively. From the spectra given in [40], a value of (≈70 dB(A)) can be calculated for the DNW. However, measurements were taken along a 12.2 m-sideline.

The total sound power generated by the facilities incorporate contributions from various components like nozzle, collector, shear layer, fan, heat exchanger etc. To rate the acoustic quality of these components, it appears desirable to estimate the lower limit below which the sound-power level cannot fall. This limit is attained in the case of a free jet without collector and any further components. Both from theory and experiments it is known that such jets are dominated by quadrupole type noise, the power of which is proportional to U^8. Experience revealed, however, that this law applies only in the Mach number range from 0.7 to 1.6 [41]. The total sound-power level $L_{W\,jet}$ is given by EQ (1):

$$L_{W_{jet}} = 153.7 + 10\log\frac{A_N}{A_{ref}} + 80\log Ma \quad dB \qquad (1)$$

Generally, at lower Mach numbers the noise level is dominated by dipole noise generated at the nozzle exit, i.e the so-called lip-noise. This leads to the U^6-law which is shown in EQ (2) for the case of low approach-flow turbulence level:

$$L_{W_{jet}} = 142.7 + 10\log\frac{A_N}{A_{ref}} + 60\log Ma \quad dB \qquad (2)$$

EQ (1) and (2) are derived with the help of acoustic efficiency factors η=acoustic power/jet power given in [41] and a set of reference quantities like speed of sound a=340 m/s, air density ρ=1.2 kg/m^3, reference area A_{ref}=1 m^2 and reference acoustic power $W_{ref}=10^{-12}$ W. For high turbulence levels the constant in EQ (2) may be increased by 10 dB due to a change of the acoustic efficiency [41].

In conjunction with the present geometry (A_N=11 m^2, L_T=10 m) EQ (1) yields a linear sound-power level of $L_{W\,jet}$=91 dB at U=150 km/h, whereas the U^6-law, EQ (2), leads to 98 dB. The difference can, at least partially, be attributed to the lip noise. Interestingly, in [42] open-jet sound-power measurements are reported, showing the U^8-law to be valid down to Mach-numbers as low as 0.23. Consequently, an eventual potential for lip-noise optimization may be present even at low Mach numbers.

The BMW-acoustic wind tunnel, which can be considered a very silent tunnel, provides a linear out-of-flow OASPL of 107 dB at 150 km/h along its 5.2 m-sideline. This follows from integration of the spectral data presented in [35]. By adding 10 log A_{jet}/A_{ref}=24 dB (A_{jet} is defined by the surface of a semi-cylinder of R=5.2 m and L_T=10 m enveloping the jet) a total sound-power level of about 131 dB is estimated. As can be concluded from the spectra presented in [35], the 33 dB-gap relative to L_W=98 dB according to EQ (2), is mainly due to low-frequency contributions, which almost vanish when A-weighted. In this frequency range however, sound-power levels cannot simply be calculated by integrating the sound power, because the plenum-wall absorption is decreasing with frequency, and the plenum characteristic may be far from anechoic. Nevertheless, this demonstrates that also a tunnel of good acoustic quality may generate more noise than what would be expected from a pure open jet, even at low Mach numbers. This indicates further areas where optimization may be promising.

As mentioned above, the turbulence of the approach flow affects the sound generation of an open jet, and low turbulence levels should be aimed for. Usually, turbulence is damped by screens and honeycombs located in the settling chamber. However, these devices may well generate additional sound. EQ (3) gives the total sound-power level of such inserts according to results presented in [41].

$$L_{W_{sc}} = 161.7 + 30\log\zeta + 10\log\frac{A_{sc}}{A_{ref}} + 60\log Ma_{sc} \qquad (3)$$

As can be deduced from EQ (3), a low pressure-loss coefficient ζ as well as a low Mach-number in the settling chamber is beneficial for low sound generation. In the present design, there will be two or three screens of $\zeta \approx 0.9$. According to the contraction ratio of 5.5, the screen area will amount to $A_{sc}=61$ m^2. At a free-stream velocity U=150 km/h in the test section, the Mach number in the settling chamber is $Ma_{sc}=0.022$. Therefore, from EQ (3) a sound-power level due to each screen of 78.3 dB is assessed. This figure, even when added up for three screens, is still lower than the values evaluated for the open jet. However, this example indicates, that such contributions may not be negligible, especially if spectral aspects are taken into account. According to [35,41,42] the acoustic power generated by a jet attains a maximum at a Strouhal number of about S=0.2-0.4. In the present case this corresponds to a frequency below 10 Hz, if the hydraulic nozzle diameter $D_h=3.26$ m is used as a reference length. In contrast, typical frequencies of the screen noise are of the order of 1000 Hz.

Another important potential for optimization of wind-tunnel noise is provided by the fan. Generally, each manufacturer will predict the sound-power level of his fan. Nevertheless, a rough estimate within ±4 dB accuracy can be made with the so-called Beranek formula [44]. In EQ (4) this formula was rewritten for the volume flux Q in m^3/s and the total-pressure increase Δp in Pascal.

$$L_{W fan} = 40.2 + 10\log\frac{Q}{m^3/s} + 20\log\frac{\Delta p}{Pascal}\, dB \qquad (4)$$

In the present design stage a pressure loss of $\Delta p=1100$ Pa is estimated for a Volume flux Q=458 m^3/s (U=150 km/h). From EQ (1) the total sound-power level of the fan can be assessed to be $L_{W\,fan}=127.6$ dB. Naturally, the spectral distribution of this power is essential. Apart from broad band noise, there may be strong contributions from shaft order tones. They originate from the interaction of a rotor with B blades and a stator with V vanes, which yields rotating acoustic pressure patterns. These patterns contain m circumferential wavelengths or lobes (modes) rotating with nB/m times shaft speed. According to [45] the mode number m is defined by:

$$m = |nB + kV\;|, \qquad k = 0, \pm 1, \pm 2, \ldots \qquad (5)$$

n : *harmonic number*

Herein, the term "mode" refers to the simplest ways in which the pressure can be distributed. From the infinite variety of modes which can be obtained from EQ (5) for any given combination of B and V, only a very limited number is of practical interest. A second condition determines, whether a specific mode can propagate along the fan duct as a spiral wave, or whether its pressure amplitudes decay exponentially with axial distance. The transition from decay to propagation occurs when the circumferential Mach number M_m of a spinning mode with which the m-lobe pattern sweeps the outer duct wall, exceeds the value of $k'^{\sigma}_{m\mu}/m$, which is the solution of a characteristic function represented by a combination of Bessel functions. Its value depends on the hub-tip ratio (σ), the number of circumferential lobes (m) and the number (μ) of nodes across the annulus in the radial direction. $k'^{\sigma}_{m\mu}/m$ Is tabulated in [45]. In conclusion, the fan design has to make sure, that only those m-modes are activated which cannot propagate along the duct. Fig. 2 illustrates the situation at U=162 km/h for a fan featuring $\sigma=0.5$, B=20 rotor blades and V=14 stator vanes. According to EQ (5), for the first harmonic (n=1) only mode numbers m>5 are activated. However, according to the tabulated functions given in [45] only the numbers m<5 can propagate (curve denoted by "$\mu=0$" inf Fig. 2). This reveals one of the main goals of fan design, i.e., by selection of an appropriate number of blades and vanes, excitation of modes that are able to propagate in the duct has to be avoided. One such mode which is usually not cut-off, is the plane-wave case m=0. It is excited by an equal number of blades and vanes, or if the former is a multiple of the latter, as can be deduced straightforwardly from EQ (5). In order to extent the cut-off range to higher harmonics, i.e. higher frequencies, the pressure pattern can be provided with a specific radial waviness. Thus, higher radial (μ) modes are introduced. Fig. 2 illustrates that for a given harmonic number n, the higher the μ number the more duct modes will be cut-off. Physically, the higher radial modes are established by an angular phase shift between hub and tip, e.g. due to vane lean, vane sweep or rotor twist [46]. In the present example more than n=5 harmonics were considered. The computations plotted in Fig. 2 revealed that up to this limit all m modes are cut off, if $\mu=5$ radial modes are present. This can be achieved with an angular phase shift between hub and tip of $\Delta\Theta=9^0$.

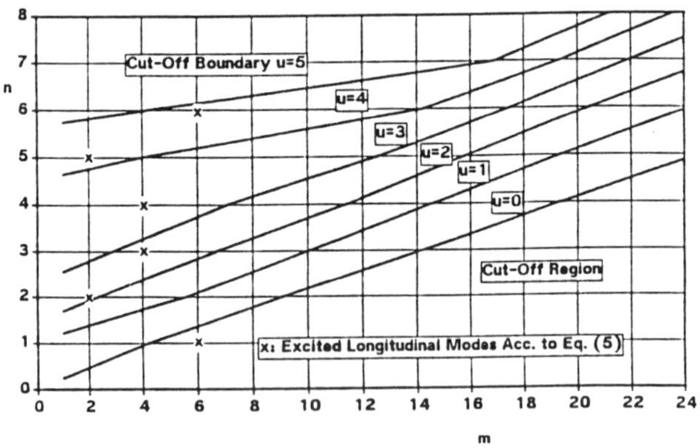

Fig. 2: Radial and Axial Mode Distribution with Cut-Off Criteria for a Fan with R=2.5 m, $\sigma=0.5$, B=20 and V=14 rotating at 350 rpm (U=162 km/h).

Nevertheless, even after tonal noise contributions are minimized, the axial fan still radiates too much noise to be placed into a "silent tunnel" without any sound absorption. In the present case, the turning vanes upstream and downstream of the fan are designed as mufflers (Fig. 1). Furthermore, noise attenuation can be enhanced with an acoustic treatment of the fan section and the collector area. Investigations which are still under progress have to clarify, whether acoustic treatment of the two remaining corners is required. The materials selected for sound absorption will be mineral wool, covered with perforated sheet metal and polyester foam. Due to the large acoustic insert losses which have to be attained especially in the low frequency range, relatively thick absorbers are required. To a certain degree, this can be compensated by selecting a somewhat higher number of rotor blades, to increase the main blade-passing frequency (BPF). The present fan design with B=20 provides a BPF of 104 Hz at a free stream velocity of 150 km/h (324 rpm).

To come to a realistic estimate of the required muffler performance, the noise radiation of the test objects has to be considered as well. For weak noise sources along the exterior contour of acoustically already optimized vehicles to become detectable outside the flow, a tunnel-noise attenuation down to the level of the open jet seems desirable. However, this level still remains to be defined. The physical limit according to EQ (2) appears unrealistic, because low-frequency noise from the collector cannot be avoided. As very limited information on low speed open-jet wind tunnels is available in the literature, 1/8-scale measurements were conducted with a very silent model of the present test-section configuration. The experiments are to be discussed in the next paragraph. An important result is displayed in Fig. (5). It reveals the A-weighted sound-pressure spectra measured at U=150 km/h, approximately in the middle of the 5.2 m sideline of the jet with an optimized collector. By integration an OASPL of 56 dB(A) can be found. As to the fan, realistic A-weighted sound-power levels amount to about 114 dB(A). With respect to the present geometry and under the assumption of semi free-field conditions in the plenum, this figure can be converted into sound pressure by subtracting $10 \log A_{sp}/A_{ref}$=25 dB. A_{sp} refers to the surface of a semi sphere, the center of which is located in the nozzle exit. Its radius is defined by the distance of the microphone position from the center of the nozzle exit, i.e. $R=\sqrt{5^2+5.2^2}$ m. These considerations lead to an A-weighted OASPL of about 89 dB(A) in the plenum due to the fan. The difference, 89 dB(A)-56 dB(A)=33 dB indicates the overall acoustic insert loss of the muffler. However, this is but a simplified estimate. In practical applications, sound attenuation has to be considered bandwise.

To achieve semi free-field conditions in the plenum, the surrounding walls and the ceiling will be treated with sound absorbing material, i.e. wedges from mineral wool. The lower cut-off frequency, where the absorption coefficient attains a value of 99 % will be of the order of 120 Hz. By linear scaling of the data published for the DNW in [40], this requirement leads to a wedge length of only 0.53 m. However, according to potential suppliers a wedge length of about 0.8 m was deemed necessary. Furthermore, all sound reflecting surfaces like nozzle, collector, traversing gear etc. have to be acoustically treated to improve the anechoic character of the plenum.

COLLECTOR DESIGN

In the following the collector design is discussed as an individual issue, because it has an influence on almost each of an aero-acoustic wind tunnel's specifications:

- it should help to minimize the power loss of the test section, since this is the largest contributor in an open-jet wind tunnel.
- it should not affect the pressure distribution of the model investigated. Especially the base pressure is not be increased.
- it should not produce significant self-noise due to its exposure to high flow velocities.
- it should help to avoid resonance effects, which might jeopardize the structure of the plenum hall.
- it should provide a possibility for fine tuning of the static pressure gradient along the test-section axis.
- it should suppress unwanted secondary flows in the plenum chamber, since the exterior noise and the location of noise sources can be investigated most conveniently by out-of-flow measurements at the sideline of the jet.

In the literature little information is available on the optimum collector design for open-jet aero-acoustic wind tunnels. Paterson et al. [38] chose a conventional collector design, which produced the well known resonance effects at low frequencies. The resonance had to be suppressed by jet exit vanes attached to the nozzle, which serve perfectly in reducing the sound power for the low frequencies, but introduce additional aerodynamic noise at high frequencies.

The same effect was experienced by Hanson [37]. He solved the problem by modifying the collector design, i.e. he added rounded lips with a diameter of nearly half the collector width. This eliminated the resonance entirely.

Ackermann [35] performed extensive experimental investigations with an 1/8-scale model of the BMW aero-acoustic wind tunnel to optimize the collector geometry. But the configuration of the BMW tunnel with the fan on axis with the test section was dictated by the limited space available in an already existing building. Thus, Ackermann's result are of limited value for an aero-acoustic tunnel of Göttingen type. However, a comparison of the pure test-section geometry shows that the optimization led to an extraordinarily large collector cross section featuring A_C/A_N=2.23. According to Lindener and Stadler [36] the benefit of the large collector is due to the circumstance, that the strong fluctuations in the shear layer enveloping the jet can not interact with the collector surface.

Ross [48] also reports on model-scale investigations, which were conducted during the design stage of the DNW aero-acoustic configuration. The aero-acoustic open-jet test section is a combination of the medium sized closed test-section's nozzle (8 m x 6 m) and the collector of the large test section (9.5 m x 9.5 m), thus limiting the costs for additional hardware. Consequently, there is a relative collector cross section of A_C/A_N=1.88. This area ratio was obviously not sufficient to avoid pumping, since additional breather slots were required. The breathers have a similar disadvantage as turning vanes; they produce a considerable increase in high-frequency noise (1-5 kHz) in the test section. To attenuate this additional background noise, Ross proposed a collar-shaped muffler for the breathers.

All these alternatives to suppress the low-frequency resonance for a conventional wind tunnel are discussed in an early paper by Seiferth [43]. He already proposed jet exit vanes, breather slots and a large collector cross section. Due to their self-noise penalty, the first two options are not an optimum solution for an aero-acoustic wind tunnel. The third measure has the disadvantage (according to Seiferth) to increase the power loss of the test section, which may result in an increase in fan noise and, not to forget, an increase in costs. Thus, at first sight none of the three possibilities appears recommendable. In addition to the three countermeasures being related only to the test-section design, Seiferth remarked, that tunnels with high aerodynamic quality, i.e. with low friction loss in the airpath, are generally more susceptible to resonance effects. That means in turn, that a design with high friction losses, for example due to mufflers, is favourable in this case. Thus, from the available literature on the subject, some ideas and hints could be picked, but an independent optimization appeared necessary.

SCALE-MODEL EXPERIMENTS

EXPERIMENTAL SET-UP - The scope of the experimental investigations reported herein comprised both, the verification of the aerodynamic as well as the acoustic test-section concept. In order to limit costs and speed up the pilot test, ways had to be found to facilitate the experimental set-up. Consequently, part of the closed circuit (Göttingen) air path was omitted. At the Technische Hochschule Darmstadt an existing open circuit, closed-wall (NPL type) tunnel was converted into an Eiffel tunnel, such that it would hold the 1/8-scale model of the AUDI plenum, including the nozzle, open-jet test section and collector (Fig. 3). All dimensions were scaled linearly, except for the acoustic treatment of the plenum walls. To model the 0.8 m-wedges which are to be used in full-scale, an appropriate commercially available substitute had to be found. This was achieved by selecting prefabricated panels of mineral wool (G+H Isover). To minimize erosion effects the surface was covered with a highly transparent fleece. According to a recommendation by G+H, two different panels were combined, thus yielding a total thickness of 150 mm. The absorption coefficient of the composite panel was claimed to be approximately 0.6 at f=500 Hz. With reference to the model-scale this corresponds to f=63 Hz in full-scale.

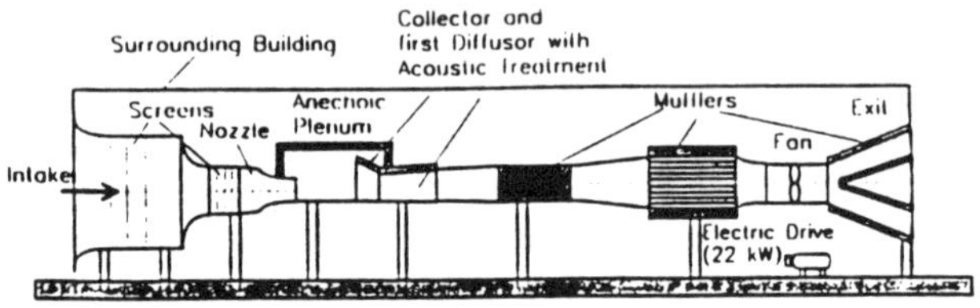

Fig. 3: Wind Tunnel at Technische Hochschule Darmstadt Housing the AUDI Pilot Tunnel.

The nozzle was fabricated at the AUDI model workshop. The computed nozzle contour was fed into the AUDI CAD-System (Computervision) and a negative shape was milled fully automatically from epowood. The final (positive) shape was obtained by laminating with epoxy resin. Overall shape accuracy was better than 0.2 mm. The 1/8-scale models were equipped with equally spaced (50 mm) pressure taps along one side-wall, top-wall and corner. To check nozzle symmetry, some control taps were bored on the opposite side.

The test-section floor was also equipped with pressure tabs spaced between 20 mm and 40 mm in the mid-section and between 40-mm and 80-mm in sections at y=-76 mm and -123 mm. In relative values based on nozzle width W, this is equivalent to y/W=-0.15 and -0.25.

The different collectors investigated were entirely designed by AUDI and fabricated from plywood in the AUDI model workshop. They feature a large bandwidth of possible entrance-area ratios, ranging from about A_C/A_N=1.0 to 5.4 (A_C=collector area; A_N=nozzle area). Whereas an area-ratio of only 1.18 was obtained as a result of open-jet computations according to the findings by Regenscheit [47], a larger collector seemed desirable for both, acoustic characteristics and for reduced liability to blockage effects [14,15]. However, only the approach by Regenscheit satisfies continuity without creating streamline curvature or requiring breathers, i.e. the collector "cuts-out" exactly the volume-flux discharged by the nozzle. With a larger collector area however, part of the entrained open-jet shear layer is collected as well. Thus breathers may become necessary to avoid a static pressure gradient due to streamline curvature, i.e. due to flow retardation. Unwanted horizontal buoyancy would be the result of such a pressure gradient. The position and size of eventual breathers was subject to extensive experimental optimizations some of which are reported by Mattern [18].

To exactly measure the power loss of the test section, the pressure taps were extended up to the end of the first diffusor. Additionally, both walls and the ceiling of the diffusor were also equipped with pressure tabs. This allowed for a measurement of the power losses without affecting the local flow profile at the collector entrance. The losses were determined with and without model in the test section, and it was a good check for the aerodynamic quality of the configuration, whether the increase of the test-section loss due to the model corresponded to the known $c_D \cdot A$ value or not.

The car was also prepared for contour-pressure-distribution measurements. It was a commercially available 1/8-scale model of the Chevrolet Corvette, which had been reinforced to prevent deformation. It was equipped with an internal pressure transducer with mechanical scanner. Pressure measurements were conducted with two different systems, a multi-transducer PSI-System provided by THD and with the AUDI system based on Scanivalve mechanical scanners.

Acoustic measurement were performed using a Brüel & Kjaer sound-pressure meter connected to a Scientific Atlanta spectrum analyzer. Thus, both 1/3-octave-band analysis to measure aerodynamic noise levels and narrow-band analysis to check for unwanted resonance peaks could be performed. With the described set-up all relevant quantities could be measured simultaneously for a given configuration, and an optimization with respect to all the above listed specifications was possible.

RESULTS OF COLLECTOR OPTIMIZATION - For a given cross section of the duct with an area ratio A_D/A_N=1.6 downstream of the collector, the collector area was varied between A_C/A_N=1.1 and 2.0. It was found, that the acoustic properties were improved when coming to larger collectors, although the specific loss coefficient of the test section, k_T was nearly constant. The results for the specific loss as a function of area ratio are given in Fig. 4. For comparison, the k-value of 0.23 for a conventional design with A_C/A_N=1.2 followed by a duct of the same cross section has been added. This figure was also measured in the model tunnel. The predicted value for the test-section loss, deduced from the theory by Regenscheit [47] for turbulent jets, was k_T=0.21. This theoretical value is but a lower limit for the real power loss, and the relatively small increase in the experimental value can well be explained by local friction losses at the collector entry. The strong increase from 0.23 (for the conventional design) to approximately 0.29 (for the different collectors combined with a duct of area ratio A_D/A_N=1.6, see Fig. 4) is, at least partly, to be attributed to further flow retardation without full pressure recovery inside the large cross section of the duct.

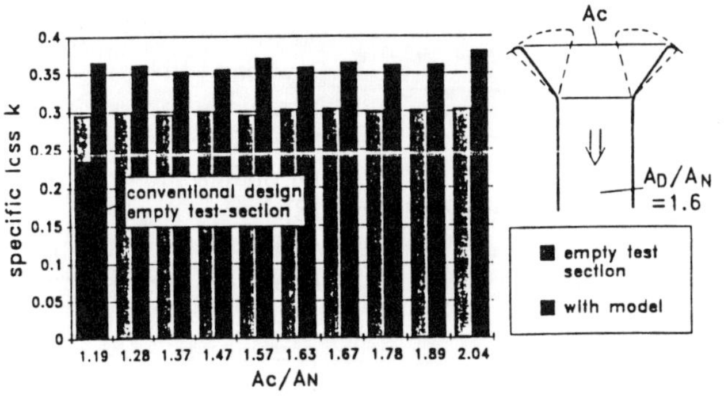

Fig. 4: Power-Loss Coefficient of Test Section as Function of Collector-Entrance Width.

Acoustic measurements were not conducted in this early design stage, since at that time, there was no acoustic lining in the plenum chamber. Thus reliable acoustic measurements were not possible. Anyway, the reduction of self-noise with large collector cross-sections was obvious and the benefits of this design were corroborated by later measurements, which will be reported in the following.

A large cross section is desirable to decrease the tunnel's overall pressure loss. Therefore, at first glance, a reduction in cross section from the large collector area to a bottle neck followed by a diffusor to retain a large cross section at the first corner again, doesn't appear meaningful. But it turned out, that a considerable increase in power loss up to a value of k_T=0.47 was obtained, when the bottle neck between collector and first corner was omitted. By reducing the cross section at the bottle neck until the flow was completely attached along its circumference, a power loss was obtained which was only slightly higher, when compared to a conventional collector in combination with a diffuser. If the additional losses were completely attributed to the diffusor, the calculated upper diffusor efficiency would be 82%. The k-value for the collector-diffusor combination will be approximately 0.31. The cross section at the bottle neck is now A_{BN}/A_N=1.58 and the cross section at the corner is A_{CO}/A_N=1.98.

With the bottle-neck cross-section taken as fixed again, now various collector designs were studied. As mentioned earlier, the collector influences most of an aero-acoustic wind tunnel's specifications. Therefore, suitable measuring techniques had to be applied and from the measured data, key values had to be derived, which characterize the different configurations with respect to the various criteria. The first such key-value is of course the specific loss k, which was taken for the complete collector and diffusor unit.

A uniform static pressure distribution in a sufficiently long part of the empty test section is a basic requirement. Thus the length, where c_p on the centerline remains below 0.01 is taken as a measure, although this gives but a rough estimate of the quality of the pressure distribution. The complete curve shape has to be checked at a later design stage. The final criterion for the pressure distribution will be $|c_p|<0.003$.

The generation of self-noise was measured using microphones and a spectrum analyser. The microphone position was outside the flow, 500 mm behind the nozzle-exit plane, at 650 mm sideline and 500 mm above the ground. 1/3-octave-band analysis was chosen, which is most suitable for rating the aerodynamic noise. Due to the 1/8-scale-model investigation a factor is introduced in the frequency distribution of the noise generated. The frequencies measured in the model tunnel are 8 times higher than those to be expected in the future full-scale tunnel. Therefore, the measured 1/3-octave-band analysis has to be downshifted by 3 octaves. Now, with the correct frequency scales for the full-scale tunnel, A-weighting was applied to obtain a measure for the relevance of the measured sound-pressure levels to the human ear. This is possible, since the sound pressure levels at least theoretically are not affected by the model-scale (see Heckl and Müller [41]). Typical results are shown in Fig. 5. Apparently the extreme low and high frequencies (below 100 Hz and above 2000 Hz) do not contribute to the A-weighted OASPL. Thus, the chosen frequency range 25-20000 Hz for the model-scale seems to be sufficient to calculate the correct A-weighted OASPL. This result is corroborated by similar findings by Ackermann [35], who compared model-scale and full-scale experiments in the BMW aero-acoustic wind tunnel. Thus, the OASPL predicted from the model experiment can be used as a key-value for self-noise generation.

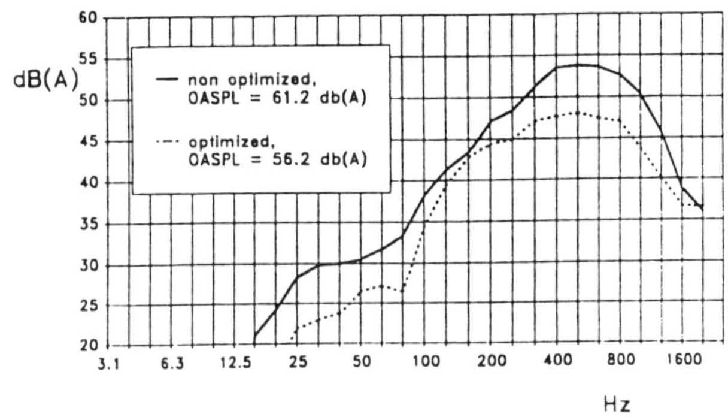

Fig. 5: 1/3-Octave-Band Analysis of Wind Tunnel Self-Noise for Different Collectors (U=150 km/h).

The liability to resonance effects was also checked by the microphone measurements, using the same set-up as described above. However, non-weighted signals were processed and narrow-band FFT-analysis was applied in the frequency range 0-100 Hz. A typical result is shown in Fig. 6. None of the various configurations investigated showed the typical resonance peaks. By contrast, Ackermann [35] found a 20 dB resonance peak, in the whole velocity range for a Strouhal number of 0.4. Indeed, in Fig. 6 for the optimized configuration a broad maximum is present at a slightly lower value of the Strouhal number, but it stays well below the SPL at lower frequencies. Thus it was concluded, that in our case the non-weighted OASPL should be a reasonable measure for possible excitations of resonant frequencies of the air duct and the plenum chamber. Due to limitations of the spectrum analyser the lowest measured 1/3-OBSPL center frequency was 25 Hz. Nevertheless, higher SPLs may exist at frequencies below 25 Hz, as revealed by control measurements. But 25 Hz in the model-scale correspond to a full-scale frequency of 3.15 Hz, whereas the calculated room resonances of the plenum chamber are above 7 Hz. Additionally, the Strouhal frequency of the jet (5 Hz at 40 m/s) is within the measuring range, and for most of the configurations the difference in the measured 1/3-OBSPLs appears as a parallel offset over the whole frequency range below 630 Hz. For the first evaluation it was thus accepted to cut off all frequencies below 25 Hz (3.15 Hz).

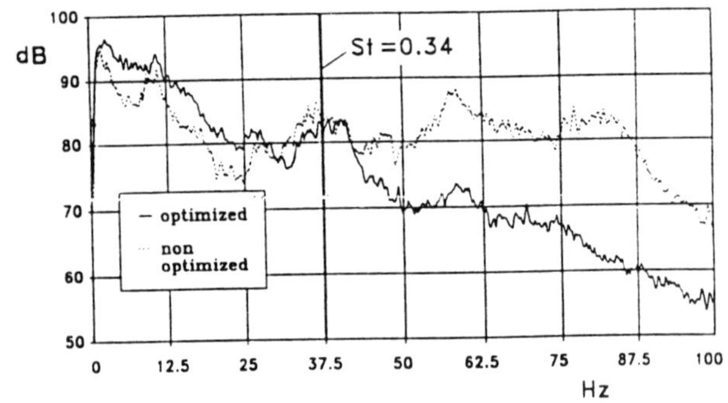

Fig. 6: Narrow-Band Frequency Analysis of Non-Weighted Signals for Different Collectors (U=150 km/h).

With the present collector geometry, no fine tuning of the pressure gradient with respect to the reproduction of known pressure distributions along models has been conducted, by the time of writing this paper. Nevertheless, in the next chapter some experiments carried out with earlier collector configurations will be discussed. One has to keep in mind however, that collector tuning always requires modifications like flaps or breathers. Consequently, the modified configurations are also subject to all the above mentioned criteria and tests.

Finally a test procedure for the strength of the secondary flow on the sideline of the jet has to be defined. But this criterion is not of primary importance, and therefore the treatment of this topic is postponed to a later design stage.

Now, by reducing the originally discussed aerodynamic properties to a set of 4 key-values (the power-loss coefficient k, the non-dimensional length of uniform pressure distribution $x(c_p<0.01)/L_T$ and the non-weighted and A-weighted OASPLs), the aero-acoustic quality of different collector designs can be easily compared.

In Fig. 7 some of the various collectors investigated in this study are sketched. As a standard case, a configuration equipped with simple rounded lips was chosen. Due to the enlarged cross-section of A_C/A_N=1.58 and the lips with R/d_h=0.1 it features already much better acoustic properties than the conventional design. Then, a collar directed perpendicular to the flow direction was added to obtain the second configuration. The surface of the collar as well as the inner surface of the collector was covered with sound absorbing material. For configuration 3 the collar was modified to form a 45°-funnel shape. To retain the original test-section length, the funnel was shifted backward by an amount of L/L_T= 0.1 (conf. 4). The influence of the funnel size was studied with configuration 5. To clarify the different effects of lip and funnel design, the last configuration was build up as a combination of both (conf. 6).

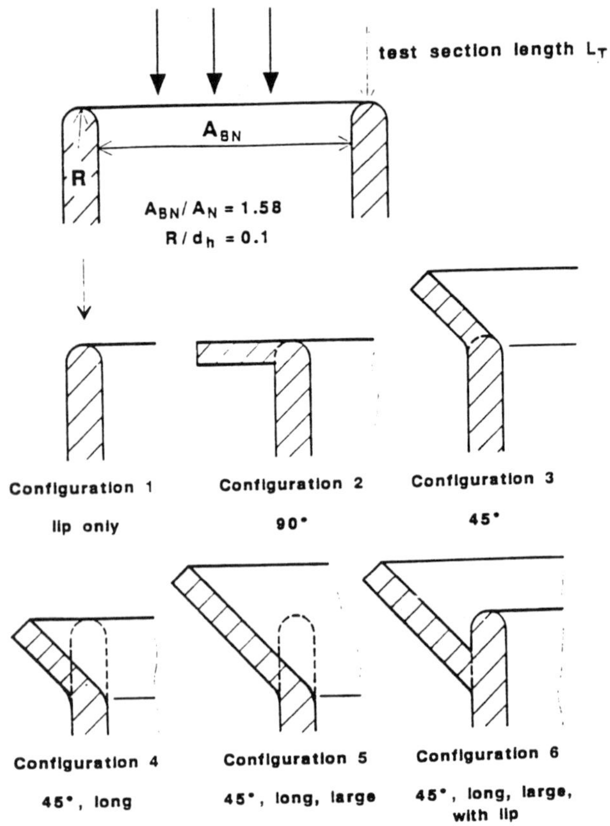

Fig. 7: Geometry of Investigated Collector Configurations.

The results for these 6 configurations are listed in Table 1. It can be seen clearly, that the funnel type collector is advantageous with respect to the power-loss coefficient as well as for the non-weighted OASPL. The only disadvantage of the funnel shape is the relatively short range of uniform pressure distribution. Thus, in configuration 4 the funnel was shifted backwards and the length of the uniform pressure distribution was increased by almost exactly the same amount. The increase in length is paid with a higher power loss and a slightly increased linear OASPL. A larger funnel (conf. 5) is better than a small one, as far as the power loss is concerned, but the benefit of the larger collector cannot compensate the losses due to the increased test-section length. In addition, more than half of the gain in extent of uniform pressure is lost again due to the stronger upstream influence.

The funnel-lip combination denoted configuration 6 demonstrates the funnel effect. The pressure distribution is almost identical to conf. 1 without funnel, whereas power loss and linear OASPL closely approximate the values of the funnel configuration (3).

The results suggest that the collector-to-nozzle area A_C/A_N may not be an adequate parameter to fully describe the collector performance with respect to all relevant criteria.

The weighted OASPL is almost constant for the listed configurations. A significantly higher weighted OASPL is only found for configurations with A_C/A_N much smaller than 1.6 (e.g. Fig. 5). For large collector cross-sections, the weighted OASPL is dominated by lip noise emanating from the nozzle. The frequency spectra of the generated self-noise differ only for frequencies below 100 Hz, as can be seen in

Fig. 8. These results justify the use of the linear OASPL as a key-value for the low frequency excitations, since the difference in the OASPL is almost exactly the offset between the curves.

Conf.	Collector Geometry	Power Loss Coefficient	$x(c_p < 0.01)$ $/L_T$	OASPL [dB]	OASPL [dB(A)]
1	lip only	0.328	0.689	101.7	57.0
2	90°	0.326	0.701	99.7	57.2
3	45°	0.310	0.681	96.2	57.3
4	45°, long	0.339	0.789	96.7	57.2
5	45°, long, large	0.324	0.712	97.0	56.2
6	45°, long, large, lip	0.314	0.687	98.8	56.9

Tab. 1: Comparison of Aero-Acoustic Quality of Different Collector Designs by Predefined Key-Values.

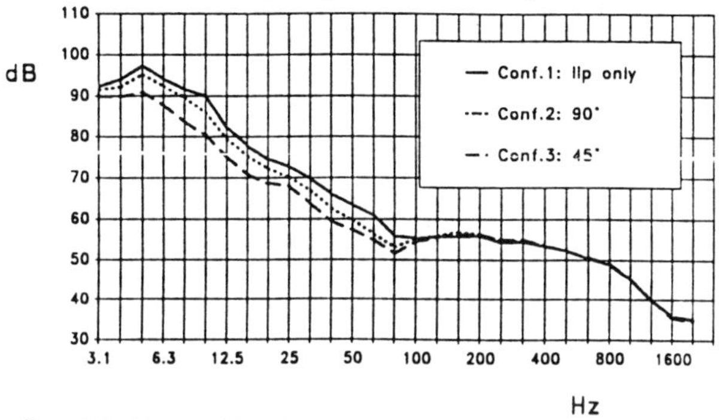

Fig. 8: 1/3-Octave-Band Analysis of Wind-Tunnel Self-Noise for Different Collectors (U=150 km/h).

An additional problem with the shape of the c_p-distribution can be deduced from Fig. 9. A weak negative pressure gradient, which is almost invariant to the collector design, is experienced with all configurations. The reason for this is not known up to now, although there is some evidence, that the gradient might be introduced by the nozzle design.

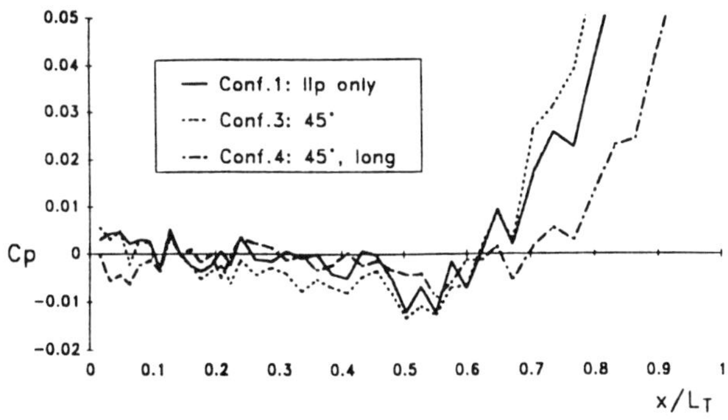

Fig. 9: Pressure Distribution along Centerline of Test Section.

An important criterion, which has not yet been mentioned, are the costs for manufacturing the collector. From Table 1 it can be deduced, that large funnels (conf. 5, conf. 6) make no sense. The aero-acoustic properties are not improved, and the costs, which are as a first approximation proportional to the collector surface, are considerably higher. For the small funnel, the situation is different. The improvement in the aero-acoustic quality is obvious, and there is only a slight increase in surface area compared with configuration 1. In addition, a large percentage of the funnel surface is plane, which is simpler to manufacture than the all-curved surface of the rounded lips. Thus, the small funnel with the elongated test section (conf. 4) was chosen as the basic configuration for further optimization.

A certain disadvantage of this configuration is the relatively high power-loss coefficient. Therefore, an additional increase of the power loss due to a breather system is not acceptable. Nevertheless, breathers are necessary, as discussed above, to provide a means for fine tuning of the pressure gradient along the test section. Various breather configurations were tested, but effective breather openings increased the loss by at least Δk=0.02. A solution was found by traversing the funnel part of the collector upstream. Thus the funnel and the duct open a variable breather slot (Fig. 10).

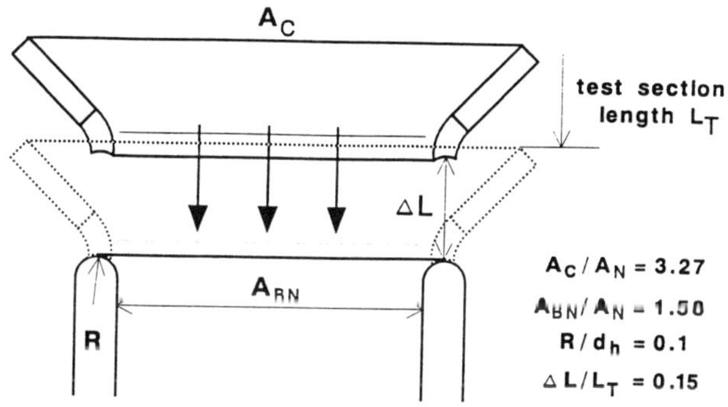

Fig. 10: Geometry of Variable Breather Slot Opened by Traversing the Funnel.

In the frontmost position the funnel was shifted by $\Delta L/L_T$=0.15. For this funnel position the c_p-distribution shown in Fig. 11 was measured. The pressure rise at the end of the test section is considerably steeper in case without breather, and the weak negative pressure gradient along the test section, which was discussed in Fig. 9, is completely eliminated by the breather.

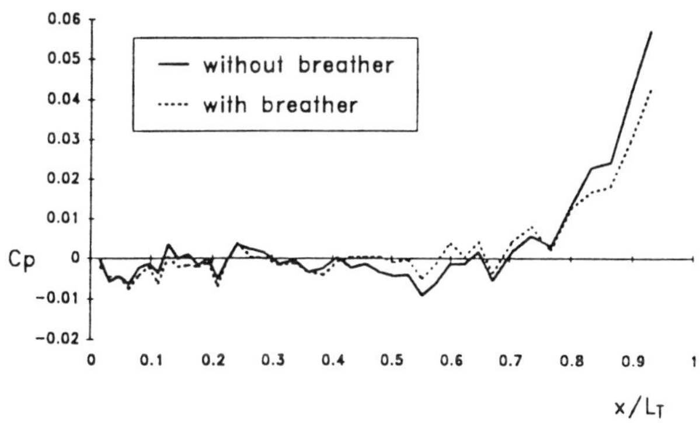

Fig. 11: Influence of the Breather System on Pressure Distribution.

Fig. 12 shows the increase in self-noise due to the breather slot. As already discussed in Chapter 3, breather systems always increase the wind tunnel self-noise, and only the amount of growth can be reduced. In this case it is merely a difference of 1.5 dB(A), whereas for cases with perforated sheet metal covering the breather slot, the increase can easily attain values of 6 dB(A). More significant is the rise in the linear OASPL. With breather slot it is 100.6 dB compared to 96.7 dB.

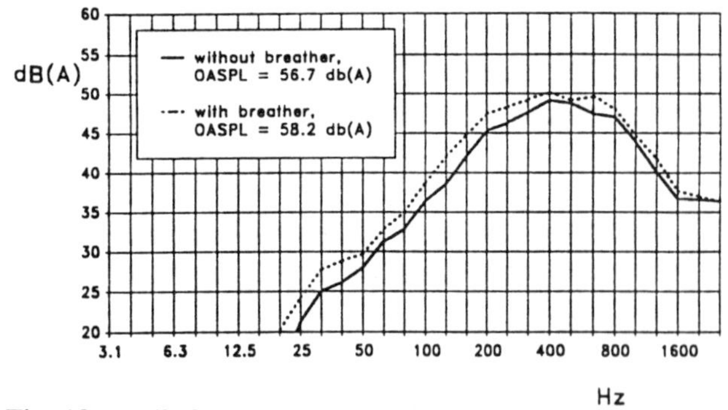

Fig. 12: 1/3-Octave-Band Analysis of Wind-Tunnel Self-Noise With and Without Breather (U=150 km/h).

The power-loss coefficient, which amounted to k_T=0.339 in the case without breather slot, increases up to k_T=0.351 when the collector is shifted into the frontmost position. In a medium position however, with a shift of $\Delta L/L_T$ = 0.075, k_T attains a value of 0.341. Furthermore, if the breather slot is sealed with the funnel in the frontmost position, for example by a special cover, a power-loss coefficient as low as k_T=0.30 can be retained. Nevertheless further optimization with respect to a reduction of the power loss is required.

INFLUENCE OF MODEL POSITION- From recent publications on the subject [14,16] it can be deduced that in open-jet wind tunnels, there is a decisive effect of longitudinal model position on the pressure distribution along the model. This effect may be expected from the static-pressure distribution in the empty test section, which generally feature a positive gradient in the streamwise direction, see e.g. Fig. 9 and Fig. 11. However, this pressure gradient represents only one of several parameters which may affect the force coefficients of a car. In addition, interactions of the vehicle and the nozzle, the collector and the free-jet boundary have to be considered. In the past, the errors due to these composite effects were simply termed "blockage effects," and in the literature they were exclusively ascribed to the interaction of the model and the free-jet boundary analogous to solid-wall effects in closed test sections [49].

All experiments to be discussed in the following, were carried out at an early time, when there was no acoustic lining of the plenum, and instead of a funnel-type collector a configuration according to Fig. 13 was used. The test were conducted with a square model (Fig. 14) featuring basic 1/8-dimensions of passenger cars [17,18], with quadratic flat plates [18] and with the aforementioned 1/8-scale model of the Corvette. For each model, surface-pressure data was taken at various model positions along the test-section axis. Furthermore, the effect of additional parameters like collector area A_C, breathers and, in the case of flat plates, blockage ratios ranging from 2 % to 25 %, was studied. For the square models as well as for the Corvette a set of "reference data" was taken in the 1.65 m^2-open-jet test section at the University of Stuttgart (IVK) at geometrical blockage ratios below 1.9 %, whereas in the present experimental set-up a blockage of 18 % was attained with the two models.

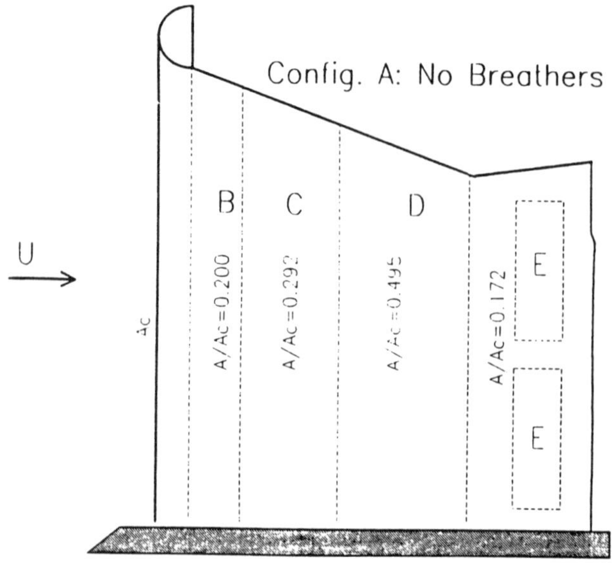

Fig. 13: Collector Configuration and Breathers Used for the Experiments on the Influence of Model Position. The Adjustable Intake Lip Allows for Variations of A_C/A_N Ranging from 1.0 to 2.035.

As to the flat plates it turned out, that the average base pressure and thus drag was only weakly dependant on the blockage ratio, if the model was positioned relatively close to the nozzle, i.e at 20 % of the test-section length (x/LT=0.2). With closed breathers, the drag coefficient which was calculated according to EQ (6), see also [49], remained between 93.5 % and 98.4 % of its extrapolated zero-blockage value [18].

$$\frac{c_D}{1-c_{p\,base}} = const. = 0.837 \qquad (6)$$

The flat plate results which contradict classical "blockage theories" may not appear relevant with respect to the test objects to be investigated in an automotive tunnel. They provide evidence however, that there are different blockage effects in the present test section which, at least partially, tend to compensate each other.

The results taken with the square model and with the Corvette were processed in a slightly different way. The c_p-figures measured along surface sections like base area or sideline were also averaged but no drag coefficient was determined, because the constant in EQ (6) was unknown or may not even be defined for the present geometries.

Fig. 14a and Fig. 14b display averaged c_p-values as a function of longitudinal position of the square model. As a parameter, the collector-to-nozzle area ratio A_C/A_N is added. A_C/A_N could be varied by changing the lateral position of the collector-intake lip (Fig. 13). The breather configuration for the results depicted in Fig. 14 is denoted by "E" in Fig. 13.

Apparently the pressure level along the model increases, when the model is shifted downstream, except for the configuration with A_C/A_N=1.18 in Fig. 14a, which will be discussed in the next paragraph. On the one hand, as discussed above, this increase may be ascribed to the horizontal buoyancy effect analogous to Fig. 9, on the other hand however, the nozzle effect is decreasing. The latter is such, that to satisfy continuity, the flow along the outer parts of the nozzle is accelerated, when the upstream stagnation effect of the model slows down the flow in the central section of the nozzle-exit area. This feature is analogous to closed-wall blockage effects, which are of opposite sign as pure open-jet blockage [49]. Therefore, the drag reading will attain a maximum, if the car is positioned as close to the nozzle as possible.

With respect to the base pressure a decisive effect of collector size A_C/A_N is discernible in Fig. 14a. With decreasing A_C/A_N the base pressure also decreases. This finding can be attributed to continuity requirements at the collector entrance. Due to wake blockage, the flow captured by the collector has to be accelerated, if the collector-entrance area is reduced. This causes the static pressure in the wake to decrease, see also [15]. In principle, the same result can be expected, if the model is continuously shifted downstream. The more the collector is blocked by the model wake, the more the flow speed has to increase to satisfy continuity. However as mentioned above, only for the smallest collector with A_C/A_N=1.18 a decrease in base pressure can be observed in Fig. 14a, whereas in all other cases the nozzle effect apparently overrules the collector influence and the pressure still increases, though with a reduced gradient. Naturally, on the side face of the model the collector effect is even weaker, as can be deduced from Fig. 14b.

In conclusion, three effects were identified, two of which (No. 1 and No. 2) produce the same phenomena, whereas one (No. 3) counteracts the others. These effects have to be differentiated from classical blockage effects ascribed to the open-jet boundary, which require but a constant correction of the dynamic pressure [49].

1) Due to the pressure gradient in the empty test section, the pressure along the model increases if the model approaches the collector (horizontal buoyancy).
2) The nozzle effect tends to accelerate the flow along the model and thus decrease the base pressure, if the nozzle is approached. Consequently, if the model is shifted towards the collector, the nozzle effect diminishes and the pressure along the model increases.
3) The collector effect causes the base pressure to decrease. This can be experienced by either moving the model towards the collector or by reducing the collector-entrance area. However, only in the case of relatively small collectors this effect becomes dominant.

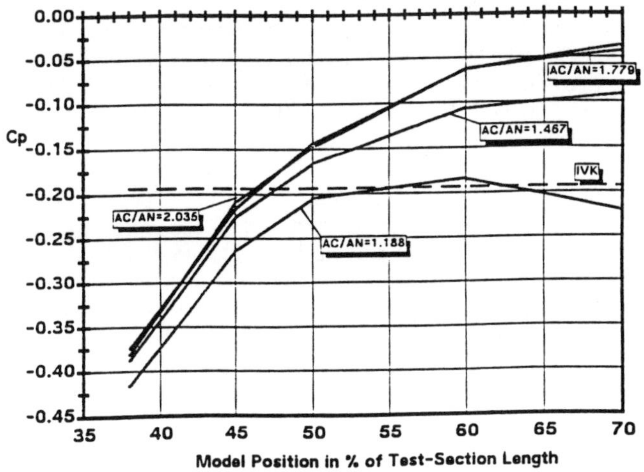

Fig. 14a: Effect of Longitudinal Model Position on the Averaged Base Pressure of a Square Model. Breather Configuration E.

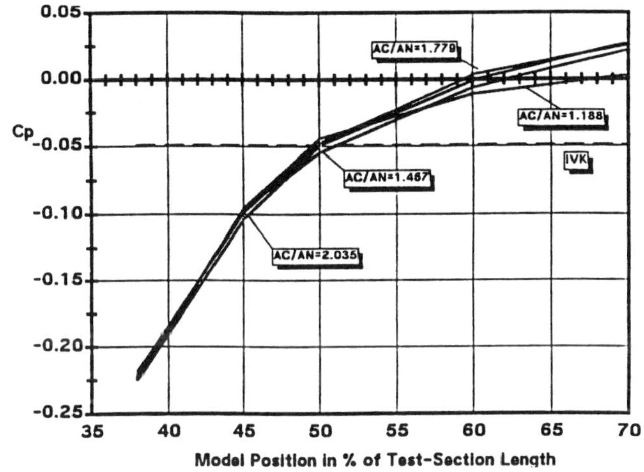

Fig. 14b: Effect of Longitudinal Model Position on the Averaged Pressure on the Side Face of a Square Model. Breather Configuration E.

Fig. 15a and Fig. 15b display results of analogous experiments with the Corvette model. Contrary to the previous tests, the breather configuration was varied according to Fig. 13, whereas the collector area remained constant at A_C/A_N=2.035.

With respect to the car position, the results are consistent with the square-model experiments. As to the breathers, Mattern [18] showed that the most effective ventilation was provided by conf. C+D, followed by conf. D, conf. C, conf. B and conf. E. Interestingly, conf. E features an inflow of mass into the collector, whereas with all other configurations mass is removed, as would be expected. The results displayed in Fig. 15 indicate that effective breathers (e.g. conf. C and conf. C+D) decrease the pressures along the car, not only in the base area. Apparently, the breathers act similar to a reduction of collector area (Fig. 14) and, analogous to the collector effect, the breather effect also results from continuity requirements:

Due to the action of entrainment the volume flux of the open jet has to increase in the streamwise direction. If there are no breathers in the collector, continuity requires the flow to divide such, that exactly the volume flux discharged by the nozzle enters the collector. Consequently, the streamlines of the jet have to curve to form a stagnation line on the inner circumference of the collector, and a positive pressure gradient will establish. If breathers are located within the collector at positions, where the pressure exceeds ambient pressure, an outwards directed flow will occur and thus, more mass has to enter the collector to satisfy continuity. Consequently, velocities at the collector entrance are increased and the positive pressure gradient is decreased. With respect to the empty test section this is corroborated by Fig. 9 and Fig. 11. Whereas according to Fig. 15, the flow along a model is affected in an analogous way. This has also been discussed in great detail in [18].

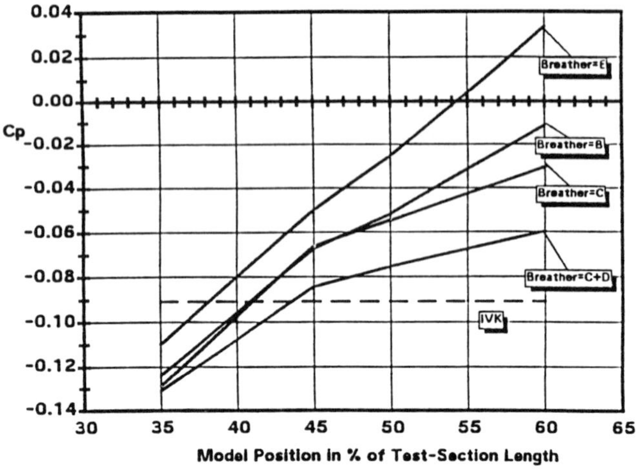

Fig. 15a: Effect of Longitudinal Model Position on the Averaged Base Pressure of a 1/8-Scale Corvette Model. A_C/A_N=2.035.

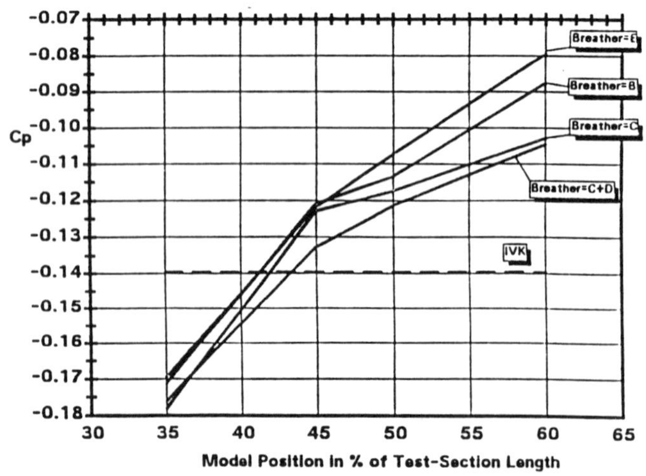

Fig. 15b: Effect of Longitudinal Model Position on the Averaged Pressure Along the Sideline of a 1/8-Scale Corvette Model. A_C/A_N=2.035.

From the results presented in Fig. 15 it can be further concluded, that the optimum car position for future testing in the AUDI tunnel will be between x/L_T=0.40 and 0.42, if a medium breather configuration is anticipated. For test objects like the square model which feature extensive regions of separated flow even near the front face, a more rearward position, e.g. x/L_T=0.47, is recommended to reduce the stronger nozzle effect (Fig. 14).

However in the near future, these tests will be extended and verified with the more recent funnel type collector.

CONCLUSIONS

The planned AUDI Aero-Acoustic Wind Tunnel in Ingolstadt is intended to serve the AUDI development engineers and designers as well as clients from other companies as a key design tool for optimization of both, full-scale cars and scale models. With its unique combination of technical features it will permit a large bandwidth of experiments to be carried out, some of which are

- acoustic inflow and out-of-flow measurements under semi-anechoic (reflecting ground plane) conditions (99 % absorption at about 120 Hz) with the plenum chamber being sufficiently large to approximate acoustic far-field conditions.
- subjective wind-noise ratings from passengers inside the car or persons positioned outside the flow along a 6.5 m-sideline. This will be possible, because of the low background noise level.
- six-component aerodynamic force measurements and pressure measurements on full-scale cars and scale models up to 60 m/s. For scale-model tests very low blockage ratios will be featured (typically 2.9 % for a 1/2.5-scale passenger car). Due to the twin-turntable concept quick changes, from model testing to full-scale tests are possible.
- full-scale and model-scale tests on racing cars and passenger cars with moving-belt ground simulation and rotating wheels in combination with the external balance. Therefore, elaborate adaptation of internal balances to the test objects are not required.

In the pre-design stage of the project extensive theoretical and 1/8-scale experimental investigations were carried out to optimize the performance of the facility.

A Göttingen type air duct with open-jet test section is considered a prerequisite to meet the aerodynamic and acoustic design goals. However, also conflicting demands have to be matched, like

- a large nozzle to minimize blockage effects and small jet dimensions to enable out-of-flow measurements and easy access even close to the car without entering the jet, as well as budget restrictions. From the aerodynamic comparison tests with models it can be concluded that the requirements can be met with a nozzle exit-area of 11 m^2.
- a long test section to decrease aerodynamic interference effects between the car, the nozzle and the collector and, on the other hand, the requirement for minimum pressure loss and jet stability have to be met. A test-section length of 10 m turned out to be a good compromise.
- the collector design has to avoid open-jet instability, which may result from a closed-loop feed-back mechanism between free shear-layer vortex generation at the nozzle and vortex impingement onto the collector. This goal has to be contrasted with the demands for streamwise pressure uniformity in the test section, with power-loss considerations and the requirement for low self-noise generation. From the acoustic and aerodynamic experiments, it is concluded that a newly developed large funnel-type collector with a bottle neck of area ratio A_{BN}/A_N=1.58 in conjunction with a variable breather slot meets the requirements best.

- collector design and streamwise position of the car have to be matched in such a way, as to optimize the tunnel's aerodynamic accuracy with respect to the reproduction of reference data for drag and pressures. As the tests revealed counteracting effects from the nozzle and the collector, it appears possible to balance these influences by an appropriate model position. From the present investigations carried out with two different models featuring a blockage ratio of 18 %, a car position at about 40 % of the test-section length can be derived.

To further verify the present design concept, additional investigations are deemed necessary, especially with respect to the acoustic absorption of the air duct and the mufflers, as well as model-collector interference and breather performance.

REFERENCES

[1] WIEDEMANN, J., DNW - Audi's Approach to the Solution of Complex Problems in Passenger-Car Aerodynamics. Colloquium Proceedings "Ten Years of Testing at DNW 1980-1990", DNW Noordoostpolder, (October, 1990).

[2] EWALD, B., Technische Planung eines Automobil-windkanals. Expert Report for AUDI AG, (1987).

[3] WALLNER,H.,R., Gutachten über den Stand der Simulationstechnik in der Kfz-Aerodynamik. Expert Report for AUDI AG, (1989).

[4] LORENZ-MEYER, W., Unterschallwindkanal für Kfz-Modell-Messungen (Maßstab 1:4). Expert Report for AUDI AG, (1989).

[5] FEUERSTEIN, A., Messung an Kfz-Modellen im Windkanal mit Untersuchung der Wirkung von Strömungsbeeinflussung auf das Meßergebnis. Studienarbeit TH Darmstadt, Fachgebiet Aerodynamik und Meßtechnik, Prof. Ewald, (1987).

[6] RETTIG, K. And WIEDEMANN, J., Untersuchung zur Wirkung der Turbulenzmanipulation auf die Reynoldszahlabhängigkeit der aerodynamischen Beiwerte eines Kfz-Modells. Report A54-88, TH Darmstadt, Fachgebiet Aerodynamik und Meßtechnik, Prof. Ewald, (1989).

[7] WIEDEMANN, J., Grenzen und Möglichkeiten der Modelltechnik innerhalb der Kraftfahrzeug - Aerodynamik. Symposium Nr. T-30-905-056-7 "Aerodynamik des Kraftfahrzeugs", Haus der Technik, Essen, (1987).

[8] WIEDEMANN, J., Windkanalvorarbeiten, Teil 1, Grundsatzuntersuchungen zur geeigneten Bodensimulation in Windkanälen. Internal Report AUDI AG, Dept. EOA, (1988).

[9] WIEDEMANN, J. and EWALD, B., Turbulence Manipulation to Increase Effective Reynolds Numbers in Vehicle Aerodynamics. AIAA Journal, Vol. 27, No. 6, pp. 763-769, June (1989).

[10] WIEDEMANN, J., Some Basic Investigations into the Principles of Ground Simulation Techniques in Automotive Wind Tunnels. SAE Paper 890369, Detroit, (1989).

[11] MERCKER, E. and WIEDEMANN, J., Comparison of Different Ground-Simulation Techniques for Use in Automotive Wind Tunnels. SAE Paper 900321, Detroit, (1990).

[12] GRUNDMANN, R. and KRAMER, C., Einheitliche Darstellung von Strömungsgeräuschen - Teil 1 und Teil 2. HLH, Vol. 42, No. 3 pp. 233-238 and No. 4 pp. 275-281, (1991).

[13] GRUNDMANN, R. and KRAMER, C., Considerations on Aero-Acoustical Wind Tunnels for Car Aero-Noise Testing. Institut für Industrie-aerodynamik GmbH, Institut an der Fachhochschule Aachen. Paper presented at meeting of SAE Standards Committee No.9 at Pininfarina Wind Tunnel, Turin, (Oct. 1991).

[14] SOCIETY OF AUTOMOTIVE ENGINEERS (SAE), Surface Vehicle Information Report J2071, "Aerodynamic Testing of Road Vehicles", Issued 1990-03. Warrendale, PA, (1990).

[15] VON SCHULZ-HAUSMANN, F.K. and VAGT., J.,D., Influence of Test-Section Length and Collector Area on Measurements in 3/4-Open-Jet Automotive Wind Tunnels. SAE Paper 880251, Detroit, (1988).

[16] GERHARDT, H.J. and KRAMER, C., Blockierungseffekte in Windkanälen mit Bodenplatte und offener Meßstrecke. Symposium No. T-30-905-056-7 "Aerodynamik des Kraftfahrzeugs", Haus der Technik, Essen, (1987).

[17] WIEDEMANN, J., AUDI Aero-Acoustic Wind Tunnel. Computational and Experimental Investigations of Aerodynamic Configurations. Results and Conclusions. Internal Report, AUDI Dept. of Aerodynamics, (1991).

[18] MATTERN, C., Experimentelle Optimierung eines Automobilwindkanals mit Freistrahlmeßstrecke. Studienarbeit TH Darmstadt, Fachgebiet Aerodynamik und Meßtechnik, Prof. Ewald, (1991).

[19] BÖRGER, G.G., Optimierung von Windkanaldüsen für den Unterschallbereich. ZFW 23, pp. 45-50, Correction: ZFW 23, pp. 282, (1975).

[20] SCHILAWA, S., Entwurf dreidimensionaler Windkanaldüsen. Fortschr.-Ber. VDI-Z., Reihe 7 Nr. 82, VDI-Verlag, Düsseldorf, (1984).

[21] VON DER VOOREN, J. and SANDERSE, A., Finite Difference Calculation of Incompressible Flow through a Straight Channel of Varying Rectangular Cross Section, with Application to Low Speed Wind Tunnels. NLR TR 77109 U, (1977).

[22] WALZ, A., Näherungstheorie für kompressible turbulente Grenzschichten. ZAMM-Sonderheft 36, pp 50-55, (1956).

[23] WALZ, A., Über Fortschritte in Näherungstheorie und Praxis der Berechnung kompressibler laminarer und turbulenter Grenzschichten mit Wärmeübergang. ZFW 13, pp 89-102, (1965).

[24] BRADSHAW, P. and UNSWORTH, K., An Inproved FORTRAN Program for the Bradshaw- Ferriss- Atwell Method of Calculating Turbulent Boundary Layers. Imperial College of Science and Technology, Dept. of Aeronautics, I.C. Aero Rept. 74-02, (1974).

[25] HACKETT, J.E., WILLIAMS, J.E., BAKER, J.B. and WALLIS, S.B., On the Influence of Ground Movement and Wheel Rotation in Tests on Modern Car Shapes. SAE Paper 870245, Detroit, (1987).

[26] BEARMAN, P.W., DE BEER, D., HAMIDY, H. and HARVEY, J.K., The Effect of a Moving Floor on Wind-Tunnel Simulation of Road Vehicles. SAE Paper 880245, Detroit, (1988).

[27] SARDOU, M., The Sensitivity of Wind-Tunnel Data to a High-Speed Moving Ground for Different Types of Road Vehicles. SAE Paper 880246, Detroit, (1988).

[28] MERCKER, E. and KNAPE, H.W., Ground Simulation with Moving Belt and Tangential Blowing for Full-Scale Automotive Testing in a Wind Tunnel, SAE Paper 890367, Detroit, (1989).

[29] LARSSON, L., HAMMAR, L., NILSSON, L.U., BERNDTSSON, A., KNUTSON, K. and DANIELSON, H., A Study of Ground Simulation - Correlation between Wind Tunnel and Water-Basin Tests of a Full-Scale Car. SAE Paper 890368, Detroit, (1989).

[30] EMMELMANN, H.-J., BERNEBURG, H. and SCHULZE, J., The Aerodynamic Development of the Opel Calibra. SAE Paper 900317, Detroit, (1990).

[31] MERCKER, E., BREUER, N., BERNEBURG, H. and EMMELMANN, H.-J., On the Aerodynamic Interference Due to the Rolling Wheels of Passenger Cars. SAE Paper 910311, Detroit, (1991).

[32] YAMAGUCHI, I., NAKATSURU, K., TAKAGI, M. and SHIMPO, Y., The Effect of a Moving Belt on Automotive Underbody Airflow in Model Wind Tunnel Tests. JSAE Review, Vol. 12, No. 3, (1991).

[33] MERCKER, E. and BERNEBURG, H., On the Simulation of Road Driving of a Passenger Car in a Wind Tunnel using a Moving Belt and Rotating Wheels. ATA & DMTI 3rd International Conference "Innovation and Reliability in Automotive Design and Testing", Florence, (1992).

[34] OGATA, N., IIDA, N. and FUJI, Y., Nissan's Low-Noise Full-Scale Wind Tunnel. SAE Paper 870250, Detroit (1987).

[35] ACKERMANN, U., Strömungsakustische Modelluntersuchungen für einen Freistrahl- Windkanal. Zeitschrift für Lärmbekämpfung 38, pp. 10-18, Springer-Verlag, (1991).

[36] LINDENER, N. and STADLER, M., Der BMW Akustik Windkanal. Tagung T-30-312-056-2 "Geräuschminderung bei Kraftfahrzeugen", Haus der Technik, Essen, (1992).

[37] HANSON, C.E., The Design and Construction of a Low-Noise, Low-Turbulence Wind Tunnel. Technical Report 79611-1, Acoustics and Vibration Laboratory, M.I.T., Cambridge, Massachusetts, (1969).

[38] PATERSON, R.W., VOGT, P.G. and FOLEY, W.M., Design and Development of the United Aircraft Laboratories Acoustic Research Tunnel. AIAA Paper 72-1005, AIAA 7th Aerodynamic Testing Conference, Palo Alto, Ca., (1972).

[39] SEIDEL, M., (ed.), Construction 1976-1980. Design, Manufacturing, Calibration of the German-Dutch Windtunnel (DNW). Published by DNW, (1982).

[40] VAN DITSHUIZEN, J.C.A., COURAGE, G.D., ROSS, R. and SCHULTZ, K.-J., Acoustic Capabilities of the German-Dutch Wind Tunnel DNW. AIAA Paper 83-0146, AIAA 21st Aerospace Sciences Meeting, Reno, Nevada, (1983).

[41] HECKL, M. and MÜLLER, H.A. (Editors), Taschenbuch der Technischen Akustik. Springer-Verlag, (1975).

[42] ACKERMANN, U., Instabilitätswellen in einem runden, turbulenten Freistrahl und ihre Schallabstrahlung. Dissertation, Georg-August-Universität Göttingen, (1976).

[43] SEIFERTH, R., Vorausberechnung und Beseitigung der Schwingungen von Freistrahl- Windkanälen. Monographien über Fortschritte der deutschen Luftfahrtforschung (Editor A. Betz), D1: Modellversuchstechnik, Teil 1: Versuchsanlagen AVA Göttingen, (1946).

[44] BERANEK, L.L. (Editor), Noise Reduction, McGraw-Hill Book Co., New York, (1960).

[45] TYLER, J.M. and SOFRIN, T.G., Axial Flow Compressor Noise Studies. Transactions of the SAE, Vol. 70, pp. 309-332, (1962).

[46] SCHULTEN, J.B.H.M., Some Remarks on Pure Tone Fan Noise Suppression of the DNW Low Speed Wind Tunnel. Memorandum AV-75-010, NLR, (1975).

[47] REGENSCHEIT, B., Isotherme Luftstrahlen, Klima + Kälteingenieur: Extra; 12, Verlag C.F. Müller, Karlsruhe, (1981).

[48] ROSS, K., Model Tunnel Measurements to Determine the DNW Aero-Acoustic Configuration. DNW-PA-79.150, (1979).

[49] GARNER, H.,C., ROGERS, E.,W.,E., ACUM, W.,E.,A. and MASKELL, E.,C., Subsonic Wind Tunnel Wall Corrections. AGARDograph 109, (1966).

950625

The Aero-Acoustic Wind Tunnel of Stuttgart University

Rudi Künstner, Jürgen Potthoff, and Ulf Essers
University of Stuttgart

ABSTRACT

The noise emission of cruising vehicles essentially consists of tire/road noise, drivetrain noise (engine with intake and exhaust system, transmission and driving axle) and aerodynamic noise due to the flow around bodywork, chassis, wheels and cooling air flow (fan). Engines and drivetrains have become quieter due to many man-years of engineering attention and tire noise has also been reduced – at least the noise reaching the passenger compartment. Consequently, the aerodynamic noise of ground vehicles has become dominant at driving speeds above 100 kph both in interior and exterior noise.

In order to determine the contribution of aerodynamic noise to the overall noise, measurements are carried out more and more in specially equipped automotive wind tunnels. Besides wind tunnels in which aeroacoustic testing was already envisaged in the specification and design phase, existing wind tunnels can be upgraded to acoustic testing by a suitable treatment of the tunnel airpath and the plenum chamber.

The existing 22.45 m^2 Open-Jet Automotive Wind Tunnel of Stuttgart University – Institute of Internal Combustion Engines and Automotive Engineering (IVK) – built in 1988 was converted into an aeroacoustic wind tunnel in 1993 by a novel silencing concept. In the return duct of this Göttingen-type wind-tunnel two U-bend silencers were installed upstream and downstream of the fan, using membrane absorbers as thin and smooth vertical splitters (for the low frequency range) in conjunction with profiling the turning vanes with coated porous Polyester-foam (for the middle and high frequency range). An anechoic plenum chamber was achieved with a new type of broadband absorbers, mounted on the plenum walls and ceiling. Due to its very low self-noise level the IVK Aero-Acoustic Wind Tunnel provides a high potential for the acoustic rating of design optimizations in the development of future vehicles.

INTRODUCTION

The research activities on vehicle aerodynamics at Stuttgart University span from the thirties of this century until today. On its former grounds in Stuttgart-Untertürkheim, adjacent to Daimler-Benz AG, the non-profit making foundation of civil law FKFS – Research Institute of Automotive Engineering and Vehicle Engines Stuttgart – operated a 32.6 m^2 open-jet automotive wind tunnel and a 1.5 m^2 open-jet model wind-tunnel, both of Göttingen-type with a single return duct.

Because of an urgent requirement to extend its factory premises, parts of the FKFS grounds, and especially the site the full-scale wind tunnel was built on, were first sold to Mercedes-Benz, followed by the wind-tunnel facility itself. The full-scale wind tunnel was operated till May 1974 and was then handed over. After renovation it has served Mercedes-Benz AG for full-scale aerodynamic testing. The 1.5 m^2 model wind-tunnel had been run until July 1987, and was then disassembled and given to the Technical Museum in Mannheim (Germany).

In 1978, the whole Institute moved to the University's new center in Stuttgart-Vaihingen on the foreseen site for Stuttgart University's Institute IVK. Since the Institute already had a full-scale automotive wind tunnel, and the willingness of German automobile manufacturers to back up and partly to sponsor the project of a new full-scale wind tunnel, led in September 1982 to a basic decision of the state government of Baden-Württemberg to establish an new wind-tunnel facility. With the financial support of the State of Baden-Württemberg, of the Federal Republic of Germany and of German car makers the project could be realized. Construction started in April 1984 and the whole wind-tunnel facility was inaugurated in September 1988 [1].

Besides a model wind-tunnel with 1.65 m^2 nozzle exit area and a top speed of 80 m/s a full-scale open-jet automotive wind tunnel of Göttingen-type was realized (Fig.1). The nozzle-exit area is 5.8 m wide by 3.87 m high giving a flow area of 22.45 m^2. The contraction ratio is 4.41. The test-section length was originally 9.5 m, but after reconstructing the adjustable collector in 1993 it was extended to 9.94 m. The ratio of the collector cross-section to the nozzle-exit area is $A_C/A_N = 1.179$. The water-cooled and variable speed DC-motor has a nominal power output of 2500

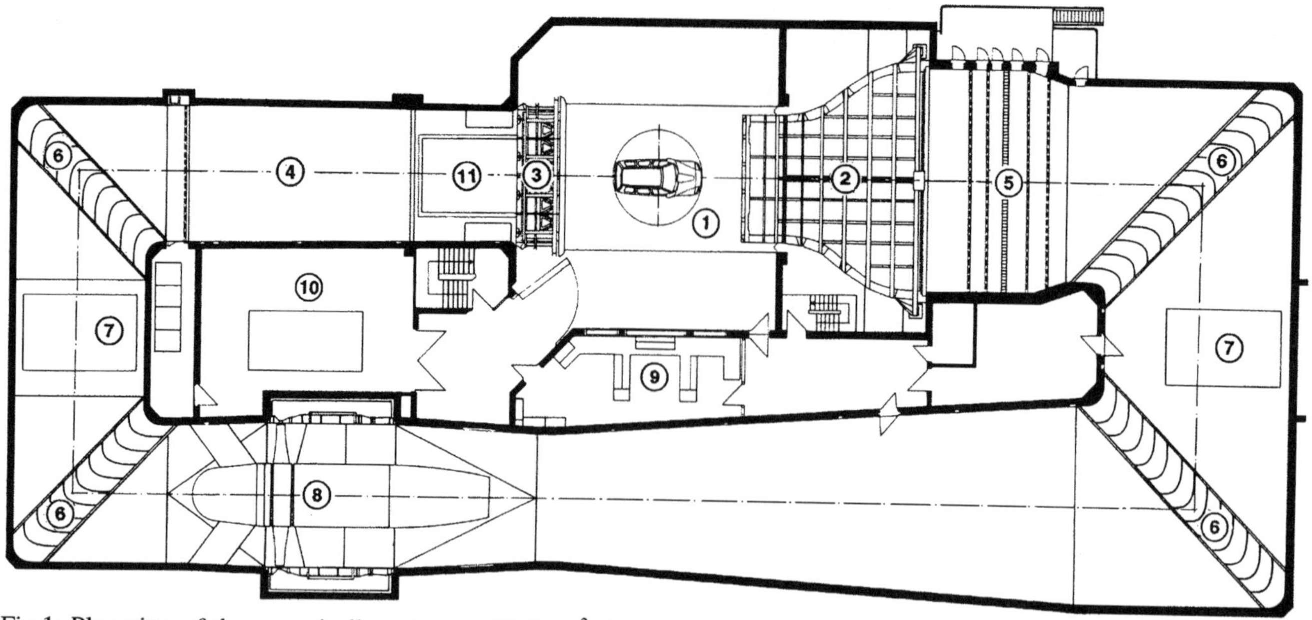

Fig.1: Plan view of the acoustically untreated 22.45 m^2 Open-Jet Automotive Wind Tunnel of Stuttgart University put into operation in 1988:

(1) Open-jet test section
(2) Nozzle
(3) Adjustable collector
(4) Test-section diffuser
(5) Honeycomb and turbulence screens
(6) Turning vanes
(7) Ventilation system
(8) Fan
(9) Control room
(10) Design studio
(11) Lift.

kW with a 20% overload capacity. It is located inside the nacelle and drives a 12 blade axial fan, which has a diameter of 7.1 m across the blade tips. The settling chamber is equipped with a honeycomb, three turbulence screens and another coarse-meshed screen in the center cross-section of the wide-angle diffuser. The four corners of the return duct were originally equipped with unprofiled sheet-metal turning vanes. The return duct itself was conservatively designed by avoiding extreme wall angles in the test section diffuser and in the fan diffuser and by avoiding extreme transitions in the shape of the return duct. The equivalent half angles of test-section diffuser and fan diffuser are 2.56 degs. and 2.73 degs., respectively.

After a couple of years of operating the wind tunnel, the Institute was faced with customer wishes to measure the aerodynamic noise of cars. The engine, power train and tire noise of newly develloped cars had been reduced, and the wind noise was now dominating at driving speeds above 100 kph. In the eighties and early nineties this annoying effect was more pronounced because of the higher top speed of the cars due to better aerodynamic shaping of the bodies and due to the increased performance of the engines. In Germany there has not yet been imposed a general speed limit on motorways. There exists only a recommended limit of 130 kph.

For the study of the engine and drivetrain noise suitable test equipment and measuring procedures have been available for many years, but the separation of wind noise from tire noise has shown to be extremely difficult and unreliable on basis of road tests and measurements in conventional wind tunnels.

Therefore, investigations were conducted to determine if the Institute's full-scale wind tunnel could be converted into an aeroacoustic facility by a suitable acoustic treatment of the tunnel airpath and the plenum chamber in order to get an acceptable signal-to-noise ratio for vehicle testing. For measuring the exterior noise emission and for detecting noise sources on vehicles, the installation of a multi-axis traversing gear was mandatory. After the design and specification phase, and after the financial support of the project was also ensured by German automotive companies, the detail planning and the realization of the acoustic conversion was started in January 1992. By end of November 1992 the IVK Wind Tunnel was closed for six months. After commissioning and recalibration it went back into operation in the beginning of June 1993.

WIND TUNNEL SELF-NOISE AND REQUIREMENTS

The noise emissions of cruising vehicles are important both to the environment and to the comfort and safety requirements of driver and passengers, because people are sensitive to the fatigue factor of excessive noise.

In the driving speed range up to 50 kph the engine and driveline noise is dominant, followed up by the tire/road noise in the velocity range up to approximately 100 kph.

With further increasing speed the wind noise gains more and more importance, because its sound power increases with the 5th to 7th power of speed, whereas the tire/road noise increases with the 3rd to 4th power.

Recent basic studies [2] have shown that the vans exterior wind-noise emission reach their tire/road noise-level at 80 kph. For current production cars it is in the speed range between 110 to 130 kph, when wind-tunnel measurements are related to the 7.5 m lateral microphone distance in the drive-by test, Fig.2. This velocity for balanced noise emissions can shift up to approximately 180 kph for cars with low aeroacoustic exterior noise. On the other hand, this balance of velocity will shift down if the engine, drivetrain and tire/road noise is further reduced.

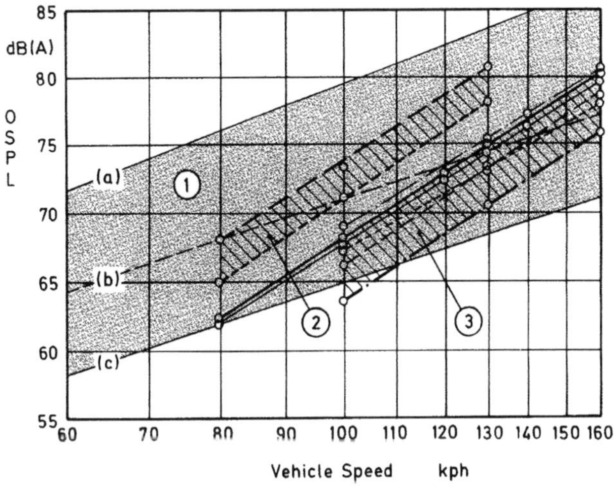

Fig.2: Exterior wind-noise emissions versus driving speed of different vehicles compared to tire/road noise:
(1) Scatterband tire/road noise of cars
(a) Upper limit: Tread-tire on concrete surface
(b) Middle: Tread-tire on ISO/drainasphalt surface
(c) Lower limit: Slick-tire on ISO/drainasphalt surface
(2) Scatterband of aerodynamic exterior noise of vans
(3) Scatterband of aerodynamic exterior noise of cars.

When several noise sources are superimposed, the determination of the share of contribution of a single sound source to the overall sound pressure level makes it necessary to either separately measure it or to suppress all the others. For the study of engine and drivetrain noise suitable test equipment and measuring procedures have been available for many years, whereas the separation of aerodynamic noise from tire/road noise has raised considerable difficulties, because the spectra of both noise sources differ only slightly.

The isolated representation of a vehicle's aerodynamic noise in conventional wind tunnels has proven to be difficult or even impossible, especially in respect to the exterior flow noise [3]. Fig.3 shows some examples from the

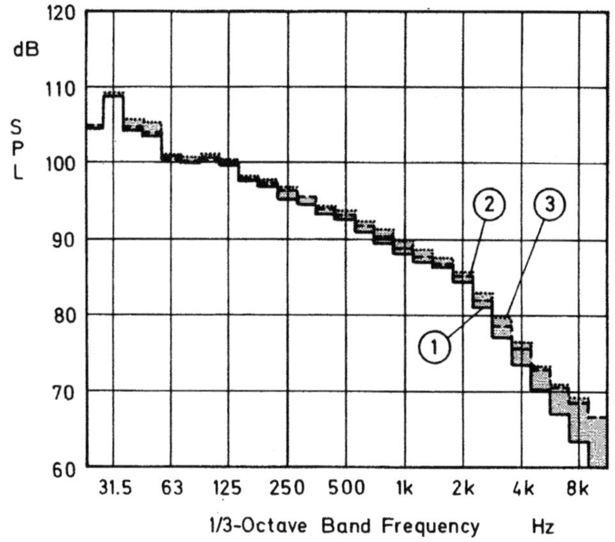

Fig.3: Out-of-flow spectra at 200 kph in the acoustically untreated IVK Wind Tunnel without vortex generators at the nozzle exit:
(1) Wind tunnel self-noise, empty test section
(2) Calibration Model BMW CC-7.2 with a smooth surface and a fastback extension of 800 mm (C_D = 0.17)
(3) Calibration Model BMW CC-7.2 as squareback with add-on spoilers at the front and rear end and at the A-pillars (C_D = 0.55).

acoustically untreated wind tunnel of the Institute before conversion. Out-of-flow measurements with a microphone position at the 6 m sideline were performed both in the empty and occupied test section. A full-scale calibration model of BMW served as a test object both as a long fastback version with a low drag coefficient (C_D = 0.17, smooth surface) and as a short squareback version with relatively big add-on spoilers at the front ends leading edge, at the rear ends trailing edge and at the A-pillars (C_D = 0.55). The graph shows that the out-of-flow measurements in the reflective plenum are dominated by the wind-tunnel self-noise. Also with the high-drag version of the calibration model there is no acceptable signal-to-noise ratio even in the middle and high frequency range.

Fig.4 shows the A-weighted overall sound pressure level (OSPL), measured out of the flow in the acoustically untreated IVK Vehicle Wind-Tunnel both with and without vortex generators ("Seiferth wings") at the nozzle exit. It is plotted versus the tunnel velocity and compares the scatterband both of conventional open-jet automotive wind tunnels and of aeroacoustic wind tunnels. The difference to the latter is between 18 and 22 dB.

In order to perform aeroacoustic investigations on a safe basis, a signal-to-noise ratio of approximately 10 dB in the frequency bandwidth from 63 Hz to 8 kHz is required. Such a reduction of the self-noise level of a conventional wind tunnel seems to be sufficient for determing the cabin noise of cars. On the assumption that the aerody-

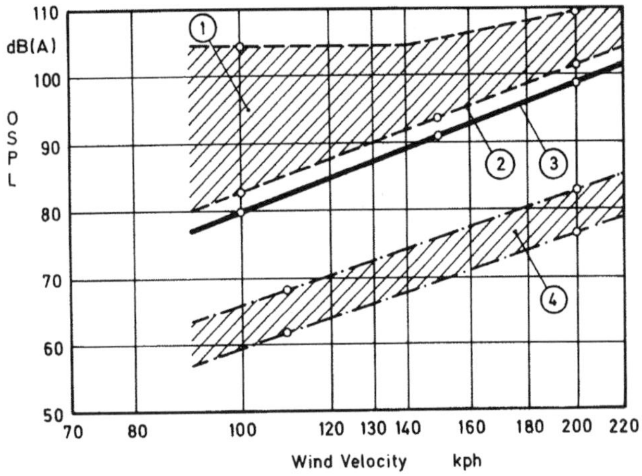

Fig.4: A-weighted overall sound pressure levels, measured out-of-flow, versus wind velocity of different vehicle wind-tunnels:
(1) Scatterband of open-jet automotive wind tunnels
(2) IVK Wind Tunnel before conversion and with vortex generators at the nozzle exit
(3) IVK Wind Tunnel before conversion and without vortex generators at the nozzle exit
(4) Scatterband of aeroacoustic vehicle wind-tunnels.

namic noise of well shaped future cars is further reduced by roughly 5 to 8 dB as against today, an additional reduction of the wind-tunnel background noise will be necessary as well, especially when the exterior flow noise is to be measured. This requirement is met by aeroacoustic wind tunnels [4,5] with a hard and reflective test-section floor which measure out-of-flow an OSPL of 72 to 77 dB(A) at a free-stream velocity of 150 kph.

Basically, the self-noise of a wind tunnel is determined by the fan, the installations in the settling chamber (turbulence screens, honeycomb and heat exchanger), the open-jet test-section with nozzle, collector, free shear layer, floor boundary-layer and the plenum's secondary flow with wall interferences. The physical limit of the wind-tunnel self-noise is given by the aerodynamic noise of the undisturbed open-jet lip-noise generated at the nozzle exit without collector.

The main noise source is the fan and its drive unit, which can only be acoustically optimized within limits imposed by the economics of building (diameter, fan speed, number of blades, blade structure and profiling, shape of stator vanes, fan efficiency). A reduction of the fan noise is feasible by installing splitter absorbers in the return duct, upstream and downstream of the fan.

When the plenum walls and ceiling are acoustically lined, a further reduction of 8 to 10 dB of the wind-tunnel self-noise will be obtained. The residual self-noise of secondary noise sources in the tunnel airpath, upstream and downstream of the fan's duct absorbers, cannot be attenuated anymore. It can only be tried, by optimizing and tuning the test section, to lower it towards the pure jet noise. Thus, the remaining self-noise of existing aeroacoustic wind tunnels is distinctly above the pure jet noise, according to a rough calculation of Wiedemann [6].

After the German automotive industry had been willing to partly financially support the conversion of the IVK Vehicle Wind-Tunnel, the Fraunhofer-Institute of Structural Physics (IBP) in Stuttgart was commissioned to work out a project study for the sound attenuation, because this Institute of the Fraunhofer Society was already experienced on this area [7]. In a cooperation of the Institute with car makers and the IBP it was agreed on the following requirements to be met:

- Reducing the wind-tunnel self-noise at least by 20 dB in order to come as close as possible to the pure jet noise. The necessary sound attenuations of the absorbers before and after the fan then has to be 30 to 35 dB, without the wind on.
- The top speed in the test section – after conversion – should not be below 245 to 250 kph. Before the conversion it was 268 kph. The non-uniformity of the velocity distribution in the jet core of ±0.3% should remain unchanged.
- All surfaces of parts to be installed in the tunnel airpath for sound attenuation should be smooth, mechanically stable, unsensitive to soiling and easy to clean.
- In order to avoid any health problems, the absorbers should be made of non-fibrous material for the sake of the wind-tunnel personnel and its clients.
- The sound-absorbing lining of the plenum walls and ceiling should be as thin as possible (<0.45 m) in order to avoid any influence on the aerodynamic coefficients of cars, because of the reduced plenum volume [8]. The absorption rate of these silencers should be higher than 0.9 in the frequency range above 125 Hz.
- The necessary changes for converting the wind tunnel's airpath and plenum should be as small as possible and the break in wind-tunnel operation should be reduced to a conceivable minimum of four months.

CONVERSION OF THE IVK WIND TUNNEL TO AN AEROACOUSTIC WIND TUNNEL

Different studies showed very soon that these requirements could only be met by a novel solution with an innovative absorber structure. After the IBP scale-model tests had shown promising results, the U-bend silencer concept for attenuating the fan noise was decided. The plan view in Fig.5 and the side view in Fig.6 give an impression of the acoustic treatment of tunnel airpath and plenum. Fig.7 shows more detailed the U-bend silencer downstream of the fan, which includes the corners 3 and 4.

The complete concept comprises the following noise-reduction measures:

- Two U-bend silencers upstream and downstream of the fan. They use thin splitter plates, which contain membrane absorbers in its steel-framework. These vertical splitter plates subdivide the return duct in three (upstream of the fan) and four sections (downstream of the fan). These sections have the same pressure loss and show the same noise reduction, respectively. The vertical splitter plates, containing membrane absorbers

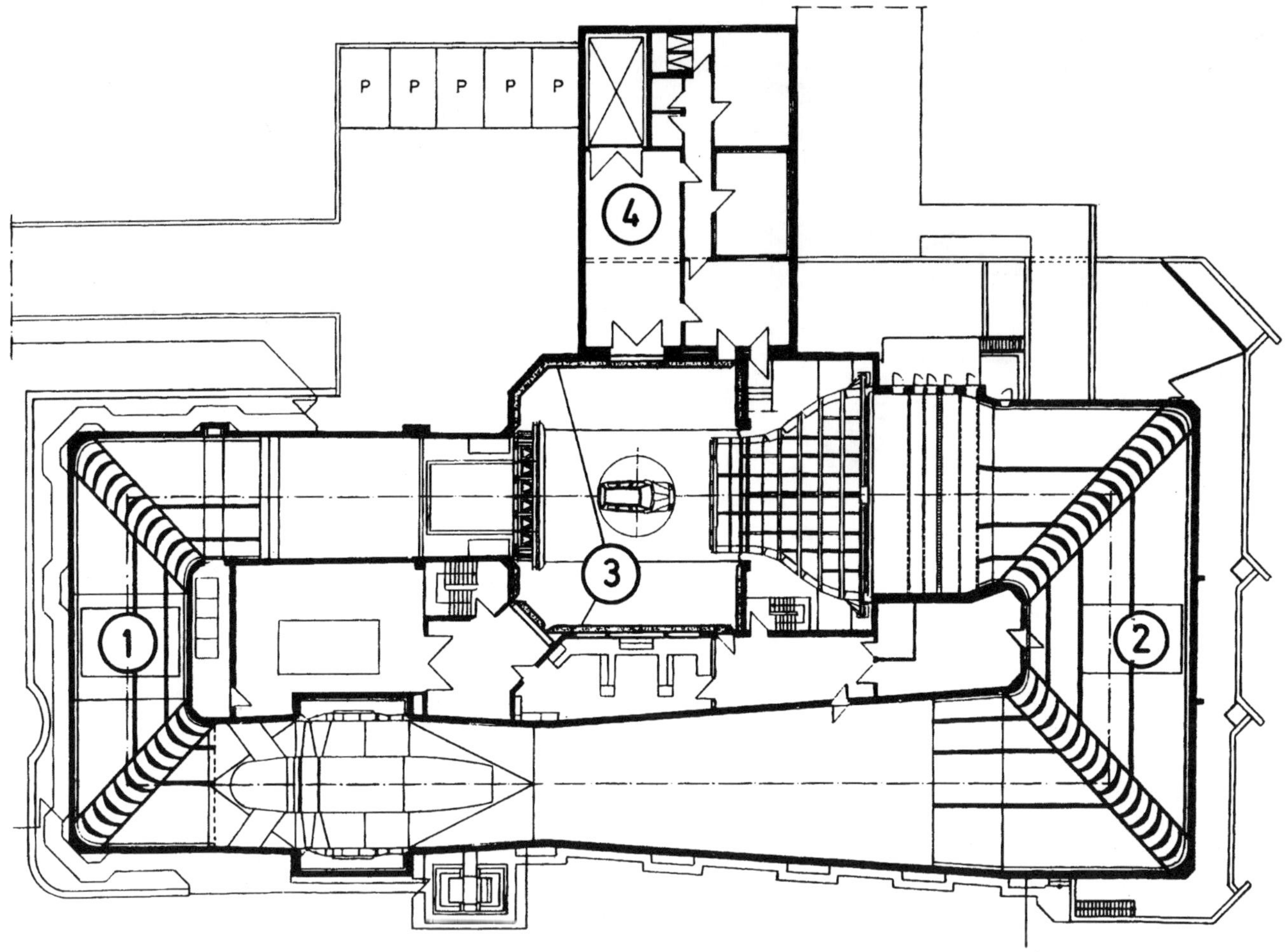

Fig.5: Plan view of the converted IVK Wind Tunnel with U-bend silencers in the return duct:
(1) U-bend silencer in the first cross-leg with 3 duct sections
(2) U bend silencer in the second cross-leg with 4 duct sections
(3) Sound-absorbing lining of the plenum chamber
(4) Outbuilding with a second access, reverberation chamber, second control room (acoustics) and an audio-laboratory.

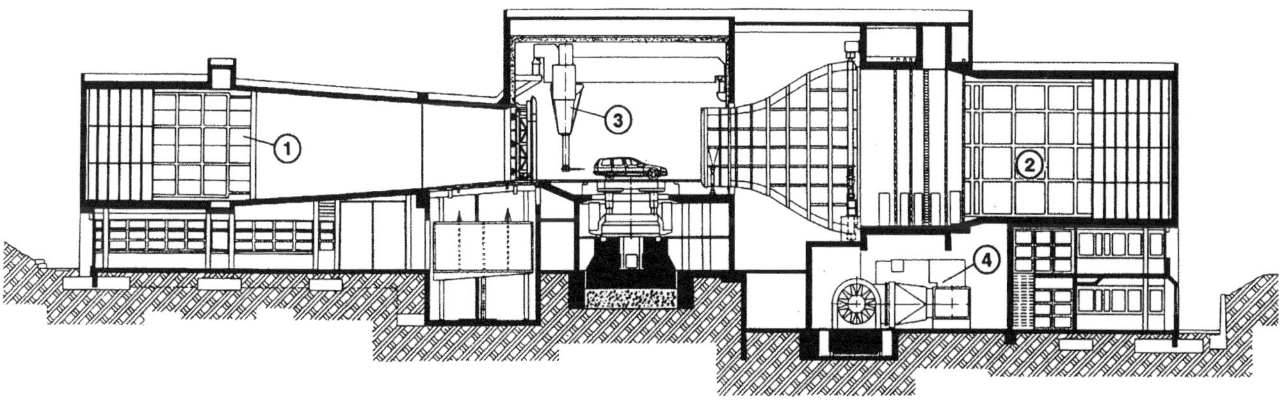

Fig.6: Side view of the converted IVK Wind Tunnel:
(1) Vertical splitter plate of U-bend silencer
(2) Vertical splitter plate of U-bend silencer
(3) Multi-axis traversing gear
(4) Model wind-tunnel.

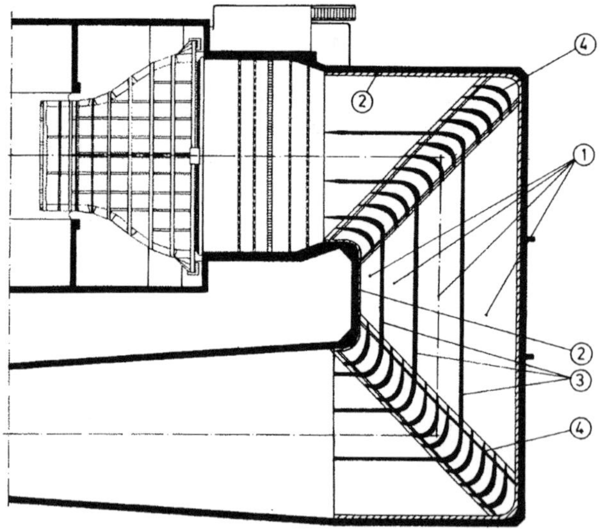

Fig.7: Innovative solution for attenuating the wind-tunnel's fan-noise by means of U-bend silencers, depicted for corners 3 and 4:
(1) Subdividing the return duct in unequal sections with the same pressure loss and acoustic transmission
(2) Acoustic lining of the duct's side-walls with membrane absorbers
(3) Vertical splitter plates containing 2 layers of membrane absorbers in back-to-back arrangement
(4) Sheet-metal turning vanes profiled with porous Polyester-foam.

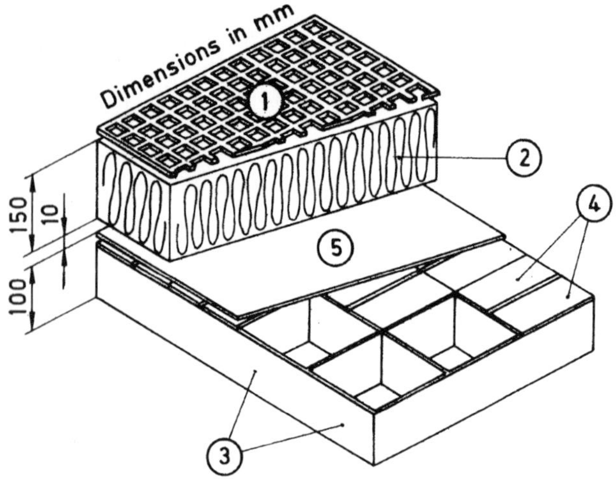

Fig.8: Schematic setup of a broadband absorber consisting of porous Polyester-foam and membrane absorbers:
(1) Perforated sheet metal
(2) Porous Polyester-foam
(3) Walls of the cavities
(4) Thin and slotted membranes enclosing the cavities
(5) Covering membrane.

that are also mounted on the inner and outer side wall of the duct, include as well the acoustically treated turning vanes in the four corners. Therewith, they form together an integral sound-attenuator unit.

- Sound-absorbing lining on the pressure and suction side of the originally unprofiled sheet-metal turning vanes in all four corners.
- Sound-absorbing lining of the plenum walls and ceiling with broad-band absorbers.

The expanded use of the wind-tunnel facility in regard to acoustics made an outbuilding necessary, adjacent to the plenum chamber, see Fig.5. It contains a second preparation room with a direct access to the wind tunnel through a door in the plenum's side-wall. This preparation room can serve as a reverberation chamber, too. Since the full-scale wind-tunnel is located on the second floor, this preparation room is accessible by a second lift. Moreover, the outbuilding contains a second control room for acoustic testing, an audio-laboratory and several offices for the vehicle-acoustics group.

The IBP silencing concept is based on dividing the necessary sound attenuation up into frequency ranges. In the low frequency range (approx. from 80 to 200 Hz) the fan noise is damped by novel membrane absorbers [9]. Fig.8 shows the main structure of such a reactive absorber. This type of absorber is made of aluminum sheet-metal and contains an array of cavities, enclosed by a thin and slotted membrane, respectively. On top, the whole membrane housing is covered by a slightly thicker membrane. Such a resonator array absorbs sound energy in the difficult and deep frequency range (<300 Hz) at cavity depths from 100 to 200 mm. In the present case, the depth was chosen to 100 mm in order to get a high absorption rate for the bandwidths of the third-octave center frequencies 100, 125 and 160 Hz.

In order to absorb the middle and high frequencies as well, a second material has to be used besides the aluminum membrane-absorbers. This non-fibrous material shall have acoustic properties similar to mineral wool. Porous Polyester-foam keeps its shape in contrast to mineral wool, but its disadvantage is its inflammability, and that it is more expensive. Eventually, porous Polyester-foam of the type GE30FR with a mass density of 30 kg/m^3 was chosen. It burns rather slowly at less than 100 mm/s and, therefore, has the fire-classification B2 according to DIN 7520 (German Industrial Standard).

Fig.9 shows how the originally unprofiled Kröber vanes in the four corners of the return duct were aerodynamically shaped with coated porous Polyester-foam and how they were integrated in the vertical splitter plates of the U-bend silencers. On the suction side of a free-standing turning vane the foam thickness is constantly 40 mm. On the pressure side it varies between 20 and 160 mm. Together with a round nose and a tailed rear end the pressure-loss coefficient is below that of the unprofiled Kröber vane. The smoothness and stability at the junction of foam with nose and rear end is ensured by overlaping sheet metal. The coated foam surface is insensitive to soiling and easy to

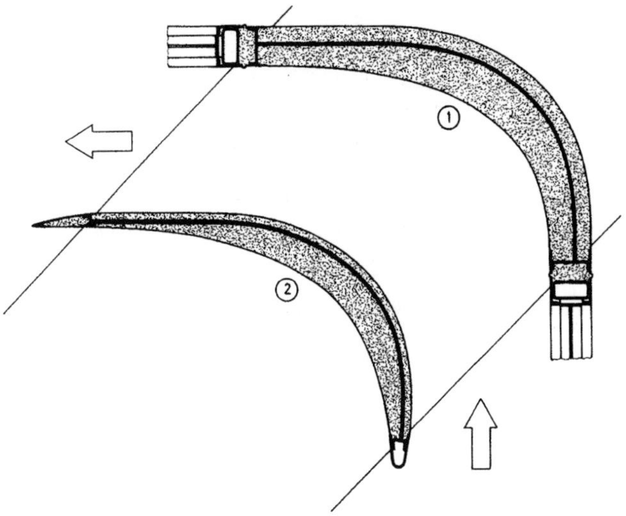

Fig.9: Profiling the sheet-metal turning vanes with coated porous Polyester-foam:
(1) Turning vane integrated in a vertical splitter plate
(2) Free-standing turning vane.

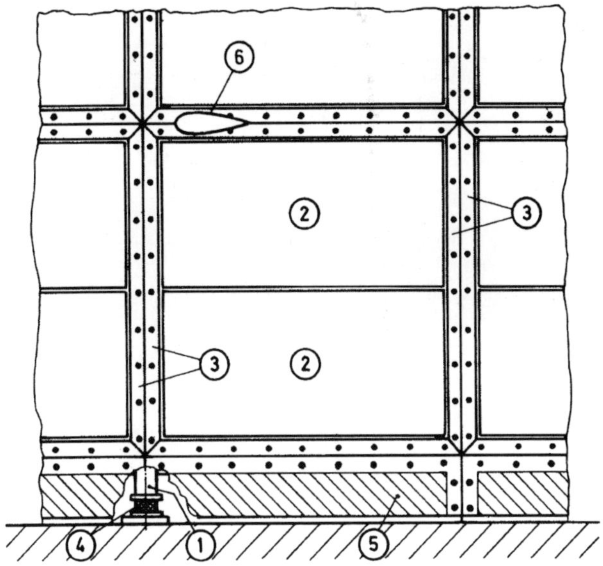

Fig.10: Constructive details of a free-standing vertical splitter plate:
(1) Bolted steel-framework
(2) Membrane absorbers in openings of the steel-framework
(3) Bolted angle irons for fixing the membrane absorbers
(4) Elastic bearing of the steel-framework
(5) Gap to floor and ceiling filled with porous Polyester-foam and covered over with perforated sheet metal
(6) Profiled horizontal support struts.

clean. The reduced pressure-loss coefficient of the profiled turning vanes partly compensates the additional pressure loss and skin-friction drag of the vertical splitter plates.

The free-standing steel-framework of the vertical splitter plates consists of rectangular steel-girders with a thickness of 210 mm. It contains membrane absorbers with a thickness of 100 mm of each in back-to-back arrangement with a 10 mm gap. The framework itself is elastically suspended with rubber bearings on floor and ceiling of the return duct, which is made of reinforced concrete. The maximum dimensions of the steel-framework in corners 3 and 4 (downstream of the fan) are 16 m in length and 8 m in height. Figs.10 – 12 show the constructive details.

The elastic suspension of the steel-framework compensates the different thermal expansion of the steel-structure and the concrete walls and prevents any structureborne noise from being transferred to the latter. The steel-framework of the U-bend silencer's inner and outer side wall is made of C-profile steel-girders and contains only one layer of membrane absorbers. This steel-structure is elastically suspended on the enclosing concrete walls, too (Fig.11, Detail A). This figure also displays details of fixing and elastically suspending the membrane absorbers in the openings of the steel-framework of the splitter plates. The membrane absorbers are clamped with bolted angle-irons, which are flush with the absorbers covering membrane. By means of special rubber profiles at the end of the angle-iron, the membrane absorber is both elastically suspended on its circumference and pressed against its small aluminum flange. Thus, a thermal expansion is enabled and the required structureborne noise insulation is ensured besides the necessary electrolytic separation of the different materials. Two membrane absorbers, one on top of the other respectively, are mounted on the same time in the

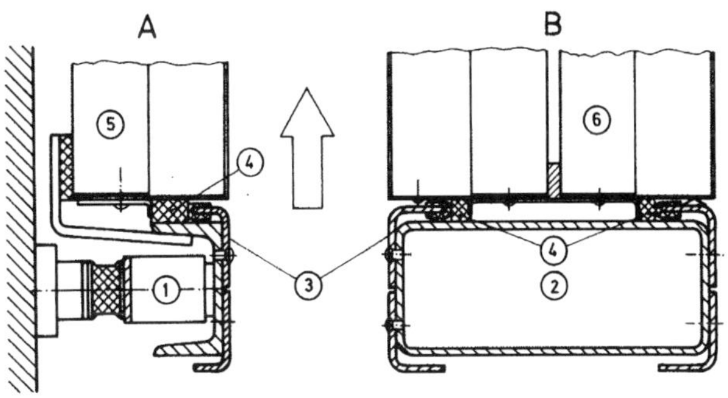

Fig.11: Constructive details of elastic fixing the membrane absorbers in the steel-framework at the duct's side-wall (A) and in a free-standing splitter plate (B):
(1) Elastic suspension
(2) Rectangular steel girder
(3) Bolted angle irons
(4) Rubber profiles
(5) Membrane absorbers in a single layer
(6) Membrane absorbers in back-to-back arrangement

Fig.12: Section of a U-bend silencer.

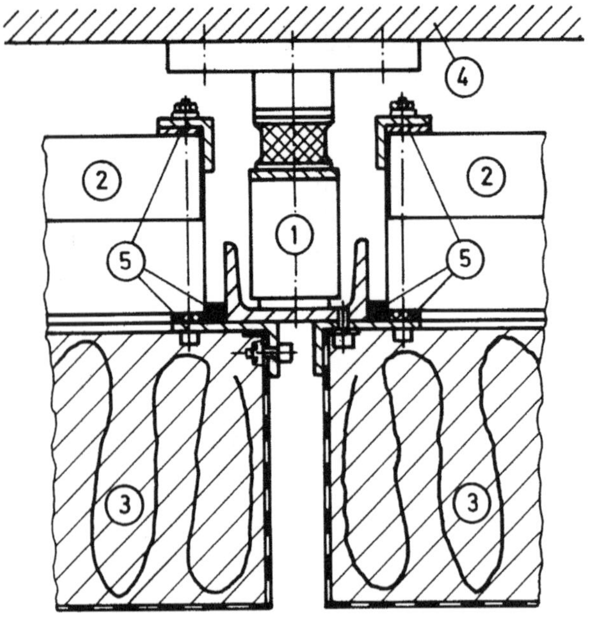

Fig.13: Constructive details of the broadband absorbers on the walls of the plenum chamber:
(1) Elastic suspension
(2) Single layer of membrane absorbers
(3) Plates of porous Polyester-foam protected by perforated sheet metal
(4) Plenum wall
(5) Rubber profiles.

steel-framework openings. The remaining gap to the concrete floor and ceiling is filled with porous Polyester-foam and covered flush with perforated sheet metal. Aerodynamically shaped horizontal struts increase the lateral stiffness of the vertical splitter plates. Fig.12 shows a section of the U-bend silencer downstream of the fan.

An anechoic plenum chamber was achieved with so-called broadband absorbers, mounted on the plenum walls and ceiling. Fig.13 shows constructive details. Again a steel-framework was elastically fixed to the plenum walls, because of the thermal expansion and the insulation from structureborne noise. The framework again contains a layer of 100 mm thick membrane absorbers, intended for attenuating the low frequency noise. Above, there are 150 mm thick plates of porous Polyester-foam with a 10 mm gap to the membrane absorbers. With the low stiffness of the foam plates it is necessary to cover them with perforated sheet metal with square holes of $20*20$ mm^2 and an open-area ratio from 52.6 to 64.2%. Altogether, with a necessary gap of 50 mm of the membrane absorbers to the concrete wall, the plenum dimensions were only reduced by 0.31 m on each wall. The remaining plenum dimensions of 12.9 m in length, 14.4 m in width and 8.2 m in height ensure that the plenum volume still is big enough to avoid any tremendous influences on the aerodynamic coefficients of cars, as stated by Deutenbach [8]. The control room's window and the access door in the plenum's side-wall can be covered by electrically driven acoustic curtains. Fig.14 gives an impression of the acoustic lining of the plenum.

Besides silencing the wind tunnel, additional measures were taken to aerodynamically optimize the open jet and to enable acoustic tests on cars. The collector flaps were shortened by 0.55 m and the annular breather gap between the collector flaps and the test-section diffuser was increased from 0.3 m to 0.4 m. It resulted in an increase of the test-section length from 9.5 m to 9.94 m. By means of an added electromechanical drive unit the collector flaps

Fig.14: View on the acoustic lining of the plenum chamber, on the nozzle with vortex generators and on the traversing gear.

Fig.15: View on the traversing gear with inflow microphone, on the acoustically lined collector and on the vertical splitter plates of the U-bend silencer.

Fig.16: View on the traversing gear with inflow microphone and on the adjustable collector with an acoustic lining and with an annular breather gap.

can be independently adjusted in the angle range from -2.5 to 34 degrees. According to the requirement of a low reflection the bellmouth and inner surface of the adjustable collector were lined with 80 mm thick coated porous Polyester-foam (Figs.15 and 16).

Although the adjustable collector was altered, the ratio of the collector cross-sectional area to the nozzle exit area (A_C/A_N = 1.179) remained unchanged after the conversion, because it turned out to be the best compromise in respect to aerodynamic and acoustic testing.

For the measurement of the exterior flow noise of vehicles an installation of a multi-axis traversing gear below the plenum ceiling was mandatory. The Dantec construction has a X-beam, which drives laterally on two fixed Y-beams, with a telescopic, stiff and aerodynamically profiled Z-arm (Figs.15 and 16). The two stiffening struts of the Z-arm have to be removed when measuring inflow. The traversing range of 7.95 m, 10.15 m and 3.75 m in X, Y and Z-direction does not only cover the inflow measuring-volume of a test object, but also big parts of the out-of-flow one. In context with different multi-axis add-on manipulating units, the construction is suitable to acoustic, aerodynamic and video measuring-tasks. A special manipulator allows – within close limits – an alignment of the inflow microphone to the local velocity vector.

ANECHOIC CHARACTERISTIC OF THE PLENUM

The anechoic characteristic of the plenum chamber was checked out with the draw-away procedure according to ISO 3745/DIN 45635. The nozzle exit and the collector entrance were not acoustically sealed during these measurements. The draw-away procedure itself requires a noise source to be put on the center of the wind-tunnel floor. From this point steel-wires were pulled to the four upper plenum corners (room diagonals) and also three lines parallel to the floor to the control-room window. The decrease of the sound pressure levels is then measured with increasing distance from the sound source. Fig.17 displays exemplarily the measured and calculated results for line 3 (room diagonal) and for different third-octave center frequencies. The graph shows tolerable deviations from the sound-propagation law according to DIN 45635 and for measuring distances of 7.5 m (50 Hz) and 9.5 m (10 kHz) in the free field.

Fig.18 shows the absorption coefficient of the plenum as a mean value of the draw-away measurements on all seven lines (curve 1). The absorption rate for frequencies higher than 63 Hz is 0.9 and increases to 0.98 until 6.3

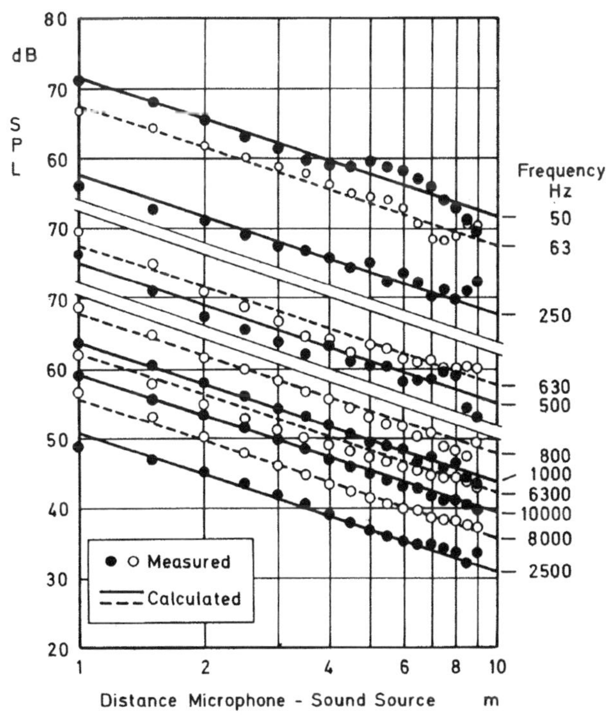

Fig.17: Checking of the anechoic characteristic of the plenum chamber by draw-away measurements on line 3 (room diagonal) according to ISO 3745.

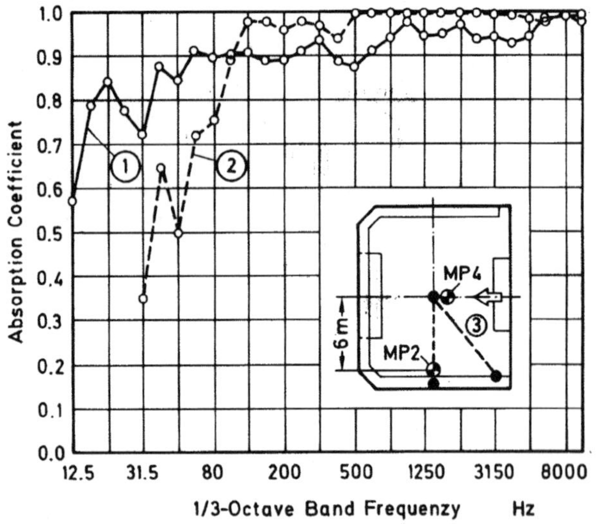

Fig.18: Comparison of absorption coefficients obtained with different acoustic materials:
(1) Acoustically treated plenum chamber (thickness of broadband absorbers 0.31m) of the IVK Wind Tunnel with a hard floor
(2) Anechoic chamber of the IBP with a hard floor and with conventional mineral wool (wedges with a thickness of 0.65 m).

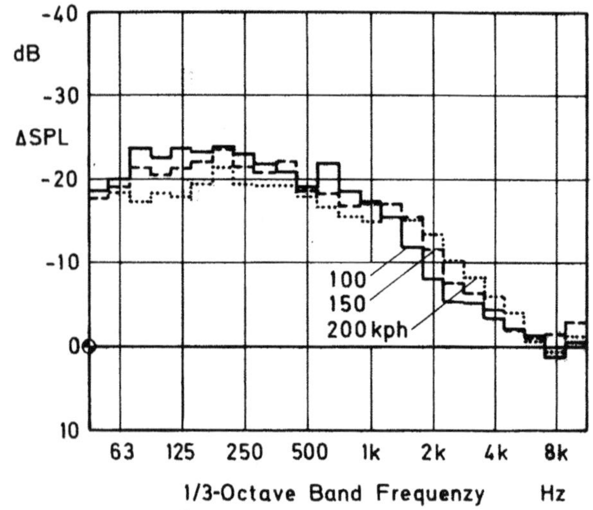

Fig.19: Inflow difference spectra of the IVK Wind Tunnel before and after conversion at different free-stream velocities.

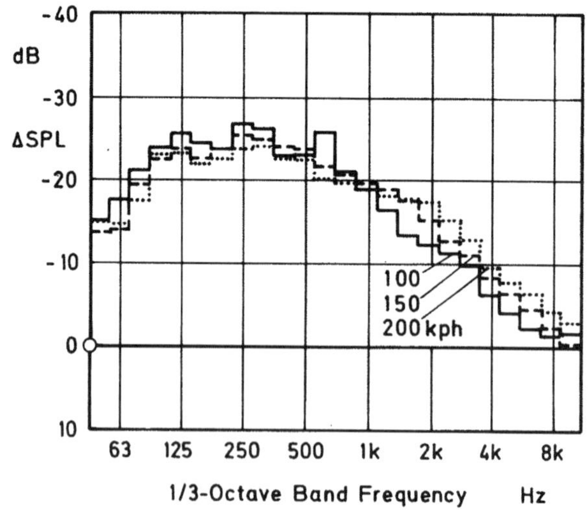

Fig.20: Out-of-flow difference spectra of the IVK Wind Tunnel before and after conversion at different free-stream velocities.

kHz. In the frequency range below 63 Hz the absorption rate only decreases to 0.8. Curve 2 relates to measurements in the anechoic chamber of IBP with a hard and reflective floor. Conventional fibrous material (wedges) with a thickness of 0.65 m was used. It can be seen that this technology offers advantages in the middle frequency range with absorption rates between 0.95 and 0.99. In the low frequency bandwidth between 31.5 Hz and 100 Hz, however, it shows significantly worse results.

WIND TUNNEL SELF-NOISE AFTER CONVERSION

Fig.19 displays the difference of two third-octave band spectra, measured inflow – microphone position 4 – in the IVK Wind Tunnel before and after conversion at the velocities 100, 150 and 200 kph. Fig.20 shows the out-of-flow difference spectra which are measured in microphone position 2. Both graphs show approximately a 20 dB reduction in the dominating frequency range of the exterior flow noise of a vehicle at around 700 Hz.

Fig.21 shows the wind-tunnel self-noise, measured both inflow and out-of-flow, versus the tunnel velocity and before and after the acoustic treatment. Without vortex generators at the nozzle exit the OSPL is reduced by 20 dB, measured out of the flow (curves 2 and 6). The share of the plenum's broadband absorbers in this reduction is roughly 8.5 dB as a measurement with installed U-bend silencers and an untreated plenum chamber showed. The inflow and out-of-flow OSPL-differences were with 0.3 dB very small before converting the wind tunnel. After completion of the acoustic treatment it slightly increased to 2.5 dB. Since the wind-tunnel fan as the main noise source has been silenced, the self-induced noise of the vortex generators at the nozzle exit is much more pronounced with an increase of 10 dB compared to 2.2 dB before the conversion.

The overall sound pressure levels 58.8 and 71.0 dB(A), measured out-of-flow at the 6 m sideline to the tunnel axis, of the IVK Aeroacoustic Wind Tunnel at free-stream velocities of 100 and 150 kph are significantly higher than the pure jet noise (lip-noise at the nozzle exit) [6]. It is caused by secondary noise sources in the test section and tunnel duct before and after the U-bend silencers. These secondary noise sources, which cannot be attenuated, are listed as follows:

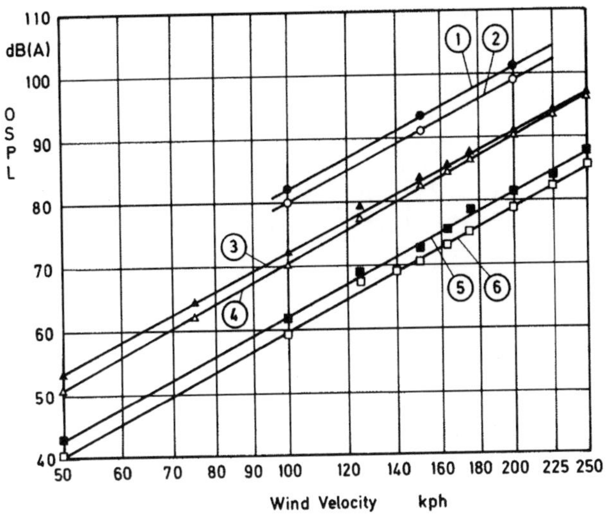

Fig.21: A-weighted inflow and out-of-flow OSPL versus wind velocity of the background noise of the IVK Wind Tunnel before and after conversion:

IVK Wind Tunnel acoustically untreated:

(1) Out-of-flow/inflow measurements with vortex generators

(2) Out-of-flow/inflow measurements without vortex generators

IVK Wind Tunnel acoustically treated:

(3) Inflow measurements with vortex generators

(4) Out-of-flow measurements with vortex generators

(5) Inflow measurements without vortex generators

(6) Out-of-flow measurements without vortex generators.

- Self-induced noise and interference noise of installations in the settling chamber like turbulence screens and combinations of it, and the honeycomb.
- Noise generated at the nozzle exit (lip noise).
- Noise from the impact of the shear layer on the collector flaps.
- Noise from the flow through the annular breather gap between the rear end of the adjustable collector and the front end of the test-section diffuser.
- Self-noise of the shear layer and the floor boundary-layer.

Fig.22 displays the A-weighted third-octave band spectra of inflow and out-of-flow measurements at 150 kph in the converted wind tunnel. The OSPL determing spectra between 800 Hz and 2 kHz is to ascribe – for the given contraction ratio of 4.41 – the installations in the settling chamber, which consist of three turbulence screens and a honeycomb. At this tunnel velocity and contraction ratio of the nozzle the contribution of the settling chamber can reach more than 60 dB in the afore mentioned frequency range [10]. The divergence of both curves in the frequency range above 2 kHz is explained by the inflow-microphone's self-induced noise at its nose cone. On the other hand, the out-of-flow microphone seems to be influenced at low frequencies by the turbulent recirculating flow in the plenum.

Fig.23 shows the A-weighted spectra of out-of-flow measurements at 150 kph of the converted wind tunnel both with and without vortex generators at the nozzle exit. The Seiferth wings influence the development of the free shear layer and cause a broadband increase of the lip noise. Superimposed is a peak at 2.5 kHz which is caused by the support bolts of the vortex generators at the given tunnel speed. This graph also elucidates that the Seiferth wings reduce the interference effects between shear layer and collector in the frequency range below 125 Hz. However, these results led to the basic decision to carry out all

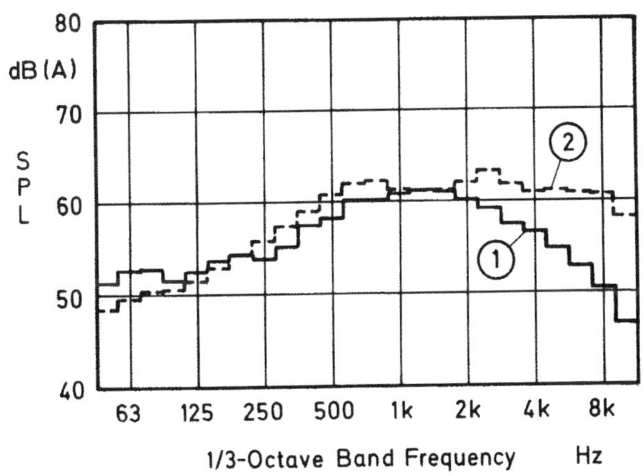

Fig.22: A-weighted self-noise spectra at 150 kph of the acoustically treated IVK Wind Tunnel without vortex generators:

(1) Out-of-flow measurement

(2) Inflow measurement.

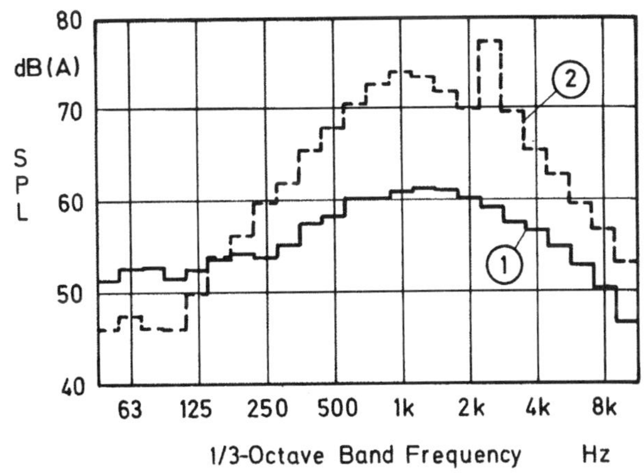

Fig.23: A-weighted self-noise spectra at 150 kph of the IVK Wind Tunnel after conversion:

(1) Out-of-flow measurement without vortex generators

(2) Out-of-flow measurement with vortex generators.

acoustic measurements without vortex generators at the nozzle exit.

COMPARISON WITH OTHER AUTOMOTIVE WIND TUNNELS

In Figs.24 and 25 the IVK Aeroacoustic Wind Tunnel is compared to the A-weighted inflow and out-of-flow OSPL of other European automotive wind tunnels. In case of the inflow measurements [11,12] the BMW Aeroacoustic Wind Tunnel shows a 10 dB higher OSPL at 150 kph than the IVK Wind Tunnel. This difference reduces to 3.5 dB in case of out-of-flow measurements [11] at the same tunnel velocity. The OSPL difference between inflow and out-of-flow measurement at 150 kph is clearly distinguished in both wind tunnels. The difference in the BMW Wind Tunnel is 9 dB, whereas in IVK it amounts only to 2.2 dB. There can be different causes for this fact, which are based on the characteristic features of both wind tunnels and need to be verified by comparative measurements.

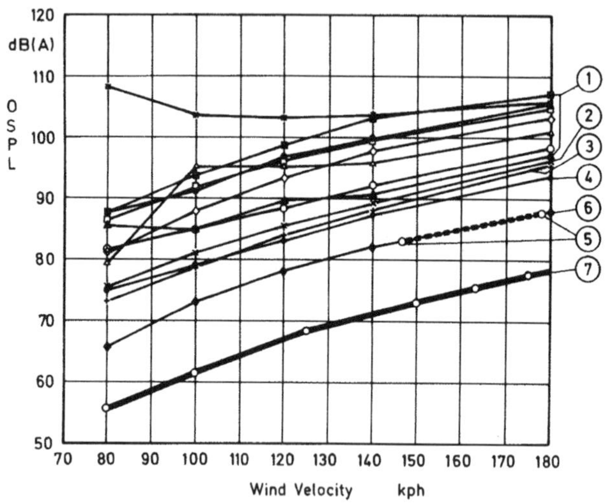

Fig.24: A-weighted inflow self-noise levels versus tunnel speed of European automotive wind tunnels compared to IVK:
(1) Acoustically untreated wind tunnels from the DNW test campaign 1992
(2) IVK Wind Tunnel before conversion
(3) Volvo Wind Tunnel with slotted-wall test-section
(4) DNW Wind Tunnel with 8*6 m^2 closed jet
(5) DNW Wind Tunnel with 8*6 m^2 acoustic open jet
(6) BMW Aeroacoustic Wind Tunnel
(7) IVK Aeroacoustic Wind Tunnel (after conversion).

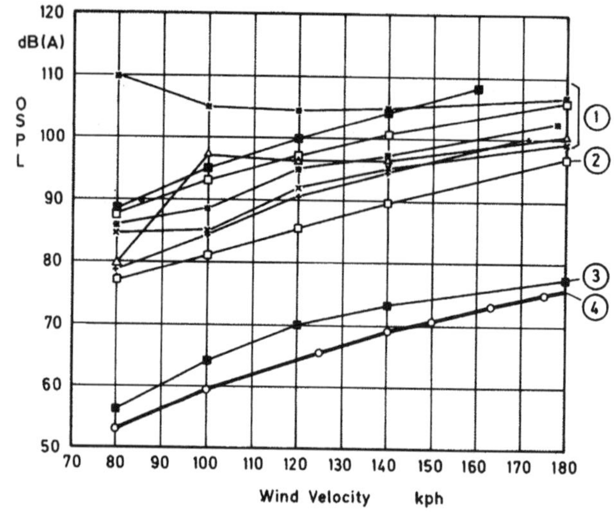

Fig.25: A-weighted out-of-flow self-noise levels versus tunnel speed of European automotive wind tunnels compared to IVK:
(1) Acoustically untreated wind tunnels from the DNW test campaign 1992
(2) IVK Wind Tunnel before conversion
(3) BMW Aeroacoustic Wind Tunnel
(4) IVK Aeroacoustic Wind Tunnel (after conversion).

INFLUENCE OF THE ACOUSTIC TREATMENT ON THE WIND TUNNEL PERFORMANCE

Table 1 compares the top speed and the power consumption of the fan's DC-motor before and after converting the wind tunnel. The top speed is reduced by roughly 10 kph for both cases in which a car with a drag area of 0.61 m^2 is in place. The additional power requirement related to the same test-section adjustments, to the same air density and to the same tunnel speed does not exceed 7.5%. There was no need to adjust the blade angle of the fan.

Figs.26 and 27 show velocity distributions in the con-

Wind Tunnel Condition	Before Conversion	After Conversion
Empty Test Section		
Top Speed	269.8 kph	262.8 kph
Fan Speed	335 min^{-1}	332.4 min^{-1}
Power Output	2835 kW	2980 kW
Air Density	1.091 kg/m^3	1.132 kg/m^3
Efficiency P_{jet} / P_{el}	1.83	1.65
Test Section Occupied with Car $C_D*A = 0.61\ m^2$		
Top Speed	267.8 kph	256.9 kph

Table 1: Performance data of the IVK Wind Tunnel before and after conversion.

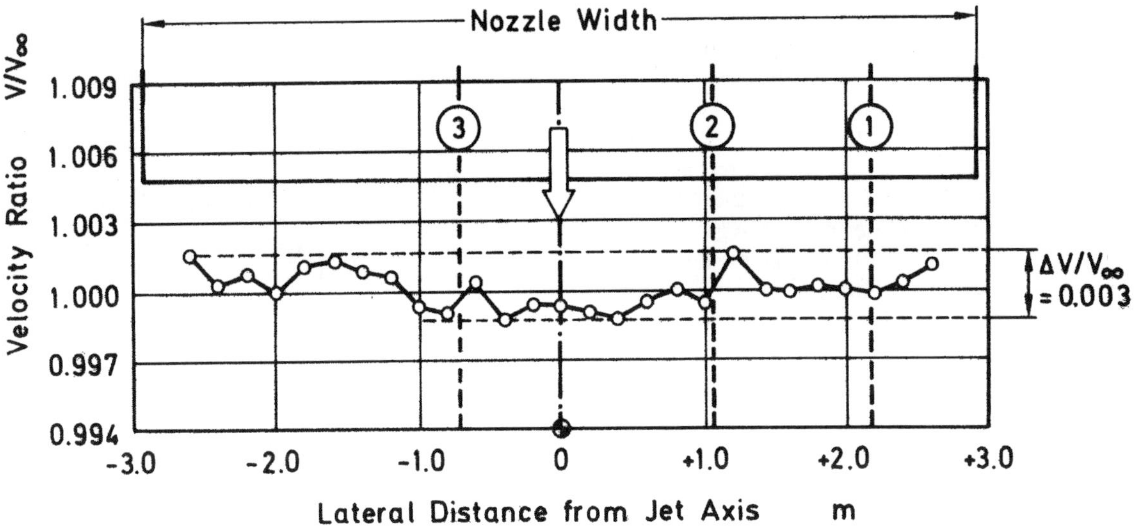

Fig.26: Velocity distribution (V/V_∞) in the open jet of the converted wind tunnel in a cross section 1 m behind the nozzle exit at an elevation of 1.3 m and for a free-stream velocity of 140 kph.
(1),(2),(3) are streamline filaments from the 3 vertical splitter plates of the U-bend silencer of corners 3 and 4.

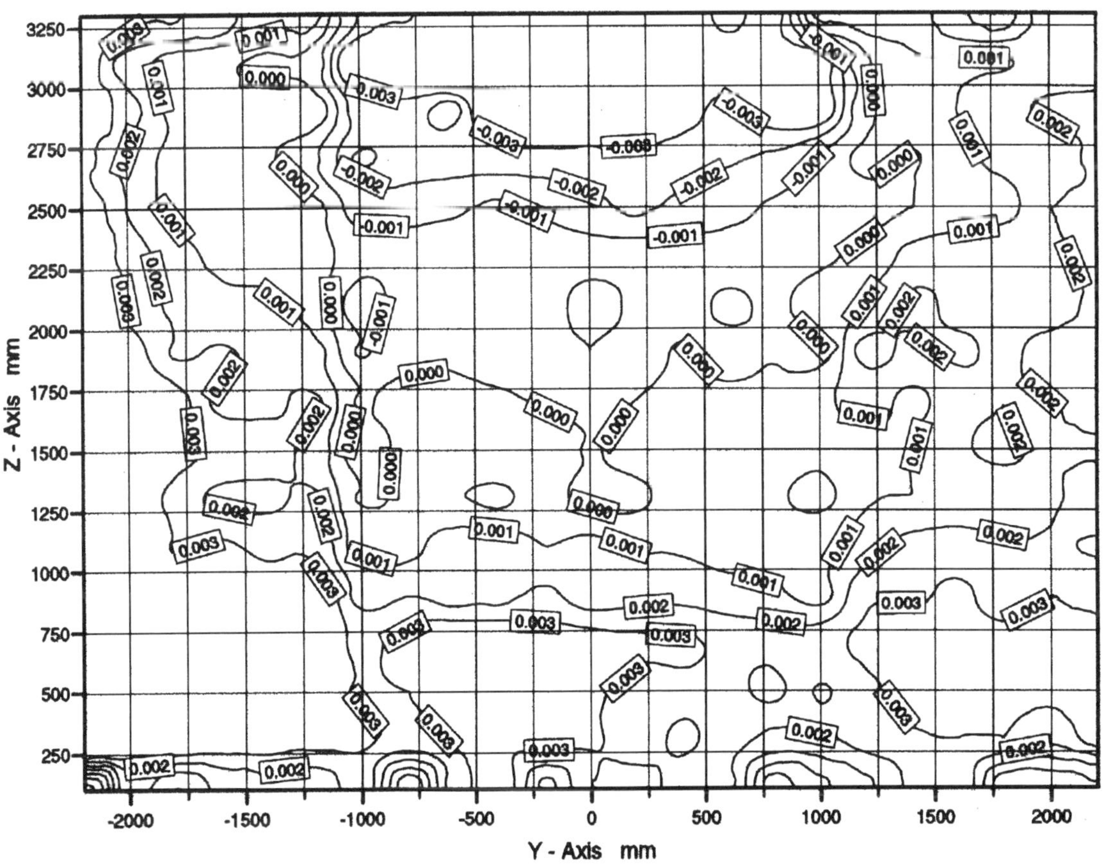

Fig.27: Velocity map ($\Delta V/V_\infty$) in the open jet of the converted wind tunnel in a cross section above the turntable's lateral axis and for a free-stream velocity of 140 kph.

verted wind tunnel in cross sections 1 m behind the nozzle exit (Fig.26) and above the turntable's lateral axis (Fig.27). The acoustic installations in the return duct have not increased the non-uniformity of velocity in the jet core with ±0.3%. Especially, no discontinuities are generated from the three free-standing vertical splitter plates of the U-bend silencer in corners 3 and 4, downstream of the fan. The according streamline filaments 1,2 and 3 in the open jet and coming from the vertical splitter plates are denoted in Fig.26. A pre-investigation in the Institute's model wind-tunnel with a vertical absorber splitter-plate with a blunt rear end, and installed between corner 4 and the first turbulence screen, showed that a wake was not detectable in the open-jet's velocity distribution. Nevertheless, all full-scale vertical splitter plates were provided with a tail. In case of the U-bend silencer downstream of the fan the tails reach until the first screen.

RESULTS OF AEROACOUSTIC VEHICLE MEASUREMENTS

Table 2 compares the exterior flow noise at 150 kph of different test vehicles before and after conversion of the wind tunnel. The data shows for the former condition a dominating background noise of the empty tunnel which leads to an unacceptable signal-to-noise ratio of vehicle measurements. The comparable data from the converted wind tunnel reveals the progress. Fig.28 shows the A-weighted third-octave band spectra of the exterior flow noise of the same test vehicles in the acoustically treated wind tunnel. The spectra were measured out-of-flow at 150 kph. Even the most silent full-scale model shows a signal-to-noise ratio from 5 to 10 dB for the frequency range from 100 Hz to 10 kHz. Thus, a safe evaluation of design optimizations is enabled.

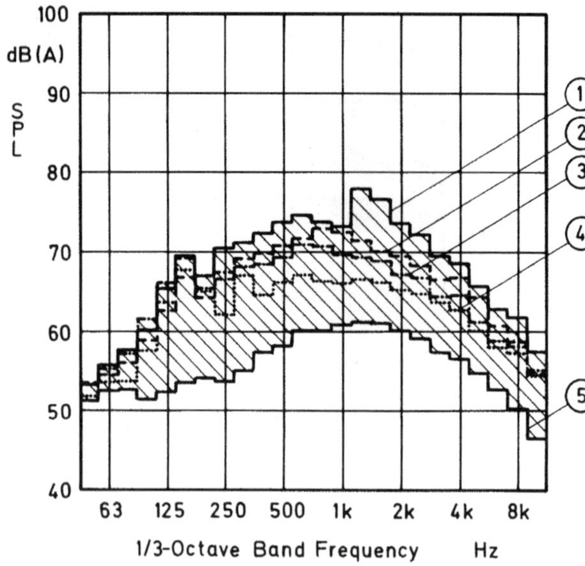

Fig.28: A-weighted out-of-flow spectra of different test vehicles at 150 kph in the IVK Aerocoustic Wind Tunnel:
(1) Mercedes-Benz 210 T1N Van with cooling air flow
(2) Audi 5000 Notchback with cooling air flow
(3) Calibration Model BMW CC-7.2 as square-back with add-on spoilers
(4) Calibration Model BMW CC-7.2 with a smooth surface and a fastback extension of 800 mm
(5) Wind tunnel self-noise.

Test Object in Open-Jet	Exterior Noise at 150 kph Out-of-Flow Measurement	
	Wind Tunnel before Conversion	*Wind Tunnel after Conversion*
Empty Test Section without Seiferth Wings	91.1 dB(A)	71.0 dB(A)
Calibration Model 1:1 BMW CC-7.2, Fastback with 800 mm Extension	91.8 dB(A)	77.9 dB(A)
Calibration Model 1:1 BMW CC-7.2, Station Wagon with Add-on Spoilers	92.2 dB(A)	80.6 dB(A)
BMW 535i Notchback	92.6 dB(A)	81.2 dB(A)
AUDI 5000 Notchback	92.8 dB(A)	81.7 dB(A)
MB210 T1N Van	94.6 dB(A)	85.2 dB(A)

Table 2: Exterior flow noise at 150 kph of different test vehicles in the IVK Wind Tunnel before and after conversion.

Disturbed flow around the bodywork, underbody and wheels of a car with vortex formations, detached flow, reattached turbulent flow and a thicker surface boundary-layer used to lead both to increased drag and increased exterior flow noise. Vice versa, the conformity of reduced drag and noise emission is a general trend, too. However, there are exceptions that depend on the kind of noise generation. The chart in Fig.29 compares the exterior flow noise to the drag area of different production cars and extreme vehicles like full-scale car models. The full-scale models of the research cars BMW CC-7.2 and UNI-CAR have a smooth underbody and have OSPL's which are in the average 3 dB below production cars with cooling air flow. If the drag area of the BMW CC-7.2 reaches nearly double the value of a production car by adding on spoilers (C_D = 0.55), the exterior flow-noise level of 80.6 dB(A) is still 0.6 dB lower than the production cars average. Note, however, that the BMW calibration model has no cooling air flow. A blanked cooling air intake can cause an increase or decrease of the exterior noise emission. For instance, it cannot be excluded that the van's relatively high noise-level of 85.2 dB(A) contains a share of unknown

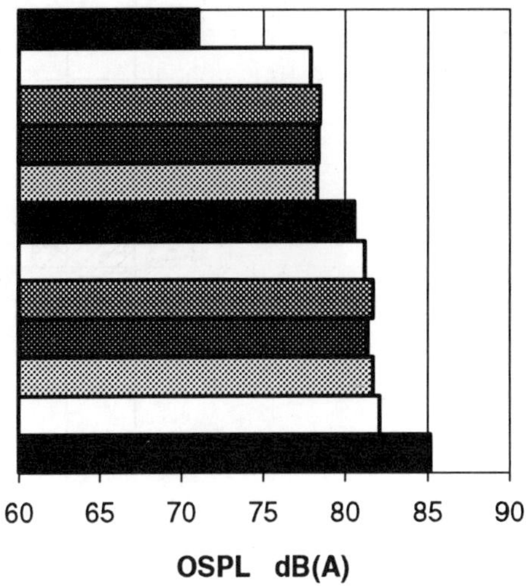

OSPL dB(A)	Test Vehicle	$A*C_D$ m²
71.0	Empty Test Section	
77.9	BMW CC-7.2 FB + 800 mm Ext.	0.345
78.5	UNI-CAR 1:1 with Cool.- Air Flow	0.492
78.4	UNI-CAR 1:1 w.o. Cool.- Air Flow	0.480
78.3	BMW CC-7.2 SW + 800 mm Ext.	0.364
80.6	BMW CC-7.2 SW + Add-on Spoilers	1.106
81.2	BMW 535i Notchback	0.645
81.7	Opel Omega A Mock-up	0.593
81.4	Opel Omega A with Cool.- Air Flow	0.608
81.7	Audi 5000 NB with Cool.- Air Flow	0.735
82.1	VW Jetta II with Cool.- Air Flow	0.680
85.2	MB 210 T1N Van	1.750

Fig.29: Signal-to-noise ratio (out-of-flow measurement) versus drag area of different test vehicles at 150 kph in the IVK Aeroacoustic Wind Tunnel.

magnitude in the OSPL, which results from the collector due to the greater jet expansion with the van.

Figs.30 and 31 show a test set-up with an acoustic mirror of 1.2 m diameter and of elliptic shape cross-section. Its microphone – positioned in the focal point – is in a 4 m sideline from the test object. This directional measuring system [14,15], suitable for sound-source location, supplies noise-radiation charts of a passenger car with and without antenna. The measurements were carried out at 140 kph and consider the frequency range from 1 kHz to 3 kHz. The distribution of the sound pressure levels of the exterior flow noise shows maxima coming from the car's front part, the wheelhouses and displays the tonal noise from the antenna.

SUMMARY AND CONCLUSIONS

The aerodynamic noise of motor vehicles becomes increasingly significant at cruising speeds above around 100 kph, after the engine, drivetrain and tire/road noise has been reduced. The aerodynamic exterior noise of cars can be determined in a wind tunnel, provided that there is a signal-to-noise ratio of at least 10 dB. For interior noise measurements the wind-tunnel's background noise-level is less decisive.

A reduction of the wind-tunnel's fan noise with conventional duct absorbers leads to high pressure losses in the tunnel airpath and to a considerable reduction of the tunnel velocity, when the power output of the fan's drive unit cannot be increased. These losses have been minimized by means of a novel sound-absorber concept which uses non-fibrous material. Two U-bend silencers were installed in the return duct of the wind tunnel, upstream and downstream of the fan. They use membrane absorbers as thin and smooth vertical splitter plates (for the low frequency range) in context with profiling the turning vanes with coated porous Polyester-foam (for the middle and high frequency range).

Fig.30: Test setup for sound-source location with an acoustic mirror mounted on the traversing gear out of flow.

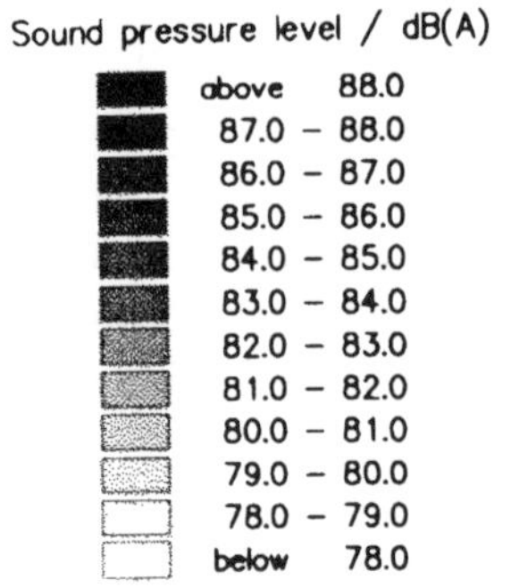

Fig.31: Noise radiation chart of a passenger car with (top) and without antenna (bottom), measured with an acoustic mirror of 1.2 m diameter at the 4 m sideline (microphone to car surface; V_∞ = 140 kph).

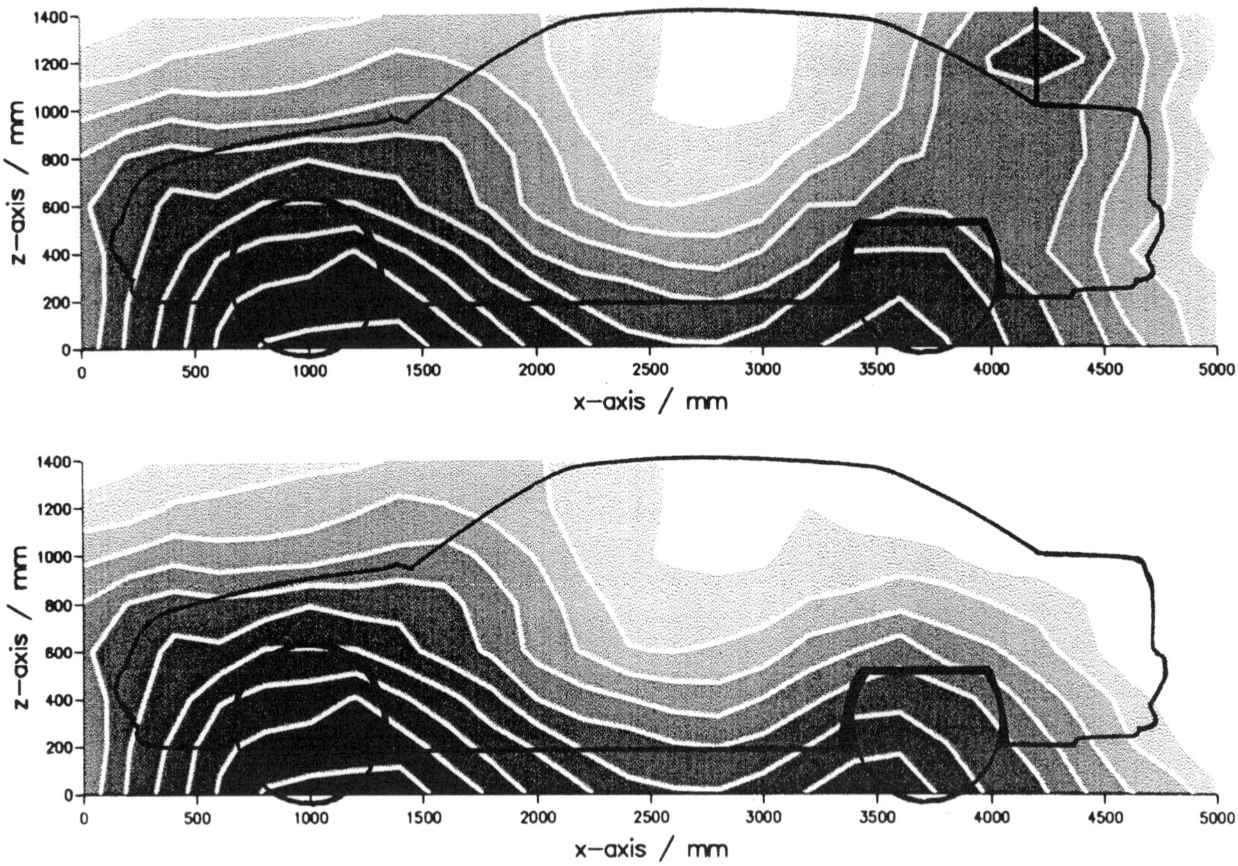

An anechoic plenum chamber was achieved with a new type of broadband absorbers which are mounted on the plenum walls and ceiling. Behind plates of porous Polyester-foam, which are protected by perforated sheet metal, there is a layer of membrane absorbers. This integral unit leads to an absorption of more than 90% for frequencies above 63 Hz, and the absorption is only reduced to 80% at low frequencies down to 16 Hz.

The acoustic treatment of tunnel airpath and plenum has reduced the wind-tunnel self-noise by 20 dB. At 150 kph an inflow overall sound-pressure level of 73.2 dB(A) is measured. Out of the flow it amounts to 71 dB(A). The top speed of the wind tunnel has only been reduced from 268 kph to 257 kph.

The reduction of the wind-tunnel's self-noise level permits the flow around a car to become the solely determing excitation mechanism of interior noise. It opens up all possibilities of subjective wind-noise ratings from passengers inside the car, and of using measurement techniques and psychoacoustic methods with regard to a higher riding comfort and a pleasant interior noise level. Even more impressive is, however, the opportunity to determine the aerodynamic exterior noise in view of its contribution to environmental noise pollution and to minimize it systematically.

The determination of the aerodynamic exterior noise under far-field conditions in the plenum in a defined sideline to the car's longitudinal axis supplies the link to drive-by/coast-by tests on road.

ACKNOWLEDGEMENTS

The Institute would like to thank Mercedes-Benz AG and Adam Opel AG as well as the State of Baden-Württemberg and the Federal Republic of Germany for generously supporting the conversion of its full-scale wind tunnel to an aeroacoustic wind tunnel and for extending its measuring equipment.

REFERENCES

[1] POTTHOFF, J.; Die IVK-Kraftfahrzeug-Windkanalanlage der Universität Stuttgart. Symposium No. T-30-905-056-7 "Aerodynamik des Kraftfahrzeugs", Haus der Technik, Essen, 1987.

[2] HELFER, M.; BUSCH, J.; Contribution of Aerodynamic Noise Sources to Interior and Exterior Vehicle Noise. DGLR-Workshop "Aeroacoustics of Cars", German-Dutch Wind Tunnel (DNW), Emmeloord/NL, 1992

[3] ACKERMANN, U.; MOHR, J.; Messung des Strömungsgeräusches von Kfz in einem schallharten Freistrahlwindkanal. Report BS 237/90, Fraunhofer-Institut für Bauphysik (IBP), Stuttgart, 1990.

[4] OGATA, N.; IDA, N.; FUJI, Y.; Nissan's Low Noise Full-Scale Wind Tunnel. SAE Paper 870250, Detroit, 1987.

[5] LINDENER, N.; STADLER, M.; Der BMW Akustik-Windkanal. Symposium No. T-30-312-056-2 "Geräuschminderung bei Kraftfahrzeugen", Haus der Technik, Essen, 1992.

[6] WIEDEMANN, J.; WICKERN, G.; EWALD, B.; MATTERN, C.; Audi Aero-Acoustic Wind Tunnel. SAE Paper 930300, Detroit, 1993.

[7] ACKERMANN, U.; Modelluntersuchungen zur Auslegung eines Akustik-Windkanals. Technisches Messen 57/1990, No. 12, pp 457-467.

[8] DEUTENBACH, K.-R.; Influence of Plenum Dimensions on Drag Measurements in ¾ Open-Jet Automotive Wind Tunnels. BMW Internal Report, 1993, and Paper planned for the 1995 SAE International Congress in Detroit.

[9] FUCHS, H.V., ACKERMANN, U.; RAMBAUSEK, N.; Membranabsorber für den technischen Schallschutz. IBP Information 14 (1987), No. 135.

[10] WIEDEMANN, J.; Written Comment to Symposium No. E-30-202-056-4 "Akustik und Aerodynamik des Kraftfahrzeugs", Haus der Technik, Essen, 1994.

[11] MERCKER, E.; PENGEL, K.; On the Induced Noise of Test Sections in Different Wind Tunnels and in the Cabin of a Passenger Car. SAE Paper 940415, Detroit, 1994.

[12] DNW Annual Report 1992; Publ. by German-Dutch Wind Tunnel, Emmeloord/NL.

[13] GEIB, W.; LINDENER, N.; Die Optimierung von Fahrzeugkomponenten im BMW Akustik-Windkanal. 4th Colloquium "Fahrzeug- und Motorentechnik", Aachen, 1993.

[14] HELFER, M.; Beurteilung von Hohlspiegelmikrofonen zur Schallquellenortung. Symposium No. E-30-202-056-4 "Akustik und Aerodynamik des Kraftfahrzeugs", Haus der Technik, Essen, 1994.

[15] BUSCH, J.; ESSERS, U.; Zur Problematik der Vergleichbarkeit von Außengeräuschmessungen auf der Straße und in Fahrzeugprüfständen. Symposium No. E-30-202-056-4 "Akustik und Aerodynamik des Kraftfahrzeugs", Haus der Technik, Essen, 1994.

930296

A Numerical Study of Wind Noise Around Front Pillar

Ming Zhu and Hideaki Miyata
University of Tokyo

Yuji Hanaoka , Kiyohira Aoki, and Ikuo Terada
Isuzu Motors, Ltd.

ABSTRACT

A numerical analysis method is developed for predicting the pressure fluctuations on the front side window surface, aiming at the elucidation of the external aerodynamic flow structure about the front pillar of a road vehicle. The simulated results are assessed by comparison with the acoustic theory and reveal fairly well the dependence of the predicted surface pressure fluctuations upon the vehicle cruising speed with the sixth power law. The features of three dimensional vortical flow are clarified from the analysis of the simulated results, indicating the strong relationship between the vortical formation and the external pressure fluctuations on the front side window surface. The external pressure fluctuations seem to be strongly related to the vortex breakdown during its interaction with the front side window and the roof-side window junction.

INTRODUCTION

It is well known that the pressure fluctuations on the front side window surface of a road vehicle are a major sound source for both the external and interior wind noises[1]. There are several methods to predict the spectrum of these pressure fluctuations empirically[2]. There are also new techniques to measure the external aerodynamic noise such as the method using near-field acoustic holography[3]. It is a fact that, except by means of measurement, there are few prediction methods today which provide information about the distribution of the unsteady pressure and the pressure fluctuations on the front side window surface. However, the experimental measurement is very costly and requires a facility free of ambient noise and instruments able to cope with high-frequency pressure fluctuation and pure aerodynamic noise. It is also difficult to obtain the detailed flow data simultaneously with the noise by means of measurement.

It is clearly noted that the unsteady pressure on the front side window surface is due to the existence of three-dimensional vortex separated from the front pillar[1]. Watanabe et al. showed the systematic structure of flow separation from the front pillar responsible for the wind noises in their experiment[4]. Also in an earlier experimental study of pressure fluctuation on a sharp edged rectangular shape by Stapleford et al.[5], it was found that the pressure spectra about the stagnation and reattachment lines of flow separation have a much higher hump. However, the mechanism of generating aerodynamic noise by vortical motion has not yet been satisfactorily explained.

It is known that the source of the aerodynamic sound due to the vortical motion is evaluated from the fluctuation of $div(\boldsymbol{u} \times \omega)$ as shown by Powell[6] and Howe[7]. Laufer explained that the vortex sound is generated by the vortex mixing[8]. However, the mechanism of generating the aerodynamic sound by the separated vortex can not be explained simply by the vortex mixing. Firstly, since there is always a vorticity component due to the shear flow of the boundary layer in the vicinity of the carbody, it may be possible that not all of the components of the vorticity over the panel take a major role of generating the unsteady pressure over the panel of the front side window. Secondly, for the aerodynamic noise about a road vehicle, the acoustic energy is generated through a kind of breakdown of flow geometry, including the separation, the reattachment, and the vortex breakdown by the flow-carbody interaction. Therefore it is necessary to elucidate the mechanism of flow breakdown for the understanding of aerodynamic noise about the front pillar.

Therefore it is important to develop an analysis method with the help of full Navier-Stokes simulation for the prediction of external pressure fluctuations on a road vehicle. With detailed information of the unsteady pressure as well as the corresponding flow data, the relationship among the carbody shape, the three-dimensional vortical flow structure and the strongest wind noise source can be elucidated so that the optimization procedure of body design for the production of low-wind-noise vehicle can be obtained.

In this paper, a numerical prediction method is proposed and developed for predicting the aerodynamic noise due to the pressure fluctuation on the front side window surface, aiming at the elucidation of relationship between the vehicle interior wind noise and the external aerodynamic flow structure about the front pillar of a road vehicle. A series of three-dimensional Navier-Stokes simulations are carried out for the vortical flow about a basic car model with different front pillar inclination and at different cruising speed. The wind noise on the side window is predicted quantitatively by using a simplified formulation of Lighthill-Curle's equation. The simulated results are assessed by comparison with the acoustic theory. Some sophisticated analyses for the simulated results are carried out to elucidate the vortical structure related to the aerodynamic noise. The distribution of the averaged pressure and the pressure fluctuation on the front side window surface are shown and the relationships between the flow-carbody interaction and the aerodynamic noise are clarified.

NUMERICAL PREDICTION METHOD

As shown in Fig.1, the whole analysis approach is designed to be composed of the numerical simulation, an analysis of the results, and both the experimental and the theoretical assessments.

NUMERICAL METHOD - In this paper, a well documented commercial code called SCRYU™ is used[9], in which the incompressible Navier-Stokes equation is discretized in the generalized boundary-fitted coordinate system and the third-order upwind TVD (MUSCL-type) scheme is employed for the convection term. Since there is no turbulence model in the computation, the partial flow cascades are dependent upon the fourth order derivative term of the upwind scheme. In the case of wind noise problem, the RANS simulation with turbulence model such as k-ε two-equation model is not suitable. Since the grid resolution of simulation is not fine enough to be able to carry out so-called large eddy simulation (LES), there are few reasonable turbulence models to cope with such unsteady three-dimensional complicated flow. The authors have verified some basic flow features by using SCRYU™, and found that reasonable result can be obtained by tuning[10] with respect to both the spatial and the temporal resolutions of simulation. The flow structure of interest in the present study is accompanied with strongly three-dimensional flow separation and reattachment. Therefore it is reasonable to choose a viscous flow past a circle as the basic flow in order to tune the numerical method. Fig.2a shows the comparison of the time-averaged pressure distribution around a two-dimensional circle for a transient Reynolds number when the grid resolution is changed, and their resultant drag force acting on the circle versus the Reynolds number of the flow past the circle. Fig.2b shows the time sequence of the streamlines past the circle for a transient Reynolds number. Both the position and its movement of separation position on the circle are quite well simulated when compared with the conventional experiments. In this paper, the know-how of the tuning is fully employed to cope with the unsteady vortical flow about the front pillar.

Although it is when the cruising speed becomes high that the wind noise is dominant, the Mach number of the road vehicle is usually lower than 0.1. Therefore, it is safe to employ a numerical method for incompressible flow to predict wind noise of a road vehicle. Actually, the assumption that the pressure fluctuation is dominant for the wind noise is also based on the low-Mach-number fact.

SIMPLIFIED LIGHTHILL - CURLE 'S EQUATION - In this paper, since the wind noise due to the pressure fluctuation is considered and the incompressible flow field at low Mach number is solved about the carbody by the numerical method, the dipole source can be considered as major one for the the sound pressure. In order to evaluate the total sound pressure effect on the front side window surface, a simplified Lighthill-Curle's equation is employed as follows[11].

$$p_f = \frac{1}{4\pi c L_0}\frac{\partial}{\partial t}\int_{\partial\Omega} p(\xi,t)\,ds(\xi) \qquad \text{(Pa)} \qquad (1)$$

where p_f is the sound pressure over the whole front side window surface $\partial\Omega$, p is the simulated unsteady pressure, c is the speed of sound and L_0 is the characteristic length and is simply set at 1.0. Eq.(1) represents the pressure fluctuation over the panel rather than the aerodynamic sound itself, aiming at extracting the pure pressure fluctuation induced by the vortical motion about the front pillar. The sound pressure level (SPL) is defined as follows.

$$SPL = 10\log_{10}\left(\frac{p_f}{p_o}\right)^2 \qquad \text{(dB)} \qquad (2)$$

where p_0 is the base sound pressure set at 2×10^{-5} (Pa).

ANALYSIS METHOD OF SIMULATED RESULT - Once the flow field is solved about the carbody, the time sequence of unsteady pressure at the respective point on the front side window surface is recorded. The sound pressure is calculated from Eq.(1) and its SPL is from Eq.(2). By Fourier transformation, SPL is transformed into frequency domain and calculated for frequencies corresponding to the preferred 1/3 octave band center frequencies. In order to extract the characteristic scale of flow structure, the frequency is nondimensionalized in a Strouhal scaling manner as follows.

$$\hat{f} = \frac{f L_0}{V_m} \qquad (3)$$

where V_m is the vehicle cruising speed.

In order to extract the component responsible for the aerodynamic noise from the vorticity, in the present paper, the vorticity is decomposed in the following two parts.

the helicity or Lamb scalar $\quad L_s=\omega\cdot\boldsymbol{u} \quad (4)$

the inertial vortex force or Lamb vector: $\quad L_v=\omega\times\boldsymbol{u} \quad (5)$

By nondimensionalization, the helicity L_s represents the vorticity component oriented with the velocity vector while the inertial vortex force L_v represents the vorticity component perpendicular to the velocity vector. The vortical sound about the front pillar may be explained as the generation and dissipation of the helicity when the flow geometry is broken down by the flow-carbody interaction.

CONDITIONS OF SIMULATION

In order to validate the method proposed in this paper, a series of three-dimensional Navier-Stokes simulations are carried out for the vortical flow around a basic car model at different cruising speed and with different front pillar inclination. Fig.3 is the basic car model and its grid. The full length of the model is 4 m and the height is 1.3 m. The degree of streamwise inclinations of the front pillar θ are set at 30, 45 and 60 degrees. The cruising speeds V_m are set at 75 km/h, 100 km/h and 150 km/h, which corresponds to the Reynolds number at $5.34x10^6$, $7.12x10^6$ and $1.07x10^7$ with the vehicle length used as the length scale. The non-slip boundary condition of velocity is imposed on the carbody. A uniform streamwise velocity of the cruising speed is set at the inflow boundary, and the pressure is set at zero at the outer boundary. Symmetric condition is set at the centerline boundary and open condition is set at the lateral boundary. The maximum pressure residual is smaller than 10^{-5} at every time step. The time increment is set at $1x10^{-4}$ second so that the maximum Courant number in the calculation is smaller than 0.4. The sampling time is about 0.5 second or 0.6 second for every case. This sampling time corresponds to more than 4 dimensionless time and seems to be enough for a separation flow about the front pillar.

Because the flow about the front pillar is the main target to be elucidated in the present study, only the fore part of the vehicles is considered. It employs only about 150,000 grid points in every case. It is known that the accuracy validation can not be easily achieved in the three-dimensional flow of interest. A half-spacing refinement of grid resolution leads to a grid with eight times points compared with the original one. On the other hand, it is clearly understood that there are few experimental data for three-dimensional, complex and unsteady flow which could be employed for the validation of numerical simulation. Therefore, in the present study, the grid resolution is considered to be achieved in the simulation of basic flow and the simulated results will be validated by comparison with the acoustic theory. As a well tuned CFD application, this amount of grid points is quite efficient in practice and shows the minimum required for the resolution of the physical phenomenon of interest.

VERIFICATION AND ANALYSIS OF NUMERICAL RESULTS

FOR VEHICLE CRUISING SPEED - Fig.4 shows the time sequences of the predicted SPL for three cruising speeds when the front pillar inclination θ is 30 degree. It denotes clearly that the sound intensities over the panel become large when the cruising speed is increased. Fig.5 is the comparison of the 1/3 octave SPL spectra between three cruising speeds. It is obvious that the spectra of three cases have a similarity when the frequency is nondimensionalized in a Strouhal scaling manner, since the flow structures are not so different between three cases in view point of Reynolds number. The humps of the spectrum correspond to the characteristic scales of the flow structure about the front pillar. From the spectra, it is also noted that the grid resolution is sufficient. Therefore it is possible to extract the characteristic scale of the flow structure about the front pillar from the predicted SPL spectrum. By integrating the sound pressure levels from the spectrum distributions in Fig.5, one can find that the simulated results reveal fairly well the dependency of the over-all SPL upon the vehicle cruising speed with sixth power-law as shown in Fig.6.

As a result, it seems that the present method is efficient enough to study the aerodynamic noise due to the pressure fluctuation. Fig.7 is the simulated time-averaged pressure contour maps on the panel of three cruising speeds (θ=30 degree). The pressure distributions almost have the same pattern between the three cases of different cruising speeds. Fig.8 is their over-all pressure fluctuation contour maps. Although the over-all SPL spectra over the panel have similarities as shown in Fig.5, the pressure fluctuation distributions on the panel are really different from each other. That is, SPL intensities about the roof-side window junction increase much more significantly than those behind the front pillar. The difference of strongest intensity at these two places is 5 dB at 75 km/h, 15 dB at 100 km/h and almost 30 dB at 150 km/h, while the maximum pressure fluctuation behind the front pillar stays at about 60 to 65 dB. Therefore these two kinds of sources show different dependence on the cruising speed and hence may be generated by a different kind of flow structure. Fig.9 is the pressure fluctuations at dimensionless frequency $\hat{f}$ = 0.904 and at $\hat{f}$ =4.352, which are the frequencies related to the first two humps of 1/3 octave SPL spectra as shown in Fig.5. It can be found that the different flow structure at the different frequency of spectrum hump, that is, the first hump is related to the aerodynamic noises due to the flow separation and reattachment behind the front pillar, and the second hump is related to those due to the interaction between the separated vortex and the roof-side window junction. Fig.10 shows the iso-helicity contour surface about the front pillar, the red portion shows the positive value and the blue shows the negative one. Firstly, it can be said that three-dimensional separated vortex (in red color) behind the front pillar is quite well described by the iso-helicity surface contour. That is, the generation of longitudinal vortex is strongly related to the three-dimensional separation about the front pillar, a kind of bifurcation of flow geometry. Secondly, it is noted that, by comparison with Fig.9, the frequency of the first hump clearly corresponds to the motions of vortex separation and reattachment. On the other hand, when the separated vortex (in red color) rises to the roof-side window junction, it looses its kinematic energy during interacting with the corner and induces the opposite vortex on the roof (in blue color). This is why the sound pressure source at the roof-side window junction is of a much higher frequency than those behind the front pillar at a high cruising speed. Therefore, it can be said that the sound pressure is generated not only by the generation of longitudinal vortex itself but also by the evidence of the vortex breakdown at the roof-side window junction.

FOR INCLINATION OF FRONT PILLAR - Fig.11 is the time sequences of predicted SPL of three inclination angles when the cruising speed V_m is 100 km/h. Both the wave heights and the unsteadiness of SPL are different between the three cases. Fig.12 is the comparison of the 1/3 octave SPL spectra between the three inclination angles. It is noted firstly that, from the spectra, in the case of 45 degrees there is only a dominant hump corresponding to the second hump of 30 degrees, which is related to the interaction between the separated vortex and the roof-side window junction in the case of 30 degrees. But there is not a hump corresponding to the separation behind the front pillar itself. Secondly, the spectrum of 60 degrees shows a clearly different flow feature from those of 30 and 45 degrees, indicating that there is a more complicated phenomenon about the front pillar in the case of

60 degrees. Fig.13 is the over-all pressure fluctuation over the panel. The above reasoning by the spectra are supported that both the strengths and the positions of SPL over the panel are different from each other cases.

In order to recognize the evidence of flow separation in the three cases, the time-averaged pressure distribution on the panel of three inclination angles are shown in Fig.14. There are negative intensities of the pressure coefficients behind the front pillar which would indicate the existence of the separated vortex from the front pillar. Fig.15 is the comparison of streamlines over the panel between the computation and the visualization measurement by means of oil film. Experiments are carried out by using the 1/5 scale models of the computation for three inclination angles. The streamwise flow speed is 30 m/s (108 km/h) which corresponds to a Reynolds number of 1.54×10^6. Although the Reynolds number is different between the computation and the experiment, the directions of simulated velocity vector show fairly good agreement with the experiment. There are stagnation parts of flow in the vicinity of roof-side pillar corner especially in the case of 60 degrees, indicating a complicated flow structure accompanied with three-dimensional flow separation. Fig.16 is the simulated flow streaks about the front pillar about the front pillar in three cases and shows a clear occurrence of flow separation.

Fig.17 is the pressure fluctuation of the frequencies corresponding to the first three humps of the 1/3 SPL spectrum in the case of 60 degree. It shows that these separated vortex-carbody interactions upon the front side window are not only at a single frequency but over the wide range of frequencies and suggests the generation of vortices in various strength, a clear evidence of vortex breakdown. Fig.18 is the time sequence of iso-helicity surface contour about the front pillar in the case of 60 degrees. The generation and dissipation of the separated vortex by the interaction with the carbody are clearly shown and are strongly related to the generation of pressure fluctuation, and the aerodynamic noise around the front pillar.

CONCLUDING REMARKS

A numerical prediction method is developed for the evaluation of the unsteady pressure over the front side window surface of a high-cruising-speed road vehicle. The simulated results demonstrate the excellent ability of the present method for the prediction of aerodynamic noise due to the pressure fluctuation over the panel and for the elucidation of the complicated three-dimensional vortical flow structure behind the front pillar. Through the analysis of the simulated results, it is shown that the sound sources of the side window surface are not only due to the separation and reattachment of the vortex behind the front pillar, but also due to the continuous interaction between the vortex and the roof-side window junction or the panel surface. Therefore one may conclude that the aerodynamic noise is strongly related to the vortex breakdown during its interaction with the front side window and the roof-side window junction.

ACKNOWLEGEMENTS

The authors would like to thank Mr. T. Iguchi and Mr. K. Okamoto of Engineering System Dept. of Isuzu Motors Ltd. for their supports when the numerical simulations have been carried out by the author Y. Hanaoka on CRAY-YMP of Isuzu Motors Ltd.

REFERENCE

[1] George, A. R., "*Automobile Aerodynamic Noise*", SAE Paper 900315, 1990.

[2] Chou, S.-T. and George, A. R., "*An Analytical Parametric Study of the Broadband Noise from Axial-Flow Fans*", Proc. Noise Con. 87, Noise Control Foundation, New York, 1987, pp.83-88.

[3] Maynard, J. D., et al., "*Nearfield Acoustic Holography: Theory of Generalized Holography and the Development of NAH*", J. Acoust. Soc. Am. 78(4), 1985, pp.2061-2068.

[4] Watanabe, M., Harita, M. and Hayashi, E., "*The Effect of Body Shapes on Wind Noise*", SAE Paper 780268, 1978.

[5] Stapleford, W. R., and Carr, G. W., "*Aerodynamic Noise in Road Vehicle, Part 1: The Relationship Between Aerodynamic Noise and the Nature of Airflow*", The Motor Industry Research Association, Report No.1971/2, November, 1970.

[6] Powell, A., "*Theory of Vortex Sound*", J. Acoust. Soc. Am. 33, 1964, 177-195.

[7] Howe, M. S., "*Contributions to the theory of aerodynamic sound, with application to excess jet noise and the theory of the flute*", J. Fluid Mech. 71, part 4, 1976, 625-673.

[8] Laufer, J., "*On the Mechanism of Noise Generation by Turbulence*", Univ. South Calif., Aerospace Engrg. 125, 1974.

[9] *SCRYU-A Package of 3D Thermal Fluid Analysis Program in Boundary-Fitted Coordinate System,* Manual of SCRYU, CRADLE, Ltd., 1992.

[10] Hanaoka, Y. and Aoki, K., "*A Verification of Simulating Some Basic Flow Features By Using SCRYU*", 1992 CRADLE User's Meeting, Yokohama, 1992.

[11] Zhu, M., Hanaoka, Y., Aoki, K. and Miyata, H., "*Numerical Prediction Method of Aerodynamic Noise Around Front Pillar*", JSAE Paper 921009, 1992. (in Japanese)

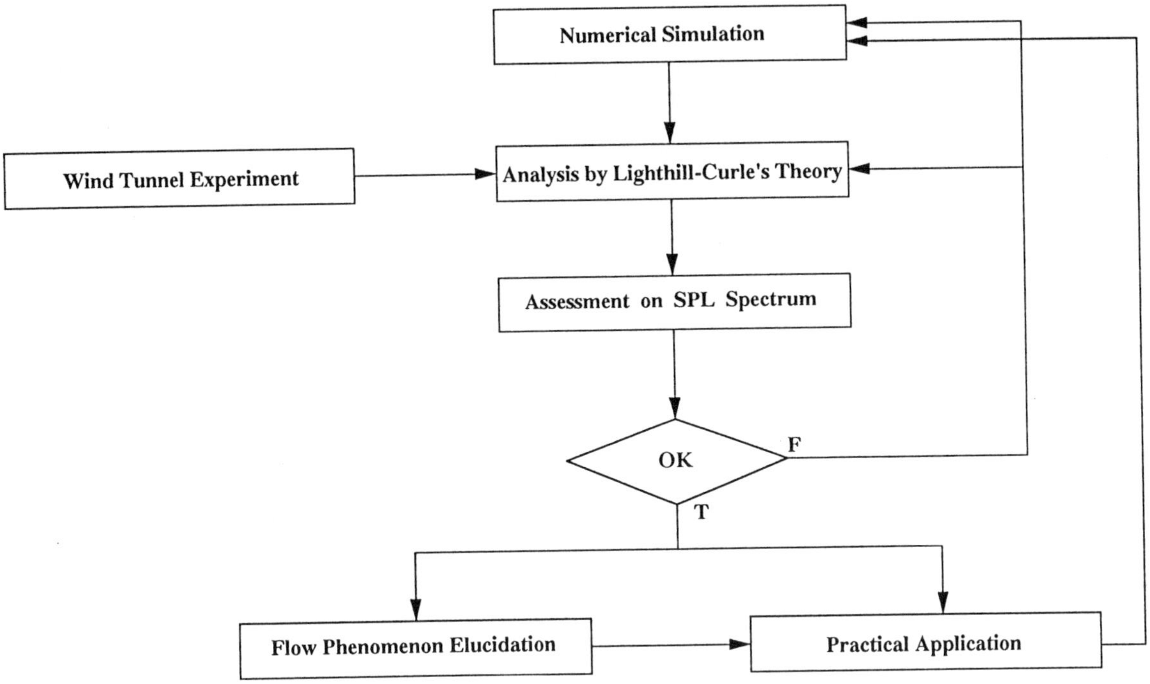

Fig.1 Analysis chart for numerical prediction of aerodynamic noise

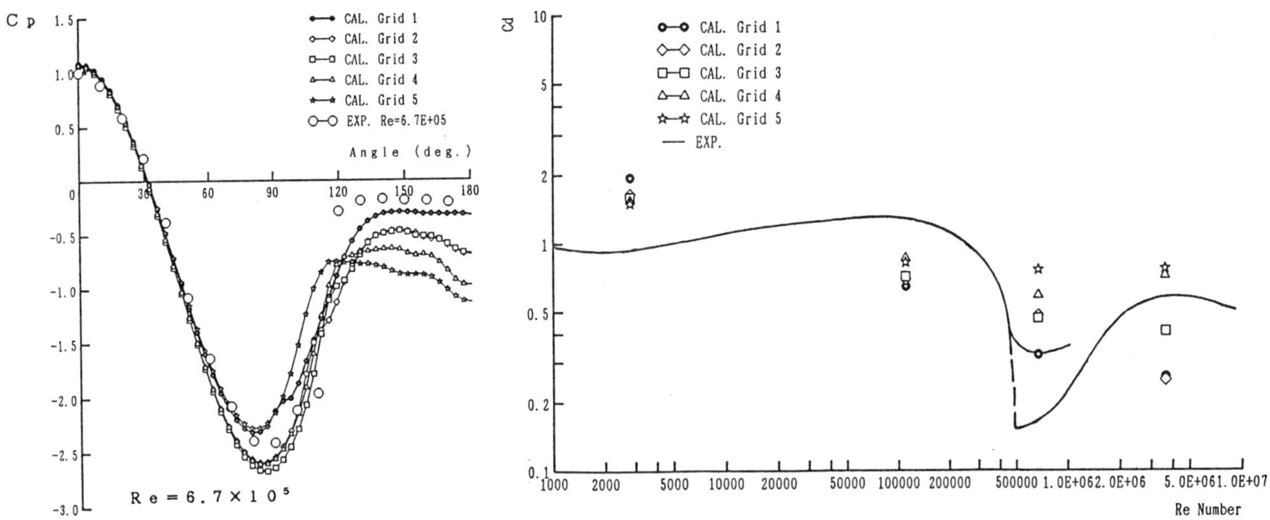

Fig.2a Verification of grid resolution in simulation by SCRYU: pressure distribution around a circle and resultant pressure force

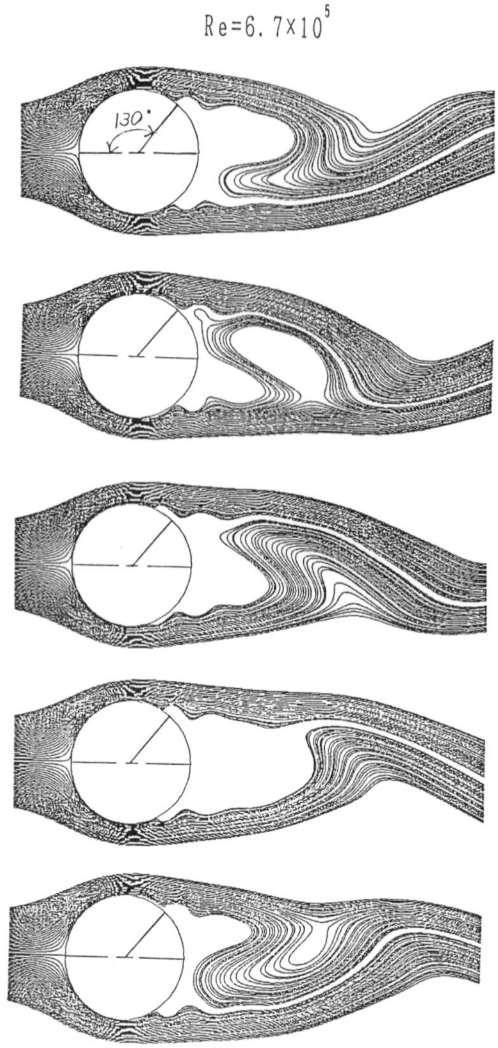

Fig.2b Continued : Time sequence of streamlines past the circle.

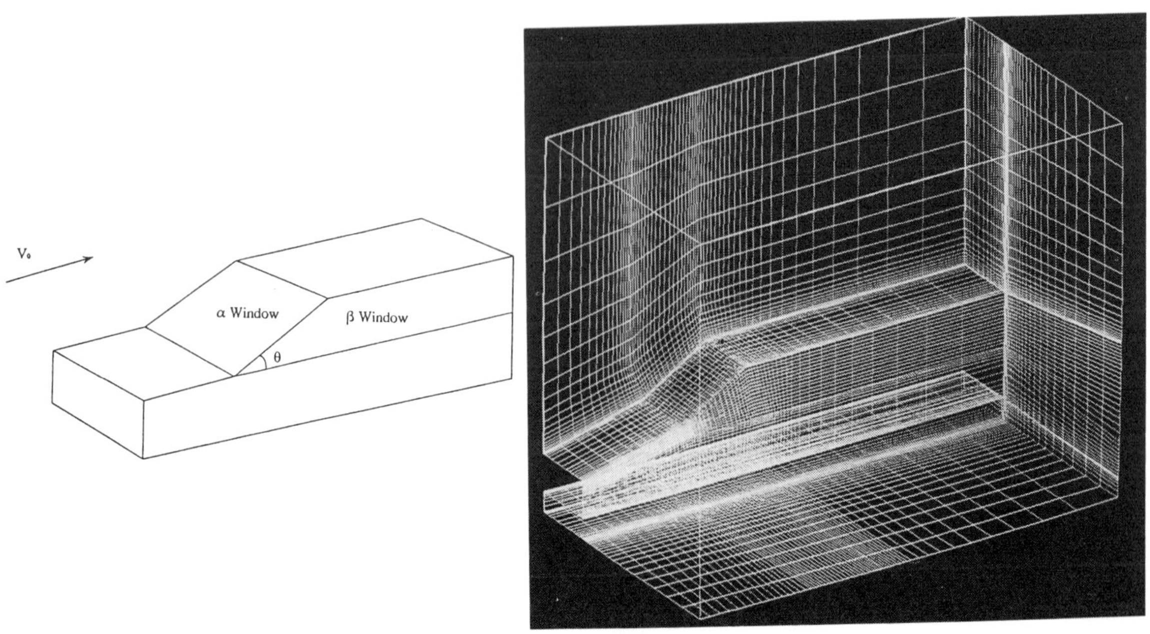

Fig.3 Base car model and its grid

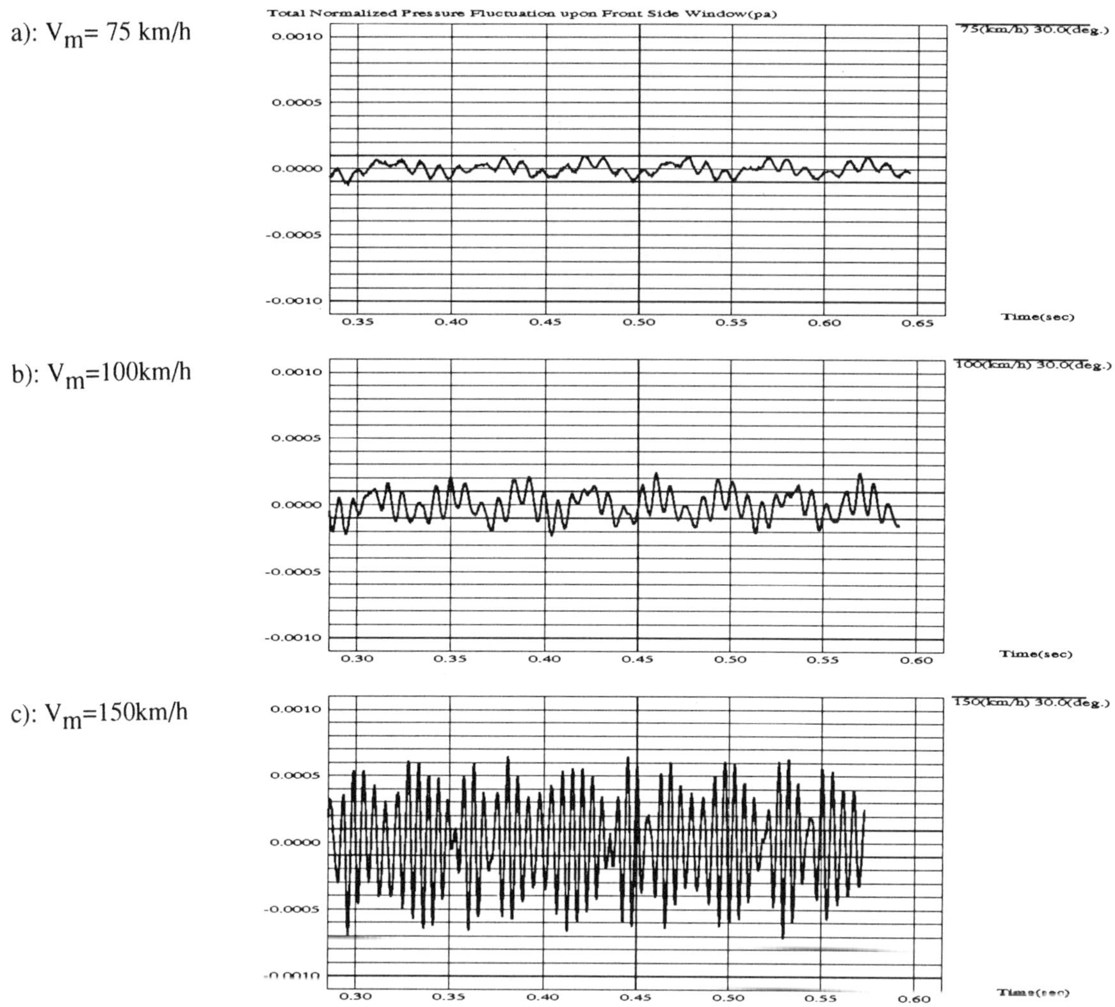

Fig.4 Time sequence of predicted sound pressure level (θ=30 degree)

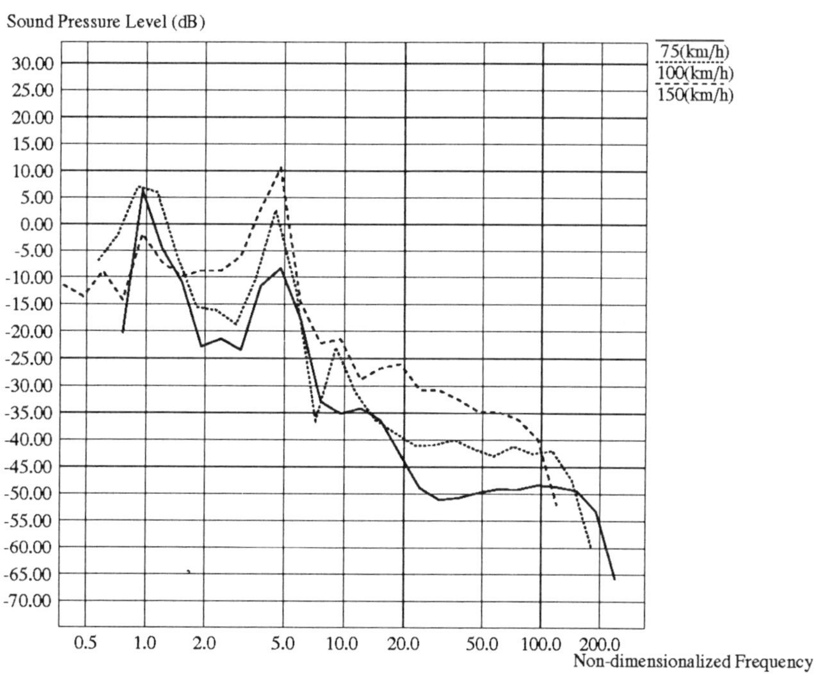

Fig.5 1/3 octave spectra of predicted sound pressure levels (θ=30 degree)

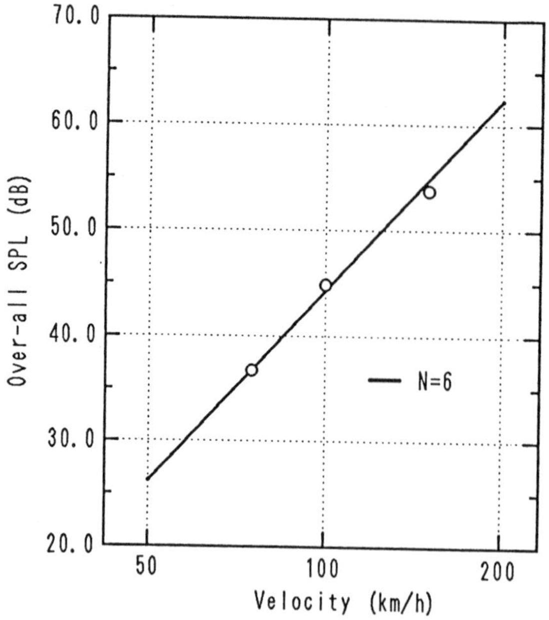

Fig.6 Over-all 1/3 octave SPL vs. the vehicle cruising speed

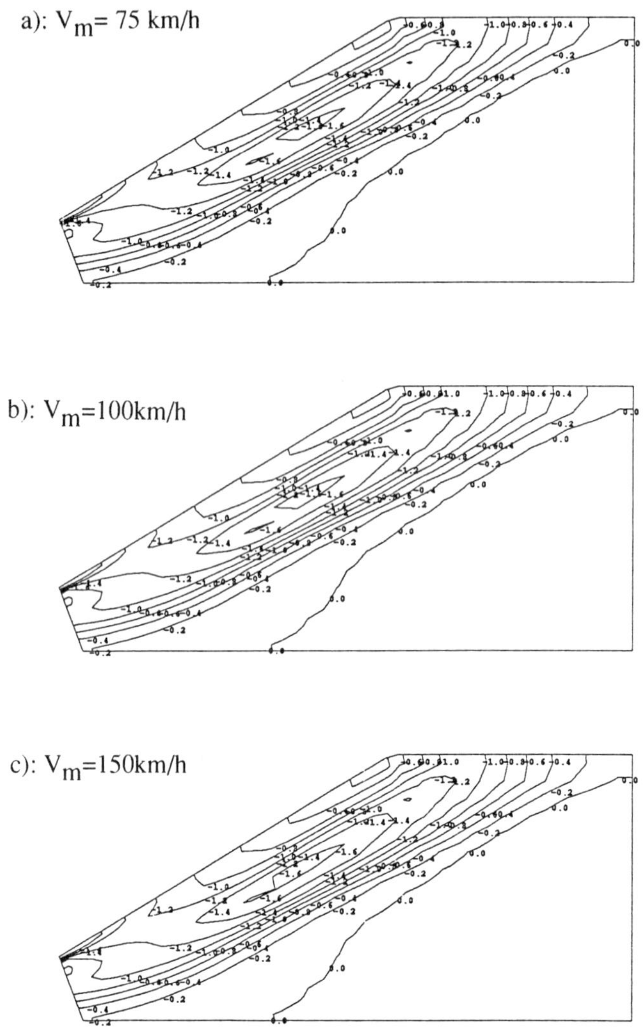

Fig.7 Time-averaged pressure distribution upon the front side window (θ=30 degree)

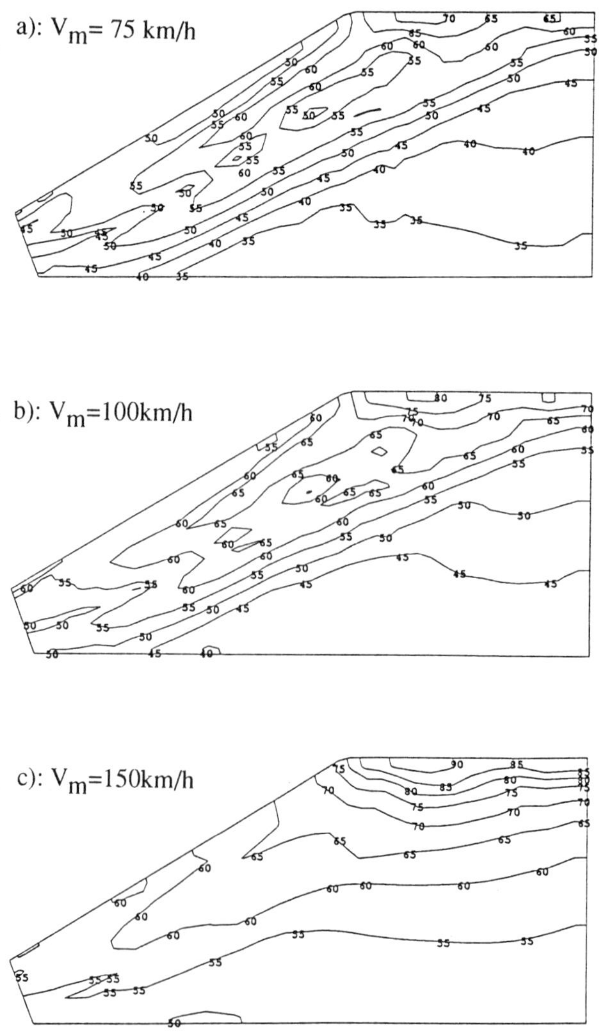

Fig.8 Over-all pressure fluctuation distribution upon the front side window (θ=30 degree)

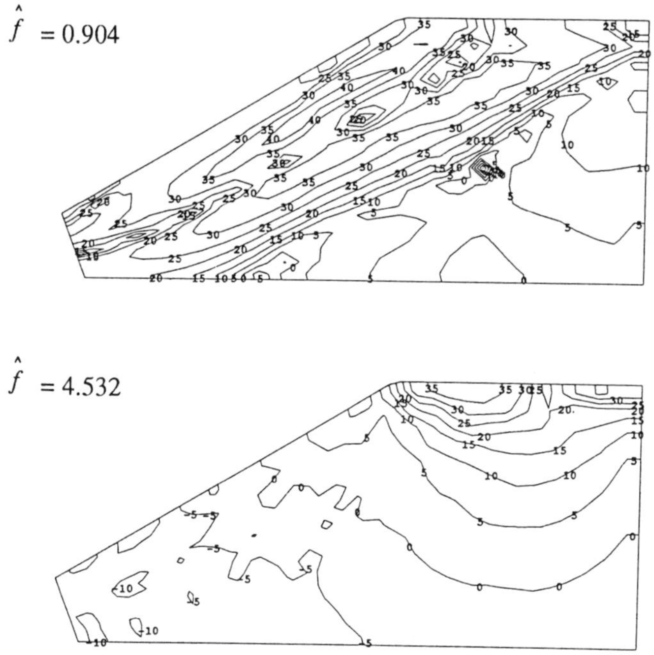

Fig.9 Pressure fluctuation distributions on the panel at $\hat{f}$ = 0.904 and $\hat{f}$ = 4.532

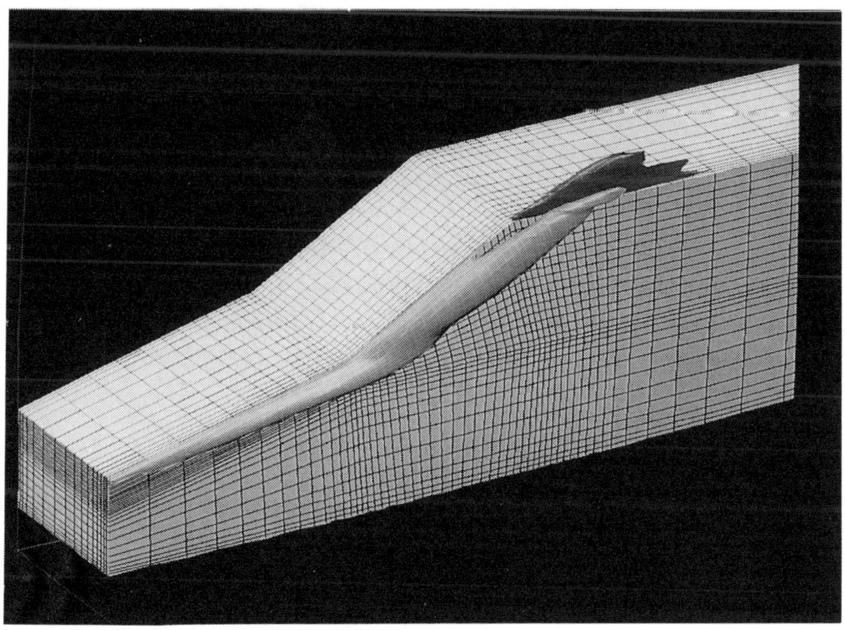

Fig.10 Iso-helicity surface contour about the front pillar (θ=30 degree)

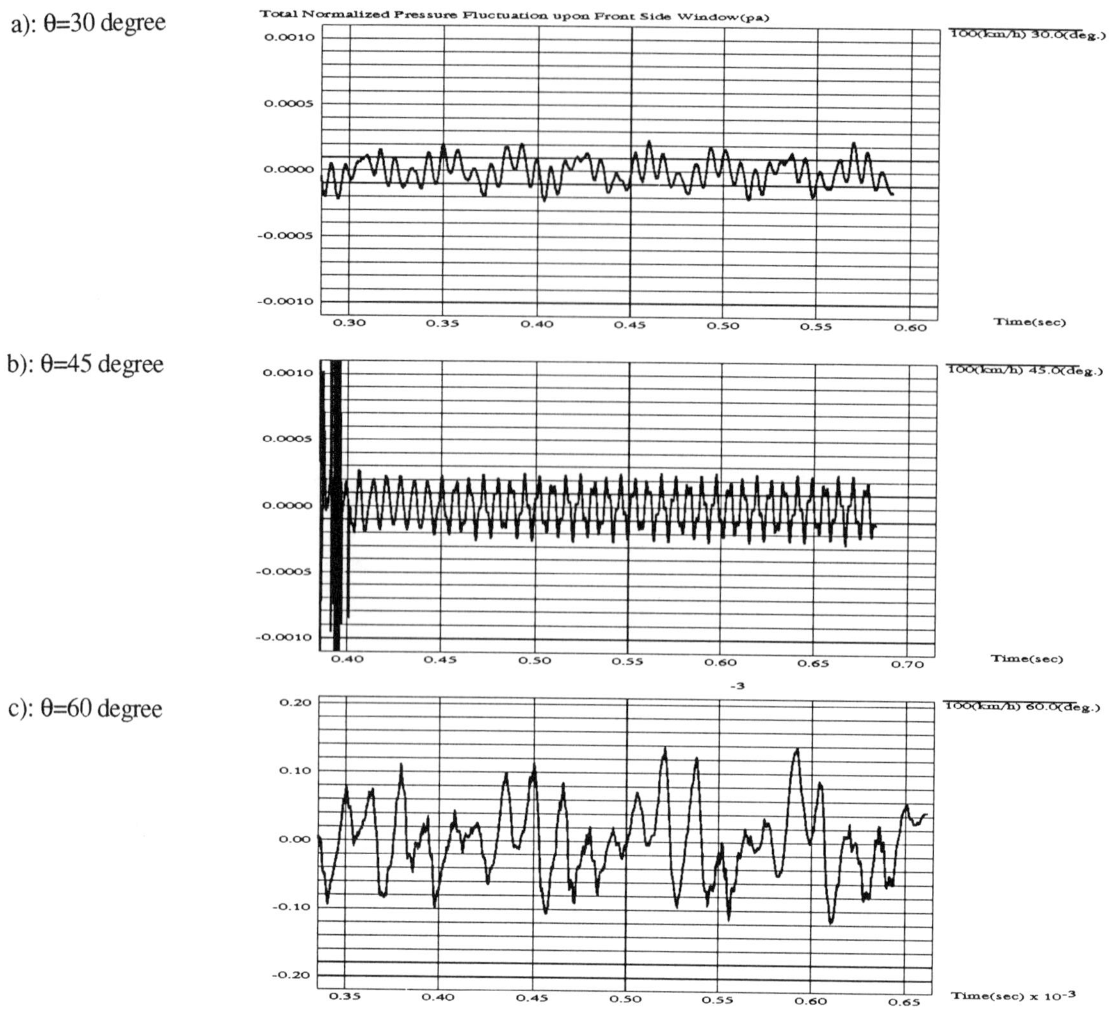

Fig.11 Time sequence of predicted sound pressure level (V_m=100km/h)

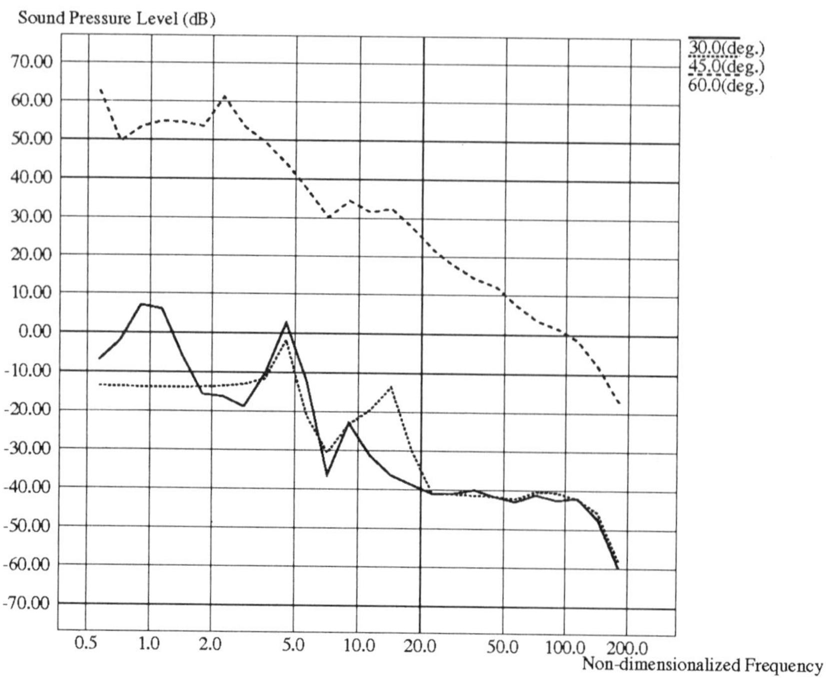

Fig.12 1/3 octave spectra of predicted sound pressure levels (V_m=100km/h)

a): θ=30 degree

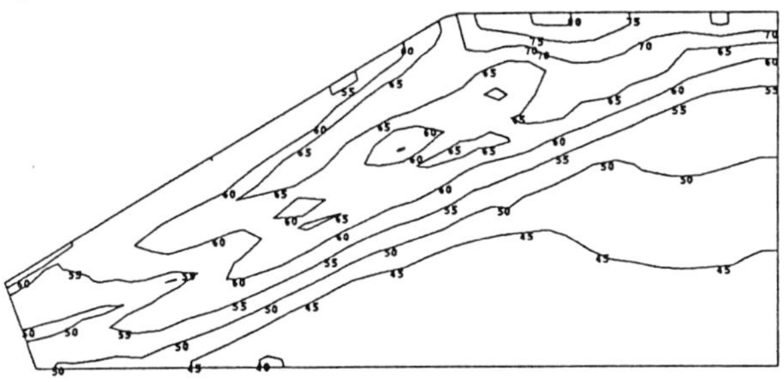

b): θ=45 degree

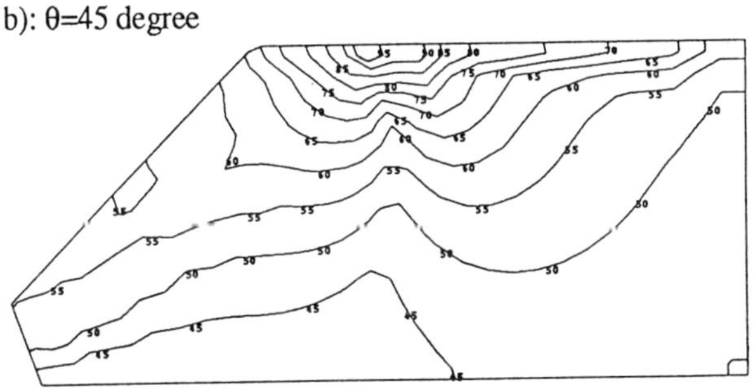

c): θ=60 degree

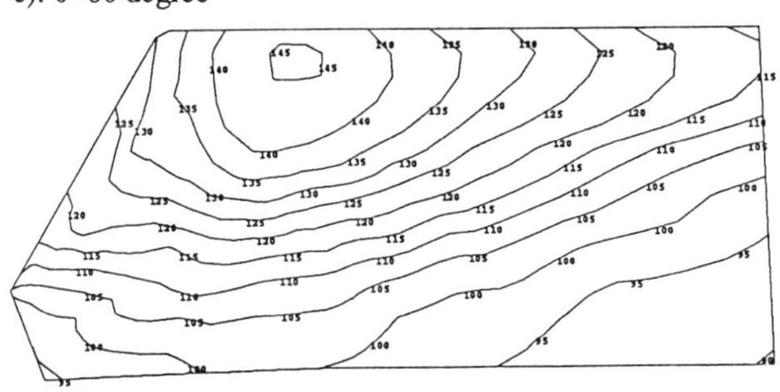

Fig.13 Comparison of computed over-all pressure fluctuation distribution upon the front side window(V_m=100km/h)

a): θ=30 degree

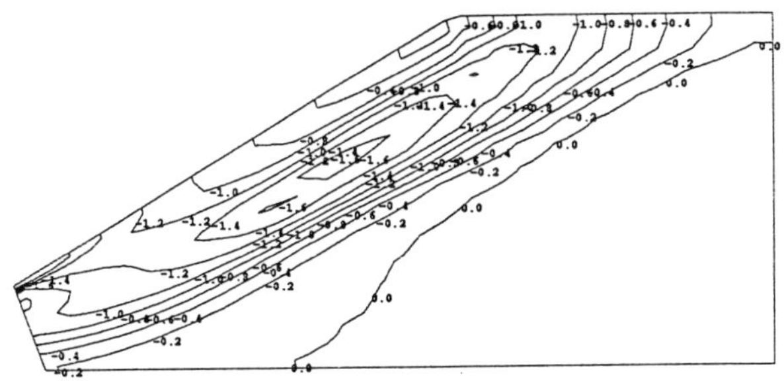

b): θ=45 degree

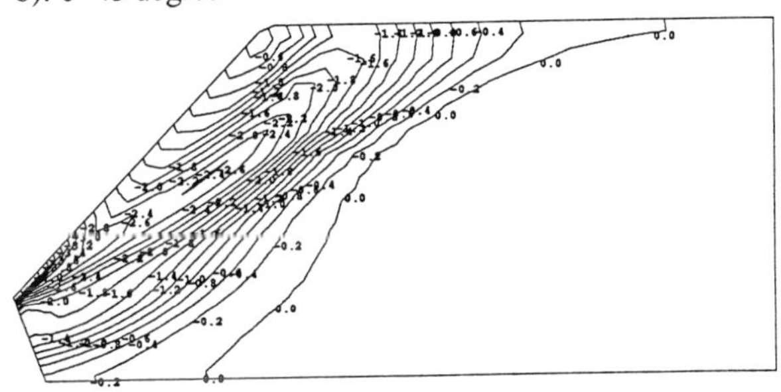

c): θ=60 degree

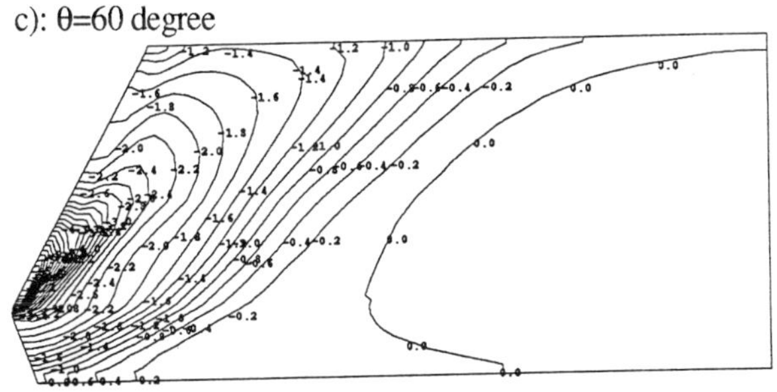

Fig.14 Time-averaged pressure distribution upon the front side window (V_m=100km/h)

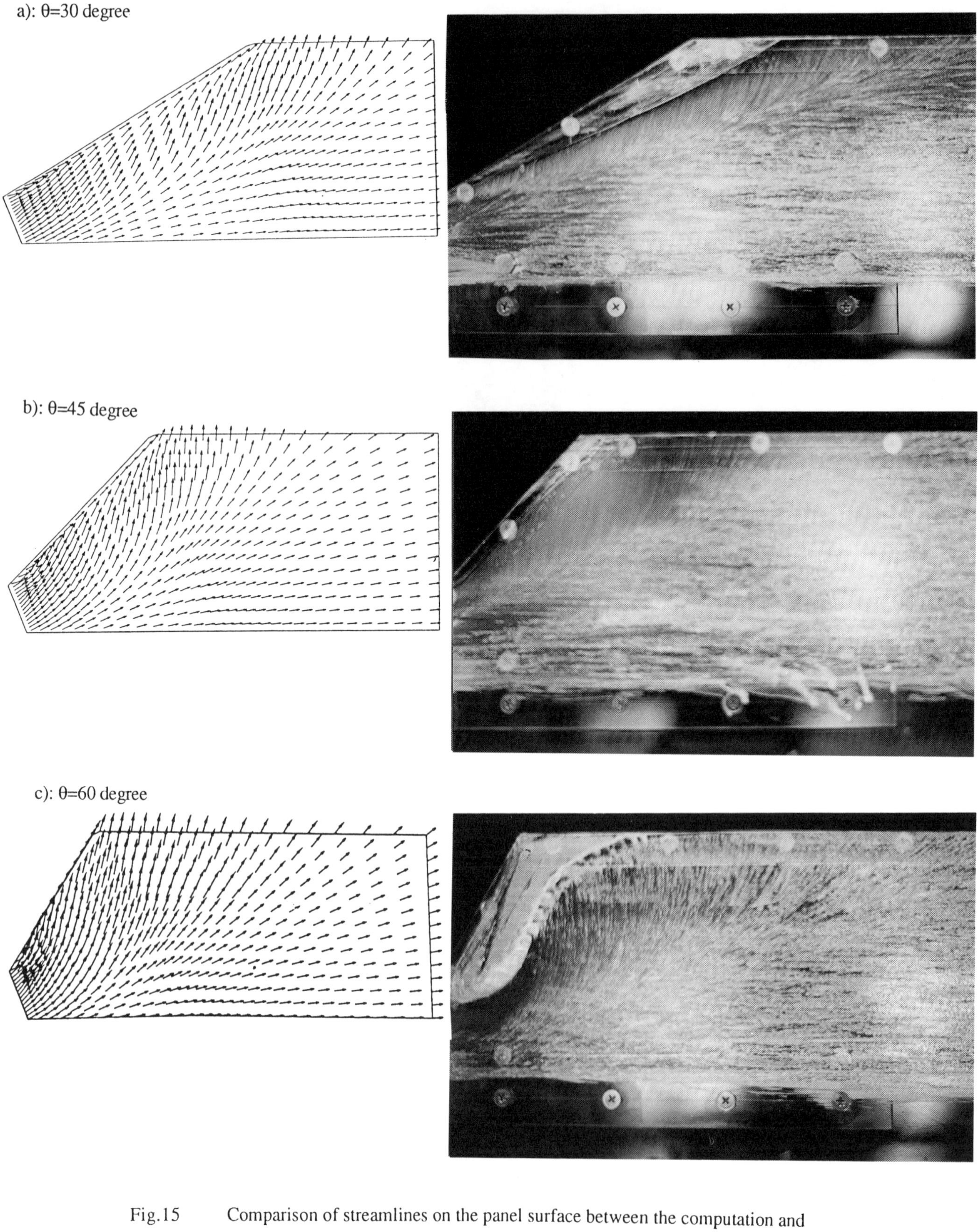

Fig.15 Comparison of streamlines on the panel surface between the computation and the visualization experiment

a): θ=30 degree

b): θ=45 degree

c): θ=60 degree

Fig.16 Flow streak about front pillar

a): $\hat{f} = 2.271$

b): $\hat{f} = 9.043$

c): $\hat{f} = 14.33$

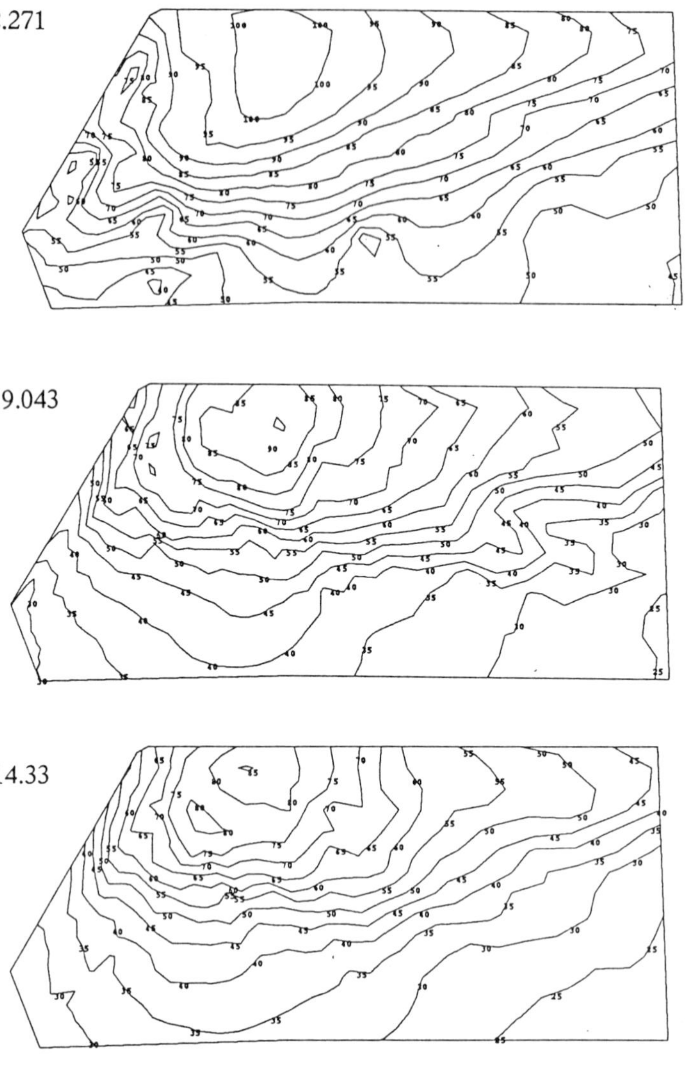

Fig.17 Comparison of computed pressure fluctuation distribution upon the front side window at three frequencies (θ=60 degree)

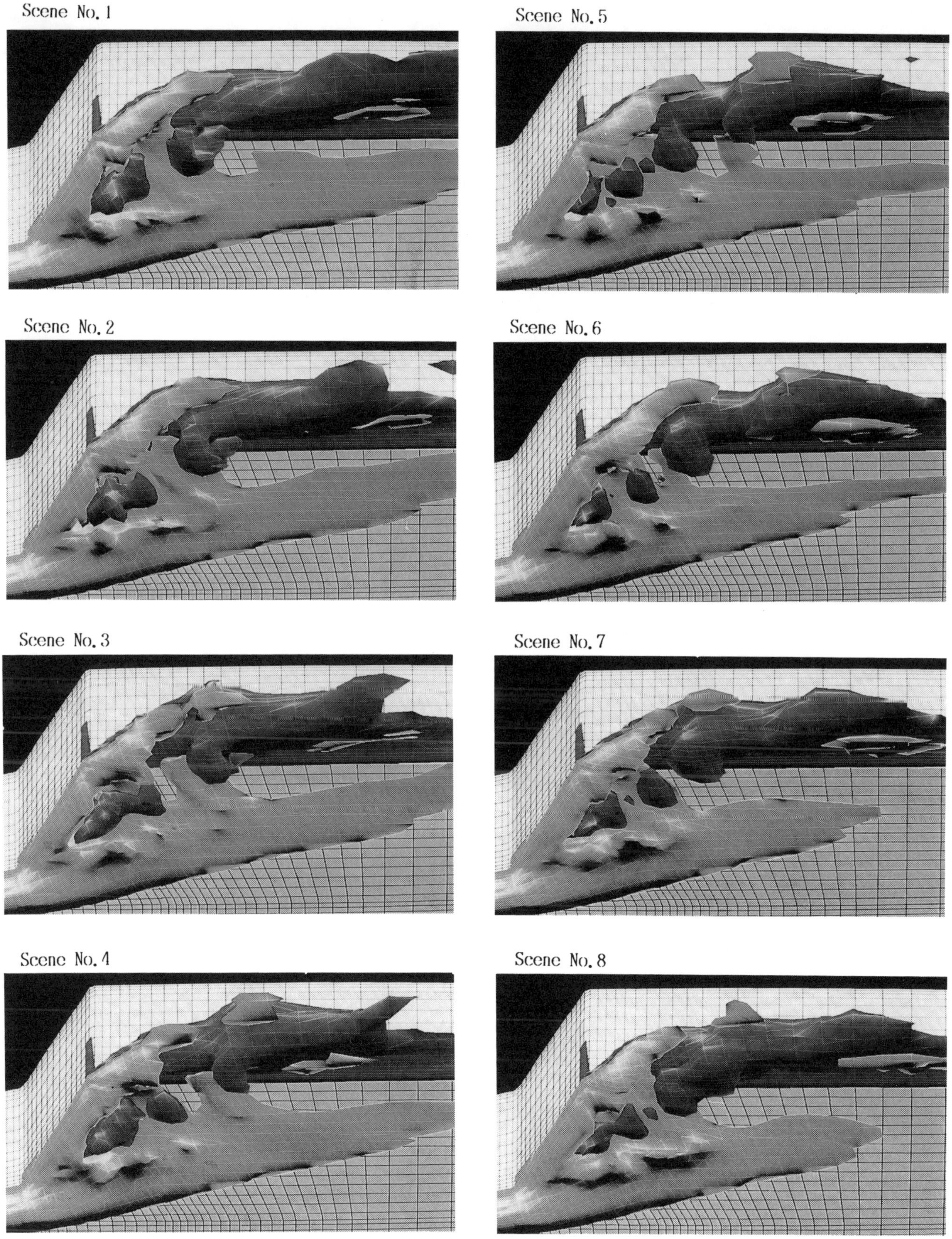

Fig.18 Time sequence of iso-helicity surface about the front pillar (θ=60 degree)

900316

An Experimental Analysis and Estimation of Aerodynamic Noise Using a Production Vehicle

Shigeru Haruna, Takahide Nouzawa, and Ichiro Kamimoto
Aerodynamics Testing & Research Group

Hiroshi Sato
Institute of Flow Research

ABSTRACT

The wind-noise-level in the cabin of a vehicle depends on the magnitude of the aerodynamically generated noise and noise isolation characteristics. Therefore, one good way to reduce the wind-noise-level in the cabin is to minimise the acoustic noise itself generated by the turbulent air flow around the vehicle at high speed cruising.

This paper describes the relation between the turbulent flow and the aerodynamic noise as well as how to estimate the magnitude of aerodynamic noise, especially around A-pillar of a production vehicle.

First, the flow visualization and the detailed measurements of flow clarify the vortex structure generated around A-pillar and side window.

Secondly, sound pressure fluctuations measured on the side window surface are discussed in relation to the vortex structure.

Lastly, in order to estimate the order of the magnitude of aerodynamic noise, we propose physical parameters given by approximating the solution of Lighthill's equation.

1. INTRODUCTION

Vehicles has come to cruise faster than before due to high performance of engine and suspension. This leads to the increase of the importance of suppressing vibration and noise of vehicles. Passenger compartment interior noise is composed mainly of the engine, tires and aerodynamic noise. Of such vehicle interior noise, aerodynamic noise becomes dominant especially at high speed cruising.

As far as aerodynamic noise is concerned, the noise level in the cabin depends on the magnitude of aerodynamically induced noise and the sealing performance for the noise. The former depends on the shape of vehicle. The latter is determined by the attenuation of transmitted sound through weatherstrips, body panel and windshield glass and so on.

Studies on the relationships between aerodynamic noise and configurations of vehicles have been performed [1, 2, 3, 4]. Stapleford [2] pointed out that the prominent noise levels occur in the front side-window area. Buchheim et al. [3] established correlations between pressure fluctuations at the outside body surface of a car and the interior noise.

However, there are few papers which relate the structure of vortex generated around vehicles to aerodynamic noise and express the estimation of aerodynamic noise generated around a vehicle.

The present paper investigates aerodynamic noise generated by the shape of a production vehicle and clarifies the mechanism of the aerodynamic noise generated by the separation of turbulent flow around A-pillar. In addition, based on the solution of Lighthill's equation, we propose physical parameters to estimate the order of the magnitude of aerodynamically induced noise around a vehicle.

2. EXPERIMENTAL EQUIPMENT AND PROCEDURES

2.1 Wind Tunnel and Test Vehicle

The experiment is performed in the test section of the MAZDA full-scale wind tunnel. The test section is 6m wide, 4m high and 12m long. This sectional area (24m^2) is large enough to neglect a blockage ratio in performing aerodynamic tests of a production car. The maximum speed is 230km/h. The turbulence intensity is approximately less than 0.3%

and the flow uniformity is about 1% at the test section.

In the present study a production car, MAZDA new Rx-7, is employed. In order to clarify the relationships between the fundamental configuration of the vehicle and aerodynamic noise, wipers and side mirrors are removed in this test.

2.2 Test Procedure

Flow visualization is performed by means of oil flow and smoke-wire techniques at the speed of 150km/h and 7km/h, respectively. Pressure fluctuations on the side window are measured by 1/4 inch condenser microphones. The diaphrams are mounted flush with the side window surface in order to avoid flow disturbances. Flow measurements are performed by I-type and X-type constant-temperature hot-wire anemometers. Then all the measured data are processed by the personal computer. All the tests except flow visualization are carried out at the speed of 50km/h (13. 9m/s). In order to investigate the effect of the cross wind to the vehicle on aerodynamic noise, experiments are performed by two yaw angles (θ), 0° and 10°

3. EXPERIMENTAL RESULTS AND DISCUSSIONS

3.1 Flow Visualization

Figure 1 shows visualized airflows on the side window surface by oil flow technique. From the results, there seems to be two different flow patterns on it. One is the rotating flow pattern and another the flow pattern along side window surface. Therefore, it follows that the flow field on the surface are classified by the reattachment line into two regions, the separated flow region and the reattached region. Compared with the reattachment line at yaw angle θ=0°, the reattachment line at θ= 10° shifts lower and as a result the separated region becomes wider. Thereby, it can be said that the separated region increases in proportion to yaw angle.

Figure 2 shows visualized airflow around A-pillar by smoke-wire technique. It is found that the flow close behind A-pillar separates in the form of a cone-shaped rotating vortex.

Based on the above test results obtained by the flow visualization, the image of airflows around A-pillar can be illustrated as shown in Figure 3. That is, airflows along windshield separate on A-pillar and flows then reattach on side window surface with the strong rotation. As a result, airflows on side window surface are composed of two different flow patterns, a cone-shaped rotating vortex and airflows along the side window.

3.2 Quantitative Analyses of Sound and Flow

(1) Sound Pressure Level

Figure 4 shows 'A' weighted sound pressure level spectra for the frequency range 25Hz-20kHz at the position of driver's ears in the cabin at the speed of 50km/h. It is found that sound level at yaw angle θ=10° is larger than that at θ=0° especially for the frequency range 100-400Hz. Generally it is known that sound pressure level is proportional to the yaw angle. Therefore, this paper investigates the cause by means of quantitative analyses of sound and flow field around A-pillar.

In order to clarify the different pressure fluctuations in the separated region and in the reattached region, pressure measurements are made on side window surface. Figure 5 schematically shows measured points and the reattachment line obtained by oil flow visualization. Pressure measurments are made by flush-mounted microphones, which are arranged in such a way that microphones cross the reattachment lines at intervals of 50mm from No. 1 to No. 17.

Figure 6 shows 'A' weighted 1/3 octave band level spectra for the frequency range 25Hz-20kHz with respect to pressure fluctations at side window surface. The difference of sound pressure level between θ=0° and 10° is that the pressure level especially for the frequency range 100-400Hz at θ=10° is much higher than that at θ=0° . This result is similar to the result obtained from sound pressure level at the position of driver's ears as shown in Figure 4. Therefore, it can be said that the pressure level on side window affects the sound pressure level in the cabin.

Figure 7 shows overall values (dBA) obtained from 'A' weighted 1/3 octave band level spectra for the frequency range 25Hz-20kHz. Compared with pressure fluctuations in the reattached flow region, the pressure level in the separated flow region is higher by approximately 10dB for respective yaw angle. Moreover, in case of yaw angle θ=10° pressure level increases in both separated and reattached flow regions in comparison with θ=0°

Since such pressure fluctuations are considered to be a reflection caused by the flow pattern around the side window, we examine closely the structure of the flows.

(2) Structure of Flows on α Plane

Cartesian co-ordinate is taken as shown in Figure 8. At z=150mm α plane is perpendicular to pressure-measuring points on the side window and parallel to the flow direction. Here structure of flows on α plane is in detail investigated by means of I-type hot-wire anemometers.

Figure 9 shows the turbulent energy distribution at 10mm from the side window surface on α plane. As has been pointed out by the flow visualization in Figures 1 and 2, in the separated flow region there exist strong spiral vortices while there are no vortices in the reattached flow region. Therefore, the turbulent energy in the separated region tends to become higher than that in the reattached flow. At yaw angle $\theta=10°$, moreover, turbulent energy becomes higher and the region spreads wider. Compared with the sound pressure level distribution, turbulent energy distribution of the flows near the side window is quite similar. Therefore, it is found that both have strong correlations.

Figure 10 shows velocity fluctuations on α plane. It is found that velocity fluctuations in the separated flow region (No. 1 and No. 3) are greater than those on the reattachment line (No. 4) as well as in the reattached flow region. Since velocity fluctuations are closely related to the generation of noise, the separated flow region plays an important role in generating aerodynamic noise. In addition, spectrum analysis of flow is performed in the two regions as shown in Figure 11. The analysis reveals that the turbulent energy in the separated region is approximately ten times as large as that in the reattached region for the every frequency.

Figure 12 shows the mean-velocity contours on α plane. Large velocity gradients in the flow field at $\theta=0°$ can be seen from the window surface to about 30mm. Therefore, it is found that the separated flow region presents turbulent shear flow especially near A-pillar. On the other hand, at $\theta=10°$ the scale of the separated flow reaches to about 70mm and the lowest velocity region exits about 30mm from the surface due to the negative velocity gradients. Figure 13 shows equi-turbulence intensity contours of flows over the side window. The intensity of the turbulence fluctuations is defined by the ratio of root mean square of the fluctuating velocity component to the mean velocity of main flow,

$$T = \frac{\sqrt{\overline{u'^2}}}{U_\infty} \times 100. \quad \cdots\cdots (1)$$

In the case of $\theta=0°$, intensity is higher near A-pillar and side window surface. As a result, the separated flow has higher intensity. For $\theta=10°$, turbulence level itself becomes much higher and its spatial scale becomes larger. In the separated flow region there appears maximum intensity at about 30mm from side window surface. Since aerodynamically induced noise has much to do with flow fluctuations, therefore, the separated flow region seems to contribute largely to the generation of aerodynamic noise.

(3) Structure of Flows on β Plane

In the previous section we have investigated the flow pattern on α plane parallel to the mean-flow-direction. As shown in Figure 14 we here attempt to examine the flow pattern on β plane vertical to the mean-flow-direction. In the figure four symbols from S_0 to P_{10} indicate points of intersection of β plane and each reattachment line and their subscripts show yaw angles.

Figure 15 shows the total pressure contours on β plane, which are obtained by means of total pressure tubes whose diameter is 5mm. Here total pressure coefficient Cp is defined as the ratio of total pressure near side window to total pressure in mean flow. From this figure it is found that the total pressure in the separated flow region is lower than that in the reattached flow region and compared with $\theta=0°$ the total pressure for $\theta=10°$ is lower and its region is also wider. In addition, the region whose total pressure is lower than that in mean flow field reaches to 50mm and 85mm from side window surface for $\theta=0°$ and $10°$, respectively. In the case of $\theta=10°$ the minimum exists at about 30mm from window surface. This figure also indicates that near the reattachment line (P_0 and P_{10}) Cp is approximately equal to 1.0 due to the recovery of total pressure. As we have seen above, the separated flow region is expressed as a region less than Cp=1.0.

In order to investigate the flow pattern on β plane in more detail, three-dimensional flow measurements are made by means of X-type hot-wire anemometers. Figure 16 shows velocity vectors of y and z components. The center of the rotating vortex is located at about 30mm from side window surface. The vector diagram also describes well how the separated flows are reattaching on side window and as a result the reattachment lines are formed. For the case of $\theta=10°$, moreover, it can be observed that in addition to a primary vortex there exists a secondary vortex.

Figure 17 shows equi-vorticity contours obtained from the velocity vector data. Vorticity ω is expressed as follows using velocity vector $\mathbf{v}$=(u, v, w)

$$\omega = \mathrm{rot}\,\mathbf{v} = \nabla \times \mathbf{v} \quad \cdots\cdots (2)$$

Here x component of vorticity is normalized by the typical length L and main-stream velocity U∞. The vorticity is calculated by

$$\omega_x = \left(\frac{\partial w}{\partial y} - \frac{\partial v}{\partial z}\right)\cdot\frac{L}{U_\infty}\,. \qquad \ldots\ldots(3)$$

Positive and negative sign indicate counterclockwise and clockwise rotation, respectively. Therefore it follows that the primary vortex rotates clockwise whereas the secondary vortex rotates counterclockwise. Figure 18 shows schematically the flow image around A-pillar.

In what follows we discuss the relationships between aerodynamic noise and the flow around A-pillar. Since vehicles running roads, say, a freeway are usually exposed to the unsteady wind, vortex generated around A-pillar is cosidered to shift the similar patterns with respect to intensity of vorticity just as shown in Figure 17. It is generally known that in the subsonic field the unsteady motion of vortex produces aerodynamic noise and moreover aerodynamic noise is generated only in the region where vorticity is nonzero [5]. Therefore it follows that as far as aerodynamic noise is concerned, the separated-flow region plays a much more important role than the reattached-flow region.

4. THE ESTIMATION OF AERODYNAMIC NOISE

In the discussion that follows we describe the method to estimate aerodynamic noise generated around a vehicle.

Lighthill's equation is expressed as follows [5, 6, 7, 8],

$$\frac{\partial^2\rho'}{\partial t^2} - c^2\nabla^2\rho' = \frac{\partial^2 T_{ij}}{\partial x_i \partial x_j} \qquad \ldots\ldots(4)$$

where

$$T_{ij} = \rho v_i v_j + P_{ij} - c^2\rho'\delta_{ij}\,.$$

Here T_{ij} is Lighthill's stress tensor, ρ density, P_{ij} stress tensor, c velocity of sound in fluid at rest, v_i component of velocity in direction x_i (i=1, 2, 3), and δ_{ij} is the Kronecker δ-function ($\delta_{ij}=1$ if i=j, $\delta_{ij}=0$ if $i\neq j$). Physically, this means that sound is generatd by a fluid flow exactly as in a uniform medium at rest which is acted upon by externally applied fluctuating stresses.

If Lighthill's acoustic analogy is applied over a limited volume, V, bounded by a stationary surface S, the most general solution of Lighthill's equation [8] is expressed in the form

$$\rho'(\mathbf{x},t) = \rho - \rho_0 = \frac{1}{4\pi c^2}\frac{\partial^2}{\partial x_i \partial x_j}\int_V \left[\frac{T_{ij}}{r}\right] dV \quad \text{(quadrupole)}$$

$$- \frac{1}{4\pi c^2}\frac{\partial}{\partial x_j}\int_S n_i\left[\frac{\rho v_i v_j + P_{ij}}{r}\right] dS \quad \text{(dipole)}$$

$$+ \frac{1}{4\pi c^2}\frac{\partial}{\partial t}\int_S \left[\frac{\rho\, \mathbf{v}\cdot\mathbf{n}}{r}\right] dS\,, \quad \text{(monopole)} \qquad \ldots\ldots(5)$$

where **n** is the unit vector whose components are the direction cosines of the outward (from the body) normal to the surface S that bounds the foreign body. Each source element at **y** radiates sound some of which travels towards the observer at **x**. The sound arriving at **x** at time t must have been launched from the source at the previous time, t-|**x**-**y**|/c, since it travels at speed c a distance |**x**-**y**| from source to observer. Here we write |**x**-**y**|=r and adopt the notation that the square brackets [ ] imply that the function they enclose is to be evaluated at position **y** at the retarded time t-r/c (Figure 19). Evidently the quardrupole field interior to the body is equivalent to the sum of a dipole and a monopole. Both these sources are concentrated on the bounding surface S and the dipole strength density is the stress applied to that surface while the monopole strength per unit surface area is mass flux crossing the surface.

If we here regard a vehicle as a foreign body, then based on the solution of Lighthill's equation (5) we can estimate the order of magnitude of aerodynamically induced noise around the vehicle. As we have seen above, the solution is composed of three terms. That is, the first term is considered as quardrupole which is generated by free turbulent flow itself. The second one expresses dipole source generated by the unsteady application of an external force to the vehicle. The third one expresses monopole source produced by mass flux crossing the surface

In case of treating the problem of the aerodynamic noise around the vehicle, the first and the third term may be neglected. This is because the first term expresses the sound generated not by the effect of the configuration of the vehicle but by the turbulent flow itself regardless of the configuration. On the other hand, the third vanishes due to **v** • **n**=0 on the motionless and rigid surface of the vehicle. Therefore it follows that only the second term, dipole source should be taken into consideration in order to evaluate the aerodynamic noise produced by the configuration of the vehicle. In addition, if the surface of the vehicle is fixed or is vibrating in its plane, normal velocity at the surface becomes zero

$$n_i v_i = 0 . \qquad \ldots\ldots (6)$$

Hence (5) reduces simply to

$$\rho'(\mathbf{x},t) \simeq -\frac{1}{4\pi c^2}\frac{\partial}{\partial x_j}\int_S \left[\frac{P_j}{r}\right] dS \qquad \ldots\ldots (7)$$

where

$$P_j \equiv n_i P_{ij} .$$

In the same way as Lighthill and Curle [6, 7] simplified the volume integral in the cases wnen $|\mathbf{x}|$ is large enough, so (7) can be simplified. When $|\mathbf{x}| \gg \lambda$, where λ is a typical wavelength of the sound generated, so that $\mathbf{x}$ lies in the radiation field of each dipole, then (7) takes simpler form

$$\rho'(\mathbf{x},t) \simeq -\frac{1}{4\pi c^2}\int_S \frac{x_j - y_j}{|\mathbf{x}-\mathbf{y}|^2}\frac{1}{c}\frac{\partial}{\partial t}[P_j]\, dS . \qquad \ldots\ldots (8)$$

Moreover, if $x \gg L$, where L is a typical dimension of a vehicle, then the above equation reduces still further to

$$\rho'(\mathbf{x},t) \simeq -\frac{1}{4\pi c^3}\frac{x_j}{x^2}\int_S \frac{\partial}{\partial t}[P_j]\, dS$$

$$= H\int_S \frac{\partial}{\partial t}[P_j]\, dS , \qquad \ldots\ldots (9)$$

where

$$H \equiv -\frac{1}{4\pi c^3}\frac{x_j}{x^2} .$$

Equation (9) physically means the force applied by the vehicle to the airflow.

Finally, the order of magnitude of $\rho'_{f_i}(\mathbf{x},t)$ radiated by the sound source of frequency f_i can be estimated by the dimensional analysis;

$$\rho'_{f_i}(\mathbf{x},t) = O\,(P_{f_i}\cdot f_i \cdot S_{f_i}) \qquad \ldots\ldots (10)$$

where suffix f_i denotes the physical quantity of the frequency f_i. This estimation shows that the magnitude of the density perturbation may be proportional to the pressure level P_{f_i}, the frequency f_i, the representative area S_{f_i}

Therefore the order of magnitude of aerodynamic noise is given by

$$\overline{\rho'^2_{f_i}(\mathbf{x},t)} = O\,((P_{f_i}\cdot f_i \cdot S_{f_i})^2) . \qquad \ldots\ldots (11)$$

According to the above equation, it is interpreted that the reduction of aerodynamic noise level can be achieved by suppressing P_{f_i}, f_i and reducing S_{f_i}, respectively. This estimated result can explain well the experimented results as we have shown in the earlier section.

Now we propose the method to estimate S_{f_i} by experimental technique: the cross correlation method with the surface pressure data P_{f_i} measured by microphones flush-mounted on side window. Figure 20 shows the typical results of iso-correlation contours on side window obtained by the present method at yaw angle $\theta = 0°$. We can estimate S_{f_i} as the area enclosed by the iso-correlation line which takes the threshold value. However, we so far don't find the best way to determine the threshold value.

Concerned with the quantitative estimation of aerodynamic noise, we will discuss in the next paper.

5. CONCLUSIONS

The following things are revealed by the experimental study of aerodynamic noise generated around A-pillar of a production car.

(1) The visualization and measurements of flow clarify the three-dimensional flow patterns, which are classified into two regions, the separated flow region and the reattached flow region. In addition, it is qualitatively and quantitatively clarified that in the separated flow region there exist the two vortices whose rotational directions are opposite.

(2) The sound pressure level in the separated flow region is much higher than that in the reattached flow region. Moreover the pressure distribution is proved to be in good correlations with the energy distribution of turbulent flow.

(3) Measurements of vorticity clarify that the flow in the separated flow region is rotational whereas that in the reattached region irrotational. Since aerodynamic noise is generated only in the region where vorticity is nonzero, it follows that the region which plays an important role in generating aerodynamic noise is not the reattached region but the separated region.

(4) It is shown that the order of magnitude of aerodynamic noise can be estimated by the products of three physical quantities; the frequency f_i the pressure level P_{f_i} and the area S_{f_i}. Therefore it seems that the reduction of aerodynamic noise level can be achieved

by minimising these three quanities. In addition, it is found that this estimated result qualitatively explains well the experimented results.

ACKNOWLEDGEMENTS

The authors would like to thank Associate Professor F. Sakao of Hiroshima University as well as Professor M. Kawamura of Hiroshima technical college for their useful advice.

REFERENCES

[1] Wolf-Heinrich Hucho et al.: "Aerodynamics of Road Vehicles", 250-256, 1987.

[2] W. R. Stapleford and G. W. Carr: "Aerodynamic Noise in Road Vehicle, Parts I and II", MIRA Reports, 1971 and 1972.

[3] R. Buchheim, W. Dobrzynski, H. Mankau, and D. schwabe: "Vehicle Interior Noise Related to External Aerodynamics", Int. J. of Vehicle Design, vol. 3, no. 4, 398-410, 1982.

[4] O. Sadakata et al.: "A Consideration of Wind Noise Reduction by Air Flow Control", 22nd FISITA No. 885115, 182-189, 1988.

[5] F. Sakao: "Sound Generation by Flow", Advance of Fluid dynamics - Turbulence in chap. 6, 221-262, 1980 (in Japanese).

[6] Lighthill, M. J.: "On Sound Generated Aerodynamically I. General Theory", Proc. R. Soc. London A211, 564-587, 1952.

[7] Curle, N.: "The Influence of Solid Boundaries upon Aerodynamic Sound", Proc. R. Soc. London A231, 505-514, 1955.

[8] A. P. Dowling and J. E. Ffowcs Williams: "Sound and Sources of Sound", 146-167, 1983.

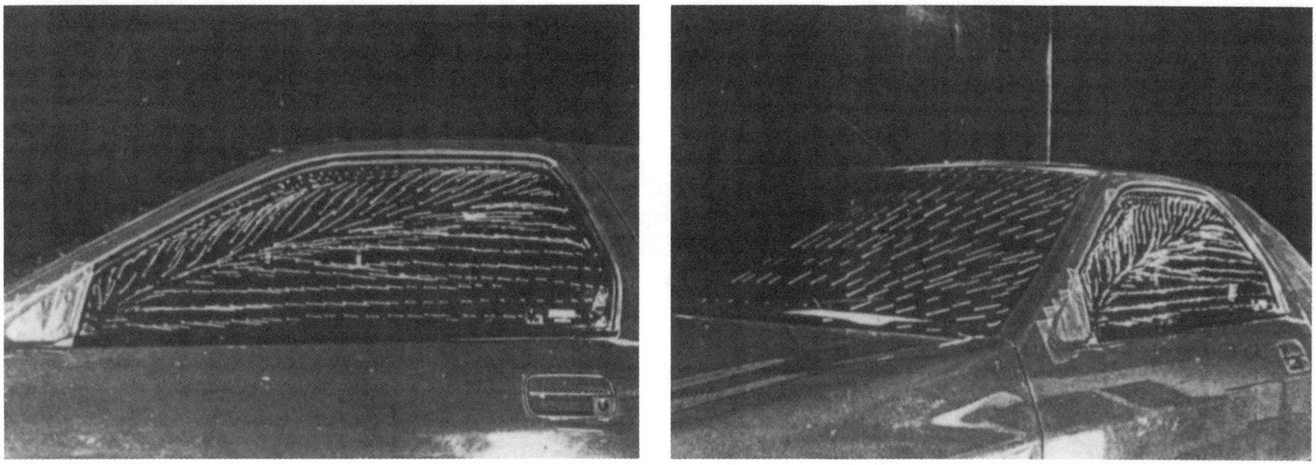

(a) $\theta = 0°$

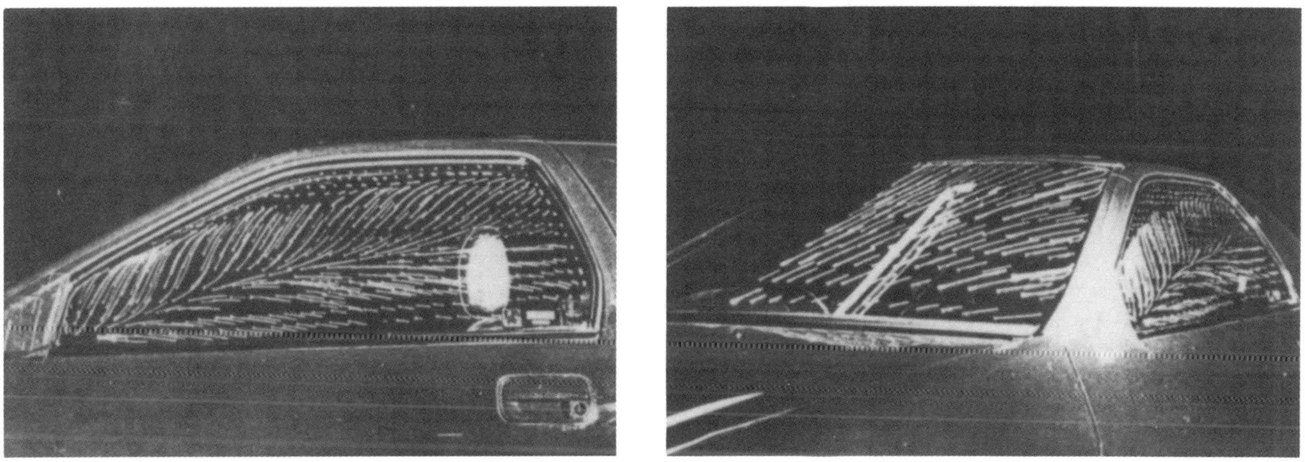

(b) $\theta = 10°$

Fig. 1 Airflows on Side-window Surface

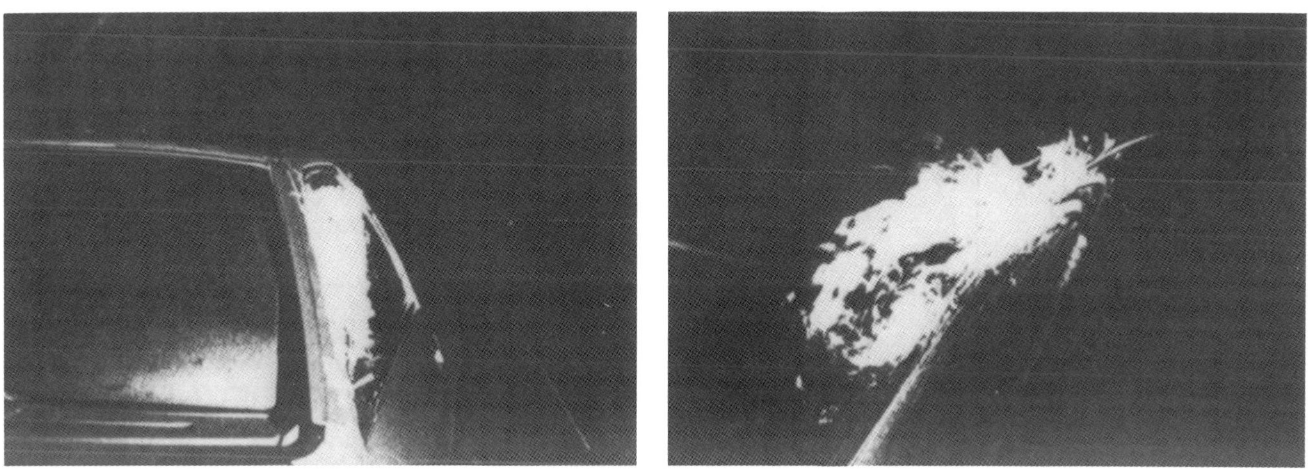

(a) Frontward View (b) Rearward View

Fig. 2 Cone-Shaped Rotating Vortex Generated Behind A-Pillar

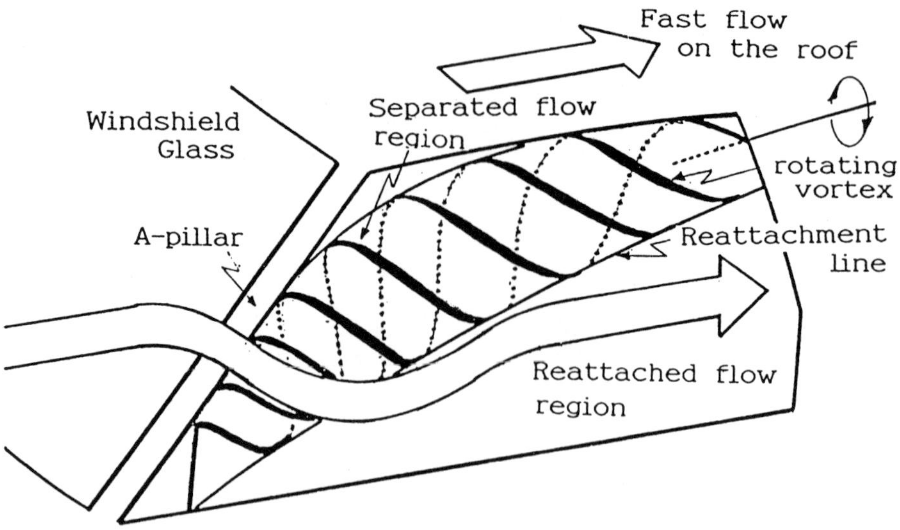

Fig. 3 Flow Pattern Around A-Pillar

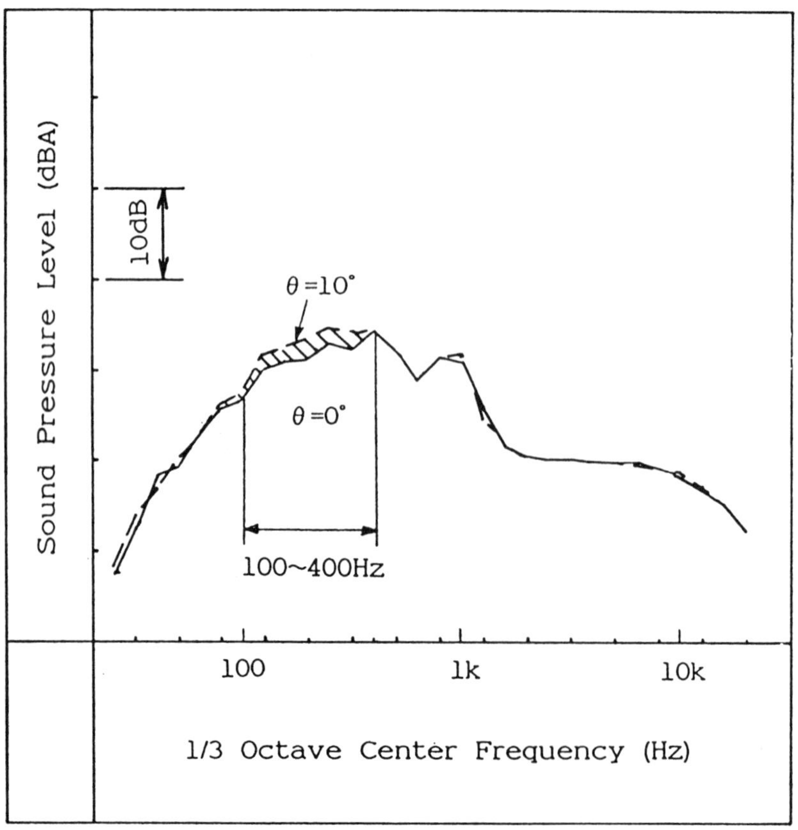

Fig. 4 Sound Pressure Level at The Position of Driver's Ear for Yaw Angle θ=0° and 10°

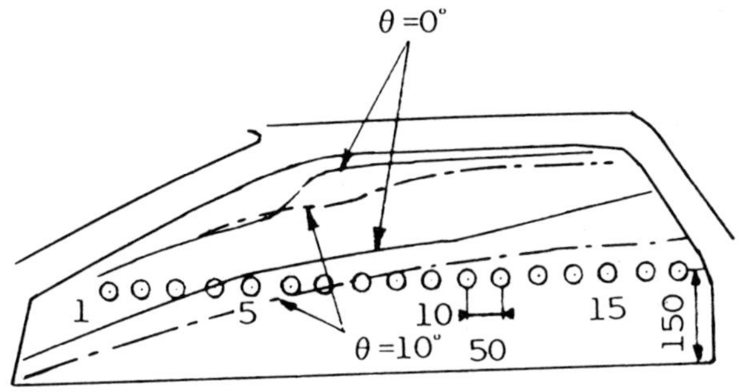

Fig. 5 Reattachment Line and Measured Points for Surface Pressure Fluctuation

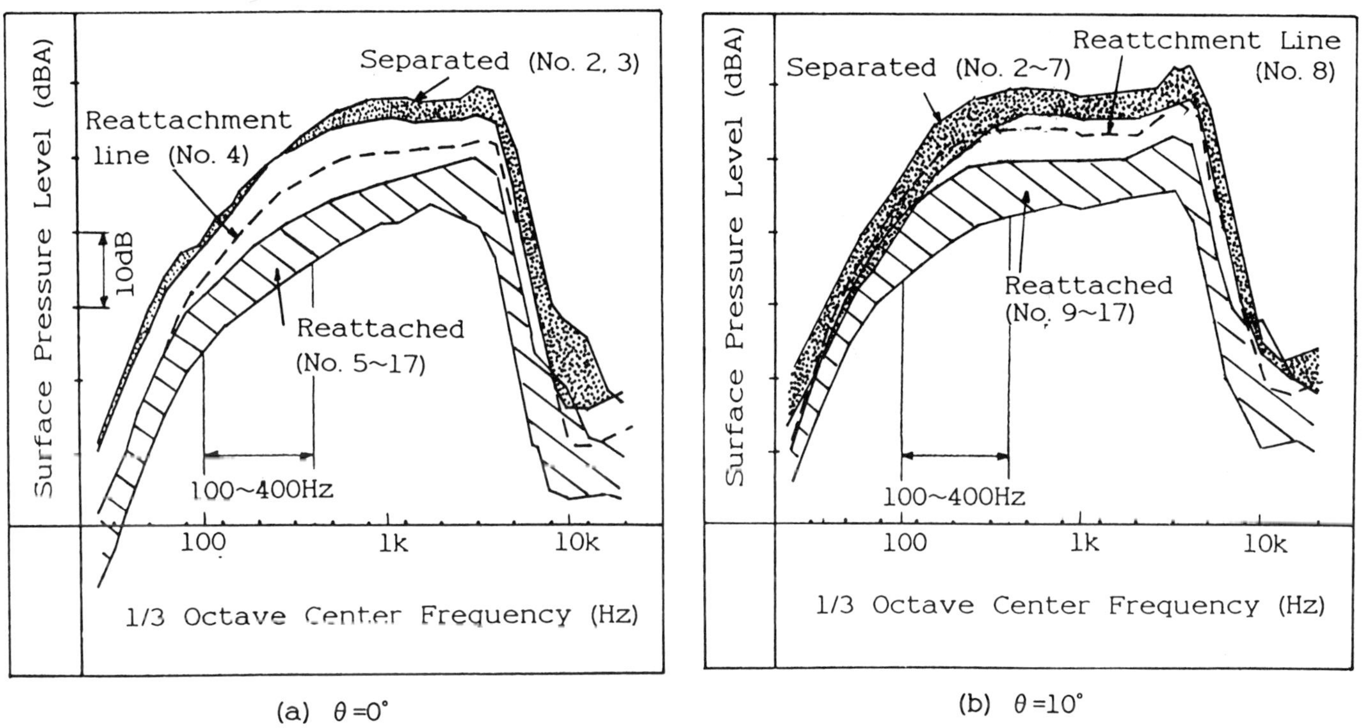

Fig. 6 'A' Weighted 1/3 Octave Band Level Spectra for Pressure Fluctuation on Side-Window Surface

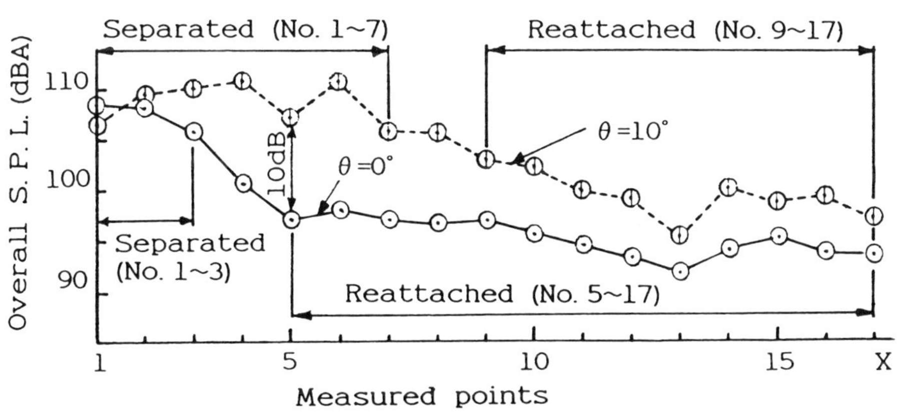

Fig. 7 Pressure Fluctuation Distribution on Side-Window Surface

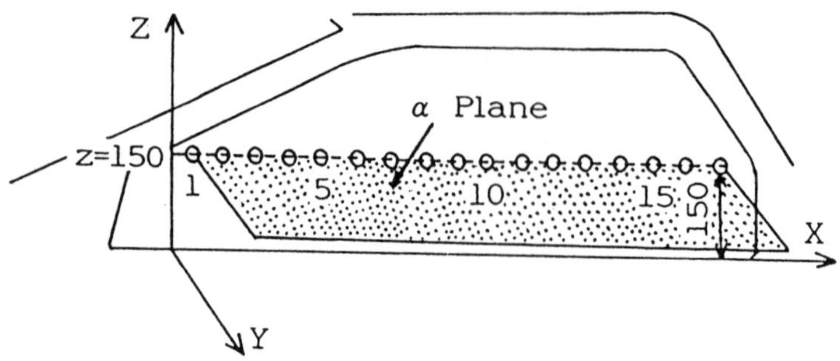

Fig. 8 α **Plane and Cartesian Co-ordinate**

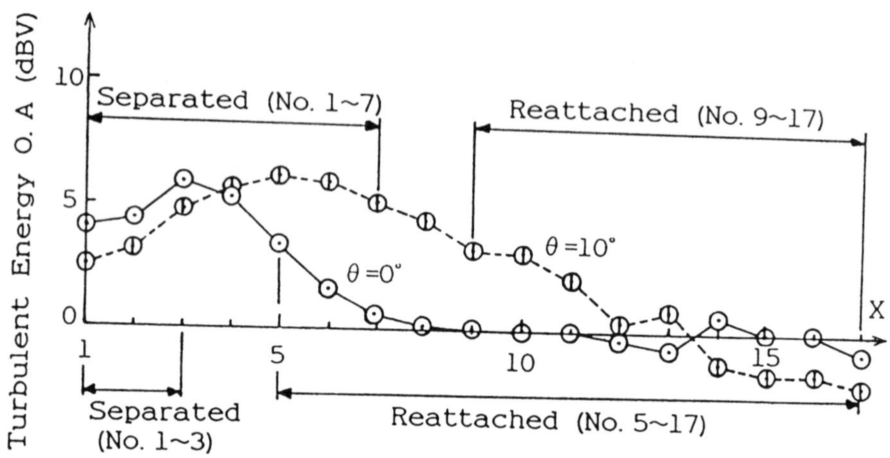

Fig. 9 Turbulent Energy Distribition on α **Plane (y=10, z=150)**

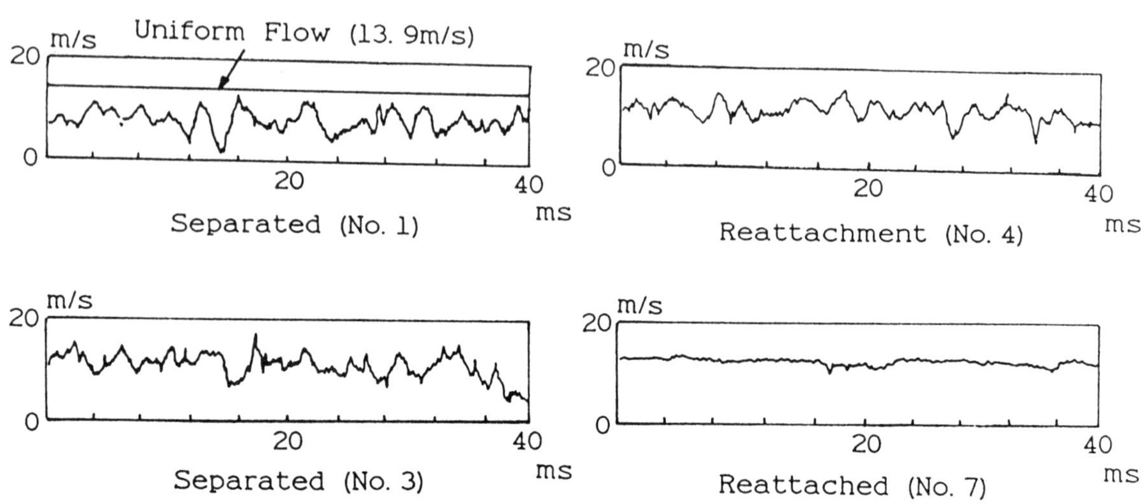

Fig. 10 Velocity Fluctuation on α **Plane (** θ **=0° , y=10, z=150)**

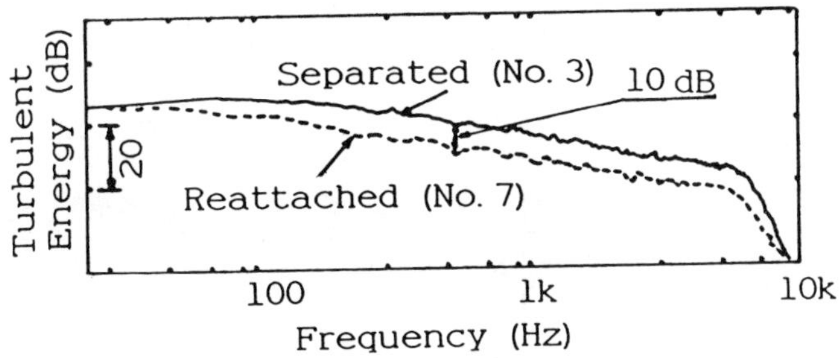

Fig. 11 Power Spectra of Flow on α Plane (θ=0°, y=10, z=150)

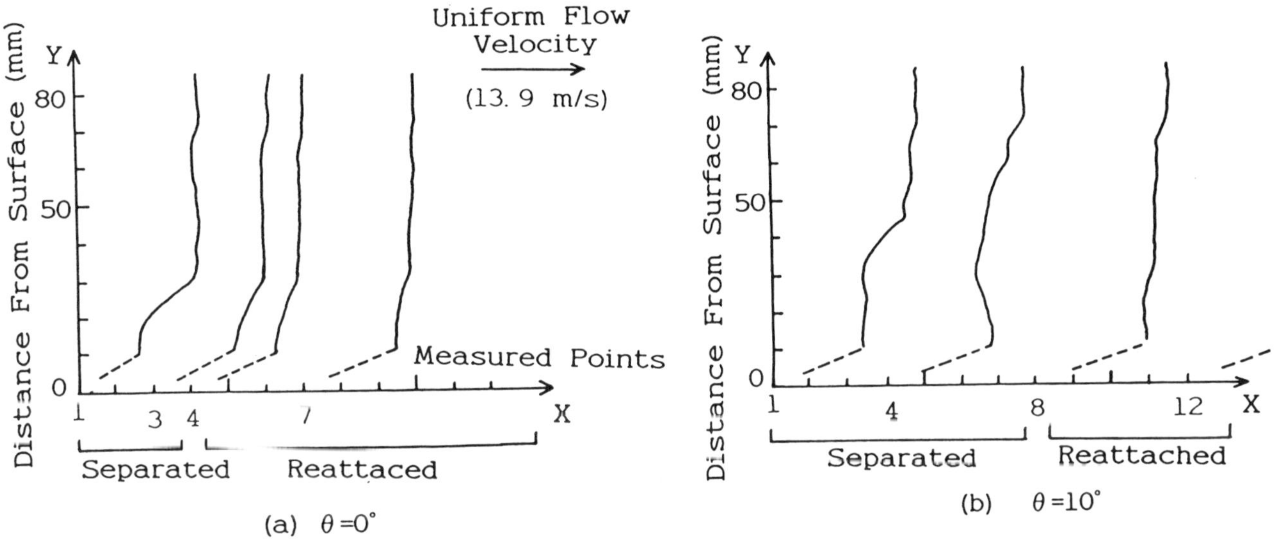

Fig. 12 Mean Flow Contours on α Plane (z=150)

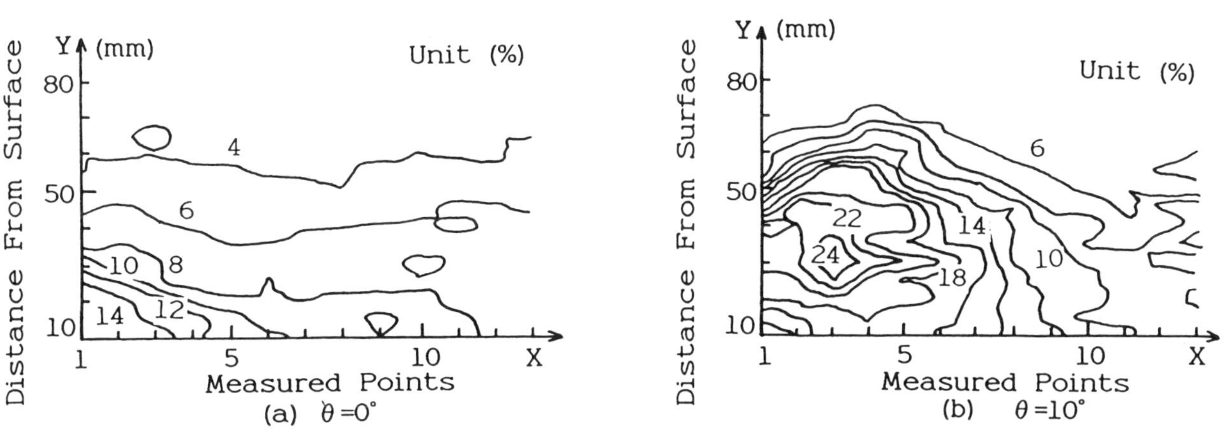

Fig. 13 Equi-Turbulent Intensity Contours on α plane (z=150)

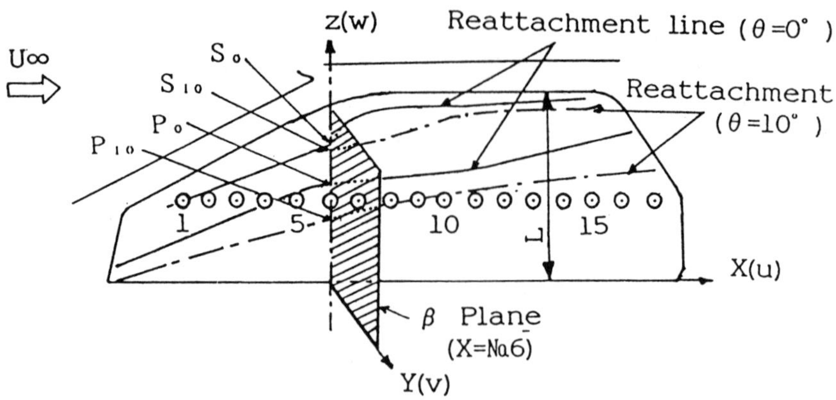

Fig. 14 β Plane and Cartesian Co-ordinate

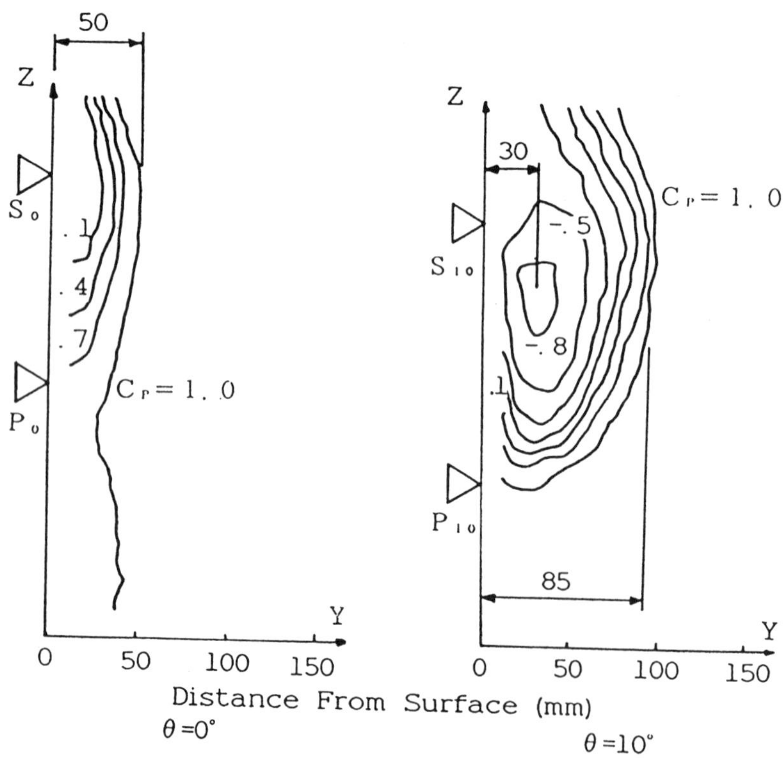

Fig. 15 Total Pressure Distribution on β Plane (x=No. 6)

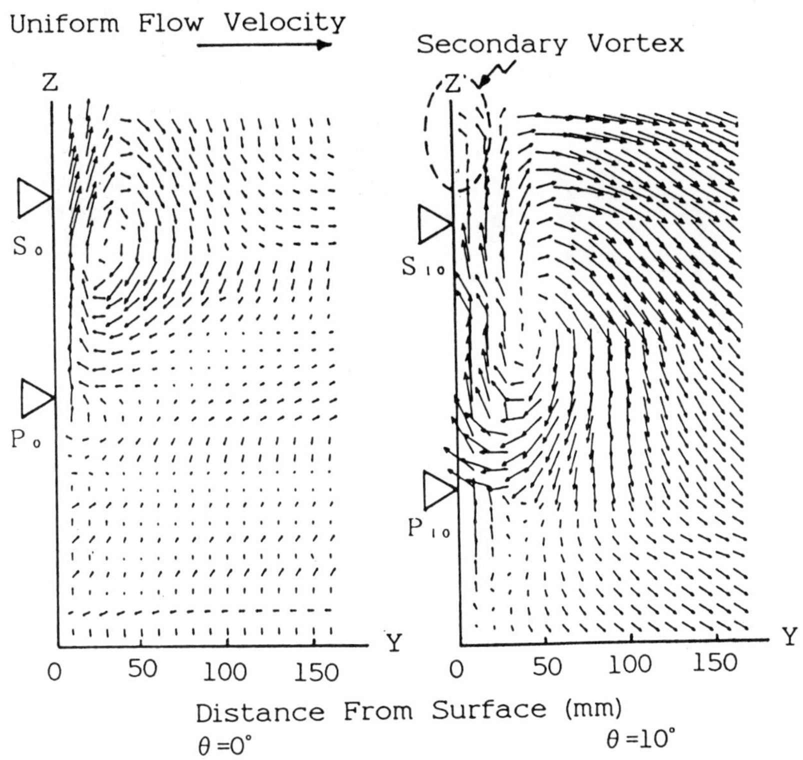

Fig. 16 Mean Velocity Vectors on β Plane (x=No. 6)

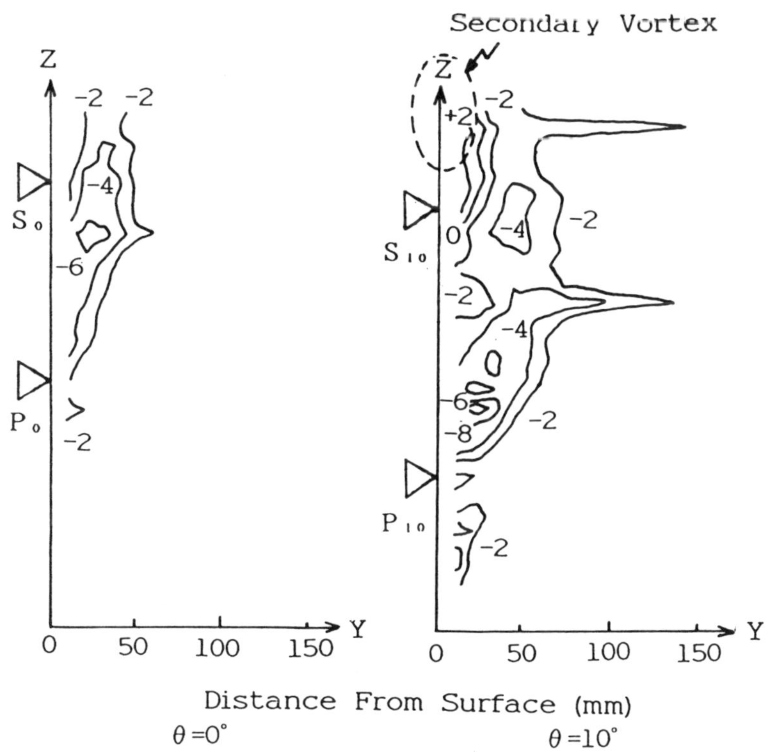

Fig. 17 Equi-Vorticity Contours on β Plane (x=No. 6)

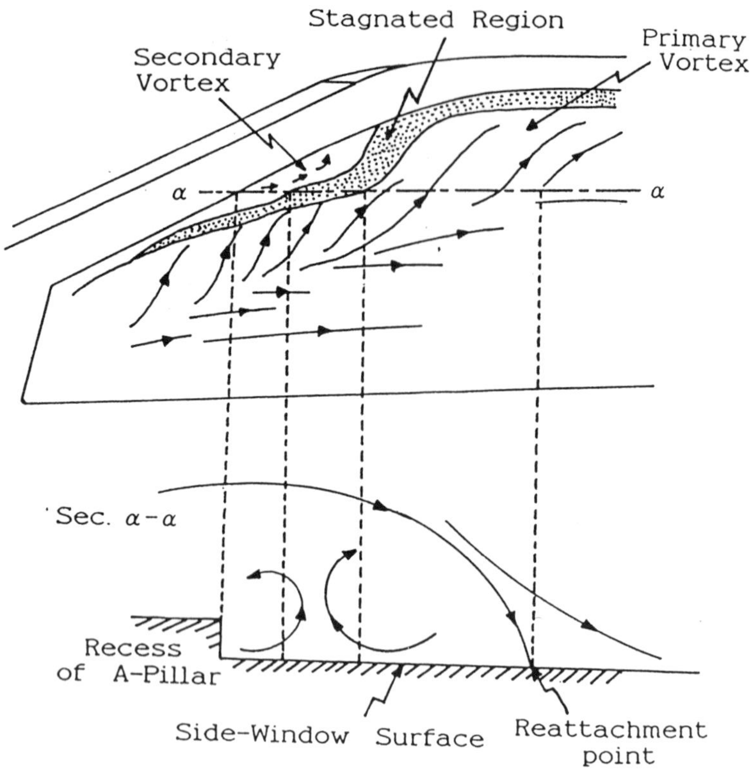

Fig. 18 Flow Pattern Close Behind A-Pillar

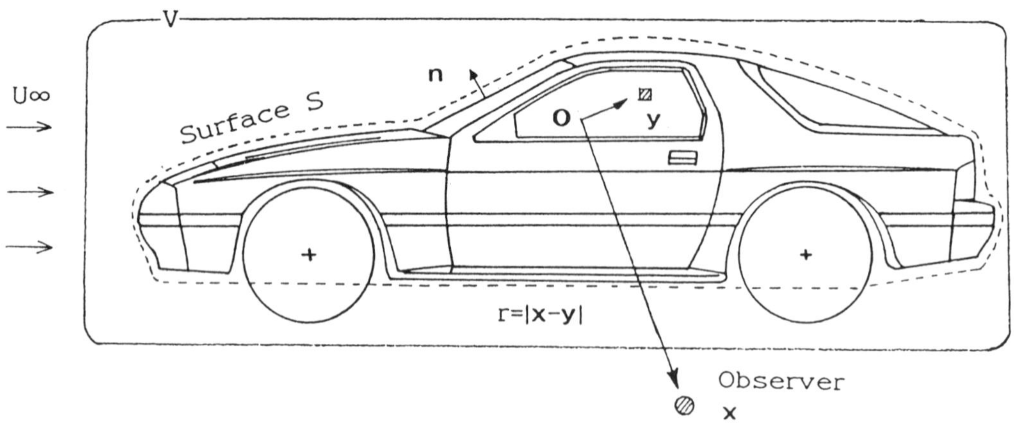

Fig. 19 A Stationary Control Surface Enclosing a Vehicle

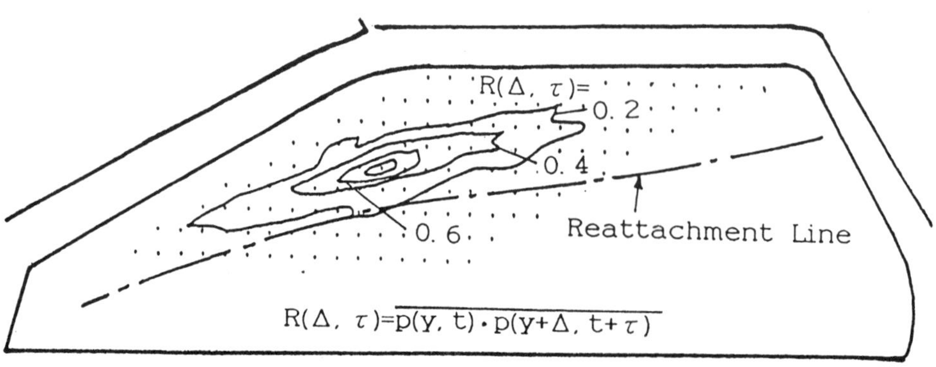

Fig. 20 Iso-correlation Contours with respect to Surface Pressure (θ=0°)